Differential and Integral
CALCULUS

PHILIP FRANKLIN, Ph.D.
PROFESSOR OF MATHEMATICS
MASSACHUSETTS INSTITUTE OF TECHNOLOGY

DOVER PUBLICATIONS, INC.
NEW YORK

Published in Canada by General Publishing Company, Ltd., 30 Lesmill Road, Don Mills, Toronto, Ontario.
Published in the United Kingdom by Constable and Company, Ltd., 10 Orange Street, London WC 2.

This Dover edition, first published in 1969, is an unabridged and unaltered republication of the work first published in 1953 by the McGraw-Hill Book Company, Inc. An 18-page section of answers to exercises, previously published separately, follows page 634.

Standard Book Number: 486-62520-6
Library of Congress Catalog Card Number: 77-96716

Manufactured in the United States of America
Dover Publications, Inc.
180 Varick Street
New York, N.Y. 10014

PREFACE

This book is an introduction to the calculus. As will be seen from the Table of Contents, it includes all the topics appropriate to a first course. This completeness enables the user of the book, by judicious omission, to select a course suitable to his taste. Thus he may place major emphasis on the acquisition of skill in formal technique, on geometric applications, on applications to physical problems, or on the theory and logical nature of the subject.

One such course would result from omitting those sections marked with an **R** or an asterisk and set in smaller type. The sections marked **R** discuss review material from analytic geometry or refresher material from trigonometry. These could be used for review in class. They also provide reference material for students lacking a fresh knowledge of these prerequisite subjects. The presence of these **R** sections makes it possible for the teacher who so desires to introduce the calculus to students with less than the traditional amount of preparation.

On the other hand the sections marked with an asterisk either treat a topic not universally studied by beginners, or contain a proof with some details a little difficult for the average class. These asterisk sections are included largely for the benefit of the thoughtful student and the student who finds it worth while to give the work a second reading to deepen his grasp of the subject.

Many students begin calculus concurrently with a physics course which makes some use of differentiation and integration. And a large number of students who wish an elementary knowledge of calculus either as an aid to understanding modern developments in natural science and economics or for its cultural value can devote only one semester to the subject. For these students it is desirable to begin the study of integration early. This book begins integration in Chap. 5, where a number of applications of integration as the inverse of differentiation are made. A full discussion of the definite integral regarded as the limit of a summation is given later in Chap. 12.

To achieve the progressive grasp of the subject with a minimum possibility of confusion, arrangement of the sections is such that the reader is never presented with more than one new term, notation, or idea at a time. And the explanations are sufficiently clear, simple, and complete so that the earnest beginning student can profitably get correct basic ideas by reading the text without having to rely entirely on the more informal class lecture presentation.

Great care has been taken to make all statements precise. The most fastidious teacher will never have to compromise his mathematical conscience. And the student who proceeds to courses in higher mathematical analysis will have nothing to unlearn. However, the introduction of subtleties beyond the grasp of the beginner has been studiously avoided.

In particular, the presentation of increasing functions and of multiple-valued functions is an improvement on the discussion customarily found in elementary texts. For many geometric quantities the algebraic sign has a significance. Instead of masking this fact by the commonly used device of taking absolute values, in each case the appropriate interpretation of the sign is brought out.

There are a large number of worked-out illustrative examples. And the solutions are given in full detail. These frequently display a model of tabular form or efficient arrangement of numerical computation or algebraic manipulation. For some problems more than one solution is given. Often the first solution describes the obvious approach, while the second solution uses a more elegant method. Instead of selecting as problems to be worked out those which can be done in minimum space, a number have been deliberately chosen because they involve apparent complications, difficulties, or paradoxes so that means of satisfactorily meeting these can be brought out in the solution.

Ample problems for the student have been provided. A set of these, listed roughly in order of difficulty, appears as an exercise after each new group of closely related topics. This arrangement facilitates the choice of suitable problems for assignments or tests.

Philip Franklin

Cambridge, Mass.,
January, 1953

CONTENTS

CHAPTER 1

LIMITS OF FUNCTIONS

The idea of limit is fundamental in calculus. In this introductory chapter we shall explain precisely what we mean when we say that a variable approaches a limit, or that a function of a variable approaches a limit. But we must begin by discussing several more elementary notions.

1. The Number Scale. The positive numbers of arithmetic, together with zero and the negative numbers introduced in elementary algebra, make up the *real number system.* It is often useful to represent these real numbers graphically by points on a straight line. As in Fig. 1, take any point O on the line as the origin, and any other point U on the line as the unit point. Then OU is a positively directed unit segment. Thus the distance from O to U is the unit of length. And the direction from O to U, or to the right in Fig. 1, is the positive direction.

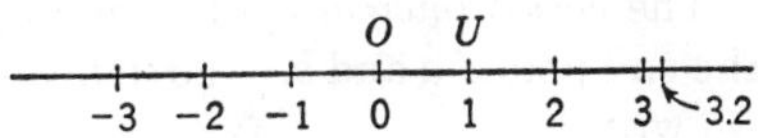

FIG. 1. A number scale.

On O mark the number 0, and on U mark the number 1. Successive intervals to the right of U, each of the same length as OU, determine the points marked 2, 3, $\cdots$. Similarly, equal intervals to the left of O determine -1, -2, -3, $\cdots$. The point for the positive number 3.2 is a point to the right of O whose distance from O is 3.2 times OU. The point for the negative number -3.2 is this same distance to the left of O. The points for any other positive or negative number may be found by a similar procedure. The line with its associated numbers is called a *number scale,* or simply a *scale.* The scale matches up real numbers with points on the line in such a way that there is one point for each number and one number for each point. In discussing numbers and points on a number scale we may identify the numbers and the points which represent them. Thus in this context we often say "3," "the number 3," or "the point 3," to mean the point with mark 3, or the point represented by the number 3. Similarly "the point a," "the number a," or "a" may mean the point whose mark is the number a. Here a is some particular real number, positive, negative, or zero.

2. Inequalities. The order of numbers is reflected by their relative position on the scale. We may indicate this by *inequalities.* Thus if the point a is to the left of the point b, we write $a < b$, read "a is less than b." In this case $b - a$ is positive so that $0 < b - a$. For example,

$$2 < 5, \qquad -5 < 2, \qquad -5 < -2,$$

and each of the differences $5 - 2 = 3$, $2 - (-5) = 7$, and $-2 - (-5) = 3$ is positive.

The relation $b > a$, read "b is greater than a," also means that the point b is to the right of the point a, or that the point a is to the left of the point b. If $b > a$, then $b - a$ is positive so that $b - a > 0$.

FIG. 2. An interval a,b.

As in Fig. 2, let a be to the left of b, and x any other point between a and b. Collectively, the interior points x make up the *open interval a,b*. To indicate that the point x or the number x belongs to the open interval, we write

$$a < x < b.$$

The *closed interval a,b* is made up of the interior points together with the end points a and b. To indicate that x belongs to the closed interval, we write

$$a \leqq x \leqq b.$$

The symbol $\leqq$ is read "is less than or equal to."

3. Absolute Value. We call 5 the *numerical value* of -5, or its *absolute value.* And we use the symbol $|-5|$ to represent it, writing $|-5| = 5$. Similarly for any negative number.

The numerical value of a positive number, or 0, is the number itself. Thus $|5| = 5$. As further examples of absolute value notation, observe that

$$|3 - 7| = |-3 + 7| = 4, \qquad |3(-7)| = |3||-7| = |-21| = 21.$$

$$\left|\frac{3}{-7}\right| = \frac{|3|}{|-7|} = \left|\frac{-3}{7}\right| = \frac{3}{7}.$$

And in general we have, for any two numbers a and b,

$$|a - b| = |b - a|, \qquad |ab| = |a||b|. \tag{1}$$

And if $b \neq 0$, for the quotient a/b, we have

$$\left|\frac{a}{b}\right| = \frac{|a|}{|b|}. \tag{2}$$

To emphasize the distinction between

$$-5 < -2 \qquad \text{and} \qquad |-5| > |-2|, \qquad \text{or } 5 > 2,$$

we say that -5 is *algebraically* less than -2 but -5 is *numerically* greater than -2.

For any point x on the number scale, the distance between 0 and x is $|x|$. And for any two numbers x_1 and x_2, the distance between the points x_1 and x_2 is $|x_2 - x_1|$. This is illustrated in Figs. 3 and 4. Note that

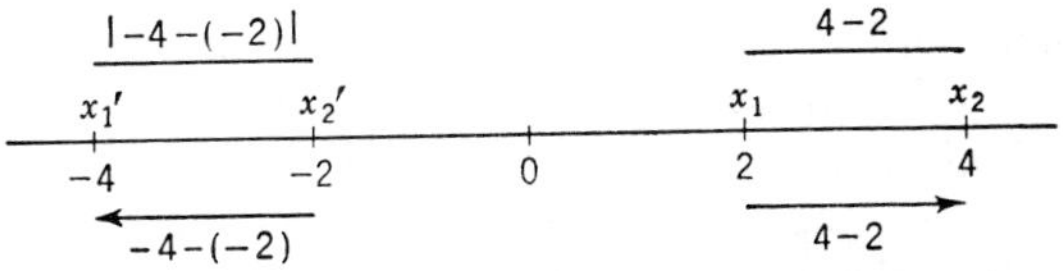

FIG. 3.

the signed quantity $x_2 - x_1$, the number for the final point minus the number for the initial point, measures the directed segment from x_1 to x_2. That is, $x_2 - x_1$ is negative when x_2 is to the left of x_1. Some directed segments are marked with arrows in Figs. 3 and 4.

An inequality involving absolute values such as

$$|x - 5| < 2 \qquad (3)$$

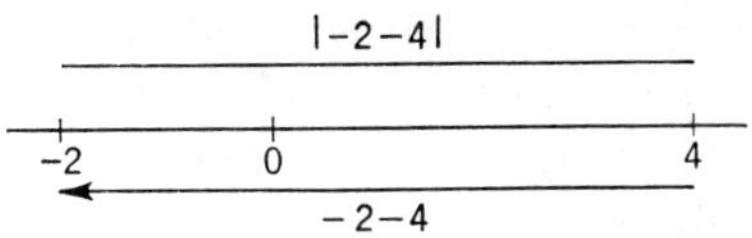

FIG. 4.

states that the distance from 5 to x is less than 2, so that the point x must lie to the right of $5 - 2 = 3$ and to the left of $5 + 2 = 7$. Thus Eq. (3) is another way of indicating that x belongs to the open interval $3 < x < 7$.

EXERCISE 1

Consider each of the following general theorems for the special case $a = 2$, $b = 5$, $c = -10$, $d = 6$. Calculate all the numbers involved, plot them on a number scale, and so verify that the hypothesis and conclusion each hold for the special case.

1. For any c, if $a < b$, then $a + c < b + c$.
2. If $a < b$, then $-a > -b$.
3. If $a < b$ and $c < d$, then $a + c < b + d$.
4. If $a < b$ and $c < 0$, then $ac > bc$.
5. If $a < b$ and $d > 0$, then $ad < bd$.
6. If $a < b$ and $c < 0$, then $a/c > b/c$.
7. If $ad > bc$ and $bd > 0$, then $a/b > c/d$.
8. For any a and b, $(a - b)^2 \geqq 0$ and $a^2 + b^2 \geqq 2ab$.
9. If $|x - a| < b$, then $a - b < x < a + b$.
10. If $|x - a| = b$, then either $x = a - b$ or $x = a + b$.
11. If $c < x < d$, then $\left| x - \dfrac{c + d}{2} \right| < \dfrac{d - c}{2}$.
12. For any a and d, $|ad| = |a| \cdot |d|$.
13. For any b and c, $|b/c| = |b|/|c|$.
14. For any b and c, $|b + c| \leqq |b| + |c|$.
15. For any b and c, $|c - b| \geqq |c| - |b|$.
16. The length of the segment with ends c and d is $|c - d|$.

4. Variables. A quantity whose value is fixed throughout a discussion is called a *constant.* A quantity which may assume different values in the course of a discussion is called a *variable.* The values which a variable can assume make up the *range* of the variable. The range may be discrete, like the sequence of positive integers 1, 2, 3, $\cdots$, or it may be continuous, like the values on an interval, $a < x < b$.

5. Functions. When two variables are so related that the value of the second variable is determined when the value of the first variable is given, the second variable is said to be a *function* of the first. The first variable, which may be given any value in a range depending on the particular problem, is called the *independent variable,* or *argument.* The second variable, whose value is determined by the choice of the independent variable, is called the *dependent variable,* or *function.*

We are sometimes given one equation connecting two variables, and we may choose either one as the independent variable. The desirable choice depends on the ease of determining the independent variable, or the application we wish to make of the functional relation.

The symbol $f(x)$, read "f of x," is used to denote a particular function of x. Here x is the independent variable. To distinguish different functions occurring in the same discussion, we write other letters in place of f. Thus $F(x)$, read "large F of x," $\phi(x)$, read "phi of x," $g(x)$, read "g of x," might represent three different functions.

Throughout any one discussion a functional symbol indicates the same relation of dependence of the function upon the independent variable. Thus if, for all values x,

$$f(x) = x^2 - 6x + 10,$$

then

$$\begin{aligned} f(0) &= 0^2 - 6 \cdot 0 + 10 = 10, \\ f(1) &= 1^2 - 6 \cdot 1 + 10 = 5, \\ f(-2) &= (-2)^2 - 6(-2) + 10 = 26, \\ f(a) &= a^2 - 6a + 10, \\ f(x+3) &= (x+3)^2 - 6(x+3) + 10 = x^2 + 1. \end{aligned}$$

Suppose that, in the same discussion, for all values of x,

$$g(x) = 3x - 5 \qquad \text{and} \qquad F(x) = f(x) + 2g(x).$$

Then we may express $F(x)$ in terms of x by observing that

$$F(x) = x^2 - 6x + 10 + 2(3x - 5) = x^2.$$

EXERCISE 2

Let $f(x) = x^2 + 4$ for all values of x. Evaluate each of the expressions

1. $f(0)$. 2. $f(2)$. 3. $f(a)$. 4. $f(a + h) - f(a)$.

Let $f(x) = 2x^3$ for all values of x. Evaluate each of the expressions

5. $f(2)$. 6. $f(-3)$. 7. $f(t - 1)$. 8. $f(3t)$. 9. $f(t^2)$.

Let $f(x) = \dfrac{x+1}{x-1}$ for all values of x except $x = 1$. Evaluate each of the expressions

10. $f(0)$. 11. $f(2)$. 12. $f(-1)$. 13. $f(t+1)$.

14. If $f(x) = x^4 + 2x^2 - 7$, prove that $f(-x) = f(x)$.
15. If $f(x) = x^3 - 3x$, prove that $f(-x) = -f(x)$.

Let $f(x) = x + \dfrac{1}{x}$ and $g(x) = x - \dfrac{1}{x}\cdot$ Prove that

16. $f(1/x) = f(x)$. 17. $g(1/x) = -g(x)$.
18. $[f(x)]^2 = f(x^2) + 2$. 19. $[g(x)]^2 = f(x^2) - 2$.
20. $f(x)\,g(x) = g(x^2)$.

21. If $f(x) = 1/x$, and $y = f(x)$, prove that $x = f(y)$.
22. If $f(x) = \dfrac{2x+5}{3x-2}$ and $y = f(x)$, prove that $x = f(y)$.
23. If $f(x) = ax + b$, prove that $f(x+h) - f(x) = ah$.
24. If $f(x) = x^2 - x$, prove that $f(x+1) - f(x) = 2x$.

Let $f(x) = 2^x$. Prove that

25. $f(x+1) - f(x) = 2^x$. 26. $f(x)\,f(z) = f(x+z)$.

R6. Graph of a Function. The functional relation between a dependent variable y and an independent variable x may be pictured to the eye by a *graph*. For *rectangular* or *Cartesian* coordinates, the method of matching a pair of real numbers x,y with a point P in the plane is as follows.

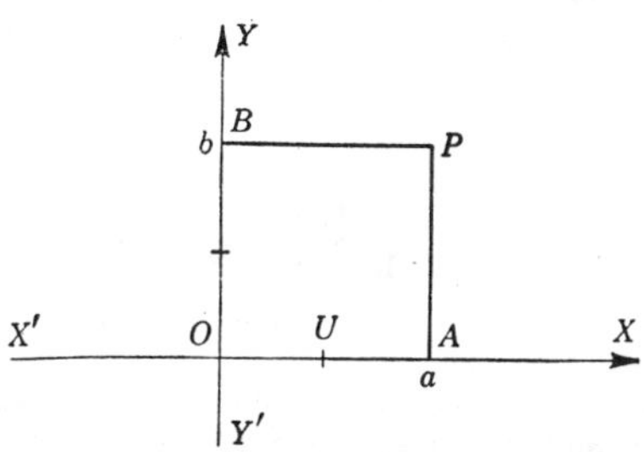

FIG. 5.

Select any point O as the origin. Take any convenient length OU as a unit. Draw a horizontal line $X'X$ through O, Fig. 5, and with OU as unit construct a number scale as in Sec. 1. We use this scale to read a value of x, or *abscissa*. And the line $X'X$ is called the x axis, or *axis of abscissas*. Next draw a vertical line $Y'Y$ through O, and on this construct a number scale by rotating the scale on $X'X$ through 90° counterclockwise. We use this scale to read a value of y, or an *ordinate*. And the line $Y'Y$ is called the y axis, or *axis of ordinates*. Taken together, with their scales, the two lines are the *coordinate axes*, or *axes of reference*. Consider any rectangle $OAPB$ such that A is on $X'X$ and B is on $Y'Y$. Then if A is marked with the number a on the $X'X$ scale, and B is marked with number b on the $Y'Y$ scale, we match the point P with the coordinates a,b. Given $x = a$ and $y = b$, we may locate A and B and hence find P. And from P and the axes of reference we may construct the rectangle $OAPB$ and hence read the coordinates a,b. We use the symbol (a,b) to mean the point with $x = a$, $y = b$. And we indicate that this is the point P by writing $P = (a,b)$ or $P(a,b)$.

To illustrate how the signs of the coordinates change in the four quadrants, in Fig. 6 we have plotted the points $P_1(2,3)$, $P_2(-4,2)$, $P_3(-1,-5)$, and $P_4(3,-2)$.

The graph of a function $f(x)$, or locus of the equation $y = f(x)$, is made up of all points $P(x,y)$ whose coordinates x and y satisfy the equation $y = f(x)$. In most cases the graph consists of one or more pieces of curves. We may draw these pieces by plotting a few points in the range of interest and joining them by a smooth curve.

For example, let $y = 4x - x^2$. We select a set of values of x in the range of interest, here the integers from -1 to 5, and tabulate the values of our function. Thus,

x	x^2	$4x - x^2 = y$
-1	1	-5
0	0	0
1	1	3
2	4	4
3	9	3
4	16	0
5	25	-5

In this case the value of y could be obtained mentally, but the tabular form with intermediate steps lessens the chance of error. And, for less simple numbers or operations this method enables us to compute each column easily from a table or slide

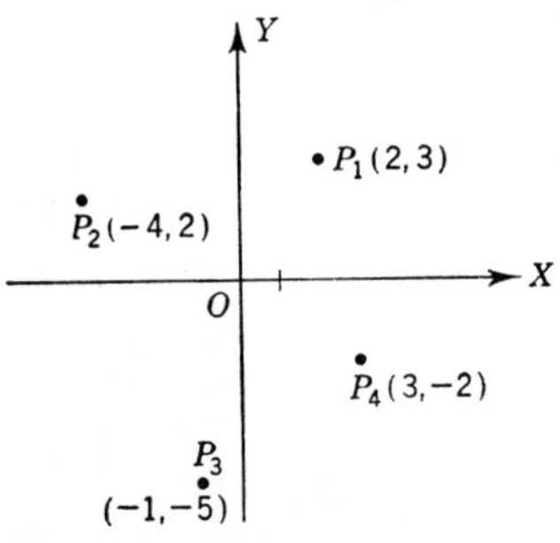

FIG. 6.

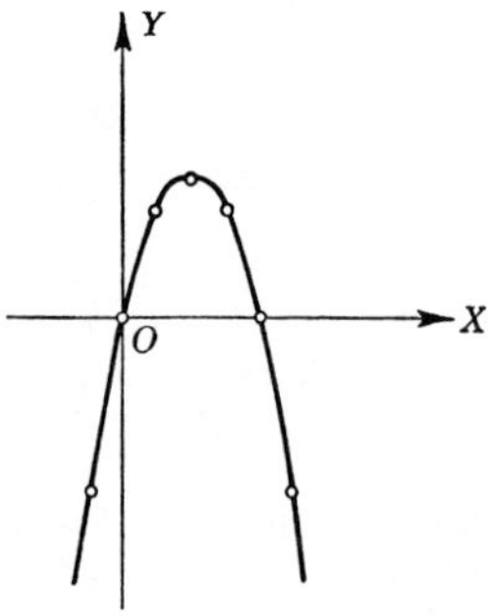

FIG. 7.

rule. For each pair of values of the table, from $(-1, -5)$ to $(5, -5)$ we plot the corresponding points, as in Fig. 7, and join them by the curve shown.

As indicated in Probs. 20 and 22 of Exercise 3, it is sometimes convenient to use a different unit on the y axis from that on the x axis.

EXERCISE 3

Consider each of the following general statements for the special case $a = 3$, $b = 4$, $c = -2$, $d = -1$. Plot all the points involved in a rectangular coordinate system and so verify that each conclusion holds for the special case.

1. The four points (a,a) $(-a,a)$, $(-a,-a)$, and $(a,-a)$ are the vertices of a square.
2. The four points $(a,0)$, $(0,a)$, $(-a,0)$, $(0,-a)$ are the vertices of a square.
3. The four points (a,b), $(-a,b)$, $(-a,-b)$, $(a,-b)$ are the vertices of a rectangle.
4. The four points (a,b), $(-b,a)$, $(-a,-b)$, $(b,-a)$ are the vertices of a square.
5. The four points $(0,0)$, (a,b), $(a + c,\ b + d)$, and (c,d) are the vertices of a parallelogram.

Plot the graph of each of the following functions.

6. $y = 2x$.
7. $y = 2x + 3$.
8. $y = 2x - 3$.
9. $y = 4$.
10. $y = -2x$.
11. $y = -2x + 3$.
12. $y = x^2/5$.
13. $y = x - x^2$.
14. $y = x^2 - 4$.
15. $y = x^3$.
16. $y = x^3 - x^2$.
17. $y = x - x^3$.

For the special case $m = 3$, $b = 2$, verify that each of the following general statements holds by plotting the graph.

18. The graph of $y = mx$ is a straight line through the origin and the point $(1,m)$.
19. The graph of $y = mx + b$ is a straight line parallel to the graph of $y = mx$, Prob. 18, and meeting the y axis at the distance b from the origin, above or below according as b is positive or negative.

20. Plot the graph of $y = x^3 + 2x - 17$ for values of x from 0 to 5, using such units that the length which represents 1 on the x axis represents 10 on the y axis.
21. A particle is moving uniformly in a straight line with a velocity of 4 ft./sec. If s is the distance in feet and t is the elapsed time in seconds, $s = 4t$. Using rectangular coordinates with the same length representing 1 sec. on the horizontal t axis, and 1 ft. on the vertical s axis, draw the graph of this relation.
22. With s and t defined as in Prob. 21, for the motion of a falling particle, $s = 16t^2$. Using rectangular coordinates with the same length representing 1 sec. on the horizontal t axis and 100 ft. on the vertical s axis, draw the graph of this relation.

***7. Limit of a Variable.** Let the variable v be a function of t, and let this function be defined for an infinite number of values of t, which we consider in order. These values may change abruptly. For example, the values of t may be $1, 2, 3, \cdots$. Or t may increase continuously from 0 through all positive values. We refer to the corresponding discrete or continuous values of v taken in order as its succession of values. We say that

The variable v approaches the constant L as a limit if beyond a certain point in its succession of values, the numerical value of the difference between v and L, $|L - v|$, becomes and stays smaller than any preassigned fixed positive quantity, however small.

To express this relation we write $\lim v = L$, read "the limit of v is L" or "v approaches L as a limit."

As an example, for either of the two t sequences mentioned, if

$$v = \frac{t}{t+1}, \qquad \lim v = 1.$$

To apply the definition to this case, we calculate

$$L - v = 1 - \frac{t}{t+1} = \frac{1}{t+1} \qquad \text{and} \qquad |L - v| = \frac{1}{t+1}.$$

Let ϵ denote a small positive number. Then $1/\epsilon$ is a large positive number. And we shall have

$$|L - v| < \epsilon \qquad \text{or} \qquad \frac{1}{t+1} < \epsilon,$$

provided that

$$t + 1 > \frac{1}{\epsilon} \qquad \text{or} \qquad t > \frac{1}{\epsilon} - 1.$$

This shows that for any fixed positive number ϵ, however small, there is a place N in the succession of values $1, 2, 3, \cdots$, or in the continuously increasing positive numbers such that for t beyond N, here $t > N$, we shall have $|L - v| < \epsilon$. We may take N as the smallest positive integer greater than $(1/\epsilon) - 1$. For example, if $\epsilon = 0.03$, $(1/\epsilon) - 1 = 32.3$, and we may take $N = 33$.

In terms of the number scale for v, Fig. 8, the definition of $\lim v = L$ means that if we choose any fixed length ϵ, *however small*, and mark the points $L - \epsilon$ and $L + \epsilon$, beyond some place in the succession of values v, all the points v will lie in the open interval $L - \epsilon < v < L + \epsilon$. The place in the succession will usually depend on the choice of the fixed number ϵ, and the smaller we select ϵ, the further out we have to pick the place in the succession.

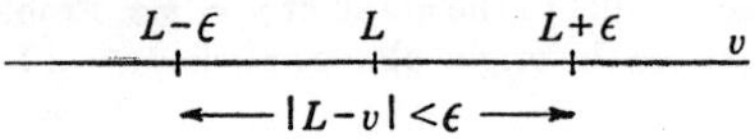

FIG. 8.

***8. Limit of a Function.** Let the dependent variable y be a given particular function of the independent variable x, so that $y = f(x)$. We consider the variable x for some succession of values, no one of which equals a, but such that, when we run through the succession as in Sec. 7, x approaches a as a limit. We write

$$x \to a, \tag{4}$$

read "x approaches a" or "x tends to a," to indicate that $\lim x = a$, while $x \neq a$.

If $f(x)$ is defined for each value of x considered, the relation $y = f(x)$ will determine a corresponding succession of values of y. Now suppose that the variable y, considered as taking on these values in order, approaches the constant A as a limit. Thus $\lim y = A$ as defined in Sec. 7. If this is the case whenever x tends to a, regardless of the particular succession of values used, we say that y approaches A when x tends to a.

To indicate that the approach of y to A is initiated by any approach of x to a, we write

$$\lim_{x \to a} y = A, \tag{5}$$

read "the limit of y, as x approaches a, is A."

We can describe the situation without reference to y by writing

$$\lim_{x \to a} f(x) = A, \tag{6}$$

read "the limit of f of x, as x approaches a, is A."

***9. The Function $f(x)$ Equals a Constant.** Let c be a constant, and $f(x) = c$. In this case, for any succession of values of x tending to a, the corresponding values of y are all equal to c. Hence they satisfy our definition of approach to c as a limit, although in a special and somewhat trivial sense. In fact here $|L - v| = |c - c| = 0$ is always less than any fixed positive quantity ϵ. Thus

$$\text{If } f(x) = c, \qquad \lim_{x \to a} f(x) = c. \tag{7}$$

A variable may, but does not necessarily, change. Here a variable always equal to c is a special case of a variable approaching the constant c as a limit. In this exceptional example,

$$\lim_{x \to a} c = c, \tag{8}$$

the variable is always equal to its limit.

***10. Operations on Limits.** We frequently derive new variables from given variables by performing arithmetic operations. In this section we state a number of principles which are helpful in evaluating the limits approached by such derived variables. These principles are plausible but may be proved as indicated in Sec. 19.

As in Sec. 8, let u, v, and w be functions of an independent variable x. And, in the notation of Eq. (5), suppose that

$$\lim_{x \to a} u = A, \qquad \lim_{x \to a} v = B, \qquad \lim_{x \to a} w = C. \tag{9}$$

Then, as a consequence of Eq. (9), in Sec. 19 it will be shown that

$$\lim_{x \to a} (u + v) = A + B, \qquad \lim_{x \to a} (u + v + w) = A + B + C, \tag{10}$$

and similarly for the sum of any finite number of variables.

As shown in Sec. 19, it also follows from Eq. (9) that

$$\lim_{x \to a} (uv) = AB, \qquad \lim_{x \to a} (uvw) = ABC, \tag{11}$$

and similarly for the product of any finite number of variables.

As in Sec. 9, we may consider a constant c as the limit of a dependent variable function always equal to c. In particular, we may take $u = c$, or $u = -1$ in the first Eq. (11) and so conclude that

$$\lim_{x \to a} (cv) = cB, \qquad \lim_{x \to a} (-v) = -B. \tag{12}$$

From Eqs. (10) and (12) we may conclude that

$$\lim_{x \to a} (u - v) = A - B, \qquad \lim_{x \to a} (u - v - w) = A - B - C. \tag{13}$$

and similarly for any finite combination with plus or minus signs.

It also follows from Eqs. (10) and (12) that

$$\begin{aligned} &\lim_{x \to a} (c_1u + c_2v) = c_1A + c_2B, \\ &\lim_{x \to a} (c_1u + c_2v + c_1w) = c_1A + c_2B + c_3C, \end{aligned} \tag{14}$$

and similarly for any finite linear combination with constant coefficients.

A further consequence of Eq. (9) is that

$$\lim_{x \to a} \frac{u}{v} = \frac{A}{B}, \qquad \text{provided } B \neq 0. \tag{15}$$

We may summarize the various consequences of Eq. (9) stated in this section in the following general rule.

Rule for the Calculation of Limits. Suppose that each of a finite number of given variables approaches a limit. Some of these variables may be constants. Let a new variable be calculated by applying a finite combination of additions, subtractions, multiplications, and divisions to the given variables. Then if no zero denominators are encountered, the limit of the new variable may be calculated by applying the same combination of additions, subtractions, multiplications, and divisions to the respective limits of the given variables.

***11. Limits of Functions.** Consider a product with one factor equal to a constant c, and one or more factors each equal to x. We may take c and x as given variables approaching limits as x tends to a with

$$\lim_{x \to a} c = c \qquad \text{and} \qquad \lim_{x \to a} x = a. \tag{16}$$

Then from the product principle of Eq. (11) it follows that

$$\lim_{x \to a} cx^2 = ca^2, \qquad \lim_{x \to a} cx^3 = ca^3, \tag{17}$$

and similarly for any positive integral power.

If we apply the sum principle of Eq. (10) to a number of terms of this type, we find that

$$\lim_{x \to a} (c_3x^3 + c_2x^2 + c_1x + c_0) = c_3a^3 + c_2a^2 + c_1a + c_0. \tag{18}$$

A similar result holds for a polynomial with constant coefficients of any degree in x.

Next take two polynomials as given variables, and apply the quotient principle of Eq. (15). It follows that

$$\lim_{x \to a} \frac{c_2x^2 + c_1x + c_0}{d_2x^2 + d_1x + d_0} = \frac{c_2a^2 + c_1a + c_0}{d_2a^2 + d_1a + d_0}, \tag{19}$$

and similarly for any rational function if we do not encounter a denominator equal to zero.

Irrational algebraic functions involve the operation u^m, with m a fraction. This and the power operation k^v, with k positive, are special cases of the general power operation u^v. When u is positive, the positive real value of u^v may be calculated and defined by the relation

$$\log (u^v) = v \log u. \tag{20}$$

From this and the property of the logarithm described in Sec. 14, it may be proved that if Eq. (9) holds, then

$$\lim_{x \to a} u^v = A^B, \qquad \text{if } A > 0. \tag{21}$$

If $A = 0$, u^v may not have a real value for some values of x near a which make u negative. But if x is restricted to tend to a through values which make u positive or zero and if B is finite, then u^v will approach $0^B = 0$.

We may summarize the various principles discussed in this section in the following general rule:

Rule for Limits of Functions. Consider a function of x which can be calculated from x and a given set of constants by applying a finite combination of algebraic and general power operations. Then if the calculation gives a definite real value for $x = a$ and for all values of x sufficiently near to a, the limit of the function as $x \to a$ is the function of a.

EXERCISE 4

Calculate each of the following limits of polynomials.

1. $\lim_{x \to 2} (x^2 - 1)$.
2. $\lim_{x \to 3} (x^3 + 2x)$.
3. $\lim_{x \to 0} (ax^4 + bx^2 + 7)$.
4. $\lim_{t \to 1} (3t^3 + t^2 - 4)$.
5. $\lim_{z \to b} (z^2 - b^2)$.
6. $\lim_{s \to 2} (bs^2 - 4bs + 4b + 5)$.

Calculate each of the following limits of rational functions.

7. $\lim_{x \to 4} \dfrac{x + 5}{x - 5}$.
8. $\lim_{x \to -1} \dfrac{x^2 - 1}{x^2 + 1}$.
9. $\lim_{t \to 0} \dfrac{at + 10}{bt^3 - 5}$.
10. $\lim_{z \to 1} \dfrac{2x^3 + 3}{3z^2 + 2}$.

Calculate each of the following limits of algebraic functions.

11. $\lim_{x \to 5} \sqrt{x^2 - 9}$.
12. $\lim_{x \to 2} \sqrt[3]{10x + 7}$.
13. $\lim_{t \to 2} \dfrac{10}{\sqrt{3t - 1}}$.
14. $\lim_{s \to 3} \sqrt{\dfrac{s + 5}{s - 1}}$.

Calculate each of the following limits of powers.

15. $\lim\limits_{x \to 2} 7^x$.

16. $\lim\limits_{x \to -1} 10^x$.

17. $\lim\limits_{x \to 9} x^{-\frac{3}{2}}$.

18. $\lim\limits_{x \to -8} x^{\frac{2}{3}}$.

Show that the general rules do not lead to a definite real finite value, and so do not apply to each of the following expressions.

19. $\lim\limits_{x \to 3} \dfrac{x}{x - 3}$.

20. $\lim\limits_{x \to 2} \dfrac{x^2 - 4}{x - 2}$.

21. $\lim\limits_{x \to 2} \sqrt{x - 4}$

22. $\lim\limits_{x \to 0} 10^{1/x}$.

***12. Infinity.** As in Sec. 7, we sometimes consider a succession of values of a variable like 1, 2, 3, $\cdots$. Or the variable may steadily increase from zero on through all positive values. In each of these cases, the variable becomes positively infinite. And in general we say that

The variable v becomes positively infinite if beyond a certain point in its succession of values v becomes and stays greater than any preassigned fixed positive quantity M, however large.

To express this behavior, we write $\lim v = +\infty$.

For a finite limit L, $L - v$ becomes numerically small. Since $+\infty$, read "plus infinity," is not a number, strictly speaking the corresponding expression in the case of an infinite limit has no meaning. Even if we regarded $+\infty$ as subject to some of the arithmetic operations in view of special definitions, for any finite value of v the definition would make $+\infty - v = +\infty$, which does not suggest a variable becoming small. Thus when $\lim v = +\infty$, v does not "approach infinity as a limit" in accordance with the definition of Sec. 7. But it is often convenient to read the relation $\lim v = +\infty$ as "limit v equals plus infinity" instead of making the more precise statement "v becomes positively infinite."

As in Sec. 8, let $y = f(x)$. We may consider the independent variable x for a succession of values such that x becomes positively infinite. We indicate this by writing $x \to +\infty$, read "x tends to plus infinity." If for all such sequences the corresponding values of y approach the constant A as a limit, we write

$$\lim_{x \to +\infty} y = A. \tag{22}$$

In a similar manner we may define the behavior of a variable v which becomes negatively infinite, and introduce the symbol $-\infty$, read "minus infinity." For example,

$$v \to -\infty \text{ and } \lim v = -\infty \qquad \text{if } \lim(-v) = +\infty. \tag{23}$$

We say that v becomes numerically infinite, to mean that $\lim |v| = +\infty$. And we introduce the symbol ∞, read "infinity," in the alternative notations:

$$\lim v = \infty, \qquad v \to \infty \qquad \text{for } \lim |v| = +\infty. \tag{24}$$

We may use any of the expressions involving infinity to describe the behavior of either the dependent variable, or of the independent, in a manner similar to that described in Sec. 8. For example,

$$\lim_{x \to \infty} \frac{1}{x} = 0, \quad \lim_{x \to 0} \frac{1}{x} = \infty, \quad \lim_{x \to 0} \frac{1}{x^2} = +\infty, \quad \lim_{x \to \infty} x^2 = +\infty, \quad \lim_{x \to +\infty} \frac{x}{x + 1} = 1.$$

The last limit is that discussed in Sec. 8 written in the new notation of this section.

More generally, we may show that

$$\lim_{x\to\infty} \frac{ax+b}{cx+d} = \frac{a}{c}, \qquad \text{if } c \neq 0. \tag{25}$$

To prove this, put $h = 1/x$. Then, when $x \to \infty$, $h \to 0$. But

$$\frac{ax+b}{cx+d} = \frac{a+b(1/x)}{c+d(1/x)} = \frac{a+bh}{c+dh}. \tag{26}$$

By the polynomial principle of Sec. 11, as $h \to 0$, the numerator $(a + bh) \to a$. And the denominator $(c + dh) \to c$. Since $c \neq 0$, we may apply the quotient principle of Eq. (15) to deduce that the limit of the last fraction in Eq. (26) is a/c.

A similar argument shows that for any two polynomials of the same degree in x, the limit of the quotient as $x \to \infty$ is equal to the quotient of the coefficients of the two terms of highest degree. For example, we have

$$\lim_{x\to\infty} \frac{c_2x^2 + c_1x + c_0}{d_2x^2 + d_1x + d_0} = \frac{c_2}{d_2}, \qquad \text{if } d_2 \neq 0. \tag{27}$$

For any two polynomials in x of different degrees, the behavior of the quotient as $x \to \infty$ is similar to the behavior of the quotient of the two terms of highest degree. For example,

$$\lim_{x\to\infty} \frac{6x+5}{3x^2+4x-2} = \lim_{x\to\infty} \frac{6x}{3x^2} = \lim_{x\to\infty} \frac{2}{x} = 0. \tag{28}$$

To prove this, note that if $h = 1/x$, $xh = 1$ and

$$\frac{6x+5}{3x^2+4x-2} = \frac{6x(1+5h/6)}{3x^2(1+4h/3-2h^2/3)}.$$

But the limit of each factor in parentheses is 1 as $h \to 0$ by the polynomial principle, and $h \to 0$ when $x \to \infty$.

The result stated may be proved in general by the same method. As a second application of the result, we have

$$\lim_{x\to\infty} \frac{6x^2-5x+3}{3x-1} = \lim_{x\to\infty} \frac{6x^2}{3x} = \lim_{x\to\infty} 2x = \infty. \tag{29}$$

A more accurate statement of the behavior in this case is

$$\lim_{x\to+\infty} \frac{6x^2-5x+3}{3x-1} = +\infty, \qquad \lim_{x\to-\infty} \frac{6x^2-5x+3}{3x-1} = -\infty. \tag{30}$$

***13. Operations and Infinity.** We have just seen that, in studying the behavior of products or quotients of polynomial *factors*, as the independent variable $x \to \infty$, we may replace each factor by its term of highest degree. We shall refer to this as the principle of the leading term for factors.

It is not so easy to predict the behavior of sums or differences. For example, consider as given limiting relations

$$\lim_{x\to\infty} (x^3+1) = \infty \qquad \text{and} \qquad \lim_{x\to\infty} (-x^3-3) = \infty.$$

In this case for the sum of the variables

$$\lim_{x\to\infty} (x^3+1) + (-x^3-3) = \lim_{x\to\infty} (-2) = -2. \tag{31}$$

For such sums or differences we must study each situation in detail, as the principles mentioned in Sec. 10 for finite limits as defined in Sec. 7 do not hold for relations involving the symbols ∞, $+\infty$, $-\infty$.

EXERCISE 5

Use the principle of the leading term to calculate each of the following limits.

1. $\lim_{x \to \infty} \frac{5x^3 - 2}{x + x^3}$.
2. $\lim_{x \to \infty} \frac{12x^2 - x}{3 + 4x^3}$.
3. $\lim_{x \to \infty} \frac{Ax + B}{C - x}$.
4. $\lim_{t \to \infty} \frac{t^2 - 1}{t^4 - 1}$.

Use the principle of the leading term to decide whether the behavior of each of the following expressions is represented by the symbol ∞, $+\infty$, or $-\infty$.

5. $\lim_{x \to \infty} \frac{5x^3 - 2}{4 - x}$.
6. $\lim_{x \to \infty} \frac{12x^2 + 4}{6x - 7}$.
7. $\lim_{x \to 0} \frac{1}{x^2}$.
8. $\lim_{x \to 0} \frac{-3}{x}$.
9. $\lim_{t \to \infty} t^4$.
10. $\lim_{x \to -\infty} \frac{-3x^2}{x + 3}$.
11. $\lim_{x \to -\infty} \frac{2x^2 + 5}{6x - 7}$.
12. $\lim_{x \to +\infty} \frac{15x^2 - 2}{2 + 5x}$.
13. $\lim_{x \to 2} \frac{5}{(x - 2)^2}$.
14. $\lim_{s \to 3} \frac{5}{3 - s}$.

If $\log u \to +\infty$, then $u \to +\infty$. Use this fact to prove that

15. $\lim_{x \to +\infty} 10^x = +\infty$.
16. $\lim_{x \to +\infty} (\frac{3}{2})^x = +\infty$.
17. $\lim_{x \to +\infty} 2^{x/100} = +\infty$.
18. $\lim_{x \to +\infty} a^x = +\infty$ if $a > 1$.

If $\log u \to -\infty$, then $u \to 0$. Use this fact to prove that

19. $\lim_{x \to -\infty} 10^x = 0$.
20. $\lim_{x \to +\infty} 10^{-x} = 0$.
21. $\lim_{x \to +\infty} (\frac{3}{5})^x = 0$.
22. $\lim_{x \to +\infty} b^x = 0$ if $0 < b < 1$.

Use Probs. 18 and 22 to show that

23. $\lim_{x \to +\infty} \frac{2 + 3^x}{3 + 2^x} = \lim_{x \to +\infty} \left(\frac{3}{2}\right)^x \frac{1 + 2(\frac{1}{3})^x}{1 + 3(\frac{1}{2})^x} = +\infty$.
24. $\lim_{x \to +\infty} \frac{7 + 5^x}{3 + 5^x} = \lim_{x \to +\infty} \frac{1 + 7(\frac{1}{5})^x}{1 + 3(\frac{1}{5})^x} = 1$.
25. $\lim_{x \to +\infty} \frac{6 + 3^x}{4 + 5^x} = \lim_{x \to +\infty} \left(\frac{3}{5}\right)^x \frac{1 + 6(\frac{1}{3})^x}{1 + 4(\frac{1}{5})^x} = 0$.

***14. Continuity.** Consider the particular function $y = x^2$ for $x = 3$, and for values of x near $x = 3$. When $x = 3$, $y = 9$. And when x is near 3, either a little less or a little greater, the value of y will be near 9. To describe this property, we say that the function $y = x^2$ is *continuous* for $x = 3$, or at $x = 3$.

Formulated in general terms, the definition is as follows:

A function $f(x)$ is continuous at $x = a$ if the function is defined for $x = a$, and if, as x tends to a, the value of $f(x)$ tends to the value assigned to the function for $x = a$.

In the notation of Sec. 8, these conditions may be written

I. A definite value is assigned to $f(a)$.

II. $\lim_{x \to a} f(x) = f(a)$.

Thus $f(x)$ is continuous at $x = a$ if both of the conditions I and II hold.

If the function $f(x)$ fails to satisfy either of the conditions I or II, the function $f(x)$ is said to be *discontinuous* at $x = a$.

Let $f(x)$ be any polynomial. Then for every value of a, $f(a)$ is defined so that condition I holds. And from Sec. 11, $f(x) \to f(a)$ as $x \to a$. Thus condition II holds. It follows that every polynomial is continuous for all values of x.

Let $f(x)$ be the quotient of two polynomials. The behavior of such a function near a value which makes the denominator zero will be discussed in the next section. For any value of x which makes the denominator different from zero, the quotient of polynomials is continuous.

Whenever the logarithmic function $\log x$ has a definite real value for $x = a$, and for all values of x near a, the function is continuous at a. This is also true of the power function b^x, the trigonometric functions, and the inverse trigonometric functions. In fact the possibility of obtaining values of these functions from a table by interpolation depends in large measure on the continuity of the functions.

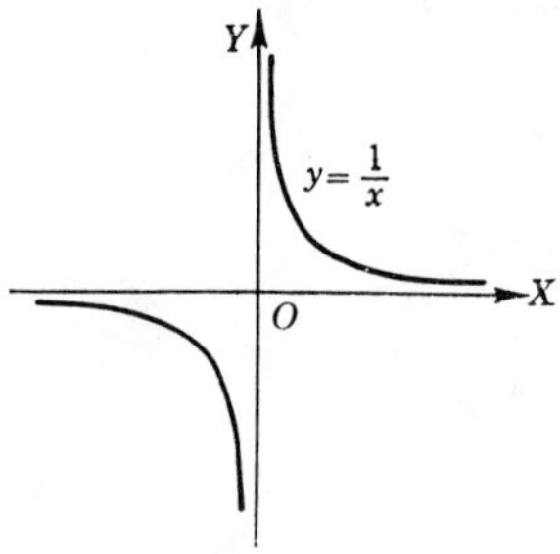

FIG. 9.

***15. Rational Functions.** Let $f(x)$ be a rational function, or quotient of two polynomials. Then $f(x)$ is continuous for any value of x which does not make the denominator zero.

For example, consider the quotient of 1 by x, or $1/x$. By the principle just stated, this is continuous for any value of x not equal to zero.

As $x \to 0$, $1/x \to \infty$. Since this is not a finite limit, the function $1/x$ is discontinuous at $x = 0$. Its graph is shown in Fig. 9. For $x = 0$, no value of $y = 1/x$ can be plotted. In fact, if $f(x) = 1/x$, $f(0)$ is not defined since 1 divided by 0 is not defined in arithmetic. In a table of values of $1/x$, it may be convenient to enter ∞ opposite 0. And one often writes $1/0 = \infty$ or says "$1/x$ is infinite at $x = 0$." But such notations and expressions are merely brief ways of indicating that no finite value is assigned to $1/x$ at $x = 0$, and that $1/x$ becomes numerically infinite as x tends to zero.

We give two additional illustrations. First, consider $y = 1/x^2$, Fig. 10. Here $y \to +\infty$ as $x \to 0$ and y is discontinuous at $x = 0$.

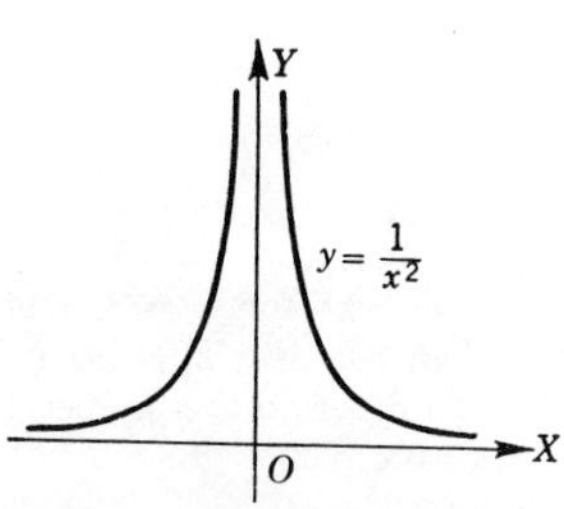

FIG. 10.

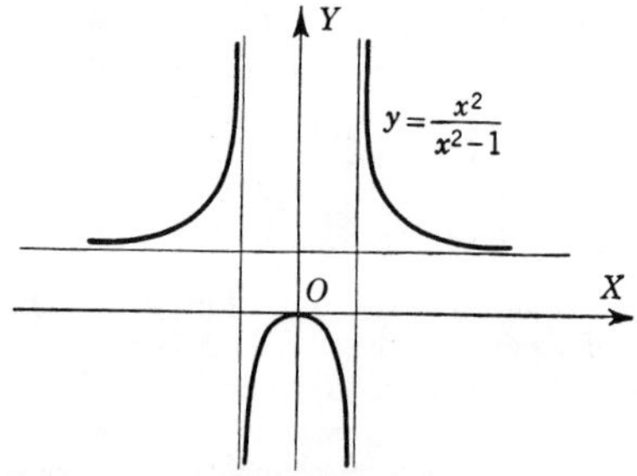

FIG. 11.

Again, let $y = x^2/(x^2 - 1)$, Fig. 11. Here the denominator is zero for $x = 1$ and $x = -1$, and the function is discontinuous at each of these points.

To sketch the graph near $x = 1$, we may put $x = 1 + h$, and consider values of h near zero. We find

$$y = \frac{x^2}{x^2 - 1} = \frac{(1 + h)^2}{(1 + h)^2 - 1} = \frac{1 + 2h + h^2}{2h + h^2} = \frac{1}{2h}\frac{1 + 2h + h^2}{1 + h/2}.$$

By Sec. 10, the last fraction tends to 1 when h tends to 0. Hence y behaves like $1/2h$ when h tends to 0, or $x = 1 + h$ tends to 1. This shows that $y \to +\infty$ when h tends to 0 through positive values, or $h \to 0+$, as we shall write in such a case. Thus

$$\lim_{x\to 1+} y = \lim_{h\to 0+} \frac{1}{2h} = +\infty.$$

For h approaching zero through negative values, we may write

$$\lim_{x\to 1-} y = \lim_{h\to 0-} \frac{1}{2h} = -\infty.$$

A simpler procedure is to write

$$y = \frac{x^2}{x^2 - 1} = \frac{x^2}{x+1}\frac{1}{x-1}.$$

Since the first factor, $x^2/(x+1)$, approaches $\frac{1}{2}$ when $x \to 1$, for x near 1, y behaves like $\frac{1}{2}\frac{1}{x-1}$. This shows that $y \to +\infty$ when $x \to 1+$, and $y \to -\infty$ when $x \to 1-$.

To study y near $x = -1$, or $k = x - (-1) = x + 1 = 0$, write

$$y = \frac{x^2}{x^2 - 1} = \frac{x^2}{x-1}\frac{1}{x+1}.$$

Since the first factor, $x^2/(x-1)$, approaches $-\frac{1}{2}$ when $x \to -1$, for x near -1, y behaves like $-\frac{1}{2}\frac{1}{x+1}$. Or, with $k = x + 1$, for k near 0, y behaves like $-\frac{1}{2}k$. This shows that as $k \to 0+$ or $x \to -1+$, $y \to -\infty$, while as $k \to 0-$, or $x \to -1-$, $y \to +\infty$.

In Fig. 11 we have drawn the horizontal line $y = 1$ and indicated that as x becomes positively or negatively large, the curve approaches this line from above. We deduced this by the following argument. We first use the principle of the leading term of Sec. 13. This shows that

$$\lim_{x\to\infty} y = \lim_{x\to\infty} \frac{x^2}{x^2 - 1} = \lim_{x\to\infty} \frac{x^2}{x^2} = 1.$$

To find out whether y is above or below 1, we calculate

$$y - 1 = \frac{x^2}{x^2 - 1} - 1 = \frac{1}{x^2 - 1}.$$

For x numerically large, this behaves like $1/x^2$ and so is near zero, but positive. Hence for large x the curve is above the line $y = 1$.

Whenever a straight line is so related to a curve that as the two are indefinitely prolonged the distance between them tends to zero, the straight line is called an *asymptote* of the curve. More precisely, let P be a point on a curve at distance OP from the origin. And let PA be the perpendicular distance from P to a fixed straight line L. Then if $PA \to 0$ as $OP \to \infty$, L is an asymptote. Thus in Fig. 11, the lines $x = -1$, $x = 1$, and $y = 1$ are all asymptotes of the curve which is the graph of $y = x^2/(x^2 - 1)$.

Consider the graph of any rational function or quotient of two polynomials. If $\lim y = b$ as $x \to \infty$, the graph will have $y = b$ as a horizontal asymptote. And if any value of a is such that $x = a$ makes the denominator equal to zero, but makes the numerator different from zero, then $\lim y = \infty$ as $x \to a$, and the graph will have $x = a$ as a vertical asymptote.

EXERCISE 6

For each of the following functions verify the statements made and sketch the graph.

1. $y = -4/x$. As $x \to 0+$, $y \to -\infty$; as $x \to 0-$, $y \to +\infty$. As $x \to +\infty$, $y \to 0-$; as $x \to -\infty$, $y \to 0+$. This y is discontinuous at $x = 0$.
2. $y = -2/x^2$. As $x \to 0$, $y \to -\infty$; as $x \to \infty$, $y \to 0-$. This y is discontinuous at $x = 0$.
3. $y = \frac{2-x}{x-3}$. As $x \to 3+$, $y \to -\infty$; as $x \to 3-$, $y \to +\infty$. As $x \to +\infty$, $y \to -1-$; as $x \to -\infty$, $y \to 1+$. This y is discontinuous at $x = 3$.
4. $y = \frac{x^3+1}{x}$. As $x \to 0+$, $y \to +\infty$; as $x \to 0-$, $y \to -\infty$. As $x \to \infty$, $y \to \infty$, and for large x the curve is near to $y = x^2$. This y is discontinuous at $x = 0$.
5. $y = \frac{4}{x^2-4}$. As $x \to 2+$, $y \to +\infty$; as $x \to 2-$, $y \to -\infty$. As $x \to -2+$, $y \to -\infty$; as $x \to -2-$, $y \to +\infty$. As $x \to \infty$, $y \to 0+$. This y is discontinuous at $x = 2$ and at $x = -2$.
6. $y = \frac{2}{x^2+1}$. As $x \to \infty$, $y \to 0+$. This y is always continuous.
7. $y = \frac{(x+1)^2}{x-1}$. As $x \to 1+$, $y \to +\infty$; as $x \to 1-$, $y \to -\infty$. Since $y = x + 3 + \frac{4}{x-1}$, for large x the curve is near the line $y = x + 3$, above for $x \to +\infty$ and below for $x \to -\infty$. Here $y = x + 3$ is an oblique asymptote. This y is discontinuous at $x = 1$.
8. $y = \frac{x^2+1}{x}$. As $x \to 0+$, $y \to +\infty$; as $x \to 0-$, $y \to -\infty$. Since $y = x + \frac{1}{x}$, for large x the curve is near the line $y = x$, above for $x \to +\infty$ and below for $x \to -\infty$. Thus $y = x$ is an oblique asymptote. This y is discontinuous at $x = 0$.
9. $y = \frac{x}{x^2+1}$. As $x \to +\infty$, $y \to 0+$; as $x \to -\infty$, $y \to 0-$. This y is always continuous.
10. $y = \frac{x^2+1}{x^2-1}$. As $x \to 1+$, $y \to +\infty$; as $x \to 1-$, $y \to -\infty$. As $x \to -1+$, $y \to -\infty$; as $x \to -1-$, $y \to +\infty$. As $x \to \infty$, $y \to 1+$. This y is discontinuous at $x = 1$ and at $x = -1$.
11. $y = \frac{2x}{x^2-9}$. As $x \to 3+$, $y \to +\infty$; as $x \to 3-$, $y \to -\infty$. As $x \to -3+$, $y \to -\infty$; as $x \to -3-$, $y \to +\infty$. As $x \to +\infty$, $y \to 0+$; as $x \to -\infty$, $y \to 0-$. This y is discontinuous at $x = 3$ and at $x = -3$.
12. $y = \frac{2x^3+1}{x^2}$. As $x \to 0$, $y \to +\infty$. Since $y = 2x + \frac{1}{x^2}$, for large x the curve is near the line $y = 2x$, but above as $x \to \infty$. Thus $y = 2x$ is an oblique asymptote. This y is discontinuous at $x = 0$.
13. $y = \frac{x^3}{1-x^2}$. As $x \to 1+$, $y \to -\infty$; as $x \to 1-$, $y \to +\infty$. As $x \to -1+$, $y \to -\infty$; as $x \to -1-$, $y \to +\infty$. Since $y = -x - \frac{x}{x^2-1}$, for large x the curve is near the line $y = -x$, below for $x \to +\infty$, and above for $x \to -\infty$. This y is discontinuous at $x = 1$ and at $x = -1$.

14. $y = \dfrac{x}{(x-2)(x-3)}$. As $x \to 2+$, $y \to -\infty$; as $x \to 2-$, $y \to +\infty$. As $x \to 3+$, $y \to +\infty$; as $x \to 3-$, $y \to -\infty$. As $x \to +\infty$, $y \to 0+$; as $x \to -\infty$, $y \to 0-$. This y is discontinuous at $x = 2$ and at $x = 3$.

***16. Discontinuous Functions.** Consider $y = f(x)$ where $f(x)$ is a rational function which is the quotient of two polynomials. If these polynomials have no common factors and hence no common roots, the fraction is said to be reduced to its lowest terms. In this case every root a of the denominator will make $y \to \infty$ as $x \to a$. Thus the fraction will be discontinuous for all the roots of the denominator, and continuous for all other values.

For a fraction not reduced to its lowest terms, there is another possibility. For example, let us consider $y = (x^2 - 1)/(x - 1)$ for the value $x = 1$. When $x = 1$, the denominator is zero. But the numerator is also zero for $x = 1$. Since zero divided by zero is not defined in arithmetic, the function is not defined at $x = 0$, so that the first condition for continuity, I of Sec. 14, is not satisfied. But for any value near, but not equal to, 1 we may write

$$y = \frac{x^2 - 1}{x - 1} = \frac{(x-1)(x+1)}{x-1} = x + 1.$$

Hence if we let x tend to 1, we have

$$\lim_{x \to 1} y = \lim_{x \to 1} (x + 1) = 2.$$

If we are willing to accept the original definition for all values of x different from 1, but at $x = 1$, where the original expression was undefined, make the special definition $y = 2$, then the new modified function is continuous for all x.

Consider a given function $f(x)$ for values near $x = a$. If as $x \to a$, $f(x) \to \infty$, then $f(x)$ is discontinuous at a. This is true even if the function tends to infinity with a fixed algebraic sign. Again, if $f(x)$ fails to approach a single limit when $x \to a$, then $f(x)$ is discontinuous at $x = a$. When $f(x)$ approaches a limit A as $x \to a$, but is undefined at $x = a$, we may take one of two points of view. We may either take the definition literally and call $f(x)$ discontinuous at a because condition I does not hold, or proceed as follows. Supplement the original definition by defining $f(a)$ as equal to A. In such a case we obtain a new continuous function which we often accept as essentially the same as the given function. The second point of view is the one which permits us to reduce any rational fraction to its lowest terms by canceling common factors and to accept the reduced fraction as essentially equivalent to the original function.

EXERCISE 7

Each of the following expressions has the form

$$\lim_{x \to a} F(x) = \lim_{x \to a} f(x) = f(a).$$

For each problem verify that $F(a)$ takes the form zero divided by zero. Hence $F(a)$ is undefined. But when x is near but not equal to a, show that $F(x) = f(x)$ and deduce from this that $F(x)$ approaches $f(a)$ as a limit when x tends to a.

1. $\displaystyle\lim_{x \to 2} \frac{x^2 - 4}{x - 2} = \lim_{x \to 2} (x + 2) = 4.$
2. $\displaystyle\lim_{x \to 3} \frac{x - 3}{x^2 - 9} = \lim_{x \to 3} \frac{1}{x + 3} = \frac{1}{6}.$

3. $\lim_{x\to 0} \dfrac{5x^2 + 2x}{x} = \lim_{x\to 0} (5x + 2) = 2.$

4. $\lim_{x\to -3} \dfrac{x^2 + 3x}{x + 3} = \lim_{x\to -3} x = -3.$

5. $\lim_{x\to 6} \dfrac{2x - 12}{x - 6} = \lim_{x\to 6} 2 = 2.$

6. $\lim_{x\to a} \dfrac{x^4 - a^4}{x^2 - a^2} = \lim_{x\to a} (x^2 + a^2) = 2a^2.$

7. $\lim_{x\to 2} \dfrac{x^2 - 3x + 2}{x^2 - 4} = \lim_{x\to 2} \dfrac{x - 1}{x + 2} = \dfrac{1}{4}.$

8. $\lim_{x\to b} \dfrac{x^3 - b^3}{x - b} = \lim_{x\to b} (x^2 + bx + b^2) = 3b^2.$

9. $\lim_{x\to 0} \dfrac{(b + x)^2 - b^2}{x} = \lim_{x\to 0} (2b + x) = 2b.$

10. $\lim_{x\to 0} \dfrac{\sqrt{b + x} - \sqrt{b}}{x} = \lim_{x\to 0} \dfrac{1}{\sqrt{b + x} + \sqrt{b}} = \dfrac{1}{2\sqrt{b}}.$

11. $\lim_{x\to 0} \dfrac{x^3 + 4x^2}{x^5 - 2x^2} = \lim_{x\to 0} \dfrac{x + 4}{x^3 - 2} = -2.$

12. $\lim_{x\to 0} \dfrac{\sin 2x}{\sin x} = \lim_{x\to 0} 2 \cos x = 2.$

Apply a procedure like that indicated for Probs. 1 to 12 to verify each of the following statements.

13. $\lim_{t\to 5} \dfrac{t - 5}{t^2 - 25} = \lim_{t\to 5} \dfrac{1}{t + 5} = \dfrac{1}{10}.$

14. $\lim_{s\to 0} \dfrac{s^5 + 9s^2}{s^4 - 3s^2} = \lim_{s\to 0} \dfrac{s^3 + 9}{s^2 - 3} = -3.$

15. $\lim_{z\to 1} \dfrac{z^2 - 2z + 1}{z^2 - 3z + 2} = \lim_{z\to 1} \dfrac{z - 1}{z - 2} = 0.$

16. $\lim_{h\to 0} \dfrac{(3 + h)^2 - 9}{h} = \lim_{h\to 0} (6 + h) = 6.$

17. $\lim_{h\to 0} \dfrac{(2 + h)^3 - 8}{h} = \lim_{h\to 0} (12 + 6h + h^2) = 12.$

18. Use the results of Probs. 19 and 20 of Exercise 5 to show that

$$\lim_{x\to 0+} \frac{1 - 10^{1/x}}{1 + 10^{1/x}} = \lim_{x\to 0+} \frac{10^{-1/x} - 1}{10^{-1/x} + 1} = -1,$$

and that

$$\lim_{x\to 0-} \frac{1 - 10^{1/x}}{1 + 10^{1/x}} = 1.$$

These results show that $y = f(x) = (1 - 10^{1/x})/(1 + 10^{1/x})$ is discontinuous at $x = 0$. This is an example of a function with a finite jump.

***17. Infinitesimals.** An *infinitesimal* is a variable which assumes numerically small values and which at a certain stage of the discussion approaches zero as a limit. For example, the variables h and k introduced in Sec. 15 are infinitesimals. In accordance with the definition of Sec. 7, specialized for the case $L = 0$, the definition of an infinitesimal is

The variable h is an infinitesimal if beyond a certain point in its succession of values, the numerical value of h, $|h|$, becomes and stays less than any preassigned fixed positive quantity, however small.

In the notation of Sec. 7, $\lim h = 0$.

By comparison of the statement made in Sec. 7 with the definition of an infinitesimal just given, we see that if $\lim v = L$, then $h = v - L$ is an infinitesimal, since $|h| = |v - L| = |L - v|$. That is, *the difference between a variable and the limit which it approaches is an infinitesimal.*

Conversely, let h be an infinitesimal and let L be any constant. Form the variable $v = L + h$. Then $L - v = -h$, so that $|L - v| = |h|$. And, since h approaches 0, $\lim v = L$. That is, *if the difference between a variable and a constant is an infinitesimal, then the variable approaches the constant as a limit.*

***18. Theorems on Infinitesimals.** In this and the following section, each of the variables is a function of the same independent variable x, and the behavior of each variable is being studied as $x \to a$, where a is a fixed value. Or, for all the variables, in place of $x \to a$, we may have $x \to +\infty$.

As in Sec. 7, the constant ϵ is a preassigned fixed number. This number ϵ is positive, so that it is greater than zero, but it may be as small a fixed positive number as we please.

To lead up to the theorems on limits, we first prove some basic theorems on combinations of infinitesimals.

Theorem. Let p and q each be constants, or variables each of which is always numerically less than some fixed quantity M, and let h and k be infinitesimals. Then $ph + qk$ is also an infinitesimal.

For, the numerical value of $ph + qk$ will become and stay less than ϵ when the numerical value of each of the pair h and k becomes and stays less than $\epsilon/2M$.

Corollary. A similar result holds for any finite number of infinitesimals, and coefficients each a constant or a variable remaining numerically less than some fixed quantity M.

As a first example of the theorem, let $p = 1$, $q = 1$. Then it follows that any sum of two infinitesimals

$$h + k \text{ is an infinitesimal.} \tag{32}$$

Again, let $p = 1$, $q = -1$. Then it follows that any difference of infinitesimals

$$h - k \text{ is an infinitesimal.} \tag{33}$$

We give two additional applications which will be useful in the proofs of the next section. Consider the corollary for three infinitesimals, the first of which is h, while the second and third are each equal to k. As the three coefficients take the constant B, the constant A, and the infinitesimal h. Since h approaches 0, ultimately we shall have $|h| < 1$. Restrict attention to this stage and take M the largest of $|A|$, $|B|$, and 1. Then it follows from the corollary that

$$Bh + Ak + hk \text{ is an infinitesimal.} \tag{34}$$

Next consider the corollary for one infinitesimal k, or the theorem with $p = 0$. As the coefficient take

$$q = \frac{-1}{B(B + k)}, \qquad \text{where } B \neq 0. \tag{35}$$

Since k approaches 0, ultimately we shall have $|k| < |B|/2$. This implies that

$$|B + k| > \frac{|B|}{2} \qquad \text{and} \qquad \left|\frac{-1}{B(B + k)}\right| < \frac{2}{|B|^2},$$

which we may take as M. Hence it follows that

$$\text{If } B \neq 0, \qquad \frac{-k}{B(B+k)} \text{ is an infinitesimal.} \tag{36}$$

***19. Theorems on Limits.** We are now in a position to prove the theorems stated in Sec. 10. As in that section, let

$$\lim_{x\to a} u = A, \qquad \lim_{x\to a} v = B, \qquad \lim_{x\to a} w = C. \tag{37}$$

Define h and k by the equations

$$h = u - A \qquad \text{and} \qquad k = v - B. \tag{38}$$

Then by the first theorem of Sec. 17, h and k each approach 0 as x tends to a. Hence h and k are infinitesimals.

Let us prove that

$$\lim_{x\to a} (u + v) = A + B. \tag{39}$$

We consider the difference

$$(u + v) - (A + B) = h + k, \tag{40}$$

by Eq. (38). But by Eq. (32) this is an infinitesimal. Hence by the converse theorem of Sec. 17, Eq. (39) is proved.

Let us next prove that

$$\lim_{x\to a} (u - v) = A - B. \tag{41}$$

We consider the difference

$$(u - v) - (A - B) = h - k, \tag{42}$$

by Eq. (38). But by Eq. (33) this is an infinitesimal. Hence by the converse theorem of Sec. 17, Eq. (41) is proved.

Theorems of a similar nature for any combination of a finite number of variables, with plus or minus signs, follow from the results for two variables given in Eqs. (39) and (41). For example,

$$\lim_{x\to a} (u + v + w) = \lim_{x\to a} (u + v) + \lim_{x\to a} w = A + B + C. \tag{43}$$

Let us next prove the result for a product

$$\lim_{x\to a} uv = AB. \tag{44}$$

From Eq. (38) we have $u = A + h$, $v = B + k$. Hence the product

$$uv = (A + h)(B + k) = AB + Bh + Ak + hk.$$

And the difference

$$uv - AB = Bh + Ak + hk. \tag{45}$$

But by Eq. (34) this is an infinitesimal. Hence by the converse theorem of Sec. 17, Eq. (44) is proved.

Theorems of a similar nature for the product of any finite number of variables follow from Eq. (44). For example,

$$\lim_{x\to a} (uvw) = \lim_{x\to a} (uv) \lim_{x\to a} w = ABC. \tag{46}$$

We next prove that when $B \neq 0$,

$$\lim_{x \to a} \frac{1}{v} = \frac{1}{B}. \tag{47}$$

Here we must consider the difference $(1/v) - (1/B)$, or since $v = B + k$,

$$\frac{1}{B + k} - \frac{1}{B} = \frac{-k}{B(B + k)}. \tag{48}$$

But this is an infinitesimal by Eq. (36). Hence Eq. (47) is proved.

Finally, for the quotient u/v when $B \neq 0$, we combine Eqs. (44) and (48) to deduce that

$$\lim_{x \to a} \left(\frac{u}{v}\right) = \lim_{x \to a} u \left(\frac{1}{v}\right) = \lim_{x \to a} u \lim_{x \to a} \frac{1}{v} = A \left(\frac{1}{B}\right) = \frac{A}{B}. \tag{49}$$

This proves the quotient principle.

Equations (39), (43), (44), (46), and (49) are the basic theorems stated in Sec. 10. Thus we have supplied the proofs omitted in that section.

The teacher or advanced student who wishes a complete account of the number system, the theory of limits, and the continuity properties of the basic elementary functions mentioned in Sec. 14 is referred to the first three chapters of the author's "A Treatise on Advanced Calculus."

CHAPTER 2

RATES AND DERIVATIVES

For the rectilinear motion of a particle, the velocity as well as the acceleration at a particular instant of time can each be precisely defined in terms of a limit of the type we discussed in Chap. 1. The same is true of the instantaneous, or true, rate of change in other physical situations. Likewise in geometry the slope of a curve at a point, which is useful in drawing the straight line tangent to the curve at the point, can be defined in terms of a limit. We explain these definitions and apply them to evaluate the quantities in a few simple examples.

We then consider a general functional relation, $y = f(x)$, and define the rate of change of y with respect to x, or the derivative of y with respect to x. The derivative is one of the two most important notions in calculus. All the physical and geometric examples of rates discussed earlier are special cases of derivatives. We introduce some special notation which is helpful in discussing the derivative. Finally we describe a systematic method of finding a derivative, or of differentiating any given function by a procedure based directly on the definition.

20. Velocity. Consider the motion of a particle along a straight path. As in Sec. 1, construct a number scale on the path with origin at O. And let s be the coordinate of P, the position of the particle at any time t. Thus if P is to the right of O, as in Fig. 12, OP is the distance from O to P. And t is the time elapsed since some fixed instant taken as $t = 0$. Then each t determines a value of s. Hence in the sense defined in Sec. 5, s is a function of t, or

FIG. 12.

$$s = f(t). \tag{1}$$

We wish to define the rate of change of s with respect to t or the instantaneous velocity at a particular fixed time t_1. We first define an approximation by the following procedure. Let t_2 be any time other than t_1. Thus $t_2 - t_1 \neq 0$. And we are particularly concerned with a time t_2 for which $t_2 - t_1$ is small. From Eq. (1), if s_1 is the coordinate of the particle at time t_1,

$$s_1 = f(t_1). \tag{2}$$

Similarly, if s_2 is the coordinate of the particle at time t_2,

$$s_2 = f(t_2). \tag{3}$$

Then the average rate of change, or the *average velocity* for the time interval between t_1 and t_2, is defined as the directed distance traveled divided by the elapsed time. Thus

$$\text{Average velocity} = \frac{s_2 - s_1}{t_2 - t_1} = \frac{f(t_2) - f(t_1)}{t_2 - t_1}. \tag{4}$$

For t_2 nearly equal to t_1, or $t_2 - t_1$ small, this gives an approximation to the velocity at time t_1. And the approximation improves for each further choice of $t_2 \neq t_1$, but nearer to t_1. Hence we keep t_1 fixed but think of t_2 as a variable tending to t_1 in the sense defined in Sec. 8. Under these conditions, the average velocity for the interval between t_1 and t_2 usually tends to a limit, and this limit v_1 is defined as the *instantaneous velocity* of the moving particle at time t_1, or simply the *velocity* at t_1. In symbols

$$v_1 = \lim_{t_2 \to t_1} \frac{s_2 - s_1}{t_2 - t_1} = \lim_{t_2 \to t_1} \frac{f(t_2) - f(t_1)}{t_2 - t_1}. \tag{5}$$

The last fraction in Eq. (5), like the expressions considered in Sec. 16, takes the form 0 divided by 0 when $t_2 = t_1$. But in many simple cases the limit can be found by the method of Sec. 16.

For example, let $s = t^3$ and $t_1 = 3$. Then $s_2 = t_2^3$ and $s_1 = 3^3$. Hence the average velocity of Eq. (4) is

$$\frac{s_2 - s_1}{t_2 - t_1} = \frac{t_2^3 - 3^3}{t_2 - 3} = t_2^2 + 3t_2 + 3^2, \tag{6}$$

since $t_2 \neq 3$. When $t_2 \to 3$, by the polynomial principle of Sec. 10, the last expression of Eq. (6) approaches $3^2 + 3^2 + 3^2 = 27$, so that $v = 27$ for $t = 3$.

For the same motion $s = t^3$ and any time t_1, $s_1 = t_1^3$. Hence the average velocity of Eq. (4) is

$$\frac{s_2 - s_1}{t_2 - t_1} = \frac{t_2^3 - t_1^3}{t_2 - t_1} = t_2^2 + t_2 t_1 + t_1^2, \tag{7}$$

since $t_2 \neq t_1$. When $t_2 \to t_1$, the last expression in Eq. (7), a polynomial in t_2, approaches $t_1^2 + t_1^2 + t_1^2 = 3t_1^2$, so that $v = 3t_1^2$ for $t = t_1$. Since t_1 may have any value, we may replace it by t and say that

$$\text{If } s = t^3, \qquad \text{the velocity at time } t \text{ is } v = 3t^2. \tag{8}$$

The velocity is also called the time rate of change of the distance. If s is in feet and t is in seconds, then v will be in feet per second.

21. Acceleration. For the general rectilinear motion of Eq. (1), the velocity at time t will be a new function of t, $v = F(t)$. The time rate of change of v is called the *acceleration*. To find the acceleration at time t_1, we select a time t_2 near to t_1 and note that

$$v_1 = F(t_1), \qquad v_2 = F(t_2).$$

Then the *average acceleration* for the time interval between t_1 and t_2 is defined as the algebraic increase in velocity divided by the elapsed time. Thus,

$$\text{Average acceleration} = \frac{v_2 - v_1}{t_2 - t_1} = \frac{F(t_2) - F(t_1)}{t_2 - t_1}. \tag{9}$$

This gives an approximation to the acceleration at time t_1. And the approximation improves for each further choice of $t_2 \neq t_1$, but nearer to t_1. Hence we keep t_1 fixed and let t_2 tend to t_1. The average acceleration for the interval between t_1 and t_2 usually tends to a limit as $t_2 \to t_1$. This limit a_1 is defined as the *instantaneous acceleration* of the moving particle at time t_1, or simply the *acceleration* at t_1. In symbols

$$a_1 = \lim_{t_2 \to t_1} \frac{v_2 - v_1}{t_2 - t_1} = \lim_{t_2 \to t_1} \frac{F(t_2) - F(t_1)}{t_2 - t_1}. \tag{10}$$

In particular, let us apply this to the motion with $s = t^3$. In Eq. (8) we found that for this motion $v = 3t^2$. Hence $v_2 = 3t_2^2$ and $v_1 = 3t_1^2$. And the average acceleration of Eq. (9) is

$$\frac{v_2 - v_1}{t_2 - t_1} = \frac{3(t_2^2 - t_1^2)}{t_2 - t_1} = 3(t_2 + t_1), \tag{11}$$

since $t_2 \neq t_1$. When $t_2 \to t_1$, the last expression in Eq. (11), a polynomial in t_2, approaches $3(t_1 + t_1) = 6t_1$, so that $a_1 = 6t_1$ for $t = t_1$. Since t_1 may have any value, we may replace it by t and say that

$$\text{If } v = 3t^2, \text{ the acceleration at time } t \text{ is } a = 6t. \tag{12}$$

If s is in feet and t is in seconds, v is in feet per second and a is in feet per second per second. For the motion $s = t^3$, when $t = 3$, $s = 27$. And from Eq. (8), $v = 27$, while from Eq. (12) $a = 18$. We sometimes add abbreviations for the units, and write at time $t = 3$ sec., $s = 27$ ft., $v = 27$ ft./sec., and $a = 18$ ft./sec.^{2}

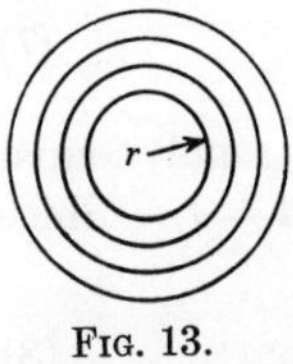

FIG. 13.

22. Rate of Change. We sometimes wish to compare the changes of two physical quantities. For example, imagine hydrogen being pumped into a small spherical balloon, causing it to expand. We assume that the balloon remains a perfect sphere as it expands. Several stages are indicated in Fig. 13. We wish to compare the changes in volume with the changes in the radius. If the volume is V and the radius is r, we have

$$V = \tfrac{4}{3}\pi r^3. \tag{13}$$

To define the instantaneous rate of change of V with respect to r when $r = r_1$, we select a second value of r, r_2, such that $r_2 \neq r_1$, but such that $r_2 - r_1$ is small. From Eq. (13), it follows that

$$V_1 = \tfrac{4}{3}\pi r_1^3 \quad \text{and} \quad V_2 = \tfrac{4}{3}\pi r_2^3. \tag{14}$$

Then the *average rate of change* for the interval between r_1 and r_2 is

$$\frac{V_2 - V_1}{r_2 - r_1} = \frac{\frac{4}{3}\pi(r_2^3 - r_1^3)}{r_2 - r_1} = \frac{4\pi}{3}(r_2^2 + r_2 r_1 + r_1^2). \tag{15}$$

This gives an approximation to the rate of change at r_1. And the approximation improves with each further choice of $r_2 \neq r_1$, but nearer to r_1. Hence we keep r_1 fixed and let r_2 tend to r_1. When $r_2 \to r_1$, the last expression in Eq. (15), a polynomial in r_2, approaches $\frac{4}{3}\pi(r_1^2 + r_1^2 + r_1^2) = 4\pi r_1^2$. This is the rate of change of V with respect to r for $r = r_1$. Since r_1 may have any value, we may write $4\pi r^2$ as the rate of change of V with

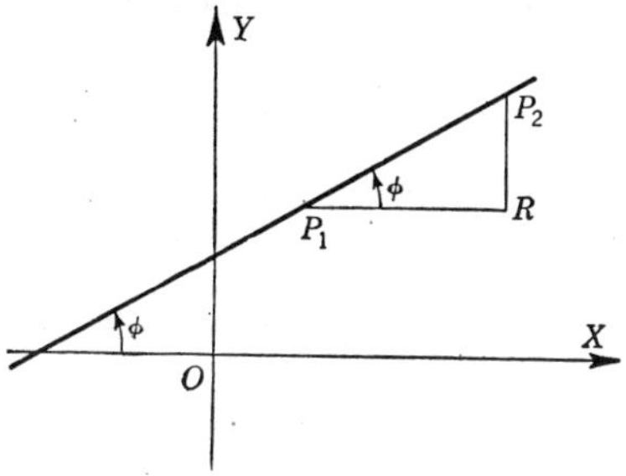

FIG. 14.

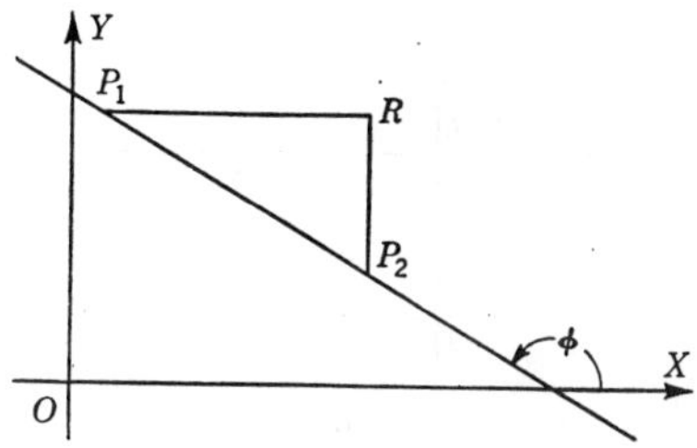

FIG. 15.

respect to r for any value of r. For example if $r = 3$ ft., the rate is 12π sq. ft. or cubic feet per foot increase in radius.

EXERCISE 8

Find the velocity at any time for each of the following given motions.

1. $s = 5t + 7$.
2. $s = 4t^2$.
3. $s = t^2 - 3t$.
4. $s = 2t^3$.

For a certain motion the velocity is a function of the time as indicated. Find the acceleration in each case.

5. $v = 5$.
6. $v = 8t$.
7. $v = 2t - 3$.
8. $v = 6t^2$.
9. If $s = At^2 + Bt + C$, show that $v = 2At + B$ and that $a = 2A$.

Find the rate of change, with respect to the radius r, of

10. The circumference of a circle, $C = 2\pi r$.
11. The area of a circle, $A = \pi r^2$.
12. The surface of a sphere, $S = 4\pi r^2$.

Find the rate of change, with respect to a side x, of

13. The area of a square, $A = x^2$.
14. The area of an equilateral triangle, $A = x^2\sqrt{3}/4$.
15. The surface of a cube, $S = 6x^2$.
16. The volume of a cube, $V = x^3$.

R23. Slope of a Straight Line. Suppose that y is such a function of x that its graph, plotted on a rectangular coordinate system as in Sec. 6, is a straight line. Let $P_1 = (x_1,y_1)$ be any point on this line, Fig. 14. And let $P_2 = (x_2,y_2)$ be any second point on the line distinct from P_1. Construct the right triangle P_1RP_2 with side P_1R parallel to OX and side RP_2 parallel to OY. When x changes from x_1 to x_2, the algebraic increase in x is $x_2 - x_1 = P_1R$. Here P_1R is a directed segment, positive when R is to the right of P_1 and negative when R is to the left of P_1. The corresponding change in y is $y_2 - y_1 = RP_2$. RP_2 is a directed segment, positive for P_2 above R and negative for P_1 below R. Thus RP_2 is negative for Fig. 15. Let ϕ be the inclination of the straight line, or angle from the direction of OX to a direction along the straight line. We may measure ϕ as in Figs. 14 and 15 so that $0 \leqq \phi < 180°$. Put $\tan \phi = m$. Then for any position of the line, or of P_1 and P_2, we have

$$m = \tan \phi = \frac{RP_2}{P_1R} = \frac{y_2 - y_1}{x_2 - x_1}. \tag{16}$$

This assumes the use of the same units on both the horizontal and the vertical axes. For the modification when different units are used, see Sec. 48.

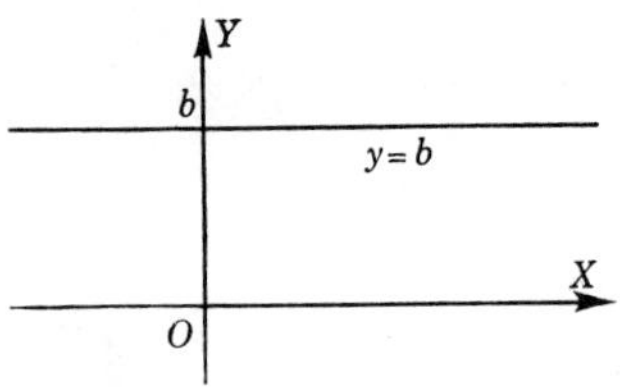

FIG. 16.

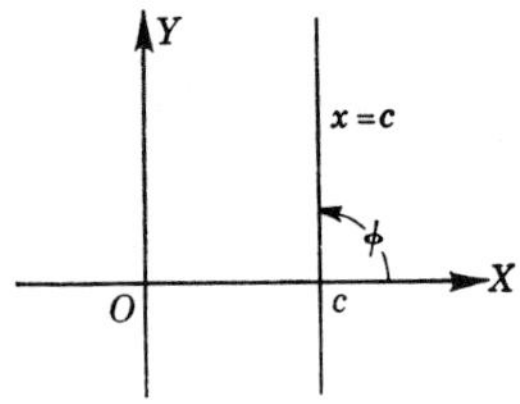

FIG. 17.

The *slope* of the straight line is defined as the tangent of the inclination to the x axis, or $\tan \phi$. Hence the slope may be found from any one of the expressions in Eq. (16). The numerical magnitude measures the steepness of the line. The slope is positive for a line slanting up to the right and ultimately extending into the first and third quadrants, as in Fig. 14. The slope is negative for a line slanting down to the right and ultimately extending into the second and fourth quadrants, as in Fig. 15.

It follows from Eq. (16) that for any P_2 not at P_1,

$$y_2 - y_1 = m(x_2 - x_1). \tag{17}$$

But this equation is true when $x_2 = x_1$ and $y_2 = y_1$. Hence we may consider $P_2 = (x_2,y_2)$ as any variable point $P = (x,y)$ which lies on the straight line. Equation (17) then becomes

$$y - y_1 = m(x - x_1). \tag{18}$$

If we solve this equation for y, and put $b = y_1 - mx_1$, we find

$$y = mx + b. \tag{19}$$

Hence the function we plotted to obtain the straight line must have been equivalent to this form. The right member of Eq. (19) is a first-degree expression when $m \neq 0$. When $m = 0$, y is a constant, and the straight line is parallel to OY as in Fig. 16. We consider the inclination ϕ as 0 in this case so that Eq. (16) is still valid.

Vertical lines would never be obtained by plotting y as a function of x, but would result from plotting x as a function of y for the case $x = c$, a constant. For such a line, as in Fig. 17, the inclination $\phi = 90°$, so that the slope $m = \tan 90° = \infty$. This is the brief notation mentioned in Sec. 15, meaning $\lim_{\phi \to 90°} \tan \phi = \infty$ in the sense defined in Sec. 12.

On the understanding that we sometimes plot x as a function of y, and sometimes plot y as a function of x, *every straight line comes from plotting some first-degree equation.* And the general first-degree equation

$$Ax + By + C = 0, \tag{20}$$

with A and B not both zero, may be written

$$y = -\frac{A}{B}x - \frac{C}{B} \quad \text{if } B \neq 0. \tag{21}$$

This has the form of Eq. (19) and so represents a straight line with slope $m = -A/B$.

If $B = 0$, Eq. (20) becomes $Ax + C = 0$ with $A \neq 0$. This may be written $x = -C/A$, which represents a straight line parallel to OY. Hence *every first-degree equation represents a straight line.*

EXERCISE 9

Find the equation of a straight line which

1. Passes through the points (2,5) and $(-1,-1)$.
2. Passes through the point (3,2) with slope 2.
3. Passes through the origin with the slope 3.
4. Passes through the point $(0,b)$ with the slope m.
5. Passes through the points $(a,0)$ and $(0,b)$.
6. Passes through the points (x_1,y_1) and (x_2,y_2).
7. Passes through the point (1,2) with an inclination of 60°.
8. Passes through the point (1,2) with an inclination of 120°.
9. Is obtained from OX by rotating OX about O through an angle equal to $-60°$.

10. Show from a diagram that two lines are parallel or coincident if their inclinations are equal, or have the same trigonometric tangents. Also, show that, conversely, if two lines are parallel, their slopes are equal.

Use Prob. 10 to find the equation of a straight line which

11. Passes through the point (3,4) and is parallel to $y = -4x$.
12. Passes through $(2,-2)$ and is parallel to the line joining (0,1) and (1,2).
13. Passes through $(3,-4)$ and is parallel to the line $y = 1$.
14. Passes through (5,3) and is parallel to the x axis.
15. Passes through (4,7) and is parallel to the y axis.
16. Passes through (x_1,y_1) and is parallel to the line representing Eq. (20).

24. Slope of a Curve. Suppose that $y = f(x)$ is any function of x whose graph is a curve, such as AB in Fig. 18. Let $P_1 = (x_1,y_1)$ be any fixed point on this curve. The straight line tangent to the curve at P_1 is defined by the following construction. Let P_2 be any point on the curve near, but not at, P_1. The segment P_1P_2 is the *chord*, and the indefinite straight line P_1S_2 obtained by producing P_1P_2 is a *secant* determined by P_1 and P_2.

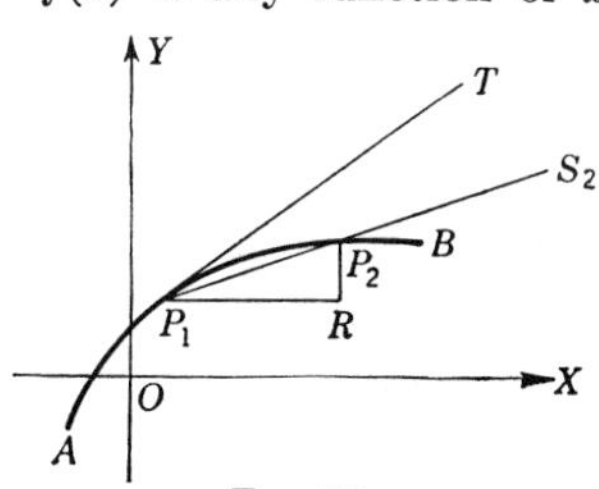

FIG. 18.

Now consider P_2 as a variable point on the curve which tends to P_1. Then the secant P_1S_2 will revolve about P_1. If there is a fixed straight line through P_1, P_1T, such that as $P_2 \to P_1$ the angle between P_1S_2 and P_1T, or $\angle S_2P_1T$ approaches zero, then the line P_1T is the *tangent* to the curve at P_1.

To indicate that a curve has a tangent at the point P_1, we say that the curve is *smooth* at P_1. The functions which we shall use have graphs which are either smooth at all points, or else smooth with the exception of a limited number of singular points. If P_1 is an interior point of a smooth arc, we may let P_2 approach P_1 from either side. In some cases, as at the end point of an arc, we still consider the line P_1T as the tangent at P_1 even though P_2 can approach P_1 from one side only.

By constructing the right triangle P_1RP_2, as in Sec. 23, we find as in Eq. (16) that the slope of a secant line is

$$m_s = \tan \phi_s = \frac{RP_2}{P_1R} = \frac{y_2 - y_1}{x_2 - x_1}. \tag{22}$$

Let $m_1 = \tan \phi_1$ be the slope of the tangent line P_1T. When x_2 tends to x_1, P_2 tends to P_1. Hence, by our definition $\phi_s - \phi_1 = \angle S_2P_1T$ approaches zero and ϕ_s approaches ϕ_1. Since, as mentioned in Sec. 15, $\tan \phi$ is a continuous function of ϕ for all ϕ which make $\tan \phi$ finite, it follows that $\tan \phi_s$ approaches $\tan \phi_1$, or m_s approaches m_1. Hence, from Eq. (22), the slope m_1 of the tangent line is

$$m_1 = \lim_{x_2 \to x_1} m_s = \lim_{x_2 \to x_1} \frac{y_2 - y_1}{x_2 - x_1}. \tag{23}$$

The *slope of the curve* at P_1 is defined as the slope of the tangent line at P_1, or m_1. The slope of the graph of $y = f(x)$ at P_1 where $x = x_1$ may be found by first noting that, since the fixed point $P_1 = (x_1,y_1)$ is on the graph, $y_1 = f(x_1)$. And since the variable point $P_2 = (x_2,y_2)$ is on the graph, $y_2 = f(x_2)$. Hence from Eq. (23), the slope of the graph of $y = f(x)$ at P_1 is

$$m_1 = \lim_{x_2 \to x_1} \frac{y_2 - y_1}{x_2 - x_1} = \lim_{x_2 \to x_1} \frac{f(x_2) - f(x_1)}{x_2 - x_1}. \tag{24}$$

For example, let $y = x^2 - x$. Then $y_1 = x_1^2 - x_1$, and $y_2 = x_2^2 - x_2$, so that in this case

$$m_s = \frac{y_2 - y_1}{x_2 - x_1} = \frac{x_2^2 - x_1^2 - (x_2 - x_1)}{x_2 - x_1} = x_2 + x_1 - 1.$$

When x_2 tends to x_1, m_s approaches the value of the last expression with $x_2 = x_1$, or $2x_1 - 1$. Hence $m_1 = 2x_1 - 1$.

To draw the tangent line at P_1, Fig. 19, we note that for any value of a, the slope of the line joining $P_1 = (x_1,y_1)$ and $P_t = (x_1 + a,\ y_1 + m_1a)$ is m_1 by Eq. (16). Hence we may draw the tangent by selecting any convenient value of a, $\neq 0$, plotting the point

$$P_t = (x_1 + a,\ y_1 + m_1a), \tag{25}$$

and joining the points P_1 and P_t by a straight line. In Fig. 19, $x_1 = 3$, $P_1 = (3,6)$, $m_1 = 5$. With $a = -2$, $P_t = (1,-4)$.

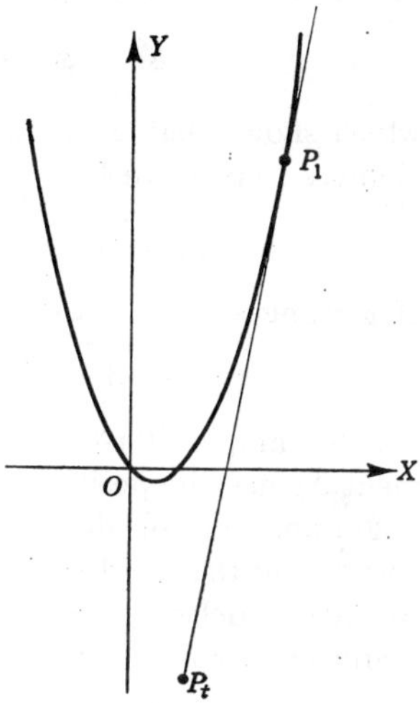

FIG. 19.

By Eq. (18) the equation of the tangent line is

$$y - y_1 = m_1(x - x_1). \tag{26}$$

EXERCISE 10

Find the slope at P_1 where $x = x_1$ for each given curve.

1. $y = x^2$.
2. $y = 2 - x^2$.
3. $y = x^3$.
4. Find the slope of the curve $y = 2x - x^2$ at the point P_1 where $x_1 = 2$. Sketch the graph and draw the tangent at P_1. Also find the equation of this tangent line.

Verify that each of the following curves passes through the point $P_1 = (1,2)$. Find the slope and the equation of the tangent line at P_1 for each curve.

5. $y = 2x^2$.
6. $y = 3 - x^2$.
7. $y = 2x^3$.
8. Show that the slope of $y = 2x^2 - 4x$ at P_1 where $x = x_1$ is $m_1 = 4x_1 - 4$.

Find the point on the curve of Prob. 8 at which

9. The slope is equal to 8.
10. The tangent is parallel to the x axis.
11. The tangent is parallel to the line $y = 12x + 3$.
12. Find the point on the curve $y = x^2/2$ at which the tangent line has an inclination of 45°.
13. Find the two points on the curve $y = 2x^3$ at which the slope is equal to 6.

25. The Increments Δx and Δy. The definition of the slope of a curve, Eq. (24), leads to the evaluation of the expression

$$\lim_{x_2 \to x_1} \frac{y_2 - y_1}{x_2 - x_1}. \tag{27}$$

And the definition of velocity, Eq. (5), of acceleration, Eq. (10), as well as the discussion of rate of change of Sec. 22 all lead to expressions of the same form as Eq. (27) with other letters in place of x and y. In dealing with such expressions, it is useful to have a special notation for the differences $y_2 - y_1$ and $x_2 - x_1$.

The *increment* of any variable which changes from a first to a second value is equal to the new value minus the old value. Thus, if x_1 is a first value of x, and x_2 a second value, the increment of x is $x_2 - x_1$. This increment is denoted by the symbol Δx, read "delta x." Thus,

$$\Delta x = x_2 - x_1 \qquad \text{and} \qquad x_2 = x_1 + \Delta x, \tag{28}$$

which shows that Δx is the algebraic increase in x. For example, if x changes from 10 to 12,

$$\Delta x = 12 - 10 = 2 \qquad \text{and} \qquad 12 = 10 + \Delta x. \tag{29}$$

If x changes from 10 to 8,

$$\Delta x = 8 - 10 = -2 \qquad \text{and} \qquad 8 = 10 + \Delta x. \tag{30}$$

These examples illustrate that, for the independent variable x, the increment Δx may be positive or negative.

For any variable dependent on x, as $y = f(x)$, the value $y_1 = f(x_1)$ will correspond to x_1 and the value $y_2 = f(x_2)$ will correspond to x_2. And by Δy, read "delta y," we mean the increment in y corresponding to the increment Δx in x. Thus

$$\Delta y = y_2 - y_1 \qquad \text{and} \qquad y_2 = y_1 + \Delta y. \tag{31}$$

From the second part of Eqs. (28) and (31) together with the fact that $y_2 = f(x_2)$, we find the relation

$$y_1 + \Delta y = f(x_1 + \Delta x). \tag{32}$$

By combining this with $y_1 = f(x_1)$, we may deduce that

$$\Delta y = f(x_1 + \Delta x) - f(x_1). \tag{33}$$

This is the expression for Δy as a function of x_1 and Δx. The increment of the dependent variable, Δy, may be positive, negative, or zero. Since x_2 does not appear in Eq. (33), we may drop the subscript 1. Thus we may write

$$\Delta y = f(x + \Delta x) - f(x), \tag{34}$$

where x is the fixed value previously denoted by x_1.

As an example, suppose that $y = 2x^2$. And let x change from 10 to 12, as in Eq. (29), so that $\Delta x = 2$. Since y changes from 200 to 288, $\Delta y = 288 - 200 = 88$. If we use Eq. (34), we find

$$\Delta y = 2(10 + 2)^2 - 2(10)^2 = 2(12^2 - 10^2) = 88. \tag{35}$$

26. Average Rate of Change. Let $y = f(x)$ be a given function of x, and consider any fixed value of x. Then if we change x by an arbitrary increment Δx, there is a corresponding change in y, Δy. The ratio of the change in y to the change in x, or quotient obtained by dividing Δy by Δx, is called the *average rate of change* of y with respect to x. Thus the

$$\text{Average rate of change} = \frac{\Delta y}{\Delta x}. \tag{36}$$

If we eliminate Δy by using Eq. (34), we obtain the relation

$$\text{Average rate of change} = \frac{f(x + \Delta x) - f(x)}{\Delta x}, \tag{37}$$

for the average rate of change of the function $f(x)$. These rates have the dimensions unit of y per unit of x.

EXERCISE 11

Find Δx and Δy for each given set of conditions.

1. $y = 5x + 2$, and x changes from 2.4 to 2.6.
2. $y = 3x^2$, and x changes from 0.9 to 0.8.
3. $y = \frac{4}{x}$, and x changes from 0.8 to 0.4.
4. $y = x^2 + 3x$, and x changes from -9 to -7.

Find Δt and Δs for each of the following motions.

5. $s = t^3$ ft., and t changes from 2 to 5 sec.
6. $s = 16t^2$ ft., and t changes from 6 to 7 sec.

Find Δt and Δv for the motion such that

7. $v = 3t^2$ ft./sec., and t changes from 4 to 6 sec.
8. $v = 15t - 2$ ft./sec., and t changes from 1.2 to 1.4 sec.

Find Δy when $\Delta x = 0.2$, if

9. $y = x^2 - 4x$, and $x_1 = 3$.
10. $y = 4x + 3$, and $x_1 = 2$.

Find the average rate of change of y with respect to x, if

11. $y = 2x + 3$, and x changes from -3 to -5.
12. $y = 5x^2 - 4$, and x changes from 5 to 4.
13. $y = 2x^2 - 3$, $\Delta x = -1$, and $x_1 = 4$.

Find the average velocity, if

14. $s = 2t^2$ ft., and t changes from 1 to 2 sec.
15. $s = t^2 + 4$ ft., $\Delta t = 0.2$ sec., and $t_1 = 2$ sec.

Find the average acceleration, if

16. $v = 4t - 3$ ft./sec., and t changes from 6 to 4 sec.
17. $v = 25t^2 + 10t$ ft./sec., $\Delta t = 0.2$ sec., and $t_1 = 1$ sec.

27. The Derivative. The true rate of change of y with respect to x is obtained by taking the limit of the average rate of change as Δx tends to zero. Thus from Eq. (36) it follows that

$$\text{True rate of change} = \lim_{\Delta x \to 0} \frac{\Delta y}{\Delta x}. \tag{38}$$

When the limit on the right exists and is the same whether $\Delta x \to 0$ through positive or negative values, this limit is called the derivative of y with respect to x. The derivative is usually denoted by the symbol dy/dx, read "dee y dee x." For the present dy/dx should be regarded in its entirety as a single symbol which suggests the limit in Eq. (38). As just defined, the derivative

$$\frac{dy}{dx} = \lim_{\Delta x \to 0} \frac{\Delta y}{\Delta x}. \tag{39}$$

Let $y = f(x)$. Then from Eq. (37) we may obtain the expression for the derivative in terms of the function

$$\frac{dy}{dx} = \lim_{\Delta x \to 0} \frac{f(x + \Delta x) - f(x)}{\Delta x}. \tag{40}$$

From Eqs. (39) and (40) or from the definitions of this and the preceding section, we may formulate the following definition of a derivative:

The derivative of a function is the limit of the ratio of the increment of the function to the increment of the independent variable, when the increment of the independent variable tends to zero.

To indicate that the limit exists, we say that the function has a derivative or is differentiable for the value in question.

If the function $f(x)$ has a derivative for any value of $x = a$, in the sense of Eq. (40), then $f(x)$ is necessarily continuous in the sense of Sec. 14. Thus differentiability implies continuity.

The converse statement is not always true. In fact, functions which are everywhere continuous but nowhere differentiable have been constructed. But such functions are rarely met in practice.

The functions which we shall use are either differentiable for all values, or else differentiable for all except a limited number of isolated values. This is in accord with the statement made in Sec. 24, since it follows from Eq. (24) that the slope of the graph of $y = f(x)$ at a point with a given value of x is equal to the derivative of y with respect to x for the same given value. Thus the graph of $y = f(x)$ is smooth for any value of x at which the function $f(x)$ is differentiable.

A notation for the derivative similar to dy/dx is used when other letters are involved. For example, let $s = f(t)$ as in Sec. 20. Then from Eq. (5) the velocity v is

$$v = \frac{ds}{dt} = \lim_{\Delta t \to 0} \frac{\Delta s}{\Delta t} = \lim_{\Delta t \to 0} \frac{f(t + \Delta t) - f(t)}{\Delta t}. \tag{41}$$

To *differentiate* a function means to find its derivative, and the process of finding a derivative is called *differentiation.*

28. Notation for Derivatives. The symbol dy/dx for the derivative of y with respect to x clearly indicates that y is the dependent variable and that x is the independent variable. But it does not indicate which fixed value of x we used.

If $y = f(x)$, the derivative of y or of $f(x)$ with respect to x may also be indicated by any one of the forms

$$\frac{df}{dx} = \frac{d\,f(x)}{dx} = \frac{d}{dx}f(x). \tag{42}$$

In the last form we may think of d/dx as the *differentiating operator* which indicates that the next following expression, here $f(x)$, is to be differentiated with respect to x.

In Sec. 24, we found from Eq. (24) that, for the graph of $y = x^2 - x$, the slope at P_1 where $x = x_1$ was $m_1 = 2x_1 - 1$. This result could be written

$$\frac{dy}{dx} = \frac{d}{dx}(x^2 - x) = 2x - 1. \tag{43}$$

As this illustrates, for a given function $y = f(x)$, the derivative depends on x and so is a new function of x. This is indicated by the notation $f'(x)$, read "f prime of x," which has the same meaning as dy/dx or the other equivalent symbols in Eq. (42).

For example, if $f(x) = x^2 - x$, from Eq. (43) we may write

$$f'(x) = \frac{d}{dx}f(x) = \frac{d}{dx}(x^2 - x) = 2x - 1. \tag{44}$$

And since $f'(x) = 2x - 1$, at x_1 and at 3 the derivatives are

$$f'(x_1) = 2x_1 - 1 \qquad \text{and} \qquad f'(3) = 5. \tag{45}$$

Thus the $f'(x)$ notation clearly indicates at what point the derivative was taken. It is particularly convenient when we have to deal with the values of the derivative of a function for several values of x in the same problem.

All these notations may be modified by replacing x, y, and f by any other letters. Thus if $v = g(u) = u^2 - u$, then

$$\frac{dv}{du} = \frac{dg}{du} = \frac{d}{du}g(u) = g'(u) = \frac{d}{du}(u^2 - u) = 2u - 1. \tag{46}$$

The symbols D or D_x are sometimes used in place of d/dx. And we sometimes use y' as a short form for dy/dx. Thus if $y = f(x)$,

$$\frac{dy}{dx} = f'(x) = y' = D_xy = Dy. \tag{47}$$

29. Differentiation. If $y = f(x)$, the derivative of y with respect to x is

$$\frac{dy}{dx} = f'(x) = \lim_{\Delta x \to 0}\frac{\Delta y}{\Delta x} = \lim_{\Delta x \to 0}\frac{f(x + \Delta x) - f(x)}{\Delta x}, \tag{48}$$

by Eqs. (39), (40), and (47). To evaluate the derivative for any given particular function $f(x)$, we must perform the operations indicated in the last expression of Eq. (48). This is conveniently done by taking the following four distinct steps.

We often write a single letter h in place of Δx wherever algebraic operations must be performed to simplify the writing.

1. Let $\Delta x = h$, replace x by $x + h$ in the given function $f(x)$, and calculate the new value of the function. The result is $y + \Delta y = f(x + h)$.
2. Subtract the given value of the function from the new value. The difference is the increment of the function, $\Delta y = f(x + h) - f(x)$.
3. Divide the increment of the function Δy by h. The result is the average rate of change $\Delta y/\Delta x$.
4. Find the limit of this quotient when h tends to zero. The result is the required derivative,

$$f'(x) = \frac{dy}{dx} = \lim_{h \to 0} \frac{f(x + h) - f(x)}{h}. \tag{49}$$

We shall illustrate this procedure for a number of examples. To facilitate reference to the description just given, in these examples we number the steps as in the description.

EXAMPLE 1. Find the derivative of $y = 4x^2 - 3x + 5$.

Solution: 1. $\Delta x = h$, $y + \Delta y = 4(x + h)^2 - 3(x + h) + 5$.

2.
$$\begin{array}{rl} y + \Delta y &= 4x^2 + 8hx + 4h^2 - 3x - 3h + 5 \\ y &= 4x^2 - 3x + 5 \\ \hline \Delta y &= 8hx + 4h^2 - 3h. \end{array}$$

3. $\Delta y/\Delta x = 8x + 4h - 3$.

4. Let h tend to zero. By the polynomial principle of Sec. 11,

$$\frac{dy}{dx} = \lim_{h \to 0} (8x + 4h - 3) = 8x - 3.$$

Thus the required derivative is $dy/dx = 8x - 3$.

EXAMPLE 2. Find the velocity if $s = t^3 - 5t^2$.

Solution: From Eq. (41), $v = ds/dt$, so we must find this derivative.

1. $\Delta t = h$, $s + \Delta s = (t + h)^3 - 5(t + h)^2$.

2.
$$\begin{array}{rl} s + \Delta s &= t^3 + 3ht^2 + 3h^2t + h^3 - 5t^2 - 10ht - 5h^2 \\ s &= t^3 - 5t^2 \\ \hline \Delta s &= 3ht^2 + 3h^2t + h^3 - 10ht - 5h^2. \end{array}$$

3. $\dfrac{\Delta s}{\Delta t} = 3t^2 + 3ht + h^2 - 10t$.

4. Let h tend to zero. By the polynomial principle of Sec. 11,

$$v = \frac{ds}{dt} = \lim_{h \to 0} (3t^2 + 3ht + h^2 - 10t) = 3t^2 - 10t.$$

Thus the required velocity is $v = 3t^2 - 10t$.

EXAMPLE 3. Find the rate of change of pressure p with respect to volume v if $pv = 100$.

Solution: $p = 100/v$, and we must find the derivative dp/dv.

1. $\Delta v = h,\ p + \Delta p = \dfrac{100}{v + h}.$

2.
$$\begin{aligned} p + \Delta p &= \frac{100}{v+h} \\ p &= \frac{100}{v} \\ \hline \Delta p &= \frac{100}{v+h} - \frac{100}{v} = \frac{-100h}{v(v+h)}. \end{aligned}$$

3. $\dfrac{\Delta p}{\Delta v} = \dfrac{-100}{v(v+h)}.$

4. Let h tend to zero. By the polynomial principle of Sec. 11, and the quotient principle of Sec. 10,

$$\frac{dp}{dv} = \lim_{h \to 0} \frac{-100}{v(v+h)} = \frac{-100}{v^2}.$$

Thus the required rate is $\dfrac{dp}{dv} = -\dfrac{100}{v^2}.$

EXAMPLE 4. Find the slope of the curve $y = x^{\frac{3}{2}}$.

Solution: From Eq. (24), the slope at a point with a given value of x is equal to $f'(x) = dy/dx$. Hence we must find this derivative.

1. $\Delta x = h,\ y + \Delta y = (x + h)^{\frac{3}{2}}.$

2.
$$\begin{aligned} y + \Delta y &= (x+h)^{\frac{3}{2}} \\ y &= x^{\frac{3}{2}} \\ \hline \Delta y &= (x+h)^{\frac{3}{2}} - x^{\frac{3}{2}}. \end{aligned}$$

3. $\dfrac{\Delta y}{\Delta x} = \dfrac{(x+h)^{\frac{3}{2}} - x^{\frac{3}{2}}}{h}.$

4. Before letting h tend to zero, it is desirable to multiply numerator and denominator by $(x + h)^{\frac{3}{2}} + x^{\frac{3}{2}}$ and to note that

$$\begin{aligned} [(x+h)^{\frac{3}{2}} - x^{\frac{3}{2}}][(x+h)^{\frac{3}{2}} + x^{\frac{3}{2}}] &= (x+h)^3 - x^3 \\ &= 3hx^2 + 3h^2x + h^3 \\ &= h(3x^2 + 3hx + h^2). \end{aligned}$$

It follows that

$$\frac{dy}{dx} = \lim_{h \to 0} \frac{3x^2 + 3hx + h^2}{(x+h)^{\frac{3}{2}} + x^{\frac{3}{2}}} = \frac{3x^2}{2x^{\frac{3}{2}}} = \frac{3}{2}x^{\frac{1}{2}}.$$

Thus the required slope is $f'(x) = \frac{3}{2}x^{\frac{1}{2}}$.

EXERCISE 12

Use the four-step procedure of Sec. 29 to find the derivative of each of the following functions.

1. $y = 3 + 4x.$
2. $y = 5x^2.$
3. $y = x^4.$
4. $y = \dfrac{1}{x}.$
5. $y = \dfrac{2 + x}{2 - x}.$
6. $y = \dfrac{4}{3x + 1}.$
7. $y = \sqrt{x}.$
8. $s = (2t + 3)^2.$
9. $v = u^3 + 2u^2.$

Find an expression for the slope at any point for each of the following given curves.

10. $y = 10 - 5x^2.$
11. $y = \dfrac{5}{x + 1}.$
12. $y = \dfrac{1}{\sqrt{x}}.$

Find the velocity at any time for each of the following given motions.

13. $s = t^3 + 2t.$ 14. $s = \dfrac{3}{4t - 5}.$ 15. $s = \sqrt{t + 1}.$

If the velocity is the given function of the time, find the acceleration. Note that, by Eq. (10), $a = dv/dt$.

16. $v = -4t^2.$ 17. $v = \sqrt{2t}.$ 18. $v = t^{\frac{3}{2}}.$

Find the rate of change with respect to r of each of the following functions.

19. $\theta = \dfrac{1}{r}.$ 20. $u = \dfrac{1}{\sqrt{r + 1}}.$ 21. $u = \dfrac{3}{r^2}.$

CHAPTER 3

BEHAVIOR OF A POLYNOMIAL. GREATEST VALUE

The use of derivatives in applications is facilitated by a few special rules, or formulas. Using these rules we can write down the derivatives of many functions of simple form by inspection, instead of having to use the lengthy procedure of Sec. 29 in each case. We begin this chapter by proving enough theorems or rules of differentiation to enable us to write down immediately the derivative of any polynomial.

We then show how the algebraic sign of the derivative may be used to find intervals in which a function of x increases when x increases, and other intervals in which the function decreases when x increases. This leads to the application of the derivative to the problem of finding for which values of the independent variable a function takes on a maximum or a minimum value.

30. The Derivative of a Constant. Let c be a constant, and $y = f(x) = c$ for all values of x. To find dy/dx we proceed as in Sec. 29.

Let $\Delta x = h$.

$$\begin{aligned} y + \Delta y &= c \\ y \quad\;\; &= c \\ \hline \Delta y &= 0. \end{aligned}$$

$$\frac{\Delta y}{\Delta x} = \frac{0}{h} = 0. \qquad \text{And} \qquad \frac{dy}{dx} = \lim_{h \to 0} 0 = 0.$$

This proves that

$$\frac{dc}{dx} = 0. \tag{1}$$

The geometric interpretation of this result is that the slope of the graph is 0. But the graph of $y = c$ is the horizontal straight line of Fig. 20. As this has the slope 0, the result of Eq. (1) might have been inferred from geometric considerations.

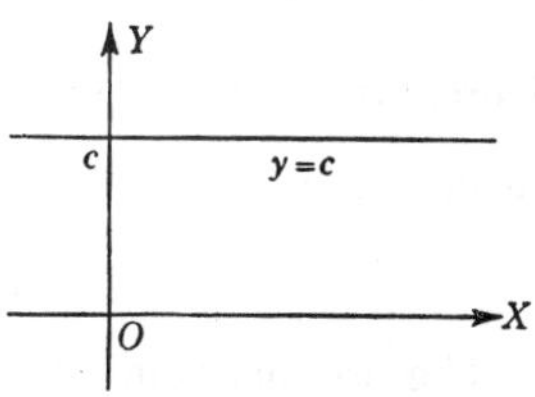

FIG. 20.

The result of Eq. (1) may be expressed in words as follows:

The derivative of a constant is zero.

31. The Derivative of x. Let $y = f(x) = x$. To find dy/dx we proceed as in Sec. 29.

Let $\Delta x = h$.
$$\begin{array}{rl} y + \Delta y &= x + h \\ y &= x \\ \hline \Delta y &= \quad h. \end{array}$$

$$\frac{\Delta y}{\Delta x} = \frac{h}{h} = 1. \qquad \text{And} \qquad \frac{dy}{dx} = \lim_{h \to 0} 1 = 1.$$

This proves that

$$\frac{dx}{dx} = 1. \tag{2}$$

This result might have been inferred from geometric considerations. For the graph of $y = x$ is the straight line of Fig. 21, with inclination 45°, which has 1 as its slope.

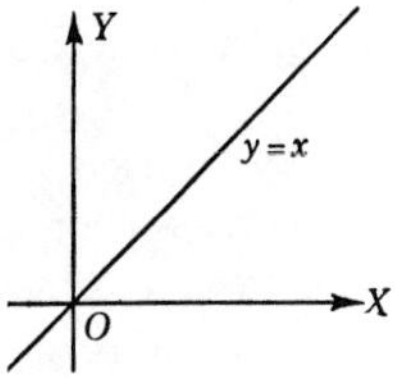

FIG. 21.

The result of Eq. (2) may be expressed in words as follows:

The derivative of any variable with respect to itself is one.

32. The Derivative of a Product. The theorems which follow deal with several functions of x, as u, v, and so forth. These functions are unspecified but are assumed to be differentiable. When x is replaced by $x + \Delta x$, u becomes $u + \Delta u$ and v becomes $v + \Delta v$. It will be convenient to retain the notation Δx, Δu, Δv to distinguish the variable whose increment is taken.

Consider the product $y = uv$. Then $y + \Delta y = (u + \Delta u)(v + \Delta v)$.

$$\begin{array}{rl} y + \Delta y &= uv + u\,\Delta v + v\,\Delta u + \Delta u\,\Delta v \\ y &= uv \\ \hline \Delta y &= \quad u\,\Delta v + v\,\Delta u + \Delta u\,\Delta v. \\ \dfrac{\Delta y}{\Delta x} &= (u + \Delta u)\dfrac{\Delta v}{\Delta x} + v\dfrac{\Delta u}{\Delta x}. \end{array}$$

By the sum and product principles for limits,

$$\frac{dy}{dx} = \lim_{\Delta x \to 0} (u + \Delta u) \lim_{\Delta x \to 0} \frac{\Delta v}{\Delta x} + v \lim_{\Delta x \to 0} \frac{\Delta u}{\Delta x} = u\frac{dv}{dx} + v\frac{du}{dx}.$$

Here, when $\Delta x \to 0$, $\Delta u \to 0$ since $\Delta x \dfrac{\Delta u}{\Delta x}$ approaches $0 \dfrac{du}{dx} = 0$. This proves that

$$\frac{d}{dx}(uv) = u\frac{dv}{dx} + v\frac{du}{dx}. \tag{3}$$

The product rule may be stated in words as follows:

The derivative of the product of two functions is equal to the first function times the derivative of the second plus the second function times the derivative of the first.

33. The Derivative of u^n. To find the derivative of u^2 with respect to x, we may put $v = u$ in Eq. (3). The result is

$$\frac{d}{dx}u^2 = \frac{d}{dx}(uu) = u\frac{du}{dx} + u\frac{du}{dx} = 2u\frac{du}{dx}.$$

This shows that if $v = u^2$, $dv/dx = 2u\,du/dx$. Now put these results in Eq. (3) and so deduce that

$$\begin{aligned}\frac{d}{dx}u^3 &= \frac{d}{dx}(uu^2) = u\frac{d}{dx}u^2 + u^2\frac{du}{dx}\\ &= u\left(2u\frac{du}{dx}\right) + u^2\frac{du}{dx} = 3u^2\frac{du}{dx}.\end{aligned}$$

The results just found

$$\frac{d}{dx}u^2 = 2u\frac{du}{dx} \quad\text{and}\quad \frac{d}{dx}u^3 = 3u^2\frac{du}{dx} \tag{4}$$

illustrate the general rule for any positive integer n,

$$\frac{d}{dx}u^n = nu^{n-1}\frac{du}{dx}. \tag{5}$$

To prove this we use mathematical induction. We assume that we have proved the result for the first k positive integers $1, 2, 3, \cdots, k$. Thus in particular $\frac{d}{dx}u^k = ku^{k-1}$. Let us put $v = u^k$, $dv/dx = ku^{k-1}$ in Eq. (3) and so deduce that

$$\begin{aligned}\frac{d}{dx}u^{k+1} &= \frac{d}{dx}(uu^k) = u\frac{d}{dx}u^k + u^k\frac{du}{dx}\\ &= u\left(ku^{k-1}\frac{du}{dx}\right) + u^k\frac{du}{dx} = (k+1)u^k\frac{du}{dx}.\end{aligned}$$

But the last expression is the result obtained from Eq. (5) by putting $n = k + 1$. This proves that Eq. (5) is true for $n = k + 1$ if it is true for $n = k$.

But Eq. (5) is true for $n = 3$, by Eq. (4). Hence it is true for $n = 3 + 1 = 4$, hence when $n = 4 + 1 = 5$, and so on through all greater positive integral values of n. This completes the proof by induction that Eq. (5) holds for n, any positive integer greater than unity.

Although it is inadvisable to use this rule when $n = 1$ or $n = 0$, it does hold for these cases also. For, let us recall from algebra that $u^0 = u^{1-1} = u/u = 1$ if $u \neq 0$. It follows that for $n = 1$, Eq. (5) becomes

$$\frac{d}{dx}u^1 = 1u^0\frac{du}{dx} \quad\text{or}\quad \frac{du}{dx} = \frac{du}{dx},$$

which is an identity. And for $n = 0$, Eq. (5) becomes

$$\frac{d}{dx} u^0 = 0u^{-1} \frac{du}{dx} \qquad \text{or} \qquad \frac{d1}{dx} = 0,$$

which is true by Eq. (1) with $c = 1$.

EXAMPLE. Let $u = x$. Then from Eq. (2), $du/dx = 1$. It follows from Eq. (5) that for n equal to any positive integer

$$\frac{d}{dx} x^n = nx^{n-1}. \tag{6}$$

34. The Derivative of cu. Consider the function cu, where c is a constant and u is any differentiable function. From the product rule of Sec. 32, we have

$$\frac{d}{dx}(cu) = c\frac{du}{dx} + u\frac{dc}{dx}.$$

But from Eq. (1), $dc/dx = 0$, so that $u\, dc/dx = 0$, and hence

$$\frac{d}{dx}(cu) = c\frac{du}{dx}. \tag{7}$$

This result may be stated in words as follows:

The derivative of a constant times a function is the constant times the derivative of the function.

EXAMPLE 1. From Eq. (6) with $n = 5$, and Eq. (7) with $c = 4$,

$$\frac{d}{dx}(4x^5) = 4(5x^4) = 20x^4.$$

EXAMPLE 2. From Eq. (6) with $n = 3$ and Eq. (7) with $c = \frac{1}{6}$,

$$\frac{d}{dx}\left(\frac{x^3}{6}\right) = \frac{1}{6}(3x^2) = \frac{x^2}{2}.$$

EXAMPLE 3. From Eqs. (6) and (7) with $u = x^n$,

$$\frac{d}{dx}(cx^n) = c(nx^{n-1}) = cnx^{n-1}. \tag{8}$$

EXAMPLE 4. Find dV/dr if $V = 4\pi r^3/3$. From Eq. (6) with $n = 3$ and Eq. (7) with $c = 4\pi/3$, and r in place of x throughout, we have

$$\frac{dV}{dr} = \frac{d}{dr}\left(\frac{4\pi r^3}{3}\right) = \frac{4\pi}{3}(3r^2) = 4\pi r^2.$$

EXAMPLE 5. Find $v = ds/dt$ if $s = 5t^8/4$. From Eqs. (6) and (7),

$$\frac{ds}{dt} = \frac{d}{dt}\left(\frac{5t^8}{4}\right) = \frac{5}{4}(8t^7) = 10t^7. \qquad \text{Hence } v = 10t^7.$$

EXERCISE 13

Find the derivative of each of the following functions of x.

1. $y = 3x^5$.
2. $y = \dfrac{x^2}{4}$.
3. $y = \frac{2}{5}x^{15}$.

Find the rate of change, with respect to a side x ft., of

4. The area of an isosceles right triangle, $A = \frac{1}{2}x^2$.
5. The perimeter of an equilateral triangle, $P = 3x$.

The altitude of a right circular cone equals the radius of its base, r. Find the rate of change with respect to r of

6. The lateral surface $L = \pi \sqrt{2}\, r^2$.
7. The volume $V = \pi r^3/3$.

For each given motion with s in feet and t in seconds find the velocity $v = ds/dt$ ft./sec.

8. $s = 16t^2$.
9. $s = 10t^3$.
10. $s = 3t^4/2$.

35. The Linearity Property. Consider the function $y = au + bv + cw$, where u, v, and w are differentiable functions of x, and a, b, and c are constants. To find the derivative dy/dx, we first notice that if x is replaced by $x + \Delta x$, then u becomes $u + \Delta u$, v becomes $v + \Delta v$, and w becomes $w + \Delta w$. It follows that since $y = au + bv + cw$,

$$y + \Delta y = a(u + \Delta u) + b(v + \Delta v) + c(w + \Delta w).$$

Thus we may write

$$\begin{array}{rl} y + \Delta y &= au + a\,\Delta u + bv + b\,\Delta v + cw + c\,\Delta w. \\ y &= au \qquad\qquad + bv \qquad\qquad + cw \\ \hline \Delta y &= \qquad a\,\Delta u \qquad + b\,\Delta v \qquad\qquad + c\,\Delta w. \end{array}$$

$$\frac{\Delta y}{\Delta x} = a\frac{\Delta u}{\Delta x} + b\frac{\Delta v}{\Delta x} + c\frac{\Delta w}{\Delta x}.$$

Hence by the linearity property of limits of Sec. 10, we have

$$\begin{aligned} \frac{dy}{dx} &= \lim_{\Delta x \to 0}\left(a\frac{\Delta u}{\Delta x} + b\frac{\Delta v}{\Delta x} + c\frac{\Delta w}{\Delta x}\right) \\ &= a\lim_{\Delta x \to 0}\frac{\Delta u}{\Delta x} + b\lim_{\Delta x \to 0}\frac{\Delta v}{\Delta x} + c\lim_{\Delta x \to 0}\frac{\Delta w}{\Delta x} \\ &= a\frac{du}{dx} + b\frac{dv}{dx} + c\frac{dw}{dx}. \end{aligned}$$

This proves that

$$\frac{d}{dx}(au + bv + cw) = a\frac{du}{dx} + b\frac{dv}{dx} + c\frac{dw}{dx}. \qquad (9)$$

A similar argument may be given for any finite number of functions. We may express the general result in words as follows:

For any linear combination of a finite number of functions with constant coefficients, the derivative is the same linear combination of the derivatives of these functions.

Equation (7) is the special case where there is just one function. As special cases for two functions, we have for the sum

$$\frac{d}{dx}(u + v) = \frac{du}{dx} + \frac{dv}{dx}. \tag{10}$$

And for the difference we have

$$\frac{d}{dx}(u - v) = \frac{du}{dx} - \frac{dv}{dx}. \tag{11}$$

For four functions, we have for the sum

$$\frac{d}{dx}(u + v + w + z) = \frac{du}{dx} + \frac{dv}{dx} + \frac{dw}{dx} + \frac{dz}{dx}. \tag{12}$$

And as an example of other possible combinations of signs,

$$\frac{d}{dx}(u - v - w + z) = \frac{du}{dx} - \frac{dv}{dx} - \frac{dw}{dx} + \frac{dz}{dx}. \tag{13}$$

EXAMPLE. Find the derivative of $y = 3x^3 - 5x^2 + 7x - 10$. By using Eq. (6) with $n = 3, 2$, Eq. (2), and Eq. (1), together with the linearity principle of this section, we find

$$\begin{aligned}\frac{dy}{dx} &= 3(3x^2) - 5(2x) + 7(1) + 0\\ &= 9x^2 - 10x + 7.\end{aligned}$$

36. The Derivative of a Polynomial. Any polynomial may be differentiated by the method used for the example at the end of Sec. 35.

We may deduce from Eqs. (6), (7), (2), and (1) that if n is a positive integer and a, b, and c are any constants,

$$\frac{d}{dx}(ax^n) = anx^{n-1}, \qquad \frac{d}{dx}(bx) = b, \qquad \frac{d}{dx}(c) = 0. \tag{14}$$

These equations may be used for the separate terms, and the derivative of a polynomial may then be obtained by applying the sum principle illustrated for four terms in Eq. (12).

For example, suppose that we wish to differentiate

$$y = ax^3 + bx^2 + cx + d. \tag{15}$$

By either of the methods described, we find that

$$\frac{dy}{dx} = 3ax^2 + 2bx + c. \tag{16}$$

And in general, for a polynomial of the mth degree,

$$y = a_m x^m + a_{m-1}x^{m-1} + \cdots + a_2x^2 + a_1x + a_0, \tag{17}$$

we find that the derivative is

$$\frac{dy}{dx} = ma_m x^{m-1} + (m-1)a_{m-1}x^{m-2} + \cdots + 2a_2x + a_1. \tag{18}$$

In particular numerical cases, we may use this equation to find the derivative. Or we may merely remember Eq. (14) and apply the sum principle.

Occasionally it is simpler to use Eq. (3) or Eq. (5) with u and v simpler polynomials. We give some examples.

EXAMPLE 1. Differentiate $y = (2x^3 - 3)^2$.
We could expand this into $y = 4x^6 - 12x^3 + 9$, and deduce that

$$\frac{dy}{dx} = 24x^5 - 36x^2. \tag{19}$$

Or we could let $2x^3 - 3 = u$, so that $du/dx = 6x^2$, and then apply Eq. (5) to $y = u^2$. This would lead to

$$\frac{dy}{dx} = \frac{d}{dx}u^2 = 2u\frac{du}{dx} = 2(2x^3 - 3)6x^2 = 12x^2(2x^3 - 3). \tag{20}$$

The last expression is equivalent to the right member of Eq. (19). In this example there is little advantage in using the second method, but for a high power, or where we wish the final result in factored form, the second method may be preferable.

EXAMPLE 2. Differentiate $y = 5(2x + 3)^6(3x - 5)^4$.
To find dy/dx, we first consider the factor $(2x + 3)^6$. From Eq. (5) with $u = 2x + 3$ and $n = 6$, we deduce that

$$\frac{d}{dx}(2x + 3)^6 = 6(2x + 3)^5 2. \tag{21}$$

The last factor, 2, which is essential, is $\frac{d}{dx}(2x + 3) = 2$, and replaces the factor du/dx of Eq. (5).

Next we consider the factor $(3x - 5)^4$. From Eq. (5) with $u = 3x - 5$ and $n = 4$, we deduce that

$$\frac{d}{dx}(3x - 5)^4 = 4(3x - 5)^3 3. \tag{22}$$

Now we apply Eq. (3) with $u = (2x + 3)^6$ and $v = (3x - 5)^4$. In view of Eqs. (21) and (22), we find

$$\begin{aligned}\frac{d}{dx}(2x + 3)^6(3x - 5)^4 &= (2x + 3)^6 4(3x - 5)^3 3 + (3x - 5)^4 6(2x + 3)^5 2\\ &= (2x + 3)^5(3x - 5)^3[12(2x + 3) + 12(3x - 5)]\\ &= 12(5x - 2)(2x + 3)^5(3x - 5)^3.\end{aligned} \tag{23}$$

Finally for the given y we treat the coefficient 5 as the c of Eq. (7). In this way we deduce from Eq. (23) that

$$\frac{dy}{dx} = \frac{d}{dx}5(2x+3)^6(3x-5)^4 = 5\frac{d}{dx}(2x+3)^6(3x-5)^4$$
$$= 60(5x-2)(2x+3)^5(3x-5)^3. \tag{24}$$

EXAMPLE 3. Differentiate $y = \dfrac{\pi(7x^2-4x+3)}{9}$. We may treat $\pi/9$ as a constant factor c of Eq. (7) and deduce that

$$\frac{dy}{dx} = \frac{\pi}{9}\frac{d}{dx}(7x^2-4x+3) = \frac{\pi}{9}(14x-4) = \frac{2\pi(7x-2)}{9}.$$

This example illustrates the advantage of carrying along a constant factor in the numerator or denominator without multiplying the factor into each term separately.

EXERCISE 14

Find the derivative of each of the following polynomials.

1. $y = 2x^2 - 3x + 7$.
2. $y = 6x^3 - 2x - 14$.
3. $y = (3-x)^3$.
4. $y = 4(2x-1)(5x+4)^2$.
5. $y = 2(x^3 - 3x^2)^5$.
6. $y = \dfrac{10x^3 - 6x^5}{15}$.

For each of the following functions find the value of the derivative dy/dx for the given value of x.

7. $y = 3x^2 + x - 6$, $x = 4$.
8. $y = 3 - 3x - x^2$, $x = -2$.
9. $y = x^6 - 2x^4$, $x = 2$.
10. $y = \dfrac{x^4}{4} - \dfrac{x^2}{2} + 12$, $x = -3$.

For each given motion find the value of the velocity $v = ds/dt$ at the time indicated.

11. $v = 2t^3 - 3t^2 + 6$, $t = 2$.
12. $s = 5t^3 - 2t + 2$, $t = 3$.
13. $v = 2(t^2 - 3t)^4$, $t = 2$.
14. $s = \dfrac{3t^2 - 6t + 7}{6}$, $t = 4$.

37. The Mean Value Theorem. It is a geometric fact that, on any arc of a smooth plane curve, there is always some intermediate point at which the straight line tangent to the curve is parallel to the chord. If the arc is part of the graph of $y = f(x)$, with end points $P_1 = (x_1,y_1)$ and $P_2 = (x_2,y_2)$ we shall have $y_1 = f(x_1)$ and $y_2 = f(x_2)$.

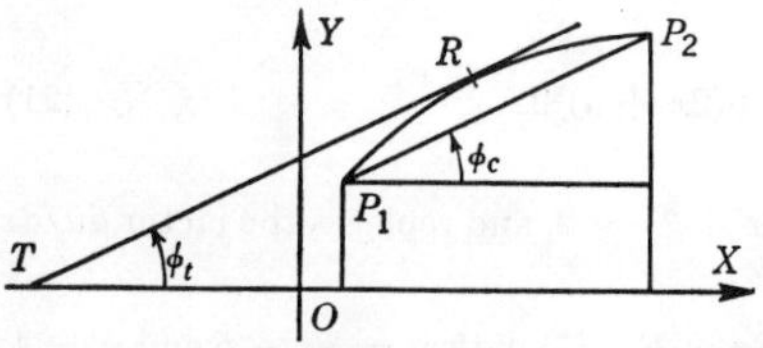

FIG. 22.

For the slope of the chord P_1P_2 in Fig. 22 we have by Sec. 23

$$\tan\phi_c = \frac{y_2-y_1}{x_2-x_1} = \frac{f(x_2)-f(x_1)}{x_2-x_1}. \tag{25}$$

But the slope of TR, tangent to the curved arc at $R = (x_0,y_0)$ with $y_0 = f(x_0)$ is the derivative dy/dx at x_0, or $f'(x_0)$, by Sec. 24. Hence

$$\tan\phi_t = f'(x_0). \tag{26}$$

But if R is a point where the tangent TR is parallel to the chord P_1P_2, the inclinations ϕ_c and ϕ_t are equal. Hence $\tan \phi_c = \tan \phi_t$ and

$$\frac{f(x_2) - f(x_1)}{x_2 - x_1} = f'(x_0). \tag{27}$$

That there always is some x_0 with $x_1 < x_0 < x_2$ satisfying Eq. (27) if $x_1 < x_2$ are any two given values and $f(x)$ is any differentiable function, is known as the *mean value theorem*.

The theorem applies to functions more general than the polynomials of this chapter. Although we based our argument on geometric intuition, the result is purely arithmetic in character. We shall treat mean value theorems more fully in Sec. 252.

38. Increasing Functions. *Definition.* The function $y = f(x)$ is said to be *increasing* in the interval a,b if for any two points x_1 and x_2 in this interval, we have

$$f(x_1) < f(x_2) \qquad \text{whenever } x_1 < x_2. \tag{28}$$

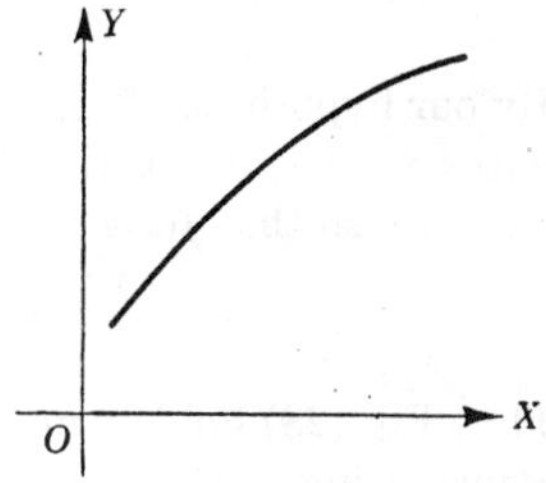

FIG. 23.

The graph of such a function, for $a < x < b$ extends up to the right, as illustrated in Fig. 23. Usually one possible choice of x_1 may be a, and one possible choice of x_2 may be b. Thus, as a particular case of Eq. (28), since $a < b$, we have

$$f(a) < f(b).$$

Again, if x is any value in the open interval a,b so that $a < x < b$, we have

$$f(a) < f(x) \qquad \text{and} \qquad f(x) < f(b). \tag{29}$$

Let x_3 be in the open interval a,b so that $a < x_3 < b$. And let p be a positive quantity so small that

$$a < x_3 - p < x_3 < x_3 + p < b.$$

Then by Eq. (28), we have

$$f(x_3 - p) < f(x_3) \qquad \text{and} \qquad f(x_3) < f(x_3 + p). \tag{30}$$

This shows that for points on the graph near $P_3 = (x_3, y_3)$ with $y_3 = f(x_3)$, those to the left are lower than P_3. And those to the right are higher than P_3. Thus if we think of the curve as being traced out by a point moving from left to right, or so that x increases from a to b, the tracing point rises. Hence we use the expression the curve representing $y = f(x)$ rises in the interval a,b to mean that $y = f(x)$ is increasing in this interval.

To say that the function $y = f(x)$ is increasing at the point where $x = x_3$ means that x_3 is inside some interval in which the function increases.

Suppose that the function $f(x)$ is differentiable and that for all points in the open interval a,b the derivative $f'(x)$ is positive. That is,

$$f'(x) > 0 \qquad \text{if } a < x < b. \tag{31}$$

Then $y = f(x)$ must be an increasing function in this interval. This is intuitively plausible, but we can prove it by using the mean value theorem as follows. For if x_1 and x_2 with $x_1 < x_2$ are any two points in the closed interval a,b, by the mean value theorem of Sec. 37, there is some point x_0 between x_1 and x_2 and hence in the open interval a,b, such that

$$\frac{f(x_2) - f(x_1)}{x_2 - x_1} = f'(x_0).$$

By our hypothesis, $f'(x_0)$ is positive, since $a < x_0 < b$. And the denominator $x_2 - x_1$ is positive, since $x_1 < x_2$. Hence the numerator must be positive, so that $f(x_2) - f(x_1) > 0$, or

$$f(x_1) < f(x_2) \qquad \text{if } x_1 < x_2. \tag{32}$$

Thus Eq. (28) holds, and hence the function $f(x)$ with positive derivative is increasing in the closed interval a,b as we stated.

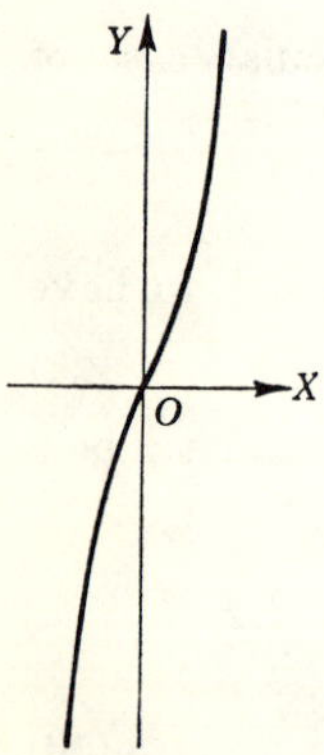

FIG. 24.

Let us next assume that $f(x)$ is a differentiable function with a continuous derivative. A continuous function, which is positive at a value x_3, is necessarily positive for all x sufficiently near x_3. Suppose that $f'(x_3) > 0$; then, since $f'(x)$ is continuous, x_3 is inside some interval throughout which $f'(x) > 0$. Hence by the theorem just proved, the function $y = f(x)$ is increasing at the point where $x = x_3$.

EXAMPLE. Consider the function $y = x^3 + 2x$. For this function $dy/dx = 3x^2 + 2$. Hence $dy/dx = f'(x) > 0$ for all real values of x, or for $-\infty < x < \infty$. The curve rises everywhere as indicated in Fig. 24.

39. Decreasing Functions. *Definition.* The function $y = f(x)$ is said to be *decreasing* in the interval a,b if for any two points x_1 and x_2 in this interval, we have

$$f(x_1) > f(x_2) \qquad \text{whenever } x_1 < x_2. \tag{33}$$

The graph of such a function extends down to the right, as illustrated in Fig. 25. If we think of a decreasing function as being traced out by a point moving from left to right, the tracing point falls. We say that the curve representing $y = f(x)$ falls in the interval a,b to mean that $y = f(x)$ is decreasing in this interval.

The function $y = f(x)$ is decreasing at the point where $x = x_3$ if x_3 is inside some interval in which the function decreases.

By reasoning as we did in Sec. 38, we can show that

The function $y = f(x)$ is decreasing in the closed interval a,b if the function $f(x)$ is differentiable, and if the derivative $f'(x)$ is negative for all values in the open interval a,b.

The function $y = f(x)$ is decreasing at the point where $x = x_3$ if $f'(x)$ is continuous at $x = x_3$ and $f'(x_3)$ is negative.

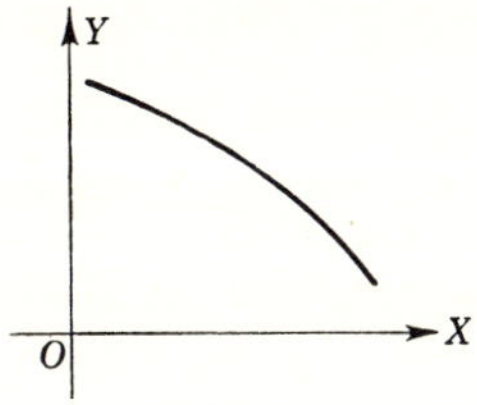

FIG. 25.

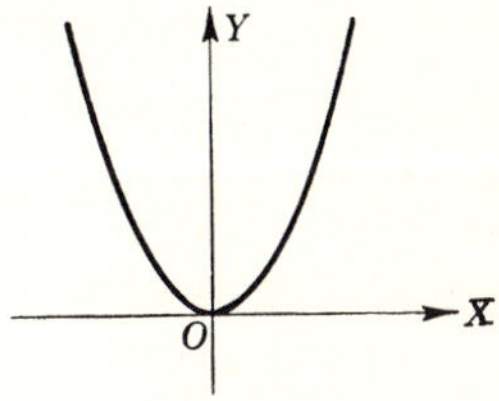

FIG. 26.

EXAMPLE. Consider the function $y = x^2$. For this function $dy/dx = 2x$. Hence $dy/dx = f'(x) > 0$ if $x > 0$. But $dy/dx = f'(x) < 0$ if $x < 0$. The graph as shown in Fig. 26 falls for all negative values $x < 0$ and rises for all positive values $x > 0$. At $x = 0$ the tangent is horizontal, and in this case we have a low point at $x = 0$ separating the decreasing from the increasing values.

EXERCISE 15

Show that each of the following functions is increasing for all values of x.

1. $y = 3x^5 + 2x - 3$.
2. $y = 4x^3 + 3x + 4$.
3. $y = x(x^2 + 1)^2$.
4. $y = (x^3 + 2x)^5$.

Show that each of the following functions is decreasing for all values of x.

5. $y = -x^3 - 3x + 10$.
6. $y = -2x^7 - x^5 + 5$.
7. $y = -x(2 + x^2)^2$.
8. $y = -(x^5 + 3x^3)^3$.

For what values of x is each of the following functions increasing, and for what values of x is it decreasing?

9. $y = 3x^4 - 5$.
10. $y = -7x^6 + 2$.
11. $y = (2x - 1)^2 + 1$.
12. $y = -(3 - x)^4 + 5$.

40. Sign of the Derivative. Most of the functions we shall use have derivatives which are positive for x in some open intervals and negative for x in some other open intervals. These intervals on the x axis may be finite or may extend to infinity in either direction. The intervals in which the sign is preserved are usually separated by isolated points. At such separation points, either there is no finite derivative or the derivative is zero.

For polynomials the separation points are points at which the derivative is zero, and we may use them to find where the function increases and where it decreases. In fact, the derivative of a polynomial is itself a polynomial and so has a factor of the form $x - a$ if it is zero at a. Since $x - a$ is negative for $x < a$ and positive for $x > a$, the sign of the product of such factors is easily determined.

EXAMPLE 1. For what values of x is the function

$$y = f(x) = 3x^4 + 2x^3 - 6x^2 - 6x + 1 \tag{34}$$

increasing and for what values is it decreasing?

Solution: We first find the derivative of y,

$$\frac{dy}{dx} = f'(x) = 12x^3 + 6x^2 - 12x - 6.$$

The right member is zero for $x = 1$ and for $x = -1$. Hence by the factor theorem† it has the factors $(x - 1)$ and $(x + 1)$ whose product is $x^2 - 1$. If we divide by this factor, we find that the remaining factor is $12x + 6 = 12(x + \frac{1}{2})$. It follows that

$$f'(x) = 12(x + 1)(x + \tfrac{1}{2})(x - 1). \tag{35}$$

Let us study the sign of $f'(x)$ from the following table:

x	12	$x + 1$	$x + \frac{1}{2}$	$x - 1$	$f'(x)$
	+	−	−	−	−
-1		0			0
	+	+	−	−	+
$-\frac{1}{2}$			0		0
	+	+	+	−	−
1				0	0
	+	+	+	+	+

The table is constructed by first noting that $f'(x)$ is zero at $1, -1$, and $-\frac{1}{2}$. Since $-1 < -\frac{1}{2} < 1$, we list these in the x column at the left in the order $-1, -\frac{1}{2}, 1$. Opposite these we put 0 in the column of the factor which vanishes, and 0 in the column for the sign of $f'(x)$, since a product of factors is zero whenever any one factor is zero. Our factors of the first degree are written with the coefficient of x positive, here $+1$. Hence each factor is minus for x less than the value which makes the factor zero. And each factor is plus for x greater than this value. We imagine x to increase algebraically as we go down the column. Thus the signs above the line for -1 apply to the open interval $-\infty < x < -1$, or values of $x < -1$. In this interval we have one plus and three minus signs, for the factors, so that the product $f'(x)$ is minus, as noted in the right-hand column.

The next row of signs apply to the interval $-1 < x < -\frac{1}{2}$, where the product has two minus signs and is plus, and so on.

From the right-hand column we may infer that, as x increases,

if $x < -1$,	$f(x)$ decreases
if $-1 < x < -\frac{1}{2}$,	$f(x)$ increases
if $-\frac{1}{2} < x < 1$,	$f(x)$ decreases
if $1 < x$,	$f(x)$ increases.

† The factor theorem states that, if $P(x)$ is a polynomial, and r is a root of $P(x) = 0$, then $P(x)$ is exactly divisible by $(x - r)$.

Thus the answer to the question of Example 1 is as follows: The function is increasing if $-1 < x < -\frac{1}{2}$ or if $x > 1$. And the function is decreasing if $x < -1$ or if $-\frac{1}{2} < x < 1$.

EXAMPLE 2. Using the results of Example 1, make a rough sketch of the graph of $y = 3x^4 + 2x^3 - 6x^2 - 6x + 1$.

Solution: To sketch the graph, we first plot the points where the derivative $f'(x)$ is zero. The table of Example 1 gives the values of x as -1, $-\frac{1}{2}$, and 1. From the expression for $f(x)$ in Eq. (34) we find that $f(-1) = 2$, $f(-\frac{1}{2}) = \frac{39}{16} = 2.44$, and $f(1) = -6$. This gives the three points $(-1,2)$, $(-0.5,2.44)$, and $(1,-6)$ which lie on the curve and at which the slope is zero. We plot these points and draw short horizontal lines through them. These lines will be tangent to the graph. We may then complete the graph of Fig. 27, making it fall for $x < -1$, rise for $-1 < x < -\frac{1}{2}$, fall for $-\frac{1}{2} < x < 1$, and rise for $x > 1$ in accord with the answer to Example 1. In drawing the outer parts of the curve it may be desirable to plot two points outside the range $-1 < x < 1$, as $(-2,21)$ and $(2,29)$.

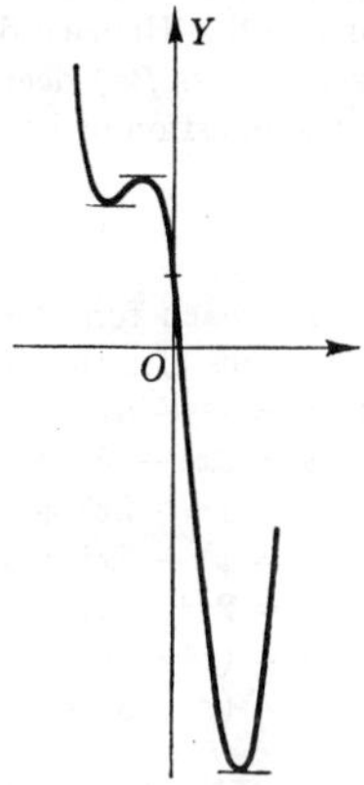

FIG. 27.

EXAMPLE 3. For what values of x is the graph of

$$y = f(x) = -2x^5 + 5x^4 - 10x^3 \tag{36}$$

rising, and for what values is it falling?

Solution: We first find the derivative of y,

$$\begin{aligned} \frac{dy}{dx} = f'(x) &= -10x^4 + 20x^3 - 30x^2 \\ &= -10x^2(x^2 - 2x + 3). \end{aligned} \tag{37}$$

To see if the second-degree factor $x^2 - 2x + 3$ has real factors, we find the roots of $x^2 - 2x + 3 = 0$. In fact, if r_1 and r_2 are the two roots of the quadratic equation

$$ax^2 + bx + c = 0, \tag{38}$$

then in accord with the factor theorem quoted in Example 1 we have the identity

$$ax^2 + bx + c = a(x - r_1)(x - r_2). \tag{39}$$

And r_1 and r_2 are given by the *quadratic formula*

$$x = \frac{-b \pm \sqrt{b^2 - 4ac}}{2a}. \tag{40}$$

In our case, for $x^2 - 2x + 3$, $a = 1$, $b = -2$, and $c = 3$ so that

$$x = \frac{-2 \pm \sqrt{-8}}{2} = 1 \pm i\sqrt{2},$$

where i is the imaginary unit, $i = \sqrt{-1}$. Hence

$$\begin{aligned} x^2 - 2x + 3 &= (x - 1 - i\sqrt{2})(x - 1 + i\sqrt{2}) \\ &= (x - 1)^2 + 2. \end{aligned}$$

This last expression, which might have been obtained directly by completing the square, shows that the real quadratic factor has no real first-degree factors and is positive for all real values of x. Thus our table in this case is

x	-10	x^2	$x^2 - 2x + 3$	$f'(x)$
	$-$	$+$	$+$	$-$
0		0		0
	$-$	$+$	$+$	$-$

Here x is negative before x increases to 0, or for $x < 0$. But x^2 is positive. From the right-hand column we may infer that as x increases, $f(x)$ decreases for $x < 0$ and also for $x > 0$. Hence it decreases as x increases through 0. Note that although $f'(0) = 0$, nevertheless $f(x)$ decreases at $x = 0$ by the definition of Sec. 39. Thus the answer to the question of Example 3 is as follows: The curve is falling for all values of x.

EXERCISE 16

For each function, find for what values of x the graph is rising, and for what values of x the graph is filling.

1. $y = x^2 - 6x + 3$.
2. $y = 2 + 4x - x^2$.
3. $y = 2x^3 - 9x^2 + 12x$.
4. $y = 4 + 6x - 2x^3$.
5. $y = x^4 - 2x^2 + 1$.
6. $y = 2 + 8x^2 - x^4$.
7. $y = x^5 - 5x^4 + 100$.
8. $y = x^4 - 32x + 3$.
9. $y = 2x^3 + 3x^2 + 14x$.
10. $y = 9x + 3x^2 - x^3$.
11. $y = (x^2 - 1)^2$.
12. $y = (x - 1)^2(x - 2)^3$.
13. $y = (x - 3)(x - 6)^2$.
14. $y = x(x + 6)^2(x - 6)^3$.

41. The Second Derivative. Let $y = f(x)$ be a given function of x. If this function is differentiable, its derivative $dy/dx = f'(x)$ will be a new function of x. Suppose that this new function is differentiable. Then the derivative of the *first* derivative is called the *second* derivative of the original function.

For example, if

$$y = 9x^5 - 7x^3 + 5x^2 + 3x - 2,$$

the first derivative is

$$\frac{dy}{dx} = 45x^4 - 21x^2 + 10x + 3.$$

And the second derivative is

$$\frac{d}{dx}\left(\frac{dy}{dx}\right) = 180x^3 - 42x + 10.$$

We usually write d^2y/dx^2 as an abbreviation for $\frac{d}{dx}\left(\frac{dy}{dx}\right)\cdot$ The reader should note carefully the position of the superscript in the numerator *between* d and y, which is intended to suggest $\frac{d}{dx}\left(\frac{d}{dx}\right)y$ or $\left(\frac{d}{dx}\right)^2 y$. Other notations analogous to $f'(x)$ and y' for the first derivative are $f''(x)$ and y''. Thus if $y = f(x)$, the second derivative is denoted by the symbols

$$\frac{d^2y}{dx^2} = f''(x) = y''. \tag{41}$$

This equation may be read "the second derivative of y with respect to x," or "dee second y dee x squared, equals f second of x equals y second."

The derivative of the second derivative is called the third derivative, and so on for derivatives of the fourth and higher order. And we use notations analogous to those of Eq. (41). Thus for the third derivative of $y = f(x)$,

$$\frac{d^3y}{dx^3} = f'''(x) = y''' = \frac{d}{dx}f''(x). \tag{42}$$

And for the nth derivative of $y = f(x)$,

$$\frac{d^ny}{dx^n} = f^{(n)}(x) = y^{(n)} = \frac{d}{dx}f^{(n-1)}(x). \tag{43}$$

EXERCISE 17

Find the second derivative, d^2y/dx^2, for each given function.

1. $y = 2x^3 - 3x^2 + 2x - 3.$
2. $y = 3x^4 - 4x^2 + 6.$
3. $y = x^5 + 5x^3 - 4x.$
4. $y = 2x^6 + 3x^4 - 7x^2.$
5. $y = 4(3 - 2x)^4.$
6. $y = (x - 1)^4(x - 2)^3.$
7. $y = x(x + 2).^5$
8. $y = (x^2 - 4)^5.$

For each of the following functions, find the value of the third derivative, d^3y/dx^3 for the given value of x.

9. $y = x^7 - x + 1,\ x = 2.$
10. $y = 2(x + 3)^4,\ x = -1.$
11. $y = 4x^3 + x^2 - x + 2,\ x = x_0.$
12. $y = ax^2 + bx + c,\ x = x_0.$
13. $y = 5x^4,\ x = -3.$
14. $y = (3x - 4)^5,\ x = 1.$

42. Velocity and Acceleration. We discussed the motion of a particle in a straight line in Secs. 20 and 21. A comparison of the definitions of velocity and acceleration found there with Sec. 27 shows that if the distance s is a function of the time t, $s = f(t)$, the velocity $v = ds/dt$ and the acceleration $a = dv/dt$. Thus, with the notation of Sec. 41 we may write for motion in a straight line

$$\text{Distance } s = f(t).$$

$$\text{Velocity } v = \frac{ds}{dt} = f'(t). \tag{44}$$

$$\text{Acceleration } a = \frac{dv}{dt} = \frac{d^2y}{dt^2} = f''(t).$$

By Sec. 38, if $f'(t)$ is positive, then $f(t)$ is an increasing function. Hence if the velocity is positive, then s increases with the time. Thus the particle is moving in the direction of positive s. On the other hand, by Sec. 39, if the velocity is negative, the particle is moving in the direction of negative s.

Similarly, if the acceleration is positive, then the velocity is algebraically increasing with time. But if the acceleration is negative, then the velocity is algebraically decreasing with time.

The absolute value of the velocity $|v|$ is called the *speed in the path*, or simply the *speed*. Whether the speed is increasing or decreasing with time depends on the sign of the velocity as well as on the sign of the acceleration. In fact, the speed $|v|$ will increase if $|v|^2 = v^2$ increases. But

$$\frac{d}{dt}v^2 = 2v\frac{dv}{dt} = 2va. \tag{45}$$

Hence, if the velocity and the acceleration have the same algebraic sign, the speed increases with time. If the velocity and the acceleration have opposite signs, the speed is a decreasing function of the time.

EXAMPLE. A particle is moving on a vertical straight line, and the upward direction is taken as the positive direction for s. Discuss the motion for which

$$s = 4t^3 - 18t^2 + 24t + 15. \tag{46}$$

Solution: We have in this case

$$\begin{aligned}\frac{ds}{dt} &= 12t^2 - 36t + 24 \\ &= 12(t^2 - 3t + 2) = 12(t - 1)(t - 2).\end{aligned} \tag{47}$$

By the method used to solve the examples of Sec. 40, we find that ds/dt is positive if $t < 1$ or if $t > 2$. But if $1 < t < 2$, then ds/dt is negative.

Hence the particle in this case is ascending for $t < 1$, descending for $1 < t < 2$, and ascending for $t > 2$.

The acceleration is found from the first expression for ds/dt to be

$$\begin{aligned}\frac{d^2s}{dt^2} &= 24t - 36 \\ &= 24(t - \tfrac{3}{2}).\end{aligned} \tag{48}$$

This is negative when $t < \frac{3}{2}$ and positive when $t > \frac{3}{2}$.

Hence the velocity is algebraically decreasing for $t < \frac{3}{2}$ and algebraically increasing for $t > \frac{3}{2}$.

To find when the speed is increasing, we may make a table of the behavior of $v = ds/dt$ and $a = d^2s/dt^2$ in the intervals where both of these preserve their sign. Thus we have

t		1		$\frac{3}{2}$		2	
v	+	0	−		−	0	+
a	−		−	0	+		+
va	−	0	+	0	−	0	+

The first line is the variation of t, with the values of t at which v or a is zero marked. In the line for v we put 0 under 1 and 2, and fill in the signs from Eq. (47). In the line for a we put 0 under $\frac{3}{2}$, and fill in the signs from Eq. (48). From the last line, which gives the sign of the product va, and the principle deduced from Eq. (45) we see that the speed increases with time if $1 < t < \frac{3}{2}$ or if $t > 2$. And the speed is a decreasing function of the time if $t < 1$ or if $\frac{3}{2} < t < 2$.

EXERCISE 18

For each of the following motions determine when the particle is moving in the direction taken as positive for s.

1. $s = 2t^2 - 6t + 5$.
2. $s = 3 + 4t - t^2$.
3. $s = t^3 - 3t^2 + 2$.
4. $s = t^3 - 9t^2 + 24t$.

For each of the following motions determine when the velocity is algebraically increasing.

5. $s = t^2 - 3t + 5$.
6. $s = t^3 - 6t^2 + 2t$.
7. $s = 2 + 12t - t^3$.
8. $s = (3 - t)^5$.

For each of the following motions determine when the speed is increasing and when the speed is decreasing.

9. $s = t^3 - 6t^2 + 9t$.
10. $s = 3t^2 - t^3 + 4$.
11. $s = t^4 - 12t^3 + 1$.
12. $s = 2 + 18t^2 - t^4$.

43. Bending of an Arc. The arcs shown in Figs. 28 and 29 have a different character. If the y axis were to extend vertically upward, and

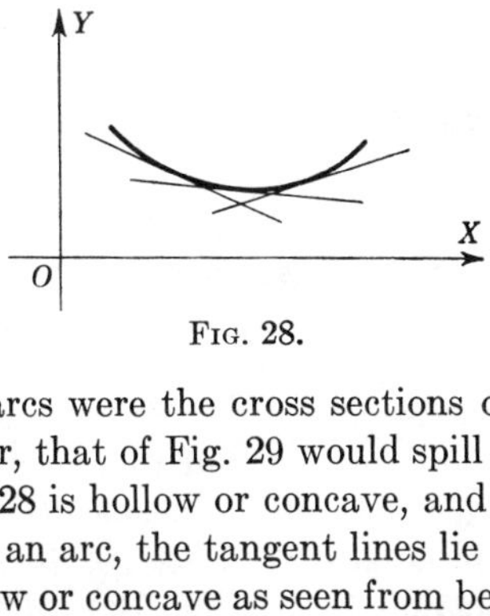

FIG. 28.

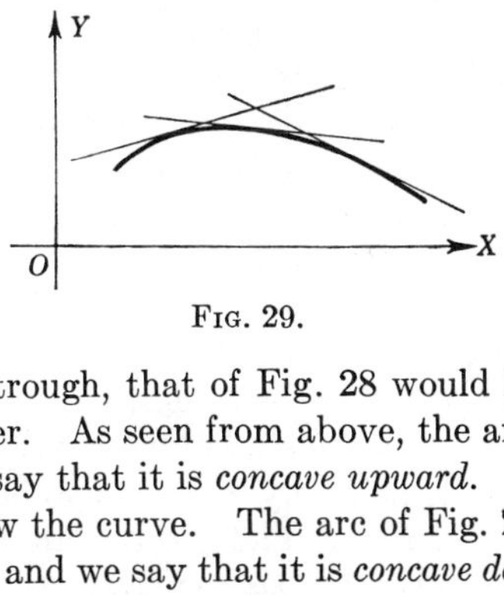

FIG. 29.

the arcs were the cross sections of a trough, that of Fig. 28 would hold water, that of Fig. 29 would spill water. As seen from above, the arc of Fig. 28 is hollow or concave, and we say that it is *concave upward.* For such an arc, the tangent lines lie below the curve. The arc of Fig. 29 is hollow or concave as seen from below, and we say that it is *concave downward.* For such an arc, the tangent lines lie above the curve.

Let us think of these arcs as portions of the graph of $y = f(x)$ being traced out by a point moving from left to right, so that x increases. For the concave upward arc of Fig. 28, the curve bends upward and the slope increases as x increases. But for the concave downward arc of Fig. 29, the curve bends downward and the slope decreases as x increases.

Since the slope of the graph of $y = f(x)$ is $f'(x)$, which has the derivative $f''(x)$, by Sec. 38 the slope will be an increasing function when $f''(x)$ is positive. And by Sec. 39 the slope will be a decreasing function when

$f''(x)$ is negative. Hence we may determine the direction of bending of an arc from the sign of the second derivative by the following principle:

The graph of $y = f(x)$ *is concave upward on any arc where the second derivative* $d^2y/dx^2 = f''(x)$ *is positive.*

And the graph of $y = f(x)$ *is concave downward on any arc where the second derivative* $d^2y/dx^2 = f''(x)$ *is negative.*

In symbols, throughout any arc

Where $f''(x)$ is $+$, the arc resembles $\cup$.
And where $f''(x)$ is $-$, the arc resembles $\cap$.

Although the result can be reasoned out from the slope behavior, one may remember the result by thinking of the arc as the edge of a trough, upright so as to hold water $(+)$, or inverted so as to spill water $(-)$.

44. Maximum. *Definition.* The function $y = f(x)$ is said to have a *maximum value* $y_1 = f(x_1)$ at x_1 if for every value x_2 sufficiently near to x_1 but not equal to x_1, $f(x_1)$ is greater than $f(x_2)$.

In particular, let p be some positive number and let $f(x)$ increase as x increases from $x_1 - p$ to x_1, and then decrease as x increases from x_1 to $x_1 + p$. Then $f(x_1)$ will necessarily be a maximum. But this situation must take place, by Secs. 38 and 39, if $f'(x)$ is positive in some interval $x_1 - p < x < x_1$ and negative in some interval $x_1 < x < x_1 + p$.

If $f'(x)$ is continuous at x_1, it can change sign only by passing through zero. This leads to one sufficient condition for a maximum:

If $f'(x_1) = 0$, *and as* x *increases through* x_1, $f'(x_1)$ *changes sign from plus to minus, then* $f(x)$ *has a maximum at* x_1.

45. Minimum. *Definition.* The function $y = f(x)$ is said to have a *minimum value* $y_1 = f(x_1)$ at x_1 if for every value x_2 sufficiently near to x_1 but not equal to x_1, $f(x_1)$ is less than $f(x_2)$.

In particular, let p be some positive number and let $f(x)$ decrease as x increases from $x_1 - p$ to x_1 and then increase as x increases from x_1 to $x_1 + p$. Then $f(x_1)$ will necessarily be a minimum. But this situation must take place, by Secs. 38 and 39, if $f'(x)$ is negative in some interval $x_1 - p < x < x_1$ and positive in some interval $x_1 < x < x_1 + p$.

Since a continuous $f'(x)$ can change sign only by passing through zero, this leads to a sufficient condition for a minimum:

If $f'(x_1) = 0$, *and as* x *increases through* x_1, $f'(x_1)$ *changes sign from minus to plus, then* $f(x)$ *has a minimum at* x_1.

46. The Second Derivative Test. Suppose that $f'(x_1) = 0$, $f''(x_1) \neq 0$, and that $f''(x)$ is continuous at x_1. Then in some interval $x_1 - p < x < x_1 + p$, $f''(x)$ preserves its sign.

First suppose that $f''(x_1)$ is positive. Then the argument used at the end of Sec. 38 shows that $f'(x)$ increases as x increases through x_1. Since $f(x_1) = 0$, the increase must be from negative values through zero to positive values. Hence from the test of Sec. 45, it follows that $f(x)$ has a minimum at x_1. And we may formulate the condition:

If $f'(x_1) = 0$ *and* $f''(x_1) > 0$, *the function* $y = f(x)$ *has a minimum value* $f(x_1)$ *at* x_1.

Similarly if $f'(x_1)$ is negative, we may conclude that $f'(x_1)$ must decrease through zero from positive values to negative values. Hence by the test of Sec. 44, $f(x)$ has a maximum at x_1. And we may formulate the condition:

If $f'(x_1) = 0$ *and* $f''(x_1) < 0$, *the function* $y = f(x)$ *has a maximum value* $f(x_1)$ *at* x_1.

47. Point of Inflection. *Definition.* A point on a smooth curve which separates two arcs having opposite directions of bending, is called a *point of inflection.*

If the graph of $y = f(x)$ has a point of inflection at x_1, there is some positive number p such that the curve is concave upward in one of the intervals† $x_1 - p, x_1$ or $x_1, x_1 + p$, and concave downward in the other of these intervals. Hence, by Sec. 43, as x increases through x_1, the second derivative will change sign. If $f''(x)$ is continuous at x_1, it can change sign only by passing through zero. This leads to the following sufficient condition for a point of inflection:

If $f''(x_1) = 0$, *and* $f''(x)$ *changes sign as* x *increases through* x_1, *then* (x_1,y_1) *with* $y_1 = f(x_1)$ *is a point of inflection on the graph of* $y = f(x)$.

FIG. 30.

A point of inflection separates an arc which is concave upward, as AP in Fig. 30, from an arc which is concave downward, as PD in Fig. 30. Near any point B on the arc AP, which bends upward, the curve lies above the tangent at B. And near any point C on the arc PD, which bends downward, the curve lies below the tangent at C. And at the point of inflection P, the tangent crosses the curve. Near any point of inflection P, the curve separates very slowly from the tangent to the curve at P, so that in practical drawing we may use a short piece of this tangent as a part of the graph.

48. Graphs. Let us illustrate the terms and results of Secs. 43 to 47 by discussing the behavior and graph of the function

$$y = x^3 - 3x^2 + 3. \tag{49}$$

The first derivative is

$$\frac{dy}{dx} = 3x^2 - 6x$$
$$= 3x(x - 2). \tag{50}$$

† The notation $x_1 - p, x_1$ for the interval $x_1 - p < x < x_1$ was discussed in Sec. 2.

By the method used to solve the examples of Sec. 40, we find that dy/dx is positive if $x < 0$ or if $x > 2$. Hence by Sec. 38 the given function is increasing in these intervals. But dy/dx is negative if $0 < x < 2$. Hence by Sec. 39 the given function decreases in this interval.

From the first expression for dy/dx in Eq. (50), we find that

$$\begin{aligned} \frac{d^2y}{dx^2} &= 6x - 6 \\ &= 6(x - 1). \end{aligned} \tag{51}$$

Thus d^2y/dx^2 is positive if $x > 1$. Hence by Sec. 43 the graph of the given function is concave upward in this interval. But d^2y/dx^2 is negative if $x < 1$. Hence in this interval the graph is concave downward.

The points where $dy/dx = 0$ are $x = 0$ and $x = 2$, from Eq. (50). At $x = 0$, y has a maximum. This may be seen from any one of the four following considerations.

1. As x increases through zero, y increases and then decreases.
2. As x increases through zero, dy/dx changes from plus to minus.
3. At $x = 0$, $dy/dx = 0$ and $d^2y/dx^2 = -6$ from Eq. (51), is negative.
4. At $x = 0$, $dy/dx = 0$ and 0 is in an interval, $x < 1$, in which the graph is concave downward.

From any one of four similar considerations, we may conclude that y has a maximum at $x = 2$. For example, analogous to item 3, at $x = 2$, $dy/dx = 0$ and $d^2y/dx^2 = 6$ from Eq. (51) is positive.

There is one point where $d^2y/dx^2 = 0$, namely, that with $x = 1$. This is point of inflection, since d^2y/dx^2 changes sign as x increases through 1.

The facts just derived are helpful in sketching the graph of the given function. From Eq. (49) we find the values of y for $x = 0, 1, 2$. They are $3, 1, -1$. We plot $(0,3)$ and $(2,-1)$ and draw short horizontal tangents at these points. We also plot the point of inflection $(1,1)$. At this point $x = 1$ and from Eq. (50) we find that $dy/dx = -3$. This is the slope of the tangent line. We draw in this tangent line, using the method of Sec. 24. For example, we may draw the line through $(1,1)$ and $(2,-2)$.

To construct an accurate graph of the curve it is useful to have two additional points with values of x outside the interval 0,2. For example, when $x = -1$, $y = -1$ from Eq. (49), and when $x = 3$, $y = 3$. At each of these points the slope, $dy/dx = 9$, from Eq. (50). Hence we may draw the tangent at $(-1,-1)$ through $(0,8)$, and the tangent at $(3,3)$ through $(2,-6)$.

The five points and tangent lines, together with the direction of bending and relation of the arc to the tangents, fairly well define the curve as sketched in Fig. 31. In particular from the graph we may read approximate values of the three real roots of $x^3 - 3x^2 + 3 = 0$.

We observe that if we change the unit on the y scale, as is suggested in some of the problems of Exercise 19, we may use all the properties above except that $m = dy/dx$ will no longer be the trigonometric tangent of the slope angle or inclination. But the value of $m = dy/dx$ will still determine the direction of the tangent line at (x_1,y_1) with $y_1 = f(x_1)$. And we may draw this tangent line through the two points (x_1,y_1) and $(x_1 + a, y_1 + ma)$, plotted on the new scales, and with a any convenient value.

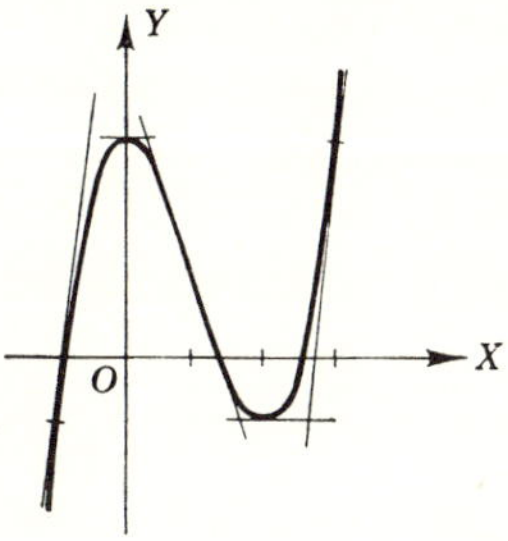

FIG. 31.

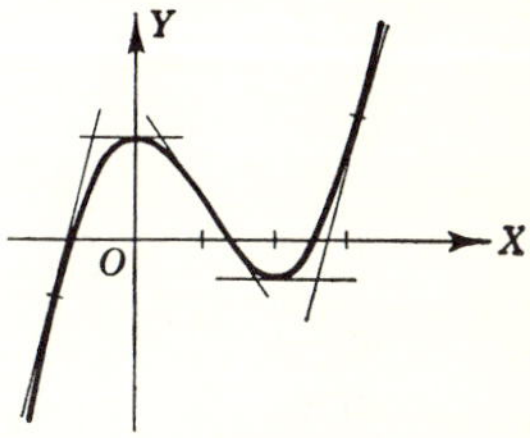

FIG. 32.

Thus in this way, Fig. 32, which is the graph of Eq. (49) with a unit on the y axis one-half as large as the unit on the x axis, may be constructed.

EXERCISE 19

Examine each of the following functions for maxima and minima. In each case plot the maximum points, minimum points, and points of inflection. Draw the tangents at these points and sketch the graph of the function. Whenever the values of y to be plotted are large compared with those of x, choose a convenient unit for y different from the unit for x.

1. $y = x^2 + 2x + 4$.
2. $y = 4 + 2x - x^2$.
3. $y = x^3 - 6x^2 + 9x$.
4. $y = 12x - x^3$.
5. $y = x^3 + 3x^2 - 4$.
6. $y = x^3 - 3x^2 + 9x$.
7. $y = x^3 + 3x^2 - 3x$.
8. $y = 12x - 3x^2 - 2x^3$.
9. $y = (x - 2)^3$.
10. $y = (x - 3)^4$.
11. $y = (x^2 - 1)^3$.
12. $y = (x^2 - 1)^2$.

***49. Roots by Newton's Method.** When approximate values of the roots of an equation $f(x) = 0$ are known, more accurate values can be computed by a process known as Newton's method of successive approximation.

Let the graph of $y = f(x)$ cross the x axis in $R = (r,0)$. Then $0 = f(r)$, so that r is a root of the equation $f(r) = 0$. We assume that $f'(r) \neq 0$ and $f''(r) \neq 0$. We also assume that $f'(x)$ and $f''(x)$ are continuous at r, so that they have fixed signs for x near r. Consider some point $P_1 = (x_1,y_1)$ with $y_1 = f(x_1)$ on the graph of $y = f(x)$ near R. We choose P_1 above OX if the arc RP is concave upward, as in Fig. 33. But we choose P_1 below OX if the arc RP is concave downard, as in Fig. 34. Then if the tangent line to $y = f(x)$ at P_1 cuts OX in a point $X_2 = (x_1,0)$, as the figures suggest, the value x_2 will be an approximation to the root of $f(x) = 0$ closer to r than x_1.

To calculate x_2, we note from triangle X_2X_1P that

$$\frac{X_1P}{X_2X_1} = \tan\phi_1 \quad \text{or} \quad \frac{y_1}{x_1 - x_2} = f'(x_1), \tag{52}$$

and this relation is true for any combination of signs of the quantities. Hence we have

$$x_1 - x_2 = \frac{y_1}{f'(x_1)} = \frac{f(x_1)}{f'(x_1)}. \tag{53}$$

This gives the formula for computing the second approximation

$$x_2 = x_1 - \frac{f(x_1)}{f'(x_1)}. \tag{54}$$

If better approximations are desired, we may compute

$$x_3 = x_2 - \frac{f(x_2)}{f'(x_2)}, \tag{55}$$

and so on. Under the conditions stated above, as we shall show in Sec. 262, when we are near a root, the error in each approximation will be of the order of magnitude of

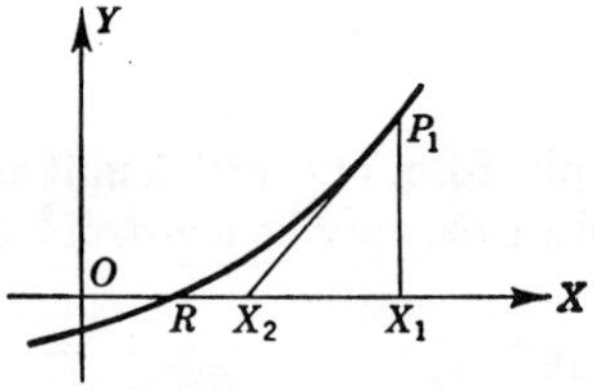

FIG. 33.

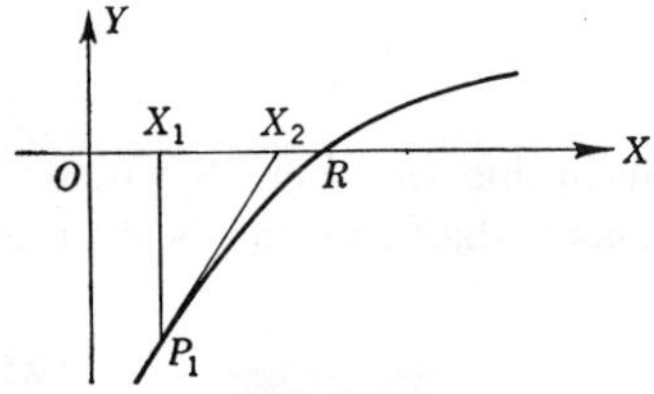

FIG. 34.

the square of the error of the preceding approximation. That is, $|r - x_2|$ will be roughly the size of $(r - x_1)^2$. The first approximation may be taken on the proper side of the root if read from a graph. Otherwise, a good approximation on the wrong side will usually lead to a second approximation on the side favorable for convergence.

EXAMPLE. Find the largest root of the equation

$$x^3 - 3x^2 + 3 = 0. \tag{56}$$

From the graph of $y = x^3 - 3x^2 + 3$ in Fig. 31, we see that the largest root is between 2 and 3. As the curve is concave upward near this root, we take $x_1 = 3$, as this makes y_1 positive. We have here

$$f(x) = x^3 - 3x^2 + 3, \qquad f(3) = 3.$$
$$f'(x) = 3x^2 - 6x, \qquad f'(3) = 9.$$

Hence, with $x_1 = 3$, from Eq. (54) the second approximation is

$$x_2 = x_1 - \frac{f(x_1)}{f'(x_1)} = 3 - \frac{3}{9} = 2.67.$$

We round this off and take $x_2 = 2.7$. Then

$$f(x_2) = f(2.7) = 0.903 \qquad \text{and} \qquad f'(x_2) = f'(2.7) = 5.58.$$

Hence with $x_2 = 2.7$, from Eq. (55) the third approximation is

$$x_3 = x_2 - \frac{f(x_2)}{f'(x_2)} = 2.7 - \frac{0.903}{5.58} = 2.54.$$

A comparison of x_2 and x_3 indicates that x_2 is about 0.2 too big, so that x_3 is about $(0.2)^2 = 0.04$ too big. Hence its second decimal place has some value, so that we use $x_3 = 2.54$ to find x_4. Then

$$f(2.54) = 0.0323, \quad f'(2.54) = 4.11.$$

Hence with $x_3 = 2.54$, the fourth approximation is

$$x_4 = x_3 - \frac{f(x_3)}{f'(x_3)} = 2.54 - \frac{0.0323}{4.11} = 2.5320.$$

This is correct to nearly four places, since a comparison of x_3 and x_4 indicates that x_3 is about 0.008 too big, so that x_4 is about $(0.008)^2 = 0.00006$, which should not seriously affect the fourth decimal place.

If $f(x_4)$ were calculated to at least eight places, and $f'(x_4)$ to at least five significant figures, they could be used to calculate a value of x_5 good to seven or eight places.

If we had not already drawn the graph, we might have obtained a first approximation 2.5. This would give as the second approximation 2.536, which would be rounded off to 2.54, and a further approximation as above would then give 2.5320. Note that 2.5 is on the unfavorable side of the root, but the later approximations are on the favorable side.

EXERCISE 20

Each of the following equations has a real root between the indicated limits. Use Newton's method to compute this root to at least two significant figures.

1. $x^3 + 2x - 6 = 0, \quad 1 < x < 2.$
2. $x^4 - 4x^3 + 4 = 0, \quad 3 < x < 4.$
3. $x^3 + x - 11 = 0, \quad 2 < x < 3.$
4. $x^4 + 11x + 5 = 0, \quad -3 < x < -2.$
5. $x^3 - 4x + 2 = 0, \quad 0 < x < 1.$
6. $x^3 - 3x - 1 = 0, \quad -1 < x < 0.$
7. $x^4 - 8x - 12 = 0, \quad 2 < x < 3.$
8. $x^5 - 3x + 1 = 0, \quad 1 < x < 2.$

50. Finding Greatest and Least Values. Many applied problems involve a relation between two variables. It is often of practical importance to determine which value of one of the variables makes the other as large as possible, or as small as possible. And we may want to calculate this greatest or least value.

It is usually advisable to begin by setting up some expression for the quantity whose extreme value is required. In the type of application discussed here, if this expression contains more than one variable, we must use other conditions of the problem to obtain one or more auxiliary relations so that all the extra variables may be expressed in terms of a single one and so eliminated from the expression first constructed. Other types of extreme value problems involving several variables will be discussed in Sec. 278.

Suppose that the single variable is x, and after elimination the expression whose extreme value is sought is $f(x)$. If, for x in the range of physical interest, $f(x)$ has a continuous derivative, $f'(x)$, any value of x *inside* this range at which $f(x)$ is a maximum or a minimum must make $f'(x) = 0$. Hence this will be true for the value which makes $f(x)$ greatest or least.

In the type of application made in this section, if a greatest value is sought, there will be only one maximum in the range of interest. And similarly, if a least value is sought, there will be only one minimum in the range of interest. Hence we may find the desired value of x among the values which make $f'(x) = 0$. Other types of maxima and minima will be discussed in Sec. 58.

In many problems it is evident from general considerations which of the roots of $f'(x) = 0$ has the desired property.

In other cases it may be useful to verify the maximum or minimum property by using the tests of Sec. 44, 45, or 46. In any case of doubt, we may draw the graph of $y = f(x)$ to check the behavior of $f(x)$ near the value in question.

After the desired value of the single variable is known, any other desired quantities may be found from the original expression, or the auxiliary relations used to find $f(x)$.

To illustrate the procedure we shall solve some examples.

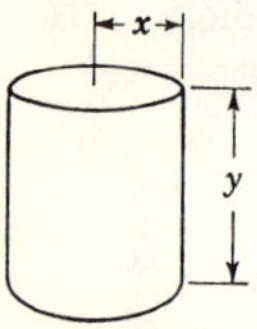

FIG. 35.

EXAMPLE 1. A closed cylindrical tin can is to have a surface of 200 sq. in. What dimensions will make its volume greatest?

Solution: Let the radius of the base be x and the height be y, as indicated in Fig. 35. Then the surface area $A = 2\pi x^2 + 2\pi xy$. And the volume $V = \pi x^2 y$. Thus we are to find the maximum of

$$V = \pi x^2 y, \qquad \text{when } 200 = 2\pi x^2 + 2\pi xy.$$

This is the auxiliary relation. Since it involves y to the first power only, we solve for y, obtaining

$$y = \frac{100 - \pi x^2}{\pi x}.$$

By substituting this value of y in $V = \pi x^2 y$, we find

$$V = 100x - \pi x^3.$$

From this we obtain

$$\frac{dV}{dx} = 100 - 3\pi x^2.$$

This is zero when $3\pi x^2 = 100$ or when $x = \pm\sqrt{100/3\pi}$. Only positive values of the radius x are admissible, so we test the plus value. The second derivative of V is $d^2V/dx^2 = -6\pi x$. This is negative when x is positive, and in particular when

$x = \sqrt{100/3\pi}$. Hence this value does make V a maximum. The value of y for this is

$$y = \frac{100 - \pi x^2}{\pi x} = \frac{200/3}{\pi \sqrt{100/3\pi}} = 2\sqrt{\frac{100}{3\pi}}.$$

Since $y = 2x$, the height equals the diameter of the base.

EXAMPLE 2. What is the volume of the right circular cone of greatest volume that can be cut from a sphere of radius a?

Solution: A cross section of the sphere and inscribed cone is shown in Fig. 36. As indicated, the distance from the base to the center is x, so that the altitude is $h = a + x$ and the radius of the base is y. Then from the right triangle, we have $x^2 + y^2 = a^2$. And the volume of the cone is $V = \frac{1}{3}\pi y^2 h$ or

$$V = \frac{\pi}{3} y^2(a + x).$$

We are to find the maximum value of this when

$$x^2 + y^2 = a^2.$$

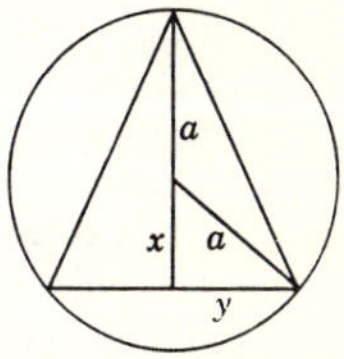

FIG. 36.

This is the auxiliary relation. Elimination of x would lead to a square root, but since V involves y^2 only, we may eliminate y^2 by solving the auxiliary equation in the form

$$y^2 = a^2 - x^2,$$

and then substituting this value in V, to obtain

$$V = \frac{\pi}{3}(a^2 - x^2)(a + x) = \frac{\pi}{3}(a + x)^2(a - x).$$

Differentiating the last product, we find

$$\frac{dV}{dx} = \frac{\pi}{3}[(a + x)^2(-1) + (a - x)(2)(a + x)(1)]$$

$$= \frac{\pi}{3}(a + x)(-a - x + 2a - 2x) = \frac{\pi}{3}(a + x)(a - 3x).$$

Thus $dV/dx = 0$ when $x = -a$, or when $a - 3x = 0$ and $x = a/3$. The second value is the only one in the range of physical interest, and leads to a maximum since, when x increases through $a/3$, the factor $\pi/3(a + x)$ of dV/dx is plus, but the factor $(a - 3x)$ changes from plus to minus.

Thus the maximum volume occurs when $x = a/3$. For this value $y^2 = a^2 - x^2 = a^2 - (a/3)^2 = 8a^2/9$, $h = a + x = a + a/3 = 4a/3$, and

$$V = \frac{\pi}{3} y^2 h = \frac{\pi}{3}\frac{8a^2}{9}\frac{4a}{3} = \frac{32\pi a^3}{81}.$$

This is reasonable, being $\frac{8}{27}$ the volume of the whole sphere.

EXAMPLE 3. A novelty salesman can sell 800 articles at \$1 each, and more if the price is lower. In fact for each 1 cent decrease in price, he estimates that 40 more articles can be sold. What price should he set initially to take in the maximum gross revenue?

Solution: At a price of x cents, the decrease of $100 - x$ means $40(100 - x)$ more sales. The total number of sales is 800 plus this or $4{,}800 - 40x$. And the revenue from these is

$$R = (4{,}800 - 40x)x = 4{,}800x - 40x^2.$$

From this we find

$$\frac{dR}{dx} = 4{,}800 - 80x.$$

This is zero when $x = 60$, which gives a maximum since the second derivative is negative. Hence the required price is 60 cents.

EXERCISE 21

1. The sum of two positive numbers is 14. For what two numbers is their product a maximum?
2. The sum of two positive numbers is 8. For what two numbers is the sum of their squares a minimum?
3. A piece of wire 24 in. in length is bent into the perimeter of a rectangle. What dimensions make the area of the rectangle a maximum?
4. A gardener has 80 ft. of wire fencing with which to fence three sides of a rectangular plot of land. The fourth side is part of a long wall already constructed. What are the dimensions for which the plot has maximum area?
5. Find the point (x,y) on the line $3x + 4y = 25$ for which the square of the distance to the origin, $x^2 + y^2$, is least.
6. Find the two points (x,y) on the curve $x^2 - y^2 + 5 = 0$ for which the square of the distance to the point (4,0), or $(x - 4)^2 + y^2$ is least.

The combined length and girth of a package sent by parcel post must not exceed 60 in. Find the dimensions of the package of maximum volume which can be mailed, if the package

7. Is rectangular with square cross section.
8. Is a right circular cylinder.
9. A piece of tin is a square 30 in. on a side. From it a box is to be constructed by cutting a small square of side x in. from each corner and bending the resulting figure. What value of x makes the volume of the box, $V = x(30 - 2x)^2$, least?
10. A printer will print 2,000 handbills for \$1 per hundred. For a larger order he will make a reduction of 1 cent per hundred on the average price for the whole lot for each hundred in excess of 2,000. Thus for an order of x handbills, the price is $100 - \frac{x - 2{,}000}{100}$ cents per hundred, or $10^{-6}(12{,}000x - x^2)$ dollars for the total price. For what size order is this price greatest?
11. A club charters bus service for an excursion. The company charges \$9 per passenger if there are 100 or fewer passengers. But for each passenger in excess of 100, the average price per passenger is reduced by 5 cents. For what number of passengers does the company take in the greatest revenue?
12. Find the dimensions of the rectangle of maximum area which can be inscribed in an equilateral triangle.
13. Find the dimensions of the cylinder of greatest volume which can be inscribed in a sphere of radius a.
14. Find the dimensions of the cylinder of greatest volume which can be inscribed in a right circular cone of altitude h and radius of base a.
15. A gardener wishes to use 30 ft. of wire fencing to bound a flower bed in the form of a sector of a circle, that is with two straight sides along two radii and one circular arc. What dimensions make the area of the plot a maximum? Note that if the central angle is θ radians and the radius is x ft., the arc is θx ft. and the area is $A = \frac{1}{2}\theta x^2$.

16. A circular filter paper 12 in. in diameter is folded into the lateral surface of a right circular cone. What is the height of the cone when its volume is a maximum?
17. Assuming that the strength of a rectangular beam is proportional to the breadth x and the square of the depth y, so that $S = kxy^2$, find the dimensions of the strongest beam which can be cut from a circular cylindrical log of radius a.
18. A window is in the form of a rectangle of dimensions $2x$ by y surmounted by a semicircle of radius x. If the perimeter of the window is a constant p, find the value of x which makes the window have maximum area.
19. A rectangular box with square base and open at the top is to have the area of its base and sides together equal to 300 sq. in. What dimensions make the volume of the box greatest?
20. The sum of two positive numbers x and y is 12. What choice of the number x makes x^3y least?
21. A manufacturer can sell x articles per week at p dollars each, where $x = 200 - p$. The cost of producing the x articles is c dollars, where $c = 125 + 120x - x^2/2$. Show that his profit is a maximum when he sells 80 articles per week and that to do this he must set the price at \$120.

CHAPTER 4

ALGEBRAIC EXPRESSIONS. RELATED RATES

Continuing the development of special rules for differentiation, we consider fractions, rational powers, inverse functions, and composite functions. This enables us to find the derivative of any algebraic function given explicitly. We then explain a method for finding the derivative of a function given implicitly. In addition to the type of application made in the preceding chapter, we treat a number of problems involving related rates.

We explain the differential notation in connection with a method of finding derivatives from geometric considerations. We then apply these notions to the arc length of a curve, and to rates related to motion in a curved path.

51. Fractions. To lead up to the derivative of negative powers and fractions, we begin by considering the function

$$y = \frac{1}{u},$$

where u is a differentiable function of x, as in Sec. 32. For $1/u$ to have a meaning, we must assume $u \neq 0$ for the value of x considered.

When x is replaced by $x + \Delta x$, u becomes $u + \Delta u$. Thus $y = 1/u$ and $y + \Delta y = 1/(u + \Delta u)$. Consequently,

$$\Delta y = \frac{1}{u + \Delta u} - \frac{1}{u} = -\frac{\Delta u}{u(u + \Delta u)}.$$

Hence

$$\frac{\Delta y}{\Delta x} = -\frac{1}{u(u + \Delta u)} \frac{\Delta u}{\Delta x}.$$

We may apply the sum and product principles to deduce that when $\Delta x \to 0$, and hence $\Delta u \to 0$, the denominator $u(u + \Delta u) \to u^2$. This is not zero since $u \neq 0$, and we may apply the quotient principle and product principle to deduce that

$$\begin{aligned} \frac{dy}{dx} &= \lim_{\Delta x \to 0} -\frac{1}{u(u + \Delta u)} \lim_{\Delta x \to 0} \frac{\Delta u}{\Delta x} \\ &= -\frac{1}{u^2} \frac{du}{dx}. \end{aligned}$$

This proves that

$$\frac{d}{dx}\left(\frac{1}{u}\right) = -\frac{1}{u^2}\frac{du}{dx}. \tag{1}$$

We may combine this with the rule for differentiating products of Sec. 32 to derive a rule for differentiating the quotient of two differentiable functions, u/v. From the rule for products, we have

$$\frac{d}{dx}\left(\frac{u}{v}\right) = \frac{d}{dx}u\left(\frac{1}{v}\right) = u\frac{d}{dx}\left(\frac{1}{v}\right) + \frac{1}{v}\frac{du}{dx}. \tag{2}$$

But by replacing u by v in Eq. (1), we obtain

$$\frac{d}{dx}\left(\frac{1}{v}\right) = -\frac{1}{v^2}\frac{dv}{dx}. \tag{3}$$

On substituting from Eq. (3) in Eq. (2), we find

$$\frac{d}{dx}\left(\frac{u}{v}\right) = u\left(-\frac{1}{v^2}\frac{dv}{dx}\right) + \frac{1}{v}\frac{du}{dx},$$

so that

$$\frac{d}{dx}\left(\frac{u}{v}\right) = \frac{v(du/dx) - u(dv/dx)}{v^2}. \tag{4}$$

This proves the rule for differentiating fractions or quotients which may be stated in words as follows:

The derivative of a fraction is equal to the denominator times the derivative of the numerator, minus the numerator times the derivative of the denominator, all divided by the square of the denominator.

Suppose that the denominator is a constant, $v = c$. Then by Sec. 30, $dv/dx = dc/dx = 0$, so that Eq. (4) reduces to

$$\frac{d}{dx}\left(\frac{u}{c}\right) = \frac{1}{c}\frac{du}{dx}. \tag{5}$$

This is more easily seen as a consequence of the rule in Sec. 34, since $u/c = (1/c)u$, and $1/c$ is also a constant. This may aid the memory to recall that the plus sign goes with the term involving the derivative of the numerator, while the minus sign comes from Eq. (3) and so goes with the term involving the derivative of the denominator.

52. The Power Rule. Consider the derivative of $y = u^n$. We showed in Sec. 33 that, for $n = 0, 1, 2$, or any positive integer,

$$\frac{d}{dx}u^n = nu^{n-1}\frac{du}{dx}. \tag{6}$$

Next let n be a negative integer, $n = -m$. Then

$$u^n = u^{-m} = \frac{1}{u^m}. \tag{7}$$

But from Eq. (3) with $v = u^m$, we have

$$\frac{d}{dx}\left(\frac{1}{u^m}\right) = -\frac{1}{(u^m)^2}\frac{d}{dx}u^m \tag{8}$$

But by Eq. (6) with n replaced by the positive integer m, we have

$$\frac{d}{dx}u^m = mu^{m-1}\frac{du}{dx}. \tag{9}$$

From Eqs. (8) and (9) we may deduce that

$$\frac{d}{dx}\left(\frac{1}{u^m}\right) = -\frac{1}{u^{2m}}mu^{m-1}\frac{du}{dx} = -mu^{-m-1}\frac{du}{dx}.$$

Now use Eq. (7) in the left member and recall that $n = -m$. Thus

$$\frac{d}{dx}u^n = nu^{n-1}\frac{du}{dx}. \tag{10}$$

This proves that the power rule, Eq. (6), holds when n is a negative integer.

Next let n be any rational number. Then $n = p/q$, a fraction in its lowest terms, is the quotient of two integers p and q, one of which may be negative. Thus

$$y = u^n = u^{p/q}. \tag{11}$$

If q is odd, we take the real value; if q is even, we assume that u is positive and take the positive real value. With this convention y is determined from the equivalent relation

$$y^q = u^p. \tag{12}$$

Since these two functions are always equal, their derivatives with respect to x are always equal. Let us assume† that y has a derivative dy/dx. Then by Eq. (6) or (10), the derivatives are found to be

$$qy^{q-1}\frac{dy}{dx} = pu^{p-1}\frac{du}{dx}. \tag{13}$$

It follows that

$$\frac{dy}{dx} = \frac{pu^{p-1}}{qy^{q-1}}\frac{du}{dx}. \tag{14}$$

† The rule will be proved for all real values of n, without this assumption, in Sec. 122.

Now recall that $y = u^{p/q}$, so that the exponent of u in u^{p-1}/y^{q-1} is $p - 1 - (q - 1)p/q = p/q - 1$. Thus

$$\frac{d}{dx}u^{p/q} = \frac{p}{q}u^{p/q-1}\frac{du}{dx}. \tag{15}$$

Finally replace p/q by n to obtain

$$\frac{d}{dx}u^n = nu^{n-1}\frac{du}{dx}. \tag{16}$$

We have now proved that the power rule, which was proved in Sec. 33 to hold for positive integral exponents, is true for any positive or negative rational exponent. We may formulate it in words as follows:

The derivative of a function raised to a constant rational power is equal to the product of this power, the function raised to this power diminished by unity, and the derivative of the function.

EXAMPLE 1. Find $\frac{dy}{dx}$ if $y = \frac{2x^2 + 1}{4x^2 - 1}$.

Solution: Using Eq. (4) with $u = 2x^2 + 1$, $du/dx = 4x$ and $v = 4x^2 - 1$, $dv/dx = 8x$, we have

$$\frac{d}{dx}\left(\frac{2x^2+1}{4x^2-1}\right) = \frac{(4x^2 - 1)4x - (2x^2 + 1)8x}{(4x^2 - 1)^2} = \frac{-12x}{(4x^2-1)^2}.$$

EXAMPLE 2. Find $\frac{dy}{dx}$ if $y = \left(\frac{2x+3}{3x+4}\right)^3$.

Solution: Here, because powers occur, the simplest procedure is to consider the expression for y as a product. Thus

$$\begin{aligned}\frac{dy}{dx} &= \frac{d}{dx}[(2x+3)^3(3x+4)^{-3}] \\ &= (2x+3)^3\frac{d}{dx}(3x+4)^{-3} + (3x+4)^{-3}\frac{d}{dx}(2x+3)^3.\end{aligned}$$

Using the power rule, Eq. (16) with $n = 3$ and $n = -3$, we find

$$\begin{aligned}\frac{dy}{dx} &= (2x+3)^3[-3(3x+4)^{-4}(3)] + (3x+4)^{-3}[3(2x+3)^2 2] \\ &= (2x+3)^2(3x+4)^{-4}(-18x - 27 + 18x + 24) \\ &= -\frac{3(2x+3)^2}{(3x+4)^4}.\end{aligned}$$

EXAMPLE 3. Find $\frac{dy}{dx}$ if $y = \frac{x^2 + 3x + 5}{3x}$.

We may write $y = \frac{1}{3}(x + 3 + 5x^{-1})$, and deduce from the linearity property and the power rule that

$$\frac{dy}{dx} = \frac{1}{3}(1 - 5x^{-2}) = \frac{x^2 - 5}{3x^2}.$$

EXERCISE 22

Find the derivative of each of the following functions.

1. $y = \sqrt{x}$.
2. $y = \dfrac{1}{x^4}$.
3. $y = \dfrac{1}{x^{25}}$.
4. $y = \sqrt{x^3}$.
5. $y = \sqrt[3]{x}$.
6. $y = \dfrac{1}{\sqrt{x}}$.
7. $y = \sqrt{3x - 5}$.
8. $y = \dfrac{5}{3 - 2x}$.
9. $y = \dfrac{3}{1 + x^2}$.
10. $y = \dfrac{2x + 5}{2x - 5}$.
11. $y = \dfrac{x^2 - 3}{x^2 + 3}$.
12. $y = \dfrac{x}{x^2 + 1}$.
13. $y = (x^2 - 2)^{-4}$.
14. $y = \sqrt{4 - x^2}$.
15. $y = x^2\sqrt{2 + 3x}$.
16. $y = \dfrac{x^2}{\sqrt{3x + 1}}$.
17. $y = \dfrac{x^2 + 4x + 3}{x}$.
18. $y = \left(\dfrac{2 - 5x}{2 + 5x}\right)^4$.
19. $y = \dfrac{x^2 - 2x + 3}{\sqrt{x}}$.
20. $y = \left(\dfrac{x^2 + 1}{x^2 - 1}\right)^3$.
21. $y = (a^{\frac{2}{3}} - x^{\frac{2}{3}})^{\frac{3}{2}}$.

53. Composite Functions. Let $y = F(u)$ be a differentiable function of u, and $u = f(x)$ be a differentiable function of x. Then each x determines a u, which in turn determines a y, so that y is a function of x. In fact $y = F[f(x)]$. When given by means of the two functions and the auxiliary variable, we call y a composite function of x. This composite function is differentiable, and

$$\frac{dy}{dx} = \frac{dy}{du}\frac{du}{dx}. \tag{17}$$

We first assume that for sufficiently small values of Δx, $\Delta u \neq 0$. This will necessarily be the case if $du/dx \neq 0$. Then since $f(x)$ is differentiable, it is continuous, and by Sec. 14, when $\Delta x \to 0$, $x + \Delta x \to x$, and $f(x + \Delta x) \to f(x)$. Hence $u + \Delta u \to u$ and $\Delta u \to 0$. Since $\Delta u \neq 0$ as $\Delta x \to 0$, we may write

$$\frac{\Delta y}{\Delta x} = \frac{\Delta y}{\Delta u}\frac{\Delta u}{\Delta x}. \tag{18}$$

And by the product rule for limits, we have

$$\lim_{\Delta x \to 0} \frac{\Delta y}{\Delta x} = \lim_{\Delta x \to 0} \frac{\Delta y}{\Delta u} \lim_{\Delta x \to 0} \frac{\Delta u}{\Delta x}$$

$$= \lim_{\Delta u \to 0} \frac{\Delta y}{\Delta u} \lim_{\Delta x \to 0} \frac{\Delta u}{\Delta x}, \tag{19}$$

since $\Delta u \to 0$ when $\Delta x \to 0$. By the assumed differentiability of $F(u)$ and $f(x)$, the limits on the right are dy/du and du/dx. Hence the left side approaches a limit, and this is dy/dx. Consequently

$$\frac{dy}{dx} = \frac{dy}{du}\frac{du}{dx}, \tag{20}$$

as we set out to prove.

When Δu can be zero for arbitrarily small values of Δx, it can be shown that Eq. (20) still holds, with each of its members equal to zero.† We omit the details.

The example of the special situation most commonly met is obtained by setting $f(x) = c$. In this case $u = c$, so that $du/dx = 0$. And $y = F(c)$, so that $dy/dx = 0$. Hence, regardless of the value of dy/du, each member of Eq. (20) is zero as we stated.

The rule for composite functions, Eq. (20), may be stated in words as follows:

If $y = F(u)$ and $u = f(x)$, the derivative of y with respect to x is equal to the product of the derivative of y with respect to u times the derivative of u with respect to x.

The letters x, y, u may be replaced by any others in applying the rule. Thus if $y = f(x)$ and $x = \phi(t)$, the rule takes the form

$$\frac{dy}{dt} = \frac{dy}{dx}\frac{dx}{dt}. \tag{21}$$

This may be written in the solved form

$$\frac{dy}{dx} = \frac{dy/dt}{dx/dt}, \tag{22}$$

provided that $dx/dt \neq 0$, and there is a derivative dy/dx.

We may have a chain of more than two relations, for example $y = f_1(u)$, $u = f_2(v)$, $v = f_3(x)$. Then from the rule as given

$$\frac{dy}{dx} = \frac{dy}{du}\frac{du}{dx} \quad \text{and} \quad \frac{du}{dx} = \frac{du}{dv}\frac{dv}{dx}. \tag{23}$$

Hence by combination of these relations we find

$$\frac{dy}{dx} = \frac{dy}{du}\frac{du}{dv}\frac{dv}{dx}, \tag{24}$$

and similarly for any number of relations. It may aid the memory of Eqs. (20) to (24) as well as that of Eq. (26) of the following section, to observe their apparent similarity to manipulations with ordinary fractions.

Example. Find dy/dx if $y = [1 - (2x - 1)^2]^2$.

Solution: We may think of this as resulting from

$$y = u^2, \qquad u = 1 - (2x - 1)^2 = 1 - v^2, \qquad v = 2x - 1.$$

† This is done by using the above argument for sequences with $\Delta u \neq 0$, and a special argument which shows that $\Delta y/\Delta x$ is always zero and so approaches zero for sequences $\Delta x \to 0$ such that Δu is always zero.

Then for the derivatives we find

$$\frac{dy}{du} = 2u = 2[1 - (2x - 1)^2],$$

$$\frac{du}{dv} = -2v = -2(2x - 1) \qquad \text{and} \qquad \frac{dv}{dx} = 2.$$

Hence from the rule for composite functions, or Eq. (24), we have

$$\begin{aligned}\frac{dy}{dx} &= \frac{dy}{du}\frac{du}{dv}\frac{dv}{dx} = 2[1 - (2x - 1)^2][-2(2x - 1)]2 \\ &= -8(2x - 1)[1 - (2x - 1)^2] = 32(2x - 1)(x - 1)x\end{aligned}$$

In this example the result may be checked by noting that $1 - (2x - 1)^2 = 4x - 4x^2$, so that $y = 16(x^2 - 2x^3 + x^4)$; and

$$\frac{dy}{dx} = 16(2x - 6x^2 + 4x^3) = 32(2x - 1)(x - 1)x.$$

If the two exponents 2 were replaced by $\frac{1}{2}$, we would have to use the first method, or the combination of the power rule and the linearity principle essentially equivalent to it.

54. Inverse Functions. Let $y = f(x)$ be a differentiable function with $dy/dx > 0$ for $a < x < b$. Then by Sec. 38 the function will be increasing in this interval. Suppose that $f(a) = c$ and $f(b) = d$. Then for any value of y such that $c < y < d$, there will be just one value of x such that $a < x < b$. See Fig. 37. Thus x is a function of y, $x = \phi(y)$. From the definition of inverse function, $y = f(x)$ and $x = \phi(y)$ are equivalent, so that

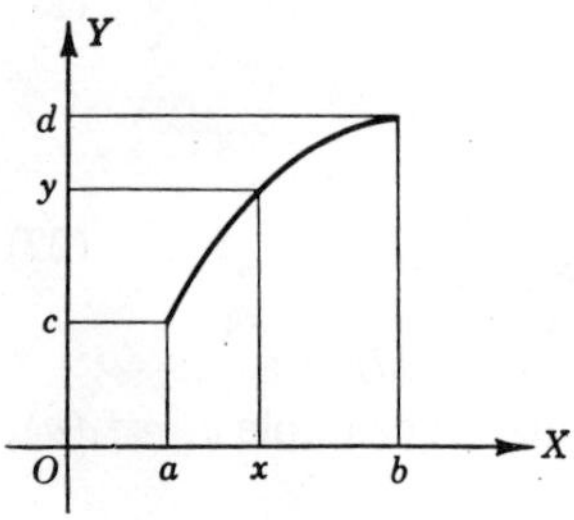

FIG. 37.

$$x = \phi[f(x)].$$

Let us differentiate this relation by the rule for composite functions. The result is

$$1 = \frac{dx}{dy}\frac{dy}{dx} = \phi'(y)f'(x). \tag{25}$$

Since we assumed that dy/dx was greater than zero, it follows from this that

$$\frac{dx}{dy} = \frac{1}{dy/dx} \qquad \text{or} \qquad \phi'(y) = \frac{1}{f'(x)}. \tag{26}$$

This equation holds whenever $dy/dx \neq 0$ in some interval. In any interval where $dy/dx < 0$, the function $f(x)$ decreases, but as in Fig. 38, there is still just one x in the a,b interval† for y in the c,d interval.

† The notation a,b for the interval $a < x < b$ was discussed in Sec. 2.

We may formulate the rule of Eq. (26) in words as follows:

The derivative of the inverse function is equal to the reciprocal of the derivative of the direct function.

As a particular example of an inverse function, consider

$$y = \frac{x^2 + 4}{2} \qquad \text{with } x > 0 \text{ and } y > 2.$$

For x and y restricted as indicated, the inverse function is $x = \sqrt{2y - 4}$. We must take y greater than 2 to make the square root real, and we must take the positive square root to get back a value of $x > 0$. Here

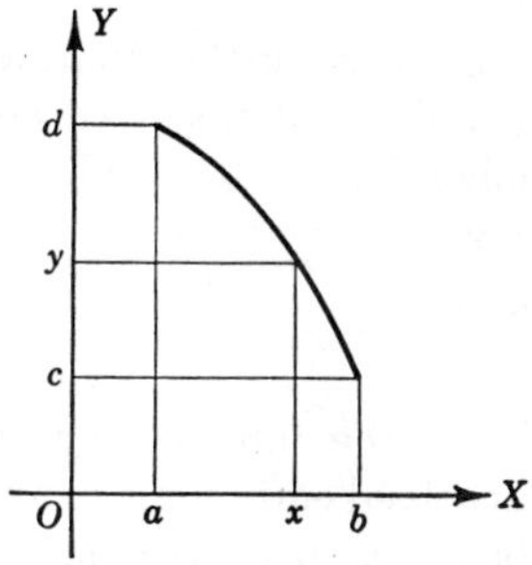

FIG. 38.

$$\frac{dy}{dx} = \frac{1}{2}\,2x = x.$$

From this by the rule for inverse functions, or Eq. (26), we find

$$\frac{dx}{dy} = \frac{1}{x} = \frac{1}{\sqrt{2y - 4}},$$

since $x = \sqrt{2y - 4}$. This may be checked by differentiating

$$x = \sqrt{2y - 4} = (2y - 4)^{\frac{1}{2}}$$

by the power rule. The result is

$$\frac{dx}{dy} = \frac{1}{2}(2y - 4)^{-\frac{1}{2}}2 = (2y - 4)^{-\frac{1}{2}} = \frac{1}{\sqrt{2y - 4}}.$$

EXERCISE 23

In each problem, first find dy/dx by considering y as a composite function. Then check by using the last form of y obtained by substitution.

1. $u = 1 - 3x.\quad y = u^3 = 1 - 9x + 27x^2 - 27x^3.$
2. $u = x^2 + 1.\quad y = u^2 - 2u = x^4 - 1.$
3. $u = 1 - x^2.\quad y = u^2 - u^3 = x^2 - 2x^4 + x^6.$
4. $u = (5 - 4x)^2.\quad y = u^3 = (5 - 4x)^6.$
5. $u = (4 + 2x)^3.\quad y = \sqrt{u} = (4 + 2x)^{\frac{3}{2}}.$

In each of the following problems two functions $y = f(x)$ and $x = \phi(y)$ are given which are inverse for x and y in suitably restricted ranges. Assuming that x and y are so restricted where necessary, for each case find dy/dx by the rule for inverse functions. Then check by differentiating the solved form.

6. $y = 2x + 3,\ x = \dfrac{y - 3}{2}.$
7. $y = (1 - 3x)^2,\ x = \dfrac{1 - \sqrt{y}}{3}.$
8. $y = \dfrac{1 + 2x}{1 - 2x},\ x = \dfrac{y - 1}{2y + 2}.$
9. $y = \sqrt{\dfrac{2 + x}{2 - x}},\ x = \dfrac{2y^2 - 2}{y^2 + 1}.$
10. $y = 2 + \sqrt{4 - x^2},\ x = \sqrt{4y - y^2}.$

55. Implicit Functions. Let y be related to x by an equation not solved for y. An example is

$$xy - 5x + 2y = 3 \qquad \text{or} \qquad xy - 5x + 2y - 3 = 0. \tag{27}$$

Suppose further that, at least with certain restrictions on the values of x and y, the unsolved equation determines y as a function of x which might be written explicitly as $y = f(x)$. In the example of Eq. (27) for $x \neq -2$, we have

$$y = \frac{5x + 3}{x + 2}. \tag{28}$$

Then the function $f(x)$ is said to be defined implicitly by the given unsolved equation. And, as given by this unsolved equation, y is said to be an *implicit function* of x.

As a second example, consider

$$y^2 - x + 2 = 0. \tag{29}$$

Here for $x > 2$, and $y > 0$, the implicit function could be written in explicit form as

$$y = \sqrt{x - 2}. \tag{30}$$

For $x > 2$ and $y < 0$, the explicit form would be

$$y = -\sqrt{x - 2}. \tag{31}$$

If we were interested in Eq. (29) for values near $x = 6$, $y = 2$, we would be concerned with the values, or branch, given by Eq. (30). For values near $x = 6$, $y = -2$, we should use the branch of Eq. (31). Without such additional information we would call the implicit function of x defined by Eq. (29) a multiple-valued function, made up of more than one, here two, single-valued functions or branches.

As another example, consider the equation

$$y^5 - 7xy - 18x^3 = 0. \tag{32}$$

This is satisfied by $(x,y) = (1,2)$ and for values of x near 1 determines values of y near 2. These could be found for any one x by the method of Sec. 49. But in this case y cannot be expressed in terms of x by any simple explicit algebraic formula.

56. Implicit Differentiation. When y is defined as an implicit function of x, it is possible to find dy/dx without solving for y. The method, called *implicit differentiation*, is as follows. We regard y as a function of x, like the u of Sec. 53, and differentiate each term of the given equation. If y is a differentiable function of x, the resulting equation is valid, since if two expressions are equal for all values of x, their derivatives with respect to x are equal. We obtain some terms by direct x differentiation, others by differentiation with respect to y and multiplication by dy/dx in accordance with Sec. 53. Thus we obtain a first-degree equation in dy/dx which may be solved for dy/dx in terms of x and y.

Let us illustrate the process for Eq. (27). From either form we find

$$x\frac{dy}{dx} + y - 5 + 2\frac{dy}{dx} = 0,$$

so that

$$\frac{dy}{dx} = \frac{-y + 5}{x + 2}. \tag{33}$$

To find the second derivative we differentiate this last equation, regarding y as a function of x, and so deduce that

$$\frac{d^2y}{dx^2} = \frac{(-dy/dx)(x + 2) - (-y + 5)}{(x + 2)^2}.$$

Since the right member contains dy/dx, we replace this by the value given in Eq. (33). The result is

$$\frac{d^2y}{dx^2} = \frac{2y - 10}{(x + 2)^2}. \tag{34}$$

Let (x_0,y_0) be any pair of values satisfying the original equation. If these values do not make any denominator zero in the expression for dy/dx and lead to a finite value, m_0, there must be just one differentiable branch of the implicit function, $y = f(x)$ which makes $y_0 = f(x_0)$. For this branch, $f'(x_0) = m_0$. And at (x_0,y_0) the value of the second and higher derivatives may also be found by implicit differentiation.

We may describe the method of implicit differentiation in words as follows:

Differentiate each term of the given equation with respect to x, considering y as a function of x like the u of a composite function. Solve the resulting equation for dy/dx.

To obtain the second derivative, differentiate the expression for dy/dx considering y and dy/dx as function of x. This gives d^2y/dx^2 in terms of x, y, and dy/dx. And dy/dx may be eliminated from this by using the expression for it in terms of x and y.

EXERCISE 24

In each of the following problems, assume that (x,y) is a given pair of values which satisfies the relation as stated. Also assume that at (x,y) the denominator of the expression for dy/dx is not zero. And find dy/dx at (x,y) by implicit differentiation.

1. $x^2 + 2xy + 3y^2 = 6$.
2. $x^2 - 3xy - 2y^2 = 4$.
3. $x^3 + y^3 = 3xy$.
4. $x^2y^2 = x^2 + y^2$.
5. $x(y - x)^2 = x + y$.
6. $xy^2 = 5$.
7. $\sqrt{x} + \sqrt{y} = \sqrt{a}$.
8. $x^{\frac{2}{3}} + y^{\frac{2}{3}} = a^{\frac{2}{3}}$.

With the same assumptions as those stated for Probs. 1 to 8, for each of the following problems find dy/dx and d^2y/dx^2 at (x,y).

9. $xy = 4$.
10. $x^2 + 4y^2 = 16$.
11. $x^2 + y^2 = a^2$.
12. $x = 4y^2$.
13. $x^3 + y^3 = a^3$.
14. $x^2 - 4y^2 = 36$.
15. $6x - 2y = 5$.
16. $(3x + y)^2 = 4$.

57. Greatest and Least Values. Applied problems in maxima and minima were discussed in Sec. 50. It often happens that the function whose extreme value we are seeking fails to increase or decrease at just one interior point of the permissible interval for the independent variable. This point separates an interval in which the function is increasing from an interval in which the function is decreasing. When this situation can be inferred from general considerations, we do not need to use the test of Secs. 44 and 45 based on sign changes, or the test of Sec. 46 based on sign of the second derivative.

Under these conditions we may locate the maximum or minimum required by finding the value of the independent variable inside the range of interest which makes the first derivative of the function equal to zero. In problems of this kind, it is sometimes more convenient to differentiate the known relations before carrying out any elimination. We illustrate the method for some examples. For other types of extreme values see Sec. 58.

EXAMPLE 1. The stiffness of a beam of rectangular cross section varies as the width and cube of the depth. Find the dimensions of the stiffest beam which can be cut out of a circular log of diameter a.

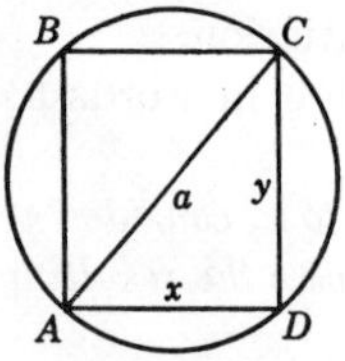

FIG. 39.

Solution: Here we must make $S = kxy^3$ a maximum, where k is a constant, x is the width, and y is the depth. And (Fig. 39) if the rectangle is inscribed in a circle, its diagonal is a diameter of the circle and so equals a. Hence

$$x^2 + y^2 = a^2. \tag{35}$$

Since S is zero when $x = 0$ and $y = a$ or when $x = a$ and $y = 0$, if $dS/dx = 0$ for just one value of x between 0 and a, this value will correspond to the maximum value of the stiffness S.

By differentiating $S = kxy^3$, regarding y as a function of x, we find $\frac{dS}{dx} = k\left(3xy^2\frac{dy}{dx} + y^3\right)$. Hence at the maximum $dS/dx = 0$ and

$$k\left(3xy^2\frac{dy}{dx} + y^3\right) = 0, \qquad \frac{dy}{dx} = -\frac{y}{3x}. \tag{36}$$

We may cancel the factor y^2, since $y \neq 0$ *inside* the interval $0 < x < a$. But from Eq. (35) we find by implicit differentiation that

$$2x + 2y\frac{dy}{dx} = 0, \qquad \frac{dy}{dx} = -\frac{x}{y}. \tag{37}$$

The values of dy/dx in Eqs. (36) and (37) can be equal only if

$$-\frac{y}{3x} = -\frac{x}{y} \quad \text{or} \quad \text{if } y^2 = 3x^2. \tag{38}$$

Hence $y = \pm\sqrt{3}\,x$. The negative value is inadmissible, so that the desired dimensions must satisfy $y = \sqrt{3}\,x$. The ratio $y/x = \sqrt{3}$ determines the shape of the beam, and we could construct the rectangle by making angle DAC in Fig. 39 equal to $\tan^{-1}\sqrt{3} = 60°$. From Eqs. (35) and (38) we find

$$4x^2 = a^2, \qquad x = \frac{a}{2}, \qquad \text{and } y = \sqrt{3}\,x = \frac{a\sqrt{3}}{2}.$$

Thus the required width is $a/2$, and the required depth is $a\sqrt{3}/2$.

EXAMPLE 2. A buoy is composed of two equal right circular cones with a common base. It is to have a given volume. For what ratio of height of cone to base is the surface of the buoy least?

Solution: A meridian section of the buoy is shown in Fig. 40. If each cone has radius of base x and height y, the volume of each cone is $\frac{1}{3}\pi x^2 y$ so that the volume of the buoy is

$$V = \tfrac{2}{3}\pi x^2 y. \tag{39}$$

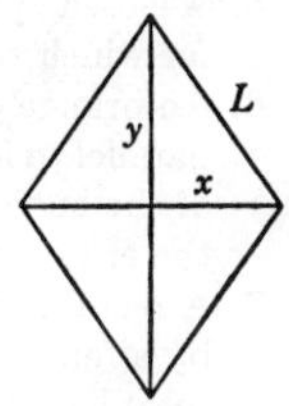

FIG. 40.

The slant height $L = \sqrt{x^2 + y^2}$, from Fig. 40, so that the lateral surface of each cone is $\pi x\sqrt{x^2 + y^2}$. Hence the surface of the buoy is

$$S = 2\pi x\sqrt{x^2 + y^2}. \tag{40}$$

Since V is fixed, if x is large, y is small. But since S exceeds $2\pi x^2$, S is large when x is large. If x is small, y is large. But since S exceeds $2\pi xy = 3V/x$, S is large when x is small, and hence $1/x$ is large. Hence if just one positive value of x makes $dS/dx = 0$, it will give the required minimum.

By differentiating Eq. (40), regarding y as a function of x, we find

$$\frac{dS}{dx} = 2\pi\left[\sqrt{x^2 + y^2} + \frac{x}{2}\,\frac{2x + 2y(dy/dx)}{\sqrt{x^2 + y^2}}\right] = 2\pi\,\frac{2x^2 + y^2 + xy(dy/dx)}{\sqrt{x^2 + y^2}}.$$

At the minimum, $dS/dx = 0$, so that

$$2x^2 + y^2 + xy\frac{dy}{dx} = 0 \quad \text{and} \quad \frac{dy}{dx} = -\frac{2x^2 + y^2}{xy}. \tag{41}$$

But from Eq. (39) we find by implicit differentiation that

$$0 = \frac{2\pi}{3}\left(x^2\frac{dy}{dx} + 2xy\right) \quad \text{and} \quad \frac{dy}{dx} = -\frac{2y}{x}. \tag{42}$$

We may cancel x since $x \neq 0$ *inside* the permissible interval $x > 0$. The values of dy/dx in Eqs. (41) and (42) can be equal only if

$$-\frac{2x^2 + y^2}{xy} = -\frac{2y}{x} \quad \text{or} \quad \text{if } 2x^3 + xy^2 = 2xy^2, \qquad 2x^3 = xy^2.$$

The value $x = 0$ has already been rejected, so that $y = \pm\sqrt{2}\,x$. As the negative value is inadmissible, $y = \sqrt{2}\,x$ and the required ratio is $\sqrt{2}$.

Since S is least when S^2 is least, we might have used $S^2 = 4\pi^2(x^4 + x^2y^2)$ in place of Eq. (40), and $2S\frac{dS}{dx} = 4\pi^2\left(4x^3 + 2xy^2 + 2x^2y\frac{dy}{dx}\right) = 0$ to derive Eq. (41) more easily.

EXERCISE 25

1. The product of two positive numbers is 25. For what two numbers is their sum least?
2. The hypotenuse of a right triangle is 10. What values of the sides make the area greatest?
3. A rectangle has two sides along the x and y axes and one vertex on the curve $x^2 + 4y^2 = 1$ in the first quadrant. Find the rectangle of maximum area.
4. A cylindrical cistern is to be constructed with open top. The total material, proportional to the sum of the area of the sides and bottom, is A. What ratio of dimensions gives the maximum volume?
5. A closed cylindrical can is to hold V cu. in. What is the ratio of height to radius for which the total area of the top, botton, and sides is least?
6. An orange crate holding V cu. ft. is divided into two equal parts by a partition parallel to its square ends of side x. If the ends are twice as thick as the material of the bottom, top, sides, and the partition, find the ratio of the length y to x for the crate whose shipping weight is least.
7. A rectangular pigpen is surrounded by a fence and divided into two equal parts by a fence parallel to one side. If its area is A, find the dimensions for which the total length of fencing is least.
8. A tank is to be constructed with square base and an open top, and is to hold 8 cu. yd. If the cost of the sides is \$1 per square yard, and the cost of the bottom is \$2 per square yard, for what ratio of dimensions does the tank cost least?
9. A page is to contain 24 sq. in. of printed material. The margins at top and bottom are each $\frac{3}{4}$ in. wide, and the margins at the sides are each $\frac{1}{2}$ in. wide. What are the dimensions of the page whose area is least?
10. A tent in the form of a square pyramid is to have a given volume. What is the ratio of height to side of base which makes the amount of canvas used for the lateral surface of the pyramid least?
11. If the slant height of a circular cone is 4 ft., what must the radius of the base be to make the volume a maximum?
12. A metal vessel, open at the top, is to be cast in the form of a prism with square base. If it is to hold 32 cu. in., and the thickness of the side and that of the bottom are each to be 1 in., what will be the inside dimensions when the least amount of metal is used?
13. If f is the coefficient of friction and x the trigonometric tangent of the pitch angle of a screw, the efficiency is $y = \frac{x - fx^2}{x + f}$. For a given value of f, what value of x makes the efficiency a maximum?
14. Two heating elements are 6 ft. apart. If the first element is eight times as powerful as the second, at a point between the two sources x ft. from the first and $(6 - x)$ ft. from the second, the intensity of heat $y = \frac{8k}{x^2} + \frac{k}{(6 - x)^2}$. At what distance x is the intensity y least?
15. The cost of fuel per hour for a vessel traveling v mi./hr. is $a + bv^3$, with a and b positive. Find the velocity for which the cost per mile, $C = \frac{a + bv^3}{v}$, is least.

16. The cost of fuel per hour for a vessel traveling with a speed of v mi. relative to the water is kv^3. Find the relative speed v for which the cost per mile of progress against a current of a mi./hr., $C = \frac{kv^3}{v - a}$, is least.
17. If three sides of a trapezoid are each 6 in. long, for what value of the third side is the area a maximum?
18. Vessel A is 41 mi. south of vessel B. If A is traveling due east at 10 mi./hr., and B is traveling south at 8 mi./hr., how much time will elapse before they are nearest to each other?
19. A man in a rowboat at a point A 4 mi. from a straight shore wishes to go to a point B which is 10 mi. up the beach from a point C opposite A. If he can row 2 mi./hr. and walk 2.5 mi./hr., how many miles from C should he land to make the total trip from A to B in the least time?
20. The banks of a river are parallel straight lines, w mi. apart. An electric power house is near the shore at A and a factory is on the opposite shore at B, s mi. downstream from C opposite A. A cable is made of one straight piece under water from A to D, between C and B, and another straight piece on land from D to B. If the part along AD cost p dollars per mile and the part along DB cost q dollars per mile, where p exceeds q, find the distance CD for which the cost of the cable is least.

***58. Other Types of Maxima and Minima.** In Secs. 50 and 57 we determined values of x which made $f(x)$ assume maximum or minimum values by setting the derivative, $f'(x)$, equal to zero. This method is sufficient whenever the extreme value is inside some interval throughout which $f(x)$ and $f'(x)$ are both continuous. Thus the method would apply to the maximum at E and the minimum at B in Fig. 41.

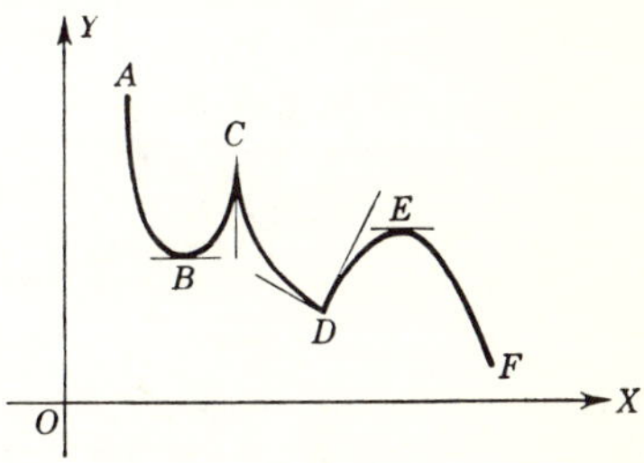

FIG. 41.

But the graph of Fig. 41 also illustrates some other types of maxima and minima. At C, the tangent is vertical and the derivative becomes infinite. At D the slope is different on the two sides, so that the derivative is not defined at this point. The curve ends at A and F, either because of imaginary values or because physical conditions of a problem restrict the values of x to be considered. The function has a maximum at C, a minimum at D, takes its greatest value for the given range at A, and takes its least value for the given range at F.

This diagram shows that in determining greatest and least values we may have to consider values where the derivative is infinite, where it is discontinuous, or where the function of physical interest fails to exist. In such cases it is usually desirable to make a graph of the function.

EXAMPLE 1. For x in the range $0 \leqq x \leqq 9$, find the greatest and least values of $y = 10 - 2(x - 1)^{\frac{2}{3}}$.

Solution: $\frac{dy}{dx} = -\frac{4}{3}(x - 1)^{-\frac{1}{3}} = \frac{-4}{3\sqrt[3]{x - 1}}$. This is infinite when $x = 1$, and changes sign from plus to minus as x increases through 1. But at $x = 1$, $y = 10$ which is finite. Hence by Sec. 44, $x = 1$ is a maximum value. At the ends of the given interval, $y = 8$ for $x = 0$ and $y = 2$ when $x = 9$. These are minimum values in the sense of being exceeded by all values near them. The graph of the function is shown in Fig. 42. Hence the greatest value is 10, assumed when $x = 1$, and the least value is 2, assumed when $x = 9$.

EXAMPLE 2. For x in the range $-2 \leqq x \leqq 3$, find the greatest and least values of $y = |x|$.

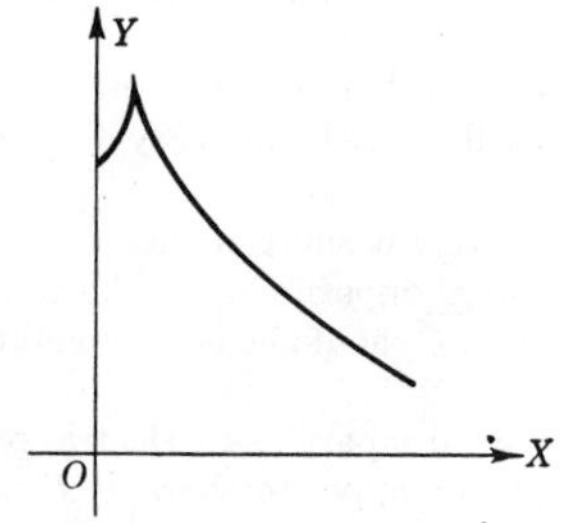

FIG. 42.

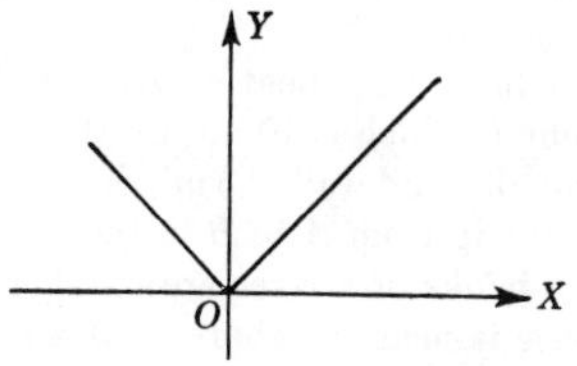

FIG. 43.

Solution: By Sec. 3, $|x| = x$ for $x \geqq 0$ and $|x| = -x$ for $x < 0$. For $x > 0$, $y = x$, $dy/dx = 1$. And for $x < 0$, $y = -x$, $dy/dx = -1$. The function is continuous at $x = 0$, where $y = 0$, and its graph is shown in Fig. 43. At the end points, $y = 2$ when $x = -2$ and $y = 3$ for $x = 3$. For the restricted interval, these are maximum values in the sense of exceeding all values near them. Hence the greatest value is 3 at $x = 3$ and the least value is 0 at $x = 0$.

EXERCISE 26

Let x be restricted to the range $-1 \leqq x \leqq 2$. Show that each given function has the indicated greatest and least values.

1. $y = x^2$. y greatest $= 4$ at $x = 2$, y least $= 0$ at $x = 0$.
2. $y = -2x + 3$. y greatest $= 5$ at $x = -1$, y least $= -1$ at $x = 2$.
3. $y = x^{\frac{2}{3}}$. y greatest $= 2^{\frac{2}{3}}$ at $x = 2$, y least $= 0$ at $x = 0$.
4. $y = -|x - 1|$. y greatest $= 0$ at $x = 1$, y least $= -2$ at $x = -1$.

For x taking any values which make y real, show that

5. $y = 4 - 3(x - 2)^{\frac{2}{3}}$ has a maximum value 4 at $x = 2$.
6. $y = 3 + 2(x - 4)^{\frac{2}{3}}$ has a minimum value 3 at $x = 4$.
7. $y = 5 + |x + 2|$ has a minimum value 5 at $x = -2$.
8. $y = 3 - |x - 4|$ has a maximum value 3 at $x = 4$.
9. $y = \sqrt{4 - x}$ has a minimum value 0 at $x = 4$.
10. $y = \sqrt{4 - x^2}$ has a minimum value 0 at $x = 2$ or -2, and a maximum value 2 for $x = 0$.

11. A man in a rowboat at a point A 3 mi. from a straight shore wishes to go to a point B which is 4 mi. up the beach from a point C opposite A. If he can row 5 mi./hr. and walk 3 mi./hr., and aims for a point on the beach x mi. up the beach from C, show that his time of making the trip from A to B in hours is $t = \frac{1}{5}\sqrt{x^2 + 9} + \frac{1}{3}|x - 4|$. Verify that $\frac{dt}{dx} = \frac{x}{5\sqrt{x^2 + 9}} + \frac{1}{3} > 0$ when $x > 4$. But $\frac{dt}{dx} = \frac{x}{5\sqrt{x^2 + 9}} - \frac{1}{3} < 0$ when $x < 4$, since $\frac{x^2}{25(x^2 + 9)} < \frac{1}{9}$ follows from $0 < 16x^2 + 225$. This shows that the least value of t is 1, for $x = 4$. That he should aim at B and row all the way is a reasonable consequence of the unrealistic assumption that he rows faster than he walks.

12. To find the point on the curve $y^2 = 4x + 2$ nearest the origin we minimize $D = \sqrt{x^2 + y^2} = \sqrt{x^2 + 4x + 2}$. Verify that dD/dx is zero for $x = -2$, that this value makes D^2 a minimum, but that for $x = -2$, y and D are imaginary. To make y real, we must have $x \geqq -\frac{1}{2}$. Show that for this range D takes its least value $\frac{1}{2}$ for $x = -\frac{1}{2}$, so that the required nearest point is $(-\frac{1}{2},0)$.

59. Related Rates. In many physical situations there are several variables, each of which is a function of the time. The time rate of change is known for some of the variables, and it is to be calculated for others. If there are just two variables, say x and y, a single relation between them will determine dy/dx as a function of x and y. We may then use this and the rule for composite functions, Eq. (21), to find a relation between dx/dt, dy/dt, x, and y which will hold at any time. Then if either rate is known for a particular value of x and y, the other rate may be found from the relation just mentioned.

In any problem of related rates of two quantities, it is necessary to find a relation between them which will hold at all times during which the conditions of the problem obtain. This is usually found by drawing a figure and using geometric considerations. We then differentiate this equation with respect to the time and solve for the rate we desire in terms of known quantities and rates. The student is warned against attempting to insert values which hold at one time only in the relation which holds at all times, until after differentiation has been completed.

EXAMPLE 1. A ladder of length 52 ft. leans against a vertical wall. How fast is the top of the ladder sliding down when the bottom is 20 ft. from the wall, if at this instant the bottom is sliding out with velocity 2 ft./sec.?

As in Fig. 44, let the bottom be x ft. from the wall and the top y ft. above the ground. Then

$$x^2 + y^2 = (52)^2. \tag{43}$$

From this by implicit differentiation, we find

$$2x + 2y\frac{dy}{dx} = 0 \quad \text{and} \quad \frac{dy}{dx} = -\frac{x}{y}. \tag{44}$$

By Eq. (21), or the rule for composite functions, we deduce that

$$\frac{dy}{dt} = \frac{dy}{dx}\frac{dx}{dt} = -\frac{x}{y}\frac{dx}{dt}. \tag{45}$$

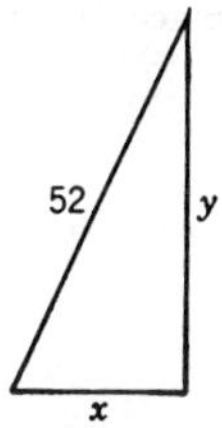

FIG. 44.

This relation holds at any time. In particular, when $x = 20$, by Eq. (43), $y = \sqrt{(52)^2 - (20)^2} = 48$. And $dx/dt = 2$. Hence at the time under consideration we have

$$\frac{dy}{dt} = -\frac{20}{48}2 = -\frac{5}{6}. \tag{46}$$

The minus sign indicates descent, and the numerical rate of descent is $\frac{5}{6}$ ft./sec.

Instead of proceeding as above, we could think of each of the variables x and y in Eq. (43) as a function of t and differentiate this equation with respect to t. This would give

$$2x\frac{dx}{dt} + 2y\frac{dy}{dt} = 0, \tag{47}$$

which holds for all values of t. Now by noting that, when $x = 20$, $y = 48$ from Eq. (43), and $dx/dt = 2$, we could deduce that

$$2 \cdot 20 \cdot 2 + 2 \cdot 48\frac{dy}{dt} = 0 \quad \text{and} \quad \frac{dy}{dt} = -\frac{5}{6}. \tag{48}$$

EXAMPLE 2. A vessel is in the shape of an inverted cone of height 12 in. and having a base of radius 5 in. Water is poured into the vessel at a rate of 50 cu. in./sec. At what rate is the wetted surface increasing when the depth of the water is 4 in.?

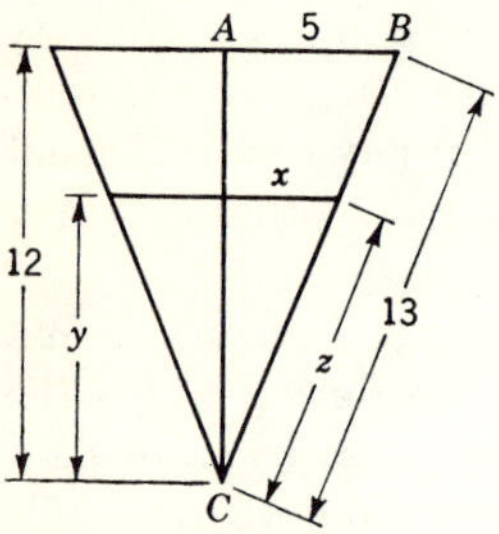

FIG. 45.

Solution: As in Fig. 45, let x be the radius of the surface of the water when the depth is y, and let z be the slant height. Then since $AB = 5$ and $AC = 12$, from the right triangle $\overline{BC}^2 = 5^2 + 12^2 = 169$ and $BC = 13$. Then from similar triangles we have

$$\frac{x}{5} = \frac{y}{12} = \frac{z}{13}, \qquad y = \frac{12}{5}x, \qquad z = \frac{13}{5}x. \tag{49}$$

The wetted surface is the lateral surface of a cone of radius x and slant height z, and so is $S = \pi xz = 13\pi x^2/5$ by Eq. (49). Hence we have

$$\frac{dS}{dt} = \frac{26\pi}{5}x\frac{dx}{dt}. \tag{50}$$

The volume of liquid is $V = \pi x^2 y/3 = 4\pi x^3/5$ by Eq. (49). Hence we have

$$\frac{dV}{dt} = \frac{12\pi}{5}x^2\frac{dx}{dt}. \tag{51}$$

Since $dV/dt = 50$, from the last equation we find

$$50 = \frac{12\pi}{5}x^2\frac{dx}{dt} \quad \text{and} \quad \frac{dx}{dt} = \frac{250}{12\pi x^2}. \tag{52}$$

On substituting this in Eq. (50), we deduce that

$$\frac{dS}{dt} = \frac{26\pi x}{5}\frac{250}{12\pi x^2} = \frac{325}{3x}. \tag{53}$$

From Eq. (49), $x = 5y/12$, so that when $y = 4$, $x = \frac{5}{3}$. Hence from Eq. (53) at this instant $dS/dt = 65$. And the wetted surface is increasing 65 sq. in./sec.

EXAMPLE 3. A locomotive is moving away from a grade crossing at the rate of 60 mi./hr. and is 400 ft. from it. An automobile is moving from the intersection at the rate of 40 mi./hr. and is 300 ft. from it. If the angle between the two directions of motion is 45°, how fast is the distance between the locomotive and the automobile increasing at the instant described?

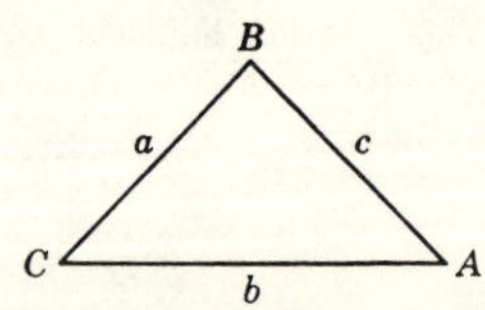

FIG. 46.

Solution: In Fig. 46, A is the position of the locomotive at any time and B is the corresponding position of the automobile. The grade crossing is at C. Then from the law of cosines† applied to triangle ABC, we have

$$c^2 = a^2 + b^2 - 2ab \cos C, \tag{54}$$

where $b = CA$, $a = BC$, and $c = AB$. As $C = 45°$, $\cos C = \sqrt{2}/2$, so that Eq. (54) becomes

$$c^2 = a^2 + b^2 - \sqrt{2}\, ab. \tag{55}$$

Let us differentiate this equation with respect to t, regarding c, a, and b as functions of t. The result is

$$2c \frac{dc}{dt} = 2a \frac{da}{dt} + 2b \frac{db}{dt} - \sqrt{2} \left(a \frac{db}{dt} + b \frac{da}{dt} \right). \tag{56}$$

We wish to find dc/dt when $da/dt = 40$ mi./hr., $db/dt = 60$ mi./hr., $a = 300$ ft., and $b = 400$ ft. From Eq. (55), we find

$$c^2 = (300)^2 + (400)^2 - \sqrt{2}\,(300)(400) = 8.03 \times 10^4, \tag{57}$$

so that $c = 283$. Since every term of Eq. (56) contains one rate and one distance, we may leave the rates in miles per hour and the distances in feet,‡ and find that

$$\frac{dc}{dt} = \frac{2 \cdot 300 \cdot 40 + 2 \cdot 400 \cdot 60 - \sqrt{2}\,(300 \cdot 60 + 400 \cdot 40)}{2 \cdot 283} = 42.3. \tag{58}$$

Hence the two vehicles are separating at the rate of 42.3 mi./hr.

EXERCISE 27

1. A ladder 50 ft. long leans against a vertical wall. If the foot of the ladder is dragged out from the wall at the rate of 1 ft./sec., how fast is the top sliding down the wall at the instant when the top is 30 ft. above the ground?
2. In Prob. 1, how high is the top of the ladder when it is sliding down at the rate of 1 ft./sec.?
3. A street light is 15 ft. above a walk on which a boy 5 ft. tall is walking. How fast is the remote end of the boy's shadow moving along the ground, if he walks at the rate of 2 ft./sec.?
4. In Prob. 3, how fast is the boy's shadow lengthening?
5. A man is walking at the rate of 2 ft./sec. toward the foot of a flagpole. At what rate is his eye approaching the top of the flagpole, 50 ft. above his eye level, when he is 40 ft. from its foot?
6. A barge is being drawn toward a dock by a cable. The end of the cable in the barge is 10 ft. below the end at dock B. If AB is being shortened at the rate of 12 ft./min., how fast is the barge moving toward the dock, when $AB = 26$ ft.?
7. A balloon is rising vertically at the rate of 20 ft./sec. Point A is directly under the balloon, and point B is on a level with A and 30 ft. from it. When the balloon is 40 ft. above A, how fast is its distance from B increasing?
8. A basin is in the form of an inverted cone of altitude 10 in. and diameter of base 20 in. If water runs into the basin at the rate of 2 cu. in./sec., how fast is the water level rising when the level is 5 in. from the top?
9. In Prob. 8, at what rate is the wetted surface increasing?

† See Eq. (51) of Sec. 94.

‡ Converting distances to miles would give the fraction in Eq. (58) with numerator and denominator each divided by 5,280, the number of feet per mile.

10. The inside of a vessel is in the form of an inverted pyramid of altitude 2 ft. The base of the pyramid, at the top of the vessel, is a square 4 ft. on a side. If water is leaking out the bottom of the vessel at the rate of 8 cu. ft./min., how fast is the level of the water falling when the water is 1 ft. deep?
11. A trough is in the form of a prism. The top is a rectangle 2 by 5 ft. and the ends are inverted isosceles right triangles, so that the trough is 1 ft. deep. If water is being poured into the tank at the rate of 20 cu. ft./min., how fast is the level rising when the depth is 6 in.?
12. In Prob. 11, at what rate is the wetted surface increasing?
13. Ship A is 50 mi. due south of ship B. If A is sailing due south at the rate of 20 mi./hr. and B is sailing due east at the rate of 25 mi./hr., how fast are they separating after 2 hr.?
14. Airplane A is 200 mi. west of airport tower T and moving west with a ground speed of 400 mi./hr. Airplane B is 100 mi. north of T and moving south with a ground speed of 300 mi./hr. At what rate is the distance AB changing at the instant described?
15. Two men A and B start walking away from point C at the uniform rate of 2 ft./sec. If A travels due east and B travels N. 30° E., how fast are they separating after 1 min.?
16. Ships A and B each start from O at the same time and travel at a speed of 15 mi./hr. But ship A travels N. 60° E., while ship B travels S. 60° E. Show that they are separating at a constant rate at all times, and find the rate.
17. Ship A is 40 mi. north of a fixed point P and is moving south at the rate of 15 mi./hr. Ship B is 10 mi. west of P and is moving east at the rate of 20 mi./hr. How fast are they approaching or separating after t hr., and when are they nearest together?
18. A viaduct crosses over a road at right angles, at a height of 110 ft. At a certain instant car A, traveling 60 mi./hr. on the viaduct, was directly above car B, traveling 30 mi./hr. on the road. How fast are they separating after 1 sec., if each travels at uniform speed?
19. Two ships A and B leave the same port on the same day, A at noon and B at 2 P.M. If A runs 10 mi./hr. on a course due north and B runs 20 mi./hr. on a course due east, how fast are they separating at 6 P.M.?
20. Planes A and B are each traveling on horizontal straight courses with ground speeds of 300 mi./hr. If the angle between their courses is 60°, and A was 2 mi. directly over B at time 0, how fast are they separating after 1 min.?

60. Geometric Differentiation. Some rate problems involve the derivative of an area or volume with respect to a length. Heretofore we have solved such problems by first finding the geometric quantity in terms of the length and by differentiating the resulting function by using the standard rules. But sometimes, as we shall now explain, we may construct an expression for the increment of the quantity from geometric considerations, and use this to find the derivative.

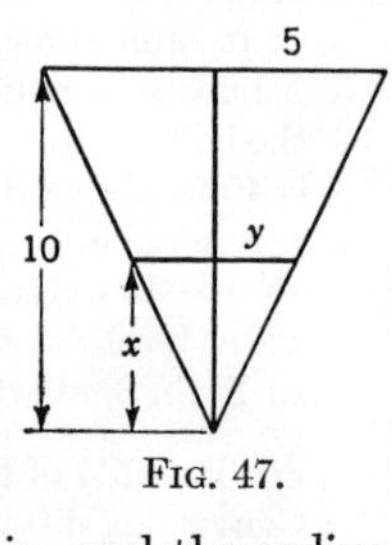

FIG. 47.

For example, consider the container in the shape of an inverted cone of revolution whose meridian section is shown in Fig. 47. Let the height of the cone be 10 in. and the radius of the base be 5 in. Suppose that water is pouring in at the rate of 8 cu. in./sec. We wish to find how fast the height of the water level is rising at the instant when the height is 6 in.

Let the height at any time be x, and the corresponding radius of the cone be y. Then from similar triangles we find that

$$\frac{y}{x} = \frac{5}{10}, \qquad y = \frac{x}{2}. \tag{59}$$

Hence the area of the section of the cone at height x is $\pi y^2 = \pi x^2/4$. Now give the height a positive increment Δx. Then the increment of the volume of the cone ΔV will be a slice between two planes Δx apart. The volume ΔV will be larger than the volume of a cylinder of height Δx and radius $x/2$, and smaller than that of a cylinder of height Δx and radius $(x + \Delta x)/2$. Hence ΔV will equal the volume of a cylinder of height x and radius $\bar{x}$, where $\bar{x}$ is a suitably chosen value between x and $x + \Delta x$. Thus we shall have

$$\Delta V = \frac{\pi}{4}\bar{x}^2\,\Delta x \qquad \text{and} \qquad \frac{\Delta V}{\Delta x} = \frac{\pi}{4}\bar{x}^2. \tag{60}$$

Now let $\Delta x \to 0$. Since $x < \bar{x} < x + \Delta x$, $\bar{x} \to x$, so that

$$\frac{dV}{dx} = \frac{\pi}{4}x^2. \tag{61}$$

We may now use the rule for composite functions to deduce that

$$\frac{dV}{dt} = \frac{dV}{dx}\frac{dx}{dt} = \frac{\pi}{4}x^2\frac{dx}{dt}. \tag{62}$$

It follows that when $x = 6$ and $dV/dt = 8$,

$$8 = \frac{\pi}{4}36\frac{dx}{dt} \qquad \text{and} \qquad \frac{dx}{dt} = \frac{8}{9\pi}. \tag{63}$$

Thus the required rate is $8/9\pi$ in./sec.

61. Differential Notation. The procedure of Sec. 60 may be shortened by the use of the notation of *differentials*. Let dx be any *fixed* number and dV be the number determined from it so that

$$dV \div dx = \frac{dV}{dx} \qquad \text{or} \qquad dV = \frac{dV}{dx} \times dx. \tag{64}$$

With y in place of V, if a is the fixed number taken as dx, and $dy/dx = m$, the corresponding definition of the differential $dy = (dy/dx)dx$ makes $dy = ma$. Thus the quantities a and ma used in Secs. 24 and 48 to draw the tangent line as the line through (x_1,y_1) and $(x_1 + a,\ y_1 + ma)$ were essentially differentials as here defined. And the present definition makes $(x_1 + dx,\ y_1 + dy)$ a point on the tangent line at (x_1,y_1).

By multiplying both sides of Eq. (61) by dx and using Eq. (64), we obtain the relation

$$dV = \frac{\pi}{4} x^2 \, dx. \tag{65}$$

This equation is equivalent to Eq. (61), because of the first part of Eq. (64).

Let us recall the geometric relation $\Delta V = \frac{\pi}{4} \bar{x}^2 \, \Delta x$ of Eq. (60). If we replace each Δ by d and omit the bar on x, this becomes $dV = \frac{\pi}{4} x^2 \, dx$, or Eq. (65). And in any similar geometric relation involving increments and an intermediate value of x, the indicated modifications lead to a correct relation between the differentials from which the derivative can be obtained by division.†

It is sometimes convenient to refer to dV or the expression for dV in terms of x and dx as the *element* of volume.

The actual written work for the deduction of dV/dx would run as follows. By noting the relation of ΔV to the cross section, from Fig. 47, we would think of $\Delta V = \frac{\pi}{4} \bar{x}^2 \, \Delta x$, but write down the expression for the element of volume,

$$dV = \frac{\pi}{4} x^2 \, dx \quad \text{and} \quad \frac{dV}{dx} = \frac{\pi}{4} x^2. \tag{66}$$

The rule for composite functions shows that in a problem involving three related variables, as x, t, and V, we may establish three *fixed* numbers for any time, dx, dt, dV which simultaneously satisfy the relations:

$$dV = \frac{dV}{dt} dt, \qquad dx = \frac{dx}{dt} dt, \qquad \text{and} \qquad dV = \frac{dV}{dx} dx. \tag{67}$$

† Let the area of the right section of a volume perpendicular to an altitude at distance x be the continuous function $A(x)$. For x between x and $x + \Delta x$, let the greatest cross section be $A(x_2)$. And let the least cross section be $A(x_1)$. Then ΔV would be increased if all the cross sections were increased. Hence $A(x_1)\Delta x < \Delta V < A(x_2)\Delta x$, or $A(x_1) < \Delta V/\Delta x < A(x_2)$. Since $\Delta V/\Delta x$ lies between $A(x_1)$ and $A(x_2)$, it must be a value assumed by the continuous function $A(x)$ for some value $\bar{x}$ between x_1 and x_2, and hence between x and $x + \Delta x$. Thus $\Delta V/\Delta x = A(\bar{x})$ and $\Delta V = A(\bar{x})\Delta x$. From this our rule gives $dV = A(x)dx$. This is correct, since taking the limit of $\Delta V/\Delta x = A(\bar{x})$ as $\Delta x \to 0$ gives $dV/dx = A(x)$ and $dV = A(x)dx$.

A similar argument applies to an area whose width perpendicular to an altitude at distance x is $L(x)$. Here ΔA would be increased if all the widths were increased. Hence $L(x_1)\Delta x < \Delta A < L(x_2)\Delta x$, or $L(x_1) < \Delta A/\Delta x < L(x_2)$. And $\Delta A/\Delta x$ must be a value assumed by the continuous function $L(x)$ for some intermediate value $\bar{x}$. Thus $\Delta A/\Delta x = L(\bar{x})$ and $\Delta A = L(\bar{x})\Delta x$. From this our rule gives $dA = L(x)dx$. This is correct, since taking the limit of $\Delta A/\Delta x = L(\bar{x})$ as $\Delta x \to 0$ gives $dA/dx = L(x)$ and $dA = L(x)dx$.

For, from the rule for composite functions, the first relation combined with the second gives

$$dV = \frac{dV}{dt}\,dt = \frac{dV}{dx}\frac{dx}{dt}\,dt = \frac{dV}{dx}\,dx. \tag{68}$$

This proves that the third relation of Eq. (67) follows from the first two. In fact, when $dx/dt \neq 0$, we may obtain the three differentials by starting with either dx or dt.

This consistency of related differentials makes it possible to avoid explicit use of the rule for composite functions, obtaining the same result by dividing by differentials. Thus from the first part of Eq. (66), on dividing by dt we find

$$\frac{dV}{dt} = \frac{\pi}{4}x^2\frac{dx}{dt}. \tag{69}$$

This is the result found in Eq. (62), and from it the problem may be solved by Eq. (63).

Although the ratio of differentials at any time is fixed, the sizes depend on one arbitrary choice. Hence as long as this is not specified, each equation in differentials will have its terms of the same degree in the differentials. Thus each term of Eqs. (66) first part, (67), and (68) is of the first degree in the differentials involved.

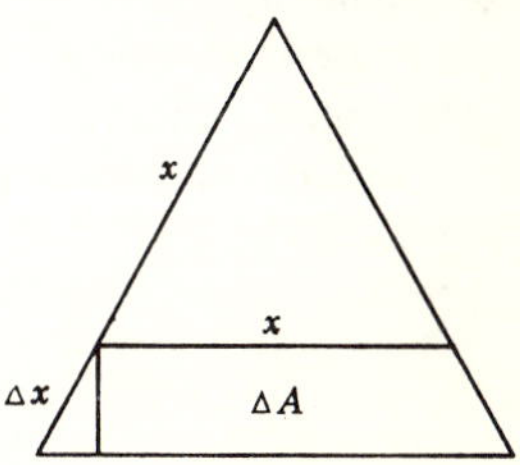

FIG. 48.

EXAMPLE. An equilateral triangle of side x is expanding. Find the time rate of change of the area in terms of x and the time rate of change of x.

Solution: If A is the area, we may consider ΔA as the trapezoid of Fig. 48. Since each of the small slant sides is Δx, inclined 60° to the base, the height is $\Delta x \sin 60° = \Delta x\,\sqrt{3}/2$. Hence we have $\Delta A = (\sqrt{3}/2)\bar{x}\,\Delta x$, so that

$$dA = \frac{\sqrt{3}}{2}x\,dx \qquad \text{and} \qquad \frac{dA}{dt} = \frac{\sqrt{3}}{2}x\frac{dx}{dt}. \tag{70}$$

EXERCISE 28

Find the time rate of change of the area of a figure in terms of x and the time rate of change of x if the figure is

1. A square of side x.
2. A circle of radius x.
3. An isosceles triangle of base $\sqrt{2}\,x$ and equal sides each x.
4. A trapezoid with bases x and $2x$ and sides each equal to x.
5. A rectangle of constant width a and height x.
6. A triangle of base x and altitude x.

Find the time rate of change of the volume of a figure in terms of x and the time rate of change of x if the figure is

7. A cube of side x.
8. A rectangular parallelepiped a by b by x.
9. A circular cone with radius of base x and altitude x.
10. A circular cone with radius of base $2x$ and altitude $3x$.
11. A square pyramid with side of base x and altitude $2x$.
12. The volume cut from a hemisphere of radius a by a plane parallel to the base of the hemisphere and distance x from it.

62. Differential of Arc. We shall discuss arc length more fully in Sec. 180. For the present we shall rest on the intuitive notion of length of a curve as obtained with a tape measure. Let the smooth curve APQ of Fig. 49 be the graph of a differentiable function $y = F(x)$. A is a fixed point and P is a variable point so moving that x increases as P moves from A to Q. If s is the arc length AP measured from A to $P = (x,y)$, s is an increasing function of x. Then if $Q = (x + \Delta x, y + \Delta y)$ is a point on the curve near P, the length of the arc PQ will be $|\Delta s|$.

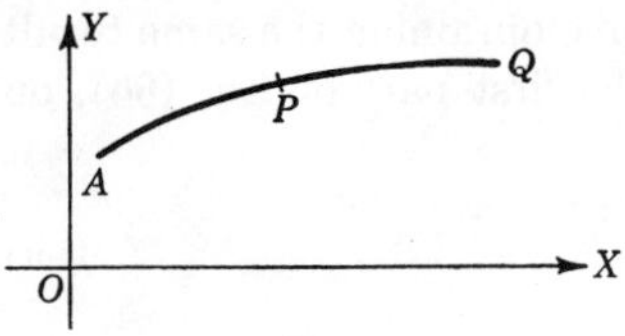

FIG. 49.

Suppose that we keep x fixed and let $\Delta x \to 0$. Then P is a fixed point on the curve, and Q is a variable nearby point on the curve tending toward P. For any smooth curve the chord $\overline{PQ}$ and the arc $\overset{\frown}{PQ}$ each tend to zero as Q tends to P in such a way that the ratio of the chord to the arc approaches one. That is†

$$\lim_{Q \to P} \frac{\text{chord } PQ}{\text{arc } PQ} = 1. \tag{71}$$

From the right triangle PRQ with $R = (x + \Delta x, y)$, we have

$$\overline{PQ}^2 = (\Delta x)^2 + (\Delta y)^2 \quad \text{or} \quad \frac{\overline{PQ}^2}{(\Delta x)^2} = 1 + \left(\frac{\Delta y}{\Delta x}\right)^2. \tag{72}$$

Since arc $PQ = |\Delta s|$, (arc $PQ)^2 = (\Delta s)^2$ and it follows that

$$\left(\frac{\text{chord } PQ}{\text{arc } PQ}\right)^2 \left(\frac{\Delta s}{\Delta x}\right)^2 = 1 + \left(\frac{\Delta y}{\Delta x}\right)^2. \tag{73}$$

Now let $Q \to P$. Then $\Delta x \to 0$ and

$$\left(\lim_{Q \to P} \frac{\text{chord } PQ}{\text{arc } PQ}\right)^2 \left(\lim_{\Delta x \to 0} \frac{\Delta s}{\Delta x}\right)^2 = 1 + \left(\lim_{\Delta x \to 0} \frac{\Delta y}{\Delta x}\right)^2. \tag{74}$$

By Eq. (71) the first factor is 1, and by the definition of a derivative the two limits of ratios of increments are each equal to a derivative. Hence we have

$$\left(\frac{ds}{dx}\right)^2 = 1 + \left(\frac{dy}{dx}\right)^2. \tag{75}$$

† We assume this result for the present. It will be proved as Eq. (82) of Sec. 182.

By multiplying by the square of the differential $(dx)^2$, we obtain the more symmetrical form in differential notation,

$$ds^2 = dx^2 + dy^2. \tag{76}$$

The last form suggests a right triangle like that in Fig. 50 with sides dx, dy, and ds. Since the inclination or slope angle ϕ has $\tan \phi = dy/dx$, the triangle has one angle equal to ϕ, as indicated. Consequently, we have

$$dx = ds \cos \phi, \qquad dy = ds \sin \phi. \tag{77}$$

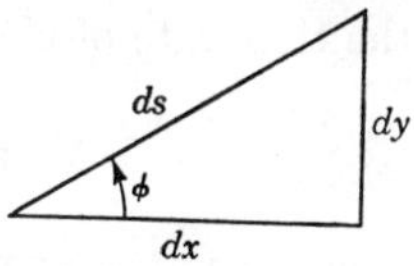

FIG. 50.

We took the direction of increasing s to be that for which x increased, so that $\cos \phi$ is positive and $-90° < \phi < 90°$.

We may take either direction on the curve as that of increasing s and still use Eq. (77), provided we take ϕ as the angle from the positive x axis to the direction on the tangent line in which s increases.

63. Motion in a Curve. It is sometimes more convenient to consider a curve as defined by relations of the form $x = f(t)$, $y = g(t)$, rather than by a single expression for y as a function of x. Thus if a point is moving on a curve, t may be the time at which the point reaches $P = (x,y)$. Or t may be a parameter, or third variable with or without some special physical or geometric meaning.

In any case, since x and y are given as functions of t, the arc $AP = s$ is also a function of t. And we may again deduce that

$$\overline{PQ}^2 = (\Delta x)^2 + (\Delta y)^2. \tag{78}$$

By dividing by $(\Delta t)^2$, we find

$$\frac{\overline{PQ}^2}{(\Delta t)^2} = \left(\frac{\Delta x}{\Delta t}\right)^2 + \left(\frac{\Delta y}{\Delta t}\right)^2. \tag{79}$$

And since $(\text{arc } PQ)^2 = (\Delta s)^2$, it follows that

$$\left(\frac{\text{chord } PQ}{\text{arc } PQ}\right)^2 \left(\frac{\Delta s}{\Delta t}\right)^2 = \left(\frac{\Delta x}{\Delta t}\right)^2 + \left(\frac{\Delta y}{\Delta t}\right)^2. \tag{80}$$

If we now take limits as $P \rightarrow Q$ or $\Delta x \rightarrow 0$, and use Eq. (71), we find

$$\left(\frac{ds}{dt}\right)^2 = \left(\frac{dx}{dt}\right)^2 + \left(\frac{dy}{dt}\right)^2. \tag{81}$$

Let t be the time, so that we are dealing with motion in a curve. Then the velocity is a vector in the direction of the tangent to the curve, whose magnitude is the speed in the path, or time rate of change of s when ds/dt is positive. Thus

$$\text{Speed in the path} = v = \frac{ds}{dt}. \tag{82}$$

The horizontal component of velocity, or component parallel to OX, is the time rate of change of x. Thus

$$\text{Horizontal component of velocity} = v_x = \frac{dx}{dt}. \tag{83}$$

And the vertical component of velocity, or component parallel to OY, is the time rate of change of y. Thus

$$\text{Vertical component of velocity} = v_y = \frac{dy}{dt}. \tag{84}$$

The relation of Eq. (81) suggests a right triangle like that in Fig. 51, with sides v, v_x, and v_y. Since

$$\frac{v_y}{v_x} = \frac{dy/dt}{dx/dt} = \frac{dy}{dx} = \tan\phi, \tag{85}$$

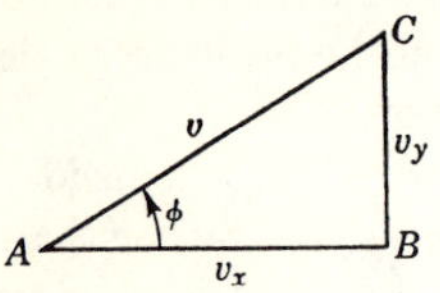

FIG. 51.

the triangle has one angle equal to the inclination or slope angle ϕ as indicated. Consequently, we have

$$v_x = v\cos\phi, \qquad v_y = v\sin\phi. \tag{86}$$

The hypotenuse of the triangle in Fig. 51 has the magnitude and direction of the velocity, and so the segment AC is one representation of the velocity vector.

We may rewrite Eq. (81) in the form

$$v^2 = v_x^2 + v_y^2. \tag{87}$$

This relation, or Eq. (81) as well as Eq. (86), can be recalled from the triangle of Fig. 51. It is also easy to read these relations from the triangle of Fig. 50, by mentally dividing each side by the differential dt.

EXAMPLE. For the motion $x = t^2$, $y = t^2 - 2t$ find the speed in the path at any time t. What is its least value? When and where is this reached? When and where is it equal to 10?

Solution: Here we have

$$v_x = \frac{dx}{dt} = 2t,$$

$$v_y = \frac{dy}{dt} = 2t - 2.$$

$$v^2 = v_x^2 + v_y^2 = (2t)^2 + (2t-2)^2 = 8t^2 - 8t + 4.$$

Hence the speed at time t is $v = 2\sqrt{2t^2 - 2t + 1}$.

Since v will be least when v^2 is least, we set the derivative of v^2 equal to zero. Thus we find

$$\frac{d}{dt}(v^2) = \frac{d}{dt}(8t^2 - 8t + 4) = 16t - 8.$$

This equals zero when $t = \frac{1}{2}$, which makes v^2 a minimum since the second derivative is 16 which is positive.

At the time of minimum speed $t = \frac{1}{2}$, we have

$$x = t^2 = (\tfrac{1}{2})^2 = \tfrac{1}{4} \qquad \text{and} \qquad y = t^2 - 2t = (\tfrac{1}{2})^2 - 2(\tfrac{1}{2}) = -\tfrac{3}{4}.$$

This gives the position. And since

$$v^2 = 8t^2 - 8t + 4 = 8(\tfrac{1}{2})^2 - 8(\tfrac{1}{2}) + 4 = 2,$$

the minimum speed in the path is $v = \sqrt{2}$.

The speed in the path will equal 10 when $v^2 = 100$. Thus

$$8t^2 - 8t + 4 = 100 \qquad \text{or} \qquad 8t^2 - 8t - 96 = 0.$$

As the left member is $8(t - 4)(t + 3)$, the roots are $t = 4$ and $t = -3$. When $t = -3$, $x = 9$, $y = 15$. And when $t = 4$, $x = 16$, $y = 8$. At each of these two positions the speed is equal to 10.

EXERCISE 29

Find v_x, v_y, and the speed in the path for each of the following motions at the given instant of time.

1. $x = 3t^2$, $y = 100 - 4t$; $t = 3$.
2. $x = 2t + 5$, $y = t^3$; $t = 2$.
3. $x = t^3$, $y = t^2$; $t = 3$.
4. $x = \sqrt{1 + 2t}$, $y = 3t$; $t = 4$.
5. $x = \dfrac{18}{1 + 2t}$, $y = 3t - 2$; $t = 1$.
6. $x = t^{\frac{3}{2}}$, $y = t^{\frac{5}{2}}$; $t = 4$.

Find the value of t for which the speed in the path is least for each of the following curvilinear motions.

7. $x = t^2 - 4t + 5$, $y = 2t - 3$.
8. $x = t^2 - 4t$, $y = t^2 - 3t + 1$.
9. $x = 2(t - 3)^4$, $y = 4t + 5$.
10. $x = (t - 2)^2$, $y = t^2$.

CHAPTER 5

INTEGRATION

The process of differentiation takes us from a function to its derivative. It is often desirable to reverse this process. That is, the derivative of a function is known, and we wish to find the function. This reversed process is called integration, and the result of the process is an integral. The branch of the subject which deals primarily with integration is known as the integral calculus. Several aspects of integration will be treated at length in later chapters.

In this chapter we can present only an introduction to integration and its applications. We begin with the definition of an indefinite integral which includes a constant of integration, and the notation for it. We then show what additional conditions are needed to determine the constant, and explain the concept of a definite integral and the notation for it. Finally we use integration to calculate a number of geometrical and physical quantities, including areas, volumes, and the force on an immersed surface due to liquid pressure.

64. Integration. Suppose that $f'(x)$, the derivative of a function with respect to x, is known. And we wish to find the function $f(x)$. This inverse process to differentiation is called *integration*. The result of the operation is called an *integral*.

For example, suppose it is known that the derivative of a function with respect to x is $3x^2$. By remembering that $\frac{d}{dx}(x^3) = 3x^2$, we see that x^3 is one possible value of $f(x)$. We say that x^3 is an integral of $3x^2$ with respect to x.

In the differential notation of Sec. 61, we may write

$$d(x^3) = 3x^2\,dx. \tag{1}$$

We also say that x^3 is an integral of the differential $3x^2\,dx$, meaning that the derivative of x^3 with respect to x is $3x^2$, and using the factor dx to indicate that the integration is with respect to x.

The operation of integration is indicated by writing the *integral sign*, $\int$, before the differential.

Thus the deduction of x^3 from $3x^2$ by integration with respect to x, or the relation of x^3 to the differential $3x^2\,dx$ is indicated by writing

$$\int 3x^2\,dx = x^3, \tag{2}$$

read "an integral of $3x^2\,dx$ equals x^3."

More generally, let $F(x)$ be the known derivative. Then

$$\int F(x)dx, \tag{3}$$

read "an integral of $F(x)dx$" means a function having $F(x)$ as its derivative with respect to x, or having $F(x)dx$ as its differential. In the expression (3), $F(x)$ is called the *integrand.*

65. Constant of Integration. We called x^3 an integral of $3x^2\,dx$ because

$$\frac{d}{dx}(x^3) = 3x^2 \qquad \text{so that } d(x^3) = 3x^2\,dx.$$

But if C is any constant,

$$\frac{d}{dx}(x^3 + C) = 3x^2 \qquad \text{and} \qquad d(x^3 + C) = 3x^2\,dx.$$

Thus for any value of the constant C, $x^3 + C$ is an integral of $3x^2\,dx$ so that

$$\int 3x^2\,dx = x^3 + C. \tag{4}$$

More generally, let the known derivative with respect to x be $F(x)$. And suppose we find some one function $f(x)$ having its derivative $f'(x) = F(x)$. Then for any constant C, we have

$$\frac{d}{dx}[f(x) + C] = f'(x) + 0 = F(x).$$

It follows that $f(x) + C$ is an integral of $F(x)dx$ or $f'(x)dx$ and

$$\int f'(x)dx = f(x) + C. \tag{5}$$

Since the constant C is unknown and to that extent is *indefinite,* the expression $f(x) + C$ is called the *indefinite* integral of $f'(x)dx$. Any constant value can be assigned to C. Hence it is sometimes called an *arbitrary constant.* It arose from an integration, and in any particular problem is a number independent of the *variable of integration.* Hence it is sometimes referred to as the *constant of integration.*

***66. The Integrals of a Function.** The determination of all the integrals of a given function from any one depends on a series of theorems which we shall next discuss.

Theorem I. *If two functions differ by a constant, they have the same derivative.*

For if $g(x) - f(x) = C$, by differentiation we find that $g'(x) - f'(x) = 0$ and so $g'(x) = f'(x)$ as we set out to prove.

Theorem II. *If $f(x)$ is an integral of $F(x)dx$, then for any constant value of C, $f(x) + C$ is also an integral of $F(x)dx$.*

Let $g(x) = f(x) + C$. Then $g(x) - f(x) = C$ and hence from theorem I, $g'(x) = f'(x)$. But $\int F(x)dx = f(x)$ implies that $F(x) = f'(x)$. It follows that $F(x) = g'(x)$, and this implies that $\int F(x)dx = g(x) = f(x) + C$, which we set out to prove.

Theorem III. *If two functions have the same derivative, their difference is a constant.*

Let $g(x)$ and $f(x)$ be two functions, having the same derivative $F(x)$. The difference $g(x) - f(x)$ is a new function. Call it $G(x)$, so that $G(x) = g(x) - f(x)$.

Then since $g'(x) = F(x)$ and $f'(x) = F(x)$, we have

$$G'(x) = g'(x) - f'(x) = F(x) - F(x) = 0. \tag{6}$$

Now apply the mean value theorem of Sec. 37 to the function $G(x)$ for the interval x_1,x_2. We have

$$G(x_2) - G(x_1) = (x_2 - x_1)G'(x_0), \tag{7}$$

where x_0 is a suitably chosen point between x_1 and x_2. But since $G'(x) = 0$ at all points, by Eq. (6), it follows that $G'(x_0) = 0$, and from Eq. (7)

$$G(x_2) - G(x_1) = 0 \qquad \text{or} \qquad G(x_2) = G(x_1). \tag{8}$$

If we keep x_1 fixed, $G(x_1)$ is a constant C. Keeping x_1 fixed, we may let x_2 vary and take any value x. Thus

$$G(x) = C \qquad \text{or} \qquad g(x) - f(x) = C, \tag{9}$$

as we set out to prove.

Theorem IV. *If $f(x)$ is an integral of $F(x)dx$, then every integral of $F(x)dx$ has the form $f(x) + C$.*

For, by theorem II, $f(x) + C$ is also an integral. And by theorem III, every other integral must be of this form.

Theorem IV enables us to determine all possible integrals of $F(x)dx$ from any given one. For example, from Eq. (2) and theorem IV, we see that Eq. (4) gives all the integrals of $3x^2\,dx$.

67. Integrals of Polynomials and Powers. Integrals of some complicated expressions will be discussed in later chapters. For the present, we merely consider integrals of polynomials or powers. We first note that integration has a linear character. For let u and v be any two functions of x. By the linear character of differentiation, we see that

$$\frac{d}{dx}\left(c_1 \int u\,dx + c_2 \int v\,dx\right) = c_1 \frac{d}{dx}\int u\,dx + c_2 \frac{d}{dx}\int v\,dx. \tag{10}$$

But, by the definition of an integral, we have

$$\frac{d}{dx}\int u\,dx = u \qquad \text{and} \qquad \frac{d}{dx}\int v\,dx = v. \tag{11}$$

It follows from Eqs. (10) and (11) that

$$\frac{d}{dx}\left(c_1 \int u\,dx + c_2 \int v\,dx\right) = c_1u + c_2v. \tag{12}$$

This shows that the expression in parentheses is an integral of $(c_1u + c_2v)dx$. Hence by theorem IV of Sec. 66 we may write

$$\int(c_1u + c_2v)dx = c_1\int u\,dx + c_2\int v\,dx + C. \tag{13}$$

This equation holds in the sense that for any particular choice of the constants of integration for the three integrals involved, there is a suitable choice of C. If either of the integrals on the right is indefinite, we may omit C from the equation.

A similar result holds for any number of functions.

For a particular choice of the constants of integration, we have

$$\int 1\,dx = \int dx = x, \qquad \int x\,dx = \frac{x^2}{2}, \qquad \int x^2\,dx = \frac{x^3}{3}, \tag{14}$$

since

$$\frac{dx}{dx} = 1, \qquad \frac{d}{dx}\left(\frac{x^2}{2}\right) = x, \qquad \frac{d}{dx}\left(\frac{x^3}{3}\right) = x^2. \tag{15}$$

It follows from this and the linearity property for three functions that

$$\int (ax^2 + bx + c)dx = \frac{ax^3}{3} + \frac{bx^2}{2} + cx + C. \tag{16}$$

More generally for any rational exponent not equal to -1, we have for a particular value of the constant of integration,

$$\int x^n\,dx = \frac{x^{n+1}}{n+1}, \qquad n \neq -1, \tag{17}$$

since

$$\frac{d}{dx}\left(\frac{x^{n+1}}{n+1}\right) = x^n. \tag{18}$$

This and the linearity property will enable us to integrate any linear combination of terms of this form, in particular a polynomial of any degree.

EXAMPLE 1. Evaluate $\int \frac{2x^4 + x^2 - 1}{3x^2}\,dx$.

Solution: From the linearity property and Eqs. (14) and (17) we find

$$\begin{aligned}\int \frac{2x^4 + x^2 - 1}{3x^2}\,dx &= \frac{1}{3}\int (2x^2 + 1 - x^{-2})dx \\ &= \frac{1}{3}\left[\frac{2x^3}{3} + x - \frac{x^{-1}}{(-1)}\right] + C \\ &= \frac{1}{9}\left(2x^3 + 3x + \frac{3}{x}\right) + C.\end{aligned}$$

EXAMPLE 2. Evaluate $\int \sqrt{2x}\,dx$.

Solution: We have

$$\int \sqrt{2x}\,dx = \sqrt{2}\int x^{\frac{1}{2}}\,dx = \sqrt{2}\,\frac{x^{\frac{3}{2}}}{\frac{3}{2}} + C = \frac{2\sqrt{2}}{3}\,x^{\frac{3}{2}} + C.$$

EXERCISE 30

Evaluate each of the following integrals.

1. $\int (4x - 3)dx.$
2. $\int (6x^2 - 8x + 5)dx.$
3. $\int (9x^2 + 2x + 4)dx.$
4. $\int (2x^2 - 3x + 6)dx.$
5. $\int 10x^{\frac{3}{2}}\, dx.$
6. $\int 6\sqrt{x}\, dx.$
7. $\int \frac{5}{\sqrt{x}}\, dx.$
8. $\int \frac{8}{x^3}\, dx.$
9. $\int \frac{10}{x^2}\, dx.$
10. $\int \frac{9}{x^4}\, dx.$
11. $\int \frac{x^2 - 2x + 3}{x^4}\, dx.$
12. $\int \frac{x^4 + x^3 + 2}{x^2}\, dx.$
13. $\int \frac{(x + 1)^2}{x^6}\, dx.$
14. $\int \frac{2x - 3}{\sqrt{x}}\, dx.$
15. $\int \sqrt{x}\,(5 - 3x)dx.$
16. $\int \frac{(x - 1)(x - 2)}{\sqrt{x}}\, dx.$

68. Curve from Given Slope. If the slope of a curve is given as a function of x, we have

$$\frac{dy}{dx} = f(x). \tag{19}$$

Hence from the definition of an integral, $y = \int f(x)dx$.

EXAMPLE. At each of its points $P(x,y)$, the slope of a curve is $2x$. Find all possible curves, also the particular one passing through the point $(-2,1)$.

Solution: Since $dy/dx = 2x$, $y = \int 2x\, dx$. But we have

$$\int 2x\, dx = 2\left(\frac{x^2}{2}\right) + C = x^2 + C.$$

Hence, with any choice of C, $y = x^2 + C$ is a possible equation of one such curve. Some of these curves are shown in Fig. 52.

If the curve is to pass through $(-2,1)$, we must have

$$1 = (-2)^2 + C \qquad \text{or} \qquad C = 1 - 4 = -3.$$

Hence the particular curve through $(-2,1)$ is $y = x^2 - 3$.

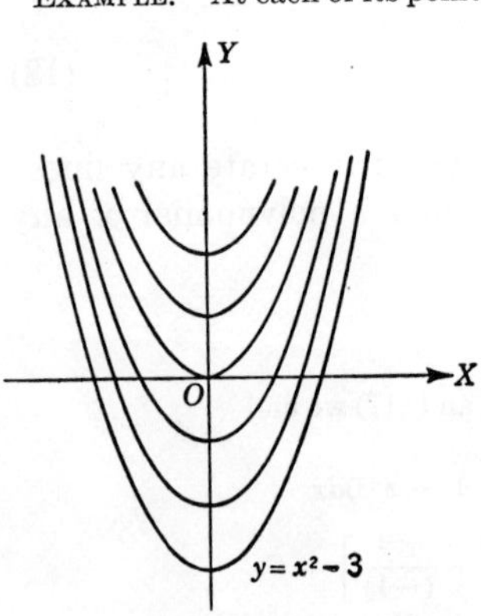

FIG. 52.

69. Motion from Given Rate. If the time rate of change of a variable x is given as a function of the time t, we have

$$\frac{dx}{dt} = f(t). \tag{20}$$

Hence from the definition of an integral, $x = \int f(t)dt$.

In particular, for straight-line motion, by Sec. 20 the velocity $v = ds/dt$, so that $s = \int v\,dt$. And the acceleration $a = dv/dt$ by Sec. 21, so that $v = \int a\,dt$.

EXAMPLE. If the acceleration $a = 12t^2$, and the particle has velocity $v_0 = 5$ and distance $s_0 = 3$ at time $t = 0$, find the velocity v and distance s at any time t.

Solution: By Eq. (16) with t in place of x we find

$$v = \int 12t^2\,dt = 12\left(\frac{t^3}{3}\right) + C = 4t^3 + C.$$

Since $v = 5$ when $t = 0$, $5 = 4(0) + C$, and $C = 5$. Hence the velocity $v = 4t^3 + 5$.

Since $v = ds/dt = 4t^3 + 5$, $s = \int(4t^3 + 5)dt$. By Eqs. (14) and (17) with t in place of x, we find

$$s = 4\int t^3\,dt + 5\int dt = 4\frac{t^4}{4} + 5t + C_1 = t^4 + 5t + C_1.$$

Since $s = 3$ when $t = 0$, $3 = 0 + 5(0) + C_1$, and $C_1 = 3$. Hence the distance $s = t^4 + 5t + 3$.

EXERCISE 31

The slope of a curve $m = dy/dx$ is the given function of the abscissa x. And the curve passes through the given point P_1. Find the equation of the curve for each set of data.

1. $m = 4x, P_1 = (2,3)$.
2. $m = \dfrac{x}{2}, P_1 = (2,-4)$.
3. $m = 1 - x^2, P_1 = (1,2)$.
4. $m = 6x^2 - 2x, P_1 = (0,2)$.
5. $m = \dfrac{2}{\sqrt{x}}, P_1 = (4,4)$.
6. $m = -\dfrac{1}{x^2}, P_1 = (3,1)$.

The quantity x is a function of the time. The time rate of change of x, $R = dx/dt$, is the given function of the time t. And the value of x at time $t = 1$ is the given value x_1. Find the expression of x in terms of t for each set of data.

7. $R = 4t, x_1 = 8$.
8. $R = 6t^2, x_1 = 2$.
9. $R = 6t^{\frac{1}{2}}, x_1 = 0$.
10. $R = 12t^3, x_1 = 1$.

For a straight-line motion, the velocity v is the given function of the time. In each problem find how far the particle will move in the given time interval t_1,t_2. Note that the required distance is the value of s at t_2 when s is taken as zero at t_1 whenever v is positive in the given interval.

11. $v = 6t - 2$; 0,4.
12. $v = 8t + 1$; 1,2.
13. $v = 8t^3 - 9t^2$; 0,2.
14. $v = \sqrt{t}$; 4,9.

For a straight-line motion, the acceleration $a = dv/dt$ is the given function of the time. The value of the velocity at time $t = 0$ is the given value v_0, while the value of the distance at time $t = 0$ is $s_0 = 0$. Find v and s in terms of t for each set of data.

15. $a = 32, v_0 = 0$.
16. $a = -32, v_0 = 100$.
17. $a = 42t^5, v_0 = 2$.
18. $a = 40t^3, v_0 = 5$.
19. $a = 6t^2 - 2t, v_0 = 3$.
20. $a = 4t + 4, v_0 = 4$.

70. Area under a Curve. Let the equation of the curve RS of Fig. 53 be $y = f(x)$, where $f(x)$ is a continuous function. For the present we assume that y is positive for values of x inside some interval under consideration, although at an end point of the interval y may be zero. Let BD be a fixed ordinate at the point where $x = OB = a$, so that $D = (a,y_a)$ with $y_a = f(a)$. And let MP be a variable ordinate at the point $P = (x,y)$ with $y = f(x)$. Let the measure of the area $BMPD$ be A, or $A(x)$, since A is a function of x.

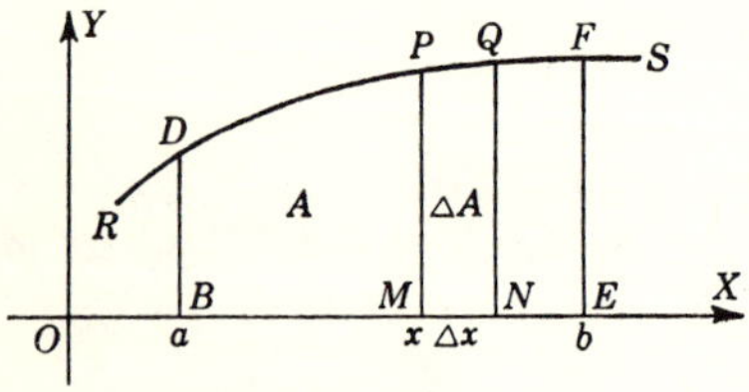

FIG. 53.

For the present, we rest on the intuitive notion of area. Area as the limit of a sum will be discussed in Sec. 168. With every bounded region R we may associate a number $A(R)$ called the area of the region. This has the property that if a region R_2 lies inside a region R_1, then $A(R_2) < A(R_1)$.

Give x a positive increment $\Delta x = MN$. Then $A(x)$ will take on an increment ΔA = area $MNQP$. Reasoning as we did in Sec. 60, we find that ΔA is greater than the area of a rectangle with base Δx and height equal to the smallest value of y on PQ, but less than the area of a rectangle with base Δx and height equal to the largest value of y on PQ. Hence there is some value of y on PQ, $\bar{y}$, for which $\Delta A = \bar{y}\,\Delta x$. This may be written

$$\frac{\Delta A}{\Delta x} = \bar{y}. \tag{21}$$

On taking the limit as $\Delta x \to 0$ and $\bar{y} \to y$, we find

$$\frac{dA}{dx} = y \qquad \text{or} \qquad dA = y\,dx. \tag{22}$$

The abbreviated procedure of Sec. 61 would have led us to write $dA = y\,dx$ directly from a consideration of the figure, and then to deduce that $dA/dx = y$.

To find the area between two fixed ordinates, BD where $x = a$ and EF where $x = b$, we may proceed as follows. Since

$$dA = y\,dx, \qquad A(x) = \int y\,dx = \int f(x)dx. \tag{23}$$

Suppose that $F(x)$ is any one function such that $F'(x) = f(x)$. Then by Secs. 65 and 66, it follows from Eq. (23) that

$$A(x) = F(x) + C. \tag{24}$$

The constant C depends on the place where we start the area. Here we began with a left-hand ordinate at $x = a$, so that $A = 0$ when $x = a$. Hence

$$0 = A(a) = F(a) + C. \tag{25}$$

It follows that $C = -F(a)$, so that

$$A(x) = F(x) - F(a). \tag{26}$$

But the area $BEFD$ which we wished to find is $A(b)$ or

$$A(b) = F(b) - F(a). \tag{27}$$

We summarize the important results of this section in the following two theorems:

Theorem I. *The area bounded above by a curve and below by the x axis, lying between a fixed left-hand ordinate and a variable right-hand ordinate through (x,y) on the curve, is a function of x, $A(x)$, whose differential is $dA = y\,dx$.*

Theorem II. *The area bounded above by the curve $y = f(x)$ and below by the x axis, lying between a left-hand ordinate at $x = a$ and a right-hand ordinate at $x = b$, is given by the difference $F(b) - F(a)$, where $F(x) = \int y\,dx$ is any function such that $F'(x) = f(x)$.*

EXAMPLE. Find the area bounded by the curve $y = -3 + 4x - x^2$ and the x axis. *Solution:* The curve, Fig. 54, meets the axis when $y = 0$, or when

$$-x^2 + 4x - 3 = 0 \qquad \text{and} \qquad x = \frac{-4 \pm \sqrt{4^2 - 4(-1)(-3)}}{2(-1)} = 3 \text{ or } 1.$$

Hence $y = -(x-1)(x-3)$ so that y is positive for $1 < x < 3$. Thus in this case the required area is bounded by the zero ordinates at $x = 1$ and $x = 3$.

By theorem I, $dA = y\,dx = (-x^2 + 4x - 3)dx$. Hence we may take

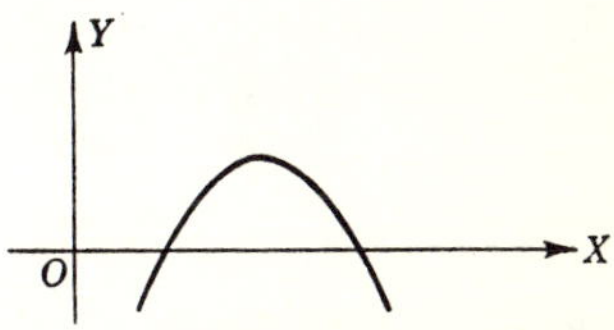

FIG. 54.

$$\int (-x^2 + 4x - 3)dx = -\frac{x^3}{3} + 4\left(\frac{x^2}{2}\right) - 3x + C,$$

with any choice of C as the $F(x)$ of theorem II. With $C = 0$, we have

$$F(x) = -\frac{x^3}{3} + 2x^2 - 3x.$$

By theorem II we then find that the area A between ordinates at $x = 1$ and at $x = 3$ is

$$A = F(3) - F(1) = \left(-\frac{3^3}{3} + 2 \cdot 3^2 - 3 \cdot 3\right) - \left(-\frac{1}{3} + 2 \cdot 1 - 3 \cdot 1\right)$$
$$= 0 - \left(-\frac{4}{3}\right) = \frac{4}{3}.$$

Thus the required area is $\frac{4}{3}$.

EXERCISE 32

For each given curve, y is positive for all values of x between the two given values. Find the area above the x axis and below the curve which lies between the two given ordinates in each case.

1. $y = 6x - 3$; $x = 3$, $x = 4$.
2. $y = 8x + 3x^2$; $x = 1$, $x = 2$.
3. $y = 8x^3 + 4x - 2$; $x = 1$, $x = 2$.
4. $y = 20x^4 + 10x^2 + 10$; $x = -2$, $x = -1$.
5. $y = 9x^2 - 6x + 2$; $x = -2$, $x = 1$.
6. $y = \sqrt{3x}$; $x = 3$, $x = 12$.
7. $y = x + \frac{10}{x^2}$; $x = 2$, $x = 5$.

Find the area in the first quadrant bounded by the x axis, the y axis, and the given curve in each of the following problems.

8. $y = 12 - 6x^2$.
9. $y = 3 - 2x - x^2$.
10. $y = 8 - x^3$.
11. $y = 1 - x^4$.
12. $y = 3 - \sqrt{x}$.
13. $y = 8 - x^{\frac{3}{2}}$.

Find the area bounded below by the x axis and above by the given curve in each of the following problems.

14. $y = 9 - x^2$.
15. $y = -3 + 4x - x^2$.
16. $y = 3 + 2x - x^2$.
17. $y = 16 - x^4$.
18. $y = x^3 - 4x$.
19. $y = 3x^2 - x^3$.

71. Integration between Limits. Theorem II of Sec. 70 states that under certain conditions an area related to the curve $y = f(x)$ and to the values $x = a$ and $x = b$ is equal to $F(b) - F(a)$, where $F(x)$ is a particular indefinite integral of $y\,dx$.

The difference $F(b) - F(a)$ as obtained from a knowledge of $F(x)$, a, and b is denoted by the symbol $[F(x)]_a^b$. And as related to a, b, and a particular indefinite integral of $y\,dx$, we represent it by the symbol

$$\int_a^b y\,dx, \tag{28}$$

read "the integral from a to b of $y\,dx$." Since $y = f(x)$, we may also represent $F(b) - F(a)$ by the symbol

$$\int_a^b f(x)dx, \tag{29}$$

read "the integral from a to b of $f(x)dx$."

The operation by which the integral (28) or (29) is obtained from the differential $y\,dx = f(x)dx$ is called *integration between limits*. In this connection a is called the *lower limit* and b is called the *upper limit* of the integral. The use of the word limit here is more closely related to its colloquial use to mean boundary than to the technical meaning explained in Chap. 1.

In terms of the symbolism just introduced, the area described in theorem II of Sec. 70 is equal to

$$\int_a^b y\,dx = \int_a^b f(x)dx = [F(x)]_a^b = F(b) - F(a). \tag{30}$$

For y positive and $a < b$, the geometric meaning of the expression proves that integration between limits leads to the same value regardless of which particular integral $F(x)$ we use.

But for any continuous function $y = f(x)$, unrestricted as to algebraic sign, and any two values a and b, let us use Eq. (30) as the definition of integration between limits, with the understanding that $F(x)$ is an indefinite integral of $f(x)dx$. A fuller discussion of integration between limits will be given in Chap. 12. If $F(x)$ is any one indefinite integral, by theorem IV of Sec. 66, any other indefinite integral $G(x)$ must have the form $F(x) + C$ where C has some constant value. It follows that

$$\begin{aligned}[G(x)]_a^b &= [F(x) + C]_a^b = [F(b) + C] - [F(a) + C] \\ &= F(b) - F(a) = [F(x)]_a^b.\end{aligned} \tag{31}$$

Thus the value of the constant of integration does not affect the result of integrating between limits as defined by Eq. (30).

Since integration between limits leads to the same *definite* value for any indefinite integral, or choice of the arbitrary constant, the resulting integral (28) or (29) is called a *definite* integral.

We note that, if y has fixed sign between a and b, Eq. (30) gives the area bounded by the x axis, the curve $y = f(x)$ and the lines $x = a$ and $x = b$ if $y(b - a)$ is positive, and Eq. (30) gives the negative of the area if $y(b - a)$ is negative.

72. Calculation of Definite Integrals. The actual calculation of a definite integral, $\int_a^b f(x)dx$, may be effected as follows.

1. Find some particular indefinite integral $F(x) = \int f(x)dx$, that is a function such that $F'(x) = f(x)$.

2. Evaluate $F(b)$ by substituting $x = b$ in $F(x)$, and evaluate $F(a)$ by substituting $x = a$ in $F(x)$.

3. Form the difference $F(b) - F(a)$ by subtracting the second result from the first.

If $F(x) = k\,G(x)$, it is sometimes convenient to factor out the constant k before subtracting and evaluate

$$[F(x)]_a^b \text{ as } k[G(x)]_a^b = k[G(b) - G(a)]. \tag{32}$$

See Example 1 below.

It is unnecessary to add an arbitrary constant C to the indefinite integral used, since Eq. (31) shows that this constant would not affect the value of the difference.

EXAMPLE 1. Find the area above the x axis and below the curve $y = (2x^2/3) + \frac{4}{9}$, lying between the lines $x = 1$ and $x = 2$.

Solution. The area is

$$A = \int_1^2 \left(\frac{2x^2}{3} + \frac{4}{9}\right) dx = \left[\frac{2x^3}{9} + \frac{4x}{9}\right]_1^2 = \frac{2}{9}[x^3 + 2x]_1^2$$
$$= \tfrac{2}{9}[(2^3 + 2 \cdot 2) - (1 + 2)] = \tfrac{2}{9}(9) = 2.$$

Hence the required area is $A = 2$.

EXAMPLE 2. Evaluate $I = \int_1^3 (-x^2 + 4x - 3)dx$.

Solution: By Sec. 67, one $F(x)$ whose derivative is the integrand is

$$\int (-x^2 + 4x - 3)dx = -\frac{x^3}{3} + 2x^2 - 3x.$$

Hence the given integral is

$$I = \left[-\frac{x^3}{3} + 2x^2 - 3x\right]_1^3 = (-9 + 18 - 9) - \left(-\frac{1}{3} + 2 - 3\right)$$
$$= 0 - (-\tfrac{4}{3}) = \tfrac{4}{3}.$$

This last chain of relations is all that need be written down, since we may insert the indefinite integral in the brackets as soon as we obtain it. Such a calculation of $I = \frac{4}{3}$ is slightly more efficient than the equivalent calculation made in the example of Sec. 70.

EXERCISE 33

Evaluate each of the following definite integrals.

1. $\int_0^2 (4x - x^3)dx.$
2. $\int_0^3 (x^2 - 4x^3)dx.$
3. $\int_1^5 \frac{3x - 4}{x^3}\,dx.$
4. $\int_1^2 \frac{x^2 - 5}{x^4}\,dx.$

For rational n not equal to -1, verify by differentiation that

$$\int (ax + b)^n\,dx = \frac{(ax + b)^{n+1}}{a(n + 1)} + C,$$

if $a \neq 0$. Use this fact to evaluate each of the following definite integrals.

5. $\int_2^1 (1 - 2x)^3\,dx.$
6. $\int_{-1}^0 (1 + 3x)^4\,dx.$
7. $\int_2^3 (3 + 2x)dx.$
8. $\int_{-1}^{-2} (2 + 3x)^2\,dx.$
9. $\int_{-1}^1 \sqrt{5 + 4x}\,dx.$
10. $\int_{-2}^1 \frac{1}{(6 + 2x)^2}\,dx.$

For rational n not equal to -1, verify by differentiation that

$$\int (ax^2 + b)^n x\,dx = \frac{(ax^2 + b)^{n+1}}{2a(n + 1)} + C,$$

if $a \neq 0$. Use this fact to evaluate each of the following definite integrals,

11. $\int_1^2 (4 + x^2)x\,dx.$
12. $\int_0^1 (1 + x^2)^4 x\,dx.$
13. $\int_0^2 x\sqrt{4 - x^2}\,dx,$
14. $\int_1^2 \frac{x}{(x^2 + 1)^2}\,dx,$

73. Areas. In Secs. 70 and 71 we found that the area under a curve, above the x axis, to the right of $x = a$ and to the left of $x = b$ has as its differential $y\,dx$ and is equal to $\int_a^b y\,dx$.

Similar reasoning can be used to express other types of area as definite integrals. Let us illustrate some of these by solving an example in several ways.

EXAMPLE 1. Find the area in the first quadrant between the curve $y = x^2$, the y axis, and the line $y = 4$.

Solution 1: From Fig. 55 we note that the area in the first quadrant bounded by the curve $y = x^2$ and the line parallel to the x axis for any y has a differential $x\,dy$. This follows from the procedure of Sec. 61 from the fact that $\Delta A = \bar{x}\,\Delta y$. As this area is zero when $y = 0$ and we wish the value when $y = 4$, the desired area is

$$A = \int_0^4 x\,dy = \int_0^4 y^{\frac{1}{2}}\,dy = \left[\frac{y^{\frac{3}{2}}}{\frac{3}{2}}\right]_0^4 = \frac{2}{3}(4^{\frac{3}{2}} - 0) = \frac{16}{3}.$$

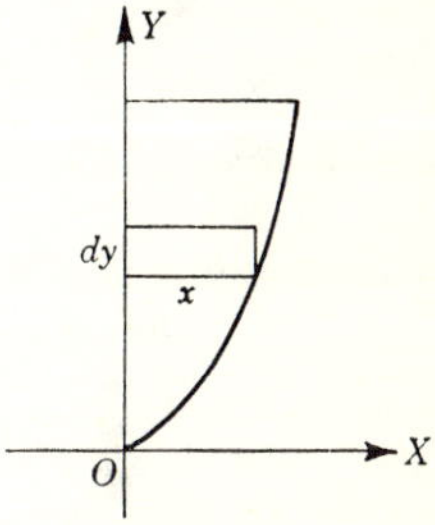

FIG. 55.

Solution 2: In the first solution we used the limits 0 and 4 for y and expressed x in terms of y.

Let us now deduce from $y = x^2$ that $dy = 2x\,dx$. Also when y increases from 0 to 4, x increases from 0 to 2, since $x = \sqrt{y}$ with a plus sign in the first quadrant. In terms of x the area is

$$A = \int_0^4 x\,dy = \int_0^2 x(2x\,dx) = \int_0^2 2x^2\,dx = \left[\frac{2x^3}{3}\right]_0^2 = \frac{2}{3}[x^3]_0^2$$
$$= \tfrac{2}{3}(2^3 - 0) = \tfrac{16}{3}.$$

Solution 3: Consider next the area in the first quadrant bounded above by $y = 4$, below by $y = x^2$, and on the right by the line parallel to the y axis at x. From Fig. 56, the ordinate from the lower curve to the upper curve at x is the length AB from $y_1 = x^2$ to $y_2 = 4$. By Sec. 3, this directed segment has length $y_2 - y_1$. Thus the element of area is

$$dA = (y_2 - y_1)dx = (4 - x^2)dx.$$

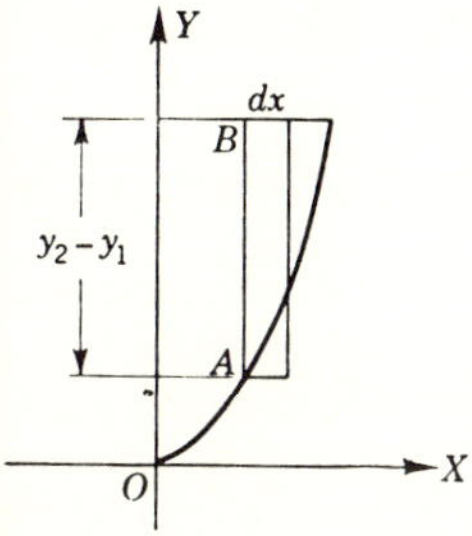

FIG. 56.

Since the variable area is zero for $x = 0$ and we wish the area for $x = 2$, the desired area is

$$A = \int_0^2 (4 - x^2)dx = \left[4x - \frac{x^3}{3}\right]_0^2 = \left(8 - \frac{8}{3}\right) - 0 = \frac{16}{3}.$$

Thus $A = \frac{16}{3}$ by any one of the three solutions.

The method used for the third solution of the example establishes the following theorem:

Theorem I. *The area bounded above by the curve $y = f_2(x)$ and below by the curve $y = f_1(x)$, lying between a left-hand ordinate at $x = a$ and a right-*

hand ordinate at $x = b$ *is given by*

$$A = \int_a^b (y_2 - y_1)dx = \int_a^b [f_2(x) - f_1(x)]dx. \tag{33}$$

This result may be recalled by reading the differential

$$dA = (y_2 - y_1)dx \tag{34}$$

from a figure.

The relations used in Secs. 70 and 71,

$$dA = y\,dx \qquad \text{and} \qquad A = \int_a^b y\,dx = \int^b f(x)dx \tag{35}$$

for an area bounded below by the x axis may be thought of as the special case of Eqs. (33) and (34) where $f_1(x) = 0$ and $y_1 = 0$, so that we may drop the subscript 2 from y_2 and $f_2(x)$.

Similarly we may establish the following theorem:

Theorem II. *The area bounded on the right by* $x = g_2(y)$ *and on the left by* $x = g_1(y)$, *lying above the line* $y = c$ *and below the line* $y = d$ *is given by*

$$A = \int_c^d (x_2 - x_1)dy = \int_c^d [g_2(y) - g_1(y)]dy. \tag{36}$$

This result may be recalled by reading the differential

$$dA = (x_2 - x_1)dy \tag{37}$$

from a diagram like Fig. 57.

For the special case of an area bounded on the right by $x = g(y)$, on the left by the y axis, lying above the line $y = c$ and below the line $y = d$, the useful relations are

$$dA = x\,dy$$

and

$$A = \int_c^d x\,dy = \int_c^d g(y)dy. \tag{38}$$

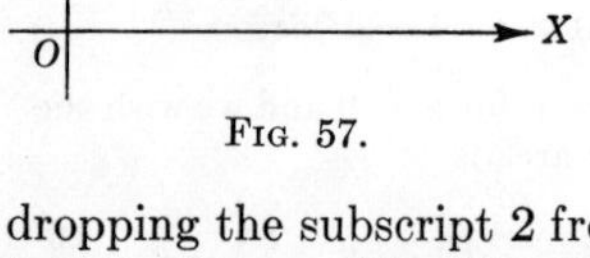

FIG. 57.

These may be obtained from Eqs. (36) and (37) by putting $g_1(y) = 0$, $x_1 = 0$, and dropping the subscript 2 from x_2 and $g_2(y)$. The first result may be read from a diagram like Fig. 55.

The first solution of Example 1 was based on a relation essentially that of Eq. (38). And as we illustrated in the second solution, it is sometimes easier to express y in terms of x, $y = f(x)$, $dy = f'(x)dx$, and use the element of area $x\,dy = xf'(x)dx$ with lower limit $g(c)$ and upper limit $g(d)$ if x is an increasing function of y. For a decreasing function, the lower limit is $g(d)$ and the upper limit $g(c)$.

EXAMPLE 2. Find the area bounded by the curve $y = x^3 - 4x^2 + 3x$ and the x axis.

Solution: We have $y = x(x^2 - 4x + 3) = x(x - 1)(x - 3)$ so that $y = 0$ when $x = 0, 1,$ or 3. By the method used in the examples of Sec. 40 we find that y is positive for $0 < x < 1$ and that y is negative for $1 < x < 3$. Hence (Fig. 58) between 0 and 1 the curve is above $y = 0$, and the element of area $(y_2 - y_1)dx = (y - 0)dx = y\,dx$. But between 1 and 3, the curve is below $y = 0$, so that the element of area $(y_2 - y_1)dx = (0 - y)dx = -y\,dx$, since we always take y_2 on the upper curve. Hence the required area is

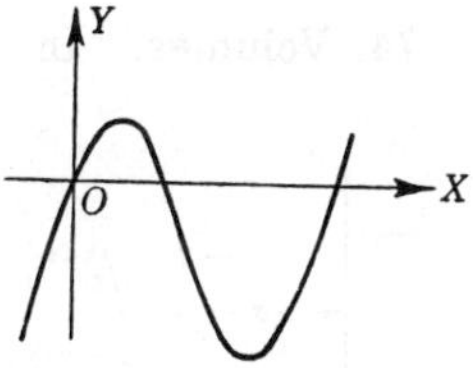

FIG. 58.

$$A = \int_0^1 (x^3 - 4x^2 + 3x)dx - \int_1^3 (x^3 - 4x^2 + 3x)dx.$$

By Sec. 67, one indefinite integral is found to be

$$\int (x^3 - 4x^2 + 3x)dx = \frac{x^4}{4} - 4\frac{x^3}{3} + 3\frac{x^2}{2} = \frac{1}{12}(3x^4 - 16x^3 + 18x^2).$$

It follows that

$$A = \tfrac{1}{12}[x^2(3x^2 - 16x + 18)]_{0,3}^{1,1},$$

meaning the result of adding the bracket evaluated for each of the two upper values and subtracting the bracket evaluated for each of the two lower values. As the two upper values are the same, and for this particular indefinite integral the factor x^2 makes the bracket zero for the lower value zero, we have

$$A = \tfrac{1}{12}[2(3 - 16 + 18) - 3^2(27 - 48 + 18)] = \tfrac{1}{12}[10 - (-27)]$$
$$= \tfrac{37}{12}.$$

Thus the required area is $A = \frac{37}{12}$.

Whenever we wish the positive area between two curves which cross, it is necessary to find the intersections and in each interval take y_2 as the upper curve in that interval to make the factor $(y_2 - y_1)$ positive.

EXERCISE 34

Find the area bounded by the x axis and the given curve in each of the following problems.

1. $y = 1 - x^4$.
2. $y = x^2 - 5x + 4$.
3. $y = 4x - x^3$.
4. $y = x^3 - 3x^2$.

Find the area bounded by the y axis and the given curve in each of the following problems.

5. $x = 9 - y^2$.
6. $x = 3 + 2y - y^2$.
7. $x = y^3 - 2y^2$.
8. $x = 9y - y^3$.

Find the area in the first quadrant bounded by the y axis, the line $y = 2$ and the curve given in each of the following problems.

9. $y = 2x^4$.
10. $y = 2x^3$.
11. $x = y^2$.
12. $2y^2 = x^3$.

Find the area bounded by each given pair of curves.

13. $y = x^2$, $y = 2 - x^2$.
14. $y = x^2 + 4$, $y = 2x^2$.
15. $y = 2x^2$, $y = x$.
16. $y^2 = x$, $x^2 = 8y$.
17. $y = x^3 - 3x$, $y = x$.
18. $y = x^3 + x^2$, $y = x^3 + 1$.

74. Volumes. In Secs. 60 and 61 we considered the volume of a solid between a fixed cross section and a variable cross section parallel to it. Let x be the perpendicular distance from some fixed cross section to the variable cross section as in Fig. 59. And let the area of the variable cross section at distance x be $A(x)$. Then we found that $\Delta V = A(\bar{x})\Delta x$ and $dV = A\,dx$. Hence by reasoning as in Secs. 70 and 71, we find that the volume between two cross sections at distances a and b is

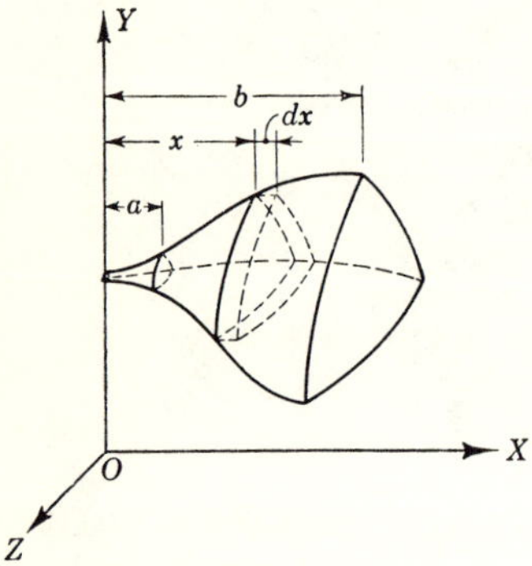

Fig. 59.

$$V = \int_a^b A\,dx. \tag{39}$$

We illustrate the application of this relation by two examples.

Example 1. In Fig. 60 let OX, OY, and OZ be three mutually perpendicular axes in space. The rectangle $PQRS$ is a typical section of a solid, with R on OX, RQ parallel to OY, and RS parallel to OZ. And when $OR = x$, $RQ = y = 2x^3$, and

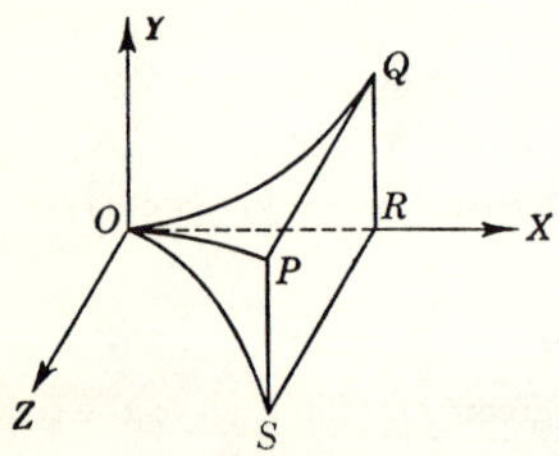

Fig. 60.

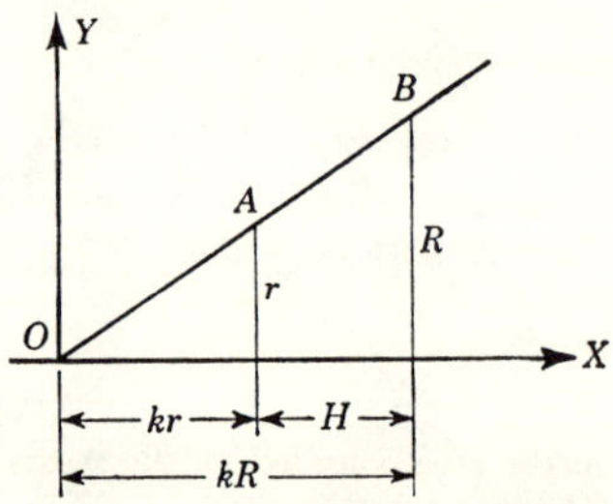

Fig. 61.

$RS = z = 3x^2$. Find the volume of that portion of the solid which lies between the planes $x = 1$ and $x = 2$, or at distances 1 and 2 from the plane containing OY and OZ.

Solution: The cross section at distance x is a rectangle of area $A = yz = (2x^3)(3x^2) = 6x^5$. Hence, from Eq. (39), the required volume is

$$V = \int_1^2 6x^5\,dx = [x^6]_1^2 = 2^6 - 1 = 63.$$

Example 2. Find the volume of a frustum of a right cone if the radius of the upper base is r, that of the lower base is R, and the height is H.

Solution: Let Fig. 61 represent a meridian section of the cone, with the axis along OX and the vertex at the origin. Let the generating line of the frustum be the segment joining the points $A = kr,r$ and $B = kR,R$. Then $kR - kr = H$, so that

$$k = \frac{H}{R - r}. \qquad \text{And} \qquad y = \frac{x}{k} \tag{40}$$

is the equation of the generating line.

The cross section at distance x is a circle of radius y. Hence its area is

$$A = \pi y^2 = \frac{\pi x^2}{k^2}.$$

It follows from Eq. (39) that the required volume is

$$V = \int_{kr}^{kR} \frac{\pi x^2}{k^2}\,dx = \left[\frac{\pi x^3}{3k^2}\right]_{kr}^{kR} = \frac{\pi}{3k^2}(k^3R^3 - k^3r^3)$$
$$= \frac{\pi k}{3}(R^3 - r^3) = \frac{\pi H}{3}\,\frac{R^3 - r^3}{R + r},$$

where the last value is found by using Eq. (40). From this we find by division that

$$V = \frac{\pi H}{3}(R^2 + rR + r^2).$$

If the area of the lower base $\pi R^2 = B$ and that of the upper base $\pi r^2 = b$, then $\pi^2R^2r^2 = bB$, and $\pi rR = \sqrt{bB}$. Hence

$$V = \frac{H}{3}(\pi R^2 + \pi rR + \pi r^2) = \frac{H}{3}(B + \sqrt{bB} + b).$$

The last form is often used in solid geometry or mensuration.

EXERCISE 35

The typical cross section of a solid is a rectangle $PQRS$ perpendicular to OX like that of Fig. 60. When $OR = x$, $RQ = y$ and $RS = z$. Find the volume of the solid between the planes $x = 0$ and $x = 1$ in each problem if y and z depend on x as given.

1. $y = 3x$, $z = 4x$.
2. $y = 4x^2$, $z = 2x$.
3. $y = 5x^3$, $z = 3x$.
4. $y = 7x^{\frac{3}{2}}$, $z = 2x$.

The area in the first quadrant bounded above by the given curve and below by the x axis, OX, is revolved about OX. Find the volume of the solid generated in each problem.

5. $y^2 = 4 - 2x$.
6. $y^2 = 4x - x^3$.
7. $y^2 = 4 - 4x^2$.
8. $y = 1 - x^2$.

Let S be the area in the first quadrant bounded above by the given curve and below by the x axis, OX. And let CD be a variable ordinate of S perpendicular to OX. Find the volume of the solid whose plane section perpendicular to OX is a square with CD as one side in each problem.

9. $y^2 = -x^2 + 3x - 2$.
10. $y = 4 - x^2$.
11. $y^2 = 16x - 4x^3$.
12. $y = 3 - x$.

Let AB be a fixed diameter of a circle of radius 4, and CD be a variable chord of this circle perpendicular to AB. Find the volume of a solid if the plane section perpendicular to AB which contains CD is the figure described in each problem.

13. A square with CD as one side.
14. An equilateral triangle with CD as one side.
15. A square with CD as one diagonal.

Find the volume of the solid generated by revolving the right triangle bounded by $x = 0$, $y = 0$, $x + y = 1$ about each of the following lines.

16. $y = 0$. 17. $y = 1$. 18. $y = -1$.

Find the volume of the solid generated when the area bounded above by the curve $y = 4 - x^2$ and below by the x axis is revolved about each of the following lines.

19. $y = 0$. 20. $y = 4$. 21. $y = -4$.

75. Pressure. Integration may be used to calculate the total force exerted by a liquid on a vertical plate such as a gate in the surface of a dam.

A liquid in contact with one side of a flat plate exerts a force perpendicular to the plate. The force per unit area, or *pressure*, at a point h units below the surface of the liquid, is $p = wh$, where w is the density, or weight of a unit volume of the liquid. Thus pressure at depth h is the same for all directions. Its value may be recalled by noting that a horizontal area A at depth h supports a column of liquid whose volume is Ah and whose weight is wAh. As this equals the force due to pressure on A, $pA = wAh$ and $p = wh$.

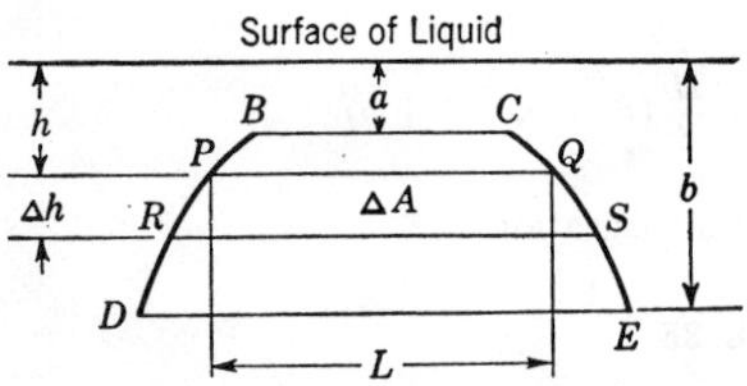

FIG. 62.

In Fig. 62 let $BCED$ be a vertical gate in a dam. Consider the force due to liquid pressure on the portion of the area $PQCB$, above a line PQ at variable depth h. This force F is a function of h. And when h is given a positive increment Δh, the increment ΔF is the force due to pressure on the element of area $\Delta A = PQSR$, where RS is a horizontal line at depth $h + \Delta h$. Since the pressure at any point of $PQSR$ is less than that at depth $h + \Delta h$, and greater than that at depth h, the total force ΔF on ΔA is equal to $wh_1\Delta A$, where h_1 is a suitable value between h and $h + \Delta h$. But if $PQ = L(h)$, by the reasoning of Sec. 60, $\Delta A = L(h_2)\Delta h$, where h_2 is another suitable value between h and $h + \Delta h$. It follows that

$$\Delta F = wh_1L(h_2)\Delta h \quad \text{and} \quad \frac{\Delta F}{\Delta h} = wh_1L(h_2). \tag{41}$$

By taking the limit as $\Delta h \to 0$ and noting that when $\Delta h \to 0$, $h_1 \to h$ and $h_2 \to h$, we find that

$$\frac{dF}{dh} = whL \quad \text{and} \quad dF = whL\,dh. \tag{42}$$

The following physical considerations may help the student to remember the result which we have just proved. First, the weight of liquid above a horizontal element of area dA at depth h is $wh\,dA$. Second, the element of area $dA = L\,dh$. Thus the element of force for the element dA in any direction is $dF = wh\,dA = whL\,dh$.

It follows from Eq. (42), by reasoning as in Secs. 70 and 71 that, if $h = a$ on BC and $h = b$ on DE, the total force due to liquid pressure on $BCED$ is given by the definite integral

$$F = \int_a^b whL\,dh. \tag{43}$$

We may replace h, L, and dh by their expressions in terms of some other variable, if more convenient, and then take as limits in place of a and b the corresponding values of the new variable.

For fresh water, the approximate value of $w = 62.5$ lb./cu. ft. or $w = \frac{1}{32}$ ton/cu. ft. may be used.

EXAMPLE 1. Find the force due to pressure on one end of a semicircular trough full of water if the diameter of the circle is 2 ft.

Solution: The vertical semicircle is shown in Fig. 63. The origin is at the center of the circle, OY is horizontal and OX is vertically downward. Then from the right triangle shown, we have

$$x^2 + y^2 = 1 \quad \text{or} \quad y = \pm\sqrt{1 - x^2}.$$

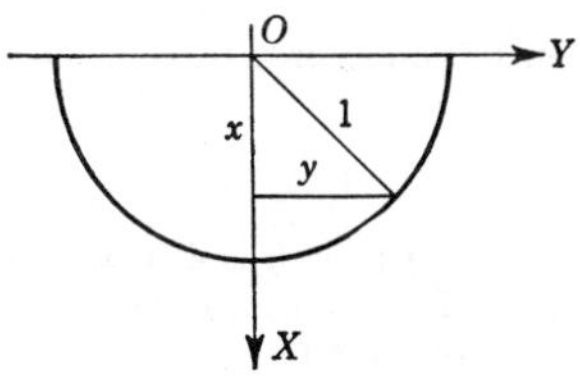

FIG. 63.

Since the depth $h = x$, $dh = dx$. And the depth $h = x$ is 0 at the top and 1 at the bottom. The width L at depth h is twice the positive y or $2\sqrt{1 - x^2}$. Hence we have

$$dF = whL\,dh = wx\,2\sqrt{1 - x^2}\,dx$$

and

$$F = 2w \int_0^1 \sqrt{1 - x^2}\,x\,dx.$$

This integral is like that of Prob. 13 of Exercise 33. And since

$$\frac{d}{dx}(1 - x^2)^{\frac{3}{2}} = \frac{3}{2}(1 - x^2)^{\frac{1}{2}}(-2x) = -3\sqrt{1 - x^2}\,x,$$

which differs from the integrand by a factor -3 only, we have

$$F = \left[-\frac{2w}{3}(1 - x^2)^{\frac{3}{2}}\right]_0^1 = -\frac{2w}{3}(0 - 1) = \frac{2w}{3}$$
$$= \frac{2(62.5)}{3} = 41.7.$$

Hence the required force is 41.7 lb.

EXAMPLE 2. Find the force due to liquid pressure on a gate in a dam in the form of a trapezoid with two sides parallel to the surface of the water, the top being 8 ft. long and 5 ft. below the surface, and the bottom being 6 ft. long and 10 ft. below the surface.

Solution: To find $L = PR$, drawn at distance x up from the bottom of the gate (Fig. 64) draw BQE parallel to AD, cutting PR in Q. Then $PQ = 6$, $BC = 8 - 6 = 2$, and from similar triangles, we have

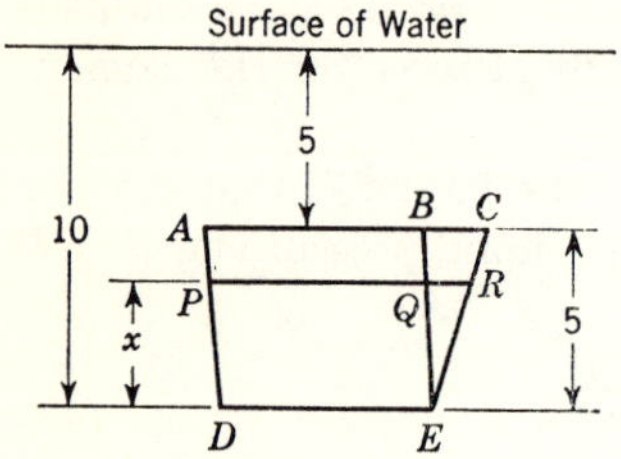

FIG. 64.

$$\frac{x}{5} = \frac{QR}{BC} = \frac{QR}{2} \qquad \text{and} \qquad QR = \frac{2x}{5}.$$

It follows that $L = PR = PQ + QR = 6 + 0.4x$. But the depth of PR, $h = 10 - x$, so that $dh = -dx$. Hence we have

$$\begin{aligned} dF &= whL\,dh = -w(10 - x)(6 + 0.4x)dx \\ &= w(0.4x^2 + 2x - 60)dx. \end{aligned}$$

Since $x = 5$ at the top and $x = 0$ at the bottom, we have

$$\begin{aligned} F &= w \int_5^0 (0.4x^2 + 2x - 60)dx = \left[w\left(0.4\frac{x^3}{3} + x^2 - 60x\right) \right]_5^0 \\ &= w\left[0 - \left(\frac{50}{3} + 25 - 300\right)\right] = \frac{775w}{3} \\ &= \tfrac{775}{3}\left(\tfrac{1}{32}\right) = 8.07. \end{aligned}$$

Hence the required force is 8.07 tons.

EXERCISE 36

A long trough is 4 ft. wide at the top and has vertical ends. It is full of water. Find the force due to pressure on one end if the end is

1. A square 4 ft. on a side.
2. An isosceles right triangle with hypotenuse 4 ft.
3. A rectangle 4 ft. wide and 2 ft. deep.
4. An isosceles trapezoid with upper base 4 ft., lower base 2 ft., and height 2 ft.
5. An equilateral triangle 4 ft. on a side.
6. One-half of a circle 2 ft. in radius.
7. A figure similar to the area bounded below by $y = x^2$ and above by $y = 4$.

Find the force due to the pressure of the water on a vertical gate in the side of a dam if the gate is in the form of

8. A square 4 ft. on a side, the upper side being parallel to and 8 ft. below the surface of the water.
9. A triangle of base 4 ft. in the surface of the water and altitude 6 ft., with altitude vertical.
10. A trapezoid of upper base 4 ft. in the surface of the water, of lower base 6 ft., and height 2 ft.
11. A trapezoid of upper base 6 ft. in the surface of the water, lower base 4 ft., and altitude 2 ft.

12. The gate in Prob. 8 is strengthened by a brace along a diagonal joining two opposite corners. Find the force due to pressure on the lower portion of the gate, below the brace.

13. Where shall a horizontal line be drawn across the gate of Prob. 8 so that the force due to pressure on the portion above the line shall be equal to the force due to pressure on the portion below the line?
14. How much higher must the water level in Prob. 8 rise to make the force due to pressure on the gate twice its original value?
15. A circular water main 4 ft. in diameter is half full of water. Find the force due to pressure on the vertical gate that closes the main.
16. An oil tank is in the form of a circular cylinder 6 ft. in diameter with the aixs of the cylinder horizontal. Find the force due to pressure on one end if the tank is half full of oil weighing 40 lb./cu. ft.

CHAPTER 6

ALGEBRAIC CURVES

We described rectangular coordinates and the graphical representation of a function $y = f(x)$ in Sec. 6. In this chapter we shall take a slightly different point of view. Here we consistently use the same unit on the x and y axes and study certain algebraic equations and the curves which represent them. The derivation of equations corresponding to curves with given geometric properties, and the solution of geometric problems by algebraic methods, constitute the branch of mathematics known as analytic geometry. In many cases the algebraic equations are simplest in their unsolved form, like those of Sec. 55. Each unsolved equation may have as its solution several branches of a multiple-valued function defined implicitly by the equation.

We begin with equations of the first degree, the straight line, and some related notions about distance and angle. Then we treat circles, parabolas, ellipses, and hyperbolas. All these have equations of the second degree. Finally we discuss a few curves having equations of degree higher than the second.

R76. Locus of an Equation. A curve, or set of points, is called the *locus* of a given equation $f(x,y) = 0$ if every point on the locus has coordinates (x,y) which satisfy the equation, and all such points are included. We say that the equation of the locus is $f(x,y) = 0$. We also say that $f(x,y) = 0$ is the equation of the locus.

R77. Straight Line. The discussion of Sec. 23 showed that the locus of every equation of the first degree, or of the form

$$Ax + By + C = 0, \tag{1}$$

with A and B not both zero, is a straight line. And the equation of every straight line is some first-degree equation. Thus the equation of every straight line is of the form of Eq. (1). And every equation of the form of Eq. (1) is the equation of some straight line.

If $B = 0$, $A \neq 0$, and Eq. (1) may be reduced to the form

$$x = -\frac{C}{A}, \tag{2}$$

a line parallel to the y axis (Fig. 65).

If $B \neq 0$, Eq. (1) may be reduced to the form

$$y = -\frac{A}{B}x - \frac{C}{B}. \tag{3}$$

The fact that $dy/dx = -A/B$ recalls that $-A/B = m$, the slope of the line as discussed in Secs. 23, 24, and 28.

Given the inclination to the x axis, ϕ, with $\tan \phi = m$, and one point on the line, $P_1 = (x_1,y_1)$, the equation of the line of Fig. 66 was found in Sec. 23 to be

$$y - y_1 = m(x - x_1). \tag{4}$$

We may check this analytically by noting that it is a first-degree equation, satisfied when $x = x_1$, $y = y_1$, and has $dy/dx = m$. Hence it is a straight line through (x_1,y_1) with slope m. If $\phi = 90°$, the equation of the line is $x = x_1$.

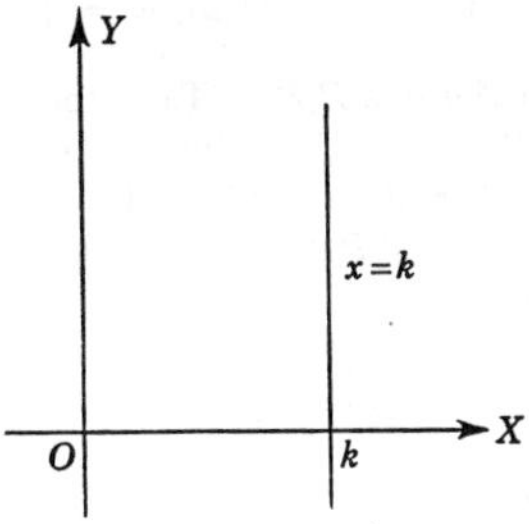

FIG. 65.

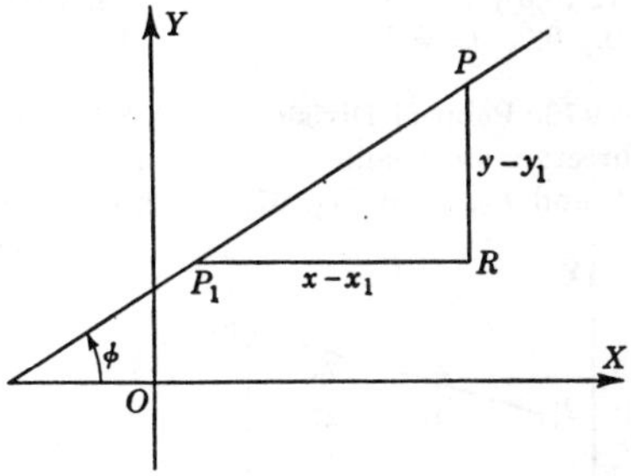

FIG. 66.

Given two distinct points on the line, $P_1 = (x_1,y_1)$ and $P_2 = (x_2,y_2)$, the equation of the line is

$$\frac{y - y_1}{x - x_1} = \frac{y_2 - y_1}{x_2 - x_1}, \tag{5}$$

if $x_2 \neq x_1$. This follows either from a consideration of slopes or from the fact that, when multiplied by $(x - x_1)$, Eq. (5) is a first-degree equation, satisfied when $x = x_1$, $y = y_1$, and also when $x = x_2$, $y = y_2$.

If $x_2 = x_1$, Eq. (5) is to be replaced by $x = x_1$.

EXAMPLE 1. Find the equation of the straight line through the points $(a,0)$ and $(0,b)$, where $a \neq 0$ and $b \neq 0$.

Solution: From Eq. (5) we have

$$\frac{y - 0}{x - a} = \frac{b - 0}{0 - a} \quad \text{or} \quad -ay = bx - ab.$$

This may be written in the more symmetrical form

$$\frac{x}{a} + \frac{y}{b} = 1, \tag{6}$$

which is the required equation.

EXAMPLE 2. Find the equation of the straight line through (2,3) and (2,5).

Solution: If we attempt to use Eq. (5), we write

$$\frac{y - 3}{x - 2} = \frac{5 - 3}{2 - 2} \quad \text{or} \quad \frac{y - 3}{x - 2} = \frac{2}{0}.$$

Taken literally, this is meaningless because of the indicated division by zero. But it suggests the alternative equations

$$\frac{x - 2}{y - 3} = \frac{0}{2} \quad \text{or} \quad (y - 3)0 = 2(x - 2).$$

Each of these is equivalent to $x = 2$, which is the required equation.

EXERCISE 37

Find the equation of the straight line through the two given points in each of the following problems.

1. (0,0), (3,5)
2. (0,0), (3,−2).
3. (−1,−2), (2,4).
4. (−4,0), (0,−1).
5. (1,2), (5,4).
6. (−1,2), (−2,1).

Find the equation of the straight line through the given point and having the given slope in each of the following problems.

7. (0,0), $m = 3$.
8. (2,0), $m = -\frac{1}{2}$.
9. (0,−3), $m = 3$.
10. (0,2), $m = \frac{1}{2}$.
11. (−3,3), $m = -1$.
12. (1,2), $m = \frac{3}{4}$.

R78. Point of Division. Let P be any point on the line P_1P_2. Take the positive direction on this line as that from $P_1 = (x_1,y_1)$ to $P_2 = (x_2,y_2)$. Thus if P is between P_1 and P_2, as in Fig. 67, the directed segments P_1P and PP_2 will both be positive. For any position of P, let k_1 and k_2 be two constants such that

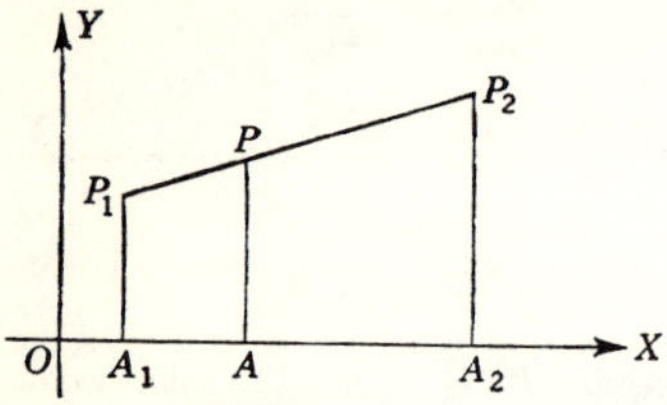

FIG. 67.

$$\frac{P_1P}{PP_2} = \frac{k_1}{k_2}. \tag{7}$$

Let us draw lines parallel to the y axis projecting the points P_1PP_2 into A_1AA_2 on the x axis. Then we have

$$\frac{A_1A}{AA_2} = \frac{x - x_1}{x_2 - x} = \frac{k_1}{k_2}.$$

We may solve this last relation for x by noting that

$$k_2(x - x_1) = k_1(x_2 - x), \qquad (k_1 + k_2)x = k_2x_1 + k_1x_2.$$

It follows from the last relation that

$$x = \frac{k_2x_1 + k_1x_2}{k_1 + k_2}. \tag{8}$$

Similarly by projecting on the y axis, we find that

$$y = \frac{k_2y_1 + k_1y_2}{k_1 + k_2}. \tag{9}$$

As P traces out the line of Fig. 68, the ratio k_1/k_2 increases from 0 to $+\infty$ as P moves from P_1 to P_2. To the right of P_2, the ratio becomes negative and varies from

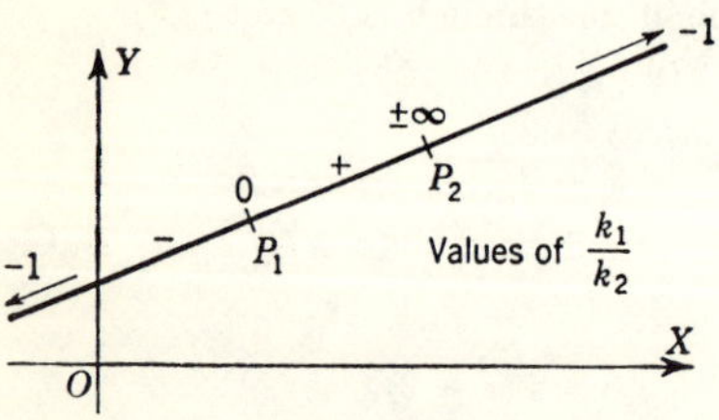

FIG. 68.

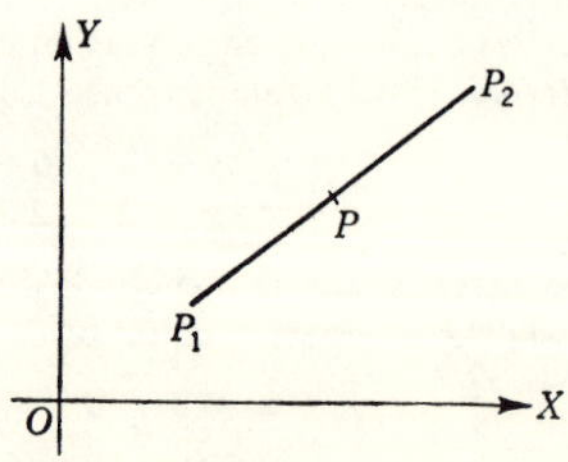

FIG. 69.

$-\infty$ to -1. That is, the ratio approaches -1 as P recedes indefinitely far out on the line. To the left of P_2, the ratio also becomes negative and varies from 0 to -1. Except for the point P_2, the ratio k_1/k_2 always increases algebraically as P moves in the positive direction, or that from P_1 to P_2.

In remembering Eqs. (8) and (9), and using them correctly, it is important for the student to notice the interchange of subscripts in the numerator, with k_2 the coefficient of x_1 or y_1, and k_1 the coefficient of x_2 or y_2.

If P is the mid-point of the segment P_1P_2 (Fig. 69) the ratio $k_1/k_2 = 1$, so that we may put $k_1 = 1$ and $k_2 = 1$. Thus, for the mid-point, Eqs. (8) and (9) reduce to

$$x = \frac{x_1 + x_2}{2}, \qquad y = \frac{y_1 + y_2}{2}. \tag{10}$$

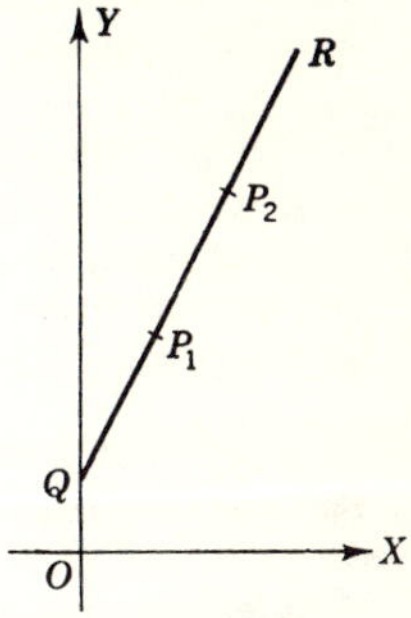

FIG. 70.

EXAMPLE, The segment QR is divided into three equal parts by the points P_1 and P_2, as in Fig. 70. If $P_1 = (1,3)$ and $P_2 = (2,5)$, find the coordinates of Q and R.

Solution: Since $P_1Q/QP_2 = -\frac{1}{2}$, we may use $k_1 = -1$ and $k_2 = 2$. Thus by Eqs. (8) and (9), for Q we have

$$x = \frac{2(1) + (-1)(2)}{2 + (-1)} = 0, \qquad y = \frac{2(3) + (-1)(5)}{2 + (-1)} = 1.$$

And $P_1R/RP_2 = 2/(-1)$, so that we may use $k_1 = 2$ and $k_2 = -1$ for P. Hence by Eqs. (8) and (9) we have

$$x = \frac{(-1)(1) + 2(2)}{-1 + 2} = 3, \qquad y = \frac{(-1)(3) + 2(5)}{-1 + 2} = 7.$$

Thus $Q = (0,1)$ and $R = (3,7)$.

R79. Angles. In Sec. 23 we defined the inclination of a straight line as the angle ϕ from the positive x axis to the line, measured counterclockwise and with $0 \leqq \phi < 180°$. Since $\tan\ \phi = m$, the slope, whenever the slope is known, we may calculate ϕ as the angle in the first or second quadrant whose tangent is m.

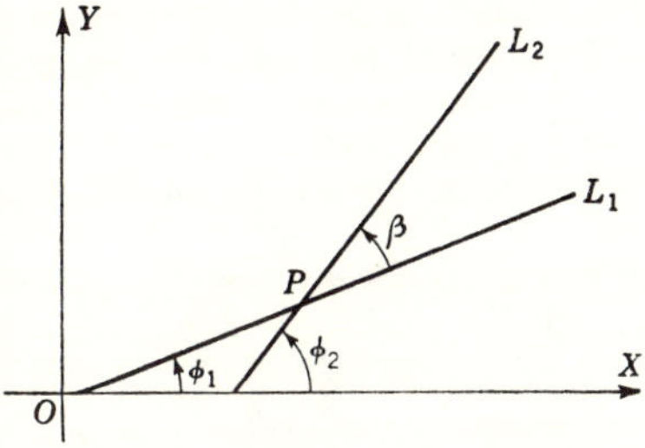

FIG. 71.

Consider the two lines L_1 and L_2 of Fig. 71, having inclinations ϕ_1 and ϕ_2, respectively, and intersecting at P. Let β, read "beta," be an angle, measured counterclockwise with $-180° < \beta < 180°$, such that a rotation of L_1 about P through angle β takes L_1 into L_2. Then either $\beta = \phi_2 - \phi_1$ or $\beta = \phi_2 - \phi_1 \pm 180°$. In either case, we have

$$\tan \beta = \tan (\phi_2 - \phi_1). \tag{11}$$

But, from trigonometry, Eq. (28) of Sec. 92,

$$\tan (\phi_2 - \phi_1) = \frac{\tan \phi_2 - \tan \phi_1}{1 + \tan \phi_1 \tan \phi_2}. \tag{12}$$

Since $\tan \phi_1 = m_1$ and $\tan \phi_2 = m_2$, it follows from Eqs. (11) and (12) that

$$\tan \beta = \frac{m_2 - m_1}{1 + m_1 m_2}. \tag{13}$$

We assumed that L_1 met L_2, and both lines met OX. By keeping one line fixed and letting the other line vary, we obtain a number of special results. First let L_2 vary so that $\phi_2 \to \phi_1$. Then $\beta \to 0$. In the limit, $m_2 = m_1$ and either L_2 coincides with L_1 or L_2 is parallel to L_1. In either case we say that the lines are "parallel" and consider the angle from L_1 to L_2 to be zero. It follows that

Two lines are "parallel" if and only if their slopes are equal.

If $1 + m_1 m_2 = 0$, or $m_2 = -1/m_1$, the expression in Eq. (13) takes the form $[(-1/m_1) - m_1]/0$. But if we let $m_2 \to -1/m_1$, since $|\tan \beta| \to \infty$, $\beta \to \pm 90°$. Conversely, if $\beta \to 90°$, $1/\tan \beta \to 0$ and in the limit $1 + m_1 m_2 = 0$. It follows that

Two lines are perpendicular if and only if the slope of one is the negative reciprocal of the other.

Next let $m_1 \to 0$ in Eq. (13). The limiting relation is now

$$\tan \beta = \frac{m_2 - 0}{1 + 0m_2} = m_2.$$

This may also be seen directly. For when $m_1 = 0$, L_1 is "parallel" to the x axis so that $\beta = \phi_2$ or $\phi_2 - 180°$, and so $\tan \beta = \tan \phi_2 = m_2$.

Again, let $m_2 \to \infty$. Then from Sec. 12 we have by the principle of the leading term,

$$\lim_{m_2 \to \infty} \tan \beta = \lim_{m_2 \to \infty} \frac{m_2 - m_1}{1 + m_1 m_2} = \lim_{m_2 \to \infty} \frac{m_2}{m_1 m_2}$$
$$= \lim_{m_2 \to \infty} \frac{1}{m_1} = \frac{1}{m_1}. \tag{14}$$

This may also be seen directly. For when $m_2 \to \infty$, L_2 in the limit is "parallel" to the y axis. Hence $\beta = 90° - \phi_1$, or $-\phi_1 - 90°$, or $270° - \phi_1$. But in each of these cases $\tan \beta = \cot \phi_1 = 1/m_1$.

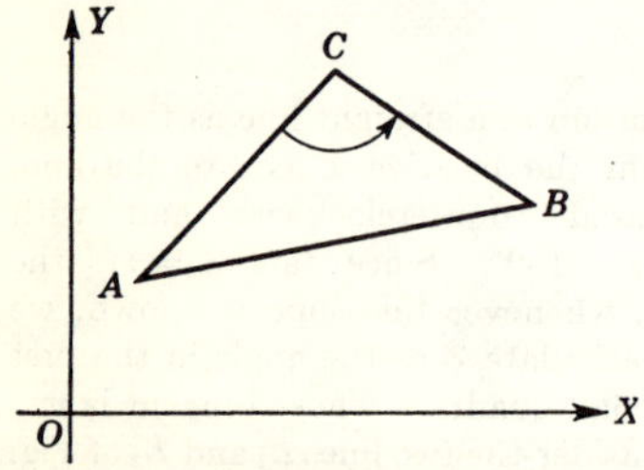

FIG. 72.

Thus, in every case, by using it either for a direct calculation or to suggest a limiting process like that of Eq. (14), the Eq. (13) leads to the value of $\tan \beta$, where β, with $-180° < \beta < 180°$, is an angle from the first line with slope m_1 to the second line with slope m_2.

EXAMPLE. Find the angle at $C = (4,5)$ of triangle ABC if $A = (1,2)$ and $B = (7,3)$.

Solution: From Fig. 72, we see that the angle C of the triangle is measured counterclockwise from CA to CB. Hence we put $m_1 = \text{slope } CA = \dfrac{2-5}{1-4} = 1$ and $m_2 = \text{slope } CB = \dfrac{3-5}{7-4} = -\dfrac{2}{3}$. With $m_1 = 1$ and $m_2 = -\frac{2}{3}$, we find from Eq. (13) that

$$\tan \angle ACB = \tan \beta = \frac{m_2 - m_1}{1 + m_1 m_2} = \frac{(-\frac{2}{3}) - 1}{1 + 1(-\frac{2}{3})} = -5.$$

The minus sign shows that the angle is obtuse, and hence

$$C = \tan^{-1}(-5) = 180° - 78.69° = 101.31°.$$

Thus the required angle C is 101.31°.

EXERCISE 38

Let $P_1 = (-3,-9)$ and $P_2 = (9,27)$. Find the coordinates of the point P on the line through P_1 and P_2 if

1. $P_1P = PP_2$.
2. $P_1P = 2PP_2$.
3. $3P_1P = 2PP_2$.
4. $P_1P_2 = P_2P$.
5. $2PP_1 = P_1P_2$.
6. $PP_1 = 3P_1P_2$.

Let $Ax + By + C = 0$ be given as the equation of a straight line L. And let $P_1 = (x_1,y_1)$ be any given point. Show that the equation of the straight line through P_1 which is

7. "Parallel" to L may be taken as $Ax + By = Ax_1 + By_1$.
8. Perpendicular to L may be taken as $Bx - Ay = Bx_1 - Ay_1$.

Use Prob. 7 to find the equation of the line which contains the given point and is parallel to the given line in each case.

9. $(1,0)$, $2x - 3y = 2$.
10. $(2,3)$, $x + 2y + 4 = 0$.
11. $(2,5)$, $x = 0$.
12. $(4,1)$, $y = 0$.

Use Prob. 8 to find the equation of the line which contains the given point and is perpendicular to the given line in each case.

13. $(2,1)$, $3x - 2y = 5$.
14. $(-3,-1)$, $x + y + 5 = 0$.
15. $(3,1)$, $x = 0$.
16. $(-2,4)$, $y = 0$.

In each of the following problems, find the tangent of the angle from the first line to the second line.

17. $y - 2x = 1$, $y = x$.
18. $2y + 3x = 4$, $5y = 2x$.
19. $y + 2x = 5$, $y + 3x = 2$.
20. $y = 3x + 2$, $y + 0$.
21. By finding the interior angles, verify that the triangle with vertices at (1,1), (2,3), (4,2) is an isosceles right triangle.

R80. Distance from a Line to a Point. Consider a straight line L whose equation is

$$Ax + By + C = 0. \tag{15}$$

And let $P_0 = (x_0,y_0)$ be a point not on L. Draw a perpendicular from P_0 to L, intersecting L in R (Fig. 73). Then the distance from the line L to P_0 is measured along RP_0. We may find an expression for this distance by the following procedure. For definiteness, in the first instance we assume that B in Eq. (15) is positive, $B > 0$, and that the inclination of the line ϕ is acute, $0 < \phi < 90°$. We also take P_0 above the line L.

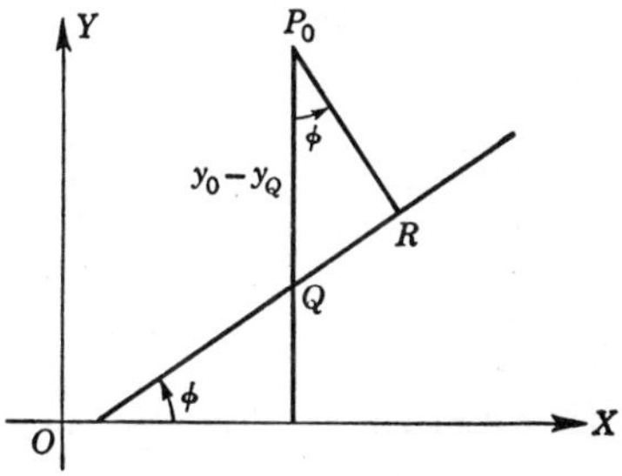

FIG. 73.

Draw a line through P_0 parallel to the y axis and let it cut L in the point $Q = (x_0,y_Q)$. Then since Q is on L, Eq. (15) is satisfied by $x = x_0$ and $y = y_Q$. It follows that

$$Ax_0 + By_Q + C = 0 \quad \text{or} \quad y_Q = -\frac{Ax_0 + C}{B}. \tag{16}$$

Hence the distance QP_0 is found to be

$$QP_0 = y_0 - y_Q = \frac{Ax_0 + By_0 + C}{B}. \tag{17}$$

Next observe that $\angle QP_0R = \phi$ so that

$$RP_0 = QP_0 \cos \phi. \tag{18}$$

But $\tan \phi = m = -A/B$, so that by trigonometry

$$\sec^2 \phi = \tan^2 \phi + 1 = \left(-\frac{A}{B}\right)^2 + 1 = \frac{A^2 + B^2}{B^2}. \tag{19}$$

Since we assumed that ϕ was acute and $B > 0$, $\sec \phi > 0$ and

$$\sec \phi = \frac{\sqrt{A^2 + B^2}}{B} \qquad \text{and} \qquad \cos \phi = \frac{B}{\sqrt{A^2 + B^2}}. \tag{20}$$

By combining Eqs. (20), (18), and (17) we find that, for the restricted case considered,

$$RP_0 = \frac{Ax_0 + By_0 + C}{\sqrt{A^2 + B^2}}. \tag{21}$$

Let us consider the distance RP_0 as positive when measured upward, or for points P_0 above L, as zero for points P_0 on L, and as negative when RP_0 points downward, or for points P_0 below L. Then Eq. (21) will hold for all cases. For, if P_0 is on the line, $y_0 - y_Q$ will be zero, and if P_0 is below the line, $y_0 - y_Q$ will be negative.

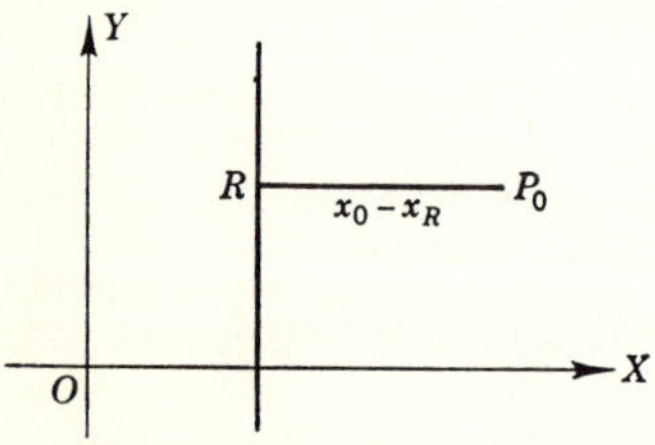

FIG. 74.

Equation (21) will also hold if B is negative and at the same time ϕ is obtuse. If B is negative and ϕ acute, or B positive and ϕ obtuse, we must insert a minus sign before the square root in Eq. (21) to make our conclusion hold as stated.

Finally, suppose that $B = 0$. Then $A \neq 0$, and if we put $B = 0$ in Eq. (21) and replace the radical by A, we obtain

$$RP_0 = \frac{Ax_0 + C}{A} \qquad \text{when } B = 0. \tag{22}$$

But in this case L is parallel to the y axis (Fig. 74). Thus RP_0 is parallel to the x axis and, if $R = (x_R, y_0)$, we have

$$Ax_R + C = 0, \qquad x_R = -\frac{C}{A}. \tag{23}$$

Hence if we measure distances along RP_0 positive to the right, we have

$$RP_0 = x_0 - x_R = x_0 - \left(-\frac{C}{A}\right) = \frac{Ax_0 + C}{A}. \tag{24}$$

This proves the result suggested by Eq. (22).

We may now formulate a rule covering all cases.

The perpendicular distance from the line L whose equation is

$$Ax + By + C = 0$$

to any point $P_0 = (x_0,y_0)$ *may be taken as*

$$p = \frac{Ax_0 + By_0 + C}{\sqrt{A^2 + B^2}} \quad \text{or}\dagger \quad p = \frac{Ax_0 + By_0 + C}{-\sqrt{A^2 + B^2}}. \tag{25}$$

Either expression is zero for P_0 *on* L, *positive for* P_0 *on one side of the line, and negative for points on the other side of the line. When the sign of the radical is chosen either way, and fixed, we call* p *the signed distance from the line to the point* $P_0 = (x_0,y_0)$.

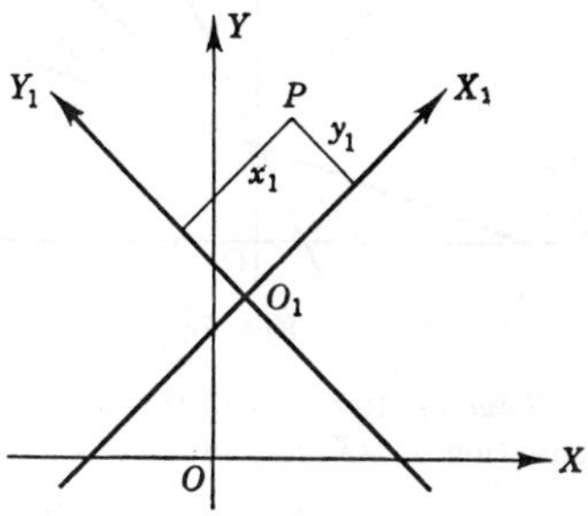

FIG. 75.

EXAMPLE 1. The lines $x - y + 2 = 0$ and $x + y - 3 = 0$ are perpendicular. They are taken as a new x axis and y axis, with positive direction upward in each case, as in Fig. 75. Find the new coordinates (x_1,y_1) of a point P in terms of its original coordinates (x,y).

Solution: The new y_1 is the distance from the new x axis. Hence by Eq. (25), y_1 is either $(x - y + 2)/\sqrt{2}$ or its negative. But the expression just written is negative for $x = 0$ and y large and positive. As y_1 should be plus in this case, we must use the other sign and

$$y_1 = \frac{x - y + 2}{-\sqrt{2}}.$$

The new x_1 is the distance from the new y axis. Hence by Eq. (25), x_1 is either $(x + y - 3)/\sqrt{2}$ or its negative. The expression just written is positive for $x = 0$ and y large and positive. As x_1 should be plus in this case, the sign is correct and

$$x_1 = \frac{x + y - 3}{\sqrt{2}}.$$

EXAMPLE 2. Find the equations of the bisectors of the angles included between the lines $y = 2x + 3$ and $x = 3y - 9$.

Solution: Any point $P(x,y)$ on either bisector is equidistant from the given lines. Hence for some combination of signs,

$$\frac{y - 2x - 3}{\sqrt{5}} = \frac{x - 3y + 9}{\pm\sqrt{10}}.$$

Thus the equations of the bisectors are

$$(1 + 2\sqrt{2})x - (3 + \sqrt{2})y + (9 + 3\sqrt{2}) = 0 \tag{26}$$

and

$$(1 - 2\sqrt{2})x - (3 - \sqrt{2})y + (9 - 3\sqrt{2}) = 0. \tag{27}$$

Since the bisector of the acute angle in Fig. 76 has positive slope, it is given by Eq. (26).

† This is more convenient for most applications than any convention which definitely fixes the sign.

EXAMPLE 3. A new set of axes through the origin OX_1 and OY_1 is obtained from the old set, OX and OY, by a positive or counterclockwise rotation through an angle ϕ about O. Find expressions for the old coordinates (x,y) in terms of the new coordinates (x_1,y_1).

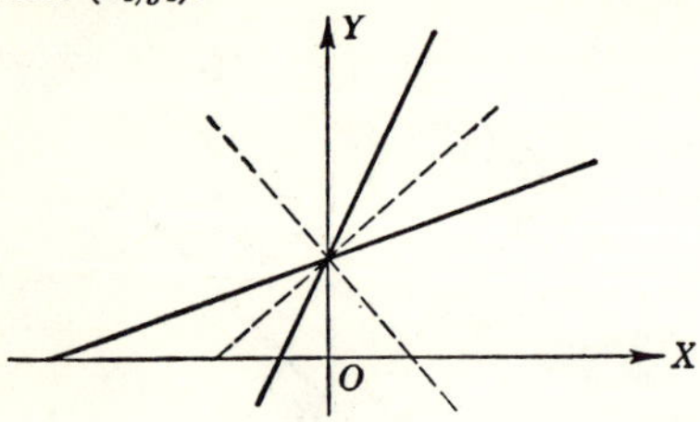

FIG. 76.

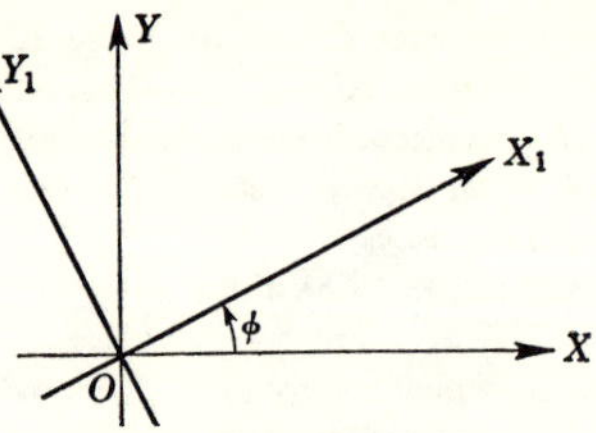

FIG. 77.

Solution: Referred to the new axes, OX in Fig. 77 has inclination $-\phi$. Hence the equation of OX may be written

$$y_1 = -x_1 \tan \phi \qquad \text{or} \qquad x_1 \sin \phi + y_1 \cos \phi = 0.$$

For the second form, we have

$$\sqrt{A^2 + B^2} = \sqrt{\sin^2 \phi + \cos^2 \phi} = 1,$$

so that the left member as it stands is plus or minus the distance from OX to $P = (x_1,y_1)$ or y since in the old coordinates $P = (x,y)$. But as $x = 0$, $y = 1$ when $x_1 = \sin \phi$, $y_1 = \cos \phi$, the plus sign is correct, and $y = x_1 \sin \phi + y_1 \cos \phi$.

Referred to the new axes, OY has inclination $-\phi + 90°$. Since $\tan (90° - \phi) = \cot \phi$, the equation of OY may be written

$$y_1 = x_1 \cot \phi \qquad \text{or} \qquad x_1 \cos \phi - y_1 \sin \phi = 0.$$

For the second form, we have

$$\sqrt{A^2 + B^2} = \sqrt{\cos^2 \phi + \sin^2 \phi} = 1,$$

so that the left member as it stands is plus or minus the distance from OY to $P = (x_1,y_1)$ or x since in the old coordinates $P = (x,y)$. But as $x = 1$, $y = 0$ when $x_1 = \cos \phi$, $y_1 = -\sin \phi$, the plus sign is correct, and $x = x_1 \cos \phi - y_1 \sin \phi$.

Thus the required expressions are

$$x = x_1 \cos \phi - y_1 \sin \phi \qquad \text{and} \qquad y = x_1 \sin \phi + y_1 \cos \phi. \tag{28}$$

EXERCISE 39

Find the positive number which measures the distance from the given line to the given point in each of the following problems.

1. $4x + 3y = 2$, $(2,3)$.
2. $12x + 5y = 15$, $(1,-2)$.
3. $2x = 3$, $(2,5)$.
4. $3y = 4$, $(7,-1)$.

Find the equations of the bisectors of the angles made by the two given lines in each of the following problems.

5. $y = x + 2$, $y = 7x + 1$.
6. $2y - 5x = 2$, $5y + 2x = 3$.
7. $y = 2x - 1$, $x = 2y + 3$.
8. $x + 3y = 2$, $3x + y = 1$.
9. Show that the perpendicular distance between the pair of parallel lines $Ax + By + C = 0$ and $Ax + By + D = 0$ is

$$p = \frac{|C - D|}{\sqrt{A^2 + B^2}}.$$

HINT: If $B \neq 0$, $(0, -D/B) = P_1$ is on the second line, while if $B = 0$, $A \neq 0$ and $(-D/A, 0) = P_2$ is on the second line. And the required distance is that from the first line to P_1 or to P_2.

Use the method or result of Prob. 9 to find the perpendicular distance between each given pair of parallel lines.

10. $3x + 4y - 1 = 0$, $3x + 4y + 21 = 0$.
11. $12x - 5y + 2 = 0$, $12x - 5y + 28 = 0$.
12. $y = 2x$, $2x - y + 3\sqrt{5} = 0$.
13. $x + 3y = 0$, $x + 3y - 4\sqrt{10} = 0$.
14. $4x - 3y + 2 = 0$, $8x - 6y = 7 = 0$.

Find the equation of the bisector of the acute angle made by each given pair of lines.

15. $y = 4x$, $x = 4y$.
16. $y + 3x = 7$, $x + 3y = 5$.

Find the new coordinates (x_1, y_1) of a point P in terms of its original coordinates (x,y) if the positive directions of the new axes point upward, and the equations of the new axes, OX_1 and OY_1 are the two given perpendicular lines.

17. $3x = 4y$, $4x = -3y$.
18. $x = 2y$, $2x = -y$.
19. $12x = 5y$, $5x = -12y$.
20. $3x = y$, $x = -3y$.

As in Example 3, a new set of axes OX_1 and OY_1 is obtained from OX and OY by a rotation through ϕ about 0.

21. Show that $x_1 = x \cos \phi + y \sin \phi$, $y_1 = -x \sin \phi + y \cos \phi$.
22. By solving the result of Prob. 21 for x and y, show that

$$x = x_1 \cos \phi - y_1 \sin \phi, \qquad y = x_1 \sin \phi + y_1 \cos \phi,$$

which checks the result found in Example 3 in the text.

23. Show from a figure that the point with $x = \cos \theta$, $y = \sin \theta$ has $x_1 = \cos (\theta - \phi)$, $y_1 = \sin (\theta - \phi)$. From this and Prob. 21, deduce one form of the addition theorem for the sine and cosine,

$$\cos (\theta - \phi) = \cos \theta \cos \phi + \sin \theta \sin \phi,$$
$$\sin (\theta - \phi) = \sin \theta \cos \phi - \cos \theta \sin \phi.$$

24. Show from a figure that the point with $x_1 = \cos \theta$, $y_1 = \sin \theta$ has $x = \cos (\theta + \phi)$, $y = \sin (\theta + \phi)$. From this and Prob. 22, deduce one form of the addition theorem for the sine and cosine,

$$\cos (\theta + \phi) = \cos \theta \cos \phi - \sin \theta \sin \phi,$$
$$\sin (\theta + \phi) = \cos \theta \sin \phi + \sin \theta \cos \phi.$$

R81. Distance between Two Points. Consider the two points $P_1 = (x_1, y_1)$ and $P_2 = (x_2, y_2)$. To find an expression for the distance P_1P_2, draw P_1R parallel to the x axis and P_2R parallel to the y axis as in Fig. 78. Then from the right triangle P_1RP_2 we have

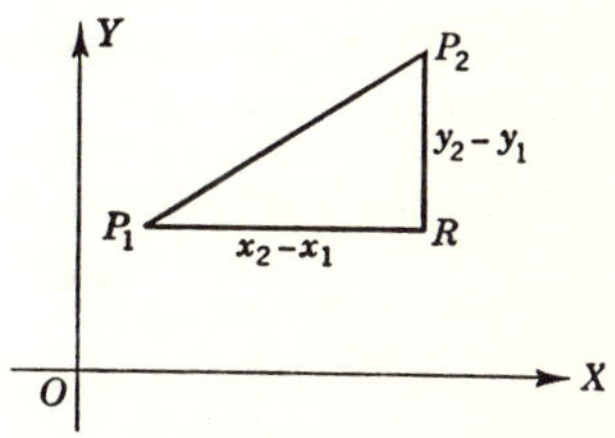

FIG. 78.

$$\overline{P_1P_2}^2 = \overline{P_1R}^2 + \overline{RP_2}^2.$$

But, by Sec. 3, $|P_1R| = |x_2 - x_1|$ and $|RP_2| = |y_2 - y_1|$. Hence we have

$$\overline{P_1P_2}^2 = (x_2 - x_1)^2 + (y_2 - y_1)^2 \tag{29}$$

and

$$P_1P_2 = \sqrt{(x_2 - x_1)^2 + (y_2 - y_1)^2}. \tag{30}$$

Example. Find the equation of the perpendicular bisector of the segment AB, joining $A = (1,3)$ and $B = (-3,-1)$, by regarding it as the locus of a point $P = (x,y)$ equidistant from A and B.

Solution: From Eq. (29) we have

$$\overline{AP}^2 = (x-1)^2 + (y-3)^2 \quad \text{and} \quad \overline{BP}^2 = [x-(-3)]^2 + [y-(-1)]^2.$$

Since $AP = BP$, $\overline{AP}^2 = \overline{BP}^2$, so that

$$(x-1)^2 + (y-3)^2 = (x+3)^2 + (y+1)^2,$$
$$x^2 - 2x + 1 + y^2 - 6y + 9 = x^2 + 6x + 9 + y^2 + 2y + 1,$$
$$-8x - 8y = 0 \quad \text{or} \quad x + y = 0.$$

Thus the required equation is $x + y = 0$, or $y = -x$.

R82. Circle. *A circle is the locus of a point whose distance from a fixed point is constant.* The fixed point is the *center* and the fixed distance is the *radius*.

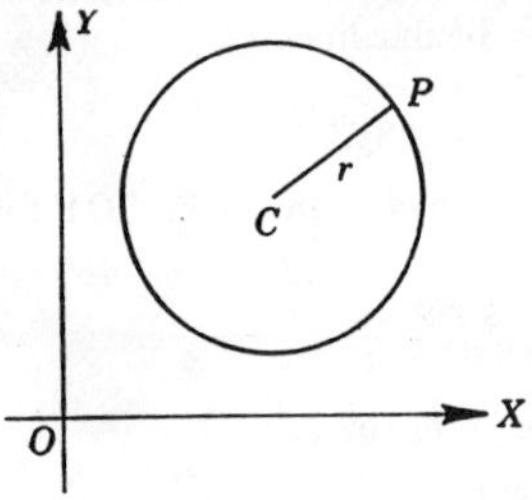

Fig. 79.

Let the center $C = (h,k)$ (Fig. 79) and the radius be r. Then if $P = (x,y)$, $CP = r$ and $\overline{CP}^2 = r^2$, so that from Eq. (29) we have

$$(x-h)^2 + (y-k)^2 = r^2. \tag{31}$$

If this equation holds, the positive square roots are equal and $|CP| = r$. Hence Eq. (31) is the equation of the circle.

The equation takes its simplest form when the center is at the origin, in which case it becomes

$$x^2 + y^2 = r^2. \tag{32}$$

For the center at $C = (h,k)$, the expanded form of Eq. (31) is

$$x^2 + y^2 - 2hx - 2ky + h^2 + k^2 - r^2 = 0. \tag{33}$$

Conversely, any equation of the form

$$Ax^2 + Ay^2 + Dx + Ey + F = 0 \tag{34}$$

with $A \neq 0$ may be transformed into the form of Eq. (33) and represents a circle, a point, or has no real locus according as the term corresponding to r^2 is positive, zero, or negative.

Example. Find the center and radius of the circle

$$2x^2 + 2y^2 + 3x - 5y - 20 = 0.$$

Solution: The equation may be written in the form

$$x^2 + \tfrac{3}{2}x \qquad + y^2 - \tfrac{5}{2}y \qquad = 10.$$

To complete the square for the x terms, add the square of one-half of $\frac{3}{2}$, or $\frac{9}{16}$. To complete the square of the y terms add the square of one-half of $-\frac{5}{2}$ or $\frac{25}{16}$. Adding these to both members, we have

$$x^2 + 2(\tfrac{3}{4})x + (\tfrac{3}{4})^2 + y^2 + 2(-\tfrac{5}{4}) + (\tfrac{5}{4})^2 = 10 + \tfrac{9}{16} + \tfrac{25}{16} = \tfrac{194}{16}.$$

This is equivalent to

$$(x + \tfrac{3}{4})^2 + (y - \tfrac{5}{4})^2 = \tfrac{194}{16},$$

from which the center and radius may be read off by thinking of

$$\left[x - \left(-\frac{3}{4}\right)\right]^2 + \left(y - \frac{5}{4}\right)^2 = \left(\frac{\sqrt{194}}{4}\right)^2,$$

as equivalent to Eq. (31). Thus the center is the point $(-\frac{3}{4},\frac{5}{4})$, and the radius is $\sqrt{194}/4$.

EXERCISE 40

Find the distance between each given pair of points.

1. (2,3), (4,7).
2. (−1,2), (3,−4).
3. (−2,−1), (−4,−3).

Find the equation of the perpendicular bisector of the segment whose end points have the given coordinates in each problem.

4. (0,2), (3,1).
5. (−5,−2), (−2,0).
6. (−3,1), (1,2).

7. Given $A = (-3,2)$, $B = (1,4)$, $C = (-2,0)$. Find the squares of the lengths of the sides of triangle ABC and deduce that it is an isosceles right triangle.

Find the equation of a circle whose center is the given point and whose radius is the given number in each problem.

8. (4,2), 4.
9. (−3,−1), 6.
10. (−2,4), 3.

Find the equation of a circle having the given pair of points as the extremity of a diameter in each problem.

11. (0,0), (0,6).
12. (4,0), (0,4).
13. (1,3), (9,9).

Find the center and radius of each of the following circles.

14. $x^2 + y^2 + 2x + 6y = 15$.
15. $x^2 + y^2 - 8x = 0$.
16. $x^2 + y^2 - 4x - 6y + 12 = 0$.
17. $2x^2 + 2y^2 + 5y = 0$.
18. $5x^2 + 5y^2 + 8x - 4y = 1$.
19. $x^2 + y^2 + 24x - 10y = 0$.

Find the equation of a circle whose center is at (3,5) if it

20. Is tangent to the x axis.
21. Is tangent to the y axis.
22. Passes through (0,0).
23. Passes through (−2,1).
24. Is tangent to $y = x$.
25. Is tangent to $y = 2x$.

R83. Parabola. *A parabola is the locus of a point whose distance from a fixed point is equal to its distance from a fixed straight line.* The fixed point is called the *focus* and the fixed straight line is called the *directrix*.

Let F, Fig. 80, be the focus and RS the directrix. Through F draw FD perpendicular to RS and intersecting it at D. Let the mid-point of DF be O. Then if $DF = 2c$, $OF = c$.

The simplest form of the equation results when we take O as the origin and DOF as the x axis. The y axis is then the perpendicular to DF through O. The focus $F = (c,0)$. Then if $P = (x,y)$ is any point on the parabola, and NP perpendicular to RS is the distance from the directrix, by the definition of the parabola, we have

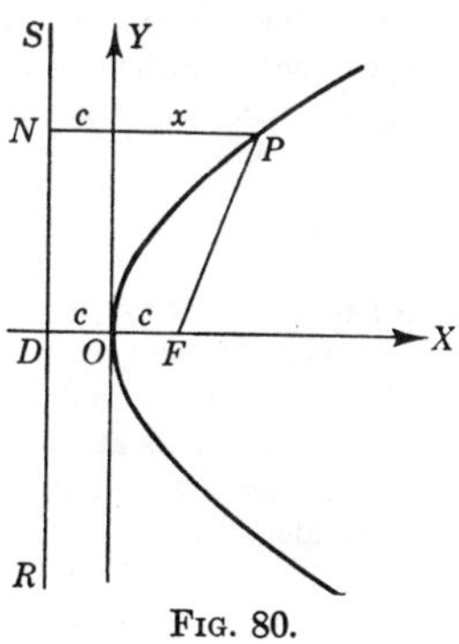

FIG. 80.

$$FP = NP. \tag{35}$$

But from Fig. 80, $NP = x + c$, in accord with Sec. 80, since $x = -c$ is the equation of RS. And from Eq. (30) we have

$$FP = \sqrt{(x - c)^2 + y^2}. \tag{36}$$

It follows that

$$\sqrt{(x - c)^2 + y^2} = x + c. \tag{37}$$

On squaring both sides, we find

$$x^2 - 2cx + c^2 + y^2 = x^2 + 2cx + c^2, \tag{38}$$

so that

$$y^2 = 4cx. \tag{39}$$

Conversely, if this equation holds, we may retrace the steps leading to Eq. (38). Since the positive square roots are equal, we may then deduce Eqs. (37) and (35). Thus Eq. (39) is the equation of the parabola having the given focus and directrix.

If we solve for y in terms of x, we have

$$y = \pm 2\sqrt{cx}.$$

Since c is positive, y is imaginary for x negative. But for x positive there are two real values of y, equal in magnitude but opposite in sign. Thus all chords parallel to OY have OX as their perpendicular bisector. Hence OX is an axis of symmetry for the parabola. We call it the *axis* of the parabola. The point O where it intersects the curve is called the *vertex*.

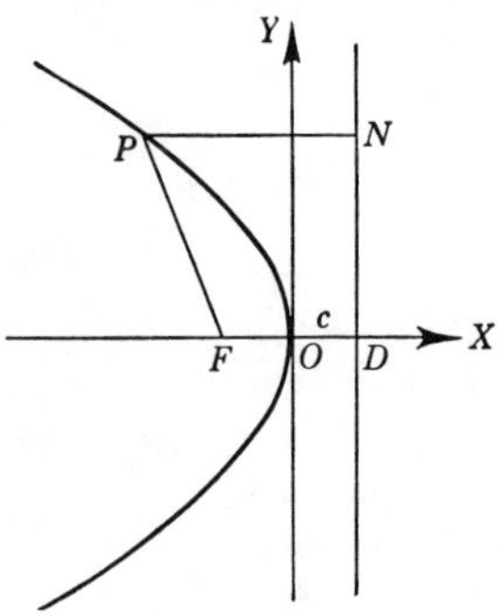

FIG. 81.

If we replace x by $-x$ in Eq. (39) it becomes

$$y^2 = -4cx,$$

which accordingly represents a parabola facing to the left, with focus at $(-c,0)$ and directrix $x = c$ (Fig. 81).

Thus $y^2 = Kx$ is a parabola facing to the right, or to the left, according as K is positive or negative.

Similarly $x^2 = Ky$ is a parabola facing up or down, according as K is positive or negative.

If p_1 is the signed distance from a fixed straight line to a variable point $P = (x,y)$, as defined in Sec. 80, $p_1 = 0$ is one form of the equation of the line. Take any second fixed straight line perpendicular to the one with equation $p_1 = 0$. Let p_2 be its signed distance to $P = (x,y)$, so that $p_2 = 0$ is one form of the equation of the second line. Then the locus of the equation

$$p_2^2 = Kp_1, \qquad K \neq 0, \tag{40}$$

is a parabola having $p_2 = 0$ as the equation of its axis and $p_1 = 0$ as the equation of the tangent at the vertex. This parabola faces to the side of $p_1 = 0$ on which Kp_1 is positive. And the distance from the vertex to the focus is $c = |K|/4$. For, if the axes had been chosen as at the beginning of this section, the y of Eq. (39) would have been $\pm p_2$, and the x would have been $\pm p_1$, and $4cx$ would have been Kp_1. Thus Eq. (40) is equivalent to Eq. (39).

In particular, if the axis of the parabola is parallel to the x axis, and the vertex is at (h,k), the equation of the parabola is

$$(y - k)^2 = K(x - h). \tag{41}$$

Conversely, any equation of the type

$$Cy^2 + Dx + Ey + F = 0 \tag{42}$$

with $C \neq 0$ and $D \neq 0$ may be transformed into an Eq. (41) and represents a parabola whose axis is parallel to OX. If $C \neq 0$ but $D = 0$, Eq. (42) corresponds to two parallel straight lines. These may be real and distinct, coincident, or imaginary, according as the two roots of $Cy^2 + Ey + F = 0$ are real and distinct, real and equal, or conjugate complex quantities.

Similarly, if the axis of the parabola is parallel to the y axis, and the vertex is at (h,k), the equation of the parabola is

$$(x - h)^2 = K(y - k). \tag{43}$$

And conversely, any equation of the type

$$Ax^2 + Dx + Ey + F = 0 \tag{44}$$

with $A \neq 0$ and $E \neq 0$ may be transformed into an Eq. (43) and represents a parabola whose axis is parallel to OY. It follows that the graphs of

$$y = ax^2 + bx + c \tag{45}$$

with $a \neq 0$, which the student plotted in connection with the quadratic equation $ax^2 + bx + c = 0$ all represented parabolas.

Example. Find the axis, vertex, and focus of the parabola

$$2y^2 + 4x + 12y = 3.$$

Solution: We may rewrite the given equation in the form

$$2(y^2 + 6y \qquad) = -4x + 3.$$

To complete the square, as in the example of Sec. 82, we add the square of one-half of 6, or 9, inside the parenthesis. Hence we add $2(9) = 18$ to each member of the equation, and so obtain

$$2(y^2 + 6y + 9) = -4x + 21 = -4(x - \tfrac{21}{4}).$$

This may be written in the form

$$(y + 3)^2 = -2(x - \tfrac{21}{4}).$$

A comparison of this equation with Eq. (41) shows that the vertex is $V = (\frac{21}{4}, -3)$ and that the equation of the axis is $y + 3 = 0$ or $y = -3$. In view of the remarks made on the meaning of K in Eq. (40), since $K = -2$, $c = |-\frac{2}{4}| = \frac{1}{2}$. The minus sign shows that the focus is to the left of the vertex, so that the focus $F = (\frac{19}{4}, -3)$. See Fig. 82.

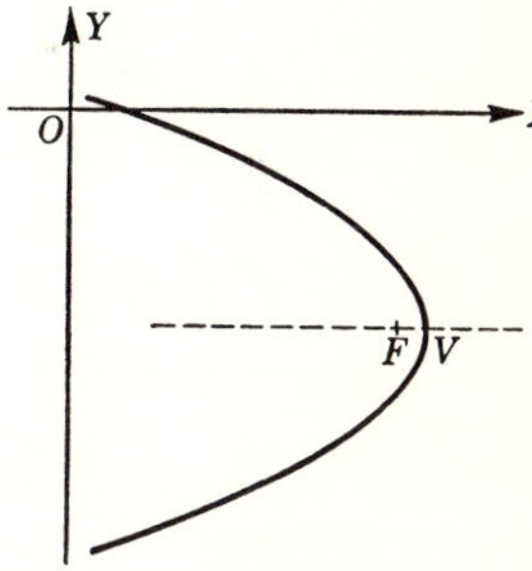

Fig. 82.

EXERCISE 41

For each given parabola find the vertex and focus. Then draw the axis and tangent at the vertex and sketch the curve.

1. $y^2 - 4x = 0.$
2. $x^2 - 8y = 0.$
3. $x^2 + 4y = 0.$
4. $y^2 + 12x = 0.$
5. $y^2 - 4y - 6x - 2 = 0.$
6. $x^2 + 6x - 12y - 3 = 0.$
7. $y^2 + 8y + 4x = 0.$
8. $x^2 - 6x + 8y = 0.$

A parabola has its vertex at the first given point and its focus at the second given point. Find its equation in each problem.

9. (0,0), (8,0).
10. (0,0), (0,−8).
11. (0,0), (−4,0).
12. (0,0), (0,4).
13. (2,3), (2,7).
14. (−3,1), (5,1).

A parabola has the first given line as its axis, the second as the tangent at the vertex, and passes through the given point. Find its equation in each problem.

15. $x = 0$, $y = 0$, (2,3).
16. $y = 0$, $x = 0$, (−2,−3).
17. $x = 1$, $y = 2$, (0,0).
18. $y = -2$, $x = 3$, (1,0).

A parabola has the given line as its directrix and the given point as its focus. Find its equation in each problem.

19. $x = 2$, (0,−2).
20. $y = -2$, (8,0).
21. $x = 2$, (6,6).
22. $y = 2$, (0,0).

R84. Ellipse. *An ellipse is the locus of a point the sum of whose distances from two fixed points is constant.* The two fixed points are called the *foci* of the ellipse.

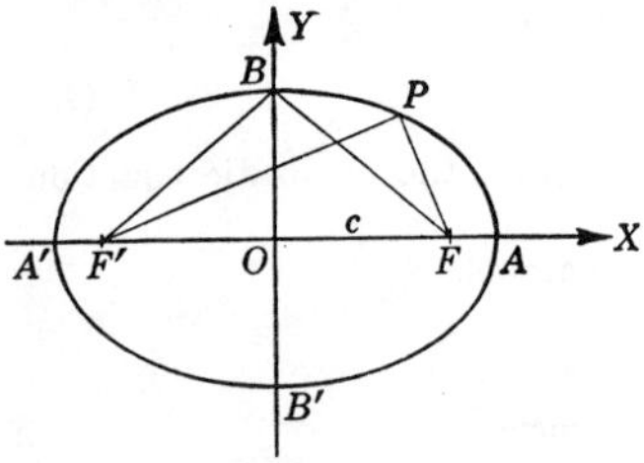

FIG. 83.

Let F' and F (Fig. 83) be the two fixed points or foci. Draw the line through $F'F$. Let the mid-point of $F'F$ be O. Then if $F'F = 2c$, $OF = c$.

The simplest form of the equation results when we take O as the origin, and $F'OF$ as the x axis. Then the y axis is the perpendicular bisector of $F'F$. And the foci are then $F' = (-c,0)$ and $F = (c,0)$. Let $P = (x,y)$ be any point on the ellipse. Then if $2a$ is the constant sum of the distances from the foci, by the definition of the ellipse, we have

$$F'P + FP = 2a. \tag{46}$$

But from Eq. (30), we have

$$F'P = \sqrt{(x+c)^2 + y^2} \quad \text{and} \quad FP = \sqrt{(x-c)^2 + y^2}. \tag{47}$$

It follows that

$$\sqrt{(x+c)^2 + y^2} + \sqrt{(x-c)^2 + y^2} = 2a. \tag{48}$$

By transposing the second radical and squaring both sides, we obtain

$$x^2 + 2cx + c^2 + y^2 = 4a^2 - 4a\sqrt{(x-c)^2 + y^2} + x^2 - 2cx + c^2 + y^2. \tag{49}$$

This may be reduced to

$$4cx - 4a^2 = -4a\sqrt{(x-c)^2 + y^2},$$

or

$$a\sqrt{(x-c)^2 + y^2} = a^2 - cx. \tag{50}$$

By again squaring both sides, we find

$$a^2(x^2 - 2cx + c^2 + y^2) = a^4 - 2a^2cx + c^2x^2,$$

or

$$(a^2 - c^2)x^2 + a^2y^2 = a^4 - a^2c^2. \tag{51}$$

In the triangle $F'PF$, the sum of two sides $F'P + FP$ is greater than the third side

$F'F$. Hence we have

$$F'P + FP > F'F \quad \text{or} \quad 2a > 2c \quad \text{and} \quad a > c. \tag{52}$$

Hence $a^2 > c^2$, and $a^2 - c^2$ is a positive quantity. It follows that $\sqrt{a^2 - c^2}$ is real. Call it b. Then we have

$$b^2 = a^2 - c^2. \tag{53}$$

We may substitute this in Eq. (51), transforming it into

$$b^2x^2 + a^2y^2 = a^2b^2,$$

so that

$$\frac{x^2}{a^2} + \frac{y^2}{b^2} = 1. \tag{54}$$

Conversely, suppose that this equation holds with $a^2 > b^2$. We assume that $a > 0$, $b > 0$. Then we may define $c = \sqrt{a^2 - b^2}$, so that $b^2 = a^2 - c^2$. We may then retrace our steps to Eq. (51). The further backward steps require square roots, which may be plus or minus. Hence we reach Eq. (48) with $\pm$ before the radicals, so that one of the relations

$$F'P + FP = 2a, \quad F'P - FP = 2a, \quad -F'P + FP = 2a, \quad -F'P - FP = 2a, \tag{55}$$

must hold. The last is impossible since a is positive. And since $a > c$, $2a > 2c$ or $2a > F'F$. Hence, if the third relation held, we would have

$$-F'P + FP > F'F \quad \text{or} \quad FP > F'F + F'P. \tag{56}$$

But this is impossible as it makes one side of triangle FPF' greater than the sum of the other two. Similarly the second relation cannot hold, and we see that Eq. (54) with $c = \sqrt{a^2 - b^2}$ implies the first relation of Eq. (55), or Eq. (46). Thus Eq. (54) is the equation of the ellipse.

Let $P_a = (x_a, y_a)$ be a point on a circle of radius a with its center at the origin. Then from Eq. (32) we have

$$x_a^2 + y_a^2 = a^2. \tag{57}$$

With each point P_a let us associate a new point $P = (x,y)$ with the same x and ordinate y shortened in the ratio b/a. Thus

$$x = x_a, \; y = \frac{b}{a} y_a$$

and

$$x_a^2 = x^2, \; y_a^2 = \frac{a^2y^2}{b^2}. \tag{58}$$

It follows from Eqs. (57) and (58) that

$$x^2 + \frac{a^2y^2}{b^2} = a^2 \quad \text{or} \quad \frac{x^2}{a^2} + \frac{y^2}{b^2} = 1. \tag{59}$$

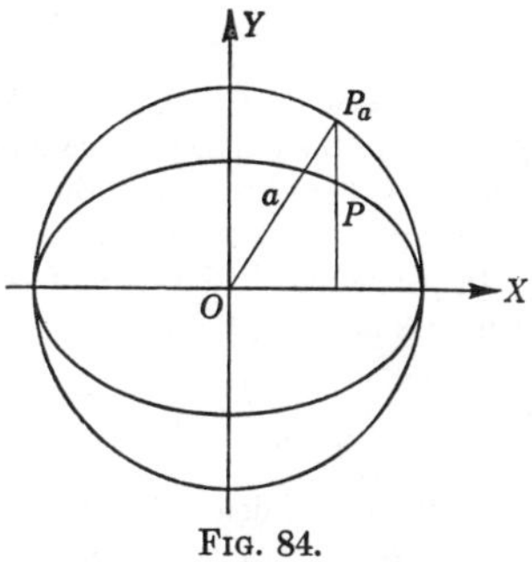

FIG. 84.

As this is the equation of the ellipse, Eq. (54), we may sketch the ellipse and visualize its shape as a foreshortened circle (Fig. 84).

We may also plot points directly from the solved form of Eq. (54) or

$$y = \pm \frac{b}{a} \sqrt{a^2 - x^2}. \tag{60}$$

The lines OX and OY are each axes of symmetry of the ellipse. We call them the *axes* of the ellipse. Their intersection O is the *center* of the ellipse. Every *diameter* of the ellipse, or chord through the center, is bisected by the center. The longest diameter $A'A = 2a$ is called the *major axis*. Its extremities A' and A, the intersections of the ellipse with the major axis, are called *vertices*. The shortest diameter $B'B = 2b$ is called the *minor axis*. The halves of $A'A$ and of $B'B$ or a and b are the *semiaxes*.

From the definition of the ellipse applied to the point B in Fig. 83, $F'B + FB = 2a$. But as B is on the perpendicular bisector of $F'F$, $F'B = FB$ and $2FB = 2a$ or $FB = a$. This fact, and the right triangle OFB help one to recall that

$$a^2 = b^2 + c^2. \tag{61}$$

This implies Eq. (53) or $c^2 = a^2 - b^2$ used to find c from Eq. (54).

The ratio $e = c/a$ is called the *eccentricity* of the ellipse. When the triangle $F'FB$ does not degenerate, we have

$$0 < c < a \qquad \text{and} \qquad 0 < \frac{c}{a} < 1, \tag{62}$$

so that e is between 0 and 1. As one extreme limiting case we have $e = 0$, $c = 0$, $b = a$. Thus $b/a = 1$, and the ellipse is the circle of radius a without foreshortening. At the other extreme we may let $e \to 1$, $c \to a$, $b \to 0$. As $b/a \to 0$, the ellipse approaches the line $A'A$ traversed twice. Each of these limiting cases satisfies our geometric definition of an ellipse. The circle having both foci coincident with the center satisfies Eq. (54) with $b = a$. Hence we admit the circle as a special case of an ellipse. But the doubled straight line having the foci coinciding with the vertices can be obtained from Eq. (54) only by a limiting process.

As in Sec. 83, consider any two perpendicular lines in the plane. Let the signed distances from them to $P = (x,y)$ (Sec. 80) be p_1 and p_2. Then $p_1 = 0$ and $p_2 = 0$ is one form of the equations of the two lines. Then the locus of the equation

$$\frac{p_1^2}{a^2} + \frac{p_2^2}{b^2} = 1 \tag{63}$$

is an ellipse with semiaxes a and b. If $a > b$, the major axis lies along the line on which $p_2 = 0$ and the minor axis lies along the line on which $p_1 = 0$. The intersection of the lines $p_1 = 0$ and $p_2 = 0$ is the center. The vertices are on $p_2 = 0$ at distance a from the center. And the foci are on $p_2 = 0$ at distance $c = \sqrt{a^2 - b^2}$ from the center.

In particular, if the axes of the ellipse are parallel to the coordinate axes, and the center is at (h,k), the equation of the ellipse is

$$\frac{(x-h)^2}{a^2} + \frac{(y-k)^2}{b^2} = 1 \qquad \text{or} \qquad \frac{(x-h)^2}{b^2} + \frac{(y-k)^2}{a^2} = 1, \tag{64}$$

since the larger denominator must be called a^2.

Conversely, any equation of the special type

$$Ax^2 + Cy^2 + Dx + Ey + F = 0, \tag{65}$$

with A and C both positive or both negative, may be transformed into

$$A(x-h)^2 + C(y-k)^2 = G. \tag{66}$$

If in Eq. (66) G has the same sign as A and C, it represents an ellipse. The locus degenerates to the point (h,k) if $G = 0$. And it becomes an imaginary locus if G has the opposite sign from that of A and C, since no real values of x and y can make $A(x-h)^2 + C(y-k)^2$ negative if A and C are each positive.

EXAMPLE. Find the vertices, foci, and eccentricity of the ellipse

$$12x^2 + 6y^2 + 48x + 36y = 5.$$

Solution: We may rewrite the given equation in the form

$$12(x^2 + 4x \qquad) + 6(y^2 + 6y \qquad) = 5.$$

To complete the square, as in the example of Sec. 82, we must add the square of half of 4, or 4, inside the first parenthesis. And we must add the square of one-half of 6, or 9, inside the second parenthesis. Hence we add $12(4) + 6(9) = 102$ to both members of the equation and so obtain

$$12(x^2 + 4x + 4) + 6(y^2 + 6y + 9) = 5 + 102 = 107$$

This may be written in the form

$$12(x+2)^2 + 6(y+3)^2 = 107,$$

or

$$\frac{(x+2)^2}{\frac{107}{12}} + \frac{(y+3)^2}{\frac{107}{6}} = 1.$$

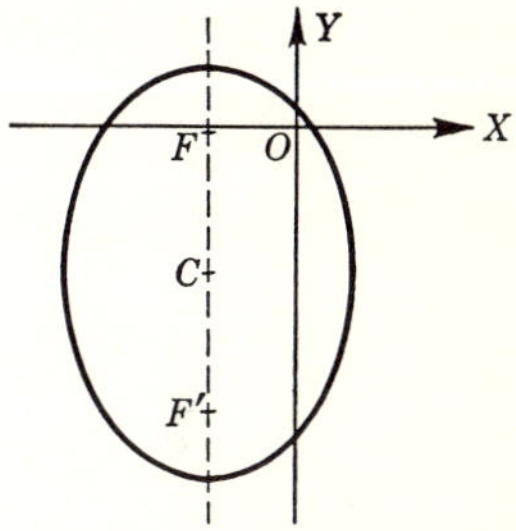

FIG. 85.

A comparison with Eq. (64) shows that the center is $(-2,-3)$. Since $\frac{107}{6} > \frac{107}{12}$, $p_1 = y + 3$ and $p_2 = x + 2$. Hence the major axis is along the line $x = -2$. And since $a^2 = \frac{107}{6}$, the vertices are $(-2, -3 \pm \sqrt{\frac{107}{6}})$. From $a^2 = \frac{107}{6}$ and $b^2 = \frac{107}{12}$ we find that $c^2 = a^2 - b^2 = \frac{107}{12}$. Hence the foci are $(-2, -3 \pm \sqrt{\frac{107}{12}})$. The eccentricity $e = c/a = \sqrt{c^2/a^2} = 1/\sqrt{2}$. See Fig. 85.

EXERCISE 42

For each given ellipse find the vertices, the foci, and the eccentricity. Also draw the major axes and sketch the curve.

1. $4x^2 + 9y^2 = 36$.
2. $25x^2 + 9y^2 = 225$.
3. $x^2 + 5y^2 = 1$.
4. $5x^2 + 2y^2 = 3$.
5. $x^2 + 25y^2 = 10x$.
6. $9x^2 + 4y^2 = 24y$.
7. $x^2 + 4y^2 - 6x - 16y + 21 = 0$.
8. $4x^2 + y^2 - 8x - 2y = 11$.

An ellipse has its foci at the two given points, and its major axis $2a$ equals the given number. Find its equation in each problem.

9. $(-3,0)$, $(3,0)$, 10.
10. $(0,-4)$, $(0,4)$, 10.
11. $(1,2)$, $(1,6)$, 6.
12. $(3,2)$, $(9,2)$, 8.

An ellipse with axes parallel to OX and OY has its center at the first given point and passes through the second and third given points. Find its equation in each problem.

13. $(0,0)$, $(5,0)$, $(0,3)$.
14. $(0,0)$, $(2,0)$, $(0,4)$.
15. $(0,0)$, $(3,4)$, $(6,2)$.
16. $(0,0)$, $(1,6)$, $(2,3)$.
17. $(2,3)$, $(2,4)$, $(5,3)$.
18. $(-2,-1)$, $(-2,3)$, $(0,-1)$.

19. Show that the locus of a point which moves so that its distance from the point $F = (4,0)$ is always $\frac{4}{5}$ of its distance from the line $4x - 25 = 0$ is an ellipse with one focus at F.

20. Show that the locus of a point which moves so that its distance from the point $F = (c,0)$ is always e times its distance from the line $cx - a^2 = 0$, where $e = c/a < 1$, is an ellipse with one focus at F.

R85. Hyperbola. *A hyperbola is the locus of a point the difference of whose distances from two fixed points is constant.* The two fixed points are called the *foci* of the hyperbola.

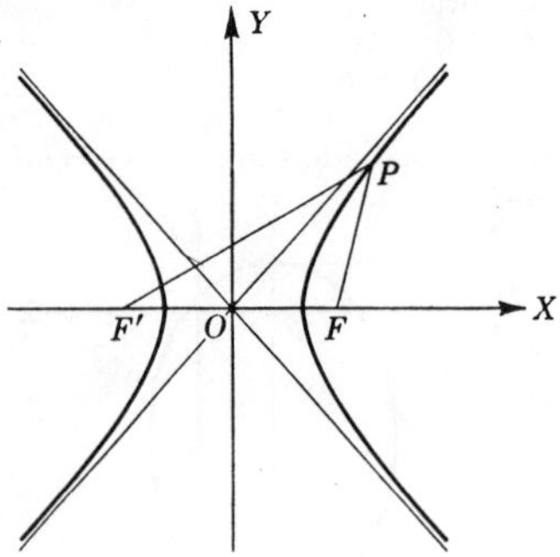

FIG. 86.

Let F' and F (Fig. 86) be the two fixed points or foci. Draw the line through $F'F$. Let the mid-point of $F'F$ be O. Then if $F'F = 2c$, $OF = c$.

The simplest form of the equation results when we take O as the origin, and $F'OF$ as the x axis. Then the y axis is the perpendicular bisector of $F'F$. And the foci are then $F' = (-c,0)$ and $F = (c,0)$. Let $P = (x,y)$ be any point on the hyperbola. Then if $2a$ is the constant difference of its distances from the foci, by the definition of the hyperbola, we have

$$F'P - FP = 2a \quad \text{or} \quad FP - F'P = 2a, \tag{67}$$

so that

$$F'P - FP = \pm 2a. \tag{68}$$

But from Eq. (30), we have

$$F'P = \sqrt{(x + c)^2 + y^2} \quad \text{and} \quad FP = \sqrt{(x - c)^2 + y^2}.$$

It follows that

$$\sqrt{(x + c)^2 + y^2} - \sqrt{(x - c)^2 + y^2} = \pm 2a. \tag{69}$$

By transposing the second radical and then squaring both sides, we obtain

$$x^2 + 2cx + c^2 = 4a^2 \pm 4a\sqrt{(x - c)^2 + y^2} + x^2 - 2cx + c^2 + y^2. \tag{70}$$

This may be reduced to

$$4cx - 4a^2 = \pm 4a\sqrt{(x - c)^2 + y^2},$$

or

$$\mp a\sqrt{(x - c)^2 + y^2} = a^2 - cx. \tag{71}$$

By again squaring both sides, we find

$$a^2(x^2 - 2cx + c^2 + y^2) = a^4 - 2a^2cx + c^2x^2,$$

or

$$(a^2 - c^2)x^2 + a^2y^2 = a^4 - a^2c^2. \tag{72}$$

If P is on the hyperbola and if $F'P > FP$, the first relation of Eq. (67) holds, and $F'P - FP = 2a$. Then in the triangle $F'PF$, the sum of two sides $F'F + FP$ is greater than the third side $F'P$. Hence we have

$$F'F + FP > F'P, \quad F'F > F'P - FP \quad \text{or} \quad 2c > 2a \text{ and } c > a. \tag{73}$$

Hence $c^2 > a^2$, and $c^2 - a^2$ is a positive quantity. It follows that $\sqrt{c^2 - a^2}$ is real. Call it b. Then we have

$$b^2 = c^2 - a^2. \tag{74}$$

We may substitute this in Eq. (72), transforming it into

$$-b^2x^2 + a^2y^2 = -a^2b^2,$$

so that

$$\frac{x^2}{a^2} - \frac{y^2}{b^2} = 1. \tag{75}$$

Conversely, suppose that this equation holds. We assume that $a > 0$, $b > 0$. Then we may define $c = \sqrt{a^2 + b^2}$, so that $b^2 = c^2 - a^2$. We may then retrace our steps to Eq. (72). The further backward steps require square roots, which may be plus or minus. Hence we reach Eq. (69) with some choice of plus or minus before the radicals, so that one of the relations

$$F'P + FP = 2a, \quad F'P - FP = 2a, \quad -F'P + FP = 2a, \quad -F'P - FP = 2a, \tag{76}$$

must hold. The last is impossible since a is positive. And since $c > a$, $2c > 2a$ or $2a < F'F$. Hence if the first relation held, we would have

$$F'P + FP < F'F. \tag{77}$$

But this is impossible as it makes one side of triangle $F'PF$ greater than the sum of the other two. Thus Eq. (75) with $c = \sqrt{a^2 + b^2}$ implies the relations of Eq. (67). Hence Eq. (75) is the equation of the hyperbola.

We may plot points on the hyperbola from the solved form of Eq. (75) or

$$y = \pm \frac{b}{a} \sqrt{x^2 - a^2}. \tag{78}$$

This shows that all chords parallel to OY have OX as their perpendicular bisector. Hence OX is an axis of symmetry. Equation (78) also shows that there are no points for $-a < x < a$, since these lead to imaginary values. And as x increases from a to $+\infty$, for the branch with the plus sign y increases from 0 to $+\infty$.

The result of solving Eq. (75) for x is

$$x = \pm \frac{a}{b} \sqrt{y^2 + b^2}. \tag{79}$$

This shows that OY is an axis of symmetry. And there are values of x for all values of y.

We call the lines OX and OY the *axes* of the hyperbola. Their intersection O is the *center* of the hyperbola. Every *diameter* of the hyperbola, or chord through the center, is bisected by the center. The shortest diameter, $A'A = 2a$, is called the *transverse axis*. Its extremities A' and A (Fig. 87), the intersections of the hyperbola with the transverse axis, are called *vertices*. The axis OY does not intersect the curve. But that part of it between $(0,-b)$ and $(0,b)$ is called the *conjugate axis*. We call a and b the *semiaxes* of the hyperbola.

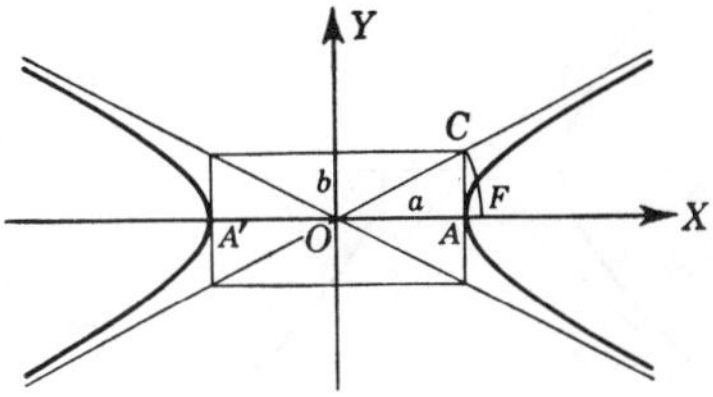

FIG. 87.

We note that the focus $F = (c,0)$ may be found by drawing a circle with center at O to pass through $C = (a,b)$. For from the right triangle OAC, we have

$$\overline{OC}^2 = \overline{OA}^2 + \overline{AC}^2 = a^2 + b^2 = c^2. \tag{80}$$

This right triangle is helpful in remembering Eq. (80), or Eq. (74).

The ratio $e = c/a$ is called the eccentricity of the hyperbola. When the triangle $F'FB$ does not degenerate, we have

$$c > a \qquad \text{and} \qquad \frac{c}{a} > 1, \tag{81}$$

so that e is greater than 1. At one extreme, we may let $e \to 1$, $c \to a$, $b \to 0$. As $b \to 0$, the hyperbola approaches the two parts of OX to the left of A' and to the right of A, each traversed trice. At the other extreme, we may let $e \to +\infty$, $a \to 0$, $b \to c$. With b fixed, as $a \to 0$ the hyperbola approaches the axis OY, traversed twice.

If we replace the 1 on the right-hand side of Eq. (75) by 0, we obtain the relation

$$\frac{x^2}{a^2} - \frac{y^2}{b^2} = 0 \qquad \text{or} \qquad \left(\frac{x}{a} - \frac{y}{b}\right)\left(\frac{x}{a} + \frac{y}{b}\right) = 0. \tag{82}$$

This will hold if and only if one of the factors is zero,

$$\frac{x}{a} - \frac{y}{b} = 0 \qquad \text{or} \qquad \frac{x}{a} + \frac{y}{b} = 0. \tag{83}$$

By Sec. 77 each of these first-degree equations represents a straight line. Hence the locus of Eq. (83), or the equivalent Eq. (82), is two straight lines. These lines have important relations to the hyperbola which we shall now study. By Sec. 80 the signed distances from these lines to $P(x,y)$ are

$$D_1 = \frac{(x/a) - (y/b)}{\sqrt{(1/a^2) + (1/b^2)}} \qquad \text{and} \qquad D_2 = \frac{(x/a) + (y/b)}{\sqrt{(1/a^2) + (1/b^2)}}, \tag{84}$$

with D_1 positive below the first line and D_2 positive above the second line. If $P(x,y)$ is on the hyperbola with Eq. (75), it follows from Eq. (84) that

$$D_1D_2 = \frac{(x^2/a^2) - (y^2/b^2)}{(1/a^2) + (1/b^2)} = \frac{1}{(1/a^2) + (1/b^2)} = \frac{a^2b^2}{a^2 + b^2} = K, \tag{85}$$

where the constant K is defined by the last equality. We may now show that the lines of Eq. (83) are asymptotes of the hyperbola in the sense defined in Sec. 15. Let the point (x,y) on the hyperbola move on the branch which runs off into the first quadrant, $x \to +\infty$, $y \to +\infty$. Then from Eq. (84), $D_2 \to +\infty$. And from Eq. (85), $D_1 = K/D_2 \to 0$. Hence $D_1 = 0$ is an asymptote approached from below by the branch in the first quadrant. Similarly it is approached from above by the branch in the third quadrant. And a similar argument shows that $D_2 = 0$ is an asymptote to the branches in the second and fourth quadrants.

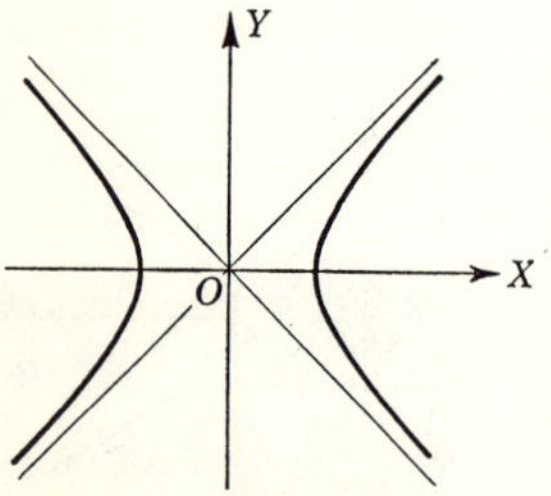

FIG. 88.

In drawing any hyperbola, it is always desirable to draw the asymptotes and then sketch the curve so as to approach them. For Eq. (75), the asymptotes may be obtained from Eq. (82).

Interpreted geometrically, Eq. (85) states that

A hyperbola is the locus of a point the product of whose distances from two fixed intersecting straight lines is constant. The two fixed lines are the *asymptotes.*

When $b = a$, the Eq. (75) reduces to

$$y^2 - x^2 = a^2, \tag{86}$$

which is called an *equilateral hyperbola* (Fig. 88). The asymptotes in this case may be found from

$$y^2 - x^2 = 0 \qquad \text{as } y - x = 0 \text{ or } y + x = 0. \tag{87}$$

These are perpendicular, so that the curve of Eq. (86) is also called a *rectangular hyperbola*. From Eq. (85), when $b = a$, $K = a^2/2$. If we take $y - x = 0$ as a new y axis, with equation $x_1 = 0$, and $y + x = 0$ as a new x axis with equation $y_1 = 0$, we shall have $D_1 = x_1$ and $D_2 = y_1$ so that the equation referred to the new axes is

$$x_1y_1 = \frac{a^2}{2}, \qquad x_1y_1 = K, \qquad \text{or } y_1 = \frac{K}{x_1}. \tag{88}$$

These are the equations of a rectangular hyperbola referred to its asymptotes as coordinate axes (Fig. 89).

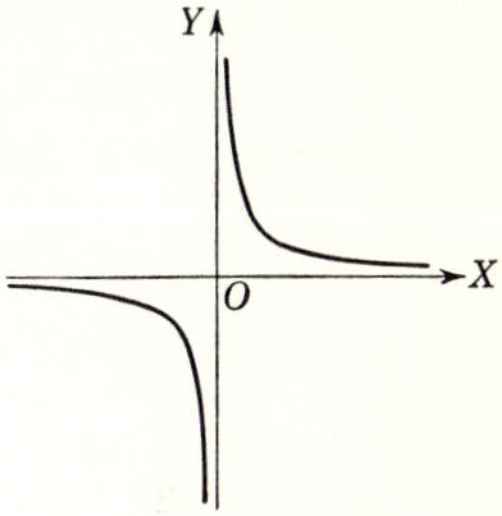

FIG. 89.

As in Sec. 83, consider any two perpendicular lines in the plane. Let the signed distances from them to $P = (x,y)$ (Sec. 78) be p_1 and p_2. Then $p_1 = 0$ and $p_2 = 0$ is one form of the equations of the two lines. Then the locus of the equation

$$\frac{p_1^2}{a^2} - \frac{p_2^2}{b^2} = 1 \tag{89}$$

is a hyperbola with semiaxes a and b. The transverse axis lies along the line on which $p_2 = 0$, and the conjugate axis lies along the line on which $p_1 = 0$. The intersection of the lines $p_1 = 0$ and $p_2 = 0$ is the center. For the vertices $p_2 = 0$ and $p_1 = \pm a$. For the foci $p_2 = 0$ and $p_1 = \pm c = \pm\sqrt{a^2 + b^2}$.

In particular, let the axes of the hyperbola be parallel to the coordinate axes, and the center be at (h,k). Then the equation of the hyperbola is

$$\frac{(x - h)^2}{a^2} - \frac{(y - k)^2}{b^2} = 1, \tag{90}$$

when the transverse axis is the line $y = k$ parallel to OX. And the equation of the hyperbola is

$$\frac{(y - k)^2}{a^2} - \frac{(x - h)^2}{b^2} = 1, \tag{91}$$

when the transverse axis is the line $x = h$ parallel to OY.

Conversely, any equation of the special type

$$Ax^2 + Cy^2 + Dx + Ey + F = 0, \tag{92}$$

with A and C having opposite signs, may be transformed into

$$A(x - h)^2 + C(y - k)^2 = G. \tag{93}$$

This is a hyperbola as in Eq. (90) if AG is plus and CG minus. It is a hyperbola as in Eq. (91) if AG is minus and CG is plus. If $G = 0$, the locus degenerates into two straight lines.

The asymptotes of the hyperbolas which are the loci of Eqs. (89) to (91) can be found by replacing the constant on the right by zero and factoring the left member into two first-degree factors. This is the same procedure we used in deducing Eqs. (82) and (83) from Eq. (75). This process may be reversed if we are given the asymptotes and one other condition, as in Example 2.

EXAMPLE 1. Find the vertices, foci, eccentricity, and asymptotes of the hyperbola

$$x^2 - 2y^2 - 2x + 8y = 0,$$

Solution: We may complete the squares of the terms in x and of the terms in y as in the example of Sec. 84. In this way we obtain in succession the relations

$$(x^2 - 2x + 1) - 2(y^2 - 4y + 4) = 1 - 8 = -7$$
$$(x - 1)^2 - 2(y - 2)^2 = -7$$

and

$$\frac{(y - 2)^2}{\frac{7}{2}} - \frac{(x - 1)^2}{7} = 1.$$

A comparison with Eq. (91) shows that the center is (1,2) and the transverse axis is along the line $x = 1$. The vertices (Fig. 90) are $(1,\ 2 \pm \sqrt{\frac{7}{2}})$. From $a^2 = \frac{7}{2}$ and $b^2 = 7$, we find that $c^2 = a^2 + b^2 = \frac{21}{2}$. Hence the foci are $(1,\ 2 \pm \sqrt{\frac{21}{2}})$. The eccentricity $e = c/a = \sqrt{c^2/a^2} = \sqrt{3}$. The asymptotes are found from

$$(x - 1)^2 - 2(y - 2)^2 = [(x - 1) - \sqrt{2}\,(y - 2)][(x - 1) + \sqrt{2}\,(y - 2)] = 0$$

to have as their equations

$$x - \sqrt{2}\,y = 1 - 2\sqrt{2}$$

and

$$x + \sqrt{2}\,y = 1 + 2\sqrt{2}.$$

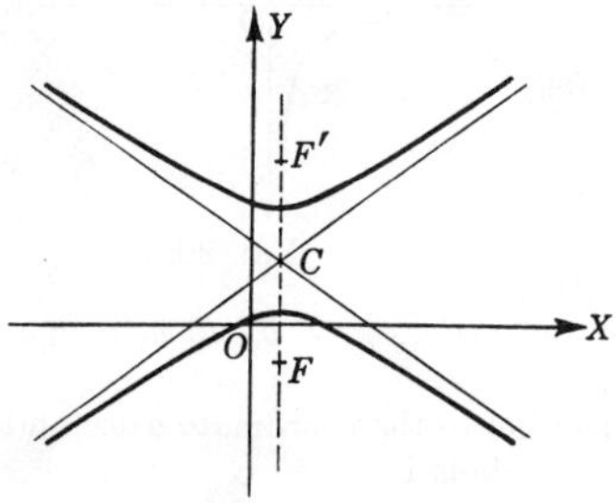

FIG. 90.

EXAMPLE 2. Find the equation of a hyperbola with asymptotes $y = x + 2$ and $x = 3$ and passing through the origin.

Solution: The equation

$$(y - x - 2)(x - 3) = 0$$

represents the asymptotes, and so may be obtained from one form of the equation of the hyperbola by replacing a constant k by 0. Hence the required equation may be taken as

$$(y - x - 2)(x - 3) = k.$$

This will pass through the origin, (0,0), if $(0 - 2)(0 - 3) = k$ or if $k = 6$. Hence the equation sought is

$$(y - x - 2)(x - 3) = 6 \quad \text{or} \quad xy - x^2 + x - 3y = 0.$$

EXERCISE 43

For each given hyperbola find the vertices, the foci, the eccentricity, and the asymptotes. Draw the asymptotes and sketch the curve.

1. $4x^2 - 9y^2 = 36$.
2. $9y^2 - 25x^2 = 225$.
3. $5x^2 - y^2 = 1$.
4. $5y^2 - 2x^2 = 3$.
5. $x^2 - 25y^2 = 10x$.
6. $4y^2 - 9x^2 = 24y$.
7. $x^2 - 4y^2 - 6x + 16y = 3$.
8. $4x^2 - y^2 - 8x + 2y = 1$.

A hyperbola has its foci at the two given points, and its major axis $2a$ equals the given number. Find its equation in each problem.

9. $(-5,0)$, $(5,0)$, 8.
10. $(0,-5)$, $(0,5)$, 6.
11. $(1,2)$, $(1,6)$, 2.
12. $(3,2)$, $(9,2)$, 4.

A hyperbola has the two given lines as its asymptotes and passes through the given point. Find its equation in each problem.

13. $x = 0$, $y = 0$, $(2,-5)$.
14. $x = 2$, $y = 3$, $(3,4)$.
15. $x = 2y$, $x = -2y$, $(1,2)$.
16. $2x = 3y$, $2x = -3y$, $(-1,-1)$.
17. $y = 3x$, $y = x$, $(1,0)$.
18. $y = x + 1$, $y = -x - 1$, $(0,0)$.

A hyperbola with axes parallel to OX and OY has its center at the first given point and passes through the second and third given point. Find its equation in each problem.

19. (0,0), (5,3), (4,0).
20. (0,0), (1,5), (0,4).
21. (0,0), (3,2), (1,0).
22. (0,0), (1,4), (0,3).
23. (−2,−1), (1,0), (0,−1)
24. (2,3), (2,4), (3,5).

25. Show that the locus of a point which moves so that its distance from the point $F = (5,0)$ is always $\frac{5}{4}$ of its distance from the line $5x - 16 = 0$ is a hyperbola with one focus at F.
26. Show that the locus of a point which moves so that its distance from the point $F = (c,0)$ is always e times its distance from the line $cx - a^2 = 0$, where $e = c/a > 1$, is a hyperbola with one focus at F.

R86. Second-degree Equation. The general equation of the second degree has the form

$$Ax^2 + Bxy + Cy^2 + Dx + Ey + F = 0, \tag{94}$$

with A, B, C not all zero.

A simple way to construct the locus of this in any given numerical case is to solve for y in terms of x by the quadratic formula. The curve may be plotted by calculating offsets parallel to OY from a diameter, as in the example below.

The locus of Eq. (94) may be entirely imaginary, have one real point only, or represent two real straight lines. But when it does not degenerate in this way, it represents either an ellipse, including the circle as a special case, a parabola, or a hyperbola. And the type of curve depends on the nature of the discriminant

$$B^2 - 4AC \tag{95}$$

which is negative for the ellipse, zero for the parabola, and positive for the hyperbola.

To outline the proof of this, we note that for a new set of rectangular axes through the origin, OX_1 and OY_1, obtained from the old set by a positive rotation through ϕ about O, we have

$$x = x_1 \cos \phi - y_1 \sin \phi \qquad \text{and} \qquad y = x_1 \sin \phi + y_1 \cos \phi, \tag{96}$$

as we found in Eq. (28). If these expressions are substituted in Eq. (94), a new equation,

$$A_1x_1^2 + B_1x_1y_1 + C_1y_1^2 + D_1x_1 + E_1y_1 + F_1 = 0, \tag{97}$$

will result. And a lengthy bit of algebra shows that

$$B_1^2 - 4A_1C_1 = B^2 - 4AC. \tag{98}$$

But if the angle of rotation ϕ satisfies

$$\tan 2\phi = \frac{B}{A - C}, \tag{99}$$

or the equivalent condition $\tan \phi = m$, a root of

$$Bm^2 + 2(A - C)m - B = 0, \tag{100}$$

the new coefficient B_1 will be zero.

Thus, if originally $B^2 - 4AC$ is negative, in the new case with $B_1 = 0$, $-4A_1C_1 < 0$, so that A_1 and C_1 have the same sign. Thus the transformed equation is essentially Eq. (65) or (66). Hence it represents an ellipse, a point, or a purely imaginary locus.

If originally $B^2 - 4AC$ is zero, in the new case with $B_1 = 0$, $-4A_1C_1 = 0$, so that A_1 or C_1 is zero. Thus the transformed equation is essentially Eq. (42) or (44). Hence it represents a parabola or a pair of parallel straight lines, which may reduce to coincidence or be imaginary.

If originally $B^2 - 4AC$ is positive, in the new case with $B_1 = 0$, $-4A_1C_1 > 0$, so that A_1 and C_1 have opposite signs. Thus the transformed equation is essentially Eq. (92) or (93). Hence it represents a hyperbola or two intersecting lines.

It may help to recall the significance of $B^2 - 4AC$ by noting that it is the discriminant of the quadratic equation in (y/x),

$$A\left(\frac{y}{x}\right)^2 + B\left(\frac{y}{x}\right) + C = 0 \quad \text{or} \quad Ax^2 + Bxy + Cy^2 = 0, \tag{101}$$

obtained from Eq. (94) by dropping all but the leading terms for large x and y. Hence the negative discriminant means no real roots or infinite branches for the ellipse. The positive discriminant means two distinct real roots or directions of infinite branches for the hyperbola, and the roots are the slopes of the asymptotes. And the zero discriminant means one infinite branch for the parabola, with the repeated root being the slope of the axis.

EXAMPLE. Sketch the graph of $2x^2 - 2xy + y^2 - 6x = 0$.

Solution: Solving the quadratic equation in y

$$y^2 - 2xy + 2x^2 - 6x = 0$$

for y, we find

$$y = \frac{2x \pm \sqrt{(-2x)^2 - 4(2x^2 - 6x)}}{2} = x \pm \sqrt{6x - x^2}.$$

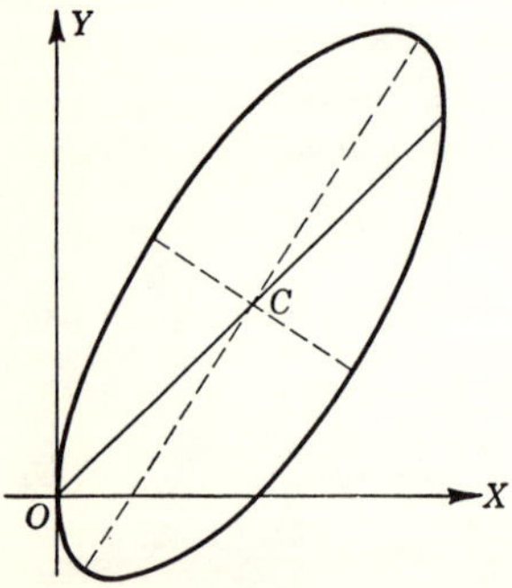

FIG. 91.

The quantity inside the radical $x(6 - x)$ is positive for $0 < x < 6$. And it is a maximum when $x = 6 - x$ or when $x = 3$. Hence the curve has vertical tangents at (0,0) and (6,6), and tangents parallel to the diameter $y = x$ at (3,0) and (3,6). The curve lies inside this parallelogram and so is an ellipse.

We would expect an actual or degenerate ellipse, since for the given equation,

$$B^2 - 4AC = (-2)^2 - 4 \cdot 1 \cdot 2 = -4 < 0.$$

The center is the mid-point of the diameter (3,3). And the axes could be found by drawing through the center (3,3) two lines with slopes equal to the roots of Eq. (100), $Bm^2 + 2(A - C)m - B = 0$. In this case $-2m^2 + 2m + 2 = 0$ or $m^2 - m - 1 = 0$ with $m = (1 \pm \sqrt{5})/2$, so that the slopes of the axes are 1.618 and -0.618. See Fig. 91.

EXERCISE 44

Sketch the graph of each given equation.

1. $x^2 - 2xy + y^2 - 4x = 0$.
2. $2xy - y^2 - 5x = 0$.
3. $2x^2 - 2xy + y^2 - 6x + 2y + 1 = 0$.
4. $x^2 + 2xy + 2y^2 = 8$.
5. $3x^2 - 4xy + y^2 + 4x = 0$.

6. $2xy - y^2 - 1 = 0$.
7. $x^2 - 2xy - 1 = 0$.
8. $2x^2 + 4xy + 4y^2 + x + 4y - 5 = 0$.
9. $4x^2 - 12xy + 9y^2 + 3x - 6y = 0$.
10. $2x^2 + 3xy - 2y^2 + x + 2y + 2 = 0$.
11. $x^2 - 4xy + 4y^2 - 4x - 2y + 8 = 0$.
12. $3x^2 + 12xy - 2y^2 - 14x = 0$.

What values of k make the graph of each given equation (a) a parabola, (b) an ellipse, (c) a hyperbola?

13. $kx^2 - 4xy + y^2 = 2y$.
14. $x^2 + kxy + 4y^2 = 2x$.
15. $9x^2 - kxy + 4y^2 = x$.
16. $9x^2 - 12xy + ky^2 = y$.

87. Tangents and Normals. Let $P_1 = (x_1,y_1)$ be a point on any given curve. And let $(dy/dx)_1$ be the value of the derivative dy/dx when $x = x_1$ and $y = y_1$. If the equation of the curve is given in the form $y = f(x)$, then $dy/dx = f'(x)$ and $(dy/dx)_1 = f'(x_1)$. If the curve is the locus of an equation of the form $F(x,y) = 0$, we may either solve for y and so obtain $y = f(x)$ as the equation of the branch through (x_1,y_1), or we may find dy/dx in terms of x and y by implicit differentiation as in Sec. 56, and so obtain $(dy/dx)_1$ in terms of x_1 and y_1.

Then $(dy/dx)_1$ as found by either method gives the slope of the curve at the point P_1 and also the slope of the straight-line tangent to the curve at P_1, by Secs. 23 and 27. That is

$$\text{Slope of tangent at } (x_1,y_1) = \left(\frac{dy}{dx}\right)_1 = \tan \phi_1. \tag{102}$$

But in Sec. 23 and in Eq. (4) the equation of a straight line through (x_1,y_1) with slope m was found to be

$$y - y_1 = m(x - x_1). \tag{103}$$

It follows from Eqs. (102) and (103) that the equation

$$y - y_1 = \left(\frac{dy}{dx}\right)_1 (x - x_1) \tag{104}$$

represents a straight line through $P_1 = (x_1,y_1)$ with slope equal to $(dy/dx)_1$ and so represents the equation of the tangent line to the given curve at P_1.

The straight line through P_1 perpendicular to the tangent at P_1 is called the *normal* to the curve at P_1. By Sec. 79 the slope of the normal is the negative reciprocal of the slope of the tangent. It follows from Eq. (102) that

$$\text{Slope of normal at } (x_1,y_1) = \frac{-1}{(dy/dx)_1}. \tag{105}$$

Hence by Eq. (103), the equation of the normal line to the given curve at P_1 is

$$y - y_1 = \frac{-1}{(dy/dx)_1}(x - x_1). \tag{106}$$

In using Eqs. (104) and (106) it should be noted that x and y are the coordinates of a variable point on the straight line, while x_1 and y_1 are the coordinates of the fixed point on the curve at which the tangent or normal is drawn. And the subscript 1 on the derivative in $(dy/dx)_1$ means that, after the derivative is found for any x and y, we must evaluate it for $x = x_1$ and $y = y_1$.

At a point where the tangent to the curve is horizontal, or parallel to OX, the derivative or slope will equal zero. Thus

$$\phi_1 = 0, \qquad \left(\frac{dy}{dx}\right)_1 = 0, \qquad \text{for a horizontal tangent.} \tag{107}$$

The equation of the tangent in this case is found from Eq. (104) to be $y = y_1$. The normal in this case has the equation $x = x_1$. Here Eq. (106) cannot be used literally but can be interpreted as indicated in Example 2 of Sec. 77.

At a point where the tangent to the curve is vertical, or parallel to OY, the derivative will not exist since the slope is infinite. But for smooth curves, as we approach such points the expression for dy/dx will become infinite as $x \to x_1$ and $y \to y_1$. Hence if we use $(dy/dx)_1$ here to mean the result of differentiating at a nearby point and letting $(x,y) \to (x_1,y_1)$ along the curve,

$$\phi_1 = 90°, \qquad \left(\frac{dy}{dx}\right)_1 = \lim_{x \to x_1} \left(\frac{dy}{dx}\right) = \infty, \tag{108}$$

for a vertical tangent, or for a vertical asymptote if $y \to \infty$ as $x \to x_1$.

Thus in practice the vertical tangents of a curve may be found by noting which values of x_1 make $(dy/dx)_1$ infinite but make y_1 finite. The equation of the tangent is $x = x_1$, and the equation of the normal is $y = y_1$, at any point $P_1 = (x_1,y_1)$ for which Eq. (108) holds.

In some cases a curve, or a given branch of a curve, is such that for all points $P_2 = (x_2,y_2)$ on the branch near $P_1 = (x_1,y_1)$, $\Delta x = x_2 - x_1$ has the same sign. In any such case, if (dy/dx) approaches a finite limit as $x \to x_1$, this limit is the slope of the tangent to the branch at P_1, where the tangent line at P_1 is defined as the limit of a revolving secant as described in Sec. 24. In any case where the limit of (dy/dx) as $x \to x_1$ exists, and a definite value may be obtained by substituting $x = x_1$ and $y = y_1$ in the expression for (dy/dx) found for points near P_1, the two values will necessarily be equal. Hence without danger of inconsistency we may denote the slope of the curve at P_1 by $(dy/dx)_1$. Thus Eqs. (104) and (106) may be used for the end points of branches.

And when the tangent at the end point of a branch is vertical so that its slope becomes infinite, Eq. (108), as well as the equations for the tangent $x = x_1$ and for the normal $y = y_1$, is applicable.

EXAMPLE 1. Find the equation of the tangent and normal to the circle $x^2 + y^2 = 25$ at the point (x_1,y_1), any point on the circle.

Solution 1: If we use implicit differentiation as in Sec. 56, we find

$$2x + 2y\frac{dy}{dx} = 0 \qquad \text{and} \qquad \frac{dy}{dx} = -\frac{x}{y}.$$

Hence from Eq. (104) the tangent at (x_1,y_1) has the equation

$$y - y_1 = -\frac{x_1}{y_1}(x - x_1).$$

This is equivalent to

$$yy_1 - y_1^2 = -xx_1 + x_1^2 \qquad \text{or} \qquad xx_1 + yy_1 = x_1^2 + y_1^2.$$

Since (x_1,y_1) must be taken on the circle, $x_1^2 + y_1^2 = 25$, so that

$$xx_1 + yy_1 = 25 \tag{109}$$

is the equation of the tangent.

For the normal, from Eq. (106), we find

$$y - y_1 = \frac{-1}{-x_1/y_1}(x - x_1).$$

This is equivalent to

$$x_1y - x_1y_1 = y_1x - x_1y_1 \qquad \text{or} \qquad x_1y - y_1x = 0. \tag{110}$$

In the forms given in Eqs. (109) and (110) the equations have meaning in all cases and represent the tangent and normal to the circle even when the slope is zero or infinite.

Solution 2: If we were to use the solved form, for a point with $y_1 > 0$, we would deduce from $x^2 + y^2 = 25$ that

$$y = \sqrt{25 - x^2} \qquad \text{and} \qquad \frac{dy}{dx} = \frac{-x}{\sqrt{25 - x^2}}. \tag{111}$$

And from Eq. (104) the equation of the tangent would be found to be

$$y - \sqrt{25 - x_1^2} = \frac{-x_1}{\sqrt{25 - x_1^2}}(x - x_1). \tag{112}$$

For a point with $y_1 < 0$, we would deduce equations similar to Eqs. (111) and (112) with a minus sign before the square root.

The corresponding equations for the normal would be

$$y - \sqrt{25 - x_1^2} = \pm\frac{\sqrt{25 - x_1^2}}{x_1}(x - x_1) \qquad \text{or} \qquad yx_1 = \pm x\sqrt{25 - x_1^2}. \tag{113}$$

Equation (112) and the second form of Eq. (113) are applicable if $x_1 = 0$, as at (0,5). For $(0,-5)$ we must put a minus sign before the radical in Eq. (112) as this is on the branch with $y_1 < 0$.

The points with $y_1 = 0$, or (5,0) and $(-5,0)$, are the end points of the branches. But the expression for the slope in Eq. (111) becomes infinite as $x \to 5$ or $x \to -5$. Thus by Eq. (108) the tangents are vertical at these points. At (5,0) the equation of the tangent is $x = 5$, while that of the normal is $y = 0$. And at $(-5,0)$ the equation of the tangent is $x = -5$, while that of the normal is $y = 0$. See Fig. 92.

EXAMPLE 2. Find the equation of the tangent and normal to the curve $y^2 = x^3$ at the point $(0,0)$.

Solution: The method of implicit differentiation gives

$$2y\frac{dy}{dx} = 3x^2 \qquad \text{and} \qquad \frac{dy}{dx} = \frac{3x^2}{2y}$$

for the slope near $(0,0)$. The fraction $3x^2/2y$ does not approach a limit if $x \to 0$ and $y \to 0$ in an unrelated way. But when (x,y) is on the curve $y^2 = x^3$,

$$y = \pm x^{\frac{3}{2}} \qquad \text{and} \qquad \frac{3x^2}{2y} = \frac{3x^2}{\pm 2x^{\frac{3}{2}}} = \pm\frac{3}{2}x^{\frac{1}{2}},$$

so that, when $x \to 0+$, $dy/dx \to 0$ for either branch.

In this problem the solved form of $y^2 = x^3$, $y = \pm x^{\frac{3}{2}}$, gives $dy/dx = \pm\frac{3}{2}x^{\frac{1}{2}}$, and this expression equals 0 when $x = 0$.

Thus either method gives the slope of the curve at 0 at $(0,0)$, so that the equation of the tangent is $y = 0$, and that of the normal is $x = 0$. The locus of the equation

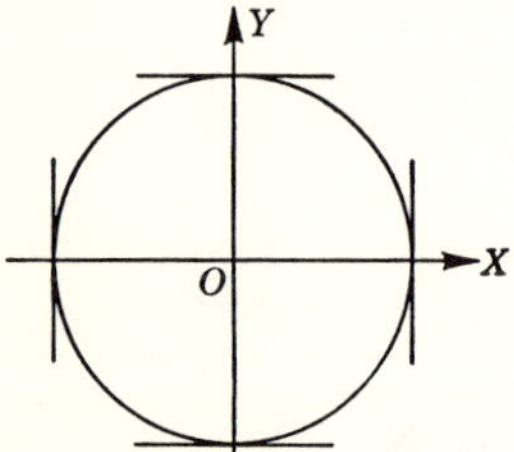

FIG. 92.

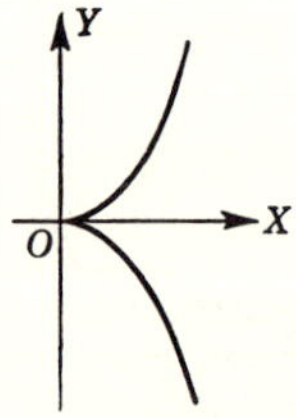

FIG. 93.

$y^2 = x^3$ is the curve shown in Fig. 93. The curve has a sharp point at the origin, since both branches lie near the tangent on the same side of the normal. Such a point is called a *cusp*.

In Example 1 the implicit method, and in Example 2 the use of the solved form, automatically gave the correct result. But as the alternative methods showed, the determination of slopes at end points of branches as limits of slopes at interior points is sometimes a convenient way of resolving apparent difficulties.

EXERCISE 45

In each problem verify that the given point is on the given curve, and find the equation of the tangent and normal to the curve at this point.

1. $(2,-3)$, $xy = -6$.
2. $(3,1)$, $xy + 2x - 4y = 5$.
3. $(4,2)$, $x^2 + y^2 = 20$.
4. $(1,2)$, $2x^2 + 3y^2 = 14$.
5. $(3,2)$, $3y^2 = 4x$.
6. $(2,1)$, $2x^2 - 3y^2 = 5$.
7. $(1,-1)$, $x^3 - x^2y = 2$.
8. $(2,-1)$, $2x^2y + y^3 = -9$.
9. $(8,1,)$, $x^{\frac{2}{3}} + y^{\frac{2}{3}} = 3$.
10. $(1,4)$, $x^{\frac{1}{2}} + y^{\frac{1}{2}} = 3$.

Find the points on each given curve where the tangent is parallel to the line $y = 24x$.

11. $y = x^3 - 3x^2$.
12. $y = 2x^3 - 15x^2 + 24x$.
13. $y = x^3 - 9x^2 + 24x$.
14. $xy + 6 = 0$.

Find the points on each given curve where the tangent is parallel to the line $y = -x$.

15. $2x^2 + y^2 - 4x = 0.$
16. $xy^2 = 32.$
17. $3x^2 + y^2 = 12.$
18. $x^2y^3 = 108.$

Find the points on each given curve where the tangent is parallel to OX, and the equations of the resulting horizontal tangents.

19. $y = x^2 - 6x.$
20. $y = x^3 - 12x.$
21. $5x^2 - 2xy + 2y^2 = 45.$
22. $x^2 + 6xy + 25y^2 = 16.$

Find the points on each given curve where the tangent is parallel to OY, and the equations of the resulting vertical tangents.

23. $y^2 + 4y - x = 0.$
24. $y^3 + 3y^2 - x = 0.$
25. $2x^2 - 2xy + y^2 = 4.$
26. $5x^2 - 4xy + y^2 = 1.$

If (x_1,y_1) is any point on the curve in question, show that the equation of the tangent to the curve at (x_1,y_1) is

27. $y_1y = 2c(x + x_1)$ for the parabola $y^2 = 4cx.$
28. $y_1x + x_1y = 2a^2$ for the hyperbola $xy = a^2.$
29. $\dfrac{x_1x}{a^2} + \dfrac{y_1y}{b^2} = 1$ for the ellipse $\dfrac{x^2}{a^2} + \dfrac{y^2}{b^2} = 1.$
30. $\dfrac{x_1x}{a^2} - \dfrac{y_1y}{b^2} = 1$ for the hyperbola $\dfrac{x^2}{a^2} - \dfrac{y^2}{b^2} = 1.$
31. $2Ax_1x + B(y_1x + x_1y) + 2Cy_1y + D(x + x_1) + E(y + y_1) + 2F = 0$, for $Ax^2 + Bxy + Cy^2 + Dx + Ey + F = 0.$

88. Geometric Properties. The angle between two curves at a point of intersection is defined as the angle between the tangents to the two curves. Hence to find this angle for two curves, we solve their equations as simultaneous to find the points of intersection. At any one such point, let m_1 be the slope of the first curve and m_2 be the slope of the second curve. Then if β, read "beta," is an angle from the first curve to the second at the point, by Eq. (13), we have

$$\tan \beta = \frac{m_2 - m_1}{1 + m_1m_2}. \tag{114}$$

When $\tan \beta$ is infinite, or one slope is infinite, we may use the limiting considerations described in Sec. 79.

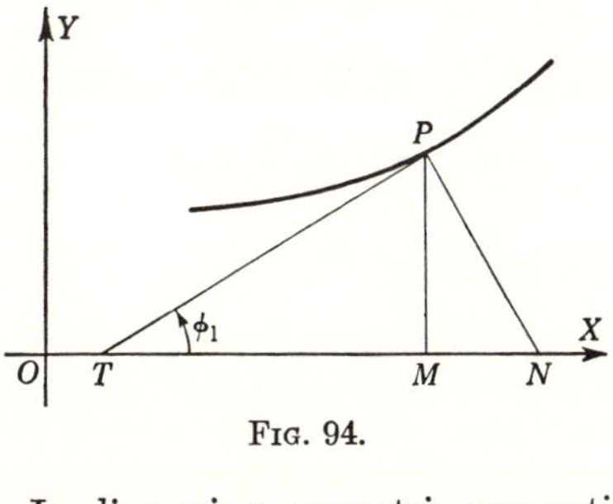

FIG. 94.

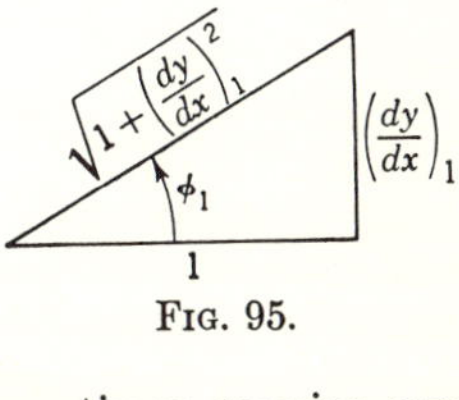

FIG. 95.

In discussing geometric properties, we sometimes require expressions for some of the lines of Fig. 94. The trigonometric functions of the inclination ϕ may be read from Fig. 95. Thus we find that

$$\text{Tangent } TP = \frac{MP}{\sin \phi_1} = \frac{y_1\sqrt{1 + (dy/dx)_1^2}}{(dy/dx)_1}. \tag{115}$$

$$\text{Subtangent } TM = \frac{MP}{\tan \phi_1} = \frac{y_1}{(dy/dx)_1}. \tag{116}$$

$$\text{Normal } PN = \frac{MP}{\cos \phi_1} = y_1\sqrt{1 + (dy/dx)_1^2}. \tag{117}$$

$$\text{Subnormal } MN = MP \tan \phi_1 = y_1\left(\frac{dy}{dx}\right)_1. \tag{118}$$

These relations make the four quantities on the right positive if y_1 and $(dy/dx)_1$ are both positive, as in Fig. 94.

We sometimes call TP the length of the tangent, and PN the length of the normal, to distinguish the segments from the indefinite straight lines whose equations were found in Sec. 87.

EXAMPLE 1. Find the angle of intersection of the curves

$$2y^2 = x \qquad \text{and} \qquad x^2 + y^2 = 5y.$$

Solution: From the first equation $x^2 = 4y^4$, and on substituting this in the second equation we obtain $4y^4 + y^2 = 5y$, or $y(y - 1)(4y^2 + 4y + 5) = 0$. Since $4y^2 + 4y + 5 = 0$ has no real roots, $y = 0$ or $y = 1$. From the first given equation, when $y = 0$, $x = 2y^2 = 0$. And when $y = 1$, $x = 2y^2 = 2$. Thus the two curves intersect at (0,0) and at (2,1).

From $2y^2 = x$, we find by implicit differentiation that

$$4y\frac{dy}{dx} = 1, \qquad \frac{dy}{dx} = \frac{1}{4y}.$$

This is ∞ when $y = 0$, and $\frac{1}{4}$ when $y = 1$.

From $x^2 + y^2 = 5y$, we find by implicit differentiation that

$$2x + 2y\frac{dy}{dx} = 5\frac{dy}{dx}, \qquad \frac{dy}{dx} = \frac{2x}{5 - 2y}.$$

This is 0 at (0,0) and is $\frac{4}{3}$ at (2,1).

It follows that at (0,0) the angle of intersection is 90°, since the first curve is tangent to OY and the second to OX.

At (2,1) the slopes are $m_1 = \frac{1}{4}$ and $m_2 = \frac{4}{3}$, so that from Eq. (114) we have

$$\tan \beta = \frac{\frac{4}{3} - \frac{1}{4}}{1 + \frac{1}{4}\frac{4}{3}} = \frac{13}{16}.$$

Hence the angle of intersection at (2,1) is $\tan^{-1} \frac{13}{16}$. See Fig. 96.

EXAMPLE 2. Find the lengths of the tangent, subtangent, normal, and subnormal at the point (1,3) on the ellipse $4x^2 + y^2 = 13$.

Solution: From $4x^2 + y^2 = 13$, we find by implicit differentiation that

$$8x + 2y\frac{dy}{dx} = 0, \qquad \frac{dy}{dx} = -\frac{4x}{y}.$$

It follows that if $(x_1y_1) = (1,3)$, we will have

$$\left(\frac{dy}{dx}\right)_1 = -\frac{4}{3}, \qquad \sqrt{1+\left(\frac{dy}{dx}\right)_1^2} = \sqrt{1+\frac{16}{9}} = \frac{5}{3}, \qquad \text{and} \qquad y_1 = 3.$$

Substitution of these values in the right member of Eqs. (115) to (118) gives $-\frac{15}{4}$, $-\frac{9}{4}$, 5, and -4, respectively. The absolute values of these quantities give the

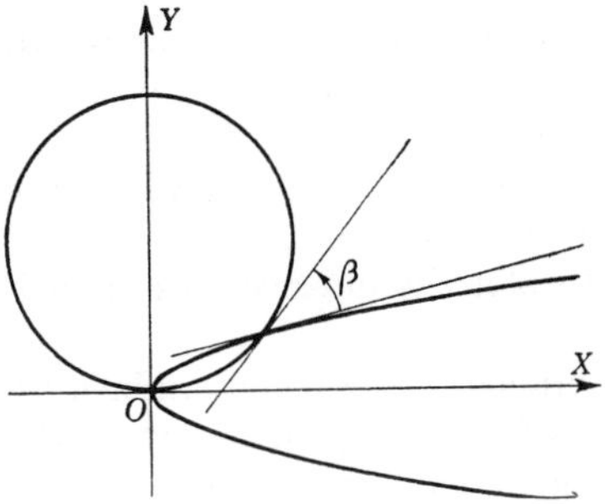

FIG. 96.

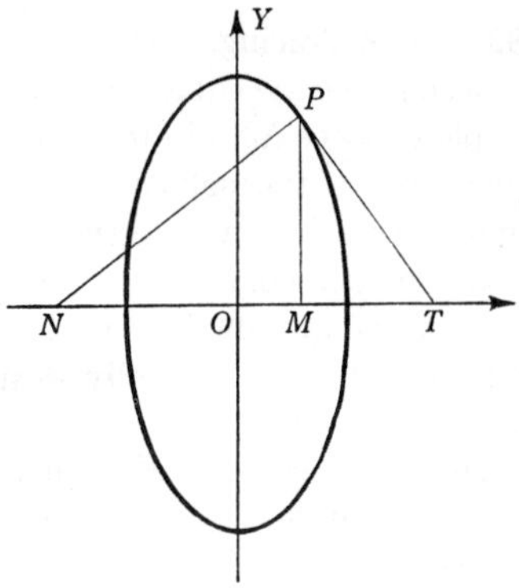

FIG. 97.

required lengths. Thus the length of the tangent $= \frac{15}{4}$, subtangent $= \frac{9}{4}$, normal $= 5$, subnormal $= 4$. For the subtangent and subnormal, the minus sign in $-\frac{9}{4}$ and -4 may be interpreted as meaning that these are on opposite sides of the ordinate in Figs. 97 and 94.

EXERCISE 46

In each problem find all the points at which the two curves intersect and the angle between the curves at each intersection.

1. $x^2 + y^2 = 4x$, $y = x$.
2. $3x^2 - 4y^2 = 12$, $x^2 + 8y^2 = 8$.
3. $y = x^2$, $y = 2x$.
4. $x^2 + y^2 = 8x$, $x^2 + y^2 = 16$.
5. $y = 2x$, $y^2 = 4x$.
6. $y^2 = x + 3$, $x^2 + y^2 = 5$.

Find the lengths of the subtangent and subnormal at the given point on the given curve in each problem.

7. (5,2), $xy = 10$.
8. (2,2), $y^3 = 2x^2$.
9. (3,1), $2x^2 - 3y^2 = 15$.
10. $(-1,1)$, $3x^2 + 4y^2 = 7$.
11. Verify analytically that, for all points on the circle $x^2 + y^2 = a^2$, the length of the normal is constant and equal to a.
12. Prove that the length of the subtangent to the ellipse $\frac{x^2}{a^2} + \frac{y^2}{b^2} = 1$ at a point with $x = x_1$ is $\frac{a^2 - x_1^2}{|x_1|}$, the same as for the circle $x^2 + y^2 = a^2$ at a point with $x = x_1$.
13. Prove that the length of the subnormal to the parabola $y^2 = 4cx$ at any point is constant and equal to $2c$.
14. Show that the subtangent to the parabola $y^2 = 4cx$ at any point is bisected by the vertex (0,0).
15. Prove that at any point P on the parabola $y^2 = 4cx$ the tangent line bisects the angle between the line through P parallel to OX, and the line drawn from the focus $F = (c,0)$ to P, produced.

16. Prove that the area of the triangle formed by OX, OY, and the tangent to the rectangular hyperbola $xy = a^2$ at any point is constant and equal to $2a^2$.
17. Prove that the sum of the intercepts on OX and OY of the tangent line at any point of the curve $x^{\frac{1}{2}} + y^{\frac{1}{2}} = a^{\frac{1}{2}}$ is constant and equal to a.
18. Prove that the part of the tangent line at any point of the curve $x^{\frac{2}{3}} + y^{\frac{2}{3}} = a^{\frac{2}{3}}$ included between OX and OY is constant and equal to a.

89. Curve Tracing. The method used in Sec. 86 to construct the locus of a second-degree equation is applicable in many other cases where it is possible to solve for y in terms of x, or for x in terms of y. In particular, regions where the solved value is imaginary will determine excluded values. And those where the solved value is real give intervals in which portions of the curve lie. Axes of symmetry, or lines from which the curve may be obtained by positive and negative offsets as in Sec. 86, are worth noting. It is usually desirable to plot the intersections of such lines with the curve.

Asymptotes or the nature of infinite branches may be determined by the methods of Sec. 15. And it is often useful to find points at which the distance from an axis of symmetry is a maximum or minimum.

EXAMPLE 1. Plot the curve $y^2 = x^2 - x^4$.

Solution: Since the curve contains only even powers of x, plus or minus values of x will give the same value of y. Hence OY is an axis of symmetry. And since

$$y = \pm \sqrt{x^2(1 - x^2)}, \tag{119}$$

OX is an axis of symmetry and the origin is a center.

To make $1 - x^2$ positive, we must have $-1 < x < 1$.

For the branch in the first quadrant, $y = 0$ when $x = 0$ and also when $x = 1$. The value of y rises to a maximum when $x^2 - x^4$ is a maximum or at a value which makes its derivative $2x - 4x^3 = 0$, which leads to $x = 0$, $x = \pm 1/\sqrt{2}$ as possibilities. The maximum y in the first quadrant occurs when $x = 1/\sqrt{2}$, and for this from Eq. (119) $y = \frac{1}{2}$. The curve is made up of two loops inside the rectangle bounded by $x = \pm 1$, $y = \pm\frac{1}{2}$.

The slope at the origin is most easily found by noting that when x is small, x^4 is small compared with x^2, so that the curve near (0,0) approximates $y^2 = x^2$ or $y = \pm x$.

The method of implicit differentiation would give

$$2y\frac{dy}{dx} = 2x - 4x^3. \tag{120}$$

This could be used to find the maximum and minimum points with $x = \pm 1/\sqrt{2}$. It also indicates the vertical tangents at $(\pm 1,0)$.

To find the tangents at (0,0) from Eq. (120), we might proceed as follows. First note that when $x \neq 0$ we have

$$\frac{dy}{dx} = \pm \frac{x - 2x^3}{\sqrt{x^2(1 - x^2)}}, \tag{121}$$

either by substituting the value of y from Eq. (119) in Eq. (120), or by direct differentiation of Eq. (119). And Eq. (121) may be reduced to

$$\frac{dy}{dx} = \pm \frac{1 - 2x^2}{\sqrt{1 - x^2}} \text{ for } x > 0 \qquad \text{and} \qquad \frac{dy}{dx} = \mp \frac{1 - 2x^2}{\sqrt{1 - x^2}} \text{ for } x < 0.$$

This shows that the tangents to the arcs with ends at the origin have as their slopes ± 1. Note that to get a smooth arc through (0,0) we must join the positive branch on one side of this point with the negative branch on the other side. See Fig. 98.

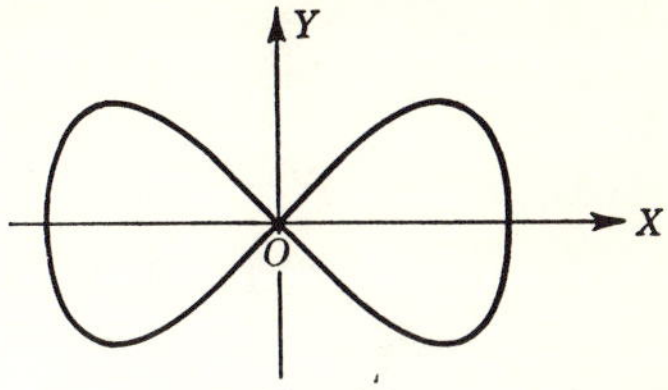

FIG. 98.

EXAMPLE 2. Discuss the graph of $y^2 = k^2(x - a)(x - b)(x - c)$.

Solution: For all values of a, b, and c, the curve has OX as an axis of symmetry, and infinite branches moving off to the right.

For $a < b < c$, the curve has a loop between a and b and an infinite branch to the right of c. Since for large x the expression for y has $kx^{\frac{3}{2}}$ as its leading term, y increases faster than x, and the curve must have a point of inflection to the right of c. See Fig. 99.

When $a < b$ but $b = c$, the loop joins the other branch as in Fig. 100. And at $x = b$, the tangents to $y^2 = k^2(x - a)(x - b)^2$ have slopes $\pm k\sqrt{b - a}$.

When $a = b$ but $b < c$, the curve is like the first case except that the loop shrinks to an isolated point.

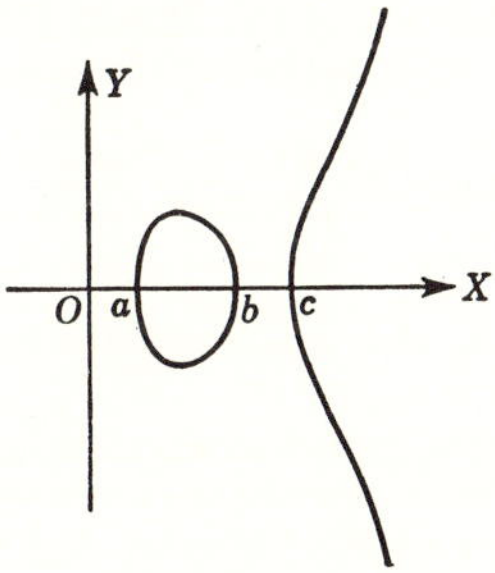

FIG. 99.

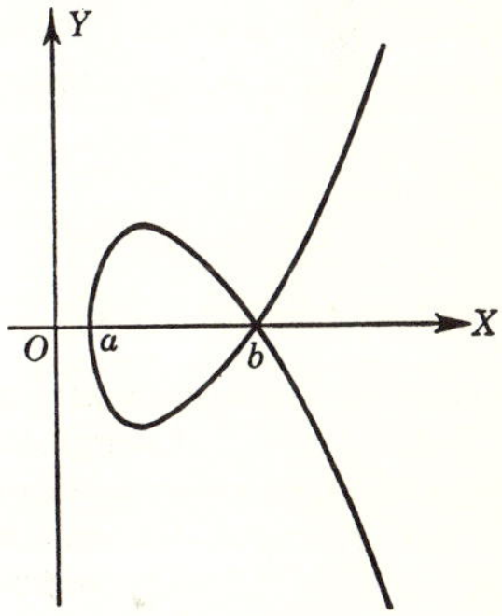

FIG. 100.

When $a = b = c$, the curve is shaped like $y^2 = x^3$ shown in Fig. 93 and discussed in Example 2 of Sec. 87. And $y^2 = k^2(x - a)^3$ has a cusp at $(a,0)$.

EXERCISE 47

Sketch the graph of each given equation.

1. $y^2 = 4(x + 2)^3$.
2. $y^2 = (x - 1)(x - 2)(x - 4)$.
3. $y^2 = x^3 - 4x$.
4. $y^2 = x^3 + 4x^2$.
5. $y^2 = x(x - 2)^2$.
6. $y^2 = x^3 - 4x^2$.
7. $y^2 = x^3 - 3x^2 + 2x$.
8. $y^2 = x^5$.
9. $y^2 = 4x^2 - x^4$.
10. $y^2 = x^4 - x^6$.
11. $xy^2 = 1 - x$.
12. $xy^2 = 1 + x$.
13. $y^2(1 - x) = x^3 + x^2$.
14. $x^2y^2 + 36 = 4y^2$.
15. $x^4 + 16y^4 = 16$.
16. $16x^4 - y^4 = 16$.
17. $y^2 = x^4 - 5x^2 + 1$.
18. $y^2 = x^5 + 5x^4$.
19. $x^{\frac{1}{2}} + y^{\frac{1}{2}} = 1$.
20. $x^{\frac{2}{3}} + y^{\frac{2}{3}} = 1$.

CHAPTER 7

TRIGONOMETRIC FUNCTIONS

The sine and the cosine are the two basic trigonometric functions. The remaining functions $\tan x$, $\cot x$, $\sec x$, and $\csc x$ may be defined in terms of $\sin x$ and $\cos x$.

There are also six inverse trigonometric functions. Each of these is the function inverse to the corresponding direct function. For example $y = \sin^{-1} x$ if $x = \sin y$.

This chapter begins with a review of the definitions and fundamental properties of trigonometric functions. We discuss radian measure of angles and derive some important limiting relations. We are then able to develop the special rules for differentiating the six direct functions and the six inverse functions.

Among other applications of the trigonometric functions we discuss angular velocity and simple harmonic motion.

R90. Angles. If two straight lines are drawn from a point, they form an *angle* having one of the lines as the initial side and the other as the terminal side. The point of intersection is the vertex. And the angle may be generated by rotating the initial line about the vertex until it coincides with the terminal line. The angle is positive if the rotation is counterclockwise. The angle is zero for the limiting case of intersecting lines when the terminal side coincides with the initial side. And the angle is negative if the rotation is clockwise. Thus in Fig. 101 angle AOB is positive while angle BOA is negative.

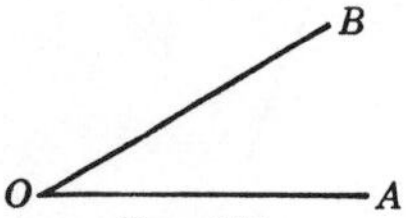

FIG. 101.

Two methods of measuring angles are in use. In *degree measure* the unit angle is $\frac{1}{360}$ of a complete revolution and is called a *degree*. Thus a right angle measures 90°, and each angle of an equilateral triangle measures 60°, and a straight angle measures 180°. The symbol ° is read "degree" or "degrees." For the usual choice of x and y axes, the angle from OX to OY, or angle XOY, is 90°.

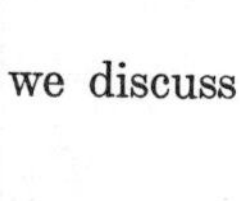

FIG. 102.

In *circular measure* the unit angle is the angle subtended at the center of a circle by an arc whose length is equal to the radius. This unit angle (Fig. 102) is called a *radian*. And the term *radian measure* is often used instead of *circular measure*. The circumference of a circle of radius r is $2\pi r$. Hence a complete revolution contains 2π radians. It follows that

$$\begin{aligned} 360° &= 2\pi \text{ radians}, \\ 180° &= \pi \text{ radians}, \end{aligned} \tag{1}$$

where, to four decimal places, $\pi = 3.1416$. Let a given angle contain $D°$ and R

radians. Then, since angles are proportional to their measures, we have

$$\frac{D}{180} = \frac{R}{\pi}. \tag{2}$$

Hence there are 57.296° in 1 radian, and

$$D = \frac{180}{\pi} R = 57.296R. \tag{3}$$

Also there is 0.017453 radian in 1°, and

$$R = \frac{\pi}{180} D = 0.017453D. \tag{4}$$

We sometimes mention radians to emphasize that circular measure is being used, but it should be understood that angles given numerically are in radians unless the degree sign is added. Thus, in this book an angle with measure $\pi/2$, 1.5708, 1.5708 radians, or 90° means a right angle. An angle with measure 2° means an angle of 0.0349 radian. And an angle with measure 2 means an angle of 2 radians or 114.6°.

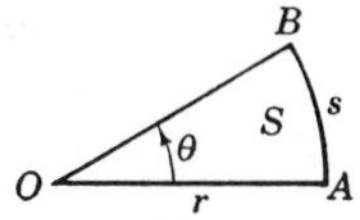

FIG. 103.

Consider an arc of length s on a circle of radius r (Fig. 103). Let the central angle be of measure θ. Then by a comparison with the whole circumference we find that

$$\frac{s}{2\pi r} = \frac{\theta}{2\pi}, \qquad s = r\theta, \qquad \theta = \frac{s}{r}. \tag{5}$$

Let the area of the sector OAB be S. Then by a comparison with the whole circle we find that

$$\frac{S}{\pi r^2} = \frac{s}{2\pi r}, \qquad S = \frac{1}{2} sr. \tag{6}$$

Thus the area of the sector S is the same as the area of a triangle whose base is equal to the arc length s and whose altitude is equal to the radius r. And from Eqs. (5) and (6) we may deduce that

$$S = \tfrac{1}{2}r^2\theta. \tag{7}$$

R91. The Sine and the Cosine. Let θ be any *number*. Then an angle of measure θ will contain θ radians or $\frac{180}{\pi}\theta°$. Construct such an angle with vertex at O, the origin of a rectangular coordinate system and initial side along the positive x axis. Choose any point $P = (x,y)$ distinct from O on the terminal side and denote the positive distance OP by r. Then the sine and cosine of θ may be *defined* by the relations

$$\sin\theta = \frac{y}{r}, \qquad \cos\theta = \frac{x}{r}. \tag{8}$$

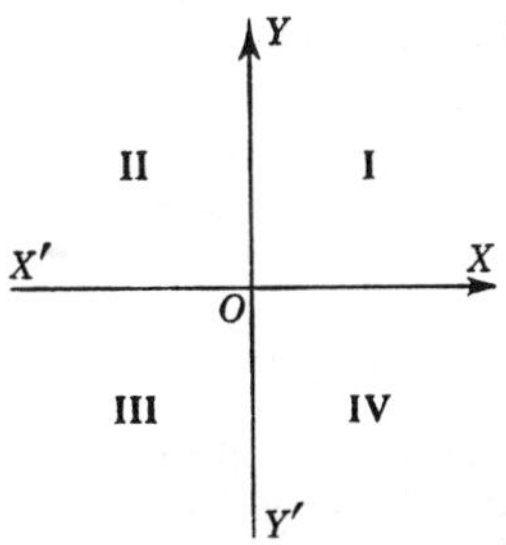

FIG. 104.

Two perpendicular coordinate axes $X'OX$ and $Y'OY$ divide the plane into four quadrants, numbered I, II, III, IV as in Fig. 104. If OX is taken along the initial line of an angle, the quadrant in which the terminal line lies is the quadrant of the angle.

Let $0 < \theta < \pi/2$ as in Fig. 105. Then x and y are positive. Hence in the first quadrant, $\sin\theta$ and $\cos\theta$ are positive.

Let $\pi/2 < \theta < \pi$ as in Fig. 106. Then x is negative and y is positive. Hence in the second quadrant $\sin\theta$ is positive but $\cos\theta$ is negative.

Let $\pi < \theta < 3\pi/2$ as in Fig. 107. Then x and y are negative. Hence in the third quadrant $\sin\theta$ is negative and $\cos\theta$ is negative.

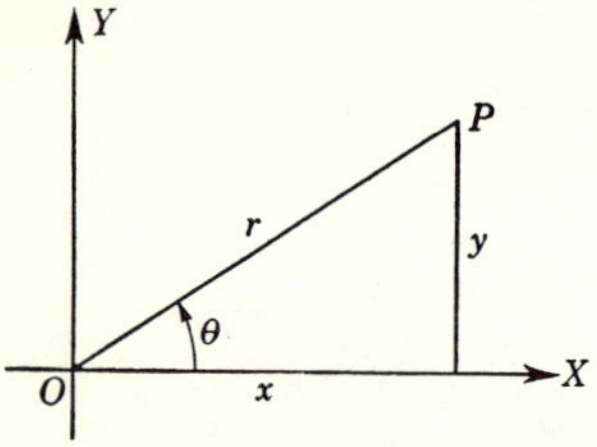

FIG. 105.

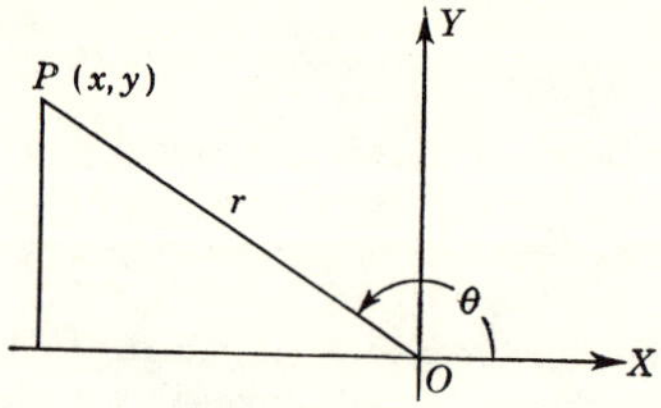

FIG. 106.

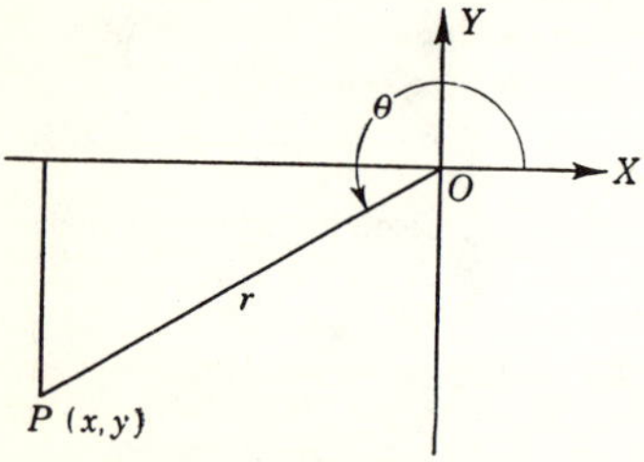

FIG. 107.

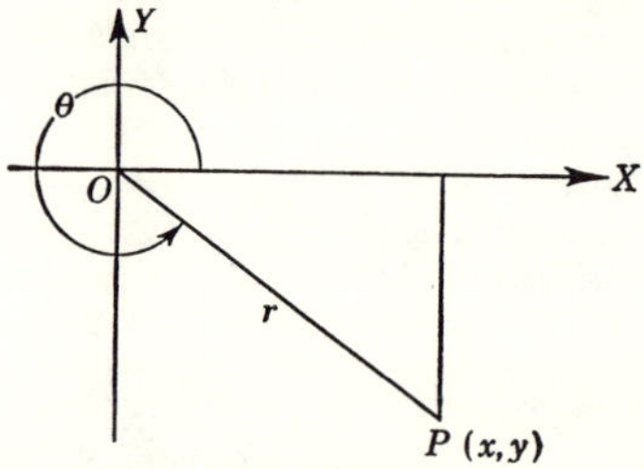

FIG. 108.

Let $3\pi/2 < \theta < 2\pi$ as in Fig. 108. Then x is positive and y is negative. Hence in the fourth quadrant $\sin\theta$ is negative and $\cos\theta$ is positive.

Since an increase or decrease of θ by 2π does not affect the position of the terminal side, we have

$$\sin(\theta \pm 2\pi) = \sin\theta \qquad \text{and} \qquad \cos(\theta \pm 2\pi) = \cos\theta. \tag{9}$$

We wrote the inequalities for the quadrants in terms of a θ between 0 and 2π. These inequalities apply equally well if all the values are changed by $2k\pi$, where k is any positive or negative integer. For example, if $2k\pi < \theta < 2k\pi + \pi/2$, θ is in the first quadrant as in Fig. 105. In particular we note that if $-\pi/2 < \theta < 0$, θ is in the fourth quadrant. And if $-\pi < \theta < -\pi/2$, θ is in the third quadrant.

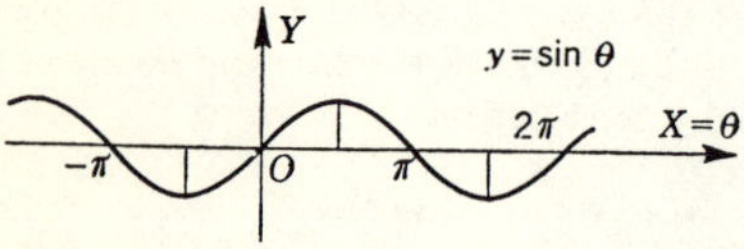

FIG. 109.

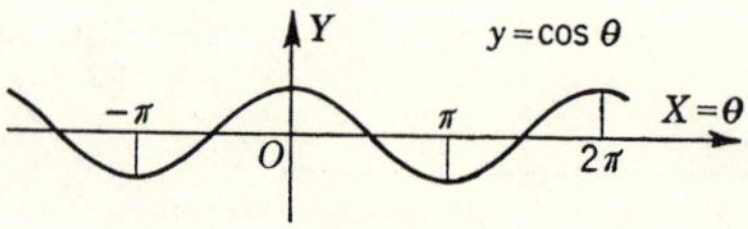

FIG. 110.

A portion of the graph of $\sin\theta$ is shown in Fig. 109. This illustrates the variation of $\sin\theta$ as θ varies over the four quadrants. The values repeat over each interval of length 2π. Similarly the graph of $\cos\theta$ is shown in Fig. 110.

Since $r^2 = x^2 + y^2$, it follows from the definition of Eq. (8) that

$$\sin^2\theta + \cos^2\theta = 1. \tag{10}$$

We recall the addition theorems proved in trigonometry. Compare Probs. 23 and 24 of Exercise 39. These are

$$\sin(A+B) = \sin A \cos B + \cos A \sin B, \tag{11}$$

$$\sin(A-B) = \sin A \cos B - \cos A \sin B, \tag{12}$$

$$\cos(A+B) = \cos A \cos B - \sin A \sin B, \tag{13}$$

$$\cos(A-B) = \cos A \cos B + \sin A \sin B. \tag{14}$$

Let k be any integer, positive or negative, or zero. Then $\sin k\pi = 0$ and $\cos k\pi = (-1)^k$. Hence it follows from Eqs. (11) and (13) with $B = k\pi$ that

$$\sin(A + k\pi) = (-1)^k \sin A, \qquad \cos(A + k\pi) = (-1)^k \cos A. \tag{15}$$

With k as just defined, $(2k+1)$ is an odd integer. And any odd integer may be written in this form. For any odd multiple of $\frac{\pi}{2}$ we have $\sin \frac{(2k+1)\pi}{2} = (-1)^k$, $\cos \frac{(2k+1)\pi}{2} = 0$. Hence it follows from Eqs. (11) and (13) with $B = \frac{(2k+1)\pi}{2}$ that

$$\sin\left[A + \frac{(2k+1)\pi}{2}\right] = (-1)^k \cos A,$$
$$\cos\left[A + \frac{(2k+1)\pi}{2}\right] = (-1)^{k+1} \sin A. \tag{16}$$

Consistent with the result of putting $A = 0$ in Eqs. (12) and (14), we may deduce directly from the definition that

$$\sin(-A) = -\sin A, \quad \cos(-A) = \cos A. \tag{17}$$

This, combined with Eq. (15), shows that

$$\sin(-A + k\pi) = (-1)^{k+1} \sin A, \qquad \cos(-A + k\pi) = (-1)^k \cos A. \tag{18}$$

And Eq. (17) combined with Eq. (16) shows that

$$\sin\left[-A + \frac{(2k+1)\pi}{2}\right] = (-1)^k \cos A,$$
$$\cos\left[-A + \frac{(2k+1)\pi}{2}\right] = (-1)^k \sin A. \tag{19}$$

Any of the Eqs. (15) through (19) are easily recalled by a reference to a figure, and use of the relationships $\pi/2$ radians equals 90° and π radians equals 180°.

EXAMPLE. Using tables, find sin 2 and cos 2.

Solution 1: Since $\pi = 3.1416$, $2 = \pi - 1.1416$. Hence from Eq. (18) with $k = 1$, we find that

$$\sin 2 = \sin 1.1416 \qquad \text{and} \qquad \cos 2 = -\cos 1.1416.$$

From tables with arguments in radian measure, we find by interpolation that $\sin 1.1416 = 0.9093$, $\cos 1.1416 = 0.4162$, so that

$$\sin 2 = 0.9093 \qquad \text{and} \qquad \cos 2 = -0.4162.$$

These are the required values.

Solution 2: Since $\pi = 3.1416$, $\pi/2 = 1.5708$ and $2 = \pi/2 + 0.4292$. Hence from Eq. (16) with $k = 0$, we find that

$$\sin 2 = \cos 0.4292 \qquad \text{and} \qquad \cos 2 = -\sin 0.4292.$$

From tables with arguments in radian measure, we find by interpolation that $\sin 0.4292 = 0.4162$ and $\cos 0.4292 = 0.9093$, so that

$$\sin 2 = 0.9093 \quad \text{and} \quad \cos 2 = -0.4162.$$

Solution 3: From Eq. (3), 2 radians $= 114.59°$. And $114.59° = 90° + 24.59° = 180° - 65.41°$. Hence from the analogues in degree measure of Eq. (16) with $k = 0$ and Eq. (18) with $k = 1$, we have

$$\sin 114.59° = \cos 24.59° = \sin 65.41°,$$
$$\cos 114.59° = -\sin 24.59° = -\cos 65.41°.$$

From tables with arguments in degrees, we find by interpolation that $\cos 24.59° = \sin 65.41° = 0.9093$, $\sin 24.59° = \cos 65.41° = 0.4161$ so that $\sin 2 = 0.9093$ and $\cos 2 = -0.4161$.

R92. The Tangent, Cotangent, Secant, and Cosecant. The trigonometric functions $\tan \theta$ and $\cot \theta$ are defined in terms of the sine and cosine by the quotients

$$\tan \theta = \frac{\sin \theta}{\cos \theta}, \qquad \cot \theta = \frac{\cos \theta}{\sin \theta}. \tag{20}$$

It follows that these functions are plus in the first and third quadrants. And they are minus in the second and fourth quadrants.

From Eq. (20) combined with Eqs. (15) through (19) we may deduce that

$$\tan (A + k\pi) = \tan A, \qquad \cot (A + k\pi) = \cot A, \tag{21}$$

$$\tan \left[A + \frac{(2k+1)\pi}{2}\right] = -\cot A, \qquad \cot \left[A + \frac{(2k+1)\pi}{2}\right] = -\tan A, \tag{22}$$

$$\tan (-A) = -\tan A, \qquad \cot (-A) = -\cot A, \tag{23}$$

$$\tan (-A + k\pi) = -\tan A, \qquad \cot (-A + k\pi) = -\cot A, \tag{24}$$

$$\tan \left[-A + \frac{(2k+1)\pi}{2}\right] = \cot A, \qquad \cot \left[-A + \frac{(2k+1)\pi}{2}\right] = \tan A. \tag{25}$$

A portion of the graph of $\tan \theta$ is shown in Fig. 111. The values repeat over each interval of length π. Similarly, the graph of $\cot \theta$ is shown in Fig. 112.

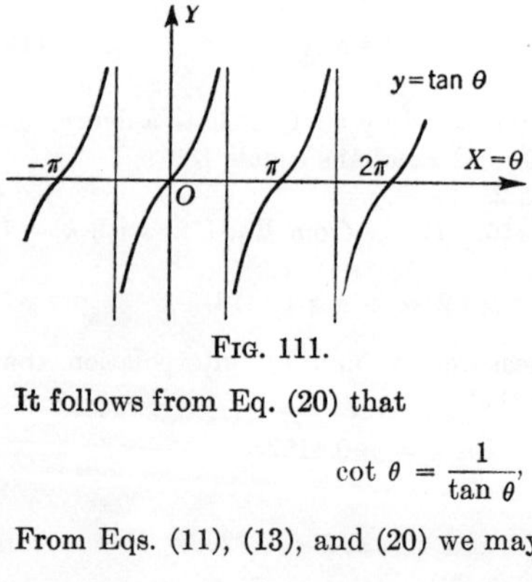

FIG. 111.

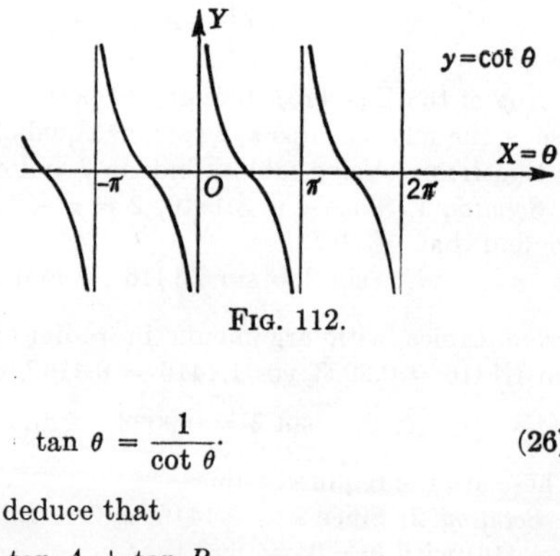

FIG. 112.

It follows from Eq. (20) that

$$\cot \theta = \frac{1}{\tan \theta}, \qquad \tan \theta = \frac{1}{\cot \theta}. \tag{26}$$

From Eqs. (11), (13), and (20) we may deduce that

$$\tan (A + B) = \frac{\tan A + \tan B}{1 - \tan A \tan B}. \tag{27}$$

And from Eqs. (12), (14), and (20) we may deduce that

$$\tan (A - B) = \frac{\tan A - \tan B}{1 + \tan A \tan B}. \tag{28}$$

The trigonometric functions sec θ and csc θ are defined in terms of the sine and cosine by the reciprocal relations

$$\sec \theta = \frac{1}{\cos \theta}, \qquad \csc \theta = \frac{1}{\sin \theta}. \tag{29}$$

These functions sec θ and csc θ satisfy relations similar to Eqs. (15) through (19) which may be obtained by taking the reciprocals of both sides of these equations. The graph of sec θ is shown in Fig. 113. And the graph of csc θ is shown in Fig. 114.

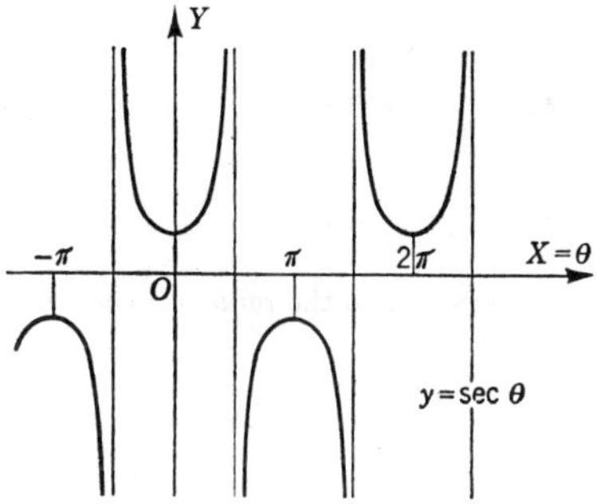

FIG. 113.

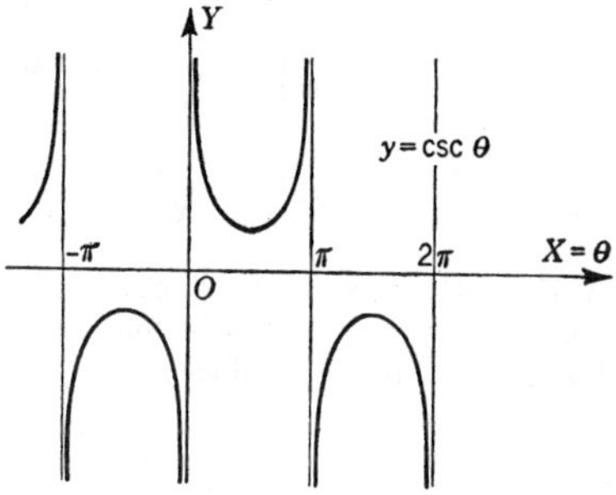

FIG. 114.

Let us divide both members of Eq. (10) by $\cos^2 \theta$. Then by using the first relation of Eqs. (20) and (29) we may deduce that

$$\tan^2 \theta + 1 = \sec^2 \theta. \tag{30}$$

Next divide both members of Eq. (10) by $\sin^2 \theta$. Then by using the second relation of Eqs. (20) and (29) we may deduce that

$$1 + \cot^2 \theta = \csc^2 \theta. \tag{31}$$

R93. Double Angles and Half Angles. Let $B = A$ in Eqs. (11), (13), and (27). This leads to the following double-angle formulas

$$\sin 2A = 2 \sin A \cos A, \tag{32}$$

$$\cos 2A = \cos^2 A - \sin^2 A, \tag{33}$$

$$\tan 2A = \frac{2 \tan A}{1 - \tan^2 A}. \tag{34}$$

It follows from Eq. (10) that

$$\sin^2 A + \cos^2 A = 1, \qquad \cos^2 A = 1 - \sin^2 A, \qquad \sin^2 A = 1 - \cos^2 A. \tag{35}$$

We may deduce from these relations and Eq. (33) that

$$\cos 2A = 1 - 2 \sin^2 A \quad \text{or} \quad 2 \sin^2 A = 1 - \cos 2A, \tag{36}$$

and

$$\cos 2A = 2 \cos^2 A - 1 \quad \text{or} \quad 2 \cos^2 A = 1 + \cos 2A. \tag{37}$$

To derive the half-angle formulas, let us put $2A = C$, $A = C/2$, in Eqs. (32), (36), and (37). This leads to

$$2 \sin \frac{C}{2} \cos \frac{C}{2} = \sin C, \tag{38}$$

$$2 \sin^2 \frac{C}{2} = 1 - \cos C, \tag{39}$$

$$2 \cos^2 \frac{C}{2} = 1 + \cos C. \tag{40}$$

It follows from Eqs. (39) and (40) that

$$\sin \frac{C}{2} = \sqrt{\frac{1 - \cos C}{2}}, \qquad \text{if } \sin \frac{C}{2} \text{ is positive,} \tag{41}$$

$$\cos \frac{C}{2} = \sqrt{\frac{1 + \cos C}{2}}, \qquad \text{if } \cos \frac{C}{2} \text{ is positive,} \tag{42}$$

$$\tan \frac{C}{2} = \sqrt{\frac{1 - \cos C}{1 + \cos C}}, \qquad \text{if } \tan \frac{C}{2} \text{ is positive.} \tag{43}$$

In Eqs. (41) through (43) a minus sign must be prefixed to the radical if the trigonometric function of $C/2$ to be found is negative.

By division, we may deduce from Eqs. (39) and (38) that

$$\tan \frac{C}{2} = \frac{1 - \cos C}{\sin C}. \tag{44}$$

And from Eqs. (40) and (38) we may deduce by division that

$$\tan \frac{C}{2} = \frac{\sin C}{1 + \cos C}. \tag{45}$$

R94. Triangles. Consider any plane triangle. As in Fig. 115 we use A, B, C to denote the vertices and a, b, c to denote the lengths of the opposite sides. We also use A to denote the interior angle at A, or $\angle BAC$, and similarly B and C for the interior angles CBA and ACB. Each of these angles is in the first or second quadrant, and

$$A + B + C = \pi, \tag{46}$$

since the sum of the three angles of a triangle is a straight angle or 180° in degree measure, and π radians in circular measure.

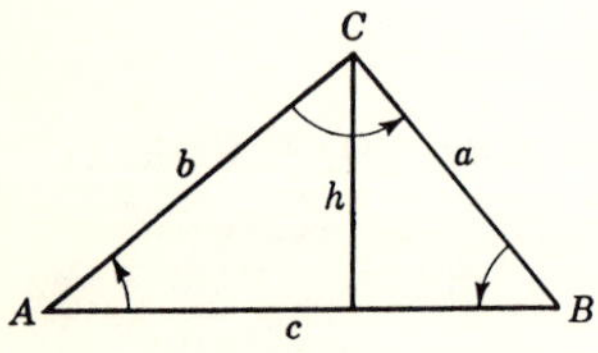

FIG. 115.

Draw the altitude h from C to side AB. Then, from the definition of the sine function, it follows that

$$\sin B = \frac{h}{a} \qquad \text{and} \qquad \sin A = \frac{h}{b}. \tag{47}$$

Thus $h = a \sin B$ and $h = b \sin A$, so that

$$a \sin B = b \sin A \qquad \text{and} \qquad \frac{a}{\sin A} = \frac{b}{\sin B}. \tag{48}$$

This is the *law of sines*.

Next take C as the origin (0,0), an x axis along CA, and the y axis perpendicular to it through C as in Fig. 116. Then

$$A = (b,0), \qquad B = (a \cos C, a \sin C). \tag{49}$$

And from the distance formula, Eq. (30) of Sec. 81, we have

$$\begin{aligned}\overline{AB}^2 &= (a \cos C - b)^2 + (a \sin C - 0)^2 \\ &= a^2(\cos^2 C + \sin^2 C) - 2ab \cos C + b^2.\end{aligned} \tag{50}$$

But $AB = c$, and the parenthesis is unity by Eq. (10). Hence

$$c^2 = a^2 + b^2 - 2ab \cos C. \tag{51}$$

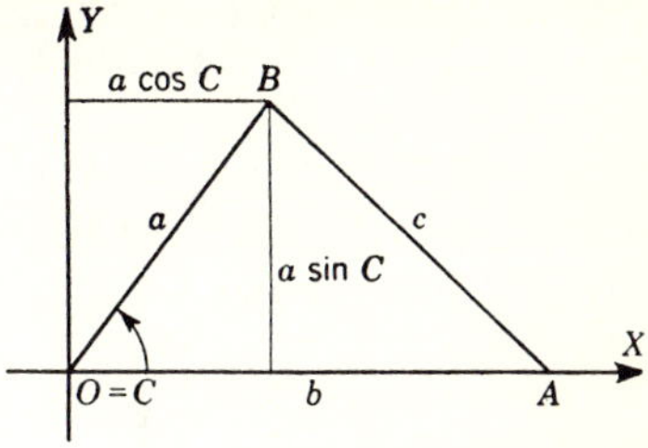

FIG. 116.

This is the *law of cosines.*

All the formulas for plane triangles in a form efficient for logarithmic computation are derived from Eqs. (46), (48), and (51). In this book, we shall use the law of cosines often, the law of sines and Eq. (46) occasionally, and have no need for the derived relations, adapted to computation.

EXERCISE 48

1. Show that $45° = \frac{\pi}{4}$ radians. And from an isosceles right triangle with sides 1, 1, $\sqrt{2}$ deduce that

$$\sin\frac{\pi}{4} = \frac{\sqrt{2}}{2}, \qquad \cos\frac{\pi}{4} = \frac{\sqrt{2}}{2}, \qquad \tan\frac{\pi}{4} = 1.$$

2. Show that $30° = \frac{\pi}{6}$ radians and that $60° = \frac{\pi}{3}$ radians. And from a right triangle with sides 1, $\sqrt{3}$, 2, deduce that

$$\sin\frac{\pi}{6} = \frac{1}{2}, \qquad \cos\frac{\pi}{6} = \frac{\sqrt{3}}{2}, \qquad \tan\frac{\pi}{6} = \frac{\sqrt{3}}{3},$$
$$\sin\frac{\pi}{3} = \frac{\sqrt{3}}{2}, \qquad \cos\frac{\pi}{3} = \frac{1}{2}, \qquad \tan\frac{\pi}{3} = \sqrt{3}.$$

3. From the result of Prob. 1 deduce that

$$\sin\frac{5\pi}{4} = -\frac{\sqrt{2}}{2}, \qquad \tan\frac{3\pi}{4} = -1, \qquad \tan\frac{5\pi}{4} = 1, \qquad \cos\frac{7\pi}{4} = \frac{\sqrt{2}}{2}.$$

4. From the result of Prob. 2 deduce that

$$\sin\frac{7\pi}{6} = -\frac{1}{2}, \qquad \sin\left(-\frac{\pi}{3}\right) = -\frac{\sqrt{3}}{2}, \qquad \tan\frac{7\pi}{6} = \frac{\sqrt{3}}{3}, \qquad \cos\left(-\frac{\pi}{3}\right) = \frac{1}{2}.$$

Use the tables to evaluate

5. sin 3.5.
6. cos 2.4
7. tan (−4).

8. If $\tan A = m$ and $\cos A$ is positive, show that

$$\cos A = \frac{1}{\sqrt{1 + m^2}}, \qquad \sin A = \frac{m}{\sqrt{1 + m^2}}.$$

9. From Eqs. (11) and (12) deduce that

$$\sin A \cos B = \tfrac{1}{2}[\sin(A + B) + \sin(A - B)],$$
$$\cos A \sin B = \tfrac{1}{2}[\sin(A + B) - \sin(A - B)].$$

10. From Eqs. (13) and (14) deduce that

$$\cos A \cos B = \tfrac{1}{2}[\cos (A + B) + \cos (A - B)],$$
$$\sin A \sin B = \tfrac{1}{2}[\cos (A - B) - \cos (A + B)].$$

If $A = \dfrac{x + y}{2}$ and $B = \dfrac{x - y}{2}$, then $A + B = x$ and $A - B = y$. From these relations and Probs. 9 and 10, deduce that

11. $\sin x + \sin y = 2 \sin \dfrac{x + y}{2} \cos \dfrac{x - y}{2}$.

12. $\sin x - \sin y = 2 \cos \dfrac{x + y}{2} \sin \dfrac{x - y}{2}$.

13. $\cos x + \cos y = 2 \cos \dfrac{x + y}{2} \cos \dfrac{x - y}{2}$.

14. $\cos x - \cos y = -2 \sin \dfrac{x + y}{2} \sin \dfrac{x - y}{2}$.

15. Let θ, x, y, r have the meanings given them in Sec. 91. Show that $\tan \theta = y/x$, $\cot \theta = x/y$, $\sec \theta = r/x$, $\csc \theta = r/y$.

Show that, in the notation described in Secs. 12 and 15,

16. $\lim\limits_{\theta \to \frac{\pi}{2}-} \tan \theta = +\infty$, $\lim\limits_{\theta \to \frac{\pi}{2}+} \tan \theta = -\infty$.

17. $\lim\limits_{\theta \to \frac{\pi}{2}-} \sec \theta = +\infty$, $\lim\limits_{\theta \to \frac{\pi}{2}+} \sec \theta = -\infty$.

18. From Probs. 16 and 17 deduce that

$$\lim_{\theta \to \frac{\pi}{2}} \tan \theta = \infty, \qquad \lim_{\theta \to \frac{\pi}{2}} \sec \theta = \infty.$$

From Eqs. (11) and (13) with $B = 2A$, combined with Eqs. (32), (33), and (35), deduce that

19. $\sin 3A = 3 \sin A - 4 \sin^3 A$.

20. $\cos 3A = 4 \cos^3 A - 3 \cos A$.

***95. The Limit of $\dfrac{\sin \theta}{\theta}$ as $\theta \to 0$.** In differentiating the sine function, we shall find it useful to know that

$$\lim_{\theta \to 0} \frac{\sin \theta}{\theta} = 1. \tag{52}$$

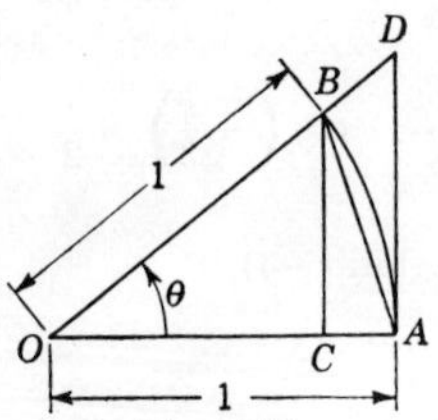

FIG. 117.

We may prove this result geometrically as follows. In Fig. 117, AB is an arc of a circle whose radius is 1, so that

$$OA = 1, \qquad OB = 1. \tag{53}$$

Let the length of arc $AB = \theta$. Then since the radius of the circle is 1, as in Eq. (5), the *radian* measure of the angle AOB is the arc AB divided by the radius, or $\theta/1 = \theta$. We assume that θ, the measure of angle AOB, is in the interval $0 < \theta < \pi/2$. Then

$$\sin \theta = \frac{CB}{OB} = CB, \qquad \tan \theta = \frac{AD}{OA} = AD. \tag{54}$$

We now consider three areas. First the area of triangle OAB,

$$K_1 = \tfrac{1}{2}OA \cdot CB = \tfrac{1}{2} \cdot 1 \cdot \sin\theta \qquad \text{and} \qquad 2K_1 = \sin\theta. \tag{55}$$

Next the area of sector OAB, as in Eq. (6), is equivalent to a triangle of altitude OA and base AB. Thus the sectorial area

$$K_2 = \tfrac{1}{2}(\text{arc } AB)OA = \tfrac{1}{2}\theta \cdot 1 \qquad \text{and} \qquad 2K_2 = \theta. \tag{56}$$

Finally, the area of triangle OAD,

$$K_3 = \tfrac{1}{2}OA \cdot AD = \tfrac{1}{2} \cdot 1 \cdot \tan\theta \qquad \text{and} \qquad 2K_3 = \tan\theta. \tag{57}$$

Since the area K_1 is a part of K_2, and K_2 is a part of K_3, we have

$$K_1 < K_2 < K_3 \qquad \text{or} \qquad 2K_1 < 2K_2 < 2K_3. \tag{58}$$

From this and the last relation of Eqs. (55), (56), and (57) we may deduce that

$$\sin\theta < \theta < \tan\theta. \tag{59}$$

This proves that in the first quadrant the radian measure of an angle exceeds its sine and is exceeded by its tangent.

For $0 < \theta < \pi/2$, $\sin\theta > 0$. Hence the inequalities of Eq. (59) will be preserved if we divide each member by $\sin\theta$. Since $\tan\theta = \sin\theta/\cos\theta$, the result of dividing Eq. (59) by $\sin\theta$ is

$$1 < \frac{\theta}{\sin\theta} < \frac{1}{\cos\theta}. \tag{60}$$

We may replace each member of this inequality by its reciprocal if we reverse the inequality signs. Thus Eq. (60) implies

$$1 > \frac{\sin\theta}{\theta} > \cos\theta \qquad \text{or} \qquad \cos\theta < \frac{\sin\theta}{\theta} < 1. \tag{61}$$

So far we have assumed that $0 < \theta < \pi/2$. However, Eq. (61) also holds for θ in the interval $-\pi/2 < \theta < 0$. For

$$\cos(-\theta) = \cos\theta \qquad \text{and} \qquad \frac{\sin(-\theta)}{-\theta} = \frac{-\sin\theta}{-\theta} = \frac{\sin\theta}{\theta}. \tag{62}$$

Hence the relations of Eq. (61) for a negative angle are equivalent to the same relations for the corresponding positive angle.

Now let θ approach zero. Then $\cos\theta$ approaches $\cos 0 = 1$, since the cosine is a continuous function of θ. Since $\sin\theta/\theta$ lies between a variable approaching 1, and one by Eq. (61), it must approach 1 also and

$$\lim_{\theta\to 0} \frac{\sin\theta}{\theta} = 1. \tag{63}$$

This is the result stated in Eq. (52).

Let $\theta = au$, where a is any constant not equal to zero. Then when u tends to zero through positive or negative values, θ tends to zero through values distinct from zero. Hence we may conclude from Eq. (63) that

$$\lim_{u\to 0} \frac{\sin au}{au} = 1 \qquad \text{and} \qquad \lim_{u\to 0} \frac{\sin au}{u} = a. \tag{64}$$

We will next show that

$$\lim_{\theta\to 0} \frac{1 - \cos \theta}{\theta} = 0. \tag{65}$$

To do this, we use Eq. (39) with $C = \theta$ to deduce that

$$\frac{1 - \cos \theta}{\theta} = \frac{2 \sin^2 \theta/2}{\theta} = \left(\frac{\sin \frac{1}{2}\theta}{\theta}\right)^2 2\theta. \tag{66}$$

By the second relation of Eq. (64) with $a = \frac{1}{2}$ and $u = \theta$, the limit of the parenthesis as θ approaches zero is $\frac{1}{2}$. Hence the right member of Eq. (66) approaches $(\frac{1}{2})^2 \cdot 2 \cdot 0 = 0$ when θ approaches zero. This proves Eq. (65).

The function $\sin x/x$ is defined for all values except $x = 0$. If we set $y = \sin x/x$ for $x \neq 0$, and $y = 1$ for $x = 0$, it follows from Eq. (52) that y is a continuous function of x for all values of x. Its graph is shown in Fig. 118.

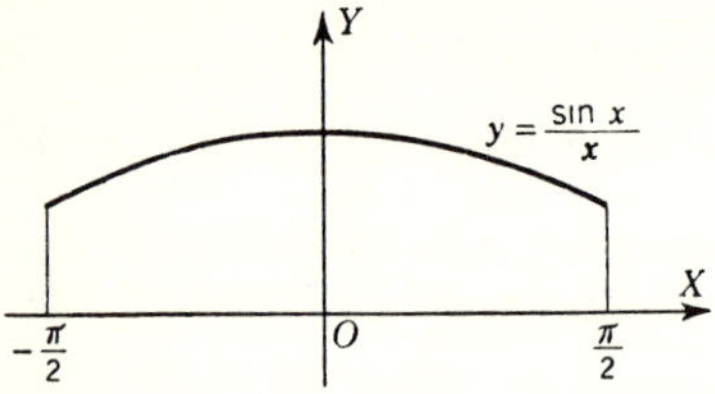

FIG. 118.

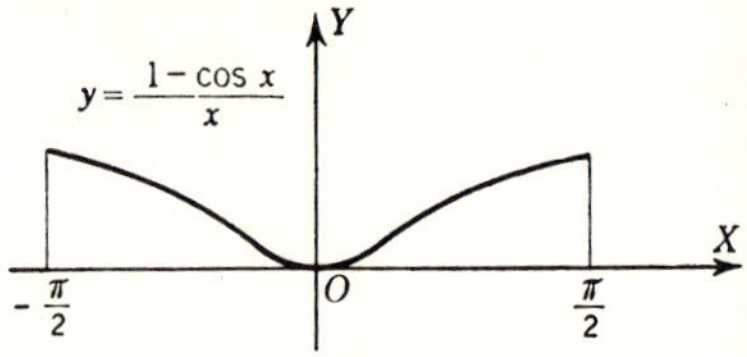

FIG. 119.

Similarly if we set $y = \dfrac{1 - \cos x}{x}$ for $x \neq 0$ and $y = 0$ for $x = 0$, it follows from Eq. (65) that y is a continuous function of x for all values of x. Its graph is shown in Fig. 119.

It is interesting to note how the numerical values conform to the limiting results of Eq. (52). Consider the values in the accompanying table.

Angle in degrees, $D°$	20°	10°	5°	4°	1°
Angle in radians, R	0.3491	0.1745	0.0873	0.0698	0.0175
Sine, $\sin R = \sin D°$	0.3420	0.1736	0.0872	0.0698	0.0175
$\sin R/R$	0.980	0.995	0.999	1.000	1.000
$\sin D°/D$	0.0171	0.0174	0.0174	0.0175	0.0175

The values of D were selected. Then the values of R and $\sin R = \sin D°$ were taken from four-place tables. And the values of $\sin R/R$ and $\sin D°/D$ were computed from the other values. The values of $\sin R/R$ are seen to be consistent with the fact that the limit of this ratio is 1 when R approaches zero. From Eqs. (3) and (4) we have

$$\frac{\sin D°}{D} = \frac{\sin R}{180R/\pi} = \frac{\pi}{180}\frac{\sin R}{R} = 0.017453 \frac{\sin R}{R}. \tag{67}$$

Hence when D approaches zero, the limit of $\sin D°/D$ to three significant figures is 0.0175 as suggested by the numerical values. When an angle approaches zero, the limiting value of the sine of the angle divided by the angle in degree measure is $\pi/180$. But the limiting value of the sine of an angle divided by the angle in radian measure is 1 when the angle approaches zero. Thus the use of radian measure avoids the appearance of a factor $\pi/180$ in a number of differentiation and integration formulas.

Because of this fact, radian measure is used almost exclusively in theoretical work involving trigonometric functions and calculus.

EXERCISE 49

Evaluate each of the following limits.

1. $\lim_{x\to 0} \frac{\sin 2x}{x}$.
2. $\lim_{x\to 0} \frac{\tan x}{x}$.
3. $\lim_{x\to 0} \frac{x}{\sin 4x}$.
4. $\lim_{x\to 0} \frac{\tan x}{\sin x}$.
5. $\lim_{x\to 0} \frac{\sin 6x}{\sin 3x}$.
6. $\lim_{x\to 0} \frac{\tan 3x}{\tan 4x}$.

Verify that, if $a \neq 0$,

7. $\lim_{x\to 0} \frac{\tan ax}{x} = \lim_{x\to 0} \frac{\sin ax}{ax} \frac{a}{\cos ax} = a$.
8. $\lim_{x\to 0} \frac{\sin bx}{\sin ax} = \lim_{x\to 0} \frac{\sin bx/x}{\sin ax/x} = \frac{b}{a}$.
9. $\lim_{x\to 0} \frac{\tan bx}{\tan ax} = \lim_{x\to 0} \frac{\tan bx/x}{\tan ax/x} = \frac{b}{a}$.
10. From Eq. (66) deduce that $\lim_{\theta\to 0} \frac{1 - \cos \theta}{\theta^2} = \frac{1}{2}$.

Use Eq. (65) and the result of Prob. 10 to evaluate each of the following limits.

11. $\lim_{x\to 0} \frac{1 - \cos 3x}{5x}$.
12. $\lim_{x\to 0} \frac{1 - \cos x}{\sin x}$.
13. $\lim_{x\to 0} \frac{1 - \cos x}{x \sin x}$.
14. $\lim_{x\to 0} \frac{1 - \cos x}{\tan x}$.
15. $\lim_{x\to 0} \frac{1 - \cos x}{\tan^2 x}$.
16. $\lim_{x\to 0} \frac{\tan x - \sin x}{x^3}$.

96. The Derivative of sin *u*. Let $y = \sin u$. Considering u as the independent variable, we may find dy/du by the procedure of Sec. 29 as follows. Give u an increment Δu. And let Δy be the corresponding increment in y. Then

$$y + \Delta y = \sin (u + \Delta u). \tag{68}$$

From this and $y = \sin u$, we find by subtraction that

$$\Delta y = \sin (u + \Delta u) - \sin u. \tag{69}$$

But from Eq. (11) with $A = u$ and $B = \Delta u$, we have

$$\sin (u + \Delta u) = \sin u \cos \Delta u + \cos u \sin \Delta u. \tag{70}$$

It follows from Eqs. (69) and (70) that

$$\begin{aligned} \Delta y &= \sin u \cos \Delta u + \cos u \sin \Delta u - \sin u \\ &= \cos u \sin \Delta u - \sin u(1 - \cos \Delta u). \end{aligned} \tag{71}$$

We may divide both sides by Δu to obtain

$$\frac{\Delta y}{\Delta u} = \cos u \frac{\sin \Delta u}{\Delta u} - \sin u \frac{1 - \cos \Delta u}{\Delta u}. \tag{72}$$

Now let Δu approach zero. Then

$$\lim_{\Delta u \to 0} \frac{\Delta y}{\Delta u} = \cos u \lim_{\Delta u \to 0} \frac{\sin \Delta u}{\Delta u} - \sin u \lim_{\Delta u \to 0} \frac{1 - \cos \Delta u}{\Delta u}. \tag{73}$$

The left number is dy/du by the definition of a derivative. And from Eqs. (52) and (65) with $\theta = \Delta u$, we have

$$\lim_{\Delta u \to 0} \frac{\sin \Delta u}{\Delta u} = 1, \qquad \lim_{\Delta u \to 0} \frac{1 - \cos \Delta u}{\Delta u} = 0. \tag{74}$$

Hence we have

$$\frac{dy}{du} = \cos u\ (1) - \sin u\ (0) = \cos u. \tag{75}$$

Next let $y = \sin u$ where u is any differentiable function of x, and consider x as the independent variable. Then by the rule for composite functions, Eq. (20) of Sec. 53, we have

$$\frac{dy}{du} = \frac{dy}{du}\frac{du}{dy}. \tag{76}$$

Replacing dy/du by $\cos u$, the value found in Eq. (75), we have

$$\frac{dy}{dx} = \cos u \frac{du}{dx}. \tag{77}$$

This proves that

$$\frac{d\ (\sin u)}{dx} = \cos u \frac{du}{dx}. \tag{78}$$

97. The Derivative of cos *u*. We wish to find the derivative of $y = \cos u$. We first put $k = 0$ and $A = u$ in Eq. (19) to deduce that

$$\sin\left(\frac{\pi}{2} - u\right) = \cos u \tag{79}$$

and

$$\cos\left(\frac{\pi}{2} - u\right) = \sin u. \tag{80}$$

It follows from Eq. (79) that

$$y = \cos u = \sin\left(\frac{\pi}{2} - u\right). \tag{81}$$

Now apply Eq. (78) with $\left(\frac{\pi}{2} - u\right)$ in place of u. The result is

$$\begin{aligned}\frac{dy}{dx} &= \frac{d}{dx}\sin\left(\frac{\pi}{2} - u\right) = \cos\left(\frac{\pi}{2} - u\right)\frac{d}{dx}\left(\frac{\pi}{2} - u\right)\\ &= \cos\left(\frac{\pi}{2} - u\right)\left(-\frac{du}{dx}\right)\\ &= -\sin u\,\frac{du}{dx},\end{aligned} \tag{82}$$

by Eq. (80). This proves that

$$\frac{d(\cos u)}{dx} = -\sin u\,\frac{du}{dx}. \tag{83}$$

EXAMPLE 1. Find dy/dx if $y = 5\sin(3x + 4)$.
Solution: By Eq. (78) with $u = 3x + 4$, we find

$$\begin{aligned}\frac{dy}{dx} &= 5\cos(3x + 4)\frac{d}{dx}(3x + 4) = 5\cos(3x + 4)3\\ &= 15\cos(3x + 4).\end{aligned}$$

This is the required derivative.

EXAMPLE 2. Find dy/dx if $y = \sqrt{\cos x^2}$.
Solution: By the power rule and Eq. (83) with $u = x^2$, we find

$$\begin{aligned}\frac{dy}{dx} &= \frac{d}{dx}(\cos x^2)^{\frac{1}{2}} = \frac{1}{2}(\cos x^2)^{-\frac{1}{2}}\frac{d}{dx}(\cos x^2)\\ &= \frac{1}{2}(\cos x^2)^{-\frac{1}{2}}(-\sin x^2)\frac{d}{dx}x^2\\ &= \tfrac{1}{2}(\cos x^2)^{-\frac{1}{2}}(-\sin x^2)\,2x\\ &= -\frac{x\sin x^2}{\sqrt{\cos x^2}}.\end{aligned}$$

This is the required derivative.

EXERCISE 50

Find dy/dx for each of the following given functions.

1. $y = \sin(4 + 2x)$.
2. $y = \cos(5 - x)$.
3. $y = 2\sin 4x$.
4. $y = 4\cos 2x$.
5. $y = 5\sin(2 - 3x)$.
6. $y = 4\cos(2 + 3x)$.
7. $y = \sin x\cos 2x$.
8. $y = \frac{1}{3}\sin^3 x$.
9. $y = \frac{1}{6}\cos^2 3x$.
10. $y = x - \sin x\cos x$.
11. $y = \frac{\sin x}{x}$.
12. $y = \sqrt{\sin 4x}$.
13. $y = \sin(x^2 - 3x)$.
14. $y = x\sin x$.
15. $x = 2\sin 3y$.
16. $x = 4\cos y$.
17. $y + x = \sin(y - x)$.
18. $y - x = \cos(2y + 3x)$.
19. $y = \sin(x + 2)\cos(x - 2)$.
20. $y = \cos(x + 3)\cos(x - 3)$.
21. Prove Eq. (83) by a direct procedure like that used to prove Eq. (78).

22. If t is the time and $x = \cos t$, $y = \sin t$, the point $P = (x,y)$ is moving around a unit circle with unit speed in the counterclockwise direction. Hence its velocity is a unit vector in the direction along the tangent with slope angle $t + (\pi/2)$. From these facts and the results of Sec. 63, deduce that

$$\frac{dx}{dt} = v_x = \cos\left(t + \frac{\pi}{2}\right) = -\sin t, \qquad \frac{dy}{dt} = v_y = \sin\left(t + \frac{\pi}{2}\right) = \cos t.$$

23. Check the results of Prob. 22 by using Eqs. (78) and (83).
24. From the result of Prob. 23 derive the formulas for the nth derivatives,

$$\frac{d^n(\cos t)}{dt^n} = \cos\left(t + \frac{n\pi}{2}\right), \qquad \frac{d^n(\sin t)}{dt^n} = \sin\left(t + \frac{n\pi}{2}\right).$$

98. The Derivatives of tan u, cot u, sec u, and csc u. Let $u = \tan u$. Then by Eq. (20) we have

$$y = \tan u = \frac{\sin u}{\cos u}. \tag{84}$$

And by the quotient rule, Eq. (4) of Sec. 51, we have

$$\frac{dy}{dx} = \frac{\cos u \dfrac{d}{dx}(\sin u) - \sin u \dfrac{d}{dx}(\cos u)}{\cos^2 u} \tag{85}$$

It follows from Eqs. (78) and (83) that

$$\begin{aligned} \frac{dy}{dx} &= \frac{\cos u(\cos u\, du/dx) - \sin u(-\sin u\, du/dx)}{\cos^2 u} \\ &= \frac{(\cos^2 u + \sin^2 u)du/dx}{\cos^2 u} \\ &= \frac{1}{\cos^2 u}\frac{du}{dx} = \sec^2 u \frac{du}{dx}, \end{aligned} \tag{86}$$

by Eqs. (10) and (29). This proves that

$$\frac{d(\tan u)}{dx} = \sec^2 u \frac{du}{dx}. \tag{87}$$

Next let $y = \cot u$. We first note that from Eq. (25) with $k = 0$,

$$\tan\left(\frac{\pi}{2} - u\right) = \cot u. \tag{88}$$

By taking reciprocals in Eq. (80) and using Eq. (29) we find that

$$\sec\left(\frac{\pi}{2} - u\right) = \csc u. \tag{89}$$

It follows from Eq. (88) that

$$y = \cot u = \tan\left(\frac{\pi}{2} - u\right). \tag{90}$$

Now apply Eq. (87) with $(\pi/2 - u)$ in place of u. The result is

$$\begin{aligned} \frac{dy}{dx} &= \frac{d}{dx}\tan\left(\frac{\pi}{2} - u\right) = \sec^2\left(\frac{\pi}{2} - u\right)\frac{d}{dx}\left(\frac{\pi}{2} - u\right) \\ &= \sec^2\left(\frac{\pi}{2} - u\right)\left(-\frac{du}{dx}\right) \\ &= -\csc^2 u\,\frac{du}{dx}, \end{aligned} \tag{91}$$

by Eq. (89). This proves that

$$\frac{d(\cot u)}{dx} = -\csc^2 u\,\frac{du}{dx}. \tag{92}$$

Again, consider $y = \csc u$. By Eq. (29) we have

$$y = \sec u = \frac{1}{\cos u} = (\cos u)^{-1}. \tag{93}$$

By the power rule, Eq. (16) of Sec. 52, we have

$$\begin{aligned} \frac{dy}{dx} &= -(\cos u)^{-2}\frac{d}{dx}(\cos u) \\ &= -(\cos u)^{-2}\left(-\sin u\,\frac{du}{dx}\right), \end{aligned} \tag{94}$$

by Eq. (83). We may deduce from this and Eqs. (20) and (29) that

$$\frac{dy}{dx} = \frac{\sin u}{\cos u}\,\frac{1}{\cos u}\,\frac{du}{dx} = \tan u \sec u\,\frac{du}{dx}. \tag{95}$$

This proves that

$$\frac{d(\sec u)}{dx} = \tan u \sec u\,\frac{du}{dx}. \tag{96}$$

Finally, let $y = \csc u$. Then from Eq. (89), we have

$$y = \csc u = \sec\left(\frac{\pi}{2} - u\right). \tag{97}$$

Now apply Eq. (96) with $(\pi/2 - u)$ in place of u. The result is

$$\begin{aligned} \frac{dy}{dx} &= \tan\left(\frac{\pi}{2} - u\right)\sec\left(\frac{\pi}{2} - u\right)\frac{d}{dx}\left(\frac{\pi}{2} - u\right) \\ &= \tan\left(\frac{\pi}{2} - u\right)\sec\left(\frac{\pi}{2} - u\right)\left(-\frac{du}{dx}\right) \\ &= -\cot u \csc u\,\frac{du}{dx}, \end{aligned} \tag{98}$$

by Eqs. (88) and (89). This proves that

$$\frac{d(\csc u)}{du} = -\cot u \csc u \frac{du}{dx}. \tag{99}$$

99. Differentiation Formulas for the Six Direct Functions. The student should memorize the six differentiation formulas of Eqs. (78), (83), (87), (92), (96), and (99), namely,

$$\begin{aligned} \frac{d(\sin u)}{dx} &= \cos u \frac{du}{dx}, & \frac{d(\cos u)}{dx} &= -\sin u \frac{du}{dx}, \\ \frac{d(\tan u)}{dx} &= \sec^2 u \frac{du}{dx}, & \frac{d(\cot u)}{dx} &= -\csc^2 u \frac{du}{dx}, \\ \frac{d(\sec u)}{dx} &= \tan u \sec u \frac{du}{dx}, & \frac{d(\csc u)}{dx} &= -\cot u \csc u \frac{du}{dx}. \end{aligned} \tag{100}$$

It is helpful to the memory to note the following facts about these six formulas. Each formula contains the factor du/dx because of the rule for composite functions. For each of the three functions sine, tangent, and secant the sign is plus. While for each of the cofunctions cosine, cotangent, and cosecant the sign is minus. Let us regard sine and cosine, tangent and cotangent, secant and cosecant as three pairs of complementary functions so that, for instance, the complement of the cotangent is the tangent. Then the six formulas consist of three complementary pairs. Either formula of each pair may be obtained from the other by one change of sign, and the replacing of each function by its complement.

EXAMPLE 1. Find dy/dx if $y = 2 \tan^2 x + 3 \sec^2 x$.

Solution: By the power rule and Eq. (100) we have

$$\begin{aligned} \frac{dy}{dx} &= 2 \cdot 2 \tan x \frac{d}{dx} (\tan x) + 3 \cdot 2 \sec x \frac{d}{dx} (\sec x) \\ &= 4 \tan x (\sec^2 x) + 6 \sec x (\sec x \tan x) \\ &= 10 \tan x \sec^2 x. \end{aligned}$$

This is the required derivative.

EXAMPLE 2. Find dy/dx if $y = \cot 2x \csc 2x$.

Solution: By the product rule and Eq. (100) we have

$$\begin{aligned} \frac{dy}{dx} &= \cot 2x \frac{d}{dx} (\csc 2x) + \csc 2x \frac{d}{dx} (\cot 2x) \\ &= \cot 2x \left[-\cot 2x \csc 2x \frac{d(2x)}{dx} \right] + \csc 2x \left[-\csc^2 2x \frac{d(2x)}{dx} \right] \\ &= -2 (\cot^2 2x + \csc^2 2x) \csc 2x. \end{aligned}$$

This is one form of the required derivative.

We may obtain a second form by deducing from Eq. (31) with $\theta = 2x$ that $\cot^2 2x = \csc^2 2x - 1$ so that

$$\frac{dy}{dx} = -2(2 \csc^2 2x - 1) \csc 2x = 2 \csc 2x - 4 \csc^3 2x.$$

EXERCISE 51

Find dy/dx for each of the following given functions.

1. $y = 6 \tan 5x$.
2. $y = 4 \cot 3x$.
3. $y = 3 \sec 7x$.
4. $y = 2 \csc 4x$.
5. $y = 4 \tan (3x + 5)$.
6. $y = 3 \cot (5 - 3x)$.
7. $y = 2 \sec (3 - 4x)$.
8. $y = 4 \csc (3 + 2x)$.
9. $y = \frac{1}{5} \sec 5x$.
10. $y = \frac{1}{3} \cot 3x$.
11. $y = 7 \tan \frac{x}{7}$.
12. $y = 2 \csc \frac{x}{2}$.
13. $y = \sqrt{\tan x}$.
14. $y = \frac{1}{\sqrt{\sec x}}$.
15. $y = \sqrt[3]{\tan 3x}$.
16. $y = \frac{1}{5} \sec^5 2x$.
17. $y = \tan^3 x - 3 \tan x + 3x$.
18. $y = \cot^3 x + 3 \cot x$.
19. $y = \tan x - x$.
20. $y = (\sec x + \tan x)^3$.
21. $\tan y = \sin 2x$.
22. $\cos y = \tan 3x$.

100. Graphs of Trigonometric Functions. For any given function $f(x)$, we may plot points on the graph of $y = f(x)$ by assigning values to x and computing the corresponding values of y, as in Sec. 6. Trigonometric functions which appear must be interpreted as applying to angles in radian measure. And their values may be found as indicated in Sec. 91. When constructing graphs, it is helpful to note intersections with coordinate axes, maxima, minima, and points of inflection as mentioned in Secs. 48 and 89. For graphs of trigonometric functions, such points often have coordinates involving π, and we recall that $\pi = 3.1416$, or 3.14 for purposes of plotting. The practical procedure of plotting is indicated in the examples which follow. When plotting $y = f(x)$, where $f(x)$ is the sum of simple terms, it may be convenient to use the method of *composition of ordinates* explained in Example 4.

EXAMPLE 1. Sketch the graph of $y = 3 \sin 2x$.

Solution: To fix the values of x which make $y = 0$, we observe that the sine is zero when the angle in radians is $0, \pi, 2\pi, 3\pi, -\pi, -2\pi$, or in general $k\pi$, where k is any positive or negative integer, or zero. Thus $y = 3 \sin 2x$ will be zero when $\sin 2x = 0$, or $2x = k\pi$. That is, if $x = -2\pi/2, -\pi/2, 0, \pi/2, 2\pi/2, 3\pi/2$, or $k\pi/2$, then $y = 0$.

The sine takes its maximum $+1$ when the angle has the value $\pi/2$, or $\pi/2 + 2k\pi$. Thus if $2x = \pi/2 + 2k\pi$, or $x = \pi/4 + k\pi$, $\sin 2x = 1$ and $y = 3 \sin 2x = 3$.

The sine takes its minimum value -1 when the angle has the value $-\pi/2$, or $-\pi/2 + 2k\pi$. Thus if $2x = -\pi/2 + 2k\pi$, or $x = -\pi/4 + k\pi$, $\sin 2x = -1$ and $y = 3 \sin 2x = -3$.

The values of x for which the sine is $+1$ or -1 lie halfway between the values for which the sine is zero.

When $y = 3 \sin x$, $dy/dx = 6 \cos 2x$, and $d^2y/dx^2 = -12 \sin 2x$. It follows that the second derivative is zero at the points $(k\pi/2, 0)$. And d^2y/dx^2 changes sign as x increases through $k\pi/2$. Hence the points $(k\pi/2, 0)$ are points of inflection by Sec. 47. At $(k\pi/2, 0)$ the slope $dy/dx = 6$ for 0 or even values of k, while $dy/dx = -6$ for 1 or odd values of k. These slopes, together with the points where $y = 0, +3$, and

-3, are sufficient to determine the general shape of the curve as shown in Fig. 120. Since all the arches are congruent, we may draw one carefully, and then make the other arches similar in shape. For example, we might use the additional points $(\pi/8, 2.12)$ and $(3\pi/8, 2.12)$ on the first arch. These points are obtained from $\sin(\pi/4) = \sin(3\pi/4) = 0.707$.

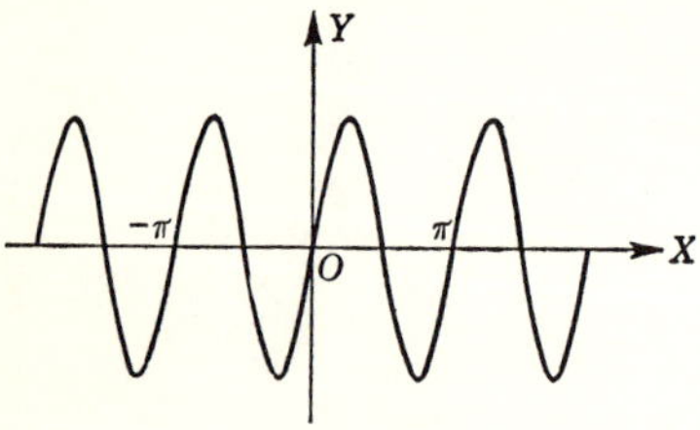

FIG. 120.

EXAMPLE 2. Sketch the graph of $y = 3\cos 2x$.

Solution: Let k denote 0, or a positive or negative integer as in Example 1. Then the cosine of an angle is zero when the angle is $\pi/2 + k\pi$, so that here $y = 0$ when $2x = \pi/2 + k\pi$ or $x = \pi/4 + k\pi/2$. The cosine takes its maximum value $+1$ when the angle is $2k\pi$. Hence here $y = 3$ when $2x = 2k\pi$ or $x = k\pi$. The cosine takes its minimum value -1 when the angle is $\pi + 2k\pi$. Hence here $y = -3$ when $2x = \pi + 2k\pi$ or $x = \pi/2 + k\pi$. The maximum and minimum points have values of x halfway between the values for which y is zero.

When $x = 3\cos 2x$, $dy/dx = -6\sin 2x$, and $d^2y/dx^2 = -12\cos 2x$. It follows that at the points $(\pi/4 + k\pi/2, 0)$, $d^2y/dx^2 = 0$. And by Sec. 47 these are points of inflection. At $(\pi/4 + k\pi/2, 0)$, $dy/dx = -6$ for $k = 0$ or any even value, while $dy/dx = 6$ for $k = 1$ or any odd value. These slopes at $(\pi/4 + k\pi/2, 0)$, together

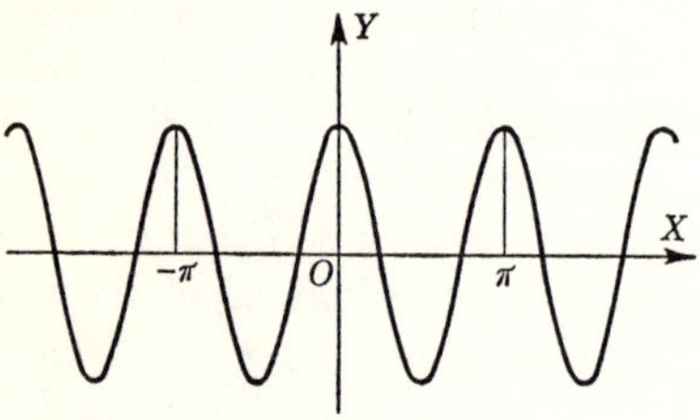

FIG. 121.

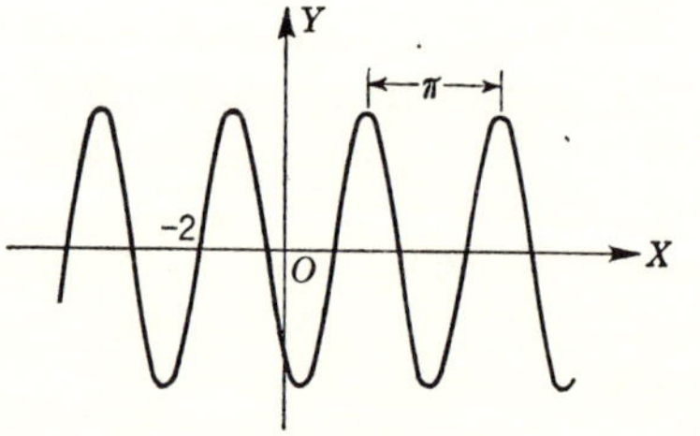

FIG. 122.

with the points $(k\pi, 3)$ and the points $(\pi/2 + k\pi, -3)$ determine the general shape of the curve as shown in Fig. 121.

EXAMPLE 3. Sketch the graph of $y = 3\sin(2x + 4)$.

Solution: Reasoning as in Example 1, we may deduce that $y = 0$ when $2x + 4 = k\pi$ or $x = -2 + k\pi/2$. Also $y = 3$ when $2x + 4 = \pi/2 + 2k\pi$, or $x = -2 + \pi/4 + k\pi$. And $y = -3$ when $2x + 4 = -\pi/2 + 2k\pi$ or $x = -2 - \pi/4 + k\pi$. This gives the maximum points $(-2 + \pi/4 + k\pi, 3)$, the minimum points $(-2 - \pi/4 + k\pi, -3)$, and the points of inflection $(-2 + k\pi/2, 0)$ with slope 6 when k is 0 or any even value and slope -6 when k is 1 or any odd value. The slopes are found from $dy/dx = 6\cos(2x + 4)$. These facts determine the required graph shown in Fig. 122.

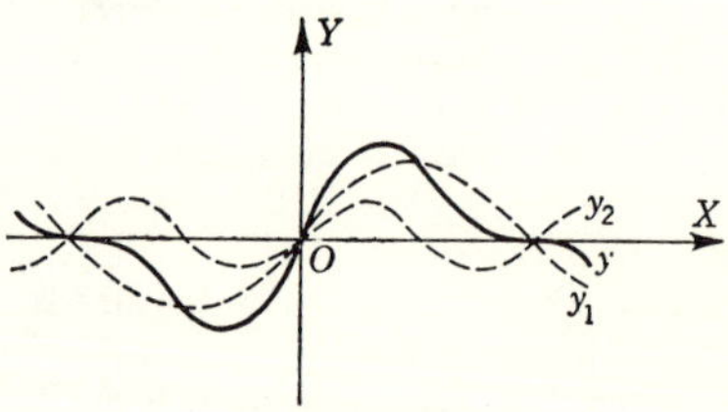

FIG. 123.

EXAMPLE 4. Sketch the graph of $y = \sin x + \frac{1}{2}\sin 2x$.

Solution: Since y is a sum of simple terms, we may use the method of *composition of ordinates*. We first sketch the graph of $y_1 = \sin x$ and $y_2 = \frac{1}{2}\sin 2x$ by the procedure used in Example 1. These are the dotted curves of Fig. 123. Then for any x we may obtain points on the full curve which is the graph of $y = y_1 + y_2 = \sin x + \frac{1}{2} \sin 2x$ by adding algebraically the ordinates of the two component curves for this x. The addition may be performed graphically. Note that, when either dotted curve is zero, the full curve crosses the other dotted curve.

EXERCISE 52

Sketch the graph of each of the following given functions.

1. $y = 2 \sin 3x$.
2. $y = 2 \cos 3x$.
3. $y = 3 \sin \frac{x}{2}$.
4. $y = 3 \cos \frac{x}{2}$.
5. $y = 4 \sin (x + 1)$.
6. $y = 5 \cos (x - 1)$.
7. $y = 2 \sin (4x - 5)$.
8. $y = 2 \cos (4x + 5)$.
9. $y = \frac{1}{2}x + \sin x$.
10. $y = \cos x - \frac{1}{2}x$.
11. $y = 1 + \sin 2x$.
12. $y = \sin x + \sin 2x$.
13. $y = 2 \sec x$.
14. $y = \csc 2x$.
15. $y = 2 \tan x$.
16. $y = \cot 2x$.
17. Let $P_1 = (x_1,y_1)$ be any point on the curve $y_1 = \sin x_1$. Show that if $y = ay_1$ and $x = (1/b)x_1 - c/b$, the point $P = (x,y)$ lies on the curve $y = a \sin (bx + c)$. Thus the graph of $y = a \sin (bx + c)$ can be obtained from the standard sine curve $y = \sin x$ by a shift of origin along the x axis to $(-c/b,0)$ and changes of scale which multiply values of x by $1/b$ and multiply values of y by a.

101. Inverse Trigonometric Functions. Let x and y be related by the equation

$$x = \sin y. \tag{101}$$

Then for any value of x in the range $-1 \leqq x \leqq 1$, a corresponding value of y can be found. Thus Eq. (101) implicitly defines y as a function of x, "the angle whose sine is x." As obtained by interpolation from a table of sines, y is the number of radians in an angle whose sine is the number x. When we wish to emphasize that y is the dependent variable, we write

$$y = \sin^{-1} x. \tag{102}$$

The symbol $\sin^{-1} x$ is read "inverse sine of x" or "angle whose sine is x." The student should observe that the superscript -1 is here part of a new symbol, with an interpretation different from that of the exponent -1 in $(\sin x)^{-1} = 1/\sin x$.

An alternative notation for $\sin^{-1} x$ is arc sin x.

In Eq. (102) consider the particular value $x = \frac{1}{2}$. Then

$$y = \sin^{-1} \tfrac{1}{2} \qquad \text{or} \qquad \tfrac{1}{2} = \sin y. \tag{103}$$

One possible value of y is $\pi/6$, since $\sin (\pi/6) = \sin 30° = \frac{1}{2}$. Another value of y is $5\pi/6$, since $\sin (5\pi/6) = \sin 150° = \frac{1}{2}$. And either $\pi/6 + 2k\pi$ or $5\pi/6 + 2k\pi$ is also a value of y, for any positive or negative integral value of k. This illustrates that $y = \sin^{-1} x$ is a multiple-valued function having infinitely many values. In general, if y_1 is any one value of

$\sin^{-1} x_1$, all possible values of $\sin^{-1} x_1$ are given by $(-1)^k y_1 + k\pi$, with $k = 0$ or any positive or negative integer.

For many purposes it is desirable to use the *principal branch* of $y = \sin^{-1} x$, for which the value lies between $-\pi/2$ and $\pi/2$. For this single-valued branch

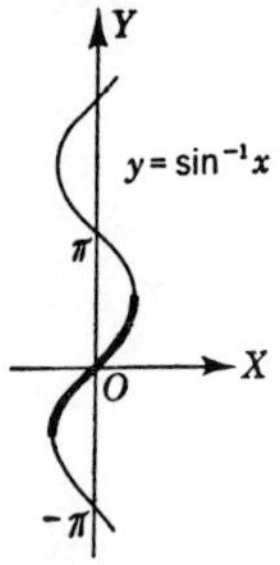

FIG. 124.

$$y = \sin^{-1} x \text{ implies } x = \sin y \qquad \text{and} \qquad -\frac{\pi}{2} \leqq x \leqq \frac{\pi}{2}. \tag{104}$$

Thus, using the principal branch we have $\sin^{-1} \frac{1}{2} = \pi/6$, $\sin^{-1}(-\frac{1}{2}) = -\pi/6, \sin^{-1} 0 = 0, \sin^{-1} 1 = \pi/2, \sin^{-1}(-1) = -\pi/2$.

To obtain the graph of $y = \sin^{-1} x$, we may plot $x = \sin y$ by the method used in Sec. 100. The graph is shown in Fig. 124. The values for the principal branch are indicated by a heavy line.

The definition of the inverse cosine of x, $\cos^{-1} x$, is similar to the one just given for the inverse sine. Thus $y = \cos^{-1} x$ if $x = \cos y$. And the principal branch of $y = \cos^{-1} x$ is such that

$$y = \cos^{-1} x \text{ implies } x = \cos y \qquad \text{and} \qquad 0 \leqq y \leqq \pi. \tag{105}$$

The graph of the multiple-valued function $y = \cos^{-1} x$ in Fig. 125 may be obtained by plotting $x = \cos y$ by the method used in Sec. 100. The values for the principal branch are indicated by a heavy line.

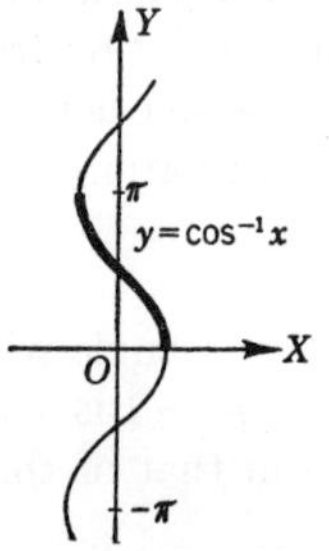

FIG. 125.

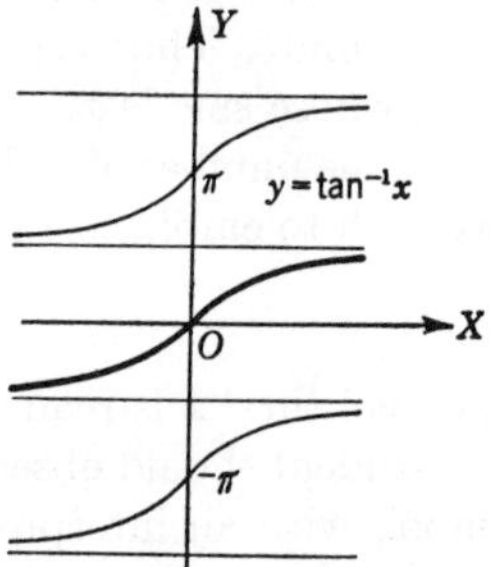

FIG. 126.

Again, to define the inverse tangent of x, $\tan^{-1} x$, we set $y = \tan^{-1} x$ if $x = \tan y$. The principal branch of $y = \tan^{-1} x$ is such that

$$y = \tan^{-1} x \text{ implies } x = \tan y \qquad \text{and} \qquad -\frac{\pi}{2} < y < \frac{\pi}{2}. \tag{106}$$

The graph of the multiple-valued function $y = \tan^{-1} x$ in Fig. 126 may be obtained by plotting $x = \tan y$ by the method used in Sec. 100. The values for the principal branch are indicated by a heavy line.

We define the other inverse trigonometric functions in a similar manner. Thus we set $y = \cot^{-1} x$ if $x = \cot y$, with $0 < y < \pi$ for the principal branch. And $y = \sec^{-1} x$ if $x = \sec y$, and $y = \csc^{-1} x$ if $x = \csc y$, with $0 < y < \pi/2$ if $x > 0$ and $-\pi < y < -\pi/2$ if $x < 0$ for the principal branch.

FIG. 127.

EXAMPLE. Sketch the graph of $y = 3 \sin^{-1} (2x - 1) + 6$.

Solution: The given relation implies $y - 6 = 3 \sin^{-1} (2x - 1)$, $\frac{y-6}{3} = \sin^{-1} (2x - 1)$, $2x - 1 = \sin \frac{y-6}{3}$, and $x = \frac{1}{2} + \frac{1}{2} \sin \frac{y-6}{3}$. Hence $\frac{dx}{dy} = \frac{1}{6} \cos \frac{y-6}{3}$ and $\frac{d^2x}{dy^2} = -\frac{1}{18} \sin \frac{y-6}{3}$. By reasoning like that used in the examples of Sec. 100, but with the roles of x and y interchanged, we may deduce from these facts that x has a maximum at the points $(1, 6 + 3\pi/2 + 6k\pi)$, that x has a mimimum at the points $(0, 6 - 3\pi/2 + 6k\pi)$, and that the curve has points of inflection at $(\frac{1}{2}, 6 + 3k\pi)$ with slope $dy/dx = 6$ for $k = 0$ or any even value and with slope $dy/dx = -6$ for $k = 1$ or any odd value. The graph is shown in Fig. 127.

EXERCISE 53

Sketch the graph of each of the following given functions.

1. $y = \sin^{-1} (x - 2)$.
2. $y = \cos^{-1} (x - 1)$.
3. $y = \sin^{-1} \frac{x}{2}$.
4. $y = \cos^{-1} \frac{x}{3}$.
5. $y = 1 + \sin^{-1} x$.
6. $y = 2 + \cos^{-1} x$.
7. $y = 2 \tan^{-1} x$.
8. $y = \frac{1}{2} \cot^{-1} x$.
9. $y = \sec^{-1} \frac{x}{2}$.
10. $y = 2 \csc^{-1} x$.
11. $y = \tan^{-1} (x - 1)$.
12. $y = \cos^{-1} (x - 2)$.
13. $y = \sin^{-1} (x - 1)$.
14. $y = \cot^{-1} (x + 2) - 1$.

102. The Derivative of $\sin^{-1} u$. Let $y = \sin^{-1} u$. Then $u = \sin y$. And by Eq. (78) the derivative of u with respect to y is

$$\frac{du}{dy} = \cos y. \tag{107}$$

If this is not zero, by Eq. (26) of Sec. 54, we have

$$\frac{dy}{du} = \frac{1}{du/dy} = \frac{1}{\cos y}. \tag{108}$$

Now let u be a differentiable function of x. Then by the rule for composite functions, Eq. (20) of Sec. 53, it follows that

$$\frac{dy}{dx} = \frac{dy}{du}\frac{du}{dx} = \frac{1}{\cos y}\frac{du}{dx}. \tag{109}$$

But by Eq. (10) we have

$$\cos^2 y + \sin^2 y = 1, \cos y = \pm \sqrt{1 - \sin^2 y} = \pm \sqrt{1 - u^2}, \quad (110)$$

since $\sin y = u$. For the principal branch of $y = \sin^{-1} u$, $-\pi/2 < y < \pi/2$, the cosine of y is positive and we must use the plus sign before the radical in Eq. (110). Hence from Eqs. (109) and (110) we may conclude that

$$\frac{d(\sin^{-1} u)}{dx} = \frac{1}{\sqrt{1 - u^2}} \frac{du}{dx} \quad (111)$$

for the principal branch where $\sin^{-1} u$ has a positive cosine. If $\sin^{-1} u$ has a negative cosine, we must use the minus sign before the radical in Eq. (110) and insert a minus sign before the radical in Eq. (111).

103. The Derivative of $\cos^{-1} u$. Let $y = \cos^{-1} u$. Then $u = \cos y$. And by Eq. (83), the derivative of u with respect to y is

$$\frac{du}{dy} = -\sin y. \quad (112)$$

If this is not zero, by Eq. (26) of Sec. 54, we have

$$\frac{dy}{du} = \frac{1}{du/dy} = -\frac{1}{\sin y}. \quad (113)$$

Now let u be a differentiable function of x. Then by the rule for composite functions, Eq. (20) of Sec. 53, it follows that

$$\frac{dy}{dx} = \frac{dy}{du}\frac{du}{dx} = -\frac{1}{\sin y}\frac{du}{dx}. \quad (114)$$

But by Eq. (10) we have

$$\sin^2 y + \cos^2 y = 1, \qquad \sin y = \pm \sqrt{1 - \cos^2 y} = \pm \sqrt{1 - u^2}, \quad (115)$$

since $\cos y = u$. For the principal branch of $y = \cos^{-1} u$, $0 < y < \pi$, the sine of y is positive, and we must use the plus sign before the radical in Eq. (115). Hence from Eqs. (114) and (115) we may conclude that

$$\frac{d(\cos^{-1} u)}{dx} = -\frac{1}{\sqrt{1 - u^2}} \frac{du}{dx} \quad (116)$$

for the principal branch where $\cos^{-1} u$ has a positive sine. If $\cos^{-1} u$ has a negative sine, we must use the minus sign before the radical in Eq. (115) and omit the minus sign before the fraction in Eq. (116).

104. The Derivative of $\tan^{-1} u$. Let $y = \tan^{-1} u$. Then $u = \tan y$. And by Eq. (87) the derivative of u with respect to y is

$$\frac{du}{dy} = \sec^2 y. \quad (117)$$

If this is not zero, by Eq. (26) of Sec. 54, we have

$$\frac{dy}{du} = \frac{1}{du/dy} = \frac{1}{\sec^2 y}. \tag{118}$$

Now let u be a differentiable function of x. Then by the rule for composite functions, Eq. (20) of Sec. 53, it follows that

$$\frac{dy}{dx} = \frac{dy}{du}\frac{du}{dx} = \frac{1}{\sec^2 y}\frac{du}{dx}. \tag{119}$$

But by Eq. (30) we have

$$\sec^2 y = 1 + \tan^2 y = 1 + u^2, \tag{120}$$

since $\tan y = u$. From Eqs. (119) and (120) we may conclude that

$$\frac{d(\tan^{-1} u)}{dx} = \frac{1}{1 + u^2}\frac{du}{dx}. \tag{121}$$

105. The Derivative of $\cot^{-1} u$. Let $y = \cot^{-1} u$. Then $u = \cot y$. And by Eq. (92) the derivative of u with respect to y is

$$\frac{du}{dy} = -\csc^2 y. \tag{122}$$

If this is not zero, by Eq. (26) of Sec. 54, we have

$$\frac{dy}{du} = \frac{1}{du/dy} = -\frac{1}{\csc^2 y}. \tag{123}$$

Now let u be a differentiable function of x. Then by the rule for composite functions, Eq. (20) of Sec. 53, it follows that

$$\frac{dy}{dx} = \frac{dy}{du}\frac{du}{dx} = -\frac{1}{\csc^2 y}\frac{du}{dx}. \tag{124}$$

But by Eq. (31) we have

$$\csc^2 y = 1 + \cot^2 y = 1 + u^2, \tag{125}$$

since $\cot y = u$. From Eqs. (124) and (125) we may conclude that

$$\frac{d(\cot^{-1} u)}{dx} = -\frac{1}{1 + u^2}\frac{du}{dx}. \tag{126}$$

***106. The Derivative of $\sec^{-1} u$.** Let $y = \sec^{-1} u$. Then $u = \sec y$. And by Eq. (96) the derivative of u with respect to y is

$$\frac{du}{dy} = \tan y \sec y. \tag{127}$$

If this is not zero, by Eq. (26) of Sec. 54, we have

$$\frac{dy}{du} = \frac{1}{du/dy} = \frac{1}{\tan y \sec y}. \tag{128}$$

Now let u be a differentiable function of x. Then by the rule for composite functions, Eq. (20) of Sec. 53, it follows that

$$\frac{dy}{dx} = \frac{dy}{du}\frac{du}{dx} = \frac{1}{\tan y \sec y}\frac{du}{dx}. \tag{129}$$

But by Eq. (30) we have

$$\tan^2 y + 1 = \sec^2 y, \qquad \tan y = \pm\sqrt{\sec^2 y - 1} = \pm\sqrt{u^2 - 1}, \tag{130}$$

since $\sec y = u$. Hence, if $\tan y$ is positive,

$$\tan y \sec y = u\sqrt{u^2 - 1}. \tag{131}$$

From Eqs. (129) and (131) we may conclude that

$$\frac{d(\sec^{-1} u)}{dx} = \frac{1}{u\sqrt{u^2 - 1}}, \tag{132}$$

if $\sec^{-1} u$ has a positive tangent. A minus sign must be prefixed to the fraction if $\tan(\sec^{-1} u)$ is negative.

***107. The Derivative of $\csc^{-1} u$.** Let $y = \csc^{-1} u$. Then $u = \csc y$. And by Eq. (99) the derivative of u with respect to y is

$$\frac{du}{dy} = -\cot y \csc y. \tag{133}$$

If this is not zero, by Eq. (26) of Sec. 54, we have

$$\frac{dy}{du} = \frac{1}{du/dy} = -\frac{1}{\cot y \csc y}. \tag{134}$$

Now let u be a differentiable function of x. Then by the rule for composite functions, Eq. (20) of Sec. 53, it follows that

$$\frac{dy}{dx} = \frac{dy}{du}\frac{du}{dx} = -\frac{1}{\cot y \csc y}\frac{du}{dx}. \tag{135}$$

But by Eq. (31) we have

$$\cot^2 y + 1 = \csc^2 y, \qquad \cot y = \pm\sqrt{\csc^2 y - 1} = \pm\sqrt{u^2 - 1}, \tag{136}$$

since $\csc y = u$. Hence, if $\cot y$ is positive,

$$\cot y \csc y = u\sqrt{u^2 - 1}. \tag{137}$$

From Eqs. (135) and (137) we may conclude that

$$\frac{d(\csc^{-1} u)}{dx} = -\frac{1}{u\sqrt{u^2 - 1}}, \tag{138}$$

if $\csc^{-1} u$ has a positive cotangent. If $\cot(\csc^{-1} u)$ is negative, we must use the minus sign before the radical in Eq. (136) and omit the minus sign before the fraction in Eq. (138).

108. Differentiation Formulas for the Six Inverse Functions. The student should memorize the six differentiation formulas of Eqs. (111), (116), (121), (126), (132), and (138), namely,

$$
\begin{aligned}
\frac{d(\sin^{-1} u)}{dx} &= \frac{1}{\sqrt{1 - u^2}} \frac{du}{dx}, & \frac{d(\cos^{-1} u)}{dx} &= -\frac{1}{\sqrt{1 - u^2}} \frac{du}{dx}, \\
\frac{d(\tan^{-1} u)}{dx} &= \frac{1}{1 + u^2} \frac{du}{dx}, & \frac{d(\cot^{-1} u)}{dx} &= -\frac{1}{1 + u^2} \frac{du}{dx}, \\
\frac{d(\sec^{-1} u)}{dx} &= \frac{1}{u\sqrt{u^2 - 1}} \frac{du}{dx}, & \frac{d(\csc^{-1} u)}{dx} &= -\frac{1}{u\sqrt{u^2 - 1}} \frac{du}{dx}.
\end{aligned} \tag{139}
$$

It is helpful to the memory to note the following facts about these six formulas. Each formula contains the factor du/dx because of the rule for composite functions. For each of the three inverse functions $\sin^{-1}$, $\tan^{-1}$, $\sec^{-1}$ the sign as written is plus. While for each of the inverse cofunctions $\cos^{-1}$, $\cot^{-1}$, $\csc^{-1}$ the sign as written is minus. We may consider $\sin^{-1}$ and $\cos^{-1}$, $\tan^{-1}$ and $\cot^{-1}$, $\sec^{-1}$ and $\csc^{-1}$ as three pairs of complementary inverse functions. Then the six formulas consist of three complementary pairs. Either formula of each pair may be obtained from the other by one change of sign, and the replacing of each inverse function by its complement.

Finally Eq. (139) may be associated with Eq. (100). Each denominator of Eq. (139) is the expression in terms of u of the trigonometric functions appearing as factors in the corresponding right member of Eq. (100).

In differentiation exercises, unless the quadrant is explicitly indicated, the student may assume that the branch of the inverse function is that for which the derivatives have the signs as given in Eq. (139). This branch is that for which the appropriate one of

$$
\begin{aligned}
\cos(\sin^{-1} u) &= \sqrt{1 - u^2}, & \sin(\cos^{-1} u) &= \sqrt{1 - u^2}, \\
\tan(\sec^{-1} u) &= \sqrt{u^2 - 1}, & \cot(\csc^{-1} u) &= \sqrt{u^2 - 1},
\end{aligned} \tag{140}
$$

is positive. When the appropriate trigonometric function which appears in a left member of Eq. (140) is negative, the sign before the fraction in Eq. (139) which contains the representative radical must be reversed.

Since the formulas for $\tan^{-1} u$ and $\cot^{-1} u$ contain no radicals, they hold as written in Eq. (139) for all branches of these functions.

EXAMPLE 1. Find dy/du if $y = 3 \sin^{-1}(1 - x^2)$ and $x > 0$.

Solution: By Eq. (139) with $u = 1 - x^2$, we find

$$\frac{dy}{dx} = 3\frac{1}{\sqrt{1 - (1 - x^2)^2}} \frac{d(1 - x^2)}{dx} = \frac{3(-2x)}{\sqrt{2x^2 - x^4}} = \frac{-6}{\sqrt{2 - x^2}},$$

since $\sqrt{x^2} = x$ when x is positive.

EXAMPLE 2. Find dy/dx if $y = 4 \cot^{-1} \sqrt{2 - 3x}$.

Solution: By Eq. (139) and the power rule, we find

$$
\begin{aligned}
\frac{dy}{dx} &= 4\frac{-1}{1 + (\sqrt{2 - 3x})^2} \frac{d}{dx}(2 - 3x)^{\frac{1}{2}} = \frac{4(-1)}{3 - 3x} \frac{1}{2}(2 - 3x)^{-\frac{1}{2}}(-3) \\
&= \frac{4(-1)(-3)}{3(1 - x)2} \frac{1}{\sqrt{2 - 3x}} = \frac{2}{(1 - x)\sqrt{2 - 3x}}.
\end{aligned}
$$

EXAMPLE 3. Find dy/dx if $y = 4 \sec^{-1} x^2$.
Solution: By Eq. (139) and the power rule, we find

$$\frac{dy}{dx} = 4 \frac{1}{x^2 \sqrt{x^4 - 1}} \frac{d(x^2)}{dx} = \frac{4(2x)}{x^2 \sqrt{x^4 - 1}} = \frac{8}{x \sqrt{x^4 - 1}}.$$

EXERCISE 54

Find dy/dx for each of the following given functions.

1. $y = \sin^{-1} 2x$.
2. $y = \cos^{-1} 3x$.
3. $y = \sin^{-1} \frac{x}{3}$.
4. $y = \cos^{-1} \frac{x}{2}$.
5. $y = 3 \sin (1 - x)$.
6. $y = 5 \cos^{-1} (1 - 2x)$.
7. $y = \sin^{-1} \sqrt{x}$.
8. $y = \cos^{-1} x^2$.
9. $y = x^2 \sin^{-1} x$.
10. $y = x \cos^{-1} x$.
11. $y = \tan^{-1} 4x$.
12. $y = \cot^{-1} 6x$.
13. $y = \tan^{-1} \frac{x}{2}$.
14. $y = \cot^{-1} \frac{x}{3}$.
15. $y = \tan^{-1} x^2$.
16. $y = \cot^{-1} x^3$.
17. $y = \sec^{-1} 2x$.
18. $y = \csc^{-1} 4x$.
19. $y = \sec^{-1} \frac{x}{4}$.
20. $y = \csc^{-1} \frac{x}{5}$.
21. $y = \sec^{-1} \frac{1}{x^2}$.
22. $y = \csc^{-1} \frac{1}{x}$.
23. $y = x \sqrt{1 - x^2} + \sin^{-1} x$.
24. $y = \frac{1}{\sqrt{1 - x^2}} + \cos^{-1} x$.

109. Angular Velocity. Consider the line OP which revolves in a plane about the fixed point O. Let OA be any fixed line in the plane through O (Fig. 128). Then the position of OP at time t is determined by the angle $AOP = \theta$. The angle θ is a function of the time, and $\theta(t_2) - \theta(t_1)$ measures the rotation of the line in time $t_2 - t_1$. The rate of change of θ with respect to the time, $d\theta/dt$, is called the *angular velocity* of the line. It is denoted by the symbol ω, read "omega," so that

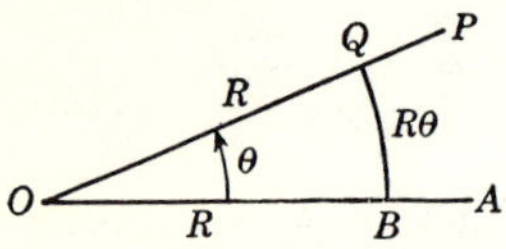

FIG. 128.

$$\text{Angular velocity} = \omega = \frac{d\theta}{dt}.$$

Similarly, the rate of change of ω with respect to the time, $d\omega/dt$, is called the *angular acceleration* of the line. It is denoted by the symbol α, read "alpha," so that

$$\text{Angular acceleration} = \alpha = \frac{d\omega}{dt} = \frac{d^2\theta}{dt^2}. \tag{141}$$

Let Q be the point on the revolving line at fixed distance $OQ = R$ from O. Then Q is moving in a circle of radius R. And if the arc $BQ = s$, by

Eq. (5), $s = R\theta$. Hence the velocity along the circle is

$$\frac{ds}{dt} = R\frac{d\theta}{dt} = R\omega. \tag{142}$$

And the acceleration along the circle is

$$\frac{d^2s}{dt^2} = R\frac{d^2\theta}{dt^2} = r\alpha. \tag{143}$$

EXAMPLE 1. A particle P traverses a circle in the positive direction at a uniform rate of n revolutions per second. The circle is of radius b, has its center at the origin, and the particle was at the point $(b,0)$ at the time $t = 0$. For the motion of the line OP find θ, ω, and α. Also find the coordinates of P in terms of t.

Solution: Since one revolution is 2π radians, the angular velocity in radians per second is $\omega = 2n\pi$. As this is $d\theta/dt$ and is constant, $\theta = 2n\pi t + C$. And $C = 0$ since $\theta = 0$ when $t = 0$. Thus for the motion of the line OP we have

$$\theta = 2n\pi t, \qquad \omega = 2n\pi, \qquad \alpha = 0.$$

And from the right triangle with hypotenuse b, angle θ, and sides x and y (Fig. 129) we have $x = b\cos\theta$, $y = b\sin\theta$. Since $\theta = 2n\pi t$,

$$x = b\cos 2n\pi t, \qquad y = b\sin 2n\pi t.$$

These are the required coordinates of P in terms of t.

It follows from these relations that

$$\frac{dx}{dt} = -2n\pi b\sin 2n\pi t, \qquad \frac{dy}{dt} = 2n\pi b\cos 2n\pi t.$$

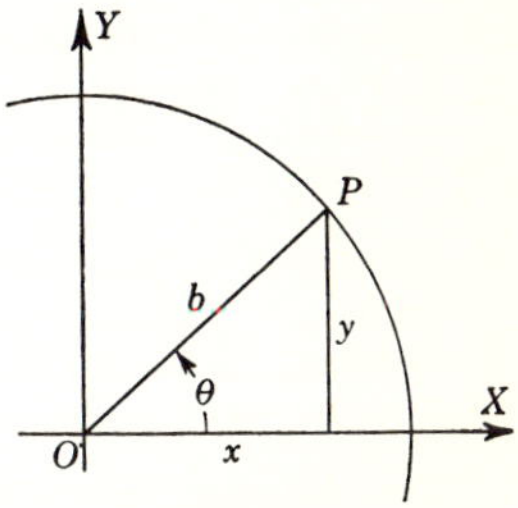

FIG. 129.

Hence from Eq. (81) of Sec. 63 we have

$$\left(\frac{ds}{dt}\right)^2 = \left(\frac{dx}{dt}\right)^2 + \left(\frac{dy}{dt}\right)^2 = (-2n\pi b\sin 2n\pi t)^2 + (2n\pi b\cos 2n\pi t)^2$$
$$= (2n\pi b)^2(\sin^2 2n\pi t + \cos^2 2n\pi t) = (2n\pi b)^2.$$

Thus $ds/dt = 2n\pi b$. This checks Eq. (142) with $R = b$ and $\omega = 2n\pi$.

EXAMPLE 2. Find ω and θ as functions of t if $\alpha = 12t^2$ and at time $t = 0$, ω was equal to 2 and θ was equal to 3.

Solution: Since $d\omega/dt = \alpha = 12t^2$, by Sec. 67, $\omega = 4t^3 + C_1$. At $t = 0$, $\omega = 2$ so that $2 = 0 + C_1$, $C_1 = 2$, and $\omega = 4t^3 + 2$. Hence $d\theta/dt = \omega = 4t^3 + 2$ and by Sec. 67, $\theta = t^4 + 2t + C_2$. But at $t = 0$, $\theta = 3$ so that $3 = 0 + 0 + C_2$, $C_2 = 3$, and $\theta = t^4 + 2t + 3$. Thus the required expressions are

$$\omega = 4t^3 + 2, \qquad \theta = t^4 + 2t + 3.$$

EXAMPLE 3. Particle A oscillates on the line $x = 1$ in such a way that at time t, A has coordinates $(1, \sin 4t)$. Particle B oscillates on the x axis in such a way that at time t, B has coordinates $(\cos 4t, 0)$. Find the angular velocity of a line through O parallel to AB at each instant.

Solution: By Eq. (16) of Sec. 23 the slope of the line AB is

$$\frac{y_B - y_A}{x_B - x_A} = \frac{0 - \sin 4t}{\cos 4t - 1} = \frac{\sin 4t}{1 - \cos 4t} = \tan\theta,$$

where θ measures the rotation angle of AB or the parallel line through O. Hence by Eq. (139) we find from $\theta = \tan^{-1} \dfrac{\sin 4t}{1 - \cos 4t}$,

$$\frac{d\theta}{dt} = \frac{\dfrac{d}{dt}\left(\dfrac{\sin 4t}{1 - \cos 4t}\right)}{1 + \left(\dfrac{\sin 4t}{1 - \cos 4t}\right)^2} = \frac{\dfrac{4 \cos 4t(1 - \cos 4t) - \sin 4t(4 \sin 4t)}{(1 - \cos 4t)^2}}{\dfrac{1 - 2 \cos 4t + \cos^2 4t + \sin^2 4t}{(1 - \cos 4t)^2}}$$

$$= \frac{4(\cos 4t - \cos^2 4t - \sin^2 4t)}{1 - 2 \cos 4t + \cos^2 4t + \sin^2 4t} = \frac{4(\cos 4t - 1)}{-2(\cos 4t - 1)} = -2.$$

Thus the required angular velocity $\omega = -2$ for any time t.

EXERCISE 55

A flywheel 10 ft. in diameter makes 4 revolutions per second about a horizontal axis. Find the horizontal and vertical components of velocity in feet per second of a point on the rim when its distance above the center of the wheel is

1. 3 ft. 2. 4 ft. 3. 5 ft. 4. 0.

The angle of rotation of a wheel θ is given as a function of the time. Find the angular velocity ω and angular acceleration α of the wheel at time t in each problem.

5. $\theta = 3t^2$. 6. $\theta = 2t^3$. 7. $\theta = 3 \sin 2t$. 8. $\theta = 4 \cos 5t$.

How many revolutions will a wheel make in the time from $t = 0$ to $t = 2$ if the angular velocity in radians per second ω is the given function of t the time in seconds?

9. $\omega = 4$. 10. $\omega = 18t$. 11. $\omega = 24t^2$. 12. $\omega = 60t^4$.

Find the angular velocity of the line OP if $O = (0,0)$, t is the time in seconds, and OP passes through the point

13. $(2,4t)$. 14. $(3t,4)$. 15. (t,t^2). 16. $(t^2,1)$.
17. $(\sin 5t, \cos 5t)$. 18. $(\cot 2t, 1)$. 19. $(1, \cot 3t)$. 20. $(\tan 4t, 1)$.

21. The line OP joins $(0,0)$ and (x,y). If the velocity components of the moving point (x,y) are $v_x = dx/dt$ and $v_y = dy/dt$, show that the angular velocity ω of the line OP is

$$\omega = \frac{xv_y - yv_x}{x^2 + y^2}.$$

110. Simple Harmonic Motion. Consider a particle P moving on a straight line. If O is some fixed point of the line, the displacement of the particle from O is a function of the time t. Let us take the line as the x axis and the fixed point O as the origin, so that the displacement $OP = x$. And in particular let

$$x = c \sin bt, \tag{144}$$

where c and b are positive constants.

For this motion the velocity is

$$v = \frac{dx}{dt} = cb \cos bt. \tag{145}$$

And the acceleration is

$$a = \frac{d^2x}{dt^2} = -cb^2 \sin bt. \tag{146}$$

When $t = 0$, $x = c \sin 0 = 0$, and the particle is at O. When $t = \frac{\pi}{2b}$, $x = c \sin \frac{\pi}{2} = c$. Hence the particle is at A (Fig. 130) where $OA = c$.

As t increases from 0 to $\pi/2b$, bt is between O and $\pi/2$. Hence by Eq. (145) v is positive, so that the particle is moving from O to A. And by Eq. (146) a is negative, so that va is negative and the particle moves with decreasing speed by Sec. 42.

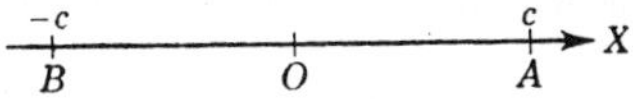

FIG. 130.

As t increases from $\pi/2b$ to π/b, bt is between $\pi/2$ and π. Hence by Eq. (145) v is negative, so that the particle is moving back from A to O. And by Eq. (146) a is negative, so that va is positive and the particle moves with increasing speed.

As t increases from π/b to $3\pi/2b$, the particle moves with decreasing speed from O to B, where $OB = -c$.

And as t increases from $3\pi/2b$ to $2\pi/b$, the particle moves back from B to O with increasing speed.

If the time increases beyond $2\pi/b$, the motion is repeated. Thus x varies periodically between $-c$ and c. And the motion is an oscillation or vibration back and forth along the line segment BA.

The motion defined by Eq. (144) is an example of *simple harmonic motion*. The maximum displacement from the center of BA, $OA = c$ is called the amplitude. And the time interval $T = 2\pi/b$, after which the motion repeats itself, is called the *period*. In any time interval of length $2\pi/b$, the particle returns to its initial position with its initial velocity. Such a portion of the motion is called a *complete vibration*. Since the time required for one complete vibration is the period $T = 2\pi/b$,

$$T = \frac{2\pi}{b} \quad \text{and} \quad n = \frac{1}{T} = \frac{b}{2\pi} \tag{147}$$

is the *frequency* of vibration, or number of complete vibrations per unit time.

A comparison of Eqs. (144) and (146) shows that

$$a = -b^2x. \tag{148}$$

This shows that the acceleration is proportional to the displacement and oppositely directed.

Consider any particle of mass m moving on a straight line under the influence of a force F. The acceleration a of the particle is determined by the relation: force equals mass times acceleration, or $F = ma$. In particular, suppose that the particle of mass m is describing the simple harmonic motion of Eq. (144). Then by Eq. (148) the force F is

$$F = ma = -mb^2x. \tag{149}$$

Thus the force is proportional to the distance from O. The negative constant of proportionality shows that the force is an attracting force toward the center O.

For the motion of Eq. (144), the time is zero when the particle is at the center O. If we measure time from some other instant, so that the time at which the particle is at O is t_0, the displacement x at time t will be

$$x = c \sin b(t - t_0) \qquad \text{or} \qquad x = c \sin (bt + h). \tag{150}$$

This also represents simple harmonic motion with amplitude c and period $2\pi/b$.

Equations (148) and (149) also hold for the motion of Eq. (150). And, as shown in Example 3, if Eq. (148) holds, the motion must be given by a relation like that of Eq. (150).

EXAMPLE 1. A particle P moves around a circle in the positive direction at a constant speed V. The circle is of radius R, has its center at the origin, and the particle was at the point $(R,0)$ at time t_0. Find the motion of its projection on a fixed diameter with slope angle A.

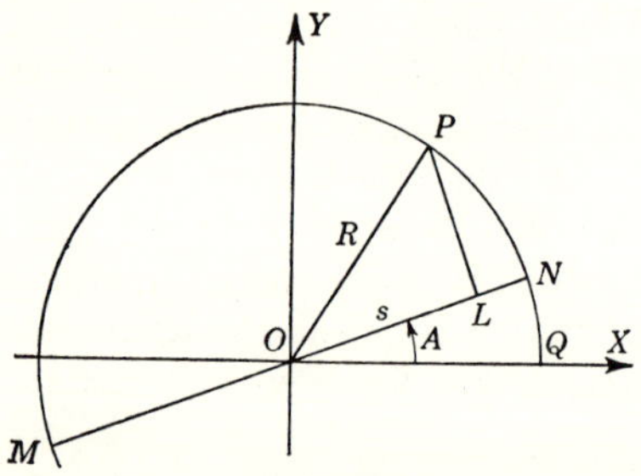

FIG. 131.

Solution: Let MON be the diameter with slope angle A (Fig. 131). And let $Q = (R,0)$. Then if the particle was at P at time t, the time elapsed after passing Q is $t - t_0$. Hence the arc $QP = V(t - t_0)$. And the angle QOP is $\frac{V}{R}(t - t_0)$ by Eq. (5). Thus the angle $NOP = \frac{V}{R}(t - t_0) - A$. Let L be the foot of the perpendicular drawn from P to the line MN. Then L is the projection of P, and the signed distance $s = OL = OP \cos NOP$. Thus

$$s = R \cos\left[\frac{V}{R}(t - t_0) - A\right] = R \sin\left[\frac{V}{R}(t - t_0) - A + \frac{\pi}{2}\right],$$

by Eq. (16). This shows that the motion of L on MN is simple harmonic with amplitude R and period $2\pi R/V$. Note that the period is the time in which P traverses one complete circumference.

EXAMPLE 2. The displacement of a particle at time t is

$$x = A \sin bt + B \cos bt.$$

Show that the motion is simple harmonic, and find the amplitude and period.

Solution: Plot the point $N = (A,B)$ (Fig. 132). Then if $C = ON$ is the distance from the origin, $C = \sqrt{A^2 + B^2}$. And, if $\theta = \angle XON$, the angle from OX to ON, $\tan \theta = B/A$. And $\theta = \tan^{-1} (B/A)$ with such a quadrant that $A = C \cos \theta$, $B = C \sin \theta$. Hence we have

$$\begin{aligned} x &= A \sin bt + B \cos bt = C(\cos \theta \sin bt + \sin \theta \cos bt) \\ &= C \sin (bt + \theta), \end{aligned}$$

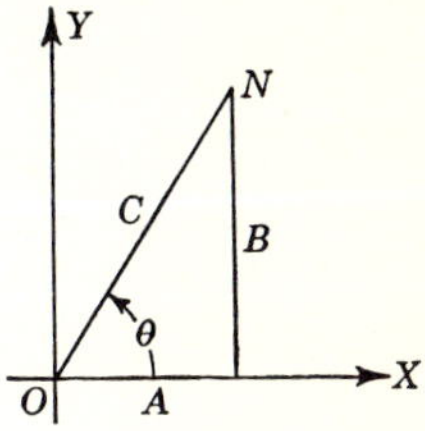

FIG. 132.

by Eq. (11). It follows that the given motion is simple harmonic with amplitude $C = \sqrt{A^2 + B^2}$ and period $T = 2\pi/b$.

EXAMPLE 3. Show that if a motion is such that Eq. (148) or (149) holds, the motion is simple harmonic.

Solution: Since $a = \dfrac{d^2x}{dt^2}$, either equation implies $\dfrac{d^2x}{dt^2} = a = -b^2x$, or $\dfrac{d^2x}{dt^2} = -b^2x$. Multiply both sides by $\dfrac{dx}{dt}$ to obtain $\dfrac{dx}{dt}\dfrac{d^2x}{dt^2} = -b^2x\dfrac{dx}{dt}$, or since $v = \dfrac{dx}{dt}$ and $\dfrac{dv}{dt} = \dfrac{d^2x}{dt^2}$, $v\dfrac{dv}{dt} = -b^2x\dfrac{dx}{dt}$. Multiply by $2\,dt$ to deduce that $2v\,dv = -b^2(2x\,dx)$. And by integrating as in Sec. 67, we find that $v^2 = -b^2x^2 + C_1$. Since v^2 is positive, C_1 must be positive. Set $\dfrac{1}{b}\sqrt{C_1} = c$, so that $C_1 = b^2c^2$. Then we may write

$$v^2 = -b^2x^2 + b^2c^2 = b^2(c^2 - x^2) \qquad \text{and} \qquad v = \frac{dx}{dt} = \pm b\sqrt{c^2 - x^2}.$$

Using the plus sign, we may deduce that

$$\frac{dx}{\sqrt{c^2 - x^2}} = b\,dt \qquad \text{or} \qquad d\left(\sin^{-1}\frac{x}{c}\right) = d(bt)$$

by Eq. (111). Hence by integration we may deduce that

$$\sin^{-1}\frac{x}{c} = bt + C_2 \qquad \text{or} \qquad \frac{x}{c} = \sin (bt + h),$$

if $C_2 = h$. With the minus sign before the radical, we would have found

$$-\sin^{-1}\frac{x}{c} = bt + C_2 \qquad \text{or} \qquad \frac{x}{c} = -\sin (bt + C_2) = \sin (bt + C_2 + \pi),$$

so that, if $C_2 + \pi = h$, we are again led to

$$x = c \sin (bt + h).$$

Thus the motion is simple harmonic, with period $2\pi/b$.

EXAMPLE 4. Use the result of Example 3 to show that the motion of Example 2 or $x = A \sin bt + B \cos bt$ is simple harmonic with period $2\pi/b$.

Solution: From the given relation $x = A \sin bt + B \cos bt$ we find the following: $dx/dt = Ab \cos bt - Bb \sin bt$, and $d^2x/dt^2 = -Ab^2 \sin bt - Bb^2 \cos bt$. Hence $a = d^2x/dt^2 = -b^2 (A \sin bt + B \cos Bt) = -b^2x$. And it follows from $a = -b^2x$ that the motion is simple harmonic with period $2\pi/b$ by the result of Example 3.

EXERCISE 56

Show that each of the following given motions is simple harmonic and find the amplitude and period of each motion.

1. $x = 8 \sin 4t$.
2. $x = 2 \sin \pi t$.
3. $x = 5 \cos 2t$.
4. $x = 4 \cos 2t + 3 \sin 2t$.
5. $x = 6 \sin (4t + 3)$.
6. $x = 8 \cos (2t - 4)$.
7. $x = 2 - 4 \sin^2 t$.
8. $x = 6 \cos^2 2t - 3$.

Find the equation of a simple harmonic motion of period π and amplitude 4 with center at the origin if

9. $x = 0$ when $t = 0$.
10. $x = 4$ when $t = 0$.
11. $x = 0$ when $t = 2$.
12. $x = 4$ when $t = 2$.

A particle P moves around a circle with constant speed in the positive direction. The circle is of radius 4 ft. and has its center at the origin O. Find the simple harmonic motion of the projection of P on the x axis if

13. The speed along the circle is 16π ft./sec.
14. The particle makes 8 revolutions per second.
15. The angular velocity of OP is 2 radians/sec.

Let $x = c \sin (bt + h)$ with c and b positive. Show that

16. x reaches its maximum c when $t = \dfrac{1}{b}\left(\dfrac{\pi}{2} - h\right) + \dfrac{2k\pi}{b}$.
17. v reaches its maximum bc when $t = -\dfrac{h}{b} + \dfrac{2k\pi}{b}$.
18. a reaches its maximum b^2c when $t = \dfrac{1}{b}\left(\dfrac{3\pi}{2} - h\right) + \dfrac{2k\pi}{b}$.
19. Show that $x = g + c \sin (bh + t)$ is a simple harmonic motion with amplitude c with center at $x = g$. Hint: Put $s = x - g$.
20. Verify that, for the motion of Prob. 19, the acceleration $a = -b^2(x - g) = -b^2x + b^2g$.

From Prob. 19 deduce that each of the following motions is simple harmonic with period $\pi/2$, amplitude 10, and center at 3.

21. $x = 3 + 10 \sin 4t$.
22. $x = 3 + 6 \cos 4t + 8 \sin 4t$.
23. $x = 3 + 10 \cos 4t$.
24. $x = 3 + 8 \cos 4t + 6 \sin 4t$.
25. $x = 20 \cos^2 2t - 7$.
26. $x = 13 - 20 \sin^2 2t$.
27. Find the amplitude and period of a simple harmonic motion for which $v = 6$ when $x = 4$, while $v = 8$ when $x = 3$.
28. If $x_1 = 3\sqrt{2} \sin (8t + \pi/4)$ and $x_2 = \sin 8t$, show that $x = x_1 + x_2$ is a simple harmonic motion.
29. If $x_i = c_i \sin (bt + h_i)$ for $i = 1, 2, 3, \cdots, n$, show that $x = x_1 + x_2 + \cdots + x_n$ is a simple harmonic motion.

111. Applied Problems. We are now able to find the derivatives of expressions involving trigonometric or inverse trigonometric functions.

Hence we can now solve applied problems in maxima and minima or rates of the kind discussed in Secs. 50, 57, and 59 which lead to trigonometric expressions.

EXAMPLE 1. For what value of x in the first quadrant is $y = 4x - \tan x$ a maximum?

Solution: We find $dy/dx = 4 - \sec^2 x$. This is zero when $4 - \sec^2 x = 0$, $\sec^2 x = 4$, $\sec x = \pm 2$. For x in the first quadrant, we must take $\sec x = 2$. Hence $\cos x = \frac{1}{2}$ and $x = \cos^{-1} \frac{1}{2} = \pi/3$. Since $d^2y/dx^2 = -2 \sec^2 x \tan x$ is negative for $x = \pi/3$, we have a maximum. Hence $\pi/3$ is the required value.

EXAMPLE 2. For what value of x in the first quadrant is $y = \dfrac{3 \cos x - 4 \sin x + 8}{\cos x}$ a minimum?

Solution: We calculate the derivative

$$\begin{aligned}\frac{dy}{dx} &= \frac{(-3 \sin x - 4 \cos x) \cos x - (-\sin x)(3 \cos x - 4 \sin x + 8)}{\cos^2 x} \\ &= \frac{-3 \sin x \cos x - 4 \cos^2 x + 3 \cos x \sin x - 4 \sin^2 x + 8 \sin x}{\cos^2 x} \\ &= \frac{-4 + 8 \sin x}{\cos^2 x}.\end{aligned}$$

This is zero when $-4 + 8 \sin x = 0$, $8 \sin x = 4$, or $\sin x = \frac{1}{2}$. And $x = \pi/6$. As x increases through $\pi/6$, $-4 + 8 \sin x$ increases through zero, or changes from minus to plus. And so does dy/dx, since $\cos^2 x$ is positive. It follows that we have a minimum. Hence $\pi/6$ is the required value of x.

EXAMPLE 3. A corridor of width 27 ft. runs at right angles to a passageway of width 8 ft. A thin pole is to be carried in a horizontal position from the corridor into the passageway. If it just clears the corner, what is its greatest permissible length?

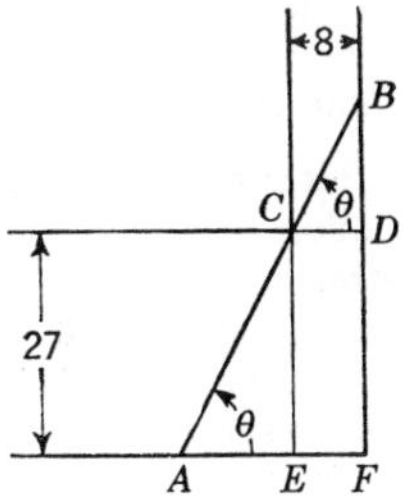

FIG. 133.

Solution: In Fig. 133, let ACB be any straight line through C. As the pole will clear if its length does not exceed AB in any position, we seek the minimum of AB. Let $\theta = \angle EAC$. Then $\angle DCB = \theta$. And $EC/AC = \sin \theta$, $AC = EC/\sin \theta = 27 \csc \theta$. Also $CD/CB = \cos \theta$, $CB = CD/\cos \theta = 8 \sec \theta$. Hence $AB = AC + CB = 27 \csc \theta + 8 \sec \theta$.

To minimize $y = 27 \csc \theta + 8 \sec \theta$, we find the derivative $dy/d\theta = -27 \cot \theta \csc \theta + 8 \tan \theta \sec \theta$. This is zero if $8 \tan \theta \sec \theta = 27 \cot \theta \csc \theta$ or $\dfrac{\tan \theta \sec \theta}{\cot \theta \csc \theta} = \dfrac{27}{8}$. Hence by Eqs. (26), (29), and (20), we have $\tan^3 \theta = \frac{27}{8}$ and $\tan \theta = \frac{3}{2}$. And $\theta = \tan^{-1} \frac{3}{2} = 0.983$. As θ increases through this value, $\tan \theta$ increases. Also $\tan^3 \theta - \frac{27}{8}$ increases through zero, or changes from minus to plus. And so does $dy/dx = 8 \cot \theta \csc \theta (\tan^3 \theta - \frac{27}{8})$, since $\cot \theta \csc \theta$ is positive. It follows that we have a minimum for y.

To find this minimum, we note that for $\tan \theta = \frac{3}{2}$, $\sin \theta = 3/\sqrt{13}$, $\cos \theta = 2/\sqrt{13}$. Hence $y = 27 \csc \theta + 8 \sec \theta = 27 (\sqrt{13}/3) + 8 (\sqrt{13}/2) = 13 \sqrt{13}$. And the required greatest permissible length is $13 \sqrt{13} = 46.9$ ft.

EXAMPLE 4. What is the volume of the right circular cone of greatest volume that can be cut from a sphere of radius a?

Solution: A cross section of the sphere and inscribed cone is shown in Fig. 134. Let $\theta = \angle EOD$. Then $ED/OD = \sin\theta$, $OE/OD = \cos\theta$. And since $OD = a$, $ED = OD\sin\theta = a\sin\theta$, $OE = OD\cos\theta = a\cos\theta$. Hence $AE = AO + OE = a + a\cos\theta$. If V is the volume of the cone generated by revolving triangle AED about AB, we have $V = \frac{1}{3}\pi\,\overline{ED}^2\overline{AE} = \frac{1}{3}\pi a^2\sin^2\theta(a + a\cos\theta)$ or $V = \frac{\pi}{3}a^3(\sin^2\theta + \sin^2\theta\cos\theta)$.

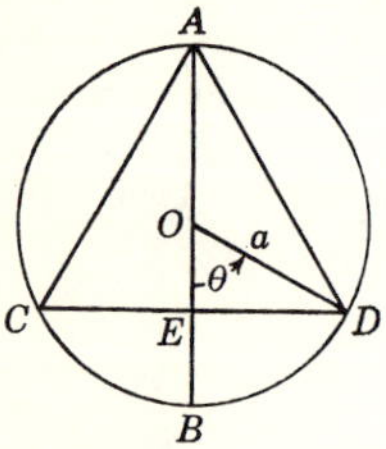

FIG. 134.

And

$$\frac{dV}{d\theta} = \frac{\pi}{3}a^3(2\sin\theta\cos\theta + 2\sin\theta\cos^2\theta - \sin^3\theta)$$

$$= \frac{\pi}{3}a^3\sin\theta(2\cos\theta + 2\cos^2\theta - \sin^2\theta).$$

This will be zero if $\sin\theta = 0$ or if $2\cos\theta + 2\cos^2\theta - \sin^2\theta = 0$. Since $\sin^2\theta = 1 - \cos^2\theta$, the second relation leads to $3\cos^2\theta + 2\cos\theta - 1 = 0$, a quadratic in $\cos\theta$ with roots -1 and $\frac{1}{3}$. Hence we may write the derivative of V in the form

$$\frac{dV}{d\theta} = \frac{\pi}{3}a^3\sin\theta(\cos\theta + 1)(3\cos\theta - 1).$$

If $\sin\theta = 0$, $V = 0$. And if $\cos\theta = -1$, $\sin\theta = 0$ and $V = 0$. Hence these values which make $dV/d\theta = 0$ cannot make V a maximum. But if $\cos\theta = \frac{1}{3}$, $\theta = \cos^{-1}\frac{1}{3} = 1.231$. As θ increases through this value, $\cos\theta$ decreases. Also $3\cos\theta - 1$ decreases through zero, or changes from plus to minus. And so does $dV/d\theta$, since $\sin\theta(\cos\theta + 1)$ is positive. It follows that we have a maximum of V.

To find this maximum, we note that if $\cos\theta = \frac{1}{3}$, $\sin^2\theta = 1 - \frac{1}{9} = \frac{8}{9}$, and $V = \frac{\pi}{3}a^3(\sin^2\theta + \sin^2\theta\cos\theta) = \frac{\pi}{3}a^3\left(\frac{8}{9} + \frac{8}{9}\frac{1}{3}\right) = \frac{32}{81}\pi a^3$. Hence the required maximum volume is $V = \frac{32}{81}\pi a^3$.

EXAMPLE 5. A searchlight in a lighthouse is a mi. offshore and revolves at the uniform rate of ω radians per second. The shore is a straight line, and the illuminated portion subtends an angle A at the light. How fast is this illuminated portion increasing when the nearest edge is x mi. from the point on the shore nearest the lighthouse?

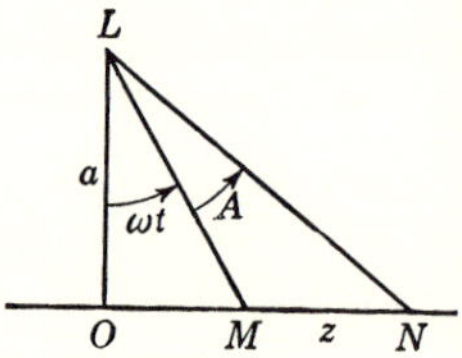

FIG. 135.

Solution: In Fig. 135, L is the lighthouse, O is the nearest point on the shore, and MN is the illuminated portion. If the nearest edge was at O at time $t = 0$, and at M at time t, we have $\angle OLM = \omega t$. Hence $OM = a\tan\omega t$. And $\angle OLN = \omega t + A$, so that $ON = a\tan(\omega t + A)$. Thus $MN = ON - OM = a\tan(\omega t + A) - a\tan\omega t$. If $z = MN$, we have

$$z = a\tan(\omega t + A) - a\tan\omega t$$

as the width of the illuminated portion of the shore. From this we find $dz/dt = a\omega\sec^2(\omega t + A) - a\omega\sec^2\omega t = a\omega[\tan^2(\omega t + A) - \tan^2\omega t]$, by Eq. (30). To evaluate this when $OM = x$, we deduce that $\tan\omega t = x/a$ and by Eq. (27) $\tan(\omega t + A) = \frac{\tan\omega t + \tan A}{1 - \tan\omega t\tan A} = \frac{(x/a) + \tan A}{1 - (x/a)\tan A} = \frac{x + a\tan A}{a - x\tan A}$. Hence, in terms of x, we have $\frac{dz}{dt} = a\omega\left[\left(\frac{x + a\tan A}{a - x\tan A}\right)^2 - \left(\frac{x}{a}\right)^2\right]$, which is the required rate.

EXERCISE 57

What value of x in the first quadrant makes each of the following given functions a minimum?

1. $y = \cot x + 2x$.
2. $y = x - 2 \sin x$.
3. $y = 2 \sec x - \tan x$.
4. $y = \sec x + \csc x$.

What value of x in the first quadrant makes each of the following given functions a maximum?

5. $y = 4x - 3 \tan x$.
6. $y = x + 2 \cos x$.
7. $y = \cot x - 2 \csc x$.
8. $y = 2 - \sec x - 8 \csc x$.

Find the greatest and least values of each of the following given functions.

9. $y = 5 \sin x + 12 \cos x$.
10. $y = \sin x - \cos x$.
11. $y = \cos 2x + 2 \cos x$.
12. $y = \sin 2x + 2 \sin x$.
13. $y = \cos x \cos (x + 1)$.
14. $y = \sin x \cos (x - 2)$.

15. The turning effect of a ship's rudder is $T = a \cos \theta \sin^2 \theta$, where θ is the angle the rudder makes with the keel. Show that E is greatest when $\theta = \tan^{-1} \sqrt{2} = 54.7°$.
16. The percentage error of a tangent galvanometer for a small error in reading is $P = a(\tan \theta + \cot \theta)$. For what value of θ will P be least?
17. A long piece of sheet metal is 18 in. wide. It is to be made into an open gutter by bending up one-third of the sheet on each side through the same angle θ. Show that the area of the trapezoidal cross section is $36 \sin \theta(1 + \cos \theta)$, and by maximizing this, find the angle θ for which the gutter has maximum capacity.
18. A weight W is dragged along the ground by a rope making an angle θ with the horizontal. If the coefficient of friction is $k = \tan A$, and the force on the rope is F, $F \cos \theta = k(W - F \sin \theta)$, so that $F = \dfrac{kW}{\cos \theta + k \sin \theta}$. Show that the force F is least when $\theta = \tan^{-1} k = A$. Also that this least force is $W \sin A$.
19. Check Prob. 18 by showing that $\theta = A$ maximizes $\dfrac{1}{F} = \dfrac{1}{W \sin A} \cos (\theta - A)$.
20. A triangle has two equal sides, each of length b. Find the angle between these sides for which the area of the triangle is greatest.

If a cylinder is inscribed in a sphere of radius a, and 2θ is the angle at the center subtended by the diameter of one base, the radius of the cylinder is $r = a \sin \theta$. And its height is $h = 2a \cos \theta$. For what value of θ does the cylinder have greatest

21. Volume, $V = \pi r^2 h = 2\pi a^3 \sin^2\theta \cos \theta$?
22. Lateral surface, $L = 2\pi rh = 4\pi a^2 \sin \theta \cos \theta = 2\pi a^2 \sin 2\theta$?
23. Total surface, $S = 2\pi r^2 + 2\pi rh = 2\pi a^2(\sin^2 \theta + 2 \sin \theta \cos \theta)$?

24. What are the dimensions of the cone of greatest lateral surface L inscribed in a sphere of radius a? With the notation of Fig. 134, the radius of the cone $r = a \sin \theta$, and the slant height $l = 2a \cos (\theta/2)$, so that $L = \pi rl = 2\pi a^2 \sin \theta \cos (\theta/2)$.
25. A girder 27 ft. long is to be moved on rollers along a corridor and into a passageway 8 ft. wide which meets the corridor at right angles. Neglect the horizontal thickness of the girder and find how wide the corridor must be in order that the girder may just clear the corner.
26. On top of a wall 27 ft. high, another parallel wall is inset 1 ft. What is the length of the shortest ladder that will just reach from the ground to the upper wall?

27. The efficiency of a screw is $E = \dfrac{\tan \theta}{\tan (\theta + A)}$, where θ is the pitch angle of the screw and $\tan A$ is the coefficient of friction. Show that E is greatest when $\theta = (\pi/4) - (A/2) = \tan^{-1} (\tan A + \sec A)$.
28. A tablet 7 ft. high is placed on a wall with its base 9 ft. above the level of an observer's eye. For what distance of his eye from the wall is the angle of vision subtended by the tablet greatest?
29. A man starts at point A and walks 30 ft. due east. He then turns and walks north at the rate of 5 ft./sec. If a spotlight at A follows him, at what rate is the beam turning 8 sec. after he turned north?
30. A revolving light in a lighthouse $\frac{1}{2}$ mi. offshore makes one revolution each minute. If the shore is a straight line, how fast is the ray of light moving along the shore when it passes a point 1 mi. from the point on shore nearest to the light?
31. The minute hand of a clock is 4 ft. long, and the hour hand is 3 ft. long. How fast are the ends of the hands separating at 9 o'clock?

CHAPTER 8

LOGARITHMS AND EXPONENTIAL FUNCTIONS

Certain numerical computations are best performed by using tables of logarithms, or their representation as scales of a slide rule. These are the *common* logarithms, or logarithms to the base 10. Thus $y = \log_{10} x$ if $10^y = x$. More generally, for any positive number a not equal to 1, $y = \log_a x$ if $a^y = x$. That is, the logarithmic function $y = \log_a x$ and the exponential function $y = a^x$ are inverse functions. The properties of these two functions are closely related.

We begin this chapter with a brief review of the definitions and fundamental properties of the exponential function as developed in algebra. And we also review the related properties of logarithms. The important limiting relation $\lim_{h\to 0} (1 + h)^{1/h} = e = 2.71828$, to five decimal places, is then discussed. We then consider *natural* logarithms, or logarithms to the base $e = 2.71828$ because with this choice of base many formulas of the calculus take their simplest form. We define $y = \ln x = \log_e x$.

We develop the special rules of differentiation for the functions $y = \ln x$ and $y = e^x$. And from these we derive the rules for $y = \log_a x$ and $y = a^x$. The process of logarithmic differentiation is explained. And this process is used to find the derivative of u^v and to prove the general power rule for irrational values of the exponent.

Among other applications of these differentiation rules we consider the compound interest law, exponential growth and decay, and damped harmonic motion.

R112. Laws of Exponents. Let a be any positive number and n be any positive integer. Then the symbol a^n means the result of raising a to the nth power by multiplication. For example,

$$a^1 = a, \qquad a^2 = a \times a, \qquad a^3 = a \times a \times a. \tag{1}$$

It follows from this definition that

$$a^m a^n = a^{m+n}, \tag{2}$$

$$\frac{a^m}{a^n} = a^{m-n}, \tag{3}$$

$$(a^m)^n = a^{mn}. \tag{4}$$

In algebra, the symbol a^x is so defined that the rules for positive integral exponents, Eqs. (2) through (4), apply in all cases. First we define $a^0 = 1$. For this makes

$$\frac{a^m}{a^m} = a^{m-m} \qquad \text{or} \qquad 1 = a^0. \tag{5}$$

Next we define $a^{-m} = 1/a^m$. For this makes

$$\frac{a^n}{a^{n+m}} = a^{n-(n+m)} \qquad \text{or} \qquad \frac{1}{a^m} = a^{-m}. \tag{6}$$

For a positive fractional exponent, p/q, we define $a^{p/q}$ as the qth root of the pth power of a. For this makes

$$(a^{p/q})^q = a^p \qquad \text{which is } a^{(p/q)q}. \tag{7}$$

It then follows as in Eq. (6) that the appropriate definition of $a^{-p/q}$ is $1/a^{p/q}$, or the reciprocal of the qth root of the pth power of a.

Any irrational number u is the limit of a sequence of rational numbers, r_n, for example its decimal approximations. And the definition

$$a^u = \lim_{m\to\infty} a^{r_m} \tag{8}$$

gives the same value of a^u for all approximating sequences.

These definitions determine the value of a^x for any real value of x. And the resulting values are such that the *laws of exponents*

$$a^x a^y = a^{x+y}, \tag{9}$$

$$\frac{a^x}{a^y} = a^{x-y}, \tag{10}$$

$$(a^x)^z = a^{xz}, \tag{11}$$

hold for all real values of x and z.

R113. The Exponential Function. Let a be any positive number other than unity. Then for every real number x, a^x has a definite value as described in Sec. 112. Thus,

$$y = a^x \tag{12}$$

defines y as a function of x. This function is continuous for all values of x. And for any a greater than 1, its graph has the general form shown in Fig. 136.

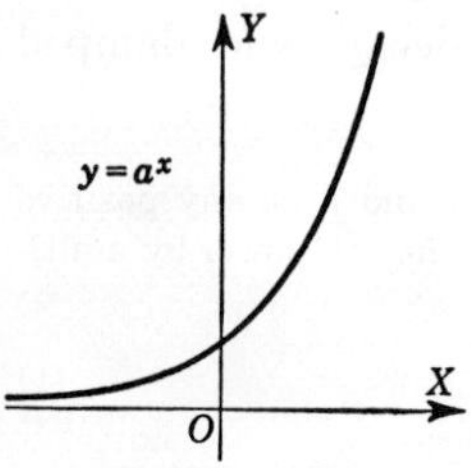

FIG. 136.

R114. The Logarithmic Function. The symbol $\log_a x$, read "logarithm of x to the base a," is defined as the power of a which equals x. That is, for any $a \neq 1$, we have

$$y = \log_a x \qquad \text{if } a^y = x. \tag{13}$$

Thus the logarithmic function is the inverse of the exponential function with the same base. And the graph of the logarithmic function may be obtained from that of the corresponding exponential function by interchanging the roles of x and y. From Fig. 136 it follows that for any $a > 1$ the graph of $y = \log_a x$ has the general form shown in Fig. 137. This function is defined and continuous for all *positive* values of x.

R115. Logarithms. If we substitute the value of y from the first relation of Eq. (13) into the second relation, we obtain

$$a^{\log_a x} = x. \tag{14}$$

This holds for any positive value of a.

And if we substitute the value of x from the second relation of Eq. (13) into the first relation, we obtain

$$y = \log_a (a^y). \tag{15}$$

This holds for any real value of y.

Each of the Eqs. (14) and (15) expresses the same inverse relation of the logarithmic to the exponential function as Eq. (13). But the alternative forms are often convenient, and the student should learn to recognize them with any letters in place of a, x, and y. We illustrate the use of Eqs. (14) and (15) by reviewing the deduction of the rules for logarithms from the law of exponents.

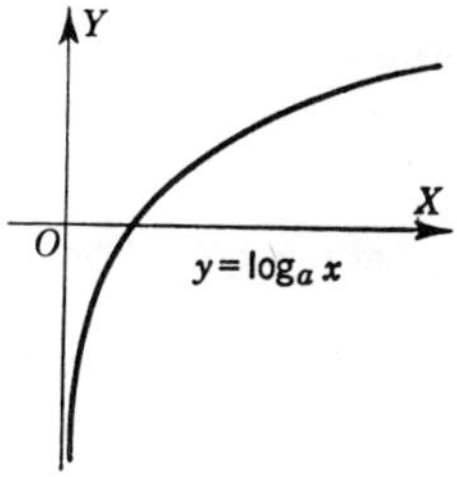

FIG. 137.

From Eq. (14), with $x = X$ and then $x = Z$, we find

$$a^{\log_a X} = X, \qquad a^{\log_a Z} = Z. \tag{16}$$

From these relations and Eq. (9), it follows that

$$XZ = a^{\log_a X} a^{\log_a Z} = a^{\log_a X + \log_a Z}. \tag{17}$$

We may deduce from this and Eq. (15) with $y = \log_a X + \log_a Z$ that

$$\log_a (XZ) = \log_a (a^{\log_a X + \log_a Z}) = \log_a X + \log_a Z. \tag{18}$$

Again, from Eqs. (16) and (10), it follows that

$$\frac{X}{Z} = \frac{a^{\log_a X}}{a^{\log_a Z}} = a^{\log_a X - \log_a Y}. \tag{19}$$

We may deduce from this and Eq. (15) that

$$\log_a \frac{X}{Z} = \log_a (a^{\log_a X - \log_a Z}) = \log_a X - \log_a Z. \tag{20}$$

Finally, from Eqs. (16) and (11) it follows that

$$X^z = (a^{\log_a X})^z = a^{z \log_a X}. \tag{21}$$

We may deduce from this and Eq. (15) that

$$\log_a (X^z) = \log_a (a^{z \log_a X}) = z \log_a X. \tag{22}$$

We summarize those properties of logarithms which are basic for their use in computation in the following equations:

$$\log_a XZ = \log_a X + \log_a Z. \tag{23}$$

$$\log_a \frac{X}{Z} = \log_a X - \log_a Z. \tag{24}$$

$$\log_a (X^n) = n \log_a X, \qquad \log_a (\sqrt[n]{X}) = \log_a (X^{1/n}) = \frac{1}{n} \log_a X. \tag{25}$$

$$\log_a 1 = 0, \qquad \log_a \frac{1}{X} = -\log_a X. \tag{26}$$

We proved Eq. (23) in Eq. (18), and Eq. (24) in Eq. (20). Also Eq. (25) is essentially Eq. (22), first with $z = n$ and then with $z = 1/n$. To verify the first relation of Eq. (26), we note from Eq. (15) that $0 = \log_a (a^0) = \log_a 1$ since $a^0 = 1$ by Eq. (5). And by Eq. (24), we have

$$\log_a \frac{1}{X} = \log_a 1 - \log_a X = 0 - \log_a X = -\log_a X. \tag{27}$$

This proves the second relation of Eq. (26).

***116. The Number *e*.** In differentiating the logarithmic function, we shall find it useful to know that $\left(1 + \frac{1}{h}\right)^h$ approaches a limit as h approaches zero. This limit is denoted by e and is approximately equal to 2.71828. Thus

$$\lim_{h\to 0} (1 + h)^{1/h} = e = 2.71828. \tag{28}$$

Let $z = 1/h$. Then $z \to \infty$ as $h \to 0$. And $h = 1/z$ so that the limit to be studied is

$$\lim_{h\to 0} (1 + h)^{1/h} = \lim_{z\to\infty} \left(1 + \frac{1}{z}\right)^z. \tag{29}$$

First let $z = n$ taking on positive integral values only. And recall the binomial theorem of algebra

$$(a + b)^n = a^n + na^{n-1}b + \frac{n(n-1)}{1\cdot 2} a^{n-2}b^2 + \frac{n(n-1)(n-2)}{1\cdot 2\cdot 3} a^{n-3}b^3 + \cdots + nab^{n-1} + b^n. \tag{30}$$

With $a = 1$ and $b = 1/n$, the binomial expansion of Eq. (30) becomes

$$\left(1 + \frac{1}{n}\right)^n = 1 + n\left(\frac{1}{n}\right) + \frac{n(n-1)}{1\cdot 2}\left(\frac{1}{n}\right)^2 + \frac{n(n-1)(n-2)}{1\cdot 2\cdot 3}\left(\frac{1}{n}\right)^3 + \cdots + \left(\frac{1}{n}\right)^n. \tag{31}$$

Using $n!$, read "factorial n," to mean $1\cdot 2\cdot 3\cdot \cdots \cdot n$, the last term may be written

$$\frac{n(n-1)\cdots 2\cdot 1}{n!}\left(\frac{1}{n}\right)^n = \frac{1}{n!}\left(1 - \frac{1}{n}\right)\left(1 - \frac{2}{n}\right)\cdots\left(1 - \frac{n-1}{n}\right). \tag{32}$$

Now let n increase to infinity through positive integral values. Then $n \to \infty$, $\frac{1}{n} \to 0$, and $\left(1 - \frac{a}{n}\right) \to 1$ for fixed a. Hence the limit of the $(k + 1)$st term in the right member of Eq. (31) is

$$\lim_{n\to\infty} \frac{1}{k!}\left(1 - \frac{1}{n}\right)\left(1 - \frac{2}{n}\right)\cdots\left(1 - \frac{k}{n}\right) = \frac{1}{k!}. \tag{33}$$

This suggests that the limit of the right member of Eq. (31) is an infinite series† whose $(k + 1)$st term is $1/k!$. It is shown in Example 1 below that this series converges. We denote its sum by e, so that

$$e = 1 + 1 + \frac{1}{2!} + \frac{1}{3!} + \cdots + \frac{1}{k!} + \cdots. \tag{34}$$

† The sum of a convergent infinite series is discussed in Sec. 233.

To evaluate this series we note that the successive terms are as tabulated on the right. These are computed successively easily, by dividing $1/(k-1)!$ by k to obtain $1/k!$. When we round these terms off to six decimal places, $1/10!$ and all the following terms are zero. Since the round-off errors may affect the last figure, we would expect 2.71828 to be a good approximation to e. And it is in fact correct to the number of decimal places written.†

$$
\begin{aligned}
1 &= 1\\
1 &= 1\\
\frac{1}{2!} &= 0.5\\
\frac{1}{3!} &= 0.166667\\
\frac{1}{4!} &= 0.041667\\
\frac{1}{5!} &= 0.008333\\
\frac{1}{6!} &= 0.001389\\
\frac{1}{7!} &= 0.000198\\
\frac{1}{8!} &= 0.000025\\
\frac{1}{9!} &= 0.000003\\
\frac{1}{10!} &= 0.000000\\
 & \overline{2.718282}
\end{aligned}
$$

It is shown in Example 2 below that as $n \to \infty$ the limit of the right member of Eq. (31) is the series of Eq. (34) which was obtained by taking the limits of the separate terms. Hence as $n \to \infty$ the left member of Eq. (31) approaches a limit. And this limit is the e of Eq. (34). It follows that

$$\lim_{n \to \infty} \left(1 + \frac{1}{n}\right)^n = e = 2.71828. \tag{35}$$

Finally, in Prob. 20 of Exercise 58 we show that the same limit is approached by $\left(1 + \frac{1}{z}\right)^z$ when $z \to \infty$ through any real values. It then follows from Eqs. (35) and (29) that Eq. (28) is true.

It is interesting to note how the numerical values obtained by the use of tables of common logarithms conform to the limiting results of Eqs. (35) and (28). From Eq. (25) we have

$$\log_{10} \left(1 + \frac{1}{n}\right)^n = n \log_{10} \left(1 + \frac{1}{n}\right) \tag{36}$$

and

$$\log_{10} (1 + h)^{1/h} = \frac{1}{h} \log_{10} (1 + h). \tag{37}$$

Let us take $n = 1{,}000$, or $h = 0.001$, and compute $(1.001)^{1{,}000}$. From a five-place table of logarithms, we find that $\log_{10} 1.001 = 0.00043$. Hence $\log_{10} (1.001)^{1{,}000} = 0.43$, and to two figures $(1.001)^{1{,}000} = 2.7$. This same value would result for any value of h less than 0.001, since if θ is any number between 0 and 1, by interpolation between $\log_{10} 1 = 0$ and $\log_{10} 1.001 = 0.00043$, we find $\log_{10} \left(1 + \frac{\theta}{1{,}000}\right) = \theta(0.00043)$. And hence by Eq. (37), $\log_{10} \left(1 + \frac{\theta}{1{,}000}\right)^{1{,}000/\theta} = 0.43$, as before.

A six-place table gives $\log_{10} (1.0001)^{10{,}000} = 0.434$ and the same value for any h less than 0.0001. This makes $(1.0001)^{10{,}000} = 2.72$ to two decimal places. A twelve-place table gives $\log_{10} (1.000001)^{1{,}000{,}000} = 0.434294$, and the same value for any h less than 10^{-6}. This shows that $(1.000001)^{1{,}000{,}000} = 2.71828$, in accord with the value found in Eq. (35).

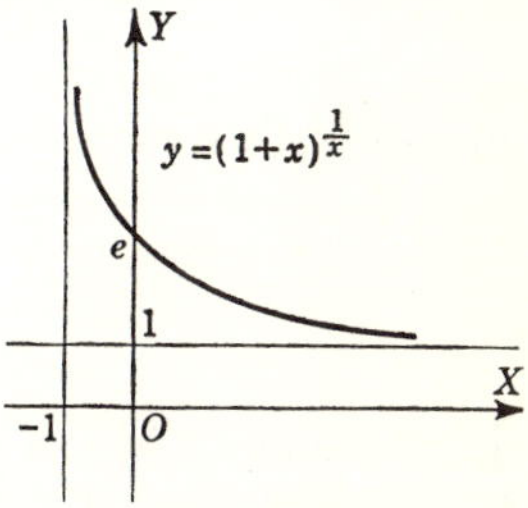

Fig. 138.

The function $(1 + x)^{1/x}$ is defined for all values of x greater than -1, except $x = 0$. If we set $y = (1 + x)^{1/x}$ for $-1 < x < 0$ and for $x > 0$, and $y = e$ for $x = 0$, it follows from Eq. (28) that y is a continuous function of x for all values of x greater than -1. The graph of this function is shown in Fig. 138. The curve has a vertical asymptote $x = -1$ by Sec. 15 since

$$\lim_{x \to -1+} (1 + x)^{1/x} = +\infty, \tag{38}$$

† Though not proved by the calculation just made. See the example in Sec. 254.

as shown in Prob. 21 of Exercise 58. And the curve has a horizontal asymptote $y = 1$ by Sec. 15, since

$$\lim_{x\to+\infty} (1 + x)^{1/x} = 1, \tag{39}$$

as shown in Prob. 22 of Exercise 58.

EXAMPLE 1. Show that the series $1 + 1 + \frac{1}{2!} + \frac{1}{3!} + \cdots + \frac{1}{k!} + \cdots$ converges.

Solution: Each term of this series is positive. Hence the sum of the first k terms increases when k increases. And this partial sum must approach a limit if no finite partial sum exceeds some fixed number.† But for the partial sum to $(k + 1)$ terms, we have

$$1 + 1 + \frac{1}{2!} + \frac{1}{3!} + \cdots + \frac{1}{k!} = 2 + \frac{1}{2} + \frac{1}{2 \cdot 3} + \cdots + \frac{1}{2 \cdot 3 \cdots k}$$
$$< 2 + \frac{1}{2} + \frac{1}{2^2} + \cdots + \frac{1}{2^{k-1}}. \tag{40}$$

And we recall from algebra that the sum of a geometric progression

$$a + ar + ar^2 + \cdots + ar^{k-1} = a\frac{r^k - 1}{r - 1} = a\frac{1 - r^k}{1 - r}. \tag{41}$$

Putting $a = 1$ and $r = \frac{1}{2}$, we find from the second form that

$$1 + \frac{1}{2} + \left(\frac{1}{2}\right)^2 + \cdots + \left(\frac{1}{2}\right)^{k-1} = \frac{1 - (\frac{1}{2})^k}{1 - \frac{1}{2}} = 2 - \frac{1}{2^{k-1}}. \tag{42}$$

It follows from this and Eq. (40) that

$$1 + 1 + \frac{1}{2!} + \frac{1}{3!} + \cdots + \frac{1}{k!} < 3 - \frac{1}{2^{k-1}} < 3. \tag{43}$$

Thus no partial finite sum exceeds 3. Hence the partial sum approaches a limit as k increases indefinitely, and the series converges.

EXAMPLE 2. Show that, as $n \to \infty$, the limit of the right member of Eq. (31) is the sum of the infinite series of Example 1.

Solution: Denote the sum of the series of Example 1 by e, so that

$$e = 1 + 1 + \frac{1}{2!} + \frac{1}{3!} + \cdots + \frac{1}{k!} + \cdots. \tag{44}$$

By Eqs. (31) and (32) we are concerned with the limit of

$$\left(1 + \frac{1}{n}\right)^n = 1 + 1 + \frac{1}{2!}\left(1 - \frac{1}{n}\right) + \frac{1}{3!}\left(1 - \frac{1}{n}\right)\left(1 - \frac{2}{n}\right)$$
$$+ \cdots + \frac{1}{n!}\left(1 - \frac{1}{n}\right)\left(1 - \frac{2}{n}\right) \cdots \left(1 - \frac{n-1}{n}\right). \tag{45}$$

As n increases, the number of terms increases. Also for every positive integer a, a/n decreases. And $(1 - a/n)$ increases but never exceeds unity. Hence each term on the right of Eq. (45) increases but remains less than the corresponding term in the series of Eq. (44). Thus the sum of the terms on the right in Eq. (45) increases, and this sum never exceeds e. It follows that the sum approaches a limit and that this limit is at most e. That is,

† See Sec. 235.

$$\lim_{n \to +\infty} \left(1 + \frac{1}{n}\right)^n = L \quad \text{and} \quad L \leqq e. \tag{46}$$

But each term on the right of Eq. (45) is positive and approaches as a limit the corresponding term in the series of Eq. (44). Hence the limit L of Eq. (46) must be greater than any partial sum of the series of Eq. (44). It follows that for any fixed value of k

$$e \geqq L > 1 + 1 + \frac{1}{2!} + \frac{1}{3!} + \cdots + \frac{1}{k!}. \tag{47}$$

But by Eq. (44), the partial sum on the right of Eq. (47) is arbitrarily close to e for sufficiently large k. Consequently, Eq. (47) implies that $L = e$. And from Eqs. (46) and (44) it then follows that

$$\lim_{n \to +\infty} \left(1 + \frac{1}{n}\right)^n = e = 1 + 1 + \frac{1}{2!} + \frac{1}{3!} + \cdots + \frac{1}{k!} + \cdots, \tag{48}$$

as was to be proved.

EXERCISE 58

Use Eqs. (28) and (29) to verify each of the following limits:

1. $\lim\limits_{n \to \infty} \left(1 + \frac{1}{n}\right)^{3n} = \left[\lim\limits_{n \to \infty} \left(1 + \frac{1}{n}\right)^n\right]^3 = e^3,$
2. $\lim\limits_{n \to 0} (1 + h)^{5/h} = [\lim\limits_{h \to 0} (1 + h)^{1/h}]^5 = e^5,$
3. $\lim\limits_{n \to \infty} \left(1 + \frac{4}{n}\right)^n = \lim\limits_{z \to \infty} \left(1 + \frac{1}{z}\right)^{4z} = e^4;$

   by putting $n = 4z$ and proceeding as in Prob. 1.
4. $\lim\limits_{x \to 0} (1 + 2x)^{1/x} = \lim\limits_{h \to 0} (1 + h)^{2/h} = e^2,$

   by putting $x = h/2$ and proceeding as in Prob. 2.
5. $\lim\limits_{n \to \infty} \left(1 + \frac{3}{n}\right)^{2n} = \lim\limits_{z \to \infty} \left(1 + \frac{1}{z}\right)^{6z} = e^6,$

   by putting $n = 3z$ and proceeding as in Prob. 1.
6. $\lim\limits_{x \to 0} (1 - 2x)^{3/x} = \lim\limits_{h \to 0} (1 + h)^{-6h} = e^{-6},$

   by putting $x = -h/2$ and proceeding as in Prob. 2.
7. $\lim\limits_{n \to \infty} \left(1 + \frac{x}{n}\right)^n = \lim\limits_{z \to \infty} \left(1 + \frac{1}{z}\right)^{zx} = e^x,$

   by putting $n = zx$ if $x \neq 0$ and proceeding as in Prob. 2. Note that if $x = 0$, each expression written equals 1.
8. From the binomial expansion of Eq. (30), deduce that

$$\begin{aligned}\left(1 + \frac{x}{n}\right)^n &= 1 + n\left(\frac{x}{n}\right) + \frac{n(n-1)}{2!}\left(\frac{x}{n}\right)^2 + \frac{n(n-1)(n-2)}{3!}\left(\frac{x}{n}\right)^3 \\ &\qquad + \cdots + \left(\frac{x}{n}\right)^n \\ &= 1 + x + \frac{x^2}{2!}\left(1 - \frac{1}{n}\right) + \frac{x^3}{3!}\left(1 - \frac{1}{n}\right)\left(1 - \frac{2}{n}\right) \\ &\qquad + \cdots + \frac{x^n}{n!}\left(1 - \frac{1}{n}\right)\left(1 - \frac{2}{n}\right) \cdots \left(1 - \frac{n-1}{n}\right).\end{aligned}$$

9. By using the results of Probs. 7 and 8 and assuming that taking the limits of the separate terms of Prob. 8 gives the correct limit of the sum, formally deduce that

$$e^x = 1 + x + \frac{x^2}{2!} + \frac{x^3}{3!} + \cdots + \frac{x^k}{k!} + \cdots.$$

Use the series of Prob. 9 to compute

10. $e^2 = 7.3892.$

11. $\sqrt{e} = e^{\frac{1}{2}} = 1.6487.$

12. $\frac{1}{e} = e^{-1} = 0.36788.$

13. $\frac{1}{\sqrt{e}} = e^{-\frac{1}{2}} = 0.60653.$

14. Deduce from Eq. (35) that

$$\lim_{n\to\infty}\left(1+\frac{1}{n}\right)^{n+1} = \left[\lim_{n\to\infty}\left(1+\frac{1}{n}\right)^n\right]\left[\lim_{n\to\infty}\left(1+\frac{1}{n}\right)\right] = e.$$

15. Deduce from Eq. (35) with n replaced by N that

$$\lim_{n\to\infty}\left(1+\frac{1}{n+1}\right)^{n+1} = \lim_{N\to\infty}\left(1+\frac{1}{N}\right)^N = e,$$

by noting that if $n \to \infty$ through positive integral values, $N = n + 1$ also takes on positive integral values, and $N \to \infty$.

16. From the result of Prob. 15 deduce that

$$\lim_{n\to\infty}\left(1+\frac{1}{n+1}\right)^n = \frac{\lim_{n\to\infty}\left(1+\frac{1}{n+1}\right)^{n+1}}{\lim_{n\to\infty}\left(1+\frac{1}{n+1}\right)} = e.$$

17. Let p be any positive number. Define a corresponding number n as 0 if $p < 1$, and otherwise the greatest positive integer not exceeding p. Then $n \leqq p < n + 1$. It follows from this that $\frac{1}{n+1} < \frac{1}{p} \leqq \frac{1}{n}$, $1 + \frac{1}{n+1} < 1 + \frac{1}{p} \leqq 1 + \frac{1}{n}$, and

$$\left(1+\frac{1}{n+1}\right)^n < \left(1+\frac{1}{p}\right)^p < \left(1+\frac{1}{n}\right)^{n+1}.$$

Suppose that $p \to +\infty$. Then $n \to \infty$. By taking the limit of the inequality just found and using the results of Probs. 16 and 14, deduce that $\lim_{p\to+\infty}\left(1+\frac{1}{p}\right)^p = e.$

18. From the result of Prob. 17 deduce that

$$\lim_{p\to+\infty}\left(1+\frac{1}{p}\right)^{p+1} = \left[\lim_{p\to+\infty}\left(1+\frac{1}{p}\right)^p\right]\left[\lim_{p\to+\infty}\left(1+\frac{1}{p}\right)\right] = e.$$

19. Let q be any negative number less than -1. Then $-q$ exceeds unity, so that $p = -q - 1$ is a positive number. Also $q = -p - 1$. Verify that this makes the quantity $1 + \frac{1}{q} = 1 - \frac{1}{p+1} = \frac{p}{p+1} = \frac{1}{1 + (1/p)} = \left(1+\frac{1}{p}\right)^{-1}$. And since $q = -(p + 1)$, $\left(1+\frac{1}{q}\right)^q = \left(1+\frac{1}{p}\right)^{p+1}$. From this fact, and the result of Prob. 18, deduce that $\lim_{q\to-\infty}\left(1+\frac{1}{q}\right)^q = e.$

20. Noting that any sequence of values of $z \to \infty$ is either a sequence $p \to +\infty$, a sequence $q \to -\infty$, or a combination of two such sequences, deduce from Probs. 17 and 19 that $\lim_{z \to \infty} \left(1 + \frac{1}{z}\right)^z = e$.

21. To evaluate $y = (1 + x)^{1/x}$ by means of common logarithms, we calculate $\log_{10} y = \frac{1}{x} \log_{10} (1 + x)$. Observing that when $x \to -1+$, $1 + x \to 0+$ and $\log_{10} (1 + x) \to -\infty$, deduce that $\log_{10} y \to +\infty$, and hence $y \to +\infty$ as $x \to -1$. This proves Eq. (38).

22. If x is large, $\log_{10} (x + 1)$ is approximately equal to the number of digits before the decimal point in the expression for x as a decimal number. This is a small fraction of x, which grows smaller as x increases. Accordingly, as $x \to +\infty$, $\frac{1}{x} \log_{10} (1 + x) \to 0$. Then, since $\frac{1}{x} \log_{10} (1 + x) = \log_{10} (1 + x)^{1/x}$, deduce that $(1 + x)^{1/x} \to 1$ as $x \to +\infty$, which is Eq. (39).

117. Natural Logarithms. Let $e = 2.71828$ be the number discussed in Sec. 116. With this number as the base, we may define the logarithm of x to the base e by Eq. (13) as

$$y = \log_e x \qquad \text{if } e^y = x. \tag{49}$$

Logarithms to the base e are called *natural logarithms.* They are widely used in theoretical work. The natural logarithm is denoted by the symbol ln. This enables us to omit the base and still distinguish between natural logarithms and common logarithms. Thus for any positive number p we write

$$\begin{aligned} \text{Natural logarithm of } p \text{ (base } e) \text{ or } \log_e p &= \ln p. \\ \text{Common logarithm of } p \text{ (base 10) or } \log_{10} p &= \log p. \end{aligned} \tag{50}$$

Using the new symbol ln, we may rewrite Eq. (49) in the form

$$y = \ln x \qquad \text{if } e^y = x \qquad \text{and} \qquad e^{\ln x} = x. \tag{51}$$

And since $\log x = \log_{10} x$, from Eq. (13) we have

$$y = \log x \qquad \text{if } 10^y = x \qquad \text{and} \qquad 10^{\log x} = x.$$

Let us find the relation between $\ln p$ and $\log p$ for any positive number p. By the last relation of Eq. (51) we have

$$e^{\ln p} = p. \tag{52}$$

Take the common logarithm of each side of this equation. Using Eq. (25) for the left side, we find

$$\log (e^{\ln p}) = (\ln p)(\log e) = \log p. \tag{53}$$

The factor $\log e$ which converts natural logarithms to common logarithms is called the *modulus* of common logarithms and is denoted by M. And from a table of common logarithms we find that

$$M = \log e = \log 2.71828 = 0.43429. \tag{54}$$

It follows from Eqs. (53) and (54) that

$$\log p = (\log e)(\ln p) = 0.43429 \ln p. \tag{55}$$

We may solve Eq. (53) for ln p and so deduce that

$$\ln p = \frac{\log p}{\log e} = 2.30259 \log p. \tag{56}$$

By using Eqs. (55) and (56), we may find values of ln x and the inverse exponential function e^x by means of a table of common logarithms. We illustrate this in Examples 1 and 2 below. We also show how these functions may be found from tables of e^x and ln x which are usually included in collections of mathematical tables.

Examples 3 and 4 illustrate how to plot graphs of functions involving powers of e or natural logarithms.

EXAMPLE 1. Find the natural logarithm of 15.75.

Solution 1: We write $15.75 = 10 \times 1.575$. From a table of natural logarithms of numbers between 1 and 10, we find $\ln 1.575 = 0.4542$, by interpolating between ln 1.57 and ln 1.58. Also $\ln 10 = 2.3026$. And since $\ln (10 \times 1.575) = \ln 10 + \ln 1.575$, $\ln 15.75 = 2.7568$.

Solution 2: Using common logarithms, we find from the tables that $\log 15.75 = 1.1973$. Hence from Eq. (56), $\ln 15.75 = 2.3026 \times 1.1973 = 2.7569$.

EXAMPLE 2. Find the value of $e^{4.82}$.

Solution 1: From tables of e^x, $e^{4.80} = 121.5$ and $e^{4.90} = 134.3$, so that by interpolation $e^{4.82} = 121.5 + (0.2)(134.3 - 121.5) = 121.5 + 2.6 = 124.1$. Hence $e^{4.82} = 124.1$ with a possible error of 1 or 2 in the last place.

Solution 2: Since 4.82 is not in the range 0 to 2.3026, the natural logarithms of numbers between 1 and 10, we subtract an integral multiple of $\ln 10 = 2.3026$ to get in this range. The multiple is 2, so we calculate $4.82 - 2(2.3026) = 4.82 - 4.6052 = 0.2148$. In the table of natural logarithms of numbers between 1 and 10, we find $\ln 1.23 = 0.2070$ and $\ln 1.24 = 0.2151$. To interpolate for 0.2148, we compute $0.2148 - 0.2070 = 0.0078$, and the difference of tabular values $0.2151 - 0.2070 = 0.0081$. This leads to the interpolation ratio $\frac{78}{81} = 1 - \frac{3}{81} = 1 - 0.04 = 0.96$ from which we conclude that $0.2184 = \ln 1.2396$. Since $2.3026 = \ln 10$, $2(2.3026) = \ln 100$, and $4.82 = 0.2148 + 2(2.3026) = \ln 1.2396 + \ln 100 = \ln 1.2396 \times 100 = \ln 123.96$. Hence from Eq. (51) we find that $e^{4.82} = 123.96$, which is the required value.

Solution 3: Since $\log e^{4.82} = 4.82 \log e$, we compute $4.82 \log e = 4.82\ (0.43429) = 2.1933$. Then from a table of common logarithms we find by interpolation that $2.1933 = \log 123.97$. It follows that $e^{4.82} = 123.97$, in good agreement with the result found in solution 2.

EXAMPLE 3. Plot the graph of $y = e^x$.

Solution: Give x a series of values, as $-2, -1, 0, 1, 2$, and read the corresponding values of $y = e^x$ from the tables of e^x. In this way we find

x	$-\infty$	$\cdots$	-2	-1	0	1	2	$\cdots$	$+\infty$
$y = e^x$	0	$\cdots$	0.14	0.37	1	2.72	7.39	$\cdots$	$+\infty$

By plotting the five finite points and joining them by a smooth curve, we obtain the graph shown in Fig. 139. The fact that as $x \to -\infty$, $y \to 0$ shows that the curve has the x axis as an asymptote.

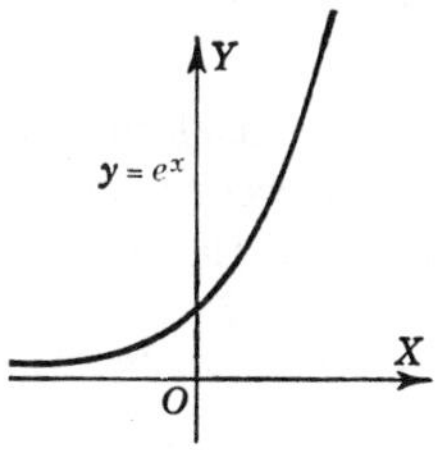

FIG. 139.

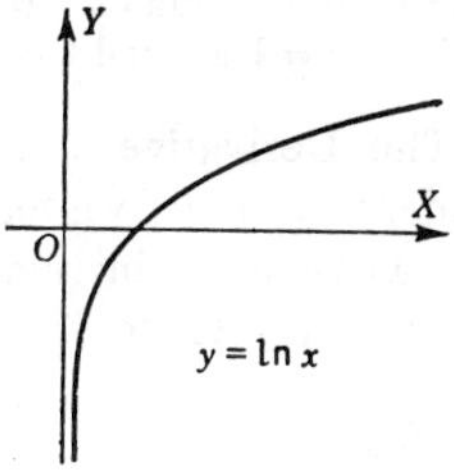

FIG. 140.

EXAMPLE 4. Plot the graph of $y = \ln x$.

Solution: Give x a series of values, as 0.5, 1, 2, 3, and read the corresponding values of $y = \ln x$ from the table of natural logarithms. In this way we find

x	0	0.5	1	2	3	. . .	$+\infty$
$y = \ln x$	$-\infty$	-0.69	0	0.69	1.10	. . .	$+\infty$

Plotting the four finite points and joining them by a smooth curve, we obtain the graph shown in Fig. 140. The fact that as $x \to 0+$, $y \to -\infty$ shows that the curve has the y axis as an asymptote.

The curve of Fig. 140 may be obtained from that of Fig. 139 by interchanging the x and y axes, since $y = \ln x$ makes $x = e^y$ by Eq. (51).

EXERCISE 59

Use any appropriate tables to evaluate

1. $e^{2.15}$.
2. $e^{-2.36}$.
3. $e^{9.8}$.
4. $\ln 2.15$.
5. $\ln 98$.
6. $\ln 0.234$.

Plot the graph of each of the following functions.

7. $y = e^{2x}$.
8. $y = e^{-x}$.
9. $y = 2e^{-2x}$.
10. $y = e^x + e^{-x}$.
11. $y = e^x - e^{-x}$.
12. $y = 2 \ln 2x$.
13. $y = \ln \frac{1}{x}$.
14. $y = e^{-x^2}$.
15. $y = e^{x+1}$.
16. For b and c positive, show that the graph of $y = ce^{bx}$ may be obtained from that of $y = e^x$ by multiplying values of y by c and multiplying values of x by $1/b$. HINT: Note that if (x_1,y_1) is a point on $y = e^x$, then $(x_1/b, cy_1)$ is a point on $y = ce^{bx}$.
17. For any base a greater than 1, show that the graph of $y = ca^x$ may be obtained from that of $y = e^x$ by multiplying values of x by $1/\ln a$ and values of y by c. HINT: Note that $a = e^{\ln a}$, so that $y = ce^{(\ln a)x}$, and use Prob. 16.
18. Derive the rules for changing logarithms from any base a to any base b, namely,

$$\log_a x = \log_a b \log_b x \qquad \text{and} \qquad \log_b x = \log_b a \log_a x.$$

Note that $x = a^{\log_a x} = b^{\log_b x}$. The first equation may be derived from this by taking logarithms to the base a. And the second equation may be derived from this by taking logarithms to the base b.

The two relations imply that $\log_a b = 1/\log_b a$. These three relations may be kept in mind by their analogy to the rules for fractions, $\frac{x}{a} = \frac{b}{a}\frac{x}{b}, \frac{x}{b} = \frac{a}{b}\frac{x}{a}, \frac{b}{a} = \frac{1}{a/b}$.

19. Deduce Eqs. (53) and (55) from Prob. 18 by taking $a = 10$ and $b = e$ and recalling that $\log_{10} x = \log x$ and $\log_e x = \ln x$.

118. The Derivative of ln *u*. Let $y = \ln u$. Considering u as the independent variable, we may find dy/du by the procedure of Sec. 29 as follows. Give u an increment Δu. And let Δy be the corresponding increment in y. Then

$$y + \Delta y = \ln (u + \Delta u). \tag{57}$$

From this and

$$y = \ln u, \tag{58}$$

we find by subtraction that

$$\Delta y = \ln (u + \Delta u) - \ln u. \tag{59}$$

But from Eq. (24) the right member is equal to

$$\ln \frac{u + \Delta u}{u} = \ln \left(1 + \frac{\Delta u}{u}\right). \tag{60}$$

It follows that

$$\frac{\Delta y}{\Delta u} = \frac{1}{\Delta u} \ln \left(1 + \frac{\Delta u}{u}\right) = \frac{1}{u}\frac{u}{\Delta u} \ln \left(1 + \frac{\Delta u}{u}\right). \tag{61}$$

Let us put $\Delta u/u = h$. We may then deduce from Eqs. (61) and (25) that

$$\frac{\Delta y}{\Delta u} = \frac{1}{u}\frac{1}{h} \ln (1 + h) = \frac{1}{u} \ln (1 + h)^{1/h}. \tag{62}$$

Now let Δu approach zero. Since u is fixed, $h = \Delta u/u \to 0$ and

$$\begin{aligned} \lim_{\Delta u \to 0} \frac{\Delta y}{\Delta u} &= \lim_{h \to 0} \frac{1}{u} \ln (1 + h)^{1/h} \\ &= \frac{1}{u} \ln [\lim_{h \to 0} (1 + h)^{1/h}], \end{aligned} \tag{63}$$

since the function $\ln u$ is continuous. But by Eq. (28), we have $\lim_{h \to 0} (1 + h)^{1/h} = e$. And since $e^1 = e$, by Eq. (51) we have $\ln e = 1$. It follows from these facts and Eq. (63) that

$$\frac{dy}{du} = \lim_{\Delta u \to 0} \frac{\Delta y}{\Delta u} = \frac{1}{u} \ln e = \frac{1}{u}. \tag{64}$$

Next let $y = \ln u$ where u is any differentiable function of x, and consider x as the independent variable. Then by the rule for composite functions, Eq. (20) of Sec. 53, we have

$$\frac{dy}{dx} = \frac{dy}{du}\frac{du}{dx}. \tag{65}$$

Replacing dy/du by $1/u$, the value found in Eq. (64), we have

$$\frac{dy}{dx} = \frac{1}{u}\frac{du}{dx}. \tag{66}$$

This proves that

$$\frac{d(\ln u)}{dx} = \frac{1}{u}\frac{du}{dx} = \frac{du/dx}{u}. \tag{67}$$

We formulate this important rule in words as follows:

The derivative of the natural logarithm of a function is equal to a fraction whose numerator is the derivative of the function and whose denominator is the function itself.

EXAMPLE 1. Find dy/dx if $y = \ln (x^4 - 2x^2)$.

Solution: From Eq. (67) with $u = x^4 - 2x^2$ and $du/dx = 4x^3 - 4x$, we have $\frac{d}{dx} \ln (x^4 - 2x^2) = \frac{4x^3 - 4x}{x^4 - 2x^2} = \frac{4x(x^2 - 1)}{x^2(x^2 - 2)} = \frac{4(x^2 - 1)}{x(x^2 - 2)}$. Hence $\frac{dy}{dx} = \frac{4(x^2 - 1)}{x(x^2 - 2)}$ is the required derivative.

EXAMPLE 2. Find dy/dx if $u = \ln x^2 \sqrt{x^2 + 3}$.

Solution: From Eqs. (23) and (25) we may write

$$y = \ln x^2 \sqrt{x^2 + 3} = 2 \ln x + \tfrac{1}{2} \ln (x^2 + 3).$$

By applying Eq. (67) to each logarithm, we may deduce from this that

$$\frac{dy}{dx} = 2\frac{1}{x} + \frac{1}{2}\frac{2x}{x^2 + 3} = \frac{2}{x} + \frac{x}{x^2 + 3} = \frac{2(x^2 + 3) + x^2}{x(x^2 + 3)} = \frac{3x^2 + 6}{x(x^2 + 3)}.$$

Hence $\frac{dy}{dx} = \frac{3(x^2 + 2)}{x(x^2 + 3)}$ is the required derivative.

EXERCISE 60

Find dy/dx for each of the following given functions.

1. $y = \ln (3x + 5)$.
2. $y = \ln (x^2 + 4x)$.
3. $y = \ln 5x^6$.
4. $y = \ln \sqrt{4 - x^2}$.
5. $y = \ln \frac{4}{x}$.
6. $y = (\ln x)^2$.
7. $y = x \ln x - x$.
8. $y = \ln (\ln x)$.
9. $y = \ln (x + \sqrt{x^2 + 4})$.
10. $y = \ln x \sqrt{5x + 3}$.
11. $y = \ln \sqrt{\frac{2 + 3x}{2 - 3x}}$.
12. $y = \sqrt{\frac{x^2 - 2}{x^2 + 2}}$.
13. $y = \ln \sin x$.
14. $y = \ln \sec x$.
15. $y = \ln \tan x$.
16. $y = \ln (\tan x + \sec x)$.
17. $y = \frac{\ln x}{x}$.
18. $y = \ln (\sqrt{x^2 + 4} - x)$.

119. The Derivative of e^u. Let $y = e^u$. Then $u = \ln y$, and by Eq. (67) the derivative of u with respect to y is

$$\frac{du}{dy} = \frac{1}{y}. \tag{68}$$

For any real value of u, $y = e^u \neq 0$. Hence by Eq. (26) of Sec. 54 we have

$$\frac{dy}{du} = \frac{1}{du/dy} = y. \tag{69}$$

Now let u be a differentiable function of x. Then by the rule for composite functions, Eq. (20) of Sec. 53, it follows that

$$\frac{dy}{dx} = \frac{dy}{du}\frac{du}{dx} = y\frac{du}{dx}. \tag{70}$$

But $y = e^u$, so that we have proved the differentiation rule

$$\frac{d(e^u)}{dx} = e^u\frac{du}{dx}. \tag{71}$$

We remind the reader that in Eq. (71) $e = 2.71828$ is the number discussed in Sec. 116. The differentiation rule corresponding to Eq. (71), expressed in words, is as follows:

The derivative of e raised to a variable power is equal to e raised to this same variable power multiplied by the derivative of the power.

EXAMPLE 1. Find dy/dx if $y = 2e^{x^3}$.

Solution: From Eq. (71) with $u = x^3$ and $du/dx = 3x^2$, we find that

$$\frac{d}{dx}(2e^{x^3}) = 2e^{x^3}(3x^2) = 6x^2e^{x^3}.$$

Hence $dy/dx = 6x^2e^{x^3}$ is the required derivative.

EXAMPLE 2. Find dy/dx if $y = e^{-2x}\sin 3x$.

Solution: Using the product rule and Eq. (71) with $u = -2x$, we have

$$\begin{aligned}\frac{d}{dx}(e^{-2x}\sin 3x) &= e^{-2x}\frac{d}{dx}(\sin 3x) + \sin 3x\frac{d}{dx}(e^{-2x})\\ &= e^{-2x}(3\cos 3x) + (\sin 3x)e^{-2x}(-2)\\ &= e^{-2x}(3\cos 3x - 2\sin 3x).\end{aligned}$$

Hence $dy/dx = e^{-2x}(3\cos 3x - 2\sin 3x)$ is the required derivative.

EXERCISE 61

Find dy/dx for each of the following given functions.

1. $y = 2e^{3x}$.
2. $y = 4e^{-2x}$.
3. $y = xe^{2x}$.
4. $y = e^{x^2}$.
5. $y = \dfrac{e^x}{x}$.
6. $y = \dfrac{4}{e^x}$.
7. $y = e^{\sqrt{x}}$.
8. $y = e^{\sin x}$.

9. $y = \frac{x}{e^x}$.
10. $y = e^{-1/x}$.
11. $y = x^2e^{-x}$.
12. $y = e^{-x} \ln x$.
13. $y = \ln (e^x + e^{-x})$.
14. $y = e^{-x} \cos 2x$.
15. $y = \frac{e^x + e^{-x}}{e^x - e^{-x}}$.
16. $y = \frac{e^{2x} - 1}{e^{2x} + 1}$.
17. $y = e^{\tan x}$.
18. $y = e^{\tan^{-1} x}$.

***120. The Derivative of log u.** Let $y = \log u$, where $\log u$ means the common logarithm to the base 10 as in Sec. 117. To find the derivative of this function, we note from Eq. (55) that

$$\log u = (\log e) \ln u. \tag{72}$$

It follows from this and Eq. (67) that

$$\frac{d(\log u)}{dx} = (\log e) \frac{1}{u} \frac{du}{dx}. \tag{73}$$

Again let $y = \log_a u$. Then from Prob. 18 of Exercise 59 we have

$$\log_a u = (\log_a e) \ln u. \tag{74}$$

It follows from this and Eq. (67) that

$$\frac{d(\log_a u)}{dx} = (\log_a e) \frac{1}{u} \frac{du}{dx}. \tag{75}$$

This relation includes Eq. (73) as the special case when $a = 10$. It also includes Eq. (67) as the special case when $a = e$, since $\log_e e = \ln e = 1$. The fact that the choice $a = e$ enables us to omit the factor $(\log_a e)$ from Eq. (75) is why natural logarithms are used almost exclusively in theoretical work involving calculus.

Since logarithms to a general base other than e or 10 are seldom used, the student will have little occasion to remember Eq. (75). And he will find it convenient to express common logarithms in terms of natural logarithms by means of Eq. (72) and so avoid the need for Eq. (73). We have included Eqs. (73) and (75) for the sake of completeness, rather than for their practical use.

Example. Find dy/dx if $y = \log (2x + 3)$.

Solution: If $y = \log (2x + 3) = (\log e) \ln (2x + 3)$, we find

$$\frac{dy}{dx} = (\log e) \frac{d}{dx} \ln (2x + 3) = (\log e) \frac{2}{2x + 3}.$$

Hence $\frac{dy}{dx} = \frac{2 \log e}{2x + 3}$ is the required derivative.

121. The Derivative of a^u. Let $y = a^u$. By Eq. (51) we have $a = e^{\ln a}$. Hence we may write

$$y = a^u = (e^{\ln a})^u = e^{u \ln a}. \tag{76}$$

It follows from Eq. (71) that

$$\frac{dy}{dx} = e^{u \ln a} \frac{d}{dx} (u \ln a) = e^{u \ln a} \ln a \frac{du}{dx}. \tag{77}$$

In view of Eq. (76) this may be written

$$\frac{d(a^u)}{dx} = (\ln a) a^u \frac{du}{dx}. \tag{78}$$

We may formulate this rule in words as follows:

The derivative of the constant a raised to a variable power is equal to the product of ln a, *a raised to this same variable power, and the derivative of the power.*

EXAMPLE. Find dy/dx if $y = 3 \cdot 2^{x^2}$.

Solution: Using Eq. (78) with $a = 2$ and $y = x^2$, $du/dx = 2x$, we have

$$\frac{d}{dx}(3 \cdot 2^{x^2}) = 3(\ln 2)2^{x^2}(2x) = 6(\ln 2)x2^{x^2}.$$

Hence $dy/dx = 6(\ln 2)x2^{x^2}$ is the required derivative.

122. Logarithmic Differentiation. Sometimes the derivative of a function $f(x)$ may be found with least effort by means of *logarithmic differentiation.* This process consists of first taking the natural logarithm of each member of the equation $y = f(x)$ and then differentiating each member of $\ln y = \ln f(x)$. Thus the steps of the process give

$$y = f(x), \quad \ln y = \ln f(x), \quad \frac{1}{y}\frac{dy}{dx} = \frac{d}{dx}[\ln f(x)], \quad \frac{dy}{dx} = y\frac{d}{dx}[\ln f(x)]. \tag{79}$$

Whenever the indicated operations can be performed by our rules for differentiation, the process proves that y is a differentiable function and enables us to find $\frac{dy}{dx}$. For, let $\ln f(x) = u$. Then the process of calculating $\frac{d}{dx}[\ln f(x)] = \frac{du}{dx}$ shows that u is a differentiable function of x. And it then follows from Eq. (71) that $y = f(x) = e^u$ is a differentiable function of x. Hence the use of Eq. (67) to calculate $\frac{d}{dx}(\ln y) = \frac{1}{y}\frac{dy}{dx}$ in Eq. (79) is legitimate.

We might have deduced the final relation of Eq. (79) by the series of steps

$$y = e^u, \qquad \frac{dy}{dx} = e^u\frac{du}{dx} = y\frac{d}{dx}[\ln f(x)], \tag{80}$$

since $y = f(x)$ makes $e^u = f(x)$ which implies that $u = \ln f(x)$.

However, in practice, the procedure of Eq. (79) is easier to remember and is usually preferable as it enables us to take advantage of the rules for logarithms to simplify $\ln f(x)$ before taking its derivative.

We shall next use logarithmic differentiation to derive some differentiation rules.

First consider $y = u^v$, where u and v are each differentiable functions of x, and u is positive for the values under consideration. Taking logarithms, we find from

$$y = u^v \qquad \text{that } \ln y = \ln(u^v) = v \ln u. \tag{81}$$

Now differentiate this by the product rule to deduce that

$$\frac{1}{y}\frac{dy}{dx} = v\left(\frac{1}{u}\frac{du}{dx}\right) + (\ln u)\frac{dv}{dx}. \tag{82}$$

Next multiply both sides by $y = u^v$. The result is

$$\frac{dy}{dx} = y\left[\frac{v}{u}\frac{du}{dx} + (\ln u)\frac{dv}{dx}\right] = vu^{v-1}\frac{du}{dx} + (\ln u)u^v\frac{dv}{dx}. \tag{83}$$

This proves the rule for differentiating a *variable base raised to a variable power*,

$$\frac{d(u^v)}{dx} = vu^{v-1}\frac{du}{dx} + (\ln u)u^v\frac{dv}{dx}. \tag{84}$$

In particular let $v = n$, any constant exponent. Then $dv/dx = 0$, and Eq. (84) reduces to

$$\frac{d(u^n)}{dx} = nu^{n-1}\frac{du}{dx}. \tag{85}$$

This is the power rule which was proved in Sec. 52 to hold for any rational value of n. The proof of Eq. (85) just given shows that the power rule holds for any value of n, whether rational or irrational.

Next let $u = a$, any positive constant, in Eq. (84). Then $du/dx = 0$, and Eq. (84) reduces to

$$\frac{d(a^v)}{dx} = (\ln a)a^v\frac{dv}{dx}. \tag{86}$$

This is equivalent to Eq. (78). A comparison of Eq. (84) with Eqs. (85) and (86) shows that the derivative of u^v consists of the sum of two terms. The first term has the same form as if v were constant and u alone were variable. And the second term has the same form as if u were constant and v alone were variable.

We give some examples of the use of logarithmic differentiation. As illustrated in Example 3, logarithmic differentiation may be advantageous even where no exponential functions appear explicitly in the function to be differentiated, if this function consists of several factors.

EXAMPLE 1. Find dy/dx if $y = x^{x^2}$.

Solution 1: Applying logarithmic differentiation, we find

$$\ln y = x^2 \ln x, \qquad \frac{1}{y}\frac{dy}{dx} = x^2\left(\frac{1}{x}\right) + (\ln x)(2x) = x(1 + 2\ln x),$$

so that $dy/dx = yx(1 + 2\ln x) = x^{x^2+1}(1 + 2\ln x)$, which is the required derivative.

Solution 2: Using Eq. (84) with $u = x$ and $v = x^2$, we obtain

$$\frac{dy}{dx} = x^2x^{x^2-1}(1) + (\ln x)x^{x^2}(2x) = x^{x^2+1}(1 + 2\ln x),$$

which is the required derivative.

EXAMPLE 2. Find dy/dx if $y = 4^{3x}5^{4x}$.

Solution: Applying logarithmic differentiation, we find $\ln y = 3x \ln 4 + 4x \ln 5$, $\frac{1}{y}\frac{dy}{dx} = 3 \ln 4 + 4 \ln 5$, so that $dy/dx = y(3 \ln 4 + 4 \ln 5) = (3 \ln 4 + 4 \ln 5)4^{3x}5^{4x}$.

EXAMPLE 3. Find dy/dx if $y = \frac{(x^5 + 4)^{\frac{2}{5}}}{(x^5 - 2)^{\frac{3}{5}}}$.

Solution: Applying logarithmic differentiation, we find

$$\ln y = \tfrac{2}{5} \ln (x^5 + 4) - \tfrac{3}{5} \ln (x^5 - 2),$$

$$\frac{1}{y}\frac{dy}{dx} = \frac{2}{5}\frac{5x^4}{x^5 + 4} - \frac{3}{5}\frac{5x^4}{x^5 - 2} = x^4\left(\frac{2}{x^5 + 4} - \frac{3}{x^5 - 2}\right) = -x^4 \frac{x^5 + 16}{(x^5 + 4)(x^5 - 2)},$$

$$\frac{dy}{dx} = -yx^4 \frac{x^5 + 16}{(x^5 + 4)(x^5 - 2)} = -x^4(x^5 + 16)(x^5 + 4)^{-\frac{3}{5}}(x^5 - 3)^{-\frac{8}{5}}.$$

Hence $\frac{dy}{dx} = \frac{-x^4(x^5 + 16)}{(x^5 + 4)^{\frac{3}{5}}(x^5 - 2)^{\frac{8}{5}}}$ is the required derivative.

EXERCISE 62

Find dy/dx for each of the following given functions.

1. $y = \log (2x + 3)$.
2. $y = \log_2 (x^2 + 1)$.
3. $y = \log \sin x$.
4. $y = \log \tan x$.
5. $y = 5^x$.
6. $y = 9 \cdot 2^{3x}$.
7. $y = 10^{-x}$.
8. $y = 3^{x^2}$.
9. $y = 3 \cdot 7^{-8x}$.
10. $y = x^2 2^x$.
11. $y = (\sqrt{e})^x$.
12. $y = \left(\frac{1}{e}\right)^x$.

For each of the following given functions, find dy/dx by using logarithmic differentiation.

13. $y = 2^{3x}3^{2x}$.
14. $y = x^{e^x}$.
15. $y = x^x$.
16. $y = x^{-x}$.
17. $y = (\sin x)^x$.
18. $y = x \sqrt{x}$.
19. $y = (2x - 3)^4(4x + 1)^3$.
20. $y = (x^3 + 2)^{\frac{1}{3}}(x^3 + 1)^{\frac{2}{3}}$.
21. $y = x^6 \sqrt{x^2 + 1}$.
22. $y = x^5 (2x + 3)^4$.
23. $y = \sqrt{\frac{x^2 - 3}{x^2 - 1}}$.
24. $y = \sqrt{\frac{1 - x^2}{1 + x^2}}$.

***123. The Compound Interest Law. Exponential Growth and Decay.** Let a sum of money be put at interest at the rate of r per cent a year. Suppose that the original amount was A dollars and that interest is compounded m times each year. Then the rate for one interest period is $r/100m$, so that in each such period \$1 becomes $\left(1 + \frac{r}{100m}\right)$ dollars. And after n interest periods \$1 becomes $\left(1 + \frac{r}{100m}\right)^n$ dollars. If we start with A dollars, after t years or mt interest periods we shall have an amount

$$x_m = A\left(1 + \frac{r}{100m}\right)^{mt}. \tag{87}$$

The principal is said to be continuously compounded† when the number of interest periods each year increases indefinitely so that the length of a single interest period approaches zero. That is, $m \to \infty$ and $r/100m \to 0$. To find the limit of the amount

† The idea of continuously compounded interest is somewhat artificial, but it is analogous to processes of wide occurrence in nature.

x_m of Eq. (87) under these conditions, put $r/100m = h$, $m = r/100h$ and write

$$x_m = A\left(1 + \frac{r}{100m}\right)^{mt} = A(1 + h)^{rt/100h} = A[(1 + h)^{1/h}]^{rt/100}. \tag{88}$$

By Eq. (28) the limit of $(1 + h)^{1/h}$ when $h \to 0$ is e. Hence for continuous compound interest, the amount x after t years is

$$x = \lim_{m\to\infty} x_m = \lim_{h\to 0} A[(1 + h)^{1/h}]^{rt/100} = Ae^{rt/100}. \tag{89}$$

This result may be derived more simply by the method of Sec. 61. For a time Δt, the rate of interest is $r\,\Delta t/100$. Hence the interest gained in time Δt, or Δx, is

$$\Delta x = \bar{x}\frac{r}{100}\Delta t \qquad \text{so that } dx = x\frac{r}{100}dt. \tag{90}$$

It follows from this relation that

$$\frac{dx}{x} = \frac{r}{100}dt. \tag{91}$$

By integrating each member as in Sec. 65 we find that

$$\int \frac{1}{x}dx = \int \frac{r}{100}dt \qquad \text{or} \qquad \ln x = \frac{r}{100}t + C, \tag{92}$$

where $\int \frac{1}{x}dx = \ln x$ because $\frac{d(\ln x)}{dx} = \frac{1}{x}$, and C is the constant of integration. If the initial amount was A, $x = A$ when $t = 0$ so that

$$\ln A = \frac{r}{100}0 + C \qquad \text{and} \qquad C = \ln A. \tag{93}$$

Hence we have

$$\ln x = \frac{r}{100}t + \ln A, \qquad e^{\ln x} = e^{(rt/100)+\ln A} = e^{rt/100}e^{\ln A}. \tag{94}$$

By using Eq. (51), we find from this that

$$x = Ae^{rt/100}. \tag{95}$$

This checks the result found by the limiting process in Eq. (89).

Many processes concerning growth or change lead us to seek a function of the time whose rate of change is proportional to the value of the function. For any such function, $x = f(t)$, we have

$$\frac{dx}{dt} = kx. \tag{96}$$

We may treat this by a process similar to that of Eqs. (90) to (95). In this way we may deduce from Eq. (96) that

$$\frac{dx}{x} = k\,dt \qquad \ln x = kt + C, \qquad x = e^C e^{kt} = Ae^{kt}. \tag{97}$$

When k is positive, we have *exponential growth*. When k is negative, we have *exponential decay*. In this case, as illustrated in Example 4, it may be convenient to write $dx/dt = -kx$, with k a positive constant.

An efficient procedure for handling specific numerical data is illustrated in the examples which follow.

EXAMPLE 1. A sum of \$1,000 is put at interest at the rate of 2 per cent per annum, continuously compounded. What is the amount after t years? After 20 years?

Solution: As in Eq. (90) the rate for time dt is $\frac{2}{100}\,dt$, so that $dx = x\,\frac{2}{100}\,dt$, $\frac{dx}{x} = \frac{dt}{50}$, $\int \frac{1}{x}\,dx = \int \frac{1}{50}\,dt$, $\ln x = \frac{t}{50} + C$. Since $x = 1{,}000$ when $t = 0$, $\ln 1{,}000 = C$, and $\ln x = \frac{t}{50} + \ln 1{,}000$. Hence $e^{\ln x} = e^{(t/50)+\ln 1{,}000} = e^{t/50}e^{\ln 1{,}000}$. This shows that $x = 1{,}000e^{t/50}$, which is the amount after t years. After 20 years the amount is $x = 1{,}000e^{20/50} = 1{,}000e^{0.4}$. From tables of e^x, $e^{0.4} = 1.4918$, so that the amount after 20 years is about \$1,492.

EXAMPLE 2. At what rate of interest, continuously compounded, will a sum double itself in 50 years?

Solution: Let the rate be r per cent. As in Eq. (90), the rate for time dt is $\frac{r}{100}\,dt$, so that $dx = x\,\frac{r}{100}\,dt$. It follows that $\frac{dx}{x} = \frac{r}{100}\,dt$, $\int \frac{1}{x}\,dx = \int \frac{r}{100}\,dt$, $\ln x = \frac{rt}{100} + C$. If the amount was A initially, it was $2A$ after 50 years. Hence $x = A$ when $t = 0$, and $x = 2A$ when $t = 50$. It follows that $\ln A = C$, and $\ln 2A = \frac{r}{100}\,50 + C = \frac{r}{2} + \ln A$. Hence $\frac{r}{2} = \ln 2A - \ln A = \ln \frac{2A}{A} = \ln 2$, and $r = 2 \ln 2$. From tables of natural logarithms, $\ln 2 = 0.693$, so that $r = 1.386$. Thus the required rate of interest is 1.39 per cent.

EXAMPLE 3. The population of the United States, in millions, was about 92 in 1910 and 106 in 1920. If a statistician had assumed the rate of growth to be proportional to the population, what would he have predicted for the population in 1950?

Solution: Let x be the population in millions t years after 1910. Then $\frac{dx}{dt} = kx$, $\frac{dx}{x} = k\,dt$, $\int \frac{1}{x}\,dx = \int k\,dt$, $\ln x = kt + C$. Since $x = 92$ when $t = 0$, $\ln 92 = C$. And since $x = 106$ when $t = 10$, $\ln 106 = 10k + C = 10k + \ln 92$. Hence $10k = \ln 106 - \ln 92 = \ln (\frac{106}{92})$. Thus $k = \frac{1}{10} \ln (\frac{106}{92})$, $\ln x = \frac{1}{10} \ln (\frac{106}{92})t + \ln 92$. For 1950, $t = 40$, so that the predicted population would have been the x with $\ln x = \frac{1}{10} \ln (\frac{106}{92})\,40 + \ln 92 = 4 \ln 106 - 3 \ln 92 = 4 \ln 1.06 - 3 \ln 9.2 + 5 \ln 10 = 4(0.0583) - 3(2.2192) + 5(2.3026) = 5.0886 = 0.4834 + 2 \ln 10 = \ln 1.622 + 2 \ln 10 = \ln 162.2$. It follows that $x = 162$ to the nearest million.

Population growth is only approximately exponential, and the actual population was 150 million in 1950.

We note that, since common logarithms are proportional to natural logarithms by Eq. (55), we might have concluded from $\ln x = 4 \ln 106 - 3 \ln 92$ that $\log x = 4 \log 106 - 3 \log 92 = 4(2.0253) - 3(1.9638) = 2.2098 = \log 162.1$. It follows that $x = 162$, to the nearest million, as before.

EXAMPLE 4. In a *first-order chemical reaction* a substance is transformed at a rate proportional to the amount of untransformed substance present. If the amount of untransformed substance x was 80 gm. initially, and was 10 gm. after 30 min., how much was there after t min.? After 10 min.?

Solution: Since transformation decreases x, with some positive constant k we have $-dx/dt = kx$, or $dx/dt = -kx$. From this we find that $\frac{dx}{x} = -k\,dt$, $\int \frac{1}{x}\,dx$

$= \int -k\,dt$, $\ln x = -kt + C$. Since $x = 80$ when $t = 0$, $\ln 80 = C$. And since $x = 10$ when $t = 30$, $\ln 10 = -30k + C = -30k + \ln 80$. Hence $30k = \ln 80 - \ln 10 = \ln 8$. Thus $k = \frac{1}{30}\ln 8$. And $\ln x = -\frac{t}{30}\ln 8 + \ln 80$. It follows that

$$e^{\ln x} = e^{-t(\ln 8)/30 + \ln 80} \quad \text{and} \quad x = 80e^{(\ln 8)(-t/30)} = 80 \cdot 8^{-t/30}.$$

Hence $x = 80(\frac{1}{8})^{t/30}$ is the required amount untransformed after t min.

When $t = 10$, $x = 80\left(\frac{1}{8}\right)^{\frac{1}{3}} = \frac{80}{\sqrt[3]{8}} = 40$. Thus after 10 min. 40 gm. remain untransformed.

EXERCISE 63

The rate of change of x with respect to t is kx, so that $dx/dt = kx$. If x equals the given amount A when $t = 0$, find the law connecting x and t in each of the following problems.

1. $A = 10$, $k = 2$.
2. $A = 100$, $k = 1$.
3. $A = 1$, $k = 0.01$.
4. $A = 5$, $k = \ln 2$.
5. $A = 10$, and $x = 20$ when $t = 1$.
6. $A = 1$, and $x = 4$ when $t = 2$.

An amount x dollars draws r per cent interest per annum, continuously compounded, so that $\frac{dx}{dt} = \frac{r}{100}x$. If x equals the given amount A when $t = 0$, find the law connecting x and t in each of the following problems.

7. $A = 1$, $r = 2$.
8. $A = 100$, $r = 4$.
9. $A = 200$, $r = 3$.
10. $A = 1{,}000$, $r = 2.5$.

At what rate of interest, r per cent, continuously compounded, will

11. \$75 increase to \$100 in 10 years?
12. \$1,000 increase to \$2,718.28 in 50 years?
13. An amount A triple itself in 100 years?

A chemical substance transforms at a rate proportional to the amount untransformed, x, so that $dx/dt = -kx$. Find the amount x after t min. for each of the following sets of conditions.

14. $x = 100$ when $t = 0$, $x = 20$ when $t = 10$.
15. $x = 40$ when $t = 0$, $x = 10$ when $t = 40$.
16. $x = 48$ when $t = 0$, $x = 3$ when $t = 80$.

Find the amount remaining untransformed after 20 min. in

17. Prob. 14.
18. Prob. 15.
19. Prob. 16.

Radioactive substances decay according to the law $dx/dt = -kx$. If an initial amount A is reduced to $A/2$ after n years, n is called the half-life. Prove that

20. $k = \frac{\ln 2}{n}$.
21. $x = Ae^{-(\ln 2)t/n} = A(\frac{1}{2})^{t/n}$.
22. For a substance with $n = 1{,}600$, 1 gm. is reduced to 0.7 gm. after 800 years.

***124. Damped Harmonic Motion.** Let a particle of mass m move on a straight line. Suppose that it is acted on by an elastic restoring force proportional to the displacement, $-Ex$. Then the equation of motion is $m(d^2x/dt^2) = -Ex$. In Example 3 of Sec. 110 we deduced from an equation of this type that the motion was simple harmonic. If in addition to the restoring force there is a friction force opposing the motion proportional to the first power of the velocity, $-fv = -f(dx/dt)$, the total force is $F = -Ex - f(dx/dt)$. And the relation: force equals mass times acceleration, or $F = ma$, leads to

$$m\frac{d^2x}{dt^2} = -Ex - f\frac{dx}{dt} \qquad \text{or} \qquad \frac{d^2x}{dt^2} + \frac{f}{m}\frac{dx}{dt} + \frac{E}{m}x = 0. \tag{98}$$

An equation of similar type is satisfied by the transient current of intensity i amp. in a single-loop electric circuit containing inductance L henrys, resistance R ohms and capacity C farads, since

$$L\frac{d^2i}{dt^2} + R\frac{di}{dt} + \frac{1}{C}i = 0 \qquad \text{or} \qquad \frac{d^2i}{dt^2} + \frac{R}{L}\frac{di}{dt} + \frac{1}{LC}i = 0. \tag{99}$$

Because of these and other physical applications, we shall briefly discuss the equation

$$\frac{d^2x}{dt^2} + p\frac{dx}{dt} + qx = 0, \tag{100}$$

where p and q are constants. This differs from Eqs. (98) and (99) in notation only.

In Eq. (100) let us† make the change of variable $x = e^{-pt/2}z$. Then we find successively that

$$\begin{aligned}\frac{dx}{dt} &= e^{-pt/2}\frac{dz}{dt} - \frac{p}{2}e^{-pt/2}z = e^{-pt/2}\left(\frac{dz}{dt} - \frac{p}{2}z\right),\\ \frac{d^2x}{dt^2} &= e^{-pt/2}\left(\frac{d^2z}{dt^2} - \frac{p}{2}\frac{dz}{dt}\right) - \frac{p}{2}e^{-pt/2}\left(\frac{dz}{dt} - \frac{p}{2}z\right)\\ &= e^{-pt/2}\left(\frac{d^2z}{dt^2} - p\frac{dz}{dt} + \frac{p^2}{4}z\right).\end{aligned} \tag{101}$$

The result of substituting the values from $x = e^{-pt/2}z$ and Eq. (101) in Eq. (100) is

$$e^{-pt/2}\left(\frac{d^2z}{dt^2} - p\frac{dz}{dt} + \frac{p^2}{4}z + p\frac{dz}{dt} - \frac{p^2}{2}z + qz\right) = 0 \tag{102}$$

so that

$$\frac{d^2z}{dt^2} - \frac{p^2}{4}z + qz = 0, \qquad \frac{d^2z}{dt^2} = \frac{p^2 - 4q}{4}z. \tag{103}$$

Let us first assume that the coefficient of z is *negative*, $p^2 < 4q$. Then $4q - p^2$ is positive, and we may put

$$b = \frac{\sqrt{4q - p^2}}{2} \qquad \text{and} \qquad \frac{p^2 - 4q}{4} = -b^2. \tag{104}$$

Hence the last relation of Eq. (103) becomes

$$\frac{d^2z}{dt^2} = -b^2z, \qquad \text{with solution } z = c\sin(bt + h), \tag{105}$$

by Example 3 of Sec. 110.

It follows that, when $p^2 < 4q$, the solution of Eq. (100) is

$$x = e^{-pt/2}z = ce^{-pt/2}\sin(bt + h). \tag{106}$$

† This substitution is unmotivated here and merely shown to be effective. It could be derived by first putting $x = uz$ and finding what function u made the coefficient of dz/dt zero in the transformed equation.

By Example 2 of Sec. 110 this may be written in the alternative form

$$x = Ae^{-pt/2} \sin bt + Be^{-pt/2} \cos bt. \tag{107}$$

A solution of the form given in Eq. (106) or Eq. (107) is called a *damped harmonic motion.*

For the sake of completeness, we shall obtain the solution of Eq. (100) in all cases. Suppose that the coefficient of z in Eq. (103) is *zero*, $p^2 = 4q$. Then Eq. (103) reduces to

$$\frac{d^2z}{dt^2} = 0. \qquad \text{Hence, } \frac{d}{dt}\left(\frac{dz}{dt}\right) = 0 \qquad \text{and} \qquad \frac{dz}{dt} = A, \tag{108}$$

a first constant of integration. Also

$$dz = A\,dt, \qquad \int dz = \int A\,dt, \qquad z = At + B, \tag{109}$$

where B is a second constant of integration. It then follows that

$$x = e^{-pt/2}z = e^{-pt/2}(At + B) = Ate^{-pt/2} + Be^{-pt/2}. \tag{110}$$

Finally the coefficient of z in Eq. (103) may be positive, $p^2 > 4q$. In this case we may put

$$k = \frac{\sqrt{p^2 - 4q}}{2} \qquad \text{and} \qquad \frac{p^2 - 4q}{4} = k^2. \tag{111}$$

Hence the last relation of Eq. (103) becomes

$$\frac{d^2z}{dt^2} = k^2z, \qquad \text{with solution } z = Ae^{kt} + Be^{-kt}, \tag{112}$$

by Example 4 below. It follows that

$$x = e^{-pt/2}z = e^{-pt/2}(Ae^{kt} + Be^{-kt}) = Ae^{r_1t} + Be^{r_2t}, \tag{113}$$

where $r_1 = -(p/2) + k$ and $r_2 = -(p/2) - k$ with k given by Eq. (111).

To put our results together in more systematic form, we note that the last solution involves terms of the form e^{rt}. Let us find directly when $x = e^{rt}$ is a solution of Eq. (100). We have

$$x = e^{rt}, \qquad \frac{dx}{dt} = re^{rt}, \qquad \frac{d^2x}{dt^2} = r^2e^{rt}. \tag{114}$$

Substitution of these results in Eq. (100) leads to

$$e^{rt}(r^2 + pr + q) = 0 \qquad \text{or} \qquad r^2 + pr + q = 0. \tag{115}$$

The roots of this equation in r as found by the quadratic formula are

$$r = \frac{-p \pm \sqrt{p^2 - 4q}}{2}. \tag{116}$$

When $p^2 > 4q$, the two roots of Eq. (116) are real and unequal and reduce to the r_1 and r_2 of Eq. (113).

When $p^2 = 4q$, the two roots are real and equal. And each equals the $-p/2$ which occurs in the exponents of Eq. (110).

Finally, when $p^2 > 4q$, the two roots are conjugate complex quantities. In this case, with b defined by Eq. (104) and $i = \sqrt{-1}$, we may write

$$r = -\frac{p}{2} \pm \sqrt{-1}\,\frac{\sqrt{4q - p^2}}{2} = -\frac{p}{2} \pm bi. \tag{117}$$

Thus the $-p/2$ which occurs in the exponent and the b which occurs in the trigonometric functions of Eqs. (106) and (107) are simply related to the complex roots of Eq. (117).

We may collect these facts in the following rule:

To find the solution of the equation

$$\frac{d^2x}{dt^2} + p\frac{dx}{dt} + qx = 0, \tag{118}$$

form the auxiliary quadratic equation,

$$r^2 + pr + q = 0, \tag{119}$$

and find its roots.

If the roots are real and distinct, r_1, r_2, *the real solution is*

$$x = Ae^{r_1t} + Be^{r_2t}, \tag{120}$$

If the roots are real and equal, $r_1 = r_2$, *the solution is*

$$x = Ate^{r_1t} + Be^{r_1t}. \tag{121}$$

If the roots are conjugate imaginaries, $a \pm ib$, *the solution is*

$$x = Ae^{at}\sin bt + Be^{at}\cos bt \qquad \text{or} \qquad x = ce^{at}\sin(bt + h). \tag{122}$$

We note that in most physical applications p is positive. Hence a, r_1, and r_2 are negative. Thus the solution of Eq. (122) represents damped harmonic motion. And the solution of Eq. (120) is a combination of exponentials decaying with time. When $p^2 - 4q$ is positive, so that the solution is given by Eq. (120), the physical system is said to be *critically damped.*

Example 1. Find the general† solution of the equation

$$\frac{d^2x}{dt^2} + 4\frac{dx}{dt} + 13x = 0.$$

Also find that particular solution which has $x = 3$ and $dx/dt = 6$ when $t = 0$.

Solution: The auxiliary equation here is $r^2 + 4r + 13 = 0$. Its roots are $r = \dfrac{-4 \pm \sqrt{4^2 - 4 \cdot 13}}{2} = -2 \pm \sqrt{-9} = -2 \pm 3i$. It follows from Eq. (122) that one form of the required general solution is

$$x = Ae^{-2t}\sin 3t + Be^{-2t}\cos 3t.$$

To find the particular solution, we first differentiate x. We have

$$\begin{aligned}\frac{dx}{dt} &= A[e^{-2t}(3\cos 3t) - 2e^{-2t}\sin 3t] + B[e^{-2t}(-3\sin 3t) - 2e^{-2t}\cos 3t]\\ &= (-2A - 3B)e^{-2t}\sin 3t + (3A - 2B)e^{-2t}\cos 3t.\end{aligned}$$

† The terms general and particular solution, illustrated in this example, are discussed in Sec. 321.

Since $e^0 = 1$, $\sin 0 = 0$, $\cos 0 = 1$; from $x = 3$ when $t = 0$, we may deduce that $3 = A1 \cdot 0 + B1 \cdot 1 = B$, so that $B = 3$. And from $dx/dt = 6$ when $t = 0$ we may deduce that $6 = (-2A - 3B)1 \cdot 0 + (3A - 2B)1 \cdot 1$, or $6 = 3A - 2B = 3A - 6$, since $B = 3$. Hence $3A = 12$ and $A = 4$. Thus the required particular solution is

$$x = 4e^{-2t} \sin 3t + 3e^{-2t} \cos 3t.$$

By the procedure used in Sec. 110, Example 2, this can be written in the form

$$x = 5e^{-2t} \sin (3t + 0.643)$$

by first computing $\sqrt{4^2 + 3^2} = 5$, and $\tan^{-1} \frac{3}{4} = \tan^{-1} 0.75 = 0.643$ radian.

The graph of the particular solution is shown in Fig 141.

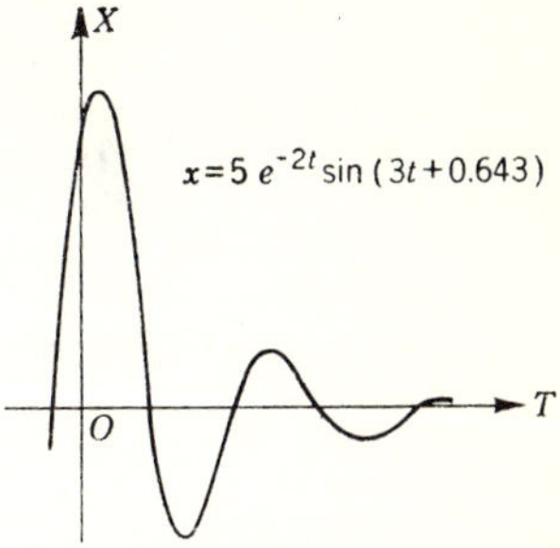

FIG. 141.

EXAMPLE 2. Find the general solution of the equation

$$\frac{d^2x}{dt^2} + 4\frac{dx}{dt} + 3x = 0.$$

Also find that particular solution which has $x = 2$ and $dx/dt = 4$ when $t = 0$.

Solution: The auxiliary equation here is $r^2 + 4r + 3 = 0$. Since $r^2 + 4r + 3 = (r + 1)(r + 3)$, the roots are $r = -1$ and $r = -3$. It follows from Eq. (120) that the required general solution is

$$x = Ae^{-t} + Be^{-3t}.$$

To find the particular solution, we first differentiate x. The result is $dx/dt = -Ae^{-t} - 3Be^{-3t}$. Since $e^0 = 1$, from $x = 2$ when $t = 0$, we may deduce that $2 = A \cdot 1 + B \cdot 1 = A + B$. And from $dx/dt = 4$ when $t = 0$, we may deduce that $4 = -A \cdot 1 - 3B \cdot 1 = -A - 3B$. We next solve the simultaneous equations $A + B = 2$, $-A - 3B = 4$ for A and B. By adding the two equations, we obtain $-2B = 6$. Hence $B = -3$. Then $A = 2 - B = 2 - (-3) = 5$. Thus $A = 5$, $B = -3$. And the required particular solution is

$$x = 5e^{-t} - 3e^{-3t}.$$

EXAMPLE 3. A body weighing 8 lb. is observed to vibrate 120 times per minute and the amplitude of the oscillation is damped to one-half in 30 sec. Evaluate f and E in Eq. (98).

Solution: Let the damped harmonic motion of the body be $x = ce^{-st} \sin (bt + h)$. If T is the period of $\sin (bt + h)$, as in Sec. 110, Eq. (147), $bT = 2\pi$. Here $T = \frac{60}{120} = \frac{1}{2}$ sec., and $b = 2\pi/T = 4\pi$. The amplitude at any time t_1 is $A_1 = ce^{-st_1}$. And at time t_2 it is $A_2 = ce^{-st_2}$. Hence $\dfrac{A_2}{A_1} = \dfrac{ce^{-st_2}}{ce^{-st_1}} = e^{-s(t_2 - t_1)}$. Since $A_2 = \frac{1}{2}A_1$ when $t_2 = t_1 + 30$, or when $t_2 - t_1 = 30$, it follows that $\frac{1}{2} = e^{-30s}$. And $-\ln 2 = -30s$, $s = \dfrac{\ln 2}{30}$. The roots of the auxiliary equation are $-s \pm bi = -\dfrac{\ln 2}{30} \pm 4\pi i$. A comparison with Eq. (117) shows that $-\dfrac{p}{2} = -\dfrac{\ln 2}{30}$, $\dfrac{\sqrt{4q - p^2}}{2} = 4\pi$. Hence $p = \dfrac{\ln 2}{15}$. And from $\dfrac{4q - p^2}{4} = 16\pi^2$, $q = 16\pi^2 + \dfrac{p^2}{4} = 16\pi^2 + \left(\dfrac{\ln 2}{30}\right)^2$.

The equivalence of Eq. (98) to Eq. (100) shows that $f/m = p$, $E/m = q$. Since the weight $w = 8$ lb. and the acceleration of gravity $g = 32$ ft./sec.^{2}, the mass $m = w/g = \frac{8}{32} = \frac{1}{4}$ slug. Consequently, we have $f = mp = \frac{1}{4}\frac{\ln 2}{15} = \frac{0.693}{60} = 0.0116$,

$$E = mq = \frac{1}{4}\left[16\pi^2 + \left(\frac{\ln 2}{30}\right)^2\right] = 4(9.87) + (0.0116)^2 = 39.48.$$

Thus $f = 0.0116$ lb.-sec./ft. and $E = 39.48$ lb./ft. are the required values.

EXAMPLE 4. Prove that the solution of $d^2z/dt^2 = k^2z$ is $z = Ae^{kt} + Be^{-kt}$ by making the transformation $z = e^{-kt}u$.

Solution: If $z = e^{-kt}u$, $dz/dt = e^{-kt}du/dt - ke^{-kt}u$, and

$$\begin{aligned}\frac{d^2z}{dt^2} &= e^{-kt}\frac{d^2u}{dt^2} - ke^{-kt}\frac{du}{dt} - ke^{-kt}\frac{du}{dt} + k^2e^{-kt}u \\ &= e^{-kt}\left(\frac{d^2u}{dt^2} - 2k\frac{du}{dt} + k^2u\right).\end{aligned}$$

The result of substituting these values in $(d^2z/dt^2) - k^2z = 0$ is

$$e^{-kt}\left(\frac{d^2u}{dt^2} - 2k\frac{du}{dt} + k^2u - k^2u\right) = 0 \qquad \text{or} \qquad \frac{d^2u}{dt^2} - 2k\frac{du}{dt} = 0.$$

It follows that $\frac{d}{dt}\left(\frac{du}{dt} - 2ku\right) = 0$ and $\frac{du}{dt} - 2ku = c_1$. And from this we may deduce $\frac{du}{dt} = 2ku + c_1$, $\frac{du}{2ku + c_1} = dt$, $\int \frac{du}{2ku + c_1} = \int dt$. But $\frac{d}{du}\ln(2ku + c_1) = \frac{2k}{2ku + c_1}$, so that $\int \frac{du}{2ku + c_1} = \frac{1}{2k}\int \frac{2k}{2k + c_1}du = \frac{1}{2k}\ln(2ku + c_1) + C_1$. And $\int dt = t + C_2$, so that if $C_2 - C_1 = c_2$, the relation $\int \frac{du}{2ku + c_1} = \int dt$ implies that $\frac{1}{2k}\ln(2ku + c_1) = t + c_2$. This leads to $\ln(2ku + c_1) = 2kt + 2kc_2$ and $2ku + c_1 = e^{2kc_2}e^{2kt}$ or $u = \left(\frac{e^{2kc_2}}{2k}\right)e^{2kt} - \frac{c_1}{2k} = Ae^{2kt} + B$, if $A = \left(\frac{e^{2kc_2}}{2k}\right)$ and $B = -\frac{c_1}{2k}$.

Finally, since $z = e^{-kt}u$, $z = e^{-kt}(Ae^{2kt} + B) = Ae^{kt} + Be^{-kt}$. This is the result we were asked to prove.

EXERCISE 64

Use the rule of the text, Eqs. (118) to (122), to verify that each of the following equations has the given expression as its general solution.

1. $\frac{d^2x}{dt^2} + 5\frac{dx}{dt} = 0$, $x = A + Be^{-5t}$.

2. $\frac{d^2x}{dt^2} + 2\frac{dx}{dt} + 5x = 0$, $x = Ae^{-t}\sin 2t + Be^{-t}\cos 2t$.

3. $\frac{d^2x}{dt^2} + 2\frac{dx}{dt} + x = 0$, $x = Ate^{-t} + Be^{-t}$.

4. $\frac{d^2x}{dt^2} + 5\frac{dx}{dt} + 6x = 0$, $x = Ae^{-2t} + Be^{-3t}$.

5. $\frac{d^2x}{dt^2} + 4\frac{dx}{dt} + 5x = 0$, $a = Ae^{-2t}\sin t + Be^{-2t}\cos t$.

6. $\frac{d^2x}{dt^2} - 9x = 0$, $x = Ae^{3t} + Be^{-3t}$.

7. $\dfrac{d^2x}{dt^2} + 9x = 0$, $x = A \sin 3t + B \cos 3t$.

8. $\dfrac{d^2x}{dt^2} + 6\dfrac{dx}{dt} + 25x = 0$, $x = Ae^{-3t} \sin 4t + Be^{-3t} \cos 4t$.

For the equation of the problem indicated, find the particular solution which has the given initial values of x and dx/dt.

9. Prob. 1, $x = 3$ and $dx/dt = -5$ when $t = 0$.
10. Prob. 2, $x = -1$ and $dx/dt = 3$ when $t = 0$.
11. Prob. 3, $x = 1$ and $dx/dt = 0$ when $t = 0$.
12. Prob. 4, $x = 5$ and $dx/dt = -12$ when $t = 0$.
13. Prob. 5, $x = 2$ and $dx/dt = -1$ when $t = 0$.
14. Prob. 7, $x = 1$ and $dx/dt = 6$ when $t = 0$.

Find the period, or time of one complete vibration, and the length of time required for the amplitude to drop to one-half its initial value for the motion determined by the equation of

15. Prob. 2. 16. Prob. 5. 17. Prob. 8.

Consider the equation $\dfrac{d^2x}{dt^2} + p\dfrac{dx}{dt} + qx = m \sin \omega t$, with $p \neq 0$.

18. Verify that $x_1 = m \dfrac{(q - \omega^2) \sin t - p\omega \cos \omega t}{p^2\omega^2 + (q - \omega^2)^2}$ is a particular solution of the given equation.

19. Let $x = x_1 + z$, where x_1 is the particular solution of Prob. 18. Show that this x is a solution of the given equation whenever

$$\frac{d^2z}{dt^2} + p\frac{dz}{dt} + qz = 0.$$

20. From Prob. 19 deduce that the general solution of the given equation is the x_1 of Prob. 18 plus an expression like the right member of the appropriate one of Eqs. (120) to (122) depending on the nature of the roots of Eq. (119).

125. Applications. The differentiation rules developed in this chapter enable us to carry out applications of the differential calculus to functions involving logarithms and exponential functions. In particular we may construct graphs of functions by methods like those of Sec. 48, and find greatest and least values by methods like those of Secs. 50 and 57.

EXAMPLE 1. Sketch the graph of $y = x^2e^{-x^2}$, after locating the maximum, minimum, and inflection points.

Solution: We find $dy/dx = x^2e^{-x^2}(-2x) + 2xe^{-x^2} = 2(x - x^3)e^{-x^2}$. And $d^2y/dx^2 = 2(x - x^3)e^{-x^2}(-2x) + 2(1 - 3x^2)e^{-x^2} = 2(1 - 5x^2 + 2x^4)e^{-x^2}$.

Thus dy/dx has the same sign changes as $(x - x^3) = -x(x - 1)(x + 1)$. This changes sign from minus to plus as x increases through 0, and changes from plus to minus as x increases through -1 and 1. At $x = 0$, $y = x^2e^{-x^2} = 0$. And at $x = \pm 1$, $y = x^2e^{-x^2} = e^{-1} = 1/e = 0.368$. Hence there is a minimum at (0,0), and a maximum at (-1,0.368) and (1,0.368).

Also d^2y/dx^2 has the same sign changes as $2x^4 - 5x^2 + 1$. By regarding $2(x^2)^2 - 5(x^2) + 1 = 0$ as a quadratic equation in x^2, we find that $x^2 = \dfrac{5 \pm \sqrt{25 - 8}}{4} = 2.282$ or 0.219. Hence $x = \pm 1.51$ or ± 0.468. As d^2y/dx^2 changes sign when x

increases through each of these four values, we therefore have points of inflection at $(\pm 1.51, 0.233)$ and at $(\pm 0.468, 0.176)$.

When $x \to \pm\infty$, $y = x^2e^{-x^2} = \dfrac{x^2}{e^{x^2}} \to 0$ since e^{x^2} increases much faster than x^2. Hence the x axis is an asymptote. The graph of the function is shown in Fig. 142.

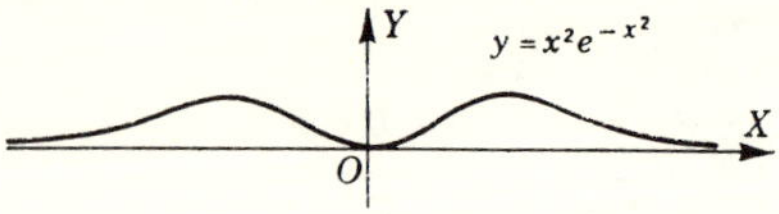

FIG. 142.

EXAMPLE 2. At what points on the curve $y^2 + \ln x = 0$ is the distance to the origin least?

Solution: At the points in question the square of the distance to the origin, $z = x^2 + y^2$, will be a minimum. From the given equation $y^2 = -\ln x$, so that $z = x^2 - \ln x$.

To minimize $z = x^2 - \ln x$, we first find $\dfrac{dz}{dx} = 2x - \dfrac{1}{x}$. This becomes infinite and changes sign when x increases through 0. But $x = 0$ makes z infinite. Again, $\dfrac{dz}{dx} = \dfrac{2x^2 - 1}{x} = 0$ when $2x^2 = 1$, $x^2 = \dfrac{1}{2}$ or $x = \pm\dfrac{1}{\sqrt{2}}$. For $x = -\dfrac{1}{\sqrt{2}}$, $\ln x$ is not real. Hence we consider $x = \dfrac{1}{\sqrt{2}}$ and note that $\dfrac{dz}{dx} = \dfrac{2}{x}\left(x + \dfrac{1}{\sqrt{2}}\right)\left(x - \dfrac{1}{\sqrt{2}}\right)$, like $\left(x - \dfrac{1}{\sqrt{2}}\right)$, changes sign from minus to plus as x increases through $\dfrac{1}{\sqrt{2}}$. Hence z is a minimum when $x = \dfrac{1}{\sqrt{2}} = 0.707$. For this value of x, $y^2 = -\ln\dfrac{1}{\sqrt{2}} = \dfrac{1}{2}\ln 2 = 0.347$ so that $y = \pm 0.589$. Hence the required points are $(0.707, \pm 0.589)$. Each of these points is at the same minimum distance 0.920 from the origin.

EXERCISE 65

Sketch the graph of each of the following functions after locating the maximum, minimum, and inflection points.

1. $y = xe^x$.
2. $y = xe^{-x}$.
3. $y = x^4e^{-x}$.
4. $y = x^2e^x$.
5. $y = x \ln x$.
6. $y = x^2 \ln x$.
7. $y = \dfrac{x}{\ln x}$.
8. $y = x(\ln x)^2$.
9. $y = e^{-x^2}$.
10. $y = xe^{-x^2}$.
11. A submarine cable has a copper core 1 cm. in diameter and an insulating layer of gutta percha x cm. thick. Its speed of transmitting signals is $S = 1.1c\dfrac{\ln x}{x^2}$, where c is the velocity of light. Show that the speed S assumes its maximum value, about 40 per cent of c, when $x = \sqrt{e}$.
12. Show that the maximum rectangle with one side on the x axis and inscribed in the curve $y = e^{-2x^2}$ has two of its vertices at the inflection points $(\pm\frac{1}{2}, e^{-\frac{1}{2}})$.
13. Show that $y = 4e^{-2x} + e^x$ assumes its minimum value, 3, when $x = \ln 2$.
14. Let $y = ae^{px} + be^{-qx}$, with a, b, p, q each greater than zero. Show that y assumes its minimum value, $\left(\dfrac{1}{p} + \dfrac{1}{q}\right)(ap)^{q/(p+q)}(bq)^{p/(p+q)}$, when $x = \dfrac{1}{p+q}\ln\dfrac{bq}{ap}$.

The rectangle $OAPB$ has O at the origin and P a point in the first quadrant on the curve $y = 4e^{-x}$. Side OA is on the x axis, and side OB is on the y axis. What point P gives the least value to

15. The area of the rectangle?
16. The volume of the cylinder generated by revolving $OAPB$ about OA?
17. The volume of the cylinder generated by revolving $AOPB$ about OB?
18. The perimeter of the rectangle, or $2(x + y)$?

The rectangle $OAPB$ has O at the origin and P a point in the first quadrant on the curve $y = \ln \dfrac{1}{\sqrt{x}} = -\dfrac{1}{2} \ln x$. Side OA is on the x axis, and side OB is on the y axis. What point P gives the least value to

19. The area of the rectangle?
20. The volume of the cylinder generated by revolving $OAPB$ about OA?
21. The volume of the cylinder generated by revolving $OAPB$ about OB?
22. The perimeter of the rectangle, or $2(x + y)$?

23. Given that $y = Ae^{ax} \sin bx + Be^{ax} \cos bx$. Let $R = \sqrt{A^2 + B^2}$, $\phi = \tan^{-1} \dfrac{B}{A}$, $r = \sqrt{a^2 + b^2}$, $\theta = \tan^{-1} \dfrac{b}{a}$, with such quadrants for ϕ and θ that $A = R \cos \phi$, $B = R \sin \phi$, $a = r \cos \theta$, $b = r \sin \theta$. Show that $y = Re^{ax} \sin (bx + \phi)$, $\dfrac{dy}{dx} = rRe^{ax} \sin (bx + \phi + \theta)$, $\dfrac{d^2y}{dx^2} = r^2Re^{ax} \sin (bx + \phi + 2\theta)$.

Use forms for the derivatives like those of Prob. 23 to locate the maximum, minimum, and inflection points for each of the following given curves with values of x in the range $0 < x < 2\pi$. Then sketch the graph for x in this range.

24. $y = e^{-x} \sin x$.
25. $y = e^{-x} \cos x$.
26. $y = e^{-0.04x} \sin x$.
27. $y = e^{-0.7x} \cos x$.

CHAPTER 9

PARAMETRIC EQUATIONS. CURVATURE

Certain problems involving curves may be solved by expressing each of the coordinates x and y of a variable point on a curve in terms of a third variable. The third variable is called a *parameter*. And the equations which express the coordinates as functions of the parameter are called *parametric equations* of the curve.

This chapter is devoted to the use of such parametric equations to study curves and to calculate quantities related to them. We first derive parametric equations for the cycloid, which is generated by a point on a circle which rolls on a straight line, and for some other curves of similar nature.

Since many questions about curves may be answered if one knows the slope dy/dx and the second derivative d^2y/dx^2, we show how to evaluate these without first eliminating the parameter. We then define curvature, which is a measure of the rapidity of bending of a curve, and express it in terms of first and second derivatives. This enables us to find the radius and center of curvature at any point of a given curve, as well as the locus of centers of curvature, or the evolute of the curve.

When t is the time, the parametric equations determine a motion. We discuss the velocity and acceleration vector at any time for such a motion. In particular we show that the normal component of acceleration equals the curvature times the square of the speed in the path.

126. Parametric Equations of a Curve. Let t be a variable parameter and $g(t)$, $h(t)$ be two given functions. Then in general the parametric equations

$$x = g(t), \qquad y = h(t) \tag{1}$$

represent a curve. To construct the curve we might assign values to t, calculate the corresponding values of x and y, and plot the resulting points. And if we can eliminate t from the two *parametric equations* (1), the resulting single equation in x and y is the *rectangular equation* of the curve.

Let us consider some examples. First, it follows from Fig. 143 that

$$x = t \cos \alpha, \qquad y = b + t \sin \alpha \tag{2}$$

are the parametric equations of a straight line with slope angle α and y

intercept b, if the distance along the line from $(0,b)$ is taken as the parameter t. We may eliminate t by noting that

$$t = \frac{x}{\cos \alpha}, \qquad y = b + \left(\frac{x}{\cos \alpha}\right) \sin \alpha = b + x \tan \alpha$$

or

$$y = (\tan \alpha)x + b. \tag{3}$$

This is the rectangular equation of the straight line.

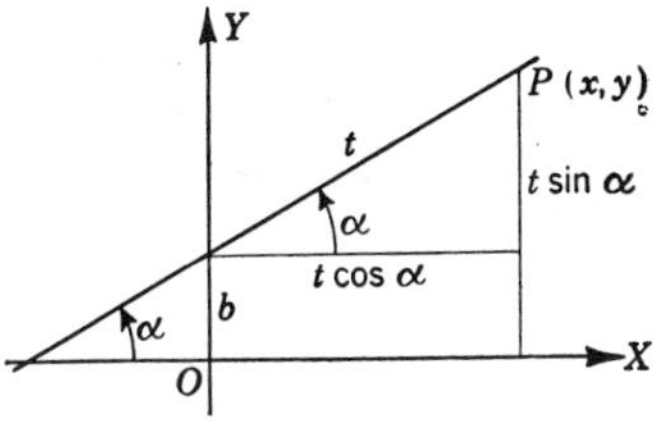

FIG. 143.

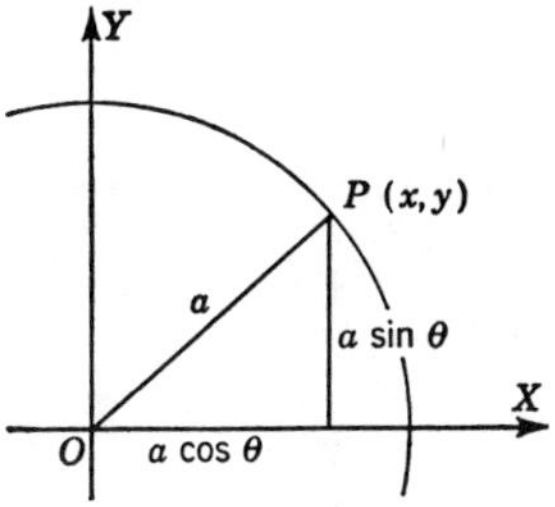

FIG. 144.

Again, from Fig. 144 it follows that

$$x = a \cos \theta, \qquad y = a \sin \theta \tag{4}$$

are the parametric equations of a circle with center at the origin and radius a, if the central angle $XOP = \theta$ is taken as the parameter. We may eliminate θ by noting that

$$\cos \theta = \frac{x}{a}, \quad \sin \theta = \frac{y}{a}, \quad \text{and hence } \left(\frac{x}{a}\right)^2 + \left(\frac{y}{a}\right)^2 = \cos^2 \theta + \sin^2 \theta = 1$$

or

$$x^2 + y^2 = a^2. \tag{5}$$

This is the rectangular equation of the circle.

Similarly if ϕ is the parameter (Fig. 145), the equations

$$x = a \cos \phi, \qquad y = b \sin \phi \tag{6}$$

represent an ellipse. For we may eliminate ϕ by noting that $\cos \phi = \frac{x}{a}$, $\sin \phi = \frac{y}{b}$, and hence $\left(\frac{x}{a}\right)^2 + \left(\frac{y}{b}\right)^2 = \cos^2 \phi + \sin^2 \phi = 1$ or

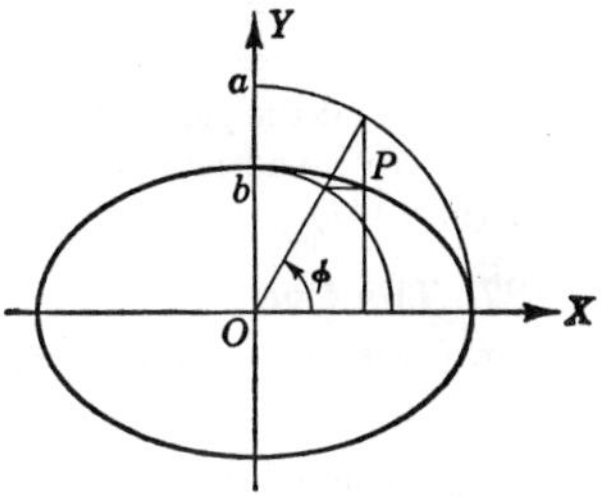

FIG. 145.

$$\frac{x^2}{a^2} + \frac{y^2}{b^2} = 1, \tag{7}$$

which is the rectangular equation of an ellipse found in Sec. 84.

EXAMPLE 1. Find the parametric equations of the curve $x^3 + y^3 = 3xy$, using as the parameter $t = y/x$.

Solution: Since $t = y/x$, $y = tx$. Substitution of this in the given rectangular equation yields $x^3 + t^3x^3 = 3tx^2$, $x + t^3x = 3t$, and $x(1 + t^3) = 3t$ so that $x = \frac{3t}{(1 + t^3)}$. Since this gives $x = 0$ for $t = 0$, nothing has been lost by our division by x^2. Also

$$y = tx = \frac{3t^2}{1 + t^3}.$$

Thus the parametric equations sought are

$$x = \frac{3t}{1 + t^3}, \qquad y = \frac{3t^2}{1 + t^3}.$$

These may be used to plot the curve (Fig. 146). When $t \to -1$, $x \to \pm\infty$, $y \to \mp\infty$, but $x + y = \frac{3t + 3t^2}{1 + t^3} = \frac{3t(1 + t)}{(1 + t)(1 - t + t^2)} = \frac{3t}{1 - t + t^2} \to -1$. In consequence, $x + y + 1 \to 0$, and $x + y + 1 = 0$ is an asymptote.

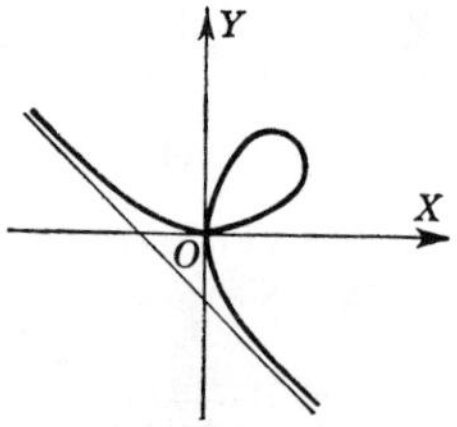

FIG. 146.

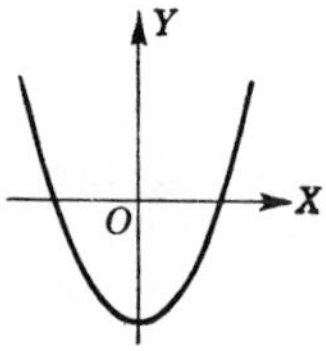

FIG. 147.

EXAMPLE 2. Find the parametric equations of a portion of the parabola $y = 2x^2 - 1$, if the parameter is such that $x = \cos t$.

Solution: Substitution of $x = \cos t$ in the given rectangular equation yields $y = 2\cos^2 t - 1 = \cos 2t$, so that the required parametric equations are

$$x = \cos t, \qquad y = \cos 2t.$$

These represent the part of the parabola for which x and y do not exceed 1 numerically, or the arc with end points $(-1,1)$ and $(1,1)$ which has the vertex $(0,-1)$ as its mid-point (Fig. 147).

127. The Cycloid. A *cycloid* is the curve generated by a point on the circumference of a circle when this circle rolls without sliding on a fixed straight line.

Let a circle of radius a roll on the x axis. In Fig. 148, C is the center of the circle, N is the point of contact with the x axis, and $P = (x,y)$ is the point on the circumference tracing the cycloid. The origin of coordinates, O, is the point found by rolling the circle to the left until P meets the x axis. This makes arc $PN = ON$. And we take as the parameter the angle $PCN = \phi$. Then the coordinates of C are $ON = a\phi$ and $NC = a$. And the coordinates of P relative to C are the projections of CP on the coordinate axes or $-a \sin \phi$ and $-a \cos \phi$. Hence the coordinates of P

relative to O are

$$x = a\phi - a \sin \phi = a(\phi - \sin \phi),$$
$$y = a - a \cos \phi = a(1 - \cos \phi). \tag{8}$$

These are the parametric equations of the cycloid with ϕ, the angle

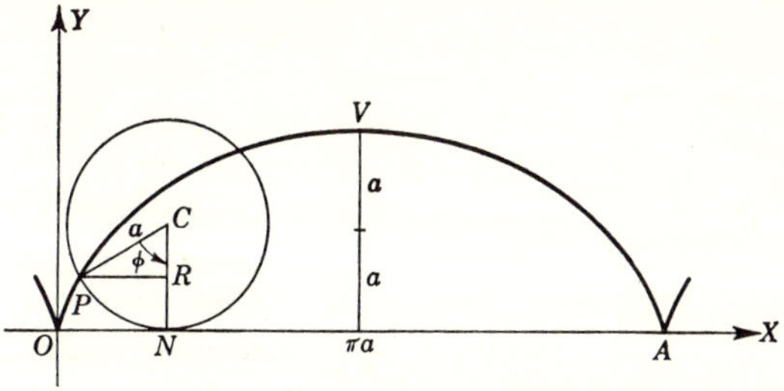

FIG. 148.

through which the radius of the rolling circle has turned, as the parameter. They are much more convenient than the rectangular equation

$$x = a \cos^{-1} \frac{a - y}{a} - \sqrt{2ay - y^2}, \tag{9}$$

which results when we substitute in the first equation the value of ϕ found by solving the second equation for $\cos \phi$.

The cycloid consists of a succession of arches, of which the first is generated by one complete revolution of the circle as ϕ varies from 0 to 2π. The base of the first arch, $OA = 2\pi a$. And the highest point $V = (a\pi, 2a)$, reached after one-half revolution when $\phi = \pi$, is called the *vertex*.

EXAMPLE. Find the parametric equations of the *trochoid* which is generated by a point rigidly attached to a circle of radius a and at distance b from the center, as the circle rolls without slipping on a fixed straight line, chosen as the x axis.

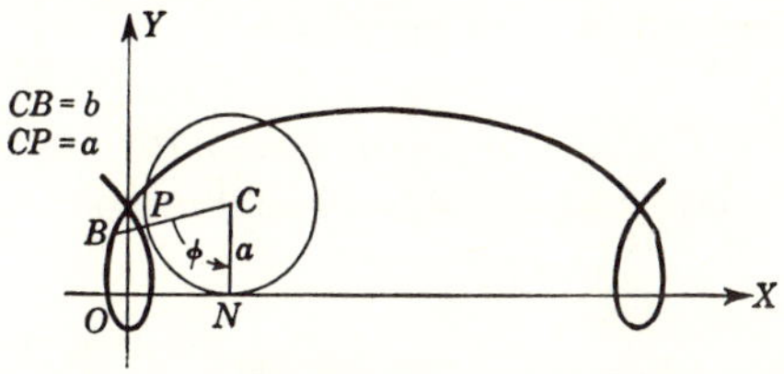

FIG. 149.

Solution: Take the parameter and origin related to the rolling circle as described for Fig. 148. And, take the tracing point $B = (x,y)$ on CP or on CP produced as in Fig. 149. Then as before $C = (a\phi, a)$ and the projections of CB are $-b \sin \phi$ and $-b \cos \phi$. Hence the coordinates of B are

$$x = a\phi - b \sin \phi, \qquad y = a - b \cos \phi. \tag{10}$$

These are the desired parametric equations of the trochoid.

128. The Epicycloid and the Hypocycloid. An *epicycloid* is the curve generated by a point on the circumference of a circle when this circle rolls without sliding on the outside of a fixed circle.

Let a circle of radius b roll on the outside of a fixed circle of radius a. Take the center of the fixed circle as the origin of coordinates, O. In Fig. 150, C is the center of the rolling circle, N is the point of contact with the fixed circle, and $P = (x,y)$ is the point tracing the epicycloid. The point A is found by rolling the circle backward until P meets the fixed circle. This makes arc NP = arc AN.

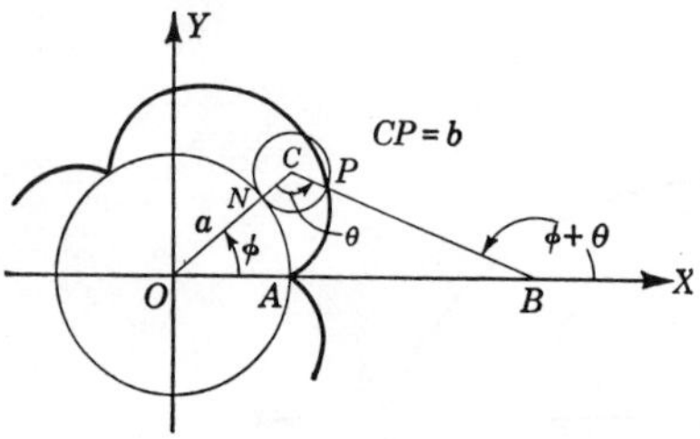

FIG. 150.

And the x axis is drawn through A. We take as the parameter the angle $AON = \phi$. Denote the angle NCP by θ. Then arc $NP = b\theta$ and arc $AN = a\phi$. Since the two arcs are equal, it follows that

$$b\theta = a\phi, \qquad \theta = \frac{a\phi}{b}, \qquad \text{and } \phi + \theta = \phi + \frac{a\phi}{b} = \frac{a+b}{b}\phi. \tag{11}$$

In triangle OBC, the exterior angle at B is the sum of the two interior angles, $\angle AON + \angle OCB = \phi + \theta$. Thus $\phi + \theta$ is the slope angle of BC, or of PC. Hence the projections of PC on the coordinate axes are $b \cos(\phi + \theta)$ and $b \sin(\phi + \theta)$. And the projections of CP are the negatives of these, or from Eq. (11),

$$-b\cos(\phi+\theta) = -b\cos\left(\frac{a+b}{b}\phi\right),$$

and

$$-b\sin(\phi+\theta) = -b\sin\left(\frac{a+b}{b}\phi\right). \tag{12}$$

The coordinates of C are $(a + b) \cos \phi$ and $(a + b) \sin \phi$. Since the coordinates of P relative to C are given by Eq. (12), it follows that the coordinates of P relative to O are

$$\begin{aligned} x &= (a+b)\cos\phi - b\cos\left(\frac{a+b}{b}\phi\right), \\ y &= (a+b)\sin\phi - b\sin\left(\frac{a+b}{b}\phi\right). \end{aligned} \tag{13}$$

These are the parametric equations of the epicycloid.

In the theory of mechanisms, it is shown that ideally designed gear teeth should be bounded by arcs of epicycloids.

A *hypocycloid* is the curve generated by a point on the circumference of a circle when this circle rolls without sliding on the inside of a fixed circle.

The notation in Fig. 151 is similar to that in Fig. 150. But in this case, the slope angle of PC is $\phi + (180 - \theta) = 180 - (\theta - \phi)$, so that the projections of PC are $-b \cos (\theta - \phi)$ and $b \sin (\theta - \phi)$. It follows by reasoning like that used to derive Eq. (13) that

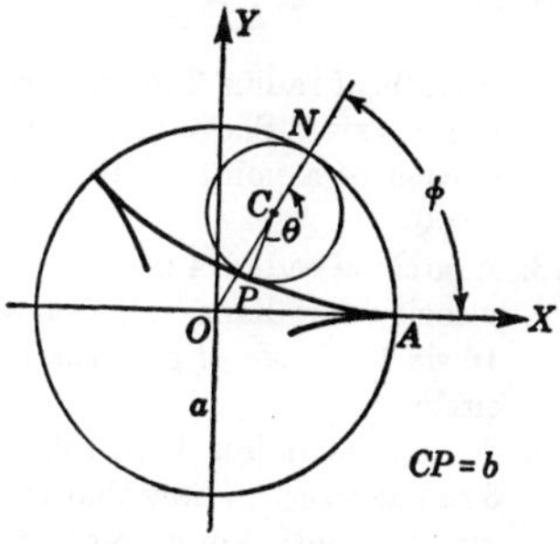

FIG. 151.

$$x = (a - b) \cos \phi + b \cos \left(\frac{a - b}{b} \phi\right),$$
$$y = (a - b) \sin \phi - b \sin \left(\frac{a - b}{b} \phi\right). \tag{14}$$

These are the parametric equations of the hypocycloid where a is the radius of the fixed circle and b is the radius of the rolling circle.

EXAMPLE. Find the rectangular equation of the hypocycloid which is described when the radius of the rolling circle is one-fourth the radius of the fixed circle.

Solution: From Eq. (14), with $b = a/4$, we find the parametric equations

$$x = \frac{3a}{4} \cos \phi + \frac{a}{4} \cos 3\phi, \qquad y = \frac{3a}{4} \sin \phi - \frac{a}{4} \sin 3\phi.$$

By combining these with the trigonometric formulas (Probs. 20 and 19 of Exercise 48),

$$\cos 3\phi = 4 \cos^3 \phi - 3 \cos \phi, \qquad \sin 3\phi = 3 \sin \phi - 4 \sin^3 \phi,$$

we may deduce the alternative form

$$x = a \cos^3 \phi, \qquad y = a \sin^3 \phi. \tag{15}$$

We may eliminate ϕ from these equations by noting that $\cos \phi = \left(\frac{x}{a}\right)^{\frac{1}{3}}$, $\sin \phi = \left(\frac{y}{a}\right)^{\frac{1}{3}}$, $\left(\frac{x}{a}\right)^{\frac{2}{3}} + \left(\frac{y}{a}\right)^{\frac{2}{3}} = \cos^2 \phi + \sin^2 \phi = 1$, so that

$$x^{\frac{2}{3}} + y^{\frac{2}{3}} = a^{\frac{2}{3}}. \tag{16}$$

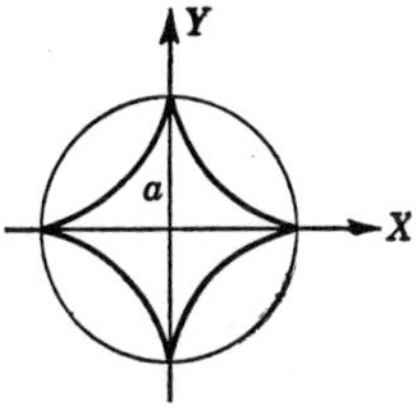

FIG. 152.

This is the required equation. The graph is shown in Fig. 152. Since the curve has four cusps, it is called the *four-cusped hypocycloid.*

EXERCISE 66

For each of the following curves, find the rectangular equation by eliminating the parameter t from the given equations.

1. $x = 6t - 12t^3$, $y = 2t$.
2. $x = t - 2$, $y = t^2 - 4t$.
3. $x = 4 \sin t$, $y = 4 \cos t$.
4. $x = \cos 2t$, $y = \sin t$.
5. $x = \sin t$, $y = \sin 2t$.
6. $x = 3 \sin t$, $y = 5 \cos t$.
7. $x = 5 - 4 \sin \omega t$, $y = -1 + 4 \cos \omega t$.

In each problem there is given the rectangular equation of a curve and an auxiliary relation involving t. Find the parametric equations of the curve, using as the parameter the variable t defined by the auxiliary equation.

8. $\frac{x^2}{a^2} + \frac{y^2}{b^2} = 1$, $x = a \sin t$.

9. $\frac{x^2}{a^2} - \frac{y^2}{b^2} = 1$, $x = a \sec t$.

10. $x^2 + y^2 = 2x$, $t = \frac{y}{x}$.

11. $x^2 - y^2 = 2y$, $t = \frac{y}{x}$.

12. A circle of radius 2 in. rolls with unit angular velocity on a straight line, taken as the x axis. Show that the equations $x = 2t - \sin t$, $y = 2 - \cos t$ give the motion of a point on the rolling circle 1 in. from the center and at (0,1) when $t = 0$.

13. A circle of radius 4 in. rolls on the outside of a circle of radius 12 in. with angular velocity 3 radians/sec. Show that the equations $x = 16 \cos t - 4 \cos 4t$, $y = 16 \sin t - 4 \sin 4t$ give the motion of a point on the circumference of the rolling circle.

14. A circle of radius 1 in. rolls inside a circle of radius 3 in. with angular velocity 3 radians/sec. Show that the equations $x = 2 \cos t + \cos 2t$, $y = 2 \sin t - \sin 2t$ give the motion of a point on the circumference of the rolling circle.

15. Show that the hypocycloid described when the radius of the rolling circle is one-half the radius of the fixed circle is the diameter $x = a \cos \phi$, $y = 0$.

Show that the equations $x = b_2 \cos b_1 t + b_1 \cos b_2 t$, $y = b_2 \sin b_1 t - b_1 \sin b_2 t$ give the motion of a point on the circumference of the rolling circle,

16. If this circle is of radius b_1 and rolls on the inside of a circle of radius $(b_1 + b_2)$ with angular velocity b_1 radians/sec.

17. If this circle is of radius b_2 and rolls on the inside of a circle of radius $(b_1 + b_2)$ with angular velocity $-b_2$ radians/sec.

18. Show that the equations

$$x = (a - b) \cos \phi + c \cos \left(\frac{a - b}{b} \phi\right),$$

$$y = (a - b) \sin \phi - c \sin \left(\frac{a - b}{b} \phi\right),$$

represent the *hypotrochoid* which is generated by a point rigidly attached to a circle of radius b and at distance c from the center, as this circle rolls without sliding on the inside of a fixed circle of radius a.

19. Show that if $a = 2b$, the hypotrochoid of Prob. 18 is the ellipse $x = (b + c) \cos \phi$, $y = (b - c) \sin \phi$.

20. The rod AB moves with its end A on the x axis and its end B on the y axis. From Fig. 153 deduce that the curve described by the point P on the rod with $AP = a$ and $PB = b$ is the ellipse

$$x = a \cos \phi, \qquad y = b \sin \phi.$$

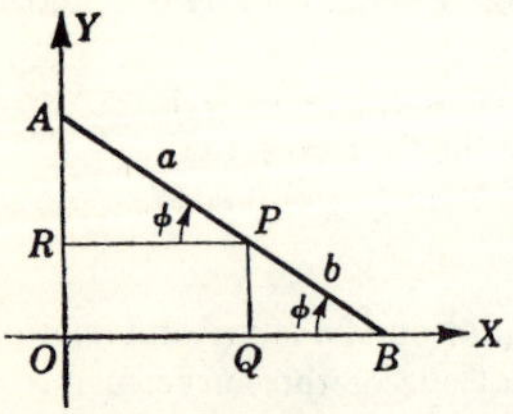

FIG. 153.

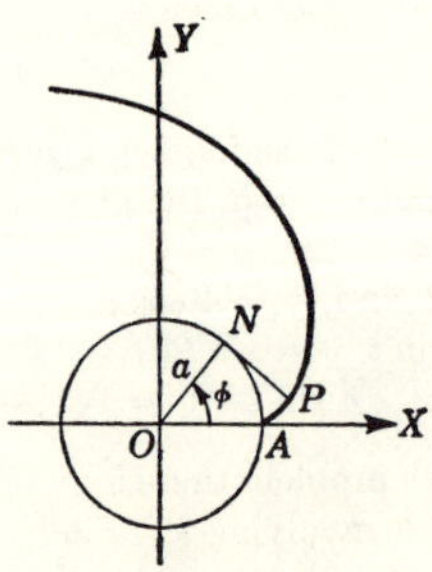

FIG. 154.

21. The *involute of the circle* is the curve generated by the end of a string which is kept taut while being unwound from a circle. From Fig. 154 deduce the parametric equations

$$x = a(\cos \phi + \phi \sin \phi), \qquad y = a(\sin \phi - \phi \cos \phi).$$

129. Slopes, Tangents, and Normals. Let a curve be given by equations in parametric form

$$x = g(t), \qquad y = h(t). \tag{17}$$

It follows from Eq. (22) of Sec. 53 that the first derivative

$$\frac{dy}{dx} = \frac{dy/dt}{dx/dt} = \frac{h'(t)}{g'(t)}. \tag{18}$$

The coordinates of P_1, the point on the curve at which $t = t_1$, are

$$x_1 = g(t_1), \qquad y_1 = h(t_1). \tag{19}$$

As in Sec. 87, let $(dy/dx)_1$ denote the value of the derivative dy/dx or slope of the curve at $P_1 = (x_1,y_1)$. Then from Eq. (18) we have

$$\tan \phi_1 = \text{slope at } P_1 = \left(\frac{dy}{dx}\right)_1 = \frac{h'(t_1)}{g'(t_1)}. \tag{20}$$

The equation of the tangent line to the curve at $P_1(x_1,y_1)$ is

$$y - y_1 = \left(\frac{dy}{dx}\right)_1 (x - x_1), \tag{21}$$

by Eq. (104) of Sec. 87. And the equation of the normal at P_1 is

$$y - y_1 = \frac{-1}{(dy/dx)_1} (x - x_1), \tag{22}$$

by Eq. (106) of Sec. 87. Hence the equations of the tangent and normal to the curve represented by Eq. (17) may be found by substituting the values of x_1, y_1, and $(dy/dx)_1$ from Eqs. (19) and (20) in Eqs. (21) and (22).

We may conclude from Eqs. (107) and (108) of Sec. 87, and Eq. (20) that

$$\phi_1 = 0, \qquad \left(\frac{dy}{dt}\right)_1 = h'(t_1) = 0, \qquad \text{for a horizontal tangent,} \tag{23}$$

and

$$\phi_1 = 90°, \quad \left(\frac{dx}{dt}\right)_1 = g'(t_1) = 0, \qquad \text{for a vertical tangent,} \tag{24}$$

unless $g'(t_1)$ and $h'(t_1)$ are zero for the same value of t_1, which requires an investigation of the limit of the quotient in Eq. (18) as $t \to t_1$.

EXAMPLE 1. Find the equations of the tangent and normal to the curve $x = e^{-t}\cos t$, $y = e^t \sin t$ at the point where $t = \pi$.

Solution: From the given equations we find that $dx/dt = -e^{-t}\sin t - e^{-t}\cos t$ and $dy/dt = e^t \cos t + e^t \sin t$. Hence, from Eq. (18), for any value of t we have

$$\frac{dy}{dx} = \frac{e^t(\cos t + \sin t)}{-e^{-t}(\cos t + \sin t)} = -e^{2t}.$$

And when $t = \pi$, $x_1 = e^{-\pi}\cos \pi = -e^{-\pi}$, $y_1 = e^{\pi}\sin\pi = 0$, $(dy/dx)_1 = -e^{2\pi}$.

By substituting these values in Eq. (21), we obtain

$$y - 0 = -e^{2\pi}[x - (-e^{-\pi})] \qquad \text{or} \qquad y + e^{2\pi}x = -e^{\pi}$$

as the equation of the tangent line.

And from Eq. (22) the equation of the normal is

$$y - 0 = \frac{-1}{-e^{2\pi}}[x - (-e^{-\pi})] \qquad \text{or} \qquad y - e^{-2\pi}x = e^{-3\pi}.$$

EXAMPLE 2. Find the points on the curve $x = 3t - t^3$, $y = 6t - t^2$ at which the tangent is parallel to a coordinate axis.

Solution: We have $dx/dt = 3 - 3t^2$, $dy/dt = 6 - 2t$. Thus $dx/dt = 0$ when $3 - 3t^2 = -3(t + 1)(t - 1) = 0$ and $t = -1$ or $t = 1$. And $dy/dt = 0$ when $6 - 2t = -2(t - 3) = 0$ and $t = 3$. Since dx/dt and dy/dt are never both zero for the same value of the parameter, it follows from Eqs. (23) and (24) that the tangent is vertical when $t = -1$ or when $t = 1$, and that the tangent is horizontal when $t = 3$. Hence the tangent to the given curve is parallel to the x axis at $(-18,9)$ and parallel to the y axis at $(-2,-7)$ and at $(2,5)$. These are the three required points.

EXAMPLE 3. Prove that the normal to the epicycloid passes through the point of contact of the rolling circle.

Solution: By Eq. (13), the coordinates of the point P tracing the epicycloid are

$$x = (a + b)\cos\phi - b\cos\left(\frac{a+b}{b}\phi\right),$$
$$y = (a + b)\sin\phi - b\sin\left(\frac{a+b}{b}\phi\right).$$

And from Fig. 150, the coordinates of the point of contact N are $a\cos\phi$ and $a\sin\phi$. Hence the slope of the line NP is

$$m_{NP} = \frac{y_P - y_N}{x_P - x_N} = \frac{b\sin\phi - b\sin\left(\frac{a+b}{b}\phi\right)}{b\cos\phi - b\cos\left(\frac{a+b}{b}\phi\right)} = \frac{\sin\phi - \sin\left(\frac{a+b}{b}\phi\right)}{\cos\phi - \cos\left(\frac{a+b}{b}\phi\right)}.$$

By differentiating the expressions for the coordinates of P we find that

$$\frac{dx}{d\phi} = -(a + b)\sin\phi + (a + b)\sin\left(\frac{a+b}{b}\phi\right),$$
$$\frac{dy}{d\phi} = (a + b)\cos\phi - (a + b)\cos\left(\frac{a+b}{b}\phi\right).$$

Hence from Eq. (18) with ϕ in place of t, we have

$$\frac{dy}{dx} = \frac{(a + b)\cos\phi - (a + b)\cos\left(\frac{a+b}{b}\phi\right)}{-(a + b)\sin\phi + (a + b)\sin\left(\frac{a+b}{b}\phi\right)} = -\frac{\cos\phi - \cos\left(\frac{a+b}{b}\phi\right)}{\sin\phi - \sin\left(\frac{a+b}{b}\phi\right)}$$

$= m$, the slope of the tangent at P.

A comparison of the expressions found for m_{NP} and m show that $m_{NP} \cdot m = -1$, or $m_{NP} = -1/m$. Thus NP has the slope of the normal, and since it passes through P, it is the normal line to the curve at P. This proves the stated property.

EXERCISE 67

Find the equations of the tangent and normal to each of the following curves at the point where t has the given value.

1. $x = t^2,\ y = t^3;\ t = 2.$
2. $x = \dfrac{2}{t},\ y = t;\ t = -2.$
3. $x = e^t,\ y = e^{-t};\ t = 1.$
4. $x = t^2,\ y = \dfrac{1}{t};\ t = -1.$
5. $x = e^{-t} \cos 5t,\ y = e^{-2t} \sin 3t;\ t = 0.$
6. $x = \dfrac{3t}{1 + t^3},\ y = \dfrac{3t^2}{1 + t^3};\ t = 1.$

In each of the following problems sketch the graph, find the rectangular equation, and find the points of contact of any horizontal or vertical tangents.

7. $x = t^3 - 12t,\ y = 2t.$
8. $x = t - 1,\ y = t^2 - 2t.$
9. $x = 5 \sin 2\phi,\ y = 5 \cos 2\phi.$
10. $x = 2 \cos 3\phi,\ y = 4 \sin 3\phi.$
11. $x = \cos 2t,\ y = \cos t.$
12. $x = \sin t,\ y = \cos 2t.$
13. $x = \cos^3 \phi,\ y = \sin^3 \phi.$
14. $x = \cos 2t,\ y = 4t.$

Prove that NP, the line from the point of contact of the rolling circle to the tracing point, is the normal at P of

15. The cycloid, Eq. (8).
16. The hypocycloid, Eq. (14).
17. The trochoid, Eq. (10).
18. The hypotrochoid, Prob. 18 of Exercise 66.

19. From Prob. 15 for the cycloid, Example 3 for the epicycloid, and Prob. 16 for the hypocycloid deduce by geometry that, for each of these curves, the tangent line at P may be constructed by extending NC to NCM, the diameter of the rolling circle through the point of contact, and drawing a straight line through P and M.
20. Show that the slope of the cycloid at any point is $\cot \phi/2$.
21. Verify from the equations of Prob. 21 of Exercise 66 that the slope of the involute to the circle at any point is $\tan \phi$.
22. Show that the slope of the tangent line to the curve

$$x = a \cos^n t, \qquad y = a \sin^n t$$

is $-\tan^{n-2} t$, the equation of the tangent line at any point is $x \sin^{n-2} t + y \cos^{n-2} t = a \cos^{n-2} t \sin^{n-2} t$, and the intercepts of the tangent line on the axes are $a \cos^{n-2} t$ and $a \sin^{n-2} t$.

Use Prob. 22 with the given value of n to show that

23. For the parabola $x^{\frac{1}{2}} + y^{\frac{1}{2}} = a^{\frac{1}{2}}$, $n = 4$, the sum of the intercepts of the tangent line on the axes is constant and equal to a.
24. For the four-cusped hypocycloid $x^{\frac{2}{3}} + y^{\frac{2}{3}} = a^{\frac{2}{3}}$, $n = 3$, the length of the tangent line cut off by the axes is constant and equal to a.
25. For the circle $x^2 + y^2 = a^2$, $n = 1$, the sum of the squares of the reciprocals of the intercepts of the tangent line on the axes is constant and equal to $1/a^2$.

130. The Second Derivative from the Parametric Form. We have seen in Secs. 43 and 47 that the direction of bending of a curve and the points of inflection may be found from an expression for the second derivative, d^2y/dx^2. We shall now find $\dfrac{d^2y}{dx^2} = \dfrac{d}{dx}\left(\dfrac{dy}{dx}\right)$ for a curve given in parametric form. We recall from Eq. (18) that

$$\frac{dy}{dx} = \frac{dy/dt}{dx/dt}. \tag{25}$$

As we may replace y by dy/dx in this relation, we find that

$$\frac{d^2y}{dx^2} = \frac{d\left(\dfrac{dy}{dx}\right)}{dx} = \frac{\dfrac{d}{dt}\left(\dfrac{dy}{dx}\right)}{dx/dt}. \tag{26}$$

From Eq. (25) and the rule for differentiating a quotient, we have

$$\frac{d}{dt}\left(\frac{dy}{dx}\right) = \frac{d}{dt}\left(\frac{dy/dt}{dx/dt}\right) = \frac{\dfrac{dx}{dt}\dfrac{d^2y}{dt^2} - \dfrac{dy}{dt}\dfrac{d^2x}{dt^2}}{(dx/dt)^2}. \tag{27}$$

By substituting this value in Eq. (26) we may deduce that

$$\frac{d^2y}{dx^2} = \frac{\dfrac{dx}{dt}\dfrac{d^2y}{dt^2} - \dfrac{dy}{dt}\dfrac{d^2x}{dt^2}}{(dx/dt)^3} = \frac{g'(t)h''(t) - h'(t)g''(t)}{[g'(t)]^3}. \tag{28}$$

In practice it is usually best to obtain dy/dx from Eq. (25) as a function of t in its simplest form, and then to use Eq. (26).

EXAMPLE 1. Given the curve $x = \tan t$, $y = 2 \sin 2t$, find d^2y/dx^2 in terms of t. Also locate the points of inflection.

Solution: From the given relations, we find by differentiation that $dx/dt = \sec^2 t$ and $dy/dt = 4 \cos 2t$. It follows from Eq. (25) that $\dfrac{dy}{dx} = \dfrac{4 \cos 2t}{\sec^2 t} = 4 \cos 2t \cos^2 t$. The derivative of this *with respect to t* is
$\dfrac{d}{dt}\left(\dfrac{dy}{dx}\right) = 4(-2 \cos 2t \cos t \sin t - 2 \sin 2t \cos^2 t) = -8 \cos t(\cos 2t \sin t + \sin 2t \cos t)$ $= -8 \cos t \sin 3t$, by the addition theorem for the sine, $\sin A \cos B + \cos A \sin B = \sin (A + B)$, with $A = t$ and $B = 2t$. Substitution of the values found in Eq. (26) gives

$$\frac{d^2y}{dx^2} = \frac{\dfrac{d}{dt}\left(\dfrac{dy}{dx}\right)}{dx/dt} = \frac{-8 \cos t \sin 3t}{\sec^2 t} = -8 \cos^3 t \sin 3t.$$

This is the required expression for the second derivative.

As $\tan t$ and $\sin 2t$ are each of period π, we need examine only values of t in the interval $0 \leqq t < \pi$. The factor $\cos t = 0$ when $t = \pi/2$, and the factor $\sin 3t = 0$ when $t = 0, \pi/3, 2\pi/3$. As t increases through any one of these values, d^2y/dx^2 changes sign. But $x = \infty$ when $t = \pi/2$. Hence we have points of inflection when $t = 0, \pi/3, 2\pi/3$ and the coordinates are $(0,0)$, $(\sqrt{3},\sqrt{3})$, and $(-\sqrt{3},-\sqrt{3})$.

EXAMPLE 2. For the epicycloid of Eq. (13), find d^2y/dx^2 in terms of ϕ.
Solution: From Example 3 of Sec. 129 we have

$$\frac{dx}{d\phi} = (a+b)\left[\sin\left(\frac{a+b}{b}\,\phi\right) - \sin\phi\right], \qquad \frac{dy}{dx} = -\frac{\cos\left(\frac{a+b}{b}\,\phi\right) - \cos\phi}{\sin\left(\frac{a+b}{b}\,\phi\right) - \sin\phi}.$$

We may apply the formulas from trigonometry, Probs. 12 and 14 of Exercise 48,

$$\sin A - \sin B = 2\cos\frac{A+B}{2}\sin\frac{A-B}{2},$$

and

$$\cos A - \cos B = -2\sin\frac{A+B}{2}\sin\frac{A-B}{2}, \qquad \text{with } A = \frac{a+b}{b}\,\phi \text{ and } B = \phi,$$

to deduce the simplified forms:

$$\frac{dx}{d\phi} = 2(a+b)\cos\left(\frac{a+2b}{2b}\,\phi\right)\sin\left(\frac{a}{2b}\,\phi\right),$$

$$\frac{dy}{dx} = \frac{2\sin\left(\frac{a+2b}{2b}\,\phi\right)\sin\left(\frac{a}{2b}\,\phi\right)}{2\cos\left(\frac{a+2b}{2b}\,\phi\right)\sin\left(\frac{a}{2b}\,\phi\right)} = \tan\left(\frac{a+2b}{2b}\,\phi\right).$$

Substitution of these simplified forms in Eq. (26) with ϕ in place of t gives

$$\frac{d^2y}{dx^2} = \frac{\frac{d}{d\phi}\left(\frac{dy}{dx}\right)}{dx/d\phi} = \frac{\frac{a+2b}{2b}\sec^2\left(\frac{a+2b}{2b}\,\phi\right)}{2(a+b)\cos\left(\frac{a+2b}{2b}\,\phi\right)\sin\left(\frac{a}{2b}\,\phi\right)}$$

$$= \frac{a+2b}{4b(a+b)\sin\left(\frac{a}{2b}\,\phi\right)\cos^3\left(\frac{a+2b}{2b}\,\phi\right)}.$$

This is the required expression.

EXERCISE 68

For each of the following curves, express d^2y/dx^2 in terms of t by the method of Sec. 130. Then eliminate the parameter, solve for y in terms of x, and express d^2y/dx^2 in terms of x and use this to check the first result.

1. $x = \frac{1}{t}$, $y = 4t$.
2. $x = t^2$, $y = 4t$.
3. $x = t^2$, $y = t^3$.
4. $x = \cos t$, $y = \sin t$.
5. $x = e^t$, $y = e^{-t}$.
6. $x = e^{2t}$, $y = e^{3t}$.

For each of the following curves find d^2y/dx^2 in terms of ϕ.

7. The cycloid, Eq. (8).
8. The hypocycloid, Eq. (14).
9. The ellipse, Eq. (6).
10. The trochoid, Eq. (10).
11. The involute of the circle, Prob. 21 of Exercise 66.
12. The curve $x = a\cos^n\phi$, $y = a\sin^n\phi$.

For each of the following curves find d^2y/dx^2 in terms of t. Also locate the points of inflection.

13. $x = \cot t,\ y = \sin^2 t.$
14. $x = 6t^2 - t^3,\ y = 3t.$
15. $x = 6t^2 - t^4,\ y = 2t.$
16. $x = (t + 3)\sqrt{t},\ y = t + 3.$
17. $x = \dfrac{t}{t^2 + 3},\ y = 4t.$
18. $x = \dfrac{1}{t},\ y = t + \dfrac{1}{t^2}.$
19. $x = \dfrac{1}{t},\ y = t^2 - \dfrac{3}{t^2}.$
20. $x = t - t^2,\ y = \dfrac{1}{t}.$

131. Curvature. The amount of curvature, or rapidity of bending, of a curved arc is measured by the rate of change of direction with respect to the distance traversed along the arc. For example, let the curve be the image of a highway on a map. Then on a hairpin turn where the curve is sharply bent, the direction changes greatly in a short distance so that the curvature is large. But on a long straightaway stretch where the curve is flat, the direction changes but little in some distance and the curvature is small.

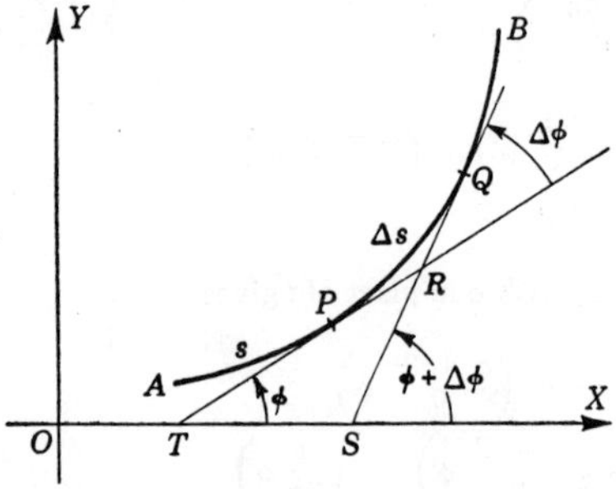

FIG. 155.

We shall now describe the concept of curvature in more precise terms. Let the arc AB be the graph of $y = f(x)$. We assume that $f(x)$ has a continuous first and second derivative, so that the arc is smooth and has a continuously turning tangent. In Fig. 155, $P = (x,y)$ is a fixed point on arc AB, and $Q = (x + \Delta x,\ y + \Delta y)$ is a variable nearby point on the curve tending toward P. The arc length s is measured from A, so that

$$\text{arc } AP = s, \qquad \text{arc } AQ = s + \Delta s, \qquad \text{arc } PQ = \Delta s. \tag{29}$$

The inclination or slope angle of a curve at any point is the angle from the positive x axis to the tangent line drawn to the curve at this point. Denote the slope angle at P by ϕ, and the slope angle at Q by $\phi + \Delta\phi$. In triangle TRS the exterior angle at S is the sum of the two interior angles at T and R. That is,

$$\angle TRS = \angle XSR - \angle STR = (\phi + \Delta\phi) - \phi = \Delta\phi. \tag{30}$$

Thus the angle from the tangent at P to the tangent at Q is $\Delta\phi$, as indicated in Fig. 155. Hence the tangent line turns through an angle $\Delta\phi$ as its point of contact moves over arc PQ and the change of the direction of motion may be measured by the change of the slope angle.

In a motion from P to Q, the change in s is arc $PQ = \Delta s$, while the change in direction is $\Delta\phi$. The average rate of change of direction with respect to arc length is $\Delta\phi/\Delta s$. We define this to be the *average curvature* of the arc PQ, so that

$$\text{Average curvature} = \frac{\Delta\phi}{\Delta s}. \tag{31}$$

When Q tends toward P, the average curvature of the arc PQ usually tends to a limit, and this limit K is defined to be the *curvature* at the point P. In symbols

$$\text{Curvature at } P = K = \lim_{\Delta s \to 0} \frac{\Delta\phi}{\Delta s} = \frac{d\phi}{ds}. \tag{32}$$

In words we may briefly define the curvature as the *rate of change of slope angle with respect to arc length*, or the derivative of ϕ with respect to s.

132. Derivative of Arc Length. The length of arc along a curve from a fixed point A to a variable point $P = (x,y)$ or s is a function of x if $y = f(x)$. And it follows from Eq. (75) of Sec. 62 that

$$\frac{ds}{dx} = \sqrt{1 + \left(\frac{dy}{dx}\right)^2}, \tag{33}$$

with the plus sign before the radical if s increases when x increases.

If $x = g(t)$ and $y = h(t)$, s is a function of the parameter t and

$$\frac{ds}{dt} = \sqrt{\left(\frac{dx}{dt}\right)^2 + \left(\frac{dy}{dt}\right)^2}, \tag{34}$$

with the plus sign before the radical if s increases when t increases, by Eq. (81) of Sec. 63.

It is easy to recall Eqs. (33) and (34) from the relation

$$ds^2 = dx^2 + dy^2, \tag{35}$$

which may be read from the right triangle with sides ds, dy, and ds. Since $\tan\phi = dy/dx$, one angle of the triangle is the slope angle ϕ, so that as noted in Eq. (77) of Sec. 62, we have

$$dx = ds\cos\phi, \qquad dy = ds\sin\phi, \tag{36}$$

if ϕ is chosen in the proper quadrant.

Let us verify that Eqs. (34) and (36) hold for the straight line of Eq. (2), or written with s in place of t,

$$x = s\cos\alpha, \qquad y = b + s\sin\alpha. \tag{37}$$

Since α is a constant, the derivatives with respect to the parameter s are

$$\frac{dx}{ds} = \cos\alpha, \qquad \frac{dy}{ds} = \sin\alpha. \tag{38}$$

Hence with s in place of t, Eq. (34) reduces to the identity

$$\frac{ds}{ds} = \sqrt{\cos^2\alpha + \sin^2\alpha} = 1. \tag{39}$$

And Eq. (36) becomes

$$ds \cos \alpha = ds \cos \phi, \qquad ds \sin \alpha = ds \sin \phi. \tag{40}$$

These relations are consistent with the fact that at all points of the straight line the slope angle ϕ is equal to the constant α.

Substitution of $\phi = \alpha$, a constant, in Eq. (32) gives

$$K = \frac{d\phi}{ds} = \frac{d\alpha}{ds} = 0. \tag{41}$$

This proves that *the curvature of a straight line at any point is zero.*

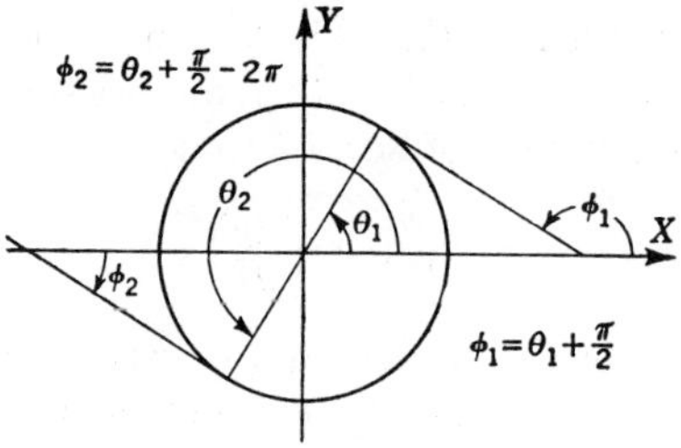

FIG. 156.

In a circle of radius a, the arc length subtended by a central angle θ is $a\theta$. Let us verify that Eqs. (34) and (36) hold for the circle of Eq. (4),

$$x = a \cos \theta, \qquad y = a \sin \theta, \qquad \text{if } s = a\theta. \tag{42}$$

The derivatives with respect to the parameter θ are

$$\frac{dx}{d\theta} = -a \sin \theta, \qquad \frac{dy}{d\theta} = a \cos \theta, \qquad \frac{ds}{d\theta} = a. \tag{43}$$

Hence with θ in place of t, Eq. (34) reduces to the identity

$$a = \frac{ds}{d\theta} = \sqrt{(-a \sin \theta)^2 + (a \cos \theta)^2} = \sqrt{a^2} = a, \tag{44}$$

and Eq. (36) becomes

$$-a \sin \theta \, d\theta = a \, d\theta \cos \phi, \qquad a \cos \theta \, d\theta = a \, d\theta \sin \phi. \tag{45}$$

But (Fig. 156) for the circle, the slope angle $\phi = \theta + \pi/2$, or $\phi = \theta + (\pi/2) - 2\pi$, so that $\cos \phi = -\sin \theta$ and $\sin \phi = \cos \theta$. And these relations are consistent with Eq. (45).

Let us substitute $\phi = \theta + \pi/2$ and $s = a\theta$ in Eq. (32), to obtain

$$K = \frac{d\phi}{ds} = \frac{d\phi/d\theta}{ds/d\theta} = \frac{1}{a}. \tag{46}$$

This proves that for a suitable direction of the arc length s, *the curvature of a circle at any point is equal to the reciprocal of the radius.*

We note that the straight line and the circle are the only curves of uniform or constant curvature, the same at all points.

EXAMPLE. Find the curvature of the cycloid $x = a(t - \sin t)$, $y = a(1 - \cos t)$, for any value of t in the interval $0 < t < 2\pi$.

Here we write t in place of the parameter of Eq. (8) to reserve ϕ for the slope angle.

Solution: The derivatives with respect to the parameter t are $dx/dt = a(1 - \cos t)$, $dy/dt = a \sin t$. Substitution in Eq. (34) gives

$$\frac{ds}{dt} = \sqrt{a^2(1 - \cos t)^2 + a^2 \sin^2 t} = a\sqrt{2 - 2\cos t} = a\sqrt{4\sin^2\frac{t}{2}}$$
$$= 2a \sin\frac{t}{2}, \qquad \text{if } 0 < t < 2\pi.$$

We could by integration deduce from $ds = 2a \sin\frac{t}{2}\,dt$ that the arc length of the cycloid measured from $t = 0$ is $s = 4a\left(1 - \cos\frac{t}{2}\right)$.

To find the slope angle ϕ, we observe that

$$\tan\phi = \frac{dy}{dx} = \frac{dy/dt}{dx/dt} = \frac{a \sin t}{a(1 - \cos t)} = \frac{2a \sin(t/2)\cos(t/2)}{2a \sin^2(t/2)} = \cot\frac{t}{2}$$
$$= \tan\left(\frac{\pi}{2} - \frac{t}{2}\right).$$

Hence one determination of $\phi = \frac{\pi}{2} - \frac{t}{2}$.

We may now substitute $\phi = \frac{\pi}{2} - \frac{t}{2}$ and $\frac{ds}{dt} = 2a \sin\frac{t}{2}$ in Eq. (32) and so deduce that for $0 < t < 2\pi$, the curvature

$$K = \frac{d\phi}{ds} = \frac{d\phi/dt}{ds/dt} = \frac{-\frac{1}{2}}{2a \sin(t/2)} = -\frac{1}{4a \sin(t/2)}.$$

This is the required value of the curvature. For $0 < t < 2\pi$, it is negative, indicating that the slope decreases as the parameter increases.

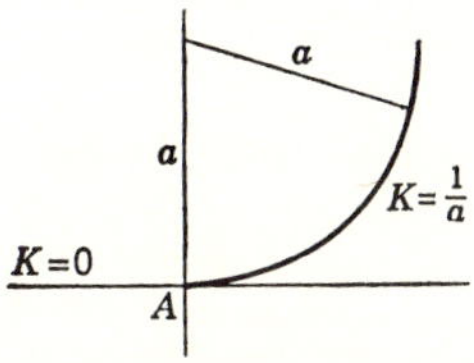

FIG. 157.

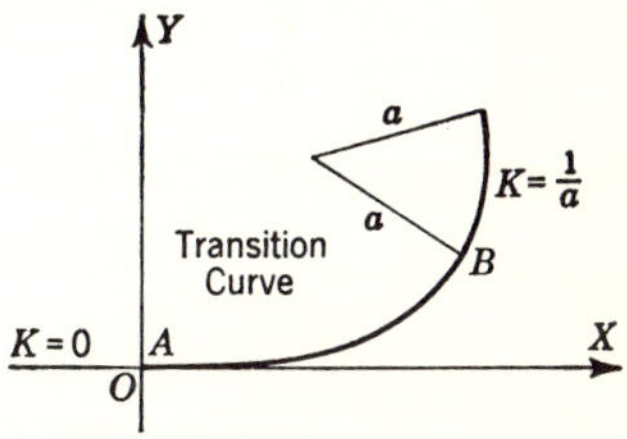

FIG. 158.

***133. Transition Curves.** For smooth riding at high speeds, railroad curves must have continuous curvature. Hence we cannot join a straight line directly to a circular arc tangent to it as in Fig. 157. For if the curves join at A and the radius of the circle is a, by Eqs. (41) and (46) the curvature is zero before A and $1/a$ after A and so suffers a sudden change. To avoid this difficulty, we must join the straight part to the circular part by an arc AB like that of Fig. 158 whose curvature is zero at A and is $1/a$ at B. Civil engineers call a curve of this type a *transition curve* or an easement curve.

One desirable form for the arc AB known as the transition spiral has the curvature at any point P proportional to the arc length $s = AP$. To find its equations, let $K = qs$. Then from Eq. (32) we have

$$\frac{d\phi}{ds} = qs, \qquad d\phi = qs\,ds = d\left(\frac{q}{2}s^2\right). \tag{47}$$

With the x axis tangent to the curve at $A = (0,0)$ as in Fig. 158, $\phi = 0$ when $s = 0$, and we may deduce from Eq. (47) that

$$\phi = \frac{q}{2}s^2 \quad \text{and} \quad s = \sqrt{\frac{2\phi}{q}}, \qquad ds = \frac{d\phi}{\sqrt{2q\phi}} = k\frac{d\phi}{\sqrt{\phi}}, \qquad \text{if } k = \frac{1}{\sqrt{2q}}. \tag{48}$$

By substitution of this value of ds in Eq. (36), we find that

$$dx = k\frac{d\phi}{\sqrt{\phi}}\cos\phi, \qquad dy = k\frac{d\phi}{\sqrt{\phi}}\sin\phi. \tag{49}$$

Since $x = 0$, $y = 0$ when $\phi = 0$, we integrate from 0 to ϕ to obtain

$$x = k\int_0^\phi \frac{\cos\phi}{\sqrt{\phi}}\,d\phi, \qquad y = k\int_0^\phi \frac{\sin\phi}{\sqrt{\phi}}\,d\phi. \tag{50}$$

These integrals cannot be expressed in closed form in terms of elementary functions. But by using the series found in Probs. 2 and 3 of Exercise 124 following Sec. 246,

$$\cos\phi = 1 - \frac{\phi^2}{2} + \frac{\phi^4}{24} - \cdots, \qquad \sin\phi = \phi - \frac{\phi^3}{6} + \frac{\phi^5}{120} - \cdots. \tag{51}$$

we may deduce from Eq. (50), by termwise integration as in Sec. 247, that

$$\begin{aligned} x &= k(2\phi^{\frac{1}{2}} - \tfrac{1}{5}\phi^{\frac{5}{2}} + \tfrac{1}{108}\phi^{\frac{9}{2}} - \cdots), \\ y &= k(\tfrac{2}{3}\phi^{\frac{3}{2}} - \tfrac{1}{21}\phi^{\frac{7}{2}} + \tfrac{1}{660}\phi^{\frac{11}{2}} - \cdots). \end{aligned} \tag{52}$$

And, with $t = \sqrt{\phi}$, the parametric equations of the transition spiral are

$$\begin{aligned} x &= k(2t - \tfrac{1}{5}t^5 + \tfrac{1}{108}t^9 - \cdots), \\ y &= k(\tfrac{2}{3}t^3 - \tfrac{1}{21}t^7 + \tfrac{1}{660}t^{11} - \cdots). \end{aligned} \tag{53}$$

By neglecting all terms of these series after the first, we obtain the relations $x = 2kt$, $y = \frac{2}{3}kt^3$, so that $t = x/(2k)$ and $y = x^3/(12k^2)$. If $c = 1/(12k^2)$, the rectangular equation is

$$y = cx^3, \tag{54}$$

and a short arc of this cubical parabola is sometimes used as a transition curve.

134. Curvature from the Rectangular Form. If $y = f(x)$, the derivatives $dy/dx = f'(x)$ and $d^2y/dx^2 = f''(x)$ may be found by direct differentiation. And if $F(x,y) = 0$, the derivatives dy/dx and d^2y/dx^2 may be expressed in terms of x and y by the method of implicit differentiation as explained in Sec. 56. Hence we may find the curvature of any curve whose equation is given in rectangular form since, as we shall show presently, the curvature may be expressed in terms of dy/dx and d^2y/dx^2.

To find the desired expression for the curvature

$$K = \frac{d\phi}{ds}, \tag{55}$$

we recall that $\tan\phi = dy/dx$ so that

$$\phi = \tan^{-1}\left(\frac{dy}{dx}\right). \tag{56}$$

By differentiation with respect to x, we find from Eq. (121) of Sec. 104 that

$$\frac{d\phi}{dx} = \frac{d}{dx}\tan^{-1}\left(\frac{dy}{dx}\right) = \frac{\frac{d}{dx}\left(\frac{dy}{dx}\right)}{1 + (dy/dx)^2} = \frac{\frac{d^2y}{dx^2}}{1 + (dy/dx)^2}. \tag{57}$$

But by Eq. (33) we have

$$\frac{ds}{dx} = \sqrt{1 + \left(\frac{dy}{dx}\right)^2}. \tag{58}$$

We may deduce from Eqs. (55), (57), and (58) that

$$K = \frac{d\phi}{ds} = \frac{\frac{d\phi}{dx}}{ds/dx} = \frac{\frac{d^2y/dx^2}{1 + (dy/dx)^2}}{\sqrt{1 + (dy/dx)^2}} = \frac{\frac{d^2y}{dx^2}}{[1 + (dy/dx)^2]^{\frac{3}{2}}}. \tag{59}$$

This proves that at any point P of a given curve, the

$$\text{Curvature} = K = \frac{d^2y/dx^2}{[1 + (dy/dx)^2]^{\frac{3}{2}}}. \tag{60}$$

EXAMPLE. Find the curvature of $y = cx^3$ at any point.

Solution: From the given equation we find by differentiation that $dy/dx = 3cx^2$, $d^2y/dx^2 = 6cx$. And by substitution in Eq. (60), we find

$$K = \frac{6cx}{(1 + 9c^2x^4)^{\frac{3}{2}}}. \tag{61}$$

This is the required curvature.

The K of Eq. (61) increases to a maximum as x increases from 0 to $1/(\sqrt{c}\,\sqrt[4]{45})$, so that an arc of Eq. (54) is useful as a transition curve only if the greatest x is less than $1/(\sqrt{c}\,\sqrt[4]{45})$ or $0.386/\sqrt{c}$.

135. Curvature from the Parametric Form. Let a curve be given in parametric form as in Eq. (17). Then we may find the derivatives dy/dx and d^2y/dx^2 in terms of the parameter as explained in Secs. 129 and 130. Hence we may find the curvature by substituting these results in Eq. (60). This is often the simplest procedure.

However, we found in Eqs. (18) and (28) that

$$\frac{dy}{dx} = \frac{dy/dt}{dx/dt}, \qquad \frac{d^2y}{dx^2} = \frac{\dfrac{dx}{dt}\dfrac{d^2y}{dt^2} - \dfrac{dy}{dt}\dfrac{d^2x}{dt^2}}{(dx/dt)^3}. \tag{62}$$

Substitution of these expressions in Eq. (60) leads to

$$\text{Curvature} = K = \frac{\dfrac{dx}{dt}\dfrac{d^2y}{dt^2} - \dfrac{dy}{dt}\dfrac{d^2x}{dt^2}}{[(dx/dt)^2 + (dy/dt)^2]^{\frac{3}{2}}}, \tag{63}$$

if dx/dt is positive.

Equation (60) is the special case of Eq. (63) with $t = x$, since in that case $dx/dt = 1$ and $d^2x/dt^2 = 0$. If $t = y$, $dy/dt = 1$ and $d^2y/dt^2 = 0$, so that Eq. (63) takes the form

$$\text{Curvature} = K = \frac{-d^2x/dy^2}{[1 + (dx/dy)^2]^{\frac{3}{2}}}, \tag{64}$$

if dx/dy is positive.

As indicated, Eqs. (60), (63), and (64) will all give the same algebraic sign to K if x, y, and t all increase together. In any case, the sign of the right member will indicate whether the slope angle ϕ increases or decreases as the independent variable increases. In particular, Eqs. (63) and (64) apply when $dx/dt = 0$ or $dx/dy = 0$.

Example. Calculate the curvature of a circle of radius a at a point in the first quadrant from each of the following four forms.

1. $y = \sqrt{a^2 - x^2}$.
2. $x = \sqrt{a^2 - y^2}$.
3. $x = a \cos t$, $y = a \sin t$.
4. $x = a \sin t$, $y = a \cos t$.

Solution 1: When $y = \sqrt{a^2 - x^2}$, $\dfrac{dy}{dx} = \dfrac{-x}{\sqrt{a^2 - x^2}}$ and

$$\frac{d^2y}{dx^2} = \frac{-\sqrt{a^2 - x^2} + x(-x/\sqrt{a^2 - x^2})}{a^2 - x^2} = \frac{-a^2}{(a^2 - x^2)^{\frac{3}{2}}}.$$

It follows from Eq. (60) that

$$K = \frac{d^2y/dx^2}{\left[1 + \left(\dfrac{dy}{dx}\right)^2\right]^{\frac{3}{2}}} = \frac{-a^2/(a^2 - x^2)^{\frac{3}{2}}}{\left(1 + \dfrac{x^2}{a^2 - x^2}\right)^{\frac{3}{2}}} = \frac{-a^2}{a^3} = -\frac{1}{a}.$$

The minus sign indicates that the slope angle decreases as x increases.

Solution 2: When $x = \sqrt{a^2 - y^2}$, $\dfrac{dx}{dy} = \dfrac{-y}{\sqrt{a^2 - y^2}}$, $\dfrac{d^2x}{dy^2} = \dfrac{-a^2}{(a^2 - y^2)^{\frac{3}{2}}}$. It follows from Eq. (64) that

$$K = \frac{-d^2x/dy^2}{\left[1 + \left(\dfrac{dx}{dy}\right)^2\right]^{\frac{3}{2}}} = \frac{a^2/(a^2 - y^2)^{\frac{3}{2}}}{\left(1 + \dfrac{y^2}{a^2 - y^2}\right)^{\frac{3}{2}}} = \frac{a^2}{a^3} = \frac{1}{a}.$$

The plus sign indicates that the slope angle increases as y increases.

Solution 3: When $x = a \cos t$, $y = a \sin t$, $dx/dt = -a \sin t$, $d^2x/dt^2 = -a \cos t$, $dy/dt = a \cos t$, $d^2y/dt^2 = -a \sin t$. It follows from Eq. (63) that

$$K = \frac{\dfrac{dx}{dt}\dfrac{d^2y}{dt^2} - \dfrac{dy}{dt}\dfrac{d^2x}{dt^2}}{[(dx/dt)^2 + (dy/dt)^2]^{\frac{3}{2}}} = \frac{a^2 \sin^2 t + a^2 \cos^2 t}{(a^2 \sin^2 t + a^2 \cos^2 t)^{\frac{3}{2}}} = \frac{a^2}{a^3} = \frac{1}{a}.$$

The plus sign indicates that the slope angle increases as t increases.

Solution 4: When $x = a \sin t$, $y = a \cos t$, $dx/dt = a \cos t$, $d^2x/dt^2 = -a \sin t$, $dy/dt = -a \sin t$, $d^2y/dt^2 = -a \cos t$. It follows from Eq. (63) that

$$K = \frac{\dfrac{dx}{dt}\dfrac{d^2y}{dt^2} - \dfrac{dy}{dt}\dfrac{d^2x}{dt^2}}{[(dx/dt)^2 + (dy/dt)^2]^{\frac{3}{2}}} = \frac{-a^2 \cos^2 t - a^2 \sin^2 t}{(a^2 \cos^2 t + a^2 \sin^2 t)^{\frac{3}{2}}} = \frac{-a^2}{a^3} = -\frac{1}{a}.$$

The minus sign indicates that the slope angle decreases as t increases.

COMMENT: From the first derivatives with respect to t in solution 3, we might have deduced that $\dfrac{dy}{dx} = \dfrac{dy/dt}{dx/dt} = \dfrac{a \cos t}{-a \sin t} = -\cot t$, so that $\dfrac{d^2y}{dx^2} = \dfrac{\dfrac{d}{dt}\left(\dfrac{dy}{dx}\right)}{dx/dt} = \dfrac{\csc^2 t}{-a \sin t} = \dfrac{\csc^3 t}{-a}$. These results, inserted in Eq. (60), give

$$K = \frac{d^2y/dx^2}{[1 + (dy/dx)^2]^{\frac{3}{2}}} = \frac{\csc^3 t/-a}{(1 + \cot^2 t)^{\frac{3}{2}}} = \frac{\csc^3 t}{-a \csc^3 t} = -\frac{1}{a},$$

as in solution 1.

136. Radius of Curvature. The *radius of curvature*, R, of any curve at a point P is the reciprocal of the curvature K at this point P. Thus from Eq. (55),

$$R = \frac{1}{K} = \frac{ds}{d\phi}. \tag{65}$$

And from Eq. (60), we may conclude that

$$R = \frac{[1 + (dy/dx)^2]^{\frac{3}{2}}}{d^2y/dx^2}. \tag{66}$$

If x and y have the dimensions of length L, dy/dx has no dimensions and $\dfrac{d^2y}{dx^2} = \dfrac{d}{dx}\left(\dfrac{dy}{dx}\right)$ has the dimensions of $1/L$. Hence the expression for the radius of curvature in Eq. (66) has the dimensions of length. These facts may help one to remember that d^2y/dx^2 is in the *denominator* of the expression for the *radius* R.

Two other expressions for R may be found from Eqs. (63) and (64). These may lead to different algebraic signs which may be interpreted as in the example of Sec. 135. And it is often best to use the first procedure described in Sec. 135.

For any point P on a given curve let us make the following construction. Draw the normal to the curve at P and measure off on the concave side of the curve a distance PC numerically equal to the radius of curvature of the curve at P. With C as a center, draw the circle which passes through P (Fig. 159). The point C is called the *center of curvature*, and the circle with C as a center and R as a radius is called the *circle of curvature* for the point P.

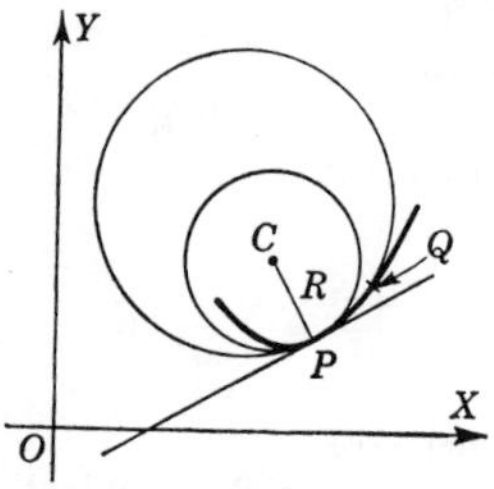

FIG. 159.

At P the circle of curvature will be tangent to the curve and will have the same curvature $K = 1/R$ as the curve has. Hence at P the values of dy/dx and d^2y/dx^2 (or of dx/dy and d^2x/dy^2 if $dx/dy = 0$) will be the same for the circle and the curve. Thus the circle of curvature is the circle which most closely fits the curve in the immediate neighborhood of P. If we pass a circle through Q, a point on the curve near to P, which is tangent to the curve at P and then let Q tend to P along the curve, the limiting position of this circle will be the circle of curvature. Because of this property, the circle of curvature is said to osculate the given curve at P. In general the circle of curvature at a point will cross the curve at that point.

If P is a point of inflection, the limiting position of the circle through Q tangent to the curve at P will be the straight line tangent to the curve at the point of inflection. The discussion of Sec. 47 shows that the slope angle ϕ has a maximum or minimum at a point of inflection. Hence $d\phi/ds = 0$, so that $K = 0$ and $R = \infty$. And we may regard the tangent straight line at a point of inflection as an osculating "circle" of infinite radius.

EXAMPLE 1. Find the radius of curvature of the curve $y = e^{-x^2}$ at the point $(1,e^{-1})$.

Solution: For any value of x, we find by differentiation that $dy/dx = -2xe^{-x^2}$, $d^2y/dx^2 = -2e^{-x^2} + 4x^2e^{-x^2}$. Hence when $x = 1$, $\dfrac{dy}{dx} = -2e^{-1}$, $\dfrac{d^2y}{dx^2} = 2e^{-1}$, and $R = \dfrac{[1 + (dy/dx)^2]^{\frac{3}{2}}}{d^2y/dx^2} = \dfrac{(1 + 4e^{-2})^{\frac{3}{2}}}{2e^{-1}} = \dfrac{(e^2 + 4)^{\frac{3}{2}}}{2e^2}$. This is the required value.

EXAMPLE 2. Find the radius of curvature of the epicycloid

$$x = (a + b)\cos t - b\cos\left(\frac{a + b}{b}t\right),$$
$$y = (a + b)\sin t - b\sin\left(\frac{a + b}{b}t\right),$$

for any value of t in the interval $0 < t < \pi b/(a + 2b)$.

Solution: From the results of Example 2 of Sec. 130, with t in place of the parameter there used, we have $\dfrac{dy}{dx} = \tan\left(\dfrac{a + 2b}{2b}t\right)$ and

$$\frac{d^2y}{dx^2} = \frac{a + 2b}{4b(a + b)\sin\left(\dfrac{a}{2b}t\right)\cos^3\left(\dfrac{a + 2b}{2b}t\right)}.$$ It follows that

$1 + \left(\frac{dy}{dx}\right)^2 = 1 + \tan^2\left(\frac{a + 2b}{2b}\,t\right) = \sec^2\left(\frac{a + 2b}{2b}\,t\right)$ and $R = \dfrac{\left[1 + \left(\frac{dy}{dx}\right)^2\right]^{\frac{3}{2}}}{\frac{d^2y}{dx^2}} =$

$\dfrac{\sec^3\left(\frac{a + 2b}{2b}\,t\right)}{\dfrac{(a + 2b)\sec^3\left(\frac{a + 2b}{2b}\,t\right)}{4b(a + b)\sin\left(\frac{a}{2b}\,t\right)}} = \dfrac{4b(a + b)}{a + 2b}\sin\left(\frac{a}{2b}\,t\right)$, since $\sec\left(\frac{a + 2b}{2b}\,t\right)$ is positive for $0 < t < \dfrac{\pi b}{a + 2b}$. Hence for t in this range, $R = \dfrac{4b(a + b)}{a + 2b}\sin\left(\frac{a}{2b}\,t\right)$ is the required value.

EXAMPLE 3. Find the curvature and radius of curvature of $y^2 = 4px$ at any point (x,y), and in particular find the radius of curvature at the point $(0,0)$.

Solution: By implicit differentiation as in Sec. 56, we find that $2y\,\dfrac{dy}{dx} = 4p$, $\dfrac{dy}{dx} = \dfrac{2p}{y}$, $\dfrac{d^2y}{dx^2} = -\dfrac{2p}{y^2}\dfrac{dy}{dx} = -\dfrac{4p^2}{y^3}$. Hence from Eq. (60),

$K = \dfrac{d^2y/dx^2}{[1 + (dy/dx)^2]^{\frac{3}{2}}} = \dfrac{-4p^2/y^3}{[1 + (4p^2/y^2)]^{\frac{3}{2}}} = \dfrac{-4p^2}{(y^2 + 4p^2)^{\frac{3}{2}}}$ if $y > 0$. Hence

$K = \dfrac{-4p^2}{(y^2 + 4p^2)^{\frac{3}{2}}}$ and $R = \dfrac{(y^2 + 4p^2)^{\frac{3}{2}}}{-4p^2}$ if y is positive. But $K = \dfrac{4p^2}{(y^2 + 4p^2)^{\frac{3}{2}}}$ and $R = \dfrac{(y^2 + 4p^2)^{\frac{3}{2}}}{4p^2}$ if y is negative. These are the required general values.

In particular, at $(0,0)$, where $y = 0$, we find by letting y tend to zero that $R = -2p$ or $R = +2p$ according as we think of $(0,0)$ as the end of an arc above or below the x axis.

Had we tried to evaluate the derivatives at $(0,0)$ we would have found $dy/dx = \infty$. This suggests that we differentiate with respect to y to deduce $2y = 4p(dx/dy)$, $dx/dy = y/2p$, $d^2x/dy^2 = 1/2p$. Hence from Eq. (64)

$$K = \frac{-d^2x/dy^2}{[1 + (dx/dy)^2]^{\frac{3}{2}}} = \frac{-1/2p}{[1 + (y^2/4p^2)]^{\frac{3}{2}}} = \frac{-4p^2}{(y^2 + 4p^2)^{\frac{3}{2}}} \qquad \text{and} \qquad R = \frac{(y^2 + 4p^2)^{\frac{3}{2}}}{-4p^2}$$

In particular when $y = 0$, $R = -2p$. This particular value could be obtained from Eqs. (65) and (64) by evaluating the derivatives at $(0,0)$ as $dx/dy = 0$ and $d^2x/dy^2 = 1/2p$. These lead to

$$R = \frac{1}{K} = \frac{[1 + (dx/dy)^2]^{\frac{3}{2}}}{-d^2x/dy^2} = \frac{1}{-1/2p} = -2p.$$

The differences in sign result from the fact that ϕ decreases as y increases for all values, making K negative in Eq. (64). But as x increases from 0, ϕ decreases on the branch with y positive, and ϕ increases on the branch with y negative, leading to opposite signs for K as found from Eq. (60) for the two branches.

EXERCISE 69

Find the radius of curvature for each of the following curves at the given point.

1. $y = x^3 - x^2$, $(2,4)$.
2. $y = x + \dfrac{2}{x}$, $(2,3)$.
3. $y = e^x$, $(0,1)$.
4. $y = \tan x$, $\left(\dfrac{\pi}{4},1\right)$.
5. $x = y^4 - 4y^2$, $(2,0)$.
6. $x = y \ln y$, (e,e).

Find the curvature of each of the following curves at any point (x,y).

7. $3y = x^3$.
8. $2y = x^2$.
9. $3y^2 = 2x^3$.
10. $y = \ln \sin x$.
11. $y = \frac{a}{2}(e^{x/a} - e^{-x/a})$.
12. $\frac{x^2}{a^2} + \frac{y^2}{b^2} = 1$.

Find the radius of curvature of each of the following curves at the point where
has the given value.

13. $x = t^2$, $y = t^3$, $t = 1$.
14. $x = \frac{2}{t}$, $y = t$, $t = 2$.
15. $x = e^{-t}$, $y = e^t$, $t = 0$.
16. $x = t^2$, $y = t$, $t = 3$.

Find the curvature of each of the following curves for any value of t.

17. $x = a \cos t$, $y = a \sin t$.
18. $x = a \cos t$, $y = b \sin t$.
19. $x = a \cos^3 t$, $y = a \sin^3 t$.
20. $x = a \cos^4 t$, $y = a \sin^4 t$.
21. $x = a(\cos t + t \sin t)$, $y = a(\sin t - t \cos t)$.
22. $x = a(t - \sin t)$, $y = a(1 - \cos t)$.

23. Find the value of x for which the radius of curvature of the curve $y = x^n$ is a minimum.

***137. Center of Curvature.** Let $P = (x,y)$ be any point on a curve, and $C = (X,Y)$ be the center of curvature for the point P. For the present, we assume that at P the derivatives dy/dx and d^2y/dx^2 are both positive. Thus the slope angle ϕ is in the first quadrant and R as found from Eq. (66) is positive. Under these conditions $PC = R$ and we may take the slope angle of the normal segment PC as $\phi + (\pi/2)$ (Fig. 160). Then the projections of PC on the coordinate axes will be

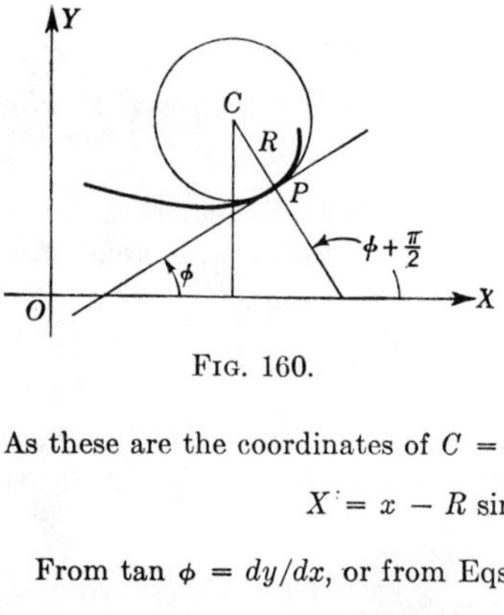

FIG. 160.

$$R \cos\left(\phi + \frac{\pi}{2}\right) = -R \sin \phi,$$

$$R \sin\left(\phi + \frac{\pi}{2}\right) = R \cos \phi. \quad (67)$$

As these are the coordinates of $C = (X,Y)$ relative to $P = (x,y)$, it follows that

$$X = x - R \sin \phi, \qquad Y = y + R \cos \phi. \quad (68)$$

From $\tan \phi = dy/dx$, or from Eqs. (36) and (35), it follows that

$$\cos \phi = \frac{1}{\sqrt{1 + (dy/dx)^2}}, \qquad \sin \phi = \frac{dy/dx}{\sqrt{1 + (dy/dx)^2}}, \quad (69)$$

when ϕ is in the first quadrant. But from Eq. (66) we have

$$R = \frac{[1 + (dy/dx)^2]^{\frac{3}{2}}}{d^2y/dx^2}. \quad (70)$$

From Eqs. (68) to (70) we may deduce that

$$X = x - \frac{dy}{dx}\frac{1 + (dy/dx)^2}{d^2y/dx^2}, \qquad Y = y + \frac{1 + (dy/dx)^2}{d^2y/dx^2}. \quad (71)$$

when dy/dx and d^2y/dx^2 are both positive.

But when d^2y/dx^2 is negative, R as found from Eq. (70) is negative. And the curve is concave downward so that the slope angle of PC is $\phi - (\pi/2)$. Hence the projections of PC are

$$(-R)\cos\left(\phi - \frac{\pi}{2}\right) = -R\sin\phi, \qquad (-R)\sin\left(\phi - \frac{\pi}{2}\right) = R\cos\phi. \tag{72}$$

These again lead to Eq. (68), which holds in all cases if we take $-\pi/2 < \phi < \pi/2$. And when ϕ is a negative acute angle, $\sin\phi$ and dy/dx are both negative so that Eq. (69) continues to hold. It follows that Eq. (71), which involves no square roots, gives the center of curvature unless dy/dx is infinite. We may deduce from Eq. (64) and a similar argument that

$$X = x + \frac{1 + (dx/dy)^2}{d^2x/dy^2}, \qquad Y = y - \frac{dx}{dy}\frac{1 + (dx/dy)^2}{d^2x/dy^2}. \tag{73}$$

This gives the center of curvature unless $dy/dx = 0$.

***138. The Evolute.** When a point $P = (x,y)$ moves along a given curve, the center of curvature $C = (X,Y)$ traces out a second curve, as in Fig. 161. This new curve is called the *evolute* of the given curve. Let the equation of the given curve be $y = f(x)$. Then $dy/dx = f'(x)$ and $d^2y/dx^2 = f''(x)$. If these values of y, dy/dx, and d^2y/dx^2 are substituted in the second members of Eq. (71), the resulting relations will be the parametric equations of the evolute, with X and Y the running coordinates and x the parameter. The rectangular equation of the evolute connecting X and Y can be found if x can be eliminated.

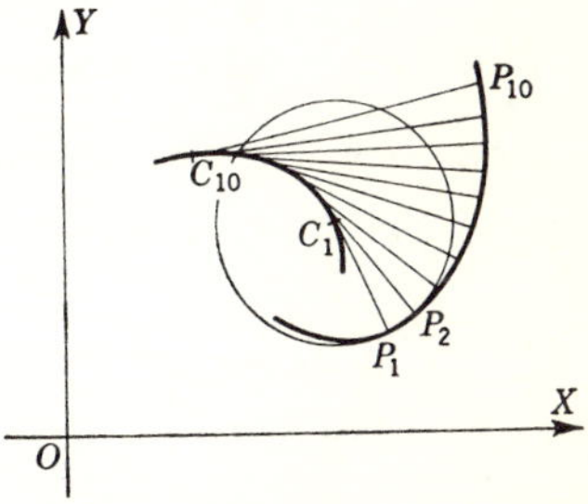

FIG. 161.

And if x and y are given in terms of a parameter t, we may express dy/dx and d^2y/dx^2 in terms of t, as in Sec. 130. Then by substituting the values of x, y, dy/dx, and d^2y/dx^2 in Eq. (71), we may find the parametric equations of the evolute in terms of the parameter t. The rectangular equation of the evolute connecting X and Y can be found if t can be eliminated from the parametric equations, as in Sec. 126.

It is sometimes convenient to find dy/dx and d^2y/dx^2 in terms of x and y by implicit differentiation, as in Sec. 56. If this is done, substitution of the values of dy/dx and d^2y/dx^2 in Eq. (71) gives X and Y in terms of x and y, which are related by the equation of the given curve. This is a form of parametric representation of the evolute. For, by substituting a series of values of (x,y) for points on the given curve, we could obtain and plot a series of points (X,Y) on the evolute. If x and y can be eliminated from the three equations, the rectangular equation of the evolute connecting X and Y can be found. In particular this may be effected by solving two of the equations for x and y and substituting these values in the third equation.

EXAMPLE 1. Find the evolute of the parabola $y = x^2$.

Solution: Since $y = x^2$, $dy/dx = 2x$, $d^2y/dx^2 = 2$. Hence, from Eq. (71),

$$X = x - \frac{dy}{dx}\frac{1 + (dy/dx)^2}{d^2y/dx^2} = x - 2x\frac{1 + 4x^2}{2} = -4x^3,$$

$$Y = y + \frac{1 + (dy/dx)^2}{d^2y/dx^2} = x^2 + \frac{1 + 4x^2}{2} = \frac{1}{2} + 3x^2.$$

From $X = -4x^3$, $x = (-X/4)^{\frac{1}{3}}$. Hence $Y = \frac{1}{2} + 3x^2 = \frac{1}{2} + 3(X^2/16)^{\frac{1}{3}}$. The evolute $X^2 = \frac{16}{27}(Y - \frac{1}{2})^3$ is a semicubical parabola with axis vertical and vertex at $(0,\frac{1}{2})$ (Fig. 162).

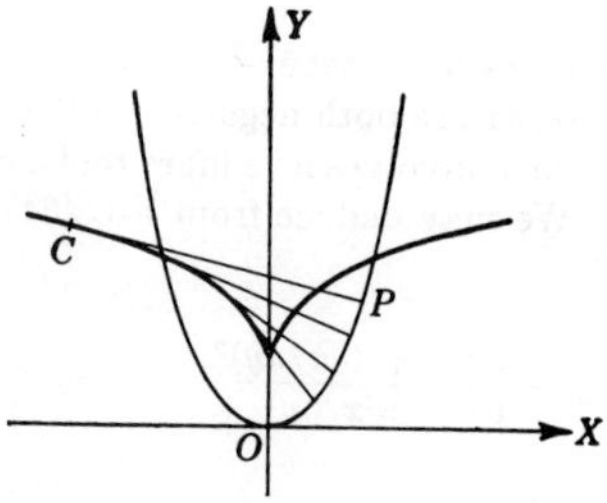

FIG. 162.

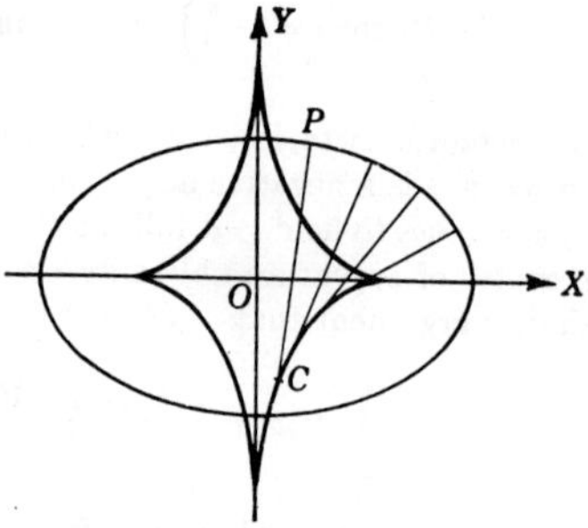

FIG. 163.

EXAMPLE 2. Find the evolute of the ellipse $\frac{x^2}{a^2} + \frac{y^2}{b^2} = 1$.

Solution: By implicit differentiation (Sec. 56), we find that $\frac{2x}{a^2} + \frac{2y}{b^2}\frac{dy}{dx} = 0$, $\frac{dy}{dx} = -\frac{b^2x}{a^2y}$, $\frac{d^2y}{dx^2} = -\frac{b^2}{a^2}\frac{y - x(dy/dx)}{y^2} = -\frac{b^2}{a^2}\frac{y - x(-b^2x/a^2y)}{y^2} = -\frac{b^2}{a^4}\frac{a^2y^2 + b^2x^2}{y^3}$. But from the given equation, $b^2x^2 + a^2y^2 = a^2b^2$, so that $\frac{d^2y}{dx^2} = -\frac{b^2}{a^4}\frac{a^2b^2}{y^3} = -\frac{b^4}{a^2y^3}$. By substituting this and $\frac{dy}{dx} = -\frac{b^2x}{a^2y}$ in Eq. (71), we find that

$$X = x - \left(-\frac{b^2x}{a^2y}\right)\frac{1 + (b^4x^2/a^4y^2)}{-b^4/a^2y^3} = x\left(1 - \frac{y^2}{b^2} - \frac{b^2x^2}{a^4}\right).$$

But from the given equation, $1 - \frac{y^2}{b^2} = \frac{x^2}{a^2}$, so that if $c^2 = a^2 - b^2$, $X = x\left(\frac{x^2}{a^2} - \frac{b^2x^2}{a^4}\right) = \frac{x^3}{a^4}(a^2 - b^2) = \frac{c^2x^3}{a^4}$. And for Y we find that $Y = y + \frac{1 + (b^4x^2/a^4y^2)}{-b^4/a^2y^3} = y\left(1 - \frac{a^2y^2}{b^4} - \frac{x^2}{a^2}\right)$. But from the given equation, $1 - \frac{x^2}{a^2} = \frac{y^2}{b^2}$, so that $Y = y\left(\frac{y^2}{b^2} - \frac{a^2y^2}{b^4}\right) = \frac{y^3}{b^4}(b^2 - a^2) = -\frac{c^2y^3}{b^4}$.

Thus the evolute of the ellipse could be plotted from the equations $X = \frac{c^2x^3}{a^4}$, $Y = -\frac{c^2y^3}{b^4}$, $\frac{x^2}{a^2} + \frac{y^2}{b^2} = 1$. We may solve the first two relations for x and y in the form $x = a\left(\frac{aX}{c^2}\right)^{\frac{1}{3}}$, $y = -b\left(\frac{bY}{c^2}\right)^{\frac{1}{3}}$. Substitution of these values in the third or given equation leads to $\left(\frac{aX}{c^2}\right)^{\frac{2}{3}} + \left(\frac{bY}{c^2}\right)^{\frac{2}{3}} = 1$ or $(aX)^{\frac{2}{3}} + (bY)^{\frac{2}{3}} = c^{\frac{4}{3}} = (a^2 - b^2)^{\frac{2}{3}}$. This is the rectangular equation of the evolute (Fig. 163).

EXAMPLE 3. Find the evolute of the cycloid, $x = a(t - \sin t)$, $y = a(1 - \cos t)$.

Solution: From the given equations, we find the derivatives $\frac{dx}{dt} = a(1 - \cos t) = 2a \sin^2 \frac{t}{2}$, $\frac{dy}{dt} = a \sin t = 2a \sin \frac{t}{2} \cos \frac{t}{2}$. Hence from Eq. (25), $\frac{dy}{dx} = \frac{dy/dt}{dx/dt} = \frac{2a \sin (t/2) \cos (t/2)}{2a \sin^2 (t/2)} = \frac{\cos (t/2)}{\sin (t/2)} = \cot \frac{t}{2}$. And from Eq. (26), $\frac{d^2y}{dx^2} = \frac{\frac{d}{dt}\left(\frac{dy}{dt}\right)}{dx/dt} = \frac{-\frac{1}{2} \csc^2 (t/2)}{2a \sin^2 (t/2)} = -\frac{1}{4a} \csc^4 \frac{t}{2}$. From $\frac{dy}{dx} = \cot \frac{t}{2}$ and $\frac{d^2y}{dx^2} = -\frac{1}{4a} \csc^4 \frac{t}{2}$, we may deduce that $\frac{1 + (dy/dx)^2}{d^2y/dx^2} = \frac{1 + \cot^2 (t/2)}{-(1/4a) \csc^4 (t/2)} = \frac{\csc^2 (t/2)}{-(1/4a) \csc^4 (t/2)} = -4a \sin^2 \frac{t}{2} = -2a(1 - \cos t)$, and $\frac{dy}{dx} \frac{1 + (dy/dx)^2}{d^2y/dx^2} = \cot \frac{t}{2}\left(-4a \sin^2 \frac{t}{2}\right) = -4a \sin \frac{t}{2} \cos \frac{t}{2} = -2a \sin t$. It follows from Eq. (71) that $X = a(t - \sin t) - (-2a \sin t) = a(t + \sin t)$, $Y = a(1 - \cos t) + (-2a)(1 - \cos t) = -a(1 - \cos t)$.

Thus $X = a(t + \sin t)$, $Y = -a(1 - \cos t)$ is one form of the parametric equations of the evolute. When $t = -\pi$, $X = -a\pi$, $Y = -2a$. Let us take this point $(-a\pi, -2a)$ as a new origin O' and begin measuring the parameter as zero at this point where $t = -\pi$. Denote the new coordinates by X', Y' and denote the new parameter by t'. Then

$$X' = X - (-a\pi), \qquad Y' = Y - (-2a), \qquad t' = t - (-\pi).$$

It follows that $X = X' - a\pi$, $Y = Y' - 2a$, and $t = t' - \pi$. The result of substituting these values in $X = a(t + \sin t)$, $Y = -a(1 - \cos t)$ is $X' - a\pi = a[t' - \pi + \sin\ (t' - \pi)]$, $Y' - 2a = -a[1 - \cos\ (t' - \pi)]$, or $X' = a(t' - \sin t')$, $Y' = a(1 - \cos t')$. A comparison of these relations with the given equations shows that the evolute of a cycloid is an equal cycloid obtained from the given curve by a parallel displacement (Fig. 164).

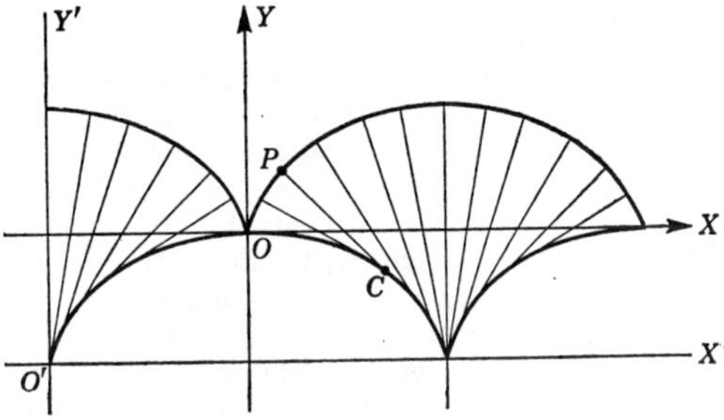

FIG. 164.

***139. Properties of the Evolute.** Let $P = (x,y)$ be any point on a given curve, and $C = (X,Y)$ be the center of curvature for this point P, or the corresponding point on the evolute. Then from Eq. (68) we have

$$X = x - R \sin \phi, \qquad Y = y + R \cos \phi. \tag{74}$$

When x and y are functions of any parameter t, R and ϕ for the given curve are also functions of t. Thus Eq. (74) determines X and Y as functions of the independent variable t. With this assumption, the differentials of X and Y are

$$dX = dx - dR \sin \phi - R \cos \phi\, d\phi, \qquad dY = dy + dR \cos \phi - R \sin \phi\, d\phi. \tag{75}$$

From Eqs. (65) and (36), or

$$R = \frac{ds}{d\phi}, \qquad dx = ds \cos \phi, \qquad dy = ds \sin \phi, \tag{76}$$

it follows that

$$ds = R\, d\phi \quad \text{and} \quad dx = R \cos \phi\, d\phi, \qquad dy = R \sin \phi\, d\phi. \tag{77}$$

This shows that in each right member of Eq. (75) the first and last terms cancel each other, so that

$$dX = -dR \sin \phi \qquad dY = dR \cos \phi. \tag{78}$$

The slope of the tangent line to the evolute at $C(X,Y)$ is

$$\frac{dY}{dX} = \frac{dR \cos \phi}{-dR \sin \phi} = \frac{-1}{\tan \phi} = \frac{-1}{dy/dx}. \tag{79}$$

Either of the last two members is an expression for the slope of the normal to the given curve at C, or of PC. This proves that

The normal PC to the given curve at P is tangent to the evolute at C, the center of curvature for P.

And the evolute of a curve is the envelope of its normals. We shall show how to find the equation of the evolute by the method of envelopes in Sec. 279.

Let S denôte arc length on the evolute. Then from Eq. (35) we have $dS^2 = dX^2 + dY^2$. It follows from this and Eq. (78) that

$$dS^2 = (-dR \sin \phi)^2 + (dR \cos \phi)^2 = dR^2 (\sin^2 \phi + \cos^2 \phi) = dR^2, \tag{80}$$

and $dS = \pm dR$. On any arc where dR/dS is of fixed algebraic sign, we may deduce from Sec. 65 that $S = \pm(R + \text{constant})$. Hence

$$S_2 - S_1 = \pm(R_2 - R_1) \tag{81}$$

if S_1 and S_2 denote the arc lengths on the evolute to C_1 and C_2 and if R_1 and R_2 are the radii of curvature at the corresponding points P_1 and P_2 of the given curve.

The length of the arc C_1C_2 is $|S_2 - S_1|$. And $|R_2| = P_2C_2$, $|R_1| = P_1C_1$ so that $|R_2 - R_1| = |P_2C_2 - P_1C_1|$ when dR preserves its sign on the arc C_1C_2. Hence for such an arc, by Eq. (81), we have the difference $P_2C_2 - P_1C_1$ numerically equal to the arc C_1C_2. This establishes the following property of the evolute:

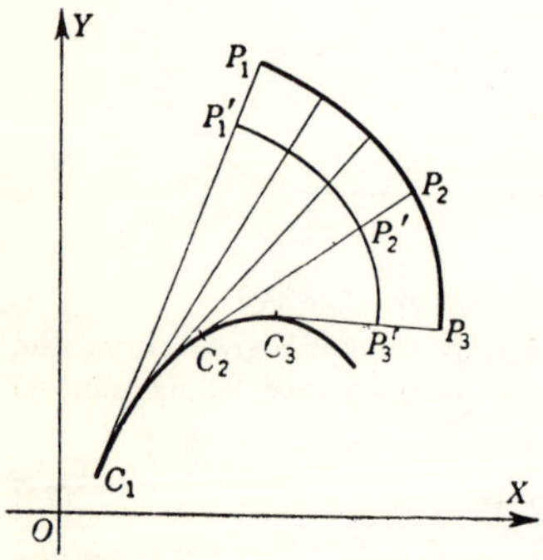

FIG. 165.

Let P_1,P_2 be two points of a given curve, and C_1,C_2 the corresponding centers of curvature or points on the evolute. Then if PC steadily increases as P moves from P_1 to P_2, arc $C_1C_2 = P_2C_2 - P_1C_1$. And if PC steadily decreases as P moves from P_1 to P_2, arc $C_1C_2 = P_1C_1 - P_2C_2$.

***140. Involutes.** Let $P_1P_2P_3$ be a given curve and $C_1C_2C_3$ be the corresponding arc of the evolute (Fig. 165). Suppose that a string of length P_1C_1 with one end fastened at C_1 is kept taut while part of it winds on the arc of the evolute. When the point of contact reaches C_2, the string will be tangent to the evolute at C_2. Hence it will extend along C_2P_2 by the first property proved in Sec. 139. And by the second property, arc $C_1C_2 = P_1C_1 - P_2C_2$, so that $P_1C_1 = \text{arc } C_1C_2 + P_2C_2$. This shows that, when the string is tangent to the evolute at C_2, it will reach along the tangent to P_2. Thus during the winding, the end of the string originally at P_1 will trace out the given arc $P_1P_2P_3$.

The curve P_1P_3 is called an *involute* of the curve C_1C_3. During the winding, any other point on the string as P_1' will trace out a curve $P_1'P_2'P_3'$ parallel to $P_1P_2P_3$ in the sense that the distance $P'P$ normal to the curve traced by P is constant. In Example 1 it is proved that the tangent at P' is parallel to that at P and that the center of curvature of $P_1'P_2'P_3'$ at P_2' is C_2. Thus the curve C_1C_3 has infinitely many involutes, and each involute has the curve C_1C_3 as its evolute. That any curve is the evolute of the involutes obtained from it by the winding process is shown in Example 2.

Example 1. For any curve traced by $P = (x,y)$, let x, y, and the slope angle ϕ be expressed in terms of some parameter. Then if b is a constant, the equations $x' = x - b \sin \phi$, $y' = y + b \cos \phi$ determine a parallel curve. Find its slope angle and radius of curvature.

Solution: From Eq. (77) we have for the differentials,

$$dx' = dx - b \cos \phi \, d\phi = R \cos \phi \, d\phi - b \cos \phi \, d\phi = (R - b) \cos \phi \, d\phi,$$
$$dy' = dy - b \sin \phi \, d\phi = R \sin \phi \, d\phi - b \sin \phi \, d\phi = (R - b) \sin \phi \, d\phi.$$

Hence $\dfrac{dy'}{dx'} = \dfrac{\sin \phi}{\cos \phi} = \dfrac{dy}{dx}$, the slope of the original curve.

From Eq. (77) applied to the curve traced by $P'(x',y')$ with radius of curvature R', slope angle $\phi' = \phi$, we have $dx' = R' \cos \phi \, d\phi$, $dy' = R' \sin \phi \, d\phi$. A comparison of these with the earlier expressions for dx' and dy' shows that $R' = R - b$. Thus the parallel curve has slope angle ϕ and radius of curvature $(R - b)$.

Example 2. For any curve traced by $C = (x,y)$, let x, y, and the slope angle ϕ be expressed in terms of some parameter which increases with s. Then if b is any constant, the equations $x' = x + (b - s) \cos \phi$, $y' = y + (b - s) \sin \phi$ determine the curve obtained by winding a string of length b on the curve. Find its slope and radius of curvature.

Solution: By using Eq. (36) we may deduce that

$$dx' = dx - ds \cos \phi - (b - s) \sin \phi \, d\phi = -(b - s) \sin \phi \, d\phi,$$
$$dy' = dy - ds \sin \phi + (b - s) \cos \phi \, d\phi = (b - s) \cos \phi \, d\phi.$$

Hence $\dfrac{dy'}{dx'} = \dfrac{\cos \phi}{-\sin \phi} = \tan\left(\phi + \dfrac{\pi}{2}\right)$, so that $\phi' = \phi + \dfrac{\pi}{2}$.

From Eq. (36) applied to the curve traced by (x', y') with arc length s' and slope angle $\phi' = \phi + \dfrac{\pi}{2}$, we have

$$dx' = ds' \cos \phi' = ds' \cos\left(\phi + \frac{\pi}{2}\right) = -ds' \sin \phi,$$
$$dy' = ds' \sin \phi' = ds' \sin\left(\phi + \frac{\pi}{2}\right) = ds' \cos \phi.$$

A comparison of these with the earlier expressions for dx' and dy' shows that $ds' = (b - s)d\phi = (b - s)d\phi'$. Hence the radius of curvature $R' = \dfrac{ds'}{d\phi'} = b - s$. Thus the involute curve obtained by the winding has the slope angle $\left(\phi + \dfrac{\pi}{2}\right)$ and radius of curvature $(b - s)$.

This shows that the curve traced by $C(x,y)$ is the evolute of the involute curve traced by (x',y').

EXERCISE 70

For each of the following curves, find the coordinates of the center of curvature at any point (x,y).

1. $4y = x^4$.
2. $3y = x^3$.
3. $2y^3 = 3x^2$.
4. $2x = y^2$.
5. $y = \ln \cos x$.
6. $2xy = a^2$.
7. $y = \frac{1}{2}(e^x - e^{-x})$.
8. $x^{\frac{2}{3}} + y^{\frac{2}{3}} = a^{\frac{2}{3}}$.
9. $x^{\frac{1}{2}} + y^{\frac{1}{2}} = a^{\frac{1}{2}}$.
10. $x^2 - y^2 = a^2$.

Find the parametric equations of the evolute of each of the following curves.

11. $x = 2t^2,\ y = 2t$.
12. $x = \dfrac{1}{t},\ y = \dfrac{a^2t}{2}$.
13. $x = a \cos t,\ y = b \sin t$.
14. $x = a \sec t,\ y = b \tan t$.
15. $x = a \cos^3 t,\ y = a \sin^3 t$.
16. $x = a \cos^4 t,\ y = a \sin^4 t$.
17. $x = a \cos t + (at - b) \sin t,\ y = a \sin t - (at - b) \cos t$.
18. $x = at + a \sin t + b \sin \dfrac{t}{2},\ y = 3a + a \cos t + b \cos \dfrac{t}{2}$.

Find the equation of the evolute in rectangular form for the curve

19. Of Prob. 17.
20. Of Prob. 13.
21. Of Prob. 14.
22. Of Prob. 4 or 11.

Show that $X + Y$ and $X - Y$ are each simply related to a perfect cube and, after calculating $\sqrt[3]{X + Y}$ and $\sqrt[3]{X - Y}$, perform the elimination and hence find the rectangular equation of the evolute for the curve of

23. Prob. 6 or 12.
24. Prob. 8 or 15.

141. Velocity and Acceleration. If t is the time as in Sec. 63, the parametric equations

$$x = g(t), \qquad y = h(t) \tag{82}$$

represent the motion of a particle along a curve. The components of the *velocity vector* along the fixed coordinate axes at any time t are

$$v_x = \frac{dx}{dt} = g'(t), \qquad v_y = \frac{dy}{dt} = h'(t). \tag{83}$$

The velocity vector lies along the tangent line with slope angle ϕ, since

$$\tan \phi = \frac{dy}{dx} = \frac{dy/dt}{dx/dt} = \frac{v_y}{v_x}. \tag{84}$$

If s increases as t increases, the speed in the path is

$$\frac{ds}{dt} = \sqrt{\left(\frac{dx}{dt}\right)^2 + \left(\frac{dy}{dt}\right)^2} = \sqrt{v_x^2 + v_y^2} = v. \tag{85}$$

Thus v, the length of the velocity vector, is the speed in the path ds/dt. The right triangle of Fig. 166 has sides v_x, v_y, and v. By Eq. (84) one angle is ϕ, as indicated. Hence we have

$$v_x = v \cos \phi, \qquad v_y = v \sin \phi. \tag{86}$$

All the above relations connecting v, ϕ, v_x, and v_y can be easily recalled from the triangle of Fig. 166.

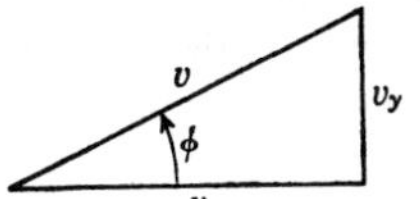

FIG. 166.

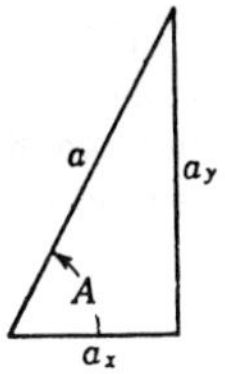

FIG. 167.

The components of the *acceleration vector* along the fixed coordinate axes at any time t are

$$a_x = \frac{dv_x}{dt} = \frac{d^2x}{dt^2} = g''(t), \qquad a_y = \frac{dv_y}{dt} = \frac{d^2y}{dt^2} = h''(t). \tag{87}$$

Let a denote the length of the acceleration vector and A the angle from the positive x axis to this vector. Then (Fig. 167) we have

$$a = \sqrt{a_x^2 + a_y^2}, \qquad \tan A = \frac{a_y}{a_x}, \qquad a_x = a \cos A, \qquad a_y = a \sin A. \tag{88}$$

EXAMPLE. Find the direction and magnitude of the velocity and acceleration vector at any time t for the motion $x = e^t \cos t$, $y = e^t \sin t$.

Solution: By differentiating the given equations we find that

$$v_x = \frac{dx}{dt} = e^t \cos t - e^t \sin t = e^t (\cos t - \sin t),$$

$$a_x = \frac{d^2x}{dt^2} = e^t (\cos t - \sin t) + e^t(-\sin t - \cos t) = -2e^t \sin t.$$

$$v_y = \frac{dy}{dt} = e^t \sin t + e^t \cos t = e^t (\sin t + \cos t),$$

$$a_y = \frac{d^2y}{dt^2} = e^t(\sin t + \cos t) + e^t(\cos t - \sin t) = 2e^t \cos t.$$

Hence

$$v^2 = v_x^2 + v_y^2 = 2e^{2t}, \qquad v = \sqrt{2}\, e^t.$$

And

$$a^2 = a_x^2 + a_y^2 = 4e^{2t}, \qquad a = 2e^t.$$

Also

$$\tan \phi = \frac{v_y}{v_x} = \frac{\sin t + \cos t}{\cos t - \sin t} = \frac{(\sin t/\cos t) + 1}{1 - (\sin t/\cos t)} = \frac{\tan t + \tan (\pi/4)}{1 - \tan t \tan (\pi/4)}$$

$$= \tan\left(t + \frac{\pi}{4}\right), \qquad \text{so that } \phi = t + \frac{\pi}{4}.$$

And

$$\tan A = \frac{a_y}{a_x} = \frac{\cos t}{-\sin t} = \tan\left(t + \frac{\pi}{2}\right), \qquad \text{so that } A = t + \frac{\pi}{2}.$$

Thus the velocity vector has magnitude $v = \sqrt{2}\, e^t$ and slope angle $\phi = t + \frac{\pi}{4}$. And the acceleration vector has magnitude $a = 2e^t$ and slope angle $A = t + \frac{\pi}{2}$. These are the required values.

Since $A - \phi = \frac{\pi}{4}$, the angle from the velocity vector to the acceleration vector is constant for this motion and always equal to 45°.

***142. Tangential and Normal Components of Acceleration.** For the motion of Eq. (82), let $P = (x,y)$ be the instantaneous position for a particular time t. If we take the slope angle in such a quadrant that Eq. (86) holds,

$$\cos\phi \qquad \text{and} \qquad \sin\phi \tag{89}$$

will be the components along the x axis and y axis of a vector of unit length along the tangent in the direction in which t and s increase. We indicate this unit vector† by **t**. And

$$\cos\left(\phi + \frac{\pi}{2}\right) = -\sin\phi, \qquad \sin\left(\phi + \frac{\pi}{2}\right) = \cos\phi \tag{90}$$

will be the components along the x axis and y axis of a unit vector **n** obtained by rotating **t** through 90° about P. Hence **n** is normal to the curve at P. We may resolve any vector along the tangential and normal directions of **t** and **n**.

The velocity vector at P has tangential component v and normal component zero. In fact, a comparison of Eq. (86) or

$$v_x = v\cos\phi, \qquad v_y = v\sin\phi \tag{91}$$

and Eq. (89) shows that the velocity vector $\mathbf{v} = v\mathbf{t}$. Hence if we use v_t and v_n to denote the tangential and normal components of **v**,

$$v_t = v, \qquad v_n = 0. \tag{92}$$

We shall now find expressions for a_t and a_n, the tangential and normal components of the acceleration vector **a**. From Eqs. (87) and (91) we find that

$$\begin{aligned} a_x &= \frac{dv_x}{dt} = \frac{dv}{dt}\cos\phi - v\sin\phi\,\frac{d\phi}{dt}, \\ a_y &= \frac{dv_y}{dt} = \frac{dv}{dt}\sin\phi + v\cos\phi\,\frac{d\phi}{dt}. \end{aligned} \tag{93}$$

A comparison of Eq. (93) with Eqs. (89) and (90) shows that the acceleration vector $\mathbf{a} = \frac{dv}{dt}\mathbf{t} + v\frac{d\phi}{dt}\mathbf{n}$. That is, a similar relation is satisfied by the corresponding x or y components. Hence the tangential and normal components of **a** are

$$a_t = \frac{dv}{dt} \qquad \text{and} \qquad a_n = v\frac{d\phi}{dt}. \tag{94}$$

From Eq. (85), $v = ds/dt$ so that $a_t = dv/dt = d^2s/dt^2$. And from Eqs. (65) and (85) we have

$$\frac{d\phi}{dt} = \frac{d\phi}{ds}\frac{ds}{dt} = \frac{1}{R}v, \quad \text{so that } a_n = v\frac{d\phi}{dt} = \frac{v^2}{R}. \tag{95}$$

† See Chap. 18 for a discussion of vectors.

Thus the tangential and normal components of acceleration are

$$a_t = \frac{dv}{dt} = \frac{d^2s}{dt^2}, \qquad a_n = \frac{v^2}{R}. \tag{96}$$

When $R = ds/d\phi$ is positive (Fig. 168), the normal direction of Eq. (90) will point toward the center of curvature. And when $R = ds/d\phi$ is negative (Fig. 169), the normal direction of Eq. (90) will point away from the center of curvature, but a_n in

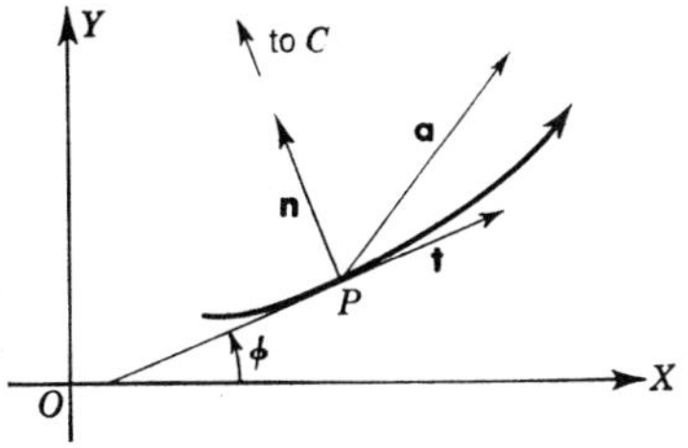

FIG. 168.

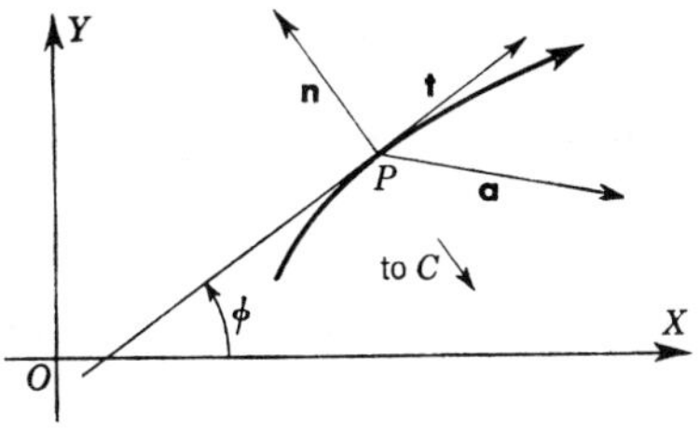

FIG. 169.

Eq. (96) will be negative, unless $v = 0$. Hence $a_n\mathbf{n}$ always points toward the center of curvature, and the acceleration vector always points toward the concave side of the path of motion, unless $a_n = 0$.

The square of the length of the acceleration vector is

$$a^2 = a_x^2 + a_y^2 = a_t^2 + a_n^2. \tag{97}$$

It follows that

$$a_n^2 = a_x^2 + a_y^2 - a_t^2 = \left(\frac{d^2x}{dt^2}\right)^2 + \left(\frac{d^2y}{dt^2}\right)^2 - \left(\frac{d^2s}{dt^2}\right)^2. \tag{98}$$

This is sometimes the simplest way to find the magnitude of a_n.

EXAMPLE 1. A point moves around a circle of radius r with constant velocity $V = r\omega$. Show that the acceleration is directed toward the center of the circle and that its magnitude is $V^2/r = r\omega^2$.

Solution: For the circle, traversed so that ϕ increases with t, the radius of curvature $R = r$. Since $v = V$, $dv/dt = 0$. Hence from Eq. (96), $a_t = 0$, and $a_n = v^2/R = V^2/r = (r\omega)^2/r = r\omega^2$. Since $a_t = 0$, the acceleration vector in this case reduces to its normal component $a_n\mathbf{n}$. Accordingly it points toward the center of curvature, or center of the circle. And its magnitude is $a_n = V^2/r = r\omega^2$.

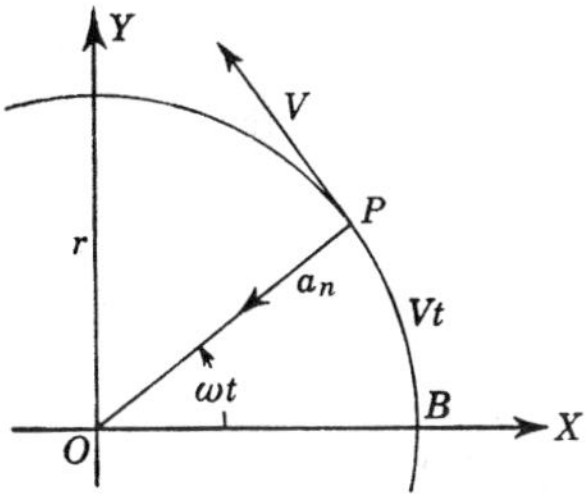

FIG. 170.

Let us check this solution by using the components along the fixed axes. Take the center of the circle at the origin, as in Fig. 170. Let arc $BP = Vt$, so that the central angle $BOP = Vt/r = r\omega t/r = \omega t$. This makes V the velocity and ω the angular velocity of OP, in accord with the data. We may then take as the equations of the motion $x = r\cos\omega t$, $y = r\sin\omega t$. By differentiation we find from these that

$$v_x = \frac{dx}{dt} = -r\omega\sin\omega t, \qquad v_y = \frac{dy}{dt} = r\omega\cos\omega t.$$

And

$$a_x = \frac{d^2x}{dt^2} = -r\omega^2\cos\omega t, \qquad a_y = \frac{d^2y}{dt^2} = -r\omega^2\sin\omega t.$$

Since $\tan A = a_y/a_x = \tan \omega t$, the acceleration vector lies along the line OP. The minus signs of a_x and a_y show that the acceleration vector points toward O. And since $a^2 = a_x^2 + a_y^2 = r^2\omega^4$, $a = r\omega^2 = (r\omega)^2/r = V^2/r$ is the magnitude of the acceleration vector.

EXAMPLE 2. Find the tangential and normal components of acceleration for the motion $x = 2t$, $y = -t^2$ at any time t.

Solution: $v_x = dx/dt = 2$, $v_y = dy/dt = -2t$, so that $v^2 = v_x^2 + v_y^2 = 4 + 4t^2$ and $v = \sqrt{4 + 4t^2} = 2\sqrt{1 + t^2}$. Hence $a_t = \dfrac{dv}{dt} = \dfrac{2t}{\sqrt{1 + t^2}}$. To use Eq. (98), we note that $a_x = \dfrac{d^2x}{dt^2} = 0$, $a_y = \dfrac{d^2y}{dt^2} = -2$, so that

$$a_n^2 = a_x^2 + a_y^2 - a_t^2 = 0 + 4 - \frac{4t^2}{1 + t^2} = \frac{4}{1 + t^2}.$$

Since $\tan \phi = \dfrac{v_y}{v_x} = \dfrac{-2t}{2} = -t$ decreases as t increases, a_n is negative and $a_n = \dfrac{-2}{\sqrt{1 + t^2}}$. Thus $a_t = \dfrac{2t}{\sqrt{1 + t^2}}$ and $a_n = \dfrac{-2}{\sqrt{1 + t^2}}$ are the required components.

We may check this solution by using Eq. (96). We have $\dfrac{dy}{dx} = \dfrac{dy/dt}{dx/dt} = \dfrac{-2t}{2} = -t$, $\dfrac{d^2y}{dx^2} = \dfrac{\dfrac{d}{dt}\left(\dfrac{dy}{dx}\right)}{dx/dt} = \dfrac{-1}{2} = -\dfrac{1}{2}$. Hence from Eq. (66), $R = \dfrac{[1 + (dy/dx)^2]^{\frac{3}{2}}}{d^2y/dx^2} = \dfrac{(1 + t^2)^{\frac{3}{2}}}{-\frac{1}{2}} = -2(1 + t^2)^{\frac{3}{2}}$. And from Eq. (96), $a_n = \dfrac{v^2}{R} = \dfrac{4(1 + t^2)}{-2(1 + t^2)^{\frac{3}{2}}} = \dfrac{-2}{\sqrt{1 + t^2}}$. This checks the earlier result.

EXERCISE 71

For each of the following motions, find the components of velocity v_x, v_y and of acceleration a_x, a_y for any t. In particular, find these components and also the speed in the path v, and the magnitude of the acceleration a, at the given instant.

1. $x = t^3 - t^2$, $y = 2t$; $t = 3$.
2. $x = 2t$, $y = t^4$; $t = 2$.
3. $x = t^2 + t$, $y = t^2 - t$; $t = 1$.
4. $x = 2t$, $y = e^{-t}$; $t = 0$.
5. $x = \sin t$, $y = \cos 3t$; $t = \pi$.
6. $x = \cos t$, $y = \cos 2t$; $t = \dfrac{\pi}{2}$.

Neglecting air resistance, the equations of motion of a projectile are

$$x = (v_0 \cos \phi_0)t, \quad y = (v_0 \sin \phi_0)t - \tfrac{1}{2}gt^2.$$

7. Verify that $v_x = v_0 \cos \phi_0$, $v_y = v_0 \sin \phi_0 - gt$, $a_x = 0$, $a_y = -g$. Show that $v = \sqrt{v_0^2 - 2gv_0t \sin \phi_0 + g^2t^2}$ and $\tan \phi = \tan \phi_0 - \dfrac{gt}{v_0} \sec \phi_0$, and deduce that v_0 and ϕ_0 are the values of v and ϕ when $t = 0$.
8. Show that the trajectory is the parabola with rectangular equation
$$y = x \tan \phi_0 - \frac{g}{2v_0^2} \sec^2 \phi_0 x^2.$$
9. The horizontal range x_F is the distance from (0,0) where the projectile rises from OX to $(x_F,0)$ where it falls on OX. Show that the range $x_F = \dfrac{v_0^2 \sin 2\phi_0}{g}$. Deduce that for a given initial velocity v_0 the range x_F is greatest when $\phi_0 = 45°$.

10. Show that, when $t = \dfrac{v_0 \sin \phi_0}{g}$, the speed in the path v is a minimum, and the height reaches its maximum, $y_M = \dfrac{v_0^2 \sin^2 \phi_0}{2g} = \dfrac{1}{8} g t_F^2$, where t_F is the time of flight for the range of Prob. 9, so that $x = x_F$ when $t = t_F$.

11. With a simple assumption about air resistance, the equations of motion of a projectile are

$$x = (V_0 \cos \phi_0) \frac{1}{k} (1 - e^{-kt}), \qquad y = (kV_0 \sin \phi_0 + g) \frac{1}{k^2} (1 - e^{-kt}) - \frac{gt}{k}.$$

Show that when $t = 0$, $x = 0$, $y = 0$, $v = v_0$, $\phi = \phi_0$ so that v_0 and ϕ_0 are the initial values of v and ϕ. Also show that at any time t,

$$a_x = -kv_x \quad \text{and} \quad a_y = -kv_y - g.$$

For each of the following motions, find the tangential and normal components of acceleration, a_t and a_n for any time t.

12. $x = t$, $y = \ln \sec t$.
13. $x = 2 \sin t$, $y = 2(1 - \cos t)$.
14. $x = \cos t^2$, $y = \sin t^2$.
15. $x = \cos e^t$, $y = \sin e^t$.
16. $x = e^t \cos t$, $y = e^t \sin t$.
17. $x = \sec t$, $y = \ln (\tan t + \sec t)$.
18. $x = b(\omega t - \sin \omega t)$, $y = b(1 - \cos \omega t)$.
19. $x = b (\cos \omega t + \omega t \sin \omega t)$, $y = b(\sin \omega t - \omega t \cos \omega t)$.

CHAPTER 10

POLAR COORDINATES

In Sec. 6 we showed how the position of a point in the plane may be determined by two signed distances, or rectangular coordinates x and y. And in Chap. 6 and elsewhere we used rectangular coordinates to solve a number of problems involving curves.

But for some purposes other types of coordinates are more convenient. Thus the problem of constructing a map from observations made at a single station suggests the location of points by means of polar coordinates. A system of polar coordinates is defined by the following procedure. We first select a fixed point in the plant, O, called the pole. And we indicate the position of any point P in the plane by giving the distance and direction of P from the pole O. To describe the direction, we take any fixed line OX through O as the initial line and measure the angle θ from OX to OP. The line segment OP is called the radius vector. Let the radial distance $OP = r$. Then the distance r and angle θ, or (r,θ), are the polar coordinates of the point P.

This chapter is devoted to the use of polar coordinates to study curves and to calculate quantities related to them. We first describe the system in detail and show how to construct curves from their equations in polar coordinates. We then discuss the relations of polar to rectangular coordinates which enable us to convert from one system to the other. And we derive the equations in polar coordinates of the curves discussed in Chap. 6.

To illustrate applications of the first derivative $dr/d\theta$, we show how to calculate the angle from the radius vector to the tangent line to a curve, the slope angle, and the angle between two curves. And we obtain an expression for the derivative of the arc length, $ds/d\theta$. As an application involving $dr/d\theta$ and $d^2r/d\theta^2$ we obtain an expression for the curvature.

If t is the time, the motion of a particle along a curve can be described by parametric equations in polar form giving r and θ in terms of t. We show how to find the velocity vector and the acceleration vector at any time for a motion described in terms of polar coordinates.

R143. Polar Coordinates. We may determine the position of any point in a plane by means of a distance and a direction from some fixed point, as follows.

Let O in Fig. 171 be the fixed point in the plane. We call O the *pole* or origin. Let OX be a fixed line through O. And take OU as the unit of length. We call OX the *initial line*, or axis. Then any point P in the plane distinct from O determines a directed line segment or vector OP. We call OP the *radius vector* for the point P. The direction of OP is determined by the angle $XOP = \theta$, read "theta," from OX to OP. And the position of P on this line is determined by the distance $r = OP$, measured with OU as the unit. Then r and θ are the *polar coordinates* of the point P.

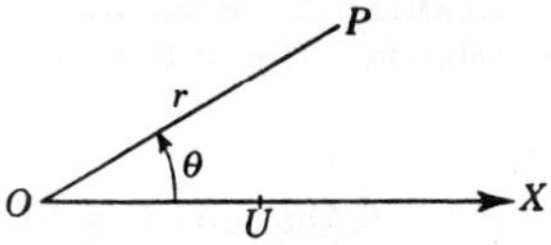

FIG. 171.

The angle θ is any angle from OX to OP, as defined in Sec. 90. Thus θ is positive when measured counterclockwise, negative when measured clockwise, and zero if P lies on OX.

We may also produce PO through O to Q, take θ as any angle from OX to OQ, and r as the negative distance $-OP$ (Fig. 172).

A given pair of polar coordinates r and θ determine a definite point in the plane. To find the point, we first construct the angle XOT having the initial side OX and the terminal side OT obtained by rotating OX through an angle of θ radians. If r is positive, we then locate the point P on OT at a distance $OP = r$ (Fig. 173). But if r is negative, we produce OT backward through O to T' and locate the point P on OT'

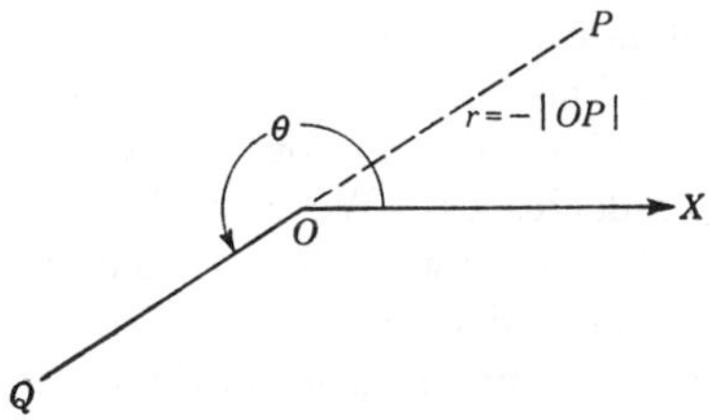

FIG. 172.

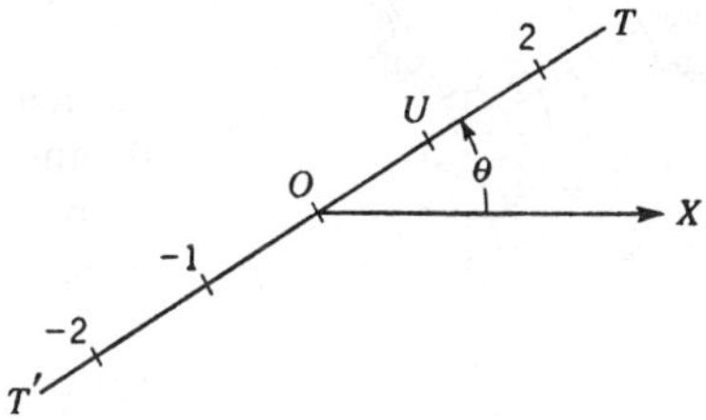

FIG 173.

at a distance $|OP| = |r|$. If r is zero, the point P is taken as O regardless of the value of θ. Thus for any value of θ, the points on the indefinite line $T'OT$ matched up with their corresponding values of r form a number scale like that of Sec. 1.

Although there is just one point for each pair r, θ, the same point may have infinitely many pairs of polar coordinates. For the angle θ may be positive or negative and make any number of complete revolutions about the origin.

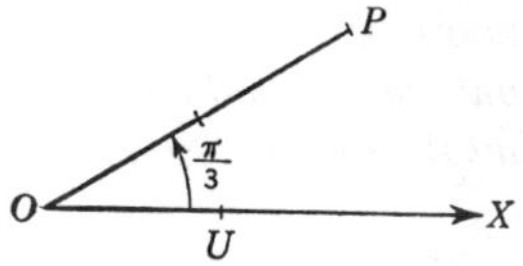

FIG. 174.

We use the symbol (a,A) to mean the point with $r = a$ and with $\theta = A$. And we indicate that this is the point P by writing $P = (a,A)$ or $P(a,A)$. If $D = (180/\pi)A$, the number of degrees in an angle of A radians, it is sometimes convenient to write $P = (a,D°)$.

EXAMPLE 1. Plot the point $P = (2,\pi/3)$ and find all possible pairs of polar coordinates for this point.

Solution: The point is located in Fig. 174 by making angle $XOP = \pi/3 = 30°$ and measuring $OP = 2$ units. To find the other pairs of coordinates, we note that if k is zero or a positive or negative integer, we may reach the terminal line OP by first making k complete revolutions, an angle of $2\pi k$ radians. This gives the coordinates $(2, \pi/3 + 2k\pi)$. And if we prolong PO through O to Q, angle XOQ has $\pi/3 + \pi$ as one value, leading to the coordinates $(-2, 4\pi/3)$. And by first making k complete revolutions, we may reach the terminal side OQ with an angle $4\pi/3 + 2k\pi$. This gives the coordinates $(-2, 4\pi/3 + 2k\pi)$. Hence any set of polar coordinates of the given point has one of the forms

$$\left(2, \frac{\pi}{3} + 2k\pi\right), \qquad \left(-2, \frac{4\pi}{3} + 2k\pi\right).$$

Example 2. What are the possible polar coordinates of $P(a,A)$?
Solution: If $a \neq 0$, as in Example 1, we find the forms

$$(a,\ A + 2k\pi), \qquad (-a,\ A + \pi + 2k\pi).$$

If $a = 0$, the point $P = (0,A)$ has coordinates $(0,\theta)$, where θ may have any value.

144. Curves. In general, the locus of points whose polar coordinates satisfy a given equation $F(r,\theta) = 0$ or $r = f(\theta)$ is a curve. We may plot the locus by tabulating pairs of values of r and θ which satisfy the equation, plotting the corresponding points and joining them by a smooth curve.

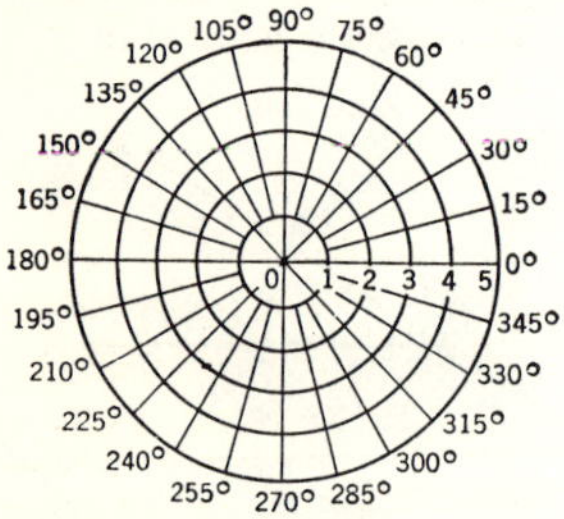

Fig. 175.

It is sometimes desirable to look for symmetry or other properties described in Sec. 89. But it is often sufficient to consider θ as increasing from some particular value, as 0, and determine intervals of θ in which r increases or decreases. We may then plot accurately the points where r is a maximum or a minimum and draw a curve through these points on which r varies in the proper direction. It is sometimes necessary to proceed in a similar way with θ decreasing from the zero value, as in Example 4.

The work is facilitated by using polar coordinate paper, ruled as in Fig. 175. The angle θ is determined from the scales at the ends of the radial straight lines. And the value of r is counted off on the concentric circles, toward the number indicating θ when r is positive, and away from the number which indicates θ when r is negative.

For the origin O we shall prove the following theorem:

A value of θ, $\theta = \theta_1$, which makes $f(\theta_1) = 0$ not only shows that the curve $r = f(\theta)$ passes through the origin, but also shows that the radius vector with $\theta = \theta_1$ is tangent to the curve at the origin.

To prove this, we first observe that since $0 = f(\theta_1)$, the coordinates $(0,\theta_1)$ satisfy $r = f(\theta)$. Hence the origin is a point on the curve. Assume that $f(\theta)$ is a continuous function of θ and that for θ_2 near θ_1, but not equal to θ_1, $f(\theta_2) \neq 0$. Then if $P_2 = (r_2,\theta_2)$ (Fig. 176), OP_2S is a secant line to the curve through O. Let θ_2 tend to θ_1. Then as $\theta_2 \to \theta_1$, $f(\theta_2) \to f(\theta_1)$ so that $r_2 \to 0$. Hence $P_2 = (r_2,\theta_2)$ tends to $O = (0,\theta_1)$. And by the discussion of Sec. 24, the limiting position of the secant OS is the tangent line to the curve at O, OT. But the secant OS is the radius vector with slope angle $\theta = \theta_2$. And $\theta_2 \to \theta_1$. Hence the radius vector with slope angle $\theta = \theta_1$ is the tangent line to the curve at O, OT.

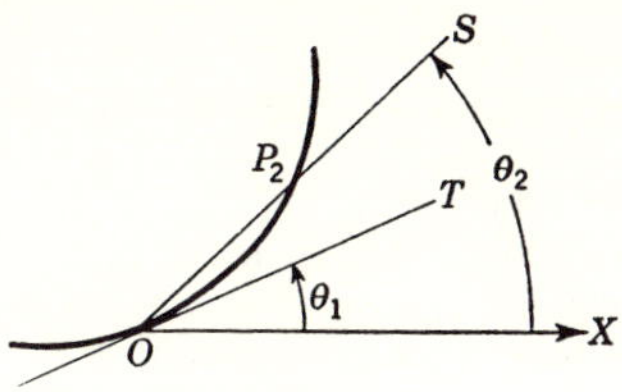

FIG. 176.

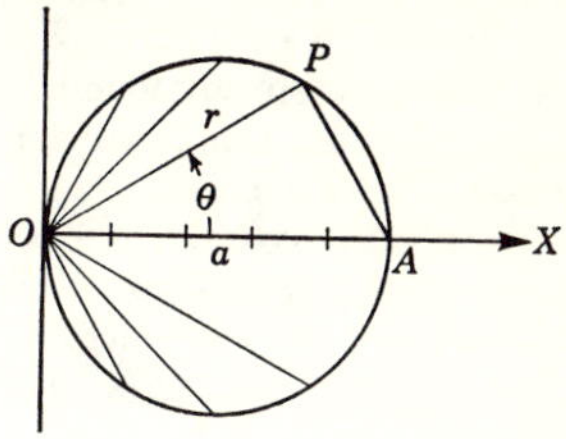

FIG. 177.

EXAMPLE 1. Plot the locus of $r = a \cos \theta$.

Solution: Let a be any† convenient length, as OA in Fig. 177. Since OA equals 5 units on the r scale, we set $a = 5$. Make the following table of values

$D°$	0°	30°	45°	60°	90°
θ	0	$\pi/6$	$\pi/4$	$\pi/3$	$\pi/2$
$\cos \theta = \cos D°$	1	0.866	0.707	0.5	0
$r = 5 \cos \theta$	5	4.33	3.54	2.5	0

Plot each of these points, using either θ or $D°$ to select the proper radial line. By joining them we obtain the upper half-loop. Its tangent at the origin is perpendicular to the initial line, since $\theta = \pi/2$ makes $r = 0$. Since $\cos(-\theta) = \cos \theta$, the values of θ from 0 to $-\pi/2$ give a symmetric half-loop below the initial line, and the curve has the initial line as an axis of symmetry. Values of θ outside the range $-\pi/2, \pi/2$ would merely give a repetition of points previously found. For example, if $\theta = 2\pi/3$, $\cos \theta = -0.5$ and $r = 5 \cos \theta = -2.5$. And the point $(-2.5, 2\pi/3)$ is the same as the point $(2.5, \pi/3)$ already plotted.

The loop looks like a circle. We may show that it actually is a circle by drawing a circle with diameter OA, and considering any point P on it. Since angle OPA is inscribed in a semicircle, it is a right angle. Let angle $AOP = \theta$, $OP = r$, and recall that $OA = a$. Then $\cos \theta = OP/OA = r/a$, so that $r = a \cos \theta$. Thus any point (r,θ) on the circle with diameter the line joining O and $A = (a,0)$ is on the locus of $r = a \cos \theta$. Hence the one loop of the locus is a circle.

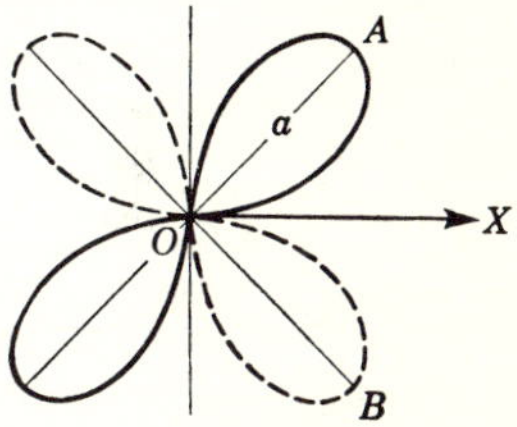

FIG. 178.

EXAMPLE 2. Plot the locus of $r = a \sin 2\theta$.

Solution: In Fig. 178, let $OA = a$. As θ increases from 0 to $\pi/4$, 2θ increases from 0 to $\pi/2$ and $r = a \sin 2\theta$ increases from 0 to a. This gives the half-loop OA. Its tangent at the origin is OX, since $\theta = 0$ makes $r = 0$. And its tangent at A is perpendicular to OA, since a is a maximum value of r. As θ increases from $\pi/4$ to $\pi/2$, another half-loop symmetric with respect to the radius vector OA is generated. This has the tangent at the origin perpendicular to the initial line, consistent with the fact that $\theta = \pi/2$ makes $r = 0$. As θ increases from $\pi/2$ to $3\pi/4$, 2θ increases from π to $3\pi/2$ and $r = a \sin 2\theta$ changes from 0 to $-a$. Thus this gives a half-loop OB in the fourth quadrant, dotted in Fig. 178, to indicate that these points correspond to

† Since $r = a \cos \theta$ makes r proportional to a, different values of a will give curves of the same shape but different sizes.

negative values of r. Continuing in this way, we find a second half-loop BO corresponding to the range of θ from $3\pi/4$ to π. The values of θ from π to $3\pi/2$ lead to the loop in the third quadrant. And the values of θ from $3\pi/2$ to 2π lead to the dotted loop in the second quadrant. Values of θ outside the range 0 to 2π merely give a repetition of loops already found. The curve is called the *rose of four leaves.*

We note that the rose $r = a \sin n\theta$ or $r = a \cos n\theta$ has $2n$ leaves when n is even, and n leaves when n is odd.

EXAMPLE 3. Plot the locus of $r = a(1 + 2 \sin \theta)$, a *limaçon.*

Solution: In Fig. 179, let $OA = a$. The curve passes through O, since $r = a(1 + 2 \sin \theta) = 0$ when $\sin \theta = -\frac{1}{2}$, which happens when $\theta = -\pi/6$ and $\theta = -5\pi/6$. And the corresponding radius vectors are each tangent to the curve

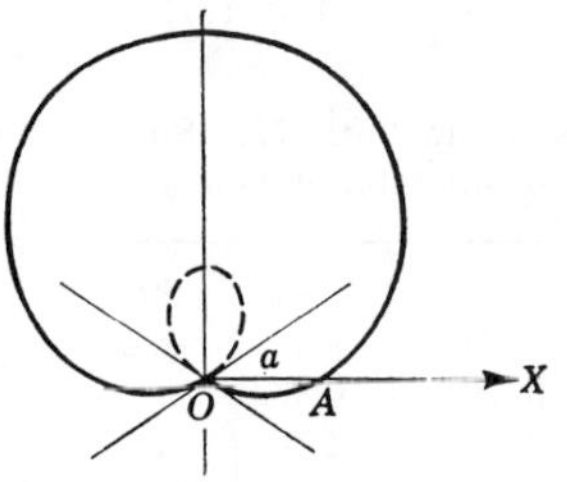

FIG. 179.

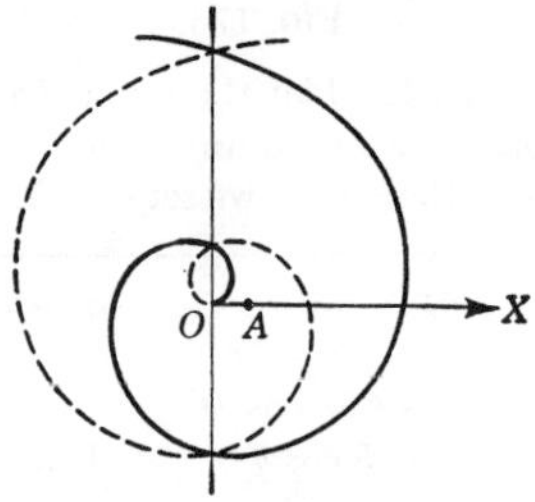

FIG. 180.

at the origin. The maximum value of r, $3a$, is reached when $\sin \theta = 1$ or when $\theta = \pi/2$. And the maximum value of $-r$, $-a$, is reached when $\sin \theta = -1$ or when $\theta = -\pi/2$. The curve consists of one large loop passing through $(3a, \pi/2)$ which may be generated by θ in the range from $-\pi/6$ to $7\pi/6$ and one small loop passing through $(-a, -\pi/2)$ which may be generated by θ in the range from $7\pi/6$ to $11\pi/6$, or $-5\pi/6$ to $-\pi/6$. The small loop is dotted to indicate that its points correspond to negative values of r.

EXAMPLE 4. Plot the locus of $r = a\theta$, the *spiral of Archimedes.*

Solution: In Fig. 180, let $OA = a$. When $\theta = 0$, $r = 0$ so that the curve passes through O and is tangent to OX at O. As θ increases, r increases, generating one coil reaching $(2\pi a, 2\pi)$ for the range of θ from 0 to 2π. As the radius vector returns to the same direction after each increase of θ by 2π, the curve consists of a succession of coils each at distance $2\pi a$ along r from the preceding coil. A symmetric series of coils, dotted in Fig. 180, is obtained by letting θ decrease from 0 through negative values.

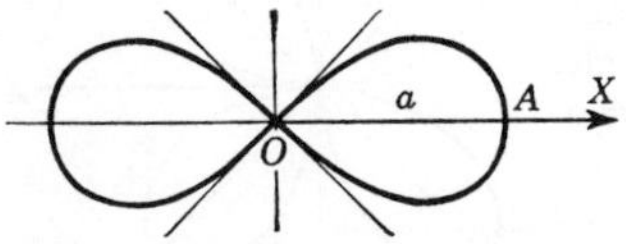

FIG. 181.

EXAMPLE 5. Plot the locus of $r^2 = a^2 \cos 2\theta$, the *lemniscate.*

Solution: In Fig. 181, let $OA = a$. By solving the equation $r^2 = a^2 \cos 2\theta$ we find that $r = \pm a \sqrt{\cos 2\theta}$. The curve passes through O, since $\cos 2\theta = 0$ when $2\theta = \pi/2$ or $-\pi/2$. Hence $r = 0$ when $\theta = \pi/4$ or $-\pi/4$. And the corresponding radius vectors are each tangent to the curve at the origin. As θ increases from 0 to $\pi/4$, the half-loop AO is generated by the positive square root. And the range $-\pi/4$ to 0 gives a half-loop symmetric about OA. For θ between $\pi/4$ and $3\pi/4$, $\cos 2\theta$ is negative so that no real values of r are obtained. A second loop results from the positive square root for θ from $3\pi/4$ to $5\pi/4$. But from $5\pi/4$ to $7\pi/4$ there are no real values of r. Further values of θ, and the negative square root, merely lead to points already plotted.

EXERCISE 72

1. Plot the points $A = (3,\pi/4)$, $B = (3,3\pi/4)$, $C = (3,5\pi/4)$, $D = (3,7\pi/4)$ and so verify that the quadrilateral $ABCD$ is a square.
2. Plot the points $A = (5,\pi/9)$, $B = (-5,\pi/9)$, $C = (5,-\pi/9)$, $D = (-5,-\pi/9)$ and so verify that the quadrilateral $ABCD$ is a rectangle.
3. Plot the points $A = (-2,0)$, $B = (-2,-\pi/3)$, $C = (0,-\pi/4)$ and so verify that the triangle ABC is equilateral.

Find all possible pairs of polar coordinates for each given point.

4. $P(4,\pi/4)$. 5. $P(-5,\pi)$. 6. $P = (0,\pi/6)$.

Plot the locus of each of the following given equations.

7. $r = a \sin \theta$.
8. $r = a \sin 3\theta$.
9. $r = a \cos 2\theta$.
10. $r = a \cos 3\theta$.
11. $r = a \sin \dfrac{\theta}{2}$.
12. $r = a \cos \dfrac{\theta}{3}$.
13. $r^2 = a^2 \sin 2\theta$.
14. $r = a(1 + \cos \theta)$.
15. $r = a(1 - \sin \theta)$.
16. $r = 2 \cos \theta + 1$.
17. $r = 2 \cos \theta + 3$.
18. $r = 2 - \cos \theta$.
19. $r = 2 \cos^2 \dfrac{\theta}{2}$.
20. $r = \sqrt{2} \sin \theta - 1$.
21. $r = -\dfrac{\theta}{\pi}$.
22. $r = e^{\theta/10}$.

145. Rectangular and Polar Coordinates. Let the point O be the pole of a system of polar coordinates and at the same time be the origin of a system of rectangular coordinates. And let OX be the initial line for polar coordinates and at the same time be the x axis. Thus the radius vector $\theta = \pi/2$ is the positive y axis.

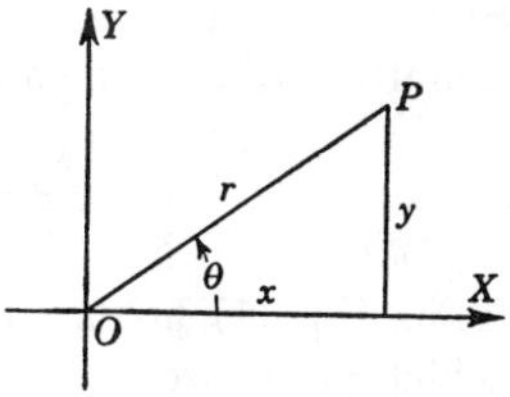

FIG. 182.

Then any point in the plane P has rectangular coordinates (x,y). And if P is distinct from O, it has polar coordinates (r,θ) with r positive (Fig. 182). Then by the definition of the sine and cosine given in Eq. (8) of Sec. 91, we have

$$\cos \theta = \frac{x}{r}, \qquad \sin \theta = \frac{y}{r}. \tag{1}$$

It follows from these relations that

$$x = r \cos \theta, \qquad y = r \sin \theta. \tag{2}$$

These relations imply that

$$x^2 + y^2 = r^2, \qquad \frac{y}{x} = \tan \theta. \tag{3}$$

And we may solve these in the form

$$r = \sqrt{x^2 + y^2}, \qquad \theta = \tan^{-1}\frac{y}{x}. \tag{4}$$

By means of Eq. (2) any expression in rectangular coordinates can be transformed into a corresponding expression in polar coordinates. And by means of Eq. (4) any expression in polar coordinates can be transformed into a corresponding expression in rectangular coordinates. But it is sometimes advisable to use Eqs. (1) and (3) at intermediate stages as illustrated in the examples.

If P is at the origin, $r = 0$, $x = 0$, and $y = 0$. Thus Eq. (2) and the first relation of Eqs. (3) and (4) still hold. The expressions 0/0 in the right members of the other relations correspond to the fact that the angle θ may be assigned any value at the origin, as illustrated in Sec. 143.

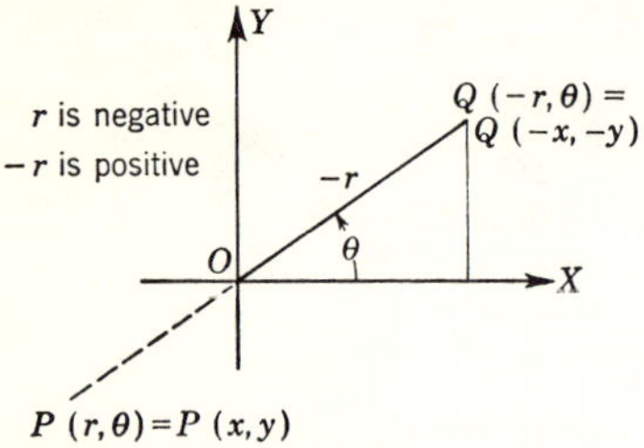

FIG. 183.

Let $P = (x,y)$ be any point distinct from O. Then we may use polar coordinates (r,θ) for P with r negative. Let us plot the point Q (Fig. 183) with polar coordinates $(-r,\theta)$ and rectangular coordinates $(-x,-y)$. Since r is negative, $-r$ is positive. Hence we may deduce from the definition of the sine and cosine that

$$\cos\theta = \frac{-x}{-r} = \frac{x}{r}, \qquad \sin\theta = \frac{-y}{-r} = \frac{y}{r}. \tag{5}$$

Thus Eq. (1) holds when r is negative. And so do Eqs. (2) and (3), which are consequences of Eq. (1). In Eq. (4) we must insert a minus sign before the radical when r is negative, and in any case we must choose such a quadrant for θ that Eq. (2) holds.

To remember Eqs. (1) to (4) we may construct a diagram like Fig. 182 with θ acute and r, x, y all positive. Any desired relation may then be read from the right triangle. In transforming equations we generally use Eq. (4) with the plus sign before the radical, as the case with the minus sign is automatically introduced when we square to eliminate the square root.

EXAMPLE 1. Find an equation in polar coordinates whose locus is the same as that of the equation $(x^2 + y^2)^3 = 4a^2x^2y^2$.

Solution: From the given equation and Eqs. (3) and (2), we find that $(r^2)^3 = 4a^2(r\cos\theta)^2(r\sin\theta)^2$ or $r^6 = a^2r^4(2\sin\theta\cos\theta)^2 = a^2r^4\sin^2 2\theta$. This is equivalent to $r^4(r + a\sin 2\theta)(r - a\sin 2\theta) = 0$ which implies that $r = 0$, $r = -a\sin 2\theta$, or $r = a\sin 2\theta$. We may omit the first two relations. For $r = 0$ represents the origin, which lies on $r = a\sin 2\theta$, since this is satisfied by $(r,\theta) = (0,0)$. And if $P = (r,\theta)$

satisfies $r = -a \sin 2\theta$, $-r = a \sin 2\theta = a \sin 2(\theta + \pi)$, so that the alternative coordinates of P, $(-r, \theta + \pi)$ satisfy $r = a \sin 2\theta$. Thus $r = a \sin 2\theta$ is an equation with the required property. See Fig. 178.

EXAMPLE 2. Find the equation in rectangular coordinates of the lemniscate, $r^2 = a^2 \cos 2\theta$. See Fig. 181.

Solution: From the given equation and Eq. (1), we find that $r^2 = a^2 \cos 2\theta = a^2(\cos^2 \theta - \sin^2 \theta) = a^2\left[\left(\frac{x}{r}\right)^2 - \left(\frac{y}{r}\right)^2\right] = \frac{a^2}{r^2}(x^2 - y^2)$. Hence $r^4 = a^2(x^2 - y^2)$. But by Eq. (3), $r^4 = (r^2)^2 = (x^2 + y^2)^2$, so that $(x^2 + y^2)^2 = a^2(x^2 - y^2)$ is the required equation.

EXAMPLE 3. Show that the locus of $r = 4 \csc \theta + 2 \sec \theta$ is a hyperbola by transforming the equation to rectangular coordinates.

Solution: From the given equation and Eq. (1), we find that $r = 4 \csc \theta + 2 \sec \theta = \frac{4}{\sin \theta} + \frac{2}{\cos \theta} = \frac{4r}{y} + \frac{2r}{x}$. It follows that $rxy = 4rx + 2ry$ or $r(xy - 4x - 2y) = 0$. We may omit the factor r, since $r = 0$ represents the origin, and $x = 0, y = 0$ makes the second factor zero. Hence the equation of the locus in rectangular form is

$$xy - 4x - 2y = 0 \qquad \text{or} \qquad (x - 2)(y - 4) = 8.$$

By Eq. (85) of Sec. 85, it follows from the second form that the locus is a hyperbola with the lines $x = 2$, $y = 4$ as asymptotes (Fig. 184).

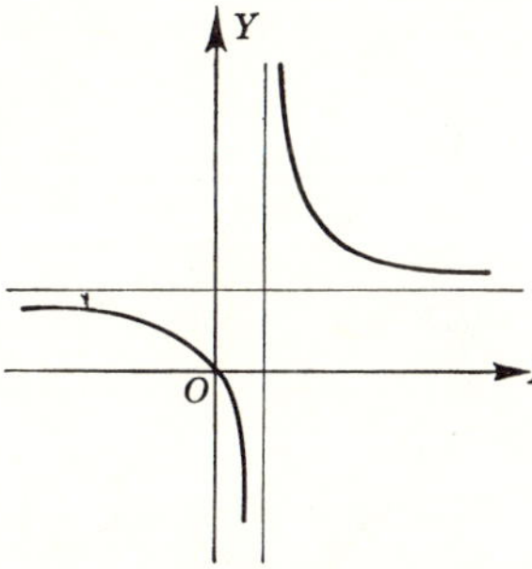

FIG. 184.

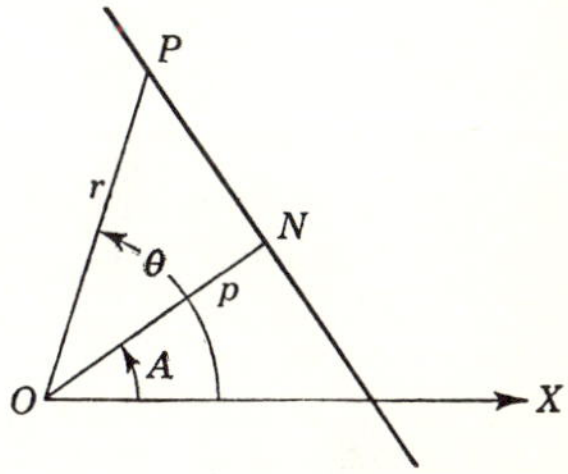

FIG. 185.

***146. Polar Equations of First- and Second-degree Curves.** The equations in rectangular form for certain types of loci were found in Chap. 6. The corresponding equations in polar coordinates could be found from these by transforming coordinates as in Sec. 145. But, as illustrated in the examples which follow, the polar equations can sometimes be found more directly by introducing geometric considerations.

EXAMPLE 1. The perpendicular from the origin to a straight line is of length p and has a slope angle A. Show that the equation of the straight line in polar coordinates is

$$r \cos (\theta - A) = p. \tag{6}$$

Solution: Let $P = (r,\theta)$ be any point on the straight line (Fig. 185) and ON be the perpendicular from O to the line. Therefore $ON = p$ and angle $XON = A$. Hence angle NOP = angle XOP − angle $XON = \theta - A$.

But by the definition of the cosine, we have

$$\frac{ON}{OP} = \cos ONP, \qquad \text{or} \qquad \frac{p}{r} = \cos (\theta - A) \text{ and } p = r \cos (\theta - A).$$

Hence Eq. (6) holds for any point $P = (r,\theta)$ on the straight line.

Example 2. A circle has its center at the point (b,B) and passes through the origin O. Show that the equation of the circle in polar coordinates is

$$r = 2b \cos(\theta - B). \tag{7}$$

Solution: Let $P = (r,\theta)$ be any point on the circle (Fig. 186). Let $C = (b,B)$. Prolong OC to D, giving the diameter OD of length $2b$. Then angle DPO is inscribed in a semicircle and therefore is a right angle. And

$$\text{Angle } DOP = \text{angle } XOP - \text{angle } XOD = \theta - B.$$

Hence from the definition of the cosine, we have

$$\frac{OP}{OD} = \cos DOP, \qquad \text{or } \frac{r}{2b} = \cos(\theta - B) \text{ and } r = 2b\cos(\theta - B).$$

Hence Eq. (7) holds for any point $P = (r,\theta)$ on the circle.

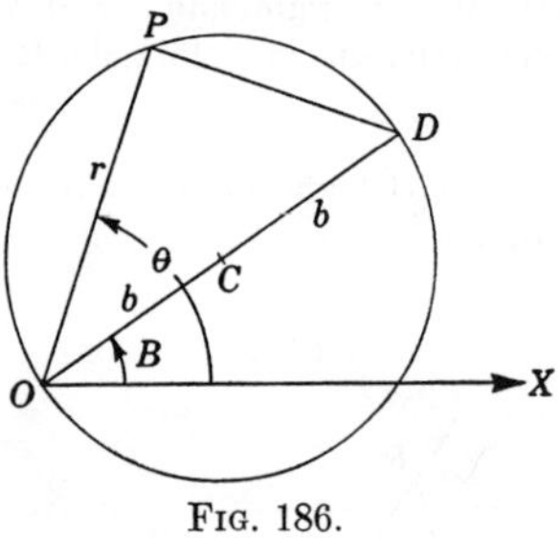

Fig. 186.

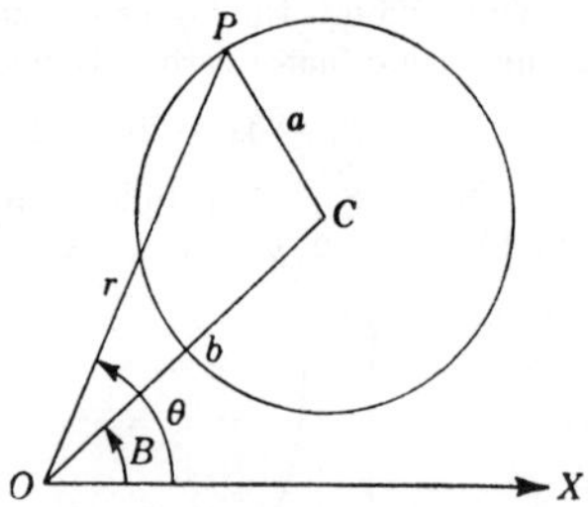

Fig. 187.

Example 3. A circle has its center at the point (b,B) and its radius equal to a. Show that the equation of the circle in polar coordinates is

$$r^2 - 2br\cos(\theta - B) + b^2 = a^2. \tag{8}$$

Solution: Let $P = (r,\theta)$ be any point on the circle (Fig. 187). Let $C = (b,B)$. Then angle COP = angle XOP − angle $XOC = \theta - B$. In triangle COP, side $OC = b$, side $OP = r$, side $CP = a$, and the angle opposite CP is angle $COP = \theta - B$. Hence by the law of cosines, Eq. (51) of Sec. 94, we have

$$a^2 = b^2 + r^2 - 2br\cos(\theta - B),$$

and Eq. (8) holds for any point $P = (r,\theta)$ on the circle.

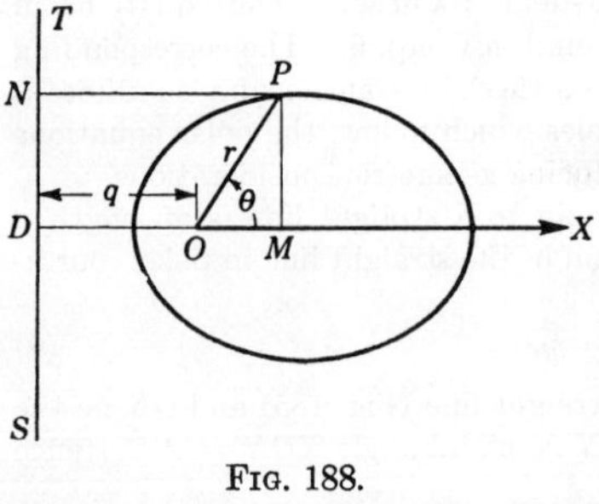

Fig. 188.

We note that if $a = b$, Eq. (8) reduces to

$$r^2 - 2br\cos(\theta - B) = 0,$$

or $r[r - 2b\cos(\theta - B)] = 0$. The factor $r - 2b\cos(\theta - B)$, set equal to zero, gives Eq. (7). And the factor r may be omitted, since $r = 0$ represents the origin, and $r = 0$, $\theta = B + (\pi/2)$ satisfies $r = 2b\cos(\theta - B)$. This checks Example 2 as a special case of Example 3.

Example 4. The straight line ST (Fig. 188), the *directrix*, has the equation $x = -q$, or $r\cos\theta = -q$. The *focus* is at the origin O. Show that the equation in polar coordinates of the locus of a point P which moves so that its distance from the focus is always e times its distance from the directrix is

$$r = \frac{eq}{1 - e\cos\theta}. \tag{9}$$

Solution: Let $P = (r,\theta)$ be any point on the locus. Then $OP = r$, $OM = r\cos\theta$, and $NP = DM = DO + OM = q + r\cos\theta$. But from the characteristic property of the locus, we have $OP = eNP$ or $r = e(q + r\cos\theta)$ and $r = \dfrac{eq}{1 - e\cos\theta}$, and Eq. (9) holds for any point $P = (r,\theta)$ on the locus.

We note that when $e = 1$ (Fig. 189) the locus is a parabola by the definition in Sec. 83. Here $e = 1$ and $q = 2c$. Hence $c = q/2$.

When $e < 1$, the locus is an ellipse by Prob. 20 of Exercise 42. Here $e = \dfrac{c}{a}$, and $q = \dfrac{a^2}{c} - c = \dfrac{a^2 - c^2}{c} = \dfrac{b^2}{c}$. And $2a$ is the sum of the values of r for $\theta = 0$ and $\theta = \pi$,

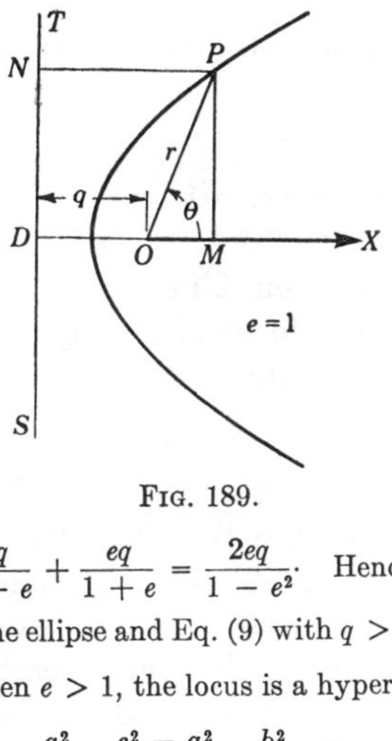

FIG. 189.

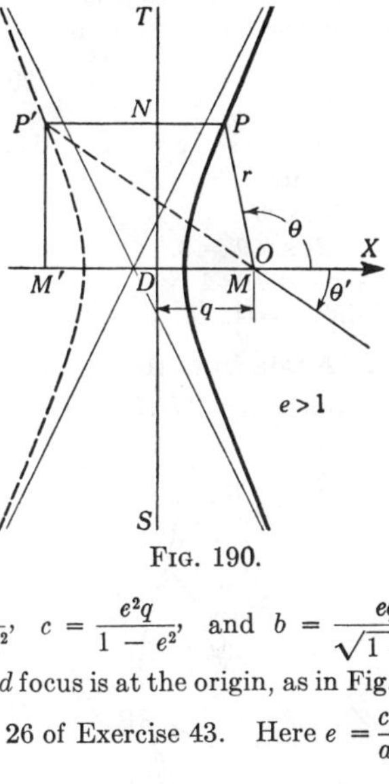

FIG. 190.

or $\dfrac{eq}{1-e} + \dfrac{eq}{1+e} = \dfrac{2eq}{1-e^2}$. Hence $a = \dfrac{eq}{1-e^2}$, $c = \dfrac{e^2q}{1-e^2}$, and $b = \dfrac{eq}{\sqrt{1-e^2}}$. For the ellipse and Eq. (9) with $q > 0$, the *left-hand* focus is at the origin, as in Fig. 188.

When $e > 1$, the locus is a hyperbola by Prob. 26 of Exercise 43. Here $e = \dfrac{c}{a}$, and $q = c - \dfrac{a^2}{c} = \dfrac{c^2 - a^2}{c} = \dfrac{b^2}{c}$. Because r is negative for $\theta = 0$ when $e > 1$, $2a$ is the algebraic sum of the values of $-r$ for $\theta = 0$ and $\theta = \pi$, or $\dfrac{-eq}{1-e} + \dfrac{-eq}{1+e} = \dfrac{2eq}{e^2-1}$. Hence $a = \dfrac{eq}{e^2-1}$, $c = \dfrac{e^2q}{e^2-1}$, and $b = \dfrac{eq}{\sqrt{e^2-1}}$. For the hyperbola and Eq. (9) with $q > 0$, the *right-hand* focus is at the origin, as in Fig. 190.

EXERCISE 73

Transform each of the following equations to polar coordinates.

1. $xy = 4$.
2. $y^2 = 4x$.
3. $x^2 - y^2 = 4$.
4. $x^2 + y^2 = 4x$.
5. $(x^2 + y^2)^2 = x^2 - y^2$.
6. $(x^2 + y^2)^2 = 2xy$.

Transform each given equation to rectangular coordinates.

7. $r = \tan\theta$.
8. $r = \cot\theta$.
9. $r^2 = \sin\theta$.
10. $r^2 = \cos\theta$.
11. $r = a\cos\theta + b$.
12. $r = a\sin\theta + b$.

As in Fig. 185, N is the foot of the perpendicular from the origin to a straight line. Find the equation in polar coordinates of the straight line for which N has the given polar coordinates.

13. $\left(4, \frac{\pi}{6}\right)$.
14. $\left(\sqrt{2}, \frac{3\pi}{4}\right)$.
15. $(3,0)$.

Find the equation in polar coordinates of a circle passing through the origin whose center has the given polar coordinates.

16. $\left(2, \frac{\pi}{2}\right)$.
17. $(6,0)$.
18. $\left(\sqrt{2}, \frac{\pi}{4}\right)$.

Find the equation in polar coordinates of a parabola with focus at the origin whose vertex has the given polar coordinates.

19. $\left(2, \frac{\pi}{2}\right)$.
20. $(5,0)$.
21. $\left(4, \frac{3\pi}{2}\right)$.

Transform each given equation to rectangular coordinates, and name the locus it represents.

22. $r = 3\sec\theta$.
23. $r = 6\sin\theta$.
24. $r^2 - 2r\sin\theta = 3$.
25. $r = -2\csc\theta$.
26. $r + r\cos\theta = 1$
27. $r(2 - \cos\theta) = 1$.
28. $r^2\sin 2\theta = 4$.
29. $r^2\cos 2\theta = 4$.

147. Angle from the Radius Vector to the Tangent Line. Suppose that the smooth arc APB of Fig. 191 is part of the locus of the equation in polar coordinates

$$r = f(\theta). \tag{10}$$

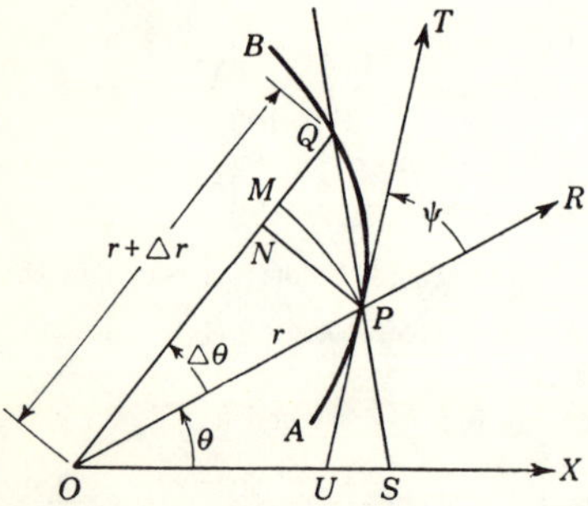

FIG. 191.

$P = (r,\theta)$ is a fixed point on the arc. The pole is O, so that OP is the radius vector. And PT is the straight line tangent to the curve at P drawn in the direction in which θ increases. If R is any point on OP produced beyond P, the angle RPT may be used to fix the direction of the tangent to the curve at P. This angle, from the direction along the radius vector in which r increases to the direction along the tangent line in which θ increases is denoted by ψ, read "psi." A formula for calculating ψ is given as the concluding equation of the following theorem.

If ψ is the angle from the radius vector through the point of contact $P = (r,\theta)$ to the tangent line drawn at P to the curve whose equation is $r = f(\theta)$, then

$$\tan\psi = \frac{r}{dr/d\theta} = \frac{f(\theta)}{f'(\theta)}. \tag{11}$$

To prove this result, make the following construction. Let $Q = (r + \Delta r, \theta + \Delta\theta)$ be a variable point near P on the curve APB. Draw the radius vector OQ and the secant line QPS. Also draw PN, the perpendicular from P to OQ. And draw PM, a circular arc with center O and radius OP intersecting OQ at M.

Now let $\Delta\theta$ approach zero as a limit. Then as $\Delta\theta \to 0$,

1. The point Q tends to P on the curve APB.

2. The radius vector OQ revolves about O and approaches OP as a limiting position.

3. The secant line QPS revolves about P and by Sec. 24 approaches the tangent line TPU as a limiting position.

It follows from 2 and 3 that angle OQP approaches angle OPU. But angle OPU equals the vertical angle RPT, or ψ, so that

$$\lim_{\Delta\theta\to 0} \angle OQP = \psi \qquad \text{and} \qquad \lim_{\Delta\theta\to 0} \tan OQP = \tan\psi, \tag{12}$$

since the trigonometric tangent is a continuous function.

From the right triangle QNP, we have

$$\tan OQP = \frac{NP}{QN} = \frac{PN}{NQ}. \tag{13}$$

We see from Fig. 191 that $PM = r\,\Delta\theta$ is a good approximation to PN, and that $MQ = OQ - OM = (r + \Delta r) - r = \Delta r$ is a good approximation to NQ. Hence we write

$$\tan OQP = \frac{PN}{NQ} = \frac{PN}{NQ}\frac{PM}{PM}\frac{MQ}{MQ} = \frac{PN}{PM}\frac{MQ}{NQ}\frac{PM}{MQ} = \frac{PN}{PM}\frac{MQ}{NQ}\frac{r\,\Delta\theta}{\Delta r}. \tag{14}$$

Since the approximation of PM to PN and of MQ to NQ is improved when we decrease the size of $\Delta\theta$, we would expect that

$$\lim_{\Delta\theta\to 0}\frac{PN}{PM} = 1 \qquad \text{and} \qquad \lim_{\Delta\theta\to 0}\frac{MQ}{NQ} = 1. \tag{15}$$

And these relations are rigorously proved to hold whenever $r \neq 0$ and $dr/d\theta \neq 0$ in Example 3 below. It follows from Eqs. (12), (14), and (15) that

$$\tan\psi = \lim_{\Delta\theta\to 0} \tan OQP = \lim_{\Delta\theta\to 0}\left(\frac{PN}{PM}\right)\left(\frac{MQ}{NQ}\right)\frac{r}{\Delta r/\Delta\theta} = \frac{r}{dr/d\theta}. \tag{16}$$

This proves Eq. (11) when $r \neq 0$ and $dr/d\theta \neq 0$. As indicated in the last member of Eq. (11), it is usually necessary to calculate $r = f(\theta)$ and $dr/d\theta = f'(\theta)$ from the given polar equation of the curve, $r = f(\theta)$.

In Fig. 191, r is positive and increases with θ, so that ψ is a positive acute angle. If r is positive and decreases when θ increases, $dr/d\theta$ is negative, and ψ is a positive obtuse angle. It is sometimes convenient to use other angles from the radius vector to the tangent line. But any such angle will equal $\psi + k\pi$ with k an integer and so have its trigonometric tangent given by $r/(dr/d\theta)$, regardless of the signs of r and of $dr/d\theta$. If $dr/d\theta = 0$ and $r \neq 0$, $\psi = \pi/2$ and $\tan \psi = \infty$, so that Eq. (11) is still valid in a limiting sense. If P is at O, strictly speaking, the radius vector and hence ψ are not defined. But suppose that $(0,\theta_1)$ is a pair of coordinates of O obtained from $f(\theta_1) = 0$, and we consider $\theta = \theta_1$ as the radius vector for this point. Then by the discussion of Fig. 176 in Sec. 144, $\theta = \theta_1$ is tangent to the curve, so that $\psi = 0$. And the value $\tan \psi = 0$ is consistent with Eq. (11), literally if $dr/d\theta \neq 0$, and in a limiting sense as $\theta \to \theta_1$ if $dr/d\theta = 0$ for $\theta = \theta_1$.

EXAMPLE 1. Find an expression for $\tan \psi$ in terms of θ for any point on the three-leaved rose $r = a \sin 3\theta$. And in particular show that $\psi = \pi/6$ when $\theta = \pi/9$ (Fig. 192).

Solution: From the given equation $r = a \sin 3\theta$, we find that $dr/d\theta = 3a \cos 3\theta$. Hence from Eq. (11) we find that

$$\tan \psi = \frac{r}{dr/d\theta} = \frac{a \sin 3\theta}{3a \cos 3\theta} = \frac{1}{3} \tan 3\theta.$$

This is the required expression for the general point.

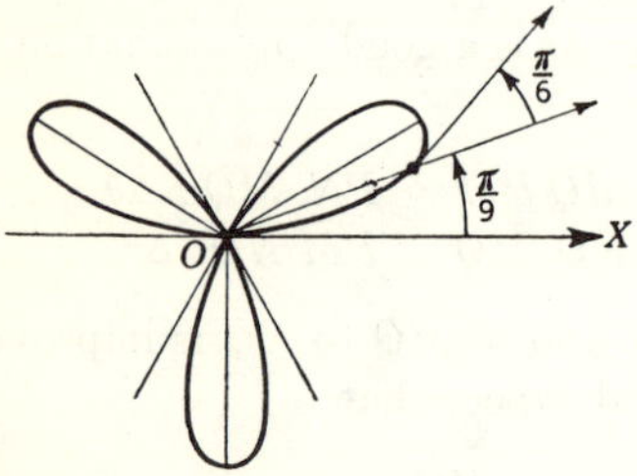

FIG. 192.

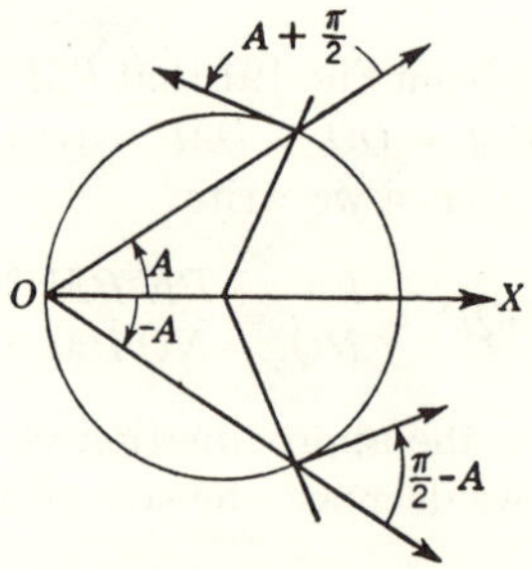

FIG. 193.

When $\theta = \pi/9$, $3\theta = \pi/3$ so that $\tan 3\theta = \sqrt{3}$. It follows that $\tan \psi = \frac{1}{3} \tan 3\theta = \frac{1}{3}\sqrt{3}$. And $\psi = \tan^{-1} \frac{1}{3}\sqrt{3} = \pi/6$ as was to be proved.

EXAMPLE 2. Use Eq. (11) to find ψ in terms of θ for the circle $r = 2a \cos \theta$, traced by θ in the range $-\pi/2$ to $\pi/2$.

Solution: From $r = 2a \cos \theta$ we find that $dr/d\theta = -2a \sin \theta$. Hence from Eq. (11), we have $\tan \psi = \dfrac{r}{dr/d\theta} = \dfrac{2a \cos \theta}{-2a \sin \theta} = -\cot \theta = \tan\left(\theta + \dfrac{\pi}{2}\right)$. If we take θ between $-\dfrac{\pi}{2}$ and $\dfrac{\pi}{2}$, $\left(\theta + \dfrac{\pi}{2}\right)$ is between 0 and π, so that $\psi = \theta + \dfrac{\pi}{2}$. This is the required result. Compare Fig. 193.

*EXAMPLE 3. Prove that Eq. (15) is true if $r \neq 0$ and $dr/d\theta \neq 0$.

Solution: We first note that, in the right triangle PON of Fig. 191, angle $PON = \Delta\theta$, $OP = r$, and $PN/OP = \sin PON$, $ON/OP = \cos PON$. Hence $PN = OP \sin PON = r \sin \Delta\theta$, $ON = OP \cos PON = r \cos \Delta\theta$.

Since $PM = r\,\Delta\theta$ by Eq. (5) of Sec. 90, and $r \neq 0$, we have $\dfrac{PN}{PM} = \dfrac{r \sin \Delta\theta}{r\,\Delta\theta} = \dfrac{\sin \Delta\theta}{\Delta\theta}$.

Hence $\lim_{\Delta\theta \to 0} \dfrac{PN}{PM} = 1$ by Eq. (52) of Sec. 95 with $\Delta\theta$ in place of θ. This proves the first relation of Eq. (15).

Also from Fig. 191 we have $MQ = OQ - OM = (r + \Delta r) - r = \Delta r$. And $NQ = OQ - ON = (r + \Delta r) - r \cos \Delta\theta$. It follows that

$$\frac{NQ}{MQ} = \frac{r + \Delta r - r \cos \Delta\theta}{\Delta r} = 1 + \frac{r}{\Delta r}(1 - \cos \Delta\theta)$$
$$= 1 + \frac{1 - \cos \Delta\theta}{\Delta\theta} \, \frac{r}{\Delta r/\Delta\theta}.$$

But

$$\lim_{\Delta\theta \to 0} \frac{\Delta r}{\Delta\theta} = \frac{dr}{d\theta} \neq 0 \qquad \text{and} \qquad \lim_{\Delta\theta \to 0} \frac{1 - \cos \Delta\theta}{\Delta\theta} = 0$$

by Eq. (65) of Sec. 95 with $\Delta\theta$ in place of θ. Hence

$$\lim_{\Delta\theta \to 0} \frac{NQ}{MQ} = 1 + 0 \cdot \frac{r}{dr/d\theta} = 1 \qquad \text{and} \qquad \lim_{\Delta\theta \to 0} \frac{MQ}{NQ} = \lim_{\Delta\theta \to 0} \frac{1}{NQ/MQ} = 1.$$

This proves the second relation of Eq. (15).

EXERCISE 74

For each given curve verify that ψ may be expressed in terms of θ as indicated, if θ is in a suitably restricted range.

1. $r = a \sin \theta,\ \psi = \theta.$
2. $r = a \sec \theta,\ \psi = \dfrac{\pi}{2} - \theta.$
3. $r = a \csc \theta,\ \psi = \pi - \theta.$
4. $r^2 = a^2 \cos 2\theta,\ \psi = \dfrac{\pi}{2} + 2\theta.$
5. $r = a(1 - \cos \theta),\ \psi = \dfrac{\theta}{2}.$
6. $r = \dfrac{a}{1 - \cos \theta},\ \psi = \pi - \dfrac{\theta}{2}.$
7. $r = a \sin^n \dfrac{\theta}{n},\ \psi = \dfrac{\theta}{n}.$
8. $r = a \csc^n \dfrac{\theta}{n},\ \psi = \pi - \dfrac{\theta}{n}.$

For each given curve, find an expression for $\tan \psi$ in terms of θ at any point. And in particular, evaluate ψ for the designated value of θ.

9. $r = a \cos 3\theta,\ \theta = \dfrac{5\pi}{18}.$
10. $r = a \sin 4\theta,\ \theta = \dfrac{\pi}{16}.$
11. $r = 2 + \cos 2\theta,\ \theta = \dfrac{\pi}{2}.$
12. $r = 2 + \sin \theta,\ \theta = 0.$
13. $r = 2\theta,\ \theta = \pi.$
14. $r + \dfrac{1}{\theta},\ \theta = \pi.$
15. For the *logarithmic spiral*, $r = ae^{b\theta}$, show that $\psi = \tan^{-1}(1/b)$. Since the curve cuts each radius vector at the same angle, the curve is called the *equiangular spiral*.
16. For the spiral of Archimedes, $r = a\theta$, show that $\psi = \tan^{-1} \theta$. Verify that $\psi = 1.004$ radians or 57.52° when $\theta = \pi/2$.
17. From Eq. (9) with $e = 1$ and $q = a$ and Prob. 6, deduce that the tangent to a parabola makes equal angles with the axis and a line from the focus to the point of contact.

148. Slope. In Fig. 194, $P = (r,\theta)$ is a point on the curve whose equation in polar coordinates is $r = f(\theta)$. The tangent line to the curve at P is UPT. And OPR is the radius vector produced so that angle $XOP = \theta$. Thus one value of angle RPT = angle $OPU = \psi$ as defined in Sec. 147. And angle $XUT = \phi$ is one value of the slope angle at P as defined in Sec. 24. In triangle OUP, by the property of an exterior angle, we have

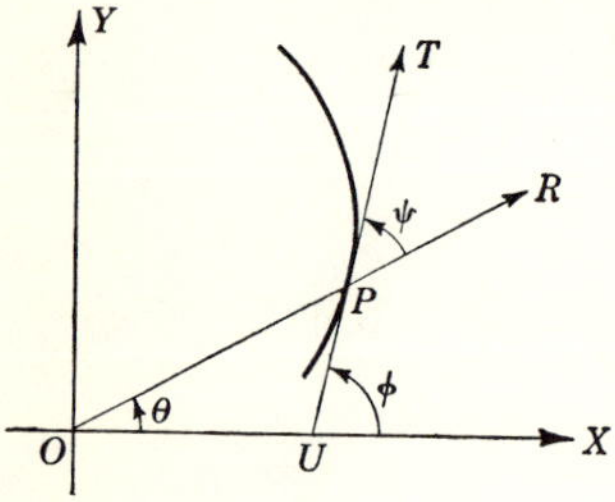

Fig. 194.

$$\angle XUP = \angle UOP + \angle OPU$$

or

$$\phi = \theta + \psi. \tag{17}$$

By Eq. (27) of Sec. 92, it follows that

$$\tan \phi = \tan (\theta + \psi) = \frac{\tan \theta + \tan \psi}{1 - \tan \theta \tan \psi}. \tag{18}$$

This holds for all admissible values of ϕ, θ, and ψ since they give the same value to $\tan \phi$, $\tan \theta$, and $\tan \psi$ as the particular choice used to derive Eq. (17). From Eqs. (18) and (11) and the relation $\tan \theta = \sin \theta/\cos \theta$, we may deduce that

$$\tan \phi = \frac{\dfrac{\sin \theta}{\cos \theta} + \dfrac{r}{dr/d\theta}}{1 - \dfrac{\sin \theta}{\cos \theta}\dfrac{r}{dr/d\theta}} = \frac{\dfrac{dr}{d\theta}\sin \theta + r \cos \theta}{\dfrac{dr}{d\theta}\cos \theta - r \sin \theta}. \tag{19}$$

We have thus proved that, for the curve $r = f(\theta)$,

$$\tan \phi = \text{slope at } P = \frac{(dr/d\theta) \sin \theta + r \cos \theta}{(dr/d\theta) \cos \theta - r \sin \theta}. \tag{20}$$

Let us check this by a purely analytic proof. We first note that Eq. (2), or

$$x = r \cos \theta, \qquad y = r \sin \theta \tag{21}$$

in combination with the relation $r = f(\theta)$, defines x and y as functions of the parameter θ. And we find by differentiation that

$$\frac{dx}{d\theta} = \frac{dr}{d\theta}\cos \theta - r \sin \theta, \qquad \frac{dy}{d\theta} = \frac{dr}{d\theta}\sin \theta + r \cos \theta. \tag{22}$$

Hence by Eq. (18) of Sec. 129 with θ in place of t, we have

$$\frac{dy}{dx} = \frac{dy/d\theta}{dx/d\theta} = \frac{(dr/d\theta) \sin \theta + r \cos \theta}{(dr/d\theta) \cos \theta - r \sin \theta}. \tag{23}$$

This checks Eq. (19) since $dy/dx = \tan \phi$.

The analytic proof holds even when $r = 0$ so that ψ is strictly speaking undefined. If $r = 0$ and $dr/d\theta \neq 0$, we find from Eq. (23) that

$$\tan \phi = \frac{dy}{dx} = \frac{\sin \theta}{\cos \theta} = \tan \theta \qquad \text{so that } \phi = \theta \text{ when } r = 0. \quad (24)$$

We observed this fact in Secs. 144 and 147.

EXAMPLE 1. Use Eq. (19) to find ϕ in terms of θ for the circle $r = 2a \cos \theta$.

Solution: From $r = 2a \cos \theta$, we find that $dr/d\theta = -2a \sin \theta$. Hence from Eq. (20) we have $\tan \phi = \dfrac{(-2a \sin \theta) \sin \theta + (2a \cos \theta) \cos \theta}{(-2a \sin \theta) \cos \theta - (2a \cos \theta) \sin \theta} = \dfrac{\cos^2 \theta - \sin^2 \theta}{-2 \sin \theta \cos \theta} = -\dfrac{\cos 2\theta}{\sin 2\theta}$ $= -\cot 2\theta = \tan\left(2\theta + \dfrac{\pi}{2}\right)$.

Hence $\phi = 2\theta + (\pi/2)$ is an expression for the slope angle, as required. For θ in the range $-\pi/2$ to $\pi/2$ this makes ϕ negative for the lower left-hand quarter and positive for the rest of the circle. But it is the choice that makes $\phi = \theta + \psi$ in accord with Eq. (17) by Example 2 of Sec. 147.

EXAMPLE 2. Find ϕ and ψ for the curve $r = a(1 + \cos \theta)$.

Solution: From $r = a(1 + \cos \theta)$, we have $dr/d\theta = -a \sin \theta$. Hence from Eq. (11) we have $\tan \psi = \dfrac{r}{dr/d\theta} = \dfrac{a(1 + \cos \theta)}{-a \sin \theta} = -\dfrac{1 + \cos \theta}{\sin \theta}$. By Eqs. (40) and (38) of Sec. 93, it follows that

$$\tan \psi = -\frac{2 \cos^2 (\theta/2)}{2 \sin (\theta/2) \cos (\theta/2)} = -\frac{\cos (\theta/2)}{\sin (\theta/2)} = -\cot \frac{\theta}{2} = \tan\left(\frac{\theta}{2} + \frac{\pi}{2}\right).$$

Hence $\psi = \theta/2 + \pi/2$ for θ in a suitable range.

From Eq. (20) we have $\tan \phi = \dfrac{(dr/d\theta) \sin \theta + r \cos \theta}{(dr/d\theta) \cos \theta - r \sin \theta}$ $= \dfrac{(-a \sin \theta) \sin \theta + a(1 + \cos \theta) \cos \theta}{(-a \sin \theta) \cos \theta - a(1 + \cos \theta) \sin \theta} = \dfrac{\cos \theta + \cos^2 \theta - \sin^2 \theta}{-\sin \theta - 2 \sin \theta \cos \theta}$. By Eqs. (33) and (32) of Sec. 93 and Probs. 11 and 13 of Exercise 48, this is $-\dfrac{\cos \theta + \cos 2\theta}{\sin \theta + \sin 2\theta} =$ $-\dfrac{2 \cos (3\theta/2) \cos (\theta/2)}{2 \sin (3\theta/2) \sin (\theta/2)} = -\cot \dfrac{3\theta}{2} = \tan\left(\dfrac{3\theta}{2} + \dfrac{\pi}{2}\right)$. Hence $\phi = (3\theta/2) + (\pi/2)$ for θ in a suitable range.

The expressions $\psi = \dfrac{\theta}{2} + \dfrac{\pi}{2}$ and $\phi = \dfrac{3\theta}{2} + \dfrac{\pi}{2}$ make $\phi = \theta + \psi$ in accord with Eq. (17).

EXERCISE 75

Use Eq. (20) to verify the expression given for ϕ in terms of θ for each given curve. Then check by using Eq. (17) and the corresponding problem of Exercise 73.

1. $r = a \sin \theta$, $\phi = 2\theta$.
2. $r = a \sec \theta$, $\phi = \dfrac{\pi}{2}$.
3. $r = a \csc \theta$, $\phi = 0$.
4. $r^2 = a^2 \cos 2\theta$, $\phi = 3\theta + \dfrac{\pi}{2}$.
5. $r = a(1 - \cos \theta)$, $\phi = \dfrac{3\theta}{2}$.
6. $r = \dfrac{a}{1 - \cos \theta}$, $\phi = \dfrac{\theta}{2}$.
7. $r = a \sin^n \dfrac{\theta}{n}$, $\phi = \dfrac{(n + 1)\theta}{n}$.
8. $r = a \csc^n \dfrac{\theta}{n}$, $\phi = \dfrac{(n - 1)\theta}{n}$.

9. Use Eq. (20) to show that, at a point on the curve $r = f(\theta)$ where the tangent line is parallel to OX,

$$\frac{dr}{d\theta}\sin\theta + r\cos\theta = 0.$$

Use the result of Prob. 9 to find the points on each of the following curves at which the tangent line is parallel to OX.

10. $r = 2 - \cos\theta$.
11. $r = 1 - 2\cos\theta$.
12. $r = \sin 2\theta$.
13. $r = \cos 2\theta$.
14. Use Eq. (20) to show that, at a point on the curve $r = f(\theta)$ where the tangent line is perpendicular to OX,

$$\frac{dr}{d\theta}\cos\theta - r\sin\theta = 0.$$

Use the result of Prob. 14 to find the points on each of the following curves at which the tangent line is perpendicular to OX.

15. $r = 2 - \cos\theta$.
16. $r = 1 - 2\cos\theta$.
17. $r = \sin 2\theta$.
18. $r = \cos 2\theta$.
19. It follows from Eq. (17) that $\tan\psi = \tan(\phi - \theta)$ so that $\tan\psi = \dfrac{\tan\phi - \tan\theta}{1 + \tan\phi\tan\theta}$. From $\tan\phi = \dfrac{dy}{dx} = \dfrac{dy/d\theta}{dx/d\theta}$ and $\tan\theta = \dfrac{\sin\theta}{\cos\theta}$, deduce that $\tan\psi = \dfrac{(dy/d\theta)\cos\theta - (dx/d\theta)\sin\theta}{(dx/d\theta)\cos\theta + (dy/d\theta)\sin\theta}$. Combine this with Eq. (22) to obtain an alternative derivation of Eq. (11).

149. Angle between Two Curves. In Fig. 195, PC_1 and PC_2 are two curves intersecting at point P. The tangent line to PC_1 at P is U_1PT_1. And the tangent line to PC_2 at P is U_2PT_2. Hence, by Sec. 88, angle $U_1PU_2 = \beta$ is one value of the angle from the first to the second curve. We wish to calculate β when the equations of PC_1 and PC_2 are given in polar coordinates.

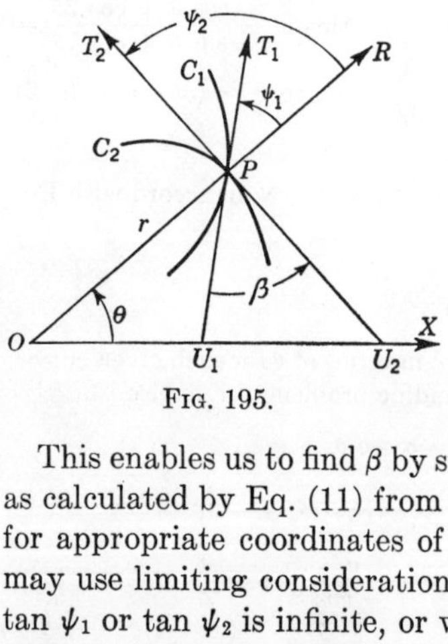

FIG. 195.

If OP is the radius vector, angle $OPU_1 = \psi_1$ and angle $OPU_2 = \psi_2$. And we may deduce from Fig. 195 that angle U_1PU_2 = angle OPU_2 − angle OPU_1, or

$$\beta = \psi_2 - \psi_1. \tag{25}$$

By Eq. (28) of Sec. 92, it follows that

$$\tan\beta = \frac{\tan\psi_2 - \tan\psi_1}{1 + \tan\psi_1\tan\psi_2}. \tag{26}$$

This enables us to find β by substituting the values of $\tan\psi_1$ and $\tan\psi_2$, as calculated by Eq. (11) from the equations of the curves and evaluated for appropriate coordinates of the point P. When $\tan\beta$ is infinite, we may use limiting considerations as described in Secs. 79 and 88. When $\tan\psi_1$ or $\tan\psi_2$ is infinite, or whenever ψ_1 and ψ_2 are easily determined, we may find β from Eq. (25).

To find the intersections of the curves whose equations are

$$r = f(\theta) \qquad \text{and} \qquad r = g(\theta), \tag{27}$$

we solve their equations as simultaneous. This frequently gives all the intersections (r_1,θ_1) as solutions of

$$f(\theta_1) = g(\theta_1), \qquad r_1 = f(\theta_1). \tag{28}$$

The curves may intersect at the origin and will, if for some two values of θ, θ_1, and θ_2,

$$f(\theta_1) = 0, \qquad g(\theta_2) = 0. \tag{29}$$

In this case, one of the angles of intersection at the origin is

$$\beta = \theta_2 - \theta_1.$$

The curves will intersect at a point $P = (r_1,\theta_1) = (-r_1, \theta + \pi)$ if

$$-f(\theta_1) = g(\theta_1 + \pi), \qquad r_1 = f(\theta_1). \tag{30}$$

For curves whose equations involve θ only through combinations of $\sin\theta$ and $\cos\theta$, all the intersections will be found by using Eqs. (28) to (30). Usually the number of intersections can be seen from a sketch of the curves. In some cases, it may be necessary to use the results of Example 2 of Sec. 143, as in Example 3 below.

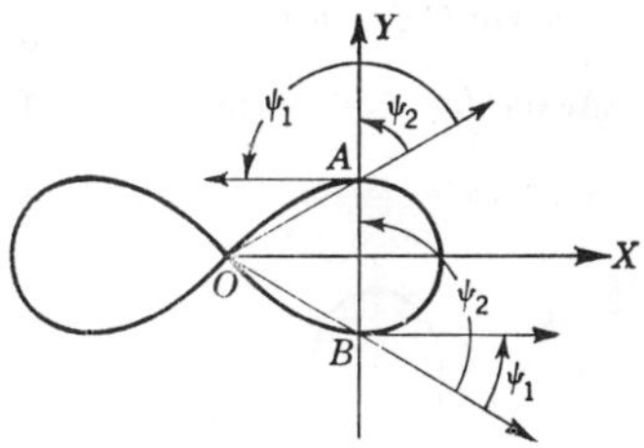

FIG. 196.

EXAMPLE 1. Find the angle of intersection of the lemniscate $r^2 = 2a^2 \cos 2\theta$ and the straight line $r \cos\theta = \frac{a}{2}\sqrt{3}$.

Solution: Elimination of r between the two equations leads to $2a^2 \cos 2\theta = \dfrac{3a^2}{4\cos^2\theta}$ or $8 \cos 2\theta \cos^2\theta = 3$. And by Eq. (37) of Sec. 93, this becomes $8(2\cos^2\theta - 1)\cos^2\theta = 3$, $16\cos^4\theta - 8\cos^2\theta - 3 = 0$, $(4\cos^2\theta - 3)(4\cos^2\theta + 1) = 0$, $\cos^2\theta = \frac{3}{4}$ and $\cos\theta = \pm\frac{1}{2}\sqrt{3}$. Hence $\theta = \pm\pi/6$, or $\pm 5\pi/6$. And $r = \dfrac{a\sqrt{3}}{2\cos\theta} = \dfrac{a\sqrt{3}}{\pm\sqrt{3}} = \pm a$. Each of the pairs $(a,\pi/6)$, $(a,-\pi/6)$, $(-a,5\pi/6)$, $(-a,-5\pi/6)$ satisfies both equations, but the first and fourth give the same point, and the second and third give the same point. As Fig. 196 shows, there are only two intersections, $A = (a,\pi/6)$ and $B = (a,-\pi/6)$ if we use the pairs with positive r.

For the lemniscate, $r^2 = 2a^2 \cos 2\theta$, $2r(dr/d\theta) = -4a^2 \sin 2\theta$, so that $\dfrac{dr}{d\theta} = -\dfrac{2}{r} a^2 \sin 2\theta$ and $\tan\psi_1 = \dfrac{r}{dr/d\theta} = \dfrac{-r^2}{2a^2 \sin 2\theta} = -\dfrac{2a^2\cos 2\theta}{2a^2 \sin 2\theta} = -\cot 2\theta = -\sqrt{3}/3$ at A where $\theta = \pi/6$, and $= \sqrt{3}/3$ at B where $\theta = -\pi/6$.

For the straight line, $r = \frac{a}{2}\sqrt{3}\sec\theta, \frac{dr}{d\theta} = \frac{a}{2}\sqrt{3}\tan\theta\sec\theta$, so that $\tan\psi_2 = \frac{r}{dr/d\theta} = \frac{\frac{a}{2}\sqrt{3}\sec\theta}{\frac{a}{2}\sqrt{3}\tan\theta\sec\theta} = \cot\theta = \sqrt{3}$ at A where $\theta = \pi/6$, and $= -\sqrt{3}$ at B where $\theta = -\pi/6$.

If we use Eq. (26), we find that at A, $\tan\beta = \frac{\sqrt{3} - (-\sqrt{3}/3)}{1 + \sqrt{3}(-\sqrt{3}/3)}$, and at B, $\tan\beta = \frac{-\sqrt{3} - (\sqrt{3}/3)}{1 + (-\sqrt{3})(\sqrt{3}/3)}$. The zero denominators, or the fact that $\tan\psi_1\tan\psi_2 = -1$ indicates that $|\beta| = \pi/2$ at each point.

Thus the given line cuts the lemniscate at right angles at $A = (a,\pi/6)$ and at $B = (a,-\pi/6)$.

We might have deduced from the values of the tangents that $\psi_1 = 5\pi/6$ at A and $\pi/6$ at B, while $\psi_2 = \pi/3$ at A and $2\pi/3$ at B. From these values and Eq. (25), we find that $\beta = \psi_2 - \psi_1 = -\pi/2$ at A and $\pi/2$ at B. This agrees with Fig. 196 and checks the perpendicularity.

EXAMPLE 2. Find the four points of intersection of the curves $r = a\cos\theta$ and $r = a\cos 2\theta$, and the corresponding angles of intersection.

Solution: As in Eq. (28), set $a\cos\theta = a\cos 2\theta$, or $\cos\theta = \cos 2\theta$. This is satisfied when $\theta = 2\theta$, or $\theta = 0$, so that $(a,0)$ is one intersection. For $r = a\cos\theta$, $dr/d\theta = -a\sin\theta$. Hence $\tan\psi_1 = \frac{a\cos\theta}{-a\sin\theta} = -\cot\theta = \infty$ for $\theta = 0$, so that $\psi_1 = \frac{\pi}{2}$.

Again for $r = a\cos 2\theta$, $dr/d\theta = -2a\sin 2\theta$. Hence $\tan\psi_2 = \frac{a\cos 2\theta}{-2a\sin 2\theta} = -\frac{1}{2}\cot 2\theta = \infty$ for $\theta = 0$, so that $\psi_2 = \frac{\pi}{2}$. It follows from Eq. (25) that at $(a,0)$,

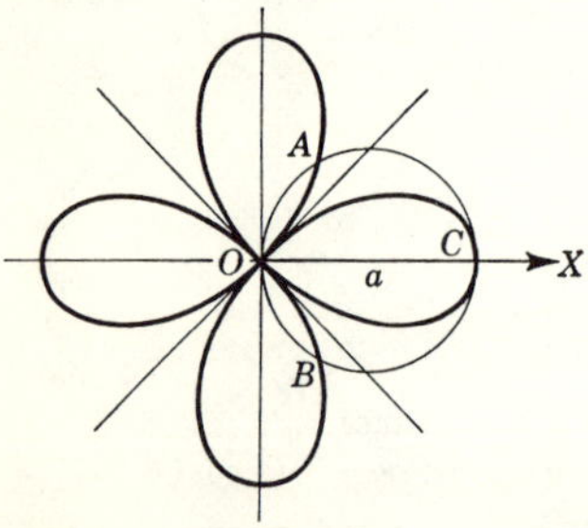

FIG. 197.

$$\beta = \psi_2 - \psi_1 = \frac{\pi}{2} - \frac{\pi}{2} = 0.$$

As in Eq. (29) set $a\cos\theta = 0$. This gives $\theta_1 = \pi/2$. And set $a\cos 2\theta = 0$ which gives $2\theta = \pi/2$ or $3\pi/2$, and $\theta_2 = \pi/4$ or $3\pi/4$. At the origin $\beta = \theta_2 - \theta_1 = -\pi/4$ or $\pi/4$.

As in Eq. (30) set $-a\cos\theta = a\cos 2(\theta + \pi)$, or $-\cos\theta = \cos 2\theta$. This is satisfied when $2\theta = \pi - \theta$ or $-\pi - \theta$. Hence $3\theta = \pm\pi$, and $\theta_1 = \pm\pi/3$. And $P = (r_1,\theta_1) = (-r_1, \theta_1 + \pi)$ is either $(a/2,\pi/3) = (-a/2,4\pi/3)$ or $(a/2,-\pi/3) = (-a/2,2\pi/3)$. For these two intersections we find $\tan\psi_1 = -\cot\theta = -\sqrt{3}/3$ at $\theta = \pi/3$ and $= \sqrt{3}/3$ at $\theta = -\pi/3$. And $\tan\psi_2 = -\frac{1}{2}\cot 2\theta = \sqrt{3}/6$ at $\theta = 4\pi/3$ and $= -\sqrt{3}/6$ at $\theta = 2\pi/3$. Hence by Eq. (26) at $(a/2,\pi/3)$,

$\tan\beta = \frac{(\sqrt{3}/6) - (-\sqrt{3}/3)}{1 + (\sqrt{3}/6)(-\sqrt{3}/3)} = \frac{\sqrt{3}/2}{\frac{5}{6}} = \frac{3}{5}\sqrt{3}$. Similarly, at $(a/2,-\pi/3)$ $\tan\beta = -\frac{3}{5}\sqrt{3}$.

Thus (Fig. 197) the two curves meet at $(a,0)$ at a zero angle, at the origin at angles of $\pm\pi/4$, at $(a/2,\pi/3)$ at an angle $\tan^{-1}\frac{3}{5}\sqrt{3} = 0.805$, and at $(a/2,-\pi/3)$ at an angle $\tan^{-1} - \frac{3}{5}\sqrt{3} = -0.805$ or 2.337.

EXAMPLE 3. Show that the two spirals $r = 3\theta$ and $r = \theta - 4$ intersect at an infinite number of points.

Solution: If (r,θ) satisfies the first equation and $(r,\theta + 2k\pi)$ satisfies the second equation, $3\theta = (\theta + 2k\pi) - 4$, so that $2\theta = 2k\pi - 4$ and $\theta = k\pi - 2$. Hence $(r,\theta) = (3k\pi - 6,\ k\pi - 2)$.

If (r,θ) satisfies the first equation and $(-r,\ \theta + 2k\pi + \pi)$ satisfies the second equation, $-3\theta = (\theta + 2k\pi + \pi) - 4$, so that $-4\theta = 2k\pi + \pi - 4$ and $\theta = -\dfrac{2k+1}{4}\pi + 1$. Hence $(-r,\ \theta + 2k\pi + \pi) = \left(\dfrac{2k+1}{4}3\pi - 3, \dfrac{2k+1}{4}3\pi + 1\right)$.

Finally, $r = 3\theta = 0$ when $\theta = 0$, and $r = \theta - 4 = 0$ when $\theta = 4$, so that the origin is a point of intersection.

Thus the intersections are $(3k\pi - 6,\ k\pi - 2)$, $\left(\dfrac{2k+1}{4}3\pi - 3, \dfrac{2k+1}{4}3\pi + 1\right)$, $(0,0)$ where k is zero or any positive or negative integer. Different values of k give different points.

EXERCISE 76

Find all the points at which each pair of curves intersect and the angle at each intersection.

1. $r \cos \theta = 6,\ r = 13 \sin \theta$.
2. $r \sin \theta = 2,\ r = \sec^2 \theta$.
3. $r = 3 \cos \theta,\ r = 1 + \cos \theta$.
4. $r = \sin \theta,\ r = \sin 2\theta$.
5. $r = 1 - \cos \theta,\ r = 1 + \cos \theta$.
6. $r \sin \theta = 2,\ r = \sec^2 \theta$.
7. $r = 2 \cos \theta,\ r = 2(1 - \cos \theta)$.
8. $r^2 = 16 \sin 2\theta,\ r^2 \sin 2\theta = 4$.

Show that each of the following pairs of curves intersect at right angles at all points of intersection.

9. $r = 4 \cos \theta,\ r = 4 \sin \theta$.
10. $r = 4 \sin \theta,\ r \sin \theta = 2$.
11. $r = 4 \sin^2 \dfrac{\theta}{2},\ r = \csc^2 \dfrac{\theta}{2}$.
12. $r = e^{\theta},\ r = e^{-\theta}$.
13. $r = a \sec^2 \dfrac{\theta}{2},\ r = b \csc^2 \dfrac{\theta}{2}$.
14. $r^2 \cos 2\theta = a^2,\ r^2 \sin 2\theta = b^2$.

Show that at any intersection with the same (r,θ) for both equations, each of the following pairs of curves cut at right angles.

15. $r = a(1 + \cos \theta),\ r = b(1 - \cos \theta)$.
16. $r = \theta,\ r = \dfrac{1}{\theta}$.
17. $r = a \sin^n \dfrac{\theta}{n},\ r = b \cos^n \dfrac{\theta}{n}$.
18. $r = ae^{k\theta},\ r = be^{-\theta/k}$.

***150. Differential of Arc Length.** Let the equation of a curve in polar coordinates be

$$r = f(\theta). \tag{31}$$

Then the length of arc along the curve from a fixed point A to a variable point $P = (r,\theta)$, or s, is a function of θ. But Eq. (2), or

$$x = r \cos \theta, \qquad y = r \sin \theta, \tag{32}$$

in combination with Eq. (31) defines x and y in terms of the parameter θ. Hence by Eq. (34) of Sec. 132 with θ in place of t, we have

$$\frac{ds}{d\theta} = \sqrt{\left(\frac{dx}{d\theta}\right)^2 + \left(\frac{dy}{d\theta}\right)^2}, \tag{33}$$

with the plus sign before the radical if s increases when θ increases.

By differentiation, we find from Eq. (32) that

$$\frac{dx}{d\theta} = \frac{dr}{d\theta}\cos\theta - r\sin\theta, \qquad \frac{dy}{d\theta} = \frac{dr}{d\theta}\sin\theta + r\cos\theta. \tag{34}$$

By squaring and adding these relations, we may deduce that

$$\left(\frac{dx}{d\theta}\right)^2 + \left(\frac{dy}{d\theta}\right)^2 = \left(\frac{dr}{d\theta}\right)^2 + r^2. \tag{35}$$

And from Eqs. (33) and (35) we may conclude that

$$\frac{ds}{d\theta} = \sqrt{\left(\frac{dr}{d\theta}\right)^2 + r^2}. \tag{36}$$

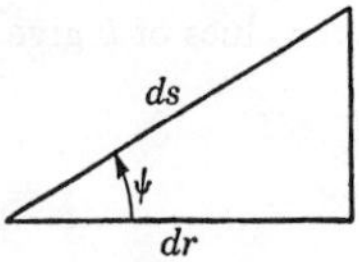

FIG. 198.

By multiplying by the differential $d\theta$, we obtain

$$ds = \sqrt{dr^2 + r^2\,d\theta^2}. \tag{37}$$

And by squaring this relation, we find that

$$ds^2 = dr^2 + r^2\,d\theta^2. \tag{38}$$

This suggests a right triangle like that in Fig. 198 with sides dr, $r\,d\theta$, and ds. Since the angle ψ of Sec. 147 has

$$\tan\psi = \frac{r}{dr/d\theta} = \frac{r\,d\theta}{dr}, \tag{39}$$

the triangle has one angle equal to ψ, as indicated. Hence

$$dr = ds\cos\psi, \qquad r\,d\theta = ds\sin\psi. \tag{40}$$

If r and $ds/d\theta$ are positive, these relations hold as written with ψ the angle defined in Sec. 147. If $ds/d\theta$ is negative, we insert a minus sign before the radical in Eqs. (33), (36), and (37). And we take ψ as the angle from the direction of the radius vector produced to the tangent line drawn in the direction of increasing s. With this convention, Eq. (40) holds whenever $r > 0$.

It is easy to recall Eqs. (36) to (40) from the triangle of Fig. 198. And this triangle is suggested by the curvilinear triangle PQM of Fig. 191 with sides $MQ = \Delta r$, $PM = r\,\Delta\theta$, and $PQ = \Delta s$, and having angle MQP approximately equal to ψ.

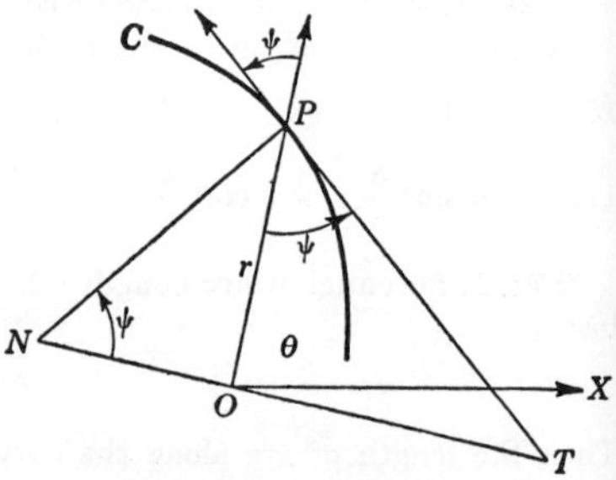

FIG. 199.

*151. Geometric Properties.

In discussing geometric properties of a curve, we sometimes require expressions for some of the lines of Fig. 199. In Fig. 199, PT is the line tangent to the curve $r = f(\theta)$ at $P = (r,\theta)$. PN is the normal line, drawn through P perpendicular to the tangent line PT. And NT is drawn through the origin O perpendicular to the radius vector OP. The trigonometric functions of ψ may be read from Fig. 198, or found from Eq. (40) combined with Eqs. (37) and (39). In this way we find that

$$\text{Polar tangent } TP = \frac{OP}{\cos\psi} = r\frac{ds}{dr} = r\frac{\sqrt{(dr/d\theta)^2 + r^2}}{dr/d\theta}. \tag{41}$$

$$\text{Polar subtangent } OT = OP \tan \psi = \frac{r^2}{dr/d\theta}. \tag{42}$$

$$\text{Polar normal } PN = \frac{OP}{\sin \psi} = \frac{ds}{d\theta} = \sqrt{(dr/d\theta)^2 + r^2}. \tag{43}$$

$$\text{Polar subnormal } ON = \frac{OP}{\tan \psi} = \frac{dr}{d\theta}. \tag{44}$$

These relations as written make the four quantities on the right positive if r and $dr/d\theta$ are both positive, as in Fig. 199.

We sometimes call TP the length of the polar tangent, and PN the length of the polar normal.

EXAMPLE 1. For the curve $r = a\theta^3$, show that the length of the polar subtangent is proportional to the square of the length of the polar subnormal.

Solution: From $r = a\theta^3$, we have $dr/d\theta = 3a\theta^2$. From Eq. (42) we find $OT = \dfrac{r^2}{dr/d\theta}$ $= \dfrac{a^2\theta^6}{3a\theta^2} = \dfrac{a}{3}\theta^4$. And from Eq. (44), $ON = \dfrac{dr}{d\theta} = 3a\theta^2$. Hence $\dfrac{OT}{\overline{ON}^2} = \dfrac{(a/3)\theta^4}{9a^2\theta^4} = \dfrac{1}{27a}$.

This shows that $OT = \dfrac{1}{27a}\overline{ON}^2$, with a constant factor of proportionality $1/(27a)$, as was to be proved.

EXAMPLE 2. Calculate the expression for the polar subtangent OT of the straight line of Eq. (6), $r \cos (\theta - A) = p$. Show that as θ approaches $\theta_1 = A + (\pi/2)$, $r \to \infty$ and $OT \to p$. And when θ approaches $\theta_2 = A - (\pi/2)$, $r \to \infty$ and $OT \to -p$.

Solution: Here $r = p \sec (\theta - A)$ and $dr/d\theta = p \tan (\theta - A) \sec (\theta - A)$. From Eq. (42), $OT = \dfrac{r^2}{dr/d\theta} = \dfrac{p^2 \sec^2 (\theta - A)}{p \tan (\theta - A) \sec (\theta - A)} = \dfrac{p \sec (\theta - A)}{\tan (\theta - A)} =$ $\dfrac{p}{\cos (\theta - A)} \dfrac{\cos (\theta - A)}{\sin (\theta - A)} = \dfrac{p}{\sin (\theta - A)}$.

When $\theta \to A + (\pi/2)$, $\theta - A \to \pi/2$, $\cos (\theta - A) \to 0$, $\sin (\theta - A) \to 1$. Hence, as θ approaches θ_1, $r = \dfrac{p}{\cos (\theta - A)} \to \infty$ and $OT = \dfrac{p}{\sin (\theta - A)} \to p$.

When $\theta \to A - (\pi/2)$, $\theta - A \to -\pi/2$, $\cos (\theta - A) \to 0$, $\sin (\theta - A) \to -1$. Hence as θ approaches θ_2, $r = \dfrac{p}{\cos (\theta - A)} \to \infty$ and $OT = \dfrac{p}{\sin (\theta - A)} \to -p$.

We note that the foot of the perpendicular from O to the straight line $(p,A) = (p, \theta_1 - \pi/2) = (-p, \theta_2 - \pi/2)$. And the equation of the line may be written as $r \cos (\theta - \theta_1 + \pi/2) = p$ or as $r \cos (\theta - \theta_2 + \pi/2) = -p$.

EXAMPLE 3. For $r = f(\theta)$, let $\lim\limits_{\theta \to \theta_1} r = \infty$ and $\lim\limits_{\theta \to \theta_1} \dfrac{r^2}{dr/d\theta} = q$. Show that the curve represented by $r = f(\theta)$ has a branch with a straight-line asymptote perpendicular to the radius vector to $(q, \theta_1 - \pi/2) = (-q, \theta_1 + \pi/2)$ at its extremity.

Solution: Since $r \to \infty$ as $\theta \to \theta_1$, there is an infinite branch. And by Eq. (42) the subtangent $OT = \dfrac{r^2}{dr/d\theta} \to q$ as $\theta \to \theta_1$. Hence the tangent line to the branch approaches a limiting position, which is an asymptote. A comparison with Example 2 shows that the asymptote will have as its equation $r \cos (\theta - \theta_1 + \pi/2) = q$ and so have $Q = (q, \theta_1 - \pi/2)$ as the foot of the perpendicular drawn to it from the origin. The remark at the end of Example 2 shows that this result holds whether $q = p$, or q is negative and $q = -p$.

EXAMPLE 4. Use the result of Example 3 to find the asymptotes of the curve $r = 4 \csc \theta + 2 \text{cse } \theta$.

Solution: Here $r = 4 \csc \theta + 2 \sec \theta$, $dr/d\theta = -4 \cot \theta \csc \theta + 2 \tan \theta \sec \theta$ The expression for r becomes infinite when $\theta \to 0$ and when $\theta \to \pi/2$.

When θ approaches 0, by the principle of the leading term for factors of Sec. 13, we have $\lim\limits_{\theta \to 0} \dfrac{r^2}{dr/d\theta} = \lim\limits_{\theta \to 0} \dfrac{(4 \csc \theta)^2}{-4 \cot \theta \csc \theta}$. But

$$\frac{4 \csc \theta}{-\cot \theta} = \frac{-4}{\sin \theta} \frac{\sin \theta}{\cos \theta} = \frac{-4}{\cos \theta} \quad \text{and} \quad \lim_{\theta \to 0} \frac{r^2}{dr/d\theta} = \lim_{\theta \to 0} \frac{-4}{\cos \theta} = -4.$$

This shows that $r \cos (\theta + \pi/2) = -4$ represents one asymptote. It is perpendicular to OQ at $Q = (q, \theta_1 - \pi/2) = (-4, -\pi/2)$.

Since $Q = (4,\pi/2)$, $r \cos (\theta - \pi/2) = 4$ is an alternative form. Either form is equivalent to $r \sin \theta = 4$.

When θ approaches $\pi/2$, by the principle of the leading term for factors of Sec. 13, we have $\lim\limits_{\theta \to \pi/2} \dfrac{r^2}{dr/d\theta} = \lim\limits_{\theta \to \pi/2} \dfrac{(2 \sec \theta)^2}{2 \tan \theta \sec \theta}$. But

$$\frac{2 \sec \theta}{\tan \theta} = \frac{2}{\cos \theta} \frac{\cos \theta}{\sin \theta} = \frac{2}{\sin \theta} \quad \text{and} \quad \lim_{\theta \to \pi/2} \frac{r^2}{dr/d\theta} = \lim_{\theta \to \pi/2} \frac{2}{\sin \theta} = 2.$$

This shows that $r \cos \theta = 2$ represents a second asymptote with $Q = (2,0)$.

The equations for the asymptotes $r \sin \theta = 4$ and $r \cos \theta = 2$ are equivalent to $y = 4$ and $x = 2$ by Eq. (2), and so check the result of Example 3 of Sec. 145 (Fig. 184).

EXAMPLE 77

Find an expression for the length of the polar tangent and of the polar normal in terms of θ for each of the following given curves.

1. $r = a \sin \theta$.
2. $r^2 = a^2 \cos 2\theta$.
3. $r = a(1 - \cos \theta)$.
4. $r = \dfrac{a}{1 - \cos \theta}$.
5. $r = a \sin^n \dfrac{\theta}{n}$.
6. $r = a \csc^n \dfrac{\theta}{n}$.
7. Show that the spiral of Archimedes $r = a\theta$ has a constant polar subnormal.
8. Show that the reciprocal spiral $r\theta = a$ has a constant polar subtangent.
9. Show that the logarithmic spiral $r = ae^{b\theta}$ has the lengths of its polar tangent, polar normal, polar subtangent, and polar subnormal each proportional to r.

Use the result of Example 3 to locate the asymptotes of each of the following curves, and sketch the curve.

10. $r = 2 \sec \theta + 3 \csc \theta$.
11. $r = \tan \theta$.
12. $r^2 \sin 2\theta = 4$.
13. $r^2 \cos 2\theta = 4$.
15. $r = \dfrac{20}{1 - 2 \cos \theta}$.
16. $r = \sqrt{\tan \theta}$.
17. $r = 6 \tan 2\theta$.
18. $r = 4 \csc 2\theta$.

***152. Curvature in Polar Form.** In Eq. (32) of Sec. 131, we defined the curvature of a curve by the relation

$$\text{Curvature at } P = K = \frac{d\phi}{ds}, \tag{45}$$

the rate of change of slope angle with respect to arc length.

We wish to calculate K for a curve whose equation is given in polar form, $r = f(\theta)$. To do this we recall Eq. (17) or

$$\phi = \theta + \psi. \tag{46}$$

From this we have for the derivative of ϕ with respect to θ,

$$\frac{d\phi}{d\theta} = 1 + \frac{d\psi}{d\theta}. \tag{47}$$

But from Eq. (11) we have

$$\tan\psi = \frac{r}{dr/d\theta}, \qquad \psi = \tan^{-1}\frac{r}{dr/d\theta}. \tag{48}$$

Differentiating this with respect to θ, using Eq. (121) of Sec. 104 and the quotient rule, we find that

$$\frac{d\psi}{d\theta} = \frac{d}{d\theta}\tan^{-1}\frac{r}{dr/d\theta} = \frac{\frac{d}{d\theta}\left(\frac{r}{dr/d\theta}\right)}{1 + \left(\frac{r}{dr/d\theta}\right)^2} = \frac{\frac{(dr/d\theta)(dr/d\theta) - r\,(d^2r/d\theta^2)}{(dr/d\theta)^2}}{\frac{(dr/d\theta)^2 + r^2}{(dr/d\theta)^2}}$$

$$= \frac{(dr/d\theta)^2 - r\,(d^2r/d\theta^2)}{(dr/d\theta)^2 + r^2}. \tag{49}$$

From Eqs. (47) and (49) we may deduce that

$$\frac{d\phi}{d\theta} = 1 + \frac{(dr/d\theta)^2 - r(d^2r/d\theta^2)}{(dr/d\theta)^2 + r^2} = \frac{r^2 + 2(dr/d\theta)^2 - r(d^2r/d\theta^2)}{(dr/d\theta)^2 + r^2}. \tag{50}$$

But by Eq. (36) we have

$$\frac{ds}{d\theta} = \sqrt{\left(\frac{dr}{d\theta}\right)^2 + r^2}. \tag{51}$$

We may deduce from Eqs. (45), (50), and (51) that

$$K = \frac{d\phi}{ds} = \frac{\frac{d\phi}{d\theta}}{ds/d\theta} = \frac{\frac{r^2 + 2(dr/d\theta)^2 - r(d^2r/d\theta^2)}{(dr/d\theta)^2 + r^2}}{\sqrt{(dr/d\theta)^2 + r^2}}. \tag{52}$$

This proves that, at any point $P = (r,\theta)$ of a given curve whose equation in polar coordinates is $r = f(\theta)$, the

$$\text{Curvature} = K = \frac{r^2 + 2(dr/d\theta)^2 - r(d^2r/d\theta^2)}{[(dr/d\theta) + r]^{\frac{3}{2}}}. \tag{53}$$

By Eq. (65) of Sec. 136, the radius of curvature $R = 1/K$. Hence

$$R = \frac{[(dr/d\theta)^2 + r^2]^{\frac{3}{2}}}{r^2 + 2(dr/d\theta)^2 - r(d^2r/d\theta^2)}. \tag{54}$$

As found from Eqs. (53) and (54), K and R will be positive if the slope angle ϕ increases when θ increases. And they will be negative if the slope angle decreases when θ increases. Compare the discussion of signs in Sec. 135.

Example. Find an expression for the radius of curvature R in terms of r for any point of the curve $r = a \sec^n (\theta/n)$.

Solution: From $r = a\sec^n\frac{\theta}{n}$, $\frac{dr}{d\theta} = an\sec^{n-1}\frac{\theta}{n}\tan\frac{\theta}{n}\sec\frac{\theta}{n}\cdot\frac{1}{n} = a\sec^n\frac{\theta}{n}\tan\frac{\theta}{n}$. And

$\dfrac{d^2r}{d\theta^2} = a\sec^n\dfrac{\theta}{n}\left(\tan^2\dfrac{\theta}{n} + \dfrac{1}{n}\sec^2\dfrac{\theta}{n}\right)$. Hence in Eq. (54), the denominator, $r^2 + 2(dr/d\theta)^2 - r(d^2r/d\theta^2)$, is

$$a^2\sec^{2n}\frac{\theta}{n}\left(1 + 2\tan^2\frac{\theta}{n} - \tan^2\frac{\theta}{n} - \frac{1}{n}\sec^2\frac{\theta}{n}\right) = \frac{n-1}{n}a^2\sec^{2n+2}\frac{\theta}{n}.$$

And $\left(\dfrac{dr}{d\theta}\right)^2 + r^2 = a^2\sec^{2n}\dfrac{\theta}{n}\left(\tan^2\dfrac{\theta}{n} + 1\right) = a^2\sec^{2n+2}\dfrac{\theta}{n}$. Hence in Eq. (54) the numerator is $\left[\left(\dfrac{dr}{d\theta}\right)^2 + r^2\right]^{\frac{3}{2}} = a^3\sec^{3n+3}\dfrac{\theta}{n}$.

It follows from Eq. (54) that, in terms of θ,

$$R = \frac{a^3\sec^{3n+3}(\theta/n)}{\dfrac{n-1}{n}a^2\sec^{2n+2}\dfrac{\theta}{n}} = \frac{an}{n-1}\sec^{n+1}\frac{\theta}{n}.$$

But from $r = a\sec^n\dfrac{\theta}{n}$, $\sec\dfrac{\theta}{n} = \left(\dfrac{r}{a}\right)^{1/n}$ so that $R = \dfrac{an}{n-1}\left(\dfrac{r}{a}\right)^{1/n}$ is the required expression in terms of r.

EXERCISE 78

For each of the following curves, verify the given expression for the radius of curvature R in terms of r for any point on the curve.

1. $r = a\cos\theta$, $R = \dfrac{a}{2}$.
2. $r = a\sin\theta$, $R = \dfrac{a}{2}$.
3. $r^2 = a^2\cos 2\theta$, $R = \dfrac{a^3}{3r^2}$.
4. $r^2\sin 2\theta = a^2$, $R = \dfrac{r^3}{a^2}$.
5. $r = a(1 - \cos\theta)$, $R = \frac{2}{3}\sqrt{2ar}$.
6. $r = \dfrac{a}{1-\cos\theta}$, $R = 2r\sqrt{\dfrac{2r}{a}}$.
7. $r = a\sin^n\dfrac{\theta}{n}$, $R = \dfrac{an}{n+1}\left(\dfrac{r}{a}\right)^{(n-1)/n}$
8. $r = a\csc^n\dfrac{\theta}{n}$, $R = \dfrac{an}{n-1}\left(\dfrac{r}{a}\right)^{(n+1)/n}$
9. $r = ae^{b\theta}$, $R = r\sqrt{1+b^2}$.
10. $r = a\theta$, $R = \dfrac{(r^2+a^2)^{\frac{3}{2}}}{r^2+2a^2}$.
11. $r = b - a\cos\theta$, $R = \dfrac{(a^2 - b^2 + 2br)^{\frac{3}{2}}}{2(a^2-b^2) + 3br}$.
12. $r = \dfrac{a}{1 - e\cos\theta}$, $R = \dfrac{[2ra + (e^2-1)r^2]^{\frac{3}{2}}}{a^2}$.
13. $r = a\theta^n$, $R = a^{(n-1)/n}\dfrac{(r^{2/n} + n^2a^{2/n})^{\frac{3}{2}}}{r^{2/n} + (n^2+n)a^{2/n}}$.
14. $r = a\sin n\theta$, $R = \dfrac{[a^2n^2 + (1-n^2)r^2]^{\frac{3}{2}}}{2a^2n^2 + (1-n^2)r^2}$.

***153. Motion in a Curve.** Let t be a variable parameter, and $G(t)$, $H(t)$ be two given functions. Then in general, the parametric equations in polar form

$$r = G(t), \qquad \theta = H(t) \tag{55}$$

determine a curve. We might construct this curve by assigning values to t, calculating the corresponding values of r and θ, and plotting the resulting points (r,θ). And if we can solve the second equation for t, substitution in the first equation leads to the equation of the curve in polar coordinates, $r = f(\theta)$.

If t is the time, Eq. (55) represents the motion of a particle along a curve. For such a motion, the components of the velocity along fixed x and y axes at any time t are

$$v_x = \frac{dx}{dt}, \qquad v_y = \frac{dy}{dt}, \tag{56}$$

by Eq. (83) of Sec. 141.

If s increases as t increases, the speed in the path is

$$\frac{ds}{dt} = v = \sqrt{v_x^2 + v_y^2}. \tag{57}$$

Let P be the instantaneous position of the particle at time t. Draw a unit vector **t** along the tangent line to the path curve in the direction in which t and s increase. And draw **n**, the unit normal to the path curve at P, obtained by rotating **t** through 90° about P. Then by Eq. (92) of Sec. 142,

$$\mathbf{v} = v\mathbf{t}, \qquad v_t = v, \qquad v_n = 0. \tag{58}$$

This is in accord with the fact that the velocity vector may be represented by a segment of length equal to the speed in the path along the tangent line in the direction in which t and s increase.

The components of the acceleration along fixed x and y axes at any time t are

$$a_x = \frac{d^2x}{dt^2}, \qquad a_y = \frac{d^2y}{dt^2}, \tag{59}$$

by Eq. (87) of Sec. 141. And by Eq. (96) of Sec. 142,

$$\mathbf{a} = \frac{dv}{dt}\mathbf{t} + \frac{v^2}{R}\mathbf{n}, \qquad a_t = \frac{dv}{dt}, \qquad a_n = \frac{v^2}{R}. \tag{60}$$

***154a Radial and Transverse Component of Velocity.** Let the motion of a particle along a curve be described by Eq. (55) in terms of polar coordinates. Then it is often useful to know the components of the velocity vector **v** in the direction of increasing r and the direction of increasing θ. To find these, we proceed as in Sec. 142. We observe that

$$\cos\theta \qquad \text{and} \qquad \sin\theta \tag{61}$$

are the components along the x axis and the y axis of a unit vector $\mathbf{u}_r$ in the direction of the radius vector OP (Fig. 200). And

$$\cos\left(\theta + \frac{\pi}{2}\right) = -\sin\theta, \qquad \sin\left(\theta + \frac{\pi}{2}\right) = \cos\theta, \tag{62}$$

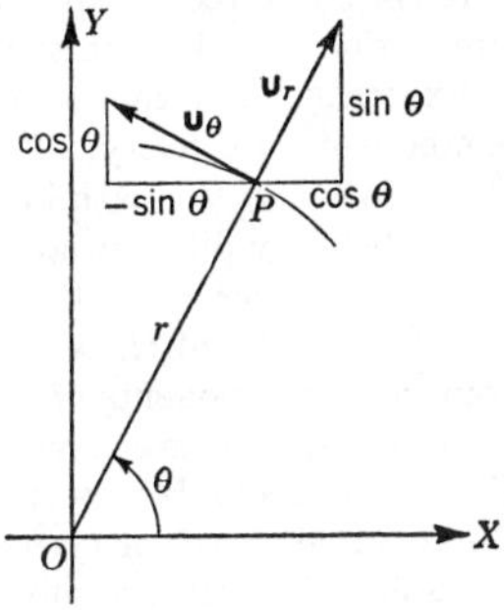

FIG. 200.

will be the components of a unit vector $\mathbf{u}_\theta$ tangent to the circle through P with center at the origin, in the direction in which θ increases. We may resolve any vector along the directions of $\mathbf{u}_r$ and $\mathbf{u}_\theta$. To do this for **v**, we recall Eq. (2) or

$$x = r\cos\theta, \qquad y = r\sin\theta. \tag{63}$$

Differentiating these with respect to t, we find that

$$\frac{dx}{dt} = \frac{dr}{dt}\cos\theta - r\sin\theta\frac{d\theta}{dt}, \qquad \frac{dy}{dt} = \frac{dr}{dt}\sin\theta + r\cos\theta\frac{d\theta}{dt}. \tag{64}$$

It follows from this and Eq. (56) that

$$v_x = \frac{dr}{dt}\cos\theta - \left(r\frac{d\theta}{dt}\right)\sin\theta, \qquad v_y = \frac{dr}{dt}\sin\theta + \left(r\frac{d\theta}{dt}\right)\cos\theta. \tag{65}$$

A comparison of Eq. (65) with Eqs. (61) and (62) shows that the velocity vector $\mathbf{v} = \frac{dr}{dt}\mathbf{u}_r + r\frac{d\theta}{dt}\mathbf{u}_\theta$. We shall use v_r to denote the *radial component* of $\mathbf{v}$ in the direction of increasing r, or the direction of $\mathbf{u}_r$. And we shall use v_θ to denote the *transverse component* of $\mathbf{v}$ in the direction of increasing θ, or the direction of $\mathbf{u}_\theta$. Then since $\mathbf{v} = v_r\mathbf{u}_r + v_\theta\mathbf{u}_\theta$, we have

$$v_r = \frac{dr}{dt}, \qquad v_\theta = r\frac{d\theta}{dt}. \tag{66}$$

Since v_r and v_θ are the components of $\mathbf{v}$ along perpendicular directions, we have

$$v^2 = v_r^2 + v_\theta^2 \quad \text{or} \quad \left(\frac{ds}{dt}\right)^2 = \left(\frac{dr}{dt}\right)^2 + r^2\left(\frac{d\theta}{dt}\right)^2. \tag{67}$$

This relation is consistent with Eq. (38). Equation (67) suggests a right triangle like that of Fig. 201 with sides $v_r = dr/dt$, $v_\theta = r(d\theta/dt)$, and $v = ds/dt$. It has one angle equal to ψ, as indicated, since

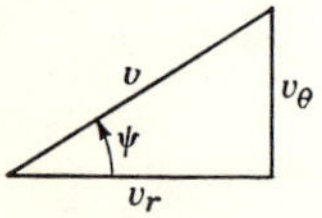

FIG. 201.

$$\frac{v_\theta}{v_r} = \frac{r(d\theta/dt)}{dr/dt} = \frac{r}{dr/d\theta} = \tan\psi. \tag{68}$$

If $r > 0$ and ψ is the angle from $\mathbf{u}_r$ to $\mathbf{t}$, as in Sec. 150, we have

$$v_r = \frac{dr}{dt} = v\cos\psi, \qquad v_\theta = r\frac{d\theta}{dt} = v\sin\psi. \tag{69}$$

This relation is consistent with Eq. (40).

It is easy to recall Eqs. (66) to (69) from the triangle of Fig. 201, whose sides are proportional to those in the triangle of Fig. 198.

EXAMPLE. For the motion $r = c\csc^2 bt$, $\theta = 2bt$, find the radial and transverse components of velocity. For t in the range 0 to π/b, find the speed in the path. Find the equation of the path in polar form. Also show that $v_x = v_r$ and $v_y = -v_\theta$.

Solution: By differentiating the given relations, we find that $dr/dt = -2bc\csc^2 bt\cot bt$, $d\theta/dt = 2b$. Hence from Eq. (66) we have $v_r = dr/dt = -2bc\csc^2 bt\cot bt$, $v_\theta = r(d\theta/dt) = 2bc\csc^2 bt$. These are the required radial and transverse components of velocity.

From Eq. (67) we have $v^2 = v_r^2 + v_\theta^2 = 4b^2c^2\csc^4 bt(\cot^2 bt + 1) = 4b^2c^2\csc^6 bt$. For $0 < t < \pi/b$, bt is between 0 and π so that $\csc bt > 0$. Hence $v = 2bc\csc^3 bt$ is the required speed for t between 0 and π/b.

To find the path, we solve $\theta = 2bt$ for t to obtain $t = \theta/2b$. Hence

$$r = c\csc^2 bt = c\csc^2\frac{\theta}{2} = \frac{c}{\sin^2(\theta/2)} = \frac{2c}{1-\cos\theta}.$$

By Example 4 of Sec. 146, the path is a parabola with focus at the origin and axis parallel to OX.

We may find v_x and v_y from Eq. (65). We have

$$\begin{aligned} v_x &= \frac{dr}{dt}\cos\theta - r\frac{d\theta}{dt}\sin\theta = -2bc\csc^2 bt(\cot bt\cos\theta + \sin\theta) \\ &= -2bc\csc^3 bt(\cos\theta\cos bt + \sin\theta\sin bt) = -2bc\csc^3 bt\cos(\theta - bt) \\ &= -2bc\csc^3 bt\cos bt, \qquad \text{since } \theta = 2bt. \end{aligned}$$

And $\csc bt \cos bt = \cot bt$, so that $v_x = -2bc \csc^2 bt \cot bt = v_r$, as was to be proved. Also,

$$\begin{aligned} v_y &= \frac{dr}{dt}\sin\theta + r\frac{d\theta}{dt}\cos\theta = -2bc\csc^2 bt\,(\cot bt \sin\theta - \cos\theta) \\ &= -2bc\csc^3 bt(\sin\theta\cos bt - \cos\theta\sin bt) = -2bc\csc^3 bt\sin(\theta - bt) \\ &= -2bc\csc^3 bt\sin bt, \text{ since } \theta = 2bt. \end{aligned}$$

And $\csc bt \sin bt = 1$, so that $v_y = -2bc \csc^2 bt = -v_\theta$, as was to be proved.

The fact that $v_x = v_r$ and $v_y = -v_\theta$ is consistent with the property of the parabola proved in Prob. 17 of Exercise 74.

***155. Radial and Transverse Components of Acceleration.** We wish to find the components of the acceleration vector in the direction of increasing r and the direction of increasing θ. Let us first seek new expressions for a_x and a_y. To do this we recall Eq. (64), or

$$\frac{dx}{dt} = \frac{dr}{dt}\cos\theta - r\sin\theta\frac{d\theta}{dt}, \qquad \frac{dy}{dt} = \frac{dr}{dt}\sin\theta + r\cos\theta\frac{d\theta}{dt}. \tag{70}$$

By differentiating these relations with respect to t, we obtain

$$\begin{aligned} \frac{d^2x}{dt^2} &= \frac{d^2r}{dt^2}\cos\theta - \frac{dr}{dt}\sin\theta\frac{d\theta}{dt} - \frac{dr}{dt}\sin\theta\frac{d\theta}{dt} - r\cos\theta\left(\frac{d\theta}{dt}\right)^2 - r\sin\theta\frac{d^2\theta}{dt^2}, \\ \frac{d^2y}{dt^2} &= \frac{d^2r}{dt^2}\sin\theta + \frac{dr}{dt}\cos\theta\frac{d\theta}{dt} + \frac{dr}{dt}\cos\theta\frac{d\theta}{dt} - r\sin\theta\left(\frac{d\theta}{dt}\right)^2 + r\cos\theta\frac{d^2\theta}{dt^2}. \end{aligned} \tag{71}$$

It follows from Eqs. (71) that

$$\begin{aligned} a_x &= \left[\frac{d^2r}{dt^2} - r\left(\frac{d\theta}{dt}\right)^2\right]\cos\theta - \left(r\frac{d^2\theta}{dt^2} + 2\frac{dr}{dt}\frac{d\theta}{dt}\right)\sin\theta, \\ a_y &= \left[\frac{d^2r}{dt^2} - r\left(\frac{d\theta}{dt}\right)^2\right]\sin\theta + \left(r\frac{d^2\theta}{dt^2} + 2\frac{dr}{dt}\frac{d\theta}{dt}\right)\cos\theta. \end{aligned} \tag{72}$$

A comparison of Eq. (72) with Eqs. (61) and (62) shows that the acceleration vector $\mathbf{a} = \left[\frac{d^2r}{dt^2} - r\left(\frac{d\theta}{dt}\right)^2\right]\mathbf{u}_r + \left(r\frac{d^2\theta}{dt^2} + 2\frac{dr}{dt}\frac{d\theta}{dt}\right)\mathbf{u}_\theta$. We shall use a_r to denote the *radial component* of $\mathbf{a}$ in the direction of increasing r, or the direction of $\mathbf{u}_r$. And we shall use a_θ to denote the transverse component of $\mathbf{a}$ in the direction of increasing θ, or the direction of $\mathbf{u}_\theta$. Since $\mathbf{a} = a_r\mathbf{u}_r + a_\theta\mathbf{u}_\theta$, we have

$$a_r = \frac{d^2r}{dt^2} - r\left(\frac{d\theta}{dt}\right)^2, \qquad a_\theta = r\frac{d^2\theta}{dt^2} + 2\frac{dr}{dt}\frac{d\theta}{dt} = \frac{1}{r}\frac{d}{dt}\left(r^2\frac{d\theta}{dt}\right), \tag{73}$$

since the expansion of the last term reduces to the first expression for a_θ.

Since a_r and a_θ are the components of a along perpendicular directions, we have

$$a^2 = a_r^2 + a_\theta^2. \tag{74}$$

EXAMPLE 1. Find the radial and transverse components of acceleration for the motion $r = 2b\cos\omega t$, $\theta = \omega t$. Show that the acceleration vector with initial point at $P = (r,\theta)$ is directed toward the point $C = (b,0)$.

Solution: By differentiating the given relations we find that $dr/dt = -2b\omega\sin\omega t$, $d^2r/dt^2 = -2b\omega^2\cos\omega t$, $d\theta/dt = \omega$, $d^2\theta/dt^2 = 0$. Hence from Eq. (73) we may deduce that

$$a_r = \frac{d^2r}{dt^2} - r\left(\frac{d\theta}{dt}\right)^2 = -2b\omega^2\cos\omega t - (2b\cos\omega t)\omega^2 = -4b\omega^2\cos\omega t.$$

$$a_\theta = r\frac{d^2\theta}{dt^2} + 2\frac{dr}{dt}\frac{d\theta}{dt} = (2b\cos\omega t)\cdot 0 + 2(-2b\omega\sin\omega t)\omega = -4b\omega^2\sin\omega t.$$

Thus $a_r = -4b\omega^2\cos\omega t$ and $a_\theta = -4b\omega^2\sin\omega t$ are the required components.

Since $\omega t = \theta$, $r = 2b \cos \omega t = 2b \cos \theta$. Hence by Eq. (7) with $B = 0$, the path $r = 2b \cos \theta$ is a circle with center at $C = (b,0)$. Draw CE perpendicular to OP, as in Fig. 202. Then angle CPO = angle COP = $\theta = \omega t$ as indicated. And $CP = b$, so that $EP = b \cos \omega t$, $EC = b \sin \omega t$. Hence the vector PC, with initial point at P, has radial and transverse components with respect to OP equal to $-EP$ and $-EC$, or $-b \cos \omega t$, $-b \sin \omega t$. As the values of a_r and a_θ are proportional to these, the acceleration vector drawn from P is directed toward C. In fact $a = 4\omega^2\overline{PC}$.

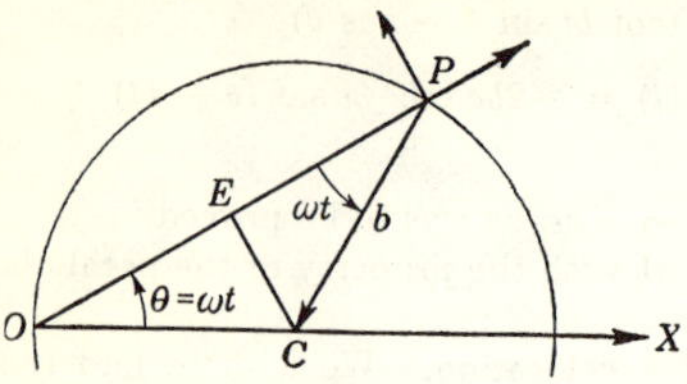

FIG. 202.

EXAMPLE 2. Show that for motion of a particle under a force directed toward O, $r^2(d\theta/dt)$ is constant.

Solution: Since $\mathbf{F}$ is directed toward O and $\mathbf{F} = m\mathbf{a}$, so is $\mathbf{a} = (1/m)\mathbf{F}$. Hence $a_\theta = 0$. And from the second form for a_θ in Eq. (73) we have $\frac{1}{r}\frac{d}{dt}\left(r^2\frac{d\theta}{dt}\right) = 0$. Hence $\frac{d}{dt}\left(r^2\frac{d\theta}{dt}\right) = 0$, so that $r^2\frac{d\theta}{dt} = h$, where h is a constant of integration.

Thus $r^2(d\theta/dt) = h$ and is constant, as was to be proved.

EXAMPLE 3. Find the radial component of acceleration of a particle moving along the curve $r = a \sec^n(\theta/n)$ if $a_\theta = 0$.

Solution: As in Example 2, if $a_\theta = 0$, then $r^2(d\theta/dt) = h$. And $\frac{d\theta}{dt} = \frac{h}{r^2} = \frac{h}{a^2}\cos^{2n}\frac{\theta}{n}$. Hence $\frac{dr}{dt} = \frac{dr}{d\theta}\frac{d\theta}{dt} = a \sec^n\frac{\theta}{n}\tan\frac{\theta}{n}\frac{h}{a^2}\cos^{2n}\frac{\theta}{n}$,
$\frac{dr}{dt} = \frac{h}{a}\tan\frac{\theta}{n}\cos^n\frac{\theta}{n}$. From this we have $\frac{d^2r}{dt^2} = \frac{d}{dt}\left(\frac{dr}{dt}\right) = \frac{d}{d\theta}\left(\frac{dr}{dt}\right)\frac{d\theta}{dt}$
$= \frac{h}{a}\left(\frac{1}{n}\sec^2\frac{\theta}{n}\cos^n\frac{\theta}{n} - \tan\frac{\theta}{n}\cos^{n-1}\frac{\theta}{n}\sin\frac{\theta}{n}\right)\frac{h}{a^2}\cos^{2n}\frac{\theta}{n}$,
$\frac{d^2r}{dt^2} = \frac{h^2}{a^3}\left(\frac{1}{n}\cos^{3n-2}\frac{\theta}{n} - \tan^2\frac{\theta}{n}\cos^{3n}\frac{\theta}{n}\right)$. And $r\left(\frac{d\theta}{dt}\right)^2 = \frac{h^2}{r^3} = \frac{h^2}{a^3}\cos^{3n}\frac{\theta}{n}$.

Hence from Eq. (73) we may deduce that

$a_r = \frac{d^2r}{dt^2} - r\left(\frac{d\theta}{dt}\right)^2 = \frac{h^2}{a^3}\cos^{3n}\frac{\theta}{n}\left(\frac{1}{n}\sec^2\frac{\theta}{n} - \tan^2\frac{\theta}{n} - 1\right)$. Since $\tan^2\frac{\theta}{n} + 1 = \sec^2\frac{\theta}{n}$,
$a_r = -\frac{h^2}{a^3}\frac{n-1}{n}\cos^{3n-2}\frac{\theta}{n} = -\frac{h^2}{a^3}\frac{n-1}{n}\left(\frac{a}{r}\right)^{(3n-2)/n}$, as required.

We may check this by finding a_t and a_n. From the first derivatives, we have by Eq. (66) $v_r = \frac{dr}{dt} = \frac{h}{a}\tan\frac{\theta}{n}\cos^n\frac{\theta}{n}$, $v_\theta = r\frac{d\theta}{dt} = \frac{h}{r} = \frac{h}{a}\cos^n\frac{\theta}{n}$. From Eq. (67) we find $v^2 = v_r^2 + v_\theta^2 = \frac{h^2}{a^2}\cos^{2n}\frac{\theta}{n}\left(\tan^2\frac{\theta}{n} + 1\right) = \frac{h^2}{a^2}\cos^{2n-2}\frac{\theta}{n}$. And from Eq. (68) we find $\tan\psi = \frac{v_\theta}{v_r} = \cot\frac{\theta}{n}$, $\psi = \frac{\pi}{2} - \frac{\theta}{n}$. Since $a_\theta = 0$, in this case $a_r\mathbf{u}_r$ is the entire acceleration, and $a_t = a_r\cos\psi$, $a_n = -a_r\sin\psi$. Thus $a_t = -\frac{h^2}{a^3}\frac{n-1}{n}\cos^{3n-2}\frac{\theta}{n}\sin\frac{\theta}{n}$, $a_n = \frac{h^2}{a^3}\frac{n-1}{n}\cos^{3n-1}\frac{\theta}{n}$. These are in accord with Eq. (60), $a_t = \frac{dv}{dt}$ and $a_n = \frac{v^2}{R}$, and the values $v = \frac{h}{a}\cos^{n-1}\frac{\theta}{n}$, $\frac{d\theta}{dt} = \frac{h}{a^2}\cos^{2n}\frac{\theta}{n}$ found above combined with the value $R = \frac{an}{n-1}\sec^{n+1}\frac{\theta}{n}$ from the example in Sec. 152.

EXERCISE 79

Find the radial and transverse components of velocity for each of the following given motions.

1. $r = 4t$, $\theta = 2t$.
2. $r = 1 - \cos 3t$, $\theta = 3t$.
3. $r = \sin 6t$, $\theta = 2t$.
4. $r = \cos 2t$, $\theta = t$.
5. $r^2 = a^2 \cos 4t$, $\theta = 2t$.
6. $r = a \sin^n t$, $\theta = nt$.
7. $r = b - a \cos t$, $\theta = t$.
8. $r = b - a \sin nt$, $\theta = t$.

For the following motions verify that the speed in the path is constant and that the magnitude of the acceleration vector is constant.

9. $r = a$, $\theta = \omega t$.
10. $r = a \sin \omega t$, $\theta = \omega t$.
11. $r = \sqrt{1 + t^2}$, $\theta = \cot^{-1} t$.
12. $r = \sqrt{p^2 + b^2t^2}$, $\theta = \tan^{-1} \dfrac{bt}{p}$.
13. For the motion $r = ae^{bt}$, $\theta = \omega t$ show that each of the components v_r, v_θ, a_r, a_θ is proportional to r.
14. For the motion $r = ae^{\omega t} + be^{-\omega t}$, $\theta = \omega t$, show that $a_r = 0$.

For the following motions, verify that a_r is proportional to r.

15. $r = A \sin bt + B \cos bt$, $\theta = \omega t$.
16. $r = Ae^{bt} + Be^{-bt}$, $\theta = \omega t$.

Verify that the radial acceleration is proportional to a power of r for each of the following curves, traversed with $a_\theta = 0$ so that $r^2\,(d\theta/dt) = h$ as in Example 2.

17. $r = a \cos \theta$.
18. $r = a \sin \theta$.
19. $r^2 = a^2 \cos 2\theta$.
20. $r^2 = a^2 \sin 2\theta$.
21. $r = a(1 - \cos \theta)$.
22. $r = a \sin^n \dfrac{\theta}{n}$.
23. $r = a \csc^n \dfrac{\theta}{n}$.
24. $r = a \cos^n \dfrac{\theta}{n}$.
25. $r = \dfrac{a}{1 - \cos \theta}$.
26. $r = \dfrac{a}{1 - e \cos \theta}$.

CHAPTER 11

DIFFERENTIALS

Differentials were defined in Sec. 61. And they proved useful in the introduction to integration presented in Chap. 5. Illustrations of the concise expression of geometric relations involving arc length were given in Secs. 62, 132, 139, and 150.

In this chapter we shall give a more systematic treatment of differentials. We first review the definition and the fundamental relation between differentials and derivatives. We translate the more useful rules of differentiation previously derived into differential form and tabulate the results. We show how the use of differentials slightly simplifies differentiation, particularly for composite functions, relations given in parametric form, and implicit functions. We explain how differentials approximate increments when the increments are small. This principle is then applied to the estimation of errors and to a new derivation of Newton's method of approximating roots of equations.

156. Differential of a Function. Let the function $y = f(x)$ have a derivative $f'(x)$ at the point x. Then the *differential* of the independent variable, dx, is any number other than zero selected arbitrarily. It is often taken as fixed throughout the discussion. But sometimes dx is itself regarded as an additional independent variable.

The *differential* of the dependent variable, dy, is then defined by the equation

$$dy = f'(x)dx. \tag{1}$$

In particular, let $y = x$ so that $f(x) = x$ and $f'(x) = 1$. Then $dy = 1 \cdot dx = dx$. This proves that

The differential of a dependent variable equal to x is equal to the differential of the independent variable x.

EXAMPLE. Show that $(x + dx, y + dy)$ is a point on the straight line tangent to the curve $y = f(x)$ at the point (x,y).

Solution: The equation of the tangent line at (x_1,y_1) to $y = f(x)$ is

$$y - y_1 = \left(\frac{dy}{dx}\right)_1 (x - x_1),$$

by Eq. (104) of Sec. 87. Since $dy/dx = f'(x)$, if $x = x_1 + dx$,

$$y - y_1 = f'(x_1)(x - x_1) = f'(x_1)dx.$$

But by Eq. (1) with $x = x_1$, $f'(x_1)dx = dy$, so that

$$y - y_1 = dy, \qquad y = y_1 + dy.$$

Hence $(x_1 + dx,\ y_1 + dy)$ is a point on the tangent to $y = f(x)$ at the point (x_1,y_1). With the subscript 1 omitted, this is the statement which was to be proved.

157. Differential of a Composite Function. When x is the independent variable, Eq. (1) holds by definition. As a consequence of that definition, we shall now prove that Eq. (1) is also valid when x is not the independent variable.

To show this, let a third variable t be taken as the independent variable. Then if $y = f(x)$ and $x = g(t)$, $y = f[g(t)] = F(t)$ is a composite function of t. By Eq. (21) of Sec. 53, if $f(x)$ is a differentiable function of x and $g(t)$ is a differentiable function of t, $y = F(t)$ is a differentiable function of t. And

$$\frac{dy}{dt} = \frac{dy}{dx}\frac{dx}{dt} \qquad \text{or} \qquad F'(t) = f'(x)g'(t). \tag{2}$$

But since t is the independent variable, in accordance with Eq. (1) with x replaced by t, we have

$$dx = g'(t)dt, \qquad dy = F'(t)dt. \tag{3}$$

We may deduce from Eqs. (3) and (2) that

$$dy = F'(t)dt = f'(x)g'(t)dt = f'(x)dx. \tag{4}$$

This shows that $dy = f'(x)dx$, as in Eq. (1).

EXAMPLE 1. If $y = x^2 + 3$ and $x = \cos t$, express dy as a function of x and dx. Also express dx and dy as functions of t and dt.

Solution: From $y = x^2 + 3$, $f'(x) = dy/dx = 2x$. And $dy = f'(x)dx = 2x\,dx$. From $x = \cos t$, $g'(t) = dx/dt = -\sin t$. And $dx = g'(t)dt = -\sin t\,dt$. From $x = \cos t$, $y = x^2 + 3 = \cos^2 t + 3$. And $F'(t) = dy/dt = -2\cos t \sin t$. And $dy = F'(t)dt = -2\cos t \sin t\,dt$. Thus the required expressions are $dy = 2x\,dx$, $dx = -\sin t\,dt$, and $dy = -2\cos t\sin t\,dt$.

Note that the expression $2x\,dx$ for dy is consistent with the other relations, combined with $x = \cos t$. This corresponds to the alternative derivation of $F'(t)$ from $F'(t) = \frac{dy}{dt} = \frac{dy}{dx}\frac{dx}{dt} = f'(x)g'(t) = 2x(-\sin t) = -2\cos t \sin t$.

EXAMPLE 2. Calculate dx and dy in Example 1 when $t = 0$.

Solution: $dx = -\sin t\,dt = 0$ and $dy = -2\cos t\sin t\,dt = 0$, the required values.

When $t = 0$, $x = \cos t = 1$, so that $dy = 2x\,dx$ becomes $0 = 2\cdot 1\cdot 0$. This illustrates that dx may be zero when t is the independent variable.

158. The Derivative as a Quotient. If $dx \neq 0$, it follows from Eq. (1) that

$$dy \div dx = f'(x) = \frac{dy}{dx}. \tag{5}$$

Hence we may think of this last notation as an actual quotient of differentials, or fraction. And when $dx \neq 0$, we may regard the first relation of Eq. (2) as equivalent to ordinary cancellation of common factors from a fraction.

159. Finding the Differential from the Derivative. Since $f'(x) = dy/dx$, we may rewrite Eq. (1) in the form

$$dy = \frac{dy}{dx}\,dx. \tag{6}$$

From the standpoint of Sec. 158 this is an algebraic identity.

To obtain the differential dy of a given function $y = f(x)$, we calculate the derivative dy/dx and multiply by dx.

EXAMPLE 1. Find dy if $y = x^4 - 2x^2 + 3$.
Solution: We have $dy/dx = 4x^3 - 4x = 4x = 4x(x^2 - 1)$. Hence

$$dy = \frac{dy}{dx}\,dx = 4x(x^2 - 1)dx.$$

EXAMPLE 2. Find dy if $y = 4e^{-3x}$.
Solution: We have $dy/dx = -12e^{-3x}$. Hence $dy = (dy/dx)\,dx = -12e^{-3x}\,dx$.
EXAMPLE 3. Find dy if $y = u/v$, where u and v are functions of x.
Solution: We have $\dfrac{dy}{dx} = \dfrac{v(du/dx) - u(dv/dx)}{v^2}$. Hence

$$dy = \frac{dy}{dx}\,dx = \frac{v\,du - u\,dv}{v^2}.$$

160. Formulas for Finding Differentials. As we observed in Sec. 159, the differential of any function may be found by multiplying its derivative with respect to x by dx. Hence any rule for finding derivatives may be converted into a corresponding rule for finding differentials by multiplying both sides of the equation expressing the rule by dx. This was illustrated for the quotient rule in Example 3 of Sec. 159. By this procedure we may deduce from Eqs. (1), (3), (5), (7), (9), (13), (16) of Chap. 3, the Eqs. (7) to (13) which follow.

$$dc = 0. \tag{7}$$
$$d(uv) = u\,dv + v\,du. \tag{8}$$
$$d(u^n) = nu^{n-1}\,du. \tag{9}$$
$$d(cu) = c\,du. \tag{10}$$
$$d(au + bv + cw) = a\,du + b\,dv + c\,dw. \tag{11}$$
$$d(u - v - w + z) = du - dv - dw + dz. \tag{12}$$
$$d(ax^3 + bx^2 + cx + d) = (3ax^2 + 2bx + c)dx. \tag{13}$$

From Eqs. (4) and (5) of Sec. 51, we may deduce Eqs. (14) and (15).

$$d\left(\frac{u}{v}\right) = \frac{v\,du - u\,dv}{v^2}. \tag{14}$$
$$d\left(\frac{u}{c}\right) = \frac{1}{c}\,du. \tag{15}$$

From Eqs. (67), (71), (78) of Chap. 8, we may deduce Eqs. (16) to (18).

$$d(\ln u) = \frac{du}{u}. \tag{16}$$
$$d(e^u) = e^u\, du. \tag{17}$$
$$d(a^u) = (\ln a)a^u\, du. \tag{18}$$

From Eqs. (100) and (139) of Chap. 7, we may deduce Eqs. (19) to (30).

$$d(\sin u) = \cos u\, du. \tag{19}$$
$$d(\cos u) = -\sin u\, du. \tag{20}$$
$$d(\tan u) = \sec^2 u\, du. \tag{21}$$
$$d(\cot u) = -\csc^2 u\, du. \tag{22}$$
$$d(\sec u) = \tan u \sec u\, du. \tag{23}$$
$$d(\csc u) = -\cot u \csc u\, du. \tag{24}$$
$$d(\sin^{-1} u) = \frac{du}{\sqrt{1-u^2}}. \tag{25}$$
$$d(\cos^{-1} u) = -\frac{du}{\sqrt{1-u^2}}. \tag{26}$$
$$d(\tan^{-1} u) = \frac{du}{1+u^2}. \tag{27}$$
$$d(\cot^{-1} u) = -\frac{du}{1+u^2}. \tag{28}$$
$$d(\sec^{-1} u) = \frac{du}{u\sqrt{u^2-1}}. \tag{29}$$
$$d(\csc^{-1} u) = -\frac{du}{u\sqrt{u^2-1}}. \tag{30}$$

161. Finding the Differential Directly. To find the differentials of functions, we may apply the formulas tabulated in Sec. 160 instead of the rules for derivatives from which they were deduced. For composite functions this involves less writing than the method of Sec. 159.

EXAMPLE 1. Find dy if $y = \sin (x^2 - 2x)$.

Solution: We use Eq. (19) with $u = x^2 - 2x$ and then Eq. (13). Thus we find $dy = \cos (x^2 - 2x)\, d(x^2 - 2x) = \cos (x^2 - 2x)\, (2x - 2)dx$. And $dy = 2(x - 1) \cos (x^2 - 2x)\, dx$ is the required differential.

EXAMPLE 2. Find dy if $y = e^{\sin^{-1} 2x}$.

Solution: We use Eq. (17) with $u = \sin^{-1} 2x$, and then Eq. (25) with $u = 2x$. Thus we find $dy = e^{\sin^{-1} 2x}\, d(\sin^{-1} 2x) = e^{\sin^{-1} 2x} \dfrac{2\, dx}{\sqrt{1-4x^2}}$. Hence $dy = \dfrac{2}{\sqrt{1-4x^2}} e^{\sin^{-1} 2x}\, dx$ is the required differential.

EXAMPLE 3. Find dy if $y = \dfrac{x}{\sqrt{4-x^2}}$.

Solution 1: We may use Eq. (14) with $u = x$ and $v = \sqrt{4 - x^2}$, and then Eq. (9) with $n = \frac{1}{2}$ and $u = 4 - x^2$. Thus we find

$$dy = \frac{\sqrt{4 - x^2}\,dx - x\,d\sqrt{4 - x^2}}{4 - x^2} = \frac{\sqrt{4 - x^2} - x\,\frac{1}{2}(4 - x^2)^{-\frac{1}{2}}(-2x)}{4 - x^2}\,dx$$
$$= \frac{4 - x^2 + x^2}{(4 - x^2)^{\frac{3}{2}}}\,dx = \frac{4\,dx}{(4 - x^2)^{\frac{3}{2}}}, \text{ the required differential.}$$

Solution 2: We may use the method of logarithmic differentiation explained in Sec. 122. We have $\ln y = \ln x - \frac{1}{2}\ln(4 - x^2)$. From this and Eq. (16) with $u = y$, $u = x$, and then with $u = (4 - x^2)$. We find that
$\frac{dy}{y} = \frac{dx}{x} - \frac{1}{2}\frac{-2x\,dx}{4 - x^2} = \frac{4 - x^2 + x^2}{x(4 - x^2)}\,dx = \frac{4\,dx}{x(4 - x^2)}$. Hence $dy = \frac{4y\,dx}{x(4 - x^2)} = \frac{x}{\sqrt{4 - x^2}}\frac{4\,dx}{x(4 - x^2)} = \frac{4\,dx}{(4 - x^2)^{\frac{3}{2}}}$.

EXERCISE 80

For each of the following functions, verify the given value of dy by using the method of Sec. 159.

1. $y = x^4 - 4x^3$. $\quad dy = 4x^2(x - 3)dx$.
2. $y = \frac{x}{2} - \frac{1}{2x}$. $\quad dy = \frac{1}{2}\left(1 + \frac{1}{x^2}\right)dx$.
3. $y = (1 - x)^3$. $\quad dy = -3(1 - x)^2\,dx$.
4. $y = \frac{3}{2 - 3x}$. $\quad dy = \frac{9\,dx}{(2 - 3x)^2}$.
5. $y = \sqrt{3 + 4x}$. $\quad dy = \frac{2\,dx}{\sqrt{3 + 4x}}$.
6. $y = ab^x$. $\quad dy = a(\ln b)b^x\,dx$.
7. $y = ae^{-bx}$. $\quad dy = -ab\,e^{-bx}\,dx$.
8. $y = a\sin bx$. $\quad dy = ab\cos bx\,dx$.
9. $y = a\sin^{-1}\frac{x}{b}$. $\quad dy = \frac{a\,dx}{\sqrt{b^2 - x^2}}$.
10. $y = a\tan^{-1}\frac{x}{b}$. $\quad dy = \frac{ab\,dx}{b^2 + x^2}$.

For each of the following functions, verify the given value of dy by using the method of Sec. 161.

11. $y = \cos(x^3 - 3x)$. $\quad dy = -3(x^2 - 1)\sin(x^3 - 3x)dx$.
12. $y = \ln\cos x$. $\quad dy = -\tan x\,dx$.
13. $y = \sin(\ln x)$. $\quad dy = \frac{\cos(\ln x)}{x}\,dx$.
14. $y = \cos^{-1} e^x$. $\quad dy = \frac{-e^x\,dx}{\sqrt{1 - e^{2x}}}$.
15. $y = e^{\tan^{-1}x}$. $\quad dy = e^{\tan^{-1}x}\frac{dx}{1 + x^2}$.
16. $y = \frac{\sin x}{x}$. $\quad dy = \frac{\sin x - x\cos x}{x^2}\,dx$.
17. $y = e^{-2x}\sin 3x$. $\quad dy = (-2\sin 3x + 3\cos 3x)e^{-2x}\sin 3x\,dx$.
18. $y = x\ln x$. $\quad dy = (1 + \ln x)dx$.
19. $y = \ln\tan x$. $\quad dy = \frac{dx}{\sin x\cos x}$.
20. $y = \ln(\tan x + \sec x)$. $\quad dy = \sec x\,dx$.

For each of the following functions, verify the given value of dy by using the method of logarithmic differentiation.

21. $y = \dfrac{\sqrt{a^2 + x^2}}{x}$. $\quad dy = \dfrac{-a^2 y\, dx}{x(a^2 + x^2)} = \dfrac{-a^2\, dx}{x^2(a^2 + x^2)^{\frac{1}{2}}}$.

22. $y = \dfrac{x}{\sqrt{a^2 + x^2}}$. $\quad dy = \dfrac{a^2 y\, dx}{x(a^2 + x^2)} = \dfrac{a^2\, dx}{(a^2 + x^2)^{\frac{3}{2}}}$.

23. $y = \sqrt{\dfrac{a + bx}{a - bx}}$. $\quad dy = \dfrac{aby\, dx}{a^2 - b^2x^2} = \dfrac{ab\, dx}{\sqrt{a + bx}(a - bx)^{\frac{3}{2}}}$.

24. $y = \left(\dfrac{a - bx}{a + bx}\right)^n$. $\quad dy = \dfrac{-2abny\, dx}{a^2 - b^2x^2} = \dfrac{-2abn(a - bx)^{n-1}\, dx}{(a + bx)^{n+1}}$.

25. $y = x^x$. $\quad dy = (1 + \ln x)y\, dx = (1 + \ln x)x^x\, dx$.

162. Implicit Differentiation. The method of implicit differentiation explained in Sec. 56 becomes a little more symmetric when we begin by finding the differential of each term of the equation which defines the function implicitly. This leads to a true equation since, if two functions are identically equal, their differentials are equal.

EXAMPLE 1. Find dy/dx if $2x^2 - 4xy + 3y^2 = 8$.

Solution: From $2x^2 - 4xy + 3y^2 = 8$, by taking the differential of each term, we find $4x\, dx - 4x\, dy - 4y\, dx + 6y\, dy = 0$. Hence $(6y - 4x)dy = (4y - 4x)\, dx$, and $\dfrac{dy}{dx} = \dfrac{2y - 2x}{3y - 2x}$.

EXAMPLE 2. Find dy/dx if $\sin 2x + \cos 2y = 2xy$.

Solution: From $\sin 2x + \cos 2y = 2xy$, by taking the differential of each term, we find $2 \cos 2x\, dx - 2 \sin 2y\, dy = 2x\, dy + 2y\, dx$. Hence $2(x + \sin 2y)dy = 2(\cos 2x - y)dx$, and $\dfrac{dy}{dx} = \dfrac{\cos 2x - y}{x + \sin 2y}$.

163. Parametric Equations. Let a curve be given by equations in parametric form

$$x = g(t), \qquad y = h(t). \tag{31}$$

Then the calculation of the derivatives dy/dx and d^2y/dx^2 may be conveniently carried out by using differentials. Thus from Eq. (31), by taking differentials, we obtain

$$dx = g'(t)dt, \qquad dy = h'(t)dt. \tag{32}$$

It follows from this by division that

$$\frac{dy}{dx} = \frac{h'(t)dt}{g'(t)dt} = \frac{h'(t)}{g'(t)}. \tag{33}$$

This is in agreement with Eq. (18) of Sec. 129.

To find the second derivative, we deduce from Eq. (33) that

$$d\left(\frac{dy}{dx}\right) = \frac{d}{dt}\left[\frac{h'(t)}{g'(t)}\right] dt. \tag{34}$$

From this and the first relation of Eq. (32) we find by division that

$$\frac{d^2y}{dx^2} = \frac{d(dy/dx)}{dx}. \tag{35}$$

This is in agreement with Eq. (26) of Sec. 130.

In practice we do not use Eq. (33) but calculate the differentials in Eqs. (32) and (34) directly, as in the examples.

EXAMPLE 1. Find dy/dx and d^2y/dx^2 if $x = a \cos^n t$, $y = b \sin^n t$.

Solution: From the given relations, we find the differentials $dx = -an \cos^{n-1} t \sin t\, dt$, $dy = bn \sin^{n-1} t \cos t$. Hence $\dfrac{dy}{dx} = \dfrac{bn \sin^{n-1} t \cos t}{-an \cos^{n-1} t \sin t} = -\dfrac{b}{a} \tan^{n-2} t$. The differential of this is $d\left(\dfrac{dy}{dx}\right) = -\dfrac{b}{a}(n-2) \tan^{n-3} t \sec^2 t\, dt$. It follows that $\dfrac{d^2y}{dx^2} = \dfrac{d(dy/dx)}{dx} = \dfrac{-(b/a)(n-2) \tan^{n-3} t \sec^2 t\, dt}{-an \cos^{n-1} t \sin t\, dt} = \dfrac{b(n-2)}{a^2n} \tan^{n-4} t \sec^{n+2} t.$

Thus the required derivatives are $\dfrac{dy}{dx} = -\dfrac{b}{a} \tan^{n-2} t$ and $\dfrac{d^2y}{d^2x} = \dfrac{b(n-2)}{a^2n} \tan^{n-4} t \sec^{n+2} t.$

EXAMPLE 2. Find dy/dx and d^2y/dx^2 if $\dfrac{x^m}{a^m} + \dfrac{y^m}{b^m} = 1$.

Solution: From $\dfrac{x^m}{a^m} + \dfrac{y^m}{b^m} = 1$, by taking differentials, we obtain $\dfrac{mx^{m-1}}{a^m}\, dx + \dfrac{my^{m-1}}{b^m}\, dy = 0$. Hence $\dfrac{dy}{dx} = -\dfrac{b^m x^{m-1}}{a^m y^{m-1}}$. The differential of this is $d\left(\dfrac{dy}{dx}\right) = -\dfrac{b^m}{a^m} \dfrac{y^{m-1}(m-1)x^{m-2}\, dx - x^{m-1}(m-1)y^{m-2}\, dy}{y^{2m-2}}$. It follows that $\dfrac{d^2y}{dx^2} = \dfrac{d(dy/dx)}{dx} = -\dfrac{b^m}{a^m} \dfrac{(m-1)x^{m-2}}{y^m}\left(y - x\dfrac{dy}{dx}\right).$

But since $\dfrac{dy}{dx} = -\dfrac{b^m x^{m-1}}{a^m y^{m-1}}$, $y - x\dfrac{dy}{dx} = y + \dfrac{b^m x^m}{a^m y^{m-1}} = \dfrac{b^m}{y^{m-1}}\left(\dfrac{y^m}{b^m} + \dfrac{x^m}{a^m}\right) = \dfrac{b^m}{y^{m-1}}$, since $\dfrac{y^m}{b^m} + \dfrac{x^m}{a^m} = 1$ by the given relation. Hence the expression found above for $\dfrac{d^2y}{dx^2}$ becomes $\dfrac{d^2y}{dx^2} = -\dfrac{b^m}{a^m} \dfrac{(m-1)x^{m-2}}{y^m} \dfrac{b^m}{y^{m-1}} = -(m-1) \dfrac{b^{2m}x^{m-2}}{a^m y^{2m-1}}$. Thus the required derivatives are $\dfrac{dy}{dx} = -\dfrac{b^m x^{m-1}}{a^m y^{m-1}}$ and $\dfrac{d^2y}{dx^2} = (1-m) \dfrac{b^{2m}x^{m-2}}{a^m y^{2m-1}}$.

The result of eliminating the parameter t from the relations of Example 1 is $(x/a)^{2/n} + (y/b)^{2/n} = 1$. And with $m = 2/n$, $x = a \cos^n t$, $y = b \sin^n t$, the result found for Example 2 reduces to that found for Example 1.

EXERCISE 81

For each given implicit function, find dy/dx.

1. $e^x \sin y = e^{-y} \cos x$.
2. $xy = \tan(x - y)$.
3. $\tan^{-1} \dfrac{y}{x} = \ln(x^2 + y^2)$.
4. $\sin^{-1} \dfrac{y}{x} = \sqrt{y^2 + x^2}$.
5. $\sin(2x - 4y) = \cos(2x + 4y)$.
6. $x \ln y + y \ln x = 1$.
7. $2x^2y^2 + xy^2 = 3$.
8. $e^{x+y} - e^{x-y} = 2$.

Find dy/dx and d^2y/dx^2 in each of the following problems.

9. $x = 3t + 5,\ y = t^2 + 1.$
10. $x = 2t^3,\ y = 4t + 1.$
11. $x = \frac{1}{2}t^2,\ y = t^3.$
12. $x = \frac{1}{3}t^3,\ y = t^2.$
13. $x = 2 \cos t,\ y = 3 \sin t.$
14. $x = 2 \tan t,\ y = 3 \sec t.$
15. $x = \sin t,\ y = \cos 2t.$
16. $x = \cos t,\ y = \sin 2t.$
17. $x = \sec t,\ y = \cos t.$
18. $x = e^{2t},\ y = e^t.$
19. $xy = 4.$
20. $xy^3 = 5.$

164. Significance of the Differential. In Fig. 203 AB is a portion of the graph of the function $y = f(x)$. Through $P(x,y)$, a point on the curve, PR is drawn parallel to OX, and PT is the tangent to the curve at P. Then

$$f'(x) = \tan \phi \qquad (36)$$

is the slope of the curve and its tangent line, so that angle $RPT = \phi$ as indicated. In passing along the curve from P to a second point Q, the increments of x and y are $\Delta x = PR$ and $\Delta y = RQ$.

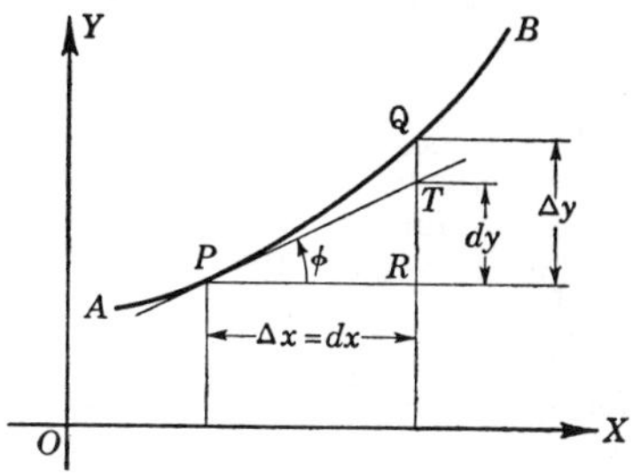

FIG. 203.

Let us take the value $\Delta x = PR$ as dx. Then by Eqs. (1) and (36) we have

$$dy = f'(x)dx = f'(x)\Delta x = (\tan \phi)\overline{PR} = RT. \qquad (37)$$

Thus if we identify the differential dx with the increment Δx, Δy is the increment in y corresponding to $\Delta x = dx$ when we move from $P(x,y)$ to $Q(x + \Delta x,\ y + \Delta y)$ along the curve. But dy is the increment in y corresponding to $\Delta x = dx$ when we move from $P(x,y)$ to $T(x + dx,\ y + dy)$ along the tangent to the curve at P.

A point describing a curve, at the instant of passing through P, is moving in the direction of the tangent to the curve at P. The differential dy is the amount y would change while x is changing to $x + dx$ if the point continued to move in the same direction as at P. In general, the motion does not continue in the same direction, so that Δy for Δx is *not* the same as dy for $dx = \Delta x$. But for a short arc, the change in direction is slight, and so Δy for a *small* Δx is *nearly equal* to dy for $dx = \Delta x$.

Again, consider the motion of a particle along a straight line such that the distance s at time t is

$$s = f(t). \qquad (38)$$

Identify the differential dt with the increment of time Δt. Then Δs is the distance the body moves during the time interval from t to $t + \Delta t = t + dt$. But the differential

$$ds = f'(t)dt = f'(t)\Delta t \qquad (39)$$

is the distance the particle would move in the time interval from t to $t + \Delta t = t + dt$ if the particle continued to move with the speed constant and equal to the same speed $f'(t)$ it had at time t. In general the speed does not remain constant so that Δs for Δt is *not* the same as ds for $dt = \Delta t$. But for a short time interval the change in speed is slight and so Δs for a *small* Δt is *nearly equal* to ds for $dt = \Delta t$.

EXAMPLE 1. Let $y = 3x^2$ and $x = 2$. Calculate Δy for $\Delta x = h$. Also calculate dy for $dx = \Delta x = h$.

Solution: To find Δy, we replace x by $x + \Delta x = x + h$ in $3x^2$. The result is $y + \Delta y = 3(x + h)^2 = 3x^2 + 6hx + 3h^2$. By subtracting $y = 3x^2$, we obtain $\Delta y = 6hx + 3h^2$.

To find dy, we have $dy = 6x\,dx = 6hx$.

Hence the required values for $x = 2$ are found to be $\Delta y = 12h + 3h^2$ and $dy = 12h$.

For this function and any value of x, Δy differs from dy by $3h^2$. And the relative error or $\dfrac{\Delta y - dy}{dy} = \dfrac{3h^2}{6hx} = \dfrac{h}{2x}$. Suppose that we accept any relative error less than 10 per cent or $\frac{1}{10}$ as leading to a fair approximation. Then Δy will approximate dy for h less than $x/5$. For example, let $x = 2$ so that $x/5 = 0.4$, and take $h = 0.2$ which is less than 0.4. Then $\Delta y = 12(0.2) + 3(0.2)^2 = 2.4 + 0.12 = 2.52$, while $dy = 12(0.2) = 2.4$.

EXAMPLE 2. Find Δy for $\Delta x = 0.2$ and dy for $dx = \Delta x = 0.2$ if $x = 2$ and $y = \ln x$.

Solution: We have $x = 2$, $\Delta x = 0.2$, $x + \Delta x = 2.2$. And from tables $\ln 2 = 0.6931$ $\ln\ 2.2 = 0.7885$. It follows that $y + \Delta y = \ln\ (x + x) = \ln\ 2.2 = 0.7885$ and $y = \ln\ x = \ln\ 2 = 0.6931$. By subtraction, we find that $\Delta y = 0.7885 - 0.693$ $= 0.0954$.

From $y = \ln x$ we have $dy = dx/x$. With $x = 2$, $dx = \Delta x = 0.2$, $dy = 0.2/2 = 0.1$. Hence the required values are

$$\Delta y = 0.0954 \qquad \text{and} \qquad dy = 0.1.$$

Here $y = 0.0954$ approximates $dy = 0.1$ to within 5 per cent.

165. Approximation of Small Increments by Differentials. In Sec. 164 we showed that for a small Δx, Δy is nearly equal to dy for $dx = \Delta x$. Hence when only an *approximate value* of the change in a function caused by a small change in the independent variable is desired, we may use the value of the corresponding differential as this is easier to calculate.

EXAMPLE 1. A hemispherical dome 80 ft. in diameter is coated with paint 0.024 in thick. Find the approximate volume of the paint.

Solution: The volume of a sphere of radius r is $\frac{4}{3}\pi r^3$. Hence for a hemisphere of radius r, $V = \frac{2}{3}\pi r^3$. The exact volume of the paint is ΔV for $r = 40$ ft. and $\Delta r = 0.002$ ft. We approximate this by $dV = 2\pi r^2\,dr$ with $r = 40$ and $dr = \Delta r = 0.002$ Thus $dV = 2\pi(40)^2(0.002) = 6.4\pi$. And the required approximate volume of the paint is 6.4π or 20.1 cu. ft.

EXAMPLE 2. Use the differential approximation to the increment to calculate $\tan 44°$ from the exact values $\tan 45° = 1$ and $\sec 45° = \sqrt{2}$.

Solution: Let $y = \tan x$. Then $45° = \pi/4$ radians and $1° = 0.01745$ radian Hence $\tan 44° = \tan\left(\dfrac{\pi}{4} - 0.01745\right) = \tan\dfrac{\pi}{4} + \Delta y = 1 + \Delta y$, where Δy is taken at $x = \pi/4$ for $\Delta x = -0.01745$. We approximate this by $dy = \sec^2 x\, dx$ with $x = \frac{\pi}{4}$ and $dx = \Delta x = -0.01745$. Thus $dy = \sec^2 \frac{\pi}{4}(-0.01745) = 2(-0.01745) = -0.0349$. And the required approximate value of $\tan 44°$ is $1 + dy = 1 - 0.0349 = 0.9651$. To three places this agrees with the tabular value 0.9657.

EXERCISE 82

In each of the following problems there is given a function $y = f(x)$, a value of x, and a value of Δx. Calculate the exact value of Δy as $f(x + \Delta x) - f(x)$. Then calculate the approximate value as the corresponding value of dy at x for $dx = \Delta x$.

1. $y = 2x^4$, $x = 1$, $\Delta x = 0.1$.
2. $y = \sin x$, $x = 0.5$, $\Delta x = 0.02$.
3. $y = e^x$, $x = 3$, $\Delta x = 0.3$.
4. $y = \sqrt{x}$, $x = 16$, $\Delta x = 0.81$.
5. $y = \log_{10} x$, $x = 50$, $\Delta x = 1$.
6. $y = \cot x$, $x = 0.4$, $\Delta x = 0.01$.

Let x increase from a to $a + h$ where h is small compared to a. Use differentials to find an approximate expression for the resulting increment in each of the following geometric quantities.

7. The area of a square of side x.
8. The area of a circle of radius x.
9. The area of an equilateral triangle of side x.
10. The surface of a sphere of radius x.
11. The volume of a sphere of radius x.
12. The lateral surface of a cylinder of radius x and altitude b.
13. The volume of a cylinder of radius x and altitude b.
14. The lateral surface of a cylinder of radius x and altitude $2x$.
15. The volume of a cylinder of radius x and altitude $2x$.

16. For $y = x^n$, verify that $dy = nx^{n-1}\,dx = nh$ where $x = 1$ and $dx = h$. From this deduce the approximate formula for h small compared with unity

$$(1 + h)^n = 1 + nh.$$

Calculate each of the following values approximately by using the formula of Prob. 16.

17. $\sqrt{102} = 10(1 + \frac{2}{100})^{\frac{1}{2}}$.
18. $\frac{1}{49} = \frac{1}{50}(1 - \frac{1}{50})^{-1}$.
19. $\dfrac{1}{\sqrt{50}} = \dfrac{1}{7}\left(1 + \dfrac{1}{49}\right)^{-\frac{1}{2}}$.
20. $\sqrt[3]{122} = 5(1 - \frac{3}{125})^{\frac{1}{3}}$.
21. $\sqrt{80} = 9(1 - \frac{1}{81})^{\frac{1}{2}}$.
22. $\dfrac{1}{\sqrt[3]{1{,}020}} = \dfrac{1}{10}\left(1 + \dfrac{2}{100}\right)^{-\frac{1}{3}}$.

Calculate each of the following values by using the given values and the differential approximation to the increment.

23. $e^{2.02}$, given $e^2 = 7.389$.
24. $\ln 20.2$, given $\ln 20 = 2.996$.
25. $\log_{10} 101$, given $\log_{10} 100 = 2$ and $\log_{10} x = 0.4343 \ln x$.
26. $\sin 61°$, given $\sin 60° = 0.8660$, $\cos 60° = 0.5$.
27. $\cot 61°$, given $\cot 60° = 0.5774$, $\csc 60° = 2$.
28. $\cos 44°$, given $\cos 45° = 0.7071$, $\sin 45° = 0.7071$.

166. Estimation of Errors. A measurement of an observed quantity x is generally slightly in error. We write $x = a \pm h$ to indicate that the error made by taking $x = a$ probably does not greatly exceed h.

If we use the value $x = a$ to compute the value of $y = f(x)$, the computed value will also be in error, so that $y = f(a) \pm k$. It is useful to know the relation of k to h. Since h and k are small and can be estimated only approximately, we may find the relation of k to h by treating h as dx and k as dy for $x = a$. Thus

$$x = a \pm dx, \qquad y = f(a) \pm dy, \qquad dy = f'(a)dx. \tag{40}$$

Let the error in the measurement of a quantity u be du. Then the

$$\text{Relative error in } u = \frac{du}{u}. \tag{41}$$

And the

$$\text{Percentage error in } u = 100\,\frac{du}{u}. \tag{42}$$

Many problems involving relative or percentage errors may be solved easily by the method of logarithmic differentiation, or by a direct application of principles similar to the results of Probs. 7, 12, and 18, of Exercise 83.

EXAMPLE 1. A point source of light is 4 ft. above a horizontal surface. Its strength is 80 candlepower so that at A, directly under the source, the intensity of illumination is 5 foot-candles. And at a point B in the surface such that $AB = x$, the intensity of illumination is $I = \dfrac{320}{(x^2 + 16)^{\frac{3}{2}}}$. Find I for $x = 3$, and estimate the error if $x = 3 \pm 0.02$.

Solution: For $x = 3$, $I = \dfrac{320}{(x^2 + 16)^{\frac{3}{2}}} = \dfrac{320}{125} = 2.56$.

We approximate the error ΔI for $\Delta x = 0.02$ by dI for $x = 3$ and $dx = 0.02$. And from $I = 320(x^2 + 16)^{-\frac{3}{2}}$, we find the differential $dI = 320(-\frac{3}{2})(x^2 + 16)^{-\frac{5}{2}}(2x\,dx) = -960x(x^2 + 16)^{-\frac{5}{2}}\,dx$. For $x = 3$ and $dx = 0.02$, $dI = -960 \cdot 3 \cdot 25^{-\frac{5}{2}}\,(0.02) = \dfrac{-57.6}{3{,}125} = -0.018$. Also for $dx = -0.02$, $dI = +0.018$. Hence the computed value of I required is written as $I = 2.56 \pm 0.018$ foot-candles.

EXAMPLE 2. The acceleration of gravity at a point above the earth's surface x mi. from the center of the earth is $y = gR^2/x^2$. Taking the radius of the earth R as 4,000 mi., estimate how high above the surface one must ascend before the decrease in gravitational acceleration is as much as 1 per cent.

Solution: For a 1 per cent decrease in y, $100\,dy/y = -1$ so that the relative error $dy/y = -0.01$. From $y = gR^2/x^2$, we find $\ln y = \ln gR^2 - 2 \ln x$, and $dy/y = -2\,dx/x$. We wish to find dx for $x = R = 4{,}000$ and $dy/y = -0.01$. Thus $-0.01 = -2\,dx/4{,}000$, and $dx = (4{,}000/2)(0.01) = 20$. Hence the required height is 20 mi.

EXAMPLE 3. Each of three lengths x, y, and z is measured with approximately the same percentage error. From them a fourth proportional, $u = yz/x$, is calculated. If the error in u must not exceed 12 per cent, what percentage error is permissible in the three measured lengths?

Solution: Let p be the desired permissible percentage error. Then the greatest error in u will occur when $100|dx/x| = p$, $100|dy/y| = p$, $100|dz/z| = p$. From

$u = yz/x$, we have $\ln u = \ln y + \ln z - \ln x$. Hence $\dfrac{du}{u} = \dfrac{dy}{y} + \dfrac{dz}{z} - \dfrac{dx}{x}$ and $100\,\dfrac{du}{u} = 100\,\dfrac{dy}{y} + 100\,\dfrac{dz}{z} - 100\,\dfrac{dx}{x}$. The right member is numerically largest when dy and dz are plus and dx is minus, or when dy and dz are minus and dx is plus. In either of these cases $100|du/u| = 100|dy/y| + 100|dz/z| + 100|dx/x|$. And $12 = p + p + p = 3p$, so that $p = 4$. Thus in each measured length the error must not exceed 4 per cent.

EXERCISE 83

Let $x = 4 \pm 0.2$. Calculate y and estimate the error in y for each of the following given functions.

1. $y = x^{\frac{3}{2}}$.
2. $y = \dfrac{1}{\sqrt{x}}$.
3. $y = \sqrt{x^2 + 9}$.
4. $y = \ln(x + 6)$.
5. $y = \sin^{-1}\dfrac{x}{5}$.
6. $y = \tan^{-1}\dfrac{x}{4}$.

7. Let $y = kx^n$. Show that the relative error of y is n times the relative error in x. Also show that the percentage error of y is n times the percentage error in x.

Let $x = 3 \pm 0.06$. For each given quantity, use Prob. 7 to find the relative error and then calculate the quantity and estimate the error.

8. The surface of a cube of side x, $S = 6x^2$. $dS/S = 0.04$, $S = 54 \pm 2.16$.
9. The volume of a cube of side x, $V = x^3$. $dV/V = 0.06$, $V = 27 \pm 1.62$.
10. The surface of a sphere of radius x, $S = 4\pi x^2$.
11. The volume of a sphere of radius x, $V = \frac{4}{3}\pi x^3$.

12. From Prob. 7 deduce that if the percentage error in $y = kx^n$ is not to exceed P, the percentage error in x must not exceed P/n.
13. If the volume of a cubical box is to be 125 ± 5 cu. ft., show that the specification for the length of an inner edge is 5 ± 0.067 ft.
14. The period of a pendulum of length L is $T = 2\pi\sqrt{L/g}$ sec. Show that if a pendulum clock is to gain or lose at most 2 min. in a day, the error in L cannot exceed 0.28 per cent.

An acute angle x is measured to within $1° = 0.01745$ radian. Show that the error in y caused by this will be numerically less than

15. 0.01 for $y = \sin x$ if x exceeds 55.1°.
16. 0.01 for $y = \cos x$ if x is less than 34.9°.
17. 0.03 for $y = \tan x$ if x is less than 40.3°.

18. Show that if $y = ku^m v^n$, the maximum relative error of y is related to the maximum relative errors of u and v by the equation $\left|\dfrac{dy}{y}\right| = \left|m\,\dfrac{du}{u}\right| + \left|n\,\dfrac{dv}{v}\right|$.

Given that measurements u and v in appropriate units are recorded as $u = 30 \pm 1.2$; $v = 40 \pm 2.4$. Use the result of Prob. 18 to find the percentage error in each of the following.

19. $y = 5u^2v$.
20. $y = 8uv^3$.
21. $y = \dfrac{u}{v^2}$.
22. $y = \sqrt{\dfrac{u}{v}}$.

***167. Newton's Method of Approximation.** Consider the problem of solving an equation of the form

$$f(x) = 0. \tag{43}$$

By graphical or other means it is often possible to find a first approximation to the desired root. Under these conditions we may use the differential approximation of Sec. 165 to obtain an improved value in the following manner.

Define the function $y = f(x)$. Use x_1 to denote the known approximate root. And let $y_1 = f(x_1)$ be the corresponding value of y. We wish to find a change in x, Δx which will change y_1 to zero and so make $x = x_1 + \Delta x$ a true root of Eq. (43). That is, Δx satisfies the equation

$$f(x_1 + \Delta x) = y_1 + \Delta y = 0. \tag{44}$$

Let us replace Δx by dx and Δy by dy. Then if $\Delta x = dx$ is small, the equation

$$y_1 + dy = f(x_1) + f'(x_1)dx = 0 \tag{45}$$

will be approximately satisfied by the value dx which makes $x_1 + dx$ a true root. Hence when $f(x)$ and $f'(x)$ are continuous, Eq. (45) will be exactly satisfied by the value dx which makes $x + dx$ an approximation to the root. Thus in general

$$x_2 = x_1 - \frac{f(x_1)}{f'(x_1)} \tag{46}$$

is the improved approximation which we were seeking.

We can continue the process with x_2 in place of x_1 to obtain

$$x_3 = x_2 - \frac{f(x_2)}{f'(x_2)}, \tag{47}$$

and so on until successive approximations agree to the number of figures desired. This process was derived from geometric considerations in Sec. 49, where it was applied to find the roots of polynomials. The method, and many of the remarks made in Sec. 49, apply to more general functions $f(x)$.

EXAMPLE 1. Solve the equation $x^2 + \ln x = 2$.

Solution: The function $x^2 + \ln x$ increases with x, is 1 for $x = 1$ and is 2.39 for $x = \sqrt{2} = 1.414$. Hence by a rough interpolation we find $x_1 = 1.3$ as a first approximation to the root. We set $f(x) = x^2 + \ln x - 2$, so that $f'(x) = 2x + (1/x)$. And $f(1.3) = (1.3)^2 + \ln 1.3 - 2 = 1.69 + 0.2624 - 2 = -0.0476$, $f'(1.3) = 2(1.3) + (1/1.3) = 2.6 + 0.769 = 3.369$. It follows that $-\frac{f(x_1)}{f'(x_1)} = -\frac{-0.0476}{3.37} = 0.014$. And from Eq. (46) we have $x_2 = 1.3 + 0.014 = 1.314$. For this value we have $f(x_2) = (1.314)^2 + \ln 1.314 - 2 = 1.7266 + 0.2731 - 2 = -0.0003$. Hence x_2 is a very good approximation to the root. We could apply the process again, finding $f'(x_2) = 2(1.31) + (1/1.31) = 3.38$. The new correction is $-\frac{f(x_2)}{f'(x_2)} = -\frac{-0.0003}{3.38} = 0.0001$. And from Eq. (47) $x_3 = 1.314 + 0.0001 = 1.3141$. This is the desired root to four decimal places.

EXAMPLE 2. Solve the equation $\tan x = 4x$ for the root near $\pi/2$.

Solution: Since $\tan \pi/2 = \infty$, we multiply the given equation to obtain $\sin x = 4x \cos x$, and set $f(x) = \sin x - 4x \cos x$. Then $f'(x) = \cos x + 4x \sin x - 4 \cos x = 4x \sin x - 3 \cos x$. Hence $\frac{f(x)}{f'(x)} = \frac{\sin x - 4x \cos x}{4x \sin x - 3 \cos x}$. Now let $x_1 = \pi/2 = 1.57$. Then $\frac{f(x_1)}{f'(x_1)} = \frac{1}{2\pi} = 0.16$. And from Eq. (46) we find $x_2 = 1.57 - 0.16 = 1.41$. We may now use $\frac{f(x)}{f'(x)} = \frac{\tan x - 4x}{4x \tan x - 3}$. For $x_2 = 1.41$, $\tan x_2 = 6.165$, and $\frac{f(x_2)}{f'(x_2)} = \frac{0.525}{31.77} = 0.016$. And from Eq. (47) $x_3 = 1.41 - 0.016 = 1.394$. For $x_3 = 1.394$, tan $x_3 = 5.597$ and $\frac{f(x_3)}{f'(x_3)} = \frac{0.021}{28.2} = 0.001$. Hence $x_4 = 1.394 - 0.001 = 1.393$. This is the desired root to three decimal places.

EXERCISE 84

For each given equation find an improved value of the root near the given approximate value x_1.

1. $x = 4 \ln x$, $x_1 = 1.3$.
2. $e^x = 26 - 2x$, $x_1 = 3$.
3. $x + \ln x = 2$, $x_1 = 1.6$.
4. $x + \sin x = 1.5$, $x_1 = 0.8$.
5. $x + \sqrt{x} = 10$, $x_1 = 7$.
6. $\sqrt[3]{x} + \sqrt{2x} = 5.9$, $x_1 = 8$.
7. $x = 1.2 \sin x$, $x_1 = 0.9$.
8. $x = \cos x$, $x_1 = 0.7$.
9. $x = 2 \cos x$, $x_1 = 1$.
10. $x = 3 \sin x$, $x_1 = 2.3$.
11. $x^2 = 2 \sin x$, $x_1 = 1.5$.
12. $2x^2 = \cos x$, $x_1 = 0.6$.
13. $x^2 = -\cot x$, $x_1 = 3$.
14. $x = 4 \sin x$, $x_1 = 1.2$.
15. $\sin x = 1 - x$, $x_1 = 0.5$.
16. $e^{-x} = x$, $x_1 = 0.6$.
17. $e^{-x} = \cos x$, $x = 1.3$.
18. $\tan x = 2e^x$, $x_1 = 1.5$.
19. $e^x = 3 - x$, $x = 0.8$.
20. $\tan x = \ln x$, $x_1 = 4.1$.

CHAPTER 12

THE DEFINITE INTEGRAL

In Chap. 5 we studied integration as the process *inverse to differentiation*. We defined integrals in terms of functions having given differentials, showed how to compute them from this definition, and used them to find certain areas, volumes, and forces due to liquid pressure whose differentials could be set up from geometrical considerations. But in a number of applications of integration to geometry and physics, it is desirable to regard an integral as the *limit* of a *sum*. In fact, this point of view is the basis of the definition of many physical concepts in mathematical terms.

To introduce the new interpretation we first observe that a simple type of area is the limit of a sum. And from this we deduce the expression for such a limit as a definite integral. This relation is known as the *fundamental theorem of the integral calculus*. We illustrate the fundamental theorem in its original and generalized form by problems in plane area,

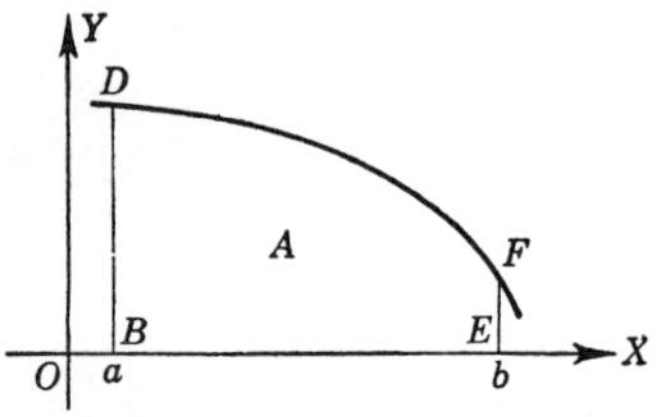

FIG. 204.

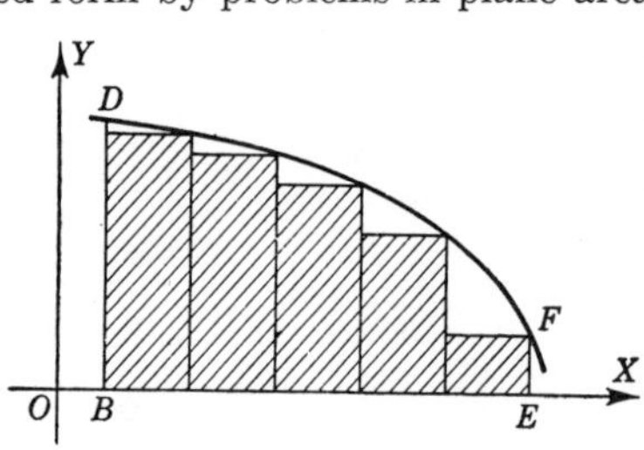

FIG. 205.

arc length, and the area of a surface of revolution. We treat these problems for curves given in rectangular coordinates, polar coordinates, or in parametric form. We also deduce several useful properties of definite integrals. And to prepare for the calculation of areas which may extend to infinity or of integrals of certain discontinuous functions, we define improper integrals and discuss their evaluation.

168. Area as a Limit. As in Sec. 70, consider an area bounded above by the curve $y = f(x)$, below by the x axis, and lying between a left-hand ordinate at $x = a$ and a right-hand ordinate at $x = b$ (Fig. 204).

For this area, make the following construction. Divide the part of the x axis between a and b into n equal intervals. At each point of division draw the ordinate from the x axis to the curve $y = f(x)$. And through the extremity of each ordinate, draw a horizontal line to the left to complete rectangles like those shown in Fig. 205. The sum of the areas of the n rectangles, or total area shaded in the figure, will be an approximation to the area $BDFE$ under the curve. Now let the number of rectangles n increase indefinitely so that the width of each rectangle $(b - a)/n$ approaches zero. Then the *limit* of the sum of the areas of the rectangles will be *equal* to the area under the curve.

We may use a more general construction, as follows. Divide the part of the x axis between a and b into n arbitrary subintervals by *any* $n - 1$ points. Select *some* point in each of the subintervals, and draw the ordinate at this point from the x axis to the curve $y = f(x)$. Using this

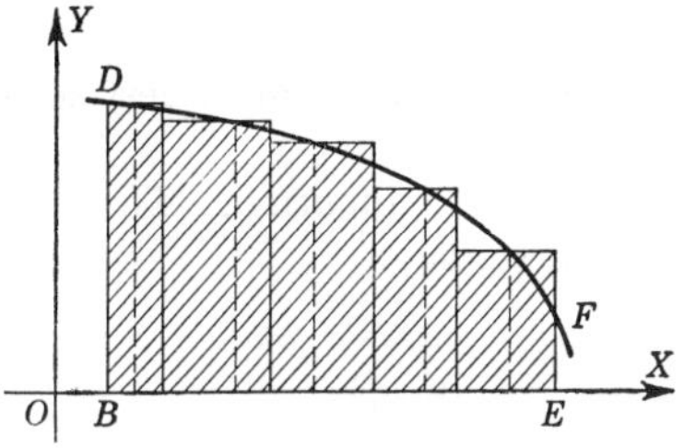

FIG. 206.

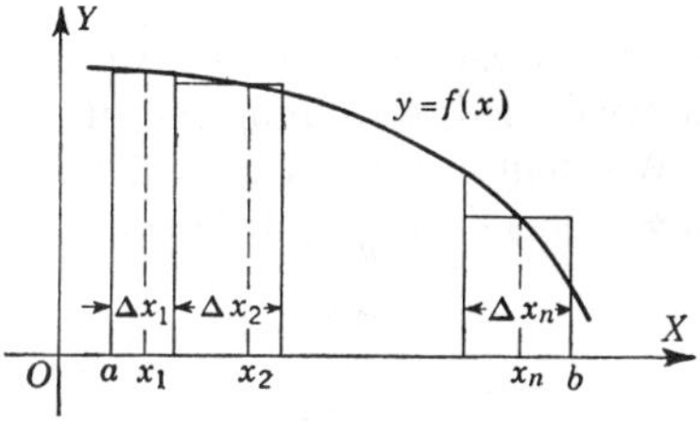

FIG. 207.

ordinate as an altitude and the subinterval as a base, construct a rectangle for each subinterval like those shown in Fig. 206. The sum of the areas of the n rectangles, or total area shaded in the figure, will be an approximation to the area $BDFE$ under the curve. Now let the number of rectangles n increase indefinitely, and the largest subinterval approach zero. Then the *limit* of the sum of the area of the rectangles will again be *equal* to the area under the curve.†

169. Notation for Sums. We shall now introduce some symbols which are useful in discussing the general construction of Sec. 168.

1. Denote the lengths of the arbitrary subintervals of a,b taken in order by

$$\Delta x_1, \Delta x_2, \Delta x_3, \cdots, \Delta x_n. \tag{1}$$

2. As in Fig. 207, denote the abscissas of the points chosen in these subintervals by

$$x_1, x_2, x_3, \cdots, x_n. \tag{2}$$

The corresponding ordinates drawn at these points to the curve $y = f(x)$ will be

$$f(x_1), f(x_2), f(x_3), \cdots, f(x_n). \tag{3}$$

† For these results we rely on geometric intuition. A purely analytic discussion involves reasoning like that used in Probs. 7 to 10 of Exercise 85. This proves that for a given bounded region R sequences have a common limit $A(R)$ with the properties of area mentioned in Sec. 70.

3. Since the area of a rectangle is equal to the product of its base by its altitude, it follows from Eqs. (1) and (3) that the areas of the shaded rectangles of Fig. 206 will be equal to

$$f(x_1)\Delta x_1, f(x_2)\Delta x_2, f(x_3)\Delta x_3, \cdots, f(x_n)\Delta x_n. \tag{4}$$

And the sum of the areas of these rectangles S_n is

$$S_n = f(x_1)\Delta x_1 + f(x_2)\Delta x_2 + f(x_3)\Delta x_3 + \cdots + f(x_n)\Delta x_n. \tag{5}$$

4. Let the area under the curve $BDFE$ be A. Then the last assertion of Sec. 168 may be written

$$A = \lim_{n\to\infty} S_n, \qquad \text{if } \lim_{n\to\infty} d_n = 0, \tag{6}$$

where for each n, d_n is the width of the widest rectangle, or largest subinterval used in forming the sum S_n.

We shall use the symbol Σ (sigma) to indicate summation. Thus for any sequence of quantities $u_1,u_2,u_3, \cdots ,u_n$, we have

$$\sum_{i=1}^{n} u_i = u_1 + u_2 + u_3 + \cdots + u_n, \tag{7}$$

where the left member is read "the sum of u_i from $i = 1$ to $i = n$" or "sigma $i = 1$ to n of u_i."

In this notation, Eq. (5) is represented by

$$S_n = \sum_{i=1}^{n} f(x_i)\Delta x_i. \tag{8}$$

And Eq. (6) may be written

$$A = \lim_{n\to\infty} \sum_{i=1}^{n} f(x_i)\Delta x_i, \qquad \text{if } \lim_{n\to\infty} d_n = 0. \tag{9}$$

170. The Fundamental Theorem of the Integral Calculus. By Eq. (30) of Sec. 71, the area under the curve $BDFE$, or A, is

$$A = \int_a^b f(x)dx = [F(x)]_a^b = F(b) - F(a), \tag{10}$$

where $F(x)$ is any particular indefinite integral of $f(x)$, or function having $f(x)$ as its derivative, so that $f(x) = F'(x)$.

A comparison of Eqs. (10) and (9) shows that

$$\lim_{n\to\infty} \sum_{i=1}^{n} f(x_i)\Delta x_i = \int_a^b f(x)dx, \qquad \text{if } \lim_{n\to\infty} d_n = 0. \tag{11}$$

In our derivation of this result, we have taken $f(x)$ as positive in the interval a,b. But if we reckon areas of rectangles or areas between curves and the x axis as negative when the ordinates extend below the x axis, an essentially similar discussion establishes the result for $f(x)$ unrestricted as to algebraic sign.

Since no geometric concepts appear explicitly in Eq. (11), this equation expresses a purely analytic result. We may express the fundamental theorem in words as follows:

Let $f(x)$ be any continuous function of x throughout the interval $x = a$ to $x = b$. Divide this interval a,b into n subintervals of respective lengths $\Delta x_1, \Delta x_2, \Delta x_3, \cdots, \Delta x_n$ and choose points $x_1, x_2, x_3, \cdots, x_n$ one in each subinterval. Consider the sum

$$S_n = f(x_1)\Delta x_1 + f(x_2)\Delta x_2 + f(x_3)\Delta x_3 + \cdots + f(x_n)\Delta x_n$$
$$= \sum_{i=1}^{n} f(x_i)\Delta x_i. \tag{12}$$

Let d_n denote the longest subinterval Δx_i used in forming the sum S_n. And take a sequence of subdivisions such that d_n approaches zero when n increases indefinitely.

Then as n becomes infinite, the sum S_n approaches a limit. And this limit is the definite integral $\int_a^b f(x)dx$.

Expressed as a formula, the *fundamental theorem* is

$$\lim_{n\to\infty} \sum_{i=1}^{n} f(x_i)\Delta x_i = \int_a^b f(x)dx. \tag{13}$$

This equation explains the origin of the notation for a definite integral. The sign of integration is a modified S, and the replacement of the Greek letters Σ and Δ by the corresponding Latin letters S and d indicates that we have performed a limiting process. This is analogous to replacing Δ by d to indicate that dy/dx is obtained from $\Delta y/\Delta x$ by a limiting process.

If any magnitude can be expressed as the limit of a sum of the form (12), the fundamental theorem, or Eq. (13), enables us to express it as a definite integral. This integral may then be calculated from an indefinite integral as indicated in Eq. (10) and as explained in detail in Sec. 72.

171. An Example. To clarify the somewhat abstract discussion of Secs. 168 to 170, we shall describe one specific sequence of sums S_n. Let the function $f(x) = x/2$, and consider it on the interval from $x = 0$ to $x = 4$. If we take all the Δx_i equal, each Δx_i will equal $4/n$. We take the x_i as the right-hand extremities of the intervals, so that

$$x_i = \frac{4i}{n}, \qquad f(x_i) = \frac{x_i}{2} = \frac{2i}{n}. \tag{14}$$

Then for the case under consideration, the sum (12) is

$$S_n = \sum_{i=1}^{n} f(x_i)\Delta x_i = \sum_{i=1}^{n} \frac{2i}{n} \cdot \frac{4}{n} = \frac{8}{n^2} \sum_{i=1}^{n} i$$
$$= \frac{8}{n^2}(1 + 2 + 3 + \cdots + n). \tag{15}$$

But the sum of an arithmetic progression is given by

$$a + (a + d) + (a + 2d) + \cdots + [a + (n - 1)d] = \frac{n}{2}[2a + (n - 1)d]. \tag{16}$$

Hence when $a = 1$ and $d = 1$, we have

$$1 + 2 + 3 + \cdots + n = \frac{n(n + 1)}{2}. \tag{17}$$

This shows that the sum S_n of Eq. (15) is

$$S_n = \frac{8}{n^2}\frac{n(n + 1)}{2} = \frac{4(n + 1)}{n}. \tag{18}$$

By the principle of the leading term of Sec. 13, it follows that

$$\lim_{n\to\infty} S_n = \lim_{n\to\infty} \frac{4(n + 1)}{n} = \lim_{n\to\infty} \frac{4n}{n} = 4. \tag{19}$$

The fundamental theorem asserts that this limit is equal to the definite integral

$$\int_0^4 f(x)dx = \int_0^4 \frac{x}{2}\,dx = \left[\frac{x^2}{4}\right]_0^4 = \frac{1}{4}(16 - 0) = 4. \tag{20}$$

The graph of $y = f(x) = x/2$ is the straight line shown in Fig. 208. The sum of the shaded rectangles is equal to the S_n of Eq. (15) with $n = 4$. The limit is the area of the right triangle BFE, or $\frac{1}{2}BE \cdot EF = \frac{1}{2}4 \cdot 2 = 4$, which checks with Eqs. (19) and (20).

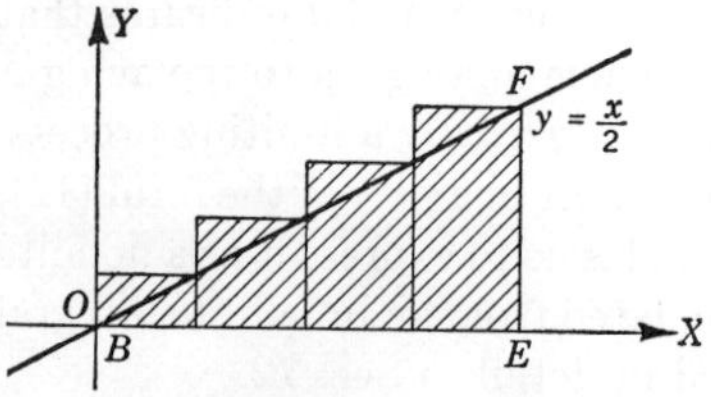

Fig. 208.

This illustrates how *exact* results are obtained as the limit of approximations.

EXERCISE 85

As in the example of the text, consider $f(x) = x/2$ on the interval 0,4 and take each subinterval equal to $4/n$. Show that

1. If x_i is the left-hand extremity, $x_i = \frac{4}{n}(i - 1)$ and $S_n = \frac{4(n - 1)}{n}$, which approaches 4 as $n \to \infty$.

2. If x_i is the mid-point, $x_i = \frac{4i - 2}{n}$ and $S_n = 4$, which approaches 4 as $n \to \infty$.
3. If x_i is any point inside the subinterval, the sum S_n will be less than that of Eq. (18) and greater than that of Prob. 1. Deduce that in any such case $S_n = \frac{4(n + \theta)}{n}$, with $-1 < \theta < 1$. In general, θ will vary with n, but since $|4 - S_n| = |4\theta/n| < 4/n$, it follows from the definition of Sec. 7 that $S_n \to 4$ as $n \to \infty$.
4. Show from a diagram that the S_n of Eq. (15) exceeds its limit by an amount equal to the area of n triangles each of base $\frac{4}{n}$ and of altitude $\frac{2}{n}$, or of total area $n\frac{1}{2}\frac{4}{n}\frac{2}{n} = \frac{4}{n}$. Deduce that $S_n = 4 + \frac{4}{n}$, which checks Eq. (18).
5. Show from a diagram that the S_n of Prob. 1 is less than its limit by an amount equal to the area of n triangles each of base $\frac{4}{n}$ and of altitude $\frac{2}{n}$, or of total area $n\frac{1}{2}\frac{4}{n}\frac{2}{n} = \frac{4}{n}$. Deduce that $S_n = 4 - \frac{4}{n}$, which checks Prob. 1.
6. Show from a diagram that, for any arc, the selection of an x_i such that $f(x_i)$ is halfway between the ordinates at the extremities of the subinterval makes the

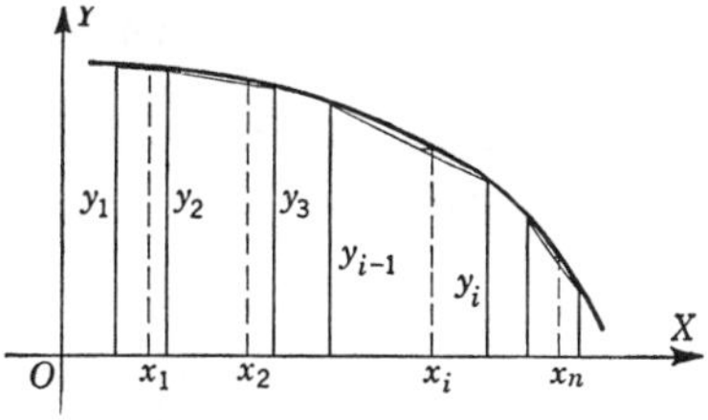

FIG. 209.

approximation S_n equal to that obtained by using inscribed polygons, or trapezoids instead of rectangles. HINT: The area of a trapezoid in Fig. 209 is $\frac{\Delta x_i(y_{i-1} + y_i)}{2}$. This equals $f(x_i)\Delta x_i$ if $f(x_i) = \frac{y_{i-1} + y_i}{2}$.

Show from a diagram that, for any increasing function $f(x)$, the selection of each x_i as

7. The left-hand extremity makes S_n less than its limit.
8. The right-hand extremity makes S_n greater than its limit.
9. A suitable intermediate value in each subinterval makes S_n equal to its limit.
10. If we take all the Δx_i equal, each Δx_i will equal $(b - a)/n$. Show from a diagram that in this case the sum S_n of Prob. 8 will exceed that of Prob. 7 by an amount equal to the area of a rectangle of base $(b - a)/n$ and altitude $f(b) - f(a)$, or equal to $\frac{1}{n}(b - a)[f(b) - f(a)]$. Thus either sum S_n will differ from its limit by an amount not exceeding this numerically.
11. Verify that Prob. 2 is a special case of Prob. 6 and also of Prob. 9.

172. Applications to Areas. In elementary geometry, areas bounded by curved arcs are often defined in terms of limits of areas of inscribed polygons. However, as Prob. 6 of Exercise 85 suggests, this definition is equivalent to that based on approximating rectangles.

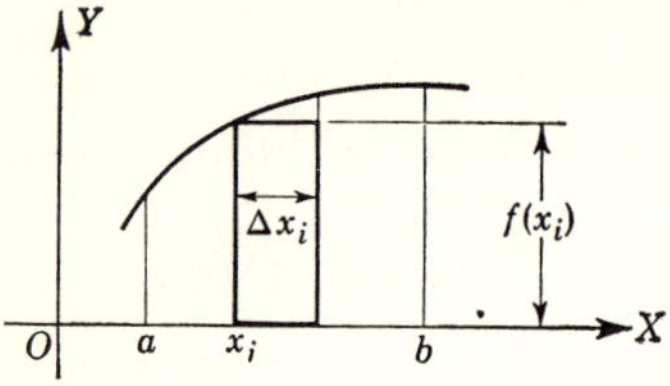

FIG. 210.

For plane areas of given type, it is usually possible to deduce from a figure a limit of a sum of rectangles equivalent to the area. From this the expression for the area as an integral may be obtained by applying the fundamental theorem.

Thus, consider the area bounded above by the curve $y = f(x)$ and below by the x axis, lying between a left-hand ordinate at $x = a$ and a right-hand ordinate at $x = b$. From the construction of Fig. 210 we see that

$$A = \lim_{n\to\infty} \sum_{i=1}^{n} f(x_i)\Delta x_i = \int_a^b f(x)dx. \tag{21}$$

This agrees with Eq. (35) of Sec. 73 where the recognition of $y\,dx$ as the element of integration corresponds to the use here of $f(x_i)\Delta x_i$ as the typical term of the sum.

Next consider the area bounded above by the curve $y = f_2(x)$ and below by the curve $y = f_1(x)$, lying between a left-hand ordinate at $x = a$ and

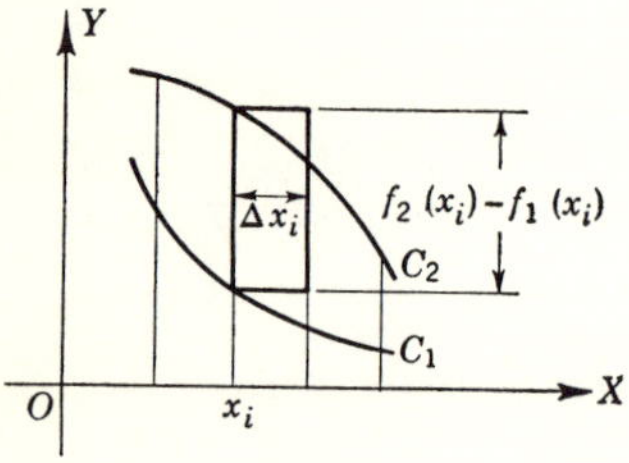

FIG. 211.

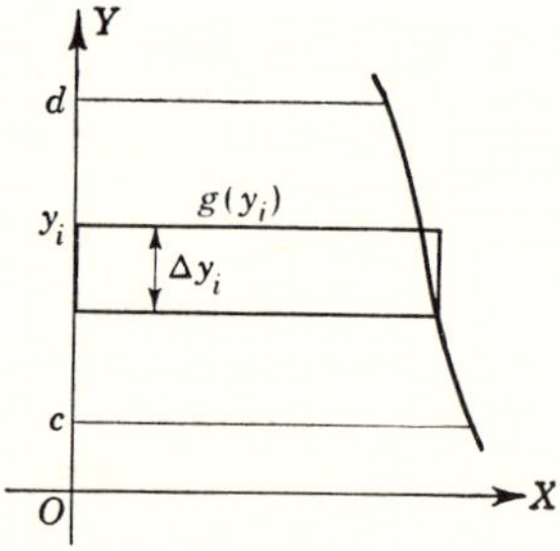

FIG. 212.

a right-hand ordinate at $x = b$. From the construction of Fig. 211, we see that

$$A = \lim_{n\to\infty} \sum_{i=1}^{n} [f_2(x_i) - f_1(x_i)]\Delta x_i = \int_a^b [f_2(x) - f_1(x)]dx. \tag{22}$$

This agrees with Eq. (33) of Sec. 73.

For an area bounded on the right by $x = g(y)$ and on the left by the y axis and lying above the line $y = c$ and below the line $y = d$, we may infer from the construction of Fig. 212 that

$$A = \lim_{n \to \infty} \sum_{i=1}^{n} g(y_i)\Delta y_i = \int_c^d g(y)dy. \tag{23}$$

This agrees with Eq. (38) of Sec. 73. And for an area bounded on the right by $x = g_2(y)$ and on the left by $x = g_1(y)$, lying above the line $y = c$ and below the line $y = d$, we may infer from the construction of Fig. 213 that

$$A = \lim_{n \to \infty} \sum_{i=1}^{n} [g_2(y_i) - g_1(y_i)]\Delta y_i = \int_c^d [g_2(y) - g_1(y)]dy. \tag{24}$$

This agrees with Eq. (36) of Sec. 73.

The formulas of this section give a clearer insight of the relation of the integral to the area but are chiefly of theoretic interest. The practical calculation of such areas is best performed as in Sec. 73, using the differential element of area to recall the appropriate formula.

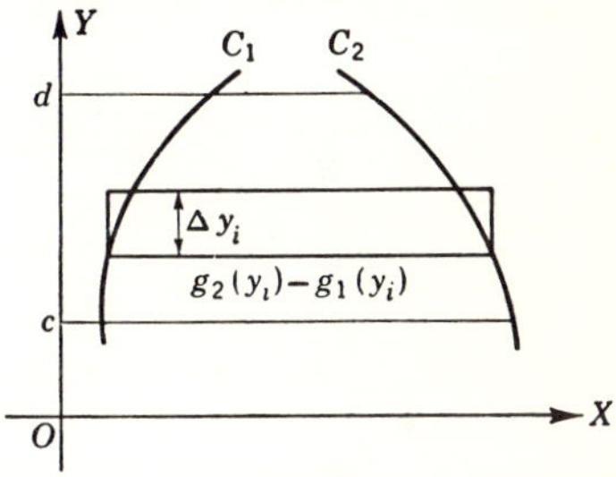

FIG. 213.

173. Integration Formulas. Although a more systematic study of integration will be made in Chap. 13, it will be convenient to collect here a few formulas for indefinite integrals for use in the examples of this chapter. The symbols c_1, c_2, and n denote constants. And C is the constant of integration of Sec. 65. The letters u and v represent functions of x, so that $du = (du/dx)dx$.

First we rewrite Eq. (13) of Sec. 67,

$$\int (c_1 u + c_2 v)dx = c_1 \int u\, dx + c_2 \int v\, dx + C. \tag{25}$$

This is the linearity property, and a similar result holds for any number of functions.

Similar to Eqs. (14) and (17) of Sec. 67, we have the formulas

$$\int 1\, du = \int du = u + C, \tag{26}$$

$$\int u\, du = \frac{u^2}{2} + C, \tag{27}$$

$$\int u^n\, du = \frac{u^{n+1}}{n+1} + C, \qquad \text{if } n \neq -1. \tag{28}$$

These may be verified by noting that

$$d(u) = du, \qquad d\left(\frac{u^2}{2}\right) = u\,du, \qquad d\left(\frac{u^{n+1}}{n+1}\right) = u^n\,du. \tag{29}$$

And for the excepted minus first power, we have

$$\int u^{-1}\,du = \int \frac{du}{u} = \ln u + C. \tag{30}$$

This follows from Sec. 160 where we showed that $d \ln u = du/u$.

For the integral of an exponential, we have

$$\int e^u\,du = e^u + C. \tag{31}$$

This follows from Sec. 160 where we showed that $de^u = e^u\,du$.

Finally for the integral of a sine or a cosine, we have

$$\int \sin u\,du = -\cos u + C, \tag{32}$$

$$\int \cos u\,du = \sin u + C. \tag{33}$$

These follow from Sec. 160 where we showed that $d \cos u = -\sin u\,du$ and $d \sin u = \cos u\,du$.

Example 1. Find $\int \sin (5x + 3)dx$.

Solution: Let $5x + 3 = u$. Then† $5\,dx = du$ and $dx = du/5$. Hence from Eqs. (25) and (32) we have

$$\int \sin (5x + 3)dx = \int \sin u \frac{du}{5} = \frac{1}{5}\int \sin u\,du = \frac{1}{5}(-\cos u + C_1).$$

Since C_1 is any constant, let $\frac{1}{5}C_1 = C$. And since $u = 5x + 3$, $\int \sin (5x + 3)dx = -\frac{1}{5}\cos (5x + 3) + C$.

Example 2. Find $\displaystyle\int \frac{x\,dx}{1 + x^2}$.

Solution: Let $1 + x^2 = u$. Then $2x\,dx = du$ and $x\,dx = du/2$. Hence from Eqs. (25) and (30) we have

$$\int \frac{x\,dx}{1 + x^2} = \int \frac{du/2}{u} = \frac{1}{2}\int \frac{du}{u} = \frac{1}{2}(\ln u + C_1).$$

Let $\frac{1}{2}C_1 = C$. And since $u = 1 + x^2$,

$$\int \frac{x\,dx}{1 + x^2} = \frac{1}{2}\ln (1 + x^2) + C.$$

Example 3. Find the area between one arch of $y = \sin (5x + 3)$ and the x axis.

Solution: Since $5x + 3 = 0$ when $x = -\frac{3}{5}$, and $5x + 3 = \pi$ when $x = (\pi - 3)/5$, these are the ends of one arch lying above the x axis. Hence the desired area is

$$\int_{-\frac{3}{5}}^{(\pi-3)/5} \sin (5x + 3)dx = -\frac{1}{5}\Big[\cos (5x + 3)\Big]_{-\frac{3}{5}}^{(\pi-3)/5} = -\frac{1}{5}(\cos \pi - \cos 0)$$
$$= -\tfrac{1}{5}(-1 - 1) = \tfrac{2}{5}.$$

† By Eq. (2) of Sec. 157. See also Eq. (56) of Sec. 178.

The integral may be set up either from Eq. (21) or by noting that $dA = y\,dx$. And the indefinite integral was found from Example 1.

EXERCISE 86

Verify each of the following integrals.

1. $\int \sin ax = -\frac{1}{a}\cos ax + C.$
2. $\int \cos ax = \frac{1}{a}\sin ax + C.$
3. $\int \tan x = -\int \frac{d(\cos x)}{\cos x} = -\ln\cos x + C.$
4. $\int e^{ax}dx = \frac{1}{a}e^{ax} + C.$
5. $\int b^x dx = \int e^{x\ln b}\,dx = \frac{b^x}{\ln b} + C.$
6. $\int \sin^2 ax\,dx = \int \frac{1-\cos 2ax}{2}\,dx = \frac{1}{2}\left(x - \frac{\sin 2ax}{2a}\right) + C.$
7. $\int \cos^2 ax\,dx = \int \frac{1+\cos 2ax}{2}\,dx = \frac{1}{2}\left(x + \frac{\sin 2ax}{2a}\right) + C.$

For each given curve, y is positive for all values of x between the two given values. Find the area above the x axis and below the curve which lies between the two given ordinates in each case.

8. $y = (3x+2)^4$, $x = -1$, $x = 0$.
9. $y = \sqrt{2x-5}$, $x = 3$, $x = 7$.
10. $y = x + \frac{1}{x}$, $x = 2$, $x = 4$.
11. $y = 4e^{2x}$, $x = 0$, $x = 1$.

Find the area in the first quadrant bounded by the x axis, the y axis, and the first arch of the curve.

12. $y = \cos x$.
13. $x = 4\cos 2y$.

Find the area bounded on the right by the line $x = 1$, bounded above by the first curve and below by the second curve.

14. $y = e^x$, $y = e^{-x}$.
15. $y = \sin x$, $y = \cos x - 1$.
16. The volume V of Sec. 74 may be approximated by a sum of slabs, of which a typical one is a cylinder (or prism) with base $A(x_i)$ and height Δx_i so that its volume is $A(x_i)\Delta x_i$. Express V as the limit of a sum, and deduce from the fundamental theorem that $V = \int_a^b A(x)dx$, which is Eq. (39) of Sec. 74.
17. Use the results of Probs. 16 and 6 to find the volume generated by revolving the first arch of the sine curve $y = \sin x$ about the x axis.

174. Area in Polar Coordinates. We sometimes wish to find the area bounded by a curve and two of its radius vectors.

Let the equation of the curve in polar coordinates be

$$r = f(\theta). \tag{34}$$

And let the two radius vectors (Fig. 214) be OD corresponding to $\theta = \alpha$, and OE corresponding to $\theta = \beta$.

Areas of this type are most conveniently approximated by sums of circular sectors. Accordingly we recall some of the properties of such

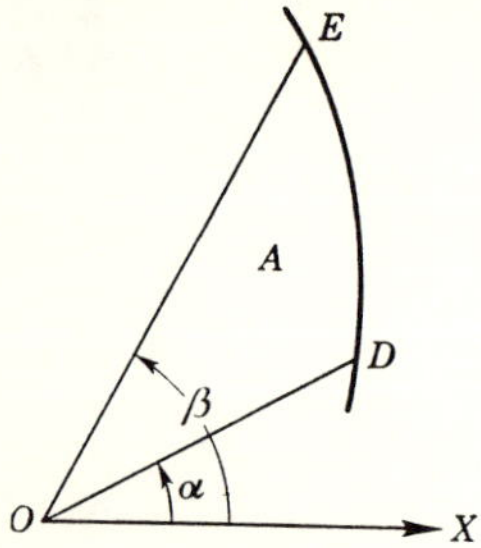

FIG. 214.

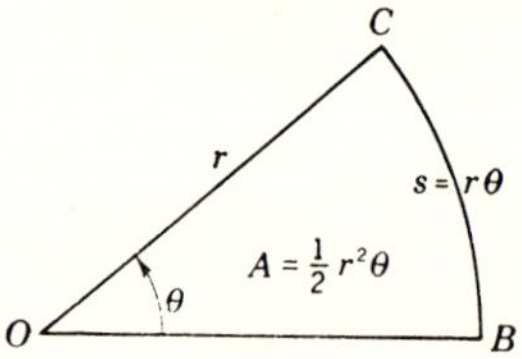

FIG. 215.

figures from plane geometry. A sector of a circle is a figure bounded by an arc of the circle and two of its radii. Let the radius be r. And let θ be the number of radians in the central angle BOC (Fig. 215). Then the length s of arc BC is $s = r\theta$. And, as in Eq. (7) of Sec. 90, the area A of the sector BOC is equivalent to that of a triangle with base s and altitude r, so that

$$A = \tfrac{1}{2}sr = \tfrac{1}{2}r^2\theta. \tag{35}$$

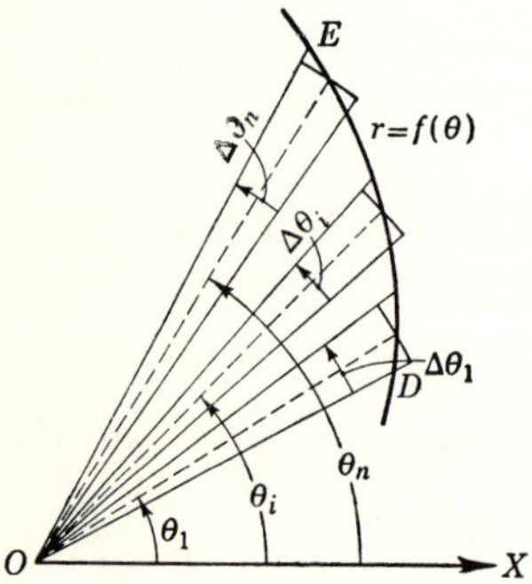

FIG. 216.

To find the area DOE of Fig. 214, we divide the angle DOE into n parts $\Delta\theta_1$, $\Delta\theta_2$, $\Delta\theta_3$, $\cdots$, $\Delta\theta_n$ by radius vectors as indicated in Fig. 216. In each interval $\Delta\theta_i$, we choose a value θ_i and construct a circular sector with radius $r_i = f(\theta_i)$ and central angle $\Delta\theta_i$. By Eq. (35), the area of the ith sector is $\frac{1}{2}r_i^2\,\Delta\theta_i$, so that the sum of the areas of all the sectors is

$$\frac{1}{2}r_1^2\,\Delta\theta_1 + \frac{1}{2}r_2^2\,\Delta\theta_2 + \frac{1}{2}r_3^2\,\Delta\theta_3 + \cdots + \frac{1}{2}r_n^2\,\Delta\theta_n$$

$$= \frac{1}{2}\sum_{i=1}^{n} r_i^2\,\Delta\theta_i = \frac{1}{2}\sum_{i=1}^{n}[f(\theta_i)]^2\,\Delta\theta_i. \tag{36}$$

If n, the number of subdivisions, increases indefinitely and the largest subdivision approaches zero, the sum of the sectors approaches the required area DOE as a limit.† Hence from Eq. (36) and the fundamental theorem, the area $A = DOE$ is

† We again rely on geometric intuition. See footnote in Sec. 168.

$$A = \frac{1}{2} \lim_{n \to \infty} \sum_{i=1}^{n} [f(\theta_i)]^2 \, \Delta\theta_i = \frac{1}{2} \int_\alpha^\beta [f(\theta)]^2 \, d\theta. \tag{37}$$

In other notation, this relation is

$$A = \frac{1}{2} \lim_{n \to \infty} \sum_{i=1}^{n} r_i^2 \, \Delta\theta_i = \frac{1}{2} \int_\alpha^\beta r^2 \, d\theta. \tag{38}$$

It follows that the sectorial area swept out by the radius vector of the curve $r = f(\theta)$ as θ increases from α to β is

$$A = \frac{1}{2} \int_\alpha^\beta r^2 \, d\theta = \frac{1}{2} \int_\alpha^\beta [f(\theta)]^2 \, d\theta. \tag{39}$$

The element of sectorial area is a circular sector with radius r and central angle $d\theta$. Hence by Eq. (35) its area is $\frac{1}{2}r^2 \, d\theta$. These facts are helpful in recalling the first integral in Eq. (39). But to evaluate this integral, we must substitute for r its value in terms of θ as found from the equation of the curve.

EXAMPLE. Find the area of one loop of the curve $r = \cos 3\theta$.

Solution: The curve is symmetrical about OX, and one-half the loop is generated when 3θ increases from 0 to $\pi/2$, or when θ increases from 0 to $\pi/6$, since this makes $\cos 3\theta$ decrease from one to zero. From Eq. (39) we have for the half loop $A = \frac{1}{2} \int_0^{\pi/6} \cos^2 3\theta \, d\theta$. By using the method or result of Prob. 7 of Exercise 86, we find that

$$\frac{1}{2} \int^{\pi/6} \cos^2 3\theta \, d\theta = \frac{1}{4} \left[\theta + \frac{\sin 6\theta}{2} \right]_0^{\pi/6} = \frac{1}{4} \left(\frac{\pi}{6} - 0 \right) = \frac{\pi}{24}.$$

Hence the required area of the whole loop is $2(\pi/24) = \pi/12$.

EXERCISE 87

Find the area bounded by the lines $\theta = 0$, $\theta = \pi/2$ and the spiral.

1. $r = e^\theta$.
2. $r = \theta$.
3. $r = \theta^2$.
4. $r = \sqrt{\theta}$.

Find the area of one loop of each of the following curves.

5. $r = \cos 5\theta$.
6. $r = \sin 5\theta$.
7. $r^2 = \cos 2\theta$.
8. $r^2 = \sin 3\theta$.

Find the total area enclosed by each of the following curves.

9. $r = 1 - \cos \theta$.
10. $r = 3 + \sin 3\theta$.

For the curve $r = 1 + 2 \cos 2\theta$, calculate

11. The area of one of the two small loops.
12. The area of one of the two large loops.
13. The total area enclosed by the curve.

For the curve $r = 1 + 2 \sin \theta$, calculate

14. The area of the small loop.
15. The area of the large loop.
16. The area between the two loops.

***175. Properties of the Definite Integral.** We shall now discuss several important properties of the definite integral which are of use in applications.

If $F(x)$ is any function having the continuous function $f(x)$ as its derivative, Eq. (10) states that

$$\int_a^b f(x)dx = F(b) - F(a). \tag{40}$$

Consequently we have

$$\int_b^a f(x)dx = F(a) - F(b) = -[F(b) - F(a)] = -\int_a^b f(x)dx. \tag{41}$$

It follows that

$$\int_a^b f(x)dx = -\int^a f(x)dx. \tag{42}$$

In discussing the fundamental theorem we assumed that a was less than b, and considered all the Δx_i positive. When a is greater than b, we must consider the increments Δx_i as negative. This reverses the sign of the sum in Eq. (13) and so makes that result consistent with Eq. (42).

For the sake of completeness, we note from Eq. (40) that

$$\int_a^a f(x)dx = 0. \tag{43}$$

The result of Eq. (42) may be formulated as a theorem:

Interchanging the limits of a definite integral is equivalent to changing its algebraic sign.

We may also deduce from Eq. (40) that

$$\int_a^k f(x)dx = F(k) - F(a) \quad \text{and} \quad \int_k^b f(x)dx = F(b) - F(k). \tag{44}$$

It follows from these relations that

$$\int_a^k f(x)dx + \int_k^b f(x)dx = F(b) - F(a). \tag{45}$$

And a comparison of Eqs. (40) and (45) shows that

$$\int_a^b f(x)dx = \int_a^k f(x)dx + \int_k^b f(x)dx. \tag{46}$$

If $a < k < b$ as in Fig. 217, this equation expresses the fact that the area between the ordinates at a and b is equal to the area from a to k plus the area from k to b.

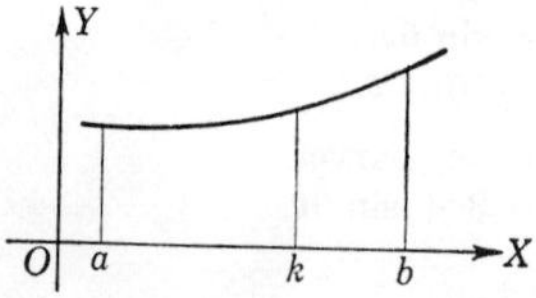

FIG. 217.

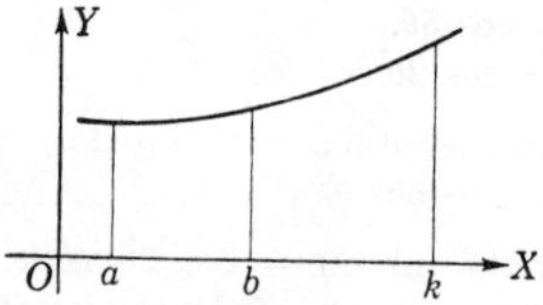

FIG. 218.

Next let $a < b < k$, as in Fig. 218. Then the integral from k to b is minus that from b to k by Eq. (42). Hence in this case Eq. (46) expresses the fact that the area between the ordinates at a and b is equal to the area from a to k minus the area from b to k. In fact, by using Eqs. (42) and (43) when necessary, we may interpret Eq. (46) geometrically for any relative order of the three letters, a, b, and k.

We may also use Eq. (40) to extend the linearity property of Eq. (25) to definite integrals and so deduce that

$$\int_a^b [c_1 f(x) + c_2 g(x)]dx = c_1 \int_a^b f(x)dx + c_2 \int_a^b g(x)dx. \tag{47}$$

It is easy to see that Eqs. (46) and (47) are consistent with the interpretation of the integrals as the limits of sums.

***176. Dependence on the Limits.** For a given continuous function $f(x)$ and a pair of numbers a and b, the construction of the limit used in the fundamental theorem determines a *number*. We have deduced the existence of the limit from geometric considerations. But this fact may be proved by purely arithmetic means. This number depends on the limits a and b, but not on the letter used to denote the variable of integration. Thus the symbol $\int_a^b f(t)dt$ means the same thing as $\int_a^b f(x)dx$, and any other letter as u, θ, or z may be used in place of t or x. For example,

$$\int_a^b f(t)dt = [F(t)]_a^b = F(b) - F(a) = \int_a^b f(x)dx. \tag{48}$$

***177. Variable Upper Limit.** Let us consider the integral of $f(x)$ over a variable interval a to x. Then for a given function $f(x)$ and a fixed number a, each value of the upper limit x determines a number. Thus the value of the integral is a function of the upper limit x, and we may write

$$\int_a^x f(t)dt = \phi(x). \tag{49}$$

We shall now evaluate the derivative of the function $\phi(x)$ by the procedure of Sec. 29. We put $\Delta x = h$ and replace x by $x + h$ in Eq. (49). This leads to

$$\int_a^{x+h} f(t)dt = \phi(x + h). \tag{50}$$

Then from Eqs. (50) and (49), by subtraction we find that

$$\begin{aligned} \Delta\phi = \phi(x + h) - \phi(x) &= \int_a^{x+h} f(t)dt - \int_a^x f(t)dt \\ &= \int_x^{x+h} f(t)dt, \end{aligned} \tag{51}$$

where the last reduction is made by using Eq. (46) with x and $x + h$ in place of k and b. Since $\Delta x = h$, it follows that

$$\frac{\Delta\phi}{\Delta x} = \frac{1}{h} \int_x^{x+h} f(t)dt. \tag{52}$$

Any sum approximating the integral on the right will lie between $hf(t_1)$ and $hf(t_2)$, where $f(t_1)$ is the smallest and $f(t_2)$ is the largest value of $f(t)$ for t between x and $x + h$. Since $f(t)$ is continuous, there will be a value t_0 between t_1 and t_2 such that the integral is equal to $hf(t_0)$. That is,

$$\frac{\Delta\phi}{\Delta x} = \frac{1}{h} [hf(t_0)] = f(t_0), \tag{53}$$

where t_0 is a suitably chosen value between x and $x + h$.

Now let $\Delta x = h \to 0$. Then since $(x + h) \to x$, any set of intermediate values t_0 will approach x, and in the limit we shall have

$$\frac{d\phi}{dx} = f(x). \tag{54}$$

This shows that the function $\phi(x)$ defined by Eq. (49) has its derivative equal to $f(x)$ and therefore is an indefinite integral of $f(x)$. We have thus sketched a proof of the theorem that

For every continuous function, indefinite integrals exist.

From Eqs. (43) and (49) we see that $\phi(a) = 0$. Thus the right member of Eq. (49), or $\phi(x) = \phi(x) - \phi(a)$, analogous to the expression $F(x) - F(a)$ in terms of any other indefinite integral $F(x)$.

And it follows from Eqs. (48), (49), and (54) that

$$\frac{d}{dx}\left[\int_a^x f(x)dx\right] = \frac{d}{dx}\left[\int_a^x f(t)dt\right] = f(x). \tag{55}$$

***178. Change of Limits for Change of Variable.** As Examples 1 and 2 of Sec. 173 illustrated, some indefinite integrals may be simplified by a substitution, or introduction of a new variable u in place of the original variable x. If $u = u(x)$ has $x = g(u)$ as its inverse function, we transform the differential by the relation

$$f(x)dx = f(x)\frac{dx}{du}du = f[g(u)]g'(u)du. \tag{56}$$

After integration, we use the relation $u = u(x)$ to translate the resulting function of u back into a function of x.

For a definite integral between the limits a and b, we may avoid this last step by changing the limits for x into proper corresponding limits for u. The new limits are

$$A = u(a) \qquad \text{and} \qquad B = u(b), \tag{57}$$

so that $a = g(A)$ and $b = g(B)$. The precise result is as follows.

Let A,B be related to a,b, by Eq. (57). Let $g'(u)$ be continuous and positive for $A < u < B$ so that to each value of x between a and b there is one and only one value of u between A and B. Then

$$\int_a^b f(x)dx = \int_A^B f[g(u)]g'(u)du. \tag{58}$$

To prove this we form the integrals with variable upper limit

$$\phi(x) = \int_a^x f(t)dt \qquad \text{and} \qquad \psi(u) = \int_A^u f[g(t)]g'(t)dt. \tag{59}$$

Then from Eq. (55) we have for the derivatives

$$\frac{d\phi}{dx} = f(x) \qquad \text{and} \qquad \frac{d\psi}{du} = f[g(u)]g'(u). \tag{60}$$

Since $x = g(u)$, $dx/du = g'(u)$. And $d\phi/dx = f(x) = f[g(u)]$ so that

$$\frac{d\phi}{du} = \frac{d\phi}{dx}\frac{dx}{du} = f[g(u)]g'(u). \tag{61}$$

From Eqs. (60) and (61) we see that $\phi(x) = \phi[g(u)]$ and $\psi(u)$ have the same derivative with respect to u. Hence by theorem III of Sec. 66, their difference is a constant.

But when $u = A$, $\psi(A) = 0$ and $\phi[g(A)] = \phi(a) = 0$, by Eqs. (59) and (43). This shows that the constant is zero, so that $\phi[g(u)] = \psi(u)$ for all values of u. In particular for $u = B$, $g(u) = g(B) = b$, so that

$$\phi(b) = \psi(B). \tag{62}$$

In view of Eq. (59), this is equivalent to Eq. (58) or the relation we wished to prove.

We have assumed that $g'(u)$ is always positive, so that A is less than B when a is less than b. If $g'(u)$ is always negative, A is greater than B when a is less than b, but the result still holds.

The essential result of this section is that, under certain restrictions, we may accomplish a change of variable in a definite integral by a simple substitution in the differential and an appropriate change of limits as in Eq. (58).

Example. Evaluate the integral $\int_0^a \sqrt{a^2 - x^2}\, dx$.

Solution: Let $x = a \sin u$. Then $u = \sin^{-1}(x/a)$, so that for the principal branch of the inverse sine, $u = 0$ when $x = 0$ and $u = \pi/2$ when $x = a$. Thus the new limits are 0 and $\pi/2$.

From $x = a \sin u$, $dx = a \cos u\, du$. And $a^2 - x^2 = a^2 - a^2 \sin^2 u = a^2(1 - \sin^2 u) = a^2 \cos^2 u$, so that $\sqrt{a^2 - x^2} = a \cos u$. Thus

$$\int_0^a \sqrt{a^2 - x^2}\, dx = \int_0^{\pi/2} a^2 \cos^2 u\, du = \frac{a^2}{2} \int_0^{\pi/2} (1 + \cos 2u) du$$
$$= \frac{a^2}{2}\left[u + \frac{\sin 2u}{2}\right]_0^{\pi/2} = \frac{a^2}{2}\left(\frac{\pi}{2} - 0\right) = \frac{\pi a^2}{4}.$$

This checks with the fact that the given integral is the expression for the area of one quarter of the circle of radius a, $x^2 + y^2 = a^2$.

179. Area Found from Parametric Equations. The area bounded above by the curve $y = f(x)$, below by the x axis, and lying between the ordinates at $x = a$ and at $x = b$ is given by a definite integral with x as the variable of integration by

$$A = \int_{x=a}^{x=b} y\, dx. \tag{63}$$

If the curve is given by equations in parametric form,

$$x = g(t), \qquad y = h(t), \tag{64}$$

it is convenient to introduce t as the variable of integration. This may be done by using the substitution rule of Eq. (58). The equation $x = g(t)$ determines the new limits t_1 and t_2 which make $a = g(t_1)$ and $b = g(t_2)$. Also $x = g(t)$ makes $dx = g'(t)dt$, so that the element of area $y\, dx = h(t)g'(t)dt$. Thus

$$A = \int_{x=a}^{x=b} y\, dx = \int_{t_1}^{t_2} h(t)g'(t)dt. \tag{65}$$

Example. Find the area of the ellipse, using the parametric form of the equations $x = a \cos t$, $y = b \sin t$, found in Sec. 126.

Solution: For the first quadrant of the ellipse (Fig. 219), the limits for x are 0 and a. The corresponding limits for t are $\pi/2$ and 0, since $0 = a \cos (\pi/2)$ and $a = a \cos 0$, where we use values for t in the first quadrant to make y positive.

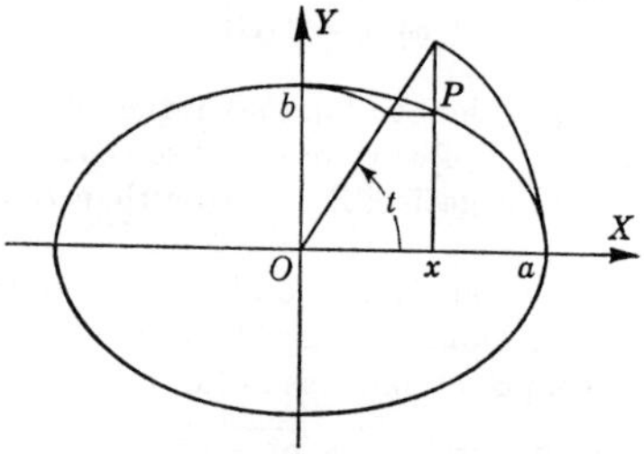

FIG. 219.

Since $x = a \cos t$, $dx = -a \sin t\, dt$ and $y\, dx = -ab \sin^2 t\, dt$. Thus the area of one quadrant is

$$A = \int_{x=0}^{x=a} y\, dx = -\int_{\pi/2}^{0} ab \cos^2 t\, dt = -\frac{ab}{2}\int_{\pi/2}^{0} (1 + \cos 2t)dt$$
$$= -\frac{ab}{2}\left[t + \frac{\sin 2t}{2}\right]_{\pi/2}^{0} = -\frac{ab}{2}\left(0 - \frac{\pi}{2}\right) = \frac{\pi ab}{4}.$$

The required total area of the ellipse is four times this, or πab.

EXERCISE 88

Verify that the indicated substitution transforms the first definite integral in x into the second one in u, and evaluate the latter integral in each case.

1. $\int_0^{\pi/2} \sin^2 x \cos x\, dx = \int_0^1 u^2\, du$, if $u = \sin x$.
2. $\int_2^7 \frac{x\, dx}{\sqrt{x+2}} = \int^3 2(u^2 - 2)du$, if $u = \sqrt{x+2}$.
3. $\int_0^2 \frac{x^2\, dx}{x+1} = \int_1^3 \left(u - 2 + \frac{1}{u}\right) du$, if $u = x + 1$.
4. $\int_0^3 x^3 \sqrt{16 + x^2}\, dx = \int_4^5 (u^4 - 16u^2)du$, if $u = \sqrt{16 + x^2}$.
5. $\int_0^{\pi} \sin^3 x \cos^2 x\, dx = -\int_1^{-1} (u^2 - u^4)du$, if $u = \cos x$.
6. $\int_0^a x^3 \sqrt{a^2 - x^2}\, dx = -\int_a^0 (a^2u^2 - u^4)du$, if $u = \sqrt{a^2 - x^2}$.
7. $\int_0^a x^2 \sqrt{a^2 - x^2}\, dx = \int_0^{\pi/2} \cos^2 u \sin^2 u\, du$, if $u = a \sin u$.
   HINT: $\cos^2 u \sin^2 u = \frac{1}{4} \sin^2 2u = \frac{1}{8}(1 - \cos 4u)$.
8. $\int_0^{\pi/4} \sec^4 dx = \int_0^1 (1 + u^2)du$, if $u = \tan x$.
9. Find the area under one arch of the cycloid $x = a(t - \sin t)$, $y = a(1 - \cos t)$ and above the x axis.

180. Length of a Curved Arc. Consider the smooth curve of Fig. 220. Any two points on this curve, as A and B, determine an arc. We wish to define its length.

Make the following construction. Divide the arc AB into n parts by intermediate points $P_1, P_2, P_3, \cdots, P_{n-1}$. Join consecutive points by straight-line segments forming the chords $AP_1, P_1P_2, P_2P_3, \cdots, P_{n-1}B$. This leads to an inscribed polygonal line, or broken line, having initial point A, terminal point B, and vertices on the curve. The length of the polygonal line is the sum of the chords or

$$L_n = AP_1 + P_1P_2 + P_2P_3 + \cdots + P_{n-1}B. \tag{66}$$

This length is one approximation to the desired arc length. Now increase the number of points of division in such a way that the length of the

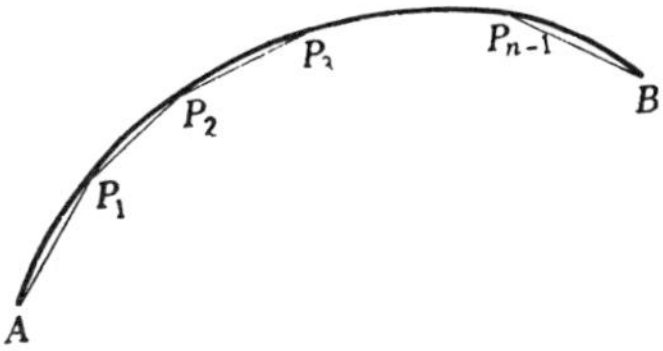

FIG. 220.

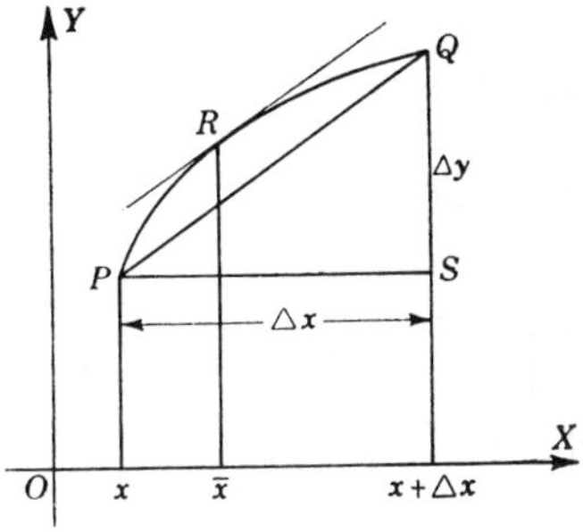

FIG. 221.

largest chord approaches zero. Then, as we shall show in Sec. 181, L_n approaches a limit L. This limit is called the length of the arc AB. Thus

The length of an arc of a curve is defined as the limit of the sum of the chords forming an inscribed polygonal line, provided that the number of chords increases indefinitely in such a way that the largest chord approaches zero as its limit.

181. Length of Arc as an Integral. We shall now use the definition of Sec. 180 to derive an expression for the length of a curved arc as an integral.

Let $y = f(x)$ be the equation of the curve. Since the curve is smooth, the derivative

$$\frac{dy}{dx} = f'(x) \tag{67}$$

must be a continuous function of x.

Let $P = (x,y)$ and $Q = (x + \Delta x, y + \Delta y)$ be any two points of this curve determining a chord PQ as in Fig. 221. Then from the right triangle PSQ we have

$$PQ = \sqrt{(\Delta x)^2 + (\Delta y)^2} = \sqrt{1 + \left(\frac{\Delta y}{\Delta x}\right)^2}\,\Delta x. \tag{68}$$

Next apply the mean value theorem, Eq. (27) of Sec. 37, with x_1, x_2, x_0

replaced by x, $x + \Delta x$, $\bar{x}$. Then $x_2 - x_1 = \Delta x$, and $f(x_2) - f(x_1) = f(x + \Delta x) - f(x) = \Delta y$, so that we have

$$\frac{\Delta y}{\Delta x} = f'(\bar{x}), \tag{69}$$

for a suitable value $\bar{x}$ between x and $x + \Delta x$, corresponding to a point R on the arc PQ where the tangent is parallel to the chord. Then from Eqs. (68) and (69) we find that

$$PQ = \sqrt{1 + [f'(\bar{x})]^2}\,\Delta x. \tag{70}$$

Consider the arc AB of the curve $y = f(x)$ (Fig. 222) lying between an initial point A with $x = a$ and a terminal point B with $x = b$. Take any intermediate points P_1, P_2, P_3, $\cdots$, P_{n-1} on this arc as in Fig. 222 and join consecutive points to form the chords

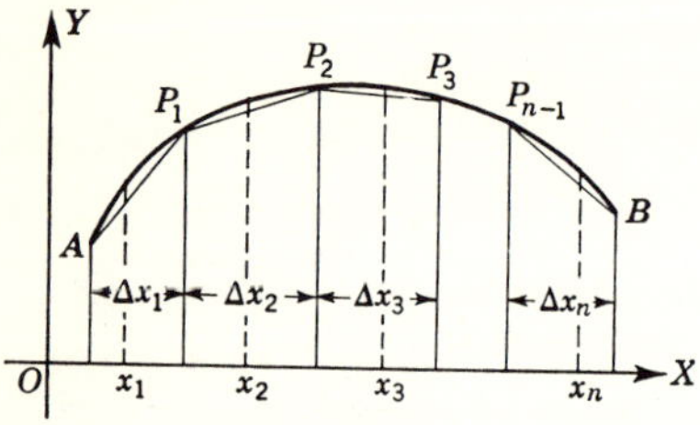

FIG. 222.

$$AP_1, P_1P_2, P_2P_3, \cdots, P_{n-1}B. \tag{71}$$

Call the projections of these chords on the x axis Δx_1, Δx_2, Δx_3, $\cdots$, Δx_n. And apply the result of Eq. (70) to each of the chords, calling the intermediate points $\bar{x}$ in order x_1, x_2, x_3, $\cdots$, x_n. In this way we find that

$$AP_1 = \sqrt{1 + [f'(x_1)]^2}\,\Delta x_1, \qquad P_1P_2 = \sqrt{1 + [f'(x_2)]^2}\,\Delta x_2,$$
$$P_2P_3 = \sqrt{1 + [f'(x_3)]^2}\,\Delta x_3, \qquad \cdots, \qquad P_{n-1}B = \sqrt{1 + [f'(x_n)]^2}\,\Delta x_n.$$

And the sum of the n chords (71) forming an inscribed polygonal line is

$$\begin{aligned} L_n &= AP_1 + P_1P_2 + P_2P_3 + \cdots + P_{n-1}B \\ &= \sqrt{1 + [f'(x_1)]^2}\,\Delta x_1 + \sqrt{1 + [f'(x_2)]^2}\,\Delta x_2 + \sqrt{1 + [f'(x_3)]^2}\,\Delta x_3 \\ &\qquad + \cdots + \sqrt{1 + [f'(x_n)]^2}\,\Delta x_n. \end{aligned} \tag{72}$$

Using the summation notation of Sec. 169, we may write

$$L_n = \sum_{i=1}^{n} \sqrt{1 + [f'(x_i)]^2}\,\Delta x_i. \tag{73}$$

The required length of arc AB or L is the limit of L_n as $n \to \infty$ in such a way that the largest chord, and hence the largest projection, $\Delta x_i \to 0$. From Eq. (73) and the fundamental theorem (13) we may conclude that

$$L = \lim_{n \to \infty} \sum_{i=1}^{n} \sqrt{1 + [f'(x_i)]^2}\,\Delta x_i = \int_a^b \sqrt{1 + [f'(x)]^2}\,dx. \tag{74}$$

We have thus proved that the length of the curve $y = f(x)$ lying between $x = a$ and $x = b$ is

$$L = \int_a^b \sqrt{1 + [f'(x)]^2}\, dx = \int_a^b \sqrt{1 + \left(\frac{dy}{dx}\right)^2}\, dx. \tag{75}$$

Similarly the length of the curve $x = g(y)$ lying between $y = c$ and $y = d$ is

$$L = \int_c^d \sqrt{1 + [g'(y)]^2}\, dy = \int_c^d \sqrt{1 + \left(\frac{dx}{dy}\right)^2}\, dy. \tag{76}$$

EXAMPLE. Find the length of the minor arc of the circle $x^2 + y^2 = 4$ lying between the points $(0,2)$ and $(\sqrt{2},\sqrt{2})$.

Solution: On the arc in question $y = \sqrt{4 - x^2}$, so that we find

$$\frac{dy}{dx} = \frac{-x}{\sqrt{4 - x^2}}, \qquad \left(\frac{dy}{dx}\right)^2 = \frac{x^2}{4 - x^2}, \qquad 1 + \left(\frac{dy}{dx}\right)^2 = \frac{4}{4 - x^2}.$$

And from Eq. (75) we have

$$L = \int_0^{\sqrt{2}} \sqrt{\frac{4}{4 - x^2}}\, dx = 2 \int_0^{\sqrt{2}} \frac{dx}{\sqrt{4 - x^2}}.$$

Proceeding as in Sec. 178, make the change of variable $x = 2 \sin u$. Then $u = \sin^{-1}(x/2)$, so that $u = 0$ when $x = 0$ and $u = \pi/4$ when $x = \sqrt{2}$. And from $x = 2 \sin u$, $dx = 2 \cos u$, and $\sqrt{4 - x^2} = 2 \cos u$. Thus

$$L = 2 \int_0^{\pi/4} \frac{2 \cos u\, du}{2 \cos u} = 2 \int_0^{\pi/4} du = 2[u]_0^{\pi/4} = \frac{\pi}{2}.$$

Thus the required length is $\pi/2$. This checks with the fact that the required arc is $\frac{1}{8}$ of a circumference of radius 2.

182. Differential of Arc. Let $P_0 = (x_0,y_0)$ be a fixed point, and $P = (x,y)$ a variable point, on a curve with equation $y = f(x)$. Let

$$s = P_0P \tag{77}$$

denote the arc from P_0 to P, considered positive for points of the curve to the right of P_0, and negative for those to the left of P_0. Then it follows from Eqs. (75) and (42) that

$$s = \int_{x_0}^{x} \sqrt{1 + \left(\frac{dy}{dx}\right)^2}\, dx. \tag{78}$$

By Eq. (55) the derivative of s with respect to x is

$$\frac{ds}{dx} = \sqrt{1 + \left(\frac{dy}{dx}\right)^2}. \tag{79}$$

Let us use this to study the ratio of a small arc $\Delta s = \overset{\frown}{PQ}$ to the chord PQ which it subtends. From Eq. (68) we have

$$\frac{\Delta s}{PQ} = \frac{\Delta s}{\Delta x \sqrt{1 + (\Delta y/\Delta x)^2}} = \frac{\Delta s/\Delta x}{\sqrt{1 + (\Delta y/\Delta x)^2}}. \tag{80}$$

Let us keep P fixed and let $Q \to P$ along the curve. Then $\Delta s \to 0$ and $\Delta x \to 0$, so that $\Delta s/\Delta x \to ds/dx$ and $\Delta y/\Delta x \to dy/dx$. Thus

$$\lim_{P\to Q} \frac{\Delta s}{PQ} = \frac{ds/dx}{\sqrt{1 + (dy/dx)^2}}. \tag{81}$$

The right member is 1, since the numerator equals the denominator by Eq. (79). Since Δs = arc PQ, this proves that

$$\lim_{P\to Q} \frac{\text{arc } PQ}{\text{chord } PQ} = 1. \tag{82}$$

The differential relation equivalent to Eq. (79) is

$$ds = \sqrt{dx^2 + dy^2}. \tag{83}$$

The results expressed in Eqs. (79), (82), and (83) are equivalent to those stated in Eqs. (75), (71), and (76) of Sec. 62, where we explained how to read Eq. (83) from a right triangle with sides dx, dy, and ds. We may recall Eqs. (75) and (76) from their relation to Eq. (83) and the right triangle (Fig. 223).

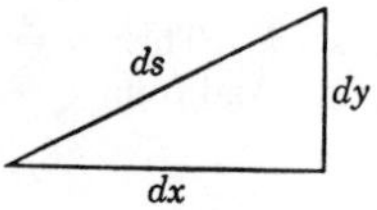

FIG. 223.

183. Length of Arc in Polar Coordinates. Let the equation of a curve in polar coordinates be

$$r = f(\theta). \tag{84}$$

And consider the arc AB of this curve lying between an initial point with $\theta = \alpha$, and a terminal point with $\theta = \beta$.

The relations of rectangular to polar coordinates are

$$x = r \cos \theta, \qquad y = r \sin \theta. \tag{85}$$

The differentials of these are

$$dx = \cos \theta \, dr - r \sin \theta \, d\theta, \qquad dy = \sin \theta \, dr + r \cos \theta \, d\theta. \tag{86}$$

By squaring and adding these, we find that

$$dx^2 + dy^2 = dr^2 + r^2 \, d\theta^2. \tag{87}$$

Thus the differential in the integral of Eq. (75) is

$$\sqrt{1 + \left(\frac{dy}{dx}\right)^2} \, dx = \sqrt{dx^2 + dy^2} = \sqrt{dr^2 + r^2 \, d\theta^2} = \sqrt{\left(\frac{dr}{d\theta}\right)^2 + r^2} \, d\theta. \tag{88}$$

Suppose first that y is a single-valued function of x on the arc considered, and that θ changes from α to β as x changes from a to b. Then in the integral of Eq. (75) we may change the variable from x to θ as in Sec. 178 and so conclude that L, the length of AB, is

$$L = \int_\alpha^\beta \sqrt{r^2 + \left(\frac{dr}{d\theta}\right)^2}\, d\theta. \qquad (89)$$

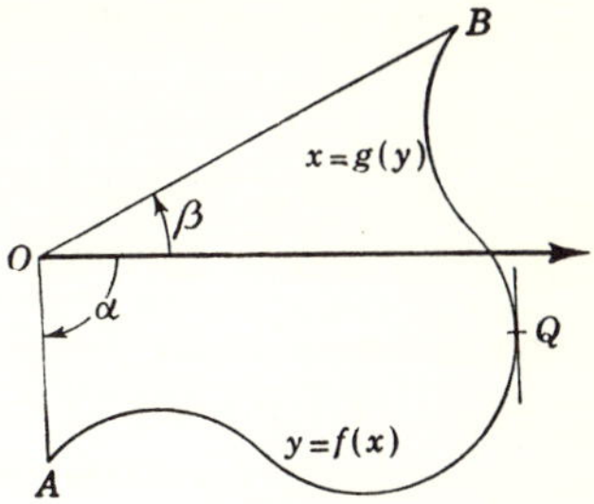

FIG. 224.

The same equation follows from Eq. (76) if x is a single-valued function of y. Suppose that an arc consists of parts of one type or the other, as AQ and QB in Fig. 224. Then by adding the integrals for the separate parts and making use of Eq. (46), we may establish Eq. (89) in this case also.

We may recall Eq. (89) from its relation to

$$ds = \sqrt{r^2\, d\theta^2 + dr^2} \qquad (90)$$

and the right triangle (Fig. 225) with sides $r\, d\theta$, dr, and ds. To evaluate the integral in Eq. (89) we must substitute for r and $dr/d\theta$ their values as found from the equation of the curve.

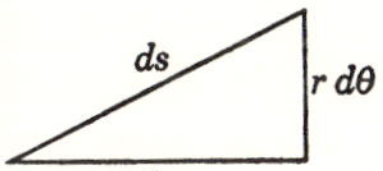

FIG. 225.

EXAMPLE. Find the total length of the curve $r = 1 + \cos\theta$.

Solution: The curve makes one single loop as θ goes from 0 to 2π, and is symmetrical about OX. Hence 0 to π gives the upper half.

From $r = 1 + \cos\theta$, we find that $dr/d\theta = -\sin\theta$, and $r^2 + (dr/d\theta)^2 = (1 + \cos\theta)^2 + (-\sin\theta)^2 = 2 + 2\cos\theta = 4\cos^2(\theta/2)$. Hence from Eq. (89) we have for the half-length

$$L = \int_0^\pi \sqrt{4\cos^2\frac{\theta}{2}}\, d\theta = 2\int_0^\pi \cos\frac{\theta}{2}\, d\theta = 4\left[\sin\frac{\theta}{2}\right]_0^\pi = 4.$$

And the required length is twice this, or 8.

The value of $2\int_0^{2\pi}\cos\frac{\theta}{2}\, d\theta = 4\left[\sin\frac{\theta}{2}\right]_0^{2\pi} = 0$, represents the upper half taken positively and the lower half taken negatively, because for $\pi < \theta < 2\pi$, $\pi/2 < \theta/2 < \pi$ and $\cos(\theta/2)$ is negative. To evaluate the length using the limits 0 and 2π we must write

$$\int_0^{2\pi}\sqrt{4\cos^2\frac{\theta}{2}}\, d\theta = \int_0^\pi \sqrt{4\cos^2\frac{\theta}{2}}\, d\theta + \int_\pi^{2\pi}\sqrt{4\cos^2\frac{\theta}{2}}\, d\theta$$
$$= 2\int_0^\pi \cos\frac{\theta}{2}\, d\theta - 2\int_0^{2\pi}\cos\frac{\theta}{2}\, d\theta = 4\left[\sin\frac{\theta}{2}\right]_{0,\,2\pi}^{\pi,\,\pi} = 8.$$

184. Length of Arc from Parametric Equations. Let a curve be given by equations in parametric form

$$x = g(t), \qquad y = h(t). \qquad (91)$$

And consider the arc AB of this curve lying between an initial point with $t = t_1$ and a terminal point with $t = t_2$.

The differentials of the x and y of Eq. (91) are

$$dx = g'(t)dt, \qquad dy = h'(t)dt. \tag{92}$$

By squaring and adding these, we find that

$$dx^2 + dy^2 = \{[g'(t)]^2 + [h'(t)]^2\}dt^2. \tag{93}$$

The differential in the integral of Eq. (75) is

$$\sqrt{1 + \left(\frac{dy}{dx}\right)^2}\, dx = \sqrt{dx^2 + dy^2} = \sqrt{[g'(t)]^2 + [h'(t)]^2}\, dt. \tag{94}$$

Hence in the integral of Eq. (75) we may change the variable of integration from x to t as in Sec. 178 and so conclude that

$$L = \int_{t_1}^{t_2} \sqrt{[g'(t)]^2 + [h'(t)]^2}\, dt = \int_{t_1}^{t_2} \sqrt{\left(\frac{dx}{dt}\right)^2 + \left(\frac{dy}{dt}\right)^2}\, dt. \tag{95}$$

With the modifications made in proving Eq. (89), this proves that Eq. (95) holds for any curve made up of parts to each of which Eq. (75) or Eq. (76) is applicable.

We may recall Eq. (95) from its relation to Eq. (83) and the right triangle of Fig. 223.

EXAMPLE. Find the length of one arch of the cycloid $x = t - \sin t$, $y = 1 - \cos t$.

Solution: The extremities are the first points where $y = 0$, or $t = 0$ and $t = 2\pi$. And from the given relations we find by differentiation that $dx/dt = 1 - \cos t$, $dy/dt = \sin t$, and $(dx/dt)^2 + (dy/dt)^2 = (1 - \cos t)^2 + \sin^2 t = 2 - 2\cos t = 4\sin^2 (t/2)$. Hence we find from Eq. (95) that

$$L = \int_0^{2\pi} \sqrt{4\sin^2 \frac{t}{2}}\, dt = 2\int_0^{2\pi} \sin \frac{t}{2}\, dt = -4\left[\cos \frac{t}{2}\right]_0^{2\pi}$$
$$= -4(\cos \pi - \cos 0) = -4(-1 - 1) = 8.$$

And the required length of one arch is 8.

EXERCISE 89

1. Find the length of the curve $y = x^{\frac{3}{2}}$ from the point (0,0) to the point (1,1).
2. Find the length of the curve $8y^2 = x^3$ from the point (0,0) to the point (2,1).
3. Find the length of the curve $9y^2 = (x^2 + 2)^3$ from the point $(1,\sqrt{3})$ to the point $(2,2\sqrt{6})$.
4. Find the length of the arc of the curve $4y = x^2 - 2 \ln x$ which lies between $x = 1$ and $x = 2$.
5. Find the length of the arc of the curve $y = \dfrac{x^3}{6} + \dfrac{1}{2x}$ which lies between $x = 1$ and $x = 2$.
6. Find the length of the catenary $y = \dfrac{a}{2}(e^{x/a} + e^{-x/a})$ which lies between $x = 0$ and $x = h$.

Verify by integration that the length of each of the following circles of radius 2 is 4π.

7. $r = 4 \sin \theta$.
8. $r = 4 \cos \theta$.
9. $r = 2\sqrt{2} (\cos \theta + \sin \theta)$.
10. $r = 2$.

11. Find the total length of the curve $r = 1 - \cos \theta$.
12. Find the total length of the curve $r = \cos^2 (\theta/2)$.
13. Find the length of one loop of the curve $r = \sin^3 (\theta/3)$.
14. Find the length of one loop of the curve $r = \cos^3 (\theta/3)$.

Find the length of arc described when θ varies from 0 to π for each of the following equiangular spirals.

15. $r = e^{-2\theta}$.
16. $r = e^{\theta/2}$.

Find the length of arc between the two points corresponding to the two given values of t for each of the following curves.

17. $x = e^t \sin t,\ y = e^t \cos t;\ t = 0,\ t = 2$.
18. $x = 2 + 4 \sin 2t,\ y = 4 - 4 \cos 2t;\ t = 0,\ t = \pi/2$.
19. $x = 4 + 3t^2,\ y = 6 - 4t^2;\ t = 0,\ t = 4$.
20. $x = \cos^3 t,\ y = \sin^3 t;\ t = 0,\ t = \pi/2$.
21. $x = a \cos^3 t,\ y = b \sin^3 t;\ t = 0,\ t = \pi/2$. HINT: After setting up the integral, make the change of variable $a^2 \cos^2 t + b^2 \sin^2 t = u^2$.
22. $x = n \cos t + \cos nt,\ y = n \sin t - \sin nt;\ t = 0,\ t = \dfrac{\pi}{n+1}$. HINT: Note that

$$1 + \sin t \sin nt - \cos t \cos nt = 1 - \cos (n+1)t = 2 \sin^2 \frac{(n+1)t}{2}.$$

***185. Theorems of Duhamel.** For some applications we need an extension of the fundamental theorem which we shall now discuss. As in Sec. 170, we are concerned with an interval from a to b on the x axis which we subdivide into n subintervals. We denote the length of the ith subinterval by Δx_i, and the largest Δx_i by d_n. And we consider a succession of subdivisions, with n becoming infinite and d_n tending to zero.

Given a continuous function $f(x)$, we may select some point x_i in the ith subdivision and form the product $f(x_i)\Delta x_i$. Then the fundamental theorem of Sec. 170 asserts that

$$\lim_{n\to\infty} \sum_{i=1}^{n} f(x_i)\Delta x_i = \int_a^b f(x)dx. \tag{96}$$

If $g(x)$ is a second continuous function, we may select some point x_i' in the ith interval, in general different from x_i, and form the products $f(x_i)g(x_i')\Delta x_i$. Then it is also true that

$$\lim_{n\to\infty} \sum_{i=1}^{n} f(x_i)\ g(x_i')\Delta x_i = \int_a^b f(x)\ g(x)dx. \tag{97}$$

In this result the product fg may be replaced by $F(f,g)$, any continuous function of the two variables f and g as defined in Sec. 269. The conclusion then is that

$$\lim_{n\to\infty} \sum_{i=1}^{n} F[f(x_i), g(x_i)']\Delta x_i = \int_a^b F[f(x), g(x)]dx. \tag{98}$$

Similar results also hold for more than two functions. These generalizations of the fundamental theorem, as well as Eqs. (97) and (98), are called *Duhamel's theorems*.

The proof of Eq. (98) depends on the following facts. Given a small positive quantity ϵ_n, for sufficiently small d_n, each term $F[f(x_i), g(x_i')]$ will differ from $F[f(x_i), g(x_i)]$ by less than ϵ_n. Hence the sum in Eq. (98) will differ from the sum

$$\sum_{i=1}^{n} F[f(x_i), g(x_i)]\, \Delta x_i \tag{99}$$

by less than $\epsilon_n(b - a)$. But as $n \to \infty$, the sum (99) approaches the integral on the right of Eq. (98) by the fundamental theorem. Hence the sum in Eq. (98) must approach the same limit.

The notion of force due to liquid pressure provides an application of Eq. (97). The discussion which led to Eq. (41) of Sec. 75 would lead us to consider the sum

$$\sum_{i=1}^{n} wh_iL(h_i')\Delta h_i \tag{100}$$

as a reasonable approximation to the force due to liquid pressure on one side of a flat submerged plate. We would expect this sum to approach a limit as $n \to \infty$, with such Δh_i that the largest tended to zero. Since h_i' and h_i are in general different, we cannot deduce this from the fundamental theorem of Sec. 170. But by Duhamel's theorem of Eq. (97), we see that the sum (100) does approach a limit as $n \to \infty$, namely,

$$F = \lim_{n\to\infty} \sum_{i=1}^{n} wh_iL(h_i')\Delta h_i = \int_a^b whL(h)dh. \tag{101}$$

And it is reasonable to use Eq. (101) as the definition of the total force on the plate due to liquid pressure.

As an application of Eq. (99), we shall derive the definite integral for the length of a curve given in parametric form,

$$x = g(t), \qquad y = h(t), \tag{102}$$

directly from the definition of arc length of Sec. 180.

For any two points $P = (x,y)$ and $Q = (x + \Delta x, y + \Delta y)$ on this curve, the chord PQ is given by

$$PQ = \sqrt{(\Delta x)^2 + (\Delta y)^2}, \tag{103}$$

as in Eq. (68).

And from the mean value theorem, by an argument like that used to derive Eq. (69), we find from Eq. (102) that

$$\Delta x = g'(\bar{t})\Delta t \qquad \text{and} \qquad \Delta y = h'(\bar{t}')\Delta t, \tag{104}$$

for a suitable t, and another suitable $\bar{t}'$ between t and $t + \Delta t$. Then from Eqs. (103) and (104), we may conclude that

$$PQ = \sqrt{[g'(\bar{t})]^2 + [h'(\bar{t}')]^2}\, \Delta t. \tag{105}$$

Now apply this result to each of the chords (71), denoting the intermediate values $\bar{t}$ and $\bar{t}'$ for the ith chord $P_{i-1}P_i$ by t_i and t_i', where $A = P_0$ and for each n, $B = P_n$. Then the sum of the chords is

$$L_n = \sum_{i=1}^{n} P_{i-1}P_i = \sum_{i=1}^{n} \sqrt{[g'(t_i)]^2 + [h'(t_1')]^2}\, \Delta t_i. \tag{106}$$

The arc length L of AB is the limit of this sum when the largest chord, and hence the largest $\Delta t_i \to 0$ as $n \to \infty$. This limit cannot be found from the fundamental theorem of Sec. 170, because t_i' and t_i are in general different. But with a change of letters, and $F(g',h') = \sqrt{g'^2 + h'^2}$, the sum in Eq. (106) is like that in Eq. (98). Hence we may conclude from Duhamel's theorem of Eq. (98) that

$$L = \lim_{n \to \infty} \sum_{i=1}^{n} \sqrt{[g'(t_i)]^2 + [h'(t_i')]^2}\,\Delta t = \int_{t_1}^{t_2} \sqrt{[g'(t)]^2 + [h'(t)]^2}\,dt. \tag{107}$$

This proves Eq. (95) for any arc on which $f'(t)$ and $g'(t)$ are continuous, without the necessity of considering partial arcs as in the proof of Sec. 184.

186. Area of a Surface of Revolution. Let a plane curve be revolved about a straight line in its plane as an axis. Then any arc of the curve generates a *surface of revolution.*

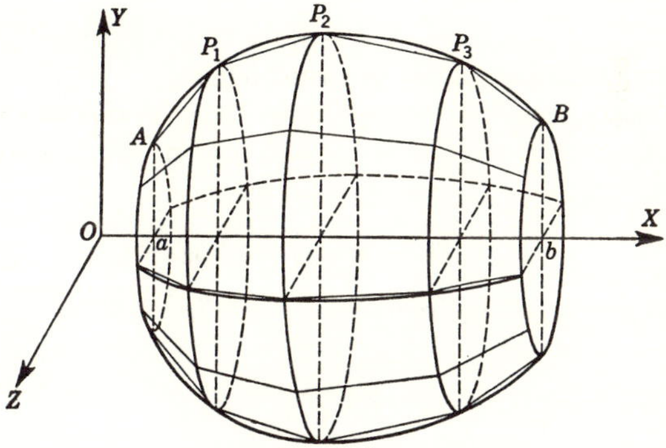

FIG. 226.

Consider in particular the smooth arc AB of the curve $y = f(x)$ (Fig. 226) lying between an initial point with $x = a$ and a terminal point with $x = b$. Assume that y is never negative for values of x between a and b. And let this arc AB revolve through a complete turn about the x axis. We wish to find S_x, the area of the surface generated.

As in Sec. 180, inscribe a polygonal line made up of n chords in the arc AB and revolve this about the x axis. Then each chord will generate the lateral surface of a frustum of a cone of revolution, so that C_i, the area generated by the ith chord, is known from solid geometry. The total area generated by the polygonal line, or sum of the n conical areas C_i, is an approximation to S_x. Now let n, the number of chords, increase indefinitely in such a way that the largest chord tends to zero as n becomes infinite. Then the sum of the n conical areas will approach a limit. This limit will be taken as our definition of S_x, the area of the surface generated by revolving the arc AB.

We recall some of the properties of a conical frustum from solid geometry. A frustum of a cone of revolution with radius of lower base R, radius of upper base r, and slant height l is shown in Fig. 227. The lateral area L is equivalent to a trapezoid whose altitude is the slant height l and whose bases are the circumferences of the bases of the frustum, namely $2\pi R$ and $2\pi r$. Hence this area is

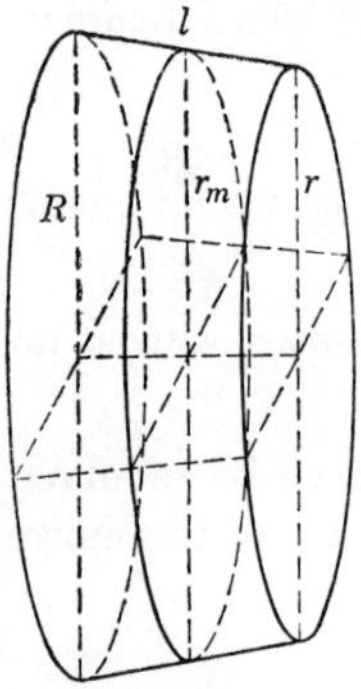

FIG. 227.

$$L = l\,\frac{2\pi R + 2\pi r}{2} = \pi l(R + r). \tag{108}$$

Let r_m be the radius of the midsection parallel to the bases. Then

$$r_m = \frac{R + r}{2}, \qquad L = 2\pi r_m l. \tag{109}$$

Thus the lateral surface of the frustum is the circumference of the midsection times the slant height.

Next let $P = (x,y)$ and $Q = (x + \Delta x, y + \Delta y)$ be any two points of the arc AB, determining a chord PQ. Then by Eq. (70) for some suitable value $\bar{x}$ between x and $x + \Delta x$, we shall have

$$PQ = \sqrt{1 + [f'(\bar{x})]^2}\,\Delta x. \tag{110}$$

The frustum generated by revolving PQ (Fig. 228) has $y = f(x)$ and

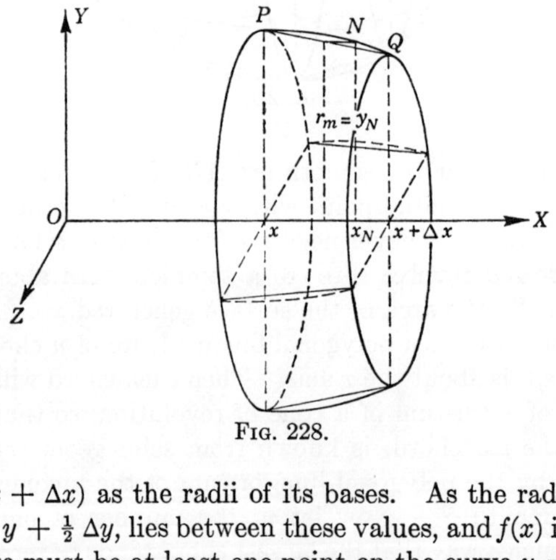

FIG. 228.

$y + \Delta y = f(x + \Delta x)$ as the radii of its bases. As the radius of its midsection, $r_m = y + \frac{1}{2}\Delta y$, lies between these values, and $f(x)$ is a continuous function, there must be at least one point on the curve $y = f(x)$ between P and Q, as $N = (x_N, y_N)$ in Fig. 228, such that

$$r_m = y_N = f(x_N). \tag{111}$$

As the slant height of the frustum $l = PQ$, from Eqs. (109) to (111) we find that C, the area generated by revolving PQ, is

$$C = 2\pi r_m l = 2\pi f(x_N)\sqrt{1 + [f'(\bar{x})]^2}\,\Delta x, \tag{112}$$

where x_N and $\bar{x}$ each lie between x and $x + \Delta x$.

Now apply this result to each of the n chords

$$AP_1 = P_0P_1,\ P_1P_2,\ P_2P_3,\ \cdots,\ P_{n-1}P_n = P_{n-1}B, \tag{113}$$

with $A = P_0$ and for each n, $B = P_n$. Then for $i = 1, 2, 3, \cdots, n$ the ith chord is $P_{i-1}P_i$. Its projection on the x axis is Δx_i. Denote the two intermediate values of x in this ith interval corresponding to x_N and $\bar{x}$ by x_i and x_i'. Then the area generated by revolving $P_{i-1}P_i$, or lateral area of the ith frustum, C_i, is

$$C_i = 2\pi f(x_i)\sqrt{1 + [f'(x_i')]^2}\,\Delta x_i. \tag{114}$$

Hence the total area generated by the polygonal line made up of the n chords, or sum of the n lateral areas, is

$$\sum_{i=1}^{n} C_i = \sum_{i=1}^{n} 2\pi f(x_i)\sqrt{1 + [f'(x'_i)]^2}\,\Delta x_i. \tag{115}$$

The area S_x, generated by revolving the arc AB, is the limit of this sum when n becomes infinite and the largest of the $\Delta x_i \to 0$ as $n \to \infty$. Hence by Duhamel's theorem, Eq. (97), we may conclude from Eq. (115) that

$$\begin{aligned} S_x &= \lim_{n\to\infty} \sum_{i=1}^{n} 2\pi f(x_i)\sqrt{1 + [f'(x'_i)]^2}\,\Delta x_i \\ &= \int_a^b 2\pi f(x)\sqrt{1 + [f'(x)]^2}\,dx. \end{aligned} \tag{116}$$

Since $y = f(x)$, $dy/dx = f'(x)$, and we may deduce from this and Eq. (79) that

$$S_x = 2\pi \int_a^b y\sqrt{1 + \left(\frac{dy}{dx}\right)^2}\,dx = 2\pi \int_a^b y\,ds. \tag{117}$$

We recall that S_x is the area generated by revolving the arc of the curve $y = f(x)$ lying between $x = a$ and $x = b$ about the x axis.

Suppose that neither a nor b is negative and that we revolve the arc AB about the y axis to generate a surface of revolution. Then its area S_y may be shown by a similar argument to be

$$S_y = 2\pi \int_a^b x\sqrt{1 + \left(\frac{dy}{dx}\right)^2}\,dx = 2\pi \int_a^b x\,ds. \tag{118}$$

The last forms in Eqs. (117) and (118), or

$$S_x = 2\pi\int y\,ds \qquad \text{and} \qquad S_y = 2\pi\int x\,ds, \tag{119}$$

may be used with appropriate limits for an arc of a curve given in polar coordinates or in parametric form, if x,y, and ds are expressed in terms of the variable of integration. For example, with θ as the independent variable as in Sec. 183, we have

$$S_x = 2\pi \int y\,ds = 2\pi \int r \sin\theta \sqrt{r^2 + \left(\frac{dr}{d\theta}\right)^2}\,d\theta. \tag{120}$$

To evaluate this integral we must substitute for r and $dr/d\theta$ their values as found from the equation of the curve in polar coordinates.

And for equations in parametric form, as in Sec. 184, we have

$$S_x = 2\pi \int y\,ds = 2\pi \int_{t_1}^{t_2} y \sqrt{\left(\frac{dx}{dt}\right)^2 + \left(\frac{dy}{dt}\right)^2}\,dt. \tag{121}$$

To evaluate this integral we must substitute for y, dx/dt, and dy/dt their values as found from the parametric equations of the curve.

The following plausibility considerations may help the student to remember the results which we have just proved. It is easy to reconstruct Eq. (117) from the first relation of Eq. (119). And this may be associated with the differential relation

$$dS_x = 2\pi y\,ds. \tag{122}$$

This may be remembered either as an expression for the lateral area of a conical frustum with radius of midsection y and slant height ds, or more simply as the area of a rectangular strip of width ds and length $2\pi y$, the circumference generated by revolving the ordinate y.

EXAMPLE 1. Find the area generated by revolving the upper half of the ellipse $x^2/a^2 + y^2/b^2 = 1$ about its major axis, OX.

Solution: From the given equation we have $y = \frac{b}{a}\sqrt{a^2 - x^2}$ on the upper half, so that $\frac{dy}{dx} = \frac{-bx}{a\sqrt{a^2 - x^2}}$. It follows that

$$\left(\frac{ds}{dx}\right)^2 = 1 + \left(\frac{dy}{dx}\right)^2 = \frac{a^4 - (a^2 - b^2)x^2}{a^2(a^2 - x^2)} = \frac{a^2 - e^2x^2}{a^2 - x^2}, \qquad \text{if } e^2 = \frac{a^2 - b^2}{a^2}.$$

Thus $e = \frac{\sqrt{a^2 - b^2}}{a} = \frac{c}{a}$ is the eccentricity of the ellipse as in Sec. 84, and

$$dS_x = 2\pi y\,ds = 2\pi \frac{b}{a}\sqrt{a^2 - x^2}\,\frac{\sqrt{a^2 - e^2x^2}}{\sqrt{a^2 - x^2}}\,dx = 2\pi\frac{b}{a}\sqrt{a^2 - e^2x^2}\,dx.$$

The area generated by one quadrant, lying between $x = 0$ and $x = a$, is

$$S_x = \frac{2\pi b}{a}\int_0^a \sqrt{a^2 - e^2x^2}\,dx = \frac{2\pi ab}{e}\int_0^{\sin^{-1} e} \cos^2 u\,du = \frac{\pi ab}{e}\left[u + \frac{\sin 2u}{2}\right]_0^{\sin^{-1} e},$$

where the change of variable $ex = a \sin u$ has been made, and the indefinite integral found as in the example of Sec. 178. When $u = \sin^{-1} e$, $\sin u = e$, $\cos^2 u = 1 - e^2 = 1 - \frac{a^2 - b^2}{a^2} = \frac{b^2}{a^2}$. Hence $\cos u = \frac{b}{a}$, and $\sin 2u = 2 \sin u \cos u = \frac{2be}{a}$. Thus $S_x = \frac{\pi ab}{e}\left(\sin^{-1} e + \frac{be}{a}\right) = \pi b^2 + \frac{\pi ab}{e} \sin^{-1} e$. And the required area is twice this, or $2\pi b\left(b + \frac{a}{e} \sin^{-1} e\right)$ with $e = \frac{\sqrt{a^2 - b^2}}{a}$.

EXAMPLE 2. Find the area generated by revolving the arc of the curve $r^2 = \cos 2\theta$ which lies in the first quadrant about the y axis.

Solution: From $r = \sqrt{\cos 2\theta}$, we find $\frac{dr}{d\theta} = \frac{-\sin 2\theta}{\sqrt{\cos 2\theta}}$. It follows that $\left(\frac{ds}{d\theta}\right)^2 = r^2 + \left(\frac{dr}{d\theta}\right)^2 = \cos 2\theta + \frac{\sin^2 2\theta}{\cos 2\theta} = \frac{1}{\cos 2\theta}$. And

$$\begin{aligned} dS_y &= 2\pi x \, ds = 2\pi r \cos \theta \left(\frac{ds}{d\theta}\right) d\theta = 2\pi \sqrt{\cos 2\theta} \cos \theta \frac{1}{\sqrt{\cos 2\theta}} d\theta \\ &= 2\pi \cos \theta \, d\theta. \end{aligned}$$

For the part in the first quadrant, θ is between 0 and $\pi/4$, so that

$$S_y = 2\pi \int_0^{\pi/4} \cos \theta \, d\theta = 2\pi \left[\sin \theta\right]_0^{\pi/4} = 2\pi \sin \frac{\pi}{4} = \pi \sqrt{2}.$$

Thus the required area is $\pi \sqrt{2}$.

EXAMPLE 3. Find the area generated by revolving the arc of the curve $x = \cos^3 t$, $y = \sin^3 t$ which lies in the first quadrant about the x axis.

Solution: From the given equations we find that $dx/dt = -3 \cos^2 t \sin t$ and $dy/dt = 3 \sin^2 t \cos t$ so that $(ds/dt)^2 = 9 \cos^4 t \sin^2 t + 9 \sin^4 t \cos^2 t = 9 \cos^2 t \sin^2 t$. And $dS_x = 2\pi y \, ds = 2\pi y (ds/dt) dt = 2\pi \sin^3 t \, (3 \cos t \sin t) dt = 6\pi \sin^4 t \cos t \, dt$. The part of the curve in the first quadrant is traced out as t varies from 0 to $\pi/2$. It follows that $S_x = 6\pi \int_0^{\pi/2} \sin^4 t \cos t \, dt = 6\pi \int_0^1 u^4 \, du = 6\pi \left[\frac{u^5}{5}\right]_0^1 = \frac{6\pi}{5}$, where we have made the change of variable $u = \sin t$. Thus the required area is $\frac{6}{5}\pi$.

EXERCISE 90

Find the area of the surface of revolution generated by revolving each of the following curved arcs about the axis indicated.

1. $y^2 = x$, from (0,0) to (4,2), about the x axis.
2. $y = x^3$, from (0,0) to (1,1) about the x axis.
3. $y = \frac{1}{2}(e^x + e^{-x})$, from $x = 0$ to $x = 1$, about the x axis.
4. $3y = 4x$, from (0,0) to (3,4), about the x axis.
5. $x^2 + y^2 = 4$, from (0,2) to $(\sqrt{2},\sqrt{2})$, about the x axis.
6. $y = x^2$, from $x = 0$ to $x = 1$, about the y axis.
7. $3y = 4x$, from (0,0) to (3,4), about the y axis.
8. $4y = x^2 - 2 \ln x$, from $x = 1$ to $x = 2$, about the y axis.
9. $y = \frac{x^3}{6} + \frac{1}{2x}$, from $x = 1$ to $x = 2$, about the y axis.
10. $x^2 + y^2 = 4$, from $(\sqrt{2},\sqrt{2})$ to (2,0), about the y axis.
11. $r^2 = \cos 2\theta$, from $\theta = 0$ to $\theta = \pi/4$, about the x axis.

12. $r = 4 \cos \theta$, from $\theta = 0$ to $\theta = \pi/2$, about the x axis.
13. $r = 4 \cos \theta$, from $\theta = 0$ to $\theta = \pi/2$, about the y axis.
14. $r = 4$, from $\theta = 0$ to $\theta = \pi$, about the x axis.
15. $x = \cos^3 t$, $y = \sin^3 t$, from $t = 0$ to $t = \pi/2$, about the y axis.
16. $x = 4 + 3t^2$, $y = 6 - 4t^2$, from $t = 0$ to $t = 1$, about the x axis.
17. $x = 4 + 3t^2$, $y = 6 - 4t^2$, from $t = 0$ to $t = 1$, about the y axis.
18. $x = 6 + 4 \sin 2t$, $y = 4 - 4 \cos 2t$, from $t = 0$ to $t = \pi$, about the x axis.
19. $x = 6 + 4 \sin 2t$, $y = 4 - 4 \cos 2t$, from $t = 0$ to $t = \pi$, about the y axis.

20. Prove that the area of any zone of height h on a sphere of radius R is $2\pi Rh$. HINT: Consider the zone as generated by the revolution of the arc of $y = \sqrt{R^2 - x^2}$ from $x = c$ to $x = c + h$ about the x axis.
21. A torus is generated by revolving a circle of radius a about a line in its plane whose distance from the center is b, where b exceeds a. Prove that its area is $4\pi^2 ab$. HINT: Consider the arc of $x = b + a \cos t$, $y = a \sin t$ from $t = 0$ to $t = 2\pi$ to be revolved about the y axis.

***187. Improper Integrals.** In our definition and discussion of the definite integral

$$\int_a^b f(x)dx, \tag{123}$$

we assumed that the interval a,b was finite, and that $f(x)$ was continuous for all values of x in this interval, including the end points a and b. Many applications lead us to consider definite integrals on infinite intervals, or integrals of functions which have discontinuities. In this section we shall define the meaning of such integrals and show how their values may be found.

We begin by considering an integral (123) of the type previously defined, so that $f(x)$ is continuous throughout the closed interval a,b. Then any indefinite integral $F(x)$ will have a derivative $F'(x) = f(x)$ for $a \leqq x \leqq b$. Hence $F(x)$ will be continuous for such values of x. Consequently we shall have

$$\lim_{x \to b} F(x) = F(b) \quad \text{and} \quad \lim_{x \to b} [F(x) - F(a)] = F(b) - F(a). \tag{124}$$

It follows that, under the conditions stated,

$$\lim_{x \to b} \int_a^x f(x)dx = \int_a^b f(x)dx. \tag{125}$$

Now suppose that $f(x)$ is continuous for all values of x with $a \leqq x < b$, but that $f(x)$ is discontinuous at b. The graph of one such function is shown in Fig. 229. If the limit on the left of Eq. (125) is finite, this limit is *defined* to be the value of the integral on the right. In this case the integral is *convergent*. If the expression on the left of Eq. (125) does not define a finite limit, the integral on the right is said to be *divergent*.

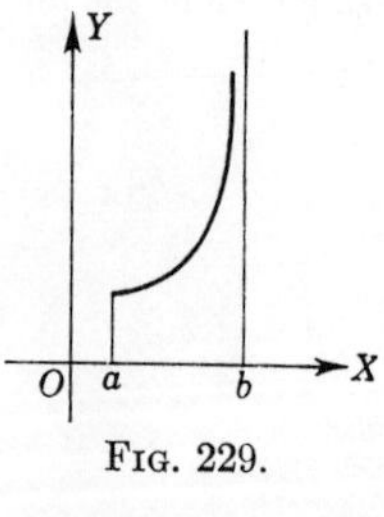

FIG. 229.

Similarly, if $f(x)$ is continuous for all values of x with $a < x \leqq b$ but is discontinuous at a, we use

$$\lim_{x \to a} \int_x^b f(x)dx = \int_a^b f(x)dx \tag{126}$$

as the definition of the right member.

Next assume that $f(x)$ is continuous for all values of x considered, but that one of the limits of the integral is infinite. The graph of one such function is shown in

Fig. 230. In this case we use as the defining equations

$$\lim_{x\to\infty} \int_a^x f(x)dx = \int_a^\infty f(x)dx, \tag{127}$$

and

$$\lim_{x\to-\infty} \int_x^b f(x)dx = \int_{-\infty}^b f(x)dx. \tag{128}$$

In the defining Eqs. (125) through (128) we have assumed that the upper limit is

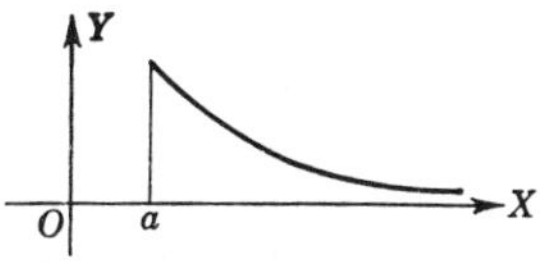

Fig. 230.

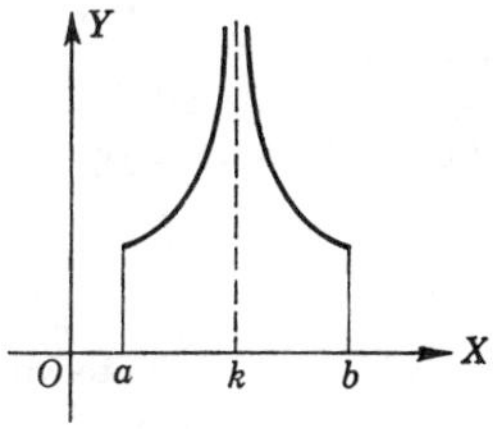

Fig. 231.

greater than the lower limit. However, similar definitions apply if this is not the case. These are such that Eq. (42), or

$$\int_a^b f(x)dx = -\int_b^a f(x)dx, \tag{129}$$

holds for the new type of integral.

Let us next consider a function $f(x)$ which is continuous throughout the interval a,b except for the value k, that is, for $a \leqq x < k$ and $k < x \leqq b$. The graph of one such function is shown in Fig. 231. Then we define

$$\int_a^b f(x)dx = \int_a^k f(x)dx + \int_k^b f(x)dx. \tag{130}$$

If both the integrals on the right of this equation are convergent when defined by relations similar in form to those of Eqs. (125) and (126), then the integral on the left of Eq. (130) is said to be *convergent*. If either or both of the integrals on the right diverge, then the integral on the left of Eq. (130) is said to be *divergent*.

If a is $-\infty$ or a point of discontinuity of $f(x)$ and b is ∞ or a point of discontinuity of $f(x)$, but $f(x)$ is continuous for all x such that $a < x < b$, we may again use Eq. (130) to define the integral on the left in terms of those on the right. And a similar decomposition may be used to define the integral when $f(x)$ has several isolated points of discontinuity.

Definite integrals that may be defined in terms of $f(x)$ by the single limiting process of the fundamental theorem, Eq. (13), are called *proper* integrals. In contradistinction, the integrals of this section which are defined in terms of the limits of proper integrals are called *improper* integrals.

The properties of Sec. 175 expressed in Eqs. (42), (43), (46), and (47) all hold for convergent improper integrals. And in any case where

$$\frac{dF}{dx} = F'(x) \text{ is finite and } = f(x) \text{ for } a < x < b, \tag{131}$$

and $F(x)$ approaches finite limits as $x \to a$ and $x \to b$, we may still use Eq. (40), or

$$\int_a^b f(x)dx = F(b) - F(a), \tag{132}$$

to evaluate the integral.

It is sometimes convenient to make a change of variable like that of Eq. (58) in an improper integral. For example, by proceeding as in the example of Sec. 181 we find that the length of the quadrant of the circle $x^2 + y^2 = 4$ from (0,2) to (2,0) is

$$2\int_0^2 \frac{dx}{\sqrt{4 - x^2}} = 2\int_0^{\pi/2} du = 2[u]_0^{\pi/2} = \pi. \tag{133}$$

Since $x = 2 \sin u$, the indefinite integral $2u = 2 \sin^{-1} (x/2)$. And the direct application of Eq. (132) would give

$$2\int_0^2 \frac{dx}{\sqrt{4 - x^2}} = 2\left[\sin^{-1}\frac{x}{2}\right]_0^2 = 2(\sin^{-1} 1 - \sin^{-1} 0) = 2\left(\frac{\pi}{2} - 0\right) = \pi. \tag{134}$$

Since $1/\sqrt{4 - x^2}$ is infinite when $x = 2$, the first integral in Eqs. (133) and (134) is improper. Either of the equations shows that this integral is convergent and has π as its value. The substitution of Eq. (133) converted the improper integral in x into the proper integral in u. That this integral in u is really that for the quadrant may be verified by using the parametric equations for the arc, $x = 2 \cos u$, $y = 2 \sin u$, from $u = 0$ to $u = \pi/2$, and Eq. (107) with u in place of t. This illustrates that, when the calculation of a geometrical or physical quantity leads to a convergent improper integral, the value of the quantity is correctly given by the definitions of this section.

Example 1. Find the area bounded above by the curve $y^3(8 - x) = 1$, below by the x axis, on the left by the y axis, and on the right by the vertical asymptote $x = 8$.

Solution: Since $y = (8 - x)^{-\frac{1}{3}}$ is positive for $0 < x < 8$, the area $A = \int y\,dx = \int_0^8 (8 - x)^{-\frac{1}{3}}\,dx$. The integral is improper, since $(8 - x)^{-\frac{1}{3}} \to \infty$ as $x \to 8$. But we may deduce from Eq. (132) that

$\int_0^8 (8 - x)^{-\frac{1}{3}}\,dx = -\frac{3}{2}[(8 - x)^{\frac{2}{3}}]_0^8 = -\frac{3}{2}(0 - 4) = 6$, since $(8 - x)^{\frac{2}{3}}$ is continuous for $0 \leqq x \leqq 8$.

Hence the required area is 6.

Example 2. Show that $\int_0^8 (8 - x)^{-1}\,dx$ is divergent.

Solution: $\int_0^8 (8 - x)^{-1}\,dx = -\int_0^8 \frac{d(8 - x)}{8 - x} = -[\ln\ (8 - x)]_0^8$ diverges, since $\ln (8 - x) \to -\infty$ as $x \to 8$.

Example 3. Show that $\int_{-1}^2 \frac{dx}{x^2}$ is divergent.

Solution: This integral is improper since $1/x^2 \to \infty$ as $x \to 0$. Hence we must write $\int_{-1}^2 \frac{dx}{x^2} = \int_{-1}^0 \frac{dx}{x^2} + \int_0^2 \frac{dx}{x^2} = \left[-\frac{1}{x}\right]_{-1}^0 + \left[-\frac{1}{x}\right]_0^2$ diverges, since $\frac{1}{x} \to \infty$ as $x \to 0$. Note that, if we overlooked the discontinuity at $x = 0$ between -1 and 2, we would obtain an incorrect result for this example, since $\left[-\frac{1}{x}\right]_{-1}^2 = -\frac{1}{2} + 1 = \frac{1}{2}$.

Example 4. Find the area bounded above by the curve $x^2y = 1$, below by the x axis, on the left by $x = 1$, and extending to infinity on the right.

Solution: Since $y = \frac{1}{x^2}$ is positive for $x > 1$, the area is $A = \int y\,dx = \int_1^\infty \frac{dx}{x^2}$

$= \left[-\frac{1}{x}\right]_1^\infty = 0 - (-1) = 1$, since $\frac{1}{x^2}$ is continuous for $x > 1$. Hence the required area is 1.

EXAMPLE 5. Find the area below the curve $x^2y + y = 1$, above the x axis, and extending to infinity in both directions.

Solution: Since $y = \frac{1}{x^2 + 1}$ is positive for all x, the area is $A = \int y\,dx = \int_{-\infty}^{\infty} \frac{dx}{x^2+1}$. Let $x = \tan u$. Then $dx = \sec^2 u\,du$, and $x^2 + 1 = \tan^2 u + 1 = \sec^2 u$. Also when u varies from $-\pi/2$ to $\pi/2$, x varies from $-\infty$ to $+\infty$. Hence, since $1/(x^2+1)$ is continuous for all x, we have

$$\int_{-\infty}^{\infty} \frac{dx}{x^2+1} = \int_{-\pi/2}^{\pi/2} \frac{\sec^2 u\,du}{\sec^2 u} = \int_{-\pi/2}^{\pi/2} du = [u]_{-\pi/2}^{\pi/2} = \frac{\pi}{2} - \left(-\frac{\pi}{2}\right) = \pi.$$

Thus the required area is π.

EXERCISE 91

Each of the following integrals is improper because the integrand is discontinuous at one of the limits. In each case, either evaluate the integral or prove that the integral diverges.

1. $\int_0^1 \frac{dx}{\sqrt{x}}$.
2. $\int_0^1 \frac{dx}{x}$.
3. $\int_2^6 \frac{dx}{\sqrt{6-x}}$.
4. $\int_2^6 \frac{dx}{(6-x)^{\frac{3}{2}}}$.
5. $\int_5^6 \frac{dx}{x-5}$.
6. $\int_5^6 \frac{dx}{(x-5)^{\frac{2}{3}}}$.

Each of the following integrals is improper because one of the limits is infinite. In each case, either evaluate the integral or prove that the integral diverges.

7. $\int_2^\infty \frac{dx}{x^3}$.
8. $\int_2^\infty \frac{dx}{x}$.
9. $\int_{-\infty}^0 \frac{dx}{(1-x)^2}$.
10. $\int_{-\infty}^0 \frac{dx}{1-x}$.
11. $\int_0^\infty \cos x\,dx$.
12. $\int_{-\infty}^0 e^x\,dx$.

Evaluate each of the following improper integrals.

13. $\int_{-1}^8 \frac{dx}{x^{\frac{1}{3}}}$.
14. $\int_0^{33} \frac{dx}{(x-1)^{\frac{4}{5}}}$.
15. $\int_{-1}^1 \frac{dx}{\sqrt{1-x^2}}$.
16. $\int_{-\infty}^\infty \frac{dx}{4+x^2}$.
17. Find the area bounded above by the curve $y = e^{-x}$, below by the x axis, on the left by the y axis, and extending to infinity on the right.
18. Find the area bounded above by the curve $y^4(1-x) = 1$, below by the x axis, on the left by the y axis, and on the right by the vertical asymptote $x = 1$.

Find the *volume* of the solid generated by revolving about the x axis the area described in

19. Prob. 17.
20. Prob. 18.
21. Find the length of the arc of the curve $r = e^{-\theta}$ from $\theta = 0$ to $\theta = \infty$.

CHAPTER 13

METHODS OF INTEGRATION

An introduction to integration as the process inverse to differentiation was given in Chap. 5. Most of the applications made there involved only polynomial functions. The definite integral as the limit of a sum was discussed in Chap. 12, and many of its properties were derived. There the applications involved integrals obtained more or less directly from the rules for differentiation or the integration formulas of Sec. 173.

In this chapter we shall give a systematic account of the principal methods of finding integrals. We begin by establishing formulas for the integrals of a number of simple types of functions of frequent occurrence. Then we discuss methods of transforming other types of integrands into forms to which the formulas for the simple types can be applied.

188. Integrals of Powers. We begin by recalling those integration formulas which enable us to integrate powers and sums of powers. In these formulas the symbols c_1, c_2, and n denote constants. And C is the constant of integration of Sec. 65. In evaluating a definite integral as in Sec. 72, the constant C may be omitted. But to give the general value of any indefinite integral one constant must be added. The letters u and v represent functions of x, so that $du = (du/dx)dx$. From Eq. (25) of Sec. 173 we have

$$\int(c_1u + c_2v)dx = c_1\int u\,dx + c_2\int v\,dx + C. \tag{1}$$

Here the C must be added if $\int u\,dx$ and $\int v\,dx$ mean particular integrals. If one or both are interpreted as indefinite integrals, the C may be omitted from Eq. (1).

The Eq. (1) expresses the *linearity* property. A similar result holds for any number of functions. This linearity property enables us to decompose the integral of a sum into a sum of integrals and to bring out *constant* factors from under the integral sign.

From Eqs. (26), (27), (28), and (30) of Sec. 173 we have

$$\int 1\,du = \int du = u + C. \tag{2}$$

$$\int u\,du = \frac{u^2}{2} + C. \tag{3}$$

$$\int u^n\,du = \frac{u^{n+1}}{n+1} + C, \text{ if } n \neq -1. \tag{4}$$

$$\int u^{-1}\,du = \int \frac{du}{u} = \ln u + C. \tag{5}$$

The Eqs. (2) to (5) enable us to integrate single powers. To integrate a sum of powers, we first use the linearity property and then apply Eqs. (2) to (5).

Example 1. Evaluate $\int \frac{ax^2 + bx + c}{x^2}\,dx$.

Solution: Divide each term in the numerator by x^2, decompose the sum as in Eq. (1), and then use Eqs. (2), (5), and (4) with $n = -2$ and $u = x$. Thus

$$\begin{aligned}\int \frac{ax^2 + bx + c}{x^2}\,dx &= a\int dx + b\int x^{-1}\,dx + c\int x^{-2}\,dx \\ &= ax + b\ln x - \frac{c}{x} + C.\end{aligned}$$

This is the required value. Although by themselves the three integrals would require separate constants, C_1, C_2, and C_3, we need write only one constant C denoting the linear combination, $C = aC_1 + bC_2 + cC_3$.

Example 2. Evaluate $\int \frac{dx}{(2-3x)^{\frac{3}{2}}}$.

Solution 1: Let $u = 2 - 3x$. Then $du = -3\,dx$, $dx = -\frac{1}{3}\,du$. And $\frac{dx}{(2-3x)^{\frac{3}{2}}} = \int \frac{-\frac{1}{3}\,du}{u^{\frac{3}{2}}} = -\frac{1}{3}\int u^{-\frac{3}{2}}\,du$. We evaluate this integral by Eq. (4) with $n = -\frac{3}{2}$ as $-\frac{1}{3}\frac{u^{-\frac{1}{2}}}{-\frac{1}{2}} + C = \frac{2}{3}u^{-\frac{1}{2}} + C = \frac{2}{3\sqrt{2-3x}} + C$. This is the required value.

Solution 2: Observe mentally that $d(2-3x) = -3\,dx$ and write $\int \frac{dx}{(2-3x)^{\frac{3}{2}}} = -\frac{1}{3}\int (2-3x)^{-\frac{3}{2}}(-3\,dx) = -\frac{1}{3}\frac{(2-3x)^{-\frac{1}{2}}}{-\frac{1}{2}} + C = \frac{2}{3\sqrt{2-3x}} + C$. This is the required value.

The method of solution 2 shows that

$$\begin{aligned}\int (ax+b)^n\,dx &= \frac{1}{a}\int (ax+b)^n(a\,dx) = \frac{(ax+b)^{n+1}}{a(n+1)} + C \qquad \text{if } n \neq -1 \text{ and} \\ &= \frac{1}{a}\ln(ax+b) \qquad \text{if } n = -1.\end{aligned}$$

Example 3. Evaluate $\int (1 + 2x_2)^{\frac{2}{3}}x\,dx$.

Solution 1: Let $u = 1 + 2x^2$. Then $du = 4x\,dx$, $x\,dx = \frac{du}{4}$. And

$$\begin{aligned}\int (1+2x^2)^{\frac{2}{3}}x\,dx &= \int u^{\frac{2}{3}}\frac{du}{4} = \frac{1}{4}\int u^{\frac{2}{3}}\,du = \frac{1}{4}\frac{u^{\frac{5}{3}}}{\frac{5}{3}} + C \\ &= \tfrac{3}{20}(1+2x^2)^{\frac{5}{3}} + C.\end{aligned}$$

We evaluated this integral by Eq. (4) with $n = \frac{2}{3}$.

Solution 2: Observe mentally that $d(1 + 2x^2) = 4x\,dx$ and write

$$\int (1 + 2x^2)^{\frac{2}{3}}\, x\, dx = \frac{1}{4}\int (1 + 2x^2)^{\frac{2}{3}}\,(4x\,dx) = \frac{1}{4}\frac{(1 + 2x^2)^{\frac{5}{3}}}{\frac{5}{3}} + C$$
$$= \tfrac{3}{20}(1 + 2x^2)^{\frac{5}{3}} + C, \text{ the required value.}$$

The method of solution 2 shows that

$$\int (ax^2 + b)^n x\,dx = \frac{1}{2a}\int (ax^2 + b)^n(2ax\,dx)$$
$$= \frac{(ax^2 + b)^{n+1}}{2a(n + 1)} + C \qquad \text{if } n \neq -1 \qquad \text{and}$$
$$= \frac{1}{2a}\ln (ax^2 + b) + C \qquad \text{if } n = -1.$$

EXAMPLE 4. Evaluate $\int \frac{e^{ax}\,dx}{b + ce^{ax}}$.

Solution: Observe mentally that $d(b + ce^{ax}) = ace^{ax}$ and write

$$\int \frac{e^{ax}\,dx}{b + ce^{ax}} = \frac{1}{ac}\int \frac{ace^{ax}\,dx}{b + ce^{ax}} = \frac{1}{ac}\ln (b + ce^{ax}) + C.$$

Here the integral was evaluated by Eq. (5) with $u = b + ce^{ax}$.

EXAMPLE 5. Evaluate $\int \sin 5x \cos^2 5x\,dx$.

Solution: Observe mentally that $d(\cos 5x) = -5 \sin 5x\,dx$ and write

$$\int \sin 5x \cos^2 5x\,dx = -\frac{1}{5}\int \cos^2 5x(-5 \sin 5x\,dx) = -\frac{1}{5}\frac{\cos^3 5x}{3} + C$$
$$= -\tfrac{1}{15}\cos^3 5x + C.$$

Here the integral was evaluated by Eq. (4) with $u = \cos 5x$ and $n = 2$.

EXAMPLE 6. Evaluate $\int \tan 3x \sec^5 3x\,dx$.

Solution: Observe mentally that $d(\sec 3x) = 3 \tan 3x \sec 3x\,dx$ and write

$$\int \tan 3x \sec^5 3x\,dx = \tfrac{1}{3}\int \sec^4 3x(3 \tan 3x \sec 3x\,dx)$$
$$= \frac{1}{3}\frac{\sec^5 3x}{5} + C = \frac{1}{15}\sec^5 3x + C.$$

Here the integral was evaluated by Eq. (4) with $u = \sec 3x$ and $n = 4$.

EXAMPLE 7. Evaluate $\int \tan^5 3x \sec^2 3x\,dx$.

Solution: Observe mentally that $d(\tan 3x) = 3 \sec^2 3x\,dx$ and write

$$\int \tan^5 3x \sec^2 3x\,dx = \frac{1}{3}\int \tan^5 3x(3 \sec^2 3x\,dx) = \frac{1}{3}\frac{\tan^6 3x}{6} + C$$
$$= \tfrac{1}{18}\tan^6 3x + C.$$

Here the integral was evaluated by Eq. (4) with $u = \tan 3x$ and $n = 5$.

EXAMPLE 8. Evaluate $\int \frac{dx}{x \ln x}$.

Solution: Let $u = \ln x$. Then $du = \frac{dx}{x}$. And $\int \frac{dx}{x \ln x} = \int \frac{du}{u}$. By Eq. (5) this is $\ln u + C = \ln (\ln x) + C$, the required value.

EXERCISE 92

Evaluate each of the following integrals.

1. $\int \frac{x^2 - 3x + 4}{x^6}\, dx.$

2. $\int \frac{x^2 + 2x - 3}{x^3}\, dx.$

3. $\int \sqrt{x}\,(5x + 6)\, dx.$

4. $\int \frac{6x - 5}{\sqrt{x}}\, dx.$

5. $\int \frac{(x - 2)^2}{x^3}\, dx$

6. $\int \left(\sqrt{x} + \frac{1}{\sqrt{x}}\right)^2 dx.$

7. $\int \sqrt{3 + 4x}\, dx.$

8. $\int \frac{dx}{\sqrt{3 - 4x}}.$

9. $\int \frac{dx}{(5 - 2x)^2}.$

10. $\int \frac{dx}{2 + 3x}.$

11. $\int \frac{x\, dx}{2x^2 + 1}.$

12. $\int \frac{x\, dx}{\sqrt{4 + 3x^2}}.$

13. $\int x \sqrt{x^2 - 1}\, dx.$

14. $\int x(1 - 2x^2)^9\, dx.$

15. $\int \frac{e^{2x}\, dx}{\sqrt{e^{2x} - 1}}.$

16. $\int \frac{e^{-x}\, dx}{(e^{-x} + 1)^2}.$

17. $\int \frac{\sqrt{\ln x}}{x}\, dx.$

18. $\int \frac{\ln x}{x}\, dx.$

19. $\int \sin^4 2x \cos 2x\, dx.$

20. $\int \sin 2x \cos^3 2x\, dx.$

21. $\int \tan \frac{x}{3} \sec^3 \frac{x}{3}\, dx.$

22. $\int \tan^3 \frac{x}{4} \sec^2 \frac{x}{4}\, dx.$

189. The Integral of *du/u*. We may use Eq. (5) as given for a range throughout which u is positive. For example, the definite integral

$$\int_2^4 \frac{dx}{x} = [\ln x]_2^4 = \ln 4 - \ln 2 = \ln \frac{4}{2} = \ln 2 = 0.6931.$$

For a range throughout which u is negative we may use

$$\int u^{-1}\, du = \int \frac{du}{u} = \ln(-u) + C. \tag{6}$$

This follows from the fact that

$$d \ln(-u) = \frac{d(-u)}{-u} = \frac{-du}{-u} = \frac{du}{u}. \tag{7}$$

For example, the definite integral

$$\int_{-4}^{-2} \frac{dx}{x} = [\ln(-x)]_{-4}^{-2} = \ln 2 - \ln 4 = \ln \frac{1}{2} = -\ln 2 = -0.6931.$$

However, it is possible to use Eq. (5) as given for the negative range if we transform the expression before trying to calculate the individual logarithms. Thus

$$\int_{-4}^{-2} \frac{dx}{x} = [\ln x]_{-4}^{-2} = \ln(-2) - \ln(-4) = \ln\frac{-2}{-4} = \ln\frac{1}{2} = -\ln 2$$
$$= -0.6931$$

While negative numbers have no real logarithms, it is possible to consider the pure imaginary $\pi i = \pi\sqrt{-1}$ as one of the logarithms of -1, as we shall show in Prob. 22 of Exercise 132. Thus Eq. (6) is equivalent to Eq. (5) with the constant C replaced by $C + \pi i$, so that

$$\ln u + C + \pi i = \ln u + C + \ln(-1) = \ln(-u) + C. \tag{8}$$

In a definite integral with integrand du/u which represents a finite real quantity, u will preserve its sign throughout the range of integration. And the result of using Eq. (5) in the way we have indicated will always give a correct result. Hence in tabulating integrals whose values involve logarithms, it is unnecessary to write out both of the possible formulas explicitly. In practice we use Eq. (6) when we realize that the range is negative sufficiently soon as in the solution 1 to Example 1 below. Otherwise we set

$$\ln(-b) - \ln(-a) = \ln\frac{b}{a} = \ln b - \ln a.$$

EXAMPLE 1. Evaluate $\int_0^2 \frac{dx}{2x - 9}$.

Solution 1: Let $u = 2x - 9$. Then $du = 2\,dx$, $dx = \frac{1}{2}\,du$. Since $u = -9$ when $x = 0$ and $u = -5$ when $x = 2$, from Eq. (58) of Sec. 178 we may deduce that

$$\int_0^2 \frac{dx}{2x - 9} = \int_{-9}^{-5} \frac{\frac{1}{2}\,du}{u} = \frac{1}{2}[\ln(-u)]_{-9}^{-5}$$
$$= \tfrac{1}{2}(\ln 5 - \ln 9) = \tfrac{1}{2}\ln\tfrac{5}{9}.$$

Solution 2: We observe that $d(2x - 9) = 2\,dx$ and write

$$\int_0^2 \frac{dx}{2x - 9} = \frac{1}{2}\int_0^2 \frac{2\,dx}{2x - 9} = \frac{1}{2}[\ln(2x - 9)]_0^2 = \frac{1}{2}[\ln(-5) - \ln(-9)] = \frac{1}{2}\ln\frac{5}{9}.$$

If we wish the result numerically, we find from the tables that $\ln 5 = 1.6094$, $\ln 9 = 2.1972$, so that $\ln\frac{5}{9} = \ln 5 - \ln 9 = -0.5878$. And $\frac{1}{2}\ln\frac{5}{9} = -0.2939$. Or, for these numbers we have more simply

$$\tfrac{1}{2}\ln\tfrac{5}{9} = -\tfrac{1}{2}\ln\tfrac{9}{5} = -\tfrac{1}{2}\ln 1.8 = -\tfrac{1}{2}(0.5878) = -0.2939.$$

EXAMPLE 2. Evaluate $\int_{3\pi/4}^{2\pi/3} \tan x\,dx$.

Solution: $\int \tan x\,dx = \int \frac{\sin x}{\cos x}\,dx$. Since $d(\cos x) = -\sin x\,dx$, we write

$$\int_{3\pi/4}^{2\pi/3} \tan x\,dx = -\int_{3\pi/4}^{2\pi/3} \frac{-\sin x\,dx}{\cos x} = -[\ln\cos x]_{3\pi/4}^{2\pi/3}$$
$$= -\ln\cos 120° + \ln\cos 135°$$
$$= -\ln\left(-\frac{1}{2}\right) + \ln\left(-\frac{1}{\sqrt{2}}\right) = \ln\frac{-2}{-\sqrt{2}}$$
$$= \ln\sqrt{2} = \tfrac{1}{2}\ln 2 = \tfrac{1}{2}(0.6931) = 0.3466, \text{ the required value.}$$

190. Integrals Leading to Logarithms. Some integrals not immediately in the form du/u may be reduced to this form by a trigonometric or algebraic substitution. In this section we illustrate this procedure for a number of integrals of such frequent occurrence that the results are worth memorizing as new formulas of integration.

To evaluate $\int \tan u\, du$, we note that

$$\tan u\, du = \frac{\sin u}{\cos u} du = -\frac{-\sin u\, du}{\cos u} = -\frac{d(\cos u)}{\cos u}. \tag{9}$$

It follows that

$$\int \tan u\, du = -\ln \cos u + C = \ln \sec u + C, \tag{10}$$

since

$$-\ln \cos u = \ln \frac{1}{\cos u} = \ln \sec u.$$

By a similar procedure we may deduce the corresponding formula

$$\int \cot u\, du = \ln \sin u + C = -\ln \csc u + C. \tag{11}$$

Consider next $\int \sec u\, du$. We have

$$\begin{aligned} \sec u\, du &= \frac{\sec u + \tan u}{\sec u + \tan u} \sec u\, du = \frac{\tan u \sec u + \sec^2 u}{\sec u + \tan u} du \\ &= \frac{d(\sec u + \tan u)}{\sec u + \tan u}. \end{aligned} \tag{12}$$

It follows that

$$\int \sec u\, du = \ln (\sec u + \tan u) + C. \tag{13}$$

By a similar procedure we may deduce the corresponding formula

$$\int \csc u\, du = -\ln (\csc u + \cot u) + C. \tag{14}$$

Or we may derive Eq. (14) by replacing u by $\pi/2 - u$ in Eq. (13).

We next consider $\int \frac{du}{\sqrt{u^2 + A}}$. We have

$$\begin{aligned} \frac{du}{\sqrt{u^2 + A}} &= \frac{u + \sqrt{u^2 + A}}{u + \sqrt{u^2 + A}} \frac{du}{\sqrt{u^2 + A}} = \frac{1 + u/\sqrt{u^2 + A}}{u + \sqrt{u^2 + A}} du \\ &= \frac{d(u + \sqrt{u^2 + A})}{u + \sqrt{u^2 + A}}. \end{aligned} \tag{15}$$

It follows that

$$\int \frac{du}{\sqrt{u^2 + A}} = \ln (u + \sqrt{u^2 + A}) + C. \tag{16}$$

We may evaluate $\int \frac{du}{u^2 - a^2}$ as follows. We first verify that

$$\frac{1}{u-a} - \frac{1}{u+a} = \frac{2a}{u^2 - a^2}. \tag{17}$$

We may deduce from this that

$$\begin{aligned}\int \frac{du}{u^2 - a^2} &= \frac{1}{2a} \int \left(\frac{du}{u-a} - \frac{du}{u+a} \right) \\ &= \frac{1}{2a} [\ln (u-a) - \ln (u+a)] + C \\ &= \frac{1}{2a} \ln \frac{u-a}{u+a} + C. \end{aligned} \tag{18}$$

EXAMPLE 1. Evaluate $\int \frac{5\,dx}{\sin 3x}$.

Solution: Write $\int \frac{5\,dx}{\sin 3x} = \frac{5}{3} \int \csc 3x\, d(3x)$. And use Eq. (14) with $u = 3x$ to deduce that this is $\frac{5}{3}[-\ln (\csc 3x + \cot 3x)] + C$. Hence

$$\int \frac{5\,dx}{\sin 3x} = -\frac{5}{3} \ln (\csc 3x + \cot 3x) + C.$$

EXAMPLE 2. Evaluate $\int \frac{dx}{\sqrt{4x^2 - 1}}$.

Solution 1: We write $\int \frac{dx}{\sqrt{4x^2 - 1}} = \frac{1}{2} \int \frac{d(2x)}{\sqrt{(2x)^2 - 1}}$. We evaluate this integral by Eq. (16) with $u = 2x$ and $A = -1$. Thus

$$\int \frac{dx}{\sqrt{4x^2 - 1}} = \frac{1}{2} \ln (2x + \sqrt{4x^2 - 1}) + C.$$

Solution 2: We write $\int \frac{dx}{\sqrt{4x^2 - 1}} = \frac{1}{2} \int \frac{dx}{\sqrt{x^2 - \frac{1}{4}}}$. We evaluate this integral by Eq. (16) with $u = x$ and $A = -\frac{1}{4}$. Thus

$$\int \frac{dx}{\sqrt{4x^2 - 1}} = \frac{1}{2} \ln \left(x + \sqrt{x^2 - \frac{1}{4}} \right) + C_1.$$

Since $x + \sqrt{x^2 - \frac{1}{4}} = \frac{2x + \sqrt{4x^2 - 1}}{2}$, the right member equals $\frac{1}{2}[\ln (2x + \sqrt{4x^2 - 1} - \ln 2] + C_1$, which agrees with the solution found in solution 1 if $C = C_1 - \frac{1}{2} \ln 2$.

EXAMPLE 3. Evaluate $\int \frac{dx}{3 - 4x^2}$.

Solution: We write $\int \frac{dx}{3 - 4x^2} = -\frac{1}{4} \int \frac{dx}{x^2 - (\sqrt{3}/2)^2}$. We evaluate this integral by Eq. (18) with $u = x$ and $a = \sqrt{3}/2$. Thus

$$\int \frac{dx}{3 - 4x^2} = -\frac{1}{4} \frac{1}{\sqrt{3}} \ln \frac{x - \sqrt{3}/2}{x + \sqrt{3}/2} + C = \frac{1}{4\sqrt{3}} \ln \frac{2x + \sqrt{3}}{2x - \sqrt{3}} + C.$$

EXERCISE 93

Evaluate each of the following integrals.

1. $\int_0^2 \frac{dx}{x+2}$.

2. $\int_2^3 \frac{dx}{1-x}$.

3. $\int \tan \frac{5x}{2}\, dx$.

4. $\int \cot (3x+2)dx$.

5. $\int_0^\pi \tan \frac{x}{3}\, dx$.

6. $\int_\pi^{3\pi/2} \cot \frac{x}{2}\, dx$.

7. $\int \csc 2x\, dx$.

8. $\int \sec (2x-3)dx$.

9. $\int_0^{\pi/4} \frac{dx}{\cos x}$.

10. $\int_{4\pi/3}^{3\pi/2} \frac{dx}{\sin x}$.

11. $\int \frac{dx}{\sqrt{x^2+2}}$.

12. $\int \frac{dx}{\sqrt{x^2-3}}$.

13. $\int \frac{dx}{x^2-4}$.

14. $\int \frac{dx}{9-x^2}$.

15. $\int \frac{dx}{\sqrt{9x^2-5}}$.

16. $\int \frac{dx}{25-4x^2}$.

17. $\int \frac{dx}{\sqrt{16x^2-8}}$.

18. $\int \frac{dx}{3x^2-6}$.

19. $\int_3^5 \frac{dx}{\sqrt{x^2-9}}$.

20. $\int_3^5 \frac{dx}{x^2-1}$.

191. Integrals Leading to Inverse Trigonometric Functions. It follows from Eq. (111) of Sec. 102 that

$$\int \frac{du}{\sqrt{1-u^2}} = \sin^{-1} u + C. \tag{19}$$

This enables us to evaluate $\int \frac{du}{\sqrt{a^2-u^2}}$. We have

$$\int \frac{du}{\sqrt{a^2-u^2}} = \int \frac{d(u/a)}{\sqrt{1-(u/a)^2}} = \sin^{-1} \frac{u}{a} + C, \tag{20}$$

by using Eq. (19) with u/a in place of u.

It follows from Eq. (121) of Sec. 104 that

$$\int \frac{du}{u^2+1} = \tan^{-1} u + C. \tag{21}$$

This enables us to evaluate $\int \frac{du}{u^2+a^2}$. We have

$$\int \frac{du}{u^2+a^2} = \frac{1}{a} \int \frac{d(u/a)}{(u/a)^2+1} = \frac{1}{a} \tan^{-1} \frac{u}{a} + C, \tag{22}$$

by using Eq. (21) with u/a in place of u.

It follows from Eq. (132) of Sec. 106 that

$$\int \frac{du}{u\sqrt{u^2-1}} = \sec^{-1} u + C. \tag{23}$$

This enables us to evaluate $\int \frac{du}{u\sqrt{u^2-a^2}}$. We have

$$\int \frac{du}{u\sqrt{u^2-a^2}} = \frac{1}{a}\int \frac{d(u/a)}{(u/a)\sqrt{(u/a)^2-1}} = \frac{1}{a}\sec^{-1}\frac{u}{a} + C. \tag{24}$$

For Eqs. (20), (22), and (24) it is easy to remember the part involving an inverse function by association with Eqs. (19), (21), (23) and the corresponding differentiation formulas. And the absence or presence of the factor $1/a$ may be related to the fact that if a and u have the same dimensions, as feet, du also has this dimension, and the integrand of Eq. (20) is of zero dimensions, while the integrands of Eq. (22) and (24) have the dimension of a minus first power, and so require a factor $1/a$ with the dimensionless inverse trigonometric function of u/a.

When applying Eq. (20), (22), or (24) to a definite integral between the limits $u = u_1$ and $u = u_2$ we must take care to choose the right values of the multiple-valued inverse functions. The method of checking the values by observing the one-to-one and continuous correspondence for u between u_1 and u_2 is illustrated in Examples 3 and 4 below.

Example 1. Evaluate $\int \frac{dx}{\sqrt{3-2x^2}}$.

Solution 1: We write $\int \frac{dx}{\sqrt{3-2x^2}} = \frac{1}{\sqrt{2}}\int \frac{dx}{\sqrt{\frac{3}{2}-x^2}}$. We evaluate this integral by Eq. (20) with $u = x$ and $a = \sqrt{\frac{3}{2}}$. Thus $\int \frac{dx}{\sqrt{3-2x^2}} = \frac{1}{\sqrt{2}}\sin^{-1}\sqrt{\frac{2}{3}}\,x + C$.

Solution 2: We write $\int \frac{dx}{\sqrt{3-2x^2}} = \frac{1}{\sqrt{2}}\int \frac{d(\sqrt{2}\,x)}{\sqrt{3-(\sqrt{2}\,x)^2}}$. We evaluate this integral by Eq. (20) with $u = \sqrt{2}\,x$ and $a = \sqrt{3}$. Thus

$$\int \frac{dx}{\sqrt{3-2x^2}} = \frac{1}{\sqrt{2}}\sin^{-1}\sqrt{\frac{2}{3}}\,x + C.$$

Example 2. Evaluate $\int \frac{dx}{x\sqrt{5x^2-4}}$.

Solution: We write $\int \frac{dx}{x\sqrt{5x^2-4}} = \int \frac{d(\sqrt{5}\,x)}{(\sqrt{5}\,x)\sqrt{(\sqrt{5}\,x)^2-4}} = \frac{1}{2}\sec^{-1}\frac{\sqrt{5}}{2}\,x + C$, by Eq. (24) with $u = \sqrt{5}\,x$ and $a = 2$.

Example 3. Evaluate $\int_{-1}^{3} \frac{dx}{x^2+3}$.

Solution: By Eq. (22) with $u = x$ and $a = \sqrt{3}$ we have

$$\int_{-1}^{3} \frac{dx}{x^2 + 3} = \frac{1}{\sqrt{3}} \left[\tan^{-1} \frac{x}{\sqrt{3}} \right]_{-1}^{3} = \frac{1}{\sqrt{3}} \left(\tan^{-1} \sqrt{3} - \tan^{-1} \frac{-1}{\sqrt{3}} \right)$$

$$= \frac{1}{\sqrt{3}} \left[\frac{\pi}{3} - \left(-\frac{\pi}{6} \right) \right] = \frac{\pi}{2\sqrt{3}}.$$

We take values between $-\pi/2$ and $\pi/2$. Thus $\tan^{-1}(x/\sqrt{3})$ increases continuously from $-\pi/6$ to $\pi/3$ as x increases continuously from -1 to 3.

Example 4. Evaluate $\displaystyle\int_{-\sqrt{2}}^{1} \frac{dx}{\sqrt{2 - x^2}}$.

Solution: By Eq. (20) with $u = x$ and $a = \sqrt{2}$, we have

$$\int_{-\sqrt{2}}^{1} \frac{dx}{\sqrt{2 - x^2}} = \left[\sin^{-1} \frac{x}{\sqrt{2}} \right]_{-\sqrt{2}}^{1} = \sin^{-1} \frac{1}{\sqrt{2}} - \sin^{-1}(-1) = \frac{\pi}{4} - \left(-\frac{\pi}{2} \right) = \frac{3\pi}{2}.$$

We take values between $-\pi/2$ and $\pi/2$. Thus $\sin^{-1}(x/\sqrt{2})$ increases continuously from $-\pi/2$ to $\pi/4$ as x increases continuously from $-\sqrt{2}$ to 1.

EXERCISE 94

Evaluate each of the following integrals.

1. $\displaystyle\int \frac{dx}{\sqrt{9 - x^2}}$.
2. $\displaystyle\int \frac{dx}{\sqrt{9 - 4x^2}}$.
3. $\displaystyle\int \frac{dx}{x^2 + 25}$.
4. $\displaystyle\int \frac{dx}{9x^2 + 4}$.
5. $\displaystyle\int \frac{dx}{x\sqrt{x^2 - 16}}$.
6. $\displaystyle\int \frac{dx}{x\sqrt{4x^2 - 9}}$.
7. $\displaystyle\int \frac{dx}{\sqrt{16 - 9x^2}}$.
8. $\displaystyle\int \frac{dx}{\sqrt{4 - 3x^2}}$.
9. $\displaystyle\int \frac{dx}{4x^2 + 9}$.
10. $\displaystyle\int \frac{dx}{3x^2 + 5}$.
11. $\displaystyle\int \frac{dx}{x\sqrt{25x^2 - 4}}$.
12. $\displaystyle\int \frac{dx}{x\sqrt{3x^2 - 4}}$.
13. $\displaystyle\int_0^1 \frac{dx}{\sqrt{4 - x^2}}$.
14. $\displaystyle\int_{\sqrt{3}}^{3} \frac{dx}{x\sqrt{4x^2 - 9}}$.
15. $\displaystyle\int_{-3}^{3} \frac{dx}{\sqrt{36 - x^2}}$.
16. $\displaystyle\int_{-3}^{3} \frac{dx}{x^2 + 9}$.
17. $\displaystyle\int_{-6}^{2} \frac{dx}{x^2 + 12}$.
18. $\displaystyle\int_{-4}^{0} \frac{dx}{x^2 + 16}$.

192. The Quadratic Expression $ax^2 + bx + c$. The Eqs. (16), (18), (20), (22), and (24) involve quadratic expressions with two terms only, one the square of the variable and the other a constant. But the general quadratic expression with three terms, $ax^2 + bx + c$, may be transformed into a quadratic expression with two terms by completing the square and regarding $(x + b/2a)$ as a new variable. In fact

$$ax^2 + bx + c = a\left[x^2 + \frac{b}{a}x + \left(\frac{b}{2a}\right)^2\right] + c - \frac{b^2}{4a}$$

$$= a\left(x + \frac{b}{2a}\right)^2 + \frac{4ac - b^2}{4a}. \qquad (25)$$

Many integrals with $ax^2 + bx + c$ or $\sqrt{ax^2 + bx + c}$ in the denominator and a constant in the numerator may be reduced to a familiar form by the use of Eq. (25).

If the numerator is a first-degree expression, it is advisable to separate a part immediately integrable by taking $ax^2 + bx + c$ as a new variable before using the transformation of Eq. (25).

Certain fractions with $(x - p)\sqrt{ax^2 + bx + c}$ in the denominator may be reduced to the forms just discussed by taking $1/(x - p)$ as a new variable.

EXAMPLE 1. Evaluate $\int \frac{dx}{3x^2 - 2x + 2}$.

Solution: By Eq. (25), we have $3x^2 - 2x + 2 = 3(x - \frac{1}{3})^2 + \frac{5}{3}$. Hence

$$\int \frac{dx}{3x^2 - 2x + 2} = \frac{1}{3}\int \frac{dx}{(x - \frac{1}{3})^2 + \frac{5}{9}} = \frac{1}{3}\frac{3}{\sqrt{5}}\tan^{-1}\frac{x - \frac{1}{3}}{\sqrt{5}/3} + C,$$

by Eq. (22) with $u = x - \frac{1}{3}$ and $a = \sqrt{5}/3$. Thus the given integral

$$\int \frac{dx}{3x^2 - 2x + 2} = \frac{1}{\sqrt{5}}\tan^{-1}\frac{3x - 1}{\sqrt{5}} + C.$$

EXAMPLE 2. Evaluate $\int \frac{dx}{\sqrt{1 + 4x - 5x^2}}$.

Solution: By Eq. (25), $-5x^2 + 4x + 1 = -5(x - \frac{2}{5})^2 + \frac{9}{5}$. Hence

$$\int \frac{dx}{\sqrt{1 + 4x - 5x^2}} = \frac{1}{\sqrt{5}}\int \frac{dx}{\sqrt{\frac{9}{25} - (x - \frac{2}{5})^2}} = \frac{1}{\sqrt{5}}\sin^{-1}\frac{x - \frac{2}{5}}{\frac{3}{5}} + C,$$

by Eq. (20) with $u = x - \frac{2}{5}$ and $a = \frac{3}{5}$. Thus the given integral

$$\int \frac{dx}{\sqrt{1 + 4x - 5x^2}} = \frac{1}{\sqrt{5}}\sin^{-1}\frac{5x - 2}{3} + C.$$

EXAMPLE 3. Evaluate $\int \frac{7x - 2}{3x^2 - 2x - 2}\,dx$.

Solution: Find $d(3x^2 - 2x - 2) = (6x - 2)dx$. Then decompose the given numerator into a multiple of $(6x - 2)$ and a constant by setting $7x - 2 = \frac{7}{6}(6x - 2) - 2 + \frac{7}{6}2 = \frac{7}{6}(6x - 2) + \frac{1}{3}$. And write

$$\int \frac{(7x - 2)dx}{3x^2 - 2x - 2} = \frac{7}{6}\int \frac{(6x - 2)dx}{3x^2 - 2x - 2} + \frac{1}{3}\int \frac{dx}{3x^2 - 2x - 2}.$$

The first integral is found from Eq. (5) with $u = 3x^2 - 2x - 2$. Thus

$$\frac{7}{6}\int \frac{(6x - 2)dx}{3x^2 - 2x - 2} = \frac{7}{6}\int \frac{d(3x^2 - 2x - 2)}{3x^2 - 2x - 2} = \frac{7}{6}\ln(3x^2 - 2x - 2) + C_1.$$

By Eq. (25) we have $3x^2 - 2x - 2 = 3(x - \frac{1}{3})^2 - \frac{7}{3}$. Hence

$$\frac{1}{3}\int \frac{dx}{3x^2 - 2x - 2} = \frac{1}{9}\int \frac{dx}{(x - \frac{1}{3})^2 - \frac{7}{9}} = \frac{1}{9}\frac{3}{2\sqrt{7}} \ln \frac{x - \frac{1}{3} - \sqrt{7}/3}{x - \frac{1}{3} + \sqrt{7}/3} + C_2,$$

by Eq. (18) with $u = x - \frac{1}{3}$ and $a = \sqrt{7}/3$. Thus the given integral

$$\int \frac{(7x - 2)dx}{3x^2 - 2x - 2} = \frac{7}{6} \ln (3x^2 - 2x - 2) + \frac{1}{6\sqrt{7}} \ln \frac{3x - 1 - \sqrt{7}}{3x - 1 + \sqrt{7}} + C.$$

EXAMPLE 4. Evaluate $\displaystyle\int \frac{dx}{(x - 1)\sqrt{x^2 + 2x + 2}}$.

Solution: Put $v = \dfrac{1}{x - 1}$. Then $x - 1 = \dfrac{1}{v}$, $x = \dfrac{1}{v} + 1$ so that $dx = -\dfrac{dv}{v^2}$,

$x^2 + 2x + 2 = \left(\dfrac{1}{v} + 1\right)^2 + 2\left(\dfrac{1}{v} + 1\right) + 2 = \dfrac{1}{v^2}(1 + 4v + 5v^2)$. Hence

$$\int \frac{dx}{(x - 1)\sqrt{x^2 + 2x + 2}} = \int \frac{-dv/v^2}{\dfrac{1}{v}\sqrt{\dfrac{1 + 4v + 5v^2}{v^2}}} = \int \frac{-dv}{\sqrt{5v^2 + 4v + 1}}.$$

By Eq. (25), $5v^2 + 4v + 1 = 5(v + \frac{2}{5})^2 + \frac{1}{5}$. Hence

$\displaystyle\int \frac{-dv}{\sqrt{5v^2 + 4v + 1}} = -\frac{1}{\sqrt{5}}\int \frac{dv}{\sqrt{(v + \frac{2}{5})^2 + \frac{1}{25}}}$. By Eq. (16) with $u = v + \frac{2}{5}$ and $A = \frac{1}{25}$, this is $-(1/\sqrt{5}) \ln [v + \frac{2}{5} + \sqrt{(v + \frac{2}{5})^2 + \frac{1}{25}}] + C_1$. But

$$v + \frac{2}{5} + \frac{1}{\sqrt{5}}\sqrt{5v^2 + 4v + 1} = \frac{v}{5}\left[5 + \frac{2}{v} + \sqrt{5}\sqrt{\frac{1}{v^2}(1 + 4v + 5v^2)}\right]$$

$= \dfrac{1}{5(x - 1)}(2x + 3 + \sqrt{5}\sqrt{x^2 + 2x + 2})$. Hence with $C = C_1 + \dfrac{\ln 5}{\sqrt{5}}$, we find

$$\int \frac{dx}{(x - 1)\sqrt{x^2 + 2x + 2}} = -\frac{1}{\sqrt{5}} \ln \frac{2x + 3 + \sqrt{5x^2 + 10x + 10}}{x - 1} + C.$$

EXERCISE 95

Evaluate each of the following integrals.

1. $\displaystyle\int \frac{dx}{x^2 - 4x + 13}$.
2. $\displaystyle\int \frac{dx}{\sqrt{7 - 6x - x^2}}$.
3. $\displaystyle\int \frac{dx}{x^2 + 6x + 8}$.
4. $\displaystyle\int \frac{dx}{\sqrt{2x^2 - 4x + 5}}$.
5. $\displaystyle\int \frac{dx}{8x - x^2 - 25}$.
6. $\displaystyle\int \frac{dx}{3x - x^2 - 2}$.
7. $\displaystyle\int \frac{dx}{\sqrt{-x^2 - 4x - 3}}$.
8. $\displaystyle\int \frac{dx}{\sqrt{5x^2 + 10x - 4}}$.
9. $\displaystyle\int \frac{dx}{x\sqrt{8 - 2x - x^2}}$.
10. $\displaystyle\int \frac{dx}{(x + 1)\sqrt{x^2 + 2x + 3}}$.
11. $\displaystyle\int_0^5 \frac{dx}{\sqrt{5x - x^2}}$.
12. $\displaystyle\int_1^4 \frac{dx}{x^2 - 2x + 10}$.
13. $\displaystyle\int_3^4 \frac{dx}{x^2 - 6x + 5}$.
14. $\displaystyle\int_1^2 \frac{dx}{\sqrt{x^2 - 3x + 2}}$.
15. $\displaystyle\int \frac{4x + 10}{x^2 + 2x + 5}\,dx$.
16. $\displaystyle\int \frac{8x - 8}{4x^2 - 4x - 3}\,dx$.

17. $\int \frac{x+3}{\sqrt{x^2+2x}}\,dx.$

18. $\int \frac{x+2}{\sqrt{4x-x^2}}\,dx.$

19. $\int \frac{x\,dx}{\sqrt{27+6x-x^2}}.$

20. $\int \frac{(2x-1)dx}{\sqrt{4x^2+4x+2}}.$

193. Integrals Leading to Trigonometric Functions. It follows from Eq. (100) of Sec. 99 that

$$\int \sin u\,du = -\cos u + C. \tag{26}$$
$$\int \cos u\,du = \sin u + C. \tag{27}$$
$$\int \sec^2 u\,du = \tan u + C. \tag{28}$$
$$\int \csc^2 u\,du = -\cot u + C. \tag{29}$$
$$\int \tan u \sec u\,du = \sec u + C. \tag{30}$$
$$\int \cot u \csc u\,du = -\csc u + C. \tag{31}$$

We recall that the integrands of Eqs. (10), (11), (13), and (14) also involved trigonometric functions. In the formulas just mentioned, as in Eqs. (26) to (31), whenever the *right* member involves one or more cofunctions (cosine, cotangent, cosecant) a minus sign follows immediately after the equality sign.

The integrals $\int \sec^2 u\,du$ and $\int \csc^2 u\,du$ are given by Eqs. (28) and (29). And the integrals $\int \tan^2 u\,du$ and $\int \cot^2 u\,du$ may be reduced to these by using the identities

$$\tan^2 u = \sec^2 u - 1 \qquad \text{and} \qquad \cot^2 u = \csc^2 u - 1. \tag{32}$$

The integrals $\int \sin^2 u\,du$ and $\int \cos^2 u\,du$ may be reduced to integrals found from Eq. (27) by using the identities

$$\sin^2 u = \tfrac{1}{2}(1 - \cos 2u) \qquad \text{and} \qquad \cos^2 u = \tfrac{1}{2}(1 + \cos 2u). \tag{33}$$

EXAMPLE 1. Evaluate $\int \sin 5x\,dx$.
Solution: By Eq. (26) with $u = 5x$, $du = 5\,dx$, we have

$$\int \sin 5x\,dx = \tfrac{1}{5}\int \sin 5x\,(5\,dx) = -\tfrac{1}{5}\cos 5x + C.$$

EXAMPLE 2: Evaluate $\int \frac{\cos(2x+3)}{\sin^2(2x+3)}\,dx.$
Solution 1: We use Eq. (31) with $u = 2x + 3$, $du = 2\,dx$. Thus

$$\begin{aligned}\int \frac{\cos(2x+3)}{\sin^2(2x+3)}\,dx &= \int \cot(2x+3)\csc(2x+3)\,dx \\ &= \tfrac{1}{2}\int \cot(2x+3)\csc(2x+3)(2\,dx) = -\tfrac{1}{2}\csc(2x+3) + C.\end{aligned}$$

Solution 2: We may use Eq. (4) with $n = -2$, $u = \sin(2x+3)$, and $du = 2\cos(2x+3)\,dx$. We write

$$\begin{aligned}\int \frac{\cos(2x+3)}{\sin^2(2x+3)}\,dx &= \frac{1}{2}\int [\sin(2x+3)]^{-2}\,[2\cos(2x+3)]\,dx \\ &= \frac{1}{2}\frac{[\sin(2x+3)]^{-1}}{-1} + C = -\frac{1}{2}\csc(2x+3) + C.\end{aligned}$$

EXAMPLE 3. Evaluate $\int(3 + \tan 2x)^2\,dx$.

Solution: $(3 + \tan 2x)^2 = 9 + 6\tan 2x + \tan^2 2x = 8 + 6\tan 2x + \sec^2 2x$, by Eq. (32). Hence the given integral

$$\begin{aligned}\int(3 + \tan 2x)^2\,dx &= 8\int dx + 3\int \tan 2x(2\,dx) + \tfrac{1}{2}\int \sec^2 2x(2\,dx)\\ &= 8x + 3\ln\sec 2x + \tfrac{1}{2}\tan 2x + C,\end{aligned}$$

by Eqs. (10) and (28) with $u = 2x$ and $du = 2\,dx$.

EXAMPLE 4. Evaluate $\int \cos^2 3x\,dx$.

Solution: $\cos^2 3x = \frac{1}{2}(1 + \cos 6x)$ by Eq. (33). Hence $\displaystyle\int \cos^2 3x\,dx = \frac{1}{2}\int dx + \frac{1}{12}\int \cos 6x\,(6\,dx) = \frac{x}{2} + \frac{1}{12}\sin 6x + C$, by Eq. (27) with $u = 6x$, $du = 6\,dx$.

EXAMPLE 5. Evaluate $\displaystyle\int \frac{dx}{\sqrt{1 - \cos 5x}}$.

Solution: By Eq. (33) with $u = 5x/2$, $\sin^2(5x/2) = \frac{1}{2}(1 - \cos 5x)$. Hence $\displaystyle\frac{1}{\sqrt{1 - \cos 5x}} = \frac{1}{\sqrt{2}\sin(5x/2)} = \frac{1}{\sqrt{2}}\csc\frac{5x}{2}$. And by Eq. (14) with $u = 5x/2$, $du = \frac{5}{2}dx$,

$$\begin{aligned}\int \frac{dx}{\sqrt{1 - \cos 5x}} &= \frac{1}{\sqrt{2}}\frac{2}{5}\int \csc\frac{5x}{2}\left(\frac{5}{2}\,dx\right)\\ &= -\frac{\sqrt{2}}{5}\ln\left(\csc\frac{5x}{2} + \cot\frac{5x}{2}\right) + C.\end{aligned}$$

EXERCISE 96

Evaluate each of the following integrals.

1. $\int \sin(4x - 7)\,dx$.
2. $\int \cos(4 - 3x)\,dx$.
3. $\int \sec(3x - 2)\tan(3x - 2)\,dx$.
4. $\int \cot\frac{x}{5}\csc\frac{x}{5}\,dx$.
5. $\int \sec^2\frac{3x}{2}\,dx$.
6. $\int \csc^2(2x - 4)\,dx$.
7. $\int \tan^2 5x\,dx$.
8. $\int \cot^2\frac{3x}{4}\,dx$.
9. $\int \tan\frac{2x}{3}\,dx$.
10. $\int \sec\frac{4x}{5}\,dx$.
11. $\int (1 - \cot x)^2\,dx$.
12. $\int (1 - \csc x)^2\,dx$.
13. $\int \sin^2 5x\,dx$.
14. $\int \cos^2\frac{5x}{8}\,dx$.
15. $\int (\sec x - \tan x)^2\,dx$.
16. $\int (\sin x - \cos x)^2\,dx$.
17. $\int \sqrt{1 - \cos x}\,dx$.
18. $\int \frac{dx}{\sqrt{1 + \cos x}}$.
19. $\int_0^{\pi} \sin\frac{x}{3}\,dx$.
20. $\int_0^{\pi} \sec^2\frac{x}{4}\,dx$.
21. $\int_0^{\pi/4} \cos^2 x\,dx$.
22. $\int_{\pi/3}^{\pi/2} \cot x\,dx$.

194. Integrals Leading to Exponential Functions. It follows from Eq. (71) of Sec. 119 that

$$\int e^u\,du = e^u + C. \tag{34}$$

And it follows from Eq. (78) of Sec. 121 that

$$\int a^u\,du = \frac{a^u}{\ln a} + C. \tag{35}$$

EXAMPLE 1. Evaluate $\int e^{-5x}\,dx$.

Solution: By Eq. (34) with $u = -5x$, $du = -5\,dx$, we have

$$\int e^{-5x}\,dx = -\tfrac{1}{5}\int e^{-5x}(-5\,dx) = -\tfrac{1}{5}e^{-5x} + C.$$

EXAMPLE 2. Evaluate $\displaystyle\int \frac{3^{1/x}}{x^2}\,dx$.

Solution: By Eq. (35) with $a = 3$, $u = 1/x$, $du = -dx/x^2$, we have

$$\int \frac{3^{1/x}}{x^2}\,dx = -\int 3^{1/x}\left(-\frac{dx}{x^2}\right) = -\frac{3^{1/x}}{\ln 3} + C.$$

EXAMPLE 3. Evaluate $\int 5^{4x-2}3^{2x+4}\,dx$.

Solution: By Eq. (35) with $a = 5^4 3^2$, $u = x$, we have

$$\int 5^{4x-2}3^{2x+4}\,dx = 5^{-2}3^4\int (5^4 3^2)^x\,dx = 5^{-2}3^4\,\frac{(5^4 3^2)^x}{\ln (5^4 3^2)} + C$$
$$= \frac{5^{4x-2}3^{2x+4}}{4\ln 5 + 3\ln 2} + C.$$

EXAMPLE 4. Evaluate $\displaystyle\int \frac{3e^{4x} + 2e^{-x}}{e^{4x} - e^{-x}}\,dx$.

Solution: $\displaystyle\frac{3e^{4x} + 2e^{-x}}{e^{4x} - e^{-x}} = \frac{3e^{5x} + 2}{e^{5x} - 1} = -2 + \frac{5e^{5x}}{e^{5x} - 1}$. Hence $\displaystyle\int \frac{3e^{4x} + 2e^{-x}}{e^{4x} - e^{-x}}\,dx =$ $\displaystyle -2\int dx + \int \frac{5e^{5x}\,dx}{e^{5x} - 1} = -2x + \ln (e^{5x} - 1) + C.$

Since $-2x = \ln e^{-2x}$, the result is equal to $\ln e^{-2x} + \ln (e^{5x} - 1) + C$, and so could be written as $\ln (e^{3x} - 2e^{-2x}) + C$.

EXERCISE 97

Evaluate each of the following integrals.

1. $\displaystyle\int e^{3x-4}\,dx.$
2. $\displaystyle\int_0^\infty e^{-2x}\,dx.$
3. $\displaystyle\int 5e^{7x}\,dx.$
4. $\displaystyle\int xe^{x^2}\,dx.$
5. $\displaystyle\int_0^\infty xe^{-x^2}\,dx.$
6. $\displaystyle\int e^{\sin x}\cos x\,dx.$
7. $\displaystyle\int e^{\tan x}\sec^2 x\,dx.$
8. $\displaystyle\int e^{x+e^x}\,dx.$
9. $\displaystyle\int_0^1 10^x\,dx.$
10. $\displaystyle\int (n^x + x^n)dx.$
11. $\displaystyle\int (e^x - x^e)dx.$
12. $\displaystyle\int_{\frac{1}{3}}^1 e^{3x-1}\,dx.$

13. $\int a^x e^x \, dx.$

14. $\int 7^{8x} 8^{7x} \, dx.$

15. $\int (e^{x/4} + e^{-x/4}) dx.$

16. $\int \frac{e^x \, dx}{e^x + 4}.$

17. $\int (e^x - e^{-x})^2 \, dx.$

18. $\int \frac{e^{2x} + e^{-2x}}{e^{2x} - e^{-2x}} \, dx.$

19. $\int \frac{e^{2x} - 1}{e^{2x} + 1} \, dx.$

20. $\int \frac{e^{3x} - 1}{e^{3x} + 2} \, dx.$

195. Integrals of the Type $\int \sin^m u \cos^n u \, du$. If n is an *odd positive integer*, we may write $n = 2p + 1$, with p either zero or some positive integer. Then

$$\begin{aligned}\int \sin^m u \cos^{2p+1} u \, du &= \int \sin^m u \, (\cos^2 u)^p \cos u \, du \\ &= \int \sin^m u \, (1 - \sin^2 u)^p \, d(\sin u). \qquad (36)\end{aligned}$$

This may be expanded into a sum of terms. And for any value of m each term can be integrated by the procedure described in Sec. 188 if $\sin u$ is regarded as a new variable.

If m is an odd positive integer, we may write $m = 2p + 1$. Then

$$\begin{aligned}\int \sin^{2p+1} u \cos^n u \, du &= \int (\sin^2 u)^p \cos^n u \sin u \, du \\ &= -\int \cos^n u \, (1 - \cos^2 u)^p \, d(\cos u). \qquad (37)\end{aligned}$$

This may be expanded into a sum of terms. And for any value of n each term can be integrated by the procedure described in Sec. 188 if $\cos u$ is regarded as a new variable.

If m and n are each either zero or an *even positive integer*, we may write $m = 2p$ and $n = 2q$, with p and q each zero or a positive integer. Then if $p \geqq q$, we may write

$$\begin{aligned}\int \sin^{2p} u \cos^{2q} u \, du &= \frac{1}{2^{p+q}} \int (2 \sin^2 u)^{p-q} (2 \sin u \cos u)^{2q} \, du \\ &= \frac{1}{2^{p+q}} \int (1 - \cos 2u)^{p-q} \sin^{2q} 2u \, du. \qquad (38)\end{aligned}$$

And if $p \leqq q$, we may write

$$\begin{aligned}\int \sin^{2p} u \cos^{2q} u \, du &= \frac{1}{2^{p+q}} \int (2 \sin u \cos u)^{2p} (2 \cos^2 u)^{q-p} \, du \\ &= \frac{1}{2^{p+q}} \int \sin^{2p} 2u \, (1 + \cos 2u)^{q-p} \, du. \qquad (39)\end{aligned}$$

The integrands in the right members of Eqs. (38) and (39) may be expanded into a sum of terms. Those terms having one odd exponent may be integrated by the process used in Eqs. (36) and (37), with $2u$ in place of u. If any term has both exponents even, we repeat the transformation used in Eqs. (38) and (39).

EXAMPLE 1. Evaluate $\int \sin^5 3x\,dx$.

Solution: By Eq. (37) with $p = 2$, $n = 0$, and $u = 3x$, we have

$$\begin{aligned}\int \sin^5 3x\,dx &= \tfrac{1}{3}\int (\sin^2 3x)^2 \sin 3x\,d(3x) = -\tfrac{1}{3}\int (1 - \cos^2 3x)^2\,d(\cos 3x)\\ &= -\tfrac{1}{3}\int (1 - 2\cos^2 3x + \cos^4 3x)\,d(\cos 3x)\\ &= -\tfrac{1}{3}(\cos 3x - \tfrac{2}{3}\cos^3 3x + \tfrac{1}{5}\cos^5 3x) + C.\end{aligned}$$

EXAMPLE 2. Evaluate $\int \sin^4 \frac{x}{2} \cos^3 \frac{x}{2}\,dx$.

Solution: By Eq. (36) with $m = 4$, $p = 1$, and $u = x/2$, we have

$$\begin{aligned}\int \sin^4 \frac{x}{2}\cos^3\frac{x}{2}\,dx &= 2\int \sin^4\frac{x}{2}\left(\cos^2\frac{x}{2}\right)\cos\frac{x}{2}\,d\left(\frac{x}{2}\right)\\ &= 2\int \sin^4\frac{x}{2}\left(1 - \sin^2\frac{x}{2}\right)d\left(\sin\frac{x}{2}\right)\\ &= 2\int\left(\sin^4\frac{x}{2} - \sin^6\frac{x}{2}\right)d\left(\sin\frac{x}{2}\right)\\ &= \frac{2}{5}\sin^5\frac{x}{2} - \frac{2}{7}\sin^7\frac{x}{2} + C.\end{aligned}$$

EXAMPLE 3. Evaluate $\int \sin^2 \frac{x}{5} \cos^4 \frac{x}{5}\,dx$.

Solution: As in Eq. (39) with $p = 1$ and $q = 2$, we may write

$$\begin{aligned}\int \sin^2\frac{x}{5}\cos^4\frac{x}{5}\,dx &= \frac{1}{2^3}\int\left(2\sin\frac{x}{5}\cos\frac{x}{5}\right)^2\left(2\cos^2\frac{x}{5}\right)dx\\ &= \frac{1}{8}\int \sin^2\frac{2x}{5}\left(1 + \cos\frac{2x}{5}\right)dx\\ &= \frac{1}{8}\int \sin^2\frac{2x}{5}\,dx - \frac{1}{8}\int \sin^2\frac{2x}{5}\cos\frac{2x}{5}\,dx.\end{aligned}$$

We treat the first integral as in Eq. (38) with $p = 1$ and $q = 0$. Thus $\frac{1}{8}\int \sin^2\frac{2x}{5}\,dx$ $= \frac{1}{8}\frac{1}{2}\int\left(1 - \cos\frac{4x}{5}\right)dx = \frac{1}{16}\int dx - \frac{1}{16}\frac{5}{4}\int \cos\frac{4x}{5}\,d\left(\frac{4x}{5}\right)$. We treat the second integral as in Eq. (36) with $p = 0$ and $m = 2$. Thus $-\frac{1}{8}\int \sin^2\frac{2x}{5}\cos\frac{2x}{5}\,dx = -\frac{1}{8}\frac{5}{2}\int \sin^2\frac{2x}{5}\,d\left(\sin\frac{2x}{5}\right)$. It follows that

$$\int \sin^2\frac{x}{5}\cos^4\frac{x}{5}\,dx = \frac{x}{16} - \frac{5}{64}\sin\frac{4x}{5} - \frac{5}{48}\sin^3\frac{2x}{5} + C.$$

EXERCISE 98

Evaluate each of the following integrals.

1. $\int \sin^3 x\,dx$.
2. $\int \cos^3 x\,dx$.
3. $\int \sin 7x \cos 7x\,dx$.
4. $\int \sin^3 x \cos^2 x\,dx$.
5. $\int \sin^2\frac{x}{4}\cos^2\frac{x}{4}\,dx$.
6. $\int \sin^4 x\,dx$.
7. $\int \cos^3\frac{x}{2}\,dx$.
8. $\int \sin^5 2x\,dx$.

9. $\int \sin^5 \frac{x}{6} \cos \frac{x}{6}\,dx.$

10. $\int \cos^3 \frac{x}{8} \sin \frac{x}{8}\,dx.$

11. $\int \sin^{\frac{2}{3}} x \cos x\,dx.$

12. $\int \sin x \sqrt{\cos x}\,dx.$

13. $\int \frac{\sin^3 x}{\cos x}\,dx.$

14. $\int \frac{\cos^3 x}{\sqrt{\sin x}}\,dx.$

15. $\int \sin^3 x \cos^3 x\,dx.$

16. $\int \cos^6 x\,dx.$

17. $\int_0^{\pi} \sin^3 \frac{x}{3}\,dx.$

18. $\int_0^{\pi} \cos^3 \frac{\pi}{6}\,dx.$

19. $\int_0^{\pi/4} \cos^4 2x\,dx.$

20. $\int_0^{\pi/8} \sin^2 2x \cos^2 2x\,dx.$

196. Products of Sines and Cosines of Different Angles. Products of sines and cosines may be expressed as sums or differences by means of the following trigonometric identities.

$$\sin A \cos B = \tfrac{1}{2} \sin (A + B) + \tfrac{1}{2} \sin (A - B). \tag{40}$$
$$\sin A \sin B = -\tfrac{1}{2} \cos (A + B) + \tfrac{1}{2} \cos (A - B). \tag{41}$$
$$\cos A \cos B = \tfrac{1}{2} \cos (A + B) + \tfrac{1}{2} \cos (A - B). \tag{42}$$

These were proved in Probs. 9 and 10 of Exercise 48.

Whenever $A = mx + a$ and $B = nx + b$, these relations reduce integrals of products of this type to a form which can be integrated by Eq. (26), (27), or (2) if $m = \pm n$.

EXAMPLE 1. Evaluate $\int \sin 3x \cos 5x\,dx$.
Solution: By Eq. (40) with $A = 3x$, $B = 5x$, we have

$$\begin{aligned}\int \sin 3x \cos 5x\,dx &= \int[\tfrac{1}{2} \sin 8x + \tfrac{1}{2} \sin (-2x)]dx \\ &= \tfrac{1}{16}\int \sin 8x\,d(8x) - \tfrac{1}{4}\int \sin 2x\,d(2x) \\ &= -\tfrac{1}{16} \cos 8x + \tfrac{1}{4} \cos 2x + C.\end{aligned}$$

EXAMPLE 2. Evaluate $\int \cos (2x + 3) \cos (1 - 2x)dx$.
Solution: By Eq. (42) with $A = 2x + 3$ and $B = 1 - 2x$, we have

$$\begin{aligned}\int \cos (2x + 3) \cos (1 - 2x)dx &= \int[\tfrac{1}{2} \cos 4 + \tfrac{1}{2} \cos (4x + 2)]dx \\ &= \tfrac{1}{2} \cos 4 \int dx + \tfrac{1}{8}\int \cos (4x + 2)\,d(4x) \\ &= \tfrac{1}{2}(\cos 4)x + \tfrac{1}{8} \sin (4x + 2) + C.\end{aligned}$$

EXAMPLE 3. Evaluate $\int \sin x \sin 3x \sin 5x\,dx$.
Solution: By Eq. (41) with $A = 3x$ and $B = x$, we have
$\sin x \sin 3x = \frac{1}{2}(-\cos 4x + \cos 2x)$. Hence by Eq. (40), we have

$$\begin{aligned}\sin x \sin 3x \sin 5x &= \tfrac{1}{2}(-\sin 5x \cos 4x + \sin 5x \cos 2x) \\ &= \tfrac{1}{4}(-\sin 9x - \sin x + \sin 7x + \sin 3x).\end{aligned}$$

It follows that the given integral $\int \sin x \sin 3x \sin 5x\,dx =$
$\frac{1}{4}(\frac{1}{9} \cos 9x + \cos x - \frac{1}{7} \cos 7x - \frac{1}{3} \cos 3x) + C.$

EXERCISE 99

Evaluate each of the following integrals.

1. $\int \sin 4x \sin 2x \, dx.$
2. $\int \sin 6x \cos 4x \, dx.$
3. $\int \cos 5x \cos 2x \, dx.$
4. $\int \sin 7x \sin x \, dx.$
5. $\int \cos x \sin 2x \, dx.$
6. $\int \cos 3x \cos 2x \, dx.$
7. $\int \sin x \cos (2x + 3) \, dx.$
8. $\int \cos 2x \cos (3x - 4) \, dx.$
9. $\int \sin 2x \sin (5x - 3) \, dx.$
10. $\int \sin \left(x - \frac{\pi}{12}\right) \cos \left(x + \frac{\pi}{12}\right) dx.$
11. $\int \cos x \cos 2x \cos 3x \, dx.$
12. $\int \sin^2 3x \sin 7x \, dx.$
13. $\int \cos^2 3x \cos 5x \, dx.$
14. $\int \sin^2 2x \sin^2 3x \, dx.$

Verify that for any positive integers p and q with $p \neq q$,

15. $\int_0^{2\pi} \sin (px + a) \sin (qx + b) \, dx = 0.$
16. $\int_0^{2\pi} \cos (px + a) \cos (qx + b) \, dx = 0.$
17. $\int_0^{2\pi} \sin (px + a) \cos (qx + b) \, dx = 0.$
18. $\int_0^{2\pi} \sin (px + a) \sin (px + b) \, dx = \pi \cos (a - b).$
19. $\int_0^{2\pi} \cos (px + a) \cos (px + b) \, dx = \pi \cos (a - b).$
20. $\int_0^{2\pi} \sin (px + a) \cos (px + b) \, dx = \pi \sin (a - b).$

197. Integrals of the Type $\int \tan^m u \sec^n u \, du$ or $\int \cot^m u \csc^n u \, du$. If n is an *even positive integer*, we may write $n = 2p + 2$, with p either zero or some positive integer. Then

$$\begin{aligned} \int \tan^m u \sec^{2p+2} u \, du &= \int \tan^m u \, (\sec^2 u)^p \sec^2 u \, du \\ &= \int \tan^m u \, (1 + \tan^2 u)^p \, d(\tan u). \end{aligned} \tag{43}$$

$$\begin{aligned} \int \cot^m u \csc^{2p+2} u \, du &= \int \cos^m u \, (\csc^2 u)^p \csc^2 u \, du \\ &= -\int \cot^m u \, (1 + \cot^2 u)^p \, d(\cot u). \end{aligned} \tag{44}$$

In each of the Eqs. (43) and (44), the last integrand can be expanded into a sum of terms. And for any value of m each of these can be integrated by the procedure of Sec. 188.

If m is an *odd positive integer*, we may write $m = 2p + 1$. Then

$$\begin{aligned} \int \tan^{2p+1} u \sec^n u \, du &= \int (\tan^2 u)^p \sec^{n-1} u \, (\tan u \sec u \, du) \\ &= \int (\sec^2 u - 1)^p \sec^{n-1} u \, d(\sec u). \end{aligned} \tag{45}$$

$$\begin{aligned} \int \cot^{2p+1} u \csc^n u \, du &= \int (\cot^2 u)^p \csc^{n-1} u \, (\cot u \csc u \, du) \\ &= -\int (\cot^2 u)^p \csc^{n-1} u \, d(\csc u). \end{aligned} \tag{46}$$

In each of the Eqs. (45) and (46), the last integrand can be expanded into a sum of terms. And for any value of n each of these can be integrated by the procedure of Sec. 188.

If $n = 0$ and m is a positive integer greater than unity, we may write

$$\int \tan^m u\,du = \int \tan^{m-2} u\,(\sec^2 u - 1)du = \frac{\tan^{m-1} u}{m-1} - \int \tan^{m-2} u\,du. \tag{47}$$

$$\int \cot^m u\,du = \int \cot^{m-2} u\,(\csc^2 u - 1)du = -\frac{\cot^{m-1} u}{m-1} - \int \cot^{m-2} u\,du. \tag{48}$$

These *reduction formulas* enable us to express the given integral in terms of a similar integral with exponent less by two. And we can continue the process if necessary. If m is even, we shall finally come to the integral of Eq. (2). And if m is odd, we shall finally come to the integral of Eq. (10) or (11).

For $m = 0$ and n a positive integer greater than unity, there is a similar reduction process which ends with the integral of Eq. (2) when n is even, and with the integral of Eq. (13) or (14) if n is odd. To derive the appropriate reduction formulas, we first verify that

$$\begin{aligned}\frac{d}{du}(\sec^{n-2} u \tan u) &= (n-2)\sec^{n-3} u\,(\sec u \tan u)\tan u + \sec^{n-2} u \sec^2 u\\ &= (n-2)\sec^{n-2} u\,(\sec^2 u - 1) + \sec^n u\\ &= (n-1)\sec^n u - (n-2)\sec^{n-2} u.\end{aligned} \tag{49}$$

By solving this relation for $\sec^n u$, we may deduce that

$$\sec^n u\,du = \frac{1}{n-1}\,d(\sec^{n-2} u \tan u) + \frac{n-2}{n-1}\sec^{n-2} u\,du. \tag{50}$$

Integration of this leads to the reduction formula

$$\int \sec^n u\,du = \frac{1}{n-1}\sec^{n-2} u \tan u + \frac{n-2}{n-1}\int \sec^{n-2} u\,du. \tag{51}$$

By a similar procedure we may show that

$$\int \csc^n u\,du = -\frac{1}{n-1}\csc^{n-2} u \cot u + \frac{n-2}{n-1}\int \csc^{n-2} u\,du. \tag{52}$$

Or we may derive Eq. (52) by replacing u by $\pi/2 - u$ in Eq. (51).

EXAMPLE 1. Evaluate $\int \tan^7 3x \sec^6 3x\,dx$.

Solution: By Eq. (43) with $m = 7$, $p = 2$, and $u = 3x$, we have

$$\begin{aligned}\int \tan^7 3x \sec^6 3x\,dx &= \tfrac{1}{3}\int \tan^7 3x\,(\sec^2 3x)^2 \sec^2 3x\,d(3x)\\ &= \tfrac{1}{3}\int \tan^7 3x\,(1 + \tan^2 3x)^2\,d(\tan 3x)\\ &= \tfrac{1}{3}\int(\tan^7 3x + 2\tan^9 3x + \tan^{11} 3x)\,d(\tan 3x)\\ &= \tfrac{1}{3}(\tfrac{1}{8}\tan^8 3x + \tfrac{1}{5}\tan^{10} 3x + \tfrac{1}{12}\tan^{12} 3x) + C.\end{aligned}$$

EXAMPLE 2. Evaluate $\int \tan^3 3x \sec^6 3x\,dx$.

Solution: By Eq. (45) with $p = 1$, $n = 6$, and $u = 3x$, we have

$$\begin{aligned}\int \tan^3 3x \sec^6 3x\,dx &= \tfrac{1}{3}\int \tan^2 3x \sec^5 3x\,(\tan 3x \sec 3x)\,d(3x)\\ &= \tfrac{1}{3}\int (\sec^2 3x - 1)\sec^5 3x\,d(\sec 3x)\\ &= \tfrac{1}{3}\int (\sec^7 3x - \sec^5 3x)\,d(\sec 3x)\\ &= \tfrac{1}{3}(\tfrac{1}{8}\sec^8 3x - \tfrac{1}{6}\sec^6 3x) + C.\end{aligned}$$

EXAMPLE 3. Evaluate $\displaystyle\int \cot^4 \frac{x}{3}\,dx$.

Solution: Use Eq. (48) with $m = 4$, $u = \dfrac{x}{3}$. Thus $\displaystyle\int \cot^4 \frac{x}{3}\,dx = 3\int \cot^4 \frac{x}{3}\,d\left(\frac{x}{3}\right)$ $\displaystyle = 3\left[-\frac{1}{3}\cot^3 \frac{x}{3} - \int \cot^2 \frac{x}{3}\,d\left(\frac{x}{3}\right)\right]$. But $\displaystyle\int \cot^2 \frac{x}{3}\,d\left(\frac{x}{3}\right) = -\cot\frac{x}{3} - \int d\left(\frac{x}{3}\right)$ from Eq. (48) with $m = 2$, so that $\displaystyle\int \cot^4 \frac{x}{3}\,dx = -\cot^3 \frac{x}{3} + 3\cot\frac{x}{3} + x + C.$

EXAMPLE 4. Evaluate $\int \tan^2 x \sec x\,dx$.

Solution: $\int \tan^2 x \sec x\,dx = \int (\sec^2 x - 1)\sec x\,dx = \int \sec^3 x\,dx - \int \sec x\,dx$. By Eq. (51) with $n = 3$ and $u = x$, $\int \sec^3 x = \frac{1}{2}\sec x \tan x + \frac{1}{2}\int \sec x\,dx$.
Hence

$$\begin{aligned}\int \tan^2 x \sec x\,dx &= \tfrac{1}{2}\tan x \sec x - \tfrac{1}{2}\int \sec x\,dx, \qquad \text{and by Eq. (13),}\\ &= \tfrac{1}{2}\tan x \sec x - \tfrac{1}{2}\ln(\sec x + \tan x) + C.\end{aligned}$$

EXERCISE 100

Evaluate each of the following integrals.

1. $\displaystyle\int \cot^4 3x \csc^2 3x\,dx.$
2. $\displaystyle\int \tan^2 \frac{x}{6} \sec^2 \frac{x}{6}\,dx.$
3. $\displaystyle\int \tan 2x \sec^3 2x\,dx.$
4. $\displaystyle\int \cot \frac{x}{5} \csc^5 \frac{x}{5}\,dx.$
5. $\displaystyle\int \tan^3 x\,dx.$
6. $\displaystyle\int \tan^3 2x \sec 2x\,dx.$
7. $\displaystyle\int \csc^4 \frac{x}{3}\,dx.$
8. $\displaystyle\int \cot^3 \frac{x}{2}\,dx.$
9. $\displaystyle\int \tan x \sec^4 x\,dx.$
10. $\displaystyle\int \tan^3 x \sec^3 x\,dx.$
11. $\displaystyle\int \tan^{\frac{2}{3}} x \sec^2 x\,dx.$
12. $\displaystyle\int \cot x \sqrt{\csc x}\,dx.$
13. $\displaystyle\int \frac{\csc^4 x}{\sqrt{\cot x}}\,dx.$
14. $\displaystyle\int \frac{\tan^3 x}{\sec x}\,dx.$
15. $\displaystyle\int \tan^4 x\,dx.$
16. $\displaystyle\int \sec^4 x\,dx.$
17. $\displaystyle\int \cot^2 x \csc x\,dx.$
18. $\displaystyle\int \sec^3 x\,dx.$
19. $\displaystyle\int_0^\pi \tan^3 \frac{x}{3} \sec \frac{x}{3}\,dx.$
20. $\displaystyle\int_0^\pi \cot^4 \frac{x}{4} \csc^4 \frac{x}{4}\,dx.$
21. $\displaystyle\int_0^{\pi/6} \cot^3 x\,dx.$
22. $\displaystyle\int_0^{\pi/4} \left(\frac{\sec x}{\tan x}\right)^4 dx.$

198. Algebraic Substitutions. An integral containing only one radical $\sqrt[q]{ax + b}$, or fractional power $(ax + b)^{p/q}$, where p is an integer and q

is a positive integer, may be freed of radicals by means of the substitution

$$(ax + b)^{1/q} = z, \qquad ax + b = z^q, \qquad dx = \frac{q}{a} z^{q-1}\,dz. \tag{53}$$

If an integral contains only one radical $\sqrt[q]{ax^n + b}$, or fractional power $(ax^n + b)^{p/q}$, and the factor $x^{n-1}\,dx$ times a simple function of x^n and $(ax^n + b)^{1/q}$, it may be simplified by means of the substitution

$$(ax^n + b)^{1/q} = z, \qquad ax^n + b = z^q, \qquad x^{n-1}\,dx = \frac{q}{an} z^{q-1}\,dz. \tag{54}$$

Whenever we apply a substitution to a definite integral, we may introduce appropriately changed limits for the new variable as in Sec. 178 and Exercise 88. This is illustrated in Example 5 below.

When the expression $ax + b$ or $ax^n + b$ appears to the first power in the denominator, the integral may be simplified by performing a division.

EXAMPLE 1. Evaluate $\displaystyle\int \frac{x^2\,dx}{(1 - 2x)^{\frac{1}{3}}}$.

Solution: As in Eq. (53) with $q = 3$, let $(1 - 2x)^{\frac{1}{3}} = z$, $1 - 2x = z^3$, $x = \frac{1}{2}(1 - z^3)$. $dx = -\frac{3}{2}z^2\,dz$. Then $x^2 = \frac{1}{4}(1 - 2z^3 + z^6)$, and

$$\int \frac{x^2\,dx}{(1 - 2x)^{\frac{1}{3}}} = \int \frac{1 - 2z^3 + z^6}{4z^4}\left(-\frac{3}{2}z^2\,dz\right) = \frac{3}{8}\int (-z^{-2} + 2z - z^4)dz$$
$$= \frac{3}{8}\left(\frac{1}{z} + z^2 - \frac{z^5}{5}\right) + C = \frac{3}{40z}(5 + 5z^3 - z^6) + C.$$

But $5 + 5z^3 - z^6 = 5 + 5(1 - 2x) - (1 - 2x)^2 = 9 - 6x - 4x^2$. Hence

$$\int \frac{x^2\,dx}{(1 - 2x)^{\frac{1}{3}}} = \frac{3}{40}(1 - 2x)^{-\frac{1}{3}}(9 - 6x - 4x^2) + C.$$

EXAMPLE 2. Evaluate $\displaystyle\int \frac{dx}{\sqrt{x} - \sqrt[3]{x}}$.

Solution: Since $\sqrt{x} = x^{\frac{1}{2}} = x^{\frac{3}{6}}$ and $\sqrt[3]{x} = x^{\frac{1}{3}} = x^{\frac{2}{6}}$, this may be thought of as involving only one radical, $x^{\frac{1}{6}}$. Let $x^{\frac{1}{6}} = z$, $z = x^6$, $dx = 6z^5\,dz$. Then $\displaystyle\int \frac{dx}{\sqrt{x} - \sqrt[3]{x}} = \int \frac{6z^5\,dz}{z^3 - z^2} = \int \frac{6z^3\,dz}{z - 1}$. Divide $6z^3$ by $z - 1$ as in algebra. $\dfrac{6z^3}{z - 1}$ $= 6z^2 + 6z + 6 + \dfrac{6}{z - 1}$. Then $\displaystyle\int \frac{6z^3\,dz}{z - 1} = \int \left(6z^2 + 6z + 6 + \frac{6}{z - 1}\right)dz = 2z^3$ $+ 3z^2 + 6z + 6\ln(z - 1) + C$. Hence

$$\int \frac{dx}{\sqrt{x} - \sqrt[3]{x}} = 2\sqrt{x} + 3\sqrt[3]{x} + 6\sqrt[6]{x} + 6\ln(\sqrt[6]{x} - 1) + C.$$

EXAMPLE 3. Evaluate $\displaystyle\int \frac{\sqrt{x^2 + a^2}}{x}\,dx$.

Solution: $\displaystyle\int \frac{\sqrt{x^2 + a^2}}{x}\,dx = \int \frac{\sqrt{x^2 + a^2}}{x^2}(x\,dx)$. Let $\sqrt{x^2 + a^2} = z$, $x^2 + a^2 = z^2$, $x^2 = z^2 - a^2$, $x\,dx = z\,dz$. Then the integral becomes $\displaystyle\int \frac{z^2\,dz}{z^2 - a^2} =$

$\int\left(1+\frac{a^2}{z^2-a^2}\right)dz = z + \frac{a}{2}\ln\frac{z-a}{z+a} + C$, where we have transformed the integrand by division as in algebra and have used Eq. (18) to integrate the second term. We note that

$$\frac{a}{2}\ln\frac{z-a}{z+a} = \frac{a}{2}\ln\frac{z-a}{z+a}\frac{z-a}{z-a} = \frac{a}{2}\ln\frac{(z-a)^2}{z^2-a^2} = a\ln\frac{z-a}{\sqrt{z^2-a^2}}.$$

Hence $\int\frac{\sqrt{x^2+a^2}}{x}\,dx = \sqrt{x^2+a^2} + a\ln\frac{\sqrt{x^2+a^2}-a}{x} + C.$

EXAMPLE 4. Evaluate $\int\frac{x^5\,dx}{\sqrt[3]{1+2x^3}}$.

Solution: $\int\frac{x^5\,dx}{\sqrt[3]{1+2x^3}} = \int\frac{x^3(x^2\,dx)}{\sqrt[3]{1+2x^3}}$. Let $\sqrt[3]{1+2x^3} = z$, $1+2x^3 = z^3$, $x^3 = \frac{1}{2}(z^3-1)$, $x^2\,dx = \frac{1}{2}z^2\,dz$. Then the integral becomes $\int\frac{z^3-1}{2z}\frac{1}{2}z^2\,dz = \frac{1}{4}\int(z^4-z)dz = \frac{1}{4}\left(\frac{z^5}{5}-\frac{z^2}{2}\right) + C = \frac{z^2}{40}(2z^3-5) + C = \frac{1}{40}(1+2x^3)^{\frac{2}{3}}(4x^3-3) + C$, the required value.

EXAMPLE 5. Evaluate $\int_0^{27}\frac{\sqrt[3]{x}+\sqrt{x}}{1+\sqrt[3]{x}}\frac{dx}{x}$.

Solution: Let $x^{\frac{1}{6}} = z$, $x = z^6$, $dx = 6z^5\,dz$, $\frac{dx}{x} = 6\frac{dz}{z}$. Then $z = 0$ when $x = 0$ and $z = \sqrt{3}$ when $x = 27$. Hence $\int_0^{27}\frac{\sqrt[3]{x}+\sqrt{x}}{1+\sqrt[3]{x}}\frac{dx}{x} = \int_0^{\sqrt{3}}\frac{z^2+z^3}{1+z^2}6\frac{dz}{z} = 6\int_0^{\sqrt{3}}\frac{z+z^2}{1+z^2}dz$. Divide z^2+z by z^2+1 as in algebra, $\frac{z^2+z}{z^2+1} = 1 + \frac{z-1}{z^2+1}$. Then $6\int_0^{\sqrt{3}}\frac{z+z^2}{1+z^2}dz = 6\int_0^{\sqrt{3}}\left(1+\frac{z}{z^2+1}-\frac{1}{z^2+1}\right)dz$

$= \Big[6z + 3\ln(z^2+1) - 6\tan^{-1}z\Big]_0^{\sqrt{3}} = 6\sqrt{3} + 3\ln 10 - 2\pi$, the required value.

EXAMPLE 6. Evaluate $\int\frac{\sqrt{x}\,dx}{\sqrt{1+\sqrt{x}}}$.

Solution: Let $\sqrt{1+\sqrt{x}} = z$, $1+\sqrt{x} = z^2$, $\sqrt{x} = z^2-1$, $x = (z^2-1)^2$, $dx = 4(z^2-1)z\,dz$. And $\int\frac{\sqrt{x}\,dx}{\sqrt{1+\sqrt{x}}} = \int\frac{4}{z}(z^2-1)^2 z\,dz = 4\int(z^4-2z^2+1)\,dz = 4\left(\frac{z^5}{5}-\frac{2}{3}z^3+z\right) + C = \frac{4z}{15}(3z^4-10z^2+15) + C$. But $3z^4-10z^2+15 = 3(1+\sqrt{x})^2 - 10(1+\sqrt{x}) + 15 = 3x - 4\sqrt{x} + 8$. Hence $\int\frac{\sqrt{x}\,dx}{\sqrt{1+\sqrt{x}}} = \frac{4}{15}(3x-4\sqrt{x}+8)\sqrt{1+\sqrt{x}} + C$.

EXERCISE 101

Evaluate each of the following integrals.

1. $\int\frac{x^2\,dx}{\sqrt{x-1}}$.

2. $\int\frac{2x+1}{2x+3}\,dx$.

3. $\int \frac{x^{\frac{1}{2}}\,dx}{1 + x^{\frac{3}{4}}}$.

4. $\int \frac{dx}{(1 + x)\sqrt{x}}$.

5. $\int \frac{x^3\,dx}{(x + 2)^2}$.

6. $\int x^3 \sqrt{2x - 3}\,dx$.

7. $\int \frac{x^3\,dx}{(x^2 + 4)^{\frac{3}{2}}}$.

8. $\int \frac{\sqrt{x^2 - 4}}{x}\,dx$.

9. $\int \sqrt{4 - \sqrt{x}}\,dx$.

10. $\int \frac{x\,dx}{(1 + 2x)^{\frac{3}{2}}}$.

11. $\int \frac{x^5\,dx}{\sqrt{x^3 + 8}}$.

12. $\int \frac{x^3\,dx}{\sqrt{x^2 - 2}}$.

13. $\int_0^4 \frac{dx}{1 + \sqrt{x}}$.

14. $\int_1^4 \frac{x\,dx}{\sqrt{1 + 2x}}$.

15. $\int_0^9 x\sqrt{9 - x^2}\,dx$.

16. $\int_3^5 \frac{x^3\,dx}{\sqrt{x^2 - 9}}$.

17. $\int_3^4 x\sqrt{x - 3}\,dx$.

18. $\int_0^4 \frac{x^3\,dx}{\sqrt{x^2 + 9}}$.

19. $\int_0^1 x^3\sqrt{x^2 + 1}\,dx$.

20. $\int_0^2 \frac{x^5\,dx}{\sqrt{x^3 + 1}}$.

199. Trigonometric Substitutions. Consider an integral which contains any one of the three radicals $\sqrt{a^2 - x^2}$, $\sqrt{x^2 + a^2}$, or $\sqrt{x^2 - a^2}$, and the factor $x^m\,dx$. If m is an odd integer, $2k + 1$, we may consider $x^m\,dx = x^{2k+1}\,dx$ as $(x^2)^k(x\,dx)$ and use Eq. (54) with $n = 2$, $q = 2$.

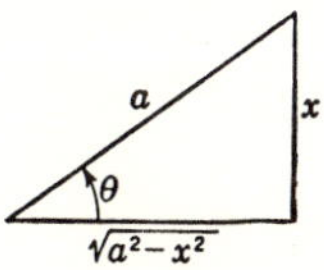

FIG. 232.

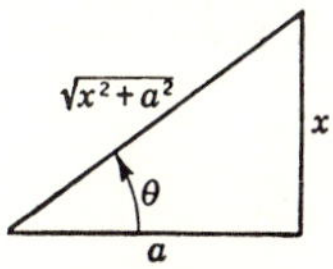

FIG. 233.

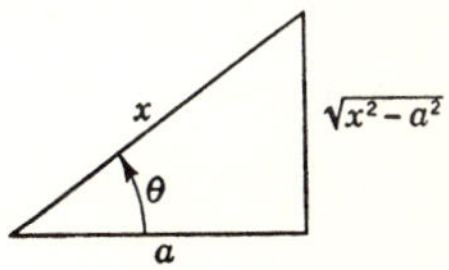

FIG. 234.

If m is an even integer, the integral may be simplified by a substitution suggested by one of the following constructions.

For $\sqrt{a^2 - x^2}$, we construct a right triangle as in Fig. 232 with sides $\sqrt{a^2 - x^2}$ and x, and hypotenuse a. Then if the angle opposite x is θ, we have

$$x = a \sin\theta, \qquad dx = a\cos\theta\,d\theta, \qquad \sqrt{a^2 - x^2} = a\cos\theta. \tag{55}$$

For $\sqrt{x^2 + a^2}$, we construct a right triangle as in Fig. 233 with sides a and x, and hypotenuse $\sqrt{x^2 + a^2}$. Then if the angle opposite x is θ, we have

$$x = a\tan\theta, \qquad dx = a\sec^2\theta\,d\theta, \qquad \sqrt{x^2 + a^2} = a\sec\theta. \tag{56}$$

For $\sqrt{x^2 - a^2}$, we construct a right triangle as in Fig. 234 with sides a and $\sqrt{x^2 - a^2}$ and hypotenuse x. Then if the angle opposite $\sqrt{x^2 - a^2}$ is θ, we have

$$x = a\sec\theta, \qquad dx = a\tan\theta\sec\theta\,d\theta, \qquad \sqrt{x^2 - a^2} = a\tan\theta. \tag{57}$$

EXAMPLE 1. Evaluate $\int \sqrt{a^2 - x^2}\, dx$.

Solution: From Eq. (55),

$$\int \sqrt{a^2 - x^2}\, dx = \int a \cos\theta (a \cos\theta\, d\theta) = \frac{a^2}{2} \int (2 \cos^2\theta) d\theta = \frac{a^2}{2} \int (1 + \cos 2\theta) d\theta$$

$= \frac{a^2}{2}\left(\theta + \frac{1}{2}\sin 2\theta\right) + C$. From Fig. 232 or Eq. (55), we have $\theta = \sin^{-1}\frac{x}{a}$, $\sin\theta = \frac{x}{a}$, $\cos\theta = \frac{1}{a}\sqrt{a^2 - x^2}$ so that $\sin 2\theta = 2 \sin\theta \cos\theta = \frac{2}{a^2} x \sqrt{a^2 - x^2}$. Hence

$$\int \sqrt{a^2 - x^2}\, dx = \frac{x}{2}\sqrt{a^2 - x^2} + \frac{a^2}{2}\sin^{-1}\frac{x}{a} + C.$$

EXAMPLE 2. Evaluate $\int_{-1}^{1} x^2 \sqrt{4 - x^2}\, dx$.

Solution: From Eq. (55) with $a = \sqrt{4} = 2$, we have $x = 2 \sin\theta$, $dx = 2 \cos\theta\, d\theta$, $\sqrt{4 - x^2} = 2\cos\theta$. And from $x = 2\sin\theta$, $\theta = \sin^{-1}(x/2)$. When $x = -1$, one value of θ is $-\pi/6$, while when $x = 1$, one value of θ is $\pi/6$. And for these values from the principal branch, x increases continuously from -1 to 1 when θ increases from $-\pi/6$ to $\pi/6$. Hence by Sec. 178,

$$\begin{aligned}\int_{-1}^{1} x^2\sqrt{4 - x^2}\, dx &= \int_{-\pi/6}^{\pi/6} (2\sin\theta)^2(2\cos\theta)(2\cos\theta\, d\theta) = 16\int_{-\pi/6}^{\pi/6} \sin^2\theta\cos^2\theta\, d\theta \\ &= 4\int_{-\pi/6}^{\pi/6} \sin^2 2\theta\, d\theta = 2\int_{-\pi/6}^{\pi/6} (1 - \cos 4\theta) d\theta \\ &= 2\left[\theta - \frac{1}{4}\sin 4\theta\right]_{-\pi/6}^{\pi/6} \\ &= 2\left\{\left[\frac{\pi}{6} - \frac{1}{4}\left(\frac{\sqrt{3}}{2}\right)\right] - \left[-\frac{\pi}{6} - \frac{1}{4}\left(-\frac{\sqrt{3}}{2}\right)\right]\right\} = \frac{2\pi}{3} - \frac{\sqrt{3}}{2},\end{aligned}$$

the required value.

EXAMPLE 3. Evaluate $\int \frac{dx}{(x^2 + a^2)^{\frac{3}{2}}}$.

Solution: From Eq. (56),

$$\begin{aligned}\int \frac{dx}{(x^2 + a^2)^{\frac{3}{2}}} &= \int \frac{a \sec^2\theta\, d\theta}{a^3 \sec^3\theta} = \frac{1}{a^2}\int \frac{d\theta}{\sec\theta} \\ &= \frac{1}{a^2}\int \cos\theta\, d\theta = \frac{1}{a^2}\sin\theta + C.\end{aligned}$$

From Fig. 233, $\sin\theta = \frac{x}{\sqrt{x^2 + a^2}}$. Hence $\int \frac{dx}{(x^2 + a^2)^{\frac{3}{2}}} = \frac{x}{a^2\sqrt{x^2 + a^2}} + C$.

EXAMPLE 4. Evaluate $\int \frac{\sqrt{x^2 - a^2}}{x^2}\, dx$.

Solution: From Eq. (57),

$$\begin{aligned}\int \frac{\sqrt{x^2 - a^2}}{x^2}\, dx &= \int \frac{a \tan\theta}{a^2 \sec^2\theta}(a \tan\theta \sec\theta\, d\theta) \\ &= \int \frac{\tan^2\theta}{\sec\theta}\, d\theta = \int \frac{\sec^2\theta - 1}{\sec\theta}\, d\theta = \int (\sec\theta - \cos\theta) d\theta,\end{aligned}$$

and by Eq. (13),

$$= \ln(\sec\theta + \tan\theta) - \sin\theta + C_1.$$

From Fig. 234, $\sec\theta = \frac{x}{a}$, $\tan\theta = \frac{\sqrt{x^2 - a^2}}{a}$, $\sin\theta = \frac{\sqrt{x^2 - a^2}}{x}$. Hence with

$C = C_1 - \ln a$, $\int \frac{\sqrt{x^2 - a^2}}{x^2}\, dx = -\frac{\sqrt{x^2 - a^2}}{x} + \ln(x + \sqrt{x^2 - a^2}) + C.$

Example 5. Evaluate $\int \sqrt{x^2 + a^2}\, dx$.

Solution: From Eq. (56), $\int \sqrt{x^2 + a^2}\, dx = \int a \sec \theta\, (a \sec^2 \theta\, d\theta) = a^2 \int \sec^3 \theta\, d\theta$. By Eq. (51) with $n = 3$ and $u = \theta$, this is $a^2 \left(\frac{1}{2} \sec \theta \tan \theta + \frac{1}{2} \int \sec \theta\, d\theta\right) = \frac{a^2}{2} [\sec \theta \tan \theta + \ln (\sec \theta + \tan \theta)] + C_1$, by Eq. (13). From Fig. 233 or Eq. (56), we have $\sec \theta = \frac{1}{a} \sqrt{x^2 + a^2}$, $\tan \theta = \frac{x}{a}$. Hence with $C = C_1 - a^2/2 \ln a$, the given integral is

$$\int \sqrt{x^2 + a^2}\, dx = \frac{x}{2} \sqrt{x^2 + a^2} + \frac{a^2}{2} \ln (x + \sqrt{x^2 + a^2}) + C.$$

EXERCISE 102

Evaluate each of the following integrals.

1. $\displaystyle\int \frac{dx}{x^2 \sqrt{4 - x^2}}$.
2. $\displaystyle\int \frac{dx}{x^2 \sqrt{3 + x^2}}$.
3. $\displaystyle\int \frac{dx}{(x^2 + 4)^2}$.
4. $\displaystyle\int \frac{x^2\, dx}{(x^2 + 9)^2}$.
5. $\displaystyle\int \frac{(x^2 - 25)^{\frac{3}{2}}}{x^6}\, dx$.
6. $\displaystyle\int \frac{dx}{(x^2 - 2)^{\frac{3}{2}}}$.
7. $\displaystyle\int \frac{dx}{x^4 \sqrt{16 - x^2}}$.
8. $\displaystyle\int \frac{dx}{x^4 \sqrt{25 + x^2}}$.
9. $\displaystyle\int \frac{dx}{x^4 \sqrt{x^2 - 4}}$.
10. $\displaystyle\int \frac{x^4\, dx}{(9 - x^2)^{\frac{5}{2}}}$.
11. $\displaystyle\int \frac{x^4\, dx}{(4 - x^2)^{\frac{7}{2}}}$.
12. $\displaystyle\int \frac{x^2\, dx}{(25 - x^2)^{\frac{3}{2}}}$.
13. $\displaystyle\int \frac{x^2\, dx}{(x^2 + 9)^{\frac{5}{2}}}$.
14. $\displaystyle\int \frac{x^2\, dx}{(9 - x^2)^{\frac{7}{2}}}$.
15. $\displaystyle\int \frac{x^2\, dx}{(x^2 + 16)^{\frac{7}{2}}}$.
16. $\displaystyle\int_0^6 \sqrt{36 - x^2}\, dx$.
17. $\displaystyle\int_3^5 \frac{dx}{x^2 \sqrt{x^2 - 9}}$.
18. $\displaystyle\int_0^2 x^2 \sqrt{4 - x^2}\, dx$.
19. $\displaystyle\int_0^4 \frac{dx}{(25 - x^2)^{\frac{3}{2}}}$.
20. $\displaystyle\int_0^3 \frac{dx}{(16 + x^2)^{\frac{3}{2}}}$.

200. Integration by Parts. Let u and v be differentiable functions of x. Then from the rule for the differential of a product we have

$$d(uv) = u\, dv + v\, du. \tag{58}$$

It follows from this that

$$u\, dv = d(uv) - v\, du. \tag{59}$$

By integrating both members of this relation, we may deduce that

$$\int u\, dv = uv - \int v\, du. \tag{60}$$

This expresses $\int u\, dv$ in terms of $\int v\, du$, which may be a simpler integral.

Suppose that a given integral contains a single inverse function like $\ln x$ or $\tan^{-1} x$ multiplied by a power of x. Then we take the inverse function as u. And the integrand $v\, du$ will contain only algebraic functions. Thus for

$$\int x^m \ln x\,dx, \qquad \int x^m \sin^{-1} ax\,dx, \qquad \int x^m \tan^{-1} ax\,dx, \tag{61}$$

we take $u = \ln x$, $\sin^{-1} ax$, or $\tan^{-1} ax$, and

$$dv = x^m\,dx \qquad \text{so that } v = \int x^m\,dx = \frac{x^{m+1}}{m+1}. \tag{62}$$

Again, suppose that a given integral contains a function of exponential type like e^x, $\sin x$, $\cos x$ multiplied by x raised to a positive integral power, n. Then we take x^n as u. And the integrand $v\,du$ will contain x to a power reduced by one, so that after n successive applications of Eq. (60) the integrand will be free of the power of x. Thus, with n a positive integer, for

$$\int x^n e^{ax}\,dx, \qquad \int x^n \cos ax\,dx, \qquad \int x^n \sin ax\,dx, \tag{63}$$

we take $u = x^n$ and $dv = e^{ax}\,dx$, $\cos ax\,dx$, or $\sin ax\,dx$ so that

$$v = \frac{1}{a} e^{ax}, \qquad \frac{1}{a} \sin ax, \qquad \text{or} \qquad -\frac{1}{a} \cos ax. \tag{64}$$

For a few integrals the application of Eq. (60) twice leads us back to the original integral, but with a coefficient different from one. In such cases the final relation may be solved for the given integral as in Examples 4 and 5 below.

To apply Eq. (60) to a given integral, the first step is the separation of the integrand into two judiciously chosen *parts*, or factors, which we identify with u and dv. For this reason integration by the use of Eq. (60) is called *integration by parts.*

When u and v are functions of x, and the integration is between the limits $x = a$ and $x = b$, we may write

$$\int_a^b u \frac{dv}{dx}\,dx = [uv]_a^b - \int_a^b v \frac{du}{dx}\,dx. \tag{65}$$

Any function having dv as its differential may be used as v. Hence, as in Eqs. (62) and (64), in finding $v = \int dv$ no constant of integration need be added.

EXAMPLE 1. Evaluate $\int x^2 \cos^{-1} 2x\,dx$.

Solution: Take $u = \cos^{-1} 2x$. Then $dv = x^2\,dx$ and $v = \int x^2\,dx = \frac{x^3}{3}$. Also $du = \frac{-2\,dx}{\sqrt{1-4x^2}}$. Hence by Eq. (60) we have

$\int x^2 \cos^{-1} 2x\,dx = \frac{x^3}{3} \cos^{-1} 2x - \int \frac{x^3}{3} \frac{(-2\,dx)}{\sqrt{1-4x^2}}$. If $z = \sqrt{1-4x^2}$, $x^2 = \frac{1-z^2}{4}$, $x\,dx = -\frac{z\,dz}{4}$. And the last integral becomes

$$\frac{2}{3}\int \frac{x^2(x\,dx)}{\sqrt{1-4x^2}} = \frac{2}{3}\int \frac{1-z^2}{4z}\left(-\frac{z\,dz}{4}\right) = \frac{1}{24}\int (z^2-1)dz = \frac{1}{24}\left(\frac{z^3}{3}-z\right)+C$$
$$= \frac{z}{72}(z^2-3)+C = -\frac{1}{36}(2x^2+1)\sqrt{1-4x^2}+C.$$

Hence $\displaystyle\int x^2 \cos^{-1} 2x\,dx = \frac{x^3}{3}\cos^{-1} 2x - \frac{1}{36}(2x^2+1)\sqrt{1-4x^2}+C.$

EXAMPLE 2. Evaluate $\int x^2 \sin 3x\,dx$.

Solution: Let $u = x^2$, $dv = \sin 3x\,dx$. Then $v = \displaystyle\int \sin 3x\,dx = -\frac{\cos 3x}{3}$. Also $du = 2x\,dx$. Hence $\displaystyle\int x^2 \sin 3x\,dx = -\frac{x^2}{3}\cos 3x - \int\left(-\frac{\cos 3x}{3}\right)2x\,dx.$

Next in $\frac{2}{3}\int x\cos 3x\,dx$, let $u = x$, $dv = \cos 3x\,dx$. Then $v = \int \cos 3x\,dx = \frac{1}{3}\sin 3x$ and $du = dx$. And we have

$$\frac{2}{3}\int x\cos 3x\,dx = \frac{2}{3}\left(\frac{x}{3}\sin 3x - \int \frac{1}{3}\sin 3x\,dx\right) = \frac{2}{3}\left(\frac{x}{3}\sin 3x + \frac{1}{9}\cos 3x\right)+C.$$

Hence $\displaystyle\int x^2\sin 3x\,dx = -\frac{x^2}{3}\cos 3x + \frac{2x}{9}\sin 3x + \frac{2}{27}\cos 3x + C.$

EXAMPLE 3. Evaluate $\int \ln(x^2+1)dx$.

Solution: Let $u = \ln(x^2+1)$, $dv = dx$. Then $v = \int dx = x$ and $du = \dfrac{2x\,dx}{x^2+1}$. Hence $\displaystyle\int \ln(x^2+1)dx = x\ln(x^2+1) - \int x\frac{2x\,dx}{x^2+1}$. $\displaystyle\int \frac{-2x^2}{x^2+1}dx = \int\left(-2+\frac{2}{x^2+1}\right)dx = -2x + 2\tan^{-1}x + C.$ Hence

$$\int \ln(x^2+1)dx = x\ln(x^2+1) - 2x + 2\tan^{-1}x + C.$$

EXAMPLE 4. Evaluate $\int e^{ax}\sin bx\,dx$ and $\int e^{ax}\cos bx\,dx$.

Solution: Let $u = e^{ax}$, $dv = \sin bx\,dx$. Then we have $v = \int \sin bx\,dx = -\dfrac{1}{b}\cos bx$ and $du = ae^{ax}\,dx$. Hence $\displaystyle\int e^{ax}\sin bx\,dx = -\frac{1}{b}e^{ax}\cos bx - \int\left(-\frac{1}{b}\cos bx\right)(ae^{ax}\,dx).$

The last term is $\displaystyle\frac{a}{b}\int e^{ax}\cos bx\,dx$. In the integral, let $u = e^{ax}$, $dv = \cos bx\,dx$. Then $v = \displaystyle\int \cos bx\,dx = \frac{1}{b}\sin bx$ and $du = e^{ax}\,dx$. Hence $\displaystyle\int e^{ax}\cos bx\,dx = \frac{1}{b}e^{ax}\sin bx - \int \frac{1}{b}\sin bx(ae^{ax}\,dx).$

Substitution of the second relation in the first shows that $\int e^{ax}\sin bx\,dx = -\dfrac{1}{b}e^{ax}\cos bx + \dfrac{a}{b^2}e^{ax}\sin bx - \dfrac{a^2}{b^2}\displaystyle\int e^{ax}\sin bx\,dx$. Replace $\int e^{ax}\sin bx\,dx$ on the left by I, and on the right by $I + C_1$. Then

$$b^2 I = e^{ax}(a\sin bx - b\cos bx) - a^2 I - a^2 C_1,$$
$$(a^2+b^2)I = e^{ax}(a\sin bx - b\cos bx) - a^2 C_1,$$

and if $C = \dfrac{-a^2C_1}{a^2+b^2}$,

$$I = \int e^{ax}\sin bx\,dx = \frac{e^{ax}(a\sin bx - b\cos bx)}{a^2+b^2} + C. \tag{66}$$

Similarly, substitution of the first relation in the second gives $\int e^{ax} \cos bx\, dx = \frac{1}{b} e^{ax} \sin bx + \frac{a}{b^2} e^{ax} \cos bx - \frac{a^2}{b^2} \int e^{ax} \cos bx\, dx$. Replace $\int e^{ax} \cos bx\, dx$ on the left by J and on the right by $J + C_2$. Then $b^2 J = e^{ax}(b \sin bx + a \cos bx) - a^2 C_2$, and if $C = \frac{-a^2 C_2}{a^2 + b^2}$,

$$J = \int e^{ax} \cos bx\, dx = \frac{e^{ax}(b \sin bx + a \cos bx)}{a^2 + b^2} + C. \tag{67}$$

EXAMPLE 5. Deduce a reduction formula for $\int \sqrt{A + Bx^2}\, dx$.

Solution: Let $u = \sqrt{A + Bx^2}$, $dv = dx$. Then $v = \int dx = x$ and $du = \frac{Bx\, dx}{\sqrt{A + Bx^2}}$. Hence $\int \sqrt{A + Bx^2}\, dx = x\sqrt{A + Bx^2} - \int \frac{Bx^2\, dx}{\sqrt{A + Bx^2}}$. But $\int \frac{-Bx^2\, dx}{\sqrt{A + Bx^2}} = \int \frac{A - (A + Bx^2)}{\sqrt{A + Bx^2}}\, dx = \int \frac{A\, dx}{\sqrt{A + Bx^2}} - \int \sqrt{A + Bx^2}\, dx$. And $\int \sqrt{A + Bx^2}\, dx = x\sqrt{A + Bx^2} + \int \frac{A\, dx}{\sqrt{A + Bx^2}} - \int \sqrt{A + Bx^2}\, dx$. Transposing the last term and dividing by 2 leads to the required reduction formula,

$$\int \sqrt{A + Bx^2}\, dx = \frac{x}{2}\sqrt{A + Bx^2} + \frac{A}{2} \int \frac{dx}{\sqrt{A + Bx^2}}. \tag{68}$$

We note that the integral on the right of Eq. (68) may be evaluated by Eq. (16) or (20), depending on the signs of A and B.

EXERCISE 103

Evaluate each of the following integrals.

1. $\int xe^{2x}\, dx$.
2. $\int \cos^{-1} x\, dx$.
3. $\int \ln x\, dx$.
4. $\int x \cos x\, dx$.
5. $\int x \sec^{-1} x\, dx$.
6. $\int x^2 e^{-x}\, dx$.
7. $\int x^3 \ln x\, dx$.
8. $\int x \sin 3x\, dx$.
9. $\int x^2 \cos x\, dx$.
10. $\int x^2 \tan^{-1} x\, dx$.
11. $\int x \cot^{-1} x\, dx$.
12. $\int x \sin^2 \frac{x}{2}\, dx$.
13. $\int_0^{\frac{1}{4}} \sin^{-1} 4x\, dx$.
14. $\int_0^{\pi/4} x \sin 2x\, dx$.
15. $\int_1^e x \ln x\, dx$.
16. $\int_{\pi/4}^{\pi/2} e^{-x} \cos x\, dx$.

Use integration by parts to derive each of the following reduction formulas.

17. $\int \sec^n x\, dx = \frac{1}{n-1} \sec^{n-2} x \tan x + \frac{n-2}{n-1} \int \sec^{n-2} x\, dx$.
18. $\int \sin^n x\, dx = -\frac{1}{n} \sin^{n-1} x \cos x + \frac{n-1}{n} \int \sin^{n-2} x\, dx$.

19. $\int \cos^n x\, dx = \frac{1}{n} \cos^{n-1} x \sin x + \frac{n-1}{n} \int \cos^{n-2} x\, dx.$

20. $\int x^n e^{ax}\, dx = \frac{1}{a} x^n e^{ax} - \frac{n}{a} \int x^{n-1} e^{ax}\, dx.$

21. Let I_n denote either $\int_0^{\pi/2} \sin^n x\, dx$ or $\int_0^{\pi/2} \cos^n x\, dx$. Deduce from Probs. 18 and 19 that, if $n > 1$, $I_n = \frac{n-1}{n} I_{n-2}$.

22. Use Prob. 21 to show that $\int_0^{\pi/2} \sin^n x\, dx = \int_0^{\pi/2} \cos^n x\, dx =$ $\frac{1 \cdot 3 \cdot 5 \cdots (n-1)}{2 \cdot 4 \cdot 6 \cdots n} \frac{\pi}{2}$ if n is an even positive integer, and $= \frac{2 \cdot 4 \cdot 6 \cdots (n-1)}{3 \cdot 5 \cdot 7 \cdots n}$ if n is an odd positive integer greater than unity.

201. Integration of Rational Fractions. A fraction whose numerator and denominator are polynomials is called a *rational fraction.* If the degree of the numerator is equal to or greater than the degree of the denominator, the fraction may be transformed by division. For example,

$$\frac{x^3}{x^2 - 3x - 2} = x + 3 + \frac{11x + 6}{x^2 - 3x - 2}.$$

A fraction having the degree of the numerator less than the degree of the denominator is called a *proper fraction.* A proper fraction can always be resolved into a sum of *partial fractions*† *whose denominators* are factors of the original denominator, $D(x)$. For each factor $(x - r)$ which occurs but once in $D(x)$ there is a single fraction of the form $A/(x - r)$. And for each factor which occurs exactly p times in $D(x)$ there is a group of fractions of the form

$$\frac{A_p}{(x - r)^p} + \cdots + \frac{A_2}{(x - r)^2} + \frac{A_1}{x - r}. \tag{69}$$

The constants in the numerator could always be obtained by assuming the complete expansion of the form stated, clearing of fractions, and equating the coefficients of the corresponding powers of x. But there is a simpler method of finding A for a simple root. To derive this, consider the general relation

$$\frac{N(x)}{D(x)} = \frac{A}{x - r} + \frac{M(x)}{E(x)}. \tag{70}$$

Here A is a constant, $N(x)$, $M(x)$, $D(x)$, $E(x)$ are all polynomials, and r is a simple root of $D(x) = 0$ so that

$$D(x) = (x - r)E(x), \qquad D(r) = 0, \qquad E(r) \neq 0. \tag{71}$$

† A proof is given in Sec. 115 of the author's "A Treatise on Advanced Calculus," John Wiley & Sons, Inc., New York, 1940 (Dover reprint).

Let us multiply both sides of Eq. (70) by $x - r$, and then let $x \to r$. The first term on the right is A, and the other term contains the factor $x - r$ after the multiplication and so approaches zero when $x \to r$. It follows that

$$A = \lim_{x \to r} (x - r) \frac{N(x)}{D(x)} = N(r) \lim_{x \to r} \frac{x - r}{D(x)}. \tag{72}$$

If we use the first relation of Eq. (71) and cancel the factor $x - r$, we see that the last limit in Eq. (72) is $1/E(r)$. Hence

$$A = \frac{N(r)}{E(r)} = \left.\frac{N(x)}{E(x)}\right|_{x=r} \tag{73}$$

This leads to the rule: In Eq. (70), *A may be found by deleting the factor $x - r$ from the denominator $D(x)$ and then evaluating the left member for $x = r$.*

The method of finding the constants for the case of multiple and complex roots will be explained by the examples which follow.

The partial fractions in Eq. (69) with exponents greater than unity may be integrated by the use of Eq. (4). When r is real and A is real, we may deduce from Eq. (5) that

$$\int \frac{A}{x - r}\, dx = A \ln (x - r) + C. \tag{74}$$

For real coefficients, with each complex root $r = s + ti$ with $A = a + bi$ where $i = \sqrt{-1}$, there is a conjugate complex root $\bar{r} = s - ti$ with $\bar{A} = a - bi$. And the integral of the sum of the two conjugate fractions is

$$\int \left(\frac{a + bi}{x - s - ti} + \frac{a - bi}{x - s + ti} \right) dx = \int \frac{2a(x - s) - 2bt}{(x - s)^2 + t^2}\, dx$$
$$= a \ln [(x - s)^2 + t^2] - 2b \tan^{-1} \frac{x - s}{t} + C. \tag{75}$$

EXAMPLE 1. Evaluate $\int \frac{2x + 1}{x^3 + x^2 - 2x}\, dx$.

Solution: Since $x^3 + x^2 - 2x = x(x - 1)(x + 2)$, we assume $\frac{2x + 1}{x(x - 1)(x + 2)} = \frac{A}{x} + \frac{B}{x - 1} + \frac{C}{x + 2}$. By Eq. (73), we find that $A = \left.\frac{2x + 1}{(x - 1)(x + 2)}\right|_{x=0} = -\frac{1}{2}$, $B = \left.\frac{2x + 1}{x(x + 2)}\right|_{x=1} = 1$, $C = \left.\frac{2x + 1}{x(x - 1)}\right|_{x=-2} = -\frac{1}{2}$. We may check these values by verifying that $\frac{-\frac{1}{2}}{x} + \frac{1}{x - 1} + \frac{-\frac{1}{2}}{x + 2} = \frac{2x + 1}{x(x - 1)(x + 2)}$. It follows from Eq. (74) that $\int \frac{2x + 1}{x^3 + x^2 - 2x}\, dx = -\frac{1}{2} \ln x + \ln (x - 1) - \frac{1}{2} \ln (x + 2) + C$.

EXAMPLE 2. Evaluate $\int \frac{x^3 + 2}{x(x - 2)^3}\, dx$.

Solution: By Eq. (69) with $r = 2$ and $p = 3$, we assume

$$\frac{x^3+2}{x(x-2)^3} = \frac{A}{x} + \frac{B}{(x-2)^3} + \frac{C}{(x-2)^2} + \frac{D}{x-2}.$$

Then $x^3 + 2 = A(x-2)^3 + Bx + Cx(x-2) + Dx(x-2)^2$ is an identity. Put $x = 0$ and $2 = A(-2)^3$, $A = -\frac{1}{4}$. Put $x = 2$ and $10 = 2B$, $B = 5$. Equate coefficients of x^3 on both sides and $1 = A + D$. Hence $D = 1 - A = 1 - (-\frac{1}{4}) = \frac{5}{4}$. Equate coefficients of x^2 on both sides and $0 = -6A + C - 4D$. Hence

$$C = 6A + 4D = 6(-\tfrac{1}{4}) + 4(\tfrac{5}{4}) = \tfrac{7}{2}.$$

Another method of finding C and D is to put any two values for x other than 0 and 2 already used in the identity. For example, we may put $x = 3$ to deduce that $29 = A + 3B + 3C + 3D$, and $x = 1$ to give $3 = -A + B - C + D$. These combined with $A = -\frac{1}{4}$ and $B = 5$ found above give $3C + 3D = \frac{57}{4}$, $-C + D = -\frac{9}{4}$ which can be solved as simultaneous equations to give $D = \frac{5}{4}$ and $C = \frac{7}{2}$ as before.

We may check our values by verifying that $\frac{-\frac{1}{4}}{x} + \frac{\frac{5}{4}}{x-2} + \frac{\frac{7}{2}}{(x-2)^2} + \frac{5}{(x-2)^3} = \frac{x^3+2}{x(x-2)^3}$. It follows that $\int \frac{x^3+2}{x(x-2)^3}\,dx = -\frac{1}{4}\ln x + \frac{5}{4}\ln(x-2) - \frac{7}{2(x-2)} - \frac{5}{2(x-2)^2} + C.$

EXAMPLE 3. Evaluate $\int \frac{x+2}{x^3+2x^2+5x}\,dx$.

Solution 1. Since the roots of $x^2 + 2x + 5$ are $-1 \pm 2i$, we assume

$$\frac{x+2}{x(x+1-2i)(x+1+2i)} = \frac{A}{x} + \frac{B}{x+1-2i} + \frac{\bar{B}}{x+1+2i}.$$

By Eq. (73) we find that $A = \left.\frac{x+2}{x^2+2x+5}\right|_{x=0} = \frac{2}{5}$, $B = \left.\frac{x+2}{x(x+1+2i)}\right|_{x=-1+2i} = \frac{1+2i}{-4(2+i)}\,\frac{2-i}{2-i} = -\frac{1}{20}(4+3i)$. Hence $\bar{B} = -\frac{1}{20}(4-3i)$. We may check these values by verifying that

$$\frac{\frac{2}{5}}{x} + \frac{-(4+3i)/20}{x+1-2i} + \frac{-(4-3i)/20}{x+1+2i} = \frac{x+2}{x^3+2x^2+5x}.$$

It follows from Eqs. (74) and (75) that

$$\int \frac{x+2}{x^3+2x^2+5x}\,dx = \frac{2}{5}\ln x - \frac{1}{5}\ln(x^2+2x+5) + \frac{3}{10}\tan^{-1}\frac{x+1}{2} + C.$$

Solution 2: To solve the problem without using i, we assume

$$\frac{x+2}{x(x^2+2x+5)} = \frac{A}{x} + \frac{Bx+C}{x^2+2x+5}.$$

Then $x + 2 = A(x^2+2x+5) + (Bx+C)x$ is an identity. Put $x = 0$ and $2 = 5A$, $A = \frac{2}{5}$. Equate coefficients of x^2 on both sides and $0 = A + B$, $B = -A = -\frac{2}{5}$. Equate coefficients of x on both sides and $1 = 2A + C$, $C = 1 - 2A = \frac{1}{5}$. Hence

$$\frac{x+2}{x(x^2+2x+5)} = \frac{\frac{2}{5}}{x} + \frac{-\frac{2}{5}x+\frac{1}{5}}{x^2+2x+5} = \frac{\frac{2}{5}}{x} + \frac{-\frac{1}{5}(2x+2)}{x^2+2x+5} + \frac{\frac{3}{5}}{(x+1)^2+2^2}.$$

It follows from Eqs. (5) and (22) that

$$\int \frac{x+2}{x^3+2x+5}\,dx = \frac{2}{5}\ln x - \frac{1}{5}\ln(x^2+2x+5) + \frac{3}{10}\tan^{-1}\frac{x+1}{2} + C.$$

EXAMPLE 4. Evaluate $\int \frac{x^3+1}{x(x^2+1)^2}\,dx.$

Solution: by Eq. (69) with $p = 2$ and $r = i$ and $-i$, we assume $\frac{x^3+1}{x(x-i)^2(x+i)^2}$ $= \frac{A}{x} + \frac{B}{(x-i)^2} + \frac{C}{x-i} + \frac{\bar{B}}{(x+i)^2} + \frac{\bar{C}}{x+i}.$ Then

$$x^3 + 1 = A(x^2+1)^2 + Bx(x+i)^2 + Cx(x-i)(x+i)^2 + \bar{B}x(x-i)^2 + \bar{C}x(x-i)^2(x+i).$$

Put $x = 0$ and $1 = A$, $A = 1$. Put $x = i$ and $-i + 1 = Bi(2i)^2 = -4Bi$, $B = (1+i)/4$ Hence $\bar{B} = (1-i)/4$. Equate coefficients of x^4 on both sides and $0 = A + C + \bar{C}$, $C + \bar{C} = -A = -1$. Equate coefficients of x^3 on both sides and $1 = B + iC + \bar{B} - i\bar{C}$, $i(C - \bar{C}) = 1 - B - \bar{B} = \frac{1}{2}$, $C - \bar{C} = -i/2$. From this and $C + \bar{C} = -1$, it follows that $C = -\frac{1}{2} - i/4$ and $\bar{C} = -\frac{1}{2} + i/4$. Hence

$$\frac{x^3+1}{x(x^2+1)^2} = \frac{1}{x} + \frac{(1+i)/4}{(x-i)^2} + \frac{(1-i)/4}{(x+i)^2} + \frac{-\frac{1}{2} - i/4}{x-i} + \frac{-\frac{1}{2} + i/4}{x+i}.$$

Since $\frac{1+i}{4}\,\frac{-1}{x-i} = -\frac{1}{4}\,\frac{1+i}{x-i}\,\frac{x+i}{x+i} = -\frac{1}{4}\,\frac{x - 1 + i(x+1)}{x^2+1}$, it follows from Eqs. (74), (75), and (4) that

$$\int \frac{x^3+1}{x(x^2+1)^2}\,dx = \ln x - \frac{1}{2}\,\frac{x-1}{x^2+1} - \frac{1}{2}\ln\,(x^2+1) + \frac{1}{2}\tan^{-1} x + C.$$

EXAMPLE 5. Evaluate $\int \frac{x^4+x}{x^3 - x^2 - 4x + 4}\,dx.$

Solution. As this is an improper fraction, we find the quotient $x + 1$, and without finding the proper fractions, assume

$$\frac{x^4+x}{(x-1)(x-2)(x+2)} = x + 1 + \frac{A}{x-1} + \frac{B}{x-2} + \frac{C}{x+2}.$$

We may then deduce from Eq. (73) that $A = \left.\frac{x^4+x}{(x+2)(x-2)}\right|_{x=1} = -\frac{2}{3}$, $B = \left.\frac{x^4+x}{(x-1)(x+2)}\right|_{x=2} = \frac{9}{2}$, $C = \left.\frac{x^4+x}{(x-1)(x-2)}\right|_{x=-2} = \frac{7}{6}$. Hence $\frac{x^4+x}{(x-1)(x-2)(x+2)} = x + 1 + \frac{-\frac{2}{3}}{x-1} + \frac{\frac{9}{2}}{x-2} + \frac{\frac{7}{6}}{x+2}.$ And

$$\int \frac{x^4+x}{x^3 - x^2 - 4x + 4}\,dx = \frac{x^2}{2} + x - \frac{2}{3}\ln\,(x-1) + \frac{9}{2}\ln\,(x-2) + \frac{7}{6}\ln\,(x+2) + C.$$

EXERCISE 104

Evaluate each of the following integrals.

1. $\int \frac{4x-6}{(x-1)(x-2)(x-3)}\,dx.$
2. $\int \frac{2x+4}{x^3-x}\,dx.$
3. $\int \frac{2x+4}{(x-1)(x-3)(x-4)}\,dx.$
4. $\int \frac{x^2\,dx}{(x+1)(x^2+4)}.$
5. $\int \frac{x^2+x-4}{x^3-x^2}\,dx.$
6. $\int \frac{2x\,dx}{(x^2+1)(x-1)}.$
7. $\int \frac{4x\,dx}{x^4+x^2}.$
8. $\int \frac{25x^3\,dx}{(x^2+4)(x-1)^2}.$
9. $\int \frac{2x^2-10x+10}{(x-1)(x-2)(x-3)}\,dx.$
10. $\int \frac{4x+2}{(x+2)(x^2-1)}\,dx.$

11. $\int \frac{2x^2 - 4x - 1}{2x^3 - x^2 - x}\,dx.$

12. $\int \frac{6x^2 - 6x - 6}{(x-1)(x^2 - x - 6)}\,dx.$

13. $\int \frac{4\,dx}{(x^2 - 1)^2}.$

14. $\int \frac{3x^2 + 15x + 6}{x^4 - 5x^2 + 4}\,dx.$

15. $\int \frac{x^2 + x}{(x-1)(x^2+1)}\,dx.$

16. $\int \frac{4\,dx}{x^4 - 1}.$

17. $\int \frac{4x^2 + 2x + 4}{x^3 - 8}\,dx.$

18. $\int \frac{8\,dx}{x^3 + 4x}.$

19. $\int \frac{2x^3 + x + 3}{(x^2+1)^2}\,dx.$

20. $\int \frac{2x^2 + x + 8}{(x^2+4)^2}\,dx.$

202. Rationalizable Integrals. Let $R(u)$ or $R(u,v)$ denote a rational function of u, or of u and v. Then the discussion of Sec. 201 shows that every integral of the form $\int R(x)dx$ is expressible in terms of known functions.

By expressing the other trigonometric functions in terms of the sine and cosine, the integral of any rational function of trigonometric functions of x may be reduced to the form $\int R(\sin x, \cos x)dx$. Let $z = \tan(x/2)$. Then $x = 2\tan^{-1} z$ and $dx = 2\,dz/(1 + z^2)$. Also $\sec^2(x/2) = 1 + \tan^2(x/2) = 1 + z^2$, $\cos^2(x/2) = 1/(1 + z^2)$. Hence $\sin x = 2\sin(x/2)\cos(x/2) = 2\tan(x/2)\cos^2(x/2) = 2z/(1 + z^2)$. And $\cos x = 2\cos^2(x/2) - 1 = (1 - z^2)/(1 + z^2)$. It follows that, if $z = \tan(x/2)$,

$$\int R(\sin x, \cos x)dx = \int R\left(\frac{2z}{1+z^2}, \frac{1-z^2}{1+z^2}\right)\frac{2\,dz}{1+z^2}. \tag{76}$$

It follows that every integral of this type can be evaluated.

Any integral of the form $\int R(x, \sqrt{ax^2 + bx + c})dx$ may be reduced to $\int R_1(u, \sqrt{\pm k^2 \pm x^2})du$ by the substitution $u = (x + b/2a)$, suggested by Eq. (25). And this in turn can be reduced to an integral of the form $\int R_2(\sin\theta, \cos\theta)d\theta$ by one of the trigonometric substitutions suggested in Sec. 199. It follows then from Eq. (76) that any integral of the form $\int R(x, \sqrt{ax^2 + bx + c})dx$ can be evaluated.

The substitution $z = e^{ax}$ makes $x = \frac{1}{a}\ln z$ and $dx = \frac{1}{a}\frac{dz}{z}$. Hence the integral

$$\int R(e^{ax})dx = \int R(z)\frac{1}{az}\,dz, \tag{77}$$

and any integral of this type can be evaluated.

The various substitutions mentioned above are occasionally practically effective. But the real purpose of our remarks is to indicate general classes of functions which can be integrated in terms of known functions and so are likely to be found worked out in integral tables.

Example. Evaluate $\int \frac{dx}{4 \sin x + 3 \cos x + 13}$.

Solution: With $z = \tan (x/2)$, by Eq (76) the integral becomes

$$\int \frac{2\,dz}{4(2z) + 3(1 - z^2) + 13(1 + z^2)} = \int \frac{2\,dz}{10z^2 + 8z + 16} = \frac{1}{5} \int \frac{dz}{(z + \frac{2}{5})^2 + \frac{36}{25}}$$
$$= \frac{1}{5}\frac{5}{6} \tan^{-1} \frac{z + \frac{2}{5}}{\frac{6}{5}} + C = \frac{1}{6} \tan^{-1} \left(\frac{5 \tan (x/2) + 2}{6}\right) + C.$$

203. Table of Integrals. The methods of integration developed in this chapter are sufficient for the solution of most problems occurring in practice which can be solved in closed form. The use of series and numerical methods will be described in Secs. 247 and 259 to 261.

Extensive tables of integrals are given as one section of many engineering handbooks and collections of mathematical tables. The student will find it convenient to become familiar with the arrangement of one such table, as its efficient use will materially reduce the work of evaluating some integrals.

To facilitate reference to the formulas of this chapter, we have collected the more important to form a brief table of integrals, Table 1 at the end of this book. A few references to sections which are added make this a key to the methods of integration which we have described.

EXERCISE 105

Evaluate each of the following integrals.

1. $\int \frac{dx}{9 - x^2}$.
2. $\int \frac{dx}{9 + x^2}$.
3. $\int \frac{dx}{\sqrt{9 - 4x^2}}$.
4. $\int \frac{dx}{\sqrt{4x^2 - 9}}$.
5. $\int \sqrt{25 - 4x^2}\, dx$.
6. $\int \sqrt{4x^2 - 25}\, dx$.
7. $\int x^2 \sqrt[3]{x^3 + 2}\, dx$.
8. $\int e^{2x} \sqrt{1 + e^{2x}}\, dx$.
9. $\int \frac{x\,dx}{x^4 + 1}$.
10. $\int \frac{x\,dx}{\sqrt{x^4 + 1}}$.
11. $\int \frac{2x\,dx}{(1 - x)^3}$.
12. $\int 5^x\, dx$.
13. $\int \cos^2 4x\, dx$.
14. $\int \sin^2 3x\, dx$.
15. $\int e^{-2x} \sin 3x\, dx$.
16. $\int e^{-3x} \cos 4x\, dx$.
17. $\int \sin^4 x\, dx$.
18. $\int_0^{\pi/4} \cos^3 2x\, dx$.
19. $\int_0^{\pi/8} \sec^3 2x\, dx$.
20. $\int_0^{\pi/2} \cos^6 x\, dx$.
21. $\int_0^{\pi/2} \sin^2 x \cos^4 x\, dx$.
22. $\int_0^{\pi/2} \sin^8 x\, dx$.

CHAPTER 14

APPLICATIONS OF INTEGRATION. MOMENTS

In Chaps. 5 and 12 we determined certain areas, volumes, and arc lengths by expressing the desired quantity in terms of an integral. With the new formulas of integration of Chap. 13 we can evaluate the integrals for these geometric quantities in many more instances. In this chapter we first review the integrals which represent area, volume, arc length, and area of a surface of revolution. Here for most of these we merely use a figure to suggest the element of integration, or differential, and so recall the integration formula which was deduced in Chap. 12 as an application of the fundamental theorem. We define the average value of a function over an interval. We then discuss the first moment and center of gravity for distributions of mass in the plane and in space, and the related notion of centroid of an arc, area, or volume. We discuss the moment of inertia for a material body about an axis and the related notion of moment of inertia of an area. As additional physical concepts defined in terms of integrals, we consider the depth of the center of pressure, the work done by a variable force, and the force due to attraction.

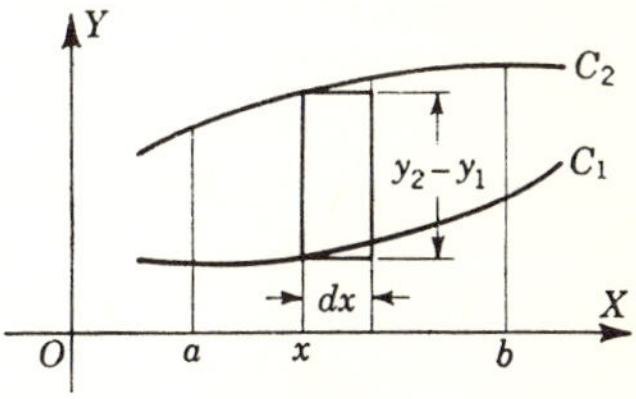

FIG. 235.

204. Area. Consider the area bounded above by the curve $y_2 = f_2(x)$ and below by the curve $y_1 = f_1(x)$, and lying between a left-hand ordinate at $x = a$ and a right-hand ordinate at $x = b$ (Fig. 235). Here the element of area is a rectangle of height $y_2 - y_1$ and of width dx. Hence the area of the element is $(y_2 - y_1)dx$. And by the argument of Sec. 172, it follows that

$$dA = (y_2 - y_1)dx, \qquad A = \int_a^b (y_2 - y_1)dx. \tag{1}$$

The other cases of plane area in rectangular coordinates described in Secs. 73 and 172 may be read from figures in a similar manner.

Next consider the area bounded by the curve in polar coordinates $r = f(\theta)$ and the two radius vectors $\theta = \alpha$ and $\theta = \beta$ (Fig. 236). Here the element of area is a circular sector of radius r and central angle $d\theta$,

Hence the area of the element is $\frac{1}{2}r^2\,d\theta$. And by the argument of Sec. 174 it follows that

$$dA = \tfrac{1}{2}r^2\,d\theta, \qquad A = \tfrac{1}{2}\int_\alpha^\beta r^2\,d\theta. \tag{2}$$

EXAMPLE 1. Let $P_1 = (a_1, k/a_1)$ and $P_2 = (a_2, k/a_2)$ with $a_2 > a_1$ be two points on the equilateral hyperbola $xy = k$ (Fig. 237). Let A be the area under arc P_1P_2

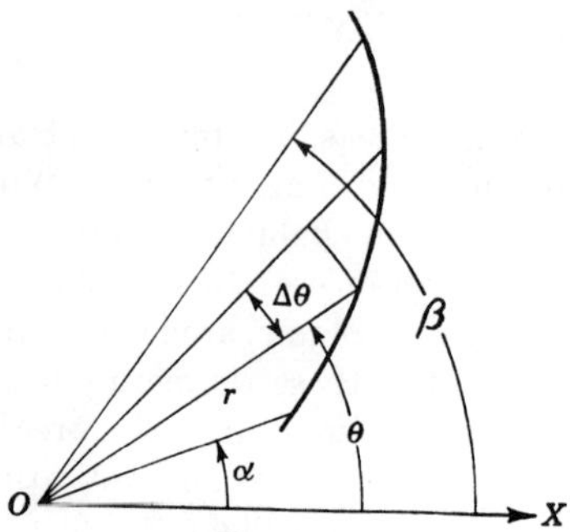

FIG. 236.

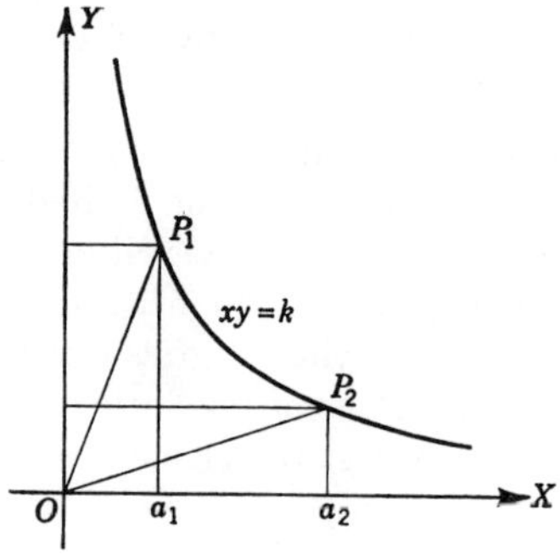

FIG. 237.

and above OX, B be the area to the left of P_2P_1 and to the right of OY, and C be the sectorial area OP_1P_2. Prove that area A = area B = area C.

Solution: We have $dA = y\,dx$, $y = k/x$, so that

$$A = \int_{a_1}^{a_2} y\,dx = \int_{a_1}^{a_2} \frac{k}{x}\,dx = [k \ln x]_{a_1}^{a_2} = k(\ln a_2 - \ln a_1) = k \ln \frac{a_2}{a_1}.$$

And $dB = x\,dy$, $x = k/y$, so that

$$B = \int_{k/a_2}^{k/a_1} \frac{k}{y}\,dy = [k \ln y]_{k/a_2}^{k/a_1} = k\left(\ln \frac{k}{a_1} - \ln \frac{k}{a_2}\right) = k \ln \frac{a_2}{a_1}.$$

And $dC = \frac{1}{2}r^2\,d\theta$. Also $x = r\cos\theta$, $y = r\sin\theta$ Hence from $xy = k$, $r^2\cos\theta\sin\theta = k$, $r^2 = \dfrac{k}{\cos\theta\sin\theta} = \dfrac{2k}{\sin 2\theta}$. We have $\tan\theta = y/x$. At P_1, $\tan\theta_1 = k/a_1^2$. And at P_2, $\tan\theta_2 = k/a_2^2$. Since $a_2 > a_1$, $\theta_2 < \theta_1$. Then $C = \dfrac{1}{2}\displaystyle\int_{\theta_2}^{\theta_1} r^2\,d\theta = \dfrac{1}{2}\int_{\theta_2}^{\theta_1} \dfrac{2k}{\sin 2\theta}\,d\theta = \dfrac{k}{2}\int_{\theta_2}^{\theta_1} \csc 2\theta\,d(2\theta) = -\dfrac{k}{2}[\ln(\csc 2\theta + \cot 2\theta)]_{\theta_2}^{\theta_1}$. We note that $\csc 2\theta + \cot 2\theta = \dfrac{1 + \cos 2\theta}{\sin 2\theta} = \dfrac{2\cos^2\theta}{2\sin\theta\cos\theta} = \cot\theta = \dfrac{1}{\tan\theta}$. Hence

$$C = -\frac{k}{2}[\ln \cot\theta]_{\theta_2}^{\theta_1} = \frac{k}{2}[\ln \tan\theta]_{\theta_2}^{\theta_1} = \frac{k}{2}(\ln \tan\theta_1 - \ln\tan\theta_2)$$
$$= \frac{k}{2}\left(\ln \frac{k}{a_1^2} - \ln \frac{k}{a_2^2}\right) = k \ln \sqrt{\frac{a_2^2}{a_1^2}} = k \ln \frac{a_2}{a_1}.$$

Thus $A = B = C = k \ln (a_2/a_1)$, as was to be proved.

EXAMPLE 2. Find the area in the first quadrant bounded by the coordinate axes and the astroid $(x/a)^{\frac{2}{3}} + (y/b)^{\frac{2}{3}} = 1$, by using parametric representation $x = a\cos^3 t$, $y = b \sin^3 t$.

Solution: Since $y = 0$ when $x = a$ (Fig. 238), the area $A = \int_0^a y\,dx$. As in Sec. 179, we may express this in terms of t. Here $x = a\cos^3 t$ shows that $x = 0$ when $t = \pi/2$ and $x = a$ when $t = 0$. And $dx = -3a\cos^2 t\sin t\,dt$, $y = b\sin^3 t$ so that, as in Sec. 195,

$$\begin{aligned} A &= -\int_{\pi/2}^{0} 3ab\cos^2 t\sin^4 t\,dt = \frac{3ab}{8}\int_0^{\pi/2}(2\cos t\sin t)^2(2\sin^2 t)dt \\ &= \frac{3ab}{8}\int_0^{\pi/2}\sin^2 2t(1-\cos 2t)dt = \frac{3ab}{8}\left[\frac{t}{2}-\frac{\sin 4t}{8}-\frac{\sin^3 2t}{6}\right]_0^{\pi/2} \\ &= \frac{3\pi ab}{32}, \qquad \text{the required area.} \end{aligned}$$

We might have used form 34 of Table 1 with $m = 4$, $n = 2$ to show that

$$\int_0^{\pi/2}\cos^2 t\sin^4 t\,dt = \frac{(1\cdot 3)1}{2\cdot 4\cdot 6}\frac{\pi}{2} = \frac{\pi}{32}.$$

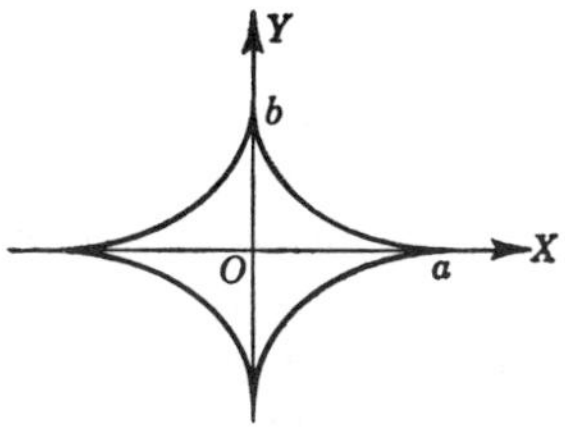

FIG. 238.

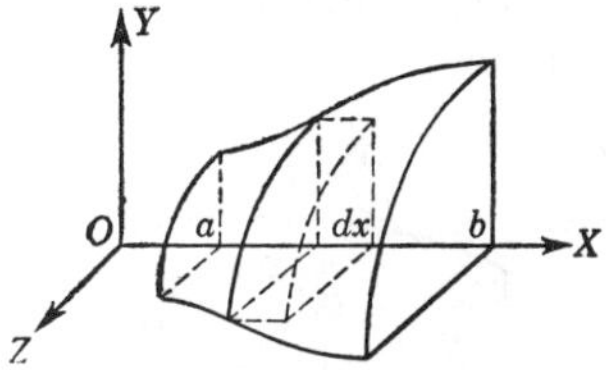

FIG. 239.

205. Volume. Consider the volume of a solid. As in Sec. 74, let x be the perpendicular distance from some fixed cross section to a variable cross section. And let the area of the variable cross section at distance x be $A(x)$. Then the element of volume (Fig. 239) is a cylinder, or prism, whose base is the cross section of area $A(x)$ and whose altitude is dx. Hence the volume of the element is $dV = A\,dx$. Thus $\Delta V = A(x')\Delta x$. And by applying the theorem of Sec. 170 to $V = \lim \Sigma\,\Delta V_i$, the volume between two cross sections at distances a and b, we may deduce that

$$dV = A\,dx, \qquad V = \int_a^b A(x)dx. \tag{3}$$

In particular, suppose that the area of Fig. 235, bounded above by the curve $y_2 = f_2(x)$, below by the curve $y_1 = f_1(x)$, and lying between a left-hand ordinate at $x = a$ and a right-hand ordinate at $x = b$, is revolved about the x axis, OX. Assume that $y_2 > y_1 \geqq 0$ for $a < x < b$, as in Fig. 240. Then we have $A(x) = \pi(y_2^2 - y_1^2)$, so that

$$dV_x = \pi(y_2^2 - y_1^2)dx, \qquad V_x = \pi\int_a^b (y_2^2 - y_1^2)dx. \tag{4}$$

A similar equation with the roles of x and y reversed holds for the volume generated by revolving an area about the y axis.

But suppose that the area of Fig. 235 is revolved about the y axis, OY. We may divide the solid generated into *cylindrical shells* by a system of

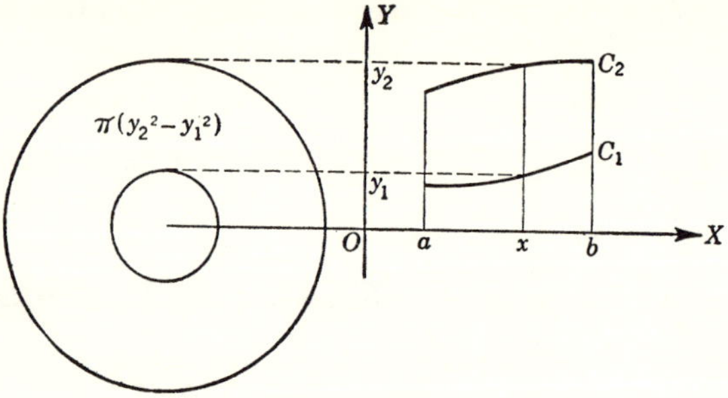

FIG. 240.

concentric circular cylinders whose common axis is OY. The volume of one of the shells such as that generated by the revolution of the area $P_1P_2Q_2Q_1$ of Fig. 241 is equal to Δx times the lateral area of a cylinder of radius x' and altitude $f_2(x'') - f_1(x'')$, where x' and x'' are suitably chosen values between x and $x + \Delta x$. Thus $\Delta V = 2\pi x'[f_2(x'') - f_1(x'')]\Delta x$. By applying the theorem of Sec. 185 to $V_y = \Sigma \lim \Delta V_i$, the volume of the solid of revolution generated, we may deduce that

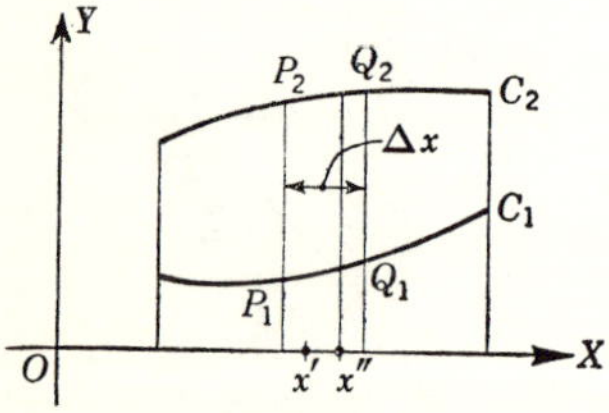

FIG. 241.

$$dV_x = 2\pi x(y_2 - y_1)dx, \qquad V_y = 2\pi \int_a^b (y_2 - y_1)x\,dx. \tag{5}$$

EXAMPLE 1. A circle of radius a is revolved about an axis in its plane b units from its center, where $b > a$, to form a solid ring, called an *anchor ring* or *torus*. Find its volume.

Solution 1: Take the x axis as the axis of revolution, and the center of the circle as $(0,b)$. Then its equation is $x^2 + (y - b)^2 = a^2$. Hence $y = b \pm \sqrt{a^2 - x^2}$. Let $y_2 = b + \sqrt{a^2 - x^2}$, $y_1 = b - \sqrt{a^2 - x^2}$. Then $y_2^2 - y_1^2 = (y_2 + y_1)(y_2 - y_1) = 4b\sqrt{a^2 - x^2}$. From symmetry we see that the volume is twice that for x from 0 to a. Hence from Eq. (4) and the substitution $x = a \sin t$, we have

$$V_x = 2\pi \int_0^a 4b\sqrt{a^2 - x^2}\,dx = 8\pi b \int_0^{\pi/2} a^2 \cos^2 t\,dt = 8\pi a^2 b \left[\frac{t}{2} + \frac{\sin 2t}{4}\right]_0^{\pi/2}$$
$$= 2\pi^2 a^2 b, \qquad \text{the required volume.}$$

Solution 2: Take the y axis as the axis of revolution, and the center of the circle as $(b,0)$. Then its equation is $y^2 + (x - b)^2 = a^2$. Hence $y = \pm\sqrt{a^2 - (x - b)^2}$. From symmetry, we see that the volume is twice that above $y_1 = 0$ and below $y_2 = \sqrt{a^2 - (x - b)^2}$. Hence from Eq (5) and the substitution $x - b = a \sin t$, we have

$$V_y = 4\pi \int_{b-a}^{b+a} x\sqrt{a^2 - (x - b)^2}\,dx = 4\pi \int_{-\pi/2}^{\pi/2} (b + a \sin t)a^2 \cos^2 t\,dt$$
$$= 4\pi a^2 b \left[\frac{t}{2} + \frac{\sin 2t}{4}\right]_{-\pi/2}^{\pi/2} + 4\pi a^3 \left[-\frac{\cos^3 t}{3}\right]_{-\pi/2}^{\pi/2} = 2\pi^2 a^2 b.$$

EXAMPLE 2. The sector of the cardioid $r = 1 + \cos\theta$ between the radius vectors $\theta = 0$ and $\theta = \pi/2$ is revolved about OX. Find the volume generated.

Solution 1: The volume $V_x = \pi\int y^2\,dx$. And $y = r\sin\theta = (1 + \cos\theta)\sin\theta$, $x = r\cos\theta = \cos\theta + \cos^2\theta$, $dx = (1 + 2\cos\theta)(-\sin\theta\,d\theta)$. Hence

$V_x = \pi\int_{\pi/2}^{0}(1 + \cos\theta)^2\sin^2\theta(1 + 2\cos\theta)(-\sin\theta\,d\theta)$. Let $\cos\theta = t$ and

$$\begin{aligned} V_x &= \pi\int_0^1 (1 + t)^2(1 - t^2)(1 + 2t)dt \\ &= \pi\int_0^1 (1 + 4t + 4t^2 - 2t^3 - 5t^4 - 2t^5)dt \\ &= \pi\left[t + 2t^2 + \frac{4}{3}t^3 - \frac{1}{2}t^4 - t^5 - \frac{1}{3}t^6\right]_0^1 = \frac{5\pi}{2}, \quad \text{the required volume.} \end{aligned}$$

Solution 2: Let a circular sector of radius r and angle $d\theta$ be revolved about OX. Then the arc $r\,d\theta$ generates a strip of area approximately $2\pi yr\,d\theta$. The conical shell with this as base and altitude r has a volume approximately $\frac{1}{3}(2\pi yr\,d\theta)r$. This suggests that $dV_x = \frac{2\pi}{3}r^3\sin\theta\,d\theta$, since $y = r\sin\theta$. Thus for $r = 1 + \cos\theta$, we may write

$$\begin{aligned} V_x &= \frac{2\pi}{3}\int_0^{\pi/2}(1 + \cos\theta)^3\sin\theta\,d\theta = -\frac{2\pi}{3}\frac{1}{4}[(1 + \cos\theta)^4]_0^{\pi/2} \\ &= -\frac{\pi}{6}(1 - 16) = \frac{5\pi}{2}, \quad \text{the required volume.} \end{aligned}$$

EXERCISE 106

Find the area between the lines $x = 0$ and $x = 1$, above the x axis and below each of the following given curves.

1. $y(x^4 + 1) = 2x$.
2. $y = xe^x$.
3. $y^2(x^2 + 1) = 1$.
4. $y^2(1 - x^2) = x^6$.

Find the area between the lines $x = 1$ and $x = 2$, above the x axis and below each of the following given curves.

5. $y = \ln x$.
6. $x^2y = x^2 - 1$.
7. $y^2(x^2 - 1) = x^2$.
8. $y^2(x^2 - 1) = 1$.

Find the area bounded by each given pair of loci.

9. $y = \dfrac{1 - x^2}{x + 2}$ and $x = 0$.
10. $y = \dfrac{x}{1 + x^2}$ and $5y = x$.
11. $x^2 = 4y$ and $y(x^2 + 4) = 8$.
12. $y(x^2 + 4) = 8$ and $y = 0$, its asymptote.

Find the area enclosed by the loop of each given curve.

13. $y^2 = x^2(1 - x)$.
14. $y^2 = x(x - 1)^2$.
15. $y^2 = x^4(1 - x)$.
16. $y^2 = x^3(1 - x)$.

Find the area enclosed by one loop of each given curve.

17. $y^2 = x^2(9 - x^2)$.
18. $y^2 = x^4(4 - x^2)$.

Find the area of the sector of each of the following curves between the lines $\theta = 0$ and $\theta = \pi/4$.

19. $r = a\sec^2(\theta/2)$.
20. $r = \tan\theta$.

Find the area of that loop of each of the following curves which is bisected by the radius vector $\theta = 0$.

21. $r^2 = \cos 2\theta\cos\theta$.
22. $r = \cos^2\theta$.

The area described in the indicated problem is revolved about the x axis. Find the volume generated.

23. Prob. 2. 24. Prob. 3. 25. Prob. 4.
26. Prob. 5 27. Prob. 6. 28. Prob. 7.

Find the volume generated when the area under the first arch of the curve $y = \sin x$ and over the x axis is revolved about

29. The y axis. 30. The line $y = 2$.

Find the volume generated when the area under the first arch of the cycloid $x = a(t - \sin t), y = a(1 - \cos t)$ is revolved about

31. The x axis. 32. The y axis.

206. Length of Arc. By Sec. 182, the basic expression for the differential of arc in rectangular coordinates is

$$ds^2 = dx^2 + dy^2, \qquad ds = \sqrt{dx^2 + dy^2}. \tag{6}$$

We may use this to recall the expression for the arc length L,

$$L = \int ds = \int_{x_1}^{x_2} \sqrt{1 + \left(\frac{dy}{dx}\right)^2}\, dx \quad \text{or} \quad L = \int_{y_1}^{y_2} \sqrt{1 + \left(\frac{dx}{dy}\right)^2}\, dy. \tag{7}$$

And when the curve is given in terms of a parameter, we have

$$L = \int_{1}^{t_2} \sqrt{\left(\frac{dx}{dt}\right)^2 + \left(\frac{dy}{dt}\right)^2}\, dt. \tag{8}$$

By Sec. 183, the relations for polar coordinates are

$$ds = \sqrt{r^2\, d\theta^2 + dr^2}, \qquad L = \int_{\theta_1}^{\theta_2} \sqrt{r^2 + \left(\frac{dr}{d\theta}\right)^2}\, d\theta. \tag{9}$$

Example 1. Find the length of arc of the parabola $y^2 = 4mx$ from the point (0,0) to the point $(k^2m, 2km)$.

Solution 1: From $x = \frac{y^2}{4m}$, $\frac{dx}{dy} = \frac{y}{2m}$. And $\left(\frac{dx}{dy}\right)^2 = \frac{y^2}{4m^2}$, $\left(\frac{dx}{dy}\right)^2 + 1 = \frac{y^2 + 4m^2}{4m^2}$, $\frac{ds}{dy} = \frac{\sqrt{y^2 + 4m^2}}{2m}$. Hence $L = \frac{1}{2m}\int_0^{2km} \sqrt{y^2 + 4m^2}\, dy$. By form 12 of Table 1 with $u = y$, $A = 4m^2$, it follows that

$$\begin{aligned} L &= \frac{1}{2m}\left[\frac{y}{2}\sqrt{y^2 + 4m^2} + 2m^2 \ln\left(y + \sqrt{y^2 + 4m^2}\right)\right]_0^{2km} \\ &= m[k\sqrt{k^2 + 1} + \ln(k + \sqrt{k^2 + 1})], \qquad \text{the required value.} \end{aligned}$$

Solution 2: Let $y = 2mt$. Then $4mx = y^2 = 4m^2t^2$ and $x = mt^2$.

$$\frac{dx}{dt} = 2mt, \qquad \frac{dy}{dt} = 2m, \qquad \left(\frac{dx}{dt}\right)^2 + \left(\frac{dy}{dt}\right)^2 = 4m^2(t^2 + 1), \qquad \frac{ds}{dt} = 2m\sqrt{t^2 + 1}.$$

Hence from form 12 of Table 1,

$$L = 2m \int_0^k \sqrt{t^2+1}\,dt = 2m\left[\frac{t}{2}\sqrt{t^2+1} + \frac{1}{2}\ln\left(t + \sqrt{t^2+1}\right)\right]_0^k$$
$$= m[k\sqrt{k^2+1} + \ln(k + \sqrt{k^2+1})], \qquad \text{the required value.}$$

Example 2. Find the length of arc of the curve $r = 1/\theta$ from the point $r = 1$, $\theta = 1$ to the point $r = 1/k$, $\theta = k$.

Solution: From $r = \dfrac{1}{\theta}$, $\dfrac{dr}{d\theta} = -\dfrac{1}{\theta^2}$. And $r^2 + \left(\dfrac{dr}{d\theta}\right)^2 = \dfrac{1}{\theta^2} + \dfrac{1}{\theta^4}$, $\dfrac{ds}{d\theta} = \dfrac{\sqrt{\theta^2+1}}{\theta^2}$.

Hence $L = \displaystyle\int_1^k \frac{\sqrt{\theta^2+1}}{\theta^2}\,d\theta$. As in Sec. 199, put $\theta = \tan t$, $\sqrt{\theta^2+1} = \sec t$, and $d\theta = \sec^2 t\,dt$. Let $\theta_k = \tan^{-1} k$. Then

$$L = \int_{\pi/4}^{\theta_k} \frac{\sec t}{\tan^2 t}\sec^2 t\,dt = \int_{\pi/4}^{\theta_k} \frac{1+\tan^2 t}{\tan^2 t}\sec t\,dt$$
$$= \int_{\pi/4}^{\theta_k}\left(\frac{\sec t}{\tan^2 t} + \sec t\right)dt = \int_{\pi/4}^{\theta_k} (\cot t \csc t + \sec t)\,dt$$
$$= [-\csc t + \ln(\sec t + \tan t)]_{\pi/4}^{\theta_k}.$$

Since $\tan\theta_k = k$, $\sec\theta_k = \sqrt{k^2+1}$, $\csc\theta_k = \dfrac{\sqrt{k^2+1}}{k}$. It follows that

$$L = \sqrt{2} - \frac{\sqrt{k^2+1}}{k} + \ln\frac{k+\sqrt{k^2+1}}{1+\sqrt{2}}, \qquad \text{the required length.}$$

207. Area of a Surface of Revolution. As shown in Sec. 186, the element of area of the surface generated by revolving a plane curve with y never negative about the x axis is equal to the area of a rectangle of width ds and length $2\pi y$, the circumference generated by revolving the ordinate y. Thus

$$dS_x = 2\pi y\,ds \qquad \text{and} \qquad S_x = 2\pi \int y\,ds. \tag{10}$$

For revolution about the y axis, with x never negative, the element of surface is

$$dS_y = 2\pi x\,ds \qquad \text{and} \qquad S_y = 2\pi \int x\,ds. \tag{11}$$

The integrals in Eqs. (10) and (11) may be used with appropriate limits for the independent variable. This may be x, y, the θ of polar coordinates, or a parameter t. In each case ds must be expressed in terms of the variable of integration as in Eqs. (7) to (9). And x or y, if present, must also be expressed in terms of this variable.

Example 1. Find the area of the surface of revolution generated by revolving the arc of the parabola $y^2 = 4mx$ from the point $(0,0)$ to the point $(k^2m, 2km)$ about the y axis.

Solution 1: From $y = 2\sqrt{mx}$, $\dfrac{dy}{dx} = \sqrt{\dfrac{m}{x}}$. And $\left(\dfrac{dy}{dx}\right)^2 = \dfrac{m}{x}$, $\left(\dfrac{dy}{dx}\right)^2 + 1 = \dfrac{x+m}{x}$, $\dfrac{ds}{dx} = \sqrt{\dfrac{x+m}{x}}$. But $S_y = 2\pi \int x\,ds = 2\pi \int x \dfrac{ds}{dx}\,dx$. $x\dfrac{ds}{dx} = x\sqrt{\dfrac{x+m}{x}}$ $= \sqrt{x^2+mx} = \sqrt{\left(x+\dfrac{m}{2}\right)^2 - \dfrac{m^2}{4}}$. Hence $S_x = 2\pi \int_0^{k^2m} \sqrt{\left(x+\dfrac{m}{2}\right)^2 - \dfrac{m^2}{4}}\,dx$. By form 12 of Table 1 with $u = x + m/2$, $A = -m^2/4$, it follows that the required area is

$$S_x = 2\pi \left[\frac{1}{2}\left(x+\frac{m}{2}\right)\sqrt{x^2+mx} - \frac{m^2}{8}\ln\left(x+\frac{m}{2}+\sqrt{x^2+mx}\right)\right]_0^{k^2m}$$
$$= \pi m^2[(k^2+\tfrac{1}{2})k\sqrt{k^2+1} - \tfrac{1}{4}\ln(2k^2+1+2k\sqrt{k^2+1})].$$

Solution 2: From $x = y^2/4m$, $dx/dy = y/2m$. And $\left(\dfrac{dx}{dy}\right)^2 = \dfrac{y^2}{4m^2}$, $\left(\dfrac{dx}{dy}\right)^2 + 1$ $= \dfrac{y^2+4m^2}{4m^2}$, $\dfrac{ds}{dy} = \dfrac{\sqrt{y^2+4m^2}}{2m}$. But $S_y = 2\pi\int x\,ds = 2\pi \int x\dfrac{ds}{dy}\,dy$. $x\dfrac{ds}{dy} =$ $\dfrac{y^2}{4m}\dfrac{\sqrt{y^2+4m^2}}{2m} = \dfrac{1}{8m^2}y^2\sqrt{y^2+4m^2}$. Hence $S_x = 2\pi\int_0^{2km}\dfrac{1}{8m^2}y^2\sqrt{y^2+4m^2}\,dy$. As in Sec. 199, let $y = 2m\tan\theta$. Let $\theta_k = \tan^{-1}k$. Then

$$S_x = \frac{\pi}{4m^2}\int_0^{\theta_k} 4m^2\tan^2\theta\, 2m\sec\theta\, 2m\sec^2\theta\,d\theta = 4\pi m^2\int_0^{\theta_k}\tan^2\theta\sec^3\theta\,d\theta.$$
$$\int\tan^2\theta\sec^3\theta\,d\theta = \int(\sec^2\theta - 1)\sec^3\theta\,d\theta = \int(\sec^5\theta - \sec^3\theta)d\theta.$$

By Eq. (51) of Sec. 197 with $n = 5$, we have $\int\sec^5\theta\,d\theta = \frac{1}{4}\sec^3\theta\tan\theta + \frac{3}{4}\int\sec^3\theta\,d\theta$. Hence by form 27 of Table 1,

$$S_x = \pi m^2[\sec^3\theta\tan\theta - \tfrac{1}{2}\sec\theta\tan\theta - \tfrac{1}{2}\ln(\sec\theta+\tan\theta)]_0^{\theta_k}$$
$$= \pi m^2[k(k^2+\tfrac{1}{2})\sqrt{k^2+1} - \tfrac{1}{2}\ln(k+\sqrt{k^2+1})], \qquad \text{the required area.}$$

This agrees with the result in solution 1, since

$$\tfrac{1}{2}\ln(k+\sqrt{k^2+1}) = \tfrac{1}{4}\ln(k+\sqrt{k^2+1})^2 = \tfrac{1}{4}\ln(2k^2+1+2k\sqrt{k^2+1}).$$

Example 2. Find the area of the surface of revolution generated by revolving the upper half of the limaçon $r = a + b\cos\theta$, $a > b$, about the x axis.

Solution: From $r = a + b\cos\theta$, $dr/d\theta = -b\sin\theta$. And

$$r^2 + (dr/d\theta)^2 = (a+b\cos\theta)^2 + b^2\sin^2\theta = a^2+b^2+2ab\cos\theta,$$

$ds/d\theta = \sqrt{a^2+b^2+2ab\cos\theta}$. But $S_x = 2\pi\int y\,ds = 2\pi\int y(ds/d\theta)d\theta$. And $y = r\sin\theta$ $= (a+b\cos\theta)\sin\theta$. Hence $S_x = 2\pi\int_0^\pi (a+b\cos\theta)\sqrt{a^2+b^2+2ab\cos\theta}\sin\theta\,d\theta$. Let $\sqrt{a^2+b^2+2ab\cos\theta} = z$, $a + b\cos\theta = \dfrac{z^2+a^2-b^2}{2a}$, $\sin\theta\,d\theta = -\dfrac{z\,dz}{ab}$.

$$S_x = -2\pi\int_{a+b}^{a-b}\frac{z^2+a^2-b^2}{2a}z\frac{z\,dz}{ab} = \frac{\pi}{a^2b}\int_{-b}^{a+b}[z^4+(a^2-b^2)z^2]\,dz$$
$$= \frac{\pi}{a^2b}\left[\frac{z^5}{5}+(a^2-b^2)\frac{z^3}{3}\right]_{a-b}^{a+b} = \frac{2\pi}{15a^2b}[(a+b)^4(4a-b)-(a-b)^4(4a+b)].$$

EXAMPLE 3. Find the area of the surface of revolution generated by revolving the arc of the cardioid $r = 1 + \cos\theta$ which lies in the first quadrant about the y axis.

Solution: From $r = 1 + \cos\theta$, $dr/d\theta = -\sin\theta$. And $r^2 + (dr/d\theta)^2 = (1 + \cos\theta)^2 + \sin^2\theta = 2(1 + \cos\theta)$, $ds/d\theta = \sqrt{2(1 + \cos\theta)}$. But $S_y = 2\pi\int x\,ds = 2\pi\int x(ds/d\theta)d\theta$. And $x = r\cos\theta = (1 + \cos\theta)\cos\theta$. Hence $S_y = 2\pi\int_0^{\pi/2}\sqrt{2}\,(1 + \cos\theta)^{\frac{3}{2}}\cos\theta\,d\theta$. Let $\sin(\theta/2) = z$, $\cos(\theta/2)d\theta = 2\,dz$. Then

$$(1 + \cos\theta)^{\frac{3}{2}}\cos\theta\,d\theta = \left(2\cos^2\frac{\theta}{2}\right)^{\frac{3}{2}}\left(1 - 2\sin^2\frac{\theta}{2}\right)d\theta$$

$$= 2\sqrt{2}\left(1 - \sin^2\frac{\theta}{2}\right)\left(1 - 2\sin^2\frac{\theta}{2}\right)\left(\cos\frac{\theta}{2}\,d\theta\right) = 4\sqrt{2}\,(1 - z^2)(1 - 2z^2)dz.$$ Then

$$S_y = 16\pi\int_0^{1/\sqrt{2}}(1 - 3z^2 + 2z^4)dz = 16\pi\left[z - z^3 + \frac{2z^5}{5}\right]_0^{1/\sqrt{2}} = \frac{24\pi\sqrt{2}}{5}.$$

EXERCISE 107

1. Find the length of the parabola $x^2 = 4y$ from the point (0,0) to the point (4,4).
2. Find the length of the arc of the curve $y = \ln\dfrac{e^x - 1}{e^x + 1}$ which lies between $x = 1$ and $x = 2$.
3. Find the length of the curve $y = \ln\sec x$ from the point (0,0) to the point $(\pi/3, \ln 2)$.
4. Find the total length of the curve $8y^2 = x^2 - x^4$.
5. Find the length of the loop of the curve $9y^2 = 3x^2 + x^3$.
6. Find the length of the curve $y = \ln x$ from the point (1,0) to the point $(e,1)$.
7. Find the length of the arc of the parabola $r = m\sec^2(\theta/2)$ which lies in the first quadrant.
8. Find the length of the curve $r = \theta$ from the point where $\theta = 0$ to the point where $\theta = 2\pi$.
9. Find the length of the curve $r = 1 - \theta$ from the point where $\theta = 1$ to the point where $\theta = 1 + 2\pi$.
10. Find the length of the curve $x = t^2 + 4t$, $y = t^2 - 4t$ from the point where $t = 0$ to the point where $t = 4$.

In each of the following problems the given arc of the curve is revolved about the indicated axis. Find the area of the surface of revolution which is thus generated.

11. $y = e^x$, (0,1) to $(1,e)$ about OX.
12. $y^2 = 9x^3$, (0,0) to (1,3) about OY.
13. $y = \sin x$, (0,0) to $(\pi,0)$ about OX.
14. $y = e^{-x}$, from $x = 0$ to $x = \infty$ about OX.
15. $r = 1 - \cos\theta$, from $\theta = 0$ to $\theta = \pi/2$ about OX.
16. $r = 1 - \cos\theta$, from $\theta = 0$ to $\theta = \pi/2$ about OY.
17. $r^2 = \cos 2\theta$, one loop, about OY.
18. $x = a\cos t$, $y = b\sin t$, $a > b$, $t = 0$ to $t = \pi/2$ about OX.
19. $x = a\cos t$, $y = b\sin t$, $a > b$, $t = 0$ to $t = \pi/2$ about OY.
20. $x = a\sec t$, $y = a\tan t$, $t = 0$ to $t = \pi/4$ about OX.
21. $x = a(t - \sin t)$, $y = a(1 - \cos t)$, $t = 0$ to $t = \pi$ about OX.

208. Average or Mean Value. Let $f(x)$ be any function of x. And consider the arc AB of Fig. 242 which is the graph of $y = f(x)$ for values of x between $x = a$ and $x = b$. Thus $OM = a$ and $ON = b$. Choose the points $M_1, M_2, \cdots, M_{n-1}$ so as to divide the interval MN into n *equal* parts, each equal to Δx. Then

$$\Delta x = \frac{b-a}{n} \qquad \text{and} \qquad n\,\Delta x = b - a. \tag{12}$$

At the points $M_1, M_2, \cdots, M_{n-1}$ erect the ordinates to the curve $y_1, y_2, \cdots, y_{n-1}$. And call the ordinate at N, y_n. Then the average value, or arithmetic mean of these n ordinates is

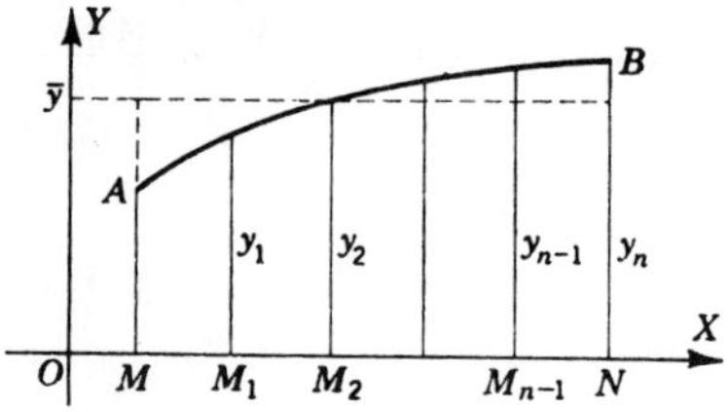

FIG. 242.

$$\frac{y_1 + y_2 + \cdots + y_n}{n}. \tag{13}$$

Multiply the numerator and denominator by Δx. Then from Eq. (12) we may deduce that the value of the fraction is

$$\frac{(y_1 + y_2 + \cdots + y_n)\Delta x}{n\,\Delta x} = \frac{y_1\,\Delta x + y_2\,\Delta x + \cdots + y_n\,\Delta x}{b-a}. \tag{14}$$

When n increases indefinitely, $\Delta x \to 0$. And it follows from Sec. 170 that the limit of the numerator is

$$\lim_{\Delta x \to 0} \sum_{i=1}^{n} y_i\,\Delta x = \int_a^b y\,dx. \tag{15}$$

Thus the limit of the fraction of Eq. (13) as $n \to \infty$ is

$$\bar{y} = \frac{1}{b-a}\int_a^b y\,dx. \tag{16}$$

This is defined as the *average* or *mean value* of y with respect to x over the interval from a to b.

It follows from Eq. (16) that

$$\int_a^b \bar{y}\,dx = \bar{y}(b-a) = \int_a^b y\,dx. \tag{17}$$

Thus the value of the integral $\int_a^b y\,dx$ is unchanged when we replace y by the *constant* average value $\bar{y}$. And $\bar{y}$ is such that for the areas between $x = a$ and $x = b$ bounded by $y = f(x)$ and $y = \bar{y}$, the portion above $y = \bar{y}$ is equal to the portion below $y = \bar{y}$. These facts are helpful in remembering Eq. (16).

Suppose next that $w(x)$ is a given weighting function. Let $w_1, w_2, \cdots, w_n$ be the values of $w(x)$ corresponding to $y_1, y_2, \cdots, y_n$. Then the *weighted average* of the y_i with weights w_i is

$$\frac{w_1y_1 + w_2y_2 + \cdots + w_ny_n}{w_1 + w_2 + \cdots + w_n}. \tag{18}$$

Multiply the numerator and denominator by Δx. Then the fraction becomes

$$\frac{w_1y_1\,\Delta x + w_2y_2\,\Delta x + \cdots + w_ny_n\,\Delta x}{w_1\,\Delta x + w_2\,\Delta x + \cdots + w_n\,\Delta x} = \frac{\sum_{i=1}^{n} w_iy_i\,\Delta x}{\sum_{i=1}^{n} w_i\,\Delta x}. \tag{19}$$

The limit of this fraction as $n \to \infty$, by Sec. 170, is

$$\bar{y}_w = \frac{\int_a^b wy\,dx}{\int_a^b w\,dx}. \tag{20}$$

This is defined as the weighted average of y with respect to x, with the weighting function w. When $w(x) = 1$, the weighted average of Eq. (20) reduces to the average of Eq. (16).

EXAMPLE 1. Find the average value of the ordinates of one arch of the cycloid $x = a(t - \sin t)$, $y = a(1 - \cos t)$ with respect to x.

Solution: When $t = 0$, $x = 0$, and when $t = 2\pi$, $x = 2\pi a$. Hence we have

$$\begin{aligned}\bar{y} &= \frac{1}{2\pi a - 0}\int_0^{2a} y\,dx = \frac{1}{2\pi a}\int_0^{2\pi} a(1 - \cos t)\,a(1 - \cos t)dt \\ &= \frac{a}{2\pi}\int_0^{2\pi}(1 - 2\cos t + \cos^2 t)dt = \frac{a}{2\pi}\left[t - 2\sin t + \frac{t}{2} + \frac{\sin 2t}{4}\right]_0^{2\pi} \\ &= \frac{3a}{2}.\end{aligned}$$

And $\bar{y} = \frac{3a}{2}$ is the required average with respect to x.

EXAMPLE 2. For the cycloidal arch of Example 1, find the average value of the ordinate with respect to t.

Solution: From $y = a(1 - \cos t)$, we have

$$\bar{y} = \frac{1}{2\pi - 0}\int_0^{2\pi} y\,dt = \frac{1}{2\pi}\int_0^{2\pi} a(1 - \cos t)dt = \frac{a}{2\pi}[t - \sin t]_0^{2\pi} = a.$$

And $\bar{y} = a$ is the required average with respect to t.

EXAMPLE 3. For the cycloidal arch of Example 1, find the average value of the ordinates with respect to the arc length s.

Solution: From $x = a(t - \sin t)$, $y = a(1 - \cos t)$, we find $\frac{dx}{dt} = a(1 - \cos t)$, $\frac{dy}{dt} = a \sin t$. And $\left(\frac{dx}{dt}\right)^2 + \left(\frac{dy}{dt}\right)^2 = a^2(1 - \cos t)^2 + a^2\sin^2 t = 2a^2(1 - \cos t) =$

$4a^2 \sin^2 \frac{t}{2}$. Hence $\frac{ds}{dt} = 2a \sin \frac{t}{2}$, for t between 0 and 2π. Hence we have

$$\int ds = \int_0^{2\pi} 2a \sin \frac{t}{2}\, dt = -4a\left[\cos \frac{t}{2}\right]_0^{2\pi} = -4a(\cos \pi - \cos 0) = 8a.$$

$$\int y\, ds = \int_0^{2\pi} a(1 - \cos t)\, 2a \sin \frac{t}{2}\, dt = 4a^2 \int_0^{2\pi} \left(1 - \cos^2 \frac{t}{2}\right) \sin \frac{t}{2}\, dt$$

$$= 8a^2\left[-\cos \frac{t}{2} + \frac{1}{3}\cos^3 \frac{t}{2}\right]_0^{2\pi} = 8a^2\left(1 - \frac{1}{3} + 1 - \frac{1}{3}\right) = \frac{32}{3}a^2.$$

It follows that $\bar{y} = \frac{\int y\, ds}{\int ds} = \frac{\frac{32}{3}a^2}{8a} = \frac{4a}{3}$. And $\bar{y} = \frac{4a}{3}$ is the required average with respect to s.

EXERCISE 108

For a freely falling body, $s = \frac{1}{2}gt^2$, $v = gt = \sqrt{2gs}$.

1. Show that the average value of v with respect to t for the interval 0 to t_1 is $\bar{v} = \frac{1}{2}gt_1$, or one-half the final velocity.
2. Show that the average value of v with respect to s for the interval 0 to s_1 is $\bar{v} = \frac{2}{3}\sqrt{2gs_1}$, or two-thirds the final velocity.

For the upper half of the circle $x = a \cos t$, $y = a \sin t$,

3. Find the average value of y with respect to x.
4. Find the average value of y with respect to the parameter t.
5. Find the average value of y with respect to the arc length s.
6. Find the average value of y^2 with respect to x.

Using $w(x) = x^2$ as a weighting function, find the weighted average of y for the interval $x = 0$ to $x = 2$ for each given function.

7. $y = 6x$.
8. $y = x^2$.
9. $y = \sqrt{x}$.

For the curve $x = t^3$, $y = t^4$ and the range $0 < t < 1$, find the average value of x

10. With respect to t.
11. With respect to y.

For the interval $x = 0$ to $x = \pi$, find the average of y with respect to x for each given function.

12. $y = \sin x$.
13. $y = \sin^2 x$.
14. $y = \cos x$.
15. $y = \sin x \sin (x + \pi/3)$.

16. In the time interval 0 to $2\pi/\omega$, the simple harmonic motion $x = A \sin \omega t$ makes one complete vibration. Find the average value of the kinetic energy $E = \frac{1}{2}mv^2 = \frac{1}{2}mA^2\omega^2 \cos^2 \omega t$.

The interval AB is 6 units in length. It is divided into two parts AC and CB by a point C chosen at random in the interval AB. Thus if $AC = x$, $CB = 6 - x$. Find the average value (with respect to AC or x) of

17. $AC - CB$.
18. $\overline{AC} \cdot \overline{CB}$.
19. $\overline{AC}^2 + \overline{CB}^2$.
20. $\overline{AC}^2 \cdot \overline{CB}$.

In statistics, for a given range of a variable x which has a frequency distribution $f(x)$, the mean value is equal to the weighted average of x with $f(x)$ as the weighting factor. For the range 0 to plus infinity, find the mean of x for each given frequency distribution.

21. $f(x) = e^{-x}$. 22. $f(x) = \dfrac{1}{x^2 + 1}$.

209. Center of Gravity of a System of Particles. Consider a system of n particles lying in a plane at points whose coordinates are (x_1,y_1), (x_2,y_2), $\cdots$, (x_n,y_n). Let their respective masses be $m_1, m_2, \cdots, m_n$. Then the point $G = (\bar{x},\bar{y})$ where

$$\begin{aligned}\bar{x} &= \frac{m_1x_1 + m_2x_2 + \cdots + m_nx_n}{m_1 + m_2 + \cdots + m_n},\\ \bar{y} &= \frac{m_1y_1 + m_2y_2 + \cdots + m_ny_n}{m_1 + m_2 + \cdots + m_n},\end{aligned} \tag{21}$$

is called the *center of gravity* of the system of n particles. The position of G is independent of the choice of coordinate axes, and for many mechanical effects the n particles may be replaced by a single particle of mass $m_1 + m_2 + \cdots + m_n$ at G.

In a plane the *first moment* of a particle about any straight line in the plane is the product of the mass of the particle times its signed distance from the straight line. Let $M = m_1 + m_2 + \cdots + m_n$. Then Eq. (21) is equivalent to the relations

$$\begin{aligned}M_x &= M\bar{x} = m_1x_1 + m_2x_2 + \cdots + m_nx_n,\\ M_y &= M\bar{y} = m_1y_1 + m_2y_2 + \cdots + m_ny_n.\end{aligned} \tag{22}$$

In this form, the equations show that the sum of the moments of the particles about either coordinate axis is equal to the moment about this axis of a single particle of mass M located at G.

210. Center of Gravity of a Plane Distribution of Mass. Consider a continuous distribution of mass in the plane. This may be a thin wire bent around a curve or a thin plate overlaying an area. Divide the mass into pieces such that each of the points in a typical piece of mass Δm_i is at a distance from the y axis between x_i and $x_i + \Delta x$. For each i, choose an x_i' which lies between x_i and $x_i + \Delta x$. Then $\Sigma x_i' \Delta m_i$ is taken as an approximation to the moment of the mass about OY. When $\Delta x \to 0$, $\Delta m \to 0$, and by Sec. 185 the sum approaches a limit $\int x\,dm$ which is defined to be the moment of the mass about OY. Thus

$$M_x = \int x\,dm. \tag{23}$$

The total mass of the body is $M = \int dm$. If this were concentrated at a point with $x = \bar{x}$, its moment about OY would be $M\bar{x}$. For the $\bar{x}$ which makes this moment equal to M_x, we have

$$M\bar{x} = M_x, \qquad \bar{x} = \frac{M_x}{M} = \frac{\int x\,dm}{\int dm}. \tag{24}$$

Similarly, the $\bar{y}$ of the point at which a concentrated mass M has the same moment about OX as the actual distributed mass is such that

$$M\bar{y} = M_y, \qquad \bar{y} = \frac{M_y}{M} = \frac{\int y\,dm}{\int dm}. \tag{25}$$

Here, in the numerator the element dm must be that approximated by a piece of mass Δm_i with all points at a distance from the x axis between y_i and $y_i + \Delta y$.

The point $G = (\bar{x},\bar{y})$, whose coordinates are given by Eqs. (24) and (25), is called the *center of gravity* of the continuously distributed mass.

If a plane mass distribution has an axis of symmetry, the point $G = (\bar{x},\bar{y})$ will lie on that axis. And if the distribution has two such axes, and hence a center of symmetry, the point $G(\bar{x},\bar{y})$ will lie at that center.

In calculating the moments M_x and M_y, we may divide the mass into parts and replace each part by the mass of the part concentrated at the center of gravity of the part. The same is true for the approximating sums. And in Eqs. (24) and (25) elements dm of a different type from that described may be used, provided that, in the integrals, x and y are replaced by appropriate functions depending on the position of the center of gravity of the elements used.

211. Centroid of a Plane Area. Consider a thin uniform plate overlaying a plane area. Let t be the thickness of the plate and D be its density. Then the mass per unit area $\rho = Dt$. And $dm = \rho\,dA$. As ρ is a constant, after substituting $dm = \rho\,dA$ in Eqs. (24) and (25) we may cancel the factor ρ and so deduce that

$$\bar{x} = \frac{\int x\,dA}{\int dA}, \qquad \bar{y} = \frac{\int y\,dA}{\int dA}. \tag{26}$$

There is no explicit reference to mass in Eq. (26). To emphasize this fact, we sometimes call $(\bar{x},\bar{y})$ the *centroid of the area A*. However, the term *center of gravity of the area* is often used.

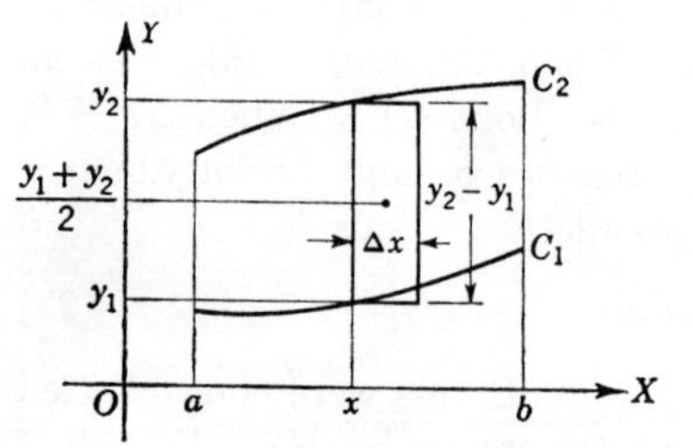

FIG. 243.

Suppose that the plane area is bounded above by the curve $y_2 = f_2(x)$, below by the curve $y_1 = f_1(x)$, and lies between a left-hand ordinate at $x = a$ and a right-hand ordinate at $x = b$ (Fig. 243). Then by Eq. (1) we have

$$dA = (y_2 - y_1)dx$$

and

$$A = \int_a^b (y_2 - y_1)dx. \tag{27}$$

The numerator of the fraction for $\bar{x}$ in Eq. (26) is

$$A\bar{x} = \int_a^b x(y_2 - y_1)dx. \tag{28}$$

For the element of area dA of Eq. (27) the points are not all at approximately the same distance from OX. But the increment ΔA corresponding to dA is a rectangle between the ordinates at x and $x + \Delta x$ bounded above by y_2 and below by y_1. Hence its center of gravity has an ordinate $\frac{1}{2}(y_1 + y_2)$ and the elementary moment is the product of this by dA or

$$\tfrac{1}{2}(y_1 + y_2)dA = \tfrac{1}{2}(y_1 + y_2)(y_1 - y_2)dx = \tfrac{1}{2}(y_2^2 - y_1^2)dx. \tag{29}$$

Thus the numerator of the fraction for $\bar{y}$ is

$$A\bar{y} = \tfrac{1}{2}\int_a^b (y_2^2 - y_1^2)dx. \tag{30}$$

Let us next consider a plane area bounded on the right by the curve $x_2 = g_2(y)$, on the left by the curve $x_1 = g_1(y)$, and lying above the line

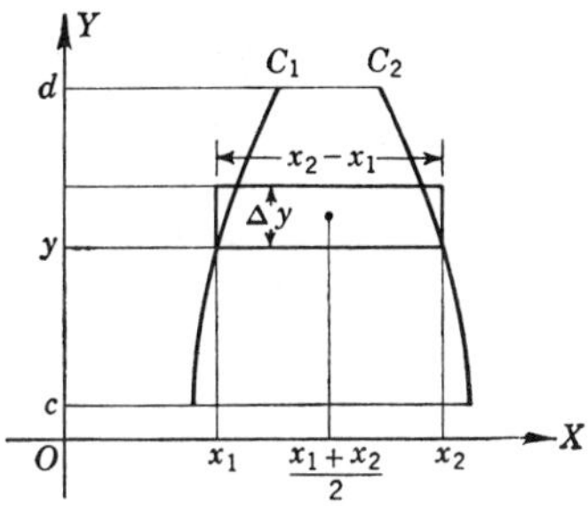

FIG. 244.

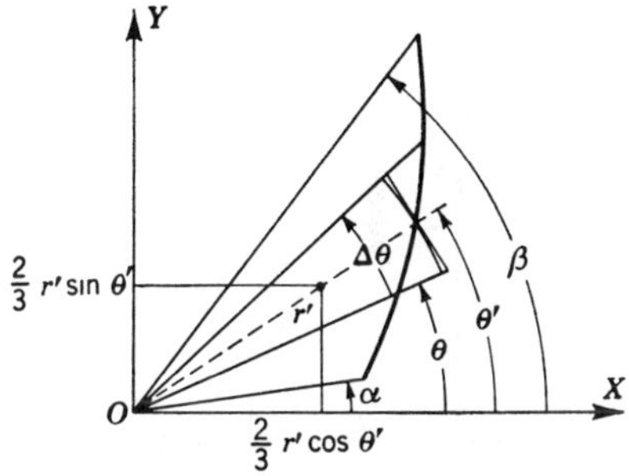

FIG. 245.

$y = c$ and below the line $y = d$ (Fig. 244). Then we may take the element of area as $dA = (x_2 - x_1)dy$, and deduce that

$$A = \int_c^d (x_2 - x_1)dy, \quad A\bar{x} = \tfrac{1}{2}\int_c^d (x_2^2 - x_1^2)dy, \quad A\bar{y} = \int_c^d (x_2 - x_1)y\,dx. \tag{31}$$

We may wish to consider the area bounded by the curve in polar coordinates $r = f(\theta)$ and the two radius vectors $\theta = \alpha$ and $\theta = \beta$ (Fig. 245). Here from Eq. (2) we have

$$dA = \tfrac{1}{2}r^2\,d\theta, \qquad A = \tfrac{1}{2}\int_\alpha^\beta r^2\,d\theta. \tag{32}$$

The increment ΔA corresponding to dA may be approximated by a circular sector of radius r' and central angle $\Delta\theta$, or an isosceles triangle of altitude r' and base $r'\,\Delta\theta$. The center of gravity of this triangle is two-thirds of the way from the vertex to the base, and so has polar coordinates $\frac{2}{3}r'$, θ'. Hence it has rectangular coordinates $\frac{2}{3}r' \cos\theta'$, $\frac{2}{3}r' \sin\theta'$. These replace x' and y' in the approximating sums for the moments. Hence in the integrals for M_x and M_y we must replace x by $\frac{2}{3}r\cos\theta$ and y by $\frac{2}{3}r\sin\theta$. This leads to the relations

$$A\bar{x} = \tfrac{1}{3}\int_\alpha^\beta r^3\cos\theta\,d\theta, \qquad A\bar{y} = \tfrac{1}{3}\int_\alpha^\beta r^3\sin\theta\,d\theta. \tag{33}$$

EXAMPLE. Find the centroid of the area of a quadrant of a circle of radius a.

Solution 1: Take the quadrant as bounded above by $y = \sqrt{a^2 - x^2}$, below by $y = 0$, and between $x = 0$ and $x = a$ (Fig. 246). Then from Eq. (27), we have $A = \int y\,dx = \int_0^a \sqrt{a^2 - x^2}\,dx = a^2 \int_0^{\pi/2} \cos^2 t\,dt = a^2\left[\frac{t}{2} + \frac{\sin 2t}{4}\right]_0^{\pi/2} = \frac{\pi a^2}{4}$. From Eq. (28), $A\bar{x} = \int xy\,dx = \int_0^a x\sqrt{a^2 - x^2}\,dx = -\frac{1}{3}[(a^2 - x^2)^{\frac{3}{2}}]_0^a = -\frac{1}{3}(0 - a^3) = \frac{a^3}{3}$. From Eq. (30), $A\bar{y} = \frac{1}{2}\int y^2\,dx = \frac{1}{2}\int_0^a (a^2 - x^2)dx = \frac{1}{2}\left[a^2x - \frac{x^3}{3}\right]_0^a = \frac{a^3}{3}$. Hence $\bar{x} = \frac{A\bar{x}}{A} = \frac{a^3/3}{\pi a^2/4} = \frac{4a}{3\pi}$, $\bar{y} = \frac{Ay}{A} = \frac{a^3/3}{\pi a^2/4} = \frac{4a}{3\pi}$. And $G = \left(\frac{4a}{3\pi}, \frac{4a}{3\pi}\right)$.

Solution 2: Take the quadrant as bounded by the lines $y = x$, $y = -x$, and on the right by $x^2 + y^2 = a^2$ (Fig. 247). From the symmetry about OX, $\bar{y} = 0$, and $\bar{x}$ is

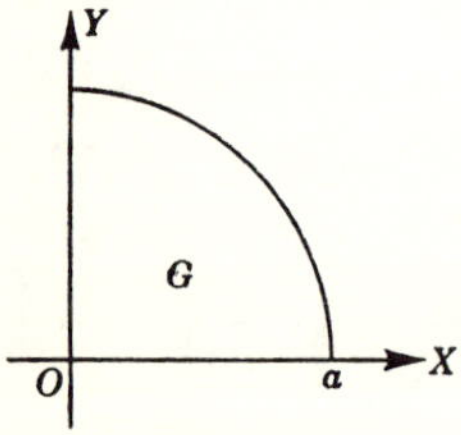

FIG. 246.

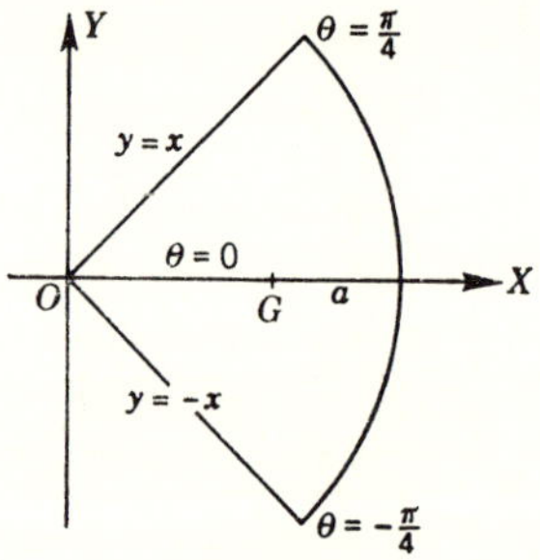

FIG. 247.

the same for the quadrant as for the upper half, bounded on the left by $x = y$, on the right by $x = \sqrt{a^2 - y^2}$, and lying between $y = 0$ and $y = a/\sqrt{2}$. From Eq. (31) we have $A = \int (x_2 - x_1)dy = \int_0^{a/\sqrt{2}} (\sqrt{a^2 - y^2} - y)\,dy = a^2\int_0^{\pi/4}\cos^2 t\,dt - \int_0^{a/\sqrt{2}} y\,dy = a^2\left[\frac{t}{2} + \frac{\sin 2t}{4}\right]_0^{\pi/4} - \left[\frac{y^2}{2}\right]_0^{a/\sqrt{2}} = a^2\left(\frac{\pi}{8} + \frac{1}{4}\right) - \frac{a^2}{4} = \frac{\pi a^2}{8}$. And $A\bar{x} = \frac{1}{2}\int (x_2^2 - x_1^2)dy = \frac{1}{2}\int_0^{a/\sqrt{2}} (a^2 - y^2 - y^2)dy = \frac{1}{2}\left[a^2y - \frac{2y^3}{3}\right]_0^{a/\sqrt{2}} = \left[\frac{y}{2}\left(a^2 - \frac{2y^2}{3}\right)\right]_0^{a/\sqrt{2}} = \frac{a^3}{3\sqrt{2}}$. It follows that $\bar{x} = \frac{A\bar{x}}{A} = \frac{a^3/(3\sqrt{2})}{\pi a^2/8} = \frac{4a}{3\pi}\sqrt{2}$.

And the required centroid $G = \left(\frac{4a}{3\pi}\sqrt{2}, 0\right)$.

Solution 3: Take the quadrant as the sector between the radius vectors $\theta = \pi/4$, $\theta = -\pi/4$ bounded by $r = a$ (Fig. 247). Then from the symmetry about $\theta = 0$, we have $\bar{y} = 0$. And $\bar{x}$ is the same for the quadrant as for the upper half, between $\theta = 0$ and $\theta = \pi/4$. From Eq. (32), $A = \frac{1}{2}\int r^2\,d\theta = \frac{1}{2}\int_0^{\pi/4} a^2\,d\theta = \frac{a^2}{2}[\theta]_0^{\pi/4} = \frac{\pi a^2}{8}$. From Eq. (33), $A\bar{x} = \frac{1}{3}\int r^3 \cos\theta\,d\theta = \frac{1}{3}\int_0^{\pi/4} a^3\cos\theta\,d\theta = \frac{a^3}{3}[\sin\theta]_0^{\pi/4} =$

$\dfrac{a^3}{3\sqrt{2}}$. Hence $\bar{x} = \dfrac{A\bar{x}}{A} = \dfrac{a^3/(3\sqrt{2})}{\pi a^2/8} = \dfrac{4a}{3\pi}\sqrt{2}$. And the required centroid has polar coordinates $r = \dfrac{4a}{3\pi}\sqrt{2}$, $\theta = 0$. Thus it is on the radius bisecting the quadrant at a distance $\dfrac{4a}{3\pi}\sqrt{2}$ from the center of the circle.

EXERCISE 109

1. Show that for a triangle the distance of the centroid from the base is one-third of the altitude.

Find the centroid of the area in the first quadrant bounded by the axes $x = 0$, $y = 0$ and each of the following given curves.

2. $\dfrac{x^2}{a^2} + \dfrac{y^2}{b^2} = 1$.
3. $\dfrac{x}{a} + \dfrac{y}{b} = 1$.
4. $y = \cos x$, $(x < \pi/2)$.
5. $y^2 = 4 - 2x$.

Find the centroid of the area bounded by the x axis and each of the following given curves.

6. $y = \sin x$ $(0 < x < \pi)$.
7. $y = x - x^3$ $(0 < x < 1)$.
8. $y = 4 - x^2$.
9. $y = 1 - x^4$.

Find the centroid of the area bounded by each of the following given sets of curves.

10. $y = x^3$, $x = 1$, $y = 0$.
11. $y^2 = 4x$, $x = 4$.
12. $a^2y = bx^2$, $y = b$.
13. $y^2 = 20x$, $x^2 = 20y$.
14. $y^2 = 10x$, $y = x$.
15. $y = x^3$, $y = x$ $(0 < x < 1)$.

Find the centroid of the area bounded by

16. $r = a(1 - \cos\theta)$.
17. $r = 3 + 2\sin\theta$.
18. $r = 2 + \cos\theta$.
19. $r = 1 - \sin\theta$.
20. Show that the centroid of a sector of a circle of radius a of central angle B lies at a distance $4a/3B \sin(B/2)$ from the center of the circle on the radius bisecting the sector.
21. Find the center of gravity of the area above OX and under the first arch of the cycloid $x = a(t - \sin t)$, $y = a(1 - \cos t)$.

212. Centroid of a Plane Arc. Consider a thin uniform wire bent along the arc of a plane curve. Let c be the area of the cross section of the wire and D be its density. Then the mass per unit length $\rho = Dc$. And $dm = \rho\, ds$. As ρ is a constant, after substituting $dm = \rho\, ds$ in Eqs. (24) and (25), we may cancel the factor ρ and so deduce that

$$\bar{x} = \frac{\int x\, ds}{\int ds}, \qquad \bar{y} = \frac{\int y\, ds}{\int ds}. \tag{34}$$

If we set $L = \int ds$ as in Sec. 206, it follows that

$$L = \int ds, \qquad L\bar{x} = \int x\, ds, \qquad L\bar{y} = \int y\, ds. \tag{35}$$

These relations determine the *centroid of a plane curve*. In applying them, we use one of the expressions for ds given in Sec. 206 and express x and y in terms of the appropriate independent variable.

EXAMPLE. Find the centroid of the arc of a quadrant of a circle of radius a.

Solution 1: Take the quadrant as the arc of $y = \sqrt{a^2 - x^2}$ between $x = 0$ and $x = a$. Then $\dfrac{dy}{dx} = \dfrac{-x}{\sqrt{a^2 - x^2}}$, $1 + \left(\dfrac{dy}{dx}\right)^2 = \dfrac{a^2}{a^2 - x^2}$, $\dfrac{ds}{dx} = \dfrac{a}{\sqrt{a^2 - x^2}}$. As in Eq. (7), $L = \int ds = \int_0^a \dfrac{a\,dx}{\sqrt{a^2 - x^2}} = a\left[\sin^{-1}\dfrac{x}{a}\right]_0^a = \dfrac{\pi a}{2}$. And from Eq. (35), $L\bar{x} = \int x\,ds = \int_0^a \dfrac{ax\,dx}{\sqrt{a^2 - x^2}} = -a\left[\sqrt{a^2 - x^2}\right]_0^a = -a(0 - a) = a^2$. Also $L\bar{y} = \int y\,ds = \int_0^a \sqrt{a^2 - x^2}\dfrac{a\,dx}{\sqrt{a^2 - x^2}} = \int_0^a a\,dx = [ax]_0^a = a^2$. Hence $\bar{x} = \dfrac{L\bar{x}}{L} = \dfrac{a^2}{\pi a/2} = \dfrac{2a}{\pi}$, $\bar{y} = \dfrac{L\bar{y}}{L} = \dfrac{a^2}{\pi a/2} = \dfrac{2a}{\pi}$. And for the centroid of arc, $\bar{x} = \dfrac{2a}{\pi}$, $\bar{y} = \dfrac{2a}{\pi}$.

Solution 2: Take the quadrant as the arc of $x = a \cos t$, $y = a \sin t$ for t from 0 to $\pi/2$. Then $dx/dt = -a \sin t$, $dy/dt = a \cos t$, $(dx/dt)^2 + (dy/dt)^2 = a^2$, $ds/dt = a$. As in Eq. (8), $L = \int ds = \int_0^{\pi/2} a\,dt = [at]_0^{\pi/2} = \pi a/2$. And from Eq. (35), $L\bar{x} = \int x\,ds = \int_0^{\pi/2} a \cos t\, a\,dt = a^2[\sin t]_0^{\pi/2} = a^2$. Also $L\bar{y} = \int y\,ds = \int_0^{\pi/2} a \sin t\, a\,dt = -a^2[\cos t]_0^{\pi/2} = -a^2(0 - 1) = a^2$. Hence $\bar{x} = \dfrac{L\bar{x}}{L} = \dfrac{a^2}{\pi a/2} = \dfrac{2a}{\pi}$, $\bar{y} = \dfrac{L\bar{y}}{L} = \dfrac{a^2}{\pi a/2} = \dfrac{2a}{\pi}$. And for the centroid of arc, $\bar{x} = \dfrac{2a}{\pi}$, $\bar{y} = \dfrac{2a}{\pi}$.

Solution 3: Take the quadrant as the arc of $r = a$ between $\theta = 0$ and $\theta = \pi/2$. Then $dr/d\theta = 0$, $r^2 + (dr/d\theta)^2 = a^2$, $ds/d\theta = a$. As in Eq. (9), $L = \int ds = \int_0^{\pi/2} a\,d\theta = [a\theta]_0^{\pi/2} = \pi a/2$. And from Eq. (35), $L\bar{x} = \int x\,ds = \int r \cos \theta\,ds = \int_0^{\pi/2} a \cos \theta\, a\,d\theta = a^2[\sin \theta]_0^{\pi/2} = a^2$. Also $L\bar{y} = \int y\,ds = \int r \sin \theta\,ds = \int_0^{\pi/2} a \sin \theta\, a\,d\theta = -a^2[\cos \theta]_0^{\pi/2} = -a^2(0 - 1) = a^2$. Hence $\bar{x} = \dfrac{L\bar{x}}{L} = \dfrac{a^2}{\pi a/2} = \dfrac{2a}{\pi}$, $\bar{y} = \dfrac{L\bar{y}}{L} = \dfrac{a^2}{\pi a/2} = \dfrac{2a}{\pi}$. And for the centroid of arc, $\bar{x} = \dfrac{2a}{\pi}$, $\bar{y} = \dfrac{2a}{\pi}$.

213. Center of Gravity of a Distribution of Mass in Space. Let x denote the signed distance from a fixed plane to any point in space, the x coordinate of the point. Then the *first moment* of a particle with respect to this fixed plane $x = 0$ is the product of the mass of the particle times its signed distance x. And for a system of n particles in space, with masses m_1 to m_n and at distances x_1 to x_n, the distance $\bar{x}$ for the center of gravity is again given by the first relation of Eqs. (21) and (22).

Next consider a continuous distribution of mass in space. Divide the mass into pieces such that each of the points in a typical piece of mass Δm_i is at a distance from the fixed plane $x = 0$ which is between x_i and $x_i + \Delta x$. For each i choose an x_i' which lies between x_i and $x_i + \Delta x$. Then the sum $\Sigma x_i' \Delta m_i$ is taken as an approximation to the moment of the mass with respect to the plane $x = 0$. When $\Delta x \to 0$, $\Delta m_i \to 0$, and by Sec. 185 this sum approaches a limit, $\int x\, dm$, which is defined to be the moment of the mass with respect to the plane $x = 0$. Thus

$$M_x = \int x\, dm. \tag{36}$$

The total mass of the body is $M = \int dm$. If this were concentrated at the center of gravity with $x = \bar{x}$, its moment with respect to $x = 0$ would be $M\bar{x}$. And this would equal M_x. Hence

$$M\bar{x} = M_x, \qquad \bar{x} = \frac{M_x}{M} = \frac{\int x\, dm}{\int dm}. \tag{37}$$

214. Centroid of a Solid of Revolution. Suppose that the area bounded above by the curve $y = f(x)$, below by the x axis, and lying between a left-hand ordinate at $x = a$ and a right-hand ordinate at $x = b$ is revolved

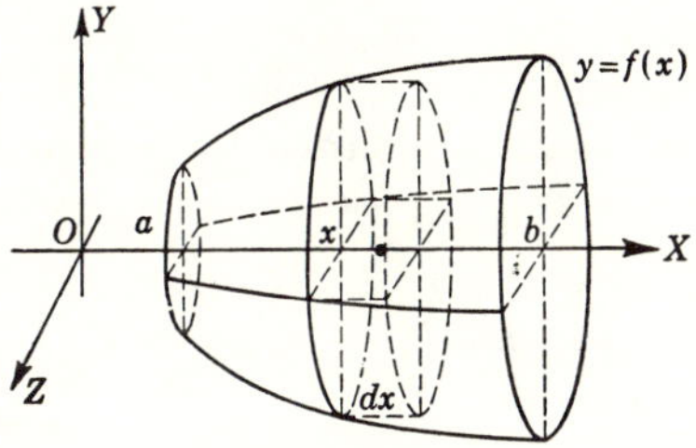

FIG. 248.

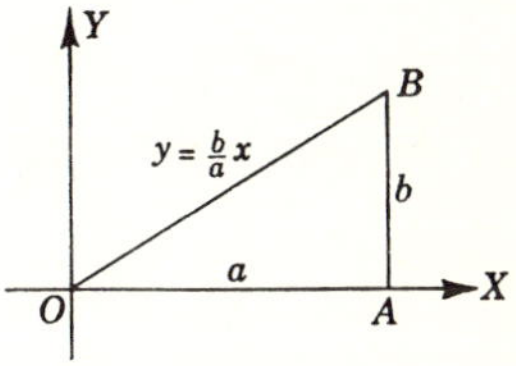

FIG. 249.

about the x axis, OX. We wish to find the center of gravity of the resulting solid of revolution (Fig. 248), assumed to be composed of matter of uniform density ρ. The center of gravity G will lie on OX. Hence its position will be determined by $\bar{x} = OG$. We take $x = 0$ as the plane through O perpendicular to OX and apply the discussion of Sec. 213 and Eq. (37). Then $dm = \rho\, dV$. And by Eq. (4), $dV = \pi y^2\, dx$. Hence

$$\bar{x} = \frac{\int x\, dm}{\int dm} = \frac{\int \rho\pi x y^2\, dx}{\int \rho\pi y^2\, dx}. \tag{38}$$

If we cancel the constant factor ρ and let $V = \pi \int y^2\, dx$, we have

$$V = \pi \int y^2\, dx, \qquad V\bar{x} = \pi \int x y^2\, dx. \tag{39}$$

This equation enables us to calculate $\bar{x} = OG$, the x coordinate of the centroid of the solid of revolution.

EXAMPLE. Show that the distance of the centroid of a solid cone of revolution from the base is one quarter of the altitude.

Solution: Let the cone be generated by revolving the right triangle OAB of Fig. 249 about OX. Let $OA = a$ and $AB = b$. Then for points on OB, $y/x = b/a$ and $y = (b/a)x$. Hence from Eq. (39) we have

$$V = \pi \int y^2\,dx = \pi \int_0^a \frac{b^2}{a^2} x^2\,dx = \frac{\pi b^2}{a^2}\left[\frac{x^3}{3}\right]_0^a = \frac{\pi ab^2}{3},$$

$$V\bar{x} = \pi \int xy^2\,dx = \pi \int_0^a x \frac{b^2}{a^2} x^2\,dx = \frac{\pi b^2}{a^2}\left[\frac{x^4}{4}\right]_0^a = \frac{\pi a^2 b^2}{4}.$$ Hence

$\bar{x} = \frac{V\bar{x}}{V} = \frac{\pi a^2 b^2/4}{\pi ab^2/3} = \frac{3a}{4}$. Since $\bar{x} = OG = \frac{3a}{4}$, $GA = OA - OG = a - \frac{3a}{4} = \frac{a}{4}$. And $GA = \frac{1}{4}OA$, as was to be proved.

215. Centroid of a Surface of Revolution. Consider a thin uniform shell overlaying the area of the surface generated by revolving a plane curve about the x axis (Fig. 250). The center of gravity of the shell G will lie on OX. Hence its position will be determined by OG. We take $x = 0$ as the plane through O perpendicular to OX, and apply the discussion of Sec. 213 and Eq. (37). Let t be the thickness of the shell and D be its density. Then the mass per unit area $\rho = Dt$ and $dm = \rho\,dS_x$. And by Eq. (10), $dS_x = 2\pi y\,ds$. Hence

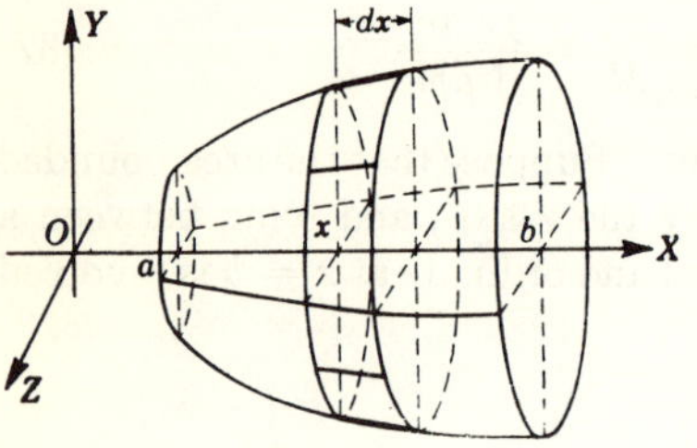

FIG. 250.

$$\bar{x} = \frac{\int x\,dm}{\int dm} = \frac{\int \rho 2\pi xy\,ds}{\int \rho 2\pi y\,ds}. \tag{40}$$

If we cancel the constant factor ρ and let $S_x = 2\pi \int y\,ds$, we have

$$S_x = 2\pi \int y\,ds, \qquad S_x\bar{x} = 2\pi \int xy\,ds. \tag{41}$$

These relations determine $\bar{x} = OG$, the x coordinate of the *centroid of the surface of revolution*. In applying them, we may use one of the expressions for ds given in Sec. 206 and express x and y in terms of the appropriate independent variable.

EXAMPLE. Show that the centroid of a hemispherical surface of radius a is the mid-point of the radial axis.

Solution: Let the surface be generated by the rotation about OX of the arc of $y = \sqrt{a^2 - x^2}$ between $x = 0$ and $x = a$. Then $\frac{dy}{dx} = \frac{-x}{\sqrt{a^2 - x^2}}$, $1 + \left(\frac{dy}{dx}\right)^2 = \frac{a^2}{a^2 - x^2}$, $\frac{ds}{dx} = \frac{a}{\sqrt{a^2 - x^2}}$. And from Eq. (41), $S_x = 2\pi \int y\,ds =$

$2\pi \int_0^a \sqrt{a^2 - x^2} \frac{a}{\sqrt{a^2 - x^2}}\,dx = 2\pi a \int_0^a dx = 2\pi a[x]_0^a = 2\pi a^2$. Also $S_x\bar{x} =$

$2\pi \int xy\,ds = 2\pi \int_0^a x\sqrt{a^2 - x^2} \frac{a}{\sqrt{a^2 - x^2}}\,dx = 2\pi a \int_0^a x\,dx = 2\pi a\left[\frac{x^2}{2}\right]_0^a = \pi a^3.$

Hence $\bar{x} = \frac{S_x\bar{x}}{S_x} = \frac{\pi a^3}{2\pi a^2} = \frac{a}{2}$.

Since $\bar{x} = a/2$, the centroid of the hemispherical surface is the mid-point of the radial axis, as was to be proved.

EXERCISE 110

1. Show that the centroid of a circular arc which subtends an angle B at the center of a circle of radius a lies at a distance $2a/B \sin (B/2)$ from the center of the circle on the radius which bisects the arc.
2. Find the centroid of the arc of $x = \cos^3 t$, $y = \sin^3 t$ for t from 0 to $\pi/2$.
3. Find the centroid of the arc of $9y^2 = 4x^3$ between the two points $(1, -\frac{2}{3})$ and $(1, \frac{2}{3})$.
4. Find the centroid of the arc of $x = t^2$, $y = t - (t^3/3)$ for t from -1 to 1.
5. Find the centroid of the arc of the catenary $y = (a/2)(e^{x/a} + e^{-x/a})$ between the points for which $x = 0$ and $x = a$.
6. Find the centroid of the arc of the cycloid $x = a(t - \sin t)$, $y = a(1 - \cos t)$ from $t = 0$ to $t = \pi$.
7. Find the centroid of the arc of the curve $r = e^\theta$ from $\theta = 0$ to $\theta = \pi/2$.
8. Find the centroid of the arc of the cardioid $r = a(1 + \cos \theta)$.
9. Show that the centroid of a solid hemisphere is on the radial axis three-eighths of the way up from the base.

The area in the first quadrant bounded by the x axis and the given curve or set of curves is revolved about OX. Find the x coordinate of the centroid of the *solid* of revolution thus generated in each of the following problems.

10. $y^2 = 4mx$, $x = m$.
11. $y^2 = 4m(m - x)$, $x = 0$.
12. $ay = x^2$, $x = a$.
13. $y = e^{-x}$, $x = 0$, $x = 1$.
14. $y = \sin x$ from $x = 0$ to $x = \pi/2$.
15. $xy = 4$, $x = 1$, $x = 4$.
16. $\dfrac{x^2}{a^2} + \dfrac{y^2}{b^2} = 1$, $x = 0$.
17. $\dfrac{x^2}{a^2} - \dfrac{y^2}{b^2} = 1$, $x = 2a$.
18. $\dfrac{x^2}{a^2} - \dfrac{y^2}{b^2} = -1$, $x = 0$, $x = a$.
19. $x^2 + y^2 = 1$, $x = 0$, $x = 1$.

In each of the following problems the given arc of the curve is revolved about OX. Find the x coordinate of the centroid of the *surface* of revolution thus generated.

20. $x = a \cos^3 t$, $y = a \sin^3 t$ from $t = 0$ to $t = \pi/2$.
21. $x^2 + y^2 = a^2$, $x = x_1$ to $x = x_2$, $-a < x_1 < x_2 < a$.
22. $y = e^{x/2} + e^{-x/2}$ from $x = 0$ to $x = 2$.
23. $y^2 = 4mx$ from $x = 0$ to $x = m$.

24. Show that the distance of the centroid of the lateral surface of a cone of revolution from the base is one-third of the altitude.

***216. Centroid of a Composite Figure.** Suppose that a body is built up of simple parts, for each of which the mass and center of gravity are known. Then the centroid of the composite body can be found by relations like those of Eq. (22). We illustrate for a body consisting of three parts and for the x coordinate. Here x is either the distance from the line $x = 0$, as in Sec. 209, or the distance from the plane $x = 0$ as in Sec. 213. Let the parts have masses M_1, M_2, M_3. And let the x coordinates of their centroids be $\bar{x}_1$, $\bar{x}_2$, and $\bar{x}_3$, respectively. Then the mass of the composite body is

$$M = M_1 + M_2 + M_3. \tag{42}$$

Let $\bar{x}$ denote the x coordinate of the centroid of the composite body. Then the moment of the body with respect to $x = 0$ is $M\bar{x}$. The moments with respect to $x = 0$ for the parts are $M_1\bar{x}_1$, $M_2\bar{x}_2$, $M_3\bar{x}_3$. Since the moment of the body equals the sum of the moments for the parts, we have

$$M\bar{x} = M_1\bar{x}_1 + M_2\bar{x}_2 + M_3\bar{x}_3. \tag{43}$$

EXAMPLE. A square of side $2a$ has a quadrant of a circle of radius a cut from one corner. Find the centroid of the area remaining (Fig. 251).

Solution: The square of side $2a$ has area $4a^2$. Its centroid is at the geometric center, or at $(\bar{x},\bar{y}) = (a,a)$ with the axes as indicated. The quadrant of the circle has area $\frac{\pi a^2}{4}$, and by the example of Sec. 211 has $(\bar{x},\bar{y}) = \left(\frac{4a}{3\pi}, \frac{4a}{3\pi}\right)$. We may replace the masses in Eq. (43) by areas so that $A\bar{x} = A_1\bar{x}_1 + A_2\bar{x}_2$. Since the quadrant is *removed*, we treat its area as *negative* and set $A_1 = 4a^2$, $A_2 = -\frac{\pi a^2}{4}$, $\bar{x}_1 = a$, $\bar{x}_2 = \frac{4a}{3\pi}$. Then $A = A_1 + A_2 = 4a^2 - \frac{\pi a^2}{4}$. And $A\bar{x} = A_1\bar{x}_1 + A_2\bar{x}_2 = 4a^2(a) + \left(-\frac{\pi a^2}{4}\right)\frac{4a}{3\pi}$ $= 4a^3 - \frac{a^3}{3} = \frac{11}{3}a^3$. Hence $\bar{x} = \frac{A\bar{x}}{A} = \frac{\frac{11}{3}a^3}{4a^2 - \pi a^2/4} = \frac{44a}{48 - 3\pi} = 1.14a$. From symmetry, $\bar{y} = \bar{x}$, so that the centroid is $(1.14a, 1.14a)$.

***217. Theorems of Pappus.** The following theorems involving the centroid may often be used to advantage in finding volumes of solids of revolution or areas of surfaces of revolution:

Theorem I. *Let a plane area be revolved about an axis in its plane not crossing the area. Then the volume generated is equal to the product of the area by the length of the circumference described by its centroid.*

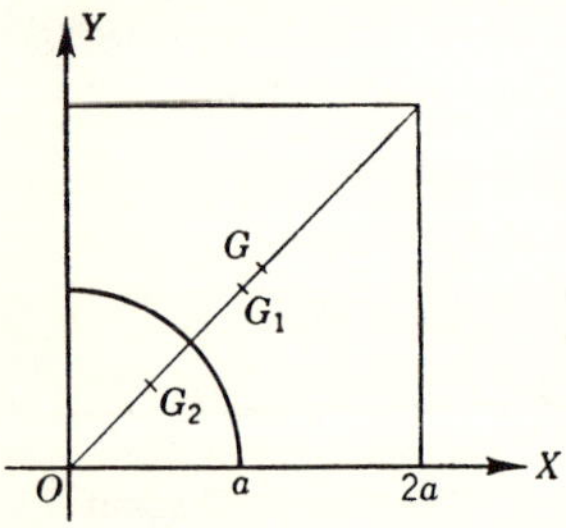

FIG. 251.

To prove this, take the axis of revolution as the y axis. And let the area be that of Fig. 243. Then from Eq. (5), the volume of the solid generated is

$$V_y = 2\pi \int_a^b (y_2 - y_1)x\,dx. \tag{44}$$

And from Eq. (28) we have

$$A\bar{x} = \int_a^b x(y_2 - y_1)dx. \tag{45}$$

A comparison of Eqs. (44) and (45) shows that

$$V_y = 2\pi \int_a^b x(y_2 - y_1)dx = 2\pi(A\bar{x}) = A(2\pi\bar{x}). \tag{46}$$

This proves theorem I, since $2\pi\bar{x}$ is the length of the circumference of radius $\bar{x}$ described by the centroid during the revolution about the y axis.

Theorem II. *Let a plane curved arc be revolved about an axis in its plane not crossing the arc. Then the area generated is equal to the product of the length of the arc by the length of the circumference described by its centroid.* To prove this, take the axis of revolution as the y axis. Then from Eq. (11), the area of the surface generated is

$$S_y = 2\pi \int x\,ds. \tag{47}$$

And from Eq. (35) we have

$$L\bar{x} = \int x\,ds. \tag{48}$$

A comparison of Eqs. (47) and (48) shows that

$$S_y = 2\pi \int x\,ds = 2\pi(L\bar{x}) = L(2\pi\bar{x}). \tag{49}$$

This proves theorem II, since $2\pi\bar{x}$ is the length of the circumference of radius $\bar{x}$ described by the centroid during the revolution about the y axis.

Example 1. A semicircular groove of radius a is cut in a cylinder of radius $4a$. Find the volume of material removed.

Solution: By the example of Sec. 211, $GD = \frac{4a}{3\pi}$ (Fig. 252). And $\bar{x} = CD - GD = 4a - \frac{4a}{3\pi}$. Also $A = \frac{1}{2}\pi a^2$. Hence by theorem I, $V_y = 2\pi A\bar{x} = 2\pi\frac{\pi a^2}{2}\left(4a - \frac{4a}{3\pi}\right) = \frac{4}{3}\pi a^3(3\pi - 1)$. Thus the required volume of material removed is $V_y = \frac{4}{3}\pi a^3(3\pi - 1)$.

EXAMPLE 2. Find the surface exposed by cutting the groove of Example 1.

Solution: If G is the centroid of the semicircular arc, by the example of Sec. 212, $GD = 2a/\pi$. And $\bar{x} = 4a - 2a/\pi$. Also $L = \frac{1}{2}(2\pi)(a) = a\pi$. Hence by theorem II, $S_y = 2\pi L\bar{x} = 2\pi(a\pi)(4a - 2a/\pi) = 4\pi a^2(2\pi - 1)$. Thus the required exposed surface is $S_y = 4\pi a^2(2\pi - 1)$.

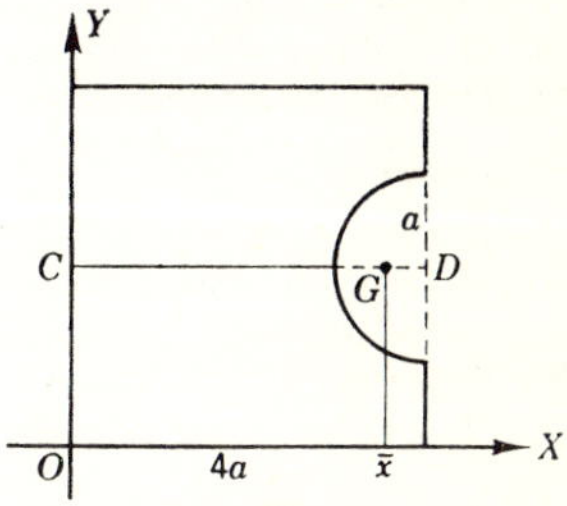

FIG. 252.

EXERCISE 111

1. A plane area is made up of two squares. Each square has one side along the x axis and one along the y axis. The first square has one corner at $(4a,4a)$. The second square has one corner at $(-2a,-2a)$. Find the centroid of the plane area.
2. A carpenter's square has each arm 12 in. on its outer edge and 10 in. on its inner edge. Assuming the outer edges along OX and OY, and 1 in. as the unit, find the centroid.
3. The square with lower corners at $(0,0)$ and $(2a,0)$ is surmounted by a semicircle with center at $(a,2a)$ and radius a. Find the centroid of the composite area.
4. Find the centroid of the perimeter of the figure of Prob. 3.
5. A circle of radius a is tangent externally to the circle $x^2 + y^2 = 4a^2$ at the point $(2a,0)$. Find the centroid of the composite area.
6. Find the centroid of the composite arc of the figure of Prob. 5.
7. Find the centroid of the area which is inside the circle $x^2 + y^2 = 4a^2$ and outside of the circle $x^2 + y^2 = 2ax$.
8. The radius of the base of a circular cylinder is 3 in. And its altitude is 8 in. A cone of revolution with the same base and altitude 4 in., with vertex at the center of the cylinder, is removed. Find the centroid of the remaining volume. See the example of Sec. 214 for the centroid of the cone.
9. From the result of Prob. 9 of Exercise 110 deduce that the centroid of a hemispherical shell of inner radius a and outer radius $a + h$ is on the axis of the shell at a distance $\frac{3}{8}\frac{(a + h)^4 - a^4}{(a + h)^3 - a^3}$ from the common base.
10. Verify that when $h \to 0$, the distance in Prob. 9 approaches $a/2$. This checks the example of Sec. 215.
11. From the result of the example of Sec. 214 deduce that the centroid of a conical shell bounded by two parallel conical surfaces of altitudes a and $a + h$, respectively, is on the axis of the shell at a distance $\frac{1}{4}\frac{(a + h)^4 - a^4}{(a + h)^3 - a^3}$ from the base.
12. Verify that, when $h \to 0$, the distance in Prob. 10 approaches $a/3$. This checks Prob. 24 of Exercise 110.

An *anchor ring,* or *torus,* is generated by the revolution of a circle of radius a about an axis in its plane b units from its center, with $b > a$. Use the appropriate theorem of Pappus to show that

13. The volume of the solid of revolution is $2\pi^2a^2b$.

14. The area of the surface of revolution is $4\pi^2ab$.

Use the appropriate theorem of Pappus to deduce that the distance of the centroid from the base for

15. The area of a triangle is one-third of the altitude.
16. The area of a semicircle of radius a is $4a/3\pi$.
17. The arc of a semicircle of radius a is $2a/\pi$.

A square of side $2a$ is revolved about an axis in its plane perpendicular to one of its diagonals at distance $3a$ from its center. Use the appropriate theorem of Pappus to find

18. The volume of the solid of revolution generated.
19. The area of the surface of revolution generated.

A triangular groove is cut in a cylinder of radius $6a$. The cross section of the groove is an isosceles triangle, with base $2a$ along an element of the cylinder and altitude $3a$ along a radius of the cylinder. Use the appropriate theorem of Pappus to find

20. The volume of the material removed.
21. The surface exposed by cutting the groove.

218. Moment of Inertia of a System of Particles. The moment of inertia of a particle about an axis is the product mr^2, where m is the mass of the particle and r is the shortest distance from the axis to the particle. The moment of inertia of a system of particles about a given axis is the sum of the moments of inertia of the separate particles about this axis. Let there be n particles of masses $m_1, m_2, \cdots, m_n$ at distances $r_1, r_2, \cdots, r_n$, respectively, from the given axis. Then their moment of inertia about this axis is

$$I = m_1r_1^2 + m_2r_2^2 + \cdots + m_nr_n^2. \tag{50}$$

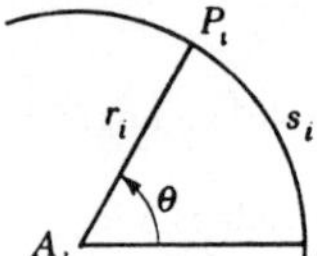

FIG. 253.

Suppose that the particles of the system are rigidly connected and rotate about the given fixed axis. Then each particle P_i describes a circle of radius $A_iP_i = r_i$ (Fig. 253). And if P_i is the position at time t, and B_i is the position at time t_0, the angle $B_iA_iP_i = \theta$ is the same for all the particles. Denote arc B_iP_i by s_i. Then we have

$$s_i = r_i\theta \quad \text{and} \quad v_i = r_i\frac{d\theta}{dt}.$$

The kinetic energy of the particle P_i is

$$\frac{1}{2}m_iv_i^2 = \frac{1}{2}m_i\left(r_i\frac{d\theta}{dt}\right)^2 = \frac{1}{2}r_i^2\left(\frac{d\theta}{dt}\right)^2. \tag{51}$$

Hence the total kinetic energy of the rigid system of particles is

$$E = \frac{1}{2}\sum m_iv_i^2 = \frac{1}{2}\left(\frac{d\theta}{dt}\right)^2\sum m_ir_i^2 = \frac{1}{2}I\left(\frac{d\theta}{dt}\right)^2. \tag{52}$$

This is typical of a number of dynamical formulas in which, for rotating rigid systems, moment of inertia I and angle of rotation θ play a role analogous to mass and distance. For example, if T is the total torque on the system about the axis, we have

$$T = I\frac{d^2\theta}{dt^2} \text{ in place of } F = m\frac{d^2s}{dt^2}. \tag{53}$$

219. Moment of Inertia of a Material Body. Consider a material body made up of a continuous distribution of mass. We may define its moment of inertia about a given axis by the following procedure. Divide the body into pieces such that each of the points in a typical piece of mass Δm_i is at a distance from the given axis between r_i and $r_i + \Delta r$. For each i choose an r_i' which lies between r_i and $r_i + \Delta r$. Then $\Sigma r_i'^2 \,\Delta m_i$ is taken as an approximation to the moment of inertia of the mass about the given axis. When $\Delta r \to 0$, $\Delta m_i \to 0$. And by Sec. 185, the sum approaches a limit, $\int r^2\, dm$, which is defined to be the moment of inertia of the body about the given axis. Thus

$$I = \int r^2\, dm. \tag{54}$$

220. Radius of Gyration. Let M denote the total mass of the body of Sec. 219, so that

$$M = \int dm. \tag{55}$$

Then we may find a constant k such that

$$k = \sqrt{\frac{I}{M}} \quad \text{and} \quad I = Mk^2. \tag{56}$$

The constant k which satisfies Eq. (56) is called the *radius of gyration* of the body about the given axis. If all the mass of the body were concentrated in a particle at distance k from the axis, the moment of inertia would be the same for this particle as for the original material body.

Let a material body be divided into n parts, of masses $m_1, m_2, \cdots, m_n$. And with respect to a given axis let the radii of gyration of the parts be $r_1, r_2, \cdots, r_n$. Then the moment of inertia of the body may be found from Eq. (50). It follows that in Eq. (54) elements dm of different type from those described may be used provided that r^2 is replaced by an appropriate function depending on the square of the radius of gyration for the type of element used.

EXAMPLE. A uniform thin plate of total mass M is in the form of a rectangle of base B and altitude H. Show that its moment of inertia about the base is $\frac{1}{3}MH^2$, so that $k = H/\sqrt{3}$.

Solution: Let the plate overlay the rectangle of Fig. 254, bounded by $x = 0$, $x = H$, $y = 0$, $y = B$. Let t be the thickness of the plate and D be its density. Then the mass per unit area $\rho = Dt$ and $dm = \rho\, dA$. Take the base on OY as the axis. As in Eq. (1), $dA = y\,dx$ or $B\,dx$, $dm = \rho B\,dx$. For this element of mass, the distance from the axis is x so that by Eq. (54) we have $I = \int r^2\,dm = \int_0^H x^2\rho B\,dx = \rho B\left[\frac{x^3}{3}\right]_0^H = \frac{\rho BH^3}{3}$. And from Eq. (55) we have $M = \int dm = \int_0^H \rho B\,dx = \rho BH$. Hence $\frac{I}{M} = \frac{\rho BH^3/3}{\rho BH} = \frac{H^2}{3}$. And from Eq. (55), $k = \sqrt{\frac{I}{M}} = \frac{H}{\sqrt{3}}$. Thus $I = \frac{MH^2}{3}$ and $k = \frac{H}{\sqrt{3}}$ as was to be proved.

221. Moment of Inertia of a Plane Area. In the theory of the bending of beams or the twisting of shafts an integral appears which is similar to

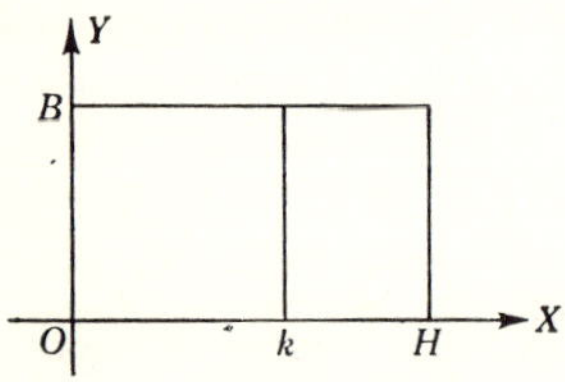

FIG. 254.

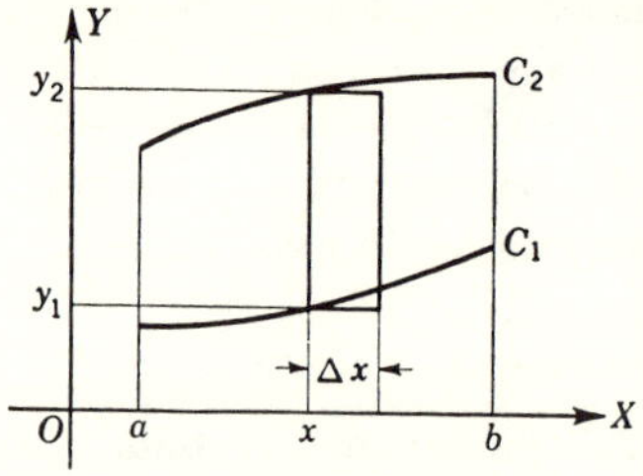

FIG. 255.

that of Eq. (54), except that mass is replaced by area in the plane of a cross section. We write

$$I = \int r^2\,dA. \tag{57}$$

We call this I the *moment of inertia of the area* about the given axis, although it has dimensions (length)^4 and has no connection with inertia. Also, if the total area $A = \int dA$, as in Eq. (56), we write

$$k = \sqrt{\frac{I}{A}} \qquad \text{and} \qquad I = Ak^2. \tag{58}$$

And we call k the *radius of gyration* of the area about the given axis.

Suppose that the plane area is bounded above by the curve $y_2 = f_2(x)$ and below by the curve $y_1 = f_1(x)$, and lies between a left-hand ordinate at $x = a$ and a right-hand ordinate at $x = b$ (Fig. 255). Then by Eq. (1) we have

$$dA = (y_2 - y_1)dx \qquad \text{and} \qquad A = \int_a^b (y_2 - y_1)dx. \tag{59}$$

For this element the distance from the y axis is x. Hence the moment of inertia about the y axis is

$$I_y = \int x^2\,dA = \int_a^b x^2(y_2 - y_1)dx. \tag{60}$$

If $y_2 > y_1 \geqq 0$, we may consider the rectangle of area $(y_2 - y_1)\Delta x$ as the figure obtained by removing a rectangle with altitude y_1 from one with altitude y_2, where both rectangles have base Δx on the x axis. The moment of inertia about the x axis is $\frac{1}{3}BH^3 = \frac{1}{3}\,\Delta x\,y_2{}^3$ for the first rectangle and $\frac{1}{3}\,\Delta x\,y_1{}^3$ for the second rectangle, by the example of Sec. 220 with the density factor ρ omitted. Hence the moment of inertia about the x axis for the rectangle of area $(y_2 - y_1)\Delta x$ is $\frac{1}{3}(y_2{}^3 - y_1{}^3)\Delta x$. And this result holds for $y_2 \geqq y_1$ regardless of the signs of y_1 and y_2. It follows that the moment of inertia of the area about the x axis is

$$I_x = \frac{1}{3}\int_a^b (y_2{}^3 - y_1{}^3)dx. \qquad (61)$$

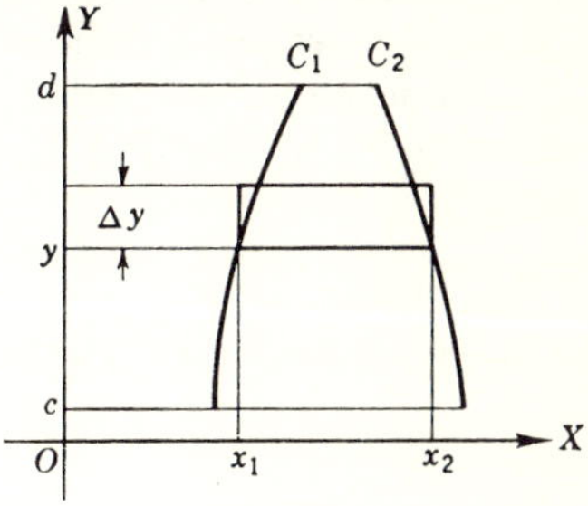

FIG. 256.

Let us next suppose that the plane area is bounded on the right by the curve $x_2 = g_2(y)$ and on the left by the curve $x_1 = g_1(y)$, and lies above the line $y = c$ and below the line $y = d$ (Fig. 256). Then we may take the element of area as $dA = (x_2 - x_1)dy$, and deduce that

$$A = \int_c^d (x_2 - x_1)dy, \qquad I_y = \frac{1}{3}\int_c^d (x_2{}^3 - x_1{}^3)dy,$$

$$I_x = \int_c^d y^2(x_2 - x_1)dy. \qquad (62)$$

EXAMPLE. Find the moment of inertia of the area of a circle of radius a about a diameter.

Solution: First consider the quadrant of the circle bounded above by $y = \sqrt{a^2 - x^2}$, below by $y = 0$, and between $x = 0$ and $x = a$. We know the area and may take the axis as OX or OY, but to illustrate our formulas we will compute all the quantities. From Eq. (59),

$$A = \int_0^a \sqrt{a^2 - x^2}\,dx = a^2\int_0^{\pi/2} \cos^2 t\,dt = a^2\left[\frac{t}{2} + \frac{\sin 2t}{4}\right]_0^{\pi/2} = \frac{\pi a^2}{4}.$$

From Eq. (60), $I_y = \int_0^a x^2\sqrt{a^2 - x^2}\,dx = a^4\int_0^{\pi/2} \sin^2 t\cos^2 t\,dt = \frac{a^4}{4}\int_0^{\pi/2} \sin^2 2t\,dt$
$= \frac{a^4}{4}\left[\frac{t}{2} - \frac{\sin 4t}{8}\right]_0^{\pi/2} = \frac{\pi a^4}{16}.$ From Eq. (61),

$$I_x = \frac{1}{3}\int_0^a (a^2 - x^2)^{\frac{3}{2}}\,dx = \frac{a^4}{3}\int_0^{\pi/2} \cos^4 t\,dt$$
$$= \frac{a^4}{3}\left[\frac{3t}{8} + \frac{\sin 2t}{4} + \frac{\sin 4t}{32}\right]_0^{\pi/2} = \frac{\pi a^4}{16}.$$

It follows that $\frac{I_y}{A} = \frac{\pi a^4/16}{\pi a^2/4} = \frac{a^2}{4}.$ Hence $I_y = \frac{Aa^2}{4}.$ And, similarly, $I_x = \frac{Aa^2}{4}.$

For the whole circle, A and I_y or I_x are each four times as large as for the quadrant. Hence the required moment of inertia of the whole circle about a diameter is $I = \pi a^4/4$ or $I = Aa^2/4$.

222. Polar Moment of Inertia of a Plane Area. For any distribution of mass in a plane, the moment of inertia about an axis perpendicular to its plane is called the *polar moment of inertia.* We usually take the point in which the axis cuts the plane of the mass as the origin. In this case we speak of the moment of inertia with respect to the origin and denote this polar moment of inertia by I_0.

Consider a single particle of mass m at $P = (x,y)$. The shortest distance from the perpendicular axis through $O = (0,0)$ to P is the distance from O to P, or $r = OP$. Thus

$$r^2 = x^2 + y^2, \qquad mr^2 = mx^2 + my^2, \qquad \text{and } I_0 = I_y + I_x. \tag{63}$$

This relation holds for systems of particles and also for continuous distributions of masses or of area in the plane.

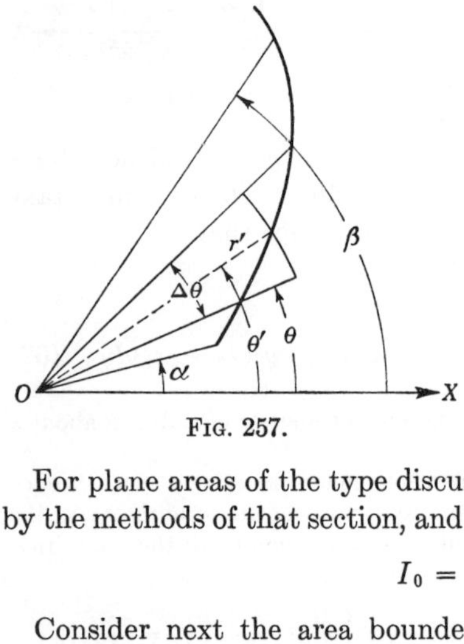

FIG. 257.

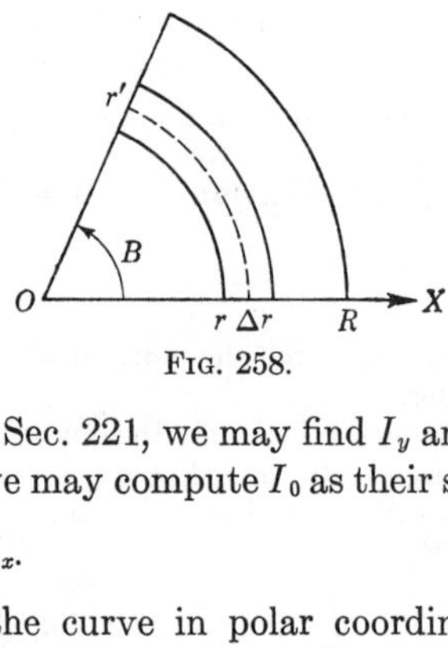

FIG. 258.

For plane areas of the type discussed in Sec. 221, we may find I_y and I_x by the methods of that section, and then we may compute I_0 as their sum:

$$I_0 = I_y + I_x. \tag{64}$$

Consider next the area bounded by the curve in polar coordinates $r = f(\theta)$ and the two radius vectors $\theta = \alpha$ and $\theta = \beta$, Fig. 257. Then by Eq. (2) we have

$$dA = \tfrac{1}{2}r^2\,d\theta, \qquad A = \tfrac{1}{2}\int_\alpha^\beta r^2\,d\theta. \tag{65}$$

Here the elementary area is approximately a circular sector of radius r' and central angle $\Delta\theta$. The polar moment of inertia of this sector is $\dfrac{\Delta\theta\, r'^4}{4}$, by Example 1 below. It follows that for the area considered

$$I_0 = \tfrac{1}{4}\int_\alpha^\beta r^4\,d\theta. \tag{66}$$

EXAMPLE 1. Show that the polar moment of inertia with respect to the origin of a circular sector with central angle B in the circle $r = R$ is $AR^2/2 = BR^4/4$.

Solution: Let the sector be bounded by $\theta = 0$, $\theta = B$, and $r = R$, as in Fig. 258. Let ΔA be the area cut out from the sector by two circles of radius r and $r + \Delta r$.

Then $\Delta A = Br'\Delta r$ so that $A = \int dA = \int_0^R Br\,dr = \left[\frac{Br^2}{2}\right]_0^R = \frac{BR^2}{2}$. And $\Delta I = r''^2\,\Delta A = Br''^2 r'\,\Delta r$, so that $I_0 = \int_0^R Br^3\,dr = \left[\frac{Br^4}{4}\right]_0^R = \frac{BR^4}{4}$. And $\frac{I_0}{A} = \frac{BR^4/4}{BR^2/2} = \frac{R^2}{2}$. Hence $I_0 = \frac{AR^2}{2} = \frac{BR^4}{4}$, as was to be proved.

EXAMPLE 2. For the area of a circle of radius a, show that the polar moment of inertia with respect to its center is $I_0 = Aa^2/2 = \pi a^4/2$.

Solution 1: Treat the circle as the area bounded by $r = a$, and use Eqs. (65) and (66) with limits 0 and 2π. This leads to $A = \frac{1}{2}\int_0^{2\pi} a^2\,d\theta = \left[\frac{a^2\theta}{2}\right]_0^{2\pi} = \pi a^2$. $I_0 = \frac{1}{4}\int_0^{2\pi} a^4\,d\theta = \left[\frac{a^4\theta}{4}\right]_0^{2\pi} = \frac{\pi a^4}{2}$. And $\frac{I_0}{A} = \frac{\pi a^4/2}{\pi a^2} = \frac{a^2}{2}$. Hence $I_0 = \frac{Aa^2}{2} = \frac{\pi a^4}{2}$.

Solution 2: From the example of Sec. 221, $I_y = I_x = \frac{\pi a^4}{4} = \frac{Aa^2}{4}$. Hence by Eq. (64), $I_0 = I_y + I_x = 2I_x = \frac{\pi a^4}{2} = \frac{Aa^2}{2}$.

Solution 3: Put $R = a$ and $B = 2\pi$ in the result of Example 1 above and so deduce that $I_0 = \frac{AR^2}{2} = \frac{BR^4}{4} = \frac{Aa^2}{2} = \frac{\pi a^4}{2}$.

EXERCISE 112

Find I_y, the moment of inertia about OY, for the area in the first quadrant bounded by each given curve and the coordinate axes.

1. $y = 4 - x^2$.
2. $y^2 = 2 - x$.
3. $y = x^2 - x^3$.
4. $y = 2x - x^2$.

Find I_x, the moment of inertia about OX, for the area in the first quadrant bounded by each given curve and ordinates and OX.

5. $y = e^x$, $x = 0$, $x = 1$.
6. $y = \sin x$, $x = 0$, $x = \pi/2$.
7. $y = 2x^3$, $x = 1$, $x = 2$.
8. $x^2 + y^2 = 4$, $x = 0$, $x = 2$.

Show that for the area of the ellipse $\frac{x^2}{a^2} + \frac{y^2}{b^2} = 1$,

9. $I_y = \frac{Aa^2}{4}$.
10. $I_x = \frac{Ab^2}{4}$.
11. $I_0 = \frac{A(a^2 + b^2)}{4}$.

A rectangle of sides $2a$ and $2b$ is bounded by $x = -a$, $x = a$, $y = -b$, $y = b$. Show that for the area of this rectangle

12. $I_y = \frac{Aa^2}{3}$.
13. $I_x = \frac{Ab^2}{3}$.
14. $I_0 = \frac{A(a^2 + b^2)}{3}$.

Find I_y, I_x, and I_0 for the right triangle whose vertices are

15. (0,0), (0,b), (a,b).
16. (0,2b), (0,b), (a,b).

Find I_y, I_x, and I_0 for the area bounded by $y = x$ and the curve

17. $y = x^2$.
18. $y = x^{\frac{1}{3}}$.
19. $y = x^4$.

Find the polar moment of inertia for the area of

20. A circle with respect to a point on its circumference.
21. A right triangle of sides a and b with respect to the vertex of the right angle.
22. A circular ring of inner radius a and outer radius b with respect to its center.

Find I_0 for the area inside one loop for each given curve.

23. $r^2 = 2\cos 2\theta$.
24. $r = \sin 2\theta$.
25. $r = \cos 3\theta$.

223. Moment of Inertia of a Solid of Revolution. Suppose that the area bounded above by the curve $y = f(x)$ and below by the x axis, which lies between a left-hand ordinate at $x = a$ and a right-hand ordinate at $x = b$, is revolved about the x axis, OX. We wish to find the moment of inertia for the resulting solid of revolution (Fig. 259), assumed to be composed of matter of uniform density ρ. Then $dm = \rho\, dV$. And by Eq. (4), $dV = \pi y^2\, dx$. By Example 2 of Sec. 222, the moment of inertia of a disk of mass M and radius a about an axis through its center perpendicular to the disk is $Ma^2/2$. Since Δm occupies a volume which is approximately a disk of radius y and thickness Δx, it has volume $\pi y^2\, \Delta x$, mass $\pi y^2 \rho\, \Delta x$, and moment of inertia about the axis OX equal to $(\rho \pi y^2\, \Delta x)\dfrac{y^2}{2}$.

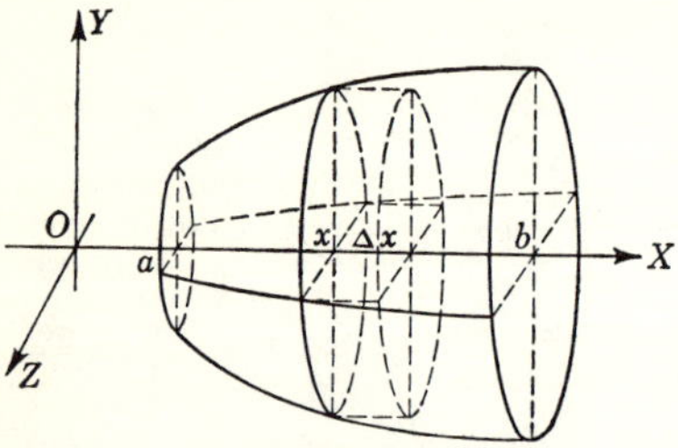

FIG. 259.

It follows that $dI = \dfrac{\pi\rho}{2} y^4\, dx$ and

$$I_x = \frac{\pi\rho}{2}\int^b y^4\, dx, \qquad M = \pi\rho \int_a^b y^2\, dx. \tag{67}$$

Next suppose that the same area is revolved about the y axis. Then $dm = \rho\, dV$. And by Eq. (5), $dV = 2\pi xy\, dx$. Here Δm occupies a volume, approximately a shell all of whose points are approximately at distance x from OY, the axis of revolution. Hence the moment of inertia of Δm about OY is approximately $x^2\, \Delta m$. It follows that $dI = x^2\, dm = x^2 \rho\, dV = x^2 \rho 2\pi xy\, dx = 2\pi\rho x^3 y\, dx$. Hence

$$I_y = 2\pi\rho \int_a^b x^3 y\, dx \qquad \text{and} \qquad M = 2\pi\rho \int_a^b xy\, dx. \tag{68}$$

EXAMPLE. Show that the moment of inertia of a homogeneous solid sphere of mass M about a diameter is $\frac{2}{5}Ma^2$.

Solution 1: Consider the hemisphere generated when the quadrant, bounded by $y = \sqrt{a^2 - x^2}$ from $x = 0$ to $x = a$ and the coordinate axes, is revolved about OX. From Eq. (67), we have $M = \pi\rho \int y^2\, dx = \pi\rho \int_0^a (a^2 - x^2)dx = \pi\rho \left[a^2 x - \dfrac{x^3}{3}\right]_0^a = \dfrac{2\pi\rho a^3}{3}$.

$$I_x = \frac{\pi\rho}{2}\int y^4\, dx = \frac{\pi\rho}{2}\int_0^a (a^2 - x^2)^2\, dx = \frac{\pi\rho}{2}\int_0^a (a^4 - 2a^2x^2 + x^4)dx$$
$$= \frac{\pi\rho}{2}\left[a^4 x - \frac{2ax^3}{3} + \frac{x^5}{5}\right]_0^a = \frac{4\pi\rho a^5}{15} \qquad \text{and} \qquad \frac{I_x}{M} = \frac{4\pi\rho a^5/15}{2\pi\rho a^5/3} = \frac{2a^2}{5}.$$

For the whole sphere, I_x and M are each doubled, so that $I_x = \frac{2}{5}Ma^2$, as was to be proved.

Solution 2: Consider the hemisphere generated when the quadrant, bounded by $y = \sqrt{a^2 - x^2}$ from $x = 0$ to $x = a$ and the coordinate axes, is revolved about OY. From Eq. (68) we have

$$M = 2\pi\rho \int xy\,dx = 2\pi\rho \int_0^a x\sqrt{a^2 - x^2}\,dx = -\frac{2}{3}\pi\rho[(a^2 - x^2)^{\frac{3}{2}}]_0^a$$

$$= -\frac{2}{3}\pi\rho(0 - a^3) = \frac{2\pi\rho a^3}{3} \quad \text{and} \quad I_y = 2\pi\rho \int x^3 y\,dx = 2\pi\rho \int_0^a x^3\sqrt{a^2 - x^2}\,dx.$$

Let $\sqrt{a^2 - x^2} = u$, $x^2 = a^2 - u^2$, $x\,dx = -u\,du$ and

$$I_y = 2\pi\rho \int_a^0 (a^2 - u^2)u(-u\,du) = -2\pi\rho \int_a^0 (a^2u^2 - u^4)du$$

$$= -2\pi\rho\left[\frac{a^2u^3}{3} - \frac{u^5}{5}\right]_a^0 = 2\pi\rho\left(\frac{a^5}{3} - \frac{a^5}{5}\right) = \frac{4\pi\rho a^5}{15}.$$

And $\dfrac{I_y}{M} = \dfrac{4\pi\rho a^5/15}{2\pi\rho a^3/3} = \dfrac{2a^2}{5}$. For the whole sphere, I_y and M are each doubled so that $I_y = \frac{2}{5}Ma^2$, as was to be proved.

***224. The Parallel-axis Theorem.** Suppose that the moment of inertia of any mass or area is known about an axis passing through its center of gravity. Then the following theorem enables us to find the moment of inertia about any parallel axis.

Theorem. Let I_g be the moment of inertia of any mass about an axis L_g which passes through the center of gravity G. And let I be the moment of inertia of the same mass about an axis L parallel to L_g. Then if the distance between the axes L_g and L is h, and the total mass is M, we have

$$I = I_g + Mh^2. \tag{69}$$

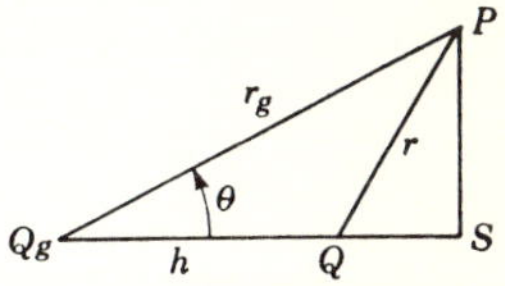

FIG. 260.

We shall prove this for a system of particles. Pass a plane perpendicular to L and L_g through P, any particle of the system. This plane is the plane of the paper in Fig. 260. Let this plane cut the axis L in Q and L_g in Q_g. Then $QQ_g = h$. Let $PQ = r$, $PQ_g = r_g$, and angle $QQ_gP = \theta$. From the law of cosines [Eq. (51) of Sec. 94] applied to triangle PQ_gQ, we find

$$r^2 = r_g^2 + h^2 - 2hr_g \cos\theta. \tag{70}$$

Multiply by m, the mass of the particle at P, and sum for all the particles. Note that $I = \Sigma r^2m$, $I_g = \Sigma r_g^2m$, and $M = \Sigma m$. Also, if we took G as the origin and the x axis parallel to Q_gQ, the distance x of P would be $Q_gS = r_g \cos\theta$. And $\bar{x}$ would be zero, so that

$$\Sigma r_g(\cos\theta)m = \Sigma mx = M\bar{x} = 0. \tag{71}$$

Hence the relation

$$\Sigma r^2m = \Sigma r_g^2m + h^2\,\Sigma m - 2h\,\Sigma r_g(\cos\theta)m \text{ becomes } I = I_g + h^2M. \tag{72}$$

This proves Eq. (69) for a system of particles.

A similar argument which replaces the sums by the multiple integrals discussed in Chap. 19 proves the results for any distribution of mass.

EXAMPLE 1. A uniform thin plate of total mass M is in the form of a rectangle of base B and altitude H. From the result of the example of Sec. 220, about the base $I = MH^2/3$. Deduce that about an axis through the center of the plate parallel to the base, $I_g = MH^2/12$.

Solution: Here we are given $I = MH^2/3$ and $h = H/2$. It follows from Eq. (69) that $MH^2/3 = I_g + M(H/2)^2$ and $I_g = M(H^2/3 - H^2/4) = MH^2/12$. Thus $I_g = MH^2/12$, as was to be proved.

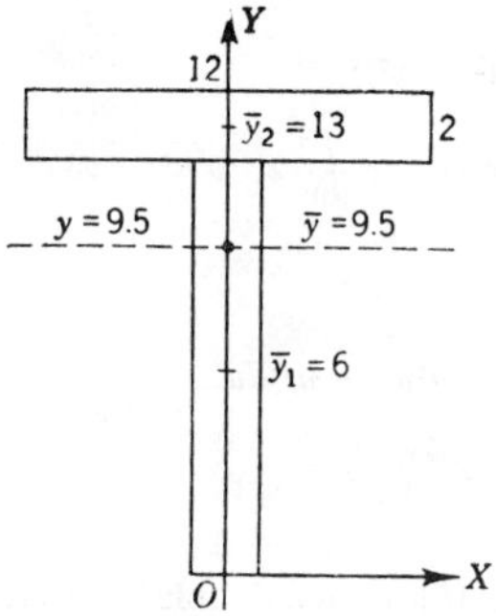

FIG. 261.

EXAMPLE 2. Find I_g, the moment of inertia of the T-shaped area of Fig. 261, about an axis parallel to the upper edge through the centroid of the area. The stem and head of the T are each rectangles 12 by 2 in.

Solution: Measure y up from the base. Then $A_1 = 2(12) = 24$ and $\bar{y}_1 = 6$. $A_2 = 2(12) = 24$ and $\bar{y}_2 = 13$. By Eq. (43) with masses replaced by areas, we have $A\bar{y} = A_1\bar{y}_1 + A_2\bar{y}_2$. Since $A = A_1 + A_2 = 48$, it follows that $48\bar{y} = 24 \times 6 + 24 \times 13$ and $\bar{y} = (24 \times 19)/48 = 9.5$. By Example 1 above, for A_1 about an axis parallel to OX through its own centroid with $y = 6$, $I_{1g} = 24(12^2)/12 = 288$. And about the line $y = 9.5$, $h = 3.5$ so that $I_1 = I_{1g} + A_1h^2 = 288 + 24(\frac{7}{2})^2 = 582$. For A_2 about an axis parallel to OX through its own centroid with $y = 13$, $I_{2g} = 24(2^2)/12 = 8$. And about the line $y = 9.5$, $h = 3.5$ so that $I_2 = I_{2g} + A_2h^2 = 8 + 24(\frac{7}{2})^2 = 302$. Since $I = I_1 + I_2 = 582 + 302 = 884$, the required $I = 884$ in.^{4}

EXERCISE 113

An area in the first quadrant is bounded by OX, OY, and the given curves. Find the moment of inertia about its axis of the homogeneous solid of revolution generated by revolving this area about OX.

1. $y = e^x$, $x = 1$.
2. $y^2 = x^3$, $x = 1$.
3. $y = \cos x$, $x = \pi/2$.
4. $y = \sec x$, $x = \pi/4$.
5. $y^2 = x - x^3$.
6. $y^4 = 4 - x^2$.

An area in the first quadrant is bounded by OX, OY, and the given curves. Find the moment of inertia about its axis of the homogeneous solid of revolution generated by revolving this area about OY.

7. $y = 4x - x^2$.
8. $y = x^2 - x^3$.
9. $y^2 = 1 - x$.
10. $4x^2 + y^2 = 4$.

Show that the moment of inertia about its axis of a homogeneous solid of altitude b in the shape of a

11. Solid cone of revolution whose base is of radius a is $I = \frac{1}{10}\pi\rho a^4 b = \frac{3}{10}Ma^2$.
12. Solid right circular cylinder of radius a is $I = \frac{1}{2}\pi\rho a^4 b = \frac{1}{2}Ma^2$.
13. Hollow right circular cylinder of inner radius a and outer radius c is $I = \frac{1}{2}\pi\rho b(c^4 - a^4) = \frac{1}{2}M(a^2 + c^2)$.

Use the parallel-axis theorem to find the moment of inertia of

14. A solid sphere of radius a about a line tangent to the sphere.
15. The cone of Prob. 11 about a line parallel to the axis through a point on the circumference of the base.
16. The cylinder of Prob. 12 about one of its elements.
17. The hollow cylinder of Prob. 13 about an element of the inner cylinder.

For each given plane area find the moment of inertia about the indicated axis in its plane.

18. The area inside a square of side $6a$ and outside a concentric and parallel square of side $3a$ about a side of the larger square.

19. The area of a carpenter's square of which each outer edge is 12 in. and each inner edge 10 in. about one of its outer edges.
20. The area inside a circle of radius $2a$ and outside a circle of radius a tangent internally to it about the common tangent.

***225. Center of Pressure.** The problem of calculating the total force F due to pressure of a liquid on an immersed vertical surface such as a gate in a dam was discussed in Sec. 75. We showed that $dF = wh\,dA = whL(h)dh$, where w is the density of the liquid, h is the depth below the surface of the area dA, and $L(h)$ is the width of wetted surface at depth h, so that $dA = L(h)dh$.

It follows from Eq. (42) of Sec. 75 and Eq. (101) of Sec. 185 that

$$dF = whL\,dh \qquad \text{and} \qquad F = w\int_a^b hL(h)dh. \tag{73}$$

Figure 262 represents a gate in a dam. Equation (73) gives the magnitude of the total force on the wetted side of the gate due to liquid pressure. The gate could be held in equilibrium by an equilibrant, or single force of this magnitude, applied to the

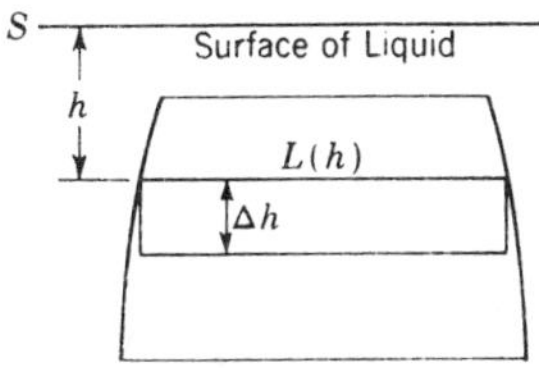

FIG. 262.

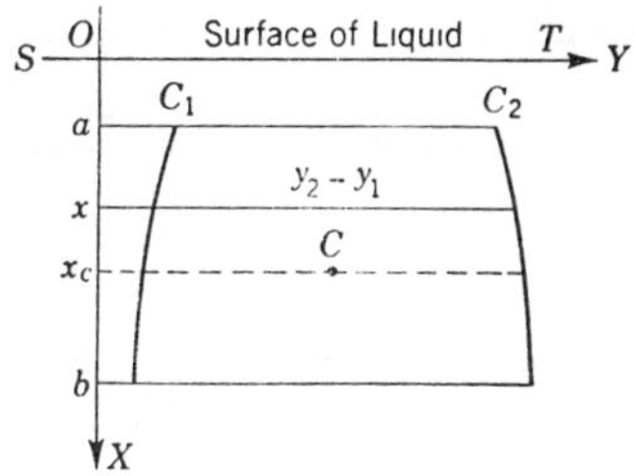

FIG. 263.

opposite side of the gate at a suitable point. This point is called the *center of pressure.* We wish to calculate its depth below the surface, h_c. Let ST be the intersection of the vertical plane through the gate and the surface of the liquid. Then the moment of the equilibrant at depth h_c about ST is Fh_c. This must equal the opposing moment of the distributed forces due to pressure. For the force $F_i = wh_i'L(h_i'')\Delta h$, the moment about ST is $h_i'''\,\Delta F_i$. Thus $wh_i'h_i'''L(h_i'')\Delta h$ is taken as an approximation to the total moment of the distributed forces due to pressure. The limit of this when $\Delta h \to 0$ is the total moment and therefore equals Fh_c. It follows from Sec. 185 that

$$Fh_c = w\int_a^b h^2L(h)dh. \tag{74}$$

By solving this relation for h_c and substituting the value of F from Eq. (73), we find for the depth of the center of pressure,

$$h_c = \frac{w}{F}\int_a^b h^2L(h)dh = \frac{\int_a^b h^2L(h)dh}{\int_a^b hL(h)dh}. \tag{75}$$

In particular, take the line ST as the y axis (Fig. 263) and take the x axis vertically downward. Suppose that the area of the gate is that between the ordinates $x = a$ and $x = b$ and between the two curves $y_1 = f_1(x)$ and $y_2 = f_2(x)$. Then $h = x$. And $L = y_2 - y_1$, so that

$$F = w \int_a^b x(y_2 - y_1)dx, \qquad Fx_c = w \int_a^b x^2(y_2 - y_1)dx. \tag{76}$$

But by Eqs. (28) and (60), we have

$$A\bar{x} = \int_a^b x(y_2 - y_1)dx, \qquad I_y = \int_a^b x^2(y_2 - y_1)dx. \tag{77}$$

A comparison of Eqs. (76) and (77) shows that

$$F = wA\bar{x}, \qquad Fx_c = wI_y, \qquad x_c = \frac{I_y}{A\bar{x}}. \tag{78}$$

Since the pressure at depth $\bar{x}$ is $w\bar{x}$, the first relation proves

Theorem I. *The total force due to pressure on a submerged vertical surface is equal to the product of the area and the pressure at the centroid of this area.*

The last relation of Eq. (78) proves

Theorem II. *The depth of the center of pressure below the line ST in the surface is equal to the moment of inertia of the area about ST divided by the first moment of the area about ST.*

We may use these theorems, instead of direct integration, whenever we know the center of gravity and the moment of inertia for the area about any horizontal axis.

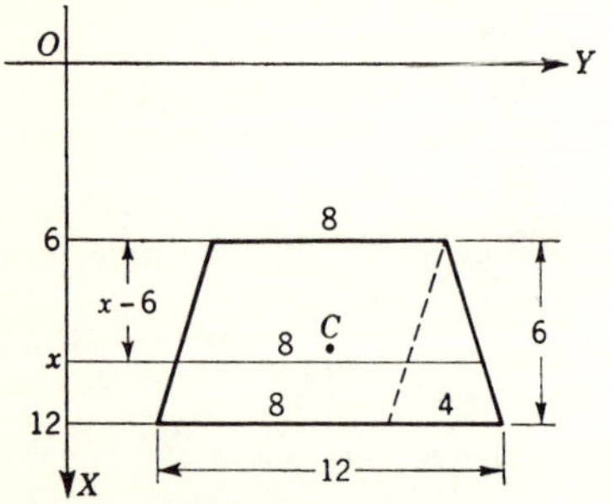

FIG. 264.

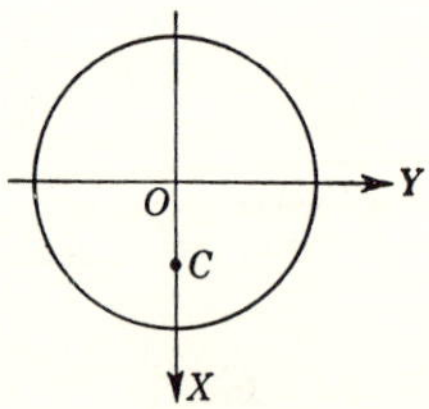

FIG. 265.

EXAMPLE 1. A vertical gate in a dam is an isosceles trapezoid with bases horizontal. The altitude is 6 ft., the lower base is 12 ft., and the upper base is 8 ft. Find the total pressure and the depth of the center of pressure if the water level is 6 ft. above the upper base.

Solution: From Fig. 264, we find that a point at depth x is $(x - 6)$ below the top of the dam and hence the width of the dam at that depth is $L = 8 + \frac{4}{6}(x - 6) = 4 + \frac{2}{3}x$. It follows from Eq. (73) that $F = w \int_6^{12} x(4 + \frac{2}{3}x)dx = w[2x^2 + \frac{2}{9}x^3]_6^{12} = w(288 + 384 - 72 - 48) = 552w$. And from Eq. (74), $Fx_c = w \int_6^{12} x^2(4 + \frac{2}{3}x)dx = w[\frac{4}{3}x^3 + \frac{1}{6}x^4]_6^{12} = w(12^2 4^2 + 2 \cdot 12^3 - 8 \cdot 6^2 - 6^3) = 72w(32 + 48 - 4 - 3) = 72 \cdot 73w$. Hence $x_c = \dfrac{Fx_c}{F} = \dfrac{72 \times 73w}{24 \times 23w} = \dfrac{219}{23} = 9.52$. And $F = 552w = \dfrac{24 \times 23}{32}$ tons $= 17.25$ tons. Thus $F = 17.25$ tons is the required total force, and $x_c = 9.52$ ft. is the required depth of the center of pressure.

EXAMPLE 2. A circular water main of radius 4 ft. is half full of water. Find the depth of the center of pressure in the gate that closes the main.

Solution: Take the y axis in the surface of the water (Fig. 265) and the x axis vertically downward. Then $A = \pi 4^2/2 = 8\pi$. By the example of Sec. 211, $\bar{x} = 4a/3\pi$

$= 16/3\pi$, so that $A\bar{x} = \frac{128}{3}$. And by the example of Sec. 221, $I_y = Aa^2/4 = 8\pi(4^2/4)$ $= 32\pi$. Hence by Eq. (78), $x_c = \frac{I_y}{A\bar{x}} = \frac{32\pi}{\frac{128}{3}} = \frac{3\pi}{4}$. And the required depth is $x_c = \frac{3\pi}{4} = 2.36$ ft.

226. Work Done by a Variable Force. The work done in moving a body by a constant force is the product of the force times the distance moved. If the force is variable, it may be expressed as some function of the distance x, $F(x)$. Divide the distance moved into parts Δx, and take $\Sigma F(x_i)\Delta x$ as an approximation to the work. Then by Sec. 170, as $\Delta x \to 0$, this sum approaches a limit $\int_a^b F(x)dx$, and that limit is defined to be the work done by the variable force in moving the body from $x = a$ to $x = b$. Thus we have

$$dW = F\,dx, \qquad W = \int_a^b F(x)dx. \tag{79}$$

Example. The force required to stretch a spring is proportional to the amount the spring has already been stretched. The force is 12 lb. when the spring has been stretched $\frac{1}{4}$ in. Find the work done in stretching the spring 2 in. from its natural, or unstretched, length.

Solution: The force F required to stretch the spring x is $F = kx$. $F = 12$ lb. when $x = \frac{1}{4}$ in. Hence $12 = k\frac{1}{4}$ and $k = 48$. Then $F = 48x$. And by Eq. (79), $W = \int F\,dx = \int_0^2 48x\,dx = [24x^2]_0^2 = 96$ in.-lb. or $\frac{96}{12} = 8$ ft.-lb. Thus $W = 8$ ft.-lb. is the required work.

227. Work Done in Emptying a Tank. Consider the total work done in emptying a tank by pumping the liquid out through the top. Let the distance x be measured down from a horizontal plane at the top of the tank. And let the cross section of the tank at distance x be $A(x)$, as in Sec. 205 (Fig. 266). Divide the volume occupied by the liquid into parts $\Delta V_i = A(x_i')\Delta x$, of weight $w\,\Delta V_i$. The work done in lifting is the weight multiplied by the vertical height. Hence the work of lifting $w\,\Delta V_i$ to the top of the tank, or through a height x_i'' is $wx_i''\,\Delta V_i$. And we take $\Sigma wx_i''\,A(x_i')\Delta x$ as an approximation to the total work done in emptying out the liquid. By Sec. 185, as $\Delta x \to 0$, this approaches a limit. And that limit is defined to be the total work done. Thus we have

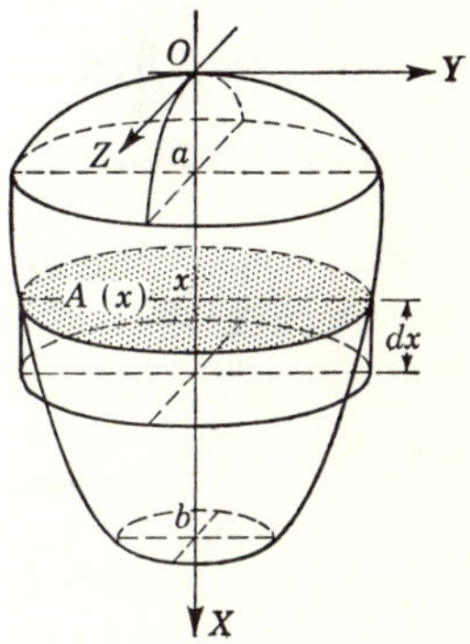

Fig. 266.

$$dW = \frac{wxA}{4}\,dx, \qquad W = w\int_a^b xA(x)dx. \tag{80}$$

As in Sec. 214 we may deduce from Eq. (37) that

$$\bar{x} = \frac{\int x\,dV}{\int dV} = \frac{1}{V}\int_a^b xA(x)dx, \qquad \int_a^b xA(x)dx = V\bar{x}. \tag{81}$$

A comparison of Eqs. (80) and (81) shows that

$$W = w(V\bar{x}) = (wV)\bar{x}. \tag{82}$$

Since wV is the weight of the liquid, this shows that the work of emptying the tank is equal to the total weight of liquid multiplied by the height through which its center of gravity is raised.

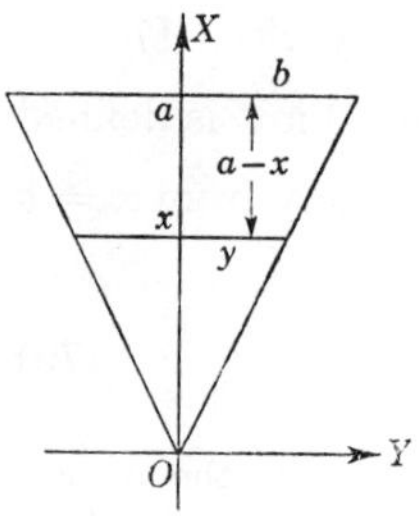

FIG. 267.

EXAMPLE. A cistern is in the form of an inverted cone of revolution of altitude a. The top of the cistern, or base of the cone, has radius b. Originally the cistern contains water up to a depth h. Find the work done in emptying the cistern by pumping this amount over the top.

Solution 1: In Fig. 267, x is the height above the vertex of the cone. Then the distance from the top is $(a - x)$. At this distance, $y/b = x/a$, $y = bx/a$. And the cross section is $A = \pi y^2 = \pi b^2x^2/a^2$. From Eq. (80) with $(a - x)$ in place of x, we have $W = w\int \pi y^2(a - x)dx = w\pi \int_0^h \frac{b^2x^2}{a^2}(a - x)dx = \frac{w\pi b^2}{a^2}\left[a\frac{x^3}{3} - \frac{x^4}{4}\right]_0^h = \frac{w\pi b^2}{a^2}\left(\frac{ah^3}{3} - \frac{h^4}{4}\right) = \frac{w\pi b^2h^3}{12a^2}(4a - 3h)$. Hence the required work is $W = \frac{w\pi b^2h^3}{12a^2}(4a - 3h)$.

Solution 2: From Fig. 267, at depth h, $y = bh/a$, and $A = \pi b^2h^2/a^2$. Hence the volume of the liquid is $\frac{1}{3}Ah = \pi b^2h^3/3a^2$. From the example of Sec. 214, the centroid of a cone is one quarter of the way up from the base, so that $\bar{x} = a - h + \frac{h}{4} = \frac{4a - 3h}{4}$. Hence from Eq. (82), $W = wV\bar{x} = w\frac{\pi b^2h^3}{3a^2}\frac{4a - 3h}{4} = \frac{w\pi b^2h}{12a^2}(4a - 3h)$. And the required work is $W = \frac{w\pi b^2h^3}{12a^2}(4a - 3h)$.

228. Work Done by an Expanding Gas. Consider the work done by a gas in a cylinder expanding against a piston head. The volume of gas is $v = Bx$, where B is the constant area of the base, or cross section, of the cylinder and x is the variable altitude. Hence $dv = B\,dx$. And the force causing the expansion is $F = pB$, where p is the pressure or force per unit area. By Eq. (79), the element of work is

$$dW = F\,dx = pB\,dx = p\,dv. \tag{83}$$

It follows that the work done by the gas in expanding from a volume v_1 to v_2 is

$$W = \int_{v_1}^{v_2} p\,dv. \tag{84}$$

If p is in pounds per square inch (psi) and v is in cubic inches, the integral gives the work as a number of inch-pounds. This may be converted to a number of foot-pounds by dividing by 12 in./ft.

To calculate W from Eq. (84) the relation between pressure and volume during the expansion must be known. In *isothermal expansion* the temperature remains constant and $pv = k$ is constant. In *adiabatic expansion*, when no heat is gained or lost, $pv^n = k$ is constant, where n is the ratio of specific heat at constant pressure to the specific heat at constant volume. For air, $n = 1.4$.

EXAMPLE 1. Air at pressure of 15 psi is compressed from 2,000 to 4 cu. in. Find the work required to compress the air isothermally so that pv is constant.

Solution: $pv = k$, $p = k/v$. And from Eq. (84), we have $W = \int p\,dx = \int_{2,000}^{4} \frac{k}{v}\,dv = k\,[\ln v]_{2,000}^{4} = k(\ln 4 - \ln 2,000) = -k \ln 500$. The minus sign indicates that work is done on the gas. Since $p_1 = 15$ when $v_1 = 2,000$, $k = p_1v_1 = 15 \cdot 2,000 = 30,000$. The work done on the gas is $-W = 30,000 \ln 500$ in.-lb. or $2,500 \ln 500$ ft.-lb.

Since $\ln 500 = \ln 5 + 2 \ln 10 = 1.6094 + 2(2.3026) = 6.2146$, $-W = 2,500(6.2146) = 15,536$ ft.-lb. Hence the work required to compress the air is 15,536 ft.-lb.

EXAMPLE 2. Find the work required to compress the air as in Example 1 adiabatically, so that $pv^{1.4}$ is constant.

Solution: $pv^{1.4} = k$, $p = kv^{-1.4}$. And from Eq. (84), we have

$$W = \int p\,dv = \int_{2,000}^{4} kv^{-1.4}dv = \left[\frac{kv^{-0.4}}{-0.4}\right]_{2,000}^{4} = -\frac{k}{0.4}(4^{-0.4} - 2,000^{-0.4}).$$

We may find the powers from a log log slide rule, or by using logarithms. For example, $\log 4^{-0.4} = -(0.4) \log 4 = -0.4(0.6021) = -0.2408 = \bar{1}.7592$ or $9.7592 - 10$. Hence $4^{-0.4} = 0.5744$. Similarly, $2,000^{-0.4} = 0.0478$. And $4^{-0.4} - 2,000^{-0.4} = 0.5266$. Since $p_1 = 15$ when $v_1 = 2,000$, $k = p_1v_1^{1.4} = 15(2,000)^{1.4}$. And the work done on the gas is $-W = \frac{15}{0.4}(2,000)^{1.4}(0.5266) = 825,900$ in.-lb. or 68,830 ft.-lb. Hence the work required to compress the air adiabatically is 68,830 ft.-lb.

***229. Attraction of Gravitation.** Let P_0 be a particle of mass m_0 at the origin (Fig. 268), and let P_1 be a particle of mass m_1 at a point whose polar coordinates are (r_1,θ_1). Then by Newton's law of gravitation, the particle P_0 is attracted by the particle P_1 with a force $\bar{F}_1$ along P_0P_1 whose intensity is

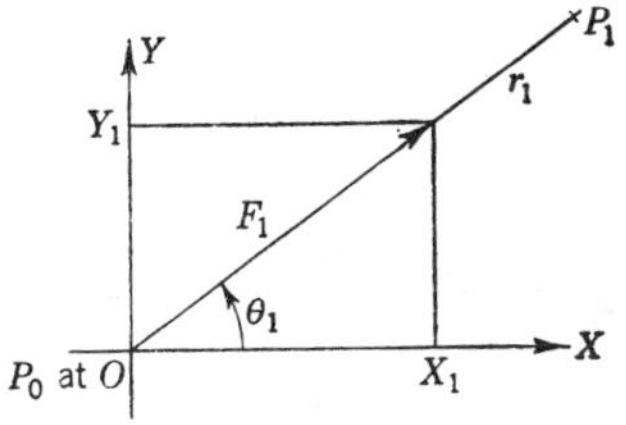

FIG. 268.

$$F_1 = K\frac{m_0m_1}{r_1^2}. \tag{85}$$

In c.g.s. units, the physical constant $K = 6.67 \times 10^{-8}$ dyne cm^{2}g^{-2}. And in m.k.s. units, $K = 6.67 \times 10^{-11}$ newton m^{2}kg^{-2}. Similarly by particles P_2 of mass m_2 at (r_2,θ_2) and P_3 of mass m_3 at (r_3,θ_3) the particle P_0 is attracted by forces $\bar{F}_2$ along P_0P_2 and $\bar{F}_3$ along P_0P_3 of respective intensities

$$F_2 = K\frac{m_0m_2}{r_2^2} \quad \text{and} \quad F_3 = K\frac{m_0m_3}{r_3^2}. \tag{86}$$

The total attraction of the three particles P_1, P_2, and P_3 on P_0 is the resultant of the three forces $\bar{F}_1$, $\bar{F}_2$, and $\bar{F}_3$. We may find this resultant by adding components. Since $\bar{F}_1$ acts along P_0P_1, or OP_1, its components are

$$X_1 = F_1 \cos \theta_1 = Km_0 \frac{m_1 \cos \theta_1}{r_1^2}, \qquad Y_1 = F_1 \sin \theta_1 = Km_0 \frac{m_1 \sin \theta_1}{r_2^2}. \tag{87}$$

The forces $\bar{F}_2$ and $\bar{F}_3$ have similar components. Hence the components X and Y of the resultant are

$$\begin{aligned} X &= Km_0 \left(\frac{m_1 \cos \theta_1}{r_1^2} + \frac{m_2 \cos \theta_2}{r_2^2} + \frac{m_3 \cos \theta_3}{r_3^2} \right), \\ Y &= Km_0 \left(\frac{m_1 \sin \theta_1}{r_1^2} + \frac{m_2 \sin \theta_2}{r_2^2} + \frac{m_3 \sin \theta_3}{r_3^2} \right). \end{aligned} \tag{88}$$

To find the attraction on P_0 of a continuous distribution of mass in the plane, we proceed as follows. Divide the mass into pieces Δm_i. Then if the attraction of each piece is approximately that of a particle of the same mass at (r_i', θ_i'), we take

$$Km_0 \sum \frac{\cos \theta_i'}{r_i'^2} \Delta m_i \qquad \text{and} \qquad Km_0 \sum \frac{\sin \theta_i'}{r_i'^2} \Delta m_i, \tag{89}$$

as approximations to the components of attraction. When the largest dimension of the pieces approaches zero, $\Delta m_i \to 0$. And by Sec. 185, the two sums approach limits. And these limits are defined to be the components of the resultant attraction. Thus

$$X = Km_0 \int \frac{\cos \theta}{r^2}\, dm, \qquad Y = Km_0 \int \frac{\sin \theta}{r^2}\, dm. \tag{90}$$

If the magnitude of the force of attraction is F and it makes an angle B with the x axis, we have

$$F = \sqrt{X^2 + Y^2} \qquad \tan B = \frac{Y}{X}. \tag{91}$$

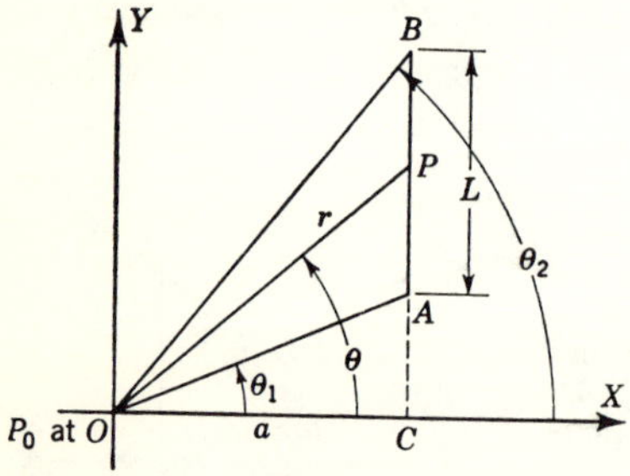

Fig. 269.

Example 1: Find the attraction exerted by a thin, straight uniform rod of length L and mass M upon a particle P_0 of mass M_0 at O if angle $COA = \theta_1$ and angle $COB = \theta_2$, where A and B are the end points of the rod, and $OC = a$ is perpendicular to BA produced (Fig. 269).

Solution: Take O as the origin and OC as the positive x axis. If $P = (r, \theta)$ is any point on the rod, $r = a \sec \theta$. And $CP = y = r \sin \theta = a \tan 6$. Since the rod has mass M for length L, it has $\frac{M}{L}$ as the mass per unit length, and $dm = \frac{M}{L}\, dy = \frac{M}{L} a \sec^2 \theta\, d\theta$. Hence by Eq. (90), $X = Km_0 \int \frac{\cos \theta}{r^2}\, dm$

$$= Km_0 \int_{\theta_1}^{\theta_2} \frac{\cos \theta}{a^2 \sec^2 \theta} \frac{M}{L} a \sec^2 \theta\, d\theta = \frac{Km_0 M}{aL} \int_{\theta_1}^{\theta_2} \cos \theta\, d\theta$$

$$= \frac{Km_0 M}{aL} [\sin \theta]_{\theta_1}^{\theta_2} = \frac{Km_0 M}{aL} (\sin \theta_2 - \sin \theta_1).$$

And $Y = Km_0 \int \frac{\sin\theta}{r^2}\,dm = Km_0 \int_{\theta_1}^{\theta_2} \frac{\sin\theta}{a^2\sec^2\theta}\frac{M}{L}\,a\sec^2\theta\,d\theta = \frac{Km_0M}{aL}\int_{\theta_1}^{\theta_2}\sin\theta\,d\theta =$ $\frac{Km_0M}{aL}[-\cos\theta]_{\theta_1}^{\theta_2} = \frac{Km_0M}{aL}(\cos\theta_1 - \cos\theta_2)$. Hence we have

$$X = \frac{Km_0M}{aL}(\sin\theta_2 - \sin\theta_1) = \frac{2Km_0M}{aL}\cos\frac{\theta_2+\theta_1}{2}\sin\frac{\theta_2-\theta_1}{2},$$

$$Y = \frac{Km_0M}{aL}(\cos\theta_1 - \cos\theta_2) = \frac{2Km_0M}{aL}\sin\frac{\theta_2+\theta_1}{2}\sin\frac{\theta_2-\theta_1}{2}.$$

It follows that $X^2 + Y^2 = \left(\frac{2Km_0M}{aL}\right)^2\sin^2\frac{\theta_2-\theta_1}{2}$ and $\frac{Y}{X} = \tan\frac{\theta_2+\theta_1}{2}$. Hence from Eq. (91) we have

$$F = \frac{2Km_0M}{aL}\sin\frac{\theta_2-\theta_1}{2}$$

and

$$B = \frac{\theta_2+\theta_1}{2}.$$

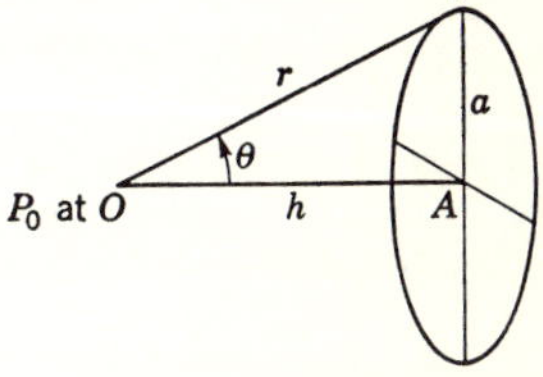

FIG. 270.

EXAMPLE 2. Find the attraction exerted by a thin, uniform, circular ring of radius a and mass M upon a particle P_0 of mass m_0 situated on the perpendicular to the plane of the ring at a distance h from the center, A.

Solution: If P is any point on the ring (Fig. 270), $r = OP = \sqrt{a^2 + h^2}$. If θ = angle AOP, $\cos\theta = \frac{h}{\sqrt{a^2+h^2}}$. By symmetry, the attraction will be along OA. And as in Eq. (90), the component along OA will be

$$x = Km_0\int\frac{\cos\theta}{r^2}\,dm = \frac{Km_0h}{(a^2+h^2)^{\frac{3}{2}}}\int dm = \frac{Km_0Mh}{(a^2+h^2)^{\frac{3}{2}}}.$$

Thus the required resultant attraction is $X = \frac{Km_0Mh}{(a^2+h^2)^{\frac{3}{2}}}$.

EXERCISE 114

For each of the following vertical wetted surfaces, find the total force due to pressure and the depth of the center of pressure.

1. A square 4 ft. on a side with the upper side parallel to and 10 ft. below the surface of the water.
2. A circle 2 ft. in radius with center 10 ft. below the surface of the water.
3. The upper half of the circle in Prob. 2.
4. The lower half of the circle in Prob. 2.
5. One end of a trough in the form of an inverted parabolic segment. The base of the segment, in the surface of the water, is 2 ft. long. And the altitude of the segment, or greatest depth, is 1 ft.
6. One end of a tank in the form of a right circular cylinder of radius 4 ft. with its axis horizontal. The tank is filled to a depth of 6 ft. with oil weighing 50 lb/ft.^{3}
7. The force required to stretch a spring is proportional to the elongation, and is 3 lb. when the extension is 1 in. Find the work done in stretching the spring from its natural length to a length 4 in. greater.
8. A negative electric charge at a fixed point O attracts a positively charged particle at P with a force inversely proportional to the square of the distance OP. If the force is 10^{-5} lb. when OP is 1 in., find the work done in moving the charged particle from a point 2 in. from O to a point 50 in. from O.

9. A wire bent into a circle of radius a with center at O carries a current of intensity I. The line OP is perpendicular to the plane of the circle. And the distance $OP = x$. Then a unit magnetic pole of polarity appropriately related to the direction of the current is repelled by a force in the direction OP whose magnitude is $F = \dfrac{2\pi I a^2}{(a^2 + x^2)^{\frac{3}{2}}}$. Find the work done in bringing the unit magnetic pole from a point P at infinity along the perpendicular to O.
10. A bucket weighing 100 lb. is to be hoisted from the bottom to the top of a shaft 50 ft. deep. If the cable weighs 0.2 lb./ft., find the work done.
11. A bag containing originally 100 lb. of sand is hoisted up a distance of 12 ft. The sand leaks out in such a way that the loss is proportional to the square of the height lifted and is 25 lb. at the maximum height of 12 ft. Find the work done in lifting the bag.

Find the work done in pumping all the water out over the top from a full tank in each of the following given forms.

12. A vertical cylinder of height 10 ft. and radius 2 ft.
13. An inverted cone of height 4 ft. and radius 3 ft.
14. A sphere of radius 3 ft.
15. The lower half of a sphere of radius 2 ft.

Find the work required to compress 200 cu. in. of air to 50 cu. in. if the original pressure was 15 psi.

16. Isothermally, $pv = k$.
17. Adiabatically, $pv^{1.4} = k$.

A gas expands from an initial pressure of 100 psi and volume 50 to 250 cu. in. Find the work done for each given relation of p to v.

18. $pv = k$.
19. $pv^{1.06} = k$.
20. $pv^{1.13} = k$.

21. Find the work done by 16 cu. in. of gas in expanding to a volume of 30 cu. in. if $pv^{1.2} = k$ and the initial pressure was 50 psi.

Determine the amount of attraction exerted on a particle P_0 of mass m_0 at the origin by a thin, uniform rod of mass M which overlays each given line.

22. The interval of $y = 0$ with $a < x < a + L$.
23. The interval of $x = a$ with $-L/2 < y < L/2$.
24. The semicircle $r = a$, $-\pi/2 < \theta < \pi/2$.
25. The semicircle $r = 2a \cos \theta$, $-\pi/4 < \theta < \pi/4$.

26. A circular disk has mass M and radius a. A particle P_0 of mass m_0 is situated on the perpendicular to the plane of the disk at a distance h from the center. By dividing the disk into rings and using Example 2 of Sec. 229, show that the attraction exerted by the disk on P_0 is
$$F = Km_0h \int_b^a \frac{M}{\pi a^2} \frac{2\pi r\, dr}{(r^2 + h^2)^{\frac{3}{2}}} = \frac{2Km_0M}{a^2}\left(1 - \frac{h}{\sqrt{h^2 + a^2}}\right).$$
27. A uniform right circular cylinder has mass M, radius a, and altitude b. A particle P_0 of mass m_0 is situated on the axis of the cylinder produced to a distance c from one face. By dividing the cylinder into disks and using Prob. 26, show that the attraction exerted by the disk on P_0 is
$$F = \frac{2Km_0}{a^2} \int_c^{b+c} \frac{M}{b}\left(1 - \frac{x}{\sqrt{x^2 + a^2}}\right) dx$$
$$= \frac{2Km_0M}{a^2b}\,[b + \sqrt{c^2 + a^2} - \sqrt{(b + c)^2 + a^2}].$$

CHAPTER 15

INFINITE SERIES

In this chapter we study infinite series, defined as the indicated sum of an infinite number of terms. For certain series a value analogous to a sum may be assigned by a limiting process. Such series are said to be convergent. To be useful in elementary calculations, a series must be convergent. Hence it becomes necessary to have methods of testing whether this property holds. We describe a number of useful tests, such as the integral test, the comparison tests, and the ratio test, which apply to series all of whose terms are positive. We describe a test for series whose terms alternate in sign. And we discuss the notion of absolute convergence, and the deduction of convergence of a series from the corresponding series of absolute values.

As types of series with variable terms, we discuss power series in x, $(x - a)$, and $1/x$. Finally we derive the binomial theorem, or power series for $(1 + x)^m$. General methods of finding power series expansions for given functions will be discussed in the next chapter.

230. Sequences. A *sequence* is a succession of numbers or *terms*. If the number of terms is limited the sequence is *finite*. For example

$$1,3,5,7,9 \tag{1}$$

and

$$\frac{1}{2}, \frac{1}{4}, \frac{1}{8}, \frac{1}{16}, \frac{1}{32} \tag{2}$$

are finite sequences.

If the number of terms is unlimited the sequence is *infinite*. Each term may be determined from the preceding by a definite rule. Or we may be given a *general term*, or nth term u_n as a function of n. For example, if the first term is 1 and each term is 2 greater than the preceding term, we have the *infinite sequence* with $u_n = 2n - 1$,

$$1,3,5, \cdots, 1 + 2(n - 1), \cdots. \tag{3}$$

And if the first term is $\frac{1}{2}$ and each term is $\frac{1}{2}$ of the preceding term, we have the *infinite sequence* with $u_n = 1/2^n$.

$$\frac{1}{2}, \frac{1}{4}, \frac{1}{8}, \cdots, \frac{1}{2^n}, \cdots. \tag{4}$$

In expressing the general term for some sequences it is convenient to use factorials. These are products of successive integers, beginning with 1. For any positive integer n, we define $n!$, read "factorial n," as

$$n! = 1 \cdot 2 \cdot 3 \cdots n. \tag{5}$$

In particular, $3! = 1 \cdot 2 \cdot 3 = 6$, $1! = 1$. And to make $n!/n = (n - 1)!$ hold for all positive integers n, including $n = 1$, we define

$$0! = 1. \tag{6}$$

EXAMPLE. In the sequence $1, \frac{3}{2}, \frac{3^2}{2 \cdot 4}, \cdots$, each term is obtained by multiplying the preceding term by 3, and dividing by twice the number of the preceding term. Find the general term.

Solution: The nth term has $(n - 1)$ factors 3, and a product of even integers in the denominator starting with 2 and ending with $2(n - 1)$. Hence the general term u_n is

$$u_n = \frac{3^{n-1}}{2 \cdot 4 \cdot 6 \cdots 2(n-1)} = \frac{3^{n-1}}{2^{n-1} 1 \cdot 2 \cdot 3 \cdots (n-1)}$$
$$= \frac{3^{n-1}}{2^{n-1}(n-1)!}.$$

Note that by the convention of Eq. (6) and the fact that $a^0 = 1$, this correctly gives $u_1 = 1$.

231. Series. A *series* is the indicated sum of the terms of a sequence. If the number of terms is limited, the series is *finite*. For example, the finite series corresponding to the sequence of Eq. (1) is

$$1 + 3 + 5 + 7 + 9 = 25. \tag{7}$$

And the finite series corresponding to the sequence of Eq. (2) is

$$\frac{1}{2} + \frac{1}{4} + \frac{1}{8} + \frac{1}{16} + \frac{1}{32} = 1 - \frac{1}{32} = \frac{31}{32}. \tag{8}$$

If the number of terms is unlimited, the series is *infinite*. For example, corresponding to the sequences of Eqs. (3) and (4) we have the infinite series

$$1 + 3 + 5 + \cdots + (2n - 1) + \cdots, \tag{9}$$

and

$$\frac{1}{2} + \frac{1}{4} + \frac{1}{8} + \cdots + \frac{1}{2^n} + \cdots. \tag{10}$$

And corresponding to the sequence with general term u_n, we may form the infinite series

$$u_1 + u_2 + u_3 + \cdots + u_n + \cdots. \tag{11}$$

This may be written in the brief form

$$\sum_{n=1}^{\infty} u_n, \tag{12}$$

read "sigma from n equals 1 to infinity of u_n." And the expression is sometimes abbreviated to Σu_n, read "sigma of u_n."

232. Partial Sums. To assign a value to the infinite series of Eq. (11), we begin by forming successively the sums

$$\begin{aligned} S_1 &= u_1 \\ S_2 &= u_1 + u_2 \\ S_3 &= u_1 + u_2 + u_3 \\ &\cdots\cdots\cdots\cdots \\ S_n &= u_1 + u_2 + u_3 + \cdots + u_n. \end{aligned} \tag{13}$$

These are called the *partial sums* of the infinite series. The nth partial sum, S_n, may be written in the brief form

$$S_n = \sum_{k=1}^{n} u_k. \tag{14}$$

We note that the nth partial sum S_n is a function of n. And the values $S_1, S_2, S_3, \cdots, S_n, \cdots$ themselves constitute an infinite sequence.

Suppose that $u_n = a + (n - 1)d$, the general term of an arithmetic progression Then the infinite series is

$$a + (a + d) + (a + 2d) + \cdots + [a + (n - 1)d] + \cdots. \tag{15}$$

And, from the sum of an arithmetic progression as found in algebra,

$$s_n = \frac{n}{2}\{a + [a + (n - 1)d]\} = \frac{n}{2}[2a + (n - 1)d]. \tag{16}$$

In particular, the sequence of Eq. (9) may be obtained from Eq. (15) by putting $a = 1$, $d = 2$, so that for that sequence

$$S_n = \frac{n}{2}[2 + (n - 1)2] = n^2. \tag{17}$$

Next suppose that $u_n = ar^{n-1}$, the general term of a geometric progression. Then the infinite series is

$$a + ar + ar^2 + \cdots + ar^{n-1} + \cdots. \tag{18}$$

And, from the sum of a geometric progression as found in algebra,

$$S_n = a\frac{r^n - 1}{r - 1} \qquad \text{if } r \neq 1. \tag{19}$$

In particular, the sequence of Eq. (4) may be obtained from Eq. (19) by putting $a = \frac{1}{2}$, $r = \frac{1}{2}$, so that for that sequence

$$S_n = \frac{1}{2}\frac{(\frac{1}{2})^n - 1}{\frac{1}{2} - 1} = 1 - \frac{1}{2^n}. \tag{20}$$

EXAMPLE. Find an expression for the partial sum S_n of the series whose general term is $\frac{3}{n(n+1)}$.

Solution: Observe that $\frac{1}{n} - \frac{1}{n+1} = \frac{1}{n(n+1)}$. Hence the partial sum

$$\frac{3}{1\cdot 2} + \frac{3}{2\cdot 3} + \frac{3}{3\cdot 4} + \cdots + \frac{3}{n(n+1)} =$$
$$3\left(1 - \frac{1}{2} + \frac{1}{2} - \frac{1}{3} + \frac{1}{3} - \frac{1}{4} + \cdots + \frac{1}{n} - \frac{1}{n+1}\right) = 3\left(1 - \frac{1}{n+1}\right).$$

And the required partial sum $S_n = \frac{3n}{n+1}$.

EXERCISE 115

For each given form of u_n, write the first three terms of the series u_n and evaluate the third partial sum $S_3 = u_1 + u_2 + u_3$.

1. $u_n = 3^n$.
2. $u_n = \frac{1}{n}$.
3. $u_n = (-2)^n$.
4. $u_n = \frac{(-1)^{n+1}}{n}$.
5. $u_n = \frac{n-2}{n+2}$.
6. $u_n = \frac{1}{n^2}$.
7. $u_n = \frac{(-1)^n}{n!}$.
8. $u_n = \frac{2^n}{n!}$.

For each given value of a and d write out the first three terms of the arithmetic series with $u_n = a + (n-1)d$. Also find the partial sums S_3 and S_n.

9. $a = 2, d = 3$.
10. $a = 4, d = 1$.
11. $a = 10, d = -2$.
12. $a = 6, d = -3$.

For each given value of a and r write out the first three terms of the geometric series with $u_n = ar^{n-1}$. Also find the partial sums S_3 and S_n.

13. $a = 1, r = 2$.
14. $a = 1, r = -\frac{1}{3}$.
15. $a = 5, r = -1$.
16. $a = 9, r = \frac{1}{3}$.
17. Deduce from Eq. (13) or (14) that $S_n - S_{n-1} = u_n$. This shows that the S_n may be obtained successively from S_1 by the rule $S_n = S_{n-1} + u_n$.

For the given value of u_n, verify that the expression for S_n is correct for $n = 1$ and $n = 2$. Also show that for any n, $S_n - S_{n-1} = u_n$ in accord with the result of Prob. 17.

18. $u_n = 3^{-n}$, $S_n = \frac{1}{2}\left(1 - \frac{1}{3^n}\right)$.
19. $u_n = e^{-n}$, $S_n = \frac{1 - e^{-n}}{e - 1}$.
20. $u_n = n^{-\frac{1}{2}} - (n+1)^{-\frac{1}{2}}$, $S_n = 1 - (n+1)^{-\frac{1}{2}}$.
21. $u_n = \frac{1}{n^2} - \frac{1}{(n+1)^2}$, $S_n = 1 - \frac{1}{(n+1)^2}$.

233. Convergence. Consider the infinite series Σu_n of Eq. (11). From it we may form the partial sums $S_1, S_2, S_3, \cdots, S_n, \cdots$ as in Eq. (13). The nth partial sum

$$S_n = u_1 + u_2 + u_3 + \cdots + u_n \tag{21}$$

is a function of n. Now let n, the number of terms used in the sum, increase indefinitely. Then if S_n approaches a finite limit S in accord with the definition of Sec. 7, we have

$$\lim_{n\to\infty} S_n = S. \tag{22}$$

Under these conditions we say that the infinite series Σu_n is *convergent.* We also say that Σu_n converges to the value or sum S. And we call S the sum, or value, of the series.

As a first illustration, recall the infinite series of Eq. (10) with $u_n = 1/2^n$. Then by Eq. (20) we have $S_n = 1 - 1/2^n$. And by Sec. 13 and Probs. 18 and 22 of Exercise 5 we have

$$\lim_{n\to\infty}\left(1 - \frac{1}{2^n}\right) = 1 - 0 = 1. \tag{23}$$

It follows that the series of Eq. (10) converges to the sum 1.

Again, by the example of Sec. 232, for the infinite series with $u_n = \dfrac{3}{n(n+1)}$, we have $S_n = \dfrac{3n}{n+1}\cdot$ By the principle of the leading term of Sec. 13, we have

$$\lim_{n\to\infty}\frac{3n}{n+1} = \lim_{n\to\infty}\frac{3n}{n} = 3. \tag{24}$$

It follows that the series $\sum \dfrac{3}{n(n+1)}$ converges to the sum 3.

Example. Show that the geometric series with $u_n = ar^{n-1}$ is convergent if $|r| < 1$.

Solution: By Eq. (19) $S_n = a\dfrac{r^n - 1}{r - 1}\cdot$ And by Prob. 22 of Exercise 5, for $|r| < 1$, as $n \to \infty$, $r^n \to 0$. Hence $S_n \to \dfrac{a}{1-r}\cdot$ It follows that the series Σar^{n-1} converges to the sum $\dfrac{a}{1-r}$ if $|r| < 1$.

234. Divergence. As in Sec. 233, let the S_n of Eq. (21) be the nth partial sum of the infinite series Σu_n. And let n, the number of terms of the sum, increase indefinitely. Then if S_n does not approach a finite limit, we say that the infinite series Σu_n is *divergent.* We do not assign any value to a divergent series.

As a first illustration, recall the infinite series of Eq. (9) with $u_n = 2n - 1$. Then by Eq. (17) we have $S_n = n^2$. And in the notation of Sec. 12,

$$\lim_{n\to\infty} n^2 = +\infty. \tag{25}$$

Since this is not a *finite* limit, the series of Eq. (9) diverges.

As a second illustration, consider the infinite series

$$1 - 1 + 1 - \cdots + (-1)^{n+1} + \cdots. \tag{26}$$

Here the partial sums are $S_1 = 1$, $S_2 = 0$, $S_3 = 1$, and so on, the odd sums being 1 and the even sums being 0. Since this sequence does not approach a limit in the sense of Sec. 7, the infinite series of Eq. (26) with $u_n = (-1)^n$ diverges.

It follows from Eq. (13) that

$$\begin{aligned} S_n - S_{n-1} &= (u_1 + u_2 + u_3 + \cdots + u_{n-1} + u_n) \\ &\quad - (u_1 + u_2 + u_3 + \cdots + u_{n-1}) \\ &= u_n. \end{aligned} \tag{27}$$

For a convergent series, by Eq. (22), we have

$$\lim_{n\to\infty} S_n = S \qquad \text{and hence} \qquad \lim_{n\to\infty} S_{n-1} = S. \tag{28}$$

Consequently, we may deduce from Eqs. (27) and (28) that

$$\lim_{n\to\infty} u_n = \lim_{n\to\infty} (S_n - S_{n-1}) = S - S = 0. \tag{29}$$

This shows that for a series to be convergent, its general term u_n must approach zero when n becomes infinite. Hence *if as n becomes infinite, u_n either approaches a nonzero limit or fails to approach a limit, the series must diverge.*

This often leads to a short proof of divergence. Thus, in the series with $u_n = 2n - 1$ of Eq. (9), $u_n \to \infty$ as $n \to \infty$, and the series diverges. This confirms our conclusion from Eq. (25).

Again, for the series of Eq. (26), $u_n = (-1)^n$ approaches no limit as $n \to \infty$, so that this series diverges. This confirms the conclusion already drawn from the sums.

Example 1. Prove that, if $a^2 + d^2 > 0$, the arithmetic series with $u_n = a + (n - 1)d$ diverges.

Solution 1: The partial sum $S_n = \frac{n}{2}[2a + (n - 1)d]$ by Eq. (16). As $n \to \infty$, this $\to \infty$ unless a and d are each zero. Hence when $a^2 + d^2 > 0$, the series diverges as was to be proved.

Solution 2: As $n \to \infty$, the individual term $u_n = a + (n - 1)d \to \infty$ if $d \neq 0$, and $\to a$ if $d = 0$. Hence if $a^2 + d^2 > 0$, u_n does not approach zero as $n \to \infty$, and the series diverges as was to be proved.

Example 2. Show that the geometric series with $u_n = ar^{n-1}$ is divergent if $|r| \geqq 1$, $a \neq 0$.

Solution 1: If $|r| > 1$, by Eq. (19), $S_n = a\dfrac{r^n - 1}{r - 1}$. And by Prob. 18 of Exercise 5, as $n \to \infty$, $r^n \to \infty$. Hence $S_n \to \infty$, and so does not approach a finite limit. If $r = 1$, $S_n = na$ and $S_n \to \infty$. If $r = -1$, $S_n = a\dfrac{(-1)^n - 1}{(-1) - 1} = \dfrac{a}{2}[1 - (-1)^n]$ which is alternately a and $-a$ and so approaches no limit when $a \neq 0$. Thus for $|r| \geqq 1$, $a \neq 0$, the series diverges as was to be proved.

Solution 2: For $a \neq 0$ and $|r| > 1$, $u_n = ar^{n-1} \to \infty$ as $n \to \infty$. For $r = 1$, $u_n = a \to a$ as $n \to \infty$. And for $r = -1$, $u_n = a(-1)^{n-1}$ is alternately a and $-a$ and so approaches no limit when $a \neq 0$. Hence u_n does not approach zero as $n \to \infty$, and the series diverges for $|r| \geqq 1$, $a \neq 0$, as was to be proved.

EXERCISE 116

For the given value of u_n, verify that the expression for S_n is correct for $n = 3$. Also prove that $\lim S_n = S$, so that the series Σu_n converges to the value S.

1. $u_n = 3^{-n}$, $S_n = \frac{1}{2}\left(1 - \frac{1}{3^n}\right)$, $S = \frac{1}{2}$.
2. $u_n = e^{-n}$, $S_n = \frac{1 - e^{-n}}{e - 1}$, $S = \frac{1}{e - 1}$.
3. $u_n = n^{-\frac{1}{2}} - (n + 1)^{-\frac{1}{2}}$, $S_n = 1 - (n + 1)^{-\frac{1}{2}}$, $S = 1$.
4. $u_n = \frac{1}{n^2} - \frac{1}{(n + 1)^2}$, $S_n = 1 - \frac{1}{(n + 1)^2}$, $S = 1$.
5. $u_n = \frac{2}{n(n + 1)}$, $S_n = \frac{2n}{n + 1}$, $S = 2$.

For the given value of u_n, show that Σu_n diverges by verifying that, as $n \to \infty$, u_n does not approach zero.

6. $u_n = n^3$.
7. $u_n = \frac{2^n}{3}$.
8. $u_n = 2 + 3^{-n}$.
9. $u_n = \frac{(-1)^n}{100}$.
10. $u_n = \frac{2^n}{n^2}$.
11. $u_n = 3 + 5(-1)^n$.
12. $u_n = n$.
13. $u_n = n(-1)^n$.
14. $u_n = \frac{2n}{n + 1}$.
15. $u_n = \frac{n^2 + 1}{2n^2}$.
16. $u_n = \frac{1}{\sqrt[n]{2}}$.
17. $u_n = \sqrt[n]{3}$.
18. $u_n = \frac{n}{2n + 1}$.
19. $u_n = \frac{3n + 1}{n}$.

235. Positive Series. We call an infinite series *positive* if each of its terms is a positive number. For such a series the partial sums S_n increase as n increases. If for some fixed number A no sum S_n ever exceeds A, we say that the sums are *bounded* and admit A as an upper bound. If every fixed number is exceeded by some S_n, we say that the sums are *unbounded*.

Let the partial sums S_n admit A as an upper bound as in Fig. 271. Then when n becomes infinite, S_n will approach a limit S not greater than A.

$$\lim_{n \to \infty} S_n = S, \qquad S \leqq A. \tag{30}$$

And the positive series will converge to the value S.

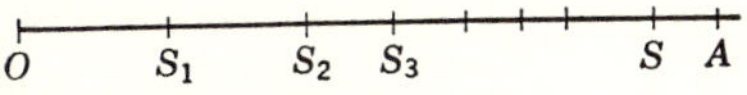

FIG. 271.

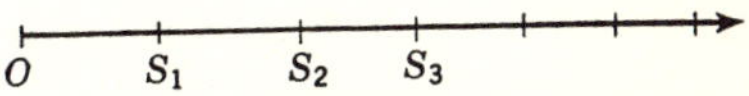

FIG. 272.

Let the partial sums S_n be unbounded as in Fig. 272. Then when n becomes infinite, S_n will become positively infinite,

$$\lim_{n \to \infty} S_n = +\infty. \tag{31}$$

And the positive series will diverge.

The simplest tests for the convergence of a positive series are indirectly based on the bounded character of the sums S_n. But they are directly based on a comparison with a convergent integral as in Sec. 236, or with another positive series known to be convergent as in Sec. 237. The form of the comparison test of Sec. 238 is often the most convenient.

We note that *a convergent series will remain convergent if we add or remove a finite number of terms either at the beginning or distributed throughout the series.* For the effect of this is a change in all S_n after a certain point by a finite constant. This can change the limit by this constant but cannot change the fact of approach to a finite limit.

Hence the conditions of the tests for convergence need hold only for values of n greater than some fixed integer.

Similar remarks apply to the corresponding tests for divergence.

***236. The Integral Test.** First consider the particular infinite series whose nth term $u_n = 1/n^p$, where p is any positive constant. The series is

$$\sum_{n=1}^{\infty} \frac{1}{n^p} = 1 + \frac{1}{2^p} + \frac{1}{3^p} + \cdots + \frac{1}{n^p} + \cdots. \tag{32}$$

In the expression for the nth term, $1/n^p$, we replace n by the continuous variable x. This leads to the function $1/x^p$ which for all values of x greater than zero is positive and decreases as x increases (Fig. 273). It follows that

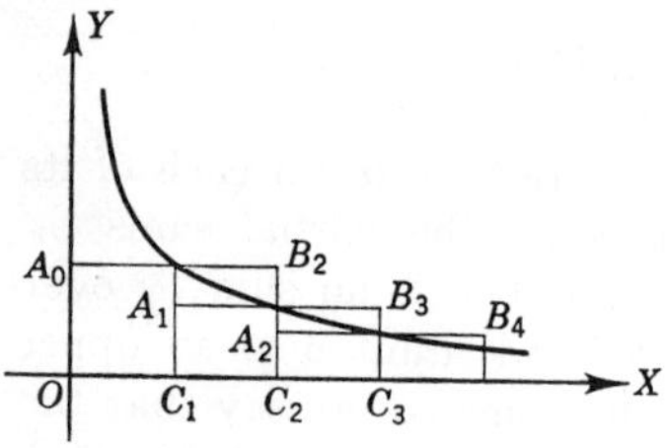

FIG. 273.

$$\begin{aligned}
\frac{1}{2^p} &< \int_1^2 \frac{1}{x^p}\,dx < 1,\\
\frac{1}{3^p} &< \int_2^3 \frac{1}{x^p}\,dx < \frac{1}{2^p},\\
\frac{1}{4^p} &< \int_3^4 \frac{1}{x^p}\,dx < \frac{1}{3^p},\\
&\cdots\cdots\cdots\cdots\\
\frac{1}{n^p} &< \int_{n-1}^{n} \frac{1}{x^p}\,dx < \frac{1}{(n-1)^p},\\
\frac{1}{(n+1)^p} &< \int_{n}^{n+1} \frac{1}{x^p}\,dx < \frac{1}{n^p}.
\end{aligned} \tag{33}$$

By adding all but the last left-hand inequalities, we may deduce that

$$S_n - 1 < \int_1^n \frac{1}{x^p}\,dx, \tag{34}$$

where S_n is the nth partial sum of the series of Eq. (32). Thus

$$S_n = 1 + \frac{1}{2^p} + \frac{1}{3^p} + \cdots + \frac{1}{n^p}. \tag{35}$$

And by adding all the right-hand inequalities of Eq. (33), we may deduce that

$$\int_1^{n+1} \frac{1}{x^p}\,dx < S_n. \tag{36}$$

The reader may check the inequalities (34) and (36) for $n = 3$ by noting in Fig. 273 that S_3 is either the sum of the rectangles $1 = A_0C_1$, A_1C_2, A_2C_3 or C_1B_2, C_2B_3, C_3B_4.

Suppose that $p > 1$. Then the integral $\int_1^n \frac{1}{x^p}\,dx$ increases with n, so that the integral of Eq. (34) is less than

$$\int_1^\infty \frac{1}{x^p}\,dx = \left[\frac{x^{-p+1}}{-p+1}\right]_1^\infty = \frac{0-1}{-p+1} = \frac{1}{p-1}, \qquad \text{if } p > 1. \tag{37}$$

It follows from the inequality of Eq. (34) that the sums S_n are bounded, and hence by Sec. 235 that the series of Eq. (32) converges if $p > 1$.

Next suppose that $p \leqq 1$. Then the integral $\int_1^{n+1} \frac{1}{x^p}\,dx$ increases indefinitely as n increases, since

$$\int_1^\infty \frac{1}{x^p}\,dx = \left[\frac{x^{-p+1}}{-p+1}\right]_1^\infty = +\infty, \qquad \text{if } p < 1, \tag{38}$$

$$\int_1^\infty \frac{1}{x}\,dx = [\ln x]_1^\infty = +\infty, \qquad \text{if } p = 1. \tag{39}$$

It follows from the inequality of Eq. (36) that the sums S_n are unbounded, and hence by Sec. 235 that the series of Eq. (32) diverges if $p \leqq 1$.

Incidentally, we now know the character of the "p series,"

$$1 + \frac{1}{2^p} + \frac{1}{3^p} + \cdots + \frac{1}{n^p} + \cdots, \tag{40}$$

for all values of p. In fact,

The p series is convergent when p is greater than unity, $p > 1$.

The p series is divergent when p is equal to or less than unity, $p \leqq 1$.

In our discussion, p was restricted to be positive, but the last statement holds if p is zero or negative, since for $p = 0$ each term is unity, and for p negative the terms increase with n. Hence in either case the nth term does not approach zero, and the series diverges by Sec. 234.

By reasoning similar to that used for the special series, we may prove the following theorem:

Let the function $f(x)$ be positive and always decrease as x increases for $x > m$, where m is some fixed positive integer. Then the series $\Sigma f(n)$, with $u_n = f(n)$, converges if the integral $\int_m^\infty f(x)dx$ *converges. And the series $\Sigma f(n)$ diverges if the integral* $\int_m^\infty f(x)dx$ *diverges.*

This is known as the *integral test.* Its principal application is to the special case of the p series.

EXAMPLE. Establish the divergence of the series

$$\frac{1}{2\ln 2} + \frac{1}{3\ln 3} + \frac{1}{4\ln 4} + \cdots + \frac{1}{(n+1)\ln(n+1)} + \cdots.$$

Solution: We may consider the series $\sum_{n=2}^{\infty} u_n$ with $u_n = \frac{1}{n\ln n}$. Then since $f(x) = \frac{1}{x\ln x}$ is positive and decreases as x increases for $x > 2$, the integral test shows that the series behaves like $\int_2^\infty \frac{1}{x\ln x}\,dx = \int_2^\infty \frac{d(\ln x)}{\ln x} = [\ln(\ln x)]_2^\infty = +\infty$. Since this integral diverges, the series diverges as was to be proved.

***237. Comparison Tests.** We may sometimes infer the convergence of a positive series from a term-by-term comparison with another positive series known to be convergent. Let

$$u_1 + u_2 + u_3 + \cdots + u_n + \cdots \tag{41}$$

be a series of positive terms to be tested for convergence. And let

$$c_1 + c_2 + c_3 + \cdots + c_n + \cdots \tag{42}$$

be a positive series which is known to be convergent to the value S'.

Then *if for every value of n each term u_n is at most as great as the corresponding term c_n, the series* (41) *is convergent*, and its value S does not exceed S'.

More generally, *if for some positive constant k and every value of n,*

$$u_n \leqq kc_n, \tag{43}$$

the series (41) *is convergent*, and its value S does not exceed kS'.

The proof of this is as follows. Denote the partial sum of the series (41) by S_n, so that

$$S_n = u_1 + u_2 + \cdots + u_n. \tag{44}$$

And denote the partial sum of the series (43) by S_n', so that

$$S_n' = c_1 + c_2 + c_3 + \cdots + c_n. \tag{45}$$

Since S' is the sum of the convergent positive series (42), $S_n' < S'$. And for k positive, $kS_n' < kS'$, so that

$$kc_1 + kc_2 + \cdots + kc_n = kS_n' < kS'. \tag{46}$$

Apply the condition of Eq. (43) with $n = 1, n = 2, n = 3, \cdots, n = n$. From this and Eqs. (44) and (46), it follows that

$$S_n \leqq kS_n' < kS'. \tag{47}$$

Hence the sums S_n are bounded. And the discussion of Sec. 235 shows that the series (41) converges. Hence as n becomes infinite, S_n approaches a limit S. And we may deduce from Eq. (47) that

$$\lim_{n\to\infty} S_n \leqq kS' \qquad \text{or} \qquad S \leqq kS'. \tag{48}$$

The first statement for the condition $u_n \leqq c_n$ is merely the important special case with $k = 1$.

There are corresponding tests for divergence. Let

$$d_1 + d_2 + d_3 + \cdots + d_n + \cdots \tag{49}$$

be a positive series which is known to be divergent.

Then, *if for every value of n, each term u_n is at least as great as the corresponding term d_n, the series* (49) *is divergent.*

More generally, *if for some positive constant k and every value of n,*

$$u_n \geqq kd_n, \tag{50}$$

the series (41) *is divergent.*

The proof of this is as follows. Denote the partial sum of the series (41) by S_n, so that Eq. (44) holds. And denote the partial sum of the series (49) by S_n'' so that

$$S_n'' = d_1 + d_2 + d_3 + \cdots + d_n. \tag{51}$$

Then

$$kS_n'' = kd_1 + kd_2 + kd_3 + \cdots + kd_n. \tag{52}$$

Apply the condition of Eq. (50) with $n = 1, n = 2, n = 3, \cdots, n = n$. From this and Eqs. (44) and (52), it follows that

$$S_n \geqq kS_n''. \tag{53}$$

Since the positive series (49) diverges, its partial sums S_n'' are unbounded. Hence as n becomes infinite, $S_n'' \to \infty$. And since k is positive, $kS_n'' \to +\infty$. It follows from Eq. (53) that the sums S_n are unbounded. And the discussion of Sec. 235 shows that the series (41) diverges.

The first statement for the condition $u_n \geqq d_n$ is merely the important special case with $k = 1$.

We have proved our results on the assumption that conditions (43) and (50) held for all values of n. By the remark made in Sec. 235, we may deduce convergence from Eq. (43) and divergence from Eq. (50) if these conditions hold for all values of n greater than some fixed integer m. However, in this case we may no longer conclude from Eq. (43) that $S \leqq kS'$, but only that $S - S_m \leqq k(S' - S_m')$.

Example 1. Use the comparison test to prove that the p series

$$1 + \frac{1}{2^p} + \frac{1}{3^p} + \cdots + \frac{1}{n^p} + \cdots$$

converges if $p > 1$.

Solution: Let $p > 1$. Consider the series

$$1 + \left(\frac{1}{2^p} + \frac{1}{3^p}\right) + \left(\frac{1}{4^p} + \frac{1}{5^p} + \frac{1}{6^p} + \frac{1}{7^p}\right) + \left(\frac{1}{8^p} + \cdots + \frac{1}{15^p}\right) + \cdots, \tag{54}$$

which reduces to the given series if we remove the parentheses. In each parenthesis the first term is $1/n^p$ with n a power of 2, $n = 2^N$. We compare this with the series

$$1 + \left(\frac{1}{2^p} + \frac{1}{2^p}\right) + \left(\frac{1}{4^p} + \frac{1}{4^p} + \frac{1}{4^p} + \frac{1}{4^p}\right) + \left(\frac{1}{8^p} + \cdots + \frac{1}{8^p}\right) + \cdots. \tag{55}$$

The result of combining the terms in parentheses in Eq. (55) is

$$1 + \frac{2}{2^p} + \frac{4}{4^p} + \frac{8}{8^p} + \cdots + \frac{2^N}{(2^N)^p} + \cdots$$
$$= 1 + \frac{1}{2^{p-1}} + \left(\frac{1}{2^{p-1}}\right)^2 + \left(\frac{1}{2^{p-1}}\right)^3 + \cdots + \left(\frac{1}{2^{p-1}}\right)^N + \cdots. \tag{56}$$

This is a geometric series with $r = 1/2^{p-1} < 1$, since $p > 1$. Hence by the example of Sec. 233, the positive series of Eq. (56) converges. It follows that the positive series of Eq. (55), with parentheses removed, has bounded partial sums and therefore converges by the discussion of Sec. 235. Let us compare a group of terms in a parenthesis in Eq. (54) with the corresponding group in Eq. (55). The first term is the same. And the later terms are smaller in Eq. (54), since they have larger denominators. It follows that, for the series of Eqs. (54) and (55) with parentheses removed, each term of Eq. (54) is at most as great as the corresponding term of Eq. (55). Hence by the first comparison test for convergence, the convergence of the series (54) follows from that of the series (55). This proves that, for $p > 1$, the p series converges.

Example 2. Use the comparison test to prove that the harmonic series, or p series with $p = 1$,

$$1 + \frac{1}{2} + \frac{1}{3} + \cdots + \frac{1}{n} + \cdots$$

diverges.

Solution: Consider the series

$$1 + \left(\frac{1}{2}\right) + \left(\frac{1}{3} + \frac{1}{4}\right) + \left(\frac{1}{5} + \frac{1}{6} + \frac{1}{7} + \frac{1}{8}\right) + \left(\frac{1}{9} + \cdots + \frac{1}{16}\right) + \cdots, \tag{57}$$

which reduces to the given series if we remove the parentheses. In each parenthesis the last term is $1/n^p$ with n a power of 2, $n = 2^N$. We compare this with the series

$$1 + \left(\frac{1}{2}\right) + \left(\frac{1}{4} + \frac{1}{4}\right) + \left(\frac{1}{8} + \frac{1}{8} + \frac{1}{8} + \frac{1}{8}\right) + \left(\frac{1}{16} + \cdots + \frac{1}{16}\right) + \cdots. \tag{58}$$

The result of combining the terms in parentheses in Eq. (58) is

$$1 + \frac{1}{2} + \frac{2}{4} + \frac{4}{8} + \frac{8}{16} + \cdots + \frac{2^{N-1}}{2^N} + \cdots$$
$$= 1 + \frac{1}{2} + \frac{1}{2} + \frac{1}{2} + \frac{1}{2} + \cdots + \frac{1}{2} + \cdots. \tag{59}$$

The partial sum of this series to N terms is $S_N'' = 1 + (N - 1)/2$, so that the partial sums are unbounded and the series diverges. It follows that the positive series of Eq. (58), with parentheses removed, has unbounded partial sums and therefore diverges by the discussion of Sec. 235. Let us compare a group of terms in a parenthesis in Eq. (57) with the corresponding group in Eq. (58). The last term is the same. And the earlier terms are larger in Eq. (57), since they have smaller denominators. It follows that, for the series of Eqs. (57) and (58) with parentheses removed, each term of Eq. (57) is at least as great as the corresponding term of Eq. (58). Hence by the first comparison test for divergence, the divergence of the series (57) follows from that of the series (58). This proves that the given harmonic series diverges.

EXAMPLE 3. Deduce from the result of Example 2 that the p series

$$1 + \frac{1}{2^p} + \frac{1}{3^p} + \cdots + \frac{1}{n^p} + \cdots$$

diverges if $p < 1$.

Solution: We compare the given positive series

$$1 + \frac{1}{2^p} + \frac{1}{3^p} + \cdots + \frac{1}{n^p} + \cdots \tag{60}$$

with the harmonic series shown to be divergent in Example 2,

$$1 + \frac{1}{2} + \frac{1}{3} + \cdots + \frac{1}{n} + \cdots. \tag{61}$$

Since $p < 1$, for $n > 1$ we have $1/n^p > 1/n$, since it has a smaller denominator. Thus each term of Eq. (60) after the first is greater than the corresponding term of Eq. (61). Hence by the first comparison test for divergence, the divergence of the series (60) follows from that of the series (61). This proves that for $p < 1$, the p series diverges.

The results of Examples 1,2, and 3 combine to check the statements made about the p series in connection with Eq. (40).

EXERCISE 117

Verify that the integral test may be applied to the series with each given u_n and use this test to prove the series convergent.

1. $u_n = \dfrac{2}{n^2}$.

2. $u_n = \dfrac{5}{n^4}$.

3. $u_n = \dfrac{3}{n^{3/2}}$.

4. $u_n = \dfrac{1}{(2n+1)^3}$.

5. $u_n = \dfrac{1}{(4n+1)^{3/2}}$.

6. $u_n = \dfrac{2n}{(n^2+1)^3}$.

7. $u_n = \dfrac{1}{n^2+1}$.

8. $u_n = \dfrac{1}{n(\ln n)^2}$.

9. $u_n = e^{-n}$.

10. $u_n = 2^{-n}$.

Verify that the integral test may be applied to the series with each given u_n and use this test to prove the series divergent.

11. $u_n = \dfrac{1}{3n}$.

12. $u_n = \dfrac{1}{2n+3}$.

13. $u_n = \dfrac{5}{\sqrt{n}}$.

14. $u_n = \dfrac{1}{\sqrt{2n+1}}$.

15. $u_n = \dfrac{2n}{n^2+1}$.

16. $u_n = \dfrac{1}{n\sqrt{\ln n}}$.

Show that each of the following series is convergent by comparing it with the series $\frac{1}{2} + \frac{1}{2^2} + \frac{1}{2^3} + \cdots + \frac{1}{2^n} + \cdots$.

17. $1 + \dfrac{1}{2^2} + \dfrac{1}{3^3} + \cdots + \dfrac{1}{n^n} + \cdots$.

18. $\dfrac{1}{2!} + \dfrac{1}{3!} + \dfrac{1}{4!} + \cdots + \dfrac{1}{(n+1)!} + \cdots$.

19. $\dfrac{1}{2!} + \dfrac{2}{4!} + \dfrac{2^2}{6!} + \cdots + \dfrac{2^{n-1}}{(2n)!} + \cdots$.

Show that each of the following series is divergent by comparing it with the series $1 + \frac{1}{2} + \frac{1}{3} + \cdots + \frac{1}{n} + \cdots$.

20. $0 + \dfrac{\ln 2}{2} + \dfrac{\ln 3}{3} + \dfrac{\ln 4}{4} + \cdots + \dfrac{\ln n}{n} + \cdots$.

21. $\dfrac{1}{\sqrt{2}} + \dfrac{1}{\sqrt{3}} + \dfrac{1}{\sqrt{4}} + \cdots + \dfrac{1}{\sqrt{n+1}} + \cdots$.

22. $\dfrac{1}{2} + \dfrac{3}{5} + \dfrac{5}{10} + \dfrac{7}{17} + \cdots + \dfrac{2n-1}{n^2+1} + \cdots$.

23. Let A be the partial sum of the harmonic series $1 + \frac{1}{2} + \frac{1}{3} + \cdots + \frac{1}{n} + \cdots$ to one million terms. From Eqs. (34) and (36) with $p = 1$, deduce that $\ln 1{,}000{,}001 < A < \ln 1{,}000{,}000 + 1$, so that A lies between 13.81 and 14.82. That the partial sums, while increasing indefinitely, have not reached 15 after one million terms indicates the slow rate of divergence.

238. Comparison Tests Using Limits. The comparison of two series may sometimes be made by using certain limiting relations. Let

$$u_1 + u_2 + u_3 + \cdots + u_n + \cdots \tag{62}$$

be a series of positive terms to be tested for convergence. And let

$$c_1 + c_2 + c_3 + \cdots + c_n + \cdots \tag{63}$$

be a positive series which is known to be convergent.

Then *if*

$$\lim_{n\to\infty} \frac{u_n}{c_n} = L, \qquad \text{a finite limit,} \tag{64}$$

the series (62) *is convergent.*

The proof of this is as follows. The relation of Eq. (64) implies that $\left|\frac{u_n}{c_n} - L\right|$ will be less than any fixed positive number for all sufficiently large n. In particular, for a suitable m, we shall have for $n > m$

$$\left|\frac{u_n}{c_n} - L\right| < 1, \qquad \frac{u_n}{c_n} - L < 1, \qquad \frac{u_n}{c_n} < L + 1. \tag{65}$$

Since the c_n are all positive, this implies that

$$u_n < (L + 1)c_n \qquad \text{for } n > m. \tag{66}$$

Thus Eq. (43) holds with $k = L + 1$, for all $n > m$. Hence by the discussion at the end of Sec. 237, the series (62) converges.

There is a corresponding form of the test for divergence. Let

$$d_1 + d_2 + d_3 + \cdots + d_n + \cdots \tag{67}$$

be a positive series which is known to be divergent.

Then *if*

$$\lim_{n\to\infty} \frac{u_n}{d_n} = L > 0, \qquad \text{or } + \infty \tag{68}$$

the series (62) *is divergent.*

The proof of this is as follows. The relation of Eq. (68) for the finite L implies that $\left|\frac{u_n}{d_n} - L\right|$ will be less than any fixed positive number for all sufficiently large n. In particular, for a suitable m, we shall have for $n > m$

$$\left|\frac{u_n}{d_n} - L\right| < \frac{L}{2}, \qquad \frac{u_n}{d_n} - L > -\frac{L}{2}, \qquad \frac{u_n}{d_n} > \frac{L}{2}. \tag{69}$$

And for the case of Eq. (68) with $+\infty$, we shall have for $n > m$,

$$\frac{u_n}{d_n} > 1. \tag{70}$$

Since the d_n are all positive, Eq. (69) or (70) implies that

$$u_n > \frac{L}{2} d_n \qquad \text{or} \qquad u_n > d_n \qquad \text{for } n > m. \tag{71}$$

Hence by the discussion at the end of Sec. 237, the series (62) diverges.

EXAMPLE 1. Test the series

$$\frac{3\sqrt{1}-5}{1\cdot 4}+\frac{3\sqrt{2}-5}{2\cdot 5}+\frac{3\sqrt{3}-5}{3\cdot 6}+\cdots+\frac{3\sqrt{n}-5}{n(n+3)}+\cdots.$$

Solution: For $n > 2$, the terms are positive. For large values of n, the leading terms in $u_n = \dfrac{3\sqrt{n}-5}{n(n+3)}$ suggest $\dfrac{3\sqrt{n}}{n^2} = \dfrac{3}{n^{\frac{3}{2}}}$. This suggests as a comparison series that whose nth term is $c_n = \dfrac{1}{n^{\frac{3}{2}}}$. This is a p series with $p = \frac{3}{2}$. Since $p > 1$, Σc_n converges by the remark made after Eq. (40). And Eq. (64) holds with $L = 3$, since $\lim\limits_{n\to\infty} \dfrac{u_n}{c_n} = \lim\limits_{n\to\infty} \dfrac{3\sqrt{n}-5}{n(n+3)} \cdot \dfrac{n^{\frac{3}{2}}}{1} = 3$ by the principle of the leading term. Hence the given series *converges.*

EXAMPLE 2. Test the series

$$1+\frac{2^q}{2\cdot 2^3-2}+\frac{3^q}{2\cdot 3^3-3}+\cdots+\frac{n^q}{2\cdot n^3-n}+\cdots.$$

Solution: For large values of n, the leading terms in $u_n = \dfrac{n^q}{2\cdot n^3 - n}$ suggest $\dfrac{n^q}{2n^3} = \dfrac{1}{2n^{3-q}}$. This suggests, as a comparison series, the p series whose nth term is $1/n^{3-q}$. By the remark made after Eq. (40), this converges if $p = 3 - q > 1$, or $q < 2$. And it diverges if $p = 3 - q \leqq 1$, or $q \geqq 2$. Also $\lim\limits_{n\to\infty} \dfrac{u_n}{1/n^{3-q}} = \dfrac{1}{2}$ by the principle of the leading term. Hence with $L = \frac{1}{2}$, by Eq. (64) when $q < 2$, the series converges. And by Eq. (68) when $q \geqq 2$, the series diverges. Thus the given series is convergent when $q < 2$ and is divergent when $q \geqq 2$.

EXERCISE 118

Prove that the series with each of the given u_n diverges by comparison with the harmonic series, or p series with $p = 1$, $1 + \dfrac{1}{2} + \dfrac{1}{3} + \cdots + \dfrac{1}{n} + \cdots$ which is known to be divergent.

1. $u_n = \dfrac{1}{7n}$.
2. $u_n = \dfrac{1}{n+5}$.
3. $u_n = \dfrac{1}{2n}$.
4. $u_n = \dfrac{1}{2n+1}$.
5. $u_n = \dfrac{n}{4(n+1)(n+2)}$.
6. $u_n = \dfrac{2}{3n-5}$.
7. $u_n = \dfrac{3n+1}{n^2-5}$.
8. $u_n = \dfrac{n^2+1}{n^3}$.
9. $u_n = \dfrac{1}{\sqrt{n(n+1)}}$.
10. $u_n = \dfrac{1}{\sqrt[3]{n(n+1)(n+2)}}$.

Prove that the series with each of the given u_n converges by comparison with the series $1 + \dfrac{1}{2^2} + \dfrac{1}{3^2} + \cdots + \dfrac{1}{n^2} + \cdots$ which is a convergent p series since $p = 2 > 1$.

11. $u_n = \dfrac{1}{n^2+5}$.
12. $u_n = \dfrac{1}{(2n+1)^2}$.
13. $u_n = \dfrac{7}{n(n+1)}$.
14. $u_n = \dfrac{1}{(2n-1)(2n)}$.

15. $u_n = \dfrac{n-1}{n^3}$.

16. $u_n = \dfrac{2}{3n^2 - 5}$.

17. $u_n = \dfrac{3n-2}{n^3 - 5n}$.

18. $u_n = \dfrac{1}{n(n+2)}$.

19. $u_n = \dfrac{1}{n\sqrt{n^2+1}}$.

20. $u_n = \dfrac{2n-1}{n(n+1)(n+2)}$.

Test the series with each of the given u_n by comparison with a p series $1 + \dfrac{1}{2^p} + \dfrac{1}{3^p} + \cdots + \dfrac{1}{n^p} + \cdots$ known to be convergent if $p > 1$ and to be divergent if $p \leqq 1$.

21. $u_n = \dfrac{1}{\sqrt{2n+1}}$.

22. $u_n = \dfrac{2}{\sqrt{n^3}}$.

23. $u_n = \dfrac{4}{\sqrt[3]{n}}$.

24. $u_n = \dfrac{4n}{1+n^4}$.

25. $u_n = \dfrac{1}{2+\sqrt{n}}$.

26. $u_n = \dfrac{3}{n^5+1}$.

Test the series with each of the given u_n by comparison with a geometric series $r + r^2 + r^3 + \cdots + r^n + \cdots$ known to be convergent if $0 < r < 1$ and divergent if $r > 1$.

27. $u_n = \dfrac{1}{5^n - 1}$.

28. $u_n = \dfrac{1}{3^n + 1}$.

29. $u_n = 5^{-(2n+1)/2}$.

30. $u_n = \dfrac{1}{\sqrt{2^n+1}}$.

31. $u_n = \dfrac{1}{n!}$.

32. $u_n = \dfrac{1}{n^n}$.

239. The Ratio Test. For many series, one of the simplest tests for convergence to apply is the ratio test. In this section we shall explain this test as applied to positive series. Let

$$u_1 + u_2 + u_3 + \cdots + u_n + \cdots \tag{72}$$

be a series of positive terms to be tested for convergence. Form the ratio of a general term to its preceding term, $\dfrac{u_{n+1}}{u_n}$. Suppose that this ratio approaches a limit, t, as n becomes infinite.

$$\lim_{n\to\infty} \frac{u_{n+1}}{u_n} = t. \tag{73}$$

Then *if t is less than one, $t < 1$, the series converges.* And *if t is greater than one, $t > 1$, the series diverges.*

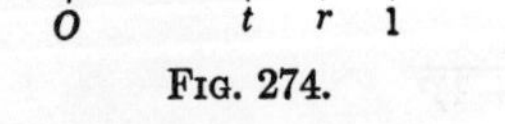

FIG. 274.

The proof of this is as follows. First suppose that $t < 1$. Then (Fig. 274) the number $r = (t+1)/2$ is such that $t < r < 1$. And $r - t > 0$. By Eq. (73), the ratio u_{n+1}/u_n approaches t as a limit when $n \to \infty$.

Hence for sufficiently large n, we shall have

$$\left|\frac{u_{n+1}}{u_n} - t\right| < r - t \qquad \text{and} \qquad \frac{u_{n+1}}{u_n} < r. \tag{74}$$

Suppose that this holds for $n > m$. Then

$$u_{m+1} < ru_m, \qquad u_{m+2} < ru_{m+1} < r^2u_m, \qquad u_{m+3} < ru_{m+2} < r^3u_m, \tag{75}$$

and so on for the following terms $u_{m+k} < r^ku_m$.

It follows that each term of the series

$$u_{m+1} + u_{m+2} + u_{m+3} + \cdots + u_{m+k} + \cdots \tag{76}$$

is less than the corresponding term of the positive series

$$u_mr + u_mr^2 + u_mr^3 + \cdots + u_mr^k + \cdots. \tag{77}$$

But the series just written is a geometric series, convergent since its positive ratio $r < 1$. Hence by the test of Sec. 237, the series of Eq. (76), and hence that of Eq. (72), converges when $t < 1$.

FIG. 275.

Next suppose that $t > 1$. Then (Fig. 275) the number $R = (t + 1)/2$ is such that $1 < R < t$. And $t - R > 0$. By Eq. (73), the ratio u_{n+1}/u_n approaches t as a limit when $n \to \infty$. Hence for sufficiently large n, we shall have

$$\left|t - \frac{u_{n+1}}{u_n}\right| < t - R \qquad \text{and} \qquad \frac{u_{n+1}}{u_n} > R > 1. \tag{78}$$

Suppose that this holds for $n > m$. Then

$$u_{m+1} > u_m, \qquad u_{m+2} > u_{m+1} > u_m, \qquad u_{m+3} > u_{m+2} > u_m, \tag{79}$$

and so on for the following terms, $u_{m+k} > u_m$. Since all the terms after u_m exceed the positive constant u_m, the term u_n cannot approach zero as n becomes infinite, and the series diverges by the test of Sec. 234. Thus the series of Eq. (72) diverges when $t > 1$.

The series of Eq. (72) also diverges if

$$\lim_{n\to\infty} \frac{u_{n+1}}{u_n} = +\infty. \tag{80}$$

For in this case, the ratio u_{n+1}/u_n will exceed any fixed number for sufficiently large n. In particular, suppose that $u_{n+1}/u_n > 1$ for $n > m$. Then we may deduce Eq. (79) and complete the argument as before.

In applying the ratio test, we may use u_{n+K+1}/u_{n+K} with K any fixed integer, in place of u_{n+1}/u_n since $n + K \to \infty$ when $n \to \infty$.

If the limit t of Eq. (73) is equal to one, $t = 1$, the theorem as stated gives no information and the simple ratio test fails. But see Prob. 15 of Exercise 119 for a more refined criterion.

EXAMPLE 1. Test the series

$$\frac{1}{100} + \frac{2!}{100^2} + \frac{3!}{100^3} + \cdots + \frac{n!}{100^n} + \cdots.$$

Solution: Here $u_n = \frac{n!}{100^n}$, so that $u_{n+1} = \frac{(n+1)!}{100^{n+1}}$. We may form u_{n+1} from u_n by multiplying $(n + 1)$ into the numerator and 100 into the denominator. It follows that $\frac{u_{n+1}}{u_n} = \frac{n+1}{100}$. And $\lim_{n\to\infty} \frac{u_{n+1}}{u_n} = \lim_{n\to\infty} \frac{n+1}{100} = +\infty$. Hence Eq. (80) holds, and the given series is divergent.

EXAMPLE 2. Test the series for which $u_n = 5^n \frac{n^2+1}{n!}$.

Solution: Here $u_n = 5^n \frac{n^2+1}{n!}$, $u_{n+1} = 5^{n+1} \frac{(n+1)^2+1}{(n+1)!}$. It follows that $\frac{u_{n+1}}{u_n} = 5 \frac{n^2+2n+2}{n^2+1} \frac{1}{n+1}$. By the principle of the leading term, the limit of this is the same as that of $\frac{5n^2}{n^3}$ or $\frac{5}{n}$. Hence $\lim_{n\to\infty} \frac{u_{n+1}}{u_n} = \lim_{n\to\infty} \frac{5}{n} = 0$. As this is less than one, the given series is convergent.

EXAMPLE 3. Test the series for which $u_n = \frac{4 \cdot 8 \cdot 12 \cdots (4n)}{8 \cdot 13 \cdots (3+5n)}$.

Solution: Here u_{n+1} is obtained from u_n by multiplying $4(n + 1)$ into the numerator and $[3 + 5(n + 1)]$ into the denominator. Hence $\frac{u_{n+1}}{u_n} = \frac{4n+4}{5n+8}$. By the principle of the leading term, the limit of this is the same as that of $4n/5n$ or $\frac{4}{5}$. Hence $\lim_{n\to\infty} \frac{u_{n+1}}{u_n} = \frac{4}{5}$. As this is less than one, the given series is convergent.

EXERCISE 119

Test the series with each of the given u_n for convergence or divergence by using the ratio test.

1. $u_n = \frac{3}{2^n}$.
2. $u_n = \frac{2^n}{3}$.
3. $u_n = \frac{3}{10^n}$.
4. $u_n = \frac{2^n}{n!}$.
5. $u_n = \frac{n}{3^n}$.
6. $u_n = \frac{n2^n}{3^n}$.
7. $u_n = \frac{3^n}{n2^n}$.
8. $u_n = \frac{9^n}{n!}$.
9. $u_n = \frac{3 \cdot 7 \cdot 11 \cdots (3+4n)}{2 \cdot 9 \cdot 16 \cdots (2+7n)}$.
10. $u_n = \frac{n!}{(2n)!}$.
11. $u_n = \frac{n^2+1}{4^n}$.
12. $u_n = \frac{2n+1}{5^n}$.
13. $u_n = \frac{(5+n)!}{3 \cdot 5 \cdot 7 \cdots (1+2n)}$.
14. $u_n = \frac{4 \cdot 7 \cdot 10 \cdots (1+3n)}{3 \cdot 5 \cdot 7 \cdots (1+2n)}$.
15. It may be proved† that, if $\frac{u_{n+1}}{u_n} = 1 - \frac{b}{n} + \frac{c}{n^2} + \cdots$, the series of Eq. (72)

† Compare the author's "A Treatise on Advanced Calculus," p. 306, John Wiley & Sons, Inc., New York, 1940 (Dover reprint).

converges if $b > 1$ and diverges if $b \leqq 1$. Verify that this gives correct results for the p series with $u_n = \frac{1}{n^p}$, by calculating $\frac{u_{n+1}}{u_n} = \frac{n^p}{(n+1)^p} = \left(1 + \frac{1}{n}\right)^{-p} = 1 - \frac{p}{n} + \frac{p(p+1)}{1 \cdot 2}\frac{1}{n^2} - \cdots$.

Use the test stated in Prob. 15 to show that the series with

16. $u_n = \frac{1 \cdot 3 \cdot 5 \cdots (2n+1)}{2 \cdot 4 \cdot 6 \cdots (2n+2)}$ is divergent.

17. $u_n = \frac{1 \cdot 3 \cdot 5 \cdots (2n+1)}{4 \cdot 6 \cdot 8 \cdots (2n+4)}$ is convergent.

18. $u_n = \frac{1}{(2n-1)(2n)}$ is convergent.

240. Alternating Series. An *alternating series* is one whose terms are alternately positive and negative. Thus if each p_n is positive, the series

$$p_1 - p_2 + p_3 - \cdots + (-1)^{n+1}p_n + \cdots \tag{81}$$

is an alternating one. For such a series we have the following test:

If each term of an alternating series is numerically less than the preceding term, and the nth term approaches zero as n becomes infinite, the series is convergent.

The proof of this is as follows. Write the series as in Eq. (81). Let m be an integer. Then $n = 2m$ is an *even* integer. We may write the partial sum of the series in Eq. (81) to $2m$ terms,

$$S_{2m} = (p_1 - p_2) + (p_3 - p_4) + (p_5 - p_6) + \cdots + (p_{2m-1} - p_{2m}). \tag{82}$$

$$S_{2m} = p_1 - (p_2 - p_3) - (p_4 - p_5) - \cdots - (p_{2m-2} - p_{2m-1}) - p_{2m}. \tag{83}$$

Since each term of the series (81) is numerically less than the preceding term,

$$p_n < p_{n-1} \qquad \text{and} \qquad (p_{n-1} - p_n) > 0. \tag{84}$$

Thus each expression in a parenthesis in Eqs. (82) and (83) is positive. It follows from Eq. (82) that, when n increases through even values $2m$, S_{2m} increases through positive values. But Eq. (83) shows that each S_{2m} is less than p_1. Hence by the argument used to derive Eq. (30), as m becomes infinite, S_{2m} will approach a limit L_1 and this positive limit L_1 will not exceed p_1. The partial sum of an odd number of terms, $S_{2m+1} = S_{2m} + p_{2m+1}$. And since p_n approaches zero as n becomes infinite,

$$\lim_{m\to\infty} S_{2m+1} = \lim_{m\to\infty} S_{2m} + \lim_{m\to\infty} p_{2m+1} = L_1 + 0 = L_1. \tag{85}$$

Hence S_n approaches L_1 whether n becomes infinite through even or through odd values, so that $S_n \to L_1$, and the alternating series converges as was to be proved.

The limit L_1 exceeds S_{2m} by the sum of the alternating series

$$p_{2m+1} - p_{2m+2} - \cdots + (-1)^{k+1}p_{2m+k} + \cdots . \tag{86}$$

As this sum is positive and less than the first term,

$$0 < L_1 - S_{2m} < p_{2m+1}, \qquad S_{2m} < L_1 < S_{2m} + p_{2m+1}. \tag{87}$$

And the sum S_{2m+1} exceeds L_1 by the sum of the alternating series

$$p_{2m+2} - p_{2m+3} + p_{2m+4} - \cdots + (-1)^k p_{2m+k} + \cdots . \tag{88}$$

As this sum is positive and less than the first term,

$$0 < S_{2m+1} - L_1 < p_{2m+2}, \qquad S_{2m+1} > L_1 > S_{2m+1} - p_{2m+2}. \tag{89}$$

It follows from Eqs. (87) and (89) that

In an alternating series satisfying the two conditions of the above theorem each even partial sum is less than the value of the series, each odd partial sum is greater than the value of the series, and the difference between the value and any partial sum is numerically less than the first of the unused terms of the series.

EXAMPLE. Show that the alternating series $1 - \frac{1}{2^6} + \frac{1}{3^6} - \cdots + (-1)^{n+1}\frac{1}{n^6} + \cdots$ is convergent, and compute its value to three decimal places.

Solution: The terms of the series alternate in sign. Also the numerical value of the nth term, $1/n^6$, decreases as n increases. And as n becomes infinite, $n^6 \to \infty$ and $\frac{1}{n^6} \to 0$. Hence the two conditions are satisfied and the series converges. Also the error in using any partial sum for the value is numerically less than the first term not used. We calculate the values of the terms by means of a tabular form and tables of powers. Thus

n	n^2	$n^6 = (n^2)^3$	$\frac{1}{n^6}$	Sign	+ terms	− terms
1	1	1	1.	+	1.	
2	4	64	0.0156	−		−0.0156
3	9	729	0.0014	+	0.0014	
4	16	4,096	0.0000	−		−0.0000
					1.0014	−0.0156
					−0.0156	
					0.9858	

It follows that 0.986 is a three-place value for the given series.

***241. Series with Positive and Negative Terms.** Most of the tests for convergence we have discussed apply only when the terms of the series, from some point onward, are positive. But we can sometimes infer the convergence of a series which contains both positive and negative terms from that of a related positive series, because of the following theorem:

A series which has both positive and negative terms,

$$u_1 + u_2 + u_3 + \cdots + u_n + \cdots \tag{90}$$

is convergent if the corresponding positive series of absolute values

$$|u_1| + |u_2| + |u_3| + \cdots + |u_n| + \cdots \tag{91}$$

is convergent.

If all but a *finite* number of terms of the series of Eq. (90) are positive, the result follows from the discussion at the end of Sec. 235. And if all but a finite number of terms are negative, we may deduce the convergence from that of the series with all signs reversed.

Next assume that there are infinitely many positive terms and infinitely many negative terms in the series of Eq. (90). Let S_n be the nth partial sum of this series. Suppose that S_n contains p positive terms, with sum P_p and q negative terms with sum $-Q_q$. Then

$$S_n = P_p - Q_q. \tag{92}$$

Let T_n be the nth partial sum of the series in Eq. (91). Then

$$T_n = P_p + Q_q. \tag{93}$$

Since the series in Eq. (91) was assumed to converge, the partial sums T_n are bounded. It follows from Eq. (93) that the positive series whose partial sum to p terms is P_p has bounded partial sums and so converges to some value P. And similarly, the positive series whose partial sum to q terms is Q_q has bounded partial sums and so converges to some value Q. Thus

$$\lim_{p\to\infty} P_p = P, \qquad \lim_{q\to\infty} Q_q = Q. \tag{94}$$

From Eqs. (92) and (94) we may deduce that

$$\lim_{n\to\infty} S_n = P - Q. \tag{95}$$

This proves that the given series (90) converges.

242. Absolute Convergence. Consider an infinite series (90) which may have both positive and negative terms. The infinite series (91), formed from it by replacing each negative term by the corresponding positive number, may be tested for convergence by any of the methods described for positive series. If the series (91) is convergent, the series (90) is said to be *absolutely convergent.*

An absolutely convergent series is necessarily convergent, as shown in Sec. 241. For example, consider the series

$$1 + \frac{1}{2} - \frac{1}{4} - \frac{1}{8} + \frac{1}{16} + \frac{1}{32} - \cdots \tag{96}$$

in which the numerical value of the nth term is $1/2^n$, and two plus signs are followed by two minus signs. The corresponding positive series of absolute values

$$1 + \frac{1}{2} + \frac{1}{2^2} + \cdots + \frac{1}{2^n} + \cdots \tag{97}$$

is a geometric series with ratio $\frac{1}{2}$ which is less than 1 and so converges. Hence the series of Eq. (96) is absolutely convergent.

A series which is convergent without being absolutely convergent is said to be *conditionally convergent.* For example, consider the series

$$1 - \frac{1}{2} + \frac{1}{3} - \frac{1}{4} + \cdots + (-1)^{n+1}\frac{1}{n} + \cdots. \tag{98}$$

This is a convergent alternating series. But the corresponding positive series of absolute values

$$1 + \frac{1}{2} + \frac{1}{3} + \cdots + \frac{1}{n} + \cdots \tag{99}$$

is the harmonic series, or p series with $p = 1$, which diverges. Hence the series of Eq. (98) is conditionally convergent.

The value of an absolutely convergent series is unaffected by a rearrangement of the order of its terms. This is not true of all rearrangements for a conditionally convergent series.

Example 1. Show that the series with each third term negative, $\frac{1}{5} + \frac{2}{5^2} - \frac{3}{5^3} + \cdots + \frac{3n-1}{5^{3n-1}} - \frac{3n}{5^{3n}} + \frac{3n+1}{5^{3n+1}} + \cdots$ is absolutely convergent.

Solution: The corresponding positive series is $\frac{1}{5} + \frac{2}{5^2} + \frac{3}{5^3} + \cdots + \frac{n}{5^n} + \cdots$. We use the ratio test of Sec. 239. Here $u_n = \frac{n}{5^n}$, $u_{n+1} = \frac{n+1}{5^{n+1}}$, $\frac{u_{n+1}}{u_n} = \frac{n+1}{5n}$. By the principle of the leading term, $\lim_{n\to\infty} \frac{u_{n+1}}{u_n} = \lim_{n\to\infty} \frac{n}{5n} = \frac{1}{5}$. As this is less than 1, the positive series converges, and the given series is absolutely convergent.

Example 2. Show that the series $\frac{1}{\sqrt{1\cdot 2}} - \frac{1}{\sqrt{2\cdot 3}} + \frac{1}{\sqrt{3\cdot 4}} - \cdots + (-1)^{n+1}\frac{1}{\sqrt{n(n+1)}} + \cdots$ is conditionally convergent.

Solution: The corresponding positive series is $\frac{1}{\sqrt{1\cdot 2}} + \frac{1}{\sqrt{2\cdot 3}} + \frac{1}{\sqrt{3\cdot 4}} + \cdots + \frac{1}{\sqrt{n(n+1)}} + \cdots$. Try the ratio test of Sec. 239. Here $u_n = \frac{1}{\sqrt{n(n+1)}}$, $u_{n+1} = \frac{1}{\sqrt{(n+1)(n+2)}}$, $\frac{u_{n+1}}{u_n} = \sqrt{\frac{n}{n+2}}$. As this approaches 1 when $n \to \infty$, we can draw no conclusion from the ratio test. However, $\sqrt{\frac{n}{n+2}} = \left(1 + \frac{2}{n}\right)^{-\frac{1}{2}} = 1 + \left(-\frac{1}{2}\right)\frac{2}{n} + \frac{(-\frac{1}{2})(-\frac{3}{2})}{1\cdot 2}\left(\frac{2}{n}\right)^2 - \cdots = 1 - \frac{1}{n} + \frac{3}{2n^2} - \cdots$ by the binomial theorem. Hence by the result stated in Prob. 15 of Exercise 119, the series diverges since here in $1 - \frac{b}{n} + \frac{c}{n^2} + \cdots$, $b = 1$.

If we wish to avoid use of the binomial theorem for fractional exponents and the unproved result, we may use the test of Sec. 238. The leading term of

$u_n = \dfrac{1}{\sqrt{n(n+1)}}$ is $\dfrac{1}{\sqrt{n \cdot n}} = \dfrac{1}{n}$. Hence we use the divergent harmonic series with general term $d_n = \dfrac{1}{n}$. Then $\lim\limits_{n \to \infty} \dfrac{u_n}{d_n} = \lim\limits_{n \to \infty} \dfrac{n}{\sqrt{n(n+1)}} = 1$ by the principle of the leading term. Hence the positive series diverges.

It follows from the divergence of the positive series of absolute values that the given series does *not* converge absolutely.

But the given series has alternating signs, and as n becomes infinite, $n(n+1)$ increases and $\to +\infty$. Hence the numerical value of the nth term, $u_n = \dfrac{1}{\sqrt{n(n+1)}}$, decreases and $\to 0$. It follows from Sec. 240 that the given series is a convergent alternating series. Since it converges, but not absolutely, it is conditionally convergent as was to be proved.

EXERCISE 120

Verify that each given series is a convergent alternating series.

1. $\dfrac{1}{3} - \dfrac{1}{3^2} + \dfrac{1}{4^4} - \cdots + (-1)^{n+1}\dfrac{1}{3^n} + \cdots$.
2. $1 - \dfrac{1}{3} + \dfrac{1}{5} - \cdots + (-1)^{n+1}\dfrac{1}{2n-1} + \cdots$.
3. $1 - \dfrac{1}{\sqrt{2}} + \dfrac{1}{\sqrt{3}} - \cdots + (-1)^{n+1}\dfrac{1}{\sqrt{n}} + \cdots$.
4. $\dfrac{1}{2} - \dfrac{2}{5} + \dfrac{3}{10} - \cdots + (-1)^{n+1}\dfrac{1}{n^2+1} + \cdots$.
5. $1 - \dfrac{1}{2^4} + \dfrac{1}{3^4} - \cdots + (-1)^{n+1}\dfrac{1}{n^4} + \cdots$.
6. $1 - \dfrac{1}{(2!)^2} + \dfrac{1}{(3!)^2} - \cdots + (-1)^{n+1}\dfrac{1}{(n!)^2} + \cdots$.

Verify that each given series with terms alternating in sign is divergent by using the test of Sec. 234.

7. $1 - \dfrac{3}{4} + \dfrac{4}{6} - \cdots + (-1)^{n+1}\dfrac{n+1}{2n} + \cdots$.
8. $1 - \dfrac{1}{\sqrt{2}} + \dfrac{1}{\sqrt[3]{2}} - \cdots + (-1)^{n+1}\dfrac{1}{\sqrt[n]{2}} + \cdots$.
9. $\dfrac{1}{2} - \dfrac{2}{5} + \dfrac{3}{8} - \cdots + (-1)^{n+1}\dfrac{n}{3n-1} + \cdots$.
10. $(\frac{2}{3}) - (\frac{2}{3})^{\frac{1}{2}} + (\frac{2}{3})^{\frac{1}{3}} - \cdots + (-1)^{n+1}(\frac{2}{3})^{1/n} + \cdots$.

Compute the value, correct to three decimal places, of the alternating series of

11. Prob. 1. 12. Prob. 5. 13. Prob. 6.

Establish the absolute convergence of the series of

14. Prob. 1. 15. Prob. 5. 16. Prob. 6.

Establish the conditional convergence of the series of

17. Prob. 2. 18. Prob. 3. 19. Prob. 4.

20. Show that the series $1 + \frac{1}{2} - \frac{1}{3} - \frac{1}{4} + \frac{1}{5} + \frac{1}{6} - \cdots$ whose general term has a numerical value $1/n$ and in which two plus signs are followed by two minus signs, converges. Hint: Note that the partial sum to $2n$ terms is the sum of the partial sum to n terms of $1 - \frac{1}{3} + \frac{1}{5} - \frac{1}{7} + \cdots$ and $\frac{1}{2} - \frac{1}{4} + \frac{1}{6} - \frac{1}{8} + \cdots$, and that each of these is a convergent alternating series.

21. In the series $1 - \frac{1}{2} + \frac{1}{3} - \frac{1}{4} - \frac{1}{6} + \frac{1}{5} - \frac{1}{8} - \frac{1}{10} - \frac{1}{12} - \frac{1}{14} + \frac{1}{7} - \frac{1}{16} - \cdots$ the terms with odd integers appear with plus signs in blocks of one, and the terms with even integers appear with plus signs in blocks doubling in length at each stage. Show that this series diverges. HINT: In the block of 2^k negative terms, the denominators which run from 2^{k+1} to $2^{k+2} - 2$ are each less than 2^{k+2}. Thus in this block each term numerically exceeds $1/2^{k+2}$. And the sum of the 2^k negative terms in the block numerically exceeds $2^k/(2^{k+2}) = \frac{1}{4}$. It follows that the partial sums to the beginning of such a block, and to the end of such a block, cannot approach the same limit. Observe that the divergent series of this problem contains the same terms, rearranged, as the convergent series of Eq. (98).

243. Power Series in x. A series whose terms contain increasing positive integral powers of a variable x, of the form

$$a_0 + a_1x + a_2x^2 + \cdots + a_nx^n + \cdots, \tag{100}$$

where the coefficients $a_0, a_1, a_2, \cdots, a_n, \cdots$ are constants, is called a *power series* in x.

Consider the absolute value of the ratio of a general coefficient to its preceding coefficient, $\left|\frac{a_{n+1}}{a_n}\right|$. Suppose that this ratio approaches a limit A as n becomes infinite.

$$\lim_{n\to\infty}\left|\frac{a_{n+1}}{a_n}\right| = A. \tag{101}$$

To test the series of Eq. (100) for absolute convergence, we consider the corresponding positive series

$$|a_0| + |a_1x| + |a_2x^2| + \cdots + |a_nx^n| + \cdots. \tag{102}$$

To apply the ratio test of Sec. 239 to this series with first term omitted, we form

$$\frac{|a_{n+1}x^{n+1}|}{|a_nx^n|} = \left|\frac{a_{n+1}}{a_n}\right| |x|. \tag{103}$$

For any fixed value of x, we may find the limit t_x,

$$t_x = \lim_{n\to\infty}\left|\frac{a_{n+1}}{a_n}\right| |x| = A|x|, \tag{104}$$

in view of Eq. (101). By Sec. 239 the positive series of Eq. (102) will converge if

$$t_x = A|x| < 1. \tag{105}$$

If $A = 0$, $A\,|x| = 0$ for all x. Hence $t_x = 0 < 1$, and the positive series of Eq. (102) converges for all x when $A = 0$.

If $A \neq 0$, $A\,|x| < 1$ when $|x| < \frac{1}{A}$, or

$$-\frac{1}{A} < x < \frac{1}{A}. \tag{106}$$

Hence for values of x inside this interval, the positive series of Eq. (102) will converge. It follows that, with A defined by Eq. (101),

The series of Eq. (100) *is absolutely convergent for values of x inside the interval of Eq.* (106) *when $A \neq 0$, and for all values of x when $A = 0$.*

Suppose that $A \neq 0$ and that x is outside of the interval of Eq. (106). Then $|x| > 1/A$. Let u_n denote the $(n + 1)$st term of the power series of Eq. (100), so that $u_n = a_n x^n$. Then it follows from Eqs. (103) and (104) that

$$\lim_{n \to \infty} \left| \frac{u_{n+1}}{u_n} \right| = A\,|x| > 1. \tag{107}$$

Since the limit exceeds 1, from a certain point on all the ratios $\left| \frac{u_{n+1}}{u_n} \right|$ will exceed 1. Beyond this point $|u_{n+1}|$ will exceed $|u_n|$, so that the terms u_n will eventually increase in numerical value. Thus the u_n cannot approach zero as n becomes infinite. And by the test of Sec. 234, the series of Eq. (100) diverges for $|x| > 1/A$. This proves that, with A defined by Eq. (101),

The series of Eq. (100) *diverges for values of x outside the interval of Eq.* (106) *when $A \neq 0$.*

The interval of Eq. (106) is called the *interval of convergence* of the power series of Eq. (100). At the end points of this interval, the series may converge or diverge. For many simple series this can be decided by using the tests of Secs. 234, 238, and 240 as illustrated in the examples of Sec. 242. At the end points of the interval of convergence, the simple ratio test can never be effective, since if $x = 1/A$ or $-1/A$, $t_x = 1$ by Eq. (104).

For some power series in x with some of their terms missing, the interval of convergence may be found by a similar consideration of u_{n+1}/u_n as illustrated in Example 3 below.

EXAMPLE 1. For what values of x does the series

$$1 - \frac{1}{3}\frac{x}{3} + \frac{1}{5}\left(\frac{x}{3}\right)^2 - \cdots + (-1)^n \frac{1}{2n+1}\left(\frac{x}{3}\right)^n + \cdots$$

converge?

Solution: Let $u_n = (-1)^n \frac{1}{2n+1}\left(\frac{x}{3}\right)^n$. Then $u_{n+1} = (-1)^{n+1} \frac{1}{2(n+1)+1}\left(\frac{x}{3}\right)^{n+1}$. And $\left|\frac{u_{n+1}}{u_n}\right| = \left|\frac{2n+1}{2n+3}\frac{x}{3}\right|$. By the principle of the leading term, the limit as $n \to \infty$ of $\frac{2n+1}{2n+3}$ is the limit of $2n/2n = 1$. Hence for this series $t_x = \lim_{n\to\infty} \left|\frac{u_{n+1}}{u_n}\right| = \frac{|x|}{3}$. Thus $t_x < 1$ for $\frac{|x|}{3} < 1$, or $|x| < 3$. And $t_x > 1$ for $\frac{|x|}{3} > 1$, or $|x| > 3$. It follows that the interval of convergence is $-3 < x < 3$. At the end point $x = 3$, the series has $u_n = (-1)^n \frac{1}{2n+1}$. Since the terms have alternate signs and numerically decrease to zero, we have convergence. At the end point $x = -3$, the series has $u_n = \frac{1}{2n+1}$. This diverges, since the ratio of its general term to that of the harmonic series with $d_n = \frac{1}{n}$ is $\frac{u_n}{d_n} = \frac{n}{2n+1}$ which approaches $\frac{1}{2}$ as $n \to \infty$. Or by the test of Prob. 15 of Exercise 119, $\frac{u_{n+1}}{u_n} = \frac{2n+1}{2n+3} = 1 - \frac{1}{n} + \frac{3}{2n^2} - \cdots$, and the coefficient of $\frac{1}{n}$ is -1, which means divergence.

The given series converges for $-3 < x \leqq 3$. This may be indicated graphically as in Fig. 276, the heavy dot at 3 indicating that the right-hand end is in the region of convergence.

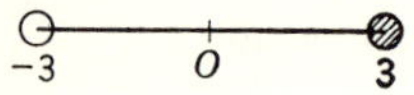

FIG. 276.

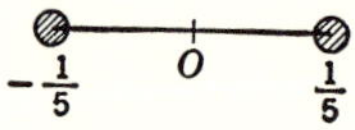

FIG. 277.

EXAMPLE 2. The $(n + 1)$st term of the series

$$1 + \tfrac{1}{2}5x - \tfrac{1}{5}(5x)^2 - \tfrac{1}{10}(5x)^3 + \cdots$$

has a coefficient of $(5x)^n$ numerically equal to $\frac{1}{n^2+1}$, and in the series two plus signs are followed by two minus signs. For what values of x does the series converge?

Solution: Let $|u_n| = \frac{1}{n^2+1}|5x|^n$. Then $|u_{n+1}| = \frac{1}{(n+1)^2+1}|5x|^{n+1}$. And $\left|\frac{u_{n+1}}{u_n}\right| = \frac{n^2+1}{n^2+2n+1}|5x|$. By the principle of the leading term, the limit as $n \to \infty$ of $\frac{n^2+1}{n^2+2n+1}$ is the limit of $\frac{n^2}{n^2} = 1$. Hence for this series $t_x = \lim_{n\to\infty}\left|\frac{u_{n+1}}{u_n}\right| = |5x|$. Thus $t_x < 1$ for $5|x| < 1$ or $|x| < \frac{1}{5}$. And $t_x > 1$ for $5|x| > 1$ or $x > \frac{1}{5}$. It follows that the interval of convergence is $-\frac{1}{5} < x < \frac{1}{5}$. At the end points, $x = -\frac{1}{5}$ or $\frac{1}{5}$, the series is absolutely convergent, since the corresponding series of positive terms has $u_n = \frac{1}{n^2+1}$. This positive series converges since the ratio of its general term to that of the convergent p series with $c_n = \frac{1}{n^2}$ is $\frac{u_n}{c_n} = \frac{n^2}{n^2+1}$ which approaches 1 as $n \to \infty$. Or by the test of Prob. 15 of Exercise 119, $\frac{u_{n+1}}{u_n} = \frac{n^2+1}{n^2+2n+2} = 1 - \frac{2}{n} + \frac{3}{n^2} - \cdots$, and the coefficient of $1/n$ is -2 which means convergence since $b = 2 > 1$.

The given series converges for $-\frac{1}{5} \leqq x \leqq \frac{1}{5}$. This may be indicated graphically as in Fig. 277, the heavy dots indicating that both ends are in the region of convergence.

EXAMPLE 3. For what values of x does the series

$$x - \frac{x^4}{3} + \frac{x^7}{3^2} - \cdots + (-1)^n\frac{x^{3n+1}}{3^n} + \cdots$$

converge?

Solution: Let $u_n = (-1)^n\frac{x^{3n+1}}{3^n}$. Then $u_{n+1} = (-1)^{n+1}\frac{x^{3(n+1)+1}}{3^{n+1}}$. And $\left|\frac{u_{n+1}}{u_n}\right| = \left|\frac{x^3}{3}\right|$. Hence $t_x = \lim_{n\to\infty}\left|\frac{u_{n+1}}{u_n}\right| = \frac{|x^3|}{3}$. Thus $t_x < 1$ for $\frac{|x^3|}{3} < 1$, $|x|^3 < 3$ or

$|x| < \sqrt[3]{3}$. And $t_x > 1$ for $\frac{|x^3|}{3} > 1$, $|x|^3 > 3$ or $|x| > \sqrt[3]{3}$. It follows that the interval of convergence is $-\sqrt[3]{3} < x < \sqrt[3]{3}$. At the end points, $x = -\sqrt[3]{3}$ or $\sqrt[3]{3}$, the series diverges, since the absolute value of u_n is $|u_n| = \frac{(\sqrt[3]{3})^{3n+1}}{3^n} = \sqrt[3]{3}$. Thus u_n cannot tend to zero as n becomes infinite.

The given series converges for $-\sqrt[3]{3} < x < \sqrt[3]{3}$, as indicated graphically in Fig. 278.

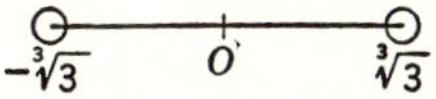

FIG. 278.

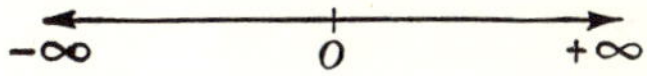

FIG. 279.

EXAMPLE 4. For what values of x does the series

$$1 - \frac{x^4}{4!} + \frac{x^8}{8!} - \cdots + (-1)^n \frac{x^{4n}}{(4n)!} + \cdots$$

converge?

Solution: Let $u_n = (-1)^n \frac{x^{4n}}{(4n)!}$. Then $u_{n+1} = (-1)^{n+1} \frac{x^{4(n+1)}}{[4(n+1)]!}$. And $\left|\frac{u_{n+1}}{u_n}\right| = \frac{|x^4|}{(4n+1)(4n+2)(4n+3)(4n+4)}$. Hence $t_x = \lim_{n\to\infty} \left|\frac{u_{n+1}}{u_n}\right| = 0$ for any fixed value of x. It follows that the given series converges for all values of x, or that the interval of convergence is $-\infty < x < +\infty$. This is indicated graphically in Fig. 279.

EXERCISE 121

Verify that each of the following series converges if $-1 < x < 1$ as indicated in Fig. 280, and diverges for all other values.

1. $1 + x + x^2 + \cdots + x^n + \cdots$.
2. $x - 2x^2 + 3x^3 - \cdots + (-1)^{n+1}nx^n + \cdots$.
3. $x - 3x^3 + 5x^5 - \cdots + (-1)^n(2n+1)x^{2n+1} + \cdots$.
4. $1 + \frac{x^2}{2} + \frac{x^4}{3} + \cdots + \frac{x^{2n}}{n+1} + \cdots$.

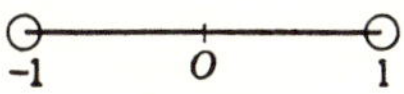

FIG. 280.

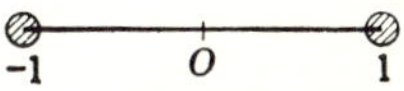

FIG. 281.

5. $x - 4x^2 + 9x^3 - \cdots + (-1)^{n+1}n^2x^n + \cdots$.

Verify that each of the following series converges if $-1 \leqq x \leqq 1$ as indicated in Fig. 281, and diverges for all other values.

6. $x - \frac{x^2}{2^2} + \frac{x^3}{3^2} - \cdots + (-1)^{n+1}\frac{x^n}{n^2} + \cdots$.
7. $x + \frac{x^2}{2^3} + \frac{x^3}{3^3} + \cdots + \frac{x^n}{n^3} + \cdots$.
8. $\frac{x}{1\cdot 2} + \frac{x^2}{2\cdot 3} + \frac{x^3}{3\cdot 4} + \cdots + \frac{x^n}{n(n+1)} + \cdots$.
9. $1 - \frac{x^2}{\sqrt{2}} + \frac{x^4}{\sqrt{4}} - \cdots + (-1)^n\frac{x^{2n}}{\sqrt{2n}} + \cdots$.

10. $\frac{x}{3} - \frac{x^2}{6} + \frac{x^3}{11} - \cdots + (-1)^{n+1} \frac{x^n}{n^2 + 2} + \cdots.$

Verify that each of the following series converges for all values of x, as indicated in Fig. 279.

11. $1 + x + \frac{x^2}{2!} + \cdots + \frac{x^n}{n!} + \cdots.$

12. $1 - \frac{x^2}{2!} + \frac{x^4}{4!} - \cdots + (-1)^n \frac{x^{2n}}{(2n)!} + \cdots.$

13. $x - \frac{x^3}{3!} + \frac{x^5}{5!} - \cdots + (-1)^n \frac{x^{2n+1}}{(2n + 1)!} + \cdots.$

14. $10x + 100 \frac{x^2}{2!} + 1{,}000 \frac{x^3}{3!} + \cdots + 10^n \frac{x^n}{n!} + \cdots.$

15. $x + \frac{x^2}{2^2} + \frac{x^3}{3^3} + \cdots + \frac{x^n}{n^n} + \cdots.$

Verify that each of the following series converges for the given values and diverges for all other values.

16. $x + \frac{x^2}{2} + \frac{x^3}{3} + \cdots + \frac{x^n}{n} + \cdots, \quad -1 \leqq x < 1.$

17. $x - \frac{x^2}{\sqrt{2}} + \frac{x^3}{\sqrt{3}} + \cdots + (-1)^{n+1} \frac{x^n}{\sqrt{n}} + \cdots, \quad -1 < x \leqq 1.$

18. $x - \frac{x^2}{2 \cdot 3^2} + \frac{x^3}{3 \cdot 3^3} - \cdots + (-1)^{n+1} \frac{x^n}{n \cdot 3^n} + \cdots, \quad -3 < x \leqq 3.$

19. $1 + \frac{2x}{5} + \frac{3x^2}{5^2} + \cdots + \frac{(n + 1)x^n}{5^n} + \cdots, \quad -5 < x < 5.$

20. $\frac{x}{6} - \frac{x^2}{2^2 \cdot 6^2} + \frac{x^3}{3^2 \cdot 6^3} - \cdots + (-1)^{n+1} \frac{x^n}{n^2 \cdot 6^n} + \cdots, \quad -6 \leqq x \leqq 6.$

***244. Other Power Series.** We sometimes consider a series of the form

$$b_0 + b_1(x - a) + b_2(x - a)^2 + \cdots + b_n(x - a)^n + \cdots, \tag{108}$$

where a and the coefficients $b_0, b_1, b_2, \cdots, b_n, \cdots$ are constants. Such a series is called a *power series* in $(x - a)$.

Suppose that

$$\lim_{n \to \infty} \left| \frac{b_{n+1}}{b_n} \right| = B, \tag{109}$$

where B is finite.

Then if $u_n = b_n(x - a)^n$, we may deduce that

$$t_x = \lim_{n \to \infty} \left| \frac{u_{n+1}}{u_n} \right| = B|x - a|. \tag{110}$$

And by an argument like that of Sec. 243, we may deduce that

The series of Eq. (108) *is absolutely convergent for all values of x such that*

$$|x - a| < \frac{1}{B}, \qquad \text{or} \qquad a - \frac{1}{B} < x < a + \frac{1}{B} \tag{111}$$

when $B \neq 0$, and for all values of x when $B = 0$.

The series of Eq. (108) *diverges for all values of x such that $|x - a| > 1/B$, or outside of the interval of* Eq. (111) *when $B \neq 0$.*

The series of negative powers of the form

$$b_0 + \frac{b_1}{x} + \frac{b^2}{x^2} + \cdots + \frac{b_n}{x^n} + \cdots \tag{112}$$

is called a power series in $1/x$.

Suppose that Eq. (109) holds. Then if $u_n = \frac{b_n}{x^n}$, we may deduce that

$$t_x = \lim_{n\to\infty} \left| \frac{u_{n+1}}{u_n} \right| = \frac{B}{|x|}. \tag{113}$$

And by an argument like that of Sec. 243, we may deduce that

The series of Eq. (112) converges for all values of x such that

$$|x| > B, \qquad \text{that is,} \qquad x < -B \text{ or } x > B \tag{114}$$

when $B \neq 0$, and for all values of x except zero when $B = 0$.

The series of Eq. (112) diverges for all values of x such that

$$|x| < B, \qquad \text{or} \qquad -B < x < B \tag{115}$$

when $B \neq 0$.

Example 1. For what values of x does the series

$$1 + \frac{2(x-2)}{3} + \frac{3(x-2)^2}{3^2} + \cdots + \frac{(n+1)(x-2)^n}{3^n} + \cdots$$

converge?

Solution: Let $u_n = \frac{(n+1)(x-2)^n}{3^n}$. Then $u_{n+1} = \frac{(n+2)(x-2)^{n+1}}{3^{n+1}}$. And $\left|\frac{u_{n+1}}{u_n}\right| = \frac{n+2}{n+1}\frac{|x-2|}{3}$. Hence $t_x = \lim_{n\to\infty}\left|\frac{u_{n+1}}{u_n}\right| = \frac{|x-2|}{3}$. Thus $t_x < 1$ if $\frac{|x-2|}{3} < 1$ or $|x-2| < 3$. And $t_x > 1$ if $\frac{|x-2|}{3} > 1$ or $|x-2| > 3$. It follows that the interval of convergence is $-3 < x - 2 < 3$ or $-1 < x < 5$. At the end points $x = -1$ and $x = 5$, the series diverges since the absolute value of u_n is $|u_n| = \frac{(n+1)3^n}{3^n} = n + 1$, which $\to \infty$, so that u_n cannot approach zero, as n becomes infinite.

The given series converges for $-1 < x < 5$, as indicated graphically in Fig. 282. Note that the center of the interval is $x = 2$, the value which makes $x - 2 = 0$, and that the series is in powers of $(x - 2)$.

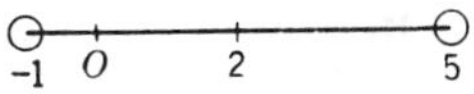

Fig. 282.

Example 2. For what values of x does the series

$$1 - \frac{5}{\sqrt{2}\,x^2} + \frac{5^2}{\sqrt{3}\,x^4} - \cdots + (-1)^n \frac{5^n}{\sqrt{n+1}\,x^{2n}} + \cdots$$

converge?

Solution: Let $u_n = (-1)^n \frac{5^n}{\sqrt{n+1}\,x^{2n}}$. Then $u_{n+1} = (-1)^{n+1}\frac{5^{n+1}}{\sqrt{n+2}\,x^{2n+2}}$. And $\left|\frac{u_{n+1}}{u_n}\right| = \frac{\sqrt{n+1}}{\sqrt{n+2}}\frac{5}{x^2}$. Hence $t_x = \lim_{n\to\infty}\left|\frac{u_{n+1}}{u_n}\right| = \frac{5}{x^2}$. Thus $t_x < 1$ if $\frac{5}{x^2} < 1$, $|x|^2 > 5$, $|x| > \sqrt{5}$. And $t_x > 1$ if $\frac{5}{x^2} > 1$, $|x|^2 < 5$, $|x| < \sqrt{5}$. It follows that the series diverges in the interval $-\sqrt{5} < x < \sqrt{5}$. At the end points, $x = -\sqrt{5}$

and $x = \sqrt{5}$, $u_n = (-1)^n \dfrac{1}{\sqrt{n+1}}$. As the terms alternate in sign and numerically decrease to zero, the series converges.

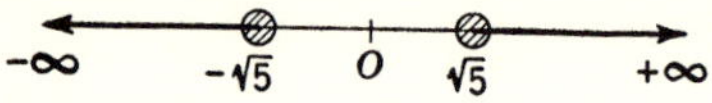

FIG. 283.

The given series converges for $x \leqq -\sqrt{5}$ or for $x \geqq \sqrt{5}$, as indicated graphically in Fig. 283.

EXERCISE 122

Verify that each of the following series converges for the given values and diverges for all other values.

1. $\dfrac{(x-2)}{3} + \dfrac{(x-2)^2}{3^2 \cdot 2} + \dfrac{(x-2)^3}{3^3 \cdot 3} + \cdots + \dfrac{(x-2)^n}{3^n \cdot n} + \cdots, \; -1 \leqq x < 5.$

2. $(x-1) - \dfrac{(x-1)^2}{2} + \dfrac{(x-1)^3}{3} - \cdots + (-1)^{n+1}\dfrac{(x-1)^n}{n} + \cdots,$
$0 < x \leqq 2.$

3. $\dfrac{(x-5)}{4} + \dfrac{(x-5)^2}{2^2 4^2} + \dfrac{(x-5)^3}{3^2 4^3} + \cdots + \dfrac{(x-5)^n}{n^2 4^n} + \cdots, \; 1 \leqq x \leqq 9.$

4. $1 + (x-3) + \dfrac{(x-3)^2}{2!} + \cdots + \dfrac{(x-3)^n}{n!} + \cdots$, all values.

5. $\dfrac{(x+3)}{2 \cdot 1} - \dfrac{(x+3)^2}{2^2 \cdot 2} + \dfrac{(x+3)^3}{2^3 \cdot 3} + \cdots + (-1)^{n+1}\dfrac{(x+3)^n}{2^n \cdot n} + \cdots,$
$-5 < x \leqq -1.$

6. $(x+1) + 2(x+1)^2 + 3(x+1)^3 + \cdots + n(x+1)^n + \cdots, \; -2 < x < 0.$

7. $\dfrac{1 \cdot 2}{x} + \dfrac{2 \cdot 3}{x^2} + \dfrac{3 \cdot 4}{x^3} + \cdots + \dfrac{n(n+1)}{x^n} + \cdots$, $x < -1$ or $x > 1$.

8. $1 + \dfrac{5}{x} + \dfrac{5^2}{x^2} + \cdots + \dfrac{5^n}{x^n} + \cdots$, $x < -5$ or $x > 5$.

9. $1 + \dfrac{7}{2^3 x} + \dfrac{7^2}{3^3 x^2} + \cdots + \dfrac{7^n}{n^3 x^n} + \cdots$, $x \leqq -7$ or $x \geqq 7$.

10. $\dfrac{1}{x} + \dfrac{1}{2^2 x^2} + \dfrac{1}{3^3 x^3} + \cdots + \dfrac{1}{n^n x^n} + \cdots, \; x \neq 0.$

11. $\dfrac{(2x+3)}{2} - \dfrac{(2x+3)^2}{2^2} + \dfrac{(2x+3)^3}{2^3} - \cdots + (-1)^{n+1}\dfrac{(2x+3)^n}{2^n} + \cdots,$
$-\dfrac{5}{2} < x < -\dfrac{1}{2}.$

12. $\dfrac{(x-3)}{1} + \dfrac{(x-3)^3}{3} + \dfrac{(x-3)^5}{3^2} + \cdots + \dfrac{(x-3)^{2n+1}}{3^n} + \cdots,$
$3 - \sqrt{3} < x < 3 + \sqrt{3}.$

13. $1 - \dfrac{(x-2)^2}{4} + \dfrac{(x-2)^4}{4^2} - \cdots + (-1)^n \dfrac{(x-2)^{2n}}{4^n} + \cdots, \; 0 < x < 4.$

14. $\dfrac{8}{3x^3} + \dfrac{8^2}{6x^6} + \dfrac{8^3}{9x^9} + \cdots + \dfrac{8^n}{3nx^{3n}} + \cdots$, $x \leqq -2$ or $x > 2$.

***245. The Binomial Series.** Let us consider the particular power series in x

$$1 + mx + \frac{m(m-1)}{1 \cdot 2}x^2 + \frac{m(m-1)(m-2)}{1 \cdot 2 \cdot 3}x^3 + \cdots + \frac{m(m-1)(m-2)\cdots(m-n+1)}{n!}x^n + \cdots, \quad (116)$$

where m is a constant.

If m is a positive integer, the term in x^{m+1} and all following terms are zero, since the numerator of their coefficients contains the factor $m - m$, which is zero.

But when m is not a positive integer, the expression in Eq. (116) is an infinite series. Since the term u_{n+1} containing x^n is obtained from the term u_n containing x^{n-1} by multiplying by $\frac{(m-n+1)x}{n}$, it follows that

$$\frac{u_{n+1}}{u_n} = \frac{m-n+1}{n}x, \qquad t_x = \lim_{n\to\infty}\left|\frac{u_{n+1}}{u_n}\right| = |x|. \tag{117}$$

Hence by Sec. 243 the series converges for $|x| < 1$ and diverges for $|x| > 1$.

By advanced analysis† it may be proved that for any value of x inside its interval of convergence, the result of differentiating a power series term by term or of integrating it from 0 to x term by term is a new power series with the same interval of convergence, representing the corresponding derivative or integral of the sum function. Accordingly, for $|x| < 1$, or $-1 < x < 1$, if

$$y = 1 + mx + \frac{m(m-1)}{2}x^2 + \frac{m(m-1)(m-2)}{3!}x^3 + \cdots \tag{118}$$

$$\frac{dy}{dx} = m + m(m-1)x + \frac{m(m-1)(m-2)}{2!}x^2 + \frac{m(m-1)(m-2)(m-3)}{3!}x^3 + \cdots$$
$$= m\left[1 + (m-1)x + \frac{(m-1)(m-2)}{2!}x^2 + \frac{(m-1)(m-2)(m-3)}{3!}x^3 + \cdots\right] \tag{119}$$

And by multiplying each member by x, we find

$$x\frac{dy}{dx} = m\left[x + (m-1)x^2 + \frac{(m-1)(m-2)}{2!}x^3 + \cdots\right]. \tag{120}$$

By addition, from Eqs. (119) and (120) we may deduce that

$$(1+x)\frac{dy}{dx} = m\left[1 + mx + \frac{m(m-1)}{2!}x^2 + \frac{m(m-1)(m-2)}{3!}x^3 + \cdots\right]$$
$$= my, \tag{121}$$

since the bracket equals the right member of Eq. (118). It follows that

$$\frac{dy}{dx} = \frac{my}{1+x}, \qquad \frac{dy}{y} = \frac{m\,dx}{1+x}, \qquad \int_0^x \frac{dy}{y} = m\int_0^x \frac{dx}{1+x}. \tag{122}$$

From Eq. (118), when $x = 0$, $y = 1$ and $\ln y = \ln 1 = 0$. Hence

$$[\ln y]_1^y = m[\ln(1+x)]_0^x, \qquad \ln y - \ln 1 = m[\ln(1+x) - \ln 1],$$
$$\ln y = m\ln(1+x) = \ln(1+x)^m, \qquad \text{and} \qquad y = (1+x)^m. \tag{123}$$

As the above calculation also holds for the finite sum when m is a positive integer, it follows that for any real value of m,

$$(1+x)^m = 1 + mx + \frac{m(m-1)}{2!}x^2 + \cdots$$
$$+ \frac{m(m-1)\cdots(m-n+1)}{n!}x^n + \cdots, \qquad -1 < x < 1. \tag{124}$$

If $x = \frac{b}{a}$, $(a+b) = a\left(1 + \frac{b}{a}\right) = a(1+x)$. Thus

$$(a+b)^m = a^m(1+x)^m, \qquad \text{for } x = \frac{b}{a}. \tag{125}$$

† Compare the author's "A Treatise on Advanced Calculus," pp. 431, 436.

Let us replace x by $\frac{b}{a}$ in the right member of Eq. (124), multiply both members by a^m, and use Eq. (125). This leads to

$$(a+b)^m = a^m + ma^{m-1}b + \frac{m(m-1)}{2!}a^{m-2}b^2 + \cdots + \frac{m(m-1)\cdots(m-n+1)}{n!}a^{m-n}b^n, + \cdots, \qquad |b| < |a|. \quad (126)$$

This is the form of the binomial theorem usually given in algebra. It is sometimes useful, but it is often more convenient to compute $(a+b)^m$ by finding $x = \frac{b}{a}$ and using Eqs. (125) and (124).

EXAMPLE 1. Expand $\sqrt{1+x}$ to five terms by the binomial theorem.

Solution: From Eq. (124) with $m = \frac{1}{2}$, we find

$$(1+x)^{\frac{1}{2}} = 1 + \frac{1}{2}x + \frac{\frac{1}{2}(-\frac{1}{2})}{1\cdot 2}x^2 + \frac{\frac{1}{2}(-\frac{1}{2})(-\frac{3}{2})}{1\cdot 2\cdot 3}x^3 + \frac{\frac{1}{2}(-\frac{1}{2})(-\frac{3}{2})(-\frac{5}{2})}{1\cdot 2\cdot 3\cdot 4}x^4 + \cdots.$$

Thus the required expansion is

$$\sqrt{1+x} = 1 + \tfrac{1}{2}x - \tfrac{1}{8}x^2 + \tfrac{1}{16}x^3 - \tfrac{5}{128}x^4 + \cdots.$$

EXAMPLE 2. Compute $\sqrt{412}$ to four decimal places.

Solution: Since 412 is near to $400 = 20^2$, we write

$$\sqrt{412} = (400+12)^{\frac{1}{2}} = 400^{\frac{1}{2}}(1 + \tfrac{12}{400})^{\frac{1}{2}} = 20(1 + \tfrac{3}{100})^{\frac{1}{2}}.$$

From the result of Example 1 with $x = \frac{3}{100} = 0.03$, we find

$$(1+0.03)^{\frac{1}{2}} = 1 + 0.015 - 0.000112 + 0.000002 - \cdots,$$

in which the terms are rounded off to six places. Since this is an alternating series with decreasing terms, the error is less than the last term used, and $(1+0.03)^{\frac{1}{2}} = 1.014890$ to within 2 in the last place. Hence $\sqrt{412} = 20(1+0.03)^{\frac{1}{2}} = 20.2978$ to four decimal places. And $\sqrt{412} = 20.2978$ is the required value.

EXAMPLE 3. Compute $\sqrt{92}$, using the five terms of the expansion of Example 1 and estimate the error made.

Solution: Since 92 is near to $100 = 10^2$, we write

$$\sqrt{92} = (100-8)^{\frac{1}{2}} = 100^{\frac{1}{2}}(1 - \tfrac{8}{100})^{\frac{1}{2}} = 10(1 - \tfrac{8}{100})^{\frac{1}{2}}.$$

From the result of Example 1 with $x = -\frac{8}{100} = -0.08$, we have

$$(1-0.08)^{\frac{1}{2}} = 1 - 0.04 - 0.0008 - 0.000032 - 0.0000016 - \cdots.$$

All the later terms are negative, but the coefficients in the expansion of Example 1 decrease numerically, since for $m = \frac{1}{2}$, $\left|\frac{m-n+1}{n}\right| = \frac{|\frac{3}{2}-n|}{n} = \frac{n-\frac{3}{2}}{n} = 1 - \frac{3}{2n} < 1.$ Hence the error E from breaking off the expansion is numerically less than the last term used times $(|x| + |x|^2 + |x|^3 + \cdots) = \frac{|x|}{1-|x|}$. That is, for the series just written for $(1-0.08)^{\frac{1}{2}}$, $|E| < 0.0000016\,\frac{0.08}{1-0.08} < 0.00000014$. Practically, we may estimate the error in the last place as $16\,\frac{0.08}{1-0.08} = \frac{16\times 8}{92} < 1.4$. It follows that $(1-0.08)^{\frac{1}{2}} = 0.9591664$ to within 2 in the last place. Thus the required value is $\sqrt{92} = 9.591664$, which is in excess by not more than 0.000002, or 2 in the last place.

EXERCISE 123

Use the binomial theorem of Eq. (124) to establish each of the following expansions, for $-1 < x < 1$.

1. $\dfrac{1}{1-x} = 1 + x + x^2 + x^3 + \cdots.$

2. $\dfrac{1}{1+x} = 1 - x + x^2 - x^3 + \cdots.$

3. $\sqrt{1+x} = 1 + \dfrac{1}{2}x - \dfrac{1}{2\cdot 4}x^2 + \dfrac{1\cdot 3}{2\cdot 4\cdot 6}x^3 - \cdots.$

4. $\dfrac{1}{\sqrt{1+x}} = 1 - \dfrac{1}{2}x + \dfrac{1\cdot 3}{2\cdot 4}x^2 - \dfrac{1\cdot 3\cdot 5}{2\cdot 4\cdot 6}x^3 + \cdots.$

5. $\sqrt{1-x^2} = 1 - \dfrac{1}{2}x^2 - \dfrac{1}{2\cdot 4}x^4 - \dfrac{1\cdot 3}{2\cdot 4\cdot 6}x^6 - \cdots.$

6. $\dfrac{1}{\sqrt{1-x^2}} = 1 + \dfrac{1}{2}x^2 + \dfrac{1\cdot 3}{2\cdot 4}x^4 + \dfrac{1\cdot 3\cdot 5}{2\cdot 4\cdot 6}x^6 + \cdots.$

7. Use the example of Sec. 233 to check Probs. 1 and 2.

Compute the value of each of the following.

8. $\frac{1}{204} = \frac{1}{200}(1 + 0.02)^{-1}$, using Prob. 2.
9. $\frac{1}{388} = \frac{1}{400}(1 - 0.03)^{-1}$, using Prob. 1.
10. $\sqrt{26} = 5(1 + 0.04)^{\frac{1}{2}}$, using Prob. 3.
11. $\dfrac{1}{\sqrt{220}} = \dfrac{1}{15}\left(1 - \dfrac{1}{45}\right)^{-\frac{1}{2}}$, using Prob. 4.
12. $\sqrt[3]{26} = 3(1 - \frac{1}{27})^{\frac{1}{3}}$, using Eq. (124).
13. $\sqrt[4]{17} = 2(1 + \frac{1}{16})^{\frac{1}{4}}$, using Eq. (124).
14. $\sqrt{2} = \frac{3}{2}(1 - \frac{1}{9})^{\frac{1}{2}}$, using Prob. 3.
15. $\sqrt[3]{2} = \frac{5}{4}(1 + 0.024)^{\frac{1}{3}}$, using Eq. (124).

16. From Prob. 1, by differentiation, deduce that

$$\frac{1}{(1-x)^2} = 1 + 2x + 3x^2 + 4x^3 + \cdots.$$

17. From $\displaystyle\int_0^x \frac{1}{1+x}\,dx = \ln(1+x)$ and Prob. 2, deduce that

$$\ln(1+x) = x - \frac{x^2}{2} + \frac{x^3}{3} - \frac{x^4}{4} + \cdots, \quad -1 < x < 1.$$

18. From $\displaystyle\int_0^x \frac{1}{1+x^2}\,dx = \tan^{-1} x$ and Prob. 2 with x replaced by x^2, deduce that

$$\tan^{-1} x = x - \frac{x^3}{3} + \frac{x^5}{5} - \frac{x^7}{7} + \cdots, \quad -1 < x < 1.$$

19. From $\displaystyle\int_0^x \frac{1}{\sqrt{1-x^2}}\,dx = \sin^{-1} x$ and Prob. 6, deduce that

$$\sin^{-1} x = x + \frac{1}{2}\frac{x^3}{3} + \frac{1\cdot 3}{2\cdot 4}\frac{x^5}{5} + \frac{1\cdot 3\cdot 5}{2\cdot 4\cdot 6}\frac{x^7}{7} + \cdots, \quad -1 < x < 1.$$

20. Let $y = 1 + x + \dfrac{x^2}{2!} + \dfrac{x^3}{3!} + \cdots + \dfrac{x^n}{n!} + \cdots.$ Verify in succession that $\dfrac{dy}{dx} = y$, $\dfrac{dy}{y} = dx$, $\displaystyle\int_1^y \frac{dy}{y} = \int_0^x dx$, $\ln y = x$, so that $y = e^x$. Thus for all real values of x, $e^x = 1 + x + \dfrac{x^2}{2!} + \dfrac{x^3}{3!} + \cdots + \dfrac{x^n}{n!} + \cdots.$

21. From the property of an alternating series, deduce that for $0 < x < 1$ in the series of Probs. 2 to 4 the error due to replacing the infinite series by a partial sum is numerically less than the first term omitted.

22. From an argument like that used in Example 3, deduce that for $-1 < x < 0$ in the series of Probs. 3 and 4 the error due to replacing the infinite series by a partial sum is numerically less than $\dfrac{1}{1 - |x|}$ times the first term omitted.

23. From an argument like that used in Example 3, deduce that for $-1 < x < 1$ in the series of Probs. 5 and 6 the error due to replacing the infinite series by a partial sum is numerically less than $\dfrac{1}{1 - x^2}$ times the first term omitted.

CHAPTER 16

TAYLOR'S SERIES. INDETERMINATE FORMS

Many simple functions may be expanded in power series in $(x - a)$. The expansion, with coefficients expressed in terms of the successive derivatives of the function, is known as Taylor's series. The special form for the expansion in a power series in x is known as Maclaurin's series. These series are first derived on the assumption that the expansions exist. Later an expression is found for the remainder by which they may be justified in certain cases. And a method of obtaining certain new series from known series by simple algebraic operations, differentiation, or integration is described.

We derive l'Hospital's rule for the evaluation of indeterminate forms. We also show how such forms are sometimes best evaluated by the use of series.

The use of series to express certain integrals and in computation is illustrated. Simpson's rule for the numerical evaluation of integrals is also discussed.

In the Maclaurin's series for e^x, $\sin x$, and $\cos x$ we replace the real variable x by the complex variable z. We use the resulting series to define e^z, $\sin z$, and $\cos z$ for a complex value of z. We then derive the Euler expressions which relate e^{iz} and e^{-iz} to $\sin z$ and $\cos z$. Finally we discuss the hyperbolic functions, whose relation to e^x is similar to the relation of the trigonometric functions to e^{ix}.

246. Maclaurin's Series. It is often possible to express a given function $f(x)$ by a power series in x, so that for appropriate values of x

$$f(x) = a_0 + a_1x + a_2x^2 + \cdots + a_nx^n + \cdots. \tag{1}$$

Suppose that the power series on the right has a sum equal to $f(x)$ for all x inside its interval of convergence. Then for these values of x, as remarked in Sec. 245, the derivative of $f(x)$ is expressed by the series obtained by differentiating the power series of Eq. (1) term by term, or as if the right member were a finite sum. Thus for x inside the interval of convergence,

$$f'(x) = a_1 + 2a_2x + 3a_3x^2 + \cdots + na_nx^{n-1} + \cdots. \tag{2}$$

This process may be applied successively to give

$$f''(x) = 2 \cdot 1a_2 + 3 \cdot 2a_3x + 4 \cdot 3a_4x^2 + \cdots + n(n-1)a_nx^{n-2} + \cdots, \tag{3}$$

$$f'''(x) = 3 \cdot 2 \cdot 1a_3 + 4 \cdot 3 \cdot 2a_4x + 5 \cdot 4 \cdot 3a_5x^2 + \cdots + n(n-1)(n-2)a_nx^{n-3} + \cdots, \quad (4)$$

and so on. For the nth derivative, $f^{(n)}(x)$, we have

$$f^{(n)}(x) = n!a_n + (n+1)n(n-1) \cdots 3 \cdot 2a_{n+1}x + (n+2)(n+1) \cdots 4 \cdot 3a_{n+2}x^2 + \cdots. \quad (5)$$

The result of substituting $x = 0$ in Eqs. (1) to (5) is

$$f(0) = a_0, \quad f'(0) = a_1, \quad f''(0) = 2!a_2, \quad f'''(0) = 3!a_3, \quad \cdots, \quad f^{(n)}(0) = n!a_n. \quad (6)$$

By solving these equations for the coefficients, we find

$$a_0 = f(0), \quad a_1 = f'(0), \quad a_2 = \frac{f''(0)}{2!}, \quad a_3 = \frac{f'''(0)}{3!}, \quad \cdots, \quad a_n = \frac{f^{(n)}(0)}{n!}. \quad (7)$$

We have now evaluated the coefficients in Eq. (1). Hence we may deduce from Eqs. (1) and (7) that

$$f(x) = f(0) + f'(0)x + \frac{f''(0)}{2!}x^2 + \frac{f'''(0)}{3!}x^3 + \cdots + \frac{f^{(n)}(0)}{n!}x^n + \cdots. \quad (8)$$

This result is known as *Maclaurin's series.* It may be used as a formula for the expansion of any given function $f(x)$ in a power series in x, provided $f(x)$ can be so expanded.

For a function to admit of a Maclaurin's series expansion, it is necessary that the function be finite and possess finite derivatives of all orders at $x = 0$. Thus $\ln x$ and $\csc x$ cannot be expanded in Maclaurin's series, since they are each infinite at $x = 0$. Again, since the expression for the first derivative of $\sqrt{x}$ and that for the second derivative of $x^{\frac{3}{2}}$ are each infinite at $x = 0$, the functions $\sqrt{x}$ and $x^{\frac{3}{2}}$ cannot be expanded in Maclaurin's series.

A method of proving that a given function admits of a Maclaurin's series expansion will be discussed in Sec. 254. In all the applications we shall make, the given function $f(x)$ will have finite derivatives of all orders at $x = 0$ which may be directly computed from the rules for differentiation. Under these conditions, the series in the right member of Eq. (8) will have a sum equal to $f(x)$ whenever the series converges. Accordingly we adopt this as a working principle.

EXAMPLE 1. Expand $\sin^2 x$ in a Maclaurin's series.

Solution: Let $f(x) = \sin^2 x$. Then $f'(x) = 2 \sin x \cos x = \sin 2x$. And we may continue the calculation in tabular form.

$$\begin{aligned} f(x) &= \sin^2 x & f(0) &= 0 \\ f'(x) &= \sin 2x & f'(0) &= 0 \\ f''(x) &= 2 \cos 2x & f''(0) &= 2 \\ f'''(x) &= -2^2 \sin 2x & f'''(0) &= 0 \\ f^{(4)}(x) &= -2^3 \cos 2x & f^{(4)}(0) &= -2^3 \\ f^{(5)}(x) &= 2^4 \sin 2x & f^{(5)}(0) &= 0 \\ f^{(6)}(x) &= 2^5 \cos 2x & f^{(6)}(0) &= 2^5 \end{aligned}$$

The values in the second column from $f'(0)$ on are alternately 0 and except for sign powers of 2. Thus $f^{(2n)}(0) = (-1)^{n+1}2^{2n-1}$, $f^{(2n+1)}(0) = 0$. By substituting these values in Eq. (8), we find that

$$\sin^2 x = \frac{2}{2!}x^2 - \frac{2^3}{4!}x^4 + \frac{2^5}{6!}x^6 - \cdots + (-1)^{n+1}\frac{2^{2n-1}}{(2n)!}x^{2n} + \cdots.$$

Let us apply the ratio test as in Sec. 243. For this series $\dfrac{u_{n+1}}{u_n} = \dfrac{2^2x^2}{(2n+1)(2n+2)}$, $t_x = \lim\limits_{n\to\infty} \dfrac{u_{n+1}}{u_n} = 0$ for any fixed x. Since $t_x = 0 < 1$, the series converges for all values of x. And by our working principle it represents $\sin^2 x$ for all values of x.

EXAMPLE 2. Expand $(1 + x)^m$ in a Maclaurin's series.

Solution: Let $f(x) = (1 + x)^m$. Then $f'(x) = m(1 + x)^{m-1}$ and

$$\begin{aligned} f(x) &= (1+x)^m & f(0) &= 1 \\ f'(x) &= m(1+x)^{m-1} & f'(0) &= m \\ f''(x) &= m(m-1)(1+x)^{m-2} & f''(0) &= m(m-1) \\ f'''(x) &= m(m-1)(m-2)(1+x)^{m-3} & f'''(0) &= m(m-1)(m-2) \end{aligned}$$

The products in the second column have one extra factor at each stage. Thus $f^{(n)}(0) = m(m-1)(m-2)\cdots(m-n+1)$. By substituting these values in Eq. (8), we find that

$$(1+x)^m = 1 + mx + \frac{m(m-1)}{2!}x^2 + \frac{m(m-1)(m-2)}{3!}x^3 \\ \cdots + \frac{m(m-1)(m-2)\cdots(m-n+1)}{n!}x^n + \cdots.$$

As we deduced from Eq. (117) of Sec. 245, this series converges for $-1 < x < 1$. Hence by our working principle, it represents the function for these values. This is in accord with the result of Sec. 245.

EXAMPLE 3. Expand $\ln(x + \sqrt{1 + x^2})$ in a Maclaurin's series.

Solution: Let $f(x) = \ln(x + \sqrt{1 + x^2})$. Then $f(0) = \ln 1 = 0$. And the derivative $f'(x) = \dfrac{1 + \frac{1}{2}(2x/\sqrt{1+x^2})}{x + \sqrt{1+x^2}} = \dfrac{1}{\sqrt{1+x^2}}\dfrac{\sqrt{1+x^2}+x}{x+\sqrt{1+x^2}} = \dfrac{1}{\sqrt{1+x^2}}$. As this is a function of x^2, the series for it would contain even powers only, and so half of our coefficients would be zero. Hence we put $u = x^2$, and $f'(x) = (1 + u)^{-\frac{1}{2}}$. The series for this could be found as in Example 2. Or, using the result of Example 2 with u in place of x and $m = -\frac{1}{2}$, we find that

$$(1+u)^{-\frac{1}{2}} = 1 - \frac{1}{2}u + \frac{1\cdot 3}{2\cdot 4}u^2 - \frac{1\cdot 3\cdot 5}{2\cdot 4\cdot 6}u^3 + \cdots \quad \text{for } -1 < u < 1.$$

It follows that for $-1 < x < 1$,

$$f'(x) = (1 + x^2)^{-\frac{1}{2}} = 1 - \frac{1}{2}x^2 + \frac{1 \cdot 3}{2 \cdot 4}x^4 - \frac{1 \cdot 3 \cdot 5}{2 \cdot 4 \cdot 6}x^6 + \cdots.$$

And by the remark on integrating power series made in Sec. 245, since $\int_0^x f'(x)dx = [f(x)]_0^x = f(x) - f(0) = \ln(x + \sqrt{1 + x^2}) - 0$, we have $\ln(x + \sqrt{1 + x^2}) = x - \frac{1}{2}\frac{x^3}{3} + \frac{1 \cdot 3}{2 \cdot 4}\frac{x^5}{5} - \frac{1 \cdot 3 \cdot 5}{2 \cdot 4 \cdot 6}\frac{x^7}{7} + \cdots$ for $-1 < x < 1$. The method used in this example is advantageous whenever a derivative of low order can be expanded by using the binomial series of Example 2, or some other series already found.

EXERCISE 124

Verify the following expansions in Maclaurin's series. And show that each series converges for all values of x.

1. $e^x = 1 + x + \frac{x^2}{2!} + \cdots + \frac{x^n}{n!} + \cdots.$
2. $\sin x = x - \frac{x^3}{3!} + \frac{x^5}{5!} - \cdots + (-1)^n \frac{x^{2n+1}}{(2n+1)!} + \cdots.$
3. $\cos x = 1 - \frac{x^2}{2!} + \frac{x^4}{4!} - \cdots + (-1)^n \frac{x^{2n}}{(2n)!} + \cdots.$
4. $\cos^2 x = 1 - \frac{2}{2!}x^2 + \frac{2^3}{4!}x^4 - \cdots + (-1)^n \frac{2^{2n-1}}{(2n)!} + \cdots.$
5. $2^x = 1 + (\ln 2)x + (\ln 2)^2 \frac{x^2}{2!} + (\ln 2)^3 \frac{x^3}{3!} + \cdots.$

Use the procedure of Example 3 to deduce that each of the following expansions in Maclaurin's series is valid for $-1 < x < 1$.

6. $\ln(1 + x) = x - \frac{x^2}{2} + \frac{x^3}{3} - \cdots + (-1)^{n+1} \frac{x^n}{n} + \cdots.$
7. $\ln\left(\frac{1 + x}{1 - x}\right) = 2\left(x + \frac{x^3}{3} + \frac{x^5}{5} + \cdots + \frac{x^{2n+1}}{2n + 1} + \cdots\right).$
8. $\sin^{-1} x = x + \frac{1}{2}\frac{x^3}{3} + \frac{1 \cdot 3}{2 \cdot 4}\frac{x^5}{5} + \frac{1 \cdot 3 \cdot 5}{2 \cdot 4 \cdot 6}\frac{x^7}{7} + \cdots.$
9. $\tan^{-1} x = x - \frac{x^3}{3} + \frac{x^5}{5} - \cdots + (-1)^n \frac{x^{2n+1}}{2n + 1} + \cdots.$
10. $\cot^{-1} x = \frac{\pi}{2} - x + \frac{x^3}{3} - \frac{x^5}{5} + \cdots + (-1)^{n+1} \frac{x^{2n+1}}{2n + 1} + \cdots.$

Verify the first three terms of each of the following series. We shall see in Sec. 247, Example 3, that these series hold for $-\pi/2 < x < \pi/2$.

11. $\sec x = 1 + \frac{x^2}{2} + \frac{5}{24}x^4 + \frac{61}{720}x^6 + \cdots.$
12. $\tan x = x + \frac{x^3}{3} + \frac{2}{15}x^5 + \frac{17}{315}x^7 + \cdots.$
13. By differentiation, deduce from the series of Prob. 11 that $\tan x \sec x = x + \frac{5}{6}x^3 + \frac{61}{120}x^5 + \cdots.$
14. By differentiation, deduce from the series of Prob. 12 that $\sec^2 x = 1 + x^2 + \frac{2}{3}x^4 + \frac{17}{45}x^6 + \cdots.$
15. By integration from 0 to x, deduce from Prob. 12 that $\ln \sec x = \frac{x^2}{2} + \frac{x^4}{12} + \frac{x^6}{45} + \frac{17}{2{,}520}x^8 + \cdots.$

Show that each of the following expansions is valid for $|x| > 1$, that is $x < -1$ or $x > 1$. HINT: Put $x = 1/u$ in the given function. And expand the appropriate power of $(1 + u)$ in increasing powers of u by using the binomial series of Example 2.

16. $\dfrac{1}{1+x} = \dfrac{1}{x} - \dfrac{1}{x^2} + \dfrac{1}{x^3} - \dfrac{1}{x^4} + \cdots.$

17. $\sqrt{1+x} = \sqrt{x}\left(1 + \dfrac{1}{2}\dfrac{1}{x} - \dfrac{1}{2\cdot 4}\dfrac{1}{x^2} + \dfrac{1\cdot 3}{2\cdot 4\cdot 6}\dfrac{1}{x^3} - \cdots\right).$

18. $(1+x)^m = x^m\left[1 + \dfrac{m}{x} + \dfrac{m(m-1)}{2!}\dfrac{1}{x^2} + \dfrac{m(m-1)(m-2)}{3!}\dfrac{1}{x^3} + \cdots\right].$

19. By replacing x by $-x^2$ in the series of Prob. 1 and then integrating, deduce that, for all values of x,

$$\int^x e^{-x^2}\,dx = x - \frac{x^3}{3} + \frac{1}{2!}\frac{x^5}{5} - \frac{1}{3!}\frac{x^7}{7} + \cdots.$$

20. By replacing x by x^2 in the series of Prob. 2 and then integrating, deduce that, for all values of x,

$$\int_0^x \sin x^2\,dx = \frac{x^3}{3} - \frac{1}{3!}\frac{x^7}{7} + \frac{1}{5!}\frac{x^{11}}{11} - \cdots.$$

21. By replacing x by $-x^4$ in the binomial series of Example 2 with $m = \frac{1}{2}$ and then integrating, deduce that, for $-1 < x < 1$,

$$\int_0^x \sqrt{1-x^4}\,dx = x - \frac{1}{2}\frac{x^5}{5} - \frac{1}{2\cdot 4}\frac{x^9}{9} - \frac{1\cdot 3}{2\cdot 4\cdot 6}\frac{x^{13}}{13} - \cdots.$$

247. Operations with Power Series. For any value of x inside its interval of convergence, the result of differentiating a power series in x term by term, or of integrating it from 0 to x term by term is a new series with the same interval of convergence, representing the corresponding derivative or integral of the sum function. This was illustrated in Sec. 245, as well as in certain problems of Exercises 123 and 124.

From the binomial series and the expansions found in the early problems of Exercise 124, a number of other series can be found by applying the fundamental algebraic operations to the series just as if they were polynomials.† Thus suppose that each of the two power series expansions holds for the indicated interval.

$$f(x) = a_0 + a_1x + a_2x^2 + \cdots + a_nx^n + \cdots, \qquad -A < x < A. \tag{9}$$

$$g(x) = b_0 + b_1x + b_2x^2 + \cdots + b_nx^n + \cdots, \qquad -B < x < B. \tag{10}$$

Then if $A \leqq B$, we may deduce by *addition* that, for the smaller interval $-A < x < A$,

$$f(x) + g(x) = (a_0 + b_0) + (a_1 + b_1)x + \cdots + (a_n + b_n)x^n + \cdots. \tag{11}$$

And by *subtraction*, for the smaller interval $-A < x < A$,

$$f(x) - g(x) = (a_0 - b_0) + (a_1 - b_1)x + \cdots + (a_n - b_n)x^n + \cdots. \tag{12}$$

† Compare the author's "A Treatise on Advanced Calculus," p. 447, John Wiley & Sons, Inc., New York, 1940 (Dover reprint).

Similarly by *multiplication*, for the smaller interval $-A < x < A$,

$$f(x)g(x) = a_0b_0 + (a_0b_1 + a_1b_0)x + (a_0b_2 + a_1b_1 + a_2b_0)x^2 + \cdots. \tag{13}$$

And if $b_0 \neq 0$, by the process of *long division* we may obtain a power series expansion for

$$\frac{f(x)}{g(x)} = \frac{a_0x + a_1x + a_2x^2 + \cdots}{b_0 + b_1x + b_2x^2 + \cdots} = \frac{a_0}{b_0} + \frac{a_1b_0 - a_0b_1}{b_0}x + \cdots. \tag{14}$$

For each *real* number r_1 such that $g(r_1) = 0$, form the absolute value $C_1 = |r_1|$. And for each *complex* number $z_2 = s_2 + it_2 = s_2 + \sqrt{-1}\,t_2$ such that $g(z_2) = 0$, form the absolute value $C_2 = \sqrt{s_2^2 + t_2^2}$. Let C be the smallest of these numbers, that is the least absolute value resulting from any root, real or complex, of the equation $g(x) = 0$. Then if K is the smallest of the three numbers C, A, and B, the series of Eq. (14) will converge for $-K < x < K$.

The process of substituting one series with constant term missing in a second is illustrated in Example 2 below.

EXAMPLE 1. From the series of Prob. 2 of Exercise 124,

$$\sin x = x - \frac{x^3}{3!} + \frac{x^5}{5!} - \cdots,$$

deduce the series for the successive powers of $\sin x$ up to the term in x^5 by multiplication.

Solution: By multiplication as in algebra, we find

$$\begin{array}{l}
x - \dfrac{x^3}{6} + \dfrac{x^5}{120} \\
x - \dfrac{x^3}{6} + \dfrac{x^5}{120} \\
\hline
x^2 - \dfrac{x^4}{6} \\
\quad - \dfrac{x^4}{6} \\
\hline
x^2 - \dfrac{x^4}{3} \quad \text{Hence } \sin^2 x = x^2 - \dfrac{x^4}{3} + \cdots. \\
x - \dfrac{x^3}{6} \\
\hline
x^3 - \dfrac{x^5}{3} \\
\quad - \dfrac{x^5}{6} \\
\hline
x^3 - \dfrac{x^5}{2} \quad \text{Hence } \sin^3 x = x^3 - \dfrac{x^5}{2} + \cdots. \\
x \\
\hline
x^4 \quad \text{Hence } \sin^4 x = x^4 - \cdots. \\
x \\
\hline
x^5 \quad \text{Hence } \sin^5 x = x^5 - \cdots.
\end{array}$$

The expansions required are those given above.

EXAMPLE 2. From the power series in x for $\sin x$ and $\cos x$, deduce the Maclaurin's series for $\cos(\sin x)$ up to terms in x^4.

Solution: From Prob. 3 of Exercise 124, with u in place of x, we have

$$\cos u = 1 - \frac{u^2}{2} + \frac{u^4}{24} - \cdots .$$

And from Prob. 2 of Exercise 124, we may deduce as in Example 1 that, if

$$u = \sin x, \qquad u^2 = x^2 - \frac{x^4}{3} + \cdots \qquad \text{and} \qquad u^4 = x^4 - \cdots .$$

By substituting these expressions in that for $\cos u$ and collecting terms, we find that

$$\cos(\sin x) = 1 - \frac{x^2}{2} + \frac{5}{24}x^4 - \cdots .$$

This is the required expansion, and it holds for all values of x.

EXAMPLE 3. From the series of Probs. 2 and 3 of Exercise 124, deduce the Maclaurin's series for $\tan x$ by division.

Solution: We have $\tan x = \dfrac{\sin x}{\cos x} = \dfrac{x - x^3/3! + \cdots}{1 - x^2/2! + \cdots}$. We may carry out the long division as in algebra.

$$\begin{array}{r|rrrr}
 & x & +\frac{1}{3}x^3 & +\frac{2}{15}x^5 & +\frac{17}{315}x^7 \\
\hline
1 - \frac{1}{2}x^2 + \frac{1}{24}x^4 - \frac{1}{720}x^6 & x & -\frac{1}{6}x^3 & +\frac{1}{120}x^5 & -\frac{1}{5040}x^7 \\
 & x & -\frac{1}{2}x^3 & +\frac{1}{24}x^5 & -\frac{1}{720}x^7 \\
\hline
 & & \frac{1}{3}x^3 & -\frac{1}{30}x^5 & +\frac{1}{840}x^7 \\
 & & \frac{1}{3}x^3 & -\frac{1}{6}x^5 & +\frac{1}{72}x^7 \\
\hline
 & & & \frac{2}{15}x^5 & -\frac{4}{315}x^7 \\
 & & & \frac{2}{15}x^5 & -\frac{1}{15}x^7 \\
\hline
 & & & & \frac{17}{315}x^7
\end{array}$$

It may be shown that $\cos x$ is not zero for any complex value of x. But it is zero for $x = \pi/2$ or $-\pi/2$. And $\pi/2$ is the smallest absolute value of a root of the denominator, $\cos x$. Also the series for $\sin x$ and $\cos x$ converge for all values. It follows that for $-\pi/2 < x < \pi/2$ we have

$$\tan x = x + \frac{x^3}{3} + \frac{2}{15}x^5 + \frac{17}{315}x^7 + \cdots .$$

EXERCISE 125

From the series of Prob. 1 of Exercise 124 and that which results when x is replaced by $-x$, deduce that, for all values of x,

1. $\dfrac{1}{2}(e^x + e^{-x}) = 1 + \dfrac{x^2}{2!} + \dfrac{x^4}{4!} + \cdots + \dfrac{x^{2n}}{(2n)!} + \cdots .$
2. $\dfrac{1}{2}(e^x - e^{-x}) = x + \dfrac{x^3}{3!} + \dfrac{x^5}{5!} + \cdots + \dfrac{x^{2n+1}}{(2n+1)!} + \cdots .$
3. Check the first three terms of the series for $\cos^2 x$ of Prob. 4 of Exercise 124 by multiplying the series of Prob. 3 of Exercise 124 by itself.
4. From the series of Prob. 6 of Exercise 124 and that which results when x is replaced by $-x$, check Prob. 7 of Exercise 124.
5. By multiplying the series of Probs. 11 and 12 of Exercise 124, check the series of Prob. 13 of Exercise 124.
6. By squaring the series of Prob. 11 of Exercise 124, check the series of Prob. 14 of Exercise 124.
7. By multiplying the series of Probs. 1 and 2 of Exercise 124, deduce that $e^x \sin x = x + x^2 + \frac{1}{3}x^3 - \frac{1}{30}x^5 + \cdots$.

From the series of Probs. 2 and 3 of Exercise 124, deduce by division

8. $\sec x = \dfrac{1}{\cos x} = 1 + \dfrac{1}{2}x^2 + \dfrac{5}{24}x^4 + \cdots, \ -\dfrac{\pi}{2} < x < \dfrac{\pi}{2}.$

9. $\cot x = \dfrac{\cos x}{\sin x} = \dfrac{1}{x} - \dfrac{1}{3}x - \dfrac{1}{45}x^3 - \cdots, \ -\pi < x < 0 \text{ or } 0 < x < \pi.$

10. $\csc x = \dfrac{1}{\sin x} = \dfrac{1}{x} + \dfrac{1}{6}x + \dfrac{7}{360}x^3 + \cdots, \ -\pi < x < 0 \text{ or } 0 < x < \pi.$

11. $\dfrac{\sin x}{x} = 1 - \dfrac{x^2}{3!} + \dfrac{x^4}{5!} - \cdots.$

12. $\dfrac{1 - \cos x}{x^2} = \dfrac{1}{2!} - \dfrac{x^2}{4!} + \dfrac{x^4}{6!} - \cdots.$

13. From the series of Prob. 3 of Exercise 124, it follows that $u = \cos x - 1 = -\dfrac{x^2}{2} + \dfrac{x^4}{24} - \dfrac{x^6}{720} + \cdots$. Deduce by multiplication that $u^2 = \dfrac{x^4}{4} - \dfrac{x^6}{24} + \cdots, \ u^3 = -\dfrac{x^6}{8} + \cdots.$

14. From the series of Prob. 1 of Exercise 124 with u in place of x, $e^u = 1 + u + \dfrac{u^2}{2} + \dfrac{u^3}{6} + \cdots$. And if $u = \cos x - 1$, $e^{\cos x} = e^{1+u} = e \cdot e^u$. By substitution from Prob. 13, deduce that

$$e^{\cos x} = e\left(1 - \frac{1}{2}x^2 + \frac{1}{6}x^4 - \frac{31}{720}x^5 + \cdots\right).$$

15. From the series found in Example 1 and the series for e^x of Prob. 1 of Exercise 124, deduce that $e^{\sin x} = 1 + x + \dfrac{1}{2}x^2 - \dfrac{1}{8}x^4 - \dfrac{1}{15}x^5 - \cdots.$

16. From the binomial series of Example 2 of Sec. 246, we have $(1 + u)^m = 1 + mu + \dfrac{m(m-1)}{2!}u^2 + \cdots$. By substitution from Prob. 13, deduce that

$$\cos^m x = 1 - \frac{m}{2}x^2 + \frac{m(3m-2)}{24}x^4 + \cdots, \ -\frac{\pi}{2} < x < \frac{\pi}{2}.$$

Use the series of Prob. 16 to check the first three terms of

17. The series for $\cos^2 x$ of Prob. 4 of Exercise 124.
18. The series for $\sec x = (\cos x)^{-1}$ of Prob. 11 of Exercise 124.
19. The series for $\sec^2 x = (\cos x)^{-2}$ of Prob. 14 of Exercise 124.

20. By integration from 0 to x, deduce from Prob. 11 that

$$\int_0^x \frac{\sin x}{x}\,dx = x - \frac{1}{3!}\frac{x^3}{3} + \frac{1}{5!}\frac{x^5}{5} - \cdots.$$

21. By integration from 0 to x, deduce from Prob. 12 that

$$\int_0^x \frac{1 - \cos x}{x^2}\,dx = \frac{1}{2!}x - \frac{1}{4!}\frac{x^3}{3} + \frac{1}{6!}\frac{x^5}{5} - \cdots.$$

248. Computation. Each partial sum of a Maclaurin's expansion is a polynomial, whose value may be calculated for a particular x. If the Maclaurin's series converges for this x, a partial sum can be found which equals the value of the series to any desired degree of accuracy. Thus in theory, a function may be computed from its Maclaurin's expansion for any value of x inside the range of validity of the expansion. For some functions, the error made by breaking off the series after the nth term, or using the partial sum S_n, could be estimated by using the theorem of Sec. 254, or the methods illustrated in Examples 2 and 3 of Sec. 245.

But a computation of this kind is convenient only when x is so small that relatively few terms of the series will give the required accuracy. Under these conditions, the practical method of finding a result accurate to within say 2 in the fourth decimal place is as follows. Compute each term used to the fifth decimal place. Include all terms that numerically exceed 0.00001. Calculate the algebraic sum of these terms and round off the result to four places.

In all the equations involving trigonometric functions of x, x is a number of radians. We note that $10° = 0.17453$ radian. Also

$$(0.17453)^2 = 0.03046, \qquad (0.17453)^3 = 0.00532, \qquad (0.17453)^4 = 0.00094,$$
$$(0.17453)^5 = 0.00016, \qquad (0.17453)^6 = 0.00003. \tag{15}$$

EXAMPLE 1. From Prob. 7 of Exercise 124, derive a series for computing $\ln (N + 1)$ from $\ln N$, where $N > 0$.

Solution: From Prob. 7 of Exercise 124, for $-1 < x < 1$, we have $\ln \left(\frac{1 + x}{1 - x}\right) = 2\left(x + \frac{x^3}{3} + \frac{x^5}{5} + \cdots\right)$. To convert this into an expression for $\ln (N + 1) - \ln N = \ln \frac{N + 1}{N}$, we let $\frac{1 + x}{1 - x} = \frac{N + 1}{N}\cdot$ Then $N(1 + x) = (N + 1)(1 - x)$, $N + Nx = N - Nx + 1 - x$, $2Nx + x = 1$, and $x = \frac{1}{2N + 1}\cdot$ This makes $0 < x < 1$ if $N > 0$. It follows that for $N > 0$, if $x = \frac{1}{2N + 1}$,

$$\ln (N + 1) - \ln N = 2\left(x + \frac{x^3}{3} + \frac{x^5}{5} + \cdots\right).$$

This leads to the required series

$$\ln (N + 1) = \ln N + 2\left[\frac{1}{2N + 1} + \frac{1}{3}\frac{1}{(2N + 1)^3} + \frac{1}{5}\frac{1}{(2N + 1)^5} + \cdots\right].$$

EXAMPLE 2. Use the result of Example 1 to compute $\ln 2$ and $\ln 5$ to four places. Also find $\ln 10$ and $\log 2$.

Solution: If $N = 1$, $\ln N = \ln 1 = 0$, $\ln (N + 1) = \ln 2$, and $x = \frac{1}{2N + 1} = \frac{1}{3}\cdot$ On substituting these values in the result of Example 1, we have $\ln 2 = 2\left(\frac{1}{3} + \frac{1}{3}\frac{1}{3^3} + \frac{1}{5}\frac{1}{3^5} + \cdots\right)\cdot$ We may compute in succession the following tabulated values.

$$\begin{array}{ll}
\frac{1}{3} = 0.33333 & \frac{1}{3} = 0.33333 \\[4pt]
\frac{1}{3^3} = 0.03704 & \frac{1}{3}\frac{1}{3^3} = 0.01235 \\[4pt]
\frac{1}{3^5} = 0.00412 & \frac{1}{5}\frac{1}{3^5} = 0.00082 \\[4pt]
\frac{1}{3^7} = 0.00046 & \frac{1}{7}\frac{1}{3^7} = 0.00007 \\[4pt]
\frac{1}{3^9} = 0.00005 & \frac{1}{9}\frac{1}{3^9} = 0.00001 \\
\hline
 & \qquad\quad 0.34658
\end{array}$$

The odd powers of $\frac{1}{3}$ in the first column are computed by dividing the previous power by 9. And from them we find the terms in the second column by division by the appropriate odd integer. It follows that $\ln 2 = 2(0.34658) = 0.69316$.

Since $\ln 4 = \ln 2^2 = 2 \ln 2$, $\ln 4 = 1.38632$.

Next let $N = 4$. Then $N + 1 = 5$ and $2N + 1 = 9$. Hence from Example 1, $\ln 5 = \ln 4 + 2\left(\frac{1}{9} + \frac{1}{3}\frac{1}{9^3} + \frac{1}{5}\frac{1}{9^5} + \cdots\right)$ We find

$$\begin{array}{ll}
\frac{1}{9} = 0.11111 & \frac{1}{9} = 0.11111 \\
\frac{1}{9^2} = 0.01235 & \\
\frac{1}{9^3} = 0.00137 & \frac{1}{3}\frac{1}{9^3} = 0.00046 \\
\frac{1}{9^4} = 0.00015 & \\
\frac{1}{9^5} = 0.00002 & \frac{1}{5}\frac{1}{9^5} = 0.00000 \\
 & \overline{\ 0.11157}
\end{array}$$

It follows that $\ln 5 = \ln 4 + 2(0.11157) = 1.38632 + 0.22314 = 1.60946$. Also $\ln 10 = \ln 5 + \ln 2 = 1.60946 + 0.69316 = 2.30260$.

And $\log 2 = \log_{10} 2 = \dfrac{\log_e 2}{\log_e 10} = \dfrac{\ln 2}{\ln 10} = \dfrac{0.69316}{2.30260} = 0.30103$.

Hence the required four-place values are

$$\ln 2 = 0.6932, \qquad \ln 5 = 1.6095, \qquad \ln 10 = 2.3026, \qquad \log 2 = 0.3010.$$

EXAMPLE 3. Use the series of Probs. 2 and 3 of Exercise 124 to compute sin 5° and cos 5° to four decimal places.

Solution: The series referred to are $\sin x = x - \dfrac{x^3}{3!} + \dfrac{x^5}{5!} - \cdots$, $\cos x = 1 - \dfrac{x^2}{2!} + \dfrac{x^4}{4!} - \cdots$. Here x is in radians. And 5° is $5\left(\dfrac{\pi}{180}\right) = 5(0.017453) = 0.08726$. And from Eq. (15), by multiplying in powers of $\frac{5}{10} = \frac{1}{2}$, we find for $x = 0.08726$, that $x^2 = 0.00762$, $x^3 = 0.00066$, $x^4 = 0.00006$. It follows that

$$\sin x = x - \frac{x^3}{6} = 0.08726 - 0.00011 = 0.08715,$$

and

$$\cos x = 1 - \frac{x^2}{2} + \frac{x^4}{24} = 1 - 0.00381 + 0.00000 = 0.99619.$$

Hence the required four-place values are

$$\sin 5° = 0.0872, \qquad \cos 5° = 0.9962.$$

249. Approximate Formulas. By using the first few nonzero terms of the Maclaurin's series, we obtain particularly simple approximate formulas. For sufficiently small values of the variable, these are accurate enough for many applications in engineering and instrumentation. The first-degree approximations found in this way are identical with those found by means of differentials in Sec. 165. But our present point of view enables us to estimate the error.

EXAMPLE 1. From the binomial series of Example 2 of Sec. 246, derive first- and second-degree approximations to $(1 + x)^m$. Estimate the error for $m = \frac{1}{2}$ and $|x| < 0.1$.

Solution: $(1 + x)^m = 1 + mx + \frac{m(m-1)}{2!}x^2 + \frac{m(m-1)(m-2)}{3!}x^3 + \cdots$.
Hence the first-degree approximation for $(1 + x)^m$ is $1 + mx$. And the second-degree approximation is $1 + mx + \frac{m(m-1)}{2}x^2$.

For $m = \frac{1}{2}$, $\sqrt{1+x} = 1 + \frac{1}{2}x - \frac{1}{8}x^2 + \frac{1}{16}x^3 - \cdots$.

For the first-degree approximation, the first term omitted is $-\frac{1}{8}x^2$. And $\frac{1}{8}(0.1)^2 = 0.00125$. Hence for $|x| < 0.1$, $|-\frac{1}{8}x^2| < 0.00125$. Thus $\sqrt{1+x} = 1 + \frac{x}{2}$ to within 0.002 if $|x| < 0.1$.

For the second-degree approximation, the first term omitted is $\frac{1}{16}x^3$. And $\frac{1}{16}(0.1)^3 = 0.0000625$. Hence for $|x| < 0.1$, $|\frac{1}{16}x^3| < 0.0000625$. Thus $\sqrt{1+x} = 1 + \frac{x}{2} - \frac{x^2}{8}$ to within 0.0001 if $|x| < 0.1$.

EXAMPLE 2. For what range of x is $\sin x = x$ to within 0.002? And for what range of x is $\sin x = x - \frac{x^3}{6}$ to within 0.002?

Solution: From Prob. 2 of Exercise 124, $\sin x = x - \frac{x^3}{3!} + \frac{x^5}{5!} - \cdots$. For the approximation $\sin x = x$, the first term omitted is $-\frac{x^3}{3!}$. The absolute value of this equals 0.002 when $\frac{x^3}{6} = 0.002$, $x^3 = 0.012$, and $x = \sqrt[3]{0.012} = 0.229$. Hence $\sin x = x$ to within 0.002 for $|x| < 0.229$ radian.

For the approximation $\sin x = x - \frac{x^3}{6}$, the first term omitted is $\frac{x^5}{5!}$. The absolute value of this equals 0.002 when $\frac{x^5}{120} = 0.002$, $x^5 = 0.24$, and $x = \sqrt[5]{0.24} = 0.752$. Hence $\sin x = x - \frac{x^3}{6}$ to within 0.002 for $|x| < 0.752$ radian.

We note that 0.229 radian equals 13.12°, and 0.752 radian equals 43.08°. Hence, since $D°$ equals $\frac{\pi}{180}D$ or $(0.01745)D$ radian,

$$\sin D° = 0.01745D \text{ to within } 0.002 \text{ for } |D| < 13.$$
$$\sin D° = 0.01745D - 0.00000089D^3 \text{ to within } 0.002 \text{ for } |D| < 43.$$

EXERCISE 126

Verify each of the following results by using the series

$$e^x = 1 + x + \frac{x^2}{2!} + \frac{x^3}{3!} + \cdots.$$

1. $e^{0.2} = 1.2214$.
2. $e^{-0.2} = 0.8188$.
3. $\sqrt{e} = e^{\frac{1}{2}} = 1.6487$.
4. $\frac{1}{e} = e^{-1} = 0.3679$.
5. $e^x = 1 + x$ to within 0.01 for $|x| < 0.14$.
6. $e^x = 1 + x + \frac{x^2}{2}$ to within 0.001 for $|x| < 0.18$.

Verify each of the following results, given that $\sin x = x - \frac{x^3}{3!} + \frac{x^5}{5!} - \cdots$, $\cos x = 1 - \frac{x^2}{2!} + \frac{x^4}{4!} - \cdots$.

7. $\sin 0.5 = 0.4794$.
8. $\cos 0.5 = 0.5463$.
9. $\sin 10° = 0.1736$.
10. $\cos 10° = 0.9848$.
11. $\sin 2° = 0.0349$.
12. $\cos 2° = 0.9994$.
13. $\sin x = x - \frac{x^3}{6} + \frac{x^5}{120}$ to within 4×10^{-5} if $|x| < \frac{\pi}{4}$.
14. $\cos x = 1 - \frac{x^2}{2} + \frac{x^4}{24}$ to within 4×10^{-4} if $|x| < \frac{\pi}{4}$.
15. $\sin (a + x) = \sin a \left(1 - \frac{x^2}{2} + \cdots\right) + \cos a \left(x - \frac{x^3}{6} + \cdots\right)$.

Given that $\sin 60° = 0.5$ and $\cos 60° = \frac{1}{2}\sqrt{3} = 0.8660$, verify that

16. $\sin 62° = 0.8829$ by using Prob. 15, or Probs. 11 and 12.
17. $\sin 58° = 0.8480$ by using Prob. 15, or Probs. 11 and 12.

18. From the series of Example 1, and $\ln 2 = 0.69315$, compute $\ln 3 = \ln (2 + 1)$. Then find $\ln 9 = 2 \ln 3$, and compute $\ln 10 = \ln (9 + 1)$.
19. Use the series $\sin^{-1} x = x + \frac{1}{2}\frac{x^3}{3} + \frac{1 \cdot 3}{2 \cdot 4}\frac{x^5}{5} + \frac{1 \cdot 3 \cdot 5}{2 \cdot 4 \cdot 6}\frac{x^7}{7} + \cdots$ to compute $\frac{\pi}{6} = \sin^{-1} \frac{1}{2} = 0.5236$.

Use the series $\tan^{-1} x = x - \frac{x^3}{3} + \frac{x^5}{5} - \cdots$ to verify that

20. $\tan^{-1} \frac{1}{3} = 0.32175$.
21. $\tan^{-1} \frac{1}{7} = 0.14190$.
22. $\tan^{-1} \frac{1}{5} = 0.19740$.
23. $\tan^{-1} \frac{1}{239} = 0.00418$.

By repeated use of $\tan (A + B) = \frac{\tan A + \tan B}{1 - \tan A \tan B}$, show that

24. $\tan (2 \tan^{-1} \frac{1}{3} + \tan^{-1} \frac{1}{7}) = 1$.
25. $\tan (4 \tan^{-1} \frac{1}{5} - \tan^{-1} \frac{1}{239}) = 1$.

26. From Prob. 24, $\pi/4 = 2 \tan^{-1} \frac{1}{3} + \tan^{-1} \frac{1}{7}$. Deduce from Probs. 20 and 21 that $\pi/4 = 0.7854$.
27. From Prob. 25, $\pi/4 = 4 \tan^{-1} \frac{1}{5} - \tan^{-1} \frac{1}{239}$. Deduce from Probs. 22 and 23 that $\pi/4 = 0.7854$.

250. Taylor's Series. It is usually possible to express a given function $f(x)$ by a power series in $(x - a)$ of the type discussed in Sec. 244. Thus, barring exceptional values of a, for values of x appropriate to the choice of a, we shall have

$$f(x) = b_0 + b_1(x - a) + b_2(x - a)^2 + \cdots + b_n(x - a)^n + \cdots. \tag{16}$$

Suppose that the power series on the right has a sum equal to $f(x)$ for all values of x inside its interval of convergence. Then for these values of x, we may differentiate the power series term by term. Thus

$$f'(x) = b_1 + 2b_2(x - a) + 3b_3(x - a)^2 + \cdots. \tag{17}$$

And a repetition of this process gives

$$f''(x) = 2 \cdot 1b_2 + 3 \cdot 2b_3(x - a) + 4 \cdot 3b_4(x - a)^2 + \cdots, \tag{18}$$

$$f'''(x) = 3 \cdot 2 \cdot 1b_3 + 4 \cdot 3 \cdot 2b_4(x - a) + 5 \cdot 4 \cdot 3b_5(x - a)^2 + \cdots, \tag{19}$$

and so on. For the nth derivative, $f^{(n)}(x)$, we have

$$f^{(n)}(x) = n!b_n + (n + 1)n(n - 1) \cdots 3 \cdot 2b_{n+2}(x - a)^2 + \cdots. \tag{20}$$

The result of substituting $x = a$ in Eqs. (16) to (20) is

$$f(a) = b_0, \quad f'(a) = b_1, \quad f''(a) = 2!b_2, \quad f'''(a) = 3!b_3, \quad \ldots, \quad f^{(n)}(a) = n!b_n. \tag{21}$$

By solving these equations for the coefficients we find

$$b_0 = f(a), \quad b_1 = f'(a), \quad b_2 = \frac{f''(a)}{2!}, \quad b_3 = \frac{f'''(a)}{3!}, \quad \ldots, \quad b_n = \frac{f^{(n)}(a)}{n!}. \tag{22}$$

We have now evaluated the coefficients in Eq. (16). Hence we may deduce from Eqs. (16) and (22) that

$$f(x) = f(a) + f'(a)(x - a) + \frac{f''(a)}{2!}(x - a)^2 + \frac{f'''(a)}{3!}(x - a)^3 + \cdots + \frac{f^{(n)}(a)}{n!}(x - a)^n + \cdots. \tag{23}$$

This result is known as *Taylor's series.* It may be used as a formula for the expansion of any given function $f(x)$ in a power series in $(x - a)$, provided $f(x)$ can be so expanded. For a function to admit of a Taylor's series expansion in powers of $(x - a)$, it is necessary that the function be finite and possess finite derivatives of all orders at $x = a$. A method of proving that the expansion exists will be discussed in Sec. 253. In the applications we shall make, the series in the right member of Eq. (23) will have a sum equal to $f(x)$ whenever the series converges. Accordingly we adopt this as a working principle.

EXAMPLE 1. Expand $1/x$ in powers of $(x - 2)$.

Solution: Let $f(x) = \frac{1}{x}\cdot$ Since $(x - 2) = (x - a)$, $a = 2$. We find

$$\begin{aligned} f(x) &= \frac{1}{x} & f(2) &= \frac{1}{2} \\ f'(x) &= -\frac{1}{x^2} & f'(2) &= -\frac{1}{2^2} \\ f''(x) &= \frac{2}{x^3} & f''(2) &= \frac{2}{2^3} \\ f'''(x) &= -\frac{2 \cdot 3}{x^4} & f'''(2) &= -\frac{2 \cdot 3}{2^4} \\ &\cdots\cdots & &\cdots\cdots \\ f^{(n)}(x) &= (-1)^n \frac{n!}{x^{n+1}} & f^{(n)}(2) &= (-1)^n \frac{n!}{2^{n+1}} \end{aligned}$$

By substituting these values in Eq. (23), we find that

$$\frac{1}{x} = \frac{1}{2} - \frac{(x-2)}{2^2} + \frac{(x-2)^2}{2^3} - \cdots + (-1)^n \frac{(x-2)^n}{2^{n+1}} + \cdots.$$

This geometric series has a ratio $-\dfrac{(x-2)}{2}$. And $\left| -\dfrac{(x-2)}{2} \right| < 1$ when $|x-2| < 2$. $-2 < x - 2 < 2$, or $0 < x < 4$. The series converges to $1/x$ for $0 < x < 4$.

EXAMPLE 2. Find the first four terms of the Taylor's series for $\tan x$ in powers of $(x - \pi/4)$.

Solution: Let $f(x) = \tan x$. Since $(x - a) = (x - \pi/4)$, $a = \pi/4$. Then $\tan (\pi/4) = 1$ and $\sec (\pi/4) = \sqrt{2}$. And we find

$$\begin{aligned}
f(x) &= \tan x & f\left(\frac{\pi}{4}\right) &= 1 \\
f'(x) &= \sec^2 x & f'\left(\frac{\pi}{4}\right) &= 2 \\
f''(x) &= 2 \sec x \,(\tan x \sec x) & & \\
&= 2 \tan x \sec^2 x & f''\left(\frac{\pi}{4}\right) &= 4 \\
f'''(x) &= 2 \sec^4 x & & \\
&\quad + 4 \tan^2 x \sec^2 x & f'''\left(\frac{\pi}{4}\right) &= 16
\end{aligned}$$

By substituting these values in Eq. (23), we find that

$$\tan x = 1 + 2\left(x - \frac{\pi}{4}\right) + 2\left(x - \frac{\pi}{4}\right)^2 + \frac{8}{3}\left(x - \frac{\pi}{4}\right)^3 + \cdots.$$

When $\cos x = 0$, $\tan x = \dfrac{\sin x}{\cos x}$ is infinite. And $\cos x = 0$ has no complex roots, while $\pi/2$ is the nearest real root to $\pi/4$. Thus for any root r, the least value of $|r - \pi/4|$ is $\pi/2 - \pi/4 = \pi/4$. As a consequence of these facts, it may be shown that the expansion found converges for $|x - \pi/4| < \pi/4$, or $0 < x < \pi/2$.

251. Incremental Form of Taylor's Series. When x changes from a to x, $(x - a)$ is the increment in x. Let $x - a = h$. Then $x = a + h$, and we may write Eq. (23) as

$$f(a+h) = f(a) + f'(a)h + \frac{f''(a)}{2!}h^2 + \frac{f'''(a)}{3!}h^3 + \cdots + \frac{f^{(n)}(a)}{n!}h^n + \cdots. \tag{24}$$

This expresses the new value of the function, $f(a + h)$, as a series in powers of h, the increment in x. This form is often easier to write in making computations by means of Taylor's series.

EXAMPLE 1. Expand ln x in a power series in h, where h is the change in x from 3 to $3 + h$.

Solution: Let $f(x) = \ln x$. Here $a = 3$. And we find

$$\begin{aligned} f(x) &= \ln x & f(3) &= \ln 3 \\ f'(x) &= \frac{1}{x} & f'(3) &= \frac{1}{3} \\ f''(x) &= -\frac{1}{x^2} & f''(3) &= -\frac{1}{3^2} \\ f'''(x) &= \frac{2}{x^3} & f'''(3) &= \frac{2}{3^3} \\ f^{(4)}(x) &= -\frac{2\cdot 3}{x^4} & f^{(4)}(3) &= -\frac{2\cdot 3}{3^4} \\ &\cdots\cdots & &\cdots\cdots \\ f^{(n)}(x) &= (-1)^{n+1}\frac{(n-1)!}{x^n} & f^{(n)}(3) &= (-1)^{n+1}\frac{(n-1)!}{3^n} \end{aligned}$$

Observing that $\frac{(n-1)!}{n!} = \frac{1}{n}$ and substituting these values in Eq. (24), we find that for $-3 < h < 3$,

$$\ln (3 + h) = \ln 3 + \frac{h}{3} - \frac{1}{2}\frac{h^2}{3^2} + \frac{1}{3}\frac{h^3}{3^3} - \cdots + (-1)^{n+1}\frac{1}{n}\frac{h^n}{3^n} + \cdots .$$

We note that, since $\ln (3 + h) = \ln 3 + \ln (1 + h/3)$, the series just found may be checked by putting $x = h/3$ in the Maclaurin's series for $\ln (1 + x)$ found in Prob. 6 of Exercise 124.

EXAMPLE 2. Compute tan 43° to four decimal places by using the power series for $\tan x$ in terms of h, where h is the change in x from $\pi/4$ to $\pi/4 + h$.

Solution: From the derivatives found in Example 2 of Sec. 250 and Eq. (24), we find that

$$\tan\left(\frac{\pi}{4} + h\right) = 1 + 2h + 2h^2 + \frac{8}{3}h^3 + \cdots .$$

This could also be obtained by placing $x = \pi/4 + h$, $x - \pi/4 = h$ in the result of Example 2 of Sec. 250.

We have $43° = 45° - 2°$ or $\frac{\pi}{4} - 2\left(\frac{\pi}{180}\right)$ radian. Hence to make $\tan\left(\frac{\pi}{4} + h\right) = \tan 43°$, we set $\frac{\pi}{4} + h = \frac{\pi}{4} - 2\left(\frac{\pi}{180}\right)$. Thus $h = -2\left(\frac{\pi}{180}\right) = -2(0.017453) = -0.03491$. And from Eq. (15), by multiplying in powers of $\frac{-2}{10} = -0.2$, we find for $h = -0.03491$, that $h^2 = 0.00122$, $h^3 = -0.00004$. It follows that

$$\begin{array}{rlrl} 1 = & 1.00000 & 2h = & -0.06982 \\ 2h^2 = & \underline{0.00244} & \frac{8}{3}h^3 = & \underline{-0.00011} \\ & 1.00244 & & -0.06993 \\ & \underline{-0.06993} & & \\ & 0.93251 & & \end{array}$$

Hence tan 43° = 0.9325 is the required four-place value.

EXERCISE 127

For each given function $f(x)$ and value of a, find the first four nonzero terms of the Taylor's series in powers of $(x - a)$.

1. $f(x) = e^x,\ a = 2.$
2. $f(x) = \sin x,\ a = \dfrac{\pi}{3}.$
3. $f(x) = \cos x,\ a = \dfrac{\pi}{4}.$
4. $f(x) = \dfrac{1}{x},\ a = 1.$
5. $f(x) = \dfrac{1}{1 + x},\ a = 2.$
6. $f(x) = \sqrt{1 + x},\ a = 3.$
7. $f(x) = \ln x,\ a = 1.$
8. $f(x) = \tan^{-1} x,\ a = 1.$
9. $f(x) = \dfrac{1}{\sqrt{x}},\ a = 1.$
10. $f(x) = x^{\frac{5}{2}},\ a = 1.$

Use Eq. (24) to verify each of the following Taylor's series.

11. $$e^{a+h} = e^a + e^a h + \frac{e^a}{2!}h^2 + \frac{e^a}{3!}h^3 + \cdots = e^a\left(1 + h + \frac{h^2}{2!} + \frac{h^3}{3!} + \cdots\right).$$

12. $$\sin(a + h) = \sin a + h\cos a - \frac{h^2}{2!}\sin a - \frac{h^3}{3!}\cos a + \cdots = \sin a\left(1 - \frac{h^2}{2!} + \cdots\right) + \cos a\left(h - \frac{h^3}{3!} + \cdots\right).$$

13. $$\cos(a + h) = \cos a - h\sin a - \frac{h^2}{2!}\cos a + \frac{h^3}{3!}\sin a + \cdots = \cos a\left(1 - \frac{h^2}{2!} + \cdots\right) - \sin a\left(h - \frac{h^3}{3!} + \cdots\right).$$

14. $$\ln(a + h) = \ln a + \frac{h}{a} - \frac{1}{2}\frac{h^2}{a^2} + \frac{1}{3}\frac{h^3}{a^3} - \cdots,\quad a \neq 0.$$

15. $$(a + h)^m = a^m + ma^{m-1}h + \frac{m(m-1)}{2!}a^{m-2}h^2 + \cdots + \frac{m(m-1)(m-2)}{3!}a^{m-3}h^3 + \cdots,\quad a \neq 0.$$

Use the series of the indicated problem to compute the value of each function to three significant figures.

16. $e^{10.2}$, given $e^{10} = 22{,}026$, from Prob. 11 with $a = 10$.
17. $e^{-5.1}$, given $e^{-5} = 0.006738$, from Prob. 11 with $a = -5$.
18. $\sin 46°$, from Prob. 12 with $a = \pi/4$.
19. $\cos 47°$, from Prob. 13 with $a = \pi/4$.
20. $\ln 2.1$, given $\ln 2 = 0.69315$ from Prob. 14 with $a = 2$.
21. $\sqrt[4]{17}$, from Prob. 15 with $a = 16$ and $m = \frac{1}{4}$.
22. $\sqrt[3]{26}$, from Prob. 15 with $a = 27$ and $m = \frac{1}{3}$.
23. $\frac{1}{98}$, from Prob. 15 with $a = 100$ and $m = -1$.

24. Check Prob. 11 by writing $e^{a+h} = e^a e^h$, and replacing x by h in the Maclaurin's series of Prob. 1 of Exercise 124.
25. Check Probs. 12 and 13 by using the addition theorems $\sin(a + h) = \sin a\cos h + \cos a\sin h$, and $\cos(a + h) = \cos a\cos h - \sin a\sin h$, and replacing x by h in the Maclaurin's series of Probs. 2 and 3 of Exercise 124.
26. Check Prob. 14 by writing $\ln(a + h) = \ln a + \ln(1 + h/a)$, and replacing x by h/a in the Maclaurin's series of Prob. 6 of Exercise 124.

27. Check Prob. 15 by writing $(a + h)^m = a^m(1 + h/a)^m$ and replacing x by h/a in the Maclaurin's series of Example 2 of Sec. 246.
28. Show that Maclaurin's series, Eq. (8), is the special case of Taylor's series, Eq. (23), with $a = 0$. Also show that if $f(a + h) = F(h)$, the Taylor's series of Eq. (24) for $f(x)$ may be obtained from the Maclaurin's series for $F(x)$ by replacing x by h.

***252. Mean Value Theorems.** In Eq. (27) of Sec. 37 we showed from geometric considerations that for any differentiable function $f(x)$ and any two values x_1 and x_2, there is always some value x_0 between x_1 and x_2 such that

$$\frac{f(x_2) - f(x_1)}{x_2 - x_1} = f'(x_0). \tag{25}$$

In a different notation, we may formulate this *mean value theorem* more precisely as follows.

Hypothesis 1. *Let $F(x)$ be a function which is continuous at all points of the interval $a \leqq x \leqq b$, and let the function $F(x)$ have a derivative $F(x)$ at each point x inside the interval $a < x < b$.*

Conclusion: Then there is a value $x = X$, inside the interval $a < X < b$, such that

$$\frac{F(b) - F(a)}{b - a} = F'(x). \tag{26}$$

An alternative statement of this conclusion is

$$F(b) - F(a) = (b - a)F'(X). \tag{27}$$

Let us set

$$b = a + h \qquad \text{and} \qquad X = a + \theta h. \tag{28}$$

Then we have

$$b - a = h > 0. \tag{29}$$

And the relation $a < X < b$ implies that $0 < \theta h < h$. Hence

$$0 < \theta < 1. \tag{30}$$

By using Eqs. (28) and (29), we may rewrite Eq. (27) in the form

$$F(a + h) - F(a) = hF'(a + \theta h). \tag{31}$$

Thus, from hypothesis 1, we may draw the following conclusion. *There is a value of θ between* 0 *and* 1, $0 < \theta < 1$, *such that Eq.* (31) *holds.*

This theorem is known as the *law of finite increments.*

If $F(a) = 0$ and $F(b) = 0$, it follows from Eq. (26) that $F'(X) = 0$. This leads to the following specialized result.

If $F(X)$ satisfies hypothesis 1, *and in addition*

$$F(a) = 0 \qquad \text{and} \qquad F(b) = 0, \tag{32}$$

then there is a value $x = X$ inside the interval $a < X < b$ such that

$$F'(X) = 0. \tag{33}$$

This special form is known as *Rolle's theorem.*

As an application of Rolle's theorem, we shall prove *Cauchy's generalized mean value theorem,* which may be formulated as follows.

As in hypothesis 1, *let each of the functions $f(x)$ and $g(x)$ be continuous for $a \leqq x \leqq b$, and differentiable for $a < x < b$. We also assume that $g(b) \neq g(a)$, and that $f'(x)$ and*

$g'(x)$ are never both zero at the same time for any value of x with $a < x < b$. Then there is a value $x = X$ inside the interval $a < X < b$ such that

$$\frac{f(b) - f(a)}{g(b) - g(a)} = \frac{f'(X)}{g'(X)}. \tag{34}$$

To prove this, consider the function

$$F(x) = [f(x) - f(a)][g(b) - g(a)] - [g(x) - g(a)][f(b) - f(a)]. \tag{35}$$

When $x = a$, the first factor of each product on the right is zero. Thus $F(a) = 0$. And when $x = b$, the two products on the right are equal and opposite in sign, so that their sum is zero. Thus $F(b) = 0$. Also $F(x)$ satisfies hypothesis 1, since $f(x)$ and $g(x)$ each satisfy a similar condition. It follows from Rolle's theorem that there is an X with $a < X < b$ for which $F'(X) = 0$ as in Eq. (33). But from Eq. (35) we find that

$$F'(x) = f'(x)[g(b) - g(a)] - g'(x)[f(b) - f(a)]. \tag{36}$$

It follows from $F'(X) = 0$ that

$$f'(X)[g(b) - g(a)] - g'(X)[f(b) - f(a)] = 0. \tag{37}$$

Suppose that $g'(X)$ were equal to 0. Then since $g(b) \neq g(a)$, it would follow from Eq. (37) that $f'(X) = 0$. Thus $f'(x)$ and $g'(x)$ would both be zero for the same value $x = X$ with $a < X < b$. As this contradicts our hypothesis, it follows that $g'(X) \neq 0$. Hence we may divide Eq. (37) by the product $g'(X)[g(b) - g(a)]$ to obtain

$$\frac{f'(X)}{g'(X)} - \frac{f(b) - f(a)}{g(b) - g(a)} = 0. \tag{38}$$

As this is equivalent to Eq. (34), the theorem is proved.

EXAMPLE. Let $f(x)$ be continuous for $a \leqq x \leqq b$ and have a first derivative $f'(x)$ and a second derivative $f''(x)$ for $a \leqq x \leqq b$. Show that there is a value of X, with $a < X < b$, such that

$$f(b) = f(a) + (b - a)f'(a) + \tfrac{1}{2}(b - a)^2 f''(X). \tag{39}$$

Solution: Let the function $F(X)$ be defined by

$$F(x) = -f(b) + f(x) + (b - x)f'(x) + \tfrac{1}{2}(b - x)^2 A,$$

where A is a constant defined by the relation

$$0 = F(a) = -f(b) + f(a) + (b - a)f'(a) + \tfrac{1}{2}(b - a)^2 A.$$

This is an equation of the first degree in A which may be solved for A. The value of the derivative of $F(x)$ is

$$\begin{aligned} F'(x) &= f'(x) - f'(x) + (b - x)f''(x) - (b - x)A \\ &= (b - x)[f''(x) - A]. \end{aligned}$$

It follows from the expression for $F(x)$ and $F'(x)$, and our assumptions, that $F(x)$ satisfies hypothesis 1. The expression for $F(x)$ shows that $F(b) = 0$. And $F(a) = 0$ by our choice of A. It follows from Rolle's theorem, Eq. (33), that there is an X, with $a < X < b$, such that $F'(X) = 0$, or

$$0 = F'(X) = (b - X)[f''(X) - A] \quad \text{and} \quad f''(X) - A = 0.$$

Hence $A = f''(X)$, and by substituting this value in the equation defining the value of A, we find

$$0 = -f(b) + f(a) + (b - a)f'(a) + \tfrac{1}{2}(b - a)^2 f''(X).$$

This is equivalent to Eq. (39), the relation which was to be proved.

***253. Taylor's Series with the Remainder Term.** Let $f(x)$ be a function possessing derivatives of all orders throughout the interval $a \leqq x \leqq b$. Then as we shall show presently, there is a value X, with $a < X < b$, such that

$$f(b) = f(a) + (b - a)f'(a) + \frac{(b-a)^2}{2!}f''(a) + \cdots + \frac{(b-a)^{n-1}}{(n-1)!}f^{(n-1)}(a) + \frac{(b-a)^n}{n!}f^{(n)}(X). \quad (40)$$

We shall prove this by applying Rolle's theorem to the function

$$F(x) = -f(b) + f(x) + (b - x)f'(x) + \frac{(b-x)^2}{2!}f''(x) + \cdots + \frac{(b-x)^{n-1}}{(n-1)!}f^{(n-1)}(x) + \frac{(b-x)^n}{n!}A, \quad (41)$$

where A is a constant defined by the relation

$$0 = F(a) = -f(b) + f(a) + (b - a)f'(a) + \frac{(b-a)^2}{2!}f''(a) + \cdots + \frac{(b-a)^{n-1}}{(n-1)!}f^{(n-1)}(a) + \frac{(b-a)^n}{n!}A. \quad (42)$$

This is an equation of the first degree in A which may be solved for A. The value of the derivative of $F(x)$ is

$$\begin{aligned} F'(x) = f'(x) - f'(x) &+ (b - x)f''(x) - (b - x)f''(x) \\ &+ \frac{(b-x)^2}{2!}f'''(x) - \frac{(b-x)^2}{2!}f'''(x) + \cdots \\ &+ \frac{(b-x)^{n-2}}{(n-2)!}f^{(n-1)}(x) - \frac{(b-x)^{n-1}}{(n-2)!}f^{(n-1)}(x) \\ &+ \frac{(b-x)^{n-1}}{(n-1)!}f^{(n)}(x) - \frac{(b-x)^{n-1}}{(n-1)}A. \end{aligned} \quad (43)$$

Since all but the last two terms cancel in pairs,

$$F'(x) = \frac{(b-x)^{n-1}}{(n-1)!}[f^{(n)}(x) - A]. \quad (44)$$

It follows from Eq. (41) and our assumptions about $f(x)$ that $F(x)$ is continuous and differentiable for $a \leqq x \leqq b$. Also $x = b$ makes all the terms in Eq. (41) zero except the first two, which cancel each other, so that $F(b) = 0$. And $F(a) = 0$, since A was chosen to make Eq. (42) hold. It follows from Rolle's theorem, Eq. (33), that there is an X, with $a < X < b$, such that $F'(X) = 0$, or from Eq. (44), such that

$$0 = F'(X) = \frac{(b-X)^{n-1}}{(n-1)!}[f^{(n)}(X) - A]. \quad (45)$$

Consequently, since $X < b$ makes $(b - X) \neq 0$,

$$f^{(n)}(X) - A = 0, \qquad A = f^{(n)}(X). \quad (46)$$

By substituting this value of A in Eq. (42), we find

$$0 = -f(b) + f(a) + (b-a)f'(a) + \frac{(b-a)^2}{2!}f''(a) + \cdots + \frac{(b-a)^{n-1}}{(n-1)!}f^{(n-1)}(a) + \frac{(b-a)^n}{n!}f^{(n)}(X). \quad (47)$$

This is equivalent to Eq. (40), which was to be proved.

Let us set $b = a + h$ and $X = a + \theta h$ as in Eq. (28). Then by using these relations and $b - a = h$, we may rewrite Eq. (40) in the form

$$f(a+h) = f(a) + f'(a)h + \frac{f''(a)}{2!}h^2 + \cdots + \frac{f^{(n-1)}(a)}{(n-1)!}h^{n-1} + \frac{f^{(n)}(a+\theta h)}{n!}h^n. \quad (48)$$

In view of Eq. (30) we have proved that there is a value of θ, with $0 < \theta < 1$, for which Eq. (48) holds.

Since $X = a + \theta h$ and $h = b - a$, we may write

$$X = a + \theta(b-a), \qquad 0 < \theta < 1 \quad (49)$$

in place of the relation $a < X < b$ in Eq. (40). By using any value x in the interval $a < x < b$ in place of b, we may conclude from Eqs. (40) and (49) that there is a value of θ, with $0 < \theta < 1$, such that

$$f(x) = f(a) + f'(a)(x-a) + \frac{f''(a)}{2!}(x-a)^2 + \frac{f'''(a)}{3!}(x-a)^3 + \cdots + \frac{f^{(n-1)}(a)}{(n-1)!}(x-a)^{n-1} + R_n, \quad (50)$$

where

$$R_n = \frac{f^{(n)}[a + \theta(x-a)]}{n!}(x-a)^n. \quad (51)$$

Any one of the relations of Eq. (40), (48), or (50) and (51) is called *Taylor's series with the remainder term*. We have established these relations on the assumption that $a < x < b$, $h > 0$. But they also hold for $b < x < a$, $h < 0$, as may be shown by a slight modification of our argument. For $b < x < a$, $b < X < a$, but we still have $0 < \theta < 1$.

The series on the right in Eq. (50) agrees with the Taylor's series of Eq. (23) up to the first n terms. For any given function $f(x)$, let S_n denote the sum of these n terms. Then if R_n is the remainder term as given by Eq. (51), we have

$$f(x) = S_n + R_n, \qquad S_n = f(x) - R_n, \qquad R_n = f(x) - S_n. \quad (52)$$

For a given function $f(x)$, a given value of a, and a fixed value of x, the remainder R_n is a function of n. It follows from the second relation of Eq. (52) that if

$$\lim_{n\to\infty} R_n = 0, \qquad \text{then } \lim_{n\to\infty} S_n = f(x). \quad (53)$$

And it follows from the third relation of Eq. (52) that if

$$\lim_{n\to\infty} S_n = f(x), \qquad \text{then } \lim_{n\to\infty} R_n = 0. \quad (54)$$

This proves that

The infinite Taylor's series of Eq. (23) *represents the function* $f(x)$ *for values of* x, *and for those values only, for which the remainder after* n *terms* R_n *of Eq.* (51) *approaches zero as* n *becomes infinite.*

***254. Maclaurin's Series with the Remainder Term.** Let $f(x)$ be a function possessing derivatives of all orders throughout the interval $|x| < b$. Then by putting $a = 0$

in Eqs. (50) and (51), we may conclude that for any x in this interval there is a value of θ, with $0 < \theta < 1$, such that

$$f(x) = f(0) + f'(0)x + \frac{f''(0)}{2!}x^2 + \frac{f'''(0)}{3!}x^3 + \cdots + \frac{f^{(n-1)}(0)}{(n-1)!}x^{n-1} + R_n \quad (55)$$

where

$$R_n = \frac{f^{(n)}(\theta x)}{n!}x^n. \quad (56)$$

The series on the right in Eq. (55) agrees with the Maclaurin's series of Eq. (8) up to the first n terms. For any given function $f(x)$, let S_n denote the sum of these terms. Then if R_n is the remainder term as given by Eq. (56), we have

$$f(x) = S_n + R_n, \qquad S_n = f(x) - R_n, \qquad R_n = f(x) - S_n. \quad (57)$$

It follows from these relations, by the reasoning used to derive Eqs. (53) and (54) from Eq. (52), that

$$\lim_{n\to\infty} S_n = f(x) \quad \text{if and only if} \quad \lim_{n\to\infty} R_n = 0. \quad (58)$$

This proves that

The infinite Maclaurin's series of Eq. (8) *represents the function $f(x)$ for values of x, and for those values only, for which the remainder after n terms R_n of Eq.* (56) *approaches zero as n becomes infinite.*

EXAMPLE. Prove that the Maclaurin's series for e^x represents the function for all values of x.

Solution: If $f(x) = e^x$, $f'(x) = e^x$, $f^{(n)}(x) = e^x$. Hence from Eq. (56), $R_n = \frac{e^{\theta x}}{n!}x^n$. Since e^x increases with x, and $0 < \theta < 1$, $e^{\theta x} < e^{|x|}$. Also, let N be a fixed positive integer greater than $2|x|$. And for $n > N$, put $n = N + k$. Then $\frac{|x|}{N+1} < \frac{1}{2}$, $\frac{|x|}{N+2} < \frac{1}{2}, \cdots, \frac{|x|}{N+k} < \frac{1}{2}$. Since $\frac{x^n}{n!} = \frac{x^N}{N!}\left(\frac{x}{N+1}\right)\left(\frac{x}{N+2}\right) \cdots \left(\frac{x}{N+k}\right)$, it follows that $\left|\frac{x^n}{n!}\right| < \frac{x^N}{N!}\frac{1}{2^k}$. And for $n > N$, $|R_n| < e^{|x|}\frac{|x|^N 2^N}{N!}\frac{1}{2^n}$. When $n \to \infty$, x and N remain fixed but $\frac{1}{2^n} \to 0$. Hence $R_n \to 0$, for any fixed x. And by the theorem, the Maclaurin's expansion

$$e^x = 1 + x + \frac{x^2}{2!} + \frac{x^3}{3!} + \cdots$$

obtained from Eq. (8) is valid for all values of x, as was to be proved.

EXERCISE 128

1. That $\lim_{n\to\infty} \frac{x^n}{n!} = 0$ was proved in Example 1. Use this fact to show that for any function $f(x)$ such that the nth derivative $f^{(n)}(x)$ is less than a fixed number, or a function of b independent of n, for $|x| < b$, the Maclaurin's series converges for all values of x with $|x| < b$.

Use the result of Prob. 1 to show that

2. $\sin x = x - \frac{x^3}{3!} + \frac{x^5}{5!} - \cdots$ for all values of x.
3. $\cos x = 1 - \frac{x^2}{2!} + \frac{x^4}{4!} - \cdots$ for all values of x.

4. The reasoning of Example 1 shows that $\lim_{n \to \infty} \frac{(x-a)^n}{n!} = 0$. From this fact and the theorem of Sec. 253, deduce that for any function $f(x)$ such that the nth derivative $f^{(n)}(x)$ is less than a fixed number, or a function of b independent of n, for $a - b < x < a + b$, the Taylor's series in powers of $(x - a)$ converges for all values of x such that $|x - a| < b$.

Use the result of Prob. 4 to show that for all values of a and x each of the following functions is represented by its Taylor's series.

5. e^x.
6. $\sin x$.
7. $\cos x$.

8. In Eq. (41), let us replace the last term by $A(b - x)$, where A is again defined by $F(a) = 0$. Show that in this case $F'(x) = \frac{(b-x)^{n-1}}{(n-1)!} f^{(n)}(x) - A$, and deduce from the reasoning of Sec. 253 that $A = \frac{(b-X)^{n-1}}{(n-1)!} f^{(n)}(X)$,

$$R_n = \frac{(x-a)(x-X)^{n-1}}{(n-1)!} f^{(n)}(X).$$

9. From Prob. 8, deduce that for a Maclaurin's series the remainder term

$$R_n = f^{(n)}(\theta x)(1 - \theta)^{n-1} \frac{x^n}{(n-1)!}.$$

10. From Prob. 9, derive the expression for the remainder in the Maclaurin's series for $\frac{1}{1-x}$, $R_n = \frac{n(1-\theta)^{n-1}x^n}{(1-\theta x)^{n+1}}$. Show that if $|x| < 1$, $0 < \theta < 1$, then $\frac{1-\theta}{1-\theta x} < 1$. Deduce that $|R_n| < \frac{n|x|^n}{(1-\theta x)^2} < \frac{n|x|^n}{(1-|x|)^2}$ which approaches zero when $n \to \infty$. Hence $\frac{1}{1-x} = 1 + x + x^2 + x^3 + \cdots$ for $-1 < x < 1$.

11. By reasoning similar to that of Prob. 10, show that $\ln(1 - x) = x + \frac{x^2}{2} + \frac{x^3}{3} + \cdots$ for $-1 < x < 1$. Here $R_n = -\frac{(1-\theta)^{n-1}x^n}{(1-\theta x)^n}$ and $|R_n| < \frac{|x|^n}{1-|x|}$ for $|x| < 1$.

12. From Prob. 9 derive the expression for the remainder in the Maclaurin's series for $(1 + x)^m$ in the form $R_n = \frac{m(m-1)\cdots(m-n+1)}{(n-1)!} x^n(1-\theta)^{n-1}(1+\theta x)^{m-n}$. And for $|x| < 1$, $|R_n| < \left|\frac{m(m-1)\cdots(m-n+1)}{(n-1)!} x^n\right| (1 + |x|)^{m-1}$ by the reasoning of Prob. 10. For $|x| < 1$, the ratio test shows that $\frac{m(m-1)\cdots(m-n+1)}{(n-1)!}|x|^n$ is the term of a convergent series, whose terms must approach zero as $n \to \infty$. It follows that $R_n \to 0$ as $n \to \infty$. This proves the validity of the binomial expansion for $-1 < x < 1$,

$$(1 + x)^m = 1 + mx + \frac{m(m-1)}{2!}x^2 + \frac{m(m-1)(m-2)}{3!}x^3 + \cdots.$$

255. Series and the Indeterminate Form 0/0. If $f(x)$ and $g(x)$ each approach zero when x approaches a, the quotient $f(x)/g(x)$ may or may not approach a limit as x approaches a. Whether there is a limit, and its value if there is one, cannot be predicted without further information about the functions. This situation is briefly described as the indeterminate form 0/0.

Let us first assume that $f(x)$ and $g(x)$ can each be expanded in Taylor's series in powers of $(x - a)$. Then we may write

$$\frac{f(x)}{g(x)} = \frac{f(a) + f'(a)(x - a) + [f''(a)/2!](x - a)^2 + \cdots}{g(a) + g'(a)(x - a) + [g''(a)/2!](x - a)^2 + \cdots}. \tag{59}$$

Since $f(x) \to 0$ and $g(x) \to 0$ when $x \to a$, in the Taylor's expansions we must have $f(a) = 0$ and $g(a) = 0$. If we omit these terms and assume that $x \neq a$, we may cancel the factor $(x - a)$ from numerator and denominator and so obtain

$$\frac{f(x)}{g(x)} = \frac{f'(a) + [f''(a)/2!](x - a) + \cdots}{g'(a) + [g''(a)/2!](x - a) + \cdots}. \tag{60}$$

It follows from this that

$$\lim_{x\to a} \frac{f(x)}{g(x)} = \frac{f'(a)}{g'(a)}, \qquad \text{if } f(a) = 0,\ g(a) = 0,\ g'(a) \neq 0. \tag{61}$$

This is one form of *l'Hospital's rule* for evaluating indeterminate forms.

If $f'(a)$ and $g'(a)$ are also both zero while $g''(a) \neq 0$, we may omit $f'(a)$ and $g'(a)$ from Eq. (60). Then by canceling the factor $(x - a)$ from numerator and denominator when $x \neq a$, we may deduce that

$$\lim_{x\to a} \frac{f(x)}{g(x)} = \frac{f''(a)}{g''(a)}, \qquad \text{if } f(a) = g(a) = f'(a) = g'(a) = 0,\ g''(a) \neq 0. \tag{62}$$

Similarly we may show that

$$\lim_{x\to a} \frac{f(x)}{g(x)} = \frac{f^{(n)}(a)}{g^{(n)}(a)}, \qquad \text{if } f(a) = f'(a) = \cdots = f^{(n-1)}(a) = 0 \quad \text{and}$$
$$g(a) = g'(a) = \cdots = g^{(n-1)}(a) = 0, \qquad \text{but } g^{(n)}(a) \neq 0. \tag{63}$$

Whenever we have to go beyond the first derivative, if the expansions of $f(x)$ and $g(x)$ are easy to obtain from known series by the methods of Sec. 247, it is practically simpler to use these series directly, as in Eq. (59).

EXAMPLE 1. Evaluate $\lim_{x\to 0} \dfrac{6^x - 3^x}{4x}$.

Solution: We cannot find the limit by direct substitution, since for $x = 0$ the numerator $6^x - 3^x = 6^0 - 3^0 = 1 - 1 = 0$, and the denominator $4x = 0$. However, we may use Eq. (61) with $f(x) = 6^x - 3^x$, $g(x) = 4x$, and $a = 0$. We have just found that $f(0) = 0$, $g(0) = 0$. And $f'(x) = 6^x \ln 6 - 3^x \ln 3$, $g'(x) = 4$.

Hence $f'(0) = 6^0 \ln 6 - 3^0 \ln 3 = \ln 6 - \ln 3 = \ln \frac{6}{3} = \ln 2$, $g'(0) = 4$. Then $\frac{f'(0)}{g'(0)} = \frac{\ln 2}{4}$. And from Eq. (61) we have $\lim_{x \to 0} \frac{6^x - 3^x}{4x} = \frac{1}{4} \ln 2$. This is the required limit.

EXAMPLE 2. Evaluate $\lim_{x \to 0} \frac{x - \tan^{-1} x}{\tan^3 x}$.

Solution 1: Let $f(x) = x - \tan^{-1} x$ and $g(x) = \tan^3 x$. Then $f(0) = 0$ and $g(0) = 0$. And $f'(x) = 1 - \frac{1}{1 + x^2}$, $g'(x) = 3 \tan^2 x \sec^2 x$ so that $f'(0) = 0$, $g'(0) = 0$. Also $f''(x) = \frac{2x}{(1 + x^2)^2}$, $g''(x) = 6 \tan x \sec^4 x + 6 \tan^3 x \sec^2 x$ so that $f''(x) = 0$, $g''(0) = 0$. But $f'''(x) = \frac{2}{(1 + x^2)^2} - \frac{8x^2}{(1 + x^2)^3}$, $g'''(x) =$ $6 \sec^6 x + 42 \tan^2 x \sec^4 x + 12 \tan^4 x \sec^2 x$ so that $f'''(0) = 2$, $g'''(0) = 6$. Then $\frac{f'''(0)}{g'''(0)} = \frac{2}{6}$. And from Eq. (63) with $n = 3$, we have $\lim_{x \to 0} \frac{x - \tan^{-1} x}{\tan^3 x} = \frac{1}{3}$. This is the required result.

Solution 2: From Probs. 9 and 12 of Exercise 124, we have $\tan^{-1} x = x - \frac{x^3}{3} + \cdots$, $\tan x = x + \cdots$. It follows that $\frac{x - \tan^{-1} x}{\tan^3 x} = \frac{x^3/3 - \cdots}{x^3 + \cdots} = \frac{1}{3} + \cdots$. Hence the required value is $\lim_{x \to 0} \frac{x - \tan^{-1} x}{\tan^3 x} = \frac{1}{3}$.

***256. The Indeterminate Form 0/0.** Let x approach a through values greater than a, indicated by writing $x \to a+$. And suppose that

$$\lim_{x \to a+} f(x) = 0, \qquad \lim_{x \to a+} g(x) = 0. \tag{64}$$

Another form of *l'Hospital's rule* for the indeterminate form 0/0 asserts that if

$$\lim_{x \to a+} \frac{f'(x)}{g'(x)} = L, \tag{65}$$

Then

$$\lim_{x \to a+} \frac{f(x)}{g(x)} = L. \tag{66}$$

Equation (65) implies that in some interval $a < x < a + h$, the functions $f(x)$ and $g(x)$ will be differentiable. And they will be continuous for $x = a$ if we define $f(a) = 0$ and $g(a) = 0$. By Cauchy's generalized mean value theorem, Eq. (34), there is a value X with $a < X < a + h$ such that

$$\frac{f(x) - f(a)}{g(x) - g(a)} = \frac{f'(X)}{g'(X)}. \tag{67}$$

When $x \to a$, $h \to 0$, and since $a < X < a + h$, $X \to a$. Also we may omit $f(a)$ and $g(a)$ from the left member, since $f(a) = 0$ and $g(a) = 0$. Hence we may conclude from Eq. (67) that

$$\lim_{x \to a} \frac{f(x)}{g(x)} = \lim_{X \to a} \frac{f'(X)}{g'(X)} = L, \tag{68}$$

by Eq. (65). This proves our conclusion, Eq. (66).

The rule of l'Hospital as stated in Eqs. (64) to (66) continues to hold if we replace $x \to a+$ by $x \to a-$ throughout, where $x \to a-$ indicates that x approaches a through values less than a. Or we may write throughout $x \to a$, with no restriction on the sign of $(x - a)$. Likewise we may apply the rule when $x \to +\infty$, $x \to -\infty$, or $x \to \infty$. See the following example. And the conclusion in any of the cases mentioned remains valid if we replace the finite limit L by $+\infty$, $-\infty$, or ∞ in Eqs. (65) and (66).

EXAMPLE. Let $\lim\limits_{x \to +\infty} f(x) = 0$ and $\lim\limits_{x \to +\infty} g(x) = 0$. Show that, if $\lim\limits_{x \to +\infty} \dfrac{f'(x)}{g'(x)} = L$, then $\lim\limits_{x \to +\infty} \dfrac{f(x)}{g(x)} = L$.

Solution: Let $x = \dfrac{1}{y}$. Then $y = \dfrac{1}{x}$, and $y \to 0+$ when $x \to +\infty$. Also

$\lim\limits_{x \to +\infty} \dfrac{f(x)}{g(x)} = \lim\limits_{y \to 0+} \dfrac{f(1/y)}{g(1/y)}$. By Eq. (66), this limit is L if $\lim\limits_{y \to 0+} \dfrac{\dfrac{d}{dy} f\left(\dfrac{1}{y}\right)}{\dfrac{d}{dy} g\left(\dfrac{1}{y}\right)} = L$.

But if $\dfrac{d}{dx} f(x) = f'(x)$, $\dfrac{d}{dy} f\left(\dfrac{1}{y}\right) = -\dfrac{1}{y^2} f'\left(\dfrac{1}{y}\right) = -x^2 f'(x)$. And similarly for $g(x)$.

It follows that $\lim\limits_{y \to 0+} \dfrac{\dfrac{d}{dy} f\left(\dfrac{1}{y}\right)}{\dfrac{d}{dy} g\left(\dfrac{1}{y}\right)} = \lim\limits_{x \to +\infty} \dfrac{-x^2 f'(x)}{-x^2 g'(x)} = \lim\limits_{x \to +\infty} \dfrac{f'(x)}{g'(x)}$. And we were given that this equals L. Thus the result is proved.

EXERCISE 129

Evaluate each of the following limits by l'Hospital's rule.

1. $\lim\limits_{x \to 2} \dfrac{x^4 - 16}{x - 2}$.
2. $\lim\limits_{x \to 3} \dfrac{e^x - e^3}{x - 3}$.
3. $\lim\limits_{x \to 1} \dfrac{\ln x}{x - 1}$.
4. $\lim\limits_{x \to 0} \dfrac{e^{-x} - 1}{x}$.
5. $\lim\limits_{x \to 0} \dfrac{\ln (1 + x)}{x}$.
6. $\lim\limits_{x \to -1} \dfrac{\sin \pi x}{1 + x}$.
7. $\lim\limits_{x \to \pi/4} \dfrac{\tan x - 1}{x - \pi/4}$.
8. $\lim\limits_{x \to \pi/3} \dfrac{\cos x - \frac{1}{2}}{x - \pi/3}$.
9. $\lim\limits_{x \to 3} \dfrac{x^2 - 5x + 6}{x^2 - 7x + 12}$.
10. $\lim\limits_{x \to 2} \dfrac{x^2 - 4}{x^3 - 8}$.
11. $\lim\limits_{x \to 0} \dfrac{e^x - e^{-x}}{\sin x}$.
12. $\lim\limits_{x \to 0} \dfrac{e^x - 1}{\tan x}$.
13. Show that each of the limits in Probs. 1 to 8 may be evaluated by a direct application of the definition of a derivative, namely,

$$\lim_{x \to a} \frac{f(x) - f(a)}{x - a} = f'(a).$$

14. Prove Eq. (61) by using the equation in Prob. 13 and the corresponding one for $g'(a)$.

Evaluate each of the following limits.

15. $\lim\limits_{x \to 0} \dfrac{\ln \cos x}{x^2}$.
16. $\lim\limits_{x \to 0} \dfrac{\cos 2x - \cos x}{x^2}$.
17. $\lim\limits_{x \to 0} \dfrac{\tan x - x}{x^3}$.
18. $\lim\limits_{x \to 0} \dfrac{\sin 2x - 2 \sin x}{\sin^3 x}$.
19. $\lim\limits_{x \to 0} \dfrac{\sin^{-1} x - x}{x^3}$.
20. $\lim\limits_{x \to 0} \dfrac{\tan x - \sin x}{\tan^3 x}$.

***257. The Indeterminate Form ∞/∞.** Suppose that

$$\lim_{x\to a+} f(x) = \infty, \qquad \lim_{x\to a+} g(x) = \infty. \tag{69}$$

Then *l'Hospital's rule* for the indeterminate form ∞/∞ asserts that if

$$\lim_{x\to a+} \frac{f'(x)}{g'(x)} = L, \tag{70}$$

Then

$$\lim_{x\to a+} \frac{f(x)}{g(x)} = L. \tag{71}$$

The deduction of this rule from Cauchy's generalized mean value theorem is outlined in the following example. By similar reasoning, the rule may be shown to hold when $x \to a+$ is replaced throughout by $x \to a-$, $x \to a$, $x \to +\infty$, $x \to -\infty$, or $x \to \infty$.

It follows from this rule that

$$\lim_{x\to+\infty} \frac{x}{e^x} = \lim_{x\to+\infty} \frac{1}{e^x} = 0. \tag{72}$$

This relation is not disturbed if we raise the functions x and e^x to any fixed positive powers. That is, if $p > 0$ and $q > 0$,

$$\lim_{x\to+\infty} \frac{x^p}{e^{qx}} = 0. \tag{73}$$

And this is true no matter how large p is, or how small q is.

To prove Eq. (73), we note that our rule gives

$$\lim_{x\to+\infty} \frac{x^p}{e^{qx}} = \lim_{x\to+\infty} \frac{px^{p-1}}{qe^{qx}}. \tag{74}$$

If $p \leqq 1$, the second limit is zero because of the finite numerator and infinite denominator. If $p > 1$, the second limit is the positive constant p/q times a limit similar to the original one with p replaced by $p - 1$. We may repeat the process, and after a sufficient number of repetitions we shall come to a fraction whose numerator contains x raised to a power which is less than 1, the case already treated.

The relation of Eq. (73), which includes that of Eq. (72) as a special case, shows that for $x \to +\infty$, e^x is an essentially higher order of infinity than x. Thus we may use the principle of the leading term to evaluate such limits as

$$\lim_{x\to+\infty} \frac{6e^{0.01x} + 100x^{50}}{2e^{0.01x} - 60x^{40}} = \lim_{x\to+\infty} \frac{6e^{0.01x}}{2e^{0.01x}} = 3. \tag{75}$$

It also follows from the rule of this section that

$$\lim_{x\to+\infty} \frac{\ln x}{x} = \lim_{x\to+\infty} \frac{1/x}{1} = 0. \tag{76}$$

This relation is not disturbed if we raise the functions $(\ln x)$ and x to any fixed positive powers. That is, if $p > 0$ and $q > 0$,

$$\lim_{x\to+\infty} \frac{(\ln x)^p}{x^q} = 0. \tag{77}$$

And this is true no matter how large p is, or how small q is.

To prove Eq. (77), we note that our rule gives

$$\lim_{x\to+\infty}\frac{(\ln x)^p}{x^q}=\lim_{x\to+\infty}\frac{p(\ln x)^{p-1}(1/x)}{qx^{q-1}}=\lim_{x\to+\infty}\frac{p(\ln x)^{p-1}}{qx^q}. \tag{78}$$

If $p \leqq 1$, the last limit is zero because of the finite numerator and infinite denominator. If $p > 1$, the last limit is the positive constant p/q times a limit similar to the original one with p replaced by $p - 1$. We may repeat the process, and after a sufficient number of repetitions we will come to a fraction whose numerator contains $(\ln x)$ raised to a power less than 1, the case already treated.

The relation of Eq. (77), which includes that of Eq. (76) as a special case, shows that for $x \to +\infty$, $\ln x$ is an essentially lower order of infinity than x. Thus we may use the principle of the leading term to evaluate such limits as

$$\lim_{x\to+\infty}\frac{80(\ln x)^{25}+10x^{0.02}}{70(\ln x)^{20}+2x^{0.02}}=\lim_{x\to+\infty}\frac{10x^{0.02}}{2x^{0.02}}=5. \tag{79}$$

Example. Prove Eq. (71).

Solution: Let x_2 be a number a little greater than a, and x_1 a number between a and x_2. Then by Eq. (34), for some X with

$$x_1 < X < x_2, \qquad \frac{f(x_2)-f(x_1)}{g(x_2)-g(x_1)}=\frac{f'(X)}{g'(X)}.$$

But

$$\frac{f(x_2)-f(x_1)}{g(x_2)-g(x_1)}=\frac{f(x_1)}{g(x_1)}\left\{\frac{1-[f(x_2)/f(x_1)]}{1-[g(x_2)/g(x_1)]}\right\}.$$

It follows from these two relations that

$$\frac{f(x_1)}{g(x_1)}=\frac{f'(X)}{g'(X)}\left\{\frac{1-[g(x_2)/g(x_1)]}{1-[f(x_2)/f(x_1)]}\right\}. \tag{80}$$

By Eq. (70), we may take x_2 so close to a that $\dfrac{f'(X)}{g'(X)}$ is close to L for X closer to a than x_2. Having fixed x_2, we may then take x_1 so much closer to a that $g(x_1)$ is large compared with $g(x_2)$, and $f(x_1)$ is large compared with $f(x_2)$, in view of Eq. (69). This will make $1-\dfrac{g(x_2)}{g(x_1)}$ close to 1 and $1-\dfrac{f(x_2)}{f(x_1)}$ close to 1. Consequently the right member of Eq. (80) will be near L when x_1 is sufficiently close to a. Hence the same is true of the left member, and this shows that $\lim\limits_{x\to a+}\dfrac{f(x)}{g(x)}=L$, or Eq. (71) which was to be proved.

258. Other Indeterminate Forms. The indeterminate forms indicated by $\infty - \infty$, $0 \cdot \infty$, 1^∞, 0^0, ∞^0 may often be reduced to cases where l'Hospital's rule applies by a simple transformation. For example, the last three types reduce to the second type $0 \cdot \infty$ if we consider the logarithm of the expression. The second type may be reduced to a fraction by putting either factor in the denominator.

As Example 1 illustrates, it is sometimes simpler to use known or easily found series expansions than it is to reduce the expression to a form that can be treated by l'Hospital's rule.

EXAMPLE 1. Evaluate $\lim_{x\to 0}\left(\csc^2 x - \frac{1}{x^2}\right)$.

Solution: This is the form $\infty - \infty$. The given expression $\csc^2 x - \frac{1}{x^2} = \frac{1}{\sin^2 x} - \frac{1}{x^2} = \frac{x^2 - \sin^2 x}{x^2 \sin^2 x}$. For $x \to 0$, this takes the form $\frac{0}{0}$. Although we could use l'Hospital's rule, it would take four applications, or differentiations. It is simpler to note that $\sin x = x - \frac{x^3}{3!} + \cdots$, $\sin^2 x = x^2 - \frac{x^4}{3} + \cdots$, so that $\frac{x^2 - \sin^2 x}{x^2 \sin^2 x} = \frac{x^4/3 - \cdots}{x^4 - \cdots}$. And $\lim_{x\to 0}\left(\csc^2 x - \frac{1}{x^2}\right) = \lim_{x\to 0}\frac{x^4/3}{x^4} = \frac{1}{3}$. Thus $\frac{1}{3}$ is the required value of the limit.

EXAMPLE 2. Evaluate $\lim_{x\to 0+}(\sec 3x)^{\cot^2 2x}$.

Solution: This is the form 1^∞. Let $u = (\sec 3x)^{\cot^2 2x}$. Then $\ln u = \cot^2 2x \ln \sec 3x = \frac{\ln \sec 3x}{\tan^2 2x}$. For $x \to 0+$, this has the form $\frac{0}{0}$. But $\frac{d}{dx}\ln \sec 3x = \frac{3 \tan 3x \sec 3x}{\sec 3x} = 3 \tan 3x$, $\frac{d}{dx}\tan^2 2x = 2 \tan 2x\,(2 \sec^2 2x)$. From Eq. (66), $\lim_{x\to 0+} \ln u = \lim_{x\to 0+}\frac{3 \tan 3x}{4 \tan 2x \sec^2 2x} = \lim_{x\to 0+}\frac{3 \tan 3x}{4 \tan 2x}$, since $\lim_{x\to 0+}\sec^2 2x = 1$. As the last limit again has the form $\frac{0}{0}$, we find $\frac{d}{dx} 3 \tan 3x = 9 \sec^2 3x$, $\frac{d}{dx} 4 \tan 2x = 8 \sec^2 2x$. Hence $\lim_{x\to 0+}\ln u = \lim_{x\to 0+}\frac{9 \sec^2 3x}{8 \sec^2 2x} = \frac{9}{8}$. It follows that $\lim_{x\to 0+}(\sec 3x)^{\cot^2 2x} = \lim_{x\to 0+} e^{\ln u} = e^{\lim_{x\to 0}\ln u} = e^{\frac{9}{8}}$. Thus $e^{\frac{9}{8}}$ is the required value of the limit.

EXERCISE 130

Evaluate each of the following limits.

1. $\lim_{x\to +\infty}\frac{x^2}{e^x}$.
2. $\lim_{x\to +\infty}\frac{\ln x}{\sqrt{x}}$.
3. $\lim_{x\to \pi/2}\frac{\csc 8x}{\csc 2x}$.
4. $\lim_{x\to 0}\frac{\tan 9x}{\tan 3x}$.
5. $\lim_{x\to \pi/4}\frac{\sec 6x}{\sec 2x}$.
6. $\lim_{x\to \pi/2}\frac{\cot 5x}{\cot 3x}$.
7. $\lim_{x\to 0}\frac{\ln \tan 2x}{\ln \sin 6x}$.
8. $\lim_{x\to \pi/2}\frac{\sec 3x}{\tan 5x}$.
9. $\lim_{x\to 0+} x \ln x$.
10. $\lim_{x\to 0} x \cot x$.
11. $\lim_{x\to +\infty}\frac{5e^x + 4x^2}{e^x + 2x^2}$.
12. $\lim_{x\to +\infty}\frac{7x + 3 \ln x}{x + 6 \ln x}$.
13. $\lim_{x\to 0}\left(\frac{1}{\sin x} - \frac{1}{x}\right)$.
14. $\lim_{x\to 0}\left(\cot x - \frac{1}{x}\right)$.
15. $\lim_{x\to 1}\left(\frac{x}{x - 1} - \frac{1}{\ln x}\right)$.
16. $\lim_{x\to 2\pi}\left(\frac{1}{x - 2\pi} - \cot x\right)$.
17. $\lim_{x\to 0}(1 + ax)^{b/x}$.
18. $\lim_{x\to \infty}\left(1 + \frac{a}{x}\right)^{bx}$.
19. $\lim_{x\to \pi/2}(\sin x)^{\tan x}$.
20. $\lim_{x\to 0+} x^x$.
21. $\lim_{x\to 1}(2 - x)^{\tan(\pi x/2)}$.
22. $\lim_{x\to 0+}(\cot x)^{\sin x}$.

259. Numerical Integration. Ordinarily, as in Sec. 72, we evaluate the definite integral

$$\int_a^b f(x)dx \tag{81}$$

by means of an indefinite integral of the function $f(x)$. But it may be impossible to express the indefinite integral of $f(x)$ exactly in terms of a finite combination of the elementary functions used in the calculus. This is true of some integrals as simple in appearance as

$$\int \sqrt{x^4+1}\,dx, \qquad \int e^{-x^2}\,dx, \qquad \int \frac{\sin x}{x}\,dx, \qquad \int \frac{dx}{\ln x}. \tag{82}$$

And, even if expressible in terms of known functions by methods like those of Chap. 13, the indefinite integral of a function $f(x)$ may be too complicated for convenient use. Moreover in some applications of the calculus, the function $f(x)$ may be known only through a table of numerical values, often obtained by reading instruments or the ordinates of a graph.

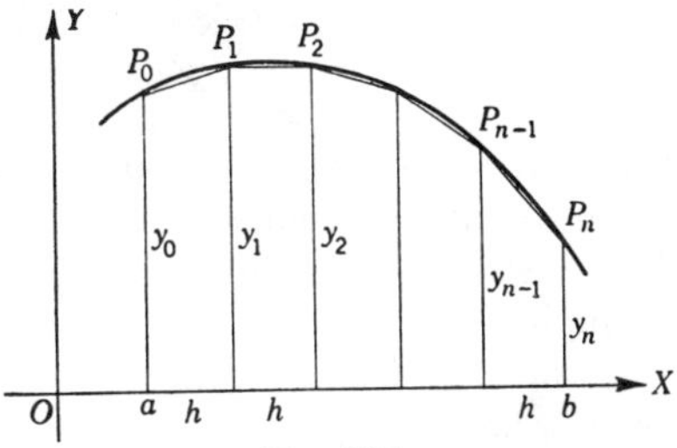

FIG. 284.

In any of these cases, by using numerical methods, it is possible to compute the definite integral (81) to an accuracy comparable with that to which the values of $f(x)$ can be found. To obtain a formula for numerical integration, we divide the interval of integration into parts and on each part replace $f(x)$ by an approximating function which can easily be integrated. We derive the trapezoidal rule in Sec. 260, and the more accurate and usually preferable Simpson's rule in Sec. 261.

260. The Trapezoidal Rule. As pointed out in Sec. 71, the value of the definite integral of Eq. (81) is equal to the measure of the area bounded above by the curve $y = f(x)$, below by the x axis, and lying between the ordinates $x = a$ and $x = b$. Let us divide the interval from a to b into n equal parts, each of length

$$h = \frac{b-a}{n}. \tag{83}$$

Call the ordinates of $y = f(x)$ at the end points and points of subdivision $y_0, y_1, y_2, \cdots, y_n$. And, as shown in Fig. 284, let these ordinates meet the curve $y = f(x)$ at $P_0, P_1, P_2, \cdots, P_n$. Then the area under the curve may be approximated by that under the polygonal line whose vertices are at these points. This area is made up of trapezoids. The area of a trapezoid is equal to one-half the altitude times the sum of the two parallel sides. It follows that

$$\text{Area of first trapezoid} = \frac{h}{2}(y_0 + y_1),$$

$$\text{Area of second trapezoid} = \frac{h}{2}(y_1 + y_2),$$

. . . .

$$\text{Area of } n\text{th trapezoid} = \frac{h}{2}(y_{n-1} + y_n).$$

Adding these expressions, we find that the approximating area is

$$T_n = h\left(\frac{y_0}{2} + y_1 + y_2 + \cdots + y_{n-1} + \frac{y_n}{2}\right). \tag{84}$$

The approximate formula

$$\int_a^b f(x)dx = T_n, \qquad h = \frac{b-a}{n} \tag{85}$$

is known as the *trapezoidal rule* for numerical integration.

Our discussion assumed that $a < b$, and $f(x) > 0$ for $a < x < b$. But these restrictions are not necessary, as shown in Example 1. And as Example 2 suggests, the error in using the trapezoidal rule is of the form Kh^2, where, for a given integral, K changes slowly with the value of n used. Hence when we change from n to $2n$, the error is multiplied by a factor that may be estimated roughly as $\left(\frac{n}{2n}\right)^2 = \frac{1}{4}$. This may be used to estimate the error made, as in Example 3.

Example 1. The points (x_1,y_1) and (x_2,y_2) are joined by a straight line. If its equation is written in the form $y = g(x)$, and $x_2 = x_1 + h$, show that $\int_{x_1}^{x_2} g(x)dx = \frac{h}{2}(y_1 + y_2)$.

Solution: From Sec. 77, the equation of the straight line is $\frac{y - y_1}{x - x_1} = m$, where $m = \frac{y_2 - y_1}{x_2 - x_1} = \frac{y_2 - y_1}{h}$, so that $mh = y_2 - y_1$. Thus $y = y_1 + m(x - x_1)$. And $\int_{x_1}^{x_2} y\,dx = \int_{x_1}^{x_2} [y_1 + m(x - x_1)]dx = \left[y_1x + \frac{m}{2}(x - x_1)^2\right]_{x_1}^{x_1+h} = y_1h + \frac{mh^2}{2} = y_1h + \frac{h}{2}(y_2 - y_1) = \frac{h}{2}(y_1 + y_2)$, as was to be proved.

*Example 2. Let $f(x_1) = y_1$, $f(x_2) = y_2$, and $x_2 = x_1 + h$. Estimate the value of Q in the equation $\int_{x_1}^{x_2} f(x)dx = \frac{h}{2}(y_1 + y_2) + Qh^3$.

Solution: Let $f(x) = F'(x)$. Then $\int_{x_1}^{x_2} f(x)dx = F(x_2) - F(x_1)$. Hence $F(x_2) - F(x_1) - \frac{h}{2}[f(x_1) + f(x_2)] - Qh^3 = 0$. Now consider the function $G(x) =$

$F\left(c+\frac{x}{2}\right) - F\left(c-\frac{x}{2}\right) - \frac{x}{2}\left[f\left(c+\frac{x}{2}\right) + f\left(c-\frac{x}{2}\right)\right] - Qx^3$, where $c = \frac{1}{2}(x_1 + x_2)$, so that $c - \frac{h}{2} = x_1$, $c + \frac{h}{2} = x_2$. The function $G(x)$ is zero for $x = h$, by the relation defining Q. And it is zero for $x = 0$. Hence by Rolle's theorem, $G'(x)$ is zero for some value X_1 between 0 and h. But since $F'(x) = f(x)$, we have

$$G'(x) = \frac{1}{2}f\left(c+\frac{x}{2}\right) + \frac{1}{2}f\left(c-\frac{x}{2}\right) - \frac{1}{2}\left[f\left(c+\frac{x}{2}\right) + f\left(c-\frac{x}{2}\right)\right] - \frac{x}{4}\left[f'\left(c+\frac{x}{2}\right) - f'\left(c-\frac{x}{2}\right)\right] - 3Qx^2.$$

The first four terms cancel, so that

$0 = G'(X_1) = -\frac{X_1}{4}\left[f'\left(c+\frac{X_1}{2}\right) - f'\left(c-\frac{X_1}{2}\right)\right] - 3QX_1^2$. And since $X_1 \neq 0$, $Q = -\frac{1}{12}\frac{f'(c + X_1/2) - f'(c - X_1/2)}{X_1}$. By Eq. (26) with $f'(x)$ in place of $F(x)$, and $b = c + \frac{X_1}{2}$, $a = c - \frac{X_1}{2}$, it follows from this that $Q = -\frac{1}{12}f''(X_2)$, where X_2 is a value between x_1 and x_2. This is the required determination.

By applying this result to each of the intervals used in deriving the trapezoidal rule, we may deduce that if

$$\int_a^b f(x)dx = T_n + E_n, \qquad E_n = -\frac{1}{12}\left[\sum f''(x_i)\right]h^3 = -\frac{1}{12}f''(X_n)(b-a)h^2,$$

where $f''(X_n)$ is the average of the n quantities $f''(x_i)$, and hence the value of the second derivative at some point X_n between a and b, and we have replaced nh by $(b - a)$. The coefficient of h^2 is the factor K mentioned in the text which varies slowly with n.

EXAMPLE 3. Calculate the integral $\int_1^5 \frac{dx}{x}$ by the trapezoidal rule, using $n = 4$ and then $n = 8$.

Solution: Since $a = 1$ and $b = 5$, $b - a = 4$ and by Eq. (85) $h = 1$ when $n = 4$ and $h = \frac{1}{2}$ when $n = 8$. Using M to denote the multiplier of y_1 in Eq. (84), we construct the accompanying table.

x	$\frac{1}{x} = y$	$n = 4$, M	$h = 1$, My	$n = 8$, M	$h = \frac{1}{2}$, My
1	1.000	$\frac{1}{2}$	0.500	$\frac{1}{2}$	0.500
1.5	0.667	..	.....	1	0.667
2	0.500	1	0.500	1	0.500
2.5	0.400	..	.....	1	0.400
3	0.333	1	0.333	1	0.333
3.5	0.286	..	.....	1	0.286
4	0.250	1	0.250	1	0.250
4.5	0.222	..	.....	1	0.222
5	0.200	$\frac{1}{2}$	0.100	$\frac{1}{2}$	0.100

Sum.............1)1.683...........2)3.258

$T = h$ (sum).....$T_4 = 1.683$........$T_8 = 1.629$

If T_8 is an improvement on T_4, T_4 is about 0.05 too large. Thus the error in T_8 is about $\frac{1}{4}$ of this or 0.01. Hence we list the required values as $T_4 = 1.68$ and $T_8 = 1.63$.

Without further calculation, using our estimate of the error, we would expect $1.63 - 0.01 = 1.62$ to be correct to within 1 or 2 in the hundredths place. This is true, as the correct value is $\ln 5 = 1.61$.

261. Simpson's Rule. As in Sec. 260, let us again consider the definite integral of Eq. (81) as the measure of the area bounded above by the curve $y = f(x)$, below by the x axis, and lying between the ordinates $x = a$ and $x = b$. Select some *even* integer n. And divide the interval from a to b into n equal parts, each of length $h = \frac{b - a}{n}$. Call the ordinates of $y = f(x)$ at the end points and points of subdivision $y_0, y_1, y_2, \cdots, y_n$.

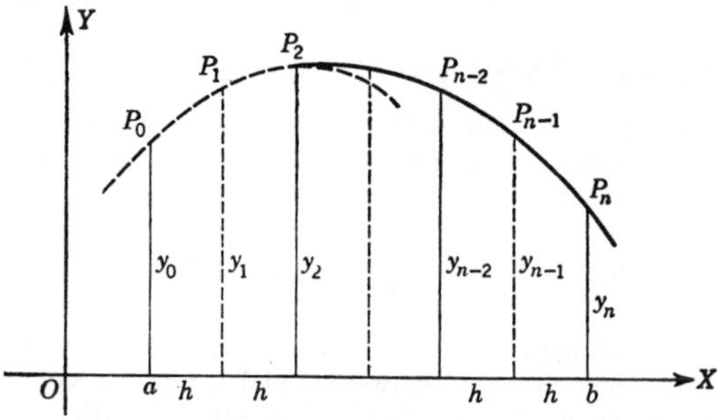

FIG. 285.

And as before let these ordinates meet the curve $y = f(x)$ at $P_0, P_1, P_2, \cdots, P_n$. But now, as in Fig. 285, let us approximate the curve $y = f(x)$ for each pair of intervals, as $P_0P_1P_2$, by the arc of a parabola with axis parallel to OY, or curve having an equation of the form $y = Ax^2 + Bx + C$. It follows from Example 1, and without regard to the sign of $f(x)$, that between the x axis and the parabolic arc

$$P_0P_1P_2, \text{ the area is } \frac{h}{3}(y_0 + 4y_1 + y_2),$$

$$P_2P_3P_4, \text{ the area is } \frac{h}{3}(y_2 + 4y_3 + y_4),$$

$$\cdots\cdots\cdots\cdots\cdots\cdots$$

$$P_{n-2}P_{n-1}P_n, \text{ the area is } \frac{h}{3}(y_{n-2} + 4y_{n-1} + y_n).$$

Adding these expressions, we find that the approximating area is

$$S_n = \frac{h}{3}(y_0 + 4y_1 + 2y_2 + 4y_3 + 2y_4 + \cdots + 2y_{n-2} + 4y_{n-1} + y_n). \quad (86)$$

The coefficients 1, 4, 2 result from the fact that each y_i with i odd is a middle ordinate once, whereas except for y_0 and y_n, each y_i with i even is an end ordinate twice.

The approximate formula, in which n is necessarily *even*,

$$\int_a^b f(x)dx = S_n, \qquad h = \frac{b-a}{n} \tag{87}$$

is known as *Simpson's rule* for numerical integration.

As Example 2 suggests, the error in using Simpson's rule is of the form Kh^4, where, for a given integral, K changes slowly with the value of n used. Hence when we change from n to $2n$, the error is multiplied by a factor which may be estimated roughly as $(n/2n)^4 = \frac{1}{16}$. This may be used to estimate the error made, as in Example 3.

Example 1. The equation of a parabola with axis parallel to OY may be written in the form $y = g(x)$, with $g(x) = A(x - x_1)^2 + B(x - x_1) + C$. If this parabola passes through the points (x_0,y_0), (x_1,y_1), (x_2,y_2), where $x_1 = x_0 + h$ and $x_2 = x_1 + h$, show that $\int_{x_0}^{x_2} g(x)dx = \frac{h}{3}(y_0 + 4y_1 + y_2)$.

Solution: $\int_{x_0}^{x_2} g(x)dx = \int_{x_1-h}^{x_1+h} [A(x - x_1)^2 + B(x - x_1) + C]dx$. With $x = x_1 + t$, this becomes

$$\int_{-h}^{h} (At^2 + Bt + C)dt = \left[\frac{A}{3}t^3 + \frac{B}{2}t^2 + Ct\right]_{-h}^{h} = \frac{2}{3}Ah^3 + 2Ch = \frac{h}{3}(2Ah^2 + 6C).$$

But $y_0 = g(x_0) = g(x_1 - h) = Ah^2 - Bh + C$, $y_1 = g(x_1) = C$, $y_2 = g(x_2) = g(x_1 + h) = Ah^2 + Bh + C$. It follows that $y_0 + y_2 = 2Ah^2 + 2C$, $y_0 + 4y_1 + y_2 = 2Ah^2 + 6C$, so that $\int_{x_0}^{x_2} g(x)dx = \frac{h}{3}(2Ah^2 + 6C) = \frac{h}{3}(y_0 + 4y_1 + y_2)$, as was to be proved.

* Example 2. Let $f(x_0) = y_0$, $f(x_1) = y_1$, $f(x_2) = y_2$. And let $x_1 = x_0 + h$, $x_2 = x_1 + h$. Estimate the value of Q in the equation

$$\int_{x_0}^{x_2} f(x)dx = \frac{h}{3}(y_0 + 4y_1 + y_2) + Qh^5.$$

Solution: Let $f(x) = F'(x)$. Then $\int_{x_0}^{x_2} f(x)dx = F(x_2) - F(x_1)$. Hence

$$F(x_2) - F(x_1) - \frac{h}{3}[f(x_0) + 4f(x_1) + f(x_2)] - Qh^5 = 0.$$

Now consider the function

$$G(x) = F(c + x) - F(c - x) - \frac{x}{3}[f(c + x) + 4f(c) + f(c - x)] - Qx^5,$$

where $c = x_1$, so that $c - h = x_0$ and $c + h = x_2$. The function $G(x)$ is zero for $x = h$, by the relation defining Q. And it is zero for $x = 0$. Hence by Rolle's theorem $G'(x)$ is zero for some value X_1 between 0 and h. But since $F'(x) = f(x)$, we have

$$G'(x) = f(c + x) + f(c - x) - \frac{1}{3}[f(c + x) + 4f(c) + f(c - x)] - \frac{x}{3}[f'(c + x) - f'(c - x)] - 5Qx^4,$$

or $G'(x) = \dfrac{2}{3}[f(c + x) + f(c - x)] - \dfrac{4}{3}f(c) - \dfrac{x}{3}[f'(c + x) - f'(c - x)] - 5Qx^4$. This is zero for $x = 0$ and $x = X_1$, so that by Rolle's theorem $G''(x)$ is zero for some value X_2 between 0 and X_1. But

$$G''(x) = \frac{2}{3}[f'(c + x) - f'(c - x)] - \frac{1}{3}[f'(c + x) - f'(c - x)] - \frac{x}{3}[f''(c + x) + f''(c - x)] - 20Qx^3,$$

or $G''(x) = \dfrac{1}{3}[f'(c + x) - f'(c - x)] - \dfrac{x}{3}[f''(c + x) + f''(x - c)] - 20Qx^3$. This is zero for $x = 0$ and $x = X_2$, so that by Rolle's theorem $G'''(x)$ is zero for some value X_3 between 0 and X_2. But

$$G'''(x) = \frac{1}{3}[f''(c + x) + f''(c - x)] - \frac{1}{3}[f''(c + x) + f''(c - x)] - \frac{x}{3}[f'''(c + x) - f'''(c - x)] - 60Qx^2.$$

It follows that $G'''(X_3) = -\dfrac{X_3}{3}[f'''(c + X_3) - f'''(c - X_3)] - 60QX_3^2 = 0$. And since $X_3 \neq 0$, $Q = -\dfrac{1}{90}\dfrac{f'''(c + X_3) - f'''(c - X_3)}{2X_3}$. By Eq. (26) with $f'''(x)$ in place of $F(x)$, and $b = c + X_3$, $a = c - X_3$, it follows from this that $Q = -\dfrac{1}{90}f^{(4)}(X_4)$, where X_4 is a value between x_0 and x_2. This is the required determination.

By applying this result to each of the doubled intervals used in deriving Simpson's rule, we may deduce that, if

$$\int_a^b f(x)dx = S_n + E_n, \qquad E_n = -\frac{1}{90}\left[\sum f^{(4)}(x_i)\right]h^5 = -\frac{1}{180}f^{(4)}(X_n)(b - a)h^4,$$

where $f^{(4)}(X_n)$ is the average of the $\dfrac{n}{2}$ quantities $f^{(4)}(x_i)$, and hence the value of the fourth derivative at some point X_n between a and b, and we have replaced nh by $(b - a)$. The coefficient of h^4 is the factor K mentioned in the text which varies slowly with n.

EXAMPLE 3. Calculate the integral $\displaystyle\int_1^5 \frac{dx}{x}$ by Simpson's rule, using $n = 4$ and $n = 8$.

Solution: Since $a = 1$ and $b = 5$, $b - a = 4$ and by Eq. (87) $h = 1$ when $n = 4$ and $h = \frac{1}{2}$ when $n = 8$. Using M to denote the multiplier of y_i in Eq. (86), we construct the accompanying table.

x	$\frac{1}{x} = y$	$n = 4$, M	$h = 1$, My	$n = 8$, M	$h = \frac{1}{2}$, My
1	1.0000	1	1.0000	1	1.0000
1.5	0.6667	..	......	4	2.6668
2	0.5000	4	2.0000	2	1.0000
2.5	0.4000	..	......	4	1.6000
3	0.3333	2	0.6666	2	0.6666
3.5	0.2857	..	......	4	1.1428
4	0.2500	4	1.0000	2	0.5000
4.5	0.2222	..	......	4	0.8888
5	0.2000	1	0.2000	1	0.2000
	Sum		3)4.8666		6)9.6650
$S = \frac{h}{3}$ (sum)			$S_4 = 1.6222$		$S_8 = 1.6108$

If S_8 is an improvement on S_4, S_4 is about 0.011 too large. Thus the error in S_8 is about $\frac{1}{16}$ of this or 0.0007. Hence we list the required values as $S_4 = 1.62$ and $S_8 = 1.611$.

Since $1.6108 - 0.0007 = 1.6101$, without further calculation, from our estimate of the error, we would expect 1.610 to be correct to within 1 or 2 in the thousandths place. This is true, as the correct value is $\ln 5 = 1.6094$.

EXERCISE 131

Compute an approximate value of each of the following integrals by using Simpson's rule, with $n = 4$. Check your result by performing the integration.

1. $\int_1^3 \frac{dx}{x}$.
2. $\int_0^4 \frac{dx}{\sqrt{64 - x^2}}$.
3. $\int_0^1 \frac{dx}{1 + x^2}$.
4. $\int_0^8 \frac{dx}{50 + x}$.
5. $\int_1^5 \frac{dx}{\sqrt{8 + x}}$.
6. $\int_0^4 \frac{dx}{4 + x^2}$.
7. $\int_4^8 \sqrt{60 + x}\, dx$.
8. $\int_2^6 \frac{x\, dx}{4 + x^2}$.
9. $\int_0^1 e^{-x}\, dx$.
10. $\int_0^1 \sin \pi x\, dx$.

Evaluate each of the following integrals by using Simpson's rule with $n = 10$. Compare your result with the numerical value found by the indicated method.

11. $\int_0^1 e^{-x^2}\, dx = 0.74682$ from the series in Prob. 19 of Exercise 124.
12. $\int_0^1 \sin x^2\, dx = 0.3103$ from the series in Prob. 20 of Exercise 124.
13. $\int_0^{0.5} \sqrt{1 - x^4}\, dx = 0.5315$ from the series in Prob. 21 of Exercise 124.
14. $\int_0^{0.5} \frac{\sin x}{x}\, dx = 0.4931$ from the series in Prob. 20 of Exercise 125.
15. $\int_0^{0.5} \frac{1 - \cos x}{x^2}\, dx = 0.2483$ from the series in Prob. 21 of Exercise 125.

For each of the following integrals, compute S_4, the approximation given by Simpson's rule with four intervals. Also compute S_8, the approximation given by Simpson's rule with eight intervals. Use these to estimate the value of the integral $I = S_8 + \frac{1}{15}(S_8 - S_4)$ as in Example 3.

16. $\int_0^2 \sqrt{1 + x^3}\, dx.$

17. $\int_2^6 \sqrt{224 - x^3}\, dx.$

18. $\int_0^4 \sqrt{256 - x^4}\, dx.$

19. $\int_0^2 \sqrt{16 + x^4}\, dx.$

20. $\int_0^2 \frac{dx}{\sqrt{8 + x^3}}.$

21. $\int_0^4 \frac{dx}{\sqrt{100 - x^3}}.$

***262. Newton's Method of Approximation.** Newton's method of finding improved approximations to the root of an equation when a value near the root is known has been described in Sec. 49, where we derived the rule from geometric considerations, and in Sec. 167, where we based the rule on the properties of differentials. We shall now deduce the rule from Taylor's series, and so obtain an estimate of the error made.

We assume that r_1 is a known approximate value of one root of the equation

$$f(x) = 0. \tag{88}$$

Write $y = f(x)$, with inverse function $x = F(y)$. Then by Eq. (23) with F in place of f and y in place of x, we have

$$x = F(a) + F'(a)(y - a) + \frac{F''(a)}{2!}(y - a)^2 + \cdots. \tag{89}$$

Let $a = f(r_1)$. Then since $y = f(x)$ defines $x = F(y)$, $x = r_1$ when $y = a$. And $F(a) = r_1$. Hence

$$a = f(r_1), \qquad F(a) = r_1. \tag{90}$$

We next find the derivatives of $F(x)$. We first note that

$$dy = f'(x)dx, \qquad dx = F'(y)dy, \qquad F'(y) = \frac{dx}{dy} = \frac{1}{dy/dx} = \frac{1}{f'(x)}. \tag{91}$$

By differentiating this relation we find

$$F''(y) = \frac{d}{dy}\left[\frac{1}{f'(x)}\right] = \frac{d}{dx}\left[\frac{1}{f'(x)}\right]\frac{dx}{dy} = -\frac{f''(x)}{[f'(x)]^2}\frac{1}{f'(x)} = -\frac{f''(x)}{[f'(x)]^3}. \tag{92}$$

Since $x = r_1$ when $y = a$ by Eq. (90), from Eqs. (91) and (92) we find

$$F'(a) = \frac{1}{f'(r_1)}, \qquad F''(a) = -\frac{f''(r_1)}{[f'(r_1)]^3}. \tag{93}$$

Substitution from Eqs. (90) and (93) in Eq. (89) leads to

$$x = r_1 + \frac{1}{f'(r_1)}[y - f(r_1)] - \frac{f''(r_1)}{2[f'(r_1)]^3}[y - f(r_1)]^2 + \cdots. \tag{94}$$

Let r denote the exact value of the root approximated by r_1. Then

$$f(r) = 0 \quad \text{and} \quad y = 0, \quad x = r \tag{95}$$

are values satisfying Eq. (94). Hence

$$r = r_1 - \frac{f(r_1)}{f'(r_1)} - \frac{f''(r_1)}{2[f'(r_1)]^3}[f(r_1)]^2 + \cdots, \tag{96}$$

Suppose that r_1 is so good an approximation to r that $f(r_1)$ is small compared to unity. Then the higher powers of $f(r_1)$ will be small compared with the first power. Neglecting these higher powers and denoting the resulting solution of Eq. (96) by r_2, we find

$$r_2 = r_1 - \frac{f(r_1)}{f'(r_1)}. \tag{97}$$

This is Newton's rule for a second approximation. As illustrated in Secs. 49 and 167, the process may be repeated for increased accuracy.

Usually, if r_1 is correct to n decimal places, $\left|\frac{f(r_1)}{f'(r_1)}\right|$ will not greatly exceed $\frac{1}{10^n}$. And $\left|\frac{f''(r_1)}{f'(r_1)}\right|$ will not greatly exceed unity. Hence for the first term neglected in Eq. (96),

$$\left| -\frac{f''(r_1)}{2[f'(r_1)]^3}[f(r_1)]^2 \right| = \frac{1}{2}\left|\frac{f''(r_1)}{f'(r_1)}\right| \left[\frac{f(r_1)}{f'(r_1)}\right]^2, \tag{98}$$

will not greatly exceed $\left(\frac{1}{10^n}\right)^2 = \frac{1}{10^{2n}}$. This explains the rule mentioned in Sec. 49 that, when r_1 is a close approximation, r_2 as found from Eq. (97) is good to nearly twice as many places as r_1.

***263. Complex Numbers.** We recall that a *complex number* is an expression of the form $a + bi$, where a and b are real numbers and i is the imaginary unit,

$$i = \sqrt{-1} \quad \text{and} \quad i^2 = -1. \tag{99}$$

For the most part, the rules for manipulating complex numbers are the same as those for real numbers. One useful principle is that, if a, b, a', and b' are all real, then the equation

$$a + bi = a' + b'i \text{ implies that } a = a' \text{ and } b = b'. \tag{100}$$

Thus, in any equation simplified to this form, we may equate the real and imaginary parts separately.

***264. Exponential and Trigonometric Functions of a Complex Number.** For a complex variable $z = x + iy$, the power function Az^n is defined by repeated multiplication. From this, by addition, we may evaluate polynomials and convergent power series with real or complex coefficients.

The functions e^z, $\sin z$, and $\cos z$ for z complex are defined by the following infinite power series.

$$e^z = 1 + z + \frac{z^2}{2!} + \frac{z^3}{3!} + \cdots, \tag{101}$$

$$\sin z = z - \frac{z^3}{3!} + \frac{z^5}{5!} - \frac{z^7}{7!} + \cdots, \tag{102}$$

$$\cos z = 1 - \frac{z^2}{2!} + \frac{z^4}{4!} - \frac{z^6}{6!} + \cdots. \tag{103}$$

These series are similar in form to the Maclaurin's series of Probs. 1 to 3 of Exercise 124 which represent the functions e^x, $\sin x$, and $\cos x$ for all real values of x. This shows that when $y = 0$, so that $z = x + iy = x$, the values obtained from the new definition will agree with those previously used for real values of the variable.

The series (101) to (103) converge for all complex values of z. Convergent series of this type may be multiplied and added together in the same way that polynomials are combined. It follows that the functions defined by the series satisfy the relation

$$e^{z_1} \cdot e^{z_2} = e^{z_1 + z_2}, \tag{104}$$

as well as the addition theorems

$$\sin(z_1 + z_2) = \sin z_1 \cos z_2 + \cos z_1 \sin z_2, \tag{105}$$

$$\cos(z_1 + z_2) = \cos z_1 \cos z_2 - \sin z_1 \sin z_2, \tag{106}$$

and the identity

$$\cos^2 z + \sin^2 z = 1. \tag{107}$$

***265. Euler's Expressions.** Let us replace z by iz in Eq. (101). In view of Eq. (99), we find that

$$\begin{aligned} e^{iz} &= 1 + (iz) + \frac{(iz)^2}{2!} + \frac{(iz)^3}{3!} + \cdots \\ &= 1 + iz - \frac{z^2}{2!} - i\frac{z^3}{3!} + \frac{z^4}{4!} + \cdots \\ &= \left(1 - \frac{z^2}{2!} + \frac{z^4}{4!} - \cdots\right) + i\left(z - \frac{z^3}{3!} + \frac{z^5}{5!} - \cdots\right). \end{aligned} \tag{108}$$

A comparison of Eq. (108) with Eqs. (102) and (103) shows that

$$e^{iz} = \cos z + i \sin z. \tag{109}$$

This is called *Euler's relation*, particularly when applied to a real value of z.

A similar calculation, which starts by replacing z by $-iz$ in Eq. (101), leads to the relation

$$e^{-iz} = \cos z - i \sin z. \tag{110}$$

To solve Eqs. (109) and (110) for $\sin z$ and $\cos z$, we proceed as follows. Subtract Eq. (110) from Eq. (109), and then divide by $2i$. The result is

$$\sin z = \frac{e^{iz} - e^{-iz}}{2i}. \tag{111}$$

Add Eq. (110) to Eq. (109) and then divide by 2. The result is

$$\cos z = \frac{e^{iz} + e^{-iz}}{2}. \tag{112}$$

The Eqs. (109) to (112) are often referred to as *Euler's expressions*. These expressions enable us to reduce any combination of sines and cosines to forms involving exponentials only. It is possible to take Eq. (101) as the definition of e^z and Eq. (104) as its fundamental property. Then Eqs. (111) and (112) may be used to define the sine and the cosine. The series (102) and (103) then follow from Eqs. (111) and (112) combined with Eqs. (101) and (99). The other trigonometric functions may be defined in terms of the sine and cosine by the usual relations. From this point of view, Eqs. (105) to (107) as well as all other trigonometric identities become a consequence of Eqs. (111), (112), and (101).

Example. Show that Eq. (106) is a consequence of Eq. (104) and Euler's expressions.

Solution: From Eq. (112), $\cos(z_1 + z_2) = \frac{1}{2}[e^{i(z_1+z_2)} + e^{-i(z_1+z_2)}]$. But Eqs. (104) and (109) imply that $e^{i(z_1+z_2)} = e^{iz_1+iz_2} = e^{iz_1}e^{iz_2} = (\cos z_1 + i \sin z_1)(\cos z_2 + i \sin z_2) = \cos z_1 \cos z_2 - \sin z_1 \sin z_2 + i(\cos z_1 \sin z_2 + \sin z_1 \cos z_2)$. And similarly we find that $e^{-i(z_1+z_2)} = \cos z_1 \cos z_2 - \sin z_1 \sin z_2 - i(\cos z_1 \sin z_2 + \sin z_1 \cos z_2)$.

By substitution of these values in the first relation found, it follows that $\cos(z_1 + z_2) = \cos z_1 \cos z_2 - \sin z_1 \sin z_2$, as was to be proved.

EXERCISE 132

1. If n is any positive integer, show that

$$\cos nz = \tfrac{1}{2}(\cos z + i \sin z)^n + \tfrac{1}{2}(\cos z - i \sin z)^n.$$

From Prob. 1, with $n = 2, 3, 4, 5$, deduce that

2. $\cos 2z = \cos^2 z - \sin^2 z$.
3. $\cos 3z = \cos^3 z - 3 \cos z \sin^2 z$.
4. $\cos 4z = \cos^4 z - 6 \cos^2 z \sin^2 z + \sin^4 z$.
5. $\cos 5z = \cos^5 z - 10 \cos^3 z \sin^2 z + 5 \cos z \sin^4 z$.

6. If n is any positive integer, show that

$$\sin nz = -\frac{i}{2}(\cos z + i \sin z)^n + \frac{i}{2}(\cos z - i \sin z)^n.$$

From Prob. 6, with $n = 2, 3, 4, 5$, deduce that

7. $\sin 2z = 2 \sin z \cos z$.
8. $\sin 3z = 3 \cos^2 z \sin z - \sin^3 z$.
9. $\sin 4z = 4 \cos^3 z \sin z - 4 \cos z \sin^3 z$.
10. $\sin 5z = 5 \cos^4 z \sin z - 10 \cos^2 z \sin^3 z + \sin^5 z$.

Evaluate each of the following integrals after transforming the integrand to the second form by means of Euler's expressions.

11. $\int_0^x 8 \cos^2 x \sin^2 x \, dx = \int_0^x (1 - \cos 4x) dx$.
12. $\int_0^x 8 \cos^4 x \, dx = \int_0^x (\cos 4x + 4 \cos 2x + 3) dx$.
13. $\int_0^x 8 \sin^4 x \, dx = \int_0^x (\cos 4x - 4 \cos 2x + 3) dx$.

The usual proof for real values shows that $\dfrac{d(az^n)}{dz} = az^{n-1}$. And convergent power series may be differentiated termwise. Deduce that

14. $\dfrac{d(e^z)}{dz} = e^z$, by using Eq. (101).
15. $\dfrac{d(e^{kz})}{dz} = ke^{kz}$, by using Eq. (101).
16. $\dfrac{d(\sin z)}{dz} = \cos z$, by using Eqs. (102) and (103).
17. $\dfrac{d(\cos z)}{dz} = -\sin z$, by using Eqs. (103) and (102).

18. The rule for composite functions $\dfrac{dw}{dz} = \dfrac{dw}{du}\dfrac{du}{dz}$ is valid in the complex case. Use this and Prob. 14 to deduce that $\dfrac{d[e^{(a+bi)z}]}{dz} = (a + bi)e^{(a+bi)z}$, for a and b real.

Use Prob. 18 and Eqs. (111) and (112)

19. To check Prob. 16.
20. To check Prob. 17.

21. From Prob. 18, deduce that one value of the indefinite integral $\int e^{(a+bi)x}\,dx$ is $\dfrac{e^{(a+bi)x}}{a+bi}$ or $\dfrac{e^{ax}e^{bix}}{a+bi} = e^{ax}\dfrac{(\cos bx + i \sin bx)}{a+bi}\dfrac{(a-bi)}{(a-bi)} = e^{ax}\dfrac{a\cos bx + b\sin bx}{a^2+b^2} + ie^{ax}\dfrac{a\sin bx - b\cos bx}{a^2+b^2}$. By equating real and imaginary parts as in Eq. (100), deduce one value of each of the indefinite integrals

$$\int e^{ax}\cos bx\,dx \qquad \text{and} \qquad \int e^{ax}\sin bx\,dx.$$

22. If k is 0 or any positive or negative integer, $m = 2k+1$ is an odd integer. Deduce from Euler's relation, Eq. (109), that $e^{im\pi} = e^{i(2k+1)\pi} = -1$. If $w = \ln z$ when $z = e^w$, it follows that $(2k+1)\pi i$ is a value of $\ln(-1)$.

***266. Computation of e^z, sin z, and cos z.** If $z = x + iy$, we find from Eqs. (104) and (109) that

$$e^z = e^{x+iy} = e^x e^{iy} = e^x(\cos y + i\sin y) = e^x\cos y + ie^x\sin y. \tag{113}$$

This enables us to compute the value of e^z from tables of values of the real functions e^x, $\cos y$, and $\sin y$ with y in radian measure.

To compute $\sin z$ and $\cos z$, we have from Eqs. (105) and (106)

$$\sin z = \sin(x+iy) = \sin x\cos iy + \cos x\sin iy, \tag{114}$$
$$\cos z = \cos(x+iy) = \cos x\cos iy - \sin x\sin iy. \tag{115}$$

But from Eqs. (111) and (112), we have

$$\sin iy = \frac{e^{-y}-e^{y}}{2i} = i\left(\frac{e^y - e^{-y}}{2}\right), \tag{116}$$
$$\cos iy = \frac{e^y + e^{-y}}{2}. \tag{117}$$

This suggests that we tabulate the real functions

$$\sinh y = \frac{e^y - e^{-y}}{2} \qquad \text{and} \qquad \cosh y = \frac{e^y + e^{-y}}{2}. \tag{118}$$

We read sinh as "hyperbolic sine" or "sinch." And we read cosh as "hyperbolic cosine" or "cosh."

The functions† $\sin z$ and $\cos z$ may then be computed from

$$\sin z = \sin x\cosh y + i\cos x\sinh y, \tag{119}$$
$$\cos z = \cos x\cosh y - i\sin x\sinh y. \tag{120}$$

*** 267. Hyperbolic Functions.** Let us study the hyperbolic functions defined by Eq. (118). With x in place of y, these are

$$\sinh x = \frac{e^x - e^{-x}}{2} \qquad \text{and} \qquad \cosh x = \frac{e^x + e^{-x}}{2}. \tag{121}$$

† A fuller discussion of exponential and trigonometric functions of complex quantities, including the computation of inverse trigonometric functions and applications to electrical and mechanical circuits, will be found in Chap. 1 of the author's "Fourier Methods," McGraw-Hill Book Company, Inc., New York, 1949.

The graph of $y = \sinh x$ is shown in Fig. 286. And that of $y = \cosh x$ is shown in Fig. 287.

It follows from Eqs. (116) and (117) that

$$\sin ix = i \sinh x \quad \text{and} \quad \cos ix = \cosh x. \tag{122}$$

These relations may be used to deduce formulas for the hyperbolic functions from

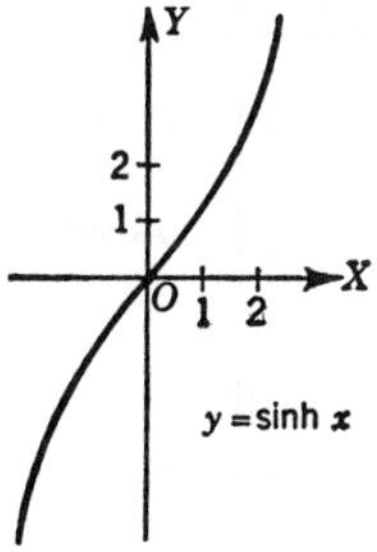

Fig. 286.

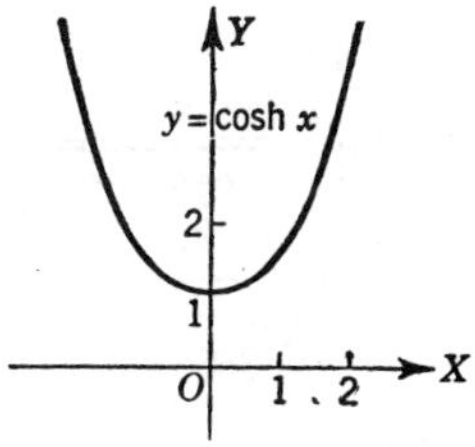

Fig. 287.

those for the trigonometric functions. For example, if we replace z_1 by ix_1 and z_2 by ix_2 in Eq. (105), we find

$$\sin i(x_1 + x_2) = \sin ix_1 \cos ix_2 + \cos ix_1 \sin ix_2. \tag{123}$$

By using the relations of Eq. (122), we may derive from this

$$i \sinh (x_1 + x_2) = i \sinh x_1 \cosh x_2 + \cosh x_1 (i \sinh x_2). \tag{124}$$

And division of this last equation by i leads to

$$\sinh (x_1 + x_2) = \sinh x_1 \cosh x_2 + \cosh x_1 \sinh x_2. \tag{125}$$

By a similar argument starting with Eq. (106), we may deduce that

$$\cosh (x_1 + x_2) = \cosh x_1 \cosh x_2 + \sinh x_1 \sinh x_2. \tag{126}$$

And from Eq. (107) we may derive the identity

$$\cosh^2 x - \sinh^2 x = 1. \tag{127}$$

By analogy with the equations which define the other trigonometric functions in terms of the sine and cosine, we may define the hyperbolic tangent, hyperbolic cotangent, hyperbolic secant, and hyperbolic cosecant by the equations

$$\tanh x = \frac{\sinh x}{\cosh x}, \qquad \coth x = \frac{\cosh x}{\sinh x} = \frac{1}{\tanh x}. \tag{128}$$

$$\operatorname{sech} x = \frac{1}{\cosh x}, \qquad \operatorname{csch} x = \frac{1}{\sinh x}. \tag{129}$$

The graph of $y = \tanh x$ is shown in Fig. 288. And that of $y = \coth x$ is shown in Fig. 289. These additional functions may be expressed in terms of exponential func-

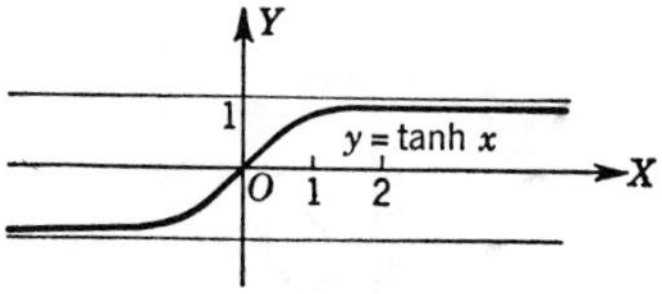

FIG. 288.

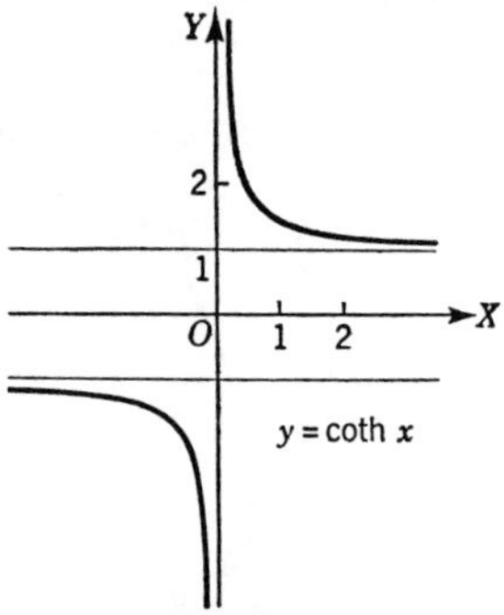

FIG. 289.

tions by using Eq. (121). For example,

$$\tanh x = \frac{e^x - e^{-x}}{e^x + e^{-x}} = \frac{e^{2x} - 1}{e^{2x} + 1}. \tag{130}$$

It follows directly from Eq. (121) that

$$\frac{d(\sinh x)}{dx} = \cosh x \quad \text{and} \quad \frac{d(\cosh x)}{dx} = \sinh x. \tag{131}$$

By putting $z = x$, and then $z = -x$ in Eq. (101), we obtain

$$e^x = 1 + x + \frac{x^2}{2!} + \frac{x^3}{3!} + \cdots, \qquad e^{-x} = 1 - x + \frac{x^2}{2!} - \frac{x^3}{3!} + \cdots. \tag{132}$$

Substitution of these in Eq. (121) leads to the Maclaurin's series

$$\sinh x = x + \frac{x^3}{3!} + \frac{x^5}{5!} + \frac{x^7}{7!} + \cdots, \tag{133}$$

$$\cosh x = 1 + \frac{x^2}{2!} + \frac{x^4}{4!} + \frac{x^6}{6!} + \cdots. \tag{134}$$

EXERCISE 133

Verify each of the following numerical evaluations.

1. $e^{\pi i} = -1$.
2. $e^{\pi i/2} = i$.
3. $e^{1-\pi i/4} = 1.967 - 1.967i$.
4. $\cos 4i = 27.31$.
5. $\sin (1 + i) = 1.298 + 0.635i$.
6. $\sin 3i = 10.02i$.

For hyperbolic functions, prove the following identities.

7. $1 - \tanh^2 x = \operatorname{sech}^2 x$.
8. $\coth^2 x - 1 = \operatorname{csch}^2 x$.
9. $\sinh 2x = 2 \sinh x \cosh x$.
10. $\cosh 2x = \cosh^2 x + \sinh^2 x$.
11. $\cosh 3x = \cosh^3 x + 3 \cosh x \sinh^2 x = 4 \cosh^3 x - 3 \cosh x$.
12. $\sinh 3x = 3 \cosh^2 x \sinh x + \sinh^3 x = 4 \sinh^3 x + 3 \sinh x$.
13. $\tanh (x_1 + x_2) = \dfrac{\tanh x_1 + \tanh x_2}{1 + \tanh x_1 \tanh x_2}$.
14. $\tanh 2x = \dfrac{2 \tanh x}{1 + \tanh^2 x}$.

In each case the value of one of the six hyperbolic functions is given. Find the value of the other functions without using tables.

15. $\sinh x = 0.75$.
16. $\cosh x = 2.6$.
17. $\tanh x = 0.8$.
18. $\operatorname{sech} x = 0.8$.

19. If $\sinh x_1 = \frac{4}{3}$, $\sinh x_2 = \frac{5}{12}$, find $\sinh (x_1 + x_2)$.
20. If $\tanh x = \frac{3}{4}$, use Prob. 14 to find $\tanh 2x$.

Prove the following rules for differentiation.

21. $\dfrac{d(\tanh x)}{dx} = \operatorname{sech}^2 x.$
22. $\dfrac{d(\operatorname{sech} x)}{dx} = -\tanh x \operatorname{sech} x.$
23. $\dfrac{d(\coth x)}{dx} = -\operatorname{csch}^2 x.$
24. $\dfrac{d(\operatorname{csch} x)}{dx} = -\coth x \operatorname{csch} x.$

25. Use Maclaurin's series, Eq. (8), and Eq. (131) to check the series of Eqs. (133) and (134).

Use the division process of Sec. 247 and Eqs. (133) and (134) to derive the following series. That they converge for $|x| < \pi/2$ corresponds to the fact that $\cosh x$ is zero for $x = -i\pi/2$ and for $x = i\pi/2$, but for no root with absolute value smaller than $\pi/2$.

26. $\operatorname{sech} x = 1 - \dfrac{x^2}{2} + \dfrac{5}{24}x^4 - \cdots, \quad -\dfrac{\pi}{2} < x < \dfrac{\pi}{2}.$
27. $\tanh x = x - \dfrac{x^3}{3} + \dfrac{2}{15}x^5 - \cdots, \quad -\dfrac{\pi}{2} < x < \dfrac{\pi}{2}.$

Use Maclaurin's series or l'Hospital's rule to show that

28. $\lim\limits_{x \to 0} \dfrac{\sinh x}{x} = 1.$
29. $\lim\limits_{x \to 0} \dfrac{\tanh x}{x} = 1.$
30. $\lim\limits_{x \to 0} \dfrac{\cosh x - 1}{x^2} = \dfrac{1}{2}.$

***268. Inverse Hyperbolic Functions.** Let $\sinh^{-1} x$ be the function inverse to the hyperbolic sine. Then

$$y = \sinh^{-1} x \quad \text{implies that} \quad x = \sinh y. \tag{135}$$

It follows from Eq. (121) that

$$x = \frac{e^y - e^{-y}}{2} \quad \text{and} \quad e^y - 2x - e^{-y} = 0. \tag{136}$$

We may rewrite this in the form

$$(e^y)^2 - 2x(e^y) - 1 = 0. \tag{137}$$

This is a quadratic equation in e^y, whose solution is

$$e^y = \frac{2x \pm \sqrt{(2x)^2 - 4(1)(-1)}}{2} = x \pm \sqrt{x^2 + 1}. \tag{138}$$

Since $x^2 + 1 > x^2$, $\sqrt{x^2 + 1} > |x|$ and the value with the minus sign is negative. But for any real y, e^y is positive. Hence we must use the plus sign. Taking natural logarithms, we find that

$$y = \sinh^{-1} x = \ln (x + \sqrt{x^2 + 1}). \tag{139}$$

By a similar procedure we may deduce from Eq. (121) that the positive value of the inverse hyperbolic cosine is

$$\cosh^{-1} x = \ln (x + \sqrt{x^2 - 1}), \qquad x \geqq 1. \tag{140}$$

And from Eq. (130) we may deduce that

$$\tanh^{-1} x = \frac{1}{2} \ln \frac{1 + x}{1 - x}, \qquad -1 < x < 1. \tag{141}$$

Let us replace x by u in Eqs. (139) to (141) and then differentiate the resulting equation with respect to x. In this way we find that

$$\frac{d(\sinh^{-1} u)}{dx} = \frac{1}{\sqrt{u^2 + 1}} \frac{du}{dx}, \tag{142}$$

$$\frac{d(\cosh^{-1} u)}{dx} = \frac{1}{\sqrt{u^2 - 1}} \frac{du}{dx}, \tag{143}$$

$$\frac{d(\tanh^{-1} u)}{dx} = \frac{1}{1 - u^2} \frac{du}{dx}. \tag{144}$$

Let us replace x by u/a in Eqs. (139) to (141). This leads to the relations

$$\ln \left(\frac{u + \sqrt{u^2 + a^2}}{a} \right) = \sinh^{-1} \frac{u}{a}, \tag{145}$$

$$\ln \left(\frac{u + \sqrt{u^2 - a^2}}{a} \right) = \cosh^{-1} \frac{u}{a}, \qquad u > a, \tag{146}$$

$$\ln \left(\frac{a + u}{a - u} \right) = 2 \tanh^{-1} \frac{u}{a}, \qquad -a < u < a. \tag{147}$$

Next replace x by a/u in Eqs. (139) to (141). This leads to the relations

$$\ln \left(\frac{a + \sqrt{a^2 - u^2}}{u} \right) = \cosh^{-1} \frac{a}{u} = \operatorname{sech}^{-1} \frac{u}{a}, \qquad u < a, \tag{148}$$

$$\ln \left(\frac{a + \sqrt{a^2 + u^2}}{u} \right) = \sinh^{-1} \frac{a}{u} = \operatorname{csch}^{-1} \frac{u}{a}, \tag{149}$$

$$\ln \left(\frac{u + a}{u - a} \right) = 2 \tanh^{-1} \frac{a}{u} = 2 \coth^{-1} \frac{u}{a}, \qquad |u| > |a|. \tag{150}$$

The relations of Eqs. (145) to (150) enable us to find alternative forms for many indefinite integrals which contain expressions like those in the left members. See Example 2 and Probs. 13 to 19 of Exercise 134.

EXAMPLE 1. Derive Eq. (142) from the properties of the direct hyperbolic functions.

Solution: Let $y = \sinh^{-1} u$. Then $u = \sinh y$. And from Eq. (131) $\dfrac{du}{dy} = \cosh y$. Hence $\dfrac{dy}{du} = \dfrac{1}{\cosh y}, \dfrac{dy}{dx} = \dfrac{dy}{du}\dfrac{du}{dx} = \dfrac{1}{\cosh y}\dfrac{du}{dx}$. But from Eq. (127), $\cosh^2 y - \sinh^2 y = 1$. Hence $\cosh^2 y = \sinh^2 y + 1 = u^2 + 1$, $\cosh y = \sqrt{u^2 + 1}$, since $\cosh y$ is positive. It follows that $\dfrac{dy}{dx} = \dfrac{1}{\cosh y}\dfrac{du}{dx} = \dfrac{1}{\sqrt{u^2 + 1}}\dfrac{du}{dx}$, which was to be proved.

EXAMPLE 2. Express the indefinite integral $\displaystyle\int \frac{du}{\sqrt{u^2 - a^2}}$ in terms of hyperbolic functions.

Solution 1: $\int \frac{du}{\sqrt{u^2 - a^2}} = \ln (u + \sqrt{u^2 - a^2}) + C_1$. By Eq. (146), $\ln \left(\frac{u + \sqrt{u^2 - a^2}}{a}\right) = \ln (u + \sqrt{u^2 - a^2}) - \ln a = \cosh^{-1} \frac{u}{a}$. Hence $\ln (u + \sqrt{u^2 - a^2}) = \cosh^{-1} \frac{u}{a} + \ln a$. And if $C = C_1 + \ln a$, $\int \frac{du}{\sqrt{u^2 - a^2}} = \cosh^{-1} \frac{u}{a} + C$, the required expression.

Solution 2: Let $u = a \cosh t$. Then from Eq. (127), $\cosh^2 t - \sinh^2 t = 1$, so that $u^2 - a^2 = a^2(\cosh^2 t - 1) = a^2 \sinh^2 t$. Also from Eq. (131), $du = a \sinh t\, dt$. It follows that $\int \frac{du}{\sqrt{u^2 - a^2}} = \int \frac{a \sinh t\, dt}{a \sinh t} = \int dt = t + C = \cosh^{-1} \frac{u}{a} + C$. This is the required expression.

EXERCISE 134

Carry out the details of the derivation of

1. Eq. (140) from Eq. (121).
2. Eq. (141) from Eq. (130).
3. Eq. (142) from Eq. (139).
4. Eq. (143) from Eq. (140).
5. Eq. (144) from Eq. (141).

6. Derive Eq. (143) from the properties of the direct hyperbolic functions of Eqs. (131) and (127).
7. Derive Eq. (144) from the properties of the direct hyperbolic functions of Probs. 21 and 7 of Exercise 133.
8. Show that $\sinh^{-1} (\tan \phi) = \ln (\tan \phi + \sec \phi)$.
9. Show that $\cosh^{-1} (\sec \phi) = \ln (\tan \phi + \sec \phi)$.
10. Show that $\tanh^{-1} (\sin \phi) = \ln (\tan \phi + \sec \phi)$.

Use the procedure of Example 3 of Sec. 246 to derive the following Maclaurin's series.

11. $\sinh^{-1} x = x - \frac{1}{2}\frac{x^3}{3} + \frac{1 \cdot 3}{2 \cdot 4}\frac{x^5}{5} - \frac{1 \cdot 3 \cdot 5}{2 \cdot 4 \cdot 6}\frac{x^7}{7} + \cdots, \qquad -1 < x < 1.$
12. $\tanh^{-1} x = x + \frac{x^3}{3} + \frac{x^5}{5} + \frac{x^7}{7} + \cdots, \qquad -1 < x < 1.$

Check the last expression for each of the following integrals.

13. $\int \frac{du}{a^2 - u^2} = \frac{1}{2a} \ln \left(\frac{a + u}{a - u}\right) + C = \frac{1}{a} \tanh^{-1} \frac{u}{a} + C, \qquad |u| < |a|.$
14. $\int \frac{du}{a^2 - u^2} = \frac{1}{2a} \ln \left(\frac{u + a}{u - a}\right) + C = \frac{1}{a} \coth^{-1} \frac{u}{a} + C, \qquad |u| > |a|.$
15. $\int \frac{du}{\sqrt{u^2 + a^2}} = \ln (u + \sqrt{u^2 + a^2}) + C_1 = \sinh^{-1} \frac{u}{a} + C.$
16. $\int \frac{du}{u \sqrt{a^2 + u^2}} = -\frac{1}{a} \ln \left(\frac{a + \sqrt{a^2 + u^2}}{u}\right) + C = -\frac{1}{a} \operatorname{csch}^{-1} \frac{u}{a} + C.$
17. $\int \frac{du}{u \sqrt{a^2 - u^2}} = -\frac{1}{a} \ln \left(\frac{a + \sqrt{a^2 - u^2}}{u}\right) + C = -\frac{1}{a} \operatorname{sech}^{-1} \frac{u}{a} + C.$
18. $\int \sqrt{u^2 + a^2}\, du = \frac{1}{2}\left[u \sqrt{u^2 + a^2} + a^2 \sinh^{-1} \frac{u}{a}\right] + C.$
19. $\int \sqrt{u^2 - a^2}\, du = \frac{1}{2}\left[u \sqrt{u^2 - a^2} - a^2 \cosh^{-1} \frac{u}{a}\right] + C.$

Verify each of the following results.

20. $\displaystyle\int_0^9 \frac{dx}{\sqrt{x^2+9}} = \sinh^{-1} 3 = 1.8184.$

21. $\displaystyle\int_3^9 \frac{dx}{\sqrt{x^2-9}} = \cosh^{-1} 3 = 1.7627.$

22. $\displaystyle\int_0^1 \frac{dx}{4-x^2} = \frac{1}{2}\tanh^{-1} 0.5 = 0.5493.$

23. $\displaystyle\int_4^\infty \frac{dx}{x^2-4} = \frac{1}{2}\coth^{-1} 2 = 0.5493.$

24. The equations for the catenary, or curve of equilibrium of a hanging chain or heavy flexible cable, take the form $d(T\cos\tau) = 0$, $d(T\sin\tau) = w\,ds$, where $\tan\tau = \frac{dy}{dx}\cdot$ Hence $T\cos\tau = H$, a constant. Let $a = \frac{H}{w}\cdot$ And put $\frac{dy}{dx} = p$. Then $a\,dp = ds = \sqrt{1+p^2}\,dx$. And $dx = \frac{a\,dp}{\sqrt{1+p^2}}\cdot$ Assume that $p = 0$ at x_0, y_0. Deduce from Prob. 15 that $x - x_0 = a\sinh^{-1} p$. Then $\frac{dy}{dx} = p = \sinh\frac{x-x_0}{a}$, so that $y = y_0 + a\cosh\left(\frac{x-x_0}{a}\right) - a.$

25. With $y_0 = a$ and $x_0 = 0$, reduce the equation of the catenary of Prob. 24 to the form $y = a\cosh\frac{x}{a} = \frac{a}{2}(e^{x/a} + e^{-x/a})$.

26. For the catenary $y = a\cosh\frac{x}{a}$ of Prob. 25, show that $\frac{dy}{dx} = \sinh\frac{x}{a}$, $\frac{ds}{dx} = \cosh\frac{x}{a}\cdot$ Deduce that $s = \int_0^x \cosh\frac{x}{a}\,dx = a\sinh\frac{x}{a}$, where s is the arc length measured from the point $(0,a)$. Thus if $p = \frac{dy}{dx}$, $s = ap$. This checks with the relation $ds = a\,dp$ of Prob. 24.

CHAPTER 17

PARTIAL DIFFERENTIATION. ENVELOPES

In many situations we must consider a geometrical or physical quantity as depending on more than one independent variable. For such cases, the notion of an ordinary derivative and the methods of the differential calculus for functions of one variable must be generalized so as to apply to functions of several variables. This extension constitutes the subject of partial differentiation, to which we devote this chapter. We define partial derivatives and total differentials and study their fundamental properties. We discuss how the partial derivatives are transformed when we change from one set of variables to another.

Finally we apply partial derivatives to the formulation of Taylor's series for functions of two variables, some problems in maxima and minima, and the problem of finding the envelope of a family of plane curves. Applications to surfaces in three space will be treated in Chap. 18.

269. Functions of Two Variables. We sometimes have a situation similar to that of Sec. 5, except that to determine one *dependent variable* we must know the value of each of two *independent variables.* For example, let

$$u = 2x^2 + 3xy + 4y^2. \tag{1}$$

Then to determine the independent variable u, we must know the value of x and the value of y. And these values may be chosen arbitrarily and independent of one another.

We say that u is a function of x and y, and use functional symbols like those of Sec. 5. Thus if

$$u = f(x,y) \tag{2}$$

denotes the particular function of Eq. (1), then when $x = 2$ and $y = 3$, $u = f(2,3) = 2 \cdot 2^2 + 3 \cdot 2 \cdot 3 + 4 \cdot 3^2 = 62.$

If for any function $f(x,y)$ and some particular pair of values $x = a$, $y = b$, $f(a,b)$ is defined and

$$\lim_{\substack{x \to a \\ y \to b}} f(x,y) = f(a,b), \tag{3}$$

where x tends to a and y tends to b as in Sec. 8 in any manner whatever, we say that the function $f(x,y)$ is *continuous* at (a,b). In this case the value of $f(x,y)$ will be nearly equal to $f(a,b)$ whenever x is sufficiently near to a and y is sufficiently near to b.

Similar definitions may be formed of continuous functions of three or more variables. Theorems similar to those of Secs. 10 and 11 hold for such functions of several variables.

270. Partial Derivatives. Let u be a function of the two independent variables x and y, so that

$$u = f(x,y). \tag{4}$$

Then for each fixed value of y, u is a function of the one variable x. If this function is differentiable, we may apply the ordinary process of differentiation to it. The result is denoted by $\partial u/\partial x$, u_x, or $f_x(x,y)$, and is called the *partial derivative* of $f(x,y)$ with respect to x. Thus from Sec. 27 we have

$$\frac{\partial u}{\partial x} = u_x = f_x(x,y) = \lim_{h \to 0} \frac{f(x + h, y) - f(x,y)}{h}. \tag{5}$$

Similarly, if we keep x fixed, we may obtain a partial derivative with respect to y denoted by $\partial u/\partial y$, u_y, or $f_y(x,y)$. And

$$\frac{\partial u}{\partial y} = u_y = f_y(x,y) = \lim_{k \to 0} \frac{f(x, y + k) - f(x,y)}{k}. \tag{6}$$

Alternative notations for the partial derivatives of Eq. (5) are

$$\frac{\partial u}{\partial x} = f_x(x,y) = f_x = \frac{\partial}{\partial x} f(x,y). \tag{7}$$

And, similarly, for the partial derivative of Eq. (6),

$$\frac{\partial u}{\partial y} = f_y(x,y) = f_y = \frac{\partial}{\partial y} f(x,y). \tag{8}$$

In the last form of Eq. (7) we may think of $\partial/\partial x$ as the differentiating operator, acting on the next following function.

Corresponding symbols and definitions are used for functions of three or more variables as illustrated in Example 3 below.

EXAMPLE 1. Find the partial derivatives of

$$u = 2x^2 + 3xy + 4y^2$$

with respect to x and y for any values, and for $(x,y) = (2,3)$.

Solution: Regarding y as a constant, from the standard rules, we find $\dfrac{\partial u}{\partial x} = 4x + 3y = f_x(x,y)$. And $f_x(2,3) = 4 \cdot 2 + 3 \cdot 3 = 17$. And regarding x as a constant, we find

$$\frac{\partial u}{\partial y} = 3x + 8y = f_y(x,y) \qquad \text{and} \qquad f_y(2,3) = 3 \cdot 2 + 8 \cdot 3 = 30.$$

EXAMPLE 2. Find the partial derivatives of $w = \ln (s^2t^3)$ with respect to s and t.

Solution: Write $w = 2 \ln s + 3 \ln t$. Then keeping t constant, we have $\dfrac{\partial w}{\partial s} = \dfrac{2}{s}$. And keeping s constant, we have $\dfrac{\partial w}{\partial t} = \dfrac{3}{t}$.

EXAMPLE 3. In a plane triangle, by the law of cosines, Sec. 94,

$$c^2 = a^2 + b^2 - 2ab \cos C.$$

Find the partial derivatives of c with respect to a, b, and C.

Solution: We could solve for c, but it is simpler to use the method of implicit differentiation of Sec. 56. Thus, keeping b and C constant, we have

$$2c \frac{\partial c}{\partial a} = 2a - 2b \cos C \qquad \text{and} \qquad \frac{\partial c}{\partial a} = \frac{a - b \cos C}{c}.$$

Next, keeping a and C constant, we have

$$2c \frac{\partial c}{\partial b} = 2b - 2a \cos C \qquad \text{and} \qquad \frac{\partial c}{\partial b} = \frac{b - a \cos C}{c}.$$

Finally, keeping a and b constant, we have

$$2c \frac{\partial c}{\partial C} = -2ab(-\sin C) \qquad \text{and} \qquad \frac{\partial c}{\partial C} = \frac{ab \sin C}{c}.$$

EXERCISE 135

Find $\partial u/\partial x$ and $\partial u/\partial y$ for each of the following functions.

1. $u = x^3 + 2xy + 3y^2$.
2. $u = (x^2 + y^2)^{\frac{1}{2}}$.
3. $u = \ln \sqrt{x^2 + y^2}$.
4. $u = e^{2x} \sin 3y$.
5. $u = e^{-x^2-y^2}$.
6. $u = e^{-2y} \cos 4x$.
7. $u = e^{3xy}$.
8. $u = \sin (2x + 3y)$.
9. $u = \cos (x^2 + xy)$.
10. $u = \tan^{-1} xy$.

Find $\partial u/\partial x$, $\partial u/\partial y$, and $\partial u/\partial z$ for each of the following functions.

11. $u = xy^2z^2$.
12. $u = 2xy + 3yz - 4xz$.
13. $u = (x^2 + y^2 + z^2)^{-\frac{1}{2}}$.
14. $u = x \sin (2y + z^2)$.
15. $u = xye^{2z}$.
16. $u = xz \ln y$.

The sides of a plane triangle are a, b, c and the angles are A, B, C. For each given formula assume that the three variables which appear in the right member are the independent variables. Find the partial derivatives with respect to a and C of

17. The area, $K = \dfrac{ab \sin C}{2}$.
18. The side, $c = \dfrac{a \sin C}{\sin A}$.
19. The side, $c = (a^2 + b^2 - 2ab \cos C)^{\frac{1}{2}}$.
20. The area, $K = \dfrac{a^2 \sin B \sin C}{\sin (B + C)}$.
21. If $u = \dfrac{2x + y}{x - y}$, show that $x \dfrac{\partial u}{\partial x} + y \dfrac{\partial u}{\partial y} = 0$.
22. If $u = \cos (x^2 + y^2)$, show that $y \dfrac{\partial u}{\partial x} - x \dfrac{\partial u}{\partial y} = 0$.
23. If $u = xe^{y/x}$, show that $x \dfrac{\partial u}{\partial x} + y \dfrac{\partial u}{\partial y} = u$.

24. If $u = e^x(x - y)$, show that $\dfrac{\partial u}{\partial x} + \dfrac{\partial u}{\partial y} = u.$

25. If $u = \ln(2x - 3y)$, show that $3\dfrac{\partial u}{\partial x} + 2\dfrac{\partial u}{\partial y} = 0.$

271. Total Differential. The law of finite increments of Sec. 252 shows that

$$f(x + h, y) - f(x,y) = h\, f_x(x + \theta_1 h, y), \tag{9}$$

for some suitably chosen value of θ_1 with $0 < \theta_1 < 1$. For, with y fixed, $f(x,y)$ becomes a function of one variable x, and with y fixed, the derivative of this function with respect to x is $f_x(x,y)$.

Similarly we may apply the law of finite increments to $f(x + h, y)$ considered to be a function of y only when we keep $x + h$ fixed. This shows that

$$f(x + h, y + k) - f(x + h, y) = k\, f_y(x + h, y + \theta_2 k), \tag{10}$$

for some suitably chosen value of θ_2 with $0 < \theta_2 < 1$.

Let Δu denote the increment of u when x and y both vary, so that

$$\Delta u = f(x + h, y + k) - f(x,y). \tag{11}$$

Then it follows from Eqs. (9) to (11) that

$$\Delta u = h\, f_x(x + \theta_1 h, y) + k\, f_y(x + h, y + \theta_2 k). \tag{12}$$

When h and k are each small, $\theta_1 h$ and $\theta_2 k$ will also be small. Thus (Fig. 290) in a plane with x and y as rectangular coordinates, the points

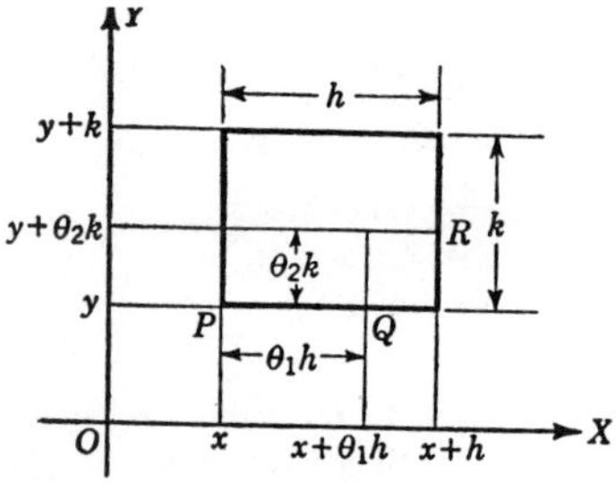

Fig. 290.

$$Q = (x + \theta_1 h, y)$$

and

$$R = (x + h, y + \theta_2 k)$$

will each be near the point $P = (x,y)$. Let us assume that each of the partial derivatives $f_x(x,y)$ and $f_y(x,y)$ is a continuous function of two variables as defined in Sec. 269. Then $f_x(x + \theta_1 h, y)$ will be nearly equal to $f_x(x,y)$ and $f_y(x + h, y + \theta_2 k)$ will be nearly equal to $f_y(x,y)$. Thus if we write

$$f_x(x + \theta_1 h, y) - f_x(x,y) = \epsilon_1, \tag{13}$$
$$f_y(x + h, y + \theta_2 k) - f_y(x,y) = \epsilon_2, \tag{14}$$

the quantities ϵ_1 and ϵ_2 will each be numerically small when h and k are small. And ϵ_1 and ϵ_2 will each approach zero when h and k approach zero. Hence we may deduce from Eqs. (12) to (14) that

$$\Delta u = h\, f_x(x,y) + k\, f_y(x,y) + \epsilon_1 h + \epsilon_2 k, \tag{15}$$

where

$$\epsilon_1 \to 0 \quad \text{and} \quad \epsilon_2 \to 0 \quad \text{when } (h,k) \to (0,0). \tag{16}$$

Since h and k are increments of x and y, we replace them by Δx and Δy and write

$$\Delta u = \Delta x\, f_x(x,y) + \Delta y\, f_y(x,y) + \epsilon_1\, \Delta x + \epsilon_2\, \Delta y. \tag{17}$$

Suppose next that x and y are each functions of a new variable t. This makes $u(x,y)$ a function of t. We may then compute the derivative du/dt from

$$\frac{\Delta u}{\Delta t} = \frac{\Delta x}{\Delta t} f_x(x,y) + \frac{\Delta y}{\Delta t} f_y(x,y) + \epsilon_1 \frac{\Delta x}{\Delta t} + \epsilon_2 \frac{\Delta y}{\Delta t}. \tag{18}$$

When $\Delta t \to 0$, $\Delta x \to 0$ and $\Delta y \to 0$, also

$$\frac{\Delta x}{\Delta t} \to \frac{dx}{dt} \quad \text{and} \quad \frac{\Delta y}{\Delta t} \to \frac{dy}{dt}. \tag{19}$$

But, by Eq. (16), when $h = \Delta x \to 0$ and $k = \Delta y \to 0$, then $\epsilon_1 \to 0$ and $\epsilon_2 \to 0$. Thus we find from Eq. (18) in the limit,

$$\frac{du}{dt} = \frac{dx}{dt} f_x(x,y) + \frac{dy}{dt} f_y(x,y). \tag{20}$$

The last two terms of Eq. (17) contributed nothing to the result just obtained in Eq. (20). Hence we define the *total differential* of u, for given dx and dy, as

$$du = dx\, f_x(x,y) + dy\, f_y(x,y). \tag{21}$$

Here dx and dy are arbitrarily chosen. We note that the limiting relation of Eq. (20) may be obtained from Eq. (21) by merely dividing by dt.

Equation (21) is consistent with the definition of a differential for a function of one variable, since, if y does not appear in the function $f(x,y)$. we shall have

$$f(x,y) = f(x), \qquad f_x(x,y) = f'(x), \qquad \text{and} \qquad f_y(x,y) = 0.$$

Hence in this case, Eq. (21) becomes

$$du = f'(x)dx \qquad \text{where } f'(x) = \frac{du}{dx} \text{ and } dx \text{ is arbitrary.} \tag{22}$$

With our first notation for the partial derivatives, Eq. (21) may be written

$$du = \frac{\partial u}{\partial x}\, dx + \frac{\partial u}{\partial y}\, dy. \tag{23}$$

And in the same notation, Eq. (20) becomes

$$\frac{du}{dt} = \frac{\partial u}{\partial x}\frac{dx}{dt} + \frac{\partial u}{\partial y}\frac{dy}{dt}. \tag{24}$$

This may be used to find the time rate of change of $u(x,y)$, a function of x and y, at any instant when the time rates of change of x and y are either known or can be found.

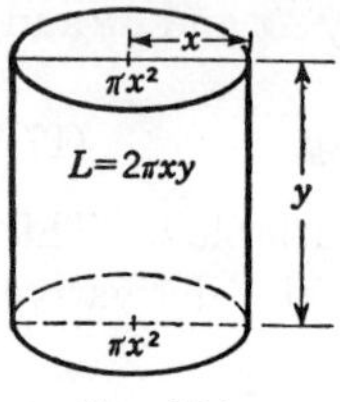

Fig. 291.

Either member of Eq. (24) is known as the *total derivative* of u with respect to t. Equation (24) is easily recalled by mentally dividing Eq. (23) by dt.

Example 1. Find the time rate of change of the surface of a right circular cylinder at the instant when the radius is 4 in. and the altitude is 6 in., if at this instant the radius is increasing at a rate of 2 in./sec. and the altitude is decreasing at a rate of 9 in./sec.

Solution: The surface of the cylinder (Fig. 291) is

$$S = 2\pi x^2 + 2\pi xy = 2\pi(x^2 + xy).$$

The partial derivatives of S are $\dfrac{\partial S}{\partial x} = 2\pi(2x + y)$ and $\dfrac{\partial S}{\partial y} = 2\pi x$. At the instant considered, $x = 4$, $y = 6$, so that

$$\frac{\partial S}{\partial x} = 28\pi \qquad \text{and} \qquad \frac{\partial S}{\partial y} = 8\pi. \qquad \text{Also } \frac{dx}{dt} = 2 \qquad \text{and} \qquad \frac{dy}{dt} = -9.$$

From Eq. (24) we have

$$\frac{\partial S}{\partial t} = \frac{\partial S}{\partial x}\frac{dx}{\partial t} + \frac{\partial S}{\partial y}\frac{dy}{dt},$$

and by substituting the values found above we obtain

$$\frac{dS}{dt} = 28\pi \cdot 2 + 8\pi \cdot (-9) = -16\pi.$$

The minus sign indicates a decrease, and the volume is decreasing at the rate of 16π cu. in./sec.

Example 2. At any time t, the coordinates of a moving point P are $(3 + 2t^2, 2 - 3t^2)$. Find the angular velocity of the line segment OP at any time t.

Solution: If $P = (x,y)$, $\tan\theta = y/x$. So that $\theta = \tan^{-1} y/x$ and

$$\frac{\partial\theta}{\partial x} = \frac{-y/x^2}{1 + (y/x)^2} = \frac{-y}{x^2 + y^2},$$
$$\frac{\partial\theta}{\partial y} = \frac{1/x}{1 + (y/x)^2} = \frac{x}{x^2 + y^2}.$$

The required angular velocity is $d\theta/dt$. And from Eq. (24) we have

$$\frac{d\theta}{dt} = \frac{\partial\theta}{\partial x}\frac{dx}{dt} + \frac{\partial\theta}{\partial y}\frac{dy}{dt}.$$

But since $x = 3 + 2t^2$, $y = 2 - 3t^2$, $dx/dt = 4t$, and $dy/dt = -6t$. So that

$$\frac{d\theta}{dt} = \frac{-y}{x^2 + y^2}4t + \frac{x}{x^2 + y^2}(-6t) = \frac{-(2 - 3t^2)4t + (3 + 2t^2)(-6t)}{(3 + 2t^2)^2 + (2 - 3t^2)^2}$$
$$= \frac{-26t}{13 + 13t^4} = \frac{-2t}{1 + t^4}.$$

The last expression gives the required rate.

EXERCISE 136

Find du/dt in each of the following problems.

1. $u = 2x^2y + 3xy^2$, $x = 2t$, $y = 4t$.
2. $u = \sin xy$, $x = t^2$, $y = t^3$.
3. $u = 5x^2y$, $x = e^{2t}$, $y = e^{-2t}$.
4. $u = 2xy$, $x = \sin t$, $y = \cos t$.
5. $u = y/x$, $y = e^{4t}$, $x = e^{2t}$.
6. $u = \ln (x^2 + y^2)$, $x = 2t$, $y = 3t$.

At a certain instant of time $x = 3$, $y = 4$. And at this time $dx/dt = 2$ and $dy/dt = -2$. Find the rate of change at the instant in question of

7. The volume of a cylinder, $V = \pi x^2 y$.
8. The hypotenuse of a right triangle, $z = \sqrt{x^2 + y^2}$.
9. The lateral surface of a cone, $L = \pi x \sqrt{x^2 + y^2}$.
10. The angle $\theta = \tan^{-1} y/x$.

A triangle has the vertex $A = (0,0)$ and the vertex $B = (4,0)$ fixed. The vertex C moves on a straight line with $x = 2t$, $y = 3 + 4t$. At time $t = 3$, find the rate of change of

11. The area of the triangle, $K = 2y$.
12. The angle $A = \tan^{-1} y/x$.
13. The angle $B = \pi - \tan^{-1} \dfrac{y}{x - 4}$.
14. The side $AC = \sqrt{x^2 + y^2}$.
15. The side $BC = \sqrt{(x - 4)^2 + y^2}$.

272. Composite Functions. Let x and y be each functions of two new variables s and t. Then for s fixed, they would be functions of t alone. Under these conditions, when we divide Eq. (23) by dt, we obtain partial derivatives. Hence we write the result as

$$\frac{\partial u}{\partial t} = \frac{\partial u}{\partial x}\frac{\partial x}{\partial t} + \frac{\partial u}{\partial y}\frac{\partial y}{\partial t}. \tag{25}$$

Similarly, by keeping t fixed and dividing by ds, we find

$$\frac{\partial u}{\partial s} = \frac{\partial u}{\partial x}\frac{\partial x}{\partial s} + \frac{\partial u}{\partial y}\frac{\partial y}{\partial s}. \tag{26}$$

If we apply the relation (23) to x and y regarded as functions of the independent variables s and t, we find

$$dx = \frac{\partial x}{\partial s}\,ds + \frac{\partial x}{\partial t}\,dt \qquad \text{and} \qquad dy = \frac{\partial y}{\partial s}\,ds + \frac{\partial y}{\partial t}\,dt. \tag{27}$$

We may now deduce from Eqs. (25) to (27) that

$$\begin{aligned}\frac{\partial u}{\partial x}\,dx + \frac{\partial u}{\partial y}\,dy &= \frac{\partial u}{\partial x}\left(\frac{\partial x}{\partial s}\,ds + \frac{\partial x}{\partial t}\,dt\right) + \frac{\partial u}{\partial y}\left(\frac{\partial y}{\partial s}\,ds + \frac{\partial y}{\partial t}\,dt\right)\\ &= \frac{\partial u}{\partial s}\,ds + \frac{\partial u}{\partial t}\,dt.\end{aligned} \tag{28}$$

But the left member defines du when x and y are the independent variables, while the right member defines du when s and t are the independent variables. Since the two are equal, it follows that the value of du obtained from an equation of the form (23) does not depend on the choice of independent variables.

In particular, formulas of differentiation, such as $df(u) = f'(u)du$, $d(uv) = u\,dv + v\,du$, $d\frac{u}{v} = \frac{v\,du - u\,dv}{u^2}$, hold good when u, v are independent or are functions of any number of other variables.

As an illustration, from the first and third formulas quoted, it follows that

$$d\tan^{-1} u = \frac{du}{1+u^2} \qquad \text{and} \qquad d\frac{y}{x} = \frac{x\,dy - y\,dx}{x^2}.$$

Hence if we put $u = y/x$, we find

$$d\tan^{-1}\frac{y}{x} = \frac{(x\,dy - y\,dx)/x^2}{1+(y/x)^2} = \frac{x\,dy - y\,dx}{x^2+y^2}.$$

This is somewhat simpler than finding the partial derivatives of $\tan^{-1}(y/x)$ directly, as we did in Example 2 of Sec. 271.

Example. Find $\frac{\partial u}{\partial s}$ and $\frac{\partial u}{\partial t}$ if $u = xy$, $x = 2s + 4t$, and $y = 2s - 4t$.

Solution: We use Eqs. (26) and (25) or

$$\frac{\partial u}{\partial s} = \frac{\partial u}{\partial x}\frac{\partial x}{\partial s} + \frac{\partial u}{\partial y}\frac{\partial y}{\partial s} \qquad \text{and} \qquad \frac{\partial u}{\partial t} = \frac{\partial u}{\partial x}\frac{\partial x}{\partial t} + \frac{\partial u}{\partial y}\frac{\partial y}{\partial t}.$$

But from the given relations we find

$$\frac{\partial u}{\partial x} = y, \quad \frac{\partial u}{\partial y} = x, \quad \frac{\partial x}{\partial s} = 2, \quad \frac{\partial x}{\partial t} = 4, \quad \frac{\partial y}{\partial s} = 2, \quad \frac{\partial y}{\partial t} = -4.$$

On substituting these values, we find that

$$\frac{\partial u}{\partial s} = y \cdot 2 + x \cdot 2 = 2(x+y) \qquad \text{and} \qquad \frac{\partial u}{\partial t} = y \cdot 4 + x(-4) = 4(y-x).$$

Since $x = 2s + 4t$ and $y = 2s - 4t$, we find from this that

$$\frac{\partial u}{\partial s} = 8s \qquad \text{and} \qquad \frac{\partial u}{\partial t} = -32t.$$

In this example it is easy to check this by substitution before differentiation. Thus $u = (2s+4t)(2s-4t) = 4s^2 - 16t^2$, from which the values $\partial u/\partial s = 8s$ and $\partial u/\partial t = -32t$ follow.

EXERCISE 137

Find $\partial u/\partial s$ and $\partial u/\partial t$ in each of the following problems.

1. $u = x^2 - y^2$, $x = 2s + t$, $y = 2s - t$.
2. $u = \ln(2x + 3y)$, $x = 3s + 2t$, $y = 3s - 2t$.
3. $u = x^3 + y^3$, $x = s + t$, $y = s - t$.

4. $u = x^2y + 3y^2$, $x = s + t$, $y = s$.
5. $u = x^2 + y^2$, $x = e^s + e^t$, $y = e^s - e^t$.
6. $u = \sin(2x + 3y)$, $x = 2s - t$, $y = t$.
7. $u = \cos(4x - y)$, $x = s + 3t$, $y = 4s$.
8. $u = e^{x+y}$, $x = s^2 + st$, $y = st - s^2$.

9. If $u = xy$, show that $df(u) = f'(u)(x\,dy + y\,dx)$.

Use Prob. 9 to find the total differential of

10. $\sin(xy)$. 11. $\cos(xy)$. 12. e^{xy}.

13. If $u = \dfrac{y}{x}$, show that $df(u) = f'(u)\dfrac{x\,dy - y\,dx}{x^2}$.

Use Prob. 13 to find the total differential of

14. $\sin\dfrac{y}{x}$. 15. $e^{y/x}$. 16. $\sin^{-1}\dfrac{y}{x}$.

17. If $u = ax + by$, show that $df(u) = f'(u)(a\,dx + b\,dy)$.

Use Prob. 17 to find the total differential of

18. $\ln(2x + 3y)$. 19. $(4x - 5y)^5$. 20. $\tan(6x - 3y)$.

273. Total Derivatives for Three or More Variables. Like the definitions of Secs. 269 and 270, the considerations of Secs. 271 and 272 apply to functions of three or more variables. For example, if $u = f(x,y,z)$, the total differential is

$$du = \frac{\partial u}{\partial x}dx + \frac{\partial u}{\partial y}dy + \frac{\partial u}{\partial z}dz. \tag{29}$$

And results like Eq. (29) hold good whether x,y,z are independent or are functions of one or more other independent variables.

If x,y,z are each functions of one variable t, we may deduce from Eq. (29) that the *total derivative* of $u = f(x,y,z)$ is

$$\frac{du}{dt} = \frac{\partial u}{\partial x}\frac{dx}{dt} + \frac{\partial u}{\partial y}\frac{dy}{dt} + \frac{\partial u}{\partial z}\frac{dz}{dt}. \tag{30}$$

And if x,y,z are functions of several new variables, of which t is one, we may deduce that the partial derivative is

$$\frac{\partial u}{\partial t} = \frac{\partial u}{\partial x}\frac{\partial x}{\partial t} + \frac{\partial u}{\partial y}\frac{\partial y}{\partial t} + \frac{\partial u}{\partial z}\frac{\partial z}{\partial t}. \tag{31}$$

Similar results hold for more than three variables.

Let $t = x$ in Eq. (30). Since $dx/dt = 1$, Eq. (30) becomes

$$\frac{du}{dx} = \frac{\partial u}{\partial x} + \frac{\partial u}{\partial y}\frac{dy}{dx} + \frac{\partial u}{\partial z}\frac{dz}{dx}. \tag{32}$$

The corresponding special case of Eq. (24) for $t = x$ is

$$\frac{du}{dx} = \frac{\partial u}{\partial x} + \frac{\partial u}{\partial y}\frac{dy}{dx}. \tag{33}$$

This is in accord with Eq. (32) since for $u(x,y)$, $\partial u/\partial z = 0$.

EXAMPLE 1. Find du/dt if $u = z \sin (y/x)$, $x = 2t$, $y = t^2$, $z = t^3$.

Solution: $\dfrac{\partial u}{\partial x} = -\dfrac{yz}{x^2} \cos \dfrac{y}{x}$, $\dfrac{\partial u}{\partial y} = \dfrac{z}{x} \cos \dfrac{y}{x}$, $\dfrac{\partial u}{\partial z} = \sin \dfrac{y}{x}$. Hence from Eq. (30) we have

$$\frac{du}{dt} = -\frac{yz}{x^2} \cos \frac{y}{x}\frac{dx}{dt} + \frac{z}{x} \cos \frac{y}{x}\frac{dy}{dt} + \sin \frac{y}{x}\frac{dz}{dt}. \tag{34}$$

This could have been found more easily from the differential

$$du = z\, d\left(\sin \frac{y}{x}\right) + \sin \frac{y}{x}\, dz = z \cos \frac{y}{x}\left(\frac{x\, dy - y\, dx}{x^2}\right) + \sin \frac{y}{x}\, dz,$$

obtained by using the product and quotient rules.

Now substitute $x = 2t$, $y = t^2$, $z = t^3$, $\dfrac{dx}{dt} = 2$, $\dfrac{dy}{dt} = 2t$, $\dfrac{dz}{dt} = 3t^2$ in Eq. (34). This gives the required result,

$$\frac{du}{dt} = -\frac{t^3}{2} \cos \frac{t}{2} + t^3 \cos \frac{t}{2} + 3t^2 \sin \frac{t}{2} = \frac{t^3}{2} \cos \frac{t}{2} + 3t^2 \sin \frac{t}{2}.$$

EXAMPLE 2. Find du/dx if $u = y \cos x + z \sin x$, $y = \sin x$, and $z = \cos x$.

Solution: $\partial u/\partial x = -y \sin x + z \cos x$, $\partial u/\partial y = \cos x$, $\partial u/\partial z = \sin x$. And $dy/dx = \cos x$, $dz/dx = -\sin x$. By substituting these values in Eq. (32), we find the required derivative

$$\frac{du}{dx} = -y \sin x + z \cos x + \cos x\,(\cos x) + \sin x\,(-\sin x).$$

In this let us replace y by $\sin x$ and z by $\cos x$, to obtain

$$\begin{aligned}\frac{du}{dx} &= -\sin^2 x + \cos^2 x + \cos^2 x - \sin^2 x = 2(\cos^2 x - \sin^2 x)\\ &= 2 \cos 2x.\end{aligned}$$

This may be checked by noting that if the values of y and z are substituted in u, we have $u = 2 \sin x \cos x = \sin 2x$.

EXERCISE 138

Find du/dt in each of the following problems.

1. $u = xyz^2$, $x = 2t^3$, $y = 3t^2$, $z = t$.
2. $u = xy + yz$, $x = e^{2t}$, $y = e^{-t}$, $z = 3e^{2t}$.
3. $u = x \cos z - y \cos z$, $x = \sin 2t$, $y = \cos 2t$, $z = 2t$.
4. $u = z \cos x + y \sin x$, $x = 3t$, $y = \cos 3t$, $z = \sin 3t$.
5. $u = 2xy + 5z$, $x = 2t$, $y = t$, $z = t^2$.
6. $u = \tan^{-1} \dfrac{x + y}{z}$, $x = t^2$, $y = 2t^2$, $z = t$.

Find du/dx in each of the following problems.

7. $u = xy \ln z$, $y = x^2$, $z = e^x$.
8. $u = x^3 + 2yz$, $y = x^2$, $z = 3x$.
9. $u = y \tan^{-1} \dfrac{z}{x}$, $y = 2x$, $z = x^2$.
10. $u = z \cos 3x - y \sin 3x$, $y = \sin 3x$, $z = \cos 3x$.
11. $u = y \cos 4x + z \sin 4x$, $y = \sin 4x$, $z = \cos 4x$.
12. $u = yze^x$, $y = e^{2x}$, $z = e^{-x}$.

274. Implicit Functions. As explained in Sec. 55, y may be defined as a function of x implicitly by an equation of the form

$$F(x,y) = 0. \tag{35}$$

The method of implicit differentiation of Sec. 56 may be expressed in terms of total and partial differentiation, as follows. Let $u = F(x,y)$. Then from Eq. (33) we find that

$$\frac{du}{dx} = \frac{dF}{dx} = \frac{\partial F}{\partial x} + \frac{\partial F}{\partial y}\frac{dy}{dx}, \tag{36}$$

if y is any particular function of x. If $y = f(x)$ is the function implicitly defined by Eq. (35), for this function $u = F(x,y) = 0$ is an identity in x. Hence $du/dx = 0$, and

$$0 = \frac{\partial F}{\partial x} + \frac{\partial F}{\partial y}\frac{dy}{dx}. \tag{37}$$

When $\partial F/\partial y \neq 0$, this equation may be solved in the form

$$\frac{dy}{dx} = -\frac{\partial F/\partial x}{\partial F/\partial y}, \qquad \frac{\partial F}{\partial y} \neq 0. \tag{38}$$

As remarked in Sec. 56, when $\partial F/\partial y \neq 0$ for any pair of values (x_1,y_1) such that $F(x_1,y_1) = 0$, there is a single branch $y = f(x)$ with $y_1 = f(x_1)$ defined by Eq. (35), and Eq. (38) gives its derivative.

Let $F(x,y) = 0$ or a constant, and x and y be functions of t. Then we may put $u = F(x,y)$, and deduce from Eq. (24) and the fact that $du/dt = 0$, that

$$\frac{\partial F}{\partial x}\frac{dx}{dt} + \frac{\partial F}{\partial y}\frac{dy}{dt} = 0. \tag{39}$$

This equation is sometimes useful in solving related rate problems.

We may have implicit functions of more than one variable. For example, z may be defined as a function of x and y implicitly by an equation of the form

$$F(x,y,z) = 0. \tag{40}$$

If $u = F(x,y,z)$, we may deduce from Eq. (29) and the fact that $du = 0$ that

$$\frac{\partial F}{\partial x}\,dx + \frac{\partial F}{\partial y}\,dy + \frac{\partial F}{\partial z}\,dz = 0, \tag{41}$$

if $z = f(x,y)$ is the function implicitly defined by Eq. (40). Now $\dfrac{\partial z}{\partial x} = \dfrac{\partial f}{\partial x}$ is the quotient of corresponding differentials dz and dx when y is constant, or $dy = 0$. Hence we may put $dy = 0$ in Eq. (41) and deduce from it that

$$\frac{\partial z}{\partial x} = \left(\frac{dz}{dx}\right)_{dy=0} = -\frac{\partial F/\partial x}{\partial F/\partial z}. \tag{42}$$

Similarly, we have

$$\frac{\partial z}{\partial y} = \left(\frac{dz}{dy}\right)_{dx=0} = -\frac{\partial F/\partial y}{\partial F/\partial z}. \tag{43}$$

Let $F(x,y,z) = 0$ or a constant, and x,y,z be functions of t. Then we may put $u = F(x,y,z)$ and deduce from Eq. (30) and the fact that $du/dt = 0$ that

$$\frac{\partial F}{\partial x}\frac{dx}{dt} + \frac{\partial F}{\partial y}\frac{dy}{dt} + \frac{\partial F}{\partial z}\frac{dz}{dt} = 0. \tag{44}$$

This equation is sometimes useful in solving related rate problems.

Similar considerations apply to implicit functions of any number of variables.

EXAMPLE 1. Find $\partial z/\partial x$ and $\partial z/\partial y$ for the function implicitly defined by the equation $x^2 + y^2 + 2z^2 = 23$ for any (x,y,z) satisfying the equation, and in particular for $(x,y,z) = (1,2,3)$.

Solution: From the given equation we find

$$2x\,dx + 2y\,dy + 4z\,dz = 0.$$

It follows from this that

$$\frac{\partial z}{\partial x} = \left(\frac{dz}{dx}\right)_{dy=0} = -\frac{x}{2z} \quad \text{and} \quad \frac{\partial z}{\partial y} = \left(\frac{dz}{dy}\right)_{dx=0} = -\frac{y}{2z}.$$

Hence for $(x,y,z) = (1,2,3)$, $\dfrac{\partial z}{\partial x} = -\dfrac{1}{6}$ and $\dfrac{\partial z}{\partial y} = -\dfrac{1}{3}$.

EXAMPLE 2. A point in space is moving on the circle of intersection of the surfaces $x^2 + y^2 + z^2 = 9$ and $x + 2y + 3z = 11$. As will be seen in Chap. 18, these are the equations of a sphere and a plane. If $dx/dt = 2$ when the point was at $(x,y,z) = (1,2,2)$, find dy/dt and dz/dt at this instant.

Solution: From Eq. (44) and the given equations we find

$$2x\frac{dx}{dt} + 2y\frac{dy}{dt} + 2z\frac{dz}{dt} = 0 \quad \text{and} \quad \frac{dx}{dt} + 2\frac{dy}{dt} + 3\frac{dz}{dt} = 0.$$

At the instant in question, $x = 1$, $y = 2$, $z = 2$, $dx/dt = 2$. By substituting these values, we find that the desired rates satisfy

$$4\frac{dy}{dt} + 4\frac{dz}{dt} = -4 \quad \text{and} \quad 2\frac{dy}{dt} + 3\frac{dz}{dt} = -2.$$

We may solve these as simultaneous in dy/dt and dz/dt. Subtracting twice the second equation from the first gives $-2\,dz/dt = 0$. And putting $dz/dt = 0$ in either of the original equations gives $dy/dt = -1$. Thus the required rates at the instant in question are

$$\frac{dy}{dt} = -1 \quad \text{and} \quad \frac{dz}{dt} = 0.$$

EXERCISE 139

In each of the following problems, assume that (x,y) is a given pair of values which satisfy the relation $F(x,y) = 0$ as stated. Also assume that for these values $\partial F/\partial y \neq 0$. Find an expression for dy/dx in terms of x and y.

1. $x + 2y - e^{xy} = 0$.
2. $xy^2 - \ln(2x + 3y) = 0$.
3. $4x - 5y - \tan xy = 0$.
4. $x^2y + \sin(2x - y) = 0$.
5. $xy - e^x \cos y = 0$.
6. $\ln(x^2 + y^2) - \tan^{-1}\dfrac{y}{x} = 0$.

In each of the following problems, assume that (x,y,z) satisfies the relation $F(x,y,z) = 0$ equivalent to the stated equation. Also assume that for these values $\partial F/\partial z \neq 0$. Find expressions for $\partial z/\partial x$ and $\partial z/\partial y$ in terms of x, y, and z.

7. $4x^2 + 5y^2 - 3z^2 = 8$.
8. $x = 4y^2 + 5z^2$.
9. $2x^2 + 3y^2 + 5z^2 = 6$.
10. $y = 4x^2 - 5z^2$.
11. $z = e^{x+2y+z}$.
12. $z = e^z \sin(x + y)$.
13. $z = \ln(x + y + z)$.
14. $z = \cos(2x - y + 3z)$.

A point is moving on the curve of intersection of two surfaces having the given equations. One of the three rates dx/dt, dy/dt, dz/dt is given for the instant when the point had a stated position. Find the other two rates at this same instant in each problem.

15. $x^2 + y^2 - z^2 = 4$, $x + y - z = 2$, $dz/dt = 4$ at $(2,1,-1)$.
16. $xyz = 6$, $x^2 + y^2 + z^2 = 13$, $dx/dt = 3$ at $(1,2,3)$.
17. $z = xy$, $z = 2x + y$, $dy/dt = 3$ at $(-1,1,-1)$.
18. $z^2 = x^2 + y^2$, $z = 5$, $dx/dt = 5$ at $(3,4,5)$.
19. $z = x^2 + y^2$, $y = x$, $dy/dt = 4$ at $(1,1,2)$.

275. Small Errors. Consider a function of two variables, $u = f(x,y)$. If we give x an increment Δx and y an increment Δy, the *increment* in u is

$$\Delta u = f(x + \Delta x, y + \Delta y) - f(x,y). \tag{45}$$

For the same value of x and y, the differential of u, du is

$$du = f_x(x,y)dx + f_y(x,y)dy, \tag{46}$$

by Eq. (21). Let us put $dx = \Delta x$ and $dy = \Delta y$ in Eq. (46). Then from Eq. (17) it follows that if

$$dx = \Delta x \text{ and } dy = \Delta y, \qquad \Delta u = du + \epsilon_1 \Delta x + \epsilon_2 \Delta y. \tag{47}$$

And from Eq. (16), ϵ_1 and ϵ_2 will be small when Δx and Δy are small. This shows that

For small values of the increments Δx and Δy, the differential du formed with $dx = \Delta x$ and $dy = \Delta y$ approximates the value of Δu.

A similar result holds for three or more variables, so that

$$du = \frac{\partial u}{\partial x}dx + \frac{\partial u}{\partial y}dy + \frac{\partial u}{\partial z}dz \tag{48}$$

formed with $dx = \Delta x$, $dy = \Delta y$, $dz = \Delta z$ approximates the value of

$$\Delta u = f(x + \Delta x, y + \Delta y, z + \Delta z) - f(x,y,z). \tag{49}$$

In practice, the error in the approximation, $|\Delta u - du|$, is of the order of magnitude of the square of the numerically largest increment of any of the independent variables. This is suggested by the development of Sec. 277.

EXAMPLE 1. Find the approximate value of the diagonal of a rectangular box of dimensions 2.02, 1.97, and 0.98.

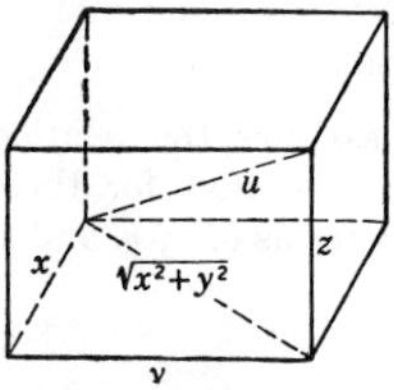

FIG. 292.

Solution: For dimensions x, y, and z, the diagonal u of Fig. 292 satisfies the relation $u^2 = x^2 + y^2 + z^2$. From this we find

$$2u\,du = 2x\,dx + 2y\,dy + 2z\,dz. \tag{50}$$

When $x = 2$, $y = 2$, $z = 1$, we have $u^2 = 9$ and $u = 3$. The given dimensions result from increments 0.02, −0.03, −0.02. Hence with differentials equal to the increments, $dx = 0.02$, $dy = -0.03$, $dz = -0.02$. Let us put these values, and $x = 2$, $y = 2$, $z = 1$, $u = 3$ in Eq. (50). This gives

$$du = \frac{x\,dx + y\,dy + z\,dz}{u} = \frac{1}{3}[2(0.02) + 2(-0.03) + 1(-0.02)]$$
$$= \frac{-0.04}{3} = -0.013.$$

Using this as an approximation to Δu, the required approximation is

$$u + du = 3 - 0.013 = 2.987.$$

We have stopped at the third decimal place, since $(-0.03)^2 = 0.0009$, so that the error of our approximation may well completely invalidate the fourth decimal place.

EXAMPLE 2. In triangle ABC, side $BC = a$ measured 40 ft., with a maximum possible error of ± 0.5 ft., while the angles at B and C each measured 60° with a maximum possible error of $\pm 1°$. Find the approximate maximum absolute error, and the approximate maximum percentage error made in calculating side $AB = c$ from these measurements.

Solution: We would calculate c from the law of sines (Sec. 94),

$$\frac{c}{\sin C} = \frac{a}{\sin A} \quad \text{or} \quad c = \frac{a \sin C}{\sin A}. \tag{51}$$

Since c involves several factors, we take natural logarithms to get

$$\ln c = \ln a + \ln \sin C - \ln \sin A.$$

From this we find the differentials

$$\frac{dc}{c} = \frac{da}{a} + \frac{\cos C}{\sin C}\,dC - \frac{\cos A}{\sin A}\,dA.$$

We have $a = 40$, $|da| = 0.5$, $C = 60°$, $|dC|$ in radians $= 1(\pi/180) = 0.01745$, $A = 180° - B - C = 60°$, $dA = \pm 1° \pm 1° = \pm 2°$ so that $|dA|$ in radians $= 2(\pi/180) = 0.03491$.

The maximum error occurs when da and dC are plus and dA is minus. Hence the maximum relative error or max $|dc/c|$ is

$$\left|\frac{da}{a}\right| + |\cot C\, dC| + |-\cot A\, dA| = \frac{0.5}{40} + \cot 60°(0.017) + \cot 60°(0.035)$$
$$= 0.012 + 0.010 + 0.020 = 0.042.$$

Hence the max $|dc/c| = 0.042$.

And, from Eq. (51) for $a = 40$, $C = 60°$, $A = 60°$, we find $c = 40$. Since $c = 40$ and max $|dc/c| = 0.042$, the max $|dc| = 40(0.042) = 1.68$. Hence the required approximate maximum absolute error is 1.68 ft. And the required approximate maximum percentage error is

$$100 \max |dc/c| = 100(0.042) = 4.2 \text{ per cent.}$$

EXERCISE 140

The two sides of a right triangle a and b were recorded as $a = 12 \pm 0.03$ ft. and $b = 5 \pm 0.02$ ft., where the terms with $\pm$ indicate maximum possible errors. Find the maximum possible error and the maximum possible percentage error in computing

1. The hypotenuse $c = \sqrt{a^2 + b^2}$.
2. The area $K = ab/2$.
3. The angle $B = \tan^{-1}(a/b)$.

Find the maximum possible relative error made in computing c by the law of sines $c = a \sin C/\sin A$ from each set of recorded data.

4. $a = 100 \pm 2$ ft., $C = 45° \pm 1°$, $A = 45° \pm 1°$.
5. $a = 60 \pm 2$ ft., $C = 50° \pm 2°$, $B = 80° \pm 2°$.

Find the maximum possible relative error made in computing c by the law of cosines $c^2 = a^2 + b^2 - 2ab \cos C$ from each set of recorded data.

6. $a = 50 \pm 2$ ft., $b = 60 \pm 3$ ft., $C = 60° \pm 1°$.
7. $a = 400 \pm 3$ ft., $b = 500 \pm 6$ ft., $C = 40° \pm 0.04°$.

Prove each of the following general principles by taking natural logarithms and forming the differentials.

8. The relative error of a product is the sum of the relative errors of the factors, $u = xy$.
9. The relative error of a quotient is the relative error of the numerator minus that of the denominator, $u = y/x$.
10. The relative error of an nth power is n times the relative error of the base, $u = x^n$.
11. For a number of factors in numerator and denominator, the maximum relative error is the sum of the maximum relative errors for the separate factors. For example, $u = xy/wz$.
12. The relative error of $u = x^n y^m/z^p$ is $n\dfrac{dx}{x} + m\dfrac{dy}{y} - p\dfrac{dz}{z}$, and the maximum relative error is $\left|n\dfrac{dx}{x}\right| + \left|m\dfrac{dy}{y}\right| + \left|p\dfrac{dz}{z}\right|$.

If measurements of x, y, z, and w in appropriate units are recorded as $x = 2 \pm 0.01$, $y = 4 \pm 0.08$, $z = 8 \pm 0.08$, and $w = 5 \pm 0.05$, use the principles or methods of Probs. 8 to 12 to find the maximum possible relative errors made in computing each of the following.

13. $u = xyz$.
14. $u = x/yz$.
15. $u = x^2yz^3$.
16. $u = x^2y/z^2$.
17. $u = \sqrt{x/y}$.
18. $u = \sqrt[3]{x^2yz}$.

276. Partial Derivatives of Higher Order. The first partial derivatives of the first partial derivatives give the second partial derivatives. Thus

$$\frac{\partial}{\partial x}\left(\frac{\partial u}{\partial x}\right) = \frac{\partial^2 u}{\partial x^2}, \qquad \frac{\partial}{\partial y}\left(\frac{\partial u}{\partial y}\right) = \frac{\partial^2 u}{\partial y^2}. \tag{52}$$

Now consider the expression

$$\Delta = f(x + h, y + k) - f(x + h, y) - f(x, y + k) + f(x,y). \tag{53}$$

We have

$$\lim_{k \to 0} \frac{\Delta}{hk} = \frac{1}{h}\left[f_y(x + h, y) - f_y(x,y)\right]. \tag{54}$$

It follows that

$$\lim_{h \to 0} \lim_{k \to 0} \frac{\Delta}{hk} = [f_y(x,y)]_x = \frac{\partial}{\partial x}\left(\frac{\partial u}{\partial y}\right) \tag{55}$$

Similarly we find

$$\lim_{k \to 0} \lim_{h \to 0} \frac{\Delta}{hk} = [f_x(x,y)]_y = \frac{\partial}{\partial y}\left(\frac{\partial u}{\partial x}\right) \tag{56}$$

We would expect that, under suitable restrictions, the repeated limit in Eq. (55) would equal that in Eq. (56), which differs from it only in the order in which the limits are taken. It would then follow that the partial derivatives on the right of these two equations are equal. In fact, it may be proved that, when all the partial derivatives that appear are continuous functions, then

$$\frac{\partial}{\partial x}\left(\frac{\partial u}{\partial y}\right) = \frac{\partial}{\partial y}\left(\frac{\partial u}{\partial x}\right) \quad \text{or} \quad \frac{\partial^2 u}{\partial x\, \partial y} = \frac{\partial^2 u}{\partial y\, \partial x}. \tag{57}$$

More generally, with similar assumptions as to continuity, any partial derivative of higher order is independent of the order in which the differentiations are performed. For example, we have

$$\frac{\partial^3 u}{\partial x^2\, \partial y} = \frac{\partial^3 u}{\partial y\, \partial x^2} = \frac{\partial^3 u}{\partial x\, \partial y\, dx}. \tag{58}$$

The definition of partial derivatives of higher order for functions of three or more variables is similar to that here given for the case of two variables. And results similar to Eqs. (57) and (58) hold for functions of several variables.

Example 1. Verify that $\frac{\partial^2 u}{\partial x\, \partial y} = \frac{\partial^2 u}{\partial y\, \partial x}$ if $u = x^2 \sin (x^2 + y^2)$.

Solution: We find, successively, that $\frac{\partial u}{\partial y} = 2x^2 y \cos (x^2 + y^2)$ and

$$\frac{\partial}{\partial x}\left(\frac{\partial u}{\partial y}\right) = 4xy \cos (x^2 + y^2) - 4x^3y \sin (x^2 + y^2),$$

$$\frac{\partial u}{\partial x} = 2x \sin (x^2 + y^2) + 2x^3 \cos (x^2 + y^2), \text{ and}$$

$$\frac{\partial}{\partial y}\left(\frac{\partial u}{\partial x}\right) = 4xy \cos (x^2 + y^2) - 4x^3y \sin (x^2 + y^2).$$

The two results agree, in accord with Eq. (57).

EXAMPLE 2. If $u = \ln (x^2 - a^2y^2)$, prove that $a^2 \frac{\partial^2 u}{\partial x^2} = \frac{\partial^2 u}{\partial y^2}$.

Solution: Write $u = \ln (x + ay) + \ln (x - ay)$. Then we find

$$\frac{\partial u}{\partial x} = \frac{1}{x + ay} + \frac{1}{x - ay}, \qquad \frac{\partial^2 u}{\partial x^2} = \frac{-1}{(x + ay)^2} + \frac{-1}{(x - ay)^2},$$

$$\frac{\partial u}{\partial y} = \frac{a}{x + ay} + \frac{-a}{x - ay}, \qquad \frac{\partial^2 u}{\partial y^2} = \frac{-a^2}{(x + ay)^2} + \frac{-a^2}{(x - ay)^2}.$$

It follows by substitution that $a^2 \frac{\partial^2 u}{\partial x^2} = \frac{\partial^2 u}{\partial y^2}$.

EXERCISE 141

Find $\frac{\partial^2 u}{\partial x^2}$, $\frac{\partial^2 u}{\partial x\, \partial y}$, and $\frac{\partial^2 u}{\partial y^2}$ for each of the following functions.

1. $u = 2x^2 + 4xy + 5y^2$.
2. $u = (2x + 3y)^3$.
3. $u = e^{xy}$.
4. $u = \sin (3x - y)$.
5. $u = \ln (x^2 - y^2)$.
6. $u = e^x \cos y$.

Verify that $\frac{\partial^2 u}{\partial x\, \partial y} = \frac{\partial^2 u}{\partial y\, \partial x}$ for each of the following functions.

7. $u = (3x - 4y)^5$.
8. $u = e^{x^2+y^2}$.
9. $u = e^y \sin (x - y)$.
10. $u = y^2 \cos (3x + 2y)$.

Verify that if

11. $u = \sqrt{x - ay}$, $a^2 \frac{\partial^2 u}{\partial x^2} - \frac{\partial^2 u}{\partial y^2} = 0$.
12. $u = \sin kx \cos akt$, $a^2 \frac{\partial^2 u}{\partial x^2} = \frac{\partial^2 u}{\partial t^2}$.
13. $u = \sin kxe^{-k^2t/a}$, $\frac{\partial^2 u}{\partial x^2} = a \frac{\partial u}{\partial t}$.
14. $u = \ln (x^2 + y^2)$, $\frac{\partial^2 u}{\partial x^2} + \frac{\partial^2 u}{\partial y^2} = 0$.
15. $u = \tan^{-1} \frac{y}{x}$, $\frac{\partial^2 u}{\partial x^2} + \frac{\partial^2 u}{\partial y^2} = 0$.
16. $u = e^{ay} \cos ax$, $\frac{\partial^2 u}{\partial x^2} + \frac{\partial^2 u}{\partial y^2} = 0$.
17. $u = x \ln \frac{y}{x}$, $x^2 \frac{\partial^2 u}{\partial x^2} + 2xy \frac{\partial^2 u}{\partial x\, \partial y} + y^2 \frac{\partial^2 u}{\partial y^2} = 0$.
18. $u = \cos xy$, $x^2 \frac{\partial^2 u}{\partial x^2} - y^2 \frac{\partial^2 u}{\partial y^2} = 0$.
19. $u = (x - y) \ln (x + y)$, $\frac{\partial^2 u}{\partial x^2} - 2 \frac{\partial^2 u}{\partial x\, \partial y} + \frac{\partial^2 u}{\partial y^2} = 0$.
20. $u = (x^2 + y^2 + z^2)^{-\frac{1}{2}}$, $\frac{\partial^2 u}{\partial x^2} + \frac{\partial^2 u}{\partial y^2} + \frac{\partial^2 u}{\partial z^2} = 0$.

21. The differential expression $M(x,y)dx + N(x,y)dy$ is said to be *exact* if it can be obtained as the total differential of some function $u(x,y)$. Thus $M\,dx + N\,dy = \frac{\partial u}{\partial x}dx + \frac{\partial u}{\partial y}dy$ and $M = \frac{\partial u}{\partial x}$, $N = \frac{\partial u}{\partial y}$. Prove that a necessary condition for $M\,dx + N\,dy$ to be exact is that $\frac{\partial M}{\partial y} = \frac{\partial N}{\partial x}$.

277. Taylor's Series. We may deduce the form of Taylor's series for functions of two variables from that for one variable as follows. Write

$$x = a + ht, \qquad y = b + kt, \tag{59}$$

so that when a, b, h, and k are constant,

$$f(x,y) = f(a + ht,\, b + kt) = F(t). \tag{60}$$

Then if $F(t)$ admits of a Maclaurin's series expansion, Eq. (8) of Sec. 246, we have

$$F(t) = F(0) + F'(0)t + F''(0)\frac{t^2}{2!} + F'''(0)\frac{t^3}{3!} + \cdots. \tag{61}$$

But we have

$$F'(t) = \frac{df}{dt} = \frac{\partial f}{\partial x}\frac{dx}{dt} + \frac{\partial f}{\partial y}\frac{dy}{dt} = h\frac{\partial f}{\partial x} + k\frac{\partial f}{\partial y} = \left(h\frac{\partial}{\partial x} + k\frac{\partial}{\partial y}\right)f, \tag{62}$$

where the last expression involves differentiating operators like those of Sec. 28. Similarly, we have

$$F''(t) = \left(h\frac{\partial}{\partial x} + k\frac{\partial}{\partial y}\right)^2 f = h^2\frac{\partial^2 f}{\partial x^2} + 2hk\frac{\partial^2 f}{\partial x\,\partial y} + k^2\frac{\partial^2 f}{\partial y^2}, \tag{63}$$

where $\left(h\frac{\partial}{\partial x} + k\frac{\partial}{\partial y}\right)^2$ means $\left(h\frac{\partial}{\partial x} + k\frac{\partial}{\partial y}\right)\left(h\frac{\partial}{\partial x} + k\frac{\partial}{\partial y}\right)$. Also

$$F'''(t) = \left(h\frac{\partial}{\partial x} + k\frac{\partial}{\partial y}\right)^3 f, \tag{64}$$

and so on for the higher derivatives. For, each application of the operator $\left(h\frac{\partial}{\partial x} + k\frac{\partial}{\partial y}\right)$ on any function of $x = a + ht$ and $y = b + kt$ is equivalent to differentiation with respect to t.

Let us next put $t = 1$ in Eq. (61) to obtain

$$F(1) = F(0) + F'(0) + \frac{1}{2!}F''(0) + \frac{1}{3!}F'''(0) + \cdots. \tag{65}$$

By Eq. (59), when $t = 1$, $x = a + h$ and $y = b + k$. Hence from Eq. (60), $F(1) = f(a + h, b + k)$. And from Eq. (59), when $t = 0$, $x = a$

and $y = b$, so that from Eq. (60) and the equations similar to Eq. (64) we have

$$F(0) = f(a,b), \qquad F^{(n)}(0) = \left(h\frac{\partial}{\partial x} + k\frac{\partial}{\partial y}\right)^n f(x,y)\bigg|_{\substack{x=a,\\ y=b}} \tag{66}$$

where the bar and subscripts mean that, after differentiation, we are to replace x, y by a, b. We may now deduce from Eqs. (65) and (66) that

$$f(a+h, y+k) = f(a,b) + \left(h\frac{\partial}{\partial x} + k\frac{\partial}{\partial y}\right) f(x,y)\bigg|_{\substack{x=a\\ y=b}} + \cdots$$
$$+ \frac{1}{n!}\left(h\frac{\partial}{\partial x} + k\frac{\partial}{\partial y}\right)^n f(x,y)\bigg|_{\substack{x=a\\ y=b}} + \cdots. \tag{67}$$

Let us denote the coefficient of h^pk^q by A_{pq}. Then if we evaluate the coefficient by expansions like that of Eq. (63), we find that

$$A_{pq} = \frac{1}{p!q!}\frac{\partial^{p+q} f}{\partial x^p\, \partial y^q}\bigg|_{\substack{x=a\\ y=b}}. \tag{68}$$

Hence if we now put

$$x = a + h, \qquad y = b + k, \qquad \text{so that } h = x - a,\ k = y - b, \tag{69}$$

we may conclude that with the A_{pq} given by Eq. (68), we have

$$f(x,y) = A_{00} + A_{10}(x - a) + A_{01}(y - b) + \cdots$$
$$+ A_{pq}(x - a)^p(y - b)^q + \cdots. \tag{70}$$

Many functions $f(x,y)$, and in particular most simple combinations of elementary functions, admit expansions of this form for general values of a,b with the exclusion of a limited number of exceptional points. For these general values of a,b the series of Eq. (70) will converge at least for sufficiently small values of $(x - a)$ and $(y - b)$.

Except for the first few terms, the formulas (68) and (70) are difficult to apply, so that if series of the type of Eq. (70) can be obtained by combining series for functions of one variable, this method should be used. The procedure is similar to that of Sec. 247.

With minor modifications, our remarks on Taylor's series for functions of two variables apply to functions of three or more variables.

EXAMPLE 1. Find the terms of the series for $\cos(x + 2y)$ in powers of x and y up to those of the second degree.

Solution: We could evaluate derivatives for $x,y = 0,0$ and use Eq. (68), but it is simpler to recall that

$$\cos u = 1 - \frac{u^2}{2!} + \cdots,$$

so that

$$\cos(x + 2y) = 1 - \tfrac{1}{2}(x + 2y)^2 + \cdots.$$

EXAMPLE 2. Find the terms of the series for $e^x \sin y$ in powers of $(x - 2)$ and $(y - 1)$ up to those of the second degree.

Solution: If $x - 2 = h$, $x = 2 + h$ and

$$e^x = e^{2+h} = e^2\left(1 + h + \frac{h^2}{2!} + \cdots\right). \tag{71}$$

And if $y - 1 = k$, $y = 1 + k$ and

$$\begin{aligned}\cos y &= \cos(1 + k) = \cos 1 \sin k + \sin 1 \cos k\\ &= \cos 1\left(k - \frac{k^3}{3!} + \cdots\right) + \sin 1\left(1 - \frac{k^2}{2!} + \cdots\right)\\ &= \sin 1 + \cos 1\, k - \sin 1\frac{k^2}{2!} + \cdots.\end{aligned} \tag{72}$$

As in Sec. 247, multiply the series in Eqs. (71) and (72) together as if they were polynomials, and omit all terms of degree higher than 2 from the product. The result is

$$e^x \cos y = e^2 \sin 1 + e^2 \sin 1\, h + e^2 \cos 1\, k + e^2 \sin 1 \frac{h^2}{2} + e^2 \cos 1\, hk - e^2 \sin 1 \frac{k^2}{2} + \cdots.$$

Now insert $h = x - 2$ and $k = y - 1$ in this result, to obtain as the required expansion

$$\begin{aligned}e^x \cos y = e^2 \sin 1 &+ e^2 \sin 1\,(x - 2) + e^2 \cos 1\,(y - 1)\\ &+ \tfrac{1}{2}e^2 \sin 1\,(x - 2)^2 + e^2 \cos 1\,(x - 2)(y - 1)\\ &+ \tfrac{1}{2}e^2 \sin 1\,(y - 1)^2 + \cdots.\end{aligned}$$

278. Maxima and Minima. Suppose that a function of two variables, $u(x,y)$, has a minimum for $x = a$, $y = b$ at an interior point of a two-dimensional region throughout which it is differentiable. Then for y fixed and equal to b, the function of a single variable, $u(x,b)$ must have a minimum at (a,b). Since this is not an end point of the region under consideration in finding the minimum, the derivative with respect to x must be zero. Thus

$$\frac{d}{dx}u(x,b)\bigg|_{x=a} = u_x(a,b) = \frac{\partial u}{\partial x}\bigg|_{a,b} = 0. \tag{73}$$

Similarly, we must have

$$\frac{d}{dy}u(a,y)\bigg|_{y=b} = u_y(a,b) = \frac{\partial u}{\partial y}\bigg|_{a,b} = 0. \tag{74}$$

Since the total differential of u as given by Eq. (23) is

$$du = \frac{\partial u}{\partial x}dx + \frac{\partial u}{\partial y}dy, \tag{75}$$

it follows that at (a,b) the total differential $du = 0$ for all values of dx and dy.

The same argument shows that the same condition must hold if $u(x,y)$

has a maximum at an interior point of a two-dimensional region throughout which it is differentiable.

It is possible to deduce tests that show in certain cases that we actually have a maximum or minimum from a study of the Taylor's expansions. In fact, if Eqs. (73) and (74) hold, in Eq. (70) we have $A_{10} = 0$ and $A_{01} = 0$, while $A_{00} = f(a,b)$. Hence

$$f(x,y) - f(a,b) = A_{20}(x-a)^2 + A_{11}(x-a)(y-b) + A_{02}(y-b)^2 + \cdots . \tag{76}$$

For (x,y) sufficiently near to (a,b), or for $x - a$ and $y - b$ sufficiently small, the sign of the right-hand side of Eq. (76) will be determined by that of the second-degree terms written out explicitly. In particular, let us suppose that these terms can be reduced to the sum of two squares by a change of axes in the x and y plane. This will be the case if† $A_{20} > 0$, $A_{02} > 0$, and

$$A_{20}(x-a)^2 + A_{11}(x-a)(y-b) + A_{02}(y-b)^2 = 1 \tag{77}$$

represents an ellipse in the x and y plane, or by Sec. 86, if

$$A_{11}^2 - 4A_{20}A_{02} < 0. \tag{78}$$

Recalling from Eq. (68) that

$$A_{20} = \frac{1}{2}\frac{\partial^2 u}{\partial x^2}, \qquad A_{11} = \frac{\partial^2 u}{\partial x\, \partial y}, \qquad A_{02} = \frac{1}{2}\frac{\partial^2 u}{\partial y^2}, \tag{79}$$

we see that the conditions assumed so far are:

$$\frac{\partial^2 u}{\partial x^2} > 0, \qquad \frac{\partial^2 u}{\partial y^2} > 0, \qquad \text{and} \qquad \left(\frac{\partial^2 u}{\partial x\, \partial y}\right)^2 - \frac{\partial^2 u}{\partial x^2}\frac{\partial^2 u}{\partial y^2} < 0. \tag{80}$$

If these relations hold, the right member of Eq. (76) is positive and $f(x,y) > f(a,b)$ for all (x,y) sufficiently near to (a,b). Hence $f(a,b)$ is a *minimum* if Eq. (80) is satisfied.

Similarly, $f(a,b)$ is a *maximum* if

$$\frac{\partial^2 u}{\partial x^2} < 0, \qquad \frac{\partial^2 u}{\partial y^2} < 0 \qquad \text{and} \left(\frac{\partial^2 u}{\partial x\, \partial y}\right)^2 - \frac{\partial^2 u}{\partial x^2}\frac{\partial^2 u}{\partial y^2} < 0. \tag{81}$$

The last condition of Eqs. (80) and (81) is that the discriminant

$$\Delta = \left(\frac{\partial^2 u}{\partial x\, \partial y}\right)^2 - \frac{\partial^2 u}{\partial x^2}\frac{\partial^2 u}{\partial y^2} \tag{82}$$

be negative. When this discriminant Δ is positive, we have neither a

† Equation (78) makes A_{20} and A_{02} have the same sign. This sign must be plus to make Eq. (77) a *real* ellipse, with the left member transformable to a *sum* of squares.

maximum nor a minimum, since the right member of Eq. (76) is then the difference of two squares and can take either sign. If the discriminant Δ is zero, there are a number of possibilities and no immediate conclusion can be drawn for all cases.

In many practical applications, we need not use tests of the kind just discussed if we may determine the sign of

$$f(x,y) - f(a,b) = f(a + h, y + k) - f(a,b) \tag{83}$$

for small h and k from other considerations.

As an example, by setting the two partial derivatives equal to zero, we find $u(0,0)$ as a possible maximum or minimum of each of the functions

$$x^2 + y^2, \qquad -x^2 - y^2, \qquad x^2 - y^2. \tag{84}$$

For values near but not equal to (0,0), the first expression is always positive and so has a minimum. For similar values, the second is always negative and so has a maximum. The third expression is positive for $y = 0$ and x near zero, but it is negative for $x = 0$ and y near zero. Consequently it has neither a maximum nor a minimum for $(x,y) = (0,0)$.

The discussion of maxima and minima of functions of three or more variables parallels that just given for two variables. In particular, a necessary condition for a function of three variables $u(x,y,z)$ to have a maximum or a minimum for $x = a$, $y = b$, $z = c$ interior to a region of differentiability is that

$$\frac{\partial u}{\partial x} = 0, \quad \frac{\partial u}{\partial y} = 0, \quad \frac{\partial u}{\partial z} = 0, \quad \text{or} \quad du = 0 \quad \text{for all } dx,\, dy,\, dz. \tag{85}$$

EXAMPLE 1. Find the maxima and minima of $u = x^3 + y^3 - 3xy$.
Solution: A necessary condition is

$$\frac{\partial u}{\partial x} = 3x^2 - 3y = 0 \qquad \text{and} \qquad \frac{\partial u}{\partial y} = 3y^2 - 3x = 0.$$

Thus $x^2 = y$, $y^2 = x$, or $y^4 = y$, and $y^4 - y = y(y^3 - 1) = 0$, so that $y = 0$ or 1, and $x = y^2 = 0$ or 1. This gives (0,0) and (1,1) as possible maxima or minima. The second derivatives are

$$\frac{\partial^2 u}{\partial x^2} = 6x, \qquad \frac{\partial^2 u}{\partial x\, \partial y} = -3, \qquad \frac{\partial^2 u}{\partial y^2} = 6y.$$

At (0,0) these reduce to 0, -3, 0 so that $\Delta = (-3)^2 - 0 = 9 > 0$. This shows that (0,0) is neither a maximum nor a minimum. We might have deduced this directly from the fact that u as given is an expansion in powers of x and y, and the terms of lowest degree or $-3xy$ may take either sign for small values of x and y.

At (1,1) the second derivatives are 6, -3, 6, and $\Delta = (-3)^2 - 6 \cdot 6 = -27 < 0$, and since $\partial^2 u/\partial x^2 = 6$ and $\partial^2 u/\partial y^2 = 6$ at (1,1) are both positive, we have a minimum there.

Thus the given u has a minimum at (1,1) for a sufficiently small neighborhood of this point.

EXAMPLE 2. Find the maxima and minima of

$$u = 2 - x^2 + 2xy - 3y^2 - 2z^2.$$

Solution: We have here as necessary conditions

$$\frac{\partial u}{\partial x} = -2x + 2y = 0, \qquad \frac{\partial u}{\partial y} = 2x - 6y = 0, \qquad \frac{\partial u}{\partial z} = -4z = 0.$$

The only solution of these simultaneous equations is $x = 0$, $y = 0$, and $z = 0$. This is a maximum, since $u(0,0,0) = 2$ and

$$u = 2 - (x - y)^2 - 2y^2 - 2z^2 < 2 \qquad \text{for } (x,y,z) \neq 0.$$

EXAMPLE 3. Find the greatest value of $u = xyz$, if x, y, and z are positive numbers such that $4x + 2y + z = 12$.

Solution 1: Solve the second equation for z, and substitute in the expression for u to obtain

$$u = xy(12 - 4x - 2y) = 12xy - 4x^2y - 2xy^2.$$

From this we may deduce the conditions

$$\frac{\partial u}{\partial x} = 12y - 8xy - 2y^2 = 0, \qquad \frac{\partial u}{\partial y} = 12x - 4x^2 - 4xy = 0.$$

Thus $2y(6 - 4x - y) = 0$ and $4x(3 - x - y) = 0$.

For a maximum, the product will be greater than zero, so that we need not consider the factors y and x, but need only test $6 - 4x - y = 0$ and $3 - x - y = 0$. Subtracting, we find that $3 - 3x = 0$ and $x = 1$, and then $y = 3 - x = 2$. Finally $z = 12 - 4x - 2y = 4$. The value $(x,y,z) = (1,2,4)$ is the required maximum. For, since $u = xyz$ is zero at the extremities of the allowable intervals, $0 < x < 3$, $0 < y < 6$, $0 < z < 12$, and is positive inside, it must be a maximum at some interior point. At this point the first derivatives of u when expressed in terms of x and y must vanish, and there is just one such point.

Solution 2: We may differentiate before eliminating. Thus, since $u = xyz$ is a maximum, its logarithm will also be a maximum. And from $\ln u = \ln x + \ln y + \ln z$, we find

$$\frac{du}{u} = \frac{dx}{x} + \frac{dy}{y} + \frac{dz}{z}.$$

This must be zero when dx, dy, dz satisfy the relation obtained from $4x + 2y + z = 12$ or $4\,dx + 2\,dy + dz = 0$.

Now eliminate $dz = -4\,dx - 2\,dy$ from the first equation, to obtain $\dfrac{du}{u} = \left(\dfrac{1}{x} - \dfrac{4}{z}\right) dx + \left(\dfrac{1}{y} - \dfrac{2}{z}\right) dy$. As this must be zero for all dx and dy, in particular $dx = 1$, $dy = 0$, or $dx = 0$, $dy = 1$, the coefficients of dx and dy must each be zero so that $x = z/4$ and $y = z/2$. Substituting these values in $4x + 2y + z = 12$, we find $4(z/4) + 2(z/2) + z = 12$, or $3z = 12$ and $z = 4$. Hence $x = z/4 = 1$ and $y = z/2 = 2$. This gives $(1,2,4)$, the values obtained in the first solution as the maximum.

The method of the second solution is particularly useful when we want only the ratios of the quantities x, y, z without their actual values.

EXERCISE 142

Find the first few terms of the Taylor's series in powers of x and y for each of the following functions. Use combinations of series in powers of one variable whenever possible.

1. $\sin(2x - y)$.
2. $\cos(x + 2y)$.
3. e^{xy}.
4. $e^x \cos y$.
5. $e^{-y} \sin x$.
6. $\cos xy + xy \sin xy$.
7. $e^x \ln(1 + y)$.
8. $\sqrt{\dfrac{1+x}{1+y}}$.

Find the value of (x,y) which makes each of the following functions a minimum or a maximum.

9. $x^4 + y^4 - 32y + 2$.
10. $x^3 + y^3 - 3x$.
11. $x^2 + 4xy + 5y^2 + y^3$.
12. $x^3 - y^3 + x^2 + y^2$.
13. $x^2 + 2xy + 2y^2 + 4x - 6y$.
14. $x^3 + y^3 + 3xy$.
15. $xy + \dfrac{8}{x} + \dfrac{27}{y}$.
16. $\sin x + \sin y + \sin(x + y)$.
17. $\sin x \sin y \sin(x + y)$.
18. $\dfrac{(2x + 3y + 4)^2}{x^2 + y^2 + 4}$.

Find the positive values of (x,y,z) which makes $x^2 + y^2 + z^2$ a minimum, subject to the condition

19. $x + 2y + 2z = 4$.
20. $2xy^4z^4 = 1$.
21. $3x + 2y + z = 12$.
22. $xy^2z^2 = 2$.

If $x + y + z = 36$, find the positive values of (x,y,z) which make each of the following functions a maximum.

23. xyz.
24. xy^2z^3.

***279. Envelope of a Family of Curves.** Let the equation of a family of curves G_c in the xy plane be

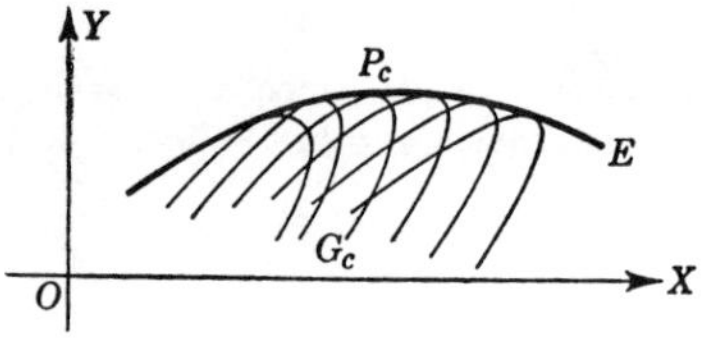

FIG. 293.

$$g(x,y,c) = 0. \tag{86}$$

Let E be a curve in the plane (Fig. 293) tangent to each curve G_c at some point P_c. Then if each point of E is a point P_c at which some curve G_c touches E, we call E the *envelope* of the curves G_c. We may use c as a parameter for the envelope E and write

$$x = x(c), \qquad y = y(c). \tag{87}$$

Since these relations make Eq. (86) an identity, we have at any P_c

$$\frac{\partial g}{\partial x}\frac{dx}{dc} + \frac{\partial g}{\partial y}\frac{dy}{dc} + \frac{\partial g}{\partial c} = 0. \tag{88}$$

On any one curve G_c, c is constant. Hence the equation

$$\frac{\partial g}{\partial x} + \frac{\partial g}{\partial y}\frac{dy}{dx} = 0, \tag{89}$$

found by differentiating Eq. (86) with c held constant, determines the slope dy/dx

at any point on G_c, and hence in particular at P_c. But at P_c, G_c has the same slope as E. Hence in Eq. (89) we have

$$\frac{dy}{dx} = \frac{dy/dc}{dx/dc} \quad \text{and} \quad \frac{\partial g}{\partial x}\frac{dx}{dc} + \frac{\partial g}{\partial y}\frac{dy}{dc} = 0. \tag{90}$$

A comparison of Eqs. (88) and (90) shows that at points P_c,

$$\frac{\partial g}{\partial c} = 0. \tag{91}$$

This leads to the rule for finding the parametric equiatons of the envelope of the family (86). Solve Eqs. (86) and (91) for x and y in terms of c.

When the functions are regular,† if the expressions for x and y in terms of c do not make $\partial g/\partial x$ and $\partial g/\partial y$ both zero, a result of this process is necessarily an envelope.

EXAMPLE 1. Find the envelope of the family of circles

$$(x - 2c)^2 + y^2 = c^2 \quad \text{or} \quad x^2 + y^2 - 4cx + 3c^2 = 0.$$

Solution: Equating the partial derivative with respect to c to zero, we find $-4x + 6c = 0$, so that $x = 3c/2$. And inserting this value in the first form gives

$$\left(-\frac{c}{2}\right)^2 + y^2 = c^2, \quad \text{so that } y^2 = \frac{3c^2}{4} \text{ and } y = \pm\frac{\sqrt{3}}{2}c.$$

Thus the envelope consists of two straight lines (Fig. 294) with parametric equations $x = \frac{3}{2}c$, $y = \frac{1}{2}\sqrt{3}\,c$. The equation in rectangular coordinates $\sqrt{3}\,y = \pm 2x$ is easily found in this case by eliminating c.

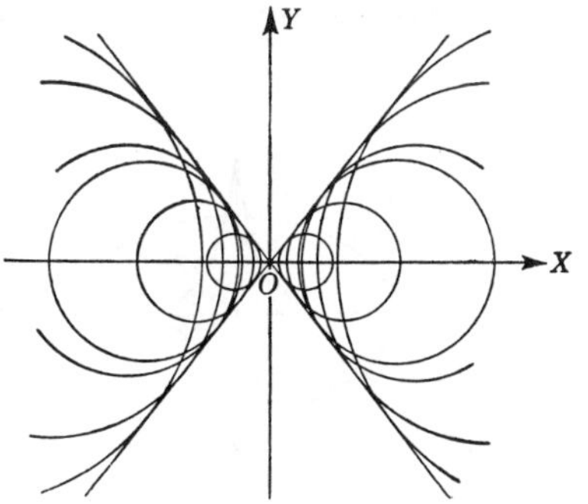

FIG. 294.

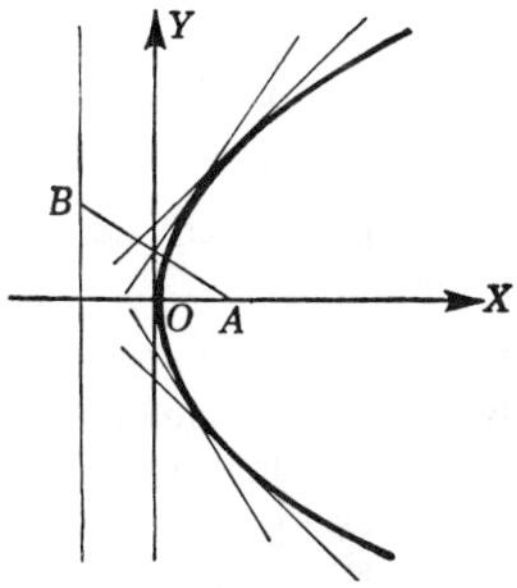

FIG. 295.

EXAMPLE 2. Let A be the fixed point (2,0) and B be any point on the line $x = -2$. Find the envelope of the straight line which is the perpendicular bisector of AB.

Solution: If $A = (2,0)$ and $B = (-2,c)$ the equation of the perpendicular bisector is found by the method used earlier in the example of Sec. 81 to be $(x - 2)^2 + y^2 = (x + 2)^2 + (y - c)^2$ or $-4x = 4x - 2cy + c^2$, $8x - 2cy + c^2 = 0$.

Equating the partial derivative with respect to c to zero, we find $-2y + 2c = 0$, $y = c$. Hence $x = \frac{1}{8}c^2$, so that $y^2 = 8x$. Some of the lines and the parabolic envelope are shown in Fig. 295.

† That is, are continuous together with all their partial derivatives of the first two orders.

EXAMPLE 3. Find the evolute of the parabola $y = x^2$ considered as the envelope of its normals.

Solution: Since $y = x^2$, $dy/dx = 2x$ and the equation of the normal, by Sec. 87, is $y - y_1 = -\dfrac{1}{(dy/dx)_1}(x - x_1)$, or in this case $y - x_1^2 = -\dfrac{1}{2x_1}(x - x_1)$. To simplify the writing, replace x_1 by c, so that the equation of the normal is $y - c^2 = -\dfrac{1}{2c}(x - c)$, or $x + 2cy - 2c^3 - c = 0$.

Equating the partial derivative with respect to c to zero, we find

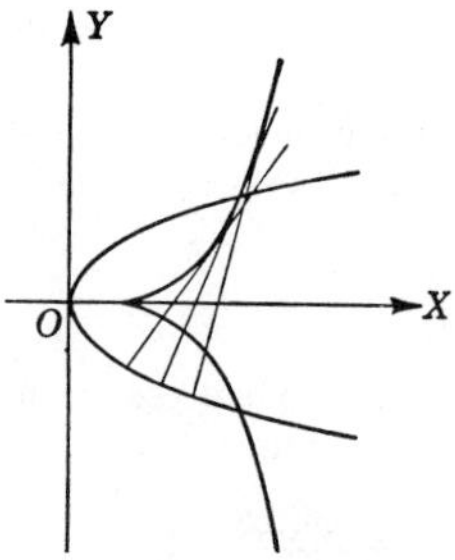

FIG. 296.

$$2y - 6c^2 - 1 = 0 \qquad \text{and} \qquad y = 3c^2 + \tfrac{1}{2}.$$

Substituting this in the equation of the normal gives

$$x + 6c^3 + c - 2c^3 - c = 0 \qquad \text{or} \qquad x = -4c^3.$$

The two expressions for x and y give the equation of the envelope, from which we may eliminate c to obtain

$$(y - \tfrac{1}{2})^3 = \tfrac{27}{16}x^2.$$

See Fig. 296.

EXAMPLE 4. Find the evolute of the ellipse,

$$x^2/a^2 + y^2/b^2 = 1,$$

considered as the envelope of its normals.

Solution: Introduce the parameter ϕ by writing $x = a \cos \phi$, $y = b \sin \phi$. Then $dx = -a \sin \phi \, d\phi$ and $dy = b \cos \phi \, d\phi$, so that $\dfrac{dy}{dx} = -\dfrac{b}{a} \cot \phi$, and the slope of the normal is $\dfrac{a}{b} \tan \phi$. Thus the equation of the normal may be written

$$y - b \sin \phi = \frac{a}{b} \tan \phi (x - a \cos \phi), \qquad \text{or}$$

$$ax \sin \phi - by \cos \phi + (b^2 - a^2) \cos \phi \sin \phi = 0.$$

Equating the partial derivative with respect to ϕ to zero, we find

$$ax \cos \phi + by \sin \phi + (b^2 - a^2)(\cos^2 \phi - \sin^2 \phi) = 0.$$

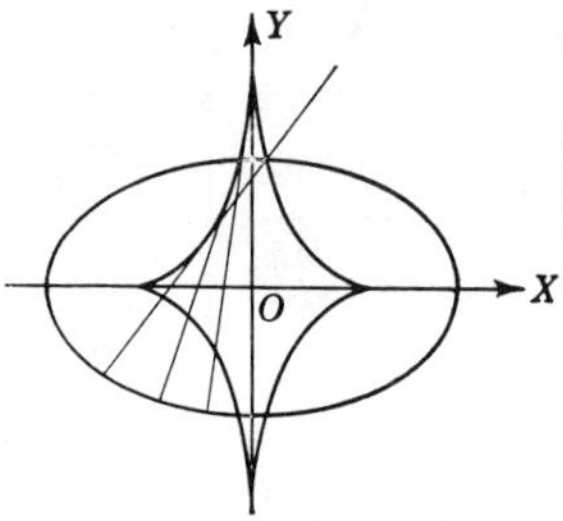

FIG. 297.

Multiplying the first equation by $\sin \phi$ and the second by $\cos \phi$ and adding gives $ax = (a^2 - b^2) \cos^3 \phi$. And multiplying the first equation by $-\cos \phi$ and the second by $\sin \phi$ and adding gives

$$by = -(a^2 - b^2) \sin^3 \phi.$$

We may eliminate ϕ from these equations by using $\cos^2 \phi + \sin^2 \phi = 1$, and so obtain

$$(ax)^{\frac{2}{3}} + (by)^{\frac{2}{3}} = (a^2 - b^2)^{\frac{2}{3}},$$

which is the equation of the required evolute (Fig. 297).

EXERCISE 143

Find the envelope of each given family of curves.

1. $x \cos c + y \sin c = 2.$
2. $(1 - c)x + cy = c - c^2.$
3. $x \sin c + y \cos c = \sin c \cos c.$
4. $x + c^2y = c.$

5. $y = cx - (1 + c^2)x^2.$

6. $y = cx + c^n.$

7. $x + y \sin c = a \cos c.$

8. $y = cx + \dfrac{4}{c}.$

9. $cx^2 + (1 - c^2)y^2 = c - c^2.$

10. $x^2 + c^2y^2 = c.$

11. $(1 - c)^2x^2 + c^2y^2 = (c - c^2)^2.$

12. $y^2 = cx + c^n.$

13. $(x - c)^2 + (y - c)^2 = c^2.$

14. $x^2 + (y - c)^2 = c.$

Find the evolute of each of the following curves as the envelope of its normals, after introducing the parameter indicated.

15. $x^{\frac{2}{3}} + y^{\frac{2}{3}} = 1,\ x = \cos^3 c,\ y = \sin^3 c.$

16. $y^2 = x^3,\ y = c^3,\ x = c^2.$

17. $x = c - \sin c,\ y = 1 - \cos c.$

18. $y^2 = 4x,\ y = 2c,\ x = c^2.$

19. $xy = 1,\ x = c,\ y = \dfrac{1}{c}.$

20. $x^{\frac{1}{2}} + y^{\frac{1}{2}} = 1,\ x = \cos^4 c,\ y = \sin^4 c.$

21. $x = e^c \cos c,\ y = e^c \sin c.$

CHAPTER 18

VECTORS AND SURFACES IN SPACE

The first part of this chapter is an introduction to solid analytic geometry. We shall treat coordinate systems, planes and straight lines, space curves, and some special classes of surfaces. This enables us to make some applications of the calculus to the geometry of curves and surfaces in space. The discussion will also provide the necessary basis for constructing spatial figures for applications of multiple integration in Chap. 19. In connection with the spatial geometry, we present some parts of vector analysis, the use of which greatly simplifies the entire discussion.

280. Coordinates in Space. We may locate points in three dimensions by using a coordinate system constructed as follows. Through any point O, selected as origin, we draw three mutually perpendicular lines, the x, y, and z axes. On each of these we take a positive direction, as indicated by the arrows in Fig. 298. We shall use right-handed systems. That is, we so choose the positive directions that a right-threaded screw along the z axis will advance in the positive direction when given the 90° turn which takes the positive x axis into the positive y axis.

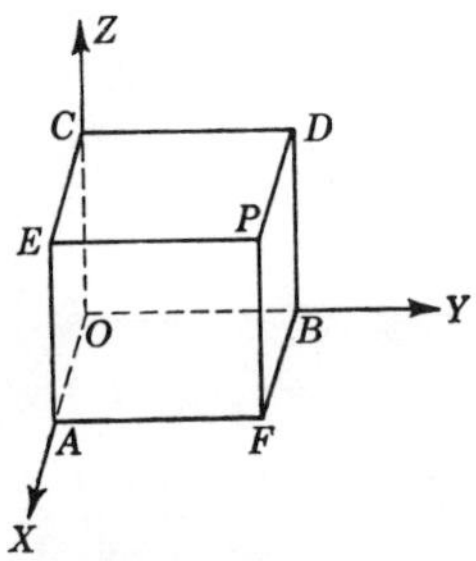

FIG. 298.

The coordinate axes determine three coordinate planes, the yz, zx, and xy planes. The signed distances of any point P from these planes are denoted by x, y, and z respectively, and are called the *coordinates* of P.

By drawing planes through P parallel to the coordinate planes, we may construct the rectangular parallelepiped shown in Fig. 298. The points D,E,F are the projections of P on the yz, zx, and xy planes. The points A,B,C are the projections of P on the x, y, and z axes.

$$x = DP = BF = CE = OA. \tag{1}$$

Thus x is the x coordinate of F, the projection of P in the xy plane, or of E, the projection of P in the zx plane. Also x is the distance from O to A, the projection of OP on the x axis.

In Fig. 298 the signs of x, y, and z are all plus since the directions from D to P, E to P, and F to P are those taken as positive on the coordinate axes. But if P were behind the yz plane, x would be minus; if P were to the left of the zx plane, y would be minus; and if P were below the xy plane, z would be minus.

If a,b,c are given values of the coordinates, the point P may be constructed either by taking $OA = a$, $AF = b$, and $FP = c$; or by taking $OA = a$, $OB = b$, and $OC = c$ on the coordinate axes and thence determining the parallelepiped having P as the vertex opposite to O.

The three coordinate planes divide space into eight regions called *octants*. The one in which all the coordinates are positive is called the *first octant*.

281. The Segment $\overline{OP}$. Let P be the point (x,y,z) and consider the directed line segment $\overline{OP}$. From the right triangles OFP and OAF (Fig. 299) we find that $OP^2 = OF^2 + FP^2$ and $OF^2 = OA^2 + AF^2$, so that

$$OP^2 = OA^2 + AF^2 + FP^2 = x^2 + y^2 + z^2. \tag{2}$$

This determines OP, the length of $\overline{OP}$.

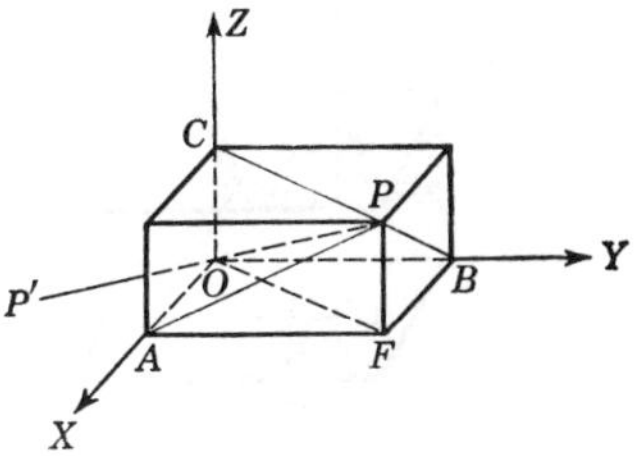

FIG. 299.

The direction of $\overline{OP}$ may be fixed by the cosines of the angles POX, POY, POZ which the segment makes with the positive coordinate axes. The cosines, which we denote by l,m,n, are called the *direction cosines* of $\overline{OP}$. The angles themselves are called the *direction angles* of $\overline{OP}$. The direction angles, denoted by α,β,γ, may be considered as always having values in the range $0 \leqq \theta \leqq \pi$. (π radians $= 180°$.) From the right triangles OAP, OBP, OCP we find that

$$l = \cos\alpha = \frac{OA}{OP} = \frac{x}{OP}.$$

Similarly, (3)

$$m = \cos\beta = \frac{y}{OP}, \qquad n = \cos\gamma = \frac{z}{OP}.$$

From these relations and Eq. (2) we may deduce that

$$l^2 + m^2 + n^2 = \cos^2\alpha + \cos^2\beta + \cos^2\gamma = 1. \tag{4}$$

If we extend the line PO through O to a point P' such that $OP' = PO$, the segment OP' will have direction angles $\pi - \alpha$, $\pi - \beta$, $\pi - \gamma$ so that its direction cosines l',m',n' will be equal to $-l, -m, -n$.

282. Direction Ratios. Let L be any straight line. Draw the line through O parallel to L, and let $\overline{P'P}$ be a segment of this line bisected by O. Then l,m,n the direction cosines of OP, or l',m',n' those of OP', are taken as the direction cosines of L. Any set of numbers a,b,c not all zero which are proportional to these cosines are called direction ratios.

Except for the distinction between $\overline{OP}$ and $\overline{OP'}$, the cosines may be found from the ratios. For if

$$l = ka, \qquad m = kb, \qquad n = kc \tag{5}$$

then

$$1 = l^2 + m^2 + n^2 = k^2(a^2 + b^2 + c^2)$$

and

$$k = \pm \frac{1}{\sqrt{a^2 + b^2 + c^2}}, \tag{6}$$

one sign giving l,m,n and the other sign giving l',m',n'.

283. Projections of Line Segments. Let L be a line through the two points A and B. And, on L, choose the positive direction as that from A to B, indicated by the arrow in Fig. 300. Also let $\overline{PQ}$ be any directed line

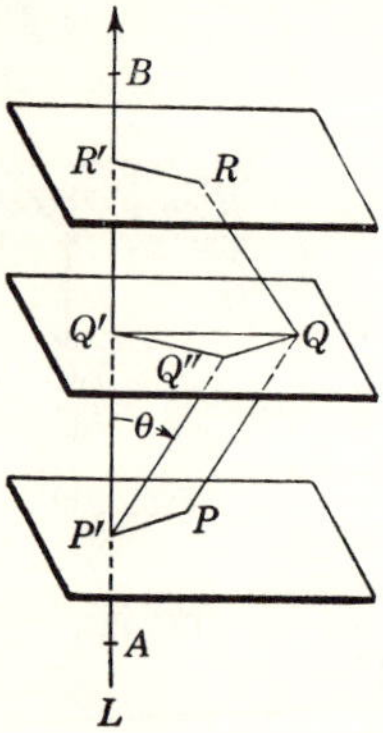

FIG. 300.

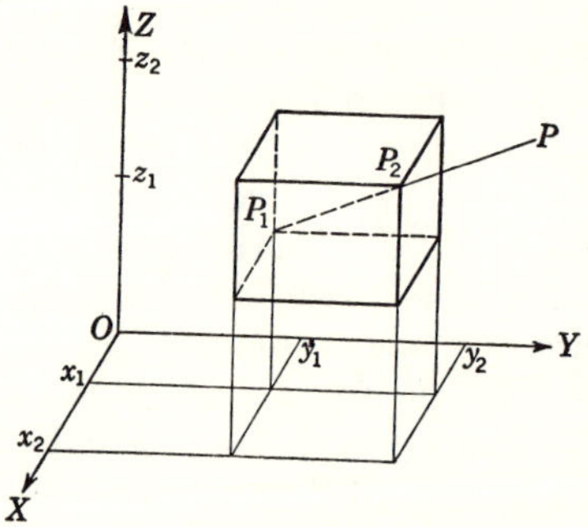

FIG. 301.

segment. The planes through P and Q perpendicular to L will determine points P' and Q' on L. The segment $\overline{P'Q'}$, or the signed distance from P' to Q', is called the *projection* of $\overline{PQ}$ on L or on $\overline{AB}$. We determine the angle θ, between L and PQ as follows. Draw a line segment parallel to $\overline{PQ}$ through any point of L, for example $\overline{P'Q''}$ through P'. Then take θ as the angle in the range $0 \leqq \theta \leqq \pi$ made by this segment with the positive direction on L, that of AB.

From the right triangle $P'Q'Q''$, we find that

$$\text{Proj}_{AB}\, \overline{PQ} = P'Q' = P'Q'' \cos\theta = PQ \cos\theta. \tag{7}$$

Consequently, we have

$$\text{Proj}_{AB}\, \overline{QP} = Q'P' = -PQ \cos\theta. \tag{8}$$

For any three points P,Q,R, we see that

$$\text{Proj}_{AB}\, \overline{PR} = \text{Proj}_{AB}\, \overline{PQ} + \text{Proj}_{AB}\, \overline{QR}, \tag{9}$$

since this is equivalent to $P'R' = P'Q' + Q'R'$ where P',Q',R' are the projections of P,Q,R, on AB.

284. The Segment $\overline{P_1P_2}$. Consider a pair of points $P_1(x_1,y_1,z_1)$ and $P_2(x_2,y_2,z_2)$. Let the segment P_1P_2 have direction cosines l,m,n and a length s. Then (Fig. 301) the projections of the segment P_1P_2 on the coordinate axes are

$$x_2 - x_1 = sl, \qquad y_2 - y_1 = sm, \qquad z_2 - z_1 = sn. \tag{10}$$

But, by Eq. (4), we have

$$l^2 + m^2 + n^2 = 1. \tag{11}$$

Consequently, from Eqs. (10) and (11), we find that

$$(x_2 - x_1)^2 + (y_2 - y_1)^2 + (z_2 - z_1)^2 = s^2(l^2 + m^2 + n^2) = s^2. \tag{12}$$

This determines the length s as

$$s = P_1P_2 = \sqrt{(x_2 - x_1)^2 + (y_2 - y_1)^2 + (z_2 - z_1)^2}. \tag{13}$$

We may also conclude from Eq. (10) that

$$l = \frac{x_2 - x_1}{s}, \qquad m = \frac{y_2 - y_1}{s} = \frac{z_2 - z_1}{s}. \tag{14}$$

Equations (13) and (14) could also be deduced from the parallelepiped with edges parallel to the axes and diagonal P_1P_2 by reasoning similar to that of Sec. 281.

Equation (14) shows that we may take $x_2 - x_1$, $y_2 - y_1$, $z_2 - z_1$ as direction ratios for the straight line through the points P_1 and P_2.

Example. Given $A = (1,-2,3)$, $B = (3,0,4)$, $C = (5,4,1)$, and $D = (1,0,-1)$. Show that the line through AB is parallel to that through CD and that segment CD is twice as long as AB and oppositely directed. Also find these lengths.

Solution: The projections of AB on the coordinate axes are $3 - 1$, $0 - (-2)$, $4 - 3$, or $2,2,1$. The corresponding projections of CD are $1 - 5$, $0 - 4$, $-1 - 1$, or $-4,-4,-2$. Since these numbers are proportional to $2,2,1$ the projections of AB, the two lines are parallel. And the factor of proportionality is -2, which shows that CD is twice as long as AB and oppositely directed.

The length of AB may be obtained either from Eq. (12), or the projections found above. From

$$AB^2 = 2^2 + 2^2 + 1^2 = 9, \qquad |AB| = 3.$$

Similarly, we may calculate

$$CD^2 = (-4)^2 + (-4)^2 + (-2)^2 = 36, \qquad |CD| = 6.$$

This agrees with $|CD| = 2|AB|$, as found above.

285. The Straight Line. Let us keep the point $P_1 = (x_1,y_1,z_1)$ fixed and take $P = (x,y,z)$ as any point on the straight line through P_1P_2.

Then if t is the (variable) signed distance from P_1 to P, we shall have the relations

$$l = \frac{x - x_1}{t}, \qquad m = \frac{y - y_1}{t}, \qquad n = \frac{z - z_1}{t}, \tag{15}$$

analogous to Eq. (14). Conversely, if the relations of Eq. (15) hold, the segment $\overline{P_1P}$ (or $\overline{PP_1}$ if t is negative) will have the same direction cosines as $\overline{P_1P_2}$, and P will be on the straight line through P_1 and P_2. It follows that the equations

$$x = x_1 + lt, \qquad y = y_1 + mt, \qquad z = z_1 + nt \tag{16}$$

determine the points on the straight line in terms of the signed distance t.

The set of equations

$$\frac{x - x_1}{l} = \frac{y - y_1}{m} = \frac{z - z_1}{n} \tag{17}$$

is equivalent to Eq. (16), since if these relations hold, we may take the common value of the ratio as the parameter t. And in this form, in place of the direction cosines l,m,n we may use any direction ratios, a,b,c proportional to them. We may then write

$$\frac{x - x_1}{a} = \frac{y - y_1}{b} = \frac{z - z_1}{c}. \tag{18}$$

These equations hold if, and only if, the segment P_1P has a fixed direction, and so are the equations of the straight line through P_1 with direction ratios a,b,c. In particular, we have

$$\frac{x - x_1}{x_2 - x_1} = \frac{y - y_1}{y_2 - y_1} = \frac{z - z_1}{z_2 - z_1} \tag{19}$$

as the equations of the straight line through the points P_1 and P_2.

We continue to write equations of the form (17), (18), and (19) even when one or two of the three denominators are zero, interpreting them as meaning that the corresponding numerators are zero. Note that it takes *two* equations to determine a line fully.

EXAMPLE 1. Find the equations of a straight line parallel to the segment joining the origin to the point (2,3,4) and passing through the point (−3,0,1).

Solution: The segment joining (0,0,0) to (2,3,4) has direction ratios 2,3,4. Hence from Eq. (18) with $a = 2$, $b = 3$, $c = 4$, and $(x_1,y_1,z_1) = (-3,0,1)$ the required equations are found to be

$$\frac{x + 3}{2} = \frac{y}{3} = \frac{z - 1}{4}.$$

EXAMPLE 2. Find the equations of a straight line through the points (3, −2,2) and (5,0,2)

Solution: With the given points as P_1 and P_2, from Eq. (19) we find

$$\frac{x-3}{2} = \frac{y+2}{2} = \frac{z-2}{0}.$$

In accordance with the stated convention on zero denominators, these equations are considered to be equivalent to

$$x = y + 5 \quad \text{and} \quad z = 2.$$

EXAMPLE 3. Find the point on the straight line

$$\frac{x-3}{3} = \frac{y-2}{4} = \frac{z-1}{6}$$

at which $x = 2$.

Solution: When $x = 2$, the first equality becomes

$$\frac{2-3}{3} = \frac{y-2}{4}, \qquad \text{or } y = 2 + 4\left(-\frac{1}{3}\right) = \frac{2}{3}.$$

And the equality of the first and third fractions becomes

$$\frac{2-3}{3} = \frac{z-1}{6}, \qquad \text{or } z = 1 + 6\left(-\frac{1}{3}\right) = -1.$$

Hence the required point is $(2,\frac{2}{3},-1)$.

EXERCISE 144

1. A rectangular parallelepiped has its sides parallel to the coordinate axes, its center at the origin, and one vertex at (3,2,5). Find the coordinates of the other seven vertices.

Find the direction angles of the segment OP, where $O = (0,0,0)$, for each of the given points P.

2. $(1,1,\sqrt{2})$. 3. $(0,1,\sqrt{3})$. 4. $(\sqrt{2},1,-1)$.

Find the length and direction cosines of OP, where $O = (0,0,0)$, for each of the given points P.

5. (2,3,6). 6. (1,2,2). 7. (2,5,14).
8. (1,4,8). 7. (3, −4,0). 10. (12,5,0).

By finding the lengths of the sides, show that the triangle having the vertices

11. (1,2,3), (0,1,2), (0,2,4) is a right triangle.
12. (1,2,3), (3,2,1), (3,0,3) is an equilateral triangle.
13. (1,2,3), (2,4,5), (3,3,1) is an isosceles right triangle.

Find the equations of an indefinite straight line which passes through the given point and has the three given numbers as direction ratios.

14. (2,1, −1) 3,4,2. 15. (0,0,0) 2,4, −1.
16. (0,0,2) 5,1,0. 17. (−2,3, −1) 2, −2,1.
18. (5,0,0) 6,0,0. 19. (3,2,1) 0,1, −1.

Find the equations of an indefinite straight line which passes through each given pair of points.

20. (3,4,2),(0,2,1). 21. (2,4,5),(3,0,0). 22. (−4, −2,5),(−3, −1,2).

Find the points on the straight line whose equations are $\frac{x-5}{1} = \frac{y-4}{2} = \frac{z-9}{3}$ at which

23. $x = 0$. 24. $y = 2$. 25. $z = 3$.

286. Equation of a Plane. We wish to find the equation of a plane through the point P_1 and perpendicular to a straight line N. Let A,B,C be any set of direction ratios for the normal N. Let $P_1 = (x_1,y_1,z_1)$ and locate the points $Q = (x_1 - A,\ y_1 - B,\ z_1 - C)$ and $R = (x_2 + A,\ y_2 + B, z_2 + C)$. Then for any point $P = (x,y,z)$ in the plane (Fig. 302), P_1P will be a perpendicular bisector of QR. And we will have $PQ = PR$. Hence $PQ^2 = PR^2$ and from Eq. (12),

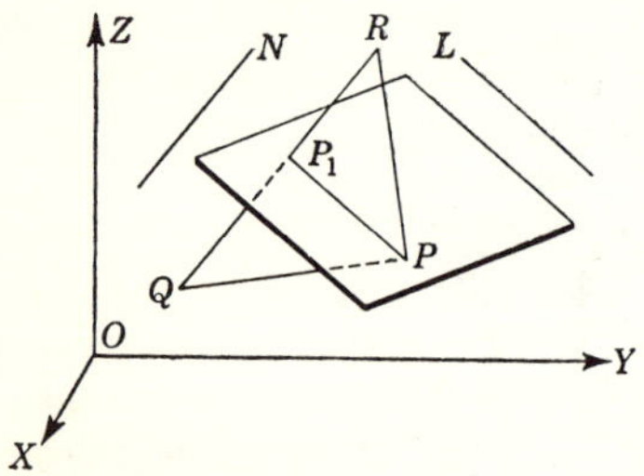

FIG. 302.

$$(x - x_1 + A)^2 + (y - y_1 + B)^2 + (z - z_1 + C)^2$$
$$= (x - x_1 - A)^2 + (y - y_1 - B)^2 + (z - z_1 - C)^2. \quad (20)$$

As the terms similar to $(x - x_1)^2$ and A^2 appear on both sides, after transposition of terms and division by 4, this reduces to

$$A(x - x_1) + B(y - y_1) + C(z - z_1) = 0. \quad (21)$$

Let us define D by the relation

$$Ax_1 + By_1 + Cz_1 = D. \quad (22)$$

Then Eq. (21) is equivalent to the first-degree equation

$$Ax + By + Cz = D. \quad (23)$$

Next consider any given first-degree equation of the form of Eq. (23). If $P_1 = (x_1,y_1,z_1)$ is any point on its locus, $x = x_1$, $y = y_1$, $z = z_1$ will satisfy Eq. (23) so that Eq. (22) holds. The preceding argument then shows that the given Eq. (23) is the equation of a plane having a normal direction with direction ratios A,B,C.

Let L be a straight line having A',B',C' as one set of direction ratios. Then the relation

$$AA' + BB' + CC' = 0 \quad (24)$$

is the condition that $x = x_1 + A'$, $y = y_1 + B'$, $z = z_1 + C'$ satisfy Eq. (21), in which case L is parallel to the plane or perpendicular to N. Hence Eq. (24) is the condition that the directions A,B,C and A',B',C' be perpendicular.

287. Vectors. In many geometrical and physical applications of mathematics we meet the concept of a *vector*, or directed magnitude. A vector is determined by its length and direction. It may be represented graphically by any line segment having this length and direction. The position of the initial point is unessential, and any two segments $\overline{P_1P_2}$ having the same length and the same direction represent the same vector. We sometimes use a particular representative segment as $\overline{P_1P_2}$ to denote a vector, but more often we find a single letter convenient. In print, we use boldface type, as **a** to indicate a vector. In writing, a dash over a letter, as $\bar{a}$, may be used. The length of a vector may be denoted by the absolute value sign or simply by the letter in ordinary type or without the dash. Thus with the notation of Sec. 284, if $\mathbf{a} = \overline{P_1P_2}$, then

$$|\mathbf{a}| \text{ or } a = |P_1P_2| = s. \tag{25}$$

We may think of a segment with coincident end points $\overline{P_1P_1}$ as representing a vector of zero length and undetermined direction. We call this the *null vector* but represent it by an ordinary zero in equations where its vector character may be inferred from the context.

For any real number C, by $C\mathbf{a}$ (or $\mathbf{a}C$) we mean the vector whose length is C times that of **a** and having the same direction as **a** if C is positive and the opposite direction if C is negative. If C is zero, $C\mathbf{a}$ is the null vector. In particular when $C = -1$, we write $-\mathbf{a}$ for $(-1)\mathbf{a}$ and if

$$\mathbf{a} = \overline{P_1P_2}, \qquad -\mathbf{a} = \overline{P_2P_1}. \tag{26}$$

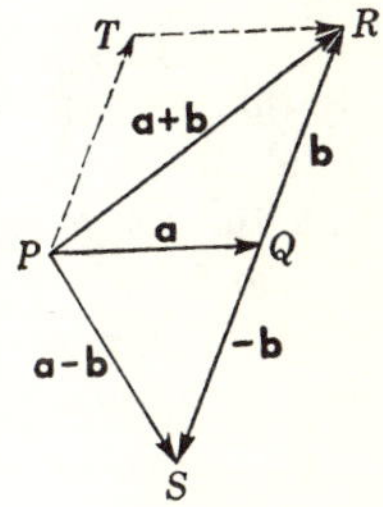

FIG. 303.

The real numbers, or numbers represented by signed coordinates on a scale like that of Sec. 1, are referred to as *scalars*. The operation that converts the vector **a** into $C\mathbf{a}$ is called *multiplication* of the vector **a** by the scalar C.

288. Addition and Subtraction of Vectors. Let $\overline{PQ}$ be a segment representing the vector **a** and $\overline{QR}$ be a segment representing the vector **b**, so that (Fig. 303),

$$\mathbf{a} = \overline{PQ}, \qquad \mathbf{b} = \overline{QR}. \qquad \text{Then } \mathbf{a} + \mathbf{b} = \overline{PR}. \tag{27}$$

Also if S is on RQ produced, so that $SQ = QR$,

$$\mathbf{b} = \overline{SQ}, \qquad -\mathbf{b} = \overline{QS}, \qquad \text{and } \mathbf{a} - \mathbf{b} = \overline{PS}. \tag{28}$$

Thus vectors are added and subtracted by the parallelogram law. That is, $\overline{PR} = \mathbf{a} + \mathbf{b}$ is the diagonal of a parallelogram $PQRT$ with $\overline{PQ} = \mathbf{a}$ and $\overline{QR} = \mathbf{b}$ as two of its sides.

It follows from the definition (Figs. 303, 304) that

$$\mathbf{a} + \mathbf{b} = \mathbf{b} + \mathbf{a}. \tag{29}$$

$$C(\mathbf{a} + \mathbf{b}) = C\mathbf{a} + C\mathbf{b}. \tag{30}$$

$$\mathbf{a} + (\mathbf{b} + \mathbf{c}) = (\mathbf{a} + \mathbf{b}) + \mathbf{c}. \tag{31}$$

$$\text{Also, if } \mathbf{a} + \mathbf{b} = \mathbf{c}, \qquad \text{then } \mathbf{c} - \mathbf{b} = \mathbf{a}. \tag{32}$$

289. Components. By drawing $\overline{OP}$ equal in magnitude and direction to a given vector $\mathbf{r}$, we obtain a representative segment for $\mathbf{r}$ with initial point at the origin. Let $P = (x,y,z)$ (Fig. 305) and construct the

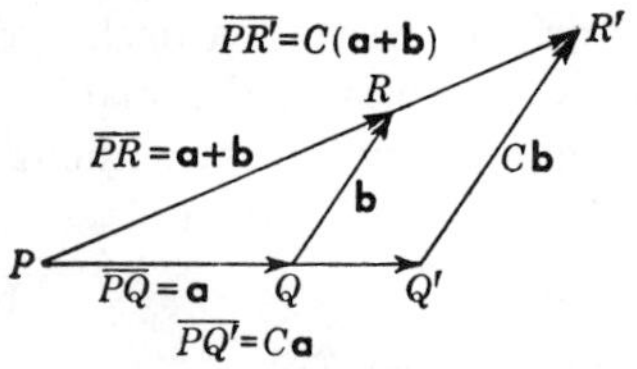

FIG. 304.

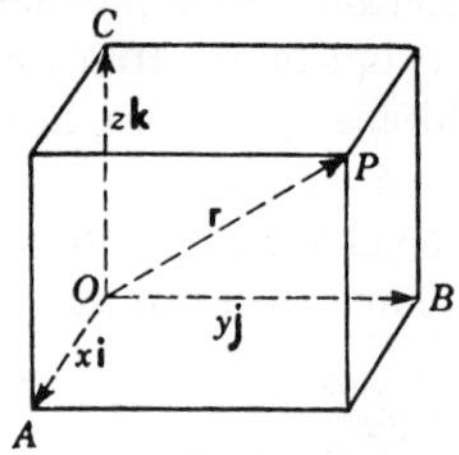

FIG. 305.

parallelepiped as in Sec. 281. Let $\mathbf{i},\mathbf{j},\mathbf{k}$ denote unit vectors† along the positive x, y, and z axes, respectively. Then we may write

$$\overline{OA} = x\mathbf{i}, \qquad \overline{OB} = y\mathbf{j}, \qquad \overline{OC} = z\mathbf{k}. \tag{33}$$

Consequently, we shall have

$$\mathbf{r} = x\mathbf{i} + y\mathbf{j} + z\mathbf{k}. \tag{34}$$

By this process we have decomposed $\mathbf{r}$ into three component vectors, each parallel to a coordinate axis. We call the scalars x,y,z which multiply $\mathbf{i},\mathbf{j},\mathbf{k}$ the *components* of the vector. Thus the vector $\mathbf{r}$ has components x, y, and z if its component vectors are $x\mathbf{i}$, $y\mathbf{j}$, and $z\mathbf{k}$, and the relation (34) holds. The components may be obtained directly from the projections of any representative segment on the coordinate axes.

By the rules of Sec. 288, we may deduce from Eq. (34) that

$$C\mathbf{r} = Cx\mathbf{i} + Cy\mathbf{j} + Cz\mathbf{k}. \tag{35}$$

Thus the components of $C\mathbf{r}$ are C times the components of $\mathbf{r}$.

Again, if

$$\mathbf{r}_1 = x_1\mathbf{i} + y_1\mathbf{j} + z_1\mathbf{k}, \tag{36}$$

we find from this and Eq. (34) that

$$\mathbf{r} + \mathbf{r}_1 = (x + x_1)\mathbf{i} + (y + y_1)\mathbf{j} + (z + z_1)\mathbf{k}, \tag{37}$$

so that the components of $\mathbf{r} + \mathbf{r}_1$ are formed by adding the corresponding components of $\mathbf{r}_1$ to those of $\mathbf{r}$.

In general, the operations of addition, subtraction, or multiplication by a scalar may be effected by applying these operations to the components of the vectors involved.

As an illustration, the two points $P_1(x_1,y_1,z_1)$ and $P_2(x_2,y_2,z_2)$ determine vectors

$$\overline{OP_1} = \mathbf{r}_1 = x_1\mathbf{i} + y_1\mathbf{j} + z_1\mathbf{k}, \qquad \overline{OP_2} = \mathbf{r}_2 = x_2\mathbf{i} + y_2\mathbf{j} + z_2\mathbf{k}. \tag{38}$$

† That is, vectors of unit magnitude.

The vector from P_1 to P_2 is

$$\overline{P_1P_2} = \mathbf{r}_2 - \mathbf{r}_1 = (x_2 - x_1)\mathbf{i} + (y_2 - y_1)\mathbf{j} + (z_2 - z_1)\mathbf{k}. \tag{39}$$

The coefficients are the projections given in Eq. (10).

If l,m,n are direction cosines, the vector

$$\mathbf{u} = l\mathbf{i} + m\mathbf{j} + n\mathbf{k} \tag{40}$$

is of unit length, since

$$|\mathbf{u}| = u = \sqrt{l^2 + m^2 + n^2} = 1. \tag{41}$$

For any vector $\mathbf{r}$ with length r and direction cosines l,m,n, we have

$$\mathbf{r} = r\mathbf{u} = rl\mathbf{i} + rm\mathbf{j} + rn\mathbf{k}. \tag{42}$$

Thus $\mathbf{r}$ has as its components rl,rm,rn and its direction cosines may be found by dividing its components by its length. Applied to the vector $\mathbf{r}_2 - \mathbf{r}_1$ of length s, this checks Eq. (14).

Similarly, Eq. (15) states that

$$\mathbf{r} - \mathbf{r}_1 = t\mathbf{u} \qquad \text{or} \qquad \mathbf{r} = \mathbf{r}_1 + t\mathbf{u}, \tag{43}$$

in which t may be positive, negative, or zero.

Example 1. Let P divide the segment $\overline{P_1P_2}$ in the ratio k_1/k_2, as in Sec. 78, so that $\frac{P_1P}{PP_2} = \frac{k_1}{k_2}$. Express $\mathbf{r} = \overline{OP}$ in terms of $\mathbf{r}_1 = \overline{OP_1}$ and $\mathbf{r}_2 = \overline{OP_2}$. Also deduce formulas for the coordinates of P in terms of those of P_1 and P_2.

Solution: The relation $\frac{P_1P}{PP_2} = \frac{k_1}{k_2}$ for the signed distances on a line implies the vector relation $k_2\overline{P_1P} = k_1\overline{PP_2}$. But $\overline{P_1P} = \overline{OP} - \overline{OP_1} = \mathbf{r} - \mathbf{r}_1$ and $\overline{PP_2} = \overline{OP_2} - \overline{OP} = \mathbf{r}_2 - \mathbf{r}$. The student should notice that each difference is the vector to the second point, or head of the vector, minus that to the first, or initial point of the vector. We may combine our relations into

$$k_2(\mathbf{r} - \mathbf{r}_1) = k_1(\mathbf{r}_2 - \mathbf{r}) \qquad \text{or} \qquad (k_1 + k_2)\mathbf{r} = k_2\mathbf{r}_1 + k_1\mathbf{r}_2.$$

This leads to the required vector expression

$$\mathbf{r} = \frac{k_2\mathbf{r}_1 + k_1\mathbf{r}_2}{k_1 + k_2}. \tag{44}$$

This implies equations for the coordinates similar to those of Sec. 78,

$$x = \frac{k_2x_1 + k_1x_2}{k_1 + k_2}, \qquad y = \frac{k_2y_1 + k_1y_2}{k_1 + k_2}, \qquad z = \frac{k_2z_1 + k_1z_2}{k_1 + k_2}. \tag{45}$$

Example 2. Find the vertex of a parallelogram P_4 opposite P_2, if the other three vertices are $P_1 = (3,4,3)$, $P_2 = (2,3,1)$, $P_3 = (7,5,8)$.

Solution: $\overline{P_1P_4} = \overline{P_2P_3} = \overline{OP_3} - \overline{OP_2} = 5\mathbf{i} + 2\mathbf{j} + 7\mathbf{k}$. Hence $\overline{OP_4} = \overline{OP_1} + \overline{P_1P_4} = 3\mathbf{i} + 4\mathbf{j} + 3\mathbf{k} + (5\mathbf{i} + 2\mathbf{j} + 7\mathbf{k}) = 8\mathbf{i} + 6\mathbf{j} + 10\mathbf{k}$. This shows that $P_4 = (8,6,10)$.

EXERCISE 145

The vectors **a**,**b**,**c** are defined by $\mathbf{a} = 2\mathbf{i} - 5\mathbf{j} + 14\mathbf{k}$, $\mathbf{b} = 6\mathbf{i} + 3\mathbf{j} - 2\mathbf{k}$, $\mathbf{c} = -\mathbf{i} - \mathbf{j} - \mathbf{k}$. Express each of the following vectors in terms of **i**,**j**,**k** and find its length.

1. $-\mathbf{a}$.
2. $5\mathbf{b}$.
3. $2\mathbf{c}$.
4. $\mathbf{a} + \mathbf{b}$.
5. $\mathbf{a} - \mathbf{b}$.
6. $2\mathbf{b} + 3\mathbf{c}$.
7. $\mathbf{a} + \mathbf{b} + \mathbf{c}$.
8. $\mathbf{b} - 2\mathbf{c}$.
9. $\mathbf{a} + 5\mathbf{b} + 2\mathbf{c}$.

The point P divides the segment $\overline{P_1P_2}$ so that $P_1P/PP_2 = k_1/k_2$. Given that $P_1 = (-8,-7,2)$ and $P_2 = (4,5,26)$, express the vector $\overline{OP}$ in terms of **i**,**j**,**k** for each indicated value of the ratio k_1/k_2.

10. 1.
11. 2.
12. $\frac{1}{3}$.
13. $\frac{1}{2}$.
14. $-\frac{1}{2}$.
15. -2.

$P_1 = (2,3,4)$, $P_2 = (14,-9,28)$, $P_3 = (1,2,2)$. Find P_4 if

16. $\overline{OP_4} = 2\overline{P_1P_2}$.
17. $\overline{P_3P_4} = \frac{1}{4}\overline{P_1P_2}$.
18. $\overline{P_1P_4}$ and $\overline{P_2P_3}$ are the diagonals of a parallelogram.
19. $\overline{P_1P_2}$ and $\overline{P_3P_4}$ are the diagonals of a parallelogram.
20. P_4 is in the xy plane and on the straight line P_1P_2.
21. P_4 is in the yz plane and on the straight line P_1P_3.

***290. The Scalar Product.** For any two vectors **a** and **b**, we may calculate the expression $ab \cos \theta$, where θ is the angle $(0 \leqq \theta \leqq \pi)$ made by their positive directions as in Sec. 283, and a and b are their lengths. We call this expression the *scalar* or *dot product*† of the two vectors and denote it by $\mathbf{a} \cdot \mathbf{b}$, read "**a** dot **b**," so that

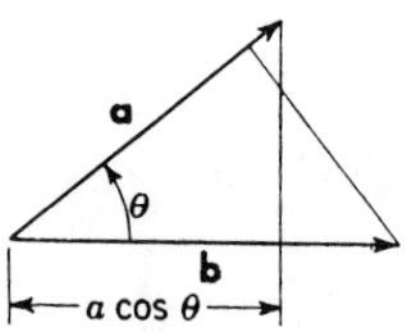

FIG. 306.

$$\mathbf{a} \cdot \mathbf{b} = ab \cos \theta. \tag{46}$$

By Sec. 283 the projection of **a** on **b** is $a \cos \theta$, and the projection of **b** on **a** is $b \cos \theta$. See Fig. 306. Consequently, we have

$$\mathbf{a} \cdot \mathbf{b} = a \operatorname{Proj}_{\mathbf{a}} \mathbf{b} = b \operatorname{Proj}_{\mathbf{b}} \mathbf{a}. \tag{47}$$

It follows from the original definition that

$$\mathbf{a} \cdot \mathbf{b} = \mathbf{b} \cdot \mathbf{a}. \tag{48}$$

Thus the scalar product is *commutative*, that is, it does not depend on the order in which the two factors are written.

Also, for any scalars p and q,

$$(p\mathbf{a}) \cdot (q\mathbf{b}) = pq\mathbf{ab} \cos \theta = pq(\mathbf{a} \cdot \mathbf{b}). \tag{49}$$

Again, by Eq. (9), we have

$$\operatorname{Proj}_{\mathbf{a}} \mathbf{b} + \operatorname{Proj}_{\mathbf{a}} \mathbf{c} = \operatorname{Proj}_{\mathbf{a}} (\mathbf{b} + \mathbf{c}). \tag{50}$$

If we multiply by a and use Eq. (47), we may deduce that

$$\mathbf{a} \cdot (\mathbf{b} + \mathbf{c}) = \mathbf{a} \cdot \mathbf{b} + \mathbf{a} \cdot \mathbf{c} \quad \text{or} \quad (\mathbf{b} + \mathbf{c}) \cdot \mathbf{a} = \mathbf{b} \cdot \mathbf{a} + \mathbf{c} \cdot \mathbf{a}. \tag{51}$$

† Here $\mathbf{a} \cdot \mathbf{b}$ does not mean algebraic multiplication, but the special combination defined by Eq. (46) or (59) which has many properties similar to those of algebraic products.

Thus the scalar product is *distributive*, and we may expand products like

$$(\mathbf{a} + \mathbf{b}) \cdot (\mathbf{c} + \mathbf{d}) = \mathbf{a} \cdot \mathbf{c} + \mathbf{a} \cdot \mathbf{d} + \mathbf{b} \cdot \mathbf{c} + \mathbf{b} \cdot \mathbf{d}. \tag{52}$$

If the two vectors **a** and **b** have the same direction, they make a zero angle. This makes $\theta = 0$ and $\cos \theta = 1$ in Eq. (46). In particular we have

$$\mathbf{a} \cdot \mathbf{a} = a^2, \tag{53}$$

and the scalar product of a vector by itself is the square of its length.

If the two vectors **a** and **b** are perpendicular, they make a right angle. This makes $\theta = 90°$ and $\cos \theta = 0$. Hence,

$$\text{If } \mathbf{a} \perp \mathbf{b}, \qquad \mathbf{a} \cdot \mathbf{b} = 0. \tag{54}$$

These facts enable us to compute all the scalar products involving the three unit vectors **i**,**j**,**k**. We have

$$\mathbf{i} \cdot \mathbf{i} = \mathbf{j} \cdot \mathbf{j} = \mathbf{k} \cdot \mathbf{k} = 1 \tag{55}$$

and

$$\mathbf{i} \cdot \mathbf{j} = \mathbf{j} \cdot \mathbf{i} = \mathbf{j} \cdot \mathbf{k} = \mathbf{k} \cdot \mathbf{j} = \mathbf{k} \cdot \mathbf{i} = \mathbf{i} \cdot \mathbf{k} = 0. \tag{56}$$

We may now express the scalar product of any two vectors in terms of their components. Let **a** have components a_x,a_y,a_z and **b** have components b_x,b_y,b_z. Then

$$\mathbf{a} = a_x\mathbf{i} + a_y\mathbf{j} + a_z\mathbf{k} \qquad \text{and} \qquad \mathbf{b} = b_x\mathbf{i} + b_y\mathbf{j} + b_z\mathbf{k}. \tag{57}$$

Under these assumptions it follows that

$$\mathbf{a} \cdot \mathbf{b} = (a_x\mathbf{i} + a_y\mathbf{j} + a_z\mathbf{k}) \cdot (b_x\mathbf{i} + b_y\mathbf{j} + b_z\mathbf{k}). \tag{58}$$

By the process used in Eqs. (51) and (52), this may be distributed into nine separate products. We may then bring out the scalar factors, as in Eq. (49), and reduce the dot products which involve **i**, **j**, and **k** only by Eqs. (55) and (56). The final result is

$$\mathbf{a} \cdot \mathbf{b} = a_xb_x + a_yb_y + a_zb_z. \tag{59}$$

In particular from Eqs. (53) and (59) with $\mathbf{b} = \mathbf{a}$, we find that

$$a^2 = \mathbf{a} \cdot \mathbf{a} = a_x^2 + a_y^2 + a_z^2. \tag{60}$$

And similarly we have

$$b^2 = \mathbf{b} \cdot \mathbf{b} = b_x^2 + b_y^2 + b_z^2. \tag{61}$$

Except for notation, these are equivalent to Eqs. (2) and (12).

We may deduce from Eq. (46), or $\mathbf{a} \cdot \mathbf{b} = ab \cos \theta$, that

$$\cos \theta = \frac{\mathbf{a} \cdot \mathbf{b}}{ab}. \tag{62}$$

And from Eqs. (59) to (62) it follows that

$$\cos \theta = \frac{a_xb_x + a_yb_y + a_zb_z}{\sqrt{a_x^2 + a_y^2 + a_z^2}\,\sqrt{b_x^2 + b_y^2 + b_z^2}}. \tag{63}$$

This determines the cosine of the angle between two vectors or segments in terms of their components.

If two vectors are perpendicular, $\cos \theta = 0$, and

$$a_xb_x + a_yb_y + a_zb_z = 0. \tag{64}$$

Conversely, if this condition holds, from Eqs. (59) and (46) we find that

$$\mathbf{a} \cdot \mathbf{b} = ab \cos \theta = 0. \tag{65}$$

Hence, either $\cos \theta = 0$ and the vectors are perpendicular, or a or b is zero, and one of the vectors is a null vector. Thus if neither vector is a null vector, condition (65) is a necessary and sufficient condition for perpendicularity. The equivalent Eq. (64) is in accord with Eq. (24).

Whereas scalar or dot multiplication of vectors may for the most part be treated by the familiar rules of algebra, there is one important exception. In ordinary algebra, if a product is zero, one of the factors must be zero. But if the scalar product of two vectors is zero, we have the alternative possibility of perpendicularity. For example, in ordinary algebra if

$$a \neq 0, \qquad \text{and} \qquad ab = ac \qquad \text{or} \qquad a(b - c) = 0, \tag{66}$$

we may conclude that

$$b - c = 0 \qquad \text{and} \qquad b = c. \tag{67}$$

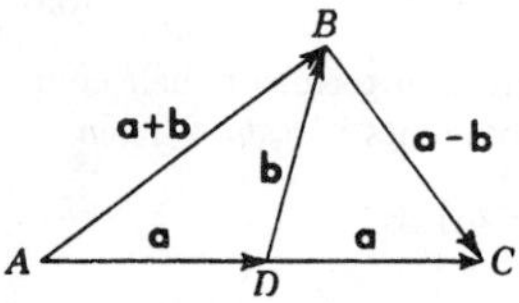

Fig. 307.

But in the algebra of vectors, if

$$\mathbf{a} \neq 0$$

and

$$\mathbf{a} \cdot \mathbf{b} = \mathbf{a} \cdot \mathbf{c}$$

or

$$\mathbf{a} \cdot (\mathbf{b} - \mathbf{c}) = 0, \tag{68}$$

we may conclude only that

$$\text{Either} \qquad \mathbf{b} - \mathbf{c} = 0 \qquad \text{or} \qquad (\mathbf{b} - \mathbf{c}) \perp \mathbf{a}. \tag{69}$$

Example 1. Find the projection of the vector $2\mathbf{i} - 3\mathbf{j} + 4\mathbf{k}$ on the line drawn from the origin to the point $(2, -2, 3)$.

Solution: Let $\mathbf{a} = 2\mathbf{i} - 3\mathbf{j} + 4\mathbf{k}$, and $\mathbf{b} = 2\mathbf{i} - 2\mathbf{j} + 3\mathbf{j}$. We wish to find $\text{Proj}_{\mathbf{b}}\, \mathbf{a} = a \cos \theta = \dfrac{\mathbf{a} \cdot \mathbf{b}}{b}$. But from Eq. (59) we have $\mathbf{a} \cdot \mathbf{b} = 2(2) - 3(-2) + 4(3) = 22$. And from Eq. (61) we have $b^2 = 2^2 + (-2)^2 + 3^2 = 17$, and $b = \sqrt{17}$. It follows that the required projection is $22/\sqrt{17}$.

Example 2. In the triangle with vertices $A = (2,4,-1)$, $B = (-1,-2,5)$, and $C = (1,-4,3)$; find the angle at vertex A.

Solution: The required angle is the angle θ between the vectors $\mathbf{c} = \overline{AB} = -3\mathbf{i} - 6\mathbf{j} + 6\mathbf{k}$ and $\mathbf{b} = \overline{AC} = -\mathbf{i} - 8\mathbf{j} + 4\mathbf{k}$. From Eq. (59), we have $cb \cos \theta = \mathbf{c} \cdot \mathbf{b} = -3(-1) - 6(-8) + 6(4) = 75$. And from Eq. (60) we find $c^2 = (-3)^2 + (-6)^2 + 6^2 = 81$, and $c = \sqrt{81} = 9$. Also $b^2 = (-1)^2 + (-8)^2 + 4^2 = 81$, and $b = \sqrt{81} = 9$. Hence $\cos \theta = \dfrac{\mathbf{c} \cdot \mathbf{b}}{cb} = \dfrac{75}{9 \cdot 9} = \dfrac{25}{27}$, and the required angle is $\theta = \cos^{-1} \dfrac{25}{27}$.

Example 3. Let A,B,C be any three distinct points in space, and D be the mid-point of $\overline{AC}$. Show that if the length of DB = the length of DC, then ABC is a right angle.

Solution: Let $\overline{AD} = \mathbf{a}$ and $\overline{DB} = \mathbf{b}$ (Fig. 307). Then $\overline{DC} = \mathbf{a}$, and $\overline{AB} = \overline{AD} + \overline{DB} = \mathbf{a} + \mathbf{b}$, while $\overline{BC} = \overline{BD} + \overline{DC} = \overline{DC} - \overline{DB} = \mathbf{a} - \mathbf{b}$. It follows that $\overline{AB} \cdot \overline{BC} = (\mathbf{a} + \mathbf{b}) \cdot (\mathbf{a} - \mathbf{b}) = \mathbf{a} \cdot \mathbf{a} - \mathbf{b} \cdot \mathbf{b} = a^2 - b^2$. But the conditions on lengths make $b^2 = a^2$, so that $\overline{AB} \cdot \overline{BC} = 0$. Since neither $\overline{AB}$ nor $\overline{BC}$ is a null vector, we may conclude that these vectors are perpendicular and ABC is a right angle, as was to be proved.

EXERCISE 146

The vectors **a**, **b**, **c**, and **d** are defined as follows. **a** = 2**i** − 3**j**, **b** = −2**i** + 4**k**, **c** = **i** + **j** + **k**, **d** = −3**i** − **j** − 2**k**. For each of the following scalar products verify the given value.

1. **a** · **c** = −1.
2. **a** · **d** = −3.
3. **b** · **c** = 2.
4. **b** · **d** = −2.

Evaluate each of the following scalar products first by performing the operations inside the parentheses, and then by distributing the products and using the results of Probs. 1 to 4.

5. (**a** + **b**) · (**c** + **d**).
6. (2**a** − **b**) · (2**c** + **d**).

In each problem find the projection of the given vector on the line from the origin to the given point.

7. 5**i** + 3**j** + **k**, (6, −2, −3).
8. 5**i** + 4**j** − 2**k**, (−1,4,−8).
9. **i** + 2**j** + 4**k**, (14, −2,5).
10. 3**i** + 4**k**, (12,15,16).

In the triangle with vertices $A = (5,3,2)$, $B = (1,1,0)$, $C = (5,1,4)$; find the angle at

11. A.
12. B.
13. C.

In the triangle with vertices $A = (1,2,3)$, $B = (-1,4,2)$, $C = (2,4,5)$, find the angle at

14. A.
15. B.
16. C.

17. Show that the projection of the vector x**i** + y**j** + z**k** on the line whose direction cosines are l,m,n is $lx + my + nz$.
18. Show that the cosine of the angle θ between a line with direction cosines l_1,m_1,n_1 and one with direction cosines l_2,m_2,n_2 is $\cos\theta = l_1l_2 + m_1m_2 + n_1n_2$.

Find the value of t for which the second vector is perpendicular to the first in each problem.

19. 2**i** − **j** + 5**k**, **i** + 2**j** + t**k**.
20. **i** + 2**j** + **k**, **i** + t**k**.
21. 5**j** + **k**, 2**i** + 3**j** + t**k**.
22. 3**i** − **k**, **i** + 2**j** + t**k**.

23. If **a** and **b** are represented by two sides of a rhombus, $a^2 = b^2$ and the diagonals of the rhombus represent **a** + **b** and **a** − **b**. From this deduce that the diagonals are perpendicular.

***291. Planes.** Consider any equation of the first degree

$$Ax + By + Cz = D. \tag{70}$$

Then A,B,C cannot all be zero, since it is of the first degree. Let $P_1(x_1,y_1,z_1)$ be one point whose coordinates satisfy the equation, so that

$$Ax_1 + By_1 + Cz_1 = D. \tag{71}$$

Then, if $P(x,y,z)$ is any point that satisfies Eq. (70), we may deduce by subtraction that

$$A(x - x_1) + B(y - y_1) + C(z - z_1) = 0. \tag{72}$$

By Eqs. (59) and (39), the left member may be considered as the scalar product of the vector $\overline{P_1P}$ and a vector **a** with components A,B,C. Since **a** ≠ 0, either $\overline{P_1P} = 0$

and P is P_1, or the vector $\overline{P_1P}$ is perpendicular to the vector $\mathbf{a}$. In either case (Fig. 308) the point P lies in the plane through P_1 perpendicular to $\mathbf{a}$. Conversely, if P is in this plane, either $\overline{P_1P} = 0$ or $\overline{P_1P}$ is perpendicular to $\mathbf{a}$, and Eq. (72) will hold. But Eqs. (72) and (71) may be added to give Eq. (70). This proves that Eq. (70) is the equation of a plane, in accord with Eq. (21).

Conversely, by taking A,B,C, any set of direction ratios for the direction perpendicular, or normal, to a given plane, and (x_1,y_1,z_1) as any point in the plane, we may write an equation of the type (72). If we define D by Eq. (71), this equation is reducible to the form of Eq. (70). This proves

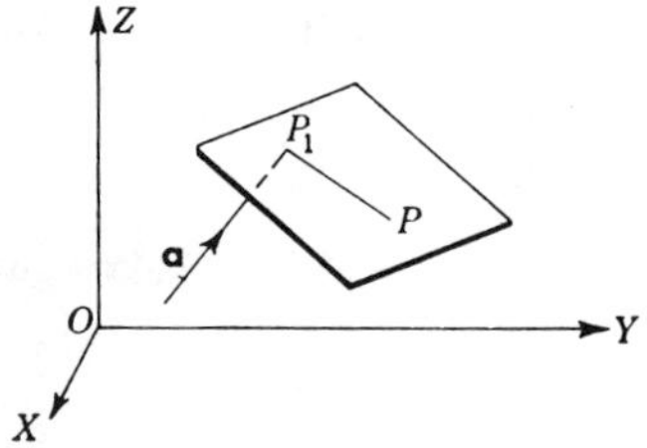

FIG. 308.

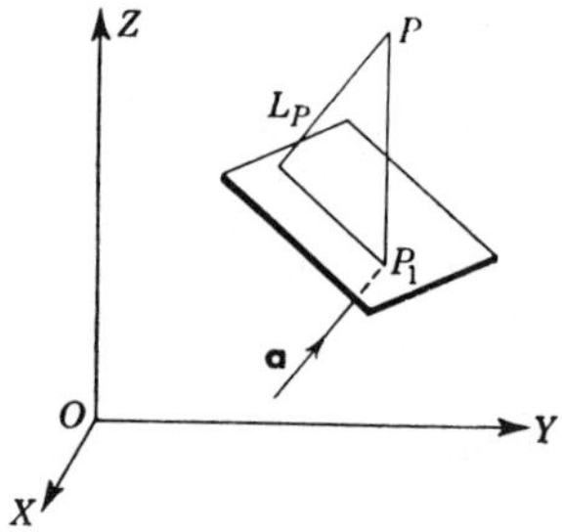

FIG. 309.

Every first-degree equation in x, y, and z represents a plane, and every plane is represented by some first-degree equation.

Consider next a point $P(x,y,z)$ *not* in the plane of Eq. (70). Let $\mathbf{a}$ (Fig. 309) be a normal vector with components A,B,C. Denote by L_P the signed perpendicular distance from the plane to P, measured positive in the direction of $\mathbf{a}$. Then L_P equals the projection of $\overline{P_1P}$ on the direction of $\mathbf{a}$, so that by Eq. (47), we have

$$\begin{aligned} aL_P &= a \operatorname{Proj}_{\mathbf{a}} \overline{P_1P} = \mathbf{a} \cdot \overline{P_1P} \\ &= A(x - x_1) + B(y - y_1) + C(z - z_1). \end{aligned} \tag{73}$$

But since $\mathbf{a} = A\mathbf{i} + B\mathbf{j} + C\mathbf{k}$, we have

$$a = \sqrt{A^2 + B^2 + C^2}, \tag{74}$$

so that

$$\begin{aligned} L_P &= \frac{A(x - x_1) + B(y - y_1) + C(z - z_1)}{\sqrt{A^2 + B^2 + C^2}} \\ &= \frac{Ax + By + Cz - D}{\sqrt{A^2 + B^2 + C^2}}. \end{aligned} \tag{75}$$

This determines the distance from the plane to P. The distance L_P is positive for (x,y,z) on the same side of the plane as $(x_1 + A,\ y_1 + B,\ z_1 + C)$ and negative for (x,y,z) on the same side of the plane as $(x_1 - A,\ y_1 - B,\ z_1 - C)$. When the point is in the plane, L_P is zero and Eq. (70) holds, so that Eq. (75) still applies in this case.

EXAMPLE 1. Find the equation of the plane through the three points $A = (a,0,0)$, $B = (0,b,0)$, $C = (0,0,c)$, where $abc \neq 0$.

Solution: Let the equation of the plane be $Ax + By + Cz = D$. Then since the coordinates of each of the given points satisfies this equation, we must have $Aa = D$, $Bb = D$, $Cc = D$, so that $A = \frac{D}{a}$, $B = \frac{D}{b}$, $C = \frac{D}{c}$. Substituting these values gives

$\frac{D}{a}x + \frac{D}{b}y + \frac{D}{c} = D$, and dividing by D gives as the required equation

$$\frac{x}{a} + \frac{y}{b} + \frac{z}{c} = 1. \tag{76}$$

Figure 310 illustrates the case where a, b, and c are all positive. Equation (76) is called the *intercept form* of the equation of a plane.

EXAMPLE 2. Find the equation of a plane passing through the point (1,3,2) and perpendicular to the line through the points $(2,-3,-3)$ and $(4,-2,-1)$.

Solution: The direction ratios of the line may be taken as the differences of the coordinates of Eq. (10), or $4 - 2$, $-2 - (-3)$, $-1 - (-3)$ which are equal to 2,1,2. Hence by Eq. (72) the equation of the plane is $2(x - 1) + 1(y - 3) + 2(z - 2) = 0$, or $2x + y + 2z = 9$. This is the required equation.

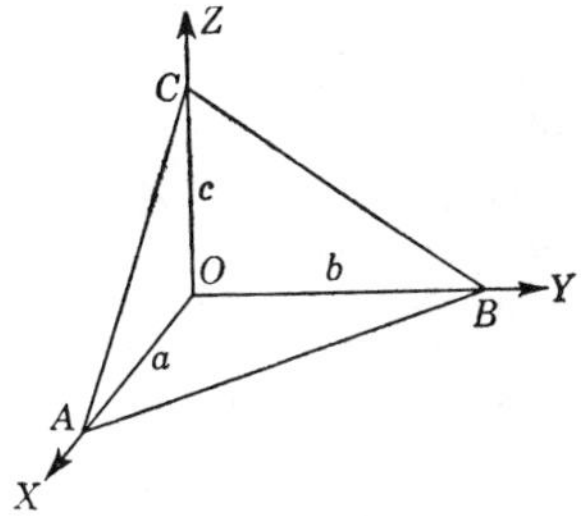

FIG. 310.

EXAMPLE 3. Find the angle between the planes whose equations are $x - 2y + 3z = 5$ and $x - 3y + 2z = 7$.

Solution: The vector $\mathbf{a} = \mathbf{i} - 2\mathbf{j} + 3\mathbf{k}$ is normal to the first plane, and the vector $\mathbf{b} = \mathbf{i} - 3\mathbf{j} + 2\mathbf{k}$ is normal to the second plane. The angle between the planes equals the angle between the normals. From Eq. (59) we have $ab \cos\theta = \mathbf{a} \cdot \mathbf{b} = 1(1) - 2(-3) + 3(2) = 13$. And from Eq. (60) $a^2 = 1^2 + (-2)^2 + 3^2 = 14$, and $b^2 = 1^2 + (-3)^2 + 2^2 = 14$. Hence $a = \sqrt{14}$, $b = \sqrt{14}$, and $\cos\theta = \frac{\mathbf{a}\cdot\mathbf{b}}{ab} = \frac{13}{\sqrt{14}\sqrt{14}} = \frac{13}{14}$.

Thus the required angle is $\cos^{-1}\frac{13}{14}$.

EXAMPLE 4. Find the equations of the planes which bisect the dihedral angles between the planes $2x + y + z = 4$ and $7x - y - 2z = 2$.

Solution: The points on each of the required planes will be equidistant from the two given planes. Hence from Eq. (75) we find that $\frac{2x + y + z - 4}{\sqrt{6}} = \pm\frac{7x - y - 2z - 2}{\sqrt{54}}$, or $3(2x + y + z - 4) = \pm(7x - y - 2z - 2)$, so that $x - 4y - 5z + 10 = 0$ and $13x + 2y + z - 14 = 0$ are the required equations.

EXERCISE 147

Find the equation of a plane through each given set of three points.

1. (2,0,0), (0,3,0), (0,0,4).
2. (4,0,0), (2,2,3), (0,4,5).
3. (1,2,3), (1,3,2), (2,3,1).
4. (3,4,2), (5,2,1), (0,2,1).

Find the equation of a plane perpendicular to the given vector which passes through the given point in each problem.

5. $2\mathbf{i} - \mathbf{j} + 2\mathbf{k}$, $(2,1,-1)$.
6. $3\mathbf{i} - \mathbf{j} + \mathbf{k}$, $(4,-4,0)$.
7. $\mathbf{i} + \mathbf{j} + \mathbf{k}$, $(3,-4,1)$.
8. $2\mathbf{i} - \mathbf{k}$, $(0,0,0)$.

Find the equation of a plane perpendicular to the line segment $\overline{AB}$ if $A = (-1,2,3)$ and $B = (3,4,7)$ which passes through

9. The point A.
10. The point $C = (0,1,-2)$.
11. The point B.
12. The mid-point of $\overline{AB}$.

Find the angle between each given pair of planes.

13. $2x + 3y - 4z = 4$, $4x + 6y - 8z = 7$.
14. $x - 2y + z = 3$, $3x + 4y + 5z = 2$.
15. $x + y = 8$, $2x + y - 2z = 6$.
16. $2x - 2y + z = 4$, $x - 2y - 2z = 7$.

Find the numerical value of the distance from the given plane to the given point in each problem.

17. $3x - 6y + 2z = 7$, $(1,-2,1)$.
18. $2x - 2y + z = 3$, $(3,2,0)$.
19. $8x - 4y - z = 3$, $(2,0,4)$.
20. $5x - 14y - 2z = 5$, $(3,1,2)$.

***292. The Vector Product.** For any two vectors **a** and **b**, *taken in this order*, we may form a new vector by the following construction. (1) Represent the vectors by two segments $\overline{OA}$ and $\overline{OB}$ having the same initial point O. (2) Determine the angle θ, $0 \leqq \theta \leqq \pi$, from **a** to **b**, and compute the value $n = ab \sin \theta$. Since $\theta = \angle AOB$, n equals the area of the parallelogram $OAPB$. (3) Draw $\overline{ON}$ of length n, perpendicular

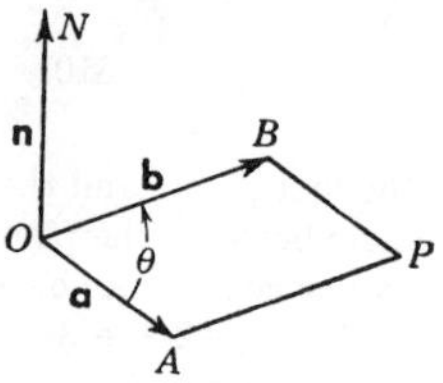

FIG. 311.

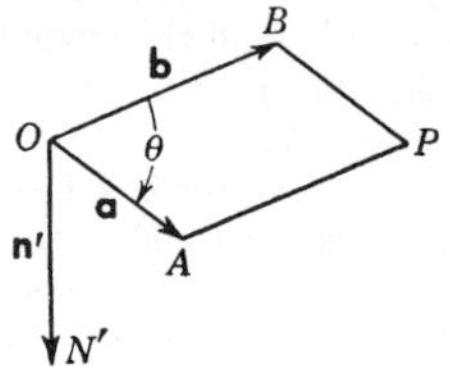

FIG. 312.

to plane OAB and in such a direction that a right-threaded screw along $\overline{ON}$ will advance when turned from $\overline{OA}$ to $\overline{OB}$ through the angle θ. Then the segment $\overline{ON}$ represents the new vector **n** to be constructed.

We call the vector **n** the *vector product*, or *cross product*, of **a** and **b** in this order, and denote it by $\mathbf{a} \times \mathbf{b}$, read "**a** cross **b**."

Thus in Fig. 311, we have

$$\mathbf{n} = \overline{ON} = \mathbf{a} \times \mathbf{b} = \overline{OA} \times \overline{OB}. \tag{77}$$

To construct the vector product $\mathbf{b} \times \mathbf{a}$, we compute

$$n' = ba \sin \theta = n,$$

as before. However, since we now rotate the screw from $\overline{OB}$ to $\overline{OA}$ through the angle θ, we must draw $\overline{ON'}$ in the opposite direction to that previously used (Fig. 312). Thus the direction of a vector product is reversed if we interchange the order of the factors, and

$$\mathbf{b} \times \mathbf{a} = -\mathbf{a} \times \mathbf{b}. \tag{78}$$

Hence vector products are *not* commutative, and we must note the order of the factors. The property expressed in Eq. (78) merits careful attention, since it makes vector products unlike the products used in arithmetic and algebra.

However, vector products are distributive. That is

$$\mathbf{a} \times (\mathbf{b} + \mathbf{c}) = \mathbf{a} \times \mathbf{b} + \mathbf{a} \times \mathbf{c}, \tag{79}$$

and

$$(\mathbf{b} + \mathbf{c}) \times \mathbf{a} = \mathbf{b} \times \mathbf{a} + \mathbf{c} \times \mathbf{a}. \tag{80}$$

These relations may be verified by a geometric interpretation of the sums and products. We omit the details.

For any two scalars p and q, it follows directly from the definition of the cross product that

$$(p\mathbf{a}) \times (q\mathbf{b}) = pq(\mathbf{a} \times \mathbf{b}). \tag{81}$$

If the two vectors **a** and **b** are parallel or lie on the same straight line, the angle between them is either a zero angle or a straight angle. This makes θ either 0° or 180° and $\sin\theta = 0$. Consequently $n = ab \sin\theta = 0$ and $\mathbf{n} = 0$. In particular,

$$\mathbf{a} \times \mathbf{a} = 0 \qquad \text{and} \qquad (p\mathbf{a}) \times (q\mathbf{a}) = 0. \tag{82}$$

If the two vectors **a** and **b** are perpendicular, they make a right angle. This makes $\theta = 90°$ and $\sin\theta = 1$. Consequently $n = ab \sin\theta = ab$, and we may conclude that

$$\text{If } \mathbf{a} \perp \mathbf{b}, \qquad |\mathbf{a} \times \mathbf{b}| = ab. \tag{83}$$

It is easy to remember Eq. (82) by noting that, when $\angle AOB$ is 0° or 180°, the parallelogram $OAPB$ degenerates into one with zero altitude and zero area. And to recall Eq. (83), observe that when $\angle AOB = 90°$, the parallelogram $OAPB$ is a rectangle with area $n = |\mathbf{a} \times \mathbf{b}| = ab$.

Using Eq. (82), we find for the three unit vectors

$$\mathbf{i} \times \mathbf{i} = 0, \qquad \mathbf{j} \times \mathbf{j} = 0, \qquad \mathbf{k} \times \mathbf{k} = 0. \tag{84}$$

And, from Fig. 313, we find that

$$\mathbf{j} \times \mathbf{k} = \mathbf{i}, \qquad \mathbf{k} \times \mathbf{i} = \mathbf{j}, \qquad \mathbf{i} \times \mathbf{j} = \mathbf{k}, \tag{85}$$

by a direct application of the definition, while

$$\mathbf{k} \times \mathbf{j} = -\mathbf{i}, \qquad \mathbf{i} \times \mathbf{k} = -\mathbf{j}, \qquad \mathbf{j} \times \mathbf{i} = -\mathbf{k}. \tag{86}$$

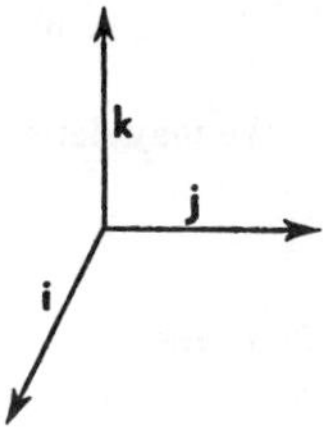

FIG. 313.

That each product has length unity is in accord with Eq. (83). Also the relation between Eqs. (86) and (85) is in accord with Eq. (78).

We may now express the vector product of two vectors **a** and **b** in terms of their components. Let

$$\mathbf{a} = a_x\mathbf{i} + a_y\mathbf{j} + a_z\mathbf{k} \qquad \text{and} \qquad \mathbf{b} = b_x\mathbf{i} + b_y\mathbf{j} + b_z\mathbf{k}. \tag{87}$$

Then the cross product

$$\mathbf{a} \times \mathbf{b} = (a_x\mathbf{i} + a_y\mathbf{j} + a_z\mathbf{k}) \times (b_x\mathbf{i} + b_y\mathbf{j} + b_z\mathbf{k}). \tag{88}$$

By the process used in Eqs. (79) and (80) this may be distributed into nine separate products. We may then bring out the scalar factors as in Eq. (81). Practically this is essentially the method of expanding a product in algebra, except that here in each final term we must write the unit vector which came from the first parenthesis in Eq. (88) before that which came from the second parenthesis. Finally, we may evaluate the cross products involving **i**, **j**, and **k** only by using Eqs. (84) to (86). The result is

$$\mathbf{a} \times \mathbf{b} = \mathbf{i}(a_yb_z - a_zb_y) + \mathbf{j}(a_zb_x - a_xb_z) + \mathbf{k}(a_xb_y - a_yb_x). \tag{89}$$

This may be written in terms of determinants:†

$$\mathbf{a} \times \mathbf{b} = \begin{vmatrix} \mathbf{i} & \mathbf{j} & \mathbf{k} \\ a_x & a_y & a_z \\ b_x & b_y & b_z \end{vmatrix} = \mathbf{i}\begin{vmatrix} a_y & a_z \\ b_y & b_z \end{vmatrix} + \mathbf{j}\begin{vmatrix} a_z & a_x \\ b_z & b_x \end{vmatrix} + \mathbf{k}\begin{vmatrix} a_x & a_y \\ b_x & b_y \end{vmatrix}. \tag{90}$$

† The reader unfamiliar with determinants will find that the discussion following Eq. (90) explains the meaning of the two- and three-rowed arrays used here and provides a convenient way of remembering Eq. (89).

In going from the three-by-three array, or third-order determinant, to the other expansion, the reader should note that the coefficient of j is either $-\begin{vmatrix} a_x & a_z \\ b_x & b_z \end{vmatrix}$ with the order *preserved* from the three-by-three array, and a *minus* sign in front, or $\begin{vmatrix} a_z & a_x \\ b_z & b_x \end{vmatrix}$ with the components in *cyclic* order, *zx* or 31, and a *plus* sign in front as in Eq. (90). Each two-by-two array, or second-order determinant, is expanded by taking the product from the principal diagonal with the plus sign, and the product from the other diagonal with the minus sign. By this process one can obtain the form of Eq. (89) from Eq. (90) and so easily remember Eq. (89).

The fact that $\mathbf{b} \times \mathbf{a} = -\mathbf{a} \times \mathbf{b}$ corresponds to the fact that interchanging the second and third rows in the third-order determinant would change its sign. In writing the third-order determinant for a cross product, we must be careful to put the components of the first vector of the product in the second row, and the components of the second vector in the third row.

We may find the length of $\mathbf{a} \times \mathbf{b}$ by using Eqs. (60) and (90) to obtain

$$|\mathbf{a} \times \mathbf{b}|^2 = (\mathbf{a} \times \mathbf{b}) \cdot (\mathbf{a} \times \mathbf{b}) = \begin{vmatrix} a_y & a_z \\ b_y & b_z \end{vmatrix}^2 + \begin{vmatrix} a_z & a_x \\ b_z & b_x \end{vmatrix}^2 + \begin{vmatrix} a_x & a_y \\ b_x & b_y \end{vmatrix}^2. \tag{91}$$

We may determine $\sin \theta$ by noting that

$$\sin^2 \theta = \frac{(ab \sin \theta)^2}{a^2b^2} = \frac{(\mathbf{a} \times \mathbf{b}) \cdot (\mathbf{a} \times \mathbf{b})}{(\mathbf{a} \cdot \mathbf{a})(\mathbf{b} \cdot \mathbf{b})}. \tag{92}$$

The expression for the numerator in terms of the components is the right member of Eq. (91), while the denominator is

$$(\mathbf{a} \cdot \mathbf{a})(\mathbf{b} \cdot \mathbf{b}) = (a_x^2 + a_y^2 + a_z^2)(b_x^2 + b_y^2 + b_z^2). \tag{93}$$

If $\mathbf{a} \times \mathbf{b}$ is the null vector, it will have zero length and conversely. That is,

$$\mathbf{a} \times \mathbf{b} = 0 \qquad \text{if, and only if, } ab \sin \theta = 0. \tag{94}$$

The last equation shows that a, b, or $\sin \theta = 0$. In the last case $\theta = 0°$ or $180°$, the direction of **a** is either the same as, or opposite to, the direction of **b**, and the vectors are parallel. Hence, if neither **a** nor **b** is a null vector, the equation

$$\mathbf{a} \times \mathbf{b} = 0 \tag{95}$$

is a necessary and sufficient condition for parallelism.

Again, from the relation between three vectors

$$\mathbf{a} \times \mathbf{b} = \mathbf{a} \times \mathbf{c} \qquad \text{or} \qquad \mathbf{a} \times (\mathbf{b} - \mathbf{c}) = 0, \tag{96}$$

we may conclude only that either

$$\mathbf{a} = 0, \qquad \mathbf{b} = 0, \qquad \text{or } (\mathbf{b} - \mathbf{c}) \text{ is parallel to } \mathbf{a}. \tag{97}$$

EXAMPLE 1. Find the area of the triangle whose vertices are $A = (2,-3,1)$, $B = (3,0,3)$, $C = (4,-2,2)$.

Solution: The area is one-half that of the parallelogram on the vectors $\mathbf{b} = \overline{AB}$ and $\mathbf{c} = \overline{AC}$, or $\frac{1}{2}|\mathbf{b} \times \mathbf{c}|$. But we find that $\mathbf{b} = \overline{AB} = \mathbf{i} + 3\mathbf{j} + 2\mathbf{k}$ and $\mathbf{c} = \overline{AC} = 2\mathbf{i} + \mathbf{j} + \mathbf{k}$. Hence from Eq. (90),

$$\mathbf{b} \times \mathbf{c} = \begin{vmatrix} \mathbf{i} & \mathbf{j} & \mathbf{k} \\ 1 & 3 & 2 \\ 2 & 1 & 1 \end{vmatrix} = \mathbf{i}\begin{vmatrix} 3 & 2 \\ 1 & 1 \end{vmatrix} + \mathbf{j}\begin{vmatrix} 2 & 1 \\ 1 & 2 \end{vmatrix} + \mathbf{k}\begin{vmatrix} 1 & 3 \\ 2 & 1 \end{vmatrix} = \mathbf{i} + 3\mathbf{j} - 5\mathbf{k}.$$

It follows that $|\mathbf{b} \times \mathbf{c}|^2 = 1^2 + 3^2 + (-5)^2 = 35$, $|\mathbf{b} \times \mathbf{c}| = \sqrt{35}$. The required area is one-half this or $\frac{1}{2}\sqrt{35}$.

EXAMPLE 2. Find the equation of a plane through the point $A = (1,2,4)$ and parallel to each of the vectors $\mathbf{b} = 2\mathbf{i} - \mathbf{j} + \mathbf{k}$ and $\mathbf{c} = 3\mathbf{i} + \mathbf{j} + \mathbf{k}$.

Solution: Since $\mathbf{n} = \mathbf{b} \times \mathbf{c}$ is perpendicular to each of the given vectors, it may be taken as the vector perpendicular to the required plane. But

$$\mathbf{n} = \mathbf{b} \times \mathbf{c} = \begin{vmatrix} \mathbf{i} & \mathbf{j} & \mathbf{k} \\ 2 & -1 & 1 \\ 3 & 1 & 1 \end{vmatrix} = -2\mathbf{i} + \mathbf{j} + 5\mathbf{k}.$$ Hence from Eq. (72),

$-2(x - 1) + 1(y - 2) + 5(z - 4) = 0$ is the equation of a plane through A perpendicular to $\mathbf{n}$. It follows that one form of the desired equation is $2x - y - 5z + 20 = 0$.

EXAMPLE 3. Find the equation of the plane through the points $A = (1,2,4)$, $B = (3,1,5)$, $C = (4,3,5)$.

Solution: The segments $\overline{AB}$ and $\overline{AC}$ in the plane determine the vectors $\mathbf{b} = \overline{AB} = 2\mathbf{i} - \mathbf{j} + \mathbf{k}$ and $\mathbf{c} = \overline{AC} = 3\mathbf{i} + \mathbf{j} + \mathbf{k}$. From these, as in Example 2, $\mathbf{n} = \mathbf{b} \times \mathbf{c} = -2\mathbf{i} + \mathbf{j} + 5\mathbf{k}$ is a normal to the plane, and the equation of the required plane through $A = (1,2,4)$ perpendicular to $\mathbf{n}$ is found as in Example 2 to be $2x - y - 5z + 20 = 0$.

EXAMPLE 4. Find the equation of the plane through the points $A = (0,1,1)$ and $B = (1,0,2)$ perpendicular to the plane $x - 3y + 2z = 1$.

Solution: The normal to the required plane is perpendicular to the vector $\overline{AB} = \mathbf{b} = \mathbf{i} - \mathbf{j} + \mathbf{k}$, and also perpendicular to the normal to the given plane, or to $\mathbf{c} = \mathbf{i} - 3\mathbf{j} + 2\mathbf{k}$. It follows that $\mathbf{n} = \mathbf{b} \times \mathbf{c} = \begin{vmatrix} \mathbf{i} & \mathbf{j} & \mathbf{k} \\ 1 & -1 & 1 \\ 1 & -3 & 2 \end{vmatrix} = \mathbf{i} - \mathbf{j} - 2\mathbf{k}$ is normal to the required plane. By Eq. (72), the plane through $A = (0,1,1)$ normal to $\mathbf{i} - \mathbf{j} - 2\mathbf{k}$ has as its equation $1(x - 0) - 1(y - 1) - 2(z - 1) = 0$, or $x - y - 2z = -3$.

EXERCISE 148

The vectors $\mathbf{a}$, $\mathbf{b}$, $\mathbf{c}$, and $\mathbf{d}$ are defined as follows: $\mathbf{a} = 3\mathbf{i} - 2\mathbf{j}$, $\mathbf{b} = -3\mathbf{i} + 2\mathbf{k}$, $\mathbf{c} = 2\mathbf{i} - \mathbf{j} + 2\mathbf{k}$, $\mathbf{d} = -\mathbf{i} + 2\mathbf{j} - \mathbf{k}$. Evaluate each of the following vector products.

1. $\mathbf{a} \times \mathbf{c}$. 2. $\mathbf{a} \times \mathbf{d}$. 3. $\mathbf{b} \times \mathbf{c}$. 4. $\mathbf{b} \times \mathbf{d}$.

Given that $A = (2,1,3)$, $B = (3,-1,4)$, $C = (1,2,1)$, $D = (4,3,2)$, find the area of each of the following triangles.

5. ABC. 6. BCD. 7. ACD. 8. ABD.

Using the data for Probs. 5 to 8, find the equation of a plane through each set of three points.

9. A,B,C. 10. B,C,D. 11. A,C,D. 12. A,B,D.

Using the data for Probs. 5 to 8, find the equation of a plane

13. Through AB and perpendicular to the plane $x - y + 2z = 7$.
14. Through BC and perpendicular to the plane $2x + 2y - z = 4$.
15. Through CD and perpendicular to the plane $z = x + 2y$.
16. Show that $(\mathbf{a} \times \mathbf{b}) \cdot \mathbf{c}$ is numerically equal to the volume of the parallelepiped having the three edges at one vertex segments which represent the vectors $\mathbf{a}$, $\mathbf{b}$, and $\mathbf{c}$. HINT: If B is the area of the parallelogram with $\mathbf{a}$ and $\mathbf{b}$ as two edges, and $\mathbf{c}$ makes an angle θ with the perpendicular to the plane of $\mathbf{a}$ and $\mathbf{b}$, the base of the parallelepiped is B and the altitude is $c \cos \theta$, so that the volume is numerically $Bc \cos \theta = (\mathbf{a} \times \mathbf{b}) \cdot \mathbf{c}$.

17. Use Prob. 16 to find the volume of the parallelepiped whose vertices are (0,0,0), (3,2,1), (4,1,2), (5,4,3).
18. Deduce from Prob. 16 that the three vectors **a**, **b**, and **c** are parallel to the same plane if and only if $(\mathbf{a} \times \mathbf{b}) \cdot \mathbf{c} = 0$.
19. Use Prob. 18 to show that the four points (2,1,3), (0,2,1), (−1,−2,−4) and (1,−3,−2) all lie in the same plane.
20. Find the volume of the tetrahedron three of whose edges at one vertex represent the vectors $\mathbf{a} = \mathbf{i} + 2\mathbf{k} - \mathbf{j}$, $\mathbf{b} = 3\mathbf{i} - 4\mathbf{j} + \mathbf{k}$, $\mathbf{c} = 2\mathbf{i} - 3\mathbf{j} + 3\mathbf{k}$. HINT: The volume is one-sixth the volume of a parallelepiped which may be found from Prob. 16.

293. Curves in Space. A set of three equations

$$x = f(t), \qquad y = g(t), \qquad z = h(t) \tag{98}$$

determine a point $P(x,y,z)$ for each value of t. In general, the point will vary continuously with t. And as t traces an interval, P will describe the arc of a curve in space. Let t_1 be a fixed value, which determines a fixed point $P_1(x_1,y_1,z_1)$. Then if $t_1 + \Delta t$ is a value near t_1, which determines P_2 with coordinates $(x_1 + \Delta x, y_1 + \Delta y, z_1 + \Delta z)$, the segment $\overline{P_1P_2}$ will have components $\Delta x, \Delta y, \Delta z$. Thus, these numbers, or

$$\frac{\Delta x}{\Delta t}, \frac{\Delta y}{\Delta t}, \frac{\Delta z}{\Delta t}, \tag{99}$$

which are proportional to them, are the direction ratios of a secant through P_1. Now let $\Delta t \to 0$. Then, if the given functions of Eq. (98) are differentiable, the ratios of Eq. (99) approach

$$\left.\frac{dx}{dt}\right|_1 = f'(t_1), \qquad \left.\frac{dy}{dt}\right|_1 = g'(t_1), \qquad \left.\frac{dz}{dt}\right|_1 = h'(t_1). \tag{100}$$

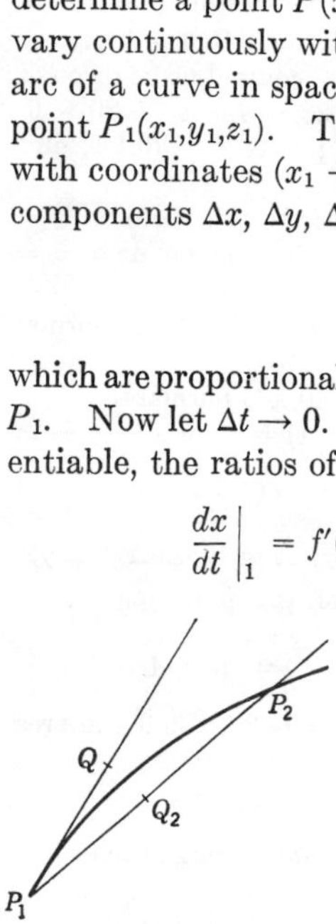

FIG. 314.

Assume that these are not all zero, so that we have

$$[f'(t_1)]^2 + [g'(t_1)]^2 + [h'(t_1)]^2 > 0. \tag{101}$$

Then the numbers in Eq. (100) may be taken as the direction ratios of a line through P_1 and $Q[x_1 + f'(t_1),\ y_1 + g'(t_1),\ z_1 + h'(t_1)]$. The equations of this line, shown in Fig. 314, may be written

$$\frac{x - x_1}{f'(t_1)} = \frac{y - y_1}{g'(t_1)} = \frac{z - z_1}{h'(t_1)}. \tag{102}$$

But the secant line through P_1 and P_2 has direction ratios (99) and therefore contains the point $Q_2\left(x_1 + \frac{\Delta x}{\Delta t}, y_1 + \frac{\Delta y}{\Delta y}, z_1 + \frac{\Delta z}{\Delta t}\right)$. When $\Delta t \to 0$, $Q_2 \to Q$, so that the secant through P_1Q_2 approaches the line through P_1Q. This limiting line is defined to be the *straight line tangent to the curve at* P_1, or the tangent at P_1. Thus the equations of the tangent are given by Eq. (102) and may be written in the alternative forms:

$$\frac{x - x_1}{\left.\dfrac{dx}{dt}\right|_1} = \frac{y - y_1}{\left.\dfrac{dy}{dt}\right|_1} = \frac{z - z_1}{\left.\dfrac{dz}{dt}\right|_1} \tag{103}$$

or

$$\frac{x - f(t_1)}{f'(t_1)} = \frac{y - g(t_1)}{g'(t_1)} = \frac{z - h(t_1)}{h'(t_1)}. \tag{104}$$

The fact that the length of the chord P_1P_2 is

$$P_1P_2 = \sqrt{\Delta x^2 + \Delta y^2 + \Delta z^2} \tag{105}$$

suggests that a definition of length of arc based on inscribed polygons like that given for plane curves in Sec. 180 would make

$$\left(\frac{ds}{dt}\right)^2 = \left(\frac{dx}{dt}\right)^2 + \left(\frac{dy}{dt}\right)^2 + \left(\frac{dz}{dt}\right)^2, \tag{106}$$

and

$$s - s_1 = \int_{t_1}^{t} \sqrt{\left(\frac{dx}{dt}\right)^2 + \left(\frac{dy}{dt}\right)^2 + \left(\frac{dz}{dt}\right)^2}\, dt. \tag{107}$$

This may be proved to be the case, and we shall accordingly use these relations as the fundamental ones for the arc length s.

EXAMPLE 1. Find the equations of the tangent line to the curve $x = \sqrt{2}\,t$, $y = e^t$, $z = e^{-t}$ at the point where $t = 2$.

Solution: From the given relations we find by differentiation that $dx/dt = \sqrt{2}$, $dy/dt = e^t$, $dz/dt = e^{-t}$. It follows from Eq. (104) that for any value of t_1, $\dfrac{x - \sqrt{2}\,t_1}{\sqrt{2}} = \dfrac{y - e^{t_1}}{e^{t_1}} = \dfrac{z - e^{-t_1}}{-e^{-t_1}}$ are the equations of the tangent. And hence when $t = 2$, the equations are

$$\frac{x - 2\sqrt{2}}{\sqrt{2}} = \frac{y - e^2}{e^2} = \frac{z - e^{-2}}{-e^{-2}}.$$

EXAMPLE 2. For the curve of Example 1, find the length of the arc between the points where $t = 1$ and $t = 2$.

Solution: Using Eq. (106) and the derivatives found in Example 1, we find $(ds/dt)^2 = (\sqrt{2})^2 + (e^t)^2 + (e^{-t})^2 = 2 + e^{2t} + e^{-2t} = (e^t + e^{-t})^2$. It follows that the required arc length is

$$s_{12} = \int_1^2 (e^t + e^{-t})dt = [e^t - e^{-t}]_1^2 = e^2 - e^{-2} - e + e^{-1}.$$

EXERCISE 149

Find the equations of the tangent line to each of the following curves in space at the point indicated.

1. $x = 3t$, $y = 4t^2$, $z = 5t^3$; $t = 2$.
2. $x = t^2$, $y = t^4$, $z = t^5$; $t = 1$.

3. $x = \cos 2t,\ y = \sin 2t,\ z = 8t;\ t = \pi/8$.
4. $x = e^{2t},\ y = e^{3t},\ z = 2t;\ t = 1$.
5. $x = t \cos t,\ y = t \sin t,\ z = t;\ t = 0$.
6. $x = \ln t,\ y = \ln (t - 1),\ z = t^2;\ t = 2$.

Find the length of arc of each of the following curves between the points where t has the two indicated values.

7. $x = t,\ y = \sqrt{3}\, t^2,\ z = 2t^3;\ t = 0,\ t = 3$.
8. $x = t,\ y = \sqrt{2} \ln t,\ z = 1/t;\ t = \frac{1}{2},\ t = 1$.
9. $x = 4 \cos t,\ y = 4 \sin t,\ z = 3t;\ t = 0,\ t = \pi$.
10. $x = 3t \cos t,\ y = 3t \sin t,\ z = 4t;\ t = 0,\ t = 1$.
11. $x = \cos t,\ y = \sin t,\ z = \ln \cos t;\ t = 0,\ t = \pi/4$.
12. $x = e^t \cos t,\ y = e^t \sin t,\ z = e^t;\ t = 0,\ t = 1$.

294. Velocity and Acceleration. If a vector has its components variable but functions of a parameter t, we may write

$$\mathbf{r}(t) = f(t)\mathbf{i} + g(t)\mathbf{j} + h(t)\mathbf{k}. \tag{108}$$

By the derivative of the vector $\mathbf{r}$ with respect to t we mean the vector

$$\mathbf{r}'(t) = \lim_{\Delta t \to 0} \frac{\mathbf{r}(t + \Delta t) - \mathbf{r}(t)}{\Delta t}. \tag{109}$$

By applying this definition to the form given in Eq. (108), we find that

$$\mathbf{r}'(t) = f'(t)\mathbf{i} + g'(t)\mathbf{j} + h'(t)\mathbf{k}. \tag{110}$$

Thus to differentiate a vector, we need merely differentiate each of its components with respect to fixed axes.

Let us take a point $P(x,y,z)$ such that

$$\overline{OP} = \mathbf{r}(t), \qquad \text{and } x = f(t), \qquad y = g(t), \qquad z = h(t). \tag{111}$$

Then, when t varies, the end point P will describe a curve in space, as in Sec. 293. The vector $\mathbf{r}'(t)$ is parallel to the tangent to this curve. And the length of $\mathbf{r}'(t)$ is

$$|\mathbf{r}'| = \sqrt{\mathbf{r}' \cdot \mathbf{r}'} = \sqrt{[f'(t)]^2 + [g'(t)]^2 + [h'(t)]^2} = \frac{ds}{dt}, \tag{112}$$

the derivative of the arc length with respect to the parameter.

When t is the time, $\mathbf{r}'(t)$ has a length equal to the speed in the path, and components equal to the velocities of x, y, and z along the fixed coordinate axes. Thus $\mathbf{r}'$ is the *velocity vector* of the moving point.

Similarly, $\mathbf{r}''(t)$ with components d^2x/dt^2, d^2y/dt^2, d^2z/dt^2 so that

$$\mathbf{r}''(t) = f''(t)\mathbf{i} + g''(t)\mathbf{j} + h''(t)\mathbf{k} \tag{113}$$

is the time derivative of the velocity vector or the *acceleration vector* of the moving point P.

EXAMPLE. Find the velocity and acceleration vector of P if $\overline{OP} = \mathbf{r}$, and $\mathbf{r} = e^t \cos t\, \mathbf{i} + e^t \sin t\, \mathbf{j} + e^t\mathbf{k}$. Also show that the length of each of these vectors is proportional to $z = e^t$.

Solution: From Eq. (110), we find that the velocity vector is

$$\mathbf{r}' = e^t(\cos t - \sin t)\mathbf{i} + e^t(\cos t + \sin t)\mathbf{j} + e^t\mathbf{k}.$$

And from Eq. (113) we find that the acceleration vector is

$$\mathbf{r}'' = e^t(-2 \sin t)\mathbf{i} + e^t(2 \cos t)\mathbf{j} + e^t\mathbf{k}.$$

These are the required vectors. And from them we find that

$$\mathbf{r}' \cdot \mathbf{r}' = e^{2t}[(\cos t - \sin t)^2 + (\cos t + \sin t)^2 + 1] = 3e^{2t},$$
$$\mathbf{r}'' \cdot \mathbf{r}'' = e^{2t}[(-2 \sin t)^2 + (2 \cos t)^2 + 1] = 5e^{2t}.$$

Hence $|\mathbf{r}'| = \sqrt{3}\, e^t = \sqrt{3}\, z$ and $|\mathbf{r}''| = \sqrt{5}\, e^t = \sqrt{5}\, z$, so that the length of each vector is proportional to z.

EXERCISE 150

Find the velocity and acceleration vector of P at the indicated time t if $\overline{OP} = \mathbf{r}$, for each of the following motions.

1. $\mathbf{r} = t^2\mathbf{i} + 2t\mathbf{j} + t^3\mathbf{k},\ t = 2.$
2. $\mathbf{r} = t^3\mathbf{i} - t^4\mathbf{j} + 2t\mathbf{k},\ t = 3.$
3. $\mathbf{r} = \cos 2t\, \mathbf{i} - \sin 2t\, \mathbf{j} + \sin 3t\, \mathbf{k},\ t = \pi/4.$
4. $\mathbf{r} = \cos t\, \mathbf{i} + \cos 3t\, \mathbf{j} + \sin 3t\, \mathbf{k},\ t = \pi/2.$
5. $\mathbf{r} = e^t\mathbf{i} + e^{-t}\mathbf{j} + e^{2t}\mathbf{k},\ t = 1.$

With $\overline{OP} = \mathbf{r}$, the given relation defines a motion of the point P. Show that the speed in the path and magnitude of the acceleration vector are constant for each of these motions.

6. $\mathbf{r} = 6 \cos 2t\, \mathbf{i} + 6 \sin 2t\, \mathbf{j} + 5t\mathbf{k}.$
7. $\mathbf{r} = 4t\mathbf{i} + 3 \sin t\, \mathbf{j} + 3 \cos t\, \mathbf{k};$
8. $\mathbf{r} = 2\mathbf{i} + \cos t\, \mathbf{j} + \sin t\, \mathbf{k}.$
9. $\mathbf{r} = (\cos t + \sin t)\mathbf{i} + (\cos t - \sin t)\mathbf{j} + 2\mathbf{k}.$
10. $\mathbf{r} = 4 \cos t\, \mathbf{i} + 5 \sin t\, \mathbf{j} + 3 \cos t\, \mathbf{k}.$

11. If $\mathbf{r} = \sin t \cos 2t\, \mathbf{i} + \sin t \sin 2t\, \mathbf{j} + \cos t\, \mathbf{k}$, show that $d\mathbf{r}/dt$ is perpendicular to $\mathbf{r}$.

295. Surfaces. The equation

$$z = f(x,y) \tag{114}$$

determines a single point P for each value of x and y. In general, when x and y vary continuously over some two-dimensional region, the point P will vary continuously over the portion of a surface in space. Similar remarks apply to $x = g(y,z)$ and to $y = h(z,x)$. A more general equation is

$$F(x,y,z) = 0. \tag{115}$$

If the partial derivatives

$$F_x = \frac{\partial F}{\partial x}, \qquad F_y = \frac{\partial F}{\partial y}, \qquad F_z = \frac{\partial F}{\partial z} \tag{116}$$

are not all zero, Eq. (115) may be solved in some restricted region for one of the variables in terms of the other two and so represents a surface.

For example, it follows from Sec. 286 or 291 that

$$Ax + By + Cz - D = 0 \tag{117}$$

represents a plane if A, B, and C are not all zero.

Again, we may conclude from Sec. 284 that the distance of a point $P(x,y,z)$ from a fixed center $C(a,b,c)$ is always equal to the constant R if and only if

$$(x - a)^2 + (y - b)^2 + (z - c)^2 - R^2 = 0. \tag{118}$$

It follows that Eq. (118) represents a sphere with center C and radius R.

Next let $x = x(t)$, $y = y(t)$, $z = z(t)$ be the equations of any curve lying in the surface (115). Then

$$F[x(t), y(t), z(t)] = 0 \tag{119}$$

for all values of t. And by total differentiation with respect to t, as in Sec. 273, we find that

$$F_x\frac{dx}{dt} + F_y\frac{dy}{dt} + F_z\frac{dz}{dt} = 0 \quad \text{or} \quad F_x\,x'(t) + F_y\,y'(t) + F_z\,z'(t) = 0. \tag{120}$$

We shall assume that the parameter t has been so selected that $[x'(t)]^2 + [y'(t)]^2 + [z'(t)]^2 > 0$. For example, we may take $t = s$, the arc length. A comparison of Eq. (120) with Eq. (24) or (64) shows that the vector with direction ratios F_x,F_y,F_z,

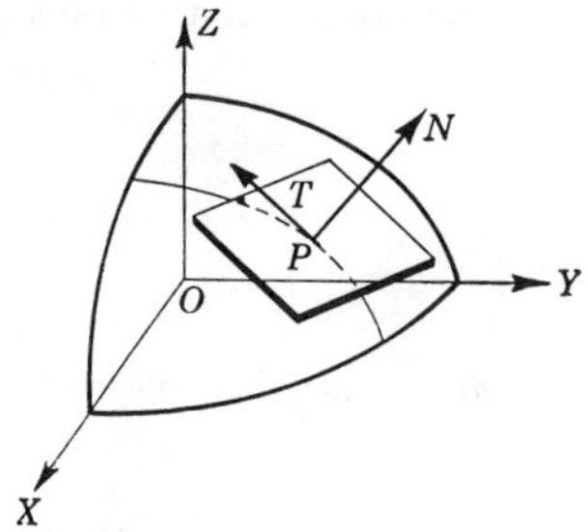

FIG. 315.

$$\mathbf{N} = F_x\,\mathbf{i} + F_y\,\mathbf{j} + F_z\,\mathbf{k} \tag{121}$$

is perpendicular to the vector with direction ratios $x'(t)$, $y'(t)$, $z'(t)$,

$$\mathbf{T} = x'(t)\,\mathbf{i} + y'(t)\,\mathbf{j} + z'(t)\,\mathbf{k}. \tag{122}$$

Thus the plane through the point $P(x,y,z)$ perpendicular to N (Fig. 315) will contain all the tangent lines drawn to curves on the surface passing through P. We call this plane the *tangent plane* to the surface at P. The direction N, which is perpendicular or normal to this plane, is called the *normal* to the surface at P.

The equation of the tangent plane to the surface at $P_1(x_1,y_1,z_1)$ is found by Eq. (21) or (72) to be

$$F_x(x_1,y_1,z_1)(x - x_1) + F_y(x_1,y_1,z_1)(y - y_1) + F_z(x_1,y_1,z_1)(z - z_1) = 0. \tag{123}$$

If the equation of the surface is given in the form

$$z = f(x,y), \tag{124}$$

we may put $F(x,y,z) = z - f(x,y)$ and write the equation of the tangent plane at $P_1(x_1,y_1,z_1)$ in the form

$$z - z_1 = \left.\frac{\partial f}{\partial x}\right|_1 (x - x_1) + \left.\frac{\partial f}{\partial y}\right|_1 (y - y_1). \tag{125}$$

296. Cylinders. A surface generated by the points on a straight line which moves so as always to remain parallel to a fixed direction while intersecting a fixed curve is called a cylindrical surface, or simply a *cylinder*. The fixed curve is called the *directing curve*. The moving line is called the *generator*. In a particular position, the generator constitutes an *element* of the cylinder (Fig. 316).

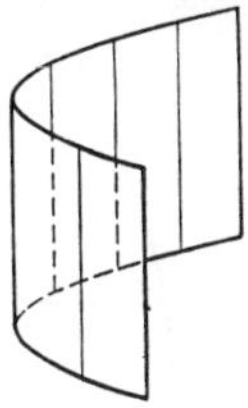

FIG. 316.

Suppose that we take the z axis parallel to the generators, and the intersection of the cylindrical surface with the xy plane as the directing curve. Regarded as a two-dimensional locus, this curve will have an equation of the form $F(x,y) = 0$. And the equation of the cylindrical surface will be this same equation, with z missing. This is because a point $P(x,y,z)$ on the cylinder with elements parallel to OZ is subject to no condition on z, but is subject to the same restrictions on x and y as the point $(x,y,0)$ which is the projection of P on the xy plane.

For example, the equation of a circle of radius 2 with center at the origin in the xy plane is

$$x^2 + y^2 = 4. \tag{126}$$

Hence, in three dimensions Eq. (126) is the equation of an indefinite right circular cylinder of radius 2, whose axis is OZ.

Similarly, $y^2 = 4x$ is the equation of an indefinite cylinder with elements parallel to OZ whose right section is a parabola.

Similar reasoning shows that any equation in x, y, and z with any one of these three variables missing is the equation of a cylinder whose generators are parallel to the coordinate axis corresponding to the missing variable. Thus $x^2 + z^2 = 4$ is the equation of an indefinite right circular cylinder of radius 2 whose axis is OY, and $z^2 = 4y$ is the equation of an indefinite cylinder with elements parallel to OX whose right section is a parabola.

Our definition makes a plane a special case of a cylindrical surface, whose directing curve may be taken as a straight line.

If the elements of a cylinder are not parallel to one of the coordinate axes, the equation will involve all three variables. For example, $(x - y)^2 + 2z^2$

= 8 is the equation of an indefinite right circular cylinder of radius 2 whose axis is the straight line along which $z = 0$ and $y = x$.

297. Surfaces of Revolution. A surface generated by the points on a curve which revolves about a fixed straight line is called a *surface of revolution*. The fixed straight line is called the axis of revolution, or simply the *axis*. Any plane section of the surface which contains the axis is called a *meridian section*. The sections perpendicular to the axis, the right sections, are circles.

Suppose that we take the z axis as the axis of revolution. The intersection of the surface with the yz plane, considered as a two-dimensional locus, will have an equation of the form $F(y,z) = 0$. We wish to find the equation satisfied by x, y, and z when $P(x,y,z)$ is any point on the surface (Fig. 317). Draw the right section through P. This will be a circle with center A and will contain a point $Q(0,y_1,z)$ on the yz meridian section. Then we have

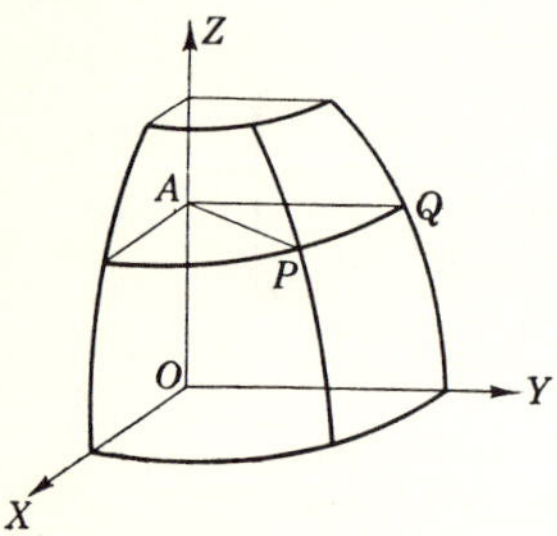

FIG. 317.

$$y_1 = AQ = AP = \sqrt{x^2 + y^2}. \quad (127)$$

But since Q is on the section with yz equation $F(y,z) = 0$, it follows that $F(y_1,z) = 0$, so that

$$F(\sqrt{x^2 + y^2}, z) = 0. \quad (128)$$

This is the equation of the surface of revolution whose axis is OZ and whose yz meridian section has the equation $F(y,z) = 0$.

By setting $y = 0$ in Eq. (128), we find that the xz meridian section has the equation $F(x,z) = 0$. And we might have derived Eq. (128) by starting with this fact.

Similar reasoning applies if the surface of revolution has OX or OY as its axis.

EXAMPLE 1. Find the equation of the surface generated by revolving the parabola $y^2 = 4z$, $x = 0$, about the z axis.

Solution: From $y_1^2 = 4z$ and $y_1 = \sqrt{x^2 + y^2}$, we find

$$x^2 + y^2 = 4z$$

as the required equation.

EXAMPLE 2. Find the equation of the surface generated by revolving the parabola $y^2 = 4z$, $x = 0$ about the line parallel to the y axis, $x = 0$, $z = -1$.

Solution: Here the distance of any point from the axis is $\sqrt{x^2 + (z+1)^2}$, which reduces to $z + 1$ when $x = 0$. Hence $z_1 + 1 = \sqrt{x^2 + (z+1)^2}$, and $y^2 = 4z_1$. It follows that $y^2/4 + 1 = \sqrt{x^2 + (z+1)^2}$, so that the required equation is

$$(y^2 + 4)^2 = 16x^2 + 16(z + 1)^2.$$

EXAMPLE 3. Find the equation of the indefinite conical surface whose vertex is at the origin, and each of whose straight line elements makes an angle α with the z axis.

Solution: The surface will have as its yz meridian section a straight line whose equation is $y = z \tan \alpha$, $x = 0$. And from $y_1 = z \tan \alpha$ and $y_1 = \sqrt{x^2 + y^2}$, we find

$$x^2 + y^2 = z^2 \tan^2 \alpha$$

as the required equation.

EXAMPLE 4. Find the equation of the tangent plane to the surface of Example 1 at the point where $x = 2$ and $y = 4$.

Solution: Since $z = \frac{1}{4}(x^2 + y^2)$, $z = \frac{1}{4}(2^2 + 4^2) = \frac{20}{4} = 5$ when $x = 2$, $y = 4$, so that the point on the surface is (2,4,5).

And if $F(x,y,z) = x^2 + y^2 - 4z$, we find that $F_x = 2x$, $F_y = 2y$, $F_z = -4$, so that at (2,4,5) we have $F_x = 4$, $F_y = 8$, $F_z = -4$.

Using these values in Eq. (123) we obtain

$$4(x - 2) + 8(y - 4) - 4(z - 5) = 0, \qquad \text{or} \qquad x + 2y - z = 5$$

as the required equation.

EXERCISE 151

For each of the following cylindrical surfaces draw a two-dimensional graph of the right section with axes correctly labeled, and then make a three-dimensional sketch of the surface.

1. $(x - 2)^2 = 9$.
2. $y^2 + z^2 = 2z$.
3. $4z^2 + 9x^2 = 36$.
4. $4y^2 - 9x^2 = 36$.
5. $z^2 = y + 2$.
6. $x^2 = 2z - 4$.

For each of the following surfaces of revolution draw a two-dimensional graph of one meridian section with the axis of revolution indicated, and then make a three-dimensional sketch of the surface.

7. $x^2 + y^2 - 4z^2 = 0$.
8. $y^2 + z^2 - x^2 = 0$.
9. $x^2 + y^2 + 4z^2 = 4$.
10. $x^2 + 4y^2 + 4z^2 = 4$.
11. $x^2 + y^2 - 8z = 0$.
12. $x^2 + y^2 - z^2 = 1$.
13. $(x - 2)^2 + y^2 = z$.
14. $z^2 + (y - 3)^2 = x$.

Find the equation of the tangent plane to the surface of the indicated problem at the given point, after verifying that the point does lie on the surface.

15. Prob. 2, (1,2,0).
16. Prob. 5, (4,7,3).
17. Prob. 6, (2,3,4).
18. Prob. 8, (5,3,4).
19. Prob. 11, (2,2,1).
20. Prob. 12, (2,1,2).
21. Prob. 13, (3,1,2).
22. Prob. 14, (8,5,2).

298. Quadric Surfaces. The general equation of the second degree in x, y, and z has the form

$$Ax^2 + By^2 + Cz^2 + Dyz + Exz + Fxy + Gx + Hy + Iz + J = 0. \qquad (129)$$

The locus of any such equation is called a *quadric surface.* All the plane sections of such a surface are curves of the type discussed in Sec. 86. We proceed to describe some general types of quadric surface.

Type 1. The equation of an *ellipsoid* is

$$\frac{x^2}{a^2} + \frac{y^2}{b^2} + \frac{z^2}{c^2} = 1. \qquad (130)$$

As indicated in Fig. 318, this surface cuts each of the coordinate planes in an ellipse. The surface may be visualized as obtained from a sphere

$$x^2 + y^2 + z^2 = a^2 \tag{131}$$

by shortening all y coordinates in the ratio b/a and all z coordinates in the ratio c/a when $c < b < a$.

We call a,b,c the *semiaxes* of the ellipsoid given by Eq. (130). They are equal to the intercepts on the coordinate axes, $a = OA$, $b = OB$, $c = OC$, from O to the *vertices* $A = (a,0,0)$, $B = (0,b,0)$, $C = (0,0,c)$. If the principal sections having as one quadrant AB, AC, and BC are drawn,

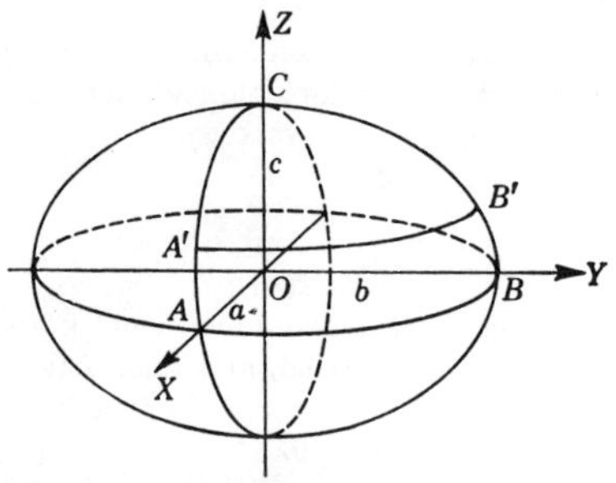

FIG. 318.

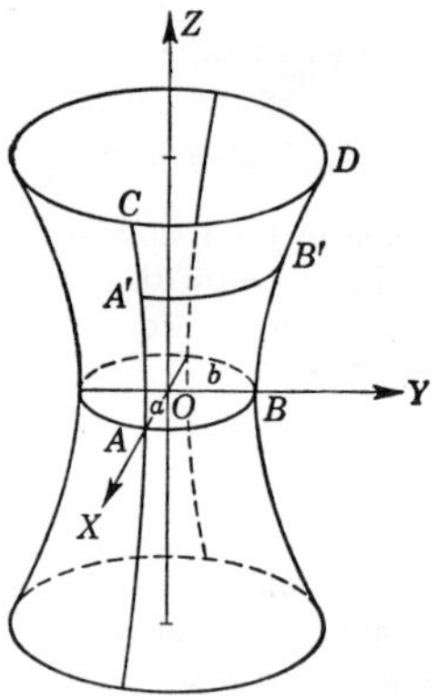

FIG. 319.

one octant of the ellipsoid may be generated by an arc similar in shape to AB and moving in a plane parallel to the xy plane with one end A' on AC and the other end B' on BC.

If two of the semiaxes are equal, the ellipsoid is called a *spheroid.* Thus a spheroid is a surface of revolution obtained by revolving an ellipse about one of its axes. Compare Probs. 9 and 10 of Exercise 151.

If the semiaxes are all equal, $b = a$ and $c = a$, the ellipsoid is the sphere given by Eq. (131).

Type 2. The equation of the *hyperboloid* of *one sheet* is

$$\frac{x^2}{a^2} + \frac{y^2}{b^2} - \frac{z^2}{c^2} = 1. \tag{132}$$

As indicated in Fig. 319, this surface cuts the xy plane in an ellipse, and the yz and zx planes in hyperbolas. One octant of the surface may be generated by a quadrant of an ellipse similar in shape to AB, moving in a plane parallel to the xy plane, with its ends A' and B' on the hyperbolic branches AC and BD.

If $a = b$, the surface given by Eq. (132) becomes a *hyperboloid of revolution* generated by revolving a hyperbola about its conjugate axis.

Type 3. The equation of the *hyperboloid of two sheets* is

$$\frac{x^2}{a^2} - \frac{y^2}{b^2} - \frac{z^2}{c^2} = 1. \tag{133}$$

As indicated in Fig. 320, this surface consists of two separate pieces, or sheets, cutting the xy and zx planes in corresponding branches of hyperbolas. One octant of the surface may be generated by a quadrant of an ellipse which moves parallel to the yz plane, remaining similar to itself with its ends B' and C' on the hyperbolic branches AD and AE.

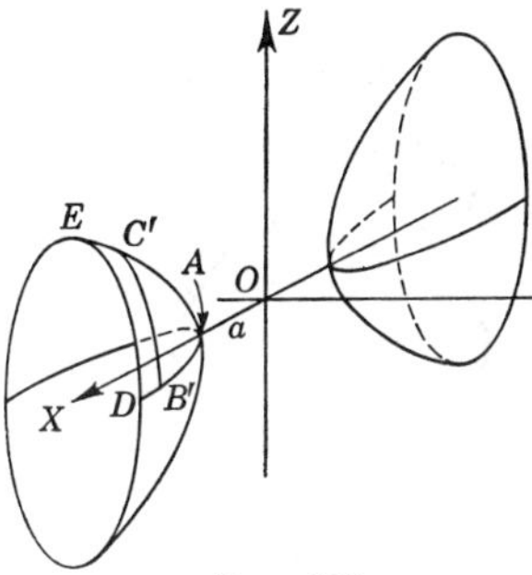

FIG. 320.

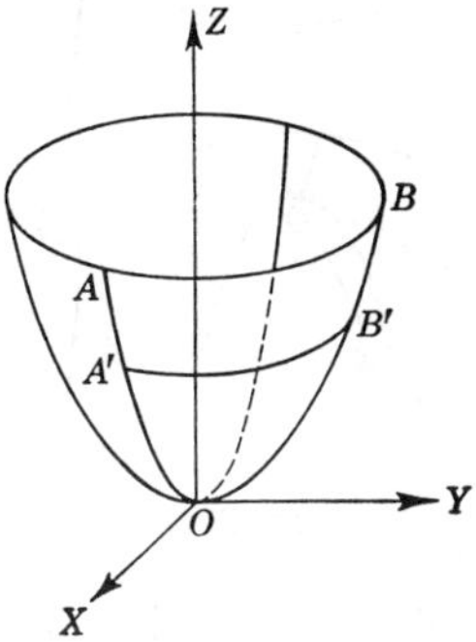

FIG. 321.

If $b = c$, the surface given by Eq. (133) becomes a *hyperboloid of revolution* generated by revolving a hyperbola about its transverse axis.

Type 4. The equation of the *elliptic paraboloid* is

$$\frac{x^2}{a^2} + \frac{y^2}{b^2} = z. \tag{134}$$

As indicated in Fig. 321, this surface cuts the yz and zx planes in parabolas. One quarter of the surface may be generated by a quadrant of an ellipse which moves parallel to the xy plane, remaining similar to itself with its ends A' and B' on the parabolic arcs OA and OB.

If $a = b$, the surface given by Eq. (134) becomes a *paraboloid of revolution* generated by revolving a parabola about its axes.

Type 5. The equation of the *hyperbolic paraboloid* is

$$\frac{x^2}{a^2} - \frac{y^2}{b^2} = z. \tag{135}$$

As indicated in Fig. 322, this surface cuts the xy plane in two straight lines, the zx plane in a parabola opening upward, and the yz plane in a parabola opening downward. The surface may be generated by a hyperbola, moving in a plane parallel to the xy plane, with its vertices on the principal parabolic sections, and its asymptotes parallel to the lines in which the surface cuts the xy plane. The portion of the surface near the origin has the general shape of a saddle.

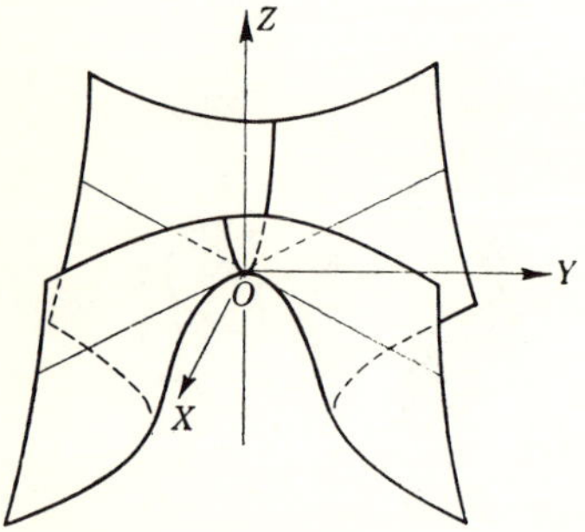

FIG. 322.

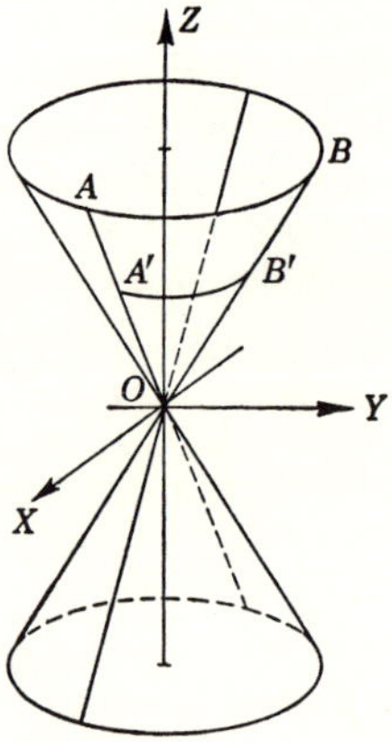

FIG. 323.

Type 6. The equation of the *elliptic cone* is

$$\frac{x^2}{a^2} + \frac{y^2}{b^2} = \frac{z^2}{c^2}. \tag{136}$$

As indicated in Fig. 323, this surface cuts the yz and zx planes in pairs of straight lines. One octant of the surface may be generated by a quadrant of an ellipse which moves parallel to the xy plane, remaining similar to itself with its ends A' and B' on the straight lines OA and OB.

We may also think of the surface as made up of all the straight lines which join the vertex of the cone O to any point on the ellipse having AB as one quadrant.

When $b = a$, the surface given by Eq. (136) becomes a *cone of revolution* like that of Example 3 of Sec. 297.

Type 7. The equation of a *quadric cylinder* with axis parallel to OZ has the form

$$Ax^2 + Bxy + Cy^2 + Dx + Ey + F = 0. \tag{137}$$

The locus will be an elliptic cylinder, parabolic cylinder, or hyperbolic cylinder whenever the corresponding plane locus of Eq. (94) of Sec. 86 is an ellipse, parabola, or hyperbola.

If the plane locus degenerates into two straight lines, one point, or no locus, the corresponding space locus will be two planes, a straight line, or no locus.

Summary. In addition to the degenerate cases of type 7, we may have a locus of just one point. For example, the origin is the only real point satisfying the equation

$$\frac{x^2}{a^2} + \frac{y^2}{b^2} + \frac{z^2}{c^2} = 0. \tag{138}$$

Except for the choice of coordinate axes, every Eq. (129) will have a locus like one of those mentioned in this section. Hence any curved surface represented by an equation of the second degree must be one of the seven types described above.

EXERCISE 152

For each of the following quadric surfaces make a three-dimensional sketch of the part in the first octant.

1. $\frac{x^2}{25} + \frac{y^2}{16} + \frac{z^2}{9} = 1.$
2. $\frac{x^2}{9} + \frac{y^2}{16} - \frac{z^2}{25} = 1.$
3. $\frac{x^2}{25} - \frac{y^2}{9} - \frac{z^2}{16} = 1.$
4. $\frac{x^2}{4} + \frac{y^2}{9} = z.$
5. $\frac{x^2}{9} - \frac{y^2}{4} = z.$
6. $\frac{x^2}{9} + \frac{y^2}{4} = z^2.$
7. $\frac{x^2}{9} + \frac{y^2}{4} = 1.$
8. $\frac{x^2}{9} - \frac{y^2}{4} = 1.$
9. $4x^2 + 9y^2 + 36z^2 = 36.$
10. $4x^2 + 9y^2 - 36z^2 = 36.$
11. $4x^2 - y^2 = 4z.$
12. $x^2 + 4y^2 = 4z.$

Find the equation of the tangent plane to the surface of the indicated problem at the given point, after verifying that the point does lie on the surface.

13. Prob. 2, (3,4,5).
14. Prob. 4, (4,3,5).
15. Prob. 5, (6,2,3).
16. Prob. 6, (9,8,5).
17. Prob. 10, (3,2,1).
18. Prob. 12, (2,1,2).

If (x_1,y_1,z_1) is any point on the surface in question, show that the equation of the tangent plane to the surface at (x_1,y_1,z_1) is

19. $\frac{x_1x}{a^2} + \frac{y_1y}{b^2} + \frac{z_1z}{c^2} = 1$ for the ellipsoid of Eq. (130).
20. $\frac{x_1x}{a^2} + \frac{y_1y}{b^2} = \frac{z_1 + z}{2}$ for the elliptic paraboloid of Eq. (134).
21. $\frac{x_1x}{a^2} + \frac{y_1y}{b^2} = \frac{z_1z}{c^2}$ for the cone of Eq. (136).

***299. Directional Derivative. Gradient.** Let $U(x,y,z)$ be any function of x,y,z. We may consider it as a potential function, determining a value of the potential U at each point of space. And we may pass an equipotential surface

$$U(x,y,z) = c \tag{139}$$

through any given point $P = (x,y,z)$. We define the vector grad U, read "the *gradient* of U," by the equation

$$\text{grad } U = \frac{\partial U}{\partial x}\mathbf{i} + \frac{\partial U}{\partial y}\mathbf{j} + \frac{\partial U}{\partial z}\mathbf{k}. \tag{140}$$

By Sec. 295, this vector is normal to the surface of Eq. (139) at P. Thus the gradient of the potential function at any point is normal to the equipotential surface through the point.

Let C be a curve in space like that of Sec. 293 which passes through P. And let s be the arc length measured along C. Then on the curve C, the coordinates x,y,z are functions of s, and hence U is a function of s. From Eq. (30) of Sec. 273, we have

$$\frac{dU}{ds} = \frac{\partial U}{\partial x}\frac{dx}{ds} + \frac{\partial U}{\partial y}\frac{dy}{ds} + \frac{\partial U}{\partial z}\frac{dz}{ds}. \tag{141}$$

If we think of a point moving along C through P, dU/ds is the rate of change of U with respect to the distance moved. We call dU/ds the *directional derivative* of U along C.

By Eq. (59), the right member of Eq. (141) may be considered as the scalar product of the vector

$$\mathbf{t} = \frac{dx}{ds}\mathbf{i} + \frac{dy}{ds}\mathbf{j} + \frac{dz}{ds}\mathbf{k} \tag{142}$$

and the vector grad U of Eq. (140). That is,

$$\frac{dU}{ds} = \mathbf{t} \cdot (\text{grad } U). \tag{143}$$

By Sec. 293, the vector $\mathbf{t}$ of Eq. (142) is directed along the tangent to the curve C at P, in the direction of increasing s. And its length is unity, since Eq. (106) or $ds^2 = dx^2 + dy^2 + dz^2$ implies that

$$\left(\frac{dx}{ds}\right)^2 + \left(\frac{dy}{ds}\right)^2 + \left(\frac{dz}{ds}\right)^2 = 1. \tag{144}$$

This shows that dx/ds, dy/ds, dz/ds are direction cosines for the tangent vector $\mathbf{t}$, or any vector $K\mathbf{t}$ with K a positive scalar constant. Any set of direction ratios a,b,c determines a vector

$$\mathbf{D} = a\mathbf{i} + b\mathbf{j} + c\mathbf{k}, \tag{145}$$

whose direction cosines may be found as in Sec. 282. For a curve C_D with tangent $\mathbf{t}$ at P in the direction of $\mathbf{D}$, we shall have

$$\frac{dx}{ds} = \frac{a}{\sqrt{a^2+b^2+c^2}}, \qquad \frac{dy}{ds} = \frac{b}{\sqrt{a^2+b^2+c^2}}, \qquad \frac{dz}{ds} = \frac{c}{\sqrt{a^2+b^2+c^2}}. \tag{146}$$

From Eqs. (141) and (146), we find that

$$\frac{dU}{ds} = \frac{1}{\sqrt{a^2+b^2+c^2}}\left(a\frac{\partial U}{\partial x} + b\frac{\partial U}{\partial y} + c\frac{\partial U}{\partial z}\right). \tag{147}$$

This shows that dU/ds depends only on the three first partial derivatives of U at P and on the direction of the tangent to the curve at P. For this reason we sometimes call dU/ds the derivative of $U(x,y,z)$ in the assigned direction. And when this direction is given by the direction ratios a,b,c, or as that of the vector of Eq. (145), the directional derivative may be found from Eq. (147).

Let θ be the angle between the gradient and the assigned direction, or that of the vector **t**. And let us put

$$|\text{grad } U| = \sqrt{\left(\frac{\partial U}{\partial x}\right)^2 + \left(\frac{\partial U}{\partial y}\right)^2 + \left(\frac{\partial U}{\partial z}\right)^2}. \tag{148}$$

Since **t** is of length 1, we find from Eqs. (143), (46), and (47) that

$$\frac{dU}{ds} = |\text{grad } U| \cos \theta = \text{Proj}_{\mathbf{t}} (\text{grad } U). \tag{149}$$

This shows how du/ds depends on the direction of the tangent to C at P, since it is equal to the projection of the gradient on this tangent. The projection is a maximum when $\cos \theta = 1$, and $\theta = 0$. That is, the vector grad U extends in the direction in which the derivative of U is a maximum, and the magnitude of the gradient, $|\text{grad } U|$ of Eq. (148), is equal to that maximum derivative.†

For a potential function $U(x,y)$ and a curve in the xy plane, Eq. (141) reduces to

$$\frac{dU}{ds} = \frac{\partial U}{\partial x}\frac{dx}{ds} + \frac{\partial U}{\partial y}\frac{dy}{ds}. \tag{150}$$

And if the plane direction is given by a slope m, we may use $1,m,0$ or $-1,-m,0$ in place of the a,b,c of Eq. (147) and so deduce that

$$\frac{dU}{ds} = \frac{1}{\pm\sqrt{1+m^2}}\left(\frac{\partial U}{\partial x} + m\frac{\partial U}{\partial y}\right), \tag{151}$$

where the sign before the radical is that of the cosine of the slope angle.

EXAMPLE 1. Find the derivative of $U = x^2 + 2yz + z^2 - x$ in the direction of $2\mathbf{i} + 2\mathbf{j} + \mathbf{k}$ at the point (1,2,2). Also find the direction at this point where the derivative is a maximum, and this maximum value.

Solution: $\partial U/\partial x = 2x - 1$, $\partial U/\partial y = 2z$, $\partial U/\partial z = 2y + 2z$. Hence at (1,2,2) $\partial U/\partial x = 1$, $\partial U/\partial y = 4$, $\partial U/\partial z = 8$ and grad $U = \mathbf{i} + 4\mathbf{j} + 8\mathbf{k}$. And in the direction $\mathbf{D} = 2\mathbf{i} + 2\mathbf{j} + \mathbf{k}$, the derivative is

$$\frac{dU}{ds} = \frac{\mathbf{D}\cdot(\text{grad } U)}{|\mathbf{D}|} = \frac{2\cdot 1 + 2\cdot 4 + 1\cdot 8}{\sqrt{2^2+2^2+1^2}} = \frac{18}{\sqrt{9}} = 6.$$

The derivative will be a maximum in the direction of grad U, or for $\mathbf{D} = \mathbf{i} + 4\mathbf{j} + 8\mathbf{k}$. And the maximum value will be

$$|\text{grad } U| = \sqrt{1^2+4^2+8^2} = \sqrt{81} = 9.$$

EXAMPLE 2. Find the derivative of $U = e^{-y}\sin x$ in the direction in the xy plane making an angle α with the positive x axis at (x,y).

Solution: $\partial U/\partial x = e^{-y}\cos x$, $\partial U/\partial y = -e^{-y}\sin x$. And the direction with $m = \tan\alpha$ has $dx/ds = \cos\alpha$ and $dy/ds = \sin\alpha$. Hence in this case

$$\frac{dU}{ds} = e^{-y}\cos x\cos\alpha - e^{-y}\sin x\sin\alpha = e^{-y}\cos(x+\alpha).$$

† The reader interested in additional applications involving vectors and partial differentiation may be referred to Chap. 8 of the author's "Methods of Advanced Calculus," McGraw-Hill Book Company, Inc., New York, 1944 (Dover reprint).

EXERCISE 153

Find the directional derivative of each given function $U(x,y,z)$ in the direction of the given vector at the given point.

1. xyz, $\mathbf{i} + \mathbf{j} + \mathbf{k}$, $(2,2,2)$.
2. $x^2 + 2y^2 - z^2$, $2\mathbf{i} - 3\mathbf{j} - 3\mathbf{k}$, $(2,1,2)$.
3. $x^2 + y^2 + z^2$, $x\mathbf{i} + y\mathbf{j} + z\mathbf{k}$, (x,y,z).
4. $\dfrac{1}{x^2 + y^2 + z^2}$, $z\mathbf{j} - y\mathbf{k}$, (x,y,z).

For each given function $U(x,y)$, find the derivative in the direction in the xy plane making the given angle with the positive x axis at the given point.

5. $\tan^{-1}(y/x)$, $30°$, $(1,2)$.
6. $\ln\sqrt{x^2 + y^2}$, $45°$, $(2,3)$.
7. $\dfrac{x}{x^2 + y^2}$, $90°$, $(2,0)$.
8. $\dfrac{1}{\sqrt{x^2 + y^2}}$, $45°$, $(2,2)$.
9. $e^x \cos y$, α, (x,y).
10. $e^{-x} \sin y$, α, (x,y).

If $U(x,y)$ is a function in a plane, show that the directional derivative dU/ds is

11. Greatest for the direction $\dfrac{\partial U}{\partial x}\mathbf{i} + \dfrac{\partial U}{\partial y}\mathbf{j}$, with slope $\dfrac{\partial U/\partial y}{\partial U/\partial x}$.
12. Least for the direction $-\dfrac{\partial U}{\partial x}\mathbf{i} - \dfrac{\partial U}{\partial y}\mathbf{j}$, with slope $\dfrac{\partial U/\partial y}{\partial U/\partial x}$.
13. Zero for the directions $\dfrac{\partial U}{\partial x}\mathbf{i} - \dfrac{\partial U}{\partial y}\mathbf{j}$ and $-\dfrac{\partial U}{\partial y}\mathbf{i} + \dfrac{\partial U}{\partial x}\mathbf{j}$ with slope $-\dfrac{\partial U/\partial x}{\partial U/\partial y}$, perpendicular to the directions of Probs. 11 and 12.
14. Check Prob. 13 by showing that the vectors of that problem are tangent to the equipotential curve $U(x,y) = c$.

CHAPTER 19

MULTIPLE INTEGRALS

The definite integral of $f(x)$ was shown to equal the limit of a summation related to an interval on a line in Chap. 12. We used this fact in Chaps. 12 and 14 to determine a number of geometrical and physical quantities by obtaining an expression for the desired quantity as a definite integral. A double integral of $f(x,y)$ can be defined which is the limit of a summation related to a two-dimensional region in a plane. Such integrals may be computed by first integrating with respect to y, keeping x constant, and then integrating with respect to x. Many problems involving quantities related to an area may be solved most efficiently by expressing the quantity as a double integral. Similarly problems involving volumes can be solved by using triple integrals, which are evaluated by three repeated integrations.

In this chapter we define double and triple integrals and develop their basic properties. We show that they are the limits of sums and may be calculated by repeated integrations. In the plane we consider double integrals expressed in terms of rectangular coordinates and polar coordinates. And in space we consider triple integrals expressed in terms of rectangle coordinates, cylindrical polar coordinates, and spherical polar coordinates. We apply these multiple integrals to a number of problems, including those of the type discussed in Chap. 14.

300. Repeated Integrals. A function of x and y may be integrated with respect to y, keeping x constant. If the limits are functions of x only, a definite integral with respect to y will equal a function of x. For example,

$$\begin{aligned}\int_x^{2x} (x^2 - 4xy + 6y^2)dy &= [x^2y - 2y^2 + 2y^3]_{y=x}^{y=2x} \\ &= (2x^3 - 8x^3 + 16x^3) - (x^3 - 2x^3 + 2x^3) \\ &= 9x^3. \qquad (1)\end{aligned}$$

The resulting function of x may be integrated with respect to x between constant limits to give a number. For instance

$$\int_0^2 9x^3 = \tfrac{9}{4}[x^4]_0^2 = \tfrac{9}{4}(16 - 0) = 36. \qquad (2)$$

The successive integrations may be indicated in one expression as

$$\int_0^2 \left[\int_x^{2x} (x^2 - 4xy + 6y^2)dy \right] dx \quad \text{or} \quad \int_0^2 dx \int_x^{2x} (x^2 - 4xy + 6y^2)dy. \tag{3}$$

Usually we leave out the brackets in the first expression. Thus we may indicate the result of Eqs. (1) and (2) by writing

$$\int_0^2 \cdot \int_x^{2x} (x^2 - 4xy + 6y^2)dy\, dx = 36. \tag{4}$$

This is a special illustration of the general expression

$$\int^b \int_{u_1}^{u_2} f(x,y)dy\, dx = \int_a^b dx \int_{u_1}^{u_2} f(x,y)dy, \tag{5}$$

in which a and b are constants and u_1 and u_2 are either constants or functions of x. Either symbol of Eq. (5) is called a *repeated integral*, or an iterated integral. It indicates that successive integrations are to be carried out as follows.

First, regarding x as a constant, we evaluate the definite integral

$$\int_{u_1}^{u_2} f(x,y)dy \tag{6}$$

as in Eq. (1). The result will be a function of x which we denote by $F(x)$.

Next, as in Eq. (2), we evaluate the ordinary definite integral

$$\int_a^b F(x)dx. \tag{7}$$

Similarly, a repeated integral may be denoted by the symbols

$$\int_c^d \int_{v_1}^{v_2} f(x,y)dx\, dy = \int_c^d dy \int_{v_1}^{v_2} f(x,y)dx, \tag{8}$$

where c and d are constants and v_1 and v_2 are either constants or functions of y. Either symbol of Eq. (8) indicates a first integration with respect to x, keeping y constant, between the limits v_1 and v_2 to be followed by a second integration with respect to y between the limits c and d. That is, if

$$\int_{v_1}^{v_2} f(x,y)dx = G(y), \qquad \int_c^d \int_{v_1}^{v_2} f(x,y)dx\, dy = \int^d G(y)dy. \tag{9}$$

EXAMPLE 1. Find the value of $\displaystyle\int_1^2 \int_x^{\sqrt{3}x} \frac{1}{x^2 + y^2}\, dy\, dx.$

Solution: The first, or inner, integral is

$$\int_x^{\sqrt{3}x} \frac{dy}{x^2 + y^2} = \frac{1}{x}\left[\tan^{-1}\frac{y}{x}\right]_{y=x}^{y=\sqrt{3}x} = \frac{1}{x}(\tan^{-1}\sqrt{3} - \tan^{-1} 1)$$

$$= \frac{1}{x}\left(\frac{\pi}{3} - \frac{\pi}{4}\right) = \frac{\pi}{12}\frac{1}{x}.$$

The second integral is

$$\int_1^2 \frac{\pi}{12}\frac{1}{x}\,dx = \frac{\pi}{12}[\ln x]_1^2 = \frac{\pi}{12}(\ln 2 - \ln 1) = \frac{\pi}{12}\ln 2, \qquad \text{the required value.}$$

EXAMPLE 2. Find the value of $\int_0^1 \int_{\sqrt{y}}^{2\sqrt{y}} x\,e^{x^2-y}\,dx\,dy.$

Solution: The first, or inner integral is

$$\int_{\sqrt{y}}^{2\sqrt{y}} x\,e^{x^2-y}\,dx = \tfrac{1}{2}\Big[e^{x^2-y}\Big]_{x=\sqrt{y}}^{x=2\sqrt{y}} = \tfrac{1}{2}(e^{3y} - 1).$$

The second integral is

$$\begin{aligned}\int_0^1 \tfrac{1}{2}(e^{3y} - 1)dy &= \tfrac{1}{2}[\tfrac{1}{3}e^{3y} - y]_0^1 = \tfrac{1}{2}[(\tfrac{1}{3}e^3 - 1) - (\tfrac{1}{3} - 0)] \\ &= \tfrac{1}{6}e^3 - \tfrac{2}{3},\end{aligned}$$

which is the required value.

EXERCISE 154

Find the value of each of the following repeated integrals.

1. $\int_1^2 \int_x^{x^2} xy^2\,dy\,dx.$
2. $\int_0^1 \int_1^{1/x} \frac{x}{y^2}\,dy\,dx.$
3. $\int_1^3 \int_x^{2x} \frac{1}{(x+y)^2}\,dy\,dx.$
4. $\int_1^3 \int_1^{y^2} \frac{y}{x^2}\,dx\,dy.$
5. $\int_1^2 \int_0^{1/x} xy\,dy\,dx.$
6. $\int_0^2 \int_0^{y^2} x\,dx\,dy.$
7. $\int_2^4 \int_y^{3y} (y - x)^2\,dx\,dy.$
8. $\int_0^\pi \int_0^{\sin y} dx\,dy.$
9. $\int_0^2 \int_0^{3-y} y^2\,dx\,dy.$
10. $\int_0^1 \int_0^{1-x} y^2\,dy\,dx.$
11. $\int_0^1 \int_0^y e^{x-y}\,dx\,dy.$
12. $\int_0^1 \int_x^{2x} e^{2x-y}\,dy\,dx.$
13. $\int_0^1 \int_0^{x^2} e^{y/x}\,dy\,dx.$
14. $\int_0^3 \int^{e^y} \frac{y}{x}\,dx\,dy.$
15. $\int_1^{\pi/2} \int_0^{x^2} \sin\frac{y}{x}\,dy\,dx.$
16. $\int_{\pi/2}^{\pi} \int_0^{y^3} \cos\frac{x}{y}\,dx\,dy.$

301. Double Integrals. Let $f(x,y)$ be a function of x and y. For any point $P = (x,y)$, we define the symbol $f(P)$ by the equation

$$f(P) = f(x,y). \tag{10}$$

Then for any two-dimensional region R of the xy plane we may make the following construction. Subdivide R into n small subregions. These may be of any shape but must completely fill R. We let d_n denote the length of the largest segment having its end points on the boundary of any one of the small subregions (Fig. 324). Let ΔR_i denote the ith subregion or its area. Select some point P_i in each subregion and form the sum

$$S_n = \sum_{i=1}^{n} f(P_i)\Delta R_i. \tag{11}$$

Then for simple types of regions R, subregions ΔR_i, and functions $f(x,y)$, any sequence of sums S_n such that $d_n \to 0$ as $n \to \infty$ will approach a limit. This limit is called the *double integral* of $f(P)$, or $f(x,y)$, over R. And we write

$$\int_R f(P)dR = \lim_{n\to\infty} \sum_{i=0}^{n} f(P_i)\Delta R_i \qquad \text{if } \lim_{n\to\infty} d_n = 0. \tag{12}$$

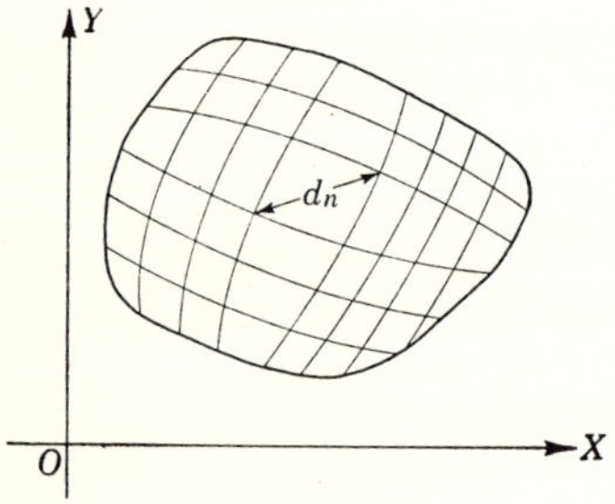

FIG. 324.

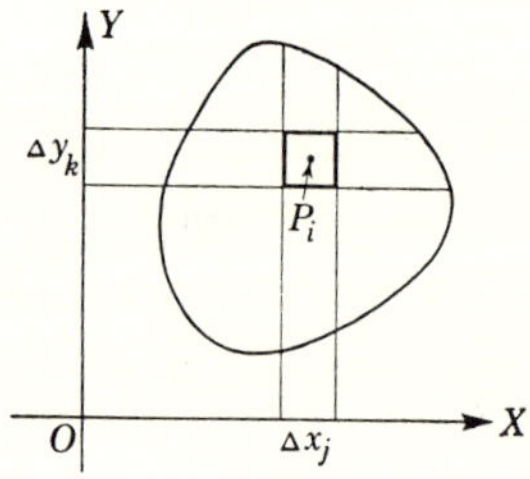

FIG. 325.

302. The Fundamental Theorem for Double Integrals. Consider the double integral over a region R, like that in Fig. 325, which is such that every straight line through an interior point and parallel to the x or y axis cuts the boundary in exactly two points. Let the subdivisions ΔR_i of Eq. (11) be rectangles formed by drawing lines parallel to the coordinate axes. Then we shall have for the area of the ith subdivision $\Delta R_i = \Delta x_j \, \Delta y_k$. For some rectangles cut by the boundary of R, only a part is a subregion of R. However, we still use the whole rectangle, as this simplifies the discussion and does not change the value of the limit in Eq. (12). We may now select x_j in Δx_j and y_k in Δy_k, and take $P_i = (x_j,y_k)$. Then

$$\begin{aligned}\sum f(P_i)\Delta R_i &= \sum_{j,k} f(x_j,y_k)\Delta x_j \, \Delta y_k \\ &= \sum_j \Delta x_j \left[\sum_k f(x_j,y_k)\Delta y_k\right] \\ &= \sum_k \Delta y_k \left[\sum_j f(x_j,y_k)\Delta x_j\right].\end{aligned} \tag{13}$$

In view of Eq. (13) of Sec. 170 and of Eq. (12), this suggests that

$$\int_R f(P)dR = \iint_R f(x,y)dy\,dx = \int_a^b dx \int_{u_1}^{u_2} f(x,y)dy = \int_c^d dy \int_{v_1}^{v_2} f(x,y)dx. \tag{14}$$

The functions and constants used as limits are so chosen that the region R is made up of those points (Fig. 326) for which

$$a < x < b \quad \text{and} \quad u_1(x) < y(x) < u_2(x), \tag{15}$$

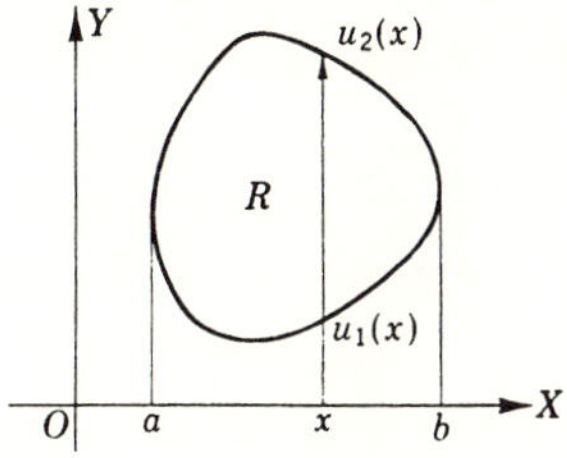

FIG. 326.

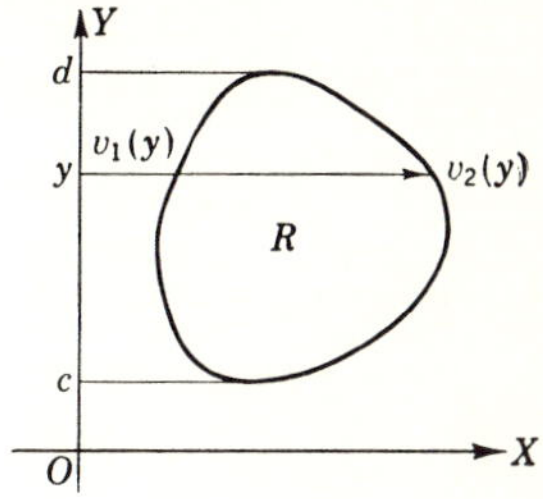

FIG. 327.

or of those points (Fig. 327) for which

$$c < y < d \quad \text{and} \quad v_1(y) < x(y) < v_2(y). \tag{16}$$

In particular, if the region R is Q, a rectangle consisting of those points (Fig. 328) for which

$$a < x < b \quad \text{and} \quad c < y < d, \tag{17}$$

the relation of Eq. (14) reduces to

$$\iint\limits_Q f(x,y)dy\,dx = \int_a^b dx \int_c^d f(x,y)dy = \int_c^d dy \int_a^b f(x,y)dx. \tag{18}$$

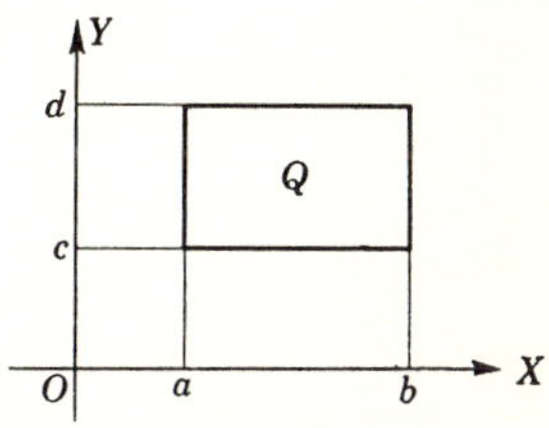

FIG. 328.

The relation (14) always holds when $f(x,y)$ is continuous in both variables. Together, Eqs. (14) and (12) express the *fundamental theorem* of the integral calculus for double integrals.† The result is also true for Duhamel modifications like those described in Sec. 185.

When $f(x,y)$ has points, or curves of discontinuity, or where some of the single integrals have infinite limits, some or all of the expressions in Eq. (14) may be meaningless. However, if either of the repeated integrals on the right involves only convergent improper integrals, it will give the value of the geometric or physical quantity represented by the double integral on the left.

The repeated integrals in Eq. (14), when thought of as expressions for the double integral, are often themselves referred to as double integrals.

Suppose that the region R of a double integral is such that some lines

† Compare the author's "A Treatise on Advanced Calculus," p. 362, John Wiley & Sons, Inc., New York, 1940 (Dover reprint).

through interior points cut the boundary in more than two points. In some cases we may still use one of the repeated integrals of Eq. (14). Thus the first one, whose limits are given by Eq. (15), is valid for the

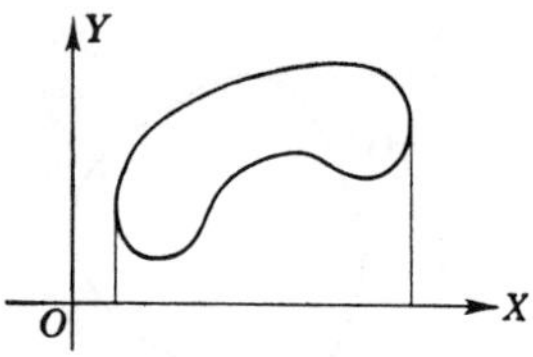

FIG. 329.

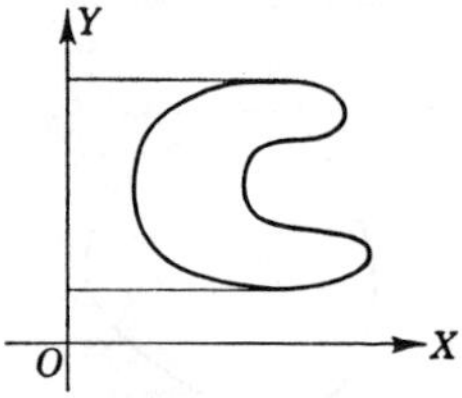

FIG. 330.

region of Fig. 329. And the second one, whose limits are given by Eq. (16), is valid for the region of Fig. 330. In other cases (Fig. 331) it is usually possible to decompose the region into parts for each of which one of the repeated integrals is valid.

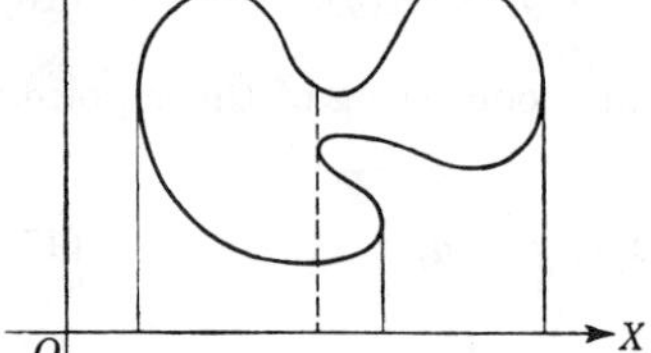

FIG. 331.

303. Area as a Double Integral in Rectangular Coordinates. Let $f(x,y) = 1$. Then $f(P_i) = 1$. And the sum $\Sigma f(P_i)\Delta R_i$ of Sec. 302 reduces to $\Sigma\, \Delta R_i$. By the convention made about boundary points, this sum $\Sigma\, \Delta R_i$ is the area of all rectangular subdivisions of the plane which lie wholly or partly in R. Hence $\Sigma\, \Delta R_i$ exceeds the area of the region R only by the parts of certain rectangles which include boundary points. For ordinary boundaries, this excess approaches zero when $d_n \to 0$. It follows from Eqs. (12) and (14) that

$$A = \text{area of } R = \int_R dR = \int dy \int dx = \int dx \int dy. \tag{19}$$

Thus *the area of any region is the value of the double integral of the function* $f(x,y) = 1$ *taken over that region.* We express this fact and the content of Eq. (19) by writing

$$A = \int dA \qquad dA = dx\, dy. \tag{20}$$

The integration may be taken in either order if limits appropriate to R are used.

Let us apply this to the area bounded above by the curve $y_2 = f_2(x)$ and below by the curve $y_1 = f_1(x)$, and lying between a left-hand ordinate at $x = a$ and a right-hand ordinate at $x = b$ (Fig. 332). Since y varies from $f_1(x)$ to $f_2(x)$ for any x between a and b, by Eq. (15) we have in this case

$$A = \int_a^b \int_{f_1(x)}^{f_2(x)} dy\, dx = \int_a^b dx \int_{f_1(x)}^{f_2(x)} dy. \tag{21}$$

The inner integral is

$$\int_{f_1(x)}^{f_2(x)} dy = [y]_{f_1(x)}^{f_2(x)} = f_2(x) - f_1(x). \tag{22}$$

It follows from Eqs. (22) and (21) that

$$A = \int_a^b [f_2(x) - f_1(x)]dx. \tag{23}$$

This agrees with Eq. (1) of Sec. 204.

Let us next apply Eq. (20) to the area bounded on the right by the curve $x = g_2(y)$, on the left by the curve $x = g_1(y)$, and lying above the

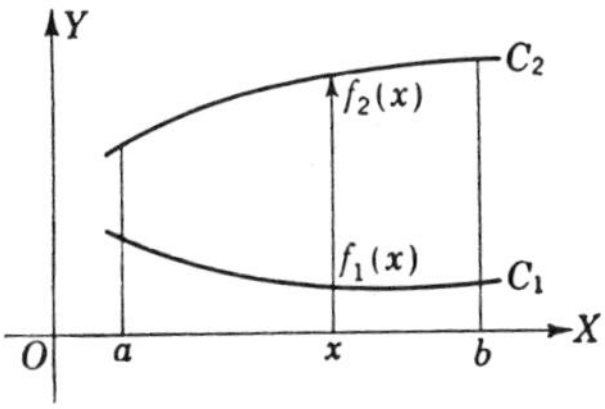

FIG. 332.

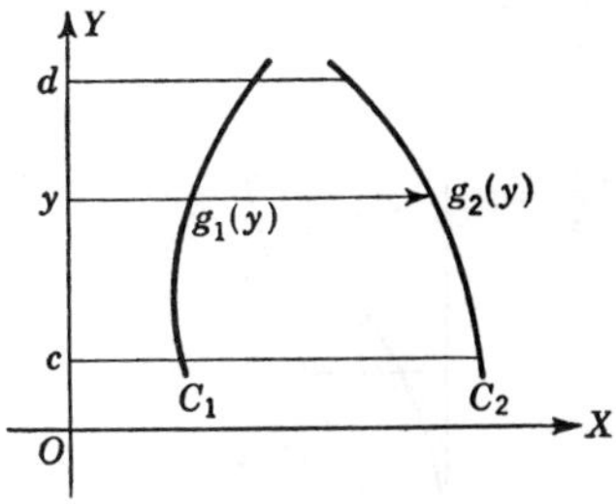

FIG. 333.

line $y = c$ and below the line $y = d$ (Fig. 333). Since x varies from $g_1(y)$ to $g_2(y)$ for any y between c and d, by Eq. (16) we have in this case

$$A = \int_c^d \int_{g_1(y)}^{g_2(y)} dx\, dy = \int_c^d dy \int_{g_1(y)}^{g_2(y)} dx. \tag{24}$$

The inner integral is

$$\int_{g_1(y)}^{g_2(y)} dx = [x]_{g_1(y)}^{g_2(y)} = g_2(y) - g_1(y). \tag{25}$$

It follows from Eqs. (25) and (24) that

$$A = \int_c^d [g_2(y) - g_1(y)]dy. \tag{26}$$

This is the result we would have obtained directly from a figure in setting up the area as a single integral.

By using Eq. (21) or (24), or the arguments which lead to them, we may express any area as a double integral. And by reversing the process, we may sketch the area over which a double integral is taken. The calculation of the area by a double integral is no more convenient than the calculation by a single integral, and in fact involves one more process. However the setting up of double integrals which represent areas gives valuable practice in the determination of the limits. And this determination is necessary for other applications of double integration.

EXAMPLE 1. Use double integration to find the area between the two parabolas $y^2 = 4x$ and $2x^2 = y$.

Solution: From $2x^2 = y$, we find $y = 2x^2$, $y^2 = 4x^4$, so that if $y^2 = 4x$, $4x = 4x^4$, $x(x^3 - 1) = 0$ and $x = 0$ or 1. When $x = 0$, $y = 2x^2 = 0$, and when $x = 1$, $y = 2x^2 = 2$. Thus the curves intersect at (0,0) and (1,2). We sketch them in Fig. 334. Since a point moving upward in a vertical line which cuts the area enters on $y = 2x^2$ and leaves on $y^2 = 4x$ or $y = 2\sqrt{x}$, the area is $A = \int_0^1 dx \int_{2x^2}^{2\sqrt{x}} dy$. The inner integral is $\int_{2x^2}^{2\sqrt{x}} dy = \Big[y\Big]_{2x^2}^{2\sqrt{x}} = 2\sqrt{x} - 2x^2$. And the outer integral is $\int_0^1 (2\sqrt{x} - 2x^2)dx = [\frac{4}{3}x^{\frac{3}{2}} - \frac{2}{3}x^3]_0^1 = \frac{2}{3}$.

Suppose we use the other order of integration. Since a point moving to the right

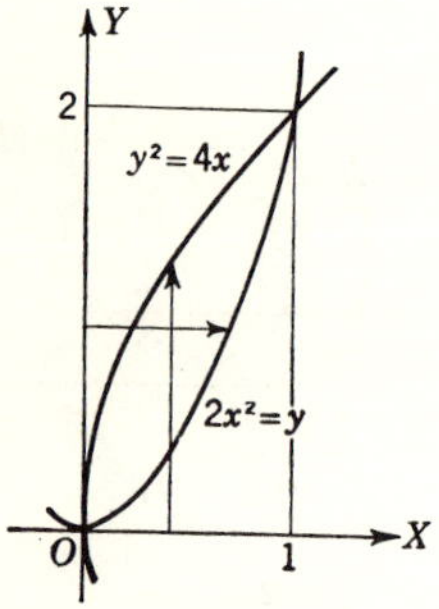

FIG. 334.

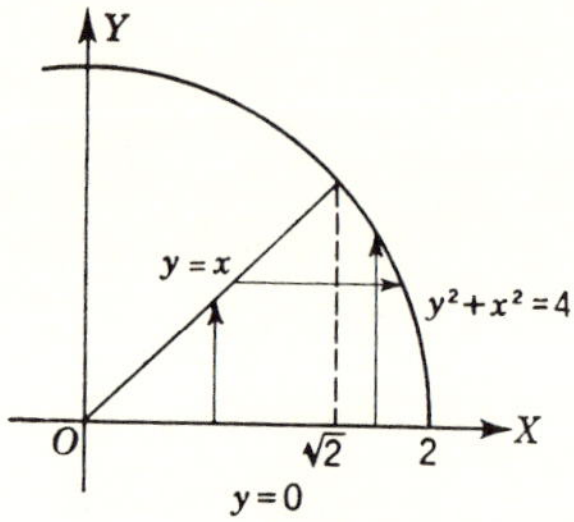

FIG. 335.

in a horizontal line which cuts the area enters on $y^2 = 4x$ or $x = y^2/4$ and leaves on $2x^2 = y$ or $x = \sqrt{y/2}$, the area is $A = \int_0^2 dy \int_{y^2/4}^{\sqrt{y/2}} dx$. The inner integral is $\int_{y^2/4}^{\sqrt{y/2}} dx = \Big[x\Big]_{y^2/4}^{\sqrt{y/2}} = \sqrt{\frac{y}{2}} - \frac{y^2}{4}$. And the outer integral is $\int_0^2 \left(\sqrt{\frac{y}{2}} - \frac{y^2}{4}\right) dy = \left[\frac{1}{\sqrt{2}}\frac{2}{3}y^{\frac{3}{2}} - \frac{y^3}{12}\right]_0^2 = \frac{4}{3} - \frac{2}{3} = \frac{2}{3}$, as before.

EXAMPLE 2. Use double integration to find the area in the first quadrant bounded by the lines $y = 0$, $y = x$, and the circle $x^2 + y^2 = 4$.

Solution: The area is shown in Fig. 335. Here vertical lines enter on $y = 0$. They leave on $y = x$ for $x < \sqrt{2}$. But they leave on $x^2 + y^2 = 4$ or $y = \sqrt{4 - x^2}$ for $x > \sqrt{2}$. Hence if we integrate with respect to y first, we must use two integrals and write $A = \int_0^{\sqrt{2}} dx \int_0^x dy + \int_{\sqrt{2}}^2 dx \int_0^{\sqrt{4-x^2}} dy$. We have $\int_0^x dy = [y]^x = x$, $\int_0^{\sqrt{2}} x\,dx = \left[\frac{x^2}{2}\right]_0^{\sqrt{2}} = 1$, the value of the first double integral. And $\int_0^{\sqrt{4-x^2}} dy = \Big[y\Big]_0^{\sqrt{4-x^2}} = \sqrt{4 - x^2}$, $\int_{\sqrt{2}}^2 \sqrt{4 - x^2}\,dx$, with $x = 2\sin t$, $= \int_{\pi/4}^{\pi/2} 4\cos^2 t\,dt = \Big[2t + \sin 2t\Big]_{\pi/4}^{\pi/2} = \frac{\pi}{2} - 1$, the value of the second double integral. Hence $A = 1 + \left(\frac{\pi}{2} - 1\right) = \frac{\pi}{2}$.

Suppose we use the other order of integration. Horizontal lines enter on $y = x$ or $x = y$ and leave on $x^2 + y^2 = 4$ or $x = \sqrt{4 - y^2}$. Thus $A = \int_0^{\sqrt{2}} dy \int_y^{\sqrt{4-y^2}} dx$. We have $\int_y^{\sqrt{4-y^2}} dx = \Big[x \Big]_y^{\sqrt{4-y^2}} = \sqrt{4 - y^2} - y$. And $A = \int_0^{\sqrt{2}} (\sqrt{4 - y^2} - y)dy$. With $y = 2 \sin t$, $\int_0^{\sqrt{2}} \sqrt{4 - y^2}\, dy = \int_0^{\pi/4} 4 \cos^2 t\, dt = \Big[2t + \sin\ 2t \Big]_0^{\pi/4} = \frac{\pi}{2} + 1$. And $-\int_0^{\sqrt{2}} y\, dy = -\Big[\frac{y^2}{2}\Big]_0^{\sqrt{2}} = -1$. Hence $A = \left(\frac{\pi}{2} + 1\right) - 1 = \frac{\pi}{2}$, as before.

EXAMPLE 3. Sketch the area which leads to the double integral $\int_1^3 dx \int^{2x} dy$, and set up the integration in the reverse order.

Solution: The area is bounded below by the straight line $y = 2$, and above by the straight line $y = 2x$, as we see from the limits for y on the inner integral. And the area lies between $x = 1$ and $x = 3$, as we see from the limits for x on the outer integral. Hence it is the triangle shown in Fig. 336. A horizontal line enters on $y = 2x$ or $x = y/2$, and leaves on $x = 3$, so that the integration in the reverse order is

$$\int_2^6 dy \int_{y/2}^3 dx.$$

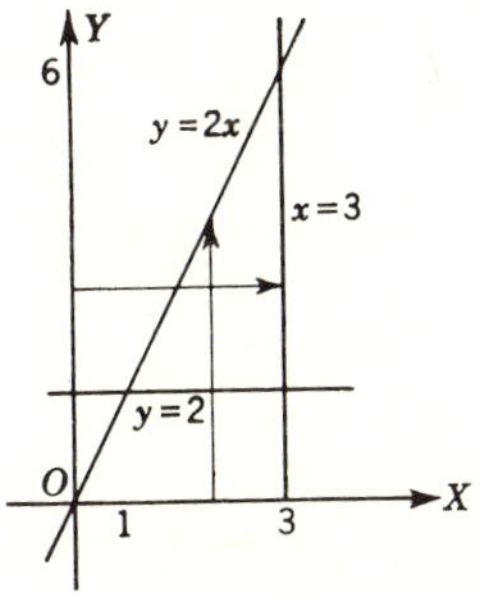

FIG. 336.

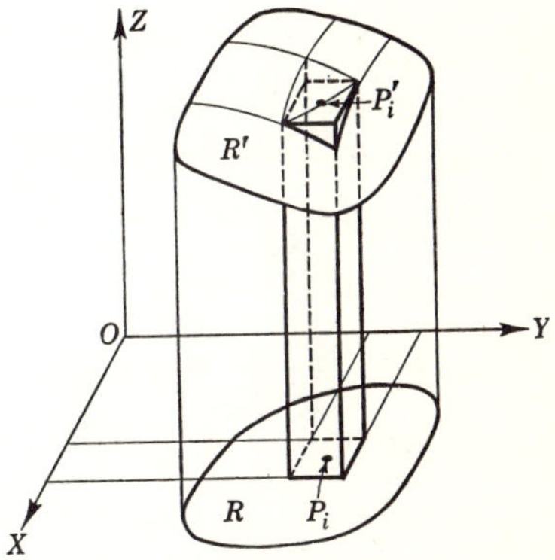

FIG. 337.

304. Volume as a Double Integral in Rectangular Coordinates. Let R be a region of the xy plane of one of the types considered in Sec. 302. On the bounding curve of R construct a right cylinder or tubular surface whose elements are parallel to the z axis. Let $f(x,y)$ be a function which is continuous, positive, and single-valued for x,y in R. Then $z = f(x,y)$ represents a surface. As indicated in Fig. 337, let R' be the portion of this surface cut out by the right cylinder. We wish to find the volume V of the solid inside the cylindrical surface bounded below by R and above by R'.

For a suitable value of $P_i = (x_i,y_i)$ in a subdivision ΔR_i wholly inside of

R, the portion of the solid over ΔR_i will have a volume equal to that of the rectangular column of base ΔR_i and height $f(P_i)$. Hence with any choice of points P_i for subdivisions which include boundary points, $\Sigma f(P_i)\Delta R_i$ will be an approximation to V. For ordinary boundaries, the error approaches zero when $d_n \to 0$. It follows from Eqs. (12) and (14) that

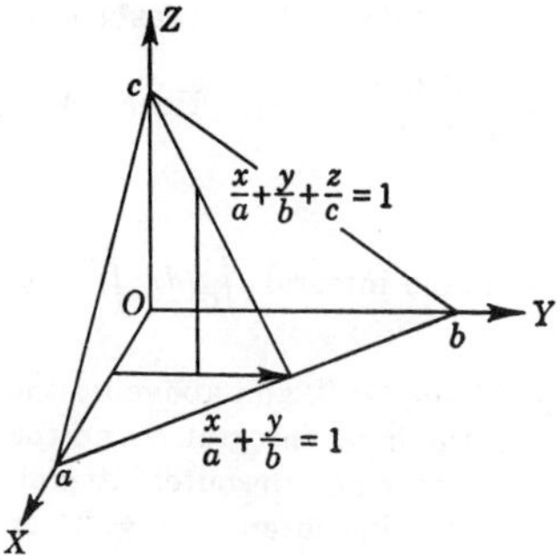

FIG. 338.

$$V = \text{volume over } R = \int_R f(P)dR$$
$$= \int dy \int f(x,y)dx = \int dx \int f(x,y)dy. \qquad (27)$$

We may express the essential content of Eq. (27) by writing

$$V = \int dV,$$
$$dV = z\,dx\,dy = f(x,y)dx\,dy. \qquad (28)$$

The limits on the repeated integrals in Eq. (27) are the same as those used in setting up a double integral for the area of R in Sec. 303.

EXAMPLE. Use double integration to find the volume bounded by the coordinate planes and the plane $\frac{x}{a} + \frac{y}{b} + \frac{z}{c} = 1$, where a, b, and c are each positive.

Solution: The volume is that of the pyramid shown in Fig. 338. Here the region R is the triangle bounded by $x = 0$, $y = 0$, and $\frac{x}{a} + \frac{y}{b} = 1$ obtained by putting $z = 0$ in the given equation. A line parallel to OY enters R on $y = 0$ and leaves on $\frac{x}{a} + \frac{y}{b} = 1$ or $y = b\left(1 - \frac{x}{a}\right)$. And from the given equation $z = c\left(1 - \frac{x}{a} - \frac{y}{b}\right)$. Hence from Eq. (27), $V = \int_0^a dx \int_0^{b(1-x/a)} c\left(1 - \frac{x}{a} - \frac{y}{b}\right) dy$. We have $\int_0^{b(1-x/a)} c\left(1 - \frac{x}{a} - \frac{y}{b}\right) dy = c\left[\left(1 - \frac{x}{a}\right)y - \frac{y^2}{2b}\right]_{y=0}^{y=b(1-x/a)} = \frac{bc}{2}\left(1 - \frac{x}{a}\right)^2$. And $\int_0^a \frac{bc}{2}\left(1 - \frac{x}{a}\right)^2 dx = -\frac{abc}{6}\left[\left(1 - \frac{x}{a}\right)^3\right]_0^a = \frac{abc}{6}$.

EXERCISE 155

Use double integration to find the area bounded by each of the following sets of curves and lines, and in the first quadrant.

1. $3y = x^2$, $3x = y^2$.
2. $y = x^2$, $y = x$.
3. $y^2 = x^3$, $y = x$.
4. $y = 0$, $x = 0$, $x^2 + y^2 = 4$.
5. $y = x^3$, $y = x$.
6. $x = 0$, $y = x$, $x^2 + y^2 = 4$.
7. $x = 0$, $y = 0$, $y = 4 - x^2$.
8. $x = 0$, $y = 0$, $x^2 + 4y^2 = 4$.

Sketch the area which leads to each of the following double integrations, and set up the integration in the reverse order.

9. $\int_0^2 dx \int_0^{\sqrt{4-x^2}} dy.$ 10. $\int_0^3 dy \int_0^{3-y} dx.$

11. $\int_1^2 dx \int_x^{2x} dy.$ 12. $\int_0^4 dy \int_y^4 dx.$

13. $\int_0^1 dx \int_{x^3}^{x} dy.$ 14. $\int_0^1 dy \int_y^{\sqrt{y}} dx.$

Use double integration to find the volume bounded below by the xy plane and above by each of the following surfaces.

15. $z = 1 - x^2 - y^2.$ 16. $36z = 72 - 9x^2 - 4y^2.$

Use double integration to find the volume over the square in the xy plane bounded by $x = 0$, $x = 1$, $y = 0$, $y = 1$ and under each of the following surfaces.

17. $z = 2x^2 + 6y^2.$ 18. $z = 4 + 8xy.$

Use double integration to find the volume bounded below by the xy plane, laterally by the cylinder $x^2 + y^2 = 1$, and above by each of the following surfaces.

19. $z = 3 - x.$ 20. $z = x^2 + y^2.$

305. Area as a Double Integral in Polar Coordinates. Let R be a plane region. And, as in Fig. 339, let a typical subdivision ΔR_i be the keystone-shaped figure bounded by two circles of radius r, $r + \Delta r$, and two straight lines θ, $\theta + \Delta\theta$. The exact area of R_i is $r_i' \, \Delta r \, \Delta\theta$, where $r_i' = r + \Delta r/2$ and so is the polar coordinate r for a point in ΔR_i. Treat subdivisions cut by the boundary of R as in Secs. 302 and 303. Then $\Sigma \, \Delta R_i$ is the area of all subdivisions of the plane which lie wholly or partly in R. Hence $\Sigma \, \Delta R_i = \Sigma r_i' \, \Delta r \, \Delta\theta$ exceeds the area of R only by the parts of certain subdivisions which include boundary points. For ordinary boundaries, this excess approaches zero when $d_n \to 0$. By the Duhamel modification of the fundamental theorem of Sec. 302, when $d_n \to 0$, $\Sigma r_i' \, \Delta r \, \Delta\theta$ approaches $\int\int r \, dr \, d\theta$ as a limit. It follows that

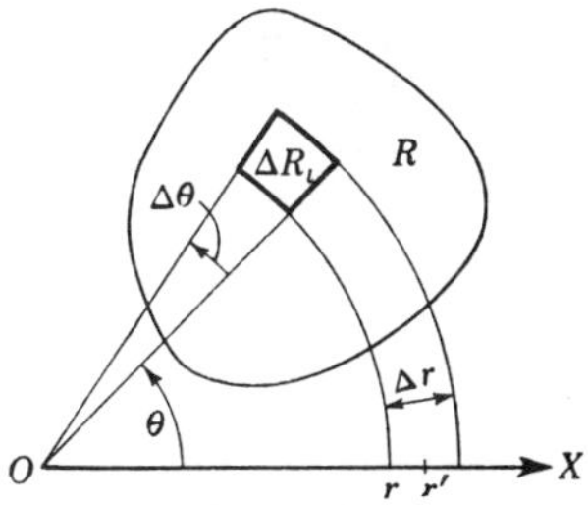

FIG. 339.

$$A = \text{area of } R = \int_R dR = \iint r \, dr \, d\theta = \int d\theta \int r \, dr. \tag{29}$$

We express the content of Eq. (29) by writing

$$A = \int dA, \qquad dA = r \, dr \, d\theta. \tag{30}$$

This may be recalled by thinking of dA as the area of a rectangle of dimensions dr and $r \, d\theta$.

Let us apply this to the area between the two radius vectors $\theta = \alpha$ and $\theta = \beta$, and the curves $r = f_1(\theta)$ and $r = f_2(\theta)$. As in Fig. 340, we assume that for $\alpha < \theta < \beta$, $f_1(\theta) < f_2(\theta)$. Since r varies from $f_1(\theta)$ to $f_2(\theta)$ for any θ between α and β, as in Eq. (15), we have in this case

$$A = \int_\alpha^\beta \int_{f_1(\theta)}^{f_2(\theta)} r\,dr\,d\theta = \int_\alpha^\beta d\theta \int_{f_1(\theta)}^{f_2(\theta)} r\,dr. \tag{31}$$

The inner integral is

$$\int_{f_1(\theta)}^{f_2(\theta)} r\,dr = \tfrac{1}{2}[r^2]_{f_1(\theta)}^{f_2(\theta)} = \tfrac{1}{2}\{[f_2(\theta)]^2 - [f_1(\theta)]^2\}. \tag{32}$$

It follows that

$$A = \tfrac{1}{2}\int \{[f_2(\theta)]^2 - [f_1(\theta)]^2\}d\theta. \tag{33}$$

With $f_1(\theta) = 0$ and $f_2(\theta) = f(\theta)$, this agrees with Eq. (2) of Sec. 204.

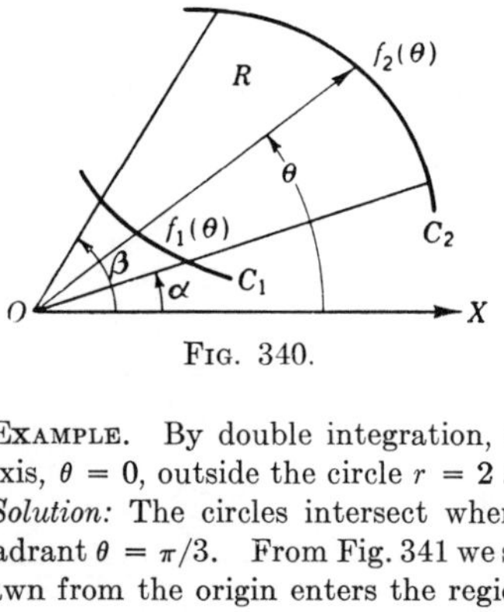

FIG. 340.

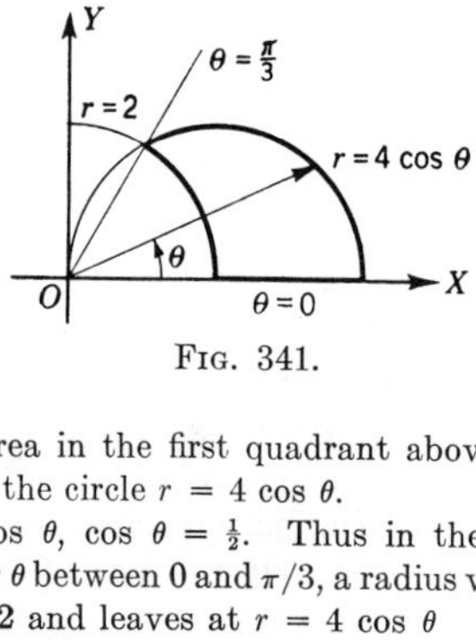

FIG. 341.

EXAMPLE. By double integration, find the area in the first quadrant above the x axis, $\theta = 0$, outside the circle $r = 2$ and inside the circle $r = 4\cos\theta$.

Solution: The circles intersect when $2 = 4\cos\theta$, $\cos\theta = \frac{1}{2}$. Thus in the first quadrant $\theta = \pi/3$. From Fig. 341 we see that for θ between 0 and $\pi/3$, a radius vector drawn from the origin enters the region at $r = 2$ and leaves at $r = 4\cos\theta$ Since $dA = r\,dr\,d\theta$, we have $A = \int_0^{\pi/3} d\theta \int_2^{4\cos\theta} r\,dr$. The inner integral is $\int_2^{4\cos\theta} r\,dr = \left[\frac{r^2}{2}\right]_2^{4\cos\theta} = 8\cos^2\theta - 2$. And $A = \int_0^{\pi/3} (8\cos^2\theta - 2)d\theta =$ $[4\theta + 2\sin 2\theta - 2\theta]_0^{\pi/3} = \frac{2\pi}{3} + \sqrt{3}$. This is the required area.

Suppose that we had been asked to find the area in the first quadrant inside both circles. Then we would have needed two integrals, since that area

$$A_1 = \int_0^{\pi/3} d\theta \int_0^2 r\,dr + \int_{\pi/3}^{\pi/2} d\theta \int_0^{4\cos\theta} r\,dr.$$

306. Cylindrical Coordinates. The location of a point in space by means of its three rectangular coordinates was described in Sec. 280. We shall now describe a second method of defining the location of a point P. We again start with a right-handed system of x, y, and z axes. Let P be any point in space. From P draw a straight line parallel to OZ until it meets the xy plane in a point F (Fig. 342). Then we retain the signed distance $FP = z$ as one coordinate. But in place of x and y, we locate F by its polar coordinates r, θ in the xy plane. The numbers (r,θ,z) are called the *cylindrical coordinates* of the point P. They are related to the rectangular coordinates (x,y,z) of P by Eqs. (1) to (4) of Sec. 145. In particular

$$x = r \cos \theta, \qquad y = r \sin \theta, \qquad x^2 + y^2 = r^2. \tag{34}$$

The discussion of Sec. 296 shows that an equation in r, θ, and z with z missing represents a cylindrical surface with elements parallel to OZ. In particular $r = a$ is a circular cylinder of radius a with OZ as its axis. $\theta = \alpha$ is a plane through OZ, a particular case of a cylinder. And the parabolic cylinder with equation $y^2 = 4x$ in rectangular coordinates has $r \sin^2 \theta = 4 \cos \theta$ as its equation in cylindrical coordinates by Eq. (34).

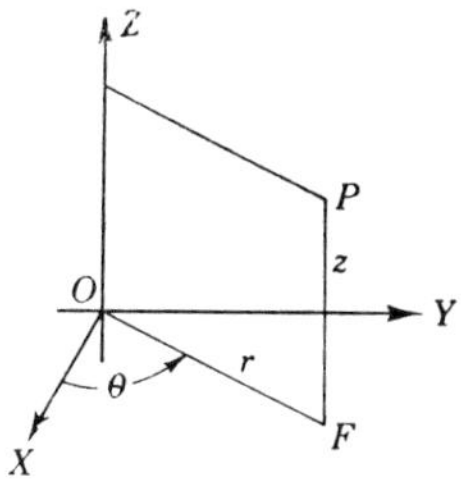

FIG. 342.

The discussion of Sec. 297 shows that an equation in r, θ, and z with θ missing represents a surface of revolution with OZ as its axis. The surface of revolution with $F(x,z) = 0$ as the equation of its meridian section in the xz plane, and $F(y,z) = 0$ as the equation of its meridian section in the yz plane has $F(\sqrt{x^2 + y^2}, z) = 0$ as its rectangular equation. Hence by Eq. (34) it has $F(r,z) = 0$ as its equation in cylindrical coordinates.

EXAMPLE 1. Find the cylindrical equation of the sphere of radius a above the xy plane and tangent to it at O.

Solution 1: The sphere is obtained by revolving the circle $x^2 + (z - a)^2 = a^2$, or $x^2 + z^2 - 2az = 0$ about the axis OZ. Replacing x by r gives $r^2 + z^2 - 2az = 0$ as the required equation.

Solution 2: The sphere has $x^2 + y^2 + (z - a)^2 = a^2$, or $x^2 + y^2 + z^2 - 2az = 0$ as its rectangular equation. Replacing $x^2 + y^2$ by r^2 gives $r^2 + z^2 - 2az = 0$ as the required equation.

EXAMPLE 2. Find the general equation of a plane in cylindrical coordinates.

Solution: In rectangular coordinates, $Ax + By + Cz = D$. From Eq. (34), in cylindrical coordinates we have $Ar \cos \theta + Br \sin \theta + Cz = 0$.

307. Volume as a Double Integral in Polar Coordinates. Let R be a region of the xy plane of the type considered in Sec. 305. On R as a base construct a right cylinder or tubular surface whose elements are parallel to OZ. Let $F(r,\theta)$ be a function which is continuous, positive, and single-valued for (r,θ) in R. Then $z = F(r,\theta)$ is the equation of a surface in cylindrical coordinates. As indicated in Fig. 343, let R' be the portion of this surface cut out by the right cylinder. We wish to find the volume V of the solid inside the cylindrical surface bounded below by R and above by R .

For a suitable value of $P_i = (r_i,\theta_i)$ in a subdivision ΔR_i wholly inside of R, the portion of the solid over ΔR_i will have a volume equal to that of the column of base $\Delta R_i = r_i' \,\Delta r\, \Delta\theta$ and height $F(r_i,\theta_i)$. Hence with any choice of points P_i for subdivisions which include boundary points,

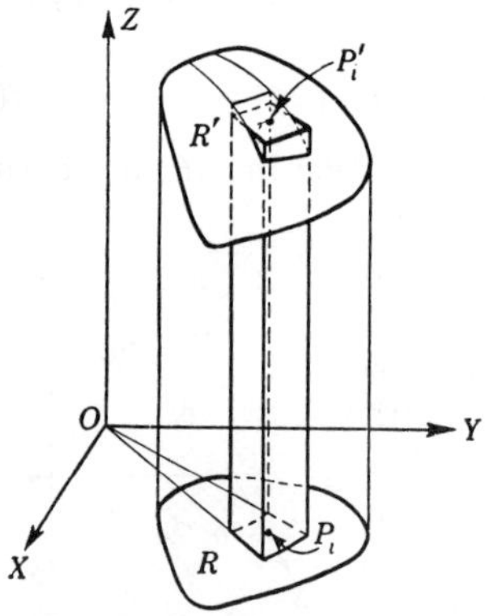

FIG. 343.

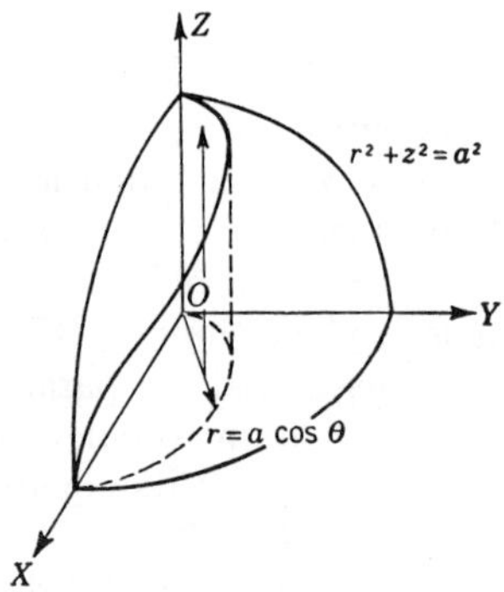

FIG. 344.

$\Sigma F(r_i,\theta_i)r_i' \,\Delta r\, \Delta\theta$ will be an approximation to V. For ordinary boundaries the error approaches zero when $d_n = 0$. It follows from the Duhamel modification of the fundamental theorem of Sec. 302 that

$$V = \text{volume over } R = \iint F(r,\theta)r\,dr\,d\theta = \int d\theta \int F(r,\theta)r\,dr. \tag{35}$$

We may express the essential content of Eq. (35) by writing

$$V = \int dV, \qquad dV = zr\,dr\,d\theta = F(r,\theta)r\,dr\,d\theta. \tag{36}$$

The limits on the repeated integral in Eq. (35) are the same as those used in setting up a double integral for the area of R in Sec. 305.

EXAMPLE 1. A sphere of radius a has its center on one element of a circular cylinder of radius $a/2$. Find the volume cut out of the sphere by the cylinder.

Solution: Let the circular cylinder have its elements parallel to OZ and cut the xy plane in a circle passing through the origin and having its center on the x axis (Fig. 344). By Sec. 146, Eq. (7), the equation of the circle, and hence of the cylinder, is $r = a\cos\theta$. If the sphere has its center at the origin, its rectangular equation is $x^2 + y^2 + z^2 = a^2$, so that its cylindrical equation is $r^2 + z^2 = a^2$. For the upper half of the sphere, $z = \sqrt{a^2 - r^2}$. The required volume is four times that in the first octant, which is over a semicircle. On this semicircle, for θ between 0 and $\pi/2$, a radius vector drawn from the origin enters R at $r = 0$, and leaves at $r = a\cos\theta$. Hence $V = 4\int_0^{\pi/2} d\theta \int_0^{a\cos\theta} \sqrt{a^2 - r^2}\, r\,dr$. The inner integral is

$$\int_0^{a\cos\theta} \sqrt{a^2 - r^2}\, r\,dr = -\frac{1}{3}\left[(a^2 - r^2)^{\frac{3}{2}}\right]_0^{a\cos\theta} = \frac{a^3}{3}(1 - \sin^3\theta).$$

Hence $V = \frac{4}{3}a^3 \int_0^{\pi/2} (1 - \sin^3\theta)d\theta = \frac{4}{3}a^3\left[\theta + \cos\theta - \frac{\cos^3\theta}{3}\right]_0^{\pi/2} =$ $\frac{4}{3}a^3\left(\frac{\pi}{2} - \frac{2}{3}\right) = \frac{2}{9}a^3(3\pi - 4).$

EXAMPLE 2. Sketch the area in the xy plane which leads to the double integration $\int_0^1 dx \int_x^{\sqrt{2x-x^2}} (x^2 + y^2)dy$. Then set up an equivalent integral in polar coordinates and evaluate it.

Solution: $y = \sqrt{2x - x^2}$ gives $y^2 = 2x - x^2$, $y^2 + (x - 1)^2 = 1$ which is the circle of Fig. 345. And $y = x$ is the line $\theta = \pi/4$. As a line parallel to OY enters the region at $y = x$ and leaves the region at $y = \sqrt{2x - x^2}$, for x between 0 and 1, the region R is the part of the circle above the line. From Eq. (34), $x^2 + y^2 = 2x$ becomes $r = 2 \cos \theta$. For θ between $\pi/4$ and $\pi/2$, a radius vector drawn from the origin enters R at $r = 0$ and leaves on the circle $r = 2 \cos \theta$. The integrand $x^2 + y^2 = r^2$, and the element of area $dx\ dy$ must be replaced by the element of area for polar coordinates, $r\ dr\ d\theta$. Hence the equivalent integral is

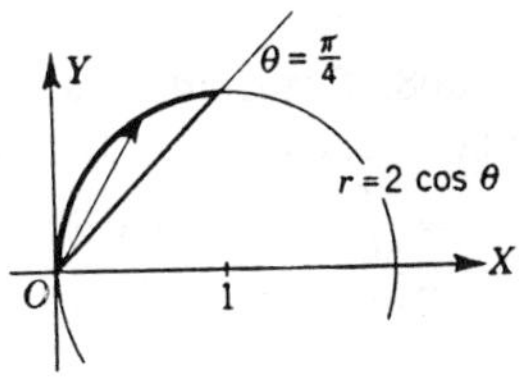

FIG. 345.

$$\iint_R r^2\, r\, dr\, d\theta = \int_{\pi/4}^{\pi/2} d\theta \int_0^{2\cos\theta} r^3\, dr.$$

The inner integral is $\int_0^{2\cos\theta} r^3\, dr = \left[\frac{r^4}{4}\right]_0^{2\cos\theta} = 4\cos^4\theta$. And $\int_{\pi/4}^{\pi/2} 4\cos^4\theta\, d\theta =$
$4\left[\frac{3}{8}\theta + \frac{\sin 2\theta}{4} + \frac{\sin 4\theta}{32}\right]_{\pi/4}^{\pi/2} = \frac{3}{8}\pi - 1.$

EXERCISE 156

Use double integration to find the area bounded by each of the following curves.

1. $r^2 = \sin 2\theta$.
2. $r = 4 \cos 3\theta$.
3. $r = 2 - \cos \theta$.
4. $r = 2 + \sin 3\theta$.

Use double integration to find the area in the first quadrant bounded by $\theta = 0$, $\theta = \pi/4$ and each of the following curves.

5. $r = 1 + \cos \theta$.
6. $r = \sec^2 \theta$.

7. Find the area inside $r = 2 \cos \theta$ and outside $r = \sqrt{2}$.
8. Find the area inside $r = 3 \cos \theta$ and outside $r = 1 + \cos \theta$.

Sketch the locus of each of the following equations.

9. The paraboloid of revolution, $z = \frac{r^2}{a^2}$.
10. The spheroid, $\frac{r^2}{a^2} + \frac{z^2}{c^2} = 1$.
11. The hyperboloid of revolution, $\frac{r^2}{a^2} - \frac{z^2}{c^2} = 1$.
12. The cone of revolution $\frac{r^2}{a^2} = \frac{z^2}{c^2}$.

Use double integration to find the volume bounded below by the plane $z = 0$ and above by each of the following surfaces.

13. $z = 4 - r^2$.
14. $z = 3 - r$.

Use double integration to find the volume inside the cylinder $r = 1$ bounded below by the plane $z = 0$ and above by each of the following surfaces.

15. $z^2 = 4 - r^2$.
16. $z = 4 + r^2$.
17. $z = 1 + r \cos \theta$.
18. $z = 2 + r^2 \sin 2\theta$.

Sketch the area in the xy plane which leads to each of the following double integrations. Then set up an equivalent integral in polar coordinates and evaluate it.

19. $\int_0^2 dx \int_0^{\sqrt{4-x^2}} dy.$

20. $\int_0^4 dy \int_0^{\sqrt{4y-y^2}} (x^2 + y^2)dx.$

21. $\int_0^1 dx \int_{x^2}^{x} \frac{1}{\sqrt{x^2 + y^2}} dy.$

22. $\int_0^{\infty} dx \int_0^{\infty} e^{-(x^2+y^2)} dy.$

308. Centroid of a Plane Area. Consider a thin plate overlaying a plane region R. Let ρ be the mass per unit area. Then the mass $M = \int_R \rho\, dR$. By Sec. 210, the moment about OY, M_x is approximated by $\Sigma x_i' \rho'' \Delta R_i$. The error approaches zero as $d_n \to 0$. It follows from Sec. 302 that $M_x = \int_R x\rho\, dR$. Thus the x coordinate of the center of gravity of the plate may be computed from

$$M = \int_R \rho\, dR, \qquad M_x = \int_R \rho x\, dR, \qquad \bar{x} = \frac{M_x}{M} = \frac{\int_R \rho x\, dR}{\int_R \rho\, dR}. \tag{37}$$

These apply for ρ any function of x and y. If ρ is constant, we may bring it out as a factor and cancel it in the last expression of Eq. (37). A similar relation holds for $\bar{y}$, so that

$$\bar{x} = \frac{\int_R x\, dR}{\int_R dR}, \qquad \bar{y} = \frac{\int_R y\, dR}{\int_R dR}. \tag{38}$$

These equations for the coordinates of the centroid of R resemble Eq. (26) of Sec. 211, but here the integrals are double integrals.

For rectangular coordinates, we replace dR by $dx\, dy$ or $dy\, dx$ and set

$$A = \iint dx\, dy, \qquad A\bar{x} = \iint x\, dx\, dy, \qquad A\bar{y} = \iint y\, dx\, dy. \tag{39}$$

These double integrals are to be evaluated as repeated integrals in either order with limits appropriate to R determined as in Sec. 303.

For polar coordinates, we replace dR by $r\, dr\, d\theta$, x by $r \cos \theta$, y by $r \sin \theta$. Thus

$$A = \iint r\, dr\, d\theta, \qquad A\bar{x} = \iint r^2 \cos\theta\, dr\, d\theta, \qquad A\bar{y} = \iint r^2 \sin\theta\, dr\, d\theta. \tag{40}$$

The limits in the equivalent repeated integrals appropriate to R are to be determined as in Sec. 305.

EXAMPLE 1. Find the centroid of the area above the x axis, bounded on the left by $y^2 = 4x$ and on the right by $y^2 = 5 - x$.

Solution: At the intersection $4x = 5 - x$, $5x = 5$, and $x = 1$. Also $y^2 = 4x = 4$, so that above the x axis, $y = 2$. From Fig. 346, for y between 0 and 2 a line parallel to OX enters the region at $y^2 = 4x$ or $x = y^2/4$ and leaves at $y^2 = 5 - x$ or $x = 5 - y^2$. Hence we have

$$A = \int_0^2 dy \int_{y^2/4}^{5-y^2} dx, \qquad A\bar{x} = \int_0^2 dy \int_{y^2/4}^{5-y^2} x\,dx, \qquad Ay = \int_0^2 dy \int_{y^2/4}^{5-y^2} y\,dx.$$

Hence $A = \int_0^2 (5 - \frac{5}{4}y^2)dy = [5y - \frac{5}{12}y^3]_0^2 = 10 - \frac{10}{3} = \frac{20}{3}$. $A\bar{x} = \frac{1}{2}\int_0^2 (25 - 10y^2 + \frac{15}{16}y^4)dy = \frac{1}{2}[25y - \frac{10}{3}y^3 + \frac{3}{16}y^5]_0^2 = 25 - \frac{40}{3} + 3 = \frac{44}{3}$. $A\bar{y} = \int_0^2 \left(5y - \frac{5}{4}y^3\right) dy = \left[\frac{5}{2}y^2 - \frac{5}{16}y^4\right]_0^2 = 10 - 5 = 5$. Hence $\bar{x} = \frac{A\bar{x}}{A} = \frac{\frac{44}{3}}{\frac{20}{3}} = \frac{11}{5}$. And $\bar{y} = \frac{A\bar{y}}{A} = \frac{5}{\frac{20}{3}} = \frac{3}{4}$. Thus the required centroid is $\left(\frac{11}{5}, \frac{3}{4}\right)$.

EXAMPLE 2. Show that the centroid of a sector of a circle of radius a with central angle B lies at a distance $4a/3B \sin (B/2)$ from the center of the circle on the radius bisecting the sector.

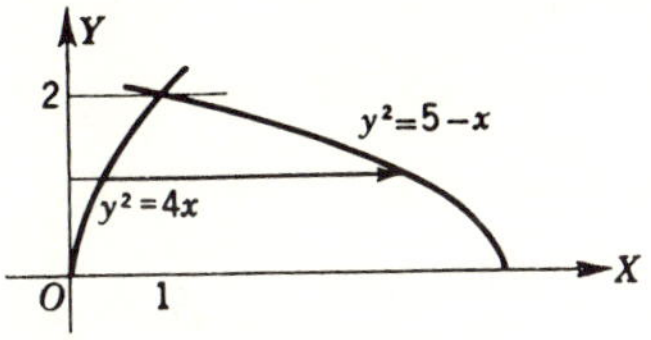

FIG. 346.

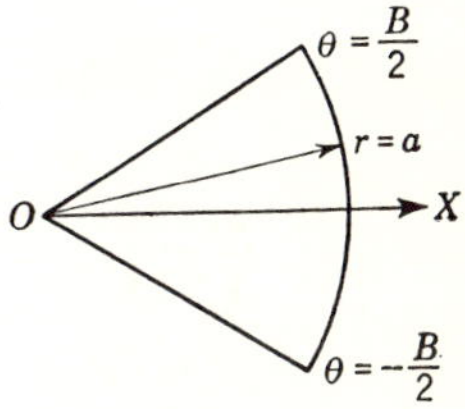

FIG. 347.

Solution: Take the sector as bounded by $r = a$ and lying between the lines $\theta = -\frac{B}{2}$ and $\theta = \frac{B}{2}$ (Fig. 347). Then from symmetry about $\theta = 0$, $\bar{y} = 0$. Also A and $A\bar{x}$ are each twice their values for the upper half between $\theta = 0$ and $\theta = \frac{B}{2}$. Hence $A = 2\int_0^{B/2} d\theta \int_0^a r\,dr = 2\left(\frac{B}{2}\right)\left(\frac{a^2}{2}\right) = \frac{1}{2}Ba^2$. Also $A\bar{x} = 2\int_0^{B/2} d\theta \int_0^a (r \cos \theta)r\,dr = 2\left(\sin\frac{B}{2}\right)\left(\frac{a^3}{3}\right) = \frac{2}{3}a^3 \sin \frac{B}{2}$. Hence $\bar{x} = \frac{A\bar{x}}{A} = \frac{\frac{2}{3}a^3 \sin (B/2)}{\frac{1}{2}Ba^2} = \frac{4a}{3B}\sin\frac{B}{2}$, as was to be proved.

309. Moment of Inertia of an Area. Let a plane region R be divided into subdivisions ΔR_i. And let r_i denote the shortest distance from some point in ΔR_i to a given axis. Then $\Sigma r_i^2\, \Delta R_i$ is an approximation to I, the moment of inertia of the area about the given axis by Secs. 219 and 221. And the error approaches zero as $d_n \to 0$. It follows from Sec. 302 that $I = \int_R r^2\, dR$. For I_y, the moment of inertia about OY or $x = 0$, $r = x$. For I_x, the moment of inertia about OX or $y = 0$, $r = y$. For I_0, the polar moment of inertia, $r^2 = x^2 + y^2$.

In rectangular coordinates, we replace dR by $dx\,dy$ or $dy\,dx$. Thus

$$I_y = \iint x^2\,dx\,dy, \qquad I_x = \iint y^2\,dx\,dy, \qquad I_0 = \iint (x^2 + y^2)dx\,dy. \quad (41)$$

These double integrals are to be evaluated as repeated integrals in either order with limits appropriate to R determined as in Sec. 303.

For polar coordinates, we replace dR by $r\,dr\,d\theta$, x by $r\cos\theta$, y by $r\sin\theta$. Thus

$$\begin{aligned} I_y &= \iint (r\cos\theta)^2\,r\,dr\,d\theta = \int d\theta \int \cos^2\theta\,r^3\,dr, \\ I_x &= \iint (r\sin\theta)^2\,r\,dr\,d\theta = \int d\theta \int \sin^2\theta\,r^3\,dr, \\ I_0 &= \iint r^2\,r\,dr\,d\theta = \int d\theta \int r^3\,dr. \end{aligned} \quad (42)$$

The limits in the repeated integrals are determined as in Sec. 305.

Any one of the moments of inertia I_x, I_y, or I_0 may be expressed in the form $I = Ak^2$. To do this we find k^2 from the relation

$$k^2 = \frac{I}{A}. \quad (43)$$

EXAMPLE 1. Find I_y for the area bounded by $y = x^2$ and $y = x$.

Solution: $I_y = \int_0^1 dx \int_{x^2}^{x} x^2\,dy$, where the limits are found from Fig. 348. Hence $I_y = \int_0^1 (x^3 - x^4)dx = \left[\frac{x^4}{4} - \frac{x^5}{5}\right]_0^1 = \frac{1}{20}$, as required.

If we wished to express this in the form Ak^2, we would compute $A = \int_0^1 dx \int_{x^2}^{x} dy$ $= \int_0^1 (x - x^2)dx = \left[\frac{x^2}{2} - \frac{x^3}{3}\right]_0^1 = \frac{1}{6}$. Then from Eq. (43), $k^2 = \frac{I}{A} = \frac{\frac{1}{20}}{\frac{1}{6}} = \frac{3}{10}$, and $I = \frac{3}{10}A$.

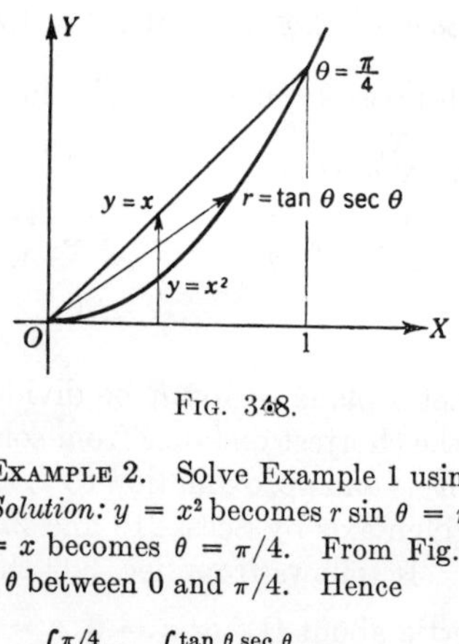

FIG. 348.

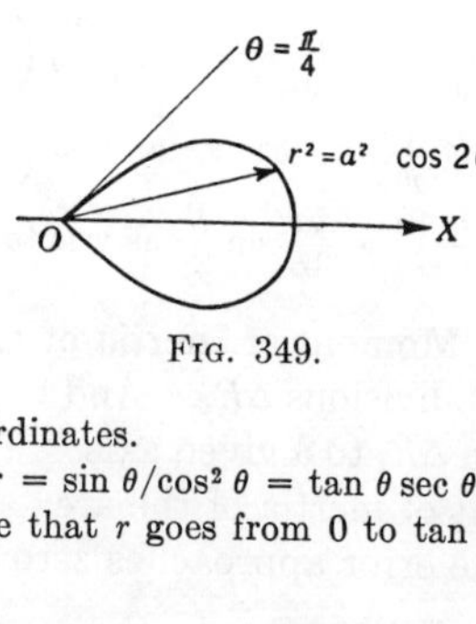

FIG. 349.

EXAMPLE 2. Solve Example 1 using polar coordinates.

Solution: $y = x^2$ becomes $r\sin\theta = r^2\cos^2\theta$, or $r = \sin\theta/\cos^2\theta = \tan\theta\sec\theta$. And $y = x$ becomes $\theta = \pi/4$. From Fig. 348, we see that r goes from 0 to $\tan\theta\sec\theta$ for θ between 0 and $\pi/4$. Hence

$$I_y = \int_0^{\pi/4} d\theta \int_0^{\tan\theta\sec\theta} \cos^2\theta\,r^3\,dr = \tfrac{1}{4}\int_0^{\pi/4} \tan^4\theta\sec^2\theta\,d\theta = \frac{1}{4}\left[\frac{\tan^5\theta}{5}\right]_0^{\pi/4} = \frac{1}{20}.$$

This is the required value.

EXAMPLE 3. Find the polar moment of inertia for the area of one loop of the lemniscate $r^2 = a^2 \cos 2\theta$.

Solution: $I_0 = 2\int_0^{\pi/4} d\theta \int_0^{a\sqrt{\cos 2\theta}} r^3\,dr$, where the limits are found from Fig. 349 for the upper half loop. Hence $I_0 = 2\int_0^{\pi/4} \frac{1}{4} a^4 \cos^2 2\theta\,d\theta = \frac{a^4}{2}\left[\frac{\theta}{2} + \frac{\sin 4\theta}{8}\right]_0^{\pi/4} = \frac{\pi a^4}{16}$, the required value.

If we wished to express this in the form Ak^2, we would compute

$$A = 2\int_0^{\pi/4} d\theta \int_0^{a\sqrt{\cos 2\theta}} r\,dr = 2\int_0^{\pi/4} \tfrac{1}{2}a^2 \cos 2\theta\,d\theta$$
$$= \frac{a^2}{2}[\sin 2\theta]_0^{\pi/4} = \frac{a^2}{2}.$$

Then $k^2 = \frac{I}{A} = \frac{\pi a^4/16}{a^2/2} = \frac{\pi a^2}{8}$ and $I = \frac{\pi a^2}{8} A$.

EXERCISE 157

Use double integration to find the centroid of the area bounded by each of the following sets of curves and lines.

1. $y = x^3$, $x = 0$, $y = 1$.
2. $2x + 3y = 6$, $y = 0$, $x = 0$.
3. $y = x^2$, $y = 1$.
4. $y = x^2$, $x = y^2$.
5. $y = x^2$, $y = x$.
6. $y = x^4$, $x = 1$.
7. Check Eq. (28) of Sec. 211, $A\bar{x} = \int_a^b x(y_2 - y_1)dx$, by working out the inner integral in $A\bar{x} = \int_a^b dx \int_{y_1}^{y_2} x\,dy$, found from Eq. (39).
8. Check Eq. (30) of Sec. 211, $A\bar{y} = \frac{1}{2}\int_a^b (y_2^2 - y_1^2)dx$, by working out the inner integral in $A\bar{y} = \int_a^b dx \int_{y_1}^{y_2} y\,dy$, found from Eq. (39).

Use double integration to find the centroid of the area bounded by each given curve.

9. $r = 1 + \cos\theta$.
10. $r = 2 - \sin\theta$.

Use double integration to find the centroid of the area bounded by the loop through $r = 1$, $\theta = 0$ for each given curve.

11. $r = \cos 2\theta$.
12. $r^2 = \cos 2\theta$.
13. Check Eq. (33) of Sec. 211, $A\bar{x} = \frac{1}{3}\int_\alpha^\beta r_1^3 \cos\theta\,d\theta$ $A\bar{y} = \frac{1}{3}\int_\alpha^\beta r_1^3 \sin\theta\,d\theta$, by working out the inner integrals in $A\bar{x} = \int_\alpha^\beta d\theta \int_0^{r_1} \cos\theta\, r^2\,dr$ and $A\bar{y} = \int_\alpha^\beta d\theta \int_0^{r_1} r^2\,dr$, found from Eq. (40), where $r_1 = f(\theta)$.

Use double integration to find I_x and I_y for the area bounded by

14. $y^2 = x$, $x = 1$.
15. $x = 4 - y^2$, $x = 0$.
16. $y = x^4$, $y = 1$,
17. $y^2 = x$, $y = x$.

Find I_0 for the area inside one loop of each given curve

18. $r^2 = 2 \cos 2\theta$. 19. $r = \sin 2\theta$. 20. $r = \cos 3\theta$.

Use double integration to find I_x, I_y, and I_0 for the area inside of the circle

21. $r = a$. 22. $r = a \cos \theta$.

23. Check Eq. (60) of Sec. 221, $I_y = \int_a^b x^2(y_2 - y_1)dx$ by working out the inner integral in $I_y = \int_a^b dx \int_{y_1}^{y_2} x^2\,dy$, found from Eq. (41).

24. Check Eq. (61) of Sec. 221, $I_x = \frac{1}{3} \int_a^b (y_2^3 - y_1^3)dx$ by working out the inner integral in $I_x = \int_a^b dx \int_{y_1}^{y_2} y^2\,dy$, found from Eq. (41).

25. Check Eq. (66), of Sec. 222, $I_0 = \frac{1}{4} \int_\alpha^\beta r_1^4\,d\theta$ by working out the inner integral in $I_0 = \int_\alpha^\beta d\theta \int_0^{r_1} r^3\,d\theta$, found from Eq. (42).

26. By working out the inner integrals in $I_y = \int_\alpha^\beta d\theta \int_0^{r_1} \cos^2\theta\, r^3\,dr$ and $I_x = \int_\alpha^\beta d\theta \int_0^{r_1} \sin^2\theta\, r^3\,dr$, found from Eq. (42), derive the single integration formulas,

$$I_y = \frac{1}{4} \int_\alpha^\beta r_1^4 \cos^2\theta\,d\theta \qquad \text{and} \qquad I_x = \frac{1}{4} \int_\alpha^\beta r_1^4 \sin^2\theta\,d\theta.$$

***310. The Area of a Curved Surface.** Let R be a region of the xy plane. On the bounding curve of R construct a right cylinder or tubular surface whose elements are parallel to the z axis. Let $f(x,y)$ be a function which is continuous and single-valued for x, y in R. Then $z = f(x,y)$ represents a surface. As indicated in Fig. 350, let R' be the portion of this surface cut out by the right cylinder. We wish to calculate S, the area of R'. Divide R into subregions ΔR_i of any convenient shape. Let $P_i = (x_i,y_i)$ be any point of ΔR_i. If $z_i = f(x_i,y_i)$, $P_i' = (x_i,y_i,z_i)$ is the point of R' whose projection is P_i. Draw the tangent plane to the surface R' at P_i'. Let ΔT_i be the area of this tangent plane whose projection is ΔR_i. The operation of projecting any area from one plane to another reduces it in the ratio of the cosine of the angle between the two planes. Thus

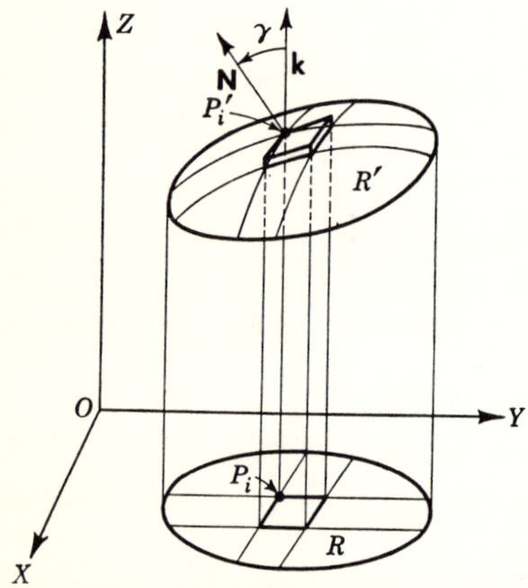

FIG. 350.

$$\Delta R_i = \cos \gamma_i\, \Delta T_i \qquad \text{and} \qquad \Delta T_i = \sec \gamma_i\, \Delta R_i, \tag{44}$$

where γ_i is the angle between the tangent plane and the xy plane. For subdivisions with d_n small, each area ΔT_i approximates a measure of the part of R' above ΔR_i. Hence we define S as the limit $\Sigma\, \Delta T_i$ when $d_n \to 0$. It then follows from Eq. (44) and Sec. 302 that

$$S = \lim_{d_n \to 0} \sum \Delta T_i = \lim_{d_n \to 0} \sum \sec \gamma_i \, \Delta R_i = \iint\limits_R \sec \gamma \, dx \, dy. \tag{45}$$

The angle γ between the tangent plane to $z = f(x,y)$ and the xy plane is the same as the angle between the normal vector $\mathbf{N}$ and OZ. By Sec. 295, for $z = f(x,y)$,

$$\mathbf{N} = -\frac{\partial f}{\partial x}\mathbf{i} - \frac{\partial f}{\partial y}\mathbf{j} + \mathbf{k} = -\frac{\partial z}{\partial x}\mathbf{i} - \frac{\partial z}{\partial y}\mathbf{j} + \mathbf{k} \tag{46}$$

is a normal vector. And since $\mathbf{k}$ is along OZ, we have from Eq. (63) of Sec. 290

$$\cos \gamma = \frac{(-\partial z/\partial x)0 + (-\partial z/\partial y)0 + 1(1)}{\sqrt{(-\partial z/\partial x)^2 + (-\partial z/\partial y)^2 + 1}\,\sqrt{0^2 + 0^2 + 1}}$$
$$= \frac{1}{\sqrt{(\partial z/\partial x)^2 + (\partial z/\partial y)^2 + 1}}. \tag{47}$$

We might have deduced this from the direction ratios of $\mathbf{N}$ by finding $\mathbf{n} = \cos \gamma$ as in Sec. 282. It follows from Eq. (47) that

$$\sec \gamma = \sqrt{1 + \left(\frac{\partial z}{\partial x}\right)^2 + \left(\frac{\partial z}{\partial y}\right)^2}. \tag{48}$$

By substitution of this value in Eq. (45), we find

$$S = \iint\limits_R \sqrt{1 + \left(\frac{\partial z}{\partial x}\right)^2 + \left(\frac{\partial z}{\partial y}\right)^2}\, dx \, dy. \tag{49}$$

This is the desired expression for the area of $z = f(x,y)$ whose projection on the xy plane is R. The double integral is to be evaluated as a repeated integral with limits determined as in Sec. 303. It is sometimes convenient to evaluate the integral by introducing polar coordinates as in Example 2, replacing $dx\,dy$ by $r\,dr\,d\theta$.

EXAMPLE 1. Find the area of the surface $z = 1 - x^2$ whose projection on the xy plane is the triangle bounded by $y = 0$, $y = x$, and $x = 1$.

Solution: From $z = 1 - x^2$, $\frac{\partial z}{\partial x} = -2x$, $\frac{\partial z}{\partial y} = 0$. It follows that $1 + \left(\frac{\partial z}{\partial x}\right)^2 + \left(\frac{\partial z}{\partial y}\right)^2 = 1 + 4x^2$. Hence from Eq. (49), we have $S = \iint\limits_R \sqrt{1 + 4x^2}\, dx\, dy = \int_0^1 dx \int_0^x \sqrt{1 + 4x^2}\, dy$. Since $\int_0^x dy = x$, $S = \int_0^1 \sqrt{1 + 4x^2}\, x\, dx = \frac{1}{12}[(1 + 4x^2)^{\frac{3}{2}}]_0^1 = \frac{1}{12}(5\sqrt{5} - 1)$.

EXAMPLE 2. Find the area of the surface $z = x^2 - y^2$ which lies inside the cylinder $x^2 + y^2 = 4$.

Solution: From $z = x^2 - y^2$, $\frac{\partial z}{\partial x} = 2x$, $\frac{\partial z}{\partial y} = -2y$. And $1 + \left(\frac{\partial z}{\partial x}\right)^2 + \left(\frac{\partial z}{\partial y}\right)^2 = 1 + 4x^2 + 4y^2$. In polar coordinates $x^2 + y^2 = r^2$, so that $\sec \gamma = \sqrt{1 + 4r^2}$. The circle $x^2 + y^2 = 4$ which bounds R becomes $r^2 = 4$ or $r = 2$. Hence $S = \iint\limits_R \sec \gamma\, r\, dr\, d\theta = \int_0^{2\pi} d\theta \int_0^2 \sqrt{1 + 4r^2}\, r\, dr$. Since $\int_0^2 \sqrt{1 + 4r^2}\, r\, dr = \frac{1}{12}[(1 + 4r^2)^{\frac{3}{2}}]_0^2 = \frac{1}{12}(17\sqrt{17} - 1)$, and $\int_0^{2\pi} d\theta = 2\pi$, $S = \frac{\pi}{6}(17\sqrt{17} - 1)$, the required area.

EXERCISE 158

1. For the upper half of the sphere $x^2 + y^2 + z^2 = a^2$, show that $\sec\gamma = \dfrac{a}{\sqrt{a^2 - x^2 - y^2}} = \dfrac{a}{\sqrt{a^2 - r^2}}$. Deduce that the part of the surface above R is $\iint\limits_R \dfrac{a}{\sqrt{a^2 - r^2}}\, r\, dr\, d\theta$.

Use the integral of Prob. 1 to verify that

2. The area of the hemisphere is $2\pi a^2$.
3. The area of the cap above $z = a\cos A$ is $2\pi a^2(1 - \cos A)$.
4. The upper area cut out by $r = a\cos\theta$ is $(\pi - 2)a^2$.
5. The area in the first octant cut out by the elliptic cylinder $z^2 + 4y^2 = a^2$ is $\pi a^2/6$. HINT: On the intersection $x = \sqrt{3}\, y$, so that R is bounded by $\theta = 0$, $\theta = \pi/6$, $r = a$.
6. The area above the first half loop of $r = a\cos n\theta$ is $\dfrac{\pi - 2}{2n}a^2$.
7. Find the area of the surface $z = 2xy$ inside the cylinder $x^2 + y^2 = a^2$.
8. Find the area of the surface $z = x^2 + y^2$ inside the cylinder $x^2 + y^2 = a^2$.
9. Show that the area over any region R is the same for each of the surfaces $az = 2xy$, $az = x^2 - y^2$, $az = x^2 + y^2$.
10. Use double integration to find the area of the triangle cut from the plane $x + 2y + 6z = 12$ by the coordinate planes.
11. Use double integration to find the area of the cone $z^2 = x^2 + y^2$ which is above the square bounded by $x = 0$, $x = 1$, $y = 0$, $y = 1$.
12. Find the part of the cylinder $z^2 + y^2 = ay$ above the xy plane which is cut out by the sphere $x^2 + y^2 + z^2 = a^2$. HINT: On the intersection $x^2 + ay = a^2$, so that R is bounded by $y = 0$, $y = \dfrac{a^2 - x^2}{a}$. And

$$S = 2\int_0^a dy \int_0^{\sqrt{a^2-ay}} \frac{a}{2\sqrt{ay - y^2}}\, dx = \int_0^a \frac{a\sqrt{a}}{\sqrt{y}}\, dy = 2a^2.$$

13. For the upper half of the conical surface $z^2 = 2xy$, show that $\sec^2\gamma = 1 + \dfrac{y}{2x} + \dfrac{x}{2y} = \dfrac{(x + y)^2}{2xy}$, and deduce that the part of the surface above R is

$$\iint\limits_R \frac{1}{\sqrt{2}}\left(\frac{\sqrt{x}}{\sqrt{y}} + \frac{\sqrt{y}}{\sqrt{x}}\right) dx\, dy = \iint\limits_R \frac{1}{\sqrt{2}} \frac{\cos\theta + \sin\theta}{\sqrt{\sin\theta\cos\theta}}\, r\, dr\, d\theta.$$

14. Use the rectangular form in Prob. 13 to find the part of $z = \sqrt{2xy}$ over the square $x = 0$, $x = 1$, $y = 0$, $y = 1$.
15. Use the polar form in Prob. 13 to find the part of $z = \sqrt{2xy}$ over the triangle bounded by $x = 0$, $y = 0$, $x + y = 2$. HINT: After the r integration, note that, if $\tan\theta = u^2$, $\sec^2\theta\, d\theta = 2u\, du$ and $\dfrac{d\theta}{(\cos\theta + \sin\theta)\sqrt{\sin\theta\cos\theta}} = \dfrac{\sec^2\theta\, d\theta}{(1 + \tan\theta)\sqrt{\tan\theta}} = \dfrac{2\, du}{1 + u^2} = d(\tan^{-1} u)$.
16. If the equation of the surface is $F(x,y,z) = 0$, from Eq. (121) of Sec. 295, $\mathbf{N} = \dfrac{\partial F}{\partial x}\mathbf{i} + \dfrac{\partial F}{\partial y}\mathbf{j} + \dfrac{\partial F}{\partial z}\mathbf{k}$. Deduce that $\sec\gamma = \dfrac{\sqrt{(\partial F/\partial x)^2 + (\partial F/\partial y)^2 + (\partial F/\partial z)^2}}{|\partial F/\partial z|}$.
17. Use Prob. 16 to check the first expression for $\sec\gamma$ in Probs. 1 and 13.

18. If $z = f(x,y) = f(r \cos \theta, r \sin \theta) = F(r,\theta)$, then $\dfrac{\partial z}{\partial r} = \dfrac{\partial z}{\partial x} \cos \theta + \dfrac{\partial z}{\partial y} \sin \theta$, $\dfrac{\partial z}{\partial \theta} = \dfrac{\partial z}{\partial x}(-r \sin \theta) + \dfrac{\partial z}{\partial y} r \cos \theta$. Deduce that $\left(\dfrac{\partial z}{\partial r}\right)^2 + \dfrac{1}{r^2}\left(\dfrac{\partial z}{\partial \theta}\right)^2 = \left(\dfrac{\partial z}{\partial x}\right)^2 + \left(\dfrac{\partial z}{\partial y}\right)^2$, so that $S = \displaystyle\iint_R \sqrt{1 + \left(\frac{\partial z}{\partial r}\right)^2 + \frac{1}{r^2}\left(\frac{\partial z}{\partial \theta}\right)^2}\, r\, dr\, d\theta$.

19. Use the expression for S in Prob. 18 to show that the area of the surface of revolution $z = F(r)$ is $S = \displaystyle\iint_R \sqrt{1 + [F'(r)]^2}\, r\, dr\, d\theta = 2\pi \int \sqrt{1 + [F'(r)]^2}\, r\, dr$. Except for notation, this is Eq. (118) of Sec. 186. And if $r = G(z)$ is the solution of $z = F(r)$, $F'(r) = \dfrac{1}{G'(z)}$, $dr = G'(z)dz$. In terms of z, $S = 2\pi \int G(z) \sqrt{1 + [G'(z)]^2}\, dz$. Except for notation, this is Eq. (116) of Sec. 186. This proves that the definition of Sec. 186 is consistent with that of Sec. 310.

311. Triple Integrals. Let $f(x,y,z)$ be a function of x, y, and z. For any point $P = (x,y,z)$, we define the symbol $f(P)$ by the equation

$$f(P) = f(x,y,z). \tag{50}$$

Then for any three-dimensional region R of space we may make the following construction. Subdivide R into n small subregions. These may be of any shape but must completely fill R. We let d_n denote the length of the largest interval having its end points on the boundary of any one of the small subregions. Let ΔR_i denote the ith subregion or its volume. Select some point P_i in each subregion and form the sum

$$S_n = \sum_{i=1}^{n} f(P_i)\Delta R_i. \tag{51}$$

Then for simple types of regions R, subregions ΔR_i, and functions $f(x,y,z)$, any sequence of sums S_n such that $d_n \to 0$ as $n \to \infty$ will approach a limit. This limit is called the *triple integral* of $f(P)$, or $f(x,y,z)$ over R. And we write

$$\int_R f(P)dR = \lim_{n\to\infty} \sum_{i=1}^{n} f(P_i)\Delta R_i \qquad \text{if } \lim_{n\to\infty} d_n = 0. \tag{52}$$

312. The Fundamental Theorem for Triple Integrals. Let the subdivisions ΔR_i of Eq. (51) be rectangular parallelepipeds formed by drawing planes parallel to the coordinate planes. Then the volume of the ith subdivision $\Delta R_i = \Delta x_j\, \Delta y_k\, \Delta z_l$. A discussion like that of Sec. 302 suggests that

$$\int_R f(P)dR = \iiint_R f(x,y,z)dz\, dy\, dx = \int_a^b dx \int_{u_1}^{u_2} dy \int_{v_1}^{v_2} f(x,y,z)dz. \tag{53}$$

The functions and constants used as limits are so chosen that the region R is made up of those points (Fig. 351) for which

$$a < x < b, \quad u_1(x) < y(x) < u_2(x), \quad w_1(x,y) < z(x,y) < w_2(x,y). \quad (54)$$

The relation (53) always holds when $f(x,y,z)$ is continuous in all three variables. Together, Eqs. (53) and (52) express the *fundamental theorem* of the integral calculus for triple integrals. The result also holds for Duhamel modifications.

The expression on the right in Eq. (53) is a repeated integral. Its evaluation by three successive single integrations is analogous to that described in Sec. 300. Such a repeated integral, when thought of as the expression for a triple integral, is often itself referred to as a triple integral.

313. Volume as a Triple Integral in Rectangular Coordinates. Let $f(x,y,z) = 1$. Then $f(P_i) = 1$. And the sum $\Sigma f(P_i)\Delta R_i$ of Sec. 311 reduces to $\Sigma\, \Delta R_i$. This approximates the volume of the region R. And

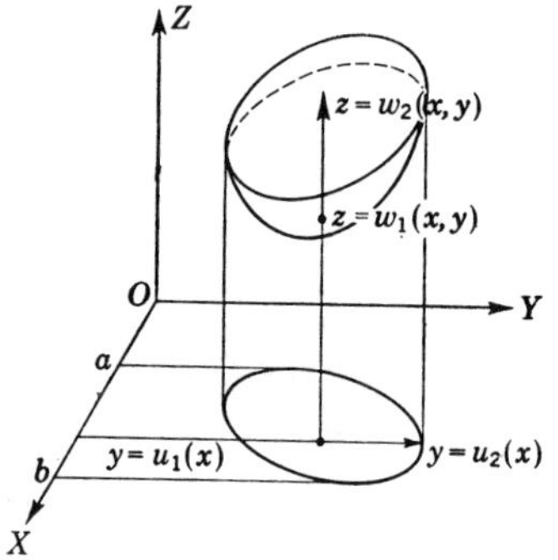

FIG. 351.

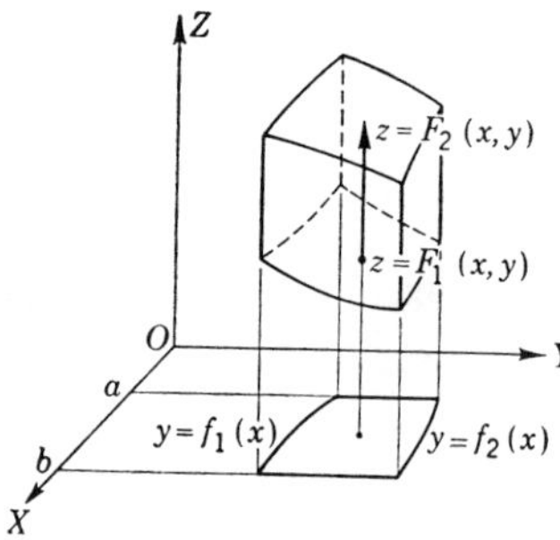

FIG. 352.

for regions bounded by ordinary surfaces, the error approaches zero when $d_n \to 0$. It follows from Eqs. (52) and (53) that

$$V = \text{volume of } R = \int_R dR = \iiint dz\, dy\, dx = \int dx \int dy \int dz. \quad (55)$$

Thus the volume of any three-dimensional region is the value of the triple integral of the function $f(x,y,z) = 1$ *taken over that region.* We express this fact and the content of Eq. (55) by writing

$$V = \int dV, \qquad dV = dx\, dy\, dz. \quad (56)$$

The integration may be taken in any one of six possible orders if limits appropriate to R are used.

Let us apply this to the volume bounded above by the surface $z_2 = F_2(x,y)$, below by the surface $z_1 = F_1(x,y)$, and laterally by the cylindrical surface whose base is the bounding curve of the area A of the xy plane of Fig. 332. Then

$$V = \int_a^b dx \int_{f_1(x)}^{f_2(x)} dy \int_{F_1(x,y)}^{F_2(x,y)} dz. \tag{57}$$

The innermost limits for z are obtained from Fig. 352, by noting that a line drawn parallel to OZ which cuts the region enters on the surface $z = F_1(x,y)$ and leaves on the surface $z = F_2(x,y)$. The remaining limits are found from Fig. 332 as in Sec. 302.

For the volume of Fig. 337, $F_1(x,y) = 0$ and $F_2(x,y) = f(x,y)$. If we make these substitutions in Eq. (57) and carry out the inner integration, we again obtain the expression found in Eq. (27).

EXAMPLE 1. Find the volume of the paraboloid $x^2 + y^2 = 2z$ which lies below the plane $z = x + 1$.

Solution: On the curve of intersection, $x^2 + y^2 = 2z$ and $z = x + 1$. Elimination of z leads to $x^2 + y^2 = 2x + 2$, the equation of the cylinder which projects this curve of intersection on to the xy plane. Thus the volume lies over the inside of the circle

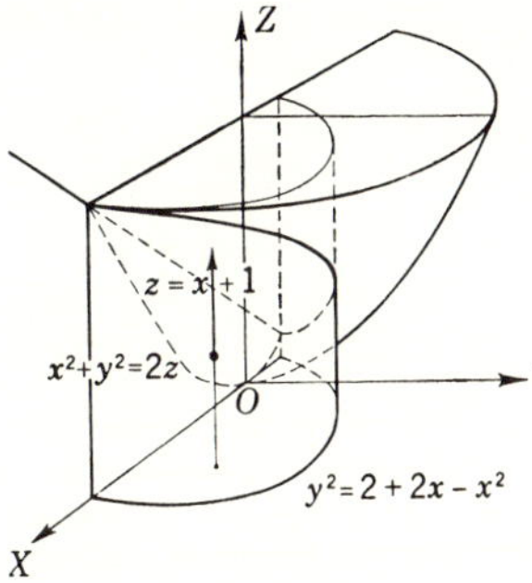

FIG. 353.

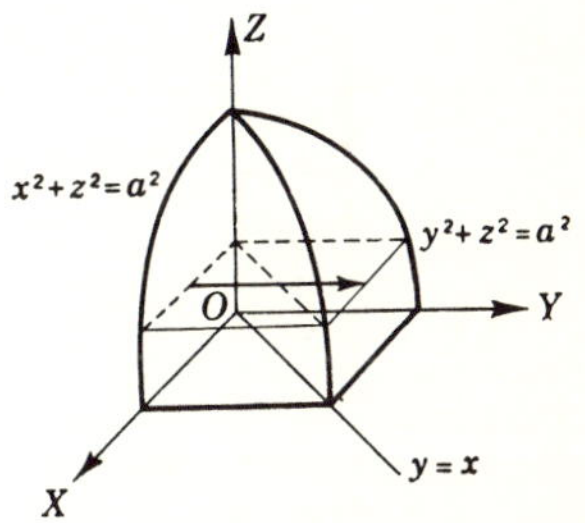

FIG. 354.

$x^2 + y^2 = 2x + 2$ or $y^2 = 2x + 2 - x^2$ in the xy plane. The circle cuts OX when $y = 0$, $x^2 - 2x - 2 = 0$, and $x = 1 \pm \sqrt{3}$. The volume is symmetrical with respect to the xz plane, so that it is twice the part with y positive, shown in Fig. 353. Thus $V = 2\int_{1-\sqrt{3}}^{1+\sqrt{3}} dx \int_0^{\sqrt{2+2x-x^2}} dy \int_{(x^2+y^2)/2}^{x+1} dz$. The inner integral is $\int_{(x^2+y^2)/2}^{x+1} dz$ $= \Big[z\Big]_{(x^2+y^2)/2}^{x+1} = \frac{1}{2}(2 + 2x - x^2 - y^2)$. And $\int_0^{\sqrt{2+2x-x^2}} \frac{1}{2}(2 + 2x - x^2 - y^2)dy$ $= \frac{1}{2}\left[(2 + 2x - x^2)y - \frac{y^3}{3}\right]_0^{\sqrt{2+2x-x^2}} = \frac{1}{3}(2 + 2x - x^2)^{\frac{3}{2}}$. $V =$ $\frac{2}{3}\int_{1-\sqrt{3}}^{1+\sqrt{3}} (2 + 2x - x^2)^{\frac{3}{2}}\, dx$. Since $(2 + 2x - x^2) = 3 - (x - 1)^2$, let $x - 1 =$ $\sqrt{3}\sin t$. Then $V = \frac{2}{3}\int_{-\pi/2}^{\pi/2} 3^2 \cos^4 t\, dt = 6\left[\frac{3t}{8} + \frac{\sin 2t}{4} + \frac{\sin 4t}{32}\right]_{-\pi/2}^{\pi/2} = \frac{9\pi}{4}$.

EXAMPLE 2. Find the volume in the first octant inside the cylinders $x^2 + z^2 = a^2$ and $y^2 + z^2 = a^2$.

Solution 1: The volume is shown in Fig. 354. A line parallel to OY enters the region on $y = 0$ and leaves on $y = \sqrt{a^2 - z^2}$. And the volume projects on the part of the

zx plane inside the quadrant of $x^2 + z^2 = a^2$ bounded by $x = 0$ and $x = \sqrt{a^2 - z^2}$, for z between 0 and a. Hence $V = \int_0^a dz \int_0^{\sqrt{a^2-z^2}} dx \int_0^{\sqrt{a^2-z^2}} dy$. $\int_0^{\sqrt{a^2-z^2}} dy = \left[y\right]_0^{\sqrt{a^2-z^2}} = \sqrt{a^2 - z^2}$. $\int_0^{\sqrt{a^2-z^2}} \sqrt{a^2 - z^2}\, dx = \sqrt{a^2 - z^2}\left[x\right]_0^{\sqrt{a^2-z^2}} = a^2 - z^2$. $V = \int_0^a (a^2 - z^2)dz = \left[a^2 z - \frac{z^3}{3}\right]_0^a = \frac{2a^3}{3}$.

Solution 2: The volume is bisected by $y = x$. Hence $V = 2\int_0^a dz \int_0^{\sqrt{a^2-z^2}} dx \int_0^x dy$. $\int_0^x dy = [y]_0^x = x$, $\int_0^{\sqrt{a^2-z^2}} x\, dx = \left[\frac{x^2}{2}\right]_0^{\sqrt{a^2-z^2}} = \frac{1}{2}(a^2 - z^2)$. And $V = \int_0^a (a^2 - z^2)dz = \left[a^2 z - \frac{z^3}{3}\right]_0^a = \frac{2a^3}{3}$.

314. Volume as a Triple Integral in Cylindrical Coordinates. Let r,θ,z be the cylindrical coordinates of Sec. 306. Then we may use the surfaces r = constant, θ = constant, z = constant to form the subdivisions ΔR_i of Eq. (51). A typical subdivision is the keystone-shaped figure bounded by two cylinders of radius r, $r + \Delta r$, two vertical planes θ, $\theta + \Delta\theta$, and two horizontal planes z, $z + \Delta z$, shown in Fig. 355. The exact volume of ΔR_i is $r_i' \, \Delta r \, \Delta\theta \, \Delta z$, where $r_i' = r + \Delta r/2$. When $d_n \to 0$, $\Sigma r_i' \, \Delta r \, \Delta\theta \, \Delta z$ approaches $\int\int\int r \, dr \, d\theta \, dz$ as a limit. It follows that, for ordinary boundaries of R,

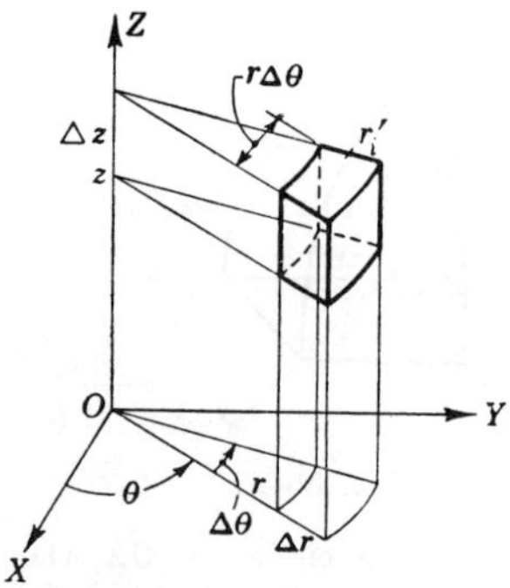

FIG. 355.

$$V = \text{volume of } R = \int\int\int r \, dz \, dr \, d\theta = \int d\theta \int dr \int r \, dz. \quad (58)$$

The integration may be taken in any one of six possible orders if limits appropriate to R are used. For solids of revolution with axis OZ, the order $\int dz \int d\theta \int r \, dr$ is often convenient.

EXAMPLE 1. Find the volume of the solid bounded above by the cone $z = 4r$ and below by the paraboloid $z = r^2 + 3$.

Solution 1: On the circles of intersection $z = 4r$ and $z = r^2 + 3$. Elimination of z leads to $4r = r^2 + 3$, $(r - 1)(r - 3) = 0$, and $r = 1,3$. Thus the solid lies over the ring between $r = 1$ and $r = 3$ in the xy plane. From Fig. 356 and Eq. (58) we have $V = \int_0^{2\pi} d\theta \int_1^3 dr \int_{r^2+3}^{4r} r\, dz$. $\int_{r^2+3}^{4r} r\, dz = r[z]_{r^2+3}^{4r} = r(4r - r^2 - 3)$. $\int_1^3 (4r^2 - r^3 - 3r)dr = \left[\frac{4}{3}r^3 - \frac{1}{4}r^4 - \frac{3}{2}r^2\right]_1^3 = \left(36 - \frac{81}{4} - \frac{27}{2}\right) - \left(\frac{4}{3} - \frac{1}{4} - \frac{3}{2}\right) = \frac{8}{3}$. $V = \frac{8}{3}\int_0^{2\pi} d\theta = \frac{8}{3}[\theta]_0^{2\pi} = \frac{16\pi}{3}$.

Solution 2: On the circles of intersection $z = 4r$ and $z = r^2 + 3$. Elimination of r leads to $z = z^2/16 + 3$, $z^2 - 16z + 48 = 0$, $(z - 4)(z - 12) = 0$ and $z = 4$, $z = 12$. Hence the solid lies between the planes $z = 4$ and $z = 12$. From Fig. 356, using the

order suggested for solids of revolution, we have $V = \int_4^{12} dz \int_0^{2\pi} d\theta \int_{z/4}^{\sqrt{z-3}} r\,dr.$
$\int_{z/4}^{\sqrt{z-3}} r\,dr = \left[\frac{r^2}{2}\right]_{z/4}^{\sqrt{z-3}} = \frac{1}{2}\left(z - 3 - \frac{z^2}{16}\right).$ Since $\int_0^{2\pi} d\theta = 2\pi,$

$$V = \pi \int_4^{12} \left(z - 3 - \frac{z^2}{16}\right) dz = \pi \left[\frac{z^2}{2} - 3z - \frac{z^3}{48}\right]_4^{12}$$
$$= \pi \left[(72 - 36 - 36) - \left(8 - 12 - \frac{4}{3}\right)\right] = \frac{16\pi}{3}.$$

Example 2. Find the volume of the paraboloid $z = r^2$ which lies below the plane $z = r \cos \theta$.

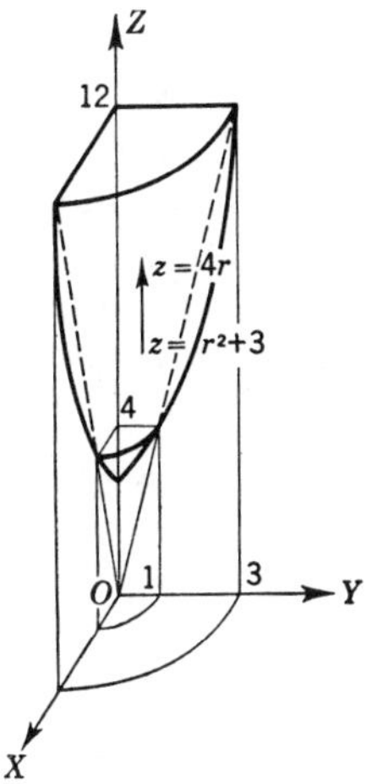

Fig. 356.

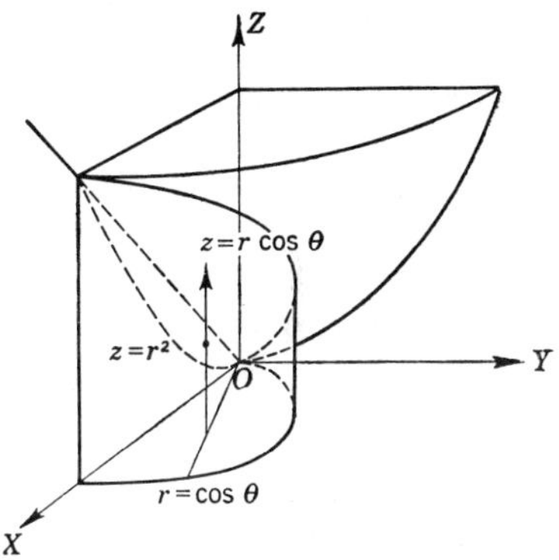

Fig. 357.

Solution: On the curve of intersection, $z = r^2$ and $z = r \cos \theta$. Elimination of z leads to $r^2 = r \cos \theta$ or $r = \cos \theta$. Thus the solid lies over the interior of the circle $r = \cos \theta$ in the xy plane. From Fig. 357 and Eq. (58) we have
$V = 2 \int_0^{\pi/2} d\theta \int_0^{\cos\theta} dr \int_{r^2}^{r\cos\theta} r\,dz.$ $\int_{r^2}^{r\cos\theta} r\,dz = [rz]_{r^2}^{r\cos\theta} = r^2 \cos\theta - r^3.$
$\int_0^{\cos\theta} (r^2 \cos\theta - r^3)dr = \left[\frac{r^3}{3}\cos\theta - \frac{r^4}{4}\right]_0^{\cos\theta} = \frac{\cos^4\theta}{12}.$ $V = \frac{1}{6}\int_0^{\pi/2} \cos^4\theta\, d\theta =$
$\frac{1}{6}\left[\frac{3\theta}{8} + \frac{\sin 2\theta}{4} + \frac{\sin 4\theta}{32}\right]_0^{\pi/2} = \frac{\pi}{32}.$

EXERCISE 159

Find the volume bounded laterally by the planes $x = 0$, $x = 1$, $y = 0$, $y = 1$, below by the plane $z = x$, and above by

1. $z = 3x + 2y$.
2. $z = 4 - x - 2y$.
3. $z = 3x + 4xy$.
4. $z = 3 + 2x^2 - y^2$.
5. Find the volume included between the planes $y = 0$ and $y = 6$, above the cylinder $z = x^2$ and below the plane $z = x$.

6. Find the volume included between the planes $x = 0$ and $x = 2$, above the cylinder $z = y^2$ and below the plane $z = y + 2$.

7. Show that the volume bounded by the coordinate planes and the plane $\frac{x}{a} + \frac{y}{b} + \frac{z}{c} = 1$, with a,b,c each positive, is $V = \int_0^a dx \int_0^{b(1-x/a)} dy \int_0^{c(1-x/a-y/b)} dz = \frac{abc}{6}$.

8. Show that the volume bounded below by the upper sheet of the elliptic cone $\frac{z^2}{c^2} = \frac{x^2}{a^2} + \frac{y^2}{b^2}$ cut off by the plane $z = c$ is $V = 4 \int_0^c dz \int_0^{az/c} dx \int_0^{b\sqrt{z^2/c^2 - x^2/a^2}} dy = \frac{\pi}{3} abc$.

9. Show that the volume of the paraboloid $\frac{z}{c} = \frac{x^2}{a^2} + \frac{y^2}{b^2}$, cut off by the plane $z = c$, is $V = 4 \int_0^c dz \int_0^{a\sqrt{z/c}} dx \int_0^{b\sqrt{z/c - x^2/a^2}} dy = \frac{\pi}{2} abc$.

10. Show that the volume of the ellipsoid $\frac{x^2}{a^2} + \frac{y^2}{b^2} + \frac{z^2}{c^2} = 1$ is

$$V = 8 \int_0^a dx \int_0^{b\sqrt{1-x^2/a^2}} dy \int_0^{c\sqrt{1-x^2/a^2-y^2/b^2}} dz = \frac{4}{3}\pi abc.$$

Find the volume bounded laterally by the cylinder $r = 2$, below by the plane $z = 1$, and above by each of the following surfaces.

11. $z = 2r \cos \theta + 3$.

12. $z = r^3 \cos^3 \theta + 2$.

13. $z = r + 2$.

14. $z = r^2 + 3$.

15. Find the volume of the solid bounded above by the sphere $r^2 + z^2 = 5$, and below by the paraboloid $z = 2r^2$.

16. Find the volume of the solid bounded above by the sphere $r^2 + z^2 = a^2$ and below by the upper sheet of the cone $z = \sqrt{3}\, r$.

Find the volume inside the sphere $r^2 + z^2 = a^2$ and included in the cylinder whose base is the curve

17. $r = a \cos \theta$.

18. One loop of $r^2 = a^2 \cos 2\theta$.

19. Show that the volume outside of $r = f_1(z)$, inside of $r = f_2(z)$, above $z = a$, and below $z = b$ is given by

$$V = \int_a^b dz \int_0^{2\pi} d\theta \int_{f_1(z)}^{f_2(z)} r\, dr = \pi \int_a^b \{[f_2(z)]^2 - [f_1(z)]^2\} dz.$$

Except for notation this is Eq. (4) of Sec. 205.

20. Show that the volume above $z = f_1(r)$, below $z = f_2(r)$, outside of $r = a$, and inside of $r = b$ is given by

$$V = \int_0^{2\pi} d\theta \int_a^b dr \int_{f_1(r)}^{f_2(r)} r\, dz = 2\pi \int_a^b [f_2(r) - f_1(r)] r\, dr.$$

Except for notation this is Eq. (5) of Sec. 205.

315. Center of Gravity of a Solid Body. Consider a solid body occupying a region of space R. Let ρ be the density or mass per unit volume.

Then the mass $M = \int_R \rho\, dR$. By Sec. 213, M_x, the first moment with respect to the yz plane with equation $x = 0$, is approximated by $\Sigma x_i' \rho'' \Delta R_i$. The error approaches zero when $d_n \to 0$. It follows from Sec. 312 that $M_x = \int_R x\rho\, dR$. Thus the x coordinate of the center of gravity of the body may be computed from the relations

$$M = \int_R \rho\, dR, \qquad M_x = \int_R \rho x\, dR, \qquad \bar{x} = \frac{M_x}{M} = \frac{\int_R \rho x\, dR}{\int_R \rho\, dR}. \tag{59}$$

These apply for ρ any function of x, y, and z. If ρ is constant, we may bring it out as a factor and cancel it in the last expression of Eq. (59). A similar relation holds for $\bar{y}$ and $\bar{z}$, so that

$$\bar{x} = \frac{\int_R x\, dR}{\int_R dR}, \qquad \bar{y} = \frac{\int_R y\, dR}{\int_R dR}, \qquad \bar{z} = \frac{\int_R z\, dR}{\int_R dR}. \tag{60}$$

For rectangular coordinates we replace dR by $dx\, dy\, dz$ and set

$$\begin{aligned} V &= \textstyle\int\int\int dz\, dy\, dx, & V\bar{x} &= \textstyle\int\int\int x\, dz\, dy\, dx, \\ V\bar{y} &= \textstyle\int\int\int y\, dz\, dy\, dx, & V\bar{z} &= \textstyle\int\int\int z\, dz\, dy\, dx. \end{aligned} \tag{61}$$

These triple integrals are to be evaluated as repeated integrals in any order with limits determined as in Sec. 313.

For cylindrical coordinates, we replace dR by $r\, dr\, d\theta\, dz$, x by $r \cos \theta$, y by $r \sin \theta$. Thus

$$\begin{aligned} V &= \textstyle\int\int\int r\, dz\, dr\, d\theta, & V\bar{x} &= \textstyle\int\int\int r^2 \cos\theta\, dz\, dr\, d\theta, \\ V\bar{y} &= \textstyle\int\int\int r^2 \sin\theta\, dz\, dr\, d\theta, & V\bar{z} &= \textstyle\int\int\int rz\, dz\, dr\, d\theta. \end{aligned} \tag{62}$$

The limits in the repeated integrals are to be determined as in Sec. 314.

EXAMPLE 1. Use triple integration to find the centroid of the volume bounded by the coordinate planes and the plane $\frac{x}{a} + \frac{y}{b} + \frac{z}{c} = 1$, where a, b, and c are each positive.

Solution: The volume is the pyramid of Fig. 338. From this and Eq. (61) we have

$$\begin{aligned} V &= \int_0^a dx \int_0^{b(1-x/a)} dy \int_0^{c(1-x/a-y/b)} dz, \\ V\bar{x} &= \int_0^a dx \int_0^{b(1-x/a)} dy \int_0^{c(1-x/a-y/b)} x\, dz, \\ V\bar{y} &= \int_0^b dy \int_0^{a(1-y/b)} dx \int_0^{c(1-x/a-y/b)} y\, dz, \\ V\bar{z} &= \int_0^c dz \int_0^{a(1-z/c)} dx \int_0^{b(1-x/a-z/c)} z\, dy. \end{aligned}$$

We have $\int_0^{c(1-x/a-y/b)} dz = [z]_0^{c(1-x/a-y/b)} = c\left(1 - \frac{x}{a} - \frac{y}{b}\right)\cdot$ And

$\int_0^{b(1-x/a)} c\left(1 - \frac{x}{a} - \frac{y}{b}\right) dy = \left[c\left(1 - \frac{x}{a}\right) y - \frac{c}{2b} y^2\right]_0^{b(1-x/a)} = \frac{bc}{2}\left(1 - \frac{x}{a}\right)^2.$

Hence $V = \frac{bc}{2} \int_0^a \left(1 - \frac{x}{a}\right)^2 dx = -\frac{abc}{6}\left[\left(1 - \frac{x}{a}\right)^3\right]_0^a = \frac{abc}{6}.$

$V\bar{x} = \frac{bc}{2} \int_0^a x\left(1 - \frac{x}{a}\right)^2 dx.$ Let $1 - \frac{x}{a} = t$. Then $x = a(1 - t)$ and

$V\bar{x} = \frac{bc}{2} \int_1^0 a(1 - t)t^2(-a\,dt) = \frac{a^2bc}{2} \int_0^1 (t^2 - t^3)dt = \frac{a^2bc}{2}\left[\frac{t^3}{3} - \frac{t^4}{4}\right]_0^1 = \frac{a^2bc}{24}.$ As here set up, $V\bar{y}$ differs from $V\bar{x}$ only in the interchange of x with y and of a with b. Hence $V\bar{y} = \frac{ab^2c}{24}.$ And $V\bar{z}$ differs from $V\bar{y}$ by the interchange of y with z and of b with c. Hence $V\bar{z} = \frac{abc^2}{24}.$ It follows that $\bar{x} = \frac{V\bar{x}}{V} = \frac{\frac{1}{24}a^2bc}{\frac{1}{6}abc} = \frac{a}{4}$, $\bar{y} = \frac{V\bar{y}}{V} = \frac{\frac{1}{24}ab^2c}{\frac{1}{6}abc} = \frac{b}{4}$, $\bar{z} = \frac{V\bar{z}}{V} = \frac{\frac{1}{24}abc^2}{\frac{1}{6}abc} = \frac{c}{4}.$ Hence the centroid is $\left(\frac{a}{4}, \frac{b}{4}, \frac{c}{4}\right).$

EXAMPLE 2. Use triple integration to find the centroid of the first octant of a solid sphere bounded by $r^2 + z^2 = a^2$.

Solution 1: From Eq. (62) we have $V = \int_0^{\pi/2} d\theta \int_0^a dr \int_0^{\sqrt{a^2-r^2}} r\,dz.$ $V\bar{x} = \int_0^{\pi/2} d\theta \int_0^a dr \int_0^{\sqrt{a^2-r^2}} r^2 \cos\theta\,dz$, $V\bar{y} = \int_0^{\pi/2} d\theta \int_0^a dr \int_0^{\sqrt{a^2-r^2}} r^2 \sin\theta\,dz$, $V\bar{z} = \int_0^{\pi/2} d\theta \int_0^a dr \int_0^{\sqrt{a^2-r^2}} rz\,dz.$ $\int_0^{\sqrt{a^2-r^2}} dz = [z]_0^{\sqrt{a^2-r^2}} = \sqrt{a^2 - r^2}.$

$\int_0^{\sqrt{a^2-r^2}} r\sqrt{a^2 - r^2}\,dr = -\frac{1}{3}[(a^2 - r^2)^{\frac{3}{2}}]_0^a = \frac{a^3}{3}$, $V = \frac{a^3}{3} \int_0^{\pi/2} d\theta = \frac{a^3}{3}[\theta]_0^{\pi/2} = \frac{\pi a^3}{6}.$

To find $\int_0^a r^2\sqrt{a^2 - r^2}\,dr$, put $r = a \sin t$, $a^4 \int_0^{\pi/2} \sin^2 t \cos^2 t\,dt =$ $\frac{a^4}{4} \int_0^{\pi/2} \sin^2 2t\,dt = \frac{a^4}{4}\left[\frac{t}{2} - \frac{\sin 4t}{8}\right]_0^{\pi/2} = \frac{\pi a^4}{16}.$ Then $V\bar{x} = \frac{\pi a^4}{16} \int_0^{\pi/2} \cos\theta\,d\theta =$ $\frac{\pi a^4}{16}[\sin\theta]_0^{\pi/2} = \frac{\pi a^4}{16}.$ And $V\bar{y} = \frac{\pi a^4}{16} \int_0^{\pi/2} \sin\theta\,d\theta = \frac{\pi a^4}{16}[-\cos\theta]_0^{\pi/2} = \frac{\pi a^4}{16}.$

$\int_0^{\sqrt{a^2-r^2}} rz\,dz = \left[\frac{rz^2}{2}\right]_0^{\sqrt{a^2-r^2}} = \frac{1}{2}(a^2r - r^3).$ $\frac{1}{2}\int_0^a (a^2r - r^3)dr = \frac{1}{2}\left[\frac{a^2r^2}{2} - \frac{r^4}{4}\right]_0^a$ $= \frac{a^4}{8}.$ $V\bar{z} = \frac{a^4}{8} \int_0^{\pi/2} d\theta = \frac{a^4}{8}[\theta]_0^{\pi/2} = \frac{\pi a^4}{16}.$ Since $V = \frac{\pi a^3}{6}$ and $V\bar{x} = V\bar{y} = V\bar{z} =$ $\frac{\pi a^4}{16}$, $\bar{x} = \bar{y} = \bar{z} = \frac{\pi a^4/16}{\pi a^3/6} = \frac{3a}{8}.$

Solution 2: $V = \int_0^a dz \int_0^{\pi/2} d\theta \int_0^{\sqrt{a^2-z^2}} r\,dr,$

$V\bar{x} = \int_0^a dz \int_0^{\pi/2} d\theta \int_0^{\sqrt{a^2-z^2}} r^2 \cos\theta\,dr$, $V\bar{y} = \int_0^a dz \int_0^{\pi/2} d\theta \int_0^{\sqrt{a^2-z^2}} r^2 \sin\theta\,dr.$

$V\bar{z} = \int_0^a dz \int_0^{\pi/2} d\theta \int_0^{\sqrt{a^2-z^2}} zr\,dr.$ $\int_0^{\sqrt{a^2-z^2}} r\,dr = \left[\frac{r^2}{2}\right]_0^{\sqrt{a^2-z^2}} = \frac{1}{2}(a^2 - z^2),$

$\int_0^{\pi/2} d\theta = [\theta]_0^{\pi/2} = \frac{\pi}{2}.$ Hence $V = \frac{\pi}{4} \int_0^a (a^2 - z^2)dz = \frac{\pi}{4}\left[a^2z - \frac{z^3}{3}\right]_0^a = \frac{\pi a^3}{6}.$ Also $V\bar{z} = \frac{\pi}{4} \int_0^a (a^2 - z^2)z\,dz = -\frac{\pi}{16}[(a^2 - z^2)^2]_0^a = \frac{\pi a^4}{16}.$ $\int_0^{\sqrt{a^2-z^2}} r^2\,dr = \left[\frac{r^3}{3}\right]_0^{\sqrt{a^2-z^2}}$ $= \frac{1}{3}(a^2 - z^2)^{\frac{3}{2}}.$ $\int_0^{\pi/2} \cos\theta\,d\theta = [\sin\theta]_0^{\pi/2} = 1,$ $\int_0^{\pi/2} \sin\theta\,d\theta = [-\cos\theta]_0^{\pi/2} = 1.$

Hence $V\bar{x} = V\bar{y} = \frac{1}{3}\int_0^a (a^2 - z^2)^{\frac{3}{2}}\,dz$. Put $z = a \sin t$ and $\frac{a^4}{3}\int_0^{\pi/2} \cos^4 t\,dt = \frac{a^4}{3}\left[\frac{3t}{8} + \frac{\sin 2t}{4} + \frac{\sin 4t}{32}\right]_0^{\pi/2} = \frac{\pi a^4}{16}$. Since $V = \frac{\pi a^3}{6}$ and $V\bar{x} = V\bar{y} = V\bar{z} = \frac{\pi a^4}{16}$, $\bar{x} = \bar{y} = \bar{z} = \frac{\pi a^4/16}{\pi a^3/6} = \frac{3a}{8}$.

316. Moment of Inertia of a Solid Body. Consider a solid body occupying a region of space R. Let its density be ρ. Divide R into subregions ΔR_i. And let r_i' denote the shortest distance from some point in ΔR_i to a given axis. Then by Sec. 219 $\Sigma r_i'^2 \rho'' \Delta R_i$ is an approximation to I, the moment of inertia of the body about the given axis. And the error approaches zero when $d_n \to 0$. It follows from Sec. 312 that

$$I = \int r^2 \rho\, dR. \tag{63}$$

Let I_x denote the moment of inertia about the x axis. The shortest distance from (x,y,z) to the x axis is $\sqrt{y^2 + z^2}$. Hence, with dR replaced by $dz\,dy\,dx$, we have from Eq. (63)

$$I_x = \iiint \rho (y^2 + z^2) dz\,dy\,dx. \tag{64}$$

Similarly for the moments of inertia about the y axis and the z axis,

$$I_y = \iiint \rho (z^2 + x^2) dz\,dy\,dx, \qquad I_z = \iiint \rho (x^2 + y^2) dz\,dy\,dx. \tag{65}$$

These triple integrals are to be evaluated as repeated integrals in any order with limits determined as in Sec. 313.

For cylindrical coordinates we replace dR by $r\,dr\,d\theta\,dz$, x by $r \cos \theta$, y by $r \sin \theta$. Thus

$$\begin{aligned} I_x &= \iiint \rho (r^2 \sin^2 \theta + z^2) r\,dz\,dr\,d\theta, \\ I_y &= \iiint \rho (r^2 \cos^2 \theta + z^2) r\,dz\,dr\,d\theta, \\ I_z &= \iiint \rho r^3\,dz\,dr\,d\theta. \end{aligned} \tag{66}$$

The limits in the repeated integrals are to be determined as in Sec. 314.

It is sometimes convenient to introduce the second moments with respect to the coordinate planes defined by

$$\begin{aligned} M_{xx} &= \iiint \rho x^2\,dz\,dy\,dx = \iiint \rho r^3 \cos^2 \theta\,dz\,dr\,d\theta, \\ M_{yy} &= \iiint \rho y^2\,dz\,dy\,dx = \iiint \rho r^3 \sin^2 \theta\,dz\,dr\,d\theta, \\ M_{zz} &= \iiint \rho z^2\,dz\,dy\,dx = \iiint \rho r z^2\,dz\,dr\,d\theta. \end{aligned} \tag{67}$$

We may then find I_x, I_y, and I_z from the relations

$$I_x = M_{yy} + M_{zz}, \qquad I_y = M_{zz} + M_{xx}, \qquad I_z = M_{xx} + M_{yy}. \tag{68}$$

Any moment of inertia may be expressed in terms of the mass M and radius of gyration k in the form $I = Mk^2$. To do this we make use of the relations

$$k^2 = \frac{I}{M}, \qquad M = \iiint \rho\, dz\, dy\, dx = \iiint \rho r\, dz\, dr\, d\theta. \tag{69}$$

EXAMPLE 1. Find I_x, I_y, and I_z for the homogeneous solid ellipsoid bounded by $\frac{x^2}{a^2} + \frac{y^2}{b^2} + \frac{z^2}{c^2} = 1$.

Solution: $M = 8\rho \int_0^a dx \int_0^{b\sqrt{1-x^2/a^2}} dy \int_0^{c\sqrt{1-x^2/a^2-y^2/b^2}} dz$. And

$M_{xz} = 8\rho \int_0^a dx \int_0^{b\sqrt{1-x^2/a^2}} dy \int_0^{c\sqrt{1-x^2/a^2-y^2/b^2}} x^2\, dz$.

$\int_0^{c\sqrt{1-x^2/a^2-y^2/b^2}} dz = \Big[z \Big]_0^{c\sqrt{1-x^2/a^2-y^2/b^2}} = c\sqrt{1 - \frac{x^2}{a^2} - \frac{y^2}{b^2}}$. In

$c \int_0^{b\sqrt{1-x^2/a^2}} \sqrt{1 - \frac{x^2}{a^2} - \frac{y^2}{b^2}}\, dy$ put $y = b\sqrt{1 - \frac{x^2}{a^2}} \sin t$.

$c \int_0^{\pi/2} b\left(1 - \frac{x^2}{a^2}\right) \cos^2 t\, dt = bc\left(1 - \frac{x^2}{a^2}\right)\left[\frac{t}{2} + \frac{\sin 2t}{4}\right]_0^{\pi/2} = \frac{\pi}{4} bc \left(1 - \frac{x^2}{a^2}\right)$.

$M = 8\rho \int_0^a \frac{\pi}{4} bc \left(1 - \frac{x^2}{a^2}\right) dx = 2\rho\pi bc \left[x - \frac{x^3}{3a^2}\right]_0^a = \frac{4}{3}\rho\pi abc$.

$M_{xz} = 8\rho \int_0^a \frac{\pi}{4} bc \left(1 - \frac{x^2}{a^2}\right) x^2\, dx = 2\rho\pi bc \left[\frac{x^3}{3} - \frac{x^5}{5a^2}\right]_0^a = \frac{4}{15}\rho\pi a^3 bc$.

$\frac{M_{xz}}{M} = \frac{\frac{4}{15}\rho\pi a^3 bc}{\frac{4}{3}\rho\pi abc} = \frac{a^2}{5}$. And $M_{xz} = \frac{Ma^2}{5}$. It follows that an interchange of letters in the calculation like that made in Example 1 of Sec. 315 would give $M_{yy} = \frac{Mb^2}{5}$ and $M_{zz} = \frac{Mc^2}{5}$. Hence from Eq. (68) we find $I_x = M_{yy} + M_{zz} = \frac{M}{5}(b^2 + c^2)$, $I_y = M_{zz} + M_{xx} = \frac{M}{5}(c^2 + a^2)$, $I_z = M_{xx} + M_{yy} = \frac{M}{5}(a^2 + b^2)$.

EXAMPLE 2. Find I_x, I_y, and I_z for the homogeneous solid bounded above by the plane $z = 2r\cos\theta$, below by the plane $z = r\cos\theta$, and laterally by the cylinder $r = a$.

Solution:

$$M = 2\rho \int_0^{\pi/2} d\theta \int_0^a dr \int_{r\cos\theta}^{2r\cos\theta} r\, dz,$$

$$M_{xx} = 2\rho \int_0^{\pi/2} d\theta \int_0^a dr \int_{r\cos\theta}^{2r\cos\theta} r^3 \cos^2\theta\, dz,$$

$$M_{yy} = 2\rho \int_0^{\pi/2} d\theta \int_0^a dr \int_{r\cos\theta}^{2r\cos\theta} r^3 \sin^2\theta\, dz,$$

$$M_{zz} = 2\rho \int_0^{\pi/2} d\theta \int_0^a dr \int_{r\cos\theta}^{2r\cos\theta} rz^2\, dz.$$

$\int_{r\cos\theta}^{2r\cos\theta} dz =]z[_{r\cos\theta}^{2r\cos\theta} = r\cos\theta$. $\int_0^a r^2\, dr = \left[\frac{r^3}{3}\right]_0^a = \frac{a^3}{3}$. Hence $M = \frac{2}{3}\rho a^3 \int_0^{\pi/2} \cos\theta\, d\theta = \frac{2}{3}\rho a^3 [\sin\theta]_0^{\pi/2} = \frac{2}{3}\rho a^3$. $\int_0^a r^4\, dr = \left[\frac{r^5}{5}\right]_0^a = \frac{a^5}{5}$. Hence $M_{xx} = \frac{2}{5}\rho a^5 \int_0^{\pi/2} \cos^3\theta\, d\theta = \frac{2}{5}\rho a^5 \left[\sin\theta - \frac{\sin^3\theta}{3}\right]_0^{\pi/2} = \frac{4}{15}\rho a^5$. Also $M_{yy} = \frac{2}{5}\rho a^5 \int_0^{\pi/2} \sin^2\theta \cos\theta\, d\theta = \frac{2}{5}\rho a^5 \left[\frac{\sin^3\theta}{3}\right]_0^{\pi/2} = \frac{2}{15}\rho a^5$. $\int_{r\cos\theta}^{2r\cos\theta} z^2\, dz = \left[\frac{z^3}{3}\right]_{r\cos\theta}^{2r\cos\theta}$

$= \frac{7}{3} r^3 \cos^3 \theta$. Hence $M_{zz} = \frac{14}{3} \rho \int_0^{\pi/2} d\theta \int_0^a r^4 \cos^3 \theta \, dr$. By the calculations made above for M_{xx}, this is $= \frac{14}{3} \rho \frac{a^5}{5} \frac{2}{3} = \frac{28}{45} \rho a^5$.

$$\frac{M_{xx}}{M} = \frac{\frac{4}{15}\rho a^5}{\frac{2}{3}\rho a^3} = \frac{2a^2}{5}. \qquad \frac{M_{yy}}{M} = \frac{\frac{2}{15}\rho a^5}{\frac{2}{3}\rho a^3} = \frac{a^2}{5}. \qquad \frac{M_{zz}}{M} = \frac{\frac{28}{45}\rho a^5}{\frac{2}{3}\rho a^3} = \frac{14a^2}{15}.$$

From $M_{xx} = \frac{2Ma^2}{5}$, $M_{yy} = \frac{Ma^2}{5}$, $M_{zz} = \frac{14Ma^2}{15}$, we find $I_x = M_{yy} + M_{zz} = \frac{17}{15} Ma^2$, $I_y = M_{zz} + M_{xx} = \frac{4}{3} Ma^2$, $I_z = M_{xx} + M_{yy} = \frac{3}{5} Ma^2$.

EXAMPLE 3. Find the moment of inertia of a solid cone of revolution about its axis if the density is proportional to the distance from the axis.

Solution: Here $\rho = Kr$. Hence $dm = \rho \, dV = Kr^2 \, dr \, d\theta \, dz$. Take the cone as inverted and bounded below by $z = \frac{hr}{a}$, above by $z = h$.

Then $M = \int_0^h dz \int_0^{2\pi} d\theta \int_0^{az/h} Kr^2 \, dr$, $I_z = \int_0^h dz \int_0^{2\pi} d\theta \int_0^{az/h} Kr^4 \, dr$.

$\int_0^{az/h} r^2 \, dr = \left[\frac{r^3}{3}\right]_0^{az/h} = \frac{a^3z^3}{3h^3}$. $\int_0^{2\pi} d\theta = [\theta]_0^{2\pi} = 2\pi$. Hence $M = \frac{2\pi Ka^3}{3h^3} \int_0^h z^3 \, dz = \frac{2\pi Ka^3}{3h^3} \left[\frac{z^4}{4}\right]_0^h = \frac{\pi Ka^3h}{6}$. $\int_0^{az/h} r^4 \, dr = \left[\frac{r^5}{5}\right]_0^{az/h} = \frac{a^5z^5}{5h^5}$. $I_z = \frac{2\pi Ka^5}{5h^5} \int_0^h z^5 \, dz = \frac{2\pi Ka^5}{5h^5} \left[\frac{z^6}{6}\right]_0^h = \frac{\pi Ka^5h}{15}$. $\frac{I_z}{M} = \frac{\frac{1}{15}\pi Ka^5h}{\frac{1}{6}\pi Ka^3h} = \frac{2}{5} a^2$, $I_z = \frac{2}{5} Ma^2$.

EXERCISE 160

1. Find the centroid of the volume bounded laterally by the planes $x = 0$, $x = 1$, $y = 0$, $y = 1$, below by the plane $z = 2x$, and above by the plane $z = 4x$.
2. Find the centroid of the volume included between the planes $y = 0$ and $y = 6$, above the cylinder $z = x^2$, and below the plane $z = x$.
3. Find the centroid of the volume in the first octant inside the cylinders $x^2 + z^2 = a^2$ and $y^2 + z^2 = a^2$.
4. Find the centroid of the volume of the paraboloid $\frac{z}{c} = \frac{x^2}{a^2} + \frac{y^2}{b^2}$ cut off by the plane $z = c$.
5. Find the centroid of the first octant of the solid ellipsoid bounded by $\frac{x^2}{a^2} + \frac{y^2}{b^2} + \frac{z^2}{c^2} = 1$.

Find the centroid of the volume bounded below by the plane $z = 0$ and above by each of the following surfaces of revolution.

6. $r^2 + z^2 = 1$.
7. $z = 1 - r^2$.
8. $z = 1 - r$.
9. $r^2 + z^2 + 2z = 3$.
10. Find the centroid of the volume inside the sphere $r^2 + z^2 = 4a^2$, and also inside the cylinder $r = a$.
11. Find the centroid of the volume bounded below by the paraboloid $az = r^2$ and above by the cone $z = 2a - r$.

Find I_x, I_y, I_z for a homogeneous solid occupying the volume of

12. Prob. 1.
13. Prob. 3.
14. Prob. 7.
15. Prob. 8.

Find I_z for a homogeneous solid occupying the volume of

16. Prob. 4.
17. Prob. 9.
18. Prob. 10.
19. Prob. 11.

A homogeneous solid right circular cylinder is of height h and radius of base a. Find its moment of inertia about

20. A diameter of its base.
21. An element of the cylinder.

22. A homogeneous solid right circular cone is of height h and radius of base a. Find its moment of inertia about an axis through the vertex parallel to the base.
23. A right cylinder of radius a and height h has its density proportional to the distance from its axis. Show that the moment of inertia about the axis is $\frac{3}{5}Ma^2$.
24. A right circular cone of radius a and height h has its density proportional to the distance from its axis. Show that the center of gravity is on the axis at distance $h/5$ up from the base.
25. A right circular cone of radius a and height h has its density proportional to the distance from the base. Show that the center of gravity is on the axis at distance $2h/5$ up from the base.

317. Spherical Coordinates. Instead of rectangular coordinates, or the cylindrical coordinates of Sec. 306, a third system is sometimes used. We again start with a right-handed system of x, y, and z axes. Let P be any point in space. From P draw a straight line parallel to OZ until it meets the xy plane in F. Let C be the projection of P on OZ. Draw the two equal and parallel lines OF and CP (Fig. 358). Let r denote the distance from O to P. Let ϕ denote the angle from OZ to OP. And let θ, as in cylindrical coordinates, be the angle from OX to the plane through P and OZ, or $OFPC$. Thus

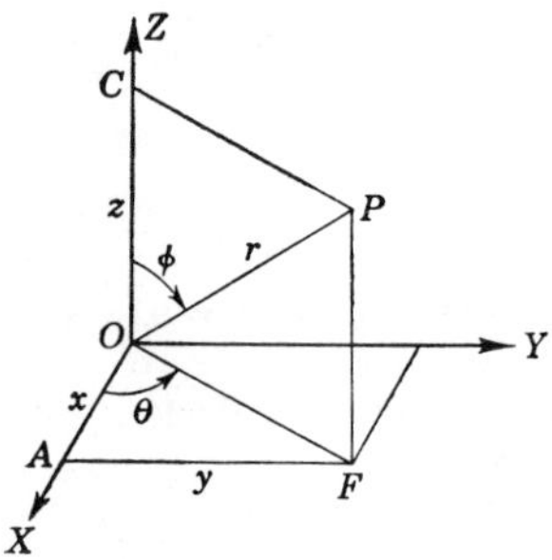

FIG. 358.

$$r = OP, \qquad \phi = \text{angle } ZOP, \qquad \theta = \text{angle } XOF. \tag{70}$$

Since θ determines plane ZOP, ϕ determines the direction of OP in that plane, and OP determines the position of P on the line, r,ϕ,θ may be used to locate the point P. The three numbers (r,ϕ,θ) are called the *spherical coordinates* of P.

From right triangle COP, we have

$$\sin\phi = \frac{CP}{OP} \quad \text{and} \quad OF = CP = OP\sin\phi = r\sin\phi. \tag{71}$$

$$\cos\phi = \frac{OC}{OP} \quad \text{and} \quad OC = OP\cos\phi \quad \text{or } z = r\cos\phi. \tag{72}$$

In right triangle AOF, angle AOF equals θ. Hence

$$x = OA = OF\cos\theta \quad \text{and} \quad y = AF = OF\sin\theta. \tag{73}$$

We might have derived Eq. (73) from Eq. (34) by noting that OF is the radius vector which was denoted by r in Sec. 306.

It follows from Eqs. (71) to (73) that

$$x = r \sin \phi \cos \theta, \qquad y = r \sin \phi \sin \theta, \qquad z = r \cos \phi. \tag{74}$$

These relations enable us to convert equations or expressions from rectangular to spherical coordinates.

We may use the surfaces r = constant, ϕ = constant, θ = constant to form the subdivisions ΔR_i of Eq. (51). A typical subdivision bounded by two spheres of radius r, $r + \Delta r$, two cones ϕ, $\phi + \Delta\phi$, and two vertical planes θ, $\theta + \Delta\theta$, is shown in Fig. 359. The exact volume of ΔR_i is $\Delta r(r' \Delta\phi)(r' \sin \phi' \Delta\theta) = r'^2 \sin \phi' \Delta r \Delta\phi \Delta\theta$ for suitably chosen values r' between r and $r + \Delta r$ and ϕ' between ϕ and $\phi + \Delta\phi$. It follows by an argument like that used in Sec. 314 that in spherical coordinates

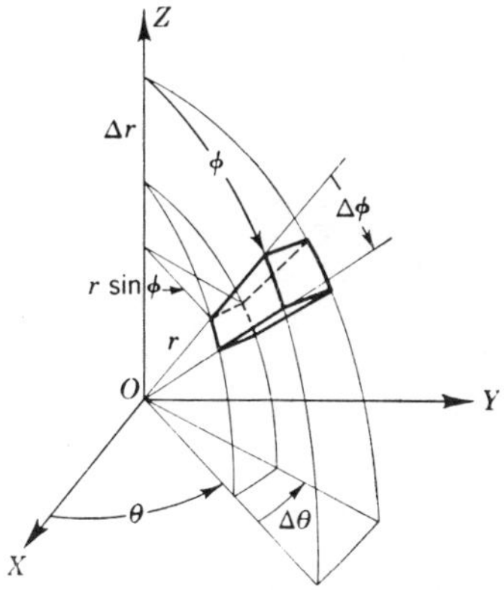

FIG. 359.

$$V = \iiint r^2 \sin \phi \, dr \, d\phi \, d\theta = \int d\theta \int d\phi \int r^2 \sin \phi \, dr. \tag{75}$$

To compute the center of gravity of a volume, we use integrals obtained from Eq. (61) by replacing $dz\, dy\, dx$ by $r^2 \sin \phi \, dr \, d\phi \, d\theta$, and x,y,z by their expressions as given in Eq. (74). For example,

$$V\bar{z} = \iiint r^3 \cos \phi \sin \phi \, dr \, d\phi \, d\theta. \tag{76}$$

A similar procedure may be applied to the integrals for finding moments of inertia given in Sec. 316. For example,

$$I_z = \iiint \rho r^4 \sin^3 \phi \, dr \, d\phi \, d\theta. \tag{77}$$

Spherical coordinates are particularly useful for problems involving spheres, or solids bounded by parts of coordinate surfaces.

EXAMPLE 1. Find the center of gravity of a solid hemisphere of radius a if the density is proportional to the distance from the center.

Solution: Here $\rho = Kr$. Hence $dm = \rho\, dV = Kr^3 \sin \phi \, dr \, d\phi \, d\theta$. And $M = \int_0^{2\pi} d\theta \int_0^{\pi/2} d\phi \int_0^a Kr^3 \sin \phi \, dr$. $M\bar{z} = \int_0^{2\pi} d\theta \int_0^{\pi/2} d\phi \int_0^a Kr^4 \cos \phi \sin \phi \, dr$. $\int_0^a r^3\, dr = \left[\frac{r^4}{4}\right]_0^a = \frac{a^4}{4}$, $\int_0^{\pi/2} \sin \phi \, d\phi = [-\cos \phi]_0^{\pi/2} = 1$, $\int_0^{2\pi} d\theta = [\theta]_0^{2\pi} = 2\pi$. It follows that $M = 2\pi(1)\frac{Ka^4}{4} = \frac{\pi}{2} Ka^4$. $\int_0^a r^4\, dr = \left[\frac{r^5}{5}\right]_0^a = \frac{a^5}{5}$, $\int_0^{\pi/2} \cos \phi \sin \phi \, d\phi = \frac{1}{2}[\sin^2 \phi]_0^{\pi/2} = \frac{1}{2}$, $\int_0^{2\pi} d\theta = [\theta]_0^{2\pi} = 2\pi$. It follows that $M\bar{z} = 2\pi\left(\frac{1}{2}\right)\frac{Ka^5}{5} = \frac{\pi}{5} Ka^5$. Hence $\bar{z} = \frac{M\bar{z}}{M} = \frac{(\pi/5)Ka^5}{(\pi/2)Ka^4} = \frac{2a}{5}$. Thus $(\bar{x},\bar{y},\bar{z}) = \left(0,0,\frac{2a}{5}\right)$.

EXAMPLE 2. Find the moment of inertia about OZ of a homogeneous solid consisting of that part of the first octant of the sphere $r = 2a$ which lies below the cone $\phi = \pi/3$ and outside of the sphere $r = a$.

Solution: Here $M = \rho \int_0^{\pi/2} d\theta \int_{\pi/3}^{\pi/2} d\phi \int_a^{2a} r^2 \sin \phi \, dr$,

$I_z = \rho \int_0^{\pi/2} d\theta \int_{\pi/3}^{\pi/2} d\phi \int_a^{2a} r^4 \sin^3 \phi \, dr$. $\int_a^{2a} r^2 \, dr = \left[\frac{r^3}{3}\right]_a^{2a} = \frac{7a^3}{3}$, $\int_{\pi/3}^{\pi/2} \sin \phi \, d\phi = [-\cos \phi]_{\pi/3}^{\pi/2} = \frac{1}{2}$, $\int_0^{\pi/2} d\theta = [\theta]_0^{\pi/2} = \frac{\pi}{2}$. It follows that $M = \rho \frac{\pi}{2}\left(\frac{1}{2}\right)\frac{7a^3}{3} = \frac{7\rho\pi a^3}{12}$. $\int_a^{2a} r^4 \, dr = \left[\frac{r^5}{5}\right]_a^{2a} = \frac{31a^5}{5}$, $\int_{\pi/3}^{\pi/2} \sin^3 \phi \, d\phi = \left[-\cos \phi + \frac{\cos^3 \phi}{3}\right]_{\pi/3}^{\pi/2} = \frac{11}{24}$, $\int_0^{\pi/2} d\theta = [\theta]_0^{\pi/2} = \frac{\pi}{2}$. It follows that $I_z = \rho \frac{\pi}{2}\left(\frac{11}{24}\right)\frac{31a^5}{5} = \frac{341\rho\pi a^5}{240}$. $\frac{I_z}{M} = \frac{\frac{341}{240}\rho\pi a^5}{\frac{7}{12}\rho\pi a^3} = \frac{341}{140} a^2$, $I_z = \frac{341}{140} Ma^2$.

318. Average or Mean Value. In Sec. 208 we defined the average value of $y = f(x)$ with respect to x over the interval from a to b as

$$\bar{y} = \frac{1}{b-a} \int_a^b y \, dx. \tag{78}$$

We may rewrite this relation in the form

$$\text{Average of } f(x) = \frac{\int_a^b f(x)dx}{\int_a^b dx}. \tag{79}$$

This suggests the average of a function of a point, $f(P)$, over a region R as

$$\text{Average of } f(P) = \frac{\int_R f(P)dR}{\int_R dR}. \tag{80}$$

For a two-dimensional plane region we may take $f(P) = f(x,y)$, $dR = dx\, dy$, and determine the limits as in Sec. 303. Or we may take $f(P) = F(r,\theta)$, $dR = r\, dr\, d\theta$, and determine the limits as in Sec. 305. For a three-dimensional region we may take $f(P) = f(x,y,z)$, $dR = dx\, dy\, dz$, and determine the limits as in Sec. 313. With cylindrical coordinates as in Sec. 314, we take $dR = r\, dr\, d\theta\, dz$. And with spherical coordinates as in Sec. 317, we take $dR = r^2 \sin \phi\, dr\, d\phi\, d\theta$.

EXAMPLE 1. A square of side a has one corner at O. Let P be any point inside the square. Find the average value of the distance OP.

Solution 1: Take two sides of the square along OX and OY. From symmetry, the average will be the same as for half of the square bounded by $y = 0$, $y = x$, $x = a$. Here $f(P) = OP = \sqrt{x^2 + y^2}$.

$A = \int_0^a dx \int_0^x dy = \int_0^a x \, dx = \left[\frac{x^2}{2}\right]_0^a = \frac{a^2}{2}$. $A\bar{f} = \int_0^a dx \int_0^x \sqrt{x^2 + y^2}\, dy$.

$$\int_0^x \sqrt{x^2+y^2}\,dy = \frac{1}{2}\left[y\sqrt{x^2+y^2} + x^2 \ln\left(y+\sqrt{x^2+y^2}\right)\right]_{y=0}^{y=x} =$$
$$\frac{1}{2}\left[\sqrt{2}\,x^2 + x^2\ln(1+\sqrt{2})x - x^2 \ln x\right] = \frac{x^2}{2}\left[\sqrt{2}+\ln(1+\sqrt{2})\right].$$

Since $\int_0^a x^2\,dx = \left[\frac{x^3}{3}\right]_0^a = \frac{a^3}{3}$, $A\bar{f} = \frac{1}{2}\left[\sqrt{2}+\ln(1+\sqrt{2})\right]\frac{a^3}{3}$. And

$$\bar{f} = \frac{A\bar{f}}{A} = \frac{\frac{1}{6}\left[\sqrt{2}+\ln(1+\sqrt{2})\right]a^3}{\frac{1}{2}a^2} = \frac{a}{3}\left[\sqrt{2}+\ln(1+\sqrt{2})\right].$$

Solution 2: In polar coordinates, take half the square as bounded by $\theta = 0$, $\theta = \pi/4$, $r = a\sec\theta$. Here $f(P) = OP = r$. $A = \int_0^{\pi/4} d\theta \int_0^{a\sec\theta} r\,dr = \int_0^{\pi/4} \frac{a^2\sec^2\theta}{2}\,d\theta$ $= \frac{a^2}{2}[\tan\theta]_0^{4/\pi} = \frac{a^2}{2}$. $A\bar{f} = \int_0^{\pi/4} d\theta \int_0^{a\sec\theta} r^2\,dr = \int_0^{\pi/4}\frac{a^3\sec^3\theta}{3}\,d\theta =$ $\frac{a^3}{3}\frac{1}{2}\left[\tan\theta\sec\theta + \ln(\tan\theta+\sec\theta)\right]_0^{\pi/4} = \frac{a^3}{6}\left[\sqrt{2}+\ln(1+\sqrt{2})\right]$. And

$$\bar{f} = \frac{A\bar{f}}{A}\;\frac{\frac{1}{6}\left[\sqrt{2}+\ln(1+\sqrt{2})\right]a^3}{\frac{1}{2}a^3} = \frac{a}{3}\left[\sqrt{2}+\ln(1+\sqrt{2})\right].$$

Example 2. Let P be any point inside of a sphere of radius a. And let B be a fixed point outside the sphere at distance b from the center of the sphere. Find the average value of $1/BP$.

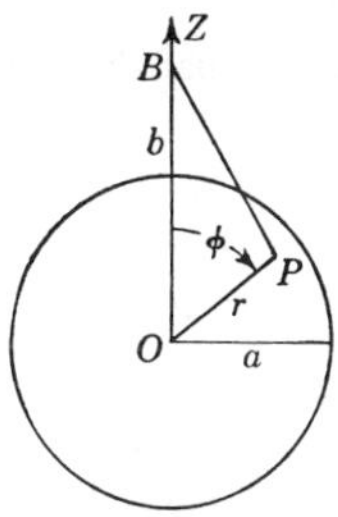

Fig. 360.

Solution: Use spherical coordinates and take B on OZ (Fig. 360). Then $\overline{BP}^2 = \overline{OP}^2 + \overline{OB}^2 - 2\overline{OP}\cdot\overline{OB}\cos\phi = r^2+b^2-2br\cos\phi$. And $f(P) = \frac{1}{BP} =$ $\frac{1}{\sqrt{r^2+b^2-2br\cos\phi}}$. $V = \int_0^{2\pi} d\theta \int_0^{\pi} d\phi \int_0^a r^2\sin\phi\,dr$. $\int_0^a r^2\,dr = \left[\frac{r^3}{3}\right]_0^a =$ $\frac{a^3}{3}$, $\int_0^{\pi}\sin\phi\,d\phi = [-\cos\phi]_0^{\pi} = 2$, $\int_0^{2\pi} d\theta = [\theta]_0^{2\pi} = 2\pi$. It follows that $V =$ $2\pi(2)\frac{a^3}{3} = \frac{4\pi a^3}{3}$. $V\bar{f} = \int_0^{2\pi} d\theta\int_0^{\pi} d\phi \int_0^a \frac{r^2\sin\phi\,dr}{\sqrt{r^2+b^2-2br\cos\phi}}$. Since the limits are all constant, $V\bar{f} = \int_0^{2\pi} d\theta \int_0^a dr \int_0^{\pi}\frac{r^2\sin\phi\,d\phi}{\sqrt{r^2+b^2-2br\cos\phi}}$.

$$\int_0^{\pi}\frac{r^2\sin\phi\,d\phi}{\sqrt{r^2+b^2-2br\cos\phi}} = \frac{r}{b}\left[\sqrt{r^2+b^2-2br\cos\phi}\right]_{\phi=0}^{\phi=\pi} = \frac{r}{b}\left[(b+r)-(b-r)\right]$$

$= \frac{2r^2}{b}$. $\int_0^a \frac{2r^2}{b}\,dr = \left[\frac{2r^3}{3b}\right]_0^a = \frac{2a^3}{3b}$. $V\bar{f} = \frac{2a^3}{3b}\int_0^{2\pi} d\theta = \frac{2a^3}{3b}[\theta]_0^{2\pi} = \frac{4\pi a^3}{3b}$. And $\bar{f} = \frac{V\bar{f}}{V} = \frac{4\pi a^3/3b}{4\pi a^3/3} = \frac{1}{b}$.

EXERCISE 161

A solid hemisphere of uniform density is bounded above by $r = a$ and below by $\phi = \pi/2$. Verify that for it

1. For the center of gravity, $\bar{z} = 3a/8$.

2. The moment of inertia $I_z = \frac{2}{5}Ma^2$.

3. The density of a solid sphere of radius a is proportional to the distance from the center. Show that its moment of inertia about a diameter is $\frac{4}{9}Ma^2$.

A solid of uniform density is bounded above by the sphere $r = a$ and below by the cone $\phi = \pi/3$. Show that for it

4. $M = \dfrac{\pi\rho a^3}{3}$. 5. $\bar{z} = \dfrac{9a}{16}$. 6. $I_z = \dfrac{Ma^2}{4}$.

Use the equations in cylindrical coordinates, $r^2 + z^2 = a^2$ and $r = \sqrt{3}\,z$, to check

7. Prob. 4. 8. Prob. 5. 9. Prob. 6.

10. The volume bounded below by $\phi = \tan^{-1}(a/h)$ and above by $r \cos \phi = h$ is an inverted cone of radius a and height h. Verify that
$$V = \int_0^{2\pi} d\theta \int_0^{\tan^{-1}(a/h)} d\phi \int_0^{h \sec \phi} r^2 \sin \phi \, dr = \frac{\pi a^2 h}{3}.$$
11. Use spherical coordinates to verify that the volume inside the sphere $r = 2a \cos \phi$ is $\frac{4}{3}\pi a^3$.

For spherical coordinates r,ϕ,θ, the projection of a point on the xy plane has polar coordinates $OF = r \sin \phi$ and θ. For a projected plane area $dA = OF\, d(OF)d\theta$. By Sec. 310, on any surface $r = g(\phi)$, $dS = dA \sec \gamma = OF\, d(OF)d\theta \sec \gamma$. Show that

12. On the sphere $r = a$, $\gamma = \phi$, and $OF = a \sin \phi$. And $dS = a \sin \phi \, d(a \sin \phi)d\theta \sec \phi = a^2 \sin \phi \, d\phi \, d\theta$.
13. On the sphere $r = 2a \cos \phi$, $\gamma = 2\phi$, and $OF = a \sin 2\phi$. And $dS = a \sin 2\phi \, d(a \sin 2\phi)d\theta \sec 2\phi = 2a^2 \sin 2\phi \, d\phi \, d\theta$.

Use the result of Prob. 12, $dS = a^2 \sin \phi \, d\phi \, d\theta$, to verify that

14. The area of the whole sphere is $S = \int_0^{2\pi} d\theta \int_0^{\pi} a^2 \sin \phi \, d\phi = 4\pi a^2$.
15. The centroid of a hemispherical surface of radius a is on its axis at distance $a/2$ from the base of the hemisphere.
16. For a thin, uniform shell covering the surface of the whole sphere, the moment of inertia about a diameter is $I_z = \frac{2}{3}Ma^2$.
17. For the right spherical triangle bounded by $\theta = 0$, $\theta = A$, $\cot \phi = \tan D \cos \theta$, the area is $S = a^2[A - \sin^{-1}(\sin D \sin A)]$. If B is the angle between the plane $\theta = A$, or $x \sin A - y \cos A = 0$ and the plane $\cot \phi = \tan D \cos \theta$, or $x \sin D - z \cos D = 0$, we have $\cos B = \sin D \sin A$. Thus $\sin^{-1}(\sin D \sin A) = \pi/2 - B$ and since $C = \pi/2$, $A - (\pi/2 - B) = A + B + C - \pi = E$, the spherical excess. It follows that $S = a^2E$.

18. Use Prob. 13 to verify that the area of the whole sphere is
$$S = \int_0^{2\pi} d\theta \int_0^{\pi/2} 2a^2 \sin 2\phi \, d\phi = 4\pi a^2.$$
19. If O is one corner of a square of side a and P is any point in the square, find the average value of $\overline{OP^2}$.

20. If O is one corner of a cube of side a and P is any point inside the cube, find the average value of $\overline{OP}^2$.

If O is a fixed point on the circumference and P is any point inside of a circle of radius a, find the average value of

21. OP. 22. $\overline{OP}^2$. 23. $1/OP$.

24. Given a sphere of radius a. Let O be a fixed point on the surface. And let P be any point on the surface. Show that the average value of $1/OP$ is $1/a$. Use spherical coordinates as in Prob. 18.
25. Show that the average value of z for points P in a volume V is the z coordinate of the centroid of V.
26. Show that the average value of the square of the distance from the origin, $x^2 + y^2$, for points P in an area A is the square of the radius of gyration, $k^2 = I_0/A$, where I_0 is the polar moment of inertia.

CHAPTER 20

DIFFERENTIAL EQUATIONS

Many physical and geometric problems lead to equations containing derivatives or differentials. Examples of such differential equations and their solution were given in Secs. 68, 69, 123, and 124. In this chapter we classify the principal useful types of solvable differential equations and describe a convenient method of solving each type.

As applications of first-order differential equations, we find curves characterized by geometric properties. And we find quantities whose time rate of change is a known function of the quantity and the time. We apply second-order differential equations to simple electrical and mechanical circuits, equilibrium curves of cables, deflection curves of beams, and to the motion of a particle in a plane due to known forces.

319. Definition of Terms. Any equation that involves differentials or derivatives is a *differential equation*. Thus

$$\frac{dy}{dx} - y = 0 \qquad \text{or} \qquad dy - y\,dx = 0, \tag{1}$$

$$\left(y - x\frac{dy}{dx}\right)^2 = \sqrt{\frac{dy}{dx}} \qquad \text{or} \qquad \left(y - x\frac{dy}{dx}\right)^4 = \frac{dy}{dx}, \tag{2}$$

$$2\frac{d^2y}{dx^2} + 7\frac{dy}{dx} + 3y = 4\sin x, \tag{3}$$

$$\frac{\partial^2 u}{\partial x^2} + \frac{\partial^2 u}{\partial y^2} = 0, \tag{4}$$

are differential equations. If the equation contains any partial derivatives, as in Eq. (4), it is a *partial differential equation*. If it does not, as in Eqs. (1) to (3), it is an *ordinary differential equation*.

The *order* of the differential equation is the same as the order of the derivative of highest order in the equation. Thus Eqs. (1) and (2) are each of the first order. But Eqs. (3) and (4) are each of the second order.

Suppose that a differential equation is reducible to a form in which each member is a polynomial in all the derivatives that occur. Then the *degree* of the equation is the largest exponent of the highest derivative in the reduced form. Thus Eq. (2) is of the fourth degree, while Eqs. (1), (3), and (4) are each of the first degree.

A differential equation is *linear* if it is a first-degree algebraic equation in the set of variables made up of the dependent variables together with all their derivatives. Thus Eqs. (1) and (3) are each linear in y, and Eq. (4) is linear in u.

320. Integral Curves. The differential equation (1) may be written in the form

$$\frac{dy}{dx} = y. \tag{5}$$

This associates a value of dy/dx, or a slope, with each point $P = (x,y)$ in the plane. Starting at any point $P_0 = (x_0,y_0)$, we may construct a polygon $P_0P_1P_2 \cdots$, as in Fig. 361, by choosing any small number h and constructing the sides successively of length h and with the slope at the first end point. A sequence of values of $h \to 0$ leads to a series of polygons approaching a curve C. The curve passes through P_0 and has at each point values of x, y, dy/dx that satisfy the differential equation. Consequently the equation of C, written explicitly as

$$y = \psi(x) \qquad \text{or implicitly as } \phi(x,y) = 0, \tag{6}$$

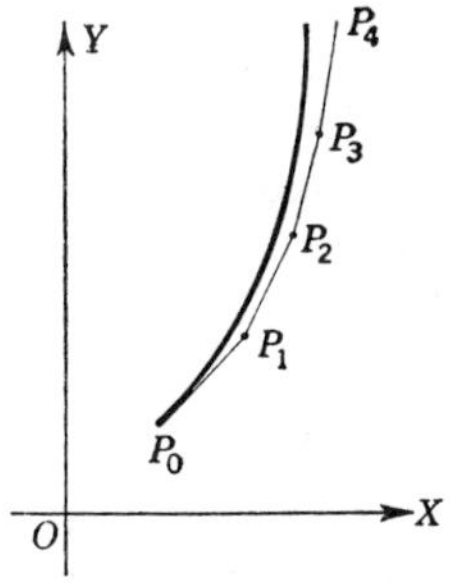

FIG. 361.

is called a *solution* of the differential equation. And C itself, which is the graph of a solution, is called an *integral curve.* The discussion suggests that the differential equation (5) has a solution, or integral curve passing through any given point.

Every differential equation of the first order, when solved for dy/dx, takes the form

$$\frac{dy}{dx} = f(x,y). \tag{7}$$

The geometric process carried out for Eq. (5) may be applied to this equation and suggests† that there is an integral curve passing through any point P_0 that is not a singular point of $f(x,y)$.

321. General Solution. Since the initial point P_0 may be varied, the differential equation (7) will have a whole family of integral curves. The equation of such a family will involve one constant c. Solved for the constant, it may be written

$$F(x,y) = c. \tag{8}$$

We may eliminate the constant from this form by differentiation.

For example, if $F(x,y)$ is $3xy - y^2$, we have

† Compare the author's "A Treatise on Advanced Calculus," p. 516, John Wiley & Sons, Inc., New York, 1940 (Dover reprint).

$$3xy - y^2 = c, \qquad 3x\frac{dy}{dx} + 3y - 2y\frac{dy}{dx} = 0, \qquad \frac{dy}{dx} = \frac{3y}{2y - 3x}. \tag{9}$$

In this way we obtain a differential equation having the given relation of Eq. (8) as its general solution.

If the family of integral curves is given in the form $g(x,y,c) = 0$, the differential equation having this as its general solution may be found by eliminating the constant c from the given relation and that found from it by differentiation.

For example, if $g(x,y,c) = y - cx - c^2$, we have

$$y = cx + c^2, \qquad \frac{dy}{dx} = c, \qquad \text{and} \qquad y = x\frac{dy}{dx} + \left(\frac{dy}{dx}\right)^2. \tag{10}$$

Suppose that a given equation contains two independent constants. Then we may obtain three equations from which they may be eliminated by differentiating twice. Thus from

$$y = c_1e^x + c_2e^{2x}, \qquad \frac{dy}{dx} = c_1e^x + 2c_2e^{2x}, \qquad \frac{d^2y}{dx^2} = c_1e^x + 4c_2e^{2x}. \tag{11}$$

The solution of the first two relations for c_1 and c_2 is

$$c_1 = e^{-x}\left(2y - \frac{dy}{dx}\right), \qquad c_2 = e^{-2x}\left(\frac{dy}{dx} - y\right). \tag{12}$$

Substitution of these expressions in the last relation leads to

$$\frac{d^2y}{dx^2} - 3\frac{dy}{dx} + 2y = 0. \tag{13}$$

Similarly, if an equation contains n independent constants, successive differentiation and elimination of the constants will lead to a differential equation of the nth order. And, conversely, an ordinary differential equation of the nth order will have a *general solution* containing n constants of integration.

Any solution obtained from the general solution by replacing the constants of integration by special numbers is called a *particular solution.*

Suppose we wish to verify that an equation having the form $y = f(x)$ is a solution of a given differential equation. We may replace y and the derivatives of y with respect to x by $f(x)$ and its derivatives in the differential equation. If this process produces an identity, $y = g(x)$ is a solution. Thus to verify that $y = c_1e^x + c_2e^{2x}$ is a solution of Eq. (13), we calculate the derivatives as in Eq. (11). Since all the terms cancel when we substitute these values in Eq. (13), the solution is verified.

EXAMPLE. Show that $y = c_1 \frac{e^x}{x} + c_2 \frac{e^{-x}}{x}$ is the general solution of the differential equation $x \frac{d^2y}{dx^2} + 2 \frac{dy}{dx} - xy = 0$.

Solution 1: Differentiating by the product rule, we find

$$\begin{array}{r|rcl} -x & y & = & \frac{1}{x}(c_1e^x + c_2e^{-x}), \\ 2 & \frac{dy}{dx} & = & -\frac{1}{x^2}(c_1e^x + c_2e^{-x}) + \frac{1}{x}(c_1e^x - c_2e^{-x}), \\ x & \frac{d^2y}{dx^2} & = & \left(\frac{1}{x} + \frac{2}{x^3}\right)(c_1e^x + c_2e^{-x}) - \frac{2}{x^2}(c_1e^x - c_2e^{-x}). \end{array}$$

We have written the coefficients of the given equation on the left. By multiplying these into the terms on the right and adding, we find that

$$-xy + 2\frac{dy}{dx} + x\frac{d^2y}{dx^2} = 0(c_1e^x + c_2e^{-x}) + 0(c_1e^x - c_2e^{-x}) = 0.$$

For a second-order differential equation, a solution containing two constants is the general solution. As $y = c_1 \frac{e^x}{x} + c_2 \frac{e^{-x}}{x}$ contains two independent constants and has been verified to be a solution, it is the general solution, as was to be proved.

Solution 2: Write the given solution as $xy = c_1e^x + c_2e^{-x}$. By differentiation we find $x\frac{dy}{dx} + y = c_1e^x - c_2e^{-x}$, and $x\frac{d^2y}{dx^2} + 2\frac{dy}{dx} = c_1e^x + c_2e^{-x}$. As the right members are the same for the first and third equations, we may eliminate the two constants by omitting the second equation and deducing that $x\frac{d^2y}{dx^2} + 2\frac{dy}{dx} = xy$. This equation, which has $xy = c_1e^x + c_2e^{-x}$ as its general solution, is equivalent to the given equation.

EXERCISE 162

In each of the following problems show that the first equation is the general solution of the given differential equation.

	General Solution	Differential Equation
1.	$y^2 = 5x^2 + c$.	$\frac{dy}{dx} = \frac{5x}{y}$.
2.	$y = cx^4$.	$\frac{dy}{dx} = \frac{4y}{x}$.
3.	$y^2 = 4cx$.	$\frac{dy}{dx} = \frac{v}{2x}$.
4.	$x^2 + y^2 = x^2y^2 + c$.	$\frac{dy}{dx} = \frac{x(y^2 - 1)}{y(1 - x^2)}$.
5.	$cy = c^2x + 1$.	$y\left(\frac{dy}{dx}\right) = x\left(\frac{dy}{dx}\right)^2 + 1$.
6.	$y^2 = cx^2 + c^2$.	$x^2y = x^3\frac{dy}{dx} + y\left(\frac{dy}{dx}\right)^2$.
7.	$x = cy + c^2xy$.	$y^3 = xy^2\frac{dy}{dx} + x^4\left(\frac{dy}{dx}\right)^2$.
8.	$y = x^2 + \frac{c}{x}$.	$x\frac{dy}{dx} + y = 3x^2$.
9.	$y = (x + c)e^{2x}$.	$\frac{dy}{dx} - 2y = e^{2x}$.

General Solution	Differential Equation
10. $y = c_1e^x + c_2e^{-x}$.	$\frac{d^2y}{dx^2} = y$.
11. $y = c_1 \sin x + c_2 \cos x$.	$\frac{d^2y}{dx^2} = -y$.
12. $y = c_1x^2 + c_2x^4$.	$x^2 \frac{d^2y}{dx^2} - 5x \frac{dy}{dx} + 8y = 0$.
13. $y = c_1 + c_2x + c_3x^2$.	$\frac{d^3y}{dx^3} = 0$.
14. $y = c(x - c)^2$.	$\left(\frac{dy}{dx}\right)^3 - 4xy \frac{dy}{dx} + 8y^2 = 0$.
15. $y = c_1e^{2x} + c_2e^{3x} + e^x$.	$\frac{d^2y}{dx^2} - 5 \frac{dy}{dx} + 6y = 2e^x$.
16. $y = \frac{x + c}{1 - cx}$.	$\frac{dy}{dx} = \frac{1 + y^2}{1 + x^2}$.
17. $y = c_1 \sin (5x + c_2)$.	$\frac{d^2y}{dx^2} = -25y$.
18. $y^2 = c_1x^2 + c_2$.	$xy \frac{d^2y}{dx^2} + x \left(\frac{dy}{dx}\right)^2 - y = 0$.
19. $y = c_1e^{-x} \cos (2x + c_2)$.	$\frac{d^2y}{dx^2} + 2 \frac{dy}{dx} + 5y = 0$.
20. $y = c_1e^{c_2x} - \frac{1}{c_2}$.	$y \frac{d^2y}{dx^2} - \left(\frac{dy}{dx}\right)^2 + \frac{dy}{dx} = 0$.

322. Equations of the First Order and First Degree. A differential equation of the first order and first degree may always be written in the form

$$M(x,y)dx + N(x,y)dy = 0 \quad \text{or} \quad N(x,y) \frac{dy}{dx} + M(x,y) = 0. \tag{14}$$

We consider the most important solvable types which are easy to recognize in Secs. 323 to 328.

323. Variables Separable. Suppose that in Eq. (14) each of the functions M and N is a product of factors, with each factor either a function of x alone, or a function of y alone. Let us divide the equation by the factor of M containing y and by the factor of N containing x. The resulting equation will then assume the form

$$f(x)dx - g(y)dy = 0 \quad \text{or} \quad f(x)dx = g(y)dy, \tag{15}$$

in which the variables are *separated*.

By direct integration of Eq. (15), we find

$$\int f(x)dx - \int g(y)dy = c \quad \text{or} \quad \int f(x)dx = \int g(y)dy + c \tag{16}$$

as the general solution.

EXAMPLE 1. Solve the equation $\frac{dy}{dx} = \frac{1 + y^3}{3(1 + x^3)xy^2}$.

Solution: Multiply the given equation by $3y^2\,dx$. And divide it by $(1 + y^3)$. This gives the separated form $\frac{3y^2\,dy}{1 + y^3} = \frac{dx}{x(1 + x^3)}$. Hence by Sec. 201, $\int \frac{3y^2\,dy}{1 + y^3} = \int \frac{dx}{x(1 + x^3)} = \int \left(\frac{1}{x} - \frac{x^2}{1 + x^3}\right) dx$. It follows that $\ln (1 + y^3) =$

$\ln x - \frac{1}{3}\ln(1+x^3) + C$. This leads to $3\ln(1+y^3) + \ln(1+x^3) = 3C + 3\ln x$, $e^{3\ln(1+y^3)+\ln(1+x^3)} = e^{3C+3\ln x}$. And by Sec. 115, $(1+y^3)^3(1+x^3) = e^{3C}x^3$. If we let $e^{3C} = c$, this becomes $(1+y^3)^3(1+x^3) = cx^3$, the required solution.

EXAMPLE 2. Solve the equation $2\left(x\dfrac{dy}{dx} - 3y\right) = xy\cot y\,\dfrac{dy}{dx}$.

Solution: From $(2x - xy\cot y)\dfrac{dy}{dx} = 6y$, by multiplying by dx and dividing by xy, we obtain $\dfrac{2 - y\cot y}{y}\,dy = 6\,\dfrac{dx}{x}$, in which the variables are separated. Hence $\displaystyle\int\left(\frac{2}{y} - \cot y\right)dy = 6\int\frac{dx}{x}$. It follows that $2\ln y - \ln\sin y = 6\ln x + C$. This leads to $e^{2\ln y - \ln\sin y} = e^{6\ln x + C}$. And by Sec. 115, $\dfrac{y^2}{\sin y} = e^{6C}x^6$. If we let $e^{6C} = c$, this becomes $\dfrac{y^2}{\sin y} = cx^6$, or $y^2 = cx^6\sin y$, the required solution.

EXERCISE 163

Solve each of the following differential equations.

1. $\dfrac{dy}{dx} = -\dfrac{x}{y}$.
2. $\dfrac{dy}{dx} = \dfrac{3+2x}{5-4y}$.
3. $y^3\,dx + x^3\,dy = 0$.
4. $\sec x\cos^2 y\,dx = \cos x\sin y\,dy$.
5. $e^{x-y}\,dx = e^{y-x}\,dy$.
6. $x\,dx + y\,dy = xy(x\,dy - y\,dx)$.
7. $\dfrac{dy}{dx} = \dfrac{y}{x}$.
8. $\dfrac{dy}{dx} = \dfrac{3+2y}{4-2x}$.
9. $(1+x^2)dy - xy\,dx = 0$.
10. $x(x+2)dy = 2y(x+1)dx$.
11. $dx = \sqrt{4-x^2}\,dy$.
12. $y\sec x\,dx = 2\sin x\,dy$.
13. $y\,dy = \sqrt{16-y^4}\,dx$.
14. $x\,dy + y\,dx = y^2\,dx$.
15. $(1+x^2)dy = -(1+y^2)dx$.
16. $\sqrt{1-x^2}\,dy = -\sqrt{1-y^2}\,dx$.

By Sec. 79, the integral curves of $\dfrac{dy}{dx} = -\dfrac{1}{f(x,y)}$ cut those of $\dfrac{dy}{dx} = f(x,y)$ at right angles. Either system of curves is called the *orthogonal trajectories* of the other. Use these facts to find the orthogonal trajectories of each given family of curves.

17. $xy = c$.
18. $x^2 + 3y^2 = c^2$.
19. $y = cx$.
20. $y = cx^2$.
21. $y = c\sin x$.
22. $y^2 + cx^2 = c$.

324. Equations Linear in *y*. Suppose that in Eq. (14) N is a function of x alone, and M is a first-degree polynomial in y. Thus the equation is

$$[B(x)y - C(x)]dx + A(x)dy = 0 \qquad \text{or} \qquad A(x)\frac{dy}{dx} + B(x)y = C(x). \tag{17}$$

Divide by $A(x)$. Then with new notation, the equation becomes

$$\frac{dy}{dx} + P(x)y = Q(x). \tag{18}$$

Multiply by $I\,dx$, where I is a function of x to be determined presently. The result is

$$I\,dy + IP(x)y\,dx = IQ(x)dx. \tag{19}$$

The first term suggests the product Iy whose differential is

$$d(Iy) = I\,dy + y\,dI. \tag{20}$$

To make the left member of Eq. (19) equal to the differential of the product IY, we must have

$$IP(x)y\,dx = y\,dI \qquad \text{or} \qquad dI = IP(x)dx. \tag{21}$$

This is a separable differential equation in the variables x and I. We may solve it for I by writing

$$\frac{dI}{I} = P(x)dx, \qquad \ln I = \int P(x)dx, \qquad \text{and } I = e^{\int P(x)dx}. \tag{22}$$

With any choice of constant in the integral, this makes

$$dI = IP(x)dx \qquad \text{and} \qquad d(Iy) = I\,dy + y\,dI = I\,dy + IP(x)y\,dx. \tag{23}$$

Hence Eq. (19) is equivalent to

$$d(Iy) = IQ(x)dx \qquad \text{and} \qquad Iy = \int IQ(x)dx + c. \tag{24}$$

To integrate a given linear equation, first reduce it to the form in Eq. (18). Then compute the *integrating factor* I from Eq. (22). Then either multiply in the factor to make the left member the differential of a product, or use Eq. (24) as a formula for the desired solution.

EXAMPLE. Solve the equation $x^2\,dy = (5x^3 + 2 - 3xy)dx$.

Solution: Divide by $x^2\,dx$ and transpose the term in y to the left member. The result is $\dfrac{dy}{dx} + \dfrac{3y}{x} = 5x + \dfrac{2}{x^2}$. Since the coefficient of y is $\dfrac{3}{x}$, by Eq. (22) we find $\ln I = \int \dfrac{3}{x}dx = 3\ln x$, $I = e^{3\ln x} = x^3$. Multiplication by $x^3\,dx$ leads to $x^3\,dy + 3x^2y\,dx = (5x^4 + 2x)dx$. Hence $x^3y = x^5 + x^2 + c$, or $y = x^2 + \dfrac{1}{x} + \dfrac{c}{x^2}$.

We may check the calculation of I, by mentally verifying $d(x^3y) = x^3\,dy + y\,d(x^3) = x^3\,dy + 3x^2y\,dx$.

325. Equations Reducible to the Linear Form. Consider the Bernoulli equation

$$A(x)\frac{dy}{dx} + B(x)y = C(x)y^n. \tag{25}$$

This differs from the linear equation only by the presence of y^n in the term on the right. Multiply by $\dfrac{(1-n)y^{-n}}{A(x)}$. Put

$$u = y^{1-n}, \qquad \frac{du}{dx} = (1-n)y^{-n}\frac{dy}{dx}. \tag{26}$$

Then the Eq. (25) becomes

$$\frac{du}{dx} + (1-n)\frac{B(x)}{A(x)}u = (1-n)\frac{C(x)}{A(x)}. \tag{27}$$

This may be solved for u by the procedure used to solve Eq. (18) for y. We may then replace u by y^{1-n} in the solution.

Again, we may interchange the roles of x and y and so solve an equation linear in x,

$$A(y)dx + [B(y)x - C(y)]dy = 0 \quad \text{or} \quad A(y)\frac{dx}{dy} + B(y)x = C(y). \tag{28}$$

More generally, we may use the procedure of Sec. 324 whenever we observe new variables $u(x,y)$ and $t(x,y)$ which reduce the given differential equation to the form

$$A(t)\frac{du}{dt} + B(t)u = C(t). \tag{29}$$

Example 1. Solve the equation $x\,dy + 2y\,dx - x^3y^3\,dx = 0$.

Solution: Divide by $x\,dx$ and transpose the term in y^3. The result is $\frac{dy}{dx} + \frac{2y}{x} = x^2y^3$, of the form of Eq. (25) with $n = 3$. Since $1 - n = -2$, put $u = y^{-2}$, $\frac{du}{dx} = -2y^{-3}\frac{dy}{dx}$. Multiplication by $-2y^{-3}$ leads to $-2y^{-3}\frac{dy}{dx} - \frac{4}{x}y^{-2} = -2x^2$ or $\frac{du}{dx} - \frac{4u}{x} = -2x^2$. Since the coefficient of u is $-\frac{4}{x}$, by Eq. (22) we find $\ln I = \int -\frac{4}{x}\,dx = -4\ln x$, $I = e^{-4\ln x} = x^{-4}$. Multiplication by $x^{-4}\,dx$ leads to $x^{-4}\,du - 4x^{-5}u\,dx = -2x^{-2}\,dx$. Hence $x^{-4}u = -2\frac{x^{-3}}{-3} + c$, $u = \frac{2}{3}x + cx^4$. Since $u = y^{-2}$, $y^{-2} = \frac{2}{3}x + cx^4$ or $y = \frac{1}{\sqrt{\frac{2}{3}x + cx^4}}$.

Example 2. Solve the equation $(xy + 1 + y^2)dy = (1 + y^2)dx$.

Solution: This is linear in x. Division by $(1 + y^2)dy$ and transposition of terms leads to $\frac{dx}{dy} - \frac{xy}{1+y^2} = 1$. Since the coefficient of x is $-\frac{y}{1+y^2}$, by Eq. (22) with y in place of x, we find $\ln I = \int \frac{-y\,dy}{1+y^2} = -\frac{1}{2}\ln(1+y^2)$, $I = e^{-\frac{1}{2}\ln(1+y^2)} = \frac{1}{\sqrt{1+y^2}}$. Multiplication by $\frac{dy}{\sqrt{1+y^2}}$ leads to $\frac{dx}{\sqrt{1+y^2}} - \frac{xy\,dy}{(1+y^2)^{\frac{3}{2}}} = \frac{dy}{\sqrt{1+y^2}}$. Hence $\frac{x}{\sqrt{1+y^2}} = \int \frac{dy}{\sqrt{1+y^2}} = \ln(y + \sqrt{y^2+1}) + c$. And $x = \sqrt{1+y^2}\,[\ln(y + \sqrt{y^2+1}) + c]$ or by Sec. 268, $x = \sqrt{1+y^2}\,(\sinh^{-1} y + c)$.

EXERCISE 164

Solve each of the following differential equations.

1. $x\dfrac{dy}{dx} + y = 3x^2.$
2. $\dfrac{dy}{dx} + y\tan x = \sec x.$
3. $x\,dy = 2(x + y)dx.$
4. $x\,dy = 3(y + 2x)dx.$
5. $\dfrac{dy}{dx} = (y - x)\cot x + 1.$
6. $x\dfrac{dy}{dx} = y + (1 - x)e^x.$
7. $(x + 1)\dfrac{dy}{dx} = 2y + (x + 1)^4.$
8. $(x^2 - 1)\dfrac{dy}{dx} + xy = x(x^2 + 1).$
9. $x^3\dfrac{dy}{dx} + 2x^2y = y^3.$
10. $x^2\,dy = y(x - y)dx.$
11. $\dfrac{dy}{dx} + \dfrac{y}{x} = y^3.$
12. $\dfrac{dy}{dx} + \dfrac{2y}{x} = 2xy^2.$
13. $5x\dfrac{dy}{dx} + 2y = xy^6.$
14. $2\dfrac{dy}{dx} + y = e^xy^3.$
15. $y\,dx = (x + 3y^4)dy.$
16. $y\,dx = 3(x + 4y)dy.$
17. $\tan y\,dx = 2(x + \sec y)dy.$
18. $y\,dx = 4(x + y)dy.$
19. $y\,dx = (\sin y - x)dy.$
20. $y^2\,dx = (1 - 2xy)dy.$

326. Homogeneous Equations. A function of any number of variables is *homogeneous* of the nth degree in these variables if, when each variable is multiplied by a scale factor k, the function is multiplied by k^n. Thus $M(x,y)$ is homogeneous of the nth degree in x and y if

$$M(kx,ky) = k^nM(x,y). \tag{30}$$

Such a function is x^n times a function of y/x, since if $k = 1/x$,

$$M\left(1,\frac{y}{x}\right) = x^{-n}M(x,y) \quad \text{and} \quad M(x,y) = x^n\,M\left(1,\frac{y}{x}\right). \tag{31}$$

The differential equation (14) is *homogeneous* if M and N are each homogeneous of the same degree, say n. Then Eq. (14) is

$$x^nM\left(1,\frac{y}{x}\right)dx + x^nN\left(1,\frac{y}{x}\right)dy = 0. \tag{32}$$

Divide by x^n. Put $y = vx$, so that $y/x = v$ and

$$dy = v\,dx + x\,dv. \tag{33}$$

Then Eq. (32) takes the form

$$M(1,v)dx + N(1,v)[v\,dx + x\,dv] = 0 \quad \text{or} \quad \frac{dx}{x} = \frac{-N(1,v)dv}{M(1,v) + vN(1,v)}. \tag{34}$$

In the last equation the variables v and x are separated. Thus it may be solved by direct integration. We may then replace v by y/x in the solution.

EXAMPLE. Solve the equation $(2x^2 + 3y^2)x \dfrac{dy}{dx} = (x^2 + 2y^2)y$.

Solution: The coefficient of dy/dx and the right member are each homogeneous polynomials of the third degree. Hence the equation is homogeneous. Multiply by dx, divide by x^3, replace y/x by v and dy by $v\,dx + x\,dv$. The result is $(2 + 3v^2)(v\,dx + x\,dv) = (1 + 2v^2)v\,dx$, $(2 + 3v^2)x\,dv = (1 + 2v^2 - 2 - 3v^2)v\,dx$, or $\dfrac{dx}{x} = -\dfrac{2 + 3v^2}{(1 + v^2)v}$ in which the variables are separated. Hence by Sec. 201, $\displaystyle\int \frac{dx}{x} = -\int \frac{2 + 3v^2}{(1 + v^2)v}\,dv = -\int \left(\frac{2}{v} + \frac{v}{1 + v^2}\right) dv$. It follows that $\ln x = -2 \ln v - \dfrac{1}{2} \ln (1 + v^2) + C$. Since $v = \dfrac{y}{x}$, this leads to

$2C = 2 \ln x + 4 \ln \dfrac{y}{x} + \ln \left(1 + \dfrac{y^2}{x^2}\right) = \ln (x^2 + y^2) + 4 \ln y - 4 \ln x$,

$e^{\ln (x^2+y^2)+4 \ln y} = e^{2C+4 \ln x}$. And by Sec. 115 with $e^{2C} = c$, $y^4(x^2 + y^2) = cx^4$.

327. Integrable Combinations. In Sec. 271 we defined the total differential of $u(x,y)$ as

$$du = \frac{\partial u}{\partial x}\,dx + \frac{\partial u}{\partial y}\,dy. \tag{35}$$

Suppose that, in Eq. (14), $M = \dfrac{\partial u}{\partial x}$ and $N = \dfrac{\partial u}{\partial y}$. Then

$$du = M(x,y)dx + N(x,y)dy = 0 \qquad \text{and} \qquad u = c. \tag{36}$$

The differential equation (14), or $M\,dx + N\,dy$, is *exact* if there is a function $u(x,y)$ whose total differential $du = M\,dx + N\,dy$. The solution of an exact differential equation can be found in the form $u(x,y) = c$ by direct integration.

In practice, we often recognize that such an equation is exact by writing $M\,dx + N\,dy$ as a sum of integrable combinations, or groups of terms each of which is a function of one variable only, or the exact differential of a simple function of x and y. From Eq. (35), or the method of Sec. 272, we may verify that

$$d(xy) = y\,dx + x\,dy, \tag{37}$$

$$d\left(\frac{y}{x}\right) = \frac{-y\,dx + x\,dy}{x^2}, \tag{38}$$

$$d(x^m y^n) = x^{m-1}y^{n-1}(my\,dx + nx\,dy). \tag{39}$$

Other combinations may be built up from the product rule, $d(uv) = u\,dv + v\,du$ with u and v each any function of x and y. And the product of du by any function of u is exact. Thus

$$\frac{-y\,dx + x\,dy}{x^2 + y^2} = \frac{-y\,dx + x\,dy}{x^2}\,\frac{x^2}{x^2 + y^2} = d\left(\frac{y}{x}\right)\frac{1}{1 + (y/x)^2}$$

$$= d \tan^{-1} \frac{y}{x}. \tag{40}$$

This is a special case of the fact that $-y\,dx + x\,dy$ becomes exact when divided by any homogeneous function of the second degree.

EXAMPLE. Solve the differential equation

$$\left(2xe^y + y^3e^x + \frac{1}{\sqrt{4 - x^2}}\right) dx + \left(x^2e^y + 3y^2e^x + \frac{1}{4 + y^2}\right) dy = 0.$$

Solution: We regroup the terms as follows:

$$(e^y\,2x\,dx + x^2e^y\,dy) + (y^3e^x\,dx + e^x\,3y^2\,dy) + \frac{dx}{\sqrt{4 - x^2}} + \frac{dy}{4 + y^2} = 0.$$

Each parenthesis contains the derivative of a product, so that

$x^2e^y + y^3e^x + \sin^{-1}\frac{x}{2} + \frac{1}{2}\tan^{-1}\frac{y}{2} = c$ is the required solution.

328. Integrating Factors. Let $u(x,y)$ and $I(x,y)$ be two functions of x and y such that $du = I(M\,dx + N\,dy)$. Then the function I is called an *integrating factor* of the differential equation (14), or $M\,dx + N\,dy = 0$, since multiplication by this factor makes the equation exact.

If the terms of an equation can be broken up into two groups, one of which is an integrable combination, it is sometimes possible to discover a factor which does not destroy the integrable character of the first group, but converts the second group into an integrable combination.

EXAMPLE 1. Solve the equation $y\,dx = (1 + y^2\sqrt{x^2 - y^2})x\,dy$.

Solution: Write the equation $(-y\,dx + x\,dy) + xy^2\sqrt{x^2 - y^2}\,dy = 0$. By the remark made after Eq. (40), $(-y\,dx + x\,dy)$ becomes exact when divided by any homogeneous function of the second degree. The term in dy will be free of x if we divide by $x\sqrt{x^2 - y^2}$, which is homogeneous of the second degree. Hence we divide by $x\sqrt{x^2 - y^2}$. The result is $\dfrac{-y\,dx + x\,dy}{x\sqrt{x^2 - y^2}} + y^2\,dy = 0$. And $\dfrac{-y\,dx + x\,dy}{x\sqrt{x^2 - y^2}} = \dfrac{-y\,dx + x\,dy}{x^2}\,\dfrac{x}{\sqrt{x^2 - y^2}} = d\left(\dfrac{y}{x}\right)\dfrac{1}{\sqrt{1 - (y/x)^2}} = d\sin^{-1}\dfrac{y}{x}$. Hence $\sin^{-1}\dfrac{y}{x} + \dfrac{y^3}{3} = c$ is the required solution.

EXAMPLE 2. Solve the equation $y^2(2y\,dx - 3x\,dy) = x^2\,dx$.

Solution: The left member suggests Eq. (39) with $m = 2$, $n = -3$, $d(x^2y^{-3}) = xy^{-4}(2y\,dx - 3x\,dy)$. Hence multiplying the given equation by xy^{-6} leads to $d(x^2y^{-3}) = x^3y^{-6}\,dx$. The left member remains exact on multiplication by any function of (x^2y^{-3}). We use $(x^2y^{-3})^{-2}$ to rid the right member of y^{-6}. This leads to

$$(x^2y^{-3})^{-2}d(x^2y^{-3}) = dx/x, \qquad -(x^2y^{-3})^{-1} = \ln x + C, \qquad \ln x + y^3/x^2 = -C,$$

or with $c = -C$, $y^3 + x^2 \ln x = cx^2$, the required solution.

EXERCISE 165

Solve each of the following homogeneous differential equations.

1. $\dfrac{dy}{dx} = \dfrac{y - x}{y + x}$.
2. $\dfrac{dy}{dx} = \dfrac{y^2 - x^2}{2xy}$.
3. $(3x + 5y)dx + (4x + 6y)dy = 0$.
4. $(x^2 + y^2)dx + xy\,dy = 0$.
5. $(8y + 10x)dx + (5y + 7x)dy = 0$.
6. $y\,dx = x\,dy + \sqrt{xy}\,dx$.
7. $(3x + y)dx + (x + y)dy = 0$.
8. $(5x + 3y)dx + 6x\,dy = 0$.
9. $y\,dx - x\,dy = (x^2 + y^2)\,dx$.
10. $y\,dx = (x + 2y)dy$.

Solve each of the following exact differential equations.

11. $(x - 3y)dx = 3(y - x)dy$.
12. $2xy\,dy = (x^2 - y^2)dx$.
13. $(x^3 + y^3)dy + 3x^2y\,dx = 0$.
14. $y\,dx = (\sin y - x)dy$.
15. $y(3x^2 - y^2)dx = x(3y^2 - x^2)dy$.
16. $2xy\,dy = (e^x - y^2)dx$.

Solve each of the following differential equations by means of an appropriately chosen integrating factor.

17. $y\,dx - x\,dy = 2x^3\,dx$.
18. $x\,dy - y\,dx = 3y^4\,dy$.
19. $y\,dx - x\,dy = (x^2 + y^2)dx$.
20. $x\,dy - y\,dx = x^2(2x\,dx + 4y\,dy)$.

329. Equations of the First Order and Higher Degree. The general differential equation of the first order is

$$F\left(x,y,\frac{dy}{dx}\right) = 0 \qquad \text{or} \qquad F(x,y,p) = 0, \tag{41}$$

if we write a single letter p in place of dy/dx.

For $F(x,y,p)$ a first-degree polynomial in p, Eq. (41) is equivalent to Eq. (14), already discussed in Secs. 322 to 328.

We now assume that $F(x,y,p)$ is a polynomial in p of degree higher than the first. And we describe methods of solving the resulting higher degree equation in Secs. 330 to 332.

330. Equations Solvable for p. It is sometimes possible to solve Eq. (41) for p. Each of the resulting first-degree equations may then be treated by one of the methods given in Secs. 323 to 328. For example,

$$y^2\left(\frac{dy}{dx}\right)^2 = 1 - y^2 \qquad \text{leads to} \qquad y\frac{dy}{dx} = \pm\sqrt{1 - y^2}. \tag{42}$$

We take the plus sign, separate the variables, and obtain

$$\frac{y\,dy}{\sqrt{1 - y^2}} = dx, \qquad -\sqrt{1 - y^2} = x - c. \tag{43}$$

Squaring both sides gives

$$y^2 + (x - c)^2 = 1, \tag{44}$$

which includes the result that would have been obtained by taking the minus sign. It is the general solution of Eq. (42) and is the family of circles shown in Fig. 362.

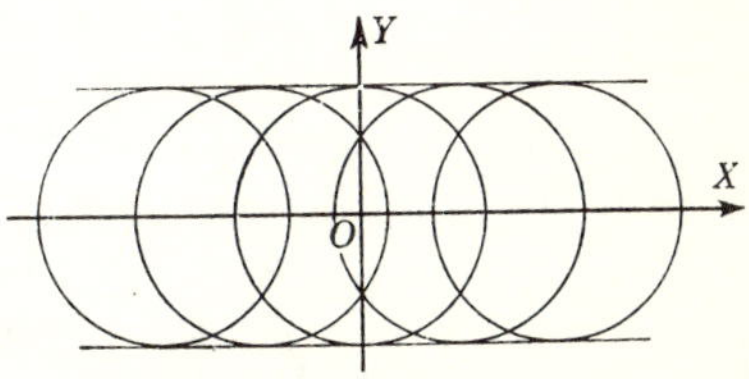

FIG. 362.

331. Singular Solutions. The circles of Eq. (44) have an envelope consisting of the two lines $y = \pm 1$. This is evident from Fig. 362 and could be found by applying the method of Sec. 279 to Eq. (44). Direct substitution in Eq. (42) shows that $y = 1$ and $y = -1$ are each solutions of this differential equation. They are called *singular solutions* because they cannot be obtained from any general solution by specializing the constant. In separating the variables, we divided by $\sqrt{1 - y^2}$, a permissible operation unless $y^2 = 1$. Thus Eq. (42) implies either Eq. (43) and the general solution of Eq. (44), or else that $y^2 = 1$, the singular solutions, since $y = \pm 1$ happens to solve the original differential equation (42).

A differential equation of the first order will always have a general solution, consisting of a one-parameter infinite family of curves. If the differential equation is of the first degree, there will be one slope and one of these curves through each point. For equations of degree higher than the first, there will be more than one slope at some points and the curves of the general solution may have an envelope at which two values of the slope coincide. Since x,y,p have the same values at each point of the envelope as they have for the tangent integral curve, the envelope will be a solution of the differential equation. Usually the envelope will not be a curve of the general solution but will be a singular solution.

Points at which Eq. (41), regarded as an equation in p, has coincident roots satisfy

$$F(x,y,p) = 0 \qquad \text{and} \qquad \frac{\partial F}{\partial p} = 0. \tag{45}$$

By Eqs. (86) and (91) of Sec. 279, the envelope of the family of curves $g(x,y,c) = 0$ which make up the general solution of Eq. (41) will satisfy

$$g(x,y,c) = 0 \qquad \text{and} \qquad \frac{\partial g}{\partial c} = 0. \tag{46}$$

Thus singular solutions may be found by testing which curves obtained from Eq. (45) or (46) satisfy the differential equation.

332. Clairaut's Form. Clairaut's differential equation is

$$y = x\frac{dy}{dx} + f\left(\frac{dy}{dx}\right) \qquad \text{or} \qquad y = px + f(p), \tag{47}$$

where f is any function of one variable. To solve it, differentiate with respect to x. The result is

$$\frac{dy}{dx} = p + x\frac{dp}{dx} + f'(p)\frac{dp}{dx} \qquad \text{or} \qquad [x + f'(p)]\frac{dp}{dx} = 0, \tag{48}$$

since we may cancel dy/dx against p. Equation (48) will hold if

$$\frac{dp}{dx} = 0, \qquad \text{which implies } p = c. \tag{49}$$

Inserting this in Eq. (47) leads to the general solution

$$y = cx + f(c). \tag{50}$$

This represents a family of straight lines. They have an envelope given in terms of the parameter c by

$$y = cx + f(c) \qquad \text{and} \qquad x + f'(c) = 0. \tag{51}$$

This is the singular solution. Replacing c by p gives

$$y = pz + f(p) \qquad \text{and} \qquad x + f'(p) = 0. \tag{52}$$

The equations in this form might have been obtained by observing that Eq. (48) would hold if the factor $x + f'(p)$ were zero, and combining $x + f'(p) = 0$ with Eq. (47). We also note that Eqs. (45) and (46), applied to the problem under discussion, lead to Eqs. (51) and (52).

Many equations of the first order and higher degree with singular solutions of simple form are reducible to Clairaut's form on introducing new variables u and t. For example, consider

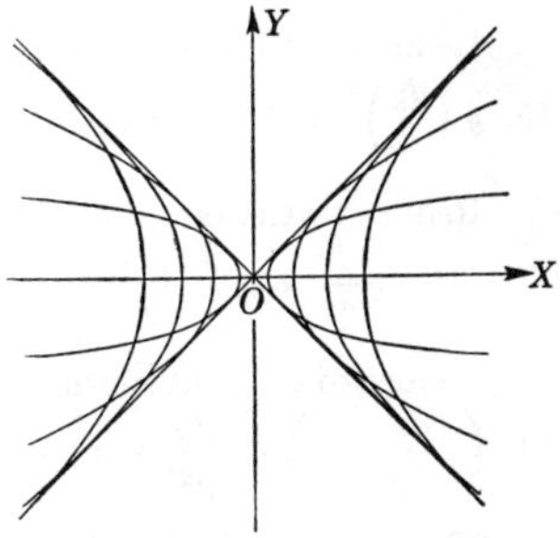

FIG. 363.

$$y\left(\frac{dy}{dx}\right)^2 - 2x\frac{dy}{dx} + y = 0. \tag{53}$$

Multiply by y and put $y^2 = u$, $2y\dfrac{dy}{dx} = \dfrac{du}{dx}$, obtaining

$$\frac{1}{4}\left(\frac{du}{dx}\right)^2 - x\frac{du}{dx} + u = 0 \qquad \text{or} \qquad u = x\frac{du}{dx} - \frac{1}{4}\left(\frac{du}{dx}\right)^2. \tag{54}$$

This is Clairaut's form, so that the general solution is

$$u = cx - \frac{c^2}{4} \qquad \text{or} \qquad y^2 = cx - \frac{c^2}{4}. \tag{55}$$

The singular solution is given by

$$y^2 = cx - \frac{c^2}{4} \qquad \text{and} \qquad 0 = x - \frac{c}{2}, \qquad \text{or} \qquad y = \pm x. \tag{56}$$

The family of parabolas of Eq. (55) and the two straight lines of Eq. (56) which make up their envelope are shown in Fig. 363.

EXERCISE 166

Solve the equivalent first-degree relations obtained by solving each of the following higher order equations for $p = dy/dx$.

1. $\left(\dfrac{dy}{dx}\right)^2 + x(y+1)\dfrac{dy}{dx} + x^2y = 0.$
2. $\left(\dfrac{dy}{dx}\right)^2 = 2x\dfrac{dy}{dx} + 3x^2.$
3. $\left(\dfrac{dy}{dx}\right)^2 + xy = (x+y)\dfrac{dy}{dx}.$
4. $y^2\left(\dfrac{dy}{dx}\right)^2 = 1 + y^2.$

5. $xy\left[\left(\frac{dy}{dx}\right)^2 + 1\right] = (x^2 + y^2)\frac{dy}{dx}.$ 6. $x^2\left(\frac{dy}{dx}\right)^2 + xy\frac{dy}{dx} = 2y^2.$

7. $\left(2x\frac{dy}{dx} - y\right)^2 = 8x^3.$ 8. $(1 + x^2)\left(\frac{dy}{dx}\right)^2 = 1.$

Solve each of the following Clairaut equations.

9. $y = x\frac{dy}{dx} + 1 + \left(\frac{dy}{dx}\right)^2.$ 10. $\left(\frac{dy}{dx}\right)^2 + y = x\frac{dy}{dx}.$

11. $\left(y - x\frac{dy}{dx}\right)^2 = \left(\frac{dy}{dx}\right)^3.$ 12. $y + \ln\frac{dy}{dx} = x\frac{dy}{dx}.$

13. $x\left(\frac{dy}{dx}\right)^2 + 1 = y\frac{dy}{dx}.$ 14. $\left(y - x\frac{dy}{dx}\right)^2 = 1 + \left(\frac{dy}{dx}\right)^2.$

Reduce to Clairaut's form by putting $y^2 = u$, and solve.

15. $y\left(\frac{dy}{dx}\right)^2 + 2x\frac{dy}{dx} = y.$ 16. $y = 2x\frac{dy}{dx} - 4y\left(\frac{dy}{dx}\right)^2.$

Reduce to Clairaut's form by putting $x^2 = t$, and solve.

17. $x + 2y\frac{dy}{dx} = x\left(\frac{dy}{dx}\right)^2.$ 18. $x\left(\frac{dy}{dx}\right)^2 + 4x = 2y\frac{dy}{dx}.$

Reduce to Clairaut's form by putting $y^2 = u$, $x^2 = t$, and solve.

19. $\left(x\frac{dy}{dx} - y\right)\left(y\frac{dy}{dx} + x\right) = \frac{dy}{dx}.$ 20. $y = x\frac{dy}{dx} + \frac{y}{x^2}\left(\frac{dy}{dx}\right)^2.$

333. Geometric Applications. A family of curves may be characterized by a geometric property which leads to an equation connecting x,y, the coordinates of any point on one of the curves, with $p = dy/dx$, the slope of that curve at (x,y). This equation will be a first-order differential equation, whose general solution represents the family of curves.

Sometimes the differential equation can be set up by reference to a figure and the use of similar triangles. But the relations of Secs. 87 and 88 often prove to be helpful. And if the expression of the property is simpler in terms of polar coordinates, we may use the relations of Secs. 148 to 151.

Example 1. The normal to a curve at the point P intersects the x axis at N, and the y axis at Q. Find the curves for which $NQ = 2QP$.

Solution: In Fig. 364, $NO/OM = NQ/QP = 2$, $NO = 2OM = 2x$, $NM = NO + OM = 3x$. Hence $\tan MPN = NM/MP = 3x/y$.

But $\tan\phi = -\tan MTP = -\tan MPN = -3x/y$. Thus $dy/dx = -3x/y$.

We might have used $\frac{MP}{NM} = \frac{y}{3x} =$ slope of normal $= -\frac{1}{dy/dx}$, by Eq. (105) of Sec. 87, which again gives $dy/dx = -3x/y$.

To solve this, we separate variables. $y\,dy = -3x\,dx$. And we integrate to obtain $y^2/2 = -3x^2/2 + C$, or with $c = 2C$, $y^2 + 3x^2 = c$, a family of similar ellipses.

Example 2. The angle from the x axis to the radius vector drawn to any point of a curve is twice the angle from the x axis to the tangent line at that point. Find the equation of the curve.

Solution 1: Let ϕ be the slope angle, and θ be the angle XOP (Fig. 365). Then $\theta = 2\phi$, $\tan\theta = \frac{2\tan\phi}{1 - \tan^2\phi}$. But $\tan\theta = \frac{y}{x}$ and $\tan\phi = \frac{dy}{dx} = p$, so that $\frac{y}{x} =$

$\dfrac{2p}{1-p^2}$ and $yp^2 + 2xp - y = 0$. Solving for p, $p = \dfrac{-2x \pm \sqrt{4x^2+4y^2}}{2y} = \dfrac{-x \pm \sqrt{x^2+y^2}}{y}$. Take the plus sign and write $\dfrac{dy}{dx} = p = \dfrac{-x + \sqrt{x^2+y^2}}{y}$. In this homogeneous equation, put $y = vx$, $\dfrac{dy}{dx} = v + x\dfrac{dv}{dx}$. $\dfrac{v\,dx + x\,dv}{dx} = \dfrac{-1+\sqrt{1+v^2}}{v}$, $(v^2 + 1 - \sqrt{1+v^2})dx = -xv\,dv$, $\dfrac{dx}{x} = \dfrac{-v\,dv}{v^2+1-\sqrt{v^2+1}}$. If $v^2 + 1 = t^2$, $\dfrac{-v\,dv}{v^2+1-\sqrt{v^2+1}} = \dfrac{-t\,dt}{t^2-t} = \dfrac{-dt}{t-1} = -d\ln(t-1)$. Hence $\ln x = -\ln(\sqrt{v^2+1}-1) + C$. Since $v = \dfrac{y}{x}$, this leads to $C = \ln x + \ln\left(\sqrt{\dfrac{y^2}{x^2}+1}-1\right) = \ln(\sqrt{x^2+y^2}-x)$. $e^C = e^{\ln(\sqrt{x^2+y^2}-x)}$. And with $e^C = c$, $c = \sqrt{x^2+y^2} - x$. Hence $\sqrt{x^2+y^2} = x + c$.

We might have deduced this from $\dfrac{dy}{dx} = \dfrac{-x+\sqrt{x^2+y^2}}{y}$ by writing it in the form $\dfrac{x\,dx + y\,dy}{\sqrt{x^2+y^2}} = dx$. This is exact, so that $\sqrt{x^2+y^2} = x + c$ follows by direct integration.

By squaring, we find $x^2 + y^2 = x^2 + 2cx + c^2$ and $y^2 = 2cx + c^2$, a family of parabolas with focus at the origin.

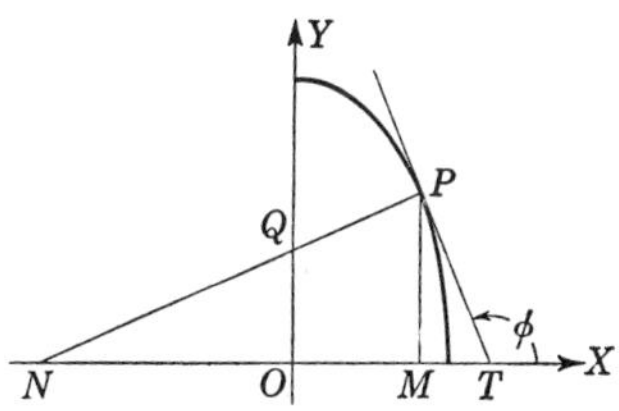

FIG. 364.

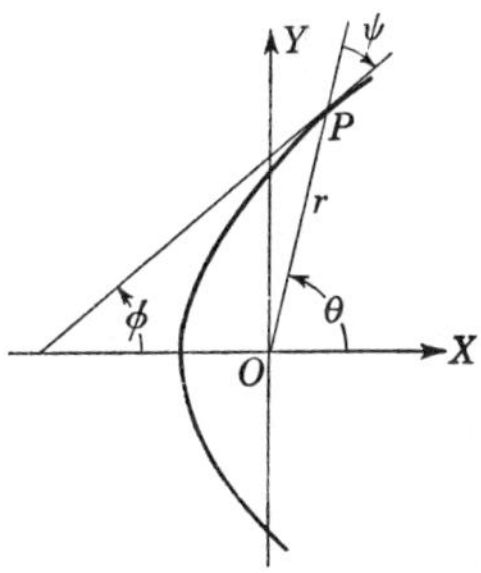

FIG. 365.

The equation $yp^2 + 2xp - y = 0$ could also be solved by the method used for Eq. (53) to give the general solution $y^2 = Cx + C^2/4$ and $x^2 + y^2 = 0$ which is satisfied by (0,0) only so that there is no singular solution.

Solution 2: Use polar coordinates (r,θ). Then if ϕ is the slope angle, $\phi = \theta/2$. But from Eq. (17) of Sec. 148, $\phi = \theta + \psi$. Hence $\psi = \phi - \theta = \theta/2 - \theta = -\theta/2$. $\tan\psi = -\tan(\theta/2)$. But by Eq. (39) of Sec. 150, $\tan\psi = \dfrac{r\,d\theta}{dr}$ so that $\dfrac{r\,d\theta}{dr} = -\tan\dfrac{\theta}{2}$. Separating the variables gives $\dfrac{dr}{r} = -\dfrac{d\theta}{\tan(\theta/2)} = \dfrac{-\cos(\theta/2)\,d\theta}{\sin(\theta/2)}$. Integrating $\ln r = -2\ln\sin\dfrac{\theta}{2} + C$, $e^{\ln r} = e^{C-2\ln\sin(\theta/2)}$, or with $e^C = c$, $r = \dfrac{c}{\sin^2(\theta/2)} = \dfrac{2c}{1-\cos\theta}$. This is a parabola with focus at the origin by Example 4 of Sec. 146.

EXAMPLE 3. The area bounded above by a given curve, below by the x axis, and between a fixed ordinate and a variable ordinate is proportional to the arc between these ordinates. Find the equation of the curve.

Solution: Let A be the area, let s be the arc length, and let a be the factor of proportionality. Then $A = as$. Hence $dA = a\,ds$. But $dA = y\,dx$ and $ds = \sqrt{dx^2 + dy^2}$ so that $y\,dx = a\sqrt{dx^2 + dy^2}$. Hence $(y^2 - a^2)dx^2 = a^2\,dy^2$, $a\,dy = \pm\sqrt{y^2 - a^2}\,dx$. Take the plus sign and separate the variables. $\dfrac{a\,dy}{\sqrt{y^2 - a^2}} = dx$ if $\sqrt{y^2 - a^2} \neq 0$.

If $\sqrt{y^2 - a^2} = 0$, $y = \pm a$ which satisfies $(y^2 - a^2)dx^2 = a^2\,dy^2$ and so gives the singular solution.

Integrating $\dfrac{a\,dy}{\sqrt{y^2 - a^2}} = dx$, $a \ln (y + \sqrt{y^2 - a^2}) = x + C$. $e^{\ln (y+\sqrt{y^2-a^2})} = e^{(x+C)/a}$. With $e^{C/a} = B$, this becomes $y + \sqrt{y^2 - a^2} = Be^{x/a}$. Hence $\sqrt{y^2 - a^2} = Be^{x/a} - y$, $y^2 - a^2 = B^2e^{2x/a} - 2Be^{x/a}y + y^2$, $2Be^{x/a}y = B^2e^{2x/a} + a^2$, $y = \frac{1}{2}\left(Be^{x/a} + \frac{a^2}{B}e^{-x/a}\right) = \frac{a}{2}\left(\frac{B}{a}e^{x/a} + \frac{a}{B}e^{-x/a}\right)$. If $\frac{B}{a} = e^{c/a}$ or $c = a \ln \frac{B}{a}$, $y = \frac{a}{2}[e^{(x+c)/a} + e^{-(x+c)/a}]$, a catenary.

We give another way of eliminating the radical from $y + \sqrt{y^2 + a^2} = Be^{x/a}$. Note that $(y + \sqrt{y^2 + a^2})(y - \sqrt{y^2 + a^2}) = y^2 - (y^2 + a^2) = -a^2$, so that $y - \sqrt{y^2 + a^2} = \dfrac{-a^2}{y + \sqrt{y^2 + a^2}} = \dfrac{-a^2}{Be^{x/a}} = -\dfrac{a^2}{B}e^{-x/a}$. It then follows by addition that $2y = Be^{x/a} - \frac{a^2}{B}e^{-x/a}$. And $y = \frac{1}{2}\left(Be^{x/a} - \frac{a^2}{B}e^{-x/a}\right)$ as before.

By Eq. (146) of Sec. 268, $\ln (y + \sqrt{y^2 - a^2}) = \cosh^{-1}\frac{y}{a} + \ln a$. Hence we could deduce from $a \ln (y + \sqrt{y^2 - a^2}) = x + C$ that $a \cosh^{-1}\frac{y}{a} + a \ln a = x + C$, or with $C - a \ln a = c$, $\cosh^{-1}\frac{y}{a} = \frac{x + c}{a}$, $\frac{y}{a} = \cosh\frac{x + c}{a}$ or $y = a \cosh\frac{x + c}{a}$.

The required curves are the family of catenaries $y = \frac{a}{2}[e^{(x+c)/a} + e^{-(x+c)/a}] = a \cosh\frac{x + c}{a}$ and the two straight lines $y = a$, $y = -a$.

Example 4. The tangent to a curve at the point P intersects the x axis at T, and the y axis at S. Find the curves for which the sum of the nth roots of the intercepts $\sqrt[n]{OT} + \sqrt[n]{OS}$ is constant, $\sqrt[n]{a}$.

Solution: By Eq. (104) of Sec. 87, the equation of the tangent line is $(y - y_1) = \left(\frac{dy}{dx}\right)_1 (x - x_1)$. Since $y = OS$ when $x = 0$, $OS = y_1 - \left(\frac{dy}{dx}\right)_1 x_1$. And $x = OT$ when $y = 0$, $-y_1 = \left(\frac{dy}{dx}\right)_1 (OT - x_1)$, $OT = x_1 - \dfrac{y_1}{(dy/dx)_1}$. Dropping the subscripts 1 and replacing $\frac{dy}{dx}$ by p, we have for the intercepts

$$OT = -\frac{y - px}{p}, \qquad OS = y - px.$$

If $\sqrt[n]{OT} + \sqrt[n]{OS} = \sqrt[n]{a}$, $\sqrt[n]{\dfrac{y - px}{-p}} + \sqrt[n]{y - px} = \sqrt[n]{a}$. Hence $\sqrt[n]{y - px}\left(1 + \dfrac{1}{\sqrt[n]{-p}}\right) = \sqrt[n]{a}$, $y - px = a\left(1 + \dfrac{1}{\sqrt[n]{-p}}\right)^{-n}$, and $y = px + a\left(1 + \dfrac{1}{\sqrt[n]{-p}}\right)^{-n}$. This is of Clairaut's form. By Eq. (50) the general

solution is $y = cx + a\left(1 + \frac{1}{\sqrt[n]{-c}}\right)^{-n}$. This merely represents the tangent straight lines to the desired curve, which is their envelope. Here $f(p) = a\left(1 + \frac{1}{\sqrt[n]{-p}}\right)^{-n}$ so that $f'(p) = -an\left(1 + \frac{1}{\sqrt[n]{-p}}\right)^{-n-1}\frac{1}{n}(-p)^{-(1/n)-1} =$
$-a(-p)^{-(n+1)/n}\left(1 + \frac{1}{\sqrt[n]{-p}}\right)^{-n-1}$. Hence by Eq. (52), $x = -f'(p) =$
$a(-p)^{-(n+1)/n}\left(1 + \frac{1}{\sqrt[n]{-p}}\right)^{-n-1}$. Substitution of this in $y = px + a\left(1 + \frac{1}{\sqrt[n]{-p}}\right)^{-n}$
gives $y = a\left[1 + \frac{1}{\sqrt[n]{-p}}\right]^{-n-1}\left[-(-p)^{-1/n} + 1 + \frac{1}{\sqrt[n]{-p}}\right] = a\left(1 + \frac{1}{\sqrt[n]{-p}}\right)^{-n-1}$.
Hence $x^{1/(n+1)} = a^{1/(n+1)}\frac{1}{\sqrt[n]{-p}}\left(1 + \frac{1}{\sqrt[n]{-p}}\right)^{-1}$, $y^{1/(n+1)} = a^{1/(n+1)}\left(1 + \frac{1}{\sqrt[n]{-p}}\right)^{-1}$.
It follows that $x^{1/(n+1)} + y^{1/(n+1)} = a^{1/(n+1)}$ is the required curve.

EXERCISE 167

The tangent to a curve at P cuts the x axis in T and the y axis in S. Find the family of curves for which, at each point,

1. $TP = PS$.
2. $ST = TP$.
3. $TS = SP$.
4. The triangle TOP has equal sides $TP = OP$.
5. In triangle TOP, angle OTP = angle $TOP + 90°$.

The normal to a curve at P cuts the x axis in N and the y axis in Q. Find the family of curves for which, at each point,

6. $NQ = QP$.
7. $QN = NP$.
8. $NP = PQ$.
9. The triangle NOP has equal sides $NP = OP$.

Through each point P of a curve, lines PD parallel to OX and PC parallel to OY are drawn to form a rectangle $OCPD$ with two sides on the axes. Let A denote the part of the rectangle below the curve, and B the part of the rectangle above the curve. Find the equation of the curve if

10. The area A, with $dA = y\,dx$, is twice the area B with $dB = x\,dy$.
11. When rotated about OY, the volume generated by A, V_A with $dV_A = 2\pi xy\,dx$, is equal to the volume generated by B, V_B with $dV_B = \pi x^2\,dy$.
12. The volume V_A of Prob. 11 is twice the volume V_B of Prob. 11.
13. Find the equation of the curve in polar coordinates which cuts all the radius vectors at the same angle A.

Let $P_1 = (r_1,\theta_1)$ be a fixed point and $P = (r,\theta)$ a variable point on a curve. Let arc $P_1P = s$, sector $P_1OP = S$. Find the equation of the curve in polar coordinates if

14. $S = k(r - r_1)$.
15. $s = k(\theta - \theta_1)$.
16. $s = k(r - r_1)$.
17. $S = ks$.

If the tangent to a curve at the point P intersects the x axis at T and the y axis at S, as in Example 4, $OT = \frac{y - px}{-p}$, $OS = y - px$. Use these to find the curve such that

18. The tangent to the curve and the coordinate axes form a triangle of constant area, a^2,

19. The sum of the intercepts $OT + OP$ is constant, a.
20. The part of the tangent cut off by the axes TS is a constant a.
21. The sum of the squares of the reciprocals of the intercepts is constant, $\frac{1}{\overline{OT}^2} + \frac{1}{\overline{OS}^2} = \frac{1}{a^2}$.

334. Applications Involving Rates. In many practical situations we seek the value of a quantity x at time t. And the conditions of the problem determine dx/dt, the rate of change of x with respect to the time, as a function of x and t. This leads to a first-order differential equation

$$\frac{dx}{dt} = f(x,t). \tag{57}$$

In setting up the differential equation, it is sometimes convenient to use differentials, as in Eqs. (37) and (39) of Sec. 164. Thus we consider dx as the change in x that *would* occur in time dt if the rate *were constant* and equal to its value at time t. This leads to the differential equation (57) in the form

$$dx = f(x,t)dt. \tag{58}$$

For example, radium decomposes at a rate proportional to the amount present. Hence if x is the amount at time t,

$$-\frac{dx}{dt} = kx \qquad \text{or} \qquad \frac{dx}{dt} = -kx. \tag{59}$$

Again, consider a sum of money put at interest at the rate of r per cent per annum, continuously compounded. Let x be the amount after t years. In dt years, the simple interest on x at $r/100$ a year would be $(rx/100)dt$. As this is the change that would occur if the rate were constantly equal to its value at time t, we have

$$dx = \frac{rx}{100}\,dt. \tag{60}$$

Equations (59) and (60) are of the type discussed in Sec. 123.

335. Determination of Particular Solutions. The general solution of Eq. (57) will determine x as a function of t and c, the constant of integration. The particular solution which satisfies the requirements of a problem may be obtained from the general solution by using an appropriately determined special value of c.

Thus if the problem requires that $x = x_1$ when $t = t_1$, c may be determined by substituting these values in the general solution.

If the differential equation contains an unknown constant like the k in Eq. (59), the solution will contain both c and k. In this case c and k can both be found if we know two pairs of corresponding values, x_1,t_1 and x_2,t_2.

After all the constants have been determined, the resulting relation between x and t may be used to find a particular value of either variable corresponding to an assigned value of the other.

In finding particular solutions, it is often desirable to integrate between limits. When this is done, as in Secs. 71 and 72, no constant of integration need be introduced.

EXAMPLE 1. If money earns 3 per cent interest per annum continuously compounded, what is the present value of an income continuously paid out at the rate of \$900 per year for 10 years?

Solution: Let x be the amount of principal remaining after t years. As in Eq. (60), in dt years the increase in x due to simple interest on x at $\frac{3}{100}$ a year would be $\frac{3x}{100}\,dt$. And in dt years the decrease due to the payment of income would be $900\,dt$. Hence we have $dx = \frac{3x}{100}\,dt - 900\,dt$ and $\frac{dx}{0.03x - 900} = dt$. Since x decreases when t increases, $\frac{dx}{dt}$ is negative, as is $0.03x - 900 = \frac{dx}{dt}\cdot$ To avoid the necessity of using the convention of Sec. 189 for logarithms of negative numbers, we rewrite the separated form as $\frac{-dx}{900 - 0.03x} = dt$.

Let $x = A$ when $t = 0$. Then $x = 0$ when $t = 10$. Hence we may integrate between these corresponding limits and deduce that $\int_A^0 \frac{-dx}{900 - 0.03x} = \int_0^{10} dt$, or $\frac{100}{3}\,[\ln\,(900 - 0.03x)]_A^0 = [t]_0^{10}$, $\frac{100}{3}\,[\ln\,900 - \ln\,(900 - 0.03A)] = 10$. It follows that $e^{\ln 900 - \ln (900 - 0.03A)} = e^{0.3}$, $\frac{900 - 0.03A}{900} = e^{-0.3}$, $A = \frac{900(1 - e^{-0.3})}{0.03} =$ $30{,}000(1 - 0.7408) = 7{,}776$. Hence the required present value is \$7,776.

EXAMPLE 2. In a *second-order chemical reaction* two substances A and B combine to form a third substance C, one molecule of A uniting with one molecule of B to yield one molecule of C. The rate at which C is formed is proportional to the product of the amounts of A and B present. Initially there were a gram-moles of A, b gram-moles of B, and no C present. And $b > a$. Find the amount of C after time t.

Solution: Let x be the number of gram-moles of C after time t. Then at that time the amount of A remaining is $(a - x)$. And that of b remaining is $(b - x)$. Hence $\frac{dx}{dt} = k(a - x)(b - x)$. $\frac{dx}{(a - x)(b - x)} = k\,dt$. But $\frac{1}{(x - a)(x - b)} =$ $\frac{1/(a - b)}{x - a} + \frac{1/(b - a)}{x - b}$, by Sec. 201. Thus $\left(\frac{1}{a - x} - \frac{1}{b - x}\right) dx = (b - a)k\,dt$.

Since $x = 0$ when $t = 0$ and $x = x$ when $t = t$, we may integrate between these corresponding limits and deduce that $[\ln\,(b - x) - \ln\,(a - x)]^x = (b - a)k[t]_0^t$. It follows that $e^{\ln (b-x) - \ln (a-x) - \ln b + \ln a} = e^{(b-a)kt}$, or $\frac{b - x}{a - x}\frac{a}{b} = \frac{e^{bkt}}{e^{akt}}$, $(ab - ax)e^{akt} =$ $(ab - bx)e^{bkt}$, and $x = \frac{ab(e^{bkt} - e^{akt})}{be^{bkt} - ae^{akt}}\cdot$ This is the required amount.

EXAMPLE 3. In Example 2, suppose that $a = 3$, $b = 5$. And that $x = 1$ when $t = 5$ min. Find the amount of C present after 10 min.

Solution: Here $\frac{dx}{dt} = k(3 - x)(5 - x)$, $\left(\frac{1}{3 - x} - \frac{1}{5 - x}\right) dx = 2k\,dt$. Integration between the limits $x = 0$, $t = 0$ and $x = 1$, $t = 5$ gives $[\ln\,(5 - x) - \ln\,(3 - x)]_0^1 =$

$2k[t]_0^5$. It follows that $\ln 4 - \ln 2 - \ln 5 + \ln 3 = 10k$, so that $k = \frac{1}{10} \ln \frac{6}{5}$. Integration between the limits $x = 0$, $t = 0$ and $x = x$, $t = 10$ gives $[\ln\,(5 - x) - \ln\,(3 - x)]_0^x = 2k[t]_0^{10}$, which leads to $e^{\ln(5-x) - \ln(3-x) - \ln 5 + \ln 3} = e^{20k}$, or $\frac{5 - x}{3 - x}\frac{3}{5} = e^{20k} = e^{2 \ln \frac{6}{5}} = \left(\frac{6}{5}\right)^2 = \frac{36}{25}$. $5(5 - x) = 12(3 - x)$, $x(12 - 5) = 36 - 25$, $x = \frac{11}{7} = 1.57$. And the required amount of C present after 10 min. is 1.57 gram-moles.

EXAMPLE 4. A tank originally contains 50 gal. of brine in which 50 lb. of salt is dissolved. Brine containing 2 lb. of salt per gallon runs into the tank at the rate of 3 gal./min. And the mixture runs out at the rate of 2 gal./min. The concentration is kept uniform by stirring. How much salt is in the tank after 10 min.?

Solution 1: Let x be the number of pounds of salt in the tank after t min. In time dt min., $2 \times 3\,dt$ lb. of salt is carried into the tank. Let V be the number of gallons of brine in the tank at time t. Then $dV = (3 - 2)dt$, $[V]_{50}^V = [t]_0^t$, $V - 50 = t$, $V = 50 + t$. Thus at time t the concentration $c = \frac{x}{V} = \frac{x}{50 + t}$. Hence in time dt min. $c \times 2\,dt = \frac{2x\,dt}{50 + t}$ lb. of salt is carried out of the tank. Thus $dx = 6\,dt - \frac{2x\,dt}{50 + t}$, or $\frac{dx}{dt} + \frac{2x}{50 + t} = 6$. This is linear in x, or of the form of Eq. (18) with x in place of y and t in place of x. The coefficient of x is $\frac{2}{50 + t}$. Hence as in Eq. (22) we find $\frac{dI}{I} = \frac{2\,dt}{50 + t}$, $\ln I = 2 \ln\,(50 + t)$, $I = e^{2 \ln(50+t)} = (50 + t)^2$. Multiplication by $(50 + t)^2\,dt$ leads to $(50 + t)^2\,dx + 2x(50 + t)dt = 6(50 + t)^2\,dt$. Integration between the limits $t = 0$, $x = 50$ and $t = 10$, $x = x$ gives $[(50 + t)^2x]_{t=0,x=50}^{t=10,x=x} = 2[(50 + t)^3]_0^{10}$ or $60^2x - 50^3 = 2(60^3 - 50^3)$, $3{,}600x = 10^3(432 - 250 + 125) = 307{,}000$. $x = \frac{307{,}000}{3{,}600} = \frac{3{,}070}{36} = 85.3$. And the required amount is 85.3 lb.

Solution 2: Let c lb./gal. be the concentration of salt in the tank after t min. Then in time dt min., $2 \times 3\,dt$ lb. of salt is carried into the tank. And in time dt min., $c \times 2\,dt$ lb. of salt is carried out of the tank. If the tank contains V gal. of brine after t min., $dV = (3 - 2)dt$, $[V]_{50}^V = [t]_0^t$, $V - 50 = t$, $V = 50 + t$. Thus the number of pounds of salt in the tank at time t is $cV = c(50 + t)$. This increases by $d\,[c(50 + t)]$ in time dt, so that $d\,[c(50 + t)] = 6\,dt - 2c\,dt$. Or $(50 + t)dc + c\,dt = 6\,dt - 2c\,dt$, $(50 + t)dc = (6 - 3c)dt$. Separating the variables, $\frac{dc}{2 - c} = \frac{3\,dt}{50 + t}$. Integration between the limits $t = 0$, $c = \frac{50}{50} = 1$ and $t = 10$, $c = c$ gives $[-\ln\,(2 - c)]_1^c = 3[\ln\,(50 + t)]_0^{10}$, $-\ln\,(2 - c) + \ln 1 = 3(\ln 60 - \ln 50)$, $e^{\ln(2-c)} = e^{3(\ln 50 - \ln 60)}$, $2 - c = (\frac{5}{6})^3$ $c = 2 - \frac{125}{216} = \frac{307}{216}$. When $t = 10$, $V = 50 + t = 60$. Hence $x = cV = \frac{307}{216}60 = \frac{3070}{36} = 85.3$. And the required amount is 85.3 lb.

EXERCISE 168

If x is the amount of money after t years, with interest at r per cent per annum continuously compounded, by Eq. (60)

$$dx = \frac{rx}{100}\,dt.$$

1. If the initial amount is A, show that the principal after t years will be $x = Ae^{rt/100}$.

2. If an initial amount A becomes B after n years, show that the rate $r = \frac{100}{n}(\ln B - \ln A) = \frac{100}{n}\ln\frac{B}{A}$.
3. A fund draws interest at r per cent per annum continuously compounded and is built up by continuous payments of P dollars per annum into the fund. If x is the number of dollars in the fund after t years, show that $dx = \frac{rx}{100}dt + P\,dt$. Deduce that if A was the initial amount in the fund, after t years, $x = \left(A + \frac{100P}{r}\right)e^{rt/100} - \frac{100P}{r}$.
4. A tank originally contains V gal. of brine in which a lb. of salt is dissolved. Water runs into the tank. And the mixture, kept uniform by stirring, runs out at the same rate. If there is s lb. of salt in the mixture after w gal. of water has run through, show that $ds = -\frac{s}{V}dw$. Deduce that $s = ae^{-w/V}$.
5. A tank originally contains V gal. of brine in which a lb. of salt is dissolved. Brine containing b lb. of salt per gallon runs into the tank at the rate of q gal./min. And the mixture, kept uniform by stirring, runs out at the same rate. If there is x lb. of salt in the tank after t min., show that $dx = bq\,dt - \frac{x}{V}q\,dt$. Deduce that $x = bV + (a - bV)e^{-qt/V}$.
6. In a *first-order chemical reaction* a substance A is transformed into a substance B. The rate is proportional to the amount of A present. The original amount of A was a, and there was no B present at the start. If x is the amount of A after time t, show that $dx/dt = k(a - x)$. Deduce that $x = a(1 - e^{-kt})$.
7. If, in the second-order reaction of Example 2, the initial values a and b are equal, deduce from $\frac{dx}{dt} = k(a - x)^2$ that $x = \frac{a^2kt}{akt + 1}$.
8. Check Prob. 7 by using l'Hospital's rule of Sec. 255 to find the limit of the result of Example 2 when $b \to a$.
9. If, in the chemical reaction of Prob. 7, $x = x_1$ when $t = t_1$, show that when $t = mt_1$, $x = \frac{amx_1}{a - x_1 + mx_1}$.
10. In Example 2, suppose that $a = 3$, $b = 4$, and $x = 2$ when $t = 20$ min. Find the amount of C present after 1 hr.
11. If, in the chemical reaction of Example 2, $x = x_1$ when $t = t_1$, show that when $t = mt_1$, $x = \frac{(b - x_1)^m a^m - (a - x_1)^m b^m}{(b - x_1)^m a^{m-1} - (a - x_1)^m b^{m-1}}$.
12. In a chemical reaction, a substance C decomposes into two substances A and B. Each of these products is formed at a rate proportional to the amount of C present. Thus if x,y,z is the number of gram-moles of A, B, C, respectively, present after time t, $\frac{dx}{dt} = k_1z$ and $\frac{dy}{dt} = k_2z$. Suppose that at $t = 0$, $x = 0$, $y = 0$, $z = z_0$. Then $x + y + z = z_0$. Show that $\frac{dz}{dt} = -(k_1 + k_2)z$, $z = z_0e^{-(k_1+k_2)t}$, $\frac{dx}{dt} = k_1z_0e^{-(k_1+k_2)t}$, $x = \frac{k_1z_0}{k_1 + k_2}[1 - e^{-(k_1+k_2)t}]$,
$y = z_0 - z - x = \frac{k_2z_0}{k_1 + k_2}[1 - e^{-(k_1+k_2)t}]$.
13. In the chemical reaction of Prob. 12, suppose that at $t = t_1$, $x = x_1$, $z = z_1$. Show

that $x = \frac{x_1 z_0}{z_0 - z_1}\left[1 - \left(\frac{z_1}{z_0}\right)^{t/t_1}\right]$, $y = \left(1 - \frac{x_1}{z_0 - z_1}\right) z_0 \left[1 - \left(\frac{z_1}{z_0}\right)^{t/t_1}\right]$, $z = z_0 \left(\frac{z_1}{z_0}\right)^{t/t_1}$.

14. In a chemical reaction between two substances A and B, 1 mole of B is produced for each mole of A consumed. The rate is proportional to the amount of A present. At the same time by a reverse reaction, B is converted into A at a rate proportional to the amount of B present. Thus if x and y are the number of gram-moles of A and B, respectively, after time t, $\frac{dx}{dt} = -\frac{dy}{dt} = -k_1 x + k_2 y$. Suppose that at $t = 0$, $x = x_0$, $y = 0$. Then $x + y = x_0$. Show that $\frac{dx}{dt} = k_2 x_0 - (k_1 + k_2)x$, $x = \frac{x_0}{k_1 + k_2}[k_2 + k_1 e^{-(k_1+k_2)t}]$, $y = \frac{x_0}{k_1 + k_2}[1 - e^{-(k_1+k_2)t}]$.

15. In the chemical reaction of Prob. 14, suppose that at $t = t_1$, $x = x_1$ and $t = \infty$, $x = X$. Show that $x = X + (x_0 - X)\left(\frac{x_1 - X}{x_0 - X}\right)^{t/t_1}$, $y = (x_0 - X)\left[1 - \left(\frac{x_1 - X}{x_0 - X}\right)^{t/t_1}\right]$.

16. A mass of insoluble material contains a soluble substance in its pores. Initially the amount is x_0. It is agitated with V gal. of liquid. The substance dissolves at a rate proportional to x, the amount of undissolved substance present. The rate is also proportional to the difference between s, the concentration of a saturated solution, and $c = \frac{x_0 - x}{V}$, the concentration actually in solution. Thus $\frac{dx}{dt} = -kx\left(s - \frac{x_0 - x}{V}\right)$. Deduce that $x = \frac{(Vs - x_0)x_0}{Vse^{kt(Vs-x_0)/V} - x_0}$.

17. In Prob. 16, suppose that $x = x_1$ when $t = t_1$. Show that

$$x = \frac{(Vs - x_0)x_0}{Vs\left[\frac{x_0(Vs + x_1 - x_0)}{Vsx_1}\right]^{t/t_1} - x_0}.$$

18. In Prob. 16 suppose that $x_0 = 12$ lb., $V = 20$ gal., and that salt is being dissolved in water so that $s = 3$ lb./gal. If 4 lb. of salt has dissolved in 5 min., when will 90 per cent be dissolved?

19. Wet wash containing 20 lb. of moisture is hung up to dry in an attic room of volume 4,000 cu. ft. The air at the beginning has a humidity of 25 per cent. Saturated air at the existent temperature contains 0.015 lb. of moisture per cubic foot. The wash dries at a rate proportional to its moisture content, x lb., and also to the difference between the moisture content of air and that of saturated air. If the humidity of the air is kept at 25 per cent by ventilation, the difference is $0.015(4{,}000) - 0.25(0.015)(4{,}000) = 60 - 15 = 45$. Thus $dx/dt = -kx(45)$. If $x = 5$ when $t = 1$ hr., deduce that $x = 20(\frac{1}{4})^t$. Also show that $k = \ln 4/45$.

20. In Prob. 19, if the room is closed, the moisture evaporated increases the moisture content of the air to $15 + (20 - x)$. Thus $\frac{dx}{dt} = -kx(25 + x)$. Using the value of k found in Prob. 19, deduce that $x = \frac{100}{9(4)^{5t/9} - 4}$. Also show that after 1 hr., $x = 6.48$ lb.

336. Equations of the Second Order. The general differential equation of the second order is

$$F\left(x,y,\frac{dy}{dx},\frac{d^2y}{dx^2}\right) = 0. \tag{61}$$

Methods of solving this equation when one of the letters x or y is absent from the function F are given in Secs. 337 and 338. And the discussion of Secs. 339 to 342 applies to certain linear equations of the second order. As we noted in Sec. 321, the general solution of a second-order equation contains two constants of integration.

337. Letter *y* Absent. If y is missing from the function F of Eq. (61), we put

$$\frac{dy}{dx} = p, \qquad \frac{d^2y}{dx^2} = \frac{dp}{dx}. \tag{62}$$

Then

$$F\left(x,\frac{dy}{dx},\frac{d^2y}{dx^2}\right) = 0 \qquad \text{becomes} \qquad F\left(x,p,\frac{dp}{dx}\right) = 0. \tag{63}$$

The solution of this first-order equation in the variables x and p may be written

$$p = G(x,c_1) \qquad \text{or} \qquad \frac{dy}{dx} = G(x,c_1). \tag{64}$$

The solution of this separable differential equation is

$$y = \int G(x,c_1)dx + c_2. \tag{65}$$

EXAMPLE 1. Solve the equation $(x + 2)\frac{d^2y}{dx^2} + 2\frac{dy}{dx} = 12x^2$.

Solution: Put $\frac{dy}{dx} = p$, $\frac{d^2y}{dx^2} = \frac{dp}{dx}$. Then $(x + 2)\frac{dp}{dx} + 2p = 12x^2$. This is linear in p. Divide by $(x + 2)$. $\frac{dp}{dx} + \frac{2p}{x+2} = \frac{12x^2}{x+2}$. The coefficient of p is $\frac{2}{x+2}$. Hence as in Eq. (22), $\frac{dI}{I} = \frac{2\,dx}{x+2}$, $\ln I = 2\ln(x + 2)$, $I = e^{2\ln(x+2)} = (x + 2)^2$. Multiplication by $(x + 2)^2\,dx$ leads to $(x + 2)^2\,dp + 2(x + 2)p\,dx = 12(x + 2)x^2\,dx$. Hence $(x + 2)^2p = \int(12x^3 + 24x^2)dx = 3x^4 + 8x^3 + C_1$. $p = \frac{dy}{dx} = \frac{3x^4 + 8x^3 + C_1}{(x + 2)^2}$ $= 3x^2 - 4x + 4 - \frac{16 + C_1}{(x + 2)^2}$, by division. And by integration

$$y = x^3 - 2x^2 + 4x + \frac{c_1}{x+2} + c_2, \text{ if } 16 + C_1 = c_1.$$

EXAMPLE 2. Solve the equation $x\frac{d^2y}{dx^2} - 6x^3 + 4x^2 - 1 = 0.$

Solution: Since $\frac{dp}{dx} = \frac{d^2y}{dx^2} = \frac{6x^3 - 4x^2 + 1}{x} = 6x^2 - 4x + \frac{1}{x}$, $p = \frac{dy}{dx} = 2x^3 - 2x^2 + \ln x + C_1$. And by a second integration $y = \frac{x^4}{2} - \frac{2}{3}x^3 + x \ln x - x + C_1x + c_2$. Thus with $C_1 - 1 = c_1$, $y = \frac{1}{2}x^4 - \frac{2}{3}x^3 + x \ln x + c_1x + c_2$.

If, as in Example 2, y and dy/dx are both missing from the function F of Eq. (61), that equation can be solved for d^2y/dx^2 and so put in the form $\frac{d^2y}{dx^2} = f(x)$. We may obtain from this $\frac{dy}{dx} = \int f(x)dx$ by recalling that $\frac{d^2y}{dx^2} = \frac{d}{dx}\left(\frac{dy}{dx}\right)$, and so need not introduce p at all.

338. Letter x Absent. Suppose that x is missing from the function F of Eq. (61). Then if

$$\frac{dy}{dx} = p, \qquad \frac{d^2y}{dx^2} = \frac{dp}{dx} = \frac{dp}{dy}\frac{dy}{dx} = p\frac{dp}{dy}. \tag{66}$$

Thus we replace $\frac{dy}{dx}$ by p and $\frac{d^2y}{dx^2}$ by $p\frac{dp}{dy}$ in

$$F\left(y, \frac{dy}{dx}, \frac{d^2y}{dx^2}\right) \text{ to obtain } F\left(y,p,p\frac{dp}{dy}\right) = 0. \tag{67}$$

The solution of this first-order equation in the variables y and p may be written

$$p = G(y,c_1) \qquad \text{or} \qquad \frac{dy}{dx} = G(y,c_1). \tag{68}$$

The solution of this separable differential equation is

$$x = \int \frac{dy}{G(y,c_1)} + c_2. \tag{69}$$

EXAMPLE 1. Solve the equation $(1 + y^2)\frac{d^2y}{dx^2} = \frac{dy}{dx} + \left(\frac{dy}{dx}\right)^3.$

Solution: Put $\frac{dy}{dx} = p$ and $\frac{d^2y}{dx^2} = p\frac{dp}{dy}$. Then $(1 + y^2)p\frac{dp}{dy} = p + p^3$. If $p \neq 0$, $\frac{dp}{1 + p^2} = \frac{dy}{1 + y^2}$, $\tan^{-1} p = \tan^{-1} y + C_1$. And $p = \tan(\tan^{-1} y + C_1) = \frac{y + \tan C_1}{1 - y \tan C_1} = \frac{y + c_1}{1 - c_1y}$ if $\tan C_1 = c_1$. $\frac{1}{p} = \frac{dx}{dy} = \frac{1 - c_1y}{y + c_1} = -c_1 + \frac{1 + c_1^2}{y + c_1}$, by division. And by integration $x = -c_1y + (1 + c_1^2)\ln(y + c_1) + c_2$.

The factor $p = 0$ or $dy/dx = 0$ gives $y = c$. But this is included in the general solution, since if $c_2 = +\infty$, $\ln(y + c_1) = -\infty$, and $y_1 + c_1 = 0$.

EXAMPLE 2. Solve the equation $\frac{d^2y}{dx^2} = \frac{1}{y^3}.$

Solution: Here $p\dfrac{dp}{dy} = \dfrac{1}{y^3}$ and $2p\,dp = 2y^{-3}\,dy$. $p^2 = y^{-2} + c_1$, $p = \pm\sqrt{y^{-2} + c_1}$. With the plus sign, $\dfrac{1}{p} = \dfrac{dx}{dy} = \dfrac{1}{\sqrt{y^{-2} + c_1}}$. And $\dfrac{c_1 y\,dy}{\sqrt{c_1y^2 + 1}} = c_1\,dx$, $\sqrt{c_1y^2 + 1} = c_1x + c_2$, $c_1y^2 + 1 = (c_1x + c_2)^2$. Thus $c_1y^2 = (c_1x + c_2)^2 - 1$, the result for both signs.

If, as in Example 2, x and dy/dx are both missing from the function F of Eq. (61), that equation can be solved for $\dfrac{d^2y}{dx^2}$ and so put in the form $\dfrac{d^2y}{dx^2} = f(y)$. If we observe that $\dfrac{d}{dx}\left(\dfrac{dy}{dx}\right)^2 = 2\dfrac{dy}{dx}\dfrac{d^2y}{dx^2} = 2f(y)\dfrac{dy}{dx}$, we may deduce that $d\left(\dfrac{dy}{dx}\right)^2 = 2f(y)dy$ and so obtain $\left(\dfrac{dy}{dx}\right)^2 = \displaystyle\int 2f(y)dy$ without introducing p.

EXERCISE 169

Solve each of the following differential equations.

1. $\dfrac{d^2y}{dx^2} + 4\cos 2x = 0.$
2. $\dfrac{d^2y}{dx^2} = x + 1.$
3. $x\dfrac{d^2y}{dx^2} = 6x^3 - 1.$
4. $e^x\dfrac{d^2y}{dx^2} = e^{2x} - 1.$
5. $x\dfrac{d^2y}{dx^2} + x\left(\dfrac{dy}{dx}\right)^2 = \dfrac{dy}{dx}.$
6. $2x\dfrac{d^2y}{dx^2} + 4\dfrac{dy}{dx} + x\left(\dfrac{dy}{dx}\right)^2 = 0.$
7. $x\dfrac{d^2y}{dx^2} = 2\left(x + \dfrac{dy}{dx}\right).$
8. $\dfrac{d^2y}{dx^2} + \tan x\dfrac{dy}{dx} = \sec x.$
9. $x\left(\dfrac{d^2y}{dx^2} + 4\right) = 3\dfrac{dy}{dx}.$
10. $x\dfrac{d^2y}{dx^2} + \dfrac{dy}{dx} = 9x^2.$
11. $x\dfrac{d^2y}{dx^2} = \dfrac{dy}{dx} + 30x^4.$
12. $x^2\dfrac{d^2y}{dx^2} = x\dfrac{dy}{dx} - \left(\dfrac{dy}{dx}\right)^2.$
13. $2y\dfrac{d^2y}{dx^2} = \left(\dfrac{dy}{dx}\right)^2.$
14. $y\dfrac{d^2y}{dx^2} + \left(\dfrac{dy}{dx}\right)^2 = 0.$
15. $y\dfrac{d^2y}{dx^2} + \left(\dfrac{dy}{dx}\right)^2 + 4 = 0.$
16. $\dfrac{d^2y}{dx^2} = \left(\dfrac{dy}{dx}\right)^2.$
17. $\dfrac{d^2y}{dx^2} + \left(\dfrac{dy}{dx}\right)^2 + 9 = 0.$
18. $\dfrac{d^2y}{dx^2} = \left[1 + \left(\dfrac{dy}{dx}\right)^2\right]^{\frac{3}{2}}.$
19. $e^{2y}\dfrac{d^2y}{dx^2} = 1.$
20. $y^3\dfrac{d^2y}{dx^2} = 2.$
21. $4\sqrt{y}\dfrac{d^2y}{dx^2} = 1.$
22. $4y^{\frac{1}{3}}\dfrac{d^2y}{dx^2} + 3 = 0.$

339. Linear Equations of Any Order. The general linear differential equation of order n is

$$a_n(x)\frac{d^ny}{dx^n} + a_{n-1}(x)\frac{d^{n-1}y}{dx^{n-1}} + a_{n-2}(x)\frac{d^{n-2}y}{dx^{n-2}} + \cdots + a_1(x)\frac{dy}{dx} + a_0(x)y = \phi(x). \quad (70)$$

The right member $\phi(x)$ and the *coefficients* $a_n(x)$, $a_{n-1}(x)$, $\cdots$, $a_1(x)$, $a_0(x)$ are given functions of x. Since the order is n, $a_n(x)$ is not identically zero. If $\phi(x) = 0$, the equation is *homogeneous.*† Thus the general homogeneous linear differential equation of the nth order is

$$L(y) = a_n(x)\frac{d^ny}{dx^n} + a_{n-1}(x)\frac{d^{n-1}y}{dx^{n-1}} + a_{n-2}(x)\frac{d^{n-2}y}{dx^{n-2}} + \cdots + a_1(x)\frac{dy}{dx} + a_0(x)y = 0. \quad (71)$$

We use $L(y)$ to denote the result of substituting any function y in the left member of Eq. (70) or (71). Since multiplying y by a constant multiplies each term by a constant,

$$L(cy_1) = cL(y_1) \quad \text{and} \quad L(cy_1) = 0 \quad \text{if } L(y_1) = 0. \quad (72)$$

Again, replacing y by $y_1 + y_2$ replaces each term in y by the sum of two similar terms, one in y_1 and one in y_2. Hence

$$L(y_1 + y_2) = L(y_1) + L(y_2), \quad (73)$$

and

$$L(y_1 + y_2) = 0 \quad \text{if } L(y_1) = 0 \text{ and } L(y_2) = 0. \quad (74)$$

Consequently the sum of two solutions, or the product of one solution by a constant, is again a solution of the homogeneous equation (71).

Suppose that $u_1, u_2, \cdots, u_n$ are n functions of x each of which is a solution of Eq. (71). Then

$$y = c_1u_1 + c_2u_2 + \cdots + c_nu_n \quad (75)$$

is also a solution, by the properties just mentioned.

It may happen that some one of the functions u_j, say u_n, can be expressed in terms of the rest by a relation of the form

$$u_n = k_1u_1 + k_2u_2 + \cdots + k_{n-1}u_{n-1}, \quad (76)$$

in which the coefficients $k_1, k_2, \cdots, k_{n-1}$ are constants. In this case the n functions u_j are *linearly dependent.* And Eq. (75) is equivalent to

$$y = (c_1 + c_nk_1)u_1 + (c_2 + c_nk_2)u_2 + \cdots + (c_{n-1} + c_nk_{n-1})u_{n-1}. \quad (77)$$

This involves at most $n - 1$ independent constants, namely, the $n - 1$ parentheses. Thus if the u_j are linearly dependent, Eq. (75) cannot be the general solution of Eq. (71).

When no relation of the form of Eq. (76) holds, the n functions u_j are *linearly independent.* We call any n linearly independent solutions of Eq. (71) a *fundamental system* of solutions u_j. Equation (75) provides the general solution of any homogeneous linear equation of order n for which a fundamental system of solutions is known.

† This new meaning of homogeneous must not be confused with the previous use of the term in Sec. 326.

Consider next the nonhomogeneous equation, Eq. (70), or

$$L(y) = \phi(x) \qquad \text{with } \phi(x) \neq 0. \tag{78}$$

Suppose that u is a solution of this equation, and u_1 is a solution of the corresponding homogeneous equation. Then

$$L(u) = \phi(x) \qquad \text{and} \qquad L(u_1) = 0. \tag{79}$$

And it follows from this and Eq. (73) that

$$L(u + u_1) = L(u) + L(u_1) = \phi(x). \tag{80}$$

Thus $u + u_1$ is a solution of Eq. (78). Hence the sum of the general solution of the homogeneous equation and any particular solution of the nonhomogeneous equation

$$y = u + c_1u_1 + c_2u_2 + \cdots + c_nu_n \tag{81}$$

is a solution of Eq. (78). There are no other solutions. For, let y be any solution of Eq. (78) and u be the particular solution just used. Then it follows from Eqs. (72) and (73) that

$$L(y - u) = L(y) - L(u) = \phi(x) - \phi(x) = 0. \tag{82}$$

Thus $y = u + (y - u)$ equals u plus a solution of the homogeneous equation.

To recapitulate; Eq. (81) provides the general solution of any linear equation of order n in terms of a particular solution of the nonhomogeneous equation and a fundamental system of solutions of the corresponding homogeneous equation. The particular solution is called the *particular integral.* The remaining terms in Eq. (81) make up the *complementary function.*

EXAMPLE. Apply the general theory to $dy/dx + 2y = 6$.

Solution: The homogeneous equation $dy/dx + 2y = 0$ is separable. $dy/y = -2\,dx$, $\ln y = -2x + C_1$. With $C_1 = 0$, $y = e^{-2x}$. Here $n = 1$ and $u_1 = e^{-2x}$ constitute a fundamental system. By inspection, we find $y = 3$ as a solution of the nonhomogeneous equation. Hence $y = 2 + c_1e^{-2x}$ is the general solution.

340. Linear Equations with Constant Coefficients and Right Member Zero. The general homogeneous linear differential equation of order n, with constant coefficients, is

$$a_n\frac{d^ny}{dx^n} + a_{n-1}\frac{d^{n-1}y}{dx^{n-1}} + a_{n-2}\frac{d^{n-2}y}{dx^{n-2}} + \cdots + a_1\frac{dy}{dx} + a_0y = 0. \tag{83}$$

The coefficients $a_n, a_{n-1}, \cdots, a_1, a_0$ are here constants, with $a_n \neq 0$. In discussing such equations, it is convenient to represent $\dfrac{d}{dx}$ by a single letter D. Thus

$$\frac{dy}{dx} = Dy, \qquad \frac{d^2y}{dx^2} = D^2y, \qquad \text{etc.} \tag{84}$$

And Eq. (83) may be written

$$(a_nD^n + a_{n-1}D^{n-1} + a_{n-2}D^{n-2} + \cdots + a_1D + a_0)y = 0. \tag{85}$$

Let us seek solutions of the form e^{rx}. Since

$$D(e^{rx}) = re^{rx}, \qquad D^2(e^{rx}) = r^2e^{rx}, \qquad \text{etc.,} \tag{86}$$

the result of putting $y = e^{rx}$ in Eq. (85) is

$$(a_nr^n + a_{n-1}r^{n-1} + a_{n-2}r^{n-2} + \cdots + a_1r + a_0)e^{rx} = 0. \tag{87}$$

Thus e^{rx} will be a solution of Eq. (85) if

$$a_nr^n + a_{n-1}r^{n-1} + a_{n-2}r^{n-2} + \cdots + a_1r + a_0 = 0. \tag{88}$$

This equation, which is easily written down by replacing D by r in the parenthesis of Eq. (85) or directly from the original form of Eq. (83), is called the *auxiliary* equation.

If the auxiliary equation has n real and distinct roots, $r_1, r_2, \cdots, r_n$, then the corresponding exponentials form a fundamental system of solutions, and the general solution of Eq. (83) is

$$y = c_1e^{r_1x} + c_2e^{r_2x} + \cdots + c_ne^{r_nx}. \tag{89}$$

When the a_j are all real, complex roots will occur in conjugate pairs. Suppose that $a \pm bi$ are a pair of conjugate complex roots, where $i = \sqrt{-1}$. To get the terms of the solution for them to add up to a real quantity, we use conjugate complex constants $p \pm qi$ as coefficients. Then by Secs. 264 and 265

$$\begin{aligned}(p + qi)e^{(a+bi)x} &+ (p - qi)e^{(a-bi)x} \\ &= 2e^{ax}\left(p\,\frac{e^{ibx} + e^{-ibx}}{2} - q\,\frac{e^{ibx} - e^{-ibx}}{2i}\right) \\ &= 2pe^{ax}\cos bx - 2qe^{ax}\sin bx.\end{aligned} \tag{90}$$

Calling $2p$ and $-2q$ new constants c_1 and c_2, we see that a pair of conjugate complex roots of the auxiliary equation,

$$a \pm bi, \qquad \text{lead to} \qquad c_1e^{ax}\cos bx + c_2e^{ax}\sin bx \tag{91}$$

in the general solution, where c_1 and c_2 are real constants.

If the auxiliary equation (88) has repeated roots, the additional terms are obtained by multiplying in powers of x. Specifically, two equal real roots of the auxiliary equation

$$s,s \text{ lead to} \qquad c_1e^{sx} + c_2xe^{sx} \qquad \text{or} \qquad e^{sx}(c_1 + c_2x). \tag{92}$$

Three equal roots of the auxiliary equation

$$s,s,s \text{ lead to } \qquad c_1e^{sx} + c_2xe^{sx} + c_2x^2e^{sx}, \tag{93}$$

and so on, if there are more than three equal roots.

Equal complex roots are treated similarly. Thus two equal pairs,

$$a \pm bi,\ a \pm bi \text{ lead to } \qquad (c_1 + c_2x)e^{ax} \cos bx + (c_3 + c_4x)e^{ax} \sin bx. \tag{94}$$

We shall verify Eq. (93). Here the auxiliary equation may be written in the form $P(r)(r - s)^3 = 0$, where $P(r)$ is a polynomial in r. The same calculation that shows this proves that the differential equation may be written

$$P(D)(D - s)^3y = 0. \tag{95}$$

But if we replace D by d/dx and expand we find that

$$(D - s)x^me^{sx} = mx^{m-1}e^{sx} \qquad \text{so that } (D - s)e^{sx} = 0. \tag{96}$$

And by repeated use of the first relation, we obtain

$$\begin{aligned} (D - s)^2xe^{sx} &= (D - s)e^{sx} = 0, \\ (D - s)^3x^2e^{sx} &= (D - s)^2 2xe^{sx} = 0. \end{aligned} \tag{97}$$

These relations show that xe^{sx} and x^2e^{sx} as well as e^{sx} are particular solutions of the differential equation (95) with three equal roots s. The verification of Eqs. (92), (94), and other cases of equal roots is similar.

EXAMPLE 1. Solve the equation $\dfrac{d^2y}{dx^2} + 2\dfrac{dy}{dx} - 15y = 0.$

Solution: The auxiliary equation is $r^2 + 2r - 15 = 0$, or $(r - 3)(r + 5) = 0$. The roots are $r = 3, -5$ Hence by Eq (89), $y = c_1e^{3x} + c_2e^{-5x}$.

EXAMPLE 2. Solve the equation $\dfrac{d^2y}{dx^2} + 8\dfrac{dy}{dx} + 20y = 0.$

Solution: The auxiliary equation is $r^2 + 8r + 20 = 0$. The roots are $r = \dfrac{-8 \pm \sqrt{8^2 - 4(20)}}{2} = -4 \pm \sqrt{-4} = -4 \pm 2i$. Hence by Eq. (91), $y = c_1e^{-4x} \cos 2x + c_2e^{-4x} \sin 2x$.

EXAMPLE 3. Solve the equation $\dfrac{d^3y}{dx^3} + 2\dfrac{d^2y}{dx^2} - 4\dfrac{dy}{dx} - 8y = 0.$

Solution: The auxiliary equation is $r^3 + 2r^2 - 4r - 8 = 0$, or $(r^2 - 4)(r + 2) = (r - 2)(r + 2)^2$. The roots are $r = 2, -2, -2$. Hence by Eqs (89) and (92), $y = c_1e^{2x} + c_2e^{-2x} + c_3xe^{-2x}$.

EXAMPLE 4. Solve the equation $\dfrac{d^3y}{dy^3} - 3\dfrac{d^2y}{dy^2} + 28\dfrac{dy}{dx} - 26y = 0.$

Solution: The auxiliary equation is $r^3 - 3r^2 + 28r - 26 = 0$. As one root is $r = 1$, we divide by $r - 1$. $(r - 1)(r^2 - 2r + 26) = 0$. The roots of $r^2 - 2r + 26 = 0$ are $r = \dfrac{2 \pm \sqrt{2^2 - 4(26)}}{2} = 1 \pm \sqrt{-25} = 1 \pm 5i$. The roots are 1, $1 \pm 5i$. Hence by Eqs (89) and (91), $y = c_1e^x + c_2e^x \cos 5x + c_3e^x \sin 5x$.

EXAMPLE 5. Solve the equation $\dfrac{d^5y}{dx^5} + 10\dfrac{d^3y}{dx^3} + 25\dfrac{dy}{dx} = 0.$

Solution: The auxiliary equation is $r^5 + 10r^3 + 25r = 0$, or $r(r^2 + 5)^2 = 0$. If $r^2 + 5 = 0$, $r = \pm\sqrt{-5} = \pm\sqrt{5}\,i$. The roots are $0, \pm\sqrt{5}\,i, \pm\sqrt{5}\,i$. Since $e^{0x} = e^0 = 1$, by Eqs (89) and (94),

$$y = c_1 + (c_2 + c_3x)\cos\sqrt{5}\,x + (c_4 + c_5x)\sin\sqrt{5}\,x.$$

EXAMPLE 6. Solve the equation $\dfrac{d^7y}{dx^7} + 4\dfrac{d^3y}{dx^3} = 0$.

Solution: The auxiliary equation $r^7 + 4r^3 = 0$, or $r^3(r^4 + 4) = 0$. By Sec 265, $-4 = 4e^{i\theta}$ if $\theta = \pm\pi, \pm 3\pi$. If $r^4 = -4 = 4e^{i\theta}$, $r = \sqrt{2}\,e^{i\theta/4} = \sqrt{2}\left(\dfrac{1}{\sqrt{2}} \pm \dfrac{1}{\sqrt{2}}i\right)$ or $\sqrt{2}\left(-\dfrac{1}{\sqrt{2}} \pm \dfrac{1}{\sqrt{2}}i\right)$. The roots are $0, 0, 0, 1 \pm i, -1 \pm i$. Since $e^{0x} = e^0 = 1$, by Eqs (93) and (91),

$$y = c_1 + c_2x + c_3x^2 + c_4e^x\cos x + c_5e^x\sin x + c_6e^{-x}\cos x + c_7e^{-x}\sin x.$$

EXERCISE 170

Solve each of the following differential equations

1. $\dfrac{dy}{dx} = 7y$.
2. $2\dfrac{dy}{dx} + 3y = 0$.
3. $\dfrac{d^2y}{dx^2} + 2\dfrac{dy}{dx} + 2y = 0$.
4. $\dfrac{d^2y}{dx^2} = \dfrac{dy}{dx} + 2y$.
5. $\dfrac{d^2y}{dx^2} + 3y = 4\dfrac{dy}{dx}$.
6. $\dfrac{d^2y}{dx^2} + y = 2\dfrac{dy}{dx}$.
7. $\dfrac{d^2y}{dx^2} + 5y = 2\dfrac{dy}{dx}$.
8. $\dfrac{d^2y}{dx^2} + 16y = 0$.
9. $\dfrac{d^2y}{dx^2} = 9y$.
10. $\dfrac{d^2y}{dx^2} = 9\dfrac{dy}{dx}$.
11. $\dfrac{d^3y}{dx^3} + 4\dfrac{dy}{dx} = 5\dfrac{d^2y}{dx^2}$.
12. $\dfrac{d^3y}{dx^3} + 9\dfrac{dy}{dx} = 6\dfrac{d^2y}{dx^2}$.
13. $\dfrac{d^3y}{dx^3} + 25\dfrac{dy}{dx} = 8\dfrac{d^2y}{dx^2}$.
14. $\dfrac{d^3y}{dx^3} = 8y$.
15. $\dfrac{d^3y}{dx^3} + 3\dfrac{dy}{dx} = 3\dfrac{d^2y}{dx^2} + y$.
16. $\dfrac{d^4y}{dx^4} + 36y = 13\dfrac{d^2y}{dx^2}$.
17. $\dfrac{d^4y}{dx^4} + 5\dfrac{d^3y}{dx^3} = 0$.
18. $\dfrac{d^4y}{dx^4} + 18\dfrac{d^2y}{dx^2} + 81y = 0$.
19. $\dfrac{d^4y}{dx^4} + 16\dfrac{d^2y}{dx^2} = 8\dfrac{d^3y}{dx^3}$.
20. $\dfrac{d^4y}{dx^4} = 3\dfrac{d^2y}{dx^2} + 4y$.

341. Constant Coefficients and Right Member a Function of x. Consider next the general nonhomogeneous linear differential equation of order n, with constant coefficients, or

$$a_n\frac{d^ny}{dx^n} + a_{n-1}\frac{d^{n-1}y}{dx^{n-1}} + a_{n-2}\frac{d^{n-2}y}{dx^{n-2}} + \cdots + a_1\frac{dy}{dx} + a_0y = \phi(x), \qquad (98)$$

with $\phi(x)$ not identically zero. By Sec. 339 we may write the general solution of this equation by adding any *particular integral* to the *complementary function*. A systematic method of finding a particular integral will be given in Sec. 342. The complementary function of Eq. (98) is the general solution of Eq. (83), obtained by replacing the right member $\phi(x)$ by zero. Hence the complementary function is the sum on the right of Eq. (89), modified when necessary as indicated in Eqs. (91) to (94).

EXAMPLE. Solve the equation $\dfrac{d^4y}{dx^4} - 6\dfrac{d^3y}{dx^3} + 13\dfrac{d^2y}{dx^2} = 9e^{3x}$.

Solution: The auxiliary equation is $r^4 - 6r^3 + 13r^2 = 0$, or $r^2(r^2 - 6r + 13) = 0$. The roots of $r^2 - 6r + 13 = 0$ are $r = \dfrac{6 \pm \sqrt{6^2 - 4(13)}}{2} = 3 \pm \sqrt{-4} = 3 \pm 2i$. The roots are $0, 0, 3 \pm 2i$. Since $e^{0x} = e^0 = 1$, by Eqs (92) and (91) the complementary function is $y_c = c_1 + c_2x + c_3e^{3x}\cos 2x + c_4e^{3x}\sin 2x$.

To find a particular integral, substitute $y = Ae^{3x}$ in the given equation. The result is $(3^4 - 6\cdot 3^3 + 13\cdot 3^2)Ae^{3x} = 36Ae^{3x} = 9e^{3x}$. This will hold if $36A = 9$ or $A = \frac{1}{4}$. Thus $y_p = \frac{1}{4}e^{3x}$ is a particular integral. Since $y = y_p + y_c$ is the complete solution, $y = \frac{1}{4}e^{3x} + c_1 + c_2x + c_3e^{3x}\cos 2x + c_4e^{3x}\sin 2x$.

342. Method of Undetermined Coefficients. In many practical cases, the right member of Eq. (98), $\phi(x)$, is a sum of terms each of which is of the type

$$kx^me^{ax}, \qquad kx^me^{ax}\cos bx, \qquad \text{or } kx^me^{ax}\sin bx. \tag{99}$$

Here m is zero or a positive integer, and a and b are any real numbers. As specific instances we have

$$3,\ 2x^3,\ 3e^{2x},\ 5xe^{2x},\ 4\cos 3x,\ 5x\sin 3x,\ 7e^x\sin x,\ 5xe^x\cos x. \tag{100}$$

In such cases a particular integral may be found by the method of undetermined coefficients. We consider separately the cases where $\phi(x)$ has a single term and where $\phi(x)$ is a sum of terms.

Case 1. *A Single Term.* With any one term T of the type under consideration, we associate the simplest polynomial in D, $Q(D)$, such that the differential relation $Q(D)T = 0$ holds. This is easily done by using in reverse the rules for forming the complementary function typified by Eqs. (89), (91) to (94). The polynomials associated with the terms of Eq. (100) are

$$D,\ D^4,\ (D-2),\ (D-2)^2,\ (D^2+9),\ (D^2+9)^2,$$
$$(D^2 - 2D + 2),\ (D^2 - 2D + 2)^2. \tag{101}$$

As these illustrate, $Q(D)$ will always be some power of a first- or second-degree factor. And we may write

$$Q(D) = F^q \qquad \text{where } F = D - a \text{ or } F = D^2 - 2aD + a^2 + b^2. \tag{102}$$

Now consider any particular integral I of Eq. (98), which we rewrite in the form $P(D)y = T(x)$. Since I is a solution of

$$P(D)y = T, \qquad P(D)I = T. \qquad \text{But } Q(D)T = 0. \tag{103}$$

It follows that I is a solution of

$$Q(D)P(D)y = 0, \qquad \text{since } Q(D)P(D)I = Q(D)T = 0. \tag{104}$$

This enables us to predict the form of I. Since any term that is in the solution of $P(D)y = 0$ is of no help in making $P(D)y = T$, we may assume I to be a linear combination with unknown coefficients of those terms in the solution of $Q(D)P(D)y = 0$ and not in the solution of $P(D)y = 0$. This leads to the following

Rule. Case 1. *A Single Term.* Let the right member of Eq. (98) consist of a single term associated with the polynomial $Q(D) = F^q$. If F is *not* a factor of $P(D)$, try

$$I = (Ax^{q-1} + Bx^{q-2} + \cdots + L)e^{ax} \qquad \text{when } F = D - a, \tag{105}$$

and try

$$\begin{aligned} I = (Ax^{q-1} + Bx^{q-2} + \cdots + L)e^{ax} \cos bx \\ + (Mx^{q-1} + Nx^{q-2} + \cdots + R)e^{ax} \sin bx \\ \text{when } F = D^2 - 2aD + a^2 + b^2. \end{aligned} \tag{106}$$

If F *is* a factor of $P(D)$ and the highest power of F which is a divisor of $P(D)$ is F^p, try the

$$I \text{ of Eq. (105) or Eq. (106) multiplied by } x^p. \tag{107}$$

Case 2. *A Sum of Terms.* With each term in $\phi(x)$, associate a polynomial $Q(D) = F^q$ as before. Arrange in one group all terms that have the same F. The particular integral of the given equation will be the sum of solutions of equations each of which has one group on the right. For any one such equation, the form of the particular integral is given by Eqs. (105) to (107) where q is the highest power of F associated with any term of the group on the right.

For example, suppose that $\phi(x)$ consists of the sum of all the terms in Eq. (100). From Eq. (101) we see that these fall into four separate groups, each made up of two consecutive terms. And if $P(D)$ is not divisible by D, $(D - 2)$, $(D^2 + 9)$, or $(D^2 - 2D + 2)$ in accordance with Eqs. (105) and (106), we try

$$Ax^3 + Bx^2 + Cx + D \qquad \text{for } 3 + 2x^3, \tag{108}$$

$$(Ax + B)e^{2x} \qquad \text{for } 2e^{2x} + 5xe^{2x}, \tag{109}$$

$$(Ax + B) \cos 3x + (Cx + D) \sin 3x \qquad \text{for } 4 \cos 3x + 5x \sin 3x, \tag{110}$$

$$(Ax + B)e^x \cos x + (Cx + D)e^x \sin x \qquad \text{for } 7e^x \sin x + 5xe^x \cos x. \tag{111}$$

On the other hand, if $P(D)$ were $D^2(D - 2)(D - 5)$ in accordance with Eq. (107), we would use

$$Ax^5 + Bx^4 + Cx^3 + Dx^2 \qquad \text{for } 3 + 2x^3, \tag{112}$$

$$(Ax^2 + Bx)e^{2x} \qquad \text{for } 2e^{2x} + 5xe^{2x}, \tag{113}$$

and Eqs. (110) and (111) as before.

By substituting the trial form in the differential equation and equating coefficients of corresponding terms, we obtain a set of first-degree equations in the unknown coefficients. A tabular form that minimizes the amount of writing is illustrated in the solution of the examples which follow.

EXAMPLE 1. Solve the equation $d^2y/dx^2 + 4y = 8 \cos 2x$.

Solution: Here $Q = F = D^2 + 4$. And $P(D) = D^2 + 4$. Thus $p = 1$ in Eq. (107), and $q = 1$, $a = 0$, $b = 2$ in Eq. (106). And we try x times $A \cos 2x + B \sin 2x$. We find

4	$y =$	A	$x \cos 2x +$ B	$x \sin 2x$		
0	$\dfrac{dy}{dx} =$	$2B$	$-2A$	$+ A$	$\cos 2x +$ B	$\sin 2x$
1	$\dfrac{d^2y}{dx^2} =$	$-4A$	$-4B$	$+4B$	$-4A$	
Totals		0	0	$4B$	$-4A$	
Should be		0	0	8	0	

The coefficients of the given equation are written on the left, and these are multiplied into each column and the results added. Under these totals, we write the coefficients in the right member, to which the totals should be equal if the equation is to hold. Here $4B = 8$, $-4A = 0$, so that $B = 2$, $A = 0$. And $y_p = 2x \sin 2x$ is a particular integral. The roots of the auxiliary equation $r^2 + 4 = 0$, $r = \pm 2i$ lead to $y_c = c_1 \cos 2x + c_2 \sin 2x$. Since $y = y_p + y_c$, $y = 2x \sin 2x + c_1 \cos 2x + c_2 \sin 2x$.

EXAMPLE 2. Solve the equation

$$\frac{d^2y}{dx^2} + 2\frac{dy}{dx} + 10y = 2e^{-2x} \cos 3x + 3e^{-2x} \sin 3x.$$

Solution: Here for both terms $Q = F = D^2 + 4D + 13$. This is not a factor of $P(D) = D^2 + 2D + 10$. Hence by Eq (106) with $q = 1$, $a = -2$, $b = 3$, we try $Ae^{-2x} \cos 3x + Be^{-2x} \sin 3x$. We find

10	$y =$	A	$e^{-2x} \cos 3x$	B	$e^{-2x} \sin 3x$
2	$\dfrac{dy}{dx} =$	$-2A$ $3B$		$-3A$ $-2B$	
1	$\dfrac{d^2y}{dx^2} =$	$-2(-2A + 3B)$ $3(-3A - 2B)$		$-3(-2A + 3B)$ $-2(-3A - 2B)$	
Totals		$A - 6B$		$6A + B$	
Should be		2		3	

The totals will be as desired if $A - 6B = 2$, $6A + B = 3$. $(A - 6B) + 6(6A + B) = 2 + 6(3)$, $37A = 20$, $A = \frac{20}{37}$. Hence $B = 3 - 6A = -\frac{9}{37}$. And

$$y_p = \tfrac{20}{37}e^{-2x} \cos 3x - \tfrac{9}{37}e^{-2x} \sin 3x.$$

The roots of the auxiliary equation $r^2 + 2r + 10 = 0$, $r = -1 \pm 3i$, lead to $y_c = c_1e^{-x} \cos 3x + c_2e^{-x} \sin 3x$. Since $y = y_p + y_c$,

$$y = \tfrac{20}{37}e^{-2x} \cos 3x - \tfrac{9}{37}e^{-2x} \sin 3x + c_1e^{-x} \cos 3x + c_2e^{-x} \sin 3x.$$

EXAMPLE 3. Solve the equation $\frac{d^3y}{dx^3} + \frac{d^2y}{dx^2} = 6e^{2x} + 6x^2 + 6x$.

Solution: Here there are two groups. For $6e^{2x}$, $Q = F = D - 2$. This is not a factor of $P(D) = D^3 + D^2$. Hence by Eq. (105) with $q = 1$, $a = 2$, we try $y = Ae^{2x}$. $(D^3 + D^2)Ae^{2x} = (2^3 + 2^2)Ae^{2x} = 12Ae^{2x} = 6e^{2x}$. This will hold if $12A = 6$, $A = \frac{1}{2}$. And $y_{p1} = \frac{1}{2}e^{2x}$ is part of the particular integral.

For $6x^2$, $Q = D^3$. And for $6x$, $Q = D^2$. These form one group, for which Eq. (105) with $q = 3$, $a = 0$ suggests $Ax^2 + Bx + C$. However, $F = D$ and D^2 is a factor of $P(D) = D^3 + D^2$. Hence by Eq. (107) we multiply by x^2 and try $Ax^4 + Bx^3 + Cx^2$. We find

0	$y = A$	$x^4 + B$	$x^3 + C$	x^2	
0	$\frac{dy}{dx} =$	$4A$	$+3B$	$+ 2C$	x
1	$\frac{d^2y}{dx^2} =$		$12A$	$+ 6B$	$+2C$
1	$\frac{d^3y}{dx^3} =$			$24A$	$+6B$
Totals	0	0	$12A$	$6B + 24A$	$2C + 6B$
Should be	0	0	6	6	0

The totals will be as desired if $12A = 6$, $6B + 24A = 6$, $2C + 6B = 0$. $A = \frac{1}{2}$, $6B = 6 - 24A = -6$, $B = -1$, $2C = -6B = 6$, $C = 3$. And

$$y_{p2} = \tfrac{1}{2}x^4 - x^3 + 3x^2.$$

The roots of the auxiliary equation $r^3 + r^2 = 0$, $r = 0, 0, -1$ lead to $y_c = c_1 + c_2x + c_3e^{-x}$. Since $y = y_{p1} + y_{p2} + y_c$,

$$y = \tfrac{1}{2}e^{2x} + \tfrac{1}{2}x^4 - x^3 + 3x^2 + c_1 + c_2x + c_3e^{-x}.$$

EXERCISE 171

1. As special cases of Eq. (105) or (106), verify that, if $\phi(x) = ke^{ax}$ and $P(a) \neq 0$, then $y_p = Ae^{ax}$; if $\phi(x) = k_1 \cos bx$, $\phi(x) = k_2 \sin bx$, or $\phi(x) = k_1 \cos bx + k_2 \sin bx$, and $P(b\sqrt{-1}) \neq 0$, then $y_p = A \cos bx + B \sin bx$; and if $P(0) \neq 0$, if $\phi(x) = k_1$, $y_p = A$ while if $\phi(x) = k_2x$ or $\phi(x) = k_1 + k_2x$, then $y_p = A + Bx$.

Solve each of the following differential equations.

2. $\frac{dy}{dx} + 2y = 2e^{3x}$.
3. $\frac{dy}{dx} + 3y = \cos x$.
4. $\frac{dy}{dx} + 2x = 2y + 1$.
5. $\frac{dy}{dx} = y + e^x$.
6. $\frac{d^2y}{dx^2} + y = 4x + 4e^x$.
7. $\frac{d^2y}{dx^2} = 4y + 4e^{2x}$.
8. $\frac{d^2y}{dx^2} - 2\frac{dy}{dx} + 5y = 10 \cos x$.
9. $\frac{d^2y}{dx^2} + \frac{dy}{dx} = 2x + 4$.
10. $\frac{d^2y}{dx^2} - 9\frac{dy}{dx} + 20y = 4x^2e^{3x}$.
11. $\frac{d^2y}{dx^2} + 9y = 9 \sin 3x$.
12. $\frac{d^2y}{dx^2} + 2\frac{dy}{dx} + y = 4e^x + 4e^{-x}$.
13. $\frac{d^2y}{dx^2} = 3\frac{dy}{dx} + 10e^{2x} \sin x$.

14. $\frac{d^3y}{dx^3} + 3\frac{dy}{dx} = 3\frac{d^2y}{dx^2} + y + e^x.$

15. $\frac{d^3y}{dx^3} = 4\frac{dy}{dx} + 12x.$

16. $\frac{d^3y}{dx^3} - 3\frac{d^2y}{dx^2} - 4y = 18xe^{-2x}.$

17. $\frac{d^3y}{dx^3} = \frac{d^2y}{dx^2} + 24x^2.$

18. $\frac{d^3y}{dx^3} - 3\frac{dy}{dx} + y = 36e^x.$

19. $\frac{d^4y}{dx^4} = y + x^3 + 3x.$

20. $\frac{d^4y}{dx^4} - 3\frac{d^2y}{dx^2} - 4y = 60e^{2x}.$

21. $\frac{d^4y}{dx^4} = \frac{d^2y}{dx^2} + 6x.$

22. Let $x = e^t$. Then $\frac{dx}{dt} = e^t$, $\frac{dt}{dx} = e^{-t}$. Deduce that $\frac{dy}{dx} = \frac{dy}{dt}\frac{dt}{dx} = e^{-t}\frac{dy}{dt}$, $\frac{d^2y}{dx^2} = \left(e^{-t}\frac{d^2y}{dt^2} - e^{-t}\frac{dy}{dt}\right)e^{-t} = e^{-2t}\left(\frac{d^2y}{dt^2} - \frac{dy}{dt}\right)$, $\frac{d^3y}{dx^3} = e^{-3t}\left(\frac{d^3y}{dt^3} - 3\frac{d^2y}{dt^2} + 2\frac{dy}{dt}\right)$. It follows that the substitution $x = e^t$ transforms the Euler-Cauchy linear differential equation $k_3x^3\frac{d^3y}{dx^3} + k_2x^2\frac{d^2y}{dx^2} + k_1x\frac{dy}{dx} + k_0y = K(x)$ into the linear equation $k_3\frac{d^3y}{dt^3} + (k_2 - 3k_3)\frac{d^2y}{dt^2} + (k_1 - k_2 + 2k_3)\frac{dy}{dt} + k_0y = K(e^t)$, which has constant coefficients. After this is solved by the method of Sec. 342, we can replace e^t by x and t by $\ln x$ in the solution.

23. In the Euler-Cauchy differential equation in x and y of Prob. 22, substitute $y = x^r$. Verify that the left member becomes $x^rP(r)$, where $P(r) = 0$ is the auxiliary equation of the transformed constant coefficient equation in t and y, $P\left(\frac{d}{dt}\right)y = K(e^t)$. This is an easily remembered way of obtaining the transformed equation.

Use Prob. 22 or 23 to solve each of the following Euler-Cauchy linear differential equations.

24. $x^2\frac{d^2y}{dx^2} + 8x\frac{dy}{dx} + 12y = 0.$

25. $x^2\frac{d^2y}{dx^2} + 2y = 2x\frac{dy}{dx}.$

26. $x^2\frac{d^2y}{dx^2} - x\frac{dy}{dx} + 6y = 12x.$

27. $x^2\frac{d^2y}{dx^2} = x\frac{dy}{dx} + 3y.$

28. $x^2\frac{d^2y}{dx^2} + 2x\frac{dy}{dx} - 2y = 4x^2.$

29. $x^2\frac{d^2y}{dx^2} - x\frac{dy}{dx} + y = x.$

30. $x^3\frac{d^3y}{dx^3} + x\frac{dy}{dx} - y = x\ln x.$

31. $x^2\frac{d^2y}{dx^2} + 5y = 0.$

32. $x^2\frac{d^3y}{dx^3} + 6x\frac{d^2y}{dx^2} + 6\frac{dy}{dx} = 0.$

33. $x\frac{d^2y}{dx^2} - \frac{dy}{dx} = 2 + 2x.$

343. Motion in a Straight Line. Consider a particle of mass m moving on a straight line under the influence of a force F. The acceleration a of the particle is determined by the relation force equals mass times acceleration, or $F = ma$. Let O be a fixed point on the line. And let P be the position of the particle at time t. Then if, as in Sec. 20, the signed distance $s = OP$, the velocity of the particle is $v = ds/dt$. And the acceleration is $a = d^2s/dt^2$. Thus

$$m\frac{d^2s}{dt^2} = F. \tag{114}$$

If we measure F in pounds, s in feet, and t in seconds, we must measure the mass m in slugs. The mass of a particle of weight w lb. is $m = w/g$ slugs, where g is the gravitational acceleration. An average value of g is 32.2 ft./sec.^{2}

Suppose that the force F is known as a function of one or more of the variables s, t, $v = ds/dt$. Then Eq. (114) becomes a second-order differential equation which may be solvable by the methods of Sec. 337, 338, or 342. The solution gives the relation between s and t, or the *equation of motion.*

EXAMPLE 1. A particle of weight 96.6 lb. is acted on by a force proportional to the square root of the time elapsed. The force is 45 lb. and toward O, after 4 sec. If the particle was initially 4 ft. to the right of O and moving away from O with a velocity of 2 ft./sec., find the equation of motion.

Solution: Since $w = 96.6$, $m = \frac{w}{g} = \frac{96.6}{32.2} = 3$. $F = k\sqrt{t}$ and $-45 = k\sqrt{4}$. Hence $k = -\frac{45}{2}$, $F = -\frac{45}{2}\sqrt{t}$. And by Eq. (114), $3\frac{d^2s}{dt^2} = -\frac{45}{2}\sqrt{t}$, or $\frac{d^2s}{dt^2} = -\frac{15}{2}\sqrt{t}$. Integration between the limits $\frac{ds}{dt} = v = 2$, $t = 0$ and $\frac{ds}{dt}$, t gives $\left[\frac{ds}{dt}\right]_2^{ds/dt} = [-5t^{\frac{3}{2}}]_0^t$, $\frac{ds}{dt} - 2 = -5t^{\frac{3}{2}}$, $\frac{ds}{dt} = 2 - 5t^{\frac{3}{2}}$. Integration between the limits $s = 4$, $t = 0$ and s,t gives $[s]_4^s = [2t - 2t^{\frac{5}{2}}]_0^t$, $s - 4 = 2t - 2t^{\frac{5}{2}}$, and $s = 4 + 2t - 2t^{\frac{5}{2}}$.

EXAMPLE 2. A particle of weight w lb. is attracted toward O by a force inversely proportional to the square of the distance. The force is $wb^2/8g$ lb. when the particle is 2 ft. from O. If the particle was initially at rest q ft. to the right of O, find the equation of motion.

Solution: $m = \frac{w}{g}$, $F = \frac{-k}{s^2}$, and $-\frac{wb^2}{8g} = \frac{-k}{2^2}$. Hence $k = \frac{wb^2}{2g}$, $F = -\frac{wb^2}{2g}\frac{1}{s^2}$. And by Eq. (114), $\frac{w}{g}\frac{d^2s}{dt^2} = -\frac{wb^2}{2g}\frac{1}{s^2}$ or $\frac{d^2s}{dt^2} = -\frac{b^2}{2s^2}$. As in Eq. (66), $\frac{d^2s}{dt^2} = \frac{dv}{dt} = \frac{dv}{ds}\frac{ds}{dt} = v\frac{dv}{ds}$. Hence $v\frac{dv}{ds} = -\frac{b^2}{2s^2}$. $2v\,dv = -b^2\frac{ds}{s^2}$. Integration between the limits $v = 0$, $s = q$ and v,s gives $[v^2]_0^v = \left[\frac{b^2}{s}\right]_q^s$, $v^2 = b^2\left(\frac{1}{s} - \frac{1}{q}\right) = \frac{b^2(q - s)}{qs}$. $v = \frac{ds}{dt} = -\frac{b}{\sqrt{q}}\frac{\sqrt{q - s}}{\sqrt{s}}$. Assuming b positive, we need the minus sign since v starts at 0 and will become negative under the negative acceleration. $\frac{b}{\sqrt{q}}dt = -\frac{\sqrt{s}}{\sqrt{q - s}}ds$. As in Sec. 192, $\frac{-\sqrt{s}}{\sqrt{q - s}} = \frac{-s}{\sqrt{qs - s^2}} = \frac{1}{2}\frac{q - 2s}{\sqrt{qs - s^2}} - \frac{q}{2}\frac{1}{\sqrt{(q/2)^2 - (s - q/2)^2}}$. Hence integration between the limits $t = 0$, $s = q$ and t,s gives $\frac{b}{\sqrt{q}}[t]_0^t = \left[\sqrt{qs - s^2} - \frac{q}{2}\sin^{-1}\frac{s - q/2}{q/2}\right]_q^s$. $\frac{bt}{\sqrt{q}} = \sqrt{qs - s^2} + \frac{q}{2}\left(-\sin^{-1}\frac{2s - q}{q} + \frac{\pi}{2}\right)$. Hence $t = \frac{\sqrt{q}}{b}\left(\sqrt{qs - s^2} + \frac{q}{2}\cos^{-1}\frac{2s - q}{q}\right)$.

EXAMPLE 3. A particle of weight 0.322 lb. is acted on by an elastic force F_e toward O proportional to the distance from O, a resisting force F_r proportional to the velocity, and a driving force $F_d = 64 \sin^2 t$ lb. The force F_e is 2 lb. when $s = 1$. And the force F_r is 2 lb. when $v = 5$ ft./sec. At time $t = 0$, $s = 2$ and $v = 4$. Find the equation of motion.

Solution: $m = \dfrac{w}{g} = \dfrac{0.322}{32.2} = \dfrac{1}{10}$. $F_e = -k_1 s$ and $-2 = -k_1$. Hence $F_e = -2s$. $F_r = -k_2 v$, and $-2 = -k_2 5$. Hence $k_2 = -\dfrac{2}{5}$ and $F_r = -\dfrac{2}{5}v$. It follows from Eq. (114) that $\dfrac{1}{10}\dfrac{d^2s}{dt^2} = F_e + F_r + F_d = -2s - \dfrac{2}{5}v + 64\sin^2 t$, or $\dfrac{d^2s}{dt^2} =$ $-20s - 4v + 640 \sin^2 t$. Since $v = \dfrac{ds}{dt}$ and $\sin^2 t = \dfrac{1 - \cos 2t}{2}$, $\dfrac{d^2s}{dt^2} + 4\dfrac{ds}{dt} + 20s =$ $320 - 320 \cos 2t$. As in Sec. 342, for 320, $Q = F = D$, which is not a factor of $P(D) = D^2 + 4D + 20$. Hence by Eq. (105) with $q = 1$, $a = 0$, we try $Ae^{0x} = A$. Substitution in the differential equation gives $20A = 320$ and $A = 16$. Hence $s_{p1} = 16$. For $-320 \cos 2t$, $Q = F = D^2 + 4$, which is not a factor of $P(D)$. Hence by Eq. (106) with $q = 1$, $a = 0$, $b = 2$, we try $A \cos 2t + B \sin 2t$. We find

20	$s =$	A	$\cos 2t$	$+ B$	$\sin 2t$
4	$\dfrac{ds}{dt} =$	$2B$		$-2A$	
1	$\dfrac{d^2s}{dt^2} =$	$-4A$		$-4B$	
Totals		$16A + 8B$		$-8A + 16B$	
Should be		-320		0	

The totals will be as desired if $16A + 8B = -320$, $-8A + 16B = 0$. $A = 2B$, $32B + 8B = -320$, $B = -8$, $A = 2B = -16$. Hence $s_{p2} = -16 \cos 2t - 8 \sin 2t$. The roots of the auxiliary equation $r^2 + 4r + 20 = 0$, $r = -2 \pm 4i$ lead to $s_c = c_1e^{-2t} \cos 4t + c_2e^{-2t} \sin 4t$. Since $s = s_{p1} + s_{p2} + s_c$, $s =$ $16 - 16 \cos 2t - 8 \sin 2t + c_1e^{-2t} \cos 4t + c_2e^{-2t} \sin 4t$. By differentiation, $v = \dfrac{ds}{dt} =$ $32 \sin 2t - 16 \cos 2t + (-2c_1 + 4c_2)e^{-2t} \cos 4t + (-4c_1 - 2c_2)e^{-2t} \sin 4t$. At $t = 0$, $s = 2$ and $v = 4$, $2 = 16 - 16 + c_1$, $4 = -16 + (-2c_1 + 4c_2)$. Hence $c_1 = 2$, $c_2 = 6$. Thus the required equation of motion is

$$s = 16 - 16 \cos 2t - 8 \sin 2t + 2e^{-2t} \cos 4t + 6e^{-2t} \sin 4t.$$

EXERCISE 172

A particle is initially at rest with $s = q$. For each given law of force, verify that the other equations follow.

1. $F = -w = -mg$. $\dfrac{d^2s}{dt^2} = -g$. $\dfrac{ds}{dt} = v = -gt$. $s = q - \dfrac{1}{2}gt^2$.

2. $F = \dfrac{m}{t+2}$. $\dfrac{d^2s}{dt^2} = \dfrac{1}{t+2}$. $\dfrac{ds}{dt} = v = \ln\left(\dfrac{t}{2} + 1\right)$.
   $s = (t + 2)\ln\left(\dfrac{t}{2} + 1\right) - t + q$.

3. $F = \dfrac{-mt}{(t^2 + 4)^{3/2}}$. $\dfrac{d^2s}{dt^2} = \dfrac{-t}{(t^2 + 4)^{3/2}}$. $\dfrac{ds}{dt} = v = \dfrac{1}{\sqrt{t^2 + 4}} - \dfrac{1}{2}$.

4. $F = -mb^2s.\quad \dfrac{d^2s}{dt^2} = -b^2s.\quad \dfrac{ds}{dt} = v = -b\sqrt{q^2 - s^2}$ or $v = -bq \sin bt$.
$s = q \cos bt$.

5. $F = mb^2s.\quad \dfrac{d^2s}{dt^2} = b^2s.\quad \dfrac{ds}{dt} = v = b\sqrt{q^2 - s^2}$ or $v = bq \sinh bt$.

$s = \dfrac{q}{2}(e^{bt} + e^{-bt})$ or $s = q \cosh bt$.

6. $F = \dfrac{mb^2}{2s^2}.\quad \dfrac{d^2s}{dt^2} = \dfrac{b^2}{2s^2}.\quad \dfrac{ds}{dt} = v = \dfrac{b}{\sqrt{q}} \dfrac{\sqrt{s - q}}{\sqrt{s}}.$

$t = \dfrac{\sqrt{q}}{b}\left(\sqrt{s^2 - qs} + \dfrac{q}{2}\ln \dfrac{2s - q + 2\sqrt{s^2 - qs}}{q}\right)$ or

$t = \dfrac{\sqrt{q}}{b}\left(s^2 - qs + \dfrac{q}{2}\cosh^{-1}\dfrac{2s - q}{q}\right).$

7. $F = \dfrac{-mb^2}{s^3}.\quad \dfrac{d^2s}{dt^2} = \dfrac{-b^2}{s^3}.\quad \dfrac{ds}{dt} = v = -\dfrac{b}{q}\dfrac{\sqrt{q^2 - s^2}}{s}.\quad t = \dfrac{q}{b}\sqrt{q^2 - s^2}.$

8. $F = \dfrac{mb^2}{s^3}.\quad \dfrac{d^2s}{dt^2} = \dfrac{b^2}{s^3}.\quad \dfrac{ds}{dt} = v = \dfrac{b}{q}\dfrac{\sqrt{s^2 - q^2}}{s}.\quad t = \dfrac{q}{b}\sqrt{s^2 - q^2}.$

A particle is initially at $s = 0$ with $v = u$. For each given law of force, verify that the other equations follow.

9. $F = -w = -mg.\quad \dfrac{d^2s}{dt^2} = -g.\quad \dfrac{ds}{dt} = v = u - gt.\quad s = ut - \dfrac{1}{2}gt^2.$

10. $F = -mb^2s.\quad \dfrac{d^2s}{dt^2} = -b^2s.\quad \dfrac{ds}{dt} = v = \sqrt{u^2 - b^2s^2}$ or $v = u \cos bt.\quad s = \dfrac{u}{b}\sin bt.$

11. $F = mb^2s.\quad \dfrac{d^2s}{dt^2} = b^2s.\quad \dfrac{ds}{dt} = v = \sqrt{b^2s^2 - u^2}$ or $v = u \cosh bt.$

$s = \dfrac{u}{2b}(e^{bt} - e^{-bt})$ or $s = \dfrac{u}{b}\sinh bt.$

12. $F = -mkv.\quad \dfrac{d^2s}{dt^2} = -k\dfrac{ds}{dt}.\quad \dfrac{ds}{dt} = v = u - ks$ or $v = ue^{-kt}.\quad s = \dfrac{u}{k}(1 - e^{-kt}).$

13. $F = mf(v).\quad \dfrac{d^2s}{dt^2} = f\left(\dfrac{ds}{dt}\right).\quad t = \displaystyle\int_u^v \dfrac{dv}{f(v)}.\quad s = \int_u^v \dfrac{v\,dv}{f(v)}.$

14. $F = -m(kv + g).\quad \dfrac{d^2s}{dt^2} = -k\dfrac{ds}{dt} - g.\quad t = \dfrac{1}{k}\ln\dfrac{ku + g}{kv + g},$

$s = \dfrac{1}{k}(u - v) - \dfrac{g}{k^2}\ln\dfrac{ku + g}{kv + g}.$ Or $s = \left(\dfrac{u}{k} + \dfrac{g}{k^2}\right)(1 - e^{-kt}) - \dfrac{g}{k}t,$

$v = \left(u + \dfrac{g}{k}\right)e^{-kt} - \dfrac{g}{k}.$

15. $F = -mkv^2.\quad \dfrac{d^2s}{dt^2} = -k\left(\dfrac{ds}{dt}\right)^2.\quad t = \dfrac{1}{k}\left(\dfrac{1}{v} - \dfrac{1}{u}\right).\quad s = \dfrac{1}{k}\ln\dfrac{u}{v}.$

16. $F = -mkv^n,\ n \neq 1,2.\quad \dfrac{d^2s}{dt^2} = -k\left(\dfrac{ds}{dt}\right)^n.\quad t = \dfrac{1}{k(n - 1)}(v^{1-n} - u^{1-n}).$

$s = \dfrac{1}{k(n - 2)}(v^{2-n} - u^{2-n}).$

17. $F = -mb(v^2 + V^2).\quad \dfrac{d^2s}{dt^2} = -b\left[\left(\dfrac{ds}{dt}\right)^2 + V^2\right].$

$t = \dfrac{1}{Vk}\left(\tan^{-1}\dfrac{u}{V} - \tan^{-1}\dfrac{v}{V}\right),\ s = \dfrac{1}{2k}\ln\dfrac{V^2 + u^2}{V^2 + v^2}.$

18. $F = mb(v^2 - V^2).\quad \dfrac{d^2s}{dt^2} = b\left[\left(\dfrac{ds}{dt}\right)^2 - V^2\right].$

If $u \neq V$, $t = \dfrac{1}{2Vk} \ln \left(\dfrac{u+V}{u-V}\dfrac{v-V}{v+V}\right)$. $\quad s = \dfrac{1}{2k} \ln \dfrac{V^2 - v^2}{V^2 - u^2}$.

If $u < V$, $t = \dfrac{1}{Vk}\left(\tanh^{-1} \dfrac{u}{V} - \tanh^{-1} \dfrac{v}{V}\right)$. If $u = V$, $v = V$, $s = Vt$.

If $u > V$, $t = \dfrac{1}{Vk}\left(\tanh^{-1} \dfrac{V}{u} - \tanh^{-1} \dfrac{V}{v}\right)$.

A particle is initially at $s = q$ with $v = u$. For each given law of force, verify that the other equations follow.

19. $F = -ks$. $m\dfrac{d^2s}{dt^2} = -ks$. $s = q \cos \sqrt{\dfrac{k}{m}}\,t + u\sqrt{\dfrac{m}{k}} \sin \sqrt{\dfrac{k}{m}}\,t$.

20. $F = -bv - ks$, where $b^2 < 4mk$. $m\dfrac{d^2s}{dt^2} = -b\dfrac{ds}{dt} - ks$. Let $\sqrt{4mk - b^2} = K$.

$$s = qe^{-bt/2m} \cos \frac{Kt}{2m} + \frac{2mu + bq}{K} e^{-bt/2m} \sin \frac{Kt}{2m}.$$

21. A rocket is projected vertically upward against gravity by firing backward P lb. of its weight per second at a constant speed S ft./sec. with respect to the rocket. Thus if the original weight was W, at time t the mass $m = \dfrac{W - Pt}{g}$ and $\dfrac{W - Pt}{g}\dfrac{d^2s}{dt^2} = \dfrac{P}{g}S - (W - Pt)$. If at time $t = 0$, $s = 0$ and $v = 0$, deduce that $\dfrac{ds}{dt} = v = S \ln \dfrac{W}{W - Pt} - gt$,

$$s = S\left(t - \frac{W - Pt}{P} \ln \frac{W}{W - Pt}\right) - \frac{1}{2}gt^2.$$

22. A rocket is projected vertically upward against gravity. During each second it takes kP lb. of air ($k > 1$) from the atmosphere, burns this with P lb. of fuel already on board, and fires the products of combustion backward at a constant speed S ft./sec. with respect to the rocket. Thus if the initial weight was W, $m = \dfrac{W - Pt}{g}$ and $\dfrac{W - Pt}{g}\dfrac{d^2s}{dt^2} = \dfrac{P}{g}S - \dfrac{kP}{g}\left(\dfrac{ds}{dt} - S\right) - (W - Pt)$. If at time $t = 0$, $s = 0$ and $v = 0$, deduce that $\dfrac{ds}{dt} = v =$

$$\frac{k+1}{k}S\left[1 - \left(\frac{W - Pt}{W}\right)^k\right] - \frac{g}{P(k-1)}\left[W - Pt - W\left(\frac{W - Pt}{W}\right)^k\right].$$

$$s = \frac{S}{k}\left[(k+1)t + \frac{W}{P}\left(\frac{W - Pt}{W}\right)^{k+1} - \frac{W}{P}\right]$$
$$- \frac{g}{(k-1)P^2}\left\{\frac{W^2 - (W - Pt)^2}{2} - \frac{W^2}{k+1}\left[1 - \left(\frac{W - Pt}{W}\right)^{k+1}\right]\right\}.$$

344. Simple Electrical and Mechanical Circuits. Suppose that a current of i amp. flows through a circuit which contains resistance R ohms, inductance L henrys, and capacitance C farads in series. Then if the impressed electromotive force is e volts at time t sec.,

$$L\frac{di}{dt} + Ri + \frac{1}{C}\int_{t_0}^{t} i\,dt = e \quad \text{and} \quad L\frac{d^2i}{dt^2} + R\frac{di}{dt} + \frac{i}{C} = \frac{de}{dt}. \tag{115}$$

Next consider a mass m attached to fixed members by a spring with stiffness constant k and a dashpot with resistance constant β. Then if F is the impressed force on the mass, and s is the linear displacement of the mass

$$m\frac{d^2s}{dt^2} + \beta\frac{ds}{dt} + ks = F. \tag{116}$$

Similarly, consider an oscillating shaft with moment of inertia I, subjected to an external torque M opposed by damping torques with constant B and elastic torques with constant K. Then

$$I\frac{d^2\theta}{dt^2} + B\frac{d\theta}{dt} + K\theta = M. \tag{117}$$

Each of the equations (115) to (117) can be solved by the method of Sec. 342. For more complicated systems, the method of Laplace transforms is often convenient.†

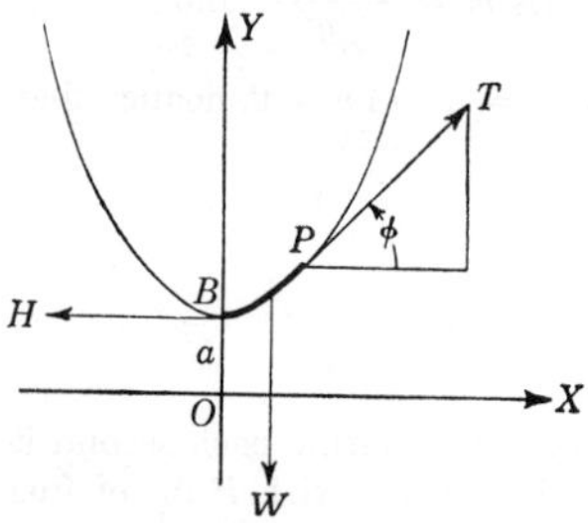

FIG. 366.

345. Equilibrium of Cables. Consider a flexible cable which is suspended at both ends and hangs in a vertical plane under the influence of a downward load. To find the curve of equilibrium, we isolate the part of the curve between the lowest point $B = (0,a)$ and $P = (x,y)$ (Fig. 366). The arc BP is then subject to the total load $W(x)$, the horizontal tension H at B, and the tension T at P making an angle with the horizontal x axis equal to ϕ, the slope angle. Since the vector sum of these forces must be zero,

$$H = T\cos\phi, \qquad W(x) = T\sin\phi. \tag{118}$$

It follows that

$$\frac{W(x)}{H} = \frac{T\sin\phi}{T\cos\phi} = \tan\phi = \frac{dy}{dx}, \qquad \frac{dy}{dx} = \frac{1}{H}W(x). \tag{119}$$

If the load can be found directly in terms of x, this is a differential equation in terms of y and x, whose integration gives the desired curve of equilibrium.

If it is easy to find the differential of $W(x)$, dW, we may use the relation found by differentiation

$$\frac{d^2y}{dx^2} = \frac{1}{H}\frac{dW}{dx}. \tag{120}$$

† The interested reader will find an elementary account of this in the author's "Fourier Methods," McGraw-Hill Book Company, Inc., New York (Dover reprint).

346. Deflection of Beams. Consider a horizontal beam under the influence of vertical forces. Take the x axis horizontal and the y axis vertical. Let E denote Young's modulus for the material of the beam when stretched or compressed longitudinally. And let I be the moment of inertia of a cross section about a horizontal axis in its plane through its center of gravity, $P = (x,y)$. Then for small deflections,

$$EI\frac{d^2y}{dx^2} = M, \tag{121}$$

where M is the bending moment on the portion of the beam to the right of $P = (x,y)$, considered positive when it would increase y, if this part were free to rotate about P.

EXAMPLE. A beam of length $2L$ is fixed at both ends. It is loaded with a weight w per unit length, and a point load W in the middle. Find the equation of the deflected curve.

Solution: Take the origin at the center of the beam (Fig. 367). Then when $x = 0$, $y = 0$ and from symmetry $dy/dx = 0$. The total weight supported is $w(2L) + W$.

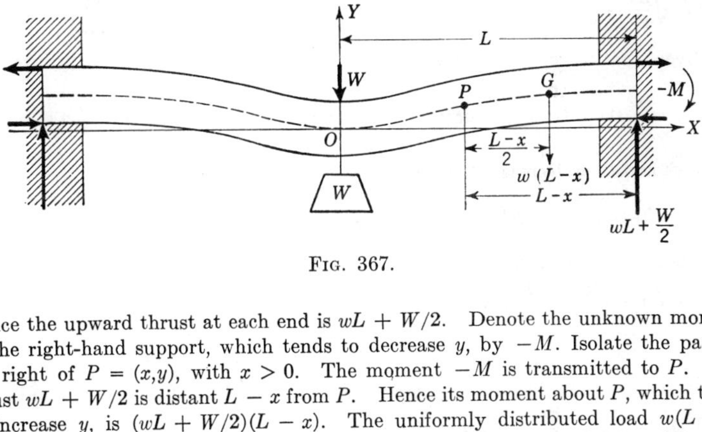

FIG. 367.

Hence the upward thrust at each end is $wL + W/2$. Denote the unknown moment at the right-hand support, which tends to decrease y, by $-M$. Isolate the part to the right of $P = (x,y)$, with $x > 0$. The moment $-M$ is transmitted to P. The thrust $wL + W/2$ is distant $L - x$ from P. Hence its moment about P, which tends to increase y, is $(wL + W/2)(L - x)$. The uniformly distributed load $w(L - x)$ can be considered as concentrated at its center of gravity G. And $PG = \frac{1}{2}(L - x)$. Hence the moment, which tends to decrease y, is $-\frac{w}{2}(L - x)^2$. By Eq. (121), $EI\frac{d^2y}{dx^2} = -M + \left(wL + \frac{W}{2}\right)(L - x) - \frac{w}{2}(L - x)^2$. Integration between the limits $\frac{dy}{dx} = 0$, $x = L$, and $\frac{dy}{dx}$, x gives

$EI\frac{dy}{dx} = M(L - x) - \frac{1}{2}\left(wL + \frac{W}{2}\right)(L - x)^2 + \frac{w}{6}(L - x)^3$. Since $\frac{dy}{dx} = 0$ when $x = 0$, $0 = ML - \frac{1}{2}\left(wL + \frac{W}{2}\right)L^2 + \frac{w}{6}L^3$, $M = \frac{WL}{4} + \frac{wL^2}{3}$.

By substituting this value we find that $EI\frac{dy}{dx} = \frac{w}{6}(L^2x - x^3) + \frac{W}{4}(Lx - x^2)$.

Integration from $x = 0$, $y = 0$ to x,y gives $EIy = \frac{w}{6}\left(L^2\frac{x^2}{2} - \frac{x^4}{4}\right) + \frac{W}{4}\left(L\frac{x^2}{2} - \frac{x^3}{3}\right)$ for $x > 0$.

By symmetry, $EIy = \frac{w}{6}\left(L^2\frac{x^2}{2} - \frac{x^4}{4}\right) + \frac{W}{4}\left(L\frac{x^2}{2} + \frac{x^3}{3}\right)$ for $x < 0$. Hence $y = \frac{x^2}{24EI}[w(2L^2 - x^2) + W(3L - 2|x|)]$, for the whole curve.

EXERCISE 173

1. Given that $L\frac{di}{dt} + Ri = E \sin \omega t$, and $i = 0$ when $t = 0$, verify that $i = \frac{E}{R^2 + L^2\omega^2}(R \sin \omega t - L \cos \omega t + L\omega e^{-Rt/L})$.

2. Given that $Ri + \frac{1}{C}\int_0^t i\,dt = E \sin \omega t$, show that $i = 0$ when $t = 0$, and that $R\frac{di}{dt} + \frac{i}{C} = E\omega \cos \omega t$. Also verify that $i = \frac{CE}{1 + C^2R^2\omega^2}(\cos \omega t + CR\omega \sin \omega t - e^{-t/RC})$.

3. Given that $L\frac{d^2i}{dt^2} + R\frac{di}{dt} + \frac{i}{C} = 0$, $i = I$ and $\frac{di}{dt} = 0$ when $t = 0$. If $R < 2\sqrt{\frac{L}{C}}$, let $\sqrt{\frac{4L}{C} - R^2} = a$, and show that $i = Ie^{-Rt/2L}\left(\cos \frac{at}{2L} + \frac{R}{a} \sin \frac{at}{2L}\right)$.

4. Given that $m\frac{d^2s}{dt^2} + \beta\frac{ds}{dt} + ks = 0$, $s = 0$, and $\frac{ds}{dt} = u$ when $t = 0$. If $\beta = 2\sqrt{mk}$, show that $s = ute^{-\sqrt{k/m}\,t}$. Thus s increases from 0 to $\frac{u}{e}\sqrt{\frac{m}{k}}$ at $t = \sqrt{\frac{m}{k}}$, and then decreases to 0 as $t \to \infty$.

5. Given that $I\frac{d^2\theta}{dt^2} + B\frac{d\theta}{dt} + K\theta = bt$, $\theta = 0$, and $\frac{d\theta}{dt} = 0$ when $t = 0$. If $B > 2\sqrt{IK}$, let $\sqrt{B^2 - 4IK} = a$, and show that
$$\theta = \frac{b}{K}t - \frac{Bb}{K^2} + \frac{b}{2aK^2}(aB + 2IK - B^2)e^{-t(B+a)/2I}$$
$$+ \frac{b}{2aK^2}(aB - 2IK + B^2)e^{-t(B-a)/2I}.$$

6. The load on the cable of a suspension bridge is $W = bx$. If the lowest point $B = (0,a)$, show that $\frac{dy}{dx} = \frac{bx}{H}$, $y = \frac{bx^2}{2H} + a$.

7. A cable supports its own weight, $W = bs$, so that $\frac{d^2y}{dx^2} = \frac{b}{H}\sqrt{1 + \left(\frac{dy}{dx}\right)^2}$. If the lowest point $B = (0,a)$, show that with $p = \frac{dy}{dx}$, $\ln (p + \sqrt{1 + p^2}) = \frac{bx}{H}$. And by one of the procedures of Example 3 of Sec. 333, $p = \frac{1}{2}(e^{bx/H} - e^{-bx/H})$ or $p = \sinh \frac{bx}{H}$. And $y = \frac{H}{2b}(e^{bx/H} + e^{-bx/H} - 2) + a$ or $y = \frac{H}{b}\cosh \frac{bx}{H} - \frac{H}{b} + a$.

8. A cable supports a curtain with lower edge horizontal. Take this edge as the x axis, and the lowest point $B = (0,a)$. Then $dW = b\,dA = by\,dx$, so that $\frac{d^2y}{dx^2} = \frac{by}{H}$. Show that $y = \frac{a}{2}(e^{\sqrt{b/H}\,x} + e^{-\sqrt{b/H}\,x}) = a \cosh \sqrt{\frac{b}{H}}\,x$.

9. When an elastic cable supports its own weight, $dW = \frac{b}{T}\,ds$, where T is the tension of Eq. (118). Thus $H = T\cos\phi = T\frac{dx}{ds}$. And $dW = b\frac{ds}{T} = b\frac{dx}{H}$, so that $\frac{d^2y}{dx^2} = \frac{b}{H^2}$. If the lowest point $B = (0,a)$, show that $y = \frac{b}{2H^2}x^2 + a$.

10. For the catenary of uniform strength, supporting its own weight, $dW = kT\,ds$, where T is the tension of Eq. (118). Thus $H = T\cos\phi = T\frac{dx}{ds}$. And $\frac{dW}{dx} = k\frac{ds}{dx}T = kH\left(\frac{ds}{dx}\right)^2$, so that $\frac{d^2y}{dx^2} = k\left[1 + \left(\frac{dy}{dx}\right)^2\right]$. If the lowest point $B = (0,a)$, show that with $p = \frac{dy}{dx}$, $kx = \tan^{-1} p$, $p = \tan kx$. And $y = 1/k \ln \sec kx + a$. Since $y = \infty$ for $kx = -\pi/2$ or $\pi/2$, the span must be less than π/k.

11. Let KA be the maximum tension a cable with cross section of area A will safely support. And let the weight of a unit volume of the cable be w. Then in Prob. 7, $b = Aw$. And for the end point of a cable of maximum span, $T = KA$. Show that $T = H\cosh\frac{bx}{H} = \frac{bK}{w}$. Let $\frac{bx}{H} = u$. Then $x = \frac{K}{w}\frac{u}{\cosh u}$. This is a maximum when $\tanh u = 1/u$, whose root is found by Newton's method (Sec. 167) to be $u = 1.1997$. For this value $\frac{u}{\cosh u} = 0.6626$. Hence the maximum span is $2x = 1.335\frac{K}{w}$.

12. With the notation of Prob. 11, in Prob. 10 the area A is variable but $dW = wA\,ds$. If $T = KA$, $dW = \frac{wT}{K}\,ds$ so that in Prob. 10, $k = \frac{w}{K}$. And the limiting span of Prob. 10, $\pi\frac{K}{w}$, is 2.353 times the maximum span found in Prob. 11. Show that for a steel cable with $K = 130{,}000$ lb./in.^{2} and $w = 480$ lb./ft.^{3}, the maximum span of a catenary is 9.86 mi., while the limiting span for a catenary of uniform strength is 23.20 mi.

A beam of length $2L$ is supported at both ends. Take the origin at the center of the beam. And show that for a

13. Load w per unit length, $EI\frac{d^2y}{dx^2} = \frac{w}{2}(L^2 - x^2)$, $y = \frac{wx^2}{24EI}(6L^2 - x^2)$.

14. Point load W at the center, $EI\frac{d^2y}{dx^2} = \frac{W}{2}(L - x)$, $y = \frac{Wx^2}{12EI}(3L - x)$.

A cantilever beam of length L is fixed at one end. Take the origin at the fixed end. And show that for a

15. Load w per unit length, $EI\frac{d^2y}{dx^2} = -\frac{w}{2}(L - x)^2$, $y = -\frac{wx^2}{24EI}(6L^2 - 4Lx + x^2)$.

16. Point load W at the free end, $EI\frac{d^2y}{dx^2} = -W(L - x)$, $y = -\frac{Wx^2}{6EI}(3L - x)$.

17. Load w per unit length and upward thrust R at the far end, $EI\frac{d^2y}{dx^2} = -\frac{w}{2}(L - x)^2 + R(L - x)$, $y = \frac{x^2}{24EI}[4R(3L - x) - w(6L^2 - 4Lx + x^2)]$.

18. In Prob. 17, the far end is supported on a level with the fixed end. Show that $R = \frac{3}{8}wL$ and $y = -\frac{wx^2}{48EI}(L - x)(3L - 2x)$.

19. A long vertical column of length L is fixed at the base and supports a weight P. Take the x axis vertical and the origin at the fixed end. If the weight causes the upper end to be at (L,a), its moment about $P = (x,y)$, which tends to increase y, is $P(a - y)$. And by Eq. (121), $EI\frac{d^2y}{dx^2} = P(a - y)$. Show that $y = a\left(1 - \cos\sqrt{\frac{P}{EI}}\,x\right)$ with $\cos\sqrt{\frac{P}{EI}}\,L = 0$. Thus the first critical load $P_1 = \frac{EI\pi^2}{4L^2}$. The column will not buckle if $P < P_1$.

347. Motion in a Plane. The acceleration of a particle of mass m moving in a plane is determined by the *vector* relation $\mathbf{F} = m\mathbf{a}$. Its component along the tangent to the path is $F_t = ma_t$. By Eq. (96) of Sec. 142, $a_t = d^2s/dt^2$. It follows that

$$m\frac{d^2s}{dt^2} = F_t. \qquad (122)$$

This may be solved as indicated in Sec. 343 whenever F_t is a known function of t, s, and $v_t = ds/dt$.

The components of $\mathbf{F} = m\mathbf{a}$ along the x and y axes are $F_x = ma_x$ and $F_y = ma_y$. By Eq. (87) of Sec. 141, $a_x = d^2x/dt^2$ and $a_y = d^2y/dt^2$. It follows that

$$m\frac{d^2x}{dt^2} = F_x, \qquad m\frac{d^2y}{dt^2} = F_y. \qquad (123)$$

These may be solved as indicated in Sec. 343 whenever F_x is a function of t, x, and $v_x = dx/dt$ while F_y is a function of t, y, and $v_y = dy/dt$.

The radial and transverse components of $F = ma$ are $F_r = ma_r$ and $F_\theta = ma_\theta$. By Eq. (73) of Sec. 155,

$$a_r = \frac{d^2r}{dt^2} - r\left(\frac{d\theta}{dt}\right)^2 \quad \text{and} \quad a_\theta = \frac{1}{r}\frac{d}{dt}\left(r^2\frac{d\theta}{dt}\right).$$

It follows that

$$m\left[\frac{d^2r}{dt^2} - r\left(\frac{d\theta}{dt}\right)^2\right] = F_r, \qquad \frac{1}{r}\frac{d}{dt}\left(r^2\frac{d\theta}{dt}\right) = F_\theta. \qquad (124)$$

For a central force, $F_\theta = 0$. In this case, from the second relation of Eq. (124), we have

$$\frac{1}{r}\frac{d}{dt}\left(r^2\frac{d\theta}{dt}\right) = 0, \qquad d\left(r^2\frac{d\theta}{dt}\right) = 0 \quad \text{and} \quad r^2\frac{d\theta}{dt} = h, \qquad (125)$$

a constant. Substitution of $\frac{d\theta}{dt} = \frac{h}{r^2}$ in the first relation of Eq. (124)

leads to

$$\frac{d^2r}{dt^2} - \frac{h^2}{r^3} = \frac{F_r}{m}. \tag{126}$$

If the force depends on the distance only, this becomes a differential equation in r and t.

EXAMPLE. Find the orbit of a planet of mass m attracted toward the sun with a force $F_r = -\frac{mk}{r^2}$. At the perihelion point, or point nearest the sun, $v_r = \frac{dr}{dt} = 0$. Let $\theta = Q$, $r = q$, and $v_\theta = r\frac{d\theta}{dt} = \frac{h}{q}$ at the perihelion point.

Solution: Since $F_\theta = 0$ as in Eq. (125), $r^2\frac{d\theta}{dt}$ is constant. When $r = q$, $r^2\frac{d\theta}{dt} = rv_\theta = q\left(\frac{h}{q}\right) = h$. Hence $\frac{d\theta}{dt} = \frac{h}{r^2}$ and by Eq. (126) $\frac{d^2r}{dt^2} - \frac{h^2}{r^3} = \frac{F_r}{m} = -\frac{k}{r^2}$. As in Eq. (66), $\frac{d^2r}{dt^2} = \frac{dv_r}{dt} = \frac{dv_r}{dr}\frac{dr}{dt} = v_r\frac{dv_r}{dr}$. Hence $2v_r\frac{dv_r}{dr} = \frac{2h^2}{r^3} - \frac{2k}{r^2}$. Integration between the limits $v_r = 0$, $r = q$ and v_r, r gives $v_r^2 = h^2\left(\frac{1}{q^2} - \frac{1}{r^2}\right) + 2k\left(\frac{1}{r} - \frac{1}{q}\right)$. Since $\frac{d\theta}{dt} = \frac{h}{r^2}$, $\frac{d\theta}{dr} = \frac{d\theta/dt}{dr/dt} = \frac{h}{r^2v_r}$. And $d\theta = \frac{(h/r^2)dr}{\sqrt{(h/q - k/h)^2 - (h/r - k/h)^2}}$. Integration between the limits $\theta = Q$, $r = q$ and θ, r gives $\theta - Q = -\sin^{-1}\frac{h/r - k/h}{h/q - k/h} + \frac{\pi}{2}$. Hence $\frac{h}{r} - \frac{k}{h} = \left(\frac{h}{q} - \frac{k}{h}\right)\cos(\theta - Q)$. $r = \frac{h^2q}{kq + (h^2 - kq)\cos(\theta - Q)}$. This is an ellipse, parabola, or hyperbola with eccentricity $e = \frac{h^2 - kq}{kq}$, since with this definition of e, and $\frac{h^2}{ek} = \bar{q}$, the equation reduces to that of Example 4 of Sec. 146, with θ replaced by $\theta - Q + \pi$.

EXERCISE 174

A projectile is fired with a velocity V and an angle of departure ϕ. Thus at $t = 0$, $x = 0$, $y = 0$, $v_x = V\cos\phi$, $v_y = V\sin\phi$. It is acted on by gravity. Show that

1. Neglecting air resistance $\frac{d^2x}{dt^2} = 0$, $\frac{d^2y}{dt^2} = -g$. And $x = Vt\cos\phi$, $y = Vt\sin\phi - \frac{1}{2}gt^2 = x\tan\phi - x^2\frac{g\sec^2\phi}{2V^2}$.
2. With air resistance equal to bmv, $\frac{d^2x}{dt^2} = -b\frac{dx}{dt}$, $\frac{d^2y}{dt^2} = -b\frac{dy}{dt} - g$. And $x = \frac{V}{b}\cos\phi(1 - e^{-bt})$, $y = \left(\frac{V}{b}\sin\phi + \frac{g}{b^2}\right)(1 - e^{-bt}) - \frac{gt}{b}$.

A particle is initially at $(q,0)$ with velocity components $v_x = 0$, $v_y = U$. A law of force is indicated. And in some problems a relation of U to q and b is given. Verify that the other equations follow.

3. No force. $\frac{d^2x}{dt^2} = 0$, $\frac{d^2y}{dt^2} = 0$. $x = q$, $y = Ut$, a straight line.
4. An attractive force toward the origin, mb^2r in magnitude. $\frac{d^2x}{dt^2} = -b^2x$, $\frac{d^2y}{dt^2} =$

$-b^2y.$ $x = q\cos bt$, $y = \dfrac{U}{b}\sin bt$, $\dfrac{x^2}{q^2} + \dfrac{b^2y^2}{U^2} = 1$, an ellipse.

5. A repulsive force from the origin, mb^2r in magnitude. $\dfrac{d^2x}{dt^2} = b^2x$, $\dfrac{d^2y}{dt^2} = b^2y$. $x = \dfrac{q}{2}(e^{bt} + e^{-bt}) = q\cosh bt$, $y = \dfrac{U}{2b}(e^{bt} - e^{-bt}) = \dfrac{U}{b}\sinh bt$. $\dfrac{x^2}{q^2} - \dfrac{b^2y^2}{U^2} = 1$, a hyperbola.

6. The force on an electron of charge e moving with velocity $\mathbf{v} = \dfrac{dx}{dt}\mathbf{i} + \dfrac{dy}{dt}\mathbf{j}$ in a field with magnetic vector $\mathbf{H} = H\mathbf{k}$ is $\dfrac{e}{c}\mathbf{v}\times\mathbf{H} = -\dfrac{eH}{c}\dfrac{dy}{dt}\mathbf{i} + \dfrac{eH}{c}\dfrac{dx}{dt}\mathbf{j}$. With $b = \dfrac{eH}{mc}$, $\dfrac{d^2x}{dt^2} = -b\dfrac{dy}{dt}$, $\dfrac{d^2y}{dt^2} = b\dfrac{dx}{dt}$. $\dfrac{dx}{dt} = -by$, $\dfrac{dy}{dt} = bx + U - bq$, $\dfrac{d^2y}{dt^2} = b\dfrac{dx}{dt} = -b^2y$. $y = \dfrac{U}{b}\sin bt$, $x = q + \dfrac{(dy/dt) - U}{b} = q - \dfrac{U}{b} + \dfrac{U}{b}\cos bt$.
$\left(x - q + \dfrac{U}{b}\right)^2 + y^2 = \dfrac{U^2}{b^2}$, a circle.

7. No force. $r^2\dfrac{d\theta}{dt} = qU$, $\dfrac{d^2r}{dt^2} - \dfrac{q^2U^2}{r^3} = 0$. $v_r^2 = \left(\dfrac{dr}{dt}\right)^2 = q^2U^2\left(\dfrac{1}{q^2} - \dfrac{1}{r^2}\right)$, $\dfrac{d\theta}{dr} = \dfrac{q}{r\sqrt{r^2 - q^2}}$, $r = q\sec\theta$, a straight line. And $t = \dfrac{1}{U}\sqrt{r^2 - q^2} = \dfrac{q}{U}\tan\theta$.

8. An attractive force toward the origin, $\dfrac{mb^2}{r^3}$ in magnitude. $r^2\dfrac{d\theta}{dt} = qU$, $\dfrac{d^2r}{dt^2} - \dfrac{q^2U^2}{r^3} = -\dfrac{b^2}{r^3}$. $v_r^2 = \left(\dfrac{dr}{dt}\right)^2 = (q^2U^2 - b^2)\left(\dfrac{1}{q^2} - \dfrac{1}{r^2}\right)$. If $Uq > b$, with $q^2U^2 - b^2 = A$, $\dfrac{d\theta}{dr} = \dfrac{q^2U^2}{A}\dfrac{1}{r\sqrt{r^2 - q^2}}$, $r = q\sec\dfrac{A\theta}{qU}$. And $t = \dfrac{q}{A}\sqrt{r^2 - q^2}$. If $Uq = b$, $r = b$ and $t = b\theta$. If $Uq < b$, with $b^2 - q^2U^2 = B$, $\dfrac{d\theta}{dr} = -\dfrac{q^2U}{B}\dfrac{1}{r\sqrt{r^2 - q^2}}$, $\theta = \dfrac{qU}{B}\ln\dfrac{q + \sqrt{q^2 - r^2}}{r} = \dfrac{qU}{B}\cosh^{-1}\dfrac{q}{r}$. And $t = \dfrac{q}{B}\sqrt{q^2 - r^2}$.

9. An attractive force toward the origin, mb^2r in magnitude. $r^2\dfrac{d\theta}{dt} = qU$, $\dfrac{d^2r}{dt^2} - \dfrac{q^2U^2}{r^3} = -b^2r$. $v_r^2 = \left(\dfrac{dr}{dt}\right)^2 = \dfrac{1}{r^2}[-q^2U^2 + r^2(b^2q^2 + U^2) - b^2r^4]$.

$$\frac{d\theta}{dr} = \frac{-1/r^3}{\sqrt{\frac{1}{4}(b^2/U^2 - 1/q^2)^2 - [1/r^2 - \frac{1}{2}(b^2/U^2 + 1/q^2)]^2}},$$

$$2\theta = \sin^{-1}\frac{1/r^2 - \frac{1}{2}(b^2/U^2 + 1/q^2)}{\frac{1}{2}(b^2/U^2 - 1/q^2)} + \frac{\pi}{2}.$$

$\left[\dfrac{1}{2}\left(\dfrac{b^2}{U^2} + \dfrac{1}{q^2}\right) - \dfrac{1}{2}\left(\dfrac{b^2}{U^2} - \dfrac{1}{q^2}\right)\cos 2\theta\right]r^2 = 1$, the ellipse $\dfrac{x^2}{q^2} + \dfrac{b^2y^2}{U^2} = 1$. And

$$\frac{dt}{dr} = -\frac{1}{b}\frac{r}{\sqrt{\frac{1}{4}(q^2 - U^2/b^2)^2 - [r^2 - \frac{1}{2}(q^2 + U^2/b^2)]^2}},$$

$2bt = -\sin^{-1}\dfrac{r^2 - \frac{1}{2}(q^2 + U^2/b^2)}{\frac{1}{2}(q^2 - U^2/b^2)} + \dfrac{\pi}{2}$. $r^2 = q^2\cos^2 bt + \dfrac{U^2}{b^2}\sin^2 bt$.

10. An attractive force toward the origin, $\dfrac{mb^2}{r^5}$ in magnitude. And $U = \dfrac{b}{q^2\sqrt{2}}$. $r^2\dfrac{d\theta}{dt} = qU$, $\dfrac{d^2r}{dt^2} - \dfrac{q^2U^2}{r^3} = -\dfrac{2q^4U^2}{r^5}$. $v_r^2 = \left(\dfrac{dr}{dt}\right)^2 = q^2U^2\left(\dfrac{q^2}{r^4} - \dfrac{1}{r^2}\right)$.
$\dfrac{d\theta}{dr} = \dfrac{-1}{\sqrt{q^2 - r^2}}$, $r = q\cos\theta$, a circle. And $t = \dfrac{q}{2u}(\theta + \sin\theta\cos\theta)$.

11. An attractive force toward the origin, $\frac{mb^2}{r^7}$ in magnitude. And $U = \frac{b}{q^3\sqrt{3}}$.

$r^2\frac{d\theta}{dt} = qU$, $\frac{d^2r}{dt^2} - \frac{q^2U^2}{r^3} = -\frac{3q^6U^2}{r^7}$. $v_r^2 = \left(\frac{dr}{dt}\right)^2 = q^2U^2\left(\frac{q^4}{r^6} - \frac{1}{r^2}\right)$.

$\frac{d\theta}{dr} = \frac{-r}{\sqrt{q^4 - r^4}}$, $r^2 = q^2 \cos 2\theta$, a lemniscate.

12. An attractive force toward the origin, $\frac{mb^2}{r^4}$ in magnitude. And $U = b\sqrt{\frac{2}{3q^3}}$.

$r^2\frac{d\theta}{dt} = qU$, $\frac{d^2r}{dt^2} - \frac{q^2U^2}{r^3} = -\frac{3q^3U^2}{2r^4}$. $v_r^2 = \left(\frac{dr}{dt}\right)^2 = q^2U^2\left(\frac{q}{r^3} - \frac{1}{r^2}\right)$.

$\frac{d\theta}{dr} = \frac{-1}{\sqrt{qr - r^2}}$, $r = \frac{q}{2}(1 + \cos\theta)$, a cardioid.

13. An attractive force toward the origin, $\frac{mb^2}{r^{2n+1}}$ in magnitude. And $U = \frac{b}{q^n\sqrt{n}}$.

$r^2\frac{d\theta}{dt} = qU$, $\frac{d^2r}{dt^2} - \frac{q^2U^2}{r^3} = -\frac{nq^{2n}U^2}{r^{2n+1}}$. $v_r^2 = \left(\frac{dr}{dt}\right)^2 = q^2U^2\left(\frac{q^{2n-2}}{r^{2n}} - \frac{1}{r^2}\right)$.

$\frac{d\theta}{dr} = \frac{-r^{n-2}}{\sqrt{q^{2n-2} - r^{2n-2}}}$, $r^{n-1} = q^{n-1}\cos(n-1)\theta$.

Let the law of force be $F_r = -\frac{mk}{r^2}$, as in the example. And let $\theta = Q$, $r = q$ at P where $v_r = 0$. Show that

14. If $v_\theta = \sqrt{\frac{k}{q}}$ and $t = 0$ at P, $r^2\frac{d\theta}{dt} = \sqrt{kq}$, $\frac{d^2r}{dt^2} - \frac{kq}{r^3} = -\frac{k}{r^2}$. $v_r^2 = -kq\left(\frac{1}{r} - \frac{1}{q}\right)^2$. Hence v_r is real only if $r = q$. But $r = q$, $v_r = \frac{dr}{dt} = 0$, $\frac{d^2r}{dt^2} = 0$ satisfy the equations in r and t. Then $\frac{d\theta}{dt} = \frac{\sqrt{kq}}{r^2} = \frac{\sqrt{kq}}{q^2}$ and $\theta = Q + \frac{\sqrt{kq}}{q^2}t$. Thus the circular orbit $r = q$ is followed with uniform speed in the path, $v_\theta = \sqrt{\frac{k}{q}}$.

15. If $v_r = 0$ at $r = \infty$, since $v_r^2 = h^2\left(\frac{1}{q^2} - \frac{1}{r^2}\right) + 2k\left(\frac{1}{r} - \frac{1}{q}\right)$ as in the example, $\frac{h^2}{q^2} - \frac{2k}{q} = 0$. Hence $h^2 = 2kq$ and $r = \frac{2q}{1 - \cos(\theta - Q)}$. This parabola is an approximation to the orbit of a comet.

16. By the example, a periodic orbit of a planet is an ellipse with the sun at one focus, which is Kepler's first law. If S is the sectorial area swept out by the radius vector, $dS = \frac{1}{2}r^2\,d\theta$. Hence $\frac{dS}{dt} = \frac{1}{2}r^2\frac{d\theta}{dt} = \frac{h}{2}$, a constant, which is Kepler's second law. It follows that, if T is the period and A the area of the whole ellipse, $\frac{A}{T} = \frac{h}{2}$. If a and b are the semiaxes of the ellipse, $A = \pi ab$. By Example 4 of Sec. 146, with $\bar{q} = \frac{h^2}{ek}$ in place of q, $a = \frac{e\bar{q}}{1 - e^2}$, $b = \frac{e\bar{q}}{\sqrt{1 - e^2}}$. Hence $b = \sqrt{a}\sqrt{e\bar{q}} = \frac{h\sqrt{a}}{\sqrt{k}}$. Thus $T = \frac{2A}{h} = \frac{2\pi ab}{h} = \frac{2\pi}{\sqrt{k}}a^{\frac{3}{2}}$. That T^2 is proportional to a^3 for any two planets is essentially Kepler's third law.

17. Verify the following alternative solution of the example, after the equations $\frac{d\theta}{dt} = \frac{h}{r^2}$ and $\frac{d^2r}{dt^2} - \frac{h^2}{r^3} = -\frac{k}{r^2}$ have been found. Let $u = \frac{1}{r}$. Then $\frac{du}{dr} = -\frac{1}{r^2} =$

$-\dfrac{1}{h}\dfrac{d\theta}{dt}$ so that $\dfrac{dr}{dt} = -h\dfrac{du}{d\theta}\cdot$ And $\dfrac{d^2r}{dt^2} = -h\dfrac{d^2u}{d\theta^2}\dfrac{d\theta}{dt} = -\dfrac{h^2}{r^2}\dfrac{d^2u}{d\theta^2}\cdot$ Hence $-\dfrac{h^2}{r^2}\dfrac{d^2u}{d\theta^2} - \dfrac{h^2}{r^3} = -\dfrac{k}{r^2}$, or $\dfrac{d^2u}{d\theta^2} + u = \dfrac{k}{h^2}\cdot$ $u = \dfrac{k}{h^2} + c_1 \cos \theta + c_2 \sin \theta$. By differentiation, $\dfrac{du}{d\theta} = -c_1 \sin \theta + c_2 \cos \theta$. When $\theta = Q$, $u = \dfrac{1}{r} = \dfrac{1}{q}$, $\dfrac{du}{d\theta} = -\dfrac{1}{h}\dfrac{dr}{dt} = 0$. $\dfrac{1}{q} = \dfrac{k}{h^2} + c_1 \cos Q + c_2 \sin Q$, $0 = -c_1 \sin Q + c_2 \cos Q$. Hence $c_1 = \left(\dfrac{1}{q} - \dfrac{k}{h^2}\right) \cos Q$, $c_2 = \left(\dfrac{1}{q} - \dfrac{k}{h^2}\right) \sin Q$ so that

$\dfrac{1}{r} = u = \dfrac{k}{h^2} + \left(\dfrac{1}{q} - \dfrac{k}{h^2}\right) \cos (\theta - Q)$ which checks the previous solution.

TABLES

Table 1. Integrals

Elementary Forms

1. $\int 1\, du = \int du = u + C.$
2. $\int u\, du = \frac{u^2}{2} + C.$
3. $\int u^n\, du = \frac{u^{n+1}}{n+1} + C$, if $n \neq -1$.
4. $\int u^{-1}\, du = \int \frac{du}{u} = \ln u + C.$ See Sec. 189.
5. $\int u\, dv = uv - \int v\, du.$ See Sec. 200.

Algebraic Forms

6. $\int \frac{du}{u^2 - a^2} = \frac{1}{2a} \ln \frac{u-a}{u+a} + C.$†
7. $\int \frac{du}{u^2 + a^2} = \frac{1}{a} \tan^{-1} \frac{u}{a} + C.$
8. $\int \frac{du}{\sqrt{a^2 - u^2}} = \sin^{-1} \frac{u}{a} + C.$
9. $\int \frac{du}{\sqrt{u^2 + A}} = \ln (u + \sqrt{u^2 + A}) + C.$†
10. $\int \frac{du}{u\sqrt{u^2 - a^2}} = \frac{1}{a} \sec^{-1} \frac{u}{a} + C.$
11. $\int \sqrt{a^2 - u^2}\, du = \frac{u}{2} \sqrt{a^2 - u^2} + \frac{a^2}{2} \sin^{-1} \frac{u}{a} + C.$
12. $\int \sqrt{u^2 + A}\, du = \frac{u}{2} \sqrt{u^2 + A} + \frac{A}{2} \ln (u + \sqrt{u^2 + A}) + C.$†

For forms containing $ax^2 + bx + c$, see Sec. 192.
For forms containing $ax + b$ or $ax^n + b$, see Sec. 198.
For forms containing $ax^2 + b$, see Sec. 199.
For forms containing rational functions, see Sec. 201.

Trigonometric Forms

13. $\int \sin u\, du = -\cos u + C.$
14. $\int \sin^2 au\, du = \frac{u}{2} - \frac{\sin 2au}{4a} + C.$
15. $\int \sin^4 au\, du = \frac{3u}{8} - \frac{\sin 2au}{4a} + \frac{\sin 4au}{32a} + C.$

† For expressions involving inverse hyperbolic functions see Sec. 268.

16. $\int \cos u \, du = \sin u + C.$

17. $\int \cos^2 au \, du = \frac{u}{2} + \frac{\sin 2au}{4a} + C.$

18. $\int \cos^4 au \, du = \frac{3u}{8} + \frac{\sin 2au}{4a} + \frac{\sin 4au}{32a} + C.$

For $\int \sin^m u \cos^n u \, du$, see Sec. 195 and items 33 and 34 below.

For $\int \sin mu \sin nu \, du$, $\int \sin mu \cos nu \, du$, etc., see Sec. 196.

19. $\int \tan u \, du = -\ln \cos u + C = \ln \sec u + C.$

20. $\int \cot u \, du = \ln \sin u + C = -\ln \csc u + C.$

21. $\int \sec u \, du = \int \frac{du}{\cos u} = \ln (\sec u + \tan u) + C.$

22. $\int \csc u \, du = \int \frac{du}{\sin u} = -\ln (\csc u + \cot u) + C.$

23. $\int \sec^2 u \, du = \int \frac{du}{\cos^2 u} = \tan u + C.$

24. $\int \csc^2 u \, du = \int \frac{du}{\sin^2 u} = -\cot u + C.$

25. $\int \tan u \sec u \, du = \sec u + C.$

26. $\int \cot u \csc u \, du = -\csc u + C.$

27. $\int \sec^3 u \, du = \frac{1}{2} \sec u \tan u + \frac{1}{2} \ln (\sec u + \tan u) + C.$

28. $\int \csc^3 u \, du = -\frac{1}{2} \csc u \cot u - \frac{1}{2} \ln (\csc u + \cot u) + C.$

For $\int \tan^m u \sec^n u \, du$, $\int \cot^m u \csc^n u \, du$, see Sec. 197.

For $\int \frac{dx}{a \sin x + b \cos x + c}$, see Sec. 202.

Exponential and Mixed Forms

29. $\int e^u \, du = e^u + C.$

30. $\int a^u \, du = \frac{a^u}{\ln a} + C.$

31. $\int e^{au} \sin bu \, du = \frac{e^{au}(a \sin bu - b \cos bu)}{a^2 + b^2} + C.$

32. $\int e^{au} \cos bu \, du = \frac{e^{au}(b \sin bu + a \cos bu)}{a^2 + b^2} + C.$

For $\int x^m e^{ax} \, dx$, $\int x^m \sin ax \, dx$, $\int x^m \cos ax \, dx$, see Sec. 200

For $\int x^m \ln x \, dx$, $\int x^m \sin^{-1} ax \, dx$, etc., see Sec. 200.

Wallis' Definite Integrals

33. $\int_0^{\pi/2} \sin^n x\, dx = \int_0^{\pi/2} \cos^n x\, dx = \dfrac{1 \cdot 3 \cdot 5 \cdots (n-1)}{2 \cdot 4 \cdot 6 \cdots n}\dfrac{\pi}{2}$, if n is an *even* positive integer, and

$= \dfrac{2 \cdot 4 \cdot 6 \cdots (n-1)}{3 \cdot 5 \cdot 7 \cdots n}$, if n is an *odd* positive integer, $n > 1$.

34. $\int_0^{\pi/2} \sin^m x \cos^n x\, dx = \dfrac{[1 \cdot 3 \cdot 5 \cdots (m-1)][1 \cdot 3 \cdot 5 \cdots (n-1)]}{2 \cdot 4 \cdot 6 \cdots (m+n)}\dfrac{\pi}{2}$, if m and n are both *even* positive integers,

$= \dfrac{2 \cdot 4 \cdot 6 \cdots (n-1)}{(m+1)(m+3)(m+5) \cdots (m+n)}$, if n is an *odd* positive integer, $n > 1$, and

$= \dfrac{2 \cdot 4 \cdot 6 \cdots (m-1)}{(n+1)(n+3)(n+5) \cdots (n+m)}$, if m is an *odd* positive integer, $m > 1$.

TABLE 2. NATURAL LOGARITHMS, BASE e

$\ln N = \log_e N$ for N a number between 1.0 and 9.9

N	.0	.1	.2	.3	.4	.5	.6	.7	.8	.9
1	0.000	0.095	0.182	0.262	0.336	0.405	0.470	0.531	0.588	0.642
2	0.693	0.742	0.788	0.833	0.875	0.916	0.956	0.993	1.030	1.065
3	1.099	1.131	1.163	1.194	1.224	1.253	1.281	1.308	1.335	1.361
4	1.386	1.411	1.435	1.459	1.482	1.504	1.526	1.548	1.569	1.589
5	1.609	1.629	1.649	1.668	1.686	1.705	1.723	1.740	1.758	1.775
6	1.792	1.808	1.825	1.841	1.856	1.872	1.887	1.902	1.917	1.932
7	1.946	1.960	1.974	1.988	2.001	2.015	2.028	2.041	2.054	2.067
8	2.079	2.092	2.104	2.116	2.128	2.140	2.152	2.163	2.175	2.186
9	2.197	2.208	2.219	2.230	2.241	2.251	2.262	2.272	2.282	2.293
	.0	.1	.2	.3	.4	.5	.6	.7	.8	.9

$\ln N = \log_e N$ for N a whole number from 10 to 99

N	0	1	2	3	4	5	6	7	8	9
1	2.303	2.398	2.485	2.565	2.639	2.708	2.773	2.833	2.890	2.944
2	2.996	3.045	3.091	3.135	3.178	3.219	3.258	3.296	3.332	3.367
3	3.401	3.434	3.466	3.497	3.526	3.555	3.584	3.611	3.638	3.664
4	3.689	3.714	3.738	3.761	3.784	3.807	3.829	3.850	3.871	3.892
5	3.912	3.932	3.951	3.970	3.989	4.007	4.025	4.043	4.060	4.078
6	4.094	4.111	4.127	4.143	4.159	4.174	4.190	4.205	4.220	4.234
7	4.248	4.263	4.277	4.290	4.304	4.317	4.331	4.344	4.357	4.369
8	4.382	4.394	4.407	4.419	4.431	4.443	4.454	4.466	4.477	4.489
9	4.500	4.511	4.522	4.533	4.543	4.554	4.564	4.575	4.585	4.595
	0	1	2	3	4	5	6	7	8	9

$\ln 10 = 2.3026$; $\ln 100 = 4.6052$; $\ln 1{,}000 = 6.9078$.

EXAMPLE: $\ln 0.43 = 3.761 - 4.605 = -0.844$.
$\ln 430 = 3.761 + 2.303 = 6.064$.

TABLE 3. EXPONENTIAL AND HYPERBOLIC FUNCTIONS

x	e^x	e^{-x}	sinh x	cosh x	tanh x	coth x
0.0	1.0000	1.0000	0.0000	1.0000	0.0000	∞
0.1	1.1052	0.9048	0.1002	1.0050	0.0997	10.033
0.2	1.2214	0.8187	0.2013	1.0201	0.1974	5.066
0.3	1.3499	0.7408	0.3045	1.0453	0.2913	3.433
0.4	1.4918	0.6703	0.4108	1.0811	0.3800	2.632
0.5	1.6487	0.6065	0.5211	1.1276	0.4621	2.1639
0.6	1.8221	0.5488	0.6367	1.1855	0.5370	1.8620
0.7	2.0138	0.4966	0.7586	1.2552	0.6044	1.6546
0.8	2.2255	0.4493	0.8881	1.3374	0.6640	1.5060
0.9	2.4596	0.4066	1.0265	1.4331	0.7163	1.3960
1.0	2.7183	0.3679	1.1752	1.5431	0.7616	1.3131
1.1	3.0042	0.3329	1.3356	1.6685	0.8005	1.2492
1.2	3.3201	0.3012	1.5095	1.8107	0.8337	1.1996
1.3	3.6693	0.2725	1.6984	1.9709	0.8617	1.1605
1.4	4.0552	0.2466	1.9043	2.1509	0.8854	1.1294
1.5	4.482	0.2231	2.129	2.352	0.9052	1.1048
1.6	4.953	0.2019	2.376	2.577	0.9217	1.0850
1.7	5.474	0.1827	2.646	2.828	0.9354	1.0691
1.8	6.050	0.1653	2.942	3.107	0.9468	1.0562
1.9	6.686	0.1496	3.268	3.418	0.9562	1.0458
2.0	7.389	0.13534	3.627	3.762	0.9640	1.0373
2.1	8.166	0.12246	4.022	4.144	0.9705	1.0304
2.2	9.025	0.11080	4.457	4.568	0.9757	1.0249
2.3	9.974	0.10026	4.937	5.037	0.9801	1.0203
2.4	11.023	0.09072	5.466	5.557	0.9837	1.0166
2.5	12.182	0.08208	6.050	6.132	0.9866	1.0136
2.6	13.464	0.07427	6.695	6.769	0.9890	1.0111
2.7	14.880	0.06721	7.406	7.473	0.9910	1.0092
2.8	16.445	0.06081	8.192	8.253	0.9926	1.0074
2.9	18.174	0.05502	9.060	9.115	0.9940	1.0061
3.0	20.086	0.04979	10.018	10.068	0.9951	1.0050
3.1	22.198	0.04505	11.076	11.122	0.9960	1.0041
3.2	24.533	0.04076	12.246	12.287	0.9967	1.0033
3.3	27.113	0.03688	13.538	13.575	0.9973	1.0027
3.4	29.964	0.03337	14.965	14.999	0.9978	1.0022
3.5	33.115	0.03020	16.543	16.573	0.9982	1.0018
4.0	54.598	0.01832	27.290	27.308	0.9993	1.0007
4.5	90.017	0.01111	45.003	45.014	0.9998	1.0002
5.0	148.41	0.00674	74.203	74.210	0.9999	1.0001
5.5	244.69	0.00409	122.34	122.35	1.0000	1.0000
6.0	403.43	0.00248	201.71	201.72	1.0000	1.0000
7.0	1096.6	0.000912	548.32	548.32	1.0000	1.0000
8.0	2981.0	0.000335	1490.5	1490.5	1.0000	1.0000
9.0	8103.1	0.000123	4051.5	4051.5	1.0000	1.0000
10.0	22026.	0.000045	11013.	11013.	1.0000	1.0000

Table 4. Common Logarithms, Base 10

N	0	1	2	3	4	5	6	7	8	9
10	0000	0043	0086	0128	0170	0212	0253	0294	0334	0374
11	0414	0453	0492	0531	0569	0607	0645	0682	0719	0755
12	0792	0828	0864	0899	0934	0969	1004	1038	1072	1106
13	1139	1173	1206	1239	1271	1303	1335	1367	1399	1430
14	1461	1492	1523	1553	1584	1614	1644	1673	1703	1732
15	1761	1790	1818	1847	1875	1903	1931	1959	1987	2014
16	2041	2068	2095	2122	2148	2175	2201	2227	2253	2279
17	2304	2330	2355	2380	2405	2430	2455	2480	2504	2529
18	2553	2577	2601	2625	2648	2672	2695	2718	2742	2765
19	2788	2810	2833	2856	2878	2900	2923	2945	2967	2989
20	3010	3032	3054	3075	3096	3118	3139	3160	3181	3201
21	3222	3243	3263	3284	3304	3324	3345	3365	3385	3404
22	3424	3444	3464	3483	3502	3522	3541	3560	3579	3598
23	3617	3636	3655	3674	3692	3711	3729	3747	3766	3784
24	3802	3820	3838	3856	3874	3892	3909	3927	3945	3962
25	3979	3997	4014	4031	4048	4065	4082	4099	4116	4133
26	4150	4166	4183	4200	4216	4232	4249	4265	4281	4298
27	4314	4330	4346	4362	4378	4393	4409	4425	4440	4456
28	4472	4487	4502	4518	4533	4548	4564	4579	4594	4609
29	4624	4639	4654	4669	4683	4698	4713	4728	4742	4757
30	4771	4786	4800	4814	4829	4843	4857	4871	4886	4900
31	4914	4928	4942	4955	4969	4983	4997	5011	5024	5038
32	5051	5065	5079	5092	5105	5119	5132	5145	5159	5172
33	5185	5198	5211	5224	5237	5250	5263	5276	5289	5302
34	5315	5328	5340	5353	5366	5378	5391	5403	5416	5428
35	5441	5453	5465	5478	5490	5502	5514	5527	5539	5551
36	5563	5575	5587	5599	5611	5623	5635	5647	5658	5670
37	5682	5694	5705	5717	5729	5740	5752	5763	5775	5786
38	5798	5809	5821	5832	5843	5855	5866	5877	5888	5899
39	5911	5922	5933	5944	5955	5966	5977	5988	5999	6010
40	6021	6031	6042	6053	6064	6075	6085	6096	6107	6117
41	6128	6138	6149	6160	6170	6180	6191	6201	6212	6222
42	6232	6243	6253	6263	6274	6284	6294	6304	6314	6325
43	6335	6345	6355	6365	6375	6385	6395	6405	6415	6425
44	6435	6444	6454	6464	6474	6484	6493	6503	6513	6522
45	6532	6542	6551	6561	6571	6580	6590	6599	6609	6618
46	6628	6637	6646	6656	6665	6675	6684	6693	6702	6712
47	6721	6730	6739	6749	6758	6767	6776	6785	6794	6803
48	6812	6821	6830	6839	6848	6857	6866	6875	6884	6893
49	6902	6911	6920	6928	6937	6946	6955	6964	6972	6981
50	6990	6998	7007	7016	7024	7033	7042	7050	7059	7067
51	7076	7084	7093	7101	7110	7118	7126	7135	7143	7152
52	7160	7168	7177	7185	7193	7202	7210	7218	7226	7235
53	7243	7251	7259	7267	7275	7284	7292	7300	7308	7316
54	7324	7332	7340	7348	7356	7364	7372	7380	7388	7396
N	**0**	**1**	**2**	**3**	**4**	**5**	**6**	**7**	**8**	**9**

TABLE 4. COMMON LOGARITHMS, BASE 10 (*Continued*)

N	0	1	2	3	4	5	6	7	8	9
55	7404	7412	7419	7427	7435	7443	7451	7459	7466	7474
56	7482	7490	7497	7505	7513	7520	7528	7536	7543	7551
57	7559	7566	7574	7582	7589	7597	7604	7612	7619	7627
58	7634	7642	7649	7657	7664	7672	7679	7686	7694	7701
59	7709	7716	7723	7731	7738	7745	7752	7760	7767	7774
60	7782	7789	7796	7803	7810	7818	7825	7832	7839	7846
61	7853	7860	7868	7875	7882	7889	7896	7903	7910	7917
62	7924	7931	7938	7945	7952	7959	7966	7973	7980	7987
63	7993	8000	8007	8014	8021	8028	8035	8041	8048	8055
64	8062	8069	8075	8082	8089	8096	8102	8109	8116	8122
65	8129	8136	8142	8149	8156	8162	8169	8176	8182	8189
66	8195	8202	8209	8215	8222	8228	8235	8241	8248	8254
67	8261	8267	8274	8280	8287	8293	8299	8306	8312	8319
68	8325	8331	8338	8344	8351	8357	8363	8370	8376	8382
69	8388	8395	8401	8407	8414	8420	8426	8432	8439	8445
70	8451	8457	8463	8470	8476	8482	8488	8494	8500	8506
71	8513	8519	8525	8531	8537	8543	8549	8555	8561	8567
72	8573	8579	8585	8591	8597	8603	8609	8615	8621	8627
73	8633	8639	8645	8651	8657	8663	8669	8675	8681	8686
74	8692	8698	8704	8710	8716	8722	8727	8733	8739	8745
75	8751	8756	8762	8768	8774	8779	8785	8791	8797	8802
76	8808	8814	8820	8825	8831	8837	8842	8848	8854	8859
77	8865	8871	8876	8882	8887	8893	8899	8904	8910	8915
78	8921	8927	8932	8938	8943	8949	8954	8960	8965	8971
79	8976	8982	8987	8993	8998	9004	9009	9015	9020	9025
80	9031	9036	9042	9047	9053	9058	9063	9069	9074	9079
81	9085	9090	9096	9101	9106	9112	9117	9122	9128	9133
82	9138	9143	9149	9154	9159	9165	9170	9175	9180	9186
83	9191	9196	9201	9206	9212	9217	9222	9227	9232	9238
84	9243	9248	9253	9258	9263	9269	9274	9279	9284	9289
85	9294	9299	9304	9309	9315	9320	9325	9330	9335	9340
86	9345	9350	9355	9360	9365	9370	9375	9380	9385	9390
87	9395	9400	9405	9410	9415	9420	9425	9430	9435	9440
88	9445	9450	9455	9460	9465	9469	9474	9479	9484	9489
89	9494	9499	9504	9509	9513	9518	9523	9528	9533	9538
90	9542	9547	9552	9557	9562	9566	9571	9576	9581	9586
91	9590	9595	9600	9605	9609	9614	9619	9624	9628	9633
92	9638	9643	9647	9652	9657	9661	9666	9671	9675	9680
93	9685	9689	9694	9699	9703	9708	9713	9717	9722	9727
94	9731	9736	9741	9745	9750	9754	9759	9763	9768	9773
95	9777	9782	9786	9791	9795	9800	9805	9809	9814	9818
96	9823	9827	9832	9836	9841	9845	9850	9854	9859	9863
97	9868	9872	9877	9881	9886	9890	9894	9899	9903	9908
98	9912	9917	9921	9926	9930	9934	9939	9943	9948	9952
99	9956	9961	9965	9969	9974	9978	9983	9987	9991	9996
N	0	1	2	3	4	5	6	7	8	9

TABLE 5. TRIGONOMETRIC FUNCTIONS

Deg	Rad	Sin	Cos	Tan	Cot	Sec	Csc		
0	0.0000	0.0000	1.0000	0.0000	∞	1.0000	∞	1.5708	90
1	0.0175	0.0175	0.9998	0.0175	57.290	1.0002	57.299	1.5533	89
2	0.0349	0.0349	0.9994	0.0349	28.636	1.0006	28.654	1.5359	88
3	0.0524	0.0523	0.9986	0.0524	19.081	1.0014	19.107	1.5184	87
4	0.0698	0.0698	0.9976	0.0699	14.301	1.0024	14.336	1.5010	86
5	0.0873	0.0872	0.9962	0.0875	11.430	1.0038	11.474	1.4835	85
6	0.1047	0.1045	0.9945	0.1051	9.5144	1.0055	9.5668	1.4661	84
7	0.1222	0.1219	0.9925	0.1228	8.1443	1.0075	8.2055	1.4486	83
8	0.1396	0.1392	0.9903	0.1405	7.1154	1.0098	7.1853	1.4312	82
9	0.1571	0.1564	0.9877	0.1584	6.3138	1.0125	6.3925	1.4137	81
10	0.1745	0.1736	0.9848	0.1763	5.6713	1.0154	5.7588	1.3963	80
11	0.1920	0.1908	0.9816	0.1944	5.1446	1.0187	5.2408	1.3788	79
12	0.2094	0.2079	0.9781	0.2126	4.7046	1.0223	4.8097	1.3614	78
13	0.2269	0.2250	0.9744	0.2309	4.3315	1.0263	4.4454	1.3439	77
14	0.2443	0.2419	0.9703	0.2493	4.0108	1.0306	4.1336	1.3265	76
15	0.2618	0.2588	0.9659	0.2679	3.7321	1.0353	3.8637	1.3090	75
16	0.2793	0.2756	0.9613	0.2867	3.4874	1.0403	3.6280	1.2915	74
17	0.2967	0.2924	0.9563	0.3057	3.2709	1.0457	3.4203	1.2741	73
18	0.3142	0.3090	0.9511	0.3249	3.0777	1.0515	3.2361	1.2566	72
19	0.3316	0.3256	0.9455	0.3443	2.9042	1.0576	3.0716	1.2392	71
20	0.3491	0.3420	0.9397	0.3640	2.7475	1.0642	2.9238	1.2217	70
21	0.3665	0.3584	0.9336	0.3839	2.6051	1.0711	2.7904	1.2043	69
22	0.3840	0.3746	0.9272	0.4040	2.4751	1.0785	2.6695	1.1868	68
23	0.4014	0.3907	0.9205	0.4245	2.3559	1.0864	2.5593	1.1694	67
24	0.4189	0.4067	0.9135	0.4452	2.2460	1.0946	2.4586	1.1519	66
25	0.4363	0.4226	0.9063	0.4663	2.1445	1.1034	2.3662	1.1345	65
26	0.4538	0.4384	0.8988	0.4877	2.0503	1.1126	2.2812	1.1170	64
27	0.4712	0.4540	0.8910	0.5095	1.9626	1.1223	2.2027	1.0996	63
28	0.4887	0.4695	0.8829	0.5317	1.8807	1.1326	2.1301	1.0821	62
29	0.5061	0.4848	0.8746	0.5543	1.8040	1.1434	2.0627	1.0647	61
30	0.5236	0.5000	0.8660	0.5774	1.7321	1.1547	2.0000	1.0472	60
31	0.5411	0.5150	0.8572	0.6009	1.6643	1.1666	1.9416	1.0297	59
32	0.5585	0.5299	0.8480	0.6249	1.6003	1.1792	1.8871	1.0123	58
33	0.5760	0.5446	0.8387	0.6494	1.5399	1.1924	1.8361	0.9948	57
34	0.5934	0.5592	0.8290	0.6745	1.4826	1.2062	1.7883	0.9774	56
35	0.6109	0.5736	0.8192	0.7002	1.4281	1.2208	1.7434	0.9599	55
36	0.6283	0.5878	0.8090	0.7265	1.3764	1.2361	1.7013	0.9425	54
37	0.6458	0.6018	0.7986	0.7536	1.3270	1.2521	1.6616	0.9250	53
38	0.6632	0.6157	0.7880	0.7813	1.2799	1.2690	1.6243	0.9076	52
39	0.6807	0.6293	0.7771	0.8098	1.2349	1.2868	1.5890	0.8901	51
40	0.6981	0.6428	0.7660	0.8391	1.1918	1.3054	1.5557	0.8727	50
41	0.7156	0.6561	0.7547	0.8693	1.1504	1.3250	1.5243	0.8552	49
42	0.7330	0.6691	0.7431	0.9004	1.1106	1.3456	1.4945	0.8378	48
43	0.7505	0.6820	0.7314	0.9325	1.0724	1.3673	1.4663	0.8203	47
44	0.7679	0.6947	0.7193	0.9657	1.0355	1.3902	1.4396	0.8209	46
45	0.7854	0.7071	0.7071	1.0000	1.0000	1.4142	1.4142	0.7854	45
		Cos	Sin	Cot	Tan	Csc	Sec	Rad	Deg

ANSWERS TO EXERCISES

Exercise 2 (Pages 4 to 5)

1. 4. **2.** 8. **3.** $a^2 + 4$ **4.** $2ah + h^2$. **5.** 16.
6. -54. **7.** $2(t - 1)^3$. **8.** $54t^3$. **9.** $2t^6$. **10.** -1.
11. 3. **12.** 0. **13.** $(t + 2)/t$.

Exercise 4 (Pages 10 to 11)

1. 3. **2.** 33. **3.** 7. **4.** 0. **5.** 0. **6.** 5.
7. -9. **8.** 0. **9.** -2. **10.** 1. **11.** 4. **12.** 3.
13. $2\sqrt{5}$. **14.** 2. **15.** 49. **16.** $\frac{1}{10}$. **17.** $\frac{1}{27}$.

Exercise 5 (Page 13)

1. 5. **2.** 0. **3.** $-A$. **4.** 0. **5.** $-\infty$. **6.** ∞.
7. $+\infty$. **8.** ∞. **9.** $+\infty$. **10.** $+\infty$. **11.** $-\infty$. **12.** $+\infty$.
13. $+\infty$. **14.** ∞.

Exercise 8 (Page 25)

1. $v = 5$. **2.** $v = 8t$. **3.** $v = 2t - 3$. **4.** $v = 6t^2$.
5. $a = 0$. **6.** $a = 8$. **7.** $a = 2$. **8.** $a = 12t$.
10. 2π. **11.** $2\pi r$. **12.** $8\pi r$. **13.** $2x$.
14. $\frac{1}{2}x\sqrt{3}$. **15.** $12x$. **16.** $3x^2$.

Exercise 9 (Page 27)

1. $y = 2x + 1$. **2.** $y = 2x - 1$. **3.** $y = 3x$. **4.** $y = mx + b$.
5. $x/a + y/b = 1$. **6.** $(y - y_1)(x_2 - x_1) = (x - x_1)(y_2 - y_1)$.
7. $y - 2 = \sqrt{3}(x - 1)$. **8.** $y - 2 = -\sqrt{3}(x - 1)$.
9. $y = -\sqrt{3}x$. **11.** $y = -4x + 16$. **12.** $y = x - 4$.
13. $y = -4$. **14.** $y = 3$. **15.** $x = 4$.
16. $Ax + By = Ax_1 + By_1$.

Exercise 10 (Page 29)

1. $2x_1$. **2.** $-2x_1$. **3.** $3x_1^2$.
4. $m_1 = -2$, $y = -2x + 4$. **5.** $m_1 = 4$, $y = 4x - 2$.
6. $m_1 = -2$, $y = -2x + 4$. **7.** $m_1 = 6$, $y = 6x - 4$.
9. (3,6). **10.** $(1,-2)$. **11.** (4,16).
12. $(1,\frac{1}{2})$. **13.** (1,2), $(-1,-2)$.

Exercise 11 (Page 31)

1. $\Delta x = 0.2$, $\Delta y = 1$. **2.** $\Delta x = -0.1$, $\Delta y = -5.1$.
3. $\Delta x = -0.4$, $\Delta y = 5$. **4.** $\Delta x = 2$, $\Delta y = 26$.
5. $\Delta t = 3$ sec., $\Delta s = 117$ ft. **6.** $\Delta t = 1$ sec., $\Delta s = 208$ ft.
7. $\Delta t = 2$ sec., $\Delta v = 60$ ft./sec. **8.** $\Delta t = 0.2$ sec., $\Delta v = 3$ ft./sec.
9. $\Delta y = 0.44$. **10.** $\Delta y = 0.8$. **11.** 2.
12. 45. **13.** 14. **14.** 6 ft./sec.
15. 4.2 ft./sec. **16.** 4 ft./sec.^{2} **17.** 65 ft./sec.^{2}

Exercise 12 (Pages 35 to 36)

1. $\frac{dy}{dx} = 4$. **2.** $\frac{dy}{dx} = 10x$. **3.** $\frac{dy}{dx} = 4x^3$. **4.** $\frac{dy}{dx} = -\frac{1}{x^2}$.
5. $\frac{dy}{dx} = \frac{4}{(2-x)^2}$. **6.** $\frac{dy}{dx} = \frac{-12}{(3x+1)^2}$. **7.** $\frac{dy}{dx} = \frac{1}{2\sqrt{x}}$.
8. $\frac{ds}{dt} = 4t + 6$. **9.** $\frac{dv}{du} = 3u^2 + 4u$. **10.** $m = -10x$.
11. $m = \frac{-5}{(x+1)^2}$. **12.** $m = -\frac{1}{2x^{\frac{3}{2}}}$. **13.** $v = 3t^2 + 2$.
14. $v = \frac{-12}{(4t-5)^2}$. **15.** $v = \frac{1}{2\sqrt{t+1}}$. **16.** $a = -12t^2$.
17. $a = \frac{1}{\sqrt{2t}}$. **18.** $a = \frac{3}{2}\sqrt{t}$. **19.** $\frac{d\theta}{dr} = -\frac{1}{r^2}$.
20. $\frac{du}{dr} = -\frac{1}{2(r+1)^{\frac{3}{2}}}$. **21.** $\frac{du}{dr} = -\frac{6}{r^3}$.

Exercise 13 (Page 41)

1. $\frac{dy}{dx} = 15x^4$. **2.** $\frac{dy}{dx} = \frac{x}{2}$. **3.** $\frac{dy}{dx} = 6x^{14}$. **4.** $\frac{dA}{dx} = x$.
5. $\frac{dP}{dx} = 3$. **6.** $\frac{dL}{dr} = 2\pi\sqrt{2}\,r$. **7.** $\frac{dV}{dr} = \pi r^2$.
8. $v = 32t$ ft./sec. **9.** $v = 30t^2$ ft./sec. **10.** $v = 6t^3$ ft./sec.

Exercise 14 (Page 44)

1. $4x - 3$. **2.** $12x^2 - 2$. **3.** $-3(3-x)^2$.
4. $24(5x+3)(5x+4)$. **5.** $30x^9(x-3)^4(x-2)$. **6.** $2x^2 - 2x^4$.
7. 25. **8.** 1. **9.** 128. **10.** -24.
11. For "v" read "s." 12. **12.** 133.
13. For "v" read "s." -64. **14.** 3.

Exercise 15 (Page 47)

9. Increasing if $x > 0$, decreasing if $x < 0$.
10. Increasing if $x < 0$, decreasing if $x > 0$.
11. Increasing if $x > \frac{1}{2}$, decreasing if $x < \frac{1}{2}$.
12. Increasing if $x < 3$, decreasing if $x > 3$.

Exercise 16 (Page 50)

1. Rising if $x > 3$, falling if $x < 3$. **2.** Rising if $x < 2$, falling if $x > 2$.
3. Rising if $x < 1$ or $x > 2$, falling if $1 < x < 2$.

4. Rising if $-1 < x < 1$, falling if $x < -1$ or $x > 1$.
5. Rising if $-1 < x < 0$ or $x > 1$, falling if $x < -1$ or $0 < x < 1$.
6. Rising if $x < -2$ or $0 < x < 2$, falling if $-2 < x < 0$ or $x > 2$.
7. Rising if $x < 0$ or $x > 4$, falling if $0 < x < 4$.
8. Rising if $x > 2$, falling if $x < 2$.
9. Rising for all x.
10. Rising if $-1 < x < 3$, falling if $x < -1$ or $x > 3$.
11. Rising if $x < -1$ or $0 < x < 1$, falling if $-1 < x < 0$ or $x > 1$.
12. Rising if $x < 1$ or $x > \frac{7}{4}$, falling if $1 < x < \frac{7}{4}$.
13. Rising if $x < 4$ or $x > 6$, falling if $4 < x < 6$.
14. Rising if $-6 < x < -3$ or $x > 2$, falling if $x < -6$ or $-3 < x < 2$.

Exercise 17 (Page 51)

1. $12x - 6$.
2. $36x^2 - 8$.
3. $20x^3 + 30x$.
4. $60x^4 + 36x^2 - 14$.
5. $192(3 - 2x)^2$.
6. $(x - 1)^2(x - 2)(42x^2 - 132x + 102)$.
7. $10(3x + 2)(x + 2)^3$.
8. $10(9x^2 - 4)(x^2 - 2)^3$.
9. 3,360.
10. 96.
11. 24.
12. 0.
13. -360.
14. 540.

Exercise 18 (Page 53)

1. $t > 3$.
2. $t < 2$.
3. $t < 0$ or $t > 2$.
4. $t < 2$ or $t > 4$.
5. All values of t.
6. $t > 2$.
7. $t < 0$.
8. $t < 3$.
9. Increasing if $1 < t < 2$ or $t > 3$, decreasing if $t < 1$ or $2 < t < 3$.
10. Increasing if $0 < t < 1$ or $t > 2$, decreasing if $t < 0$ or $1 < t < 2$.
11. Increasing if $0 < t < 6$ or $t > 9$, decreasing if $t < 0$ or $6 < t < 9$.
12. Increasing if $t < -3$ or $-\sqrt{3} < t < 0$ or $\sqrt{3} < t < 3$, decreasing if $-3 < t < -\sqrt{3}$ or $0 < t < \sqrt{3}$ or $t > 3$.

Exercise 19 (Page 57)

1. Min $x = -1$, $y = 3$.
2. Max $x = 1$, $y = 5$.
3. Max $x = 1$, $y = 4$; Min $x = 3$, $y = 0$; inflection $x = 2$, $y = 2$, $m = -3$.
4. Max $x = 2$, $y = 16$; Min $x = -2$, $y = -16$; inflection $x = 0$, $y = 0$, $m = 12$.
5. Max $x = -2$, $y = 0$; Min $x = 0$, $y = -4$; inflection $x = -1$, $y = -2$, $m = -3$.
6. Inflection $x = 1$, $y = 7$, $m = 6$.
7. Max $x = -1 - \sqrt{2} = -2.4$; $y = 10.66$; Min $x = -1 + \sqrt{2} = 0.4$, $y = -0.66$; inflection $x = -1$, $y = 5$, $m = -6$.
8. Max $x = 1$, $y = 7$; Min $x = -2$, $y = -20$; inflection $x = -\frac{1}{2}$, $y = -\frac{13}{2}$, $m = \frac{27}{2}$.
9. Inflection $x = 2$, $y = 0$, $m = 0$.
10. Min $x = 3$, $y = 0$.
11. Min $x = 0$, $y = -1$; inflections $x = \pm 1$, $y = 0$, $m = 0$ and $x = \pm \dfrac{1}{\sqrt{5}}$, $y = -\dfrac{64}{125}$, $m = \pm \dfrac{96}{25\sqrt{5}}$.
12. Max $x = 0$, $y = 1$; Min $x = \pm 1$, $y = 0$; inflections $x = \pm \dfrac{1}{\sqrt{3}}$, $y = \dfrac{4}{9}$, $m = \mp \dfrac{8}{3\sqrt{3}}$.

Exercise 20 (Page 59)

1. 1.46.
2. 3.93.
3. 2.07.
4. -2.05.
5. 0.539.
6. -0.347.
7. 2.36.
8. 1.32.

Exercise 21 (Pages 62 to 63)

1. 7, 7. 2. 4, 4. 3. 6 by 6 in. 4. 20 by 40 ft.
5. $x = 3,\ y = 4$. 6. $x = 2,\ y = \pm 3$.
7. 10 by 10 by 20 in. 8. Radius $20/\pi$ in., length 20 in.
9. For "least" read "greatest." $x = 5$. 10. 6,000.
11. $x = 140$. 12. For side a, base $a/2$, alt. $(a/4)\sqrt{3}$.
13. Radius $\sqrt{2}\,a/\sqrt{3}$, alt. $2a/\sqrt{3}$. 14. Radius $2a/3$, alt. $h/3$.
15. $x = 7.5$ ft., arc $= 15$ ft., $\theta = 2$ radians.
16. $2\sqrt{3}$ in. 17. $x = 2a/\sqrt{3},\ y = 2\sqrt{2}a/3$.
18. $x = p/(4 + \pi)$. 19. $x = 10$ in.
20. For "least" read "greatest." $x = 9,\ y = 3$.

Exercise 22 (Page 68)

1. $\dfrac{1}{2\sqrt{x}}$. 2. $-\dfrac{4}{x^5}$. 3. $-\dfrac{25}{x^{26}}$. 4. $\frac{3}{2}\sqrt{x}$. 5. $\frac{1}{3}x^{-\frac{2}{3}}$.

6. $-\frac{1}{2}x^{-\frac{3}{2}}$. 7. $\dfrac{3}{2\sqrt{3x-5}}$. 8. $\dfrac{10}{(3-2x)^2}$. 9. $-\dfrac{6x}{(1+x^2)^2}$.

10. $-\dfrac{20}{(2x-5)^2}$. 11. $\dfrac{12x}{(x^2+3)^2}$. 12. $\dfrac{1-x^2}{(x^2+1)^2}$. 13. $-8x(x^2-2)^{-5}$.

14. $\dfrac{-x}{\sqrt{4-x^2}}$. 15. $\dfrac{8x+15x^2}{2\sqrt{2+3x}}$. 16. $\dfrac{9x^2+4x}{2(3x+1)^{\frac{3}{2}}}$. 17. $1 - \dfrac{3}{x^2}$.

18. $-\dfrac{80(2-5x)^3}{(2+5x)^5}$. 19. $\dfrac{3x^2-2x-3}{x\sqrt{x}}$. 20. $-\dfrac{12x(x^2+1)^2}{(x^2-1)^4}$.

21. $-(a^{\frac{2}{3}} - x^{\frac{2}{3}})^{\frac{1}{2}}x^{-\frac{1}{3}}$.

Exercise 23 (Page 71)

1. $-9(1-3x)^2$. 2. $4x^3$. 3. $2x(1-x^2)(1-3x^2)$.

4. $-24(5-4x)^5$. 5. $3\sqrt{4+2x}$. 6. $\dfrac{dy}{dx} = 2$.

7. $\dfrac{dy}{dx} = -6\sqrt{y}$. 8. $\dfrac{dy}{dx} = (y+1)^2$. 9. $\dfrac{dy}{dx} = -\dfrac{8y}{(y^2+1)^2}$.

10. $\dfrac{dy}{dx} = \dfrac{\sqrt{4y-y^2}}{2-y}$.

Exercise 24 (Pages 73 to 74)

1. $-\dfrac{x+y}{3x+y}$. 2. $\dfrac{2x-3y}{3x+4y}$. 3. $\dfrac{y-x^2}{y^2-x}$. 4. $\dfrac{x(y^2-1)}{y(1-x^2)}$.

5. $\dfrac{(y-x)(y-3x)-1}{1-2x(y-x)}$.

6. $-\dfrac{y}{2x}$. 7. $-\dfrac{\sqrt{y}}{\sqrt{x}}$. 8. $-\dfrac{y^{\frac{1}{3}}}{x^{\frac{1}{3}}}$.

9. $\dfrac{dy}{dx} = -\dfrac{y}{x},\ \dfrac{d^2y}{dx^2} = \dfrac{2y}{x^2}$. 10. $\dfrac{dy}{dx} = -\dfrac{x}{4y},\ \dfrac{d^2y}{dx^2} = -\dfrac{1}{y^3}$.

11. $\dfrac{dy}{dx} = -\dfrac{x}{y},\ \dfrac{d^2y}{dx^2} = -\dfrac{a^2}{y^2}$. 12. $\dfrac{dy}{dx} = \dfrac{1}{8y},\ \dfrac{d^2y}{dx^2} = -\dfrac{1}{64y^3}$.

13. $\frac{dy}{dx} = -\frac{x^2}{3y^2}, \frac{d^2y}{dx^2} = -\frac{6xy^3 + 2x^4}{9y^5}$.
14. $\frac{dy}{dx} = \frac{x}{4y}, \frac{d^2y}{dx^2} = -\frac{9}{4y^3}$.
15. $\frac{dy}{dx} = 3, \frac{d^2y}{dx^2} = 0$.
16. $\frac{dy}{dx} = -3, \frac{d^2y}{dx^2} = 0$.

Exercise 25 (Pages 76 to 77)

1. 5, 5.
2. $5\sqrt{2}, 5\sqrt{2}$.
3. $x = 1/\sqrt{2}, y = 1/(2\sqrt{2})$.
4. $h = r$.
5. $h/r = 2$.
6. $y/x = \frac{5}{2}$.
7. $3\sqrt{A/6}$ by $2\sqrt{A/6}$.
8. $y/x = 1$, 2 by 2 yd.
9. 6 by 4 in.
10. $h/s = 1/\sqrt{2}$.
11. $4\sqrt{\frac{2}{3}}$ ft.
12. 4 by 4 by 2 in.
13. $x = -f + \sqrt{1 + f^2}$.
14. $x = 4$ ft.
15. $v = \sqrt[3]{a/2b}$.
16. $v = 3a/2$.
17. 12 in.
18. 2 hr.
19. $\frac{16}{3}$.
20. $w/\sqrt{p^2 - q^2}$ mi.

Exercise 27 (Pages 81 to 82)

1. $\frac{4}{3}$ ft./sec.
2. $25\sqrt{2}$ ft.
3. 3 ft./sec.
4. 1 ft./sec.
5. $8/\sqrt{41}$ ft./sec.
6. 13 ft./min.
7. 16 ft./sec.
8. $2/(25\pi)$ in./sec.
9. $4\sqrt{2}/5$ sq. ft./sec.
10. 2 ft./min.
11. 4 ft./min.
12. $(8 + 40\sqrt{2})$ sq. ft./min.
13. $305/\sqrt{106}$ mi./hr.
14. $100\sqrt{5}$ mi./hr.
15. 2 ft./sec.
16. 15 mi./hr.
17. $\frac{5(25t - 32)}{\sqrt{25t^2 - 64t + 68}}$ mi./hr., $t = \frac{32}{25}$ hr.
18. $88\sqrt{5}/3$ ft./sec. or $20\sqrt{5}$ mi./hr.
19. 22 mi./hr.
20. $1{,}500/\sqrt{29}$ mi./hr.

Exercise 28 (Pages 85 to 86)

1. $2x\frac{dx}{dt}$.
2. $2\pi x\frac{dx}{dt}$.
3. $x\frac{dx}{dt}$.
4. $\frac{3\sqrt{3}}{2}x\frac{dx}{dt}$.
5. $a\frac{dx}{dt}$.
6. $x\frac{dx}{dt}$.
7. $3x^2\frac{dx}{dt}$.
8. $ab\frac{dx}{dt}$.
9. $\pi x^2\frac{dx}{dt}$.
10. $12\pi x^2\frac{dx}{dt}$.
11. $2x^2\frac{dx}{dt}$.
12. $\pi(a^2 - x^2)\frac{dx}{dt}$.

Exercise 29 (Page 89)

1. $v_x = 18, v_y = -4, v = 2\sqrt{85}$.
2. $v_x = 2, v_y = 12, v = 2\sqrt{37}$.
3. $v_x = 27, v_y = 6, v = 3\sqrt{85}$.
4. $v_x = \frac{1}{3}, v_y = 3, v = \frac{1}{3}\sqrt{82}$.
5. $v_x = -4, v_y = 3, v = 5$.
6. $v_x = 3, v_y = 20, v = \sqrt{409}$.
7. $t = 2$.
8. $t = \frac{7}{4}$.
9. $t = 3$.
10. $t = 1$.

Exercise 30 (Page 94)

1. $2x^2 - 3x + C$.
2. $2x^3 - 4x^2 + 5x + C$.
3. $3x^3 + x^2 + 4x + C$.
4. $\frac{2}{3}x^3 - \frac{3}{2}x^2 + 6x + C$.
5. $4x^{\frac{5}{4}} + C$.
6. $4x^{\frac{3}{2}} + C$.
7. $10\sqrt{x} + C$.
8. $-\frac{4}{x^2} + C$.
9. $-\frac{10}{x} + C$.
10. $-\frac{3}{x^3} + C$.

11. $\dfrac{-x^2 + x - 1}{x^3} + C.$ 12. $\dfrac{1}{3}x^3 + \dfrac{1}{2}x^2 - \dfrac{2}{x} + C.$
13. $-\frac{1}{3}x^{-3} - \frac{1}{2}x^{-4} - \frac{1}{5}x^{-5} + C.$ 14. $\frac{4}{3}x^{\frac{3}{2}} - 6x^{\frac{1}{2}} + C.$
15. $\frac{10}{3}x^{\frac{3}{2}} - \frac{6}{5}x^{\frac{5}{2}} + C.$ 16. $\frac{2}{5}x^{\frac{5}{2}} - 2x^{\frac{3}{2}} + 4x^{\frac{1}{2}} + C.$

Exercise 31 (Page 95)

1. $y = 2x^2 - 5.$ 2. $y = \dfrac{x^2}{4} - 5.$ 3. $y = x - \dfrac{x^3}{3} + \dfrac{4}{3}.$
4. $y = 2x^3 - x^2 + 2.$ 5. $y = 4\sqrt{x} - 4.$ 6. $y = \dfrac{1}{x} + \dfrac{2}{3}.$
7. $x = 2t^2 + 6.$ 8. $x = 2t^3.$ 9. $x = \frac{12}{5}t^{\frac{5}{2}} - \frac{12}{5}.$
10. $x = 3t^4 - 2.$ 11. $40\frac{2}{3}.$ 12. 13.
13. $9\frac{139}{1024}.$ 14. $\frac{38}{3}.$ 15. $v = 32t,\ s = 16t^2.$
16. $v = 100 - 32t,\ s = 100t - 16t^2.$ 17. $v = 7t^6 + 2,\ s = t^7 + 2t.$
18. $v = 10t^4 + 5,\ s = 2t^5 + 5t.$ 19. $v = 2t^3 - t^2 + 3,\ s = \frac{1}{2}t^4 - \frac{1}{3}t^3 + 3t.$
20. $v = 2t^2 + 4t + 4,\ s = \frac{2}{3}t^3 + 2t^2 + 4t.$

Exercise 32 (Page 98)

1. 18. 2. 19. 3. 34. 4. $157\frac{1}{3}.$ 5. 32.
6. 42. 7. $13\frac{1}{2}.$ 8. $8\sqrt{2}.$ 9. $\frac{5}{3}.$ 10. 12.
11. $\frac{4}{5}.$ 12. 9. 13. 19.2. 14. 36. 15. $\frac{4}{3}.$
16. $10\frac{2}{3}.$ 17. 51.2. 18. 4. 19. $\frac{3}{4}.$

Exercise 33 (Page 100)

1. 4. 2. $-72.$ 3. $\frac{12}{25}.$ 4. $-\frac{23}{24}.$ 5. 10.
6. $\frac{11}{5}.$ 7. 8. 8. $-7.$ 9. $\frac{13}{3}.$ 10. $\frac{3}{16}.$
11. $\frac{39}{4}.$ 12. 3.1. 13. $\frac{8}{3}.$ 14. $\frac{3}{20}.$

Exercise 34 (Pages 103 to 104)

1. $\frac{8}{5}.$ 2. $\frac{11}{2}.$ 3. 8. 4. $\frac{27}{4}.$ 5. 36. 6. $\frac{32}{3}.$
7. $\frac{4}{3}.$ 8. $\frac{81}{2}.$ 9. 1.6. 10. $\frac{3}{2}.$ 11. $\frac{8}{3}.$ 12. 2.4.
13. $\frac{8}{3}.$ 14. $\frac{32}{3}.$ 15. $\frac{1}{24}.$ 16. $\frac{8}{3}.$ 17. 8. 18. $\frac{4}{3}.$

Exercise 35 (Pages 105 to 106)

1. 4. 2. 2. 3. 3. 4. 4. 5. $4\pi.$ 6. $4\pi.$
7. $\frac{8}{3}\pi.$ 8. $\frac{8}{15}\pi.$ 9. $\frac{1}{6}.$ 10. $\frac{256}{15}.$ 11. 16. 12. 9.
13. $\frac{1024}{3}.$ 14. $\frac{256}{3}\sqrt{3}.$ 15. $\frac{512}{3}.$ 16. $\pi/3.$ 17. $2\pi/3.$
18. $4\pi/3.$ 19. $\frac{512}{15}\pi.$ 20. $\frac{256}{5}\pi.$ 21. $\frac{1792}{15}\pi.$

Exercise 36 (Pages 108 to 109)

1. 1 ton. 2. $\frac{1}{12}$ ton. 3. $\frac{1}{4}$ ton. 4. $\frac{5}{24}$ ton. 5. $\frac{1}{4}$ ton.
6. $\frac{1}{6}$ ton. 7. $\frac{8}{15}$ ton. 8. 5 tons. 9. $\frac{3}{4}$ ton. 10. $\frac{1}{3}$ ton.
11. $\frac{7}{24}$ ton. 12. $\frac{8}{3}$ tons.
13. $\sqrt{104} - 8 = 2.20$ ft. below top of gate.
14. 10 ft. 15. $\pi/8$ ton. 16. 720 lb.

Exercise 37 (Page 112)

1. $5x - 3y = 0.$ 2. $2x + 3y = 0.$ 3. $y - 2x = 0.$
4. $x + 4y + 4 = 0.$ 5. $x - 2y + 3 = 0.$ 6. $x - y + 3 = 0.$
7. $y = 3x.$ 8. $x + 2y = 2.$ 9. $y = 3x - 3.$
10. $2y = x + 4.$ 11. $y = -x.$ 12. $4y - 3x = 5.$

Exercise 38 (Page 115)

1. (3,9). **2.** (5,15). **3.** $(\frac{9}{5},\frac{27}{5})$. **4.** (21,63). **5.** (−9,−27). **6.** (−39,−117). **9.** $2x - 3y = 2$. **10.** $x + 2y = 8$. **11.** $x = 2$. **12.** $y = 1$. **13.** $2x + 3y = 7$. **14.** $y = x + 2$. **15.** $y = 1$. **16.** $x = -2$. **17.** $-\frac{1}{3}$. **18.** $-\frac{11}{16}$. **19.** $-\frac{1}{7}$. **20.** For "$y + 0$" read "$y = 0$." −3.

Exercise 39 (Pages 118 to 119)

1. 3. **2.** 1. **3.** $\frac{1}{2}$. **4.** $\frac{17}{3}$. **5.** $2x + 4y = 9$, $6y = 12x + 11$. **6.** $7y = 3x + 5$, $7x + 3y = 1$. **7.** $x + y + 1 = 0$, $3x - 3y = 4$. **8.** $4x + 4y = 3$, $2y = 2x + 1$. **10.** $\frac{22}{5}$. **11.** 2. **12.** 3. **13.** 4. **14.** For "=7" read "+7." $\frac{3}{10}$. **15.** $y = x$. **16.** $x + y = 3$.

17. $x_1 = \dfrac{4x + 3y}{5}$, $y_1 = \dfrac{4y - 3x}{5}$. **18.** $x_1 = \dfrac{2x + y}{\sqrt{5}}$, $y_1 = \dfrac{2y - x}{\sqrt{5}}$.

19. $x_1 = \dfrac{5x + 12y}{13}$, $y_1 = \dfrac{5y - 12x}{13}$. **20.** $x_1 = \dfrac{x + 3y}{\sqrt{10}}$, $y_1 = \dfrac{y - 3x}{\sqrt{10}}$.

Exercise 40 (Page 121)

1. $2\sqrt{5}$. **2.** $\sqrt{52}$. **3.** $2\sqrt{2}$. **4.** $y = 3x - 3$. **5.** $6x + 4y + 25 = 0$. **6.** $8x + 2y + 5 = 0$. **7.** $25 = 20 + 5$, right but *not* isosceles. Read "$C = (-1,-2)$." $40 = 20 + 20$. **8.** $x^2 + y^2 - 8x - 4y + 16 = 0$. **9.** $x^2 + y^2 + 6x + 2y = 26$. **10.** $x^2 + y^2 + 4x - 8y + 11 = 0$. **11.** $x^2 + y^2 = 6y$. **12.** $x^2 + y^2 = 4x + 4y$. **13.** $x^2 + y^2 - 10x - 12y + 36 = 0$. **14.** (−1,−3), 5. **15.** (4,0), 4. **16.** (2,3), 2. **17.** $(0,-\frac{5}{4})$, $\frac{5}{4}$. **18.** $(-\frac{4}{5},\frac{2}{5})$, 1. **19.** (−12,5), 13. **20.** $x^2 + y^2 - 6x - 10y + 9 = 0$. **21.** $x^2 + y^2 - 6x - 10y + 25 = 0$. **22.** $x^2 + y^2 - 6x - 10y = 0$. **23.** $x^2 + y^2 - 6x - 10y = 7$. **24.** $x^2 + y^2 - 6x - 10y + 32 = 0$. **25.** $5x^2 + 5y^2 - 30x - 50y + 169 = 0$.

Exercise 41 (Pages 123 to 124)

1. $V = (0,0)$, $F = (1,0)$. **2.** $V = (0,0)$, $F = (0,2)$. **3.** $V = (0,0)$, $F = (0,-1)$. **4.** $V = (0,0)$, $F = (-3,0)$. **5.** $V = (-1,2)$, $F = (\frac{1}{2},2)$. **6.** $V = (-3,-1)$, $F = (-3,2)$. **7.** $V = (4,-4)$, $F = (3,-4)$. **8.** $V = (3,\frac{9}{8})$, $F = (3,-\frac{7}{8})$. **9.** $y^2 = 32x$. **10.** $x^2 = -32y$. **11.** $y^2 = -16x$. **12.** $x^2 = 16y$. **13.** $x^2 - 4x - 16y + 44 = 0$. **14.** $y^2 - 2y - 32x - 95 = 0$. **15.** $3x^2 = 4y$. **16.** $2y^2 = -9x$. **17.** $2x^2 + 4x + y = 0$. **18.** $y^2 + 4y + 2x - 2 = 0$. **19.** $y^2 = -8x$. **20.** $x^2 - 16x - 4y + 60 = 0$. **21.** $y^2 - 12y - 8x + 68 = 0$. **22.** $x^2 + 4y - 4 = 0$.

Exercise 42 (Pages 127 to 128)

1. $V = (\pm 3,0)$, $F = (\pm\sqrt{5},0)$, $e = \frac{1}{3}\sqrt{5}$.
2. $V = (0,\pm 5)$, $F = (0,\pm 4)$, $e = \frac{4}{5}$.
3. $V = (\pm 1,0)$, $F = \left(\pm\dfrac{2}{\sqrt{5}},0\right)$, $e = \dfrac{2}{\sqrt{5}}$.

4. $V = \left(0, \pm \sqrt{\frac{3}{2}}\right)$, $F = \left(0, \pm \frac{3}{\sqrt{10}}\right)$, $e = \frac{2}{\sqrt{10}}$.
5. $V = (0,0)$ or $(10,0)$, $F = (5 \pm 2\sqrt{6}, 0)$, $e = \frac{2}{5}\sqrt{6}$.
6. $V = (0,0)$ or $(0,6)$, $F = (0, 3 \pm \sqrt{5})$, $e = \frac{1}{3}\sqrt{5}$.
7. $V = (-2,2)$ or $(8,2)$, $F = (3 \pm \frac{5}{2}\sqrt{3}, 2)$, $e = \frac{3}{5}$.
8. $V = (1, -3)$ or $(1,5)$, $F = (1, 1 \pm 2\sqrt{3})$, $e = \frac{1}{2}\sqrt{3}$.

9. $\frac{x^2}{25} + \frac{y^2}{16} = 1$.
10. $\frac{x^2}{3} + \frac{y^2}{25} = 1$.
11. $\frac{(x-1)^2}{9} + \frac{(y-4)^2}{5} = 1$.
12. $\frac{(x-6)^2}{7} + \frac{(y-2)^2}{16} = 1$.
13. $\frac{x^2}{25} + \frac{y^2}{9} = 1$.
14. $\frac{x^2}{4} + \frac{y^2}{16} = 1$.
15. $\frac{x^2}{45} + \frac{y^2}{20} = 1$.
16. $\frac{x^2}{5} + \frac{y^2}{45} = 1$.
17. $\frac{(x-2)^2}{9} + \frac{(y-3)^2}{1} = 1$.
18. $\frac{(x+2)^2}{4} + \frac{(y+1)^2}{16} = 1$.
19. $\frac{x^2}{25} + \frac{y^2}{9} = 1$.
20. $\frac{x^2}{a^2} + \frac{y^2}{a^2 - c^2} = 1$.

Exercise 43 (Pages 132 to 133)

1. $V = (\pm 3,0)$, $F = (\pm 5,0)$, $e = \frac{5}{3}$, $3y = \pm 2x$.
2. $V = (0, \pm 5)$, $F = (0, \pm\sqrt{34})$, $e = \frac{1}{5}\sqrt{34}$, $3y = \pm 5x$.
3. $V = \left(\pm \frac{1}{\sqrt{5}}, 0\right)$, $F = \left(\pm\sqrt{\frac{6}{5}}, 0\right)$, $e = \sqrt{6}$, $y = \pm\sqrt{5}\,x$.
4. $V = (0, \pm\sqrt{\frac{3}{5}})$, $F = (0, \pm\sqrt{\frac{21}{10}})$, $e = \sqrt{\frac{7}{2}}$, $\sqrt{5}\,y = \pm\sqrt{2}\,x$.
5. $V = (0,0)$ or $(10,0)$, $F = (5 \pm \sqrt{26}, 0)$, $e = \frac{1}{5}\sqrt{26}$, $5y = \pm(x - 5)$.
6. $V = (0,0)$ or $(0,6)$, $F = (0, 3 \pm \sqrt{13})$, $e = \frac{1}{3}\sqrt{13}$, $2y = 6 \pm 3x$.
7. $V = (2,2)$ or $(2,4)$, $F = (2, 3 \pm \sqrt{5})$, $e = \sqrt{5}$, $2y = 4 \pm (x - 3)$.
8. $V = (0,1)$ or $(2,1)$, $F = (1 \pm \sqrt{5}, 1)$, $e = \sqrt{5}$, $y = 1 \pm 2(x - 1)$.

9. $\frac{x^2}{16} - \frac{y^2}{9} = 1$.
10. $\frac{y^2}{9} - \frac{x^2}{16} = 1$.
11. $\frac{(y-4)^2}{1} - \frac{(x-1)^2}{3} = 1$.
12. $\frac{(x-6)^2}{4} - \frac{(y-2)^2}{5} = 1$.
13. $xy = -10$.
14. $xy - 3x - 2y + 5 = 0$.
15. $4y^2 - x^2 = 15$.
16. $9y^2 - 4x^2 = 5$.
17. $3x^2 - 4xy + y^2 = 3$.
18. $x^2 - y^2 + 2x = 0$.
19. $x^2 - y^2 = 16$.
20. $y^2 - 9x^2 = 16$.
21. $x^2 - 2y^2 = 1$.
22. $y^2 - 7x^2 = 9$.
23. $(x + 2)^2 - 5(y + 1)^2 = 4$.
24. $(y - 3)^2 - 3(x - 2)^2 = 1$.
25. $\frac{x^2}{16} - \frac{y^2}{9} = 1$.
26. $\frac{x^2}{a^2} - \frac{y^2}{c^2 - a^2} = 1$.

Exercise 44 (Pages 134 to 135)

1. Parabola.
2. Hyperbola.
3. Ellipse.
4. Ellipse.
5. Hyperbola.
6. Hyperbola.
7. Hyperbola.
8. Ellipse.
9. Parabola.
10. Hyperbola.
11. Parabola.
12. Hyperbola.
13. (a) $k = 4$; (b) $k > 4$; (c) $k < 4$.

14. (a) $k = \pm 4$; (b) $-4 < k < 4$; (c) $k < -4$ or $k > 4$.
15. (a) $k = \pm 12$; (b) $-12 < k < 12$; (c) $k < -12$ or $k > 12$.
16. (a) $k = 4$. (b) $k > 4$. (c) $k < 4$.

Exercise 45 (Pages 138 to 139)

1. Tangent $2y - 3x + 12 = 0$, normal $2x + 3y + 5 = 0$.
2. Tangent $y - 3x + 8 = 0$, normal $3y + x - 6 = 0$.
3. Tangent $2x + y = 10$, normal $2y - x = 0$.
4. Tangent $x + 3y = 7$, normal $y - 3x + 1 = 0$.
5. Tangent $x - 3y + 3 = 0$, normal $3x + y = 11$.
6. Tangent $4x - 3y = 5$, normal $3x + 4y = 10$.
7. Tangent $y - 5x + 6 = 0$, normal $x + 5y + 4 = 0$.
8. Tangent $11y - 8x + 27 = 0$, normal $11x + 8y = 14$.
9. For "$= 3$" read "$= 5$." Tangent $x + 2y = 10$, normal $2x - y = 15$.
10. Tangent $2x + y = 6$, normal $y - 2x = 2$.
11. $(-2,-20)$ and $(4,16)$.
12. $(0,0)$ and $(5,-5)$.
13. $(0,0)$ and $(6,36)$.
14. $(\frac{1}{2},-12)$ and $(-\frac{1}{2},12)$.
15. $(1 + \sqrt{\frac{1}{3}}, 2\sqrt{\frac{1}{3}})$ and $(1 - \sqrt{\frac{1}{3}}, -2\sqrt{\frac{1}{3}})$.
16. $(2,4)$.
17. $(1,3)$ and $(-1,-3)$.
18. $(2,3)$.
19. $(3,-9)$, $y = -9$.
20. $(2,-16)$, $y = -16$ and $(-2,16)$, $y = 16$.
21. $(1,5)$, $y = 5$ and $(-1,-5)$, $y = -5$.
22. $(3,-1)$, $y = -1$ and $(-3,1)$, $y = 1$.
23. $(-4,-2)$, $x = -4$.
24. $(0,0)$, $x = 0$ and $(4,-2)$, $x = 4$.
25. $(2,2)$, $x = 2$ and $(-2,-2)$ $x = -2$.
26. $(1,2)$, $x = 1$ and $(-1,-2)$, $x = -1$.

Exercise 46 (Pages 141 to 142)

1. $(0,0)$, 45° and $(2,2)$, 45°.
2. $(\sqrt{\frac{32}{7}}, \pm\sqrt{\frac{3}{7}})$, $(-\sqrt{\frac{32}{7}}, \pm\sqrt{\frac{3}{7}})$ at $\tan^{-1}(5\sqrt{6}/12)$.
3. $(0,0)$, $\tan^{-1} 2$ and $(2,4)$, $\tan^{-1}\frac{2}{9}$.
4. $(2, \pm 2\sqrt{3})$, 60°.
5. $(0,0)$, $\tan^{-1}\frac{1}{2}$ and $(1,2)$, $\tan^{-1}\frac{1}{3}$.
6. $(1,\pm 2)$, $\tan^{-1}\frac{6}{7}$ and $(-2,\pm 1)$, $\tan^{-1}\frac{3}{4}$.
7. Subtangent 5, subnormal $\frac{4}{5}$.
8. Subtangent 3, subnormal $\frac{4}{3}$.
9. Subtangent $\frac{1}{2}$, subnormal 2.
10. Subtangent $\frac{4}{3}$, subnormal $\frac{3}{4}$.

Exercise 48 (Pages 151 to 152)

5. -0.3508.
6. -0.7374.
7. -1.1578.

Exercise 49 (Page 155)

1. 2.
2. 1.
3. $\frac{1}{4}$.
4. 1.
5. 2.
6. $\frac{3}{4}$.
11. 0.
12. 0.
13. $\frac{1}{2}$.
14. 0.
15. $\frac{1}{2}$.
16. $\frac{1}{2}$.

Exercise 50 (Pages 157 to 158)

1. $2\cos(4 + 2x)$.
2. $\sin(5 - x)$.
3. $8\cos 4x$.
4. $-8\sin 2x$.
5. $-15\cos(2 - 3x)$.
6. $-12\cos(2 + 3x)$.
7. $\cos x \cos 2x - 2\sin x \sin 2x$.
8. $\sin^2 x \cos x$.
9. $\cos 3x \sin 3x$.
10. $2\sin^2 x$.
11. $\dfrac{x\cos x - \sin x}{x^2}$.
12. $\dfrac{2\cos 4x}{\sqrt{\sin 4x}}$.
13. $(2x - 3)\cos(x^2 - 3x)$.

14. $\sin x + x \cos x.$ 15. $\frac{1}{6 \cos 3y}.$ 16. $-\frac{1}{4 \sin y}.$

17. $\frac{\cos (y - x) + 1}{\cos (y - x) - 1}.$ 18. $\frac{1 - 3 \sin (2y + 3x)}{1 + 2 \sin (2y + 3x)}.$

19. $\cos 2x.$ 20. $-\sin 2x.$

Exercise 51 (Page 161)

1. $30 \sec^2 5x.$ 2. $-12 \csc^2 3x.$ 3. $21 \tan 7x \sec 7x.$
4. $-8 \cot 4x \csc 4x.$ 5. $12 \sec^2 (3x + 5).$ 6. $9 \csc^2 (5 - 3x).$
7. $-8 \tan (3 - 4x) \sec (3 - 4x).$ 8. $-8 \cot (3 + 2x) \csc (3 + 2x).$

9. $\tan 5x \sec 5x.$ 10. $-\csc^2 3x.$ 11. $\sec^2 \frac{x}{7}.$

12. $-\cot \frac{x}{2} \csc \frac{x}{2}.$ 13. $\frac{\sec^2 x}{2 \sqrt{\tan x}}.$ 14. $-\frac{\tan x}{2 \sqrt{\sec x}}.$

15. $\frac{\sec^2 3x}{\sqrt[3]{\tan^2 3x}}.$ 16. $2 \sec^5 2x \tan 2x.$ 17. $3 \tan^4 x.$

18. $-3 \csc^4 x.$ 19. $\tan^2 x.$ 20. $3(\sec x + \tan x)^3 \sec x.$
21. $2 \cos 2x \cos^2 y.$ 22. $-3 \sec^2 3x \csc y.$

Exercise 54 (Page 170)

1. $\frac{2}{\sqrt{1 - 4x^2}}.$ 2. $\frac{-3}{\sqrt{1 - 9x^2}}.$ 3. $\frac{1}{\sqrt{9 - x^2}}.$ 4. $\frac{-1}{\sqrt{4 - x^2}}.$

5. $\frac{-3}{\sqrt{2x - x^2}}.$ 6. $\frac{5}{\sqrt{x - x^2}}.$ 7. $\frac{1}{2 \sqrt{x - x^2}}.$ 8. $\frac{-2x}{\sqrt{1 - x^4}}.$

9. $2x \sin^{-1} x + \frac{x^2}{\sqrt{1 - x^2}}.$ 10. $\cos^{-1} x - \frac{x}{\sqrt{1 - x^2}}.$

11. $\frac{4}{1 + 16x^2}.$ 12. $\frac{-6}{1 + 36x^2}.$ 13. $\frac{2}{4 + x^2}.$ 14. $\frac{-3}{9 + x^2}.$

15. $\frac{2x}{1 + x^4}.$ 16. $\frac{-3x^2}{1 + x^6}.$ 17. $\frac{1}{x \sqrt{4x^2 - 1}}.$ 18. $\frac{-1}{x \sqrt{16x^2 - 1}}.$

19. $\frac{16}{x \sqrt{x^2 - 16}}.$ 20. $\frac{-25}{x \sqrt{x^2 - 25}}.$ 21. $\frac{-2x}{\sqrt{1 - x^4}}.$

22. $\frac{1}{\sqrt{1 - x^2}}$ if $x > 0$, $\frac{-1}{\sqrt{1 - x^2}}$ if $x < 0$.

23. $2 \sqrt{1 - x^2}.$ 24. $\frac{x^2 + x - 1}{(1 - x^2)^{\frac{3}{2}}}.$

Exercise 55 (Page 172)

1. $\frac{dx}{dt} = -24\pi, \frac{dy}{dt} = \pm 32\pi.$ 2. $\frac{dx}{dt} = -32\pi, \frac{dy}{dt} = \pm 24\pi.$

3. $\frac{dx}{dt} = -40\pi, \frac{dy}{dt} = 0.$ 4. $\frac{dx}{dt} = 0, \frac{dy}{dt} = \pm 40\pi.$

5. $\omega = 6t, \alpha = 6.$ 6. $\omega = 6t^2, \alpha = 12t.$
7. $\omega = 6 \cos 2t, \alpha = -12 \sin 2t.$ 8. $\omega = -20 \sin 5t, \alpha = -100 \cos 5t.$

9. $\frac{4}{\pi}.$ 10. $\frac{18}{\pi}.$ 11. $\frac{32}{\pi}.$ 12. $\frac{192}{\pi}.$

13. $\dfrac{2}{1+4t^2}$. **14.** $\dfrac{-12}{16+9t^2}$. **15.** $\dfrac{1}{1+t^2}$. **16.** $\dfrac{-2t}{1+t^4}$.
17. -5. **18.** 2. **19.** -3. **20.** -4.

Exercise 56 (Page 176)

1. $c = 8,\ T = \pi/2$. **2.** $c = 2,\ T = 2$. **3.** $c = 5,\ T = \pi$.
4. $c = 5,\ T = \pi$. **5.** $c = 6,\ T = \pi/2$. **6.** $c = 8,\ T = \pi$.
7. $c = 2,\ T = \pi$. **8.** $c = 3,\ T = \pi/2$. **9.** $x = 4 \sin 2t$.
10. $x = 4 \cos 2t$. **11.** $x = 4 \sin (2t - 4)$. **12.** $x = 4 \cos (2t - 4)$.
13. $x = 4 \sin (4\pi t - A)$. **14.** $x = 4 \sin (16\pi t - A)$.
15. $x = 4 \sin (2t - A)$. **27.** $x^2 + \dfrac{v^2}{b^2} = c^2,\ c = 5,\ T = \pi$.
28. $x = 5 \sin (8t + \tan^{-1} \frac{3}{4})$.

Exercise 57 (Pages 179 to 180)

1. $\pi/6$. **2.** $\pi/3$. **3.** $\pi/6$. **4.** $\pi/4$. **5.** $\pi/6$. **6.** $\pi/6$. **7.** $\pi/3$.
8. $\tan^{-1} 2$. **9.** $13,\ -13$. **10.** $\sqrt{2},\ -\sqrt{2}$. **11.** $3,\ -1$.
12. $\frac{3}{2}\sqrt{3},\ -\frac{3}{2}\sqrt{3}$. **13.** $\frac{1}{2}(1 + \cos 1),\ -\frac{1}{2}(1 + \cos 1)$.
14. $\frac{1}{2}(1 + \sin 2),\ -\frac{1}{2}(1 + \sin 2)$. **16.** $\pm(\pi/4)$.
17. $\pi/3$. **20.** $\pi/2$. **21.** $\tan^{-1} \sqrt{2}$. **22.** $\pi/4$.
23. $\pi/2 - \frac{1}{2} \tan^{-1} 2$. **24.** $2 \tan^{-1} (1/\sqrt{2})$. **25.** $5\sqrt{5}$ ft.
26. $10\sqrt{10}$ ft. **28.** 12 ft. **29.** 0.06 radian/sec.
30. 5π mi./min. **31.** $\frac{22}{5}\pi$ radians/hr.

Exercise 59 (Pages 191 to 192)

1. 8.585. **2.** 0.09442. **3.** 18,034.
4. 0.7655. **5.** 4.5850. **6.** -1.4524.

Exercise 60 (Page 193)

1. $\dfrac{3}{3x+5}$. **2.** $\dfrac{2x+4}{x^2+4x}$. **3.** $\dfrac{6}{x}$. **4.** $\dfrac{x}{x^2-4}$. **5.** $-\dfrac{1}{x}$.
6. $\dfrac{2 \ln x}{x}$. **7.** $\ln x$. **8.** $\dfrac{1}{x \ln x}$. **9.** $\dfrac{1}{\sqrt{x^2+4}}$.
10. $\dfrac{10x+11}{2x(5x+3)}$. **11.** $\dfrac{24}{4-9x^2}$. **12.** $\dfrac{4x}{x^2-4}$.
13. $\cot x$. **14.** $\tan x$. **15.** $\sec x \csc x$.
16. $\sec x$. **17.** $\dfrac{1 - \ln x}{x^2}$. **18.** $\dfrac{-1}{\sqrt{x^2+4}}$.

Exercise 61 (Pages 194 to 195)

1. $6e^{3x}$. **2.** $-8e^{-2x}$. **3.** $(2x+1)e^{2x}$. **4.** $2xe^{x^2}$.
5. $\dfrac{x-1}{x^2} e^x$. **6.** $-4e^{-x}$. **7.** $\dfrac{1}{2\sqrt{x}} e^{\sqrt{x}}$. **8.** $\cos x\, e^{\sin x}$.
9. $(1-x)e^{-x}$. **10.** $\dfrac{1}{x^2} e^{-1/x}$. **11.** $(2-x)xe^{-x}$.
12. $\dfrac{1}{x}(1 - x \ln x)e^{-x}$. **13.** $\dfrac{e^{2x}-1}{e^{2x}+1}$.

14. $-e^{-x}(\cos 2x + 2 \sin 2x)$. **15.** $\dfrac{-4}{(e^x - e^{-x})^2}$.

16. $\dfrac{4e^{2x}}{(e^{2x} + 1)^2}$. **17.** $\sec^2 x \, e^{\tan x}$. **18.** $\dfrac{1}{1 + x^2} e^{\tan^{-1}} e^{\tan^{-1} x}$.

Exercise 62 (Page 198)

1. $\dfrac{2(\log e)}{2x + 3}$. **2.** $\dfrac{2x(\log_2 e)}{x^2 + 1}$. **3.** $(\log e) \tan x$.

4. $(\log e) \sec x \csc x$. **5.** $(\ln 5)5^x$. **6.** $27(\ln 2)2^{3x}$.

7. $-(\ln 10)10^{-x}$. **8.** $6(\ln 3)x3^{x^2}$. **9.** $-24(\ln 7)7^{-8x}$.

10. $(2 + x \ln 2)x2^x$. **11.** $\frac{1}{2}e^{x/2}$. **12.** $-e^{-x}$.

13. $(3 \ln 2 + 2 \ln 3)2^{3x}3^{2x}$. **14.** $\dfrac{1}{x}(1 + x \ln x)e^x x^{e^x}$.

15. $(1 + \ln x)x^x$. **16.** $-(1 + \ln x)x^{-x}$.

17. $(\tan x + \ln \sin x)(\sin x)^x$. **18.** $\dfrac{1}{2\sqrt{x}}(2 + \ln x)x^{\sqrt{x}}$.

19. $4(12x - 7)(2x - 3)^3(4x + 1)^2$. **20.** $x^2(3x^3 + 5)(x^3 + 2)^{-\frac{2}{3}}(x^3 + 1)^{-\frac{1}{3}}$.

21. $\dfrac{(7x^2 + 6)x^5}{\sqrt{x^2 + 1}}$. **22.** $3(6x + 5)x^4(2x + 3)^3$.

23. $2x(x^2 - 3)^{-\frac{1}{2}}(x^2 - 1)^{-\frac{3}{2}}$. **24.** $-2x(1 - x^2)^{-\frac{1}{2}}(1 + x^2)^{-\frac{3}{2}}$.

Exercise 63 (Page 201)

1. $x = 10e^{2t}$. **2.** $x = 100e^t$. **3.** $x = e^{0.01t}$. **4.** $x = 5 \cdot 2^t$.

5. $x = 10 \cdot 2^t$. **6.** $x = 2^t$. **7.** $x = e^{0.02t}$.

8. $x = 100e^{0.04t}$. **9.** $x = 200e^{0.03t}$. **10.** $x = 1{,}000e^{0.025t}$.

11. $r = 2.877$. **12.** $r = 2$. **13.** $r = 1.0986$.

14. $x = 100 \cdot 5^{-t/10}$. **15.** $x = 40 \cdot 2^{-t/20}$. **16.** $x = 48 \cdot 2^{-t/20}$.

17. $x_1 = 4$. **18.** $x_1 = 20$. **19.** $x_1 = 24$.

Exercise 64 (Pages 206 to 207)

9. $x = 2 + e^{-5t}$. **10.** $x = e^{-t} \sin 2t - e^{-t} \cos 2t$.

11. $x = te^{-t} + e^{-t}$. **12.** $x = 3e^{-2t} + 2e^{-3t}$.

13. $x = 3e^{-2t} \sin t + 2e^{-2t} \cos t$. **14.** $x = \frac{3}{2}e^{3t} - \frac{1}{2}e^{-3t}$.

15. $T = \pi$, $t_1 = \ln 2$. **16.** $T = 2\pi$, $t_1 = \frac{1}{2} \ln 2$.

17. $T = \pi/2$, $t_1 = \frac{1}{3} \ln 2$.

Exercise 65 (Pages 208 to 209)

1. Min $(-1, -e^{-1})$, inflection $(-2, -2e^{-2})$.
2. Max $(1,e^{-1})$, inflection $(2,2e^{-2})$.
3. Max $(4, 256e^{-4})$, Min $(0,0)$, inflections $(2,16e^{-2})$, $(6,1296e^{-6})$.
4. Max $(-2,4e^{-2})$, Min $(0,0)$, inflections $[-2 \pm \sqrt{2}, (6 \mp 4\sqrt{2})e^{-2\pm\sqrt{2}}]$.
5. Min $(e^{-1}, -e^{-1})$.
6. Min $(e^{-\frac{1}{2}}, -\frac{1}{2}e^{-1})$, inflection $(e^{-\frac{3}{2}}, -\frac{3}{2}e^{-3})$.
7. Min (e,e), inflections $\left(e^{\pm\sqrt{2}}, \pm \dfrac{1}{\sqrt{2}} e^{\pm\sqrt{2}}\right)$.
8. Max $(e^{-2},4e^{-2})$, Min $(1,0)$, inflection (e^{-1},e^{-1})
9. Max $(0,1)$, inflections $\left(\pm \dfrac{1}{\sqrt{2}}, e^{-\frac{1}{2}}\right)$.

10. Max $\left(\frac{1}{\sqrt{2}}, \frac{1}{\sqrt{2}} e^{-\frac{1}{2}}\right)$, Min $\left(-\frac{1}{\sqrt{2}}, -\frac{1}{\sqrt{2}} e^{-\frac{1}{2}}\right)$, inflections $\left(\pm\sqrt{\frac{3}{2}}, \pm\sqrt{\frac{3}{2}} e^{-3}\right)$.

15. $(1, 4e^{-1})$, greatest area.

16. $(\frac{1}{2}, 4e^{-\frac{1}{2}})$, greatest volume.

17. $(2, 4e^{-2})$, greatest volume.

18. $(\ln 4, 1)$, least value.

19. $(e^{-1}, \frac{1}{2})$, greatest area.

20. $(e^{-2}, 1)$, greatest volume.

21. $(e^{-\frac{1}{2}}, \frac{1}{4})$, greatest volume.

22. $(\frac{1}{2}, \frac{1}{2} \ln 2)$, least value.

24. $b = 1$, $\phi = 0$, $\theta = 3\pi/4$. Max $x = \pi/4$, Min $x = 5\pi/4$, inflections $x = \pi/2$, $x = 3\pi/2$.

25. $b = 1$, $\phi = \pi/2$, $\theta = 3\pi/4$. Max $x = 7\pi/4$, Min $x = 3\pi/4$, inflections 0, π, 2π.

26. $b = 1$, $\phi = 0$, $\theta = \pi/2 + \tan^{-1}(0.04)$. Max $x = \pi/2 - \tan^{-1}(0.04)$, Min $x = 3\pi/2 - \tan^{-1}(0.04)$, inflections $x = \pi - 2\tan^{-1}(0.04)$, $x = 2\pi - 2\tan^{-1}(0.04)$.

27. $b = 1$, $\phi = \pi/2$, $\theta = \pi/2 + \tan^{-1}(0.07)$. Max $x = 2\pi - \tan^{-1}(0.07)$, Min $x = \pi - \tan^{-1}(0.07)$, inflections $x = \pi/2 - 2\tan^{-1}(0.07)$, $x = 3\pi/2 - 2\tan^{-1}(0.07)$.

Exercise 66 (Pages 215 to 217)

1. $2x = 6y - 3y^3$.

2. $y = x^2 - 4$.

3. $x^2 + y^2 = 16$.

4. $x = 1 - 2y^2$, $-1 \leqq x \leqq 1$, $-1 \leqq y \leqq 1$.

5. $y^2 = 4x^2(1 - x^2)$.

6. $x^2/9 + y^2/25 = 1$.

7. $(x - 5)^2 + (y + 1)^2 = 16$.

8. $x = a \sin t$, $y = b \cos t$.

9. $x = a \sec t$, $y = b \tan t$.

10. $x = \frac{2}{1 + t^2}$, $y = \frac{2t}{1 + t^2}$.

11. $x = \frac{2t}{1 - t^2}$, $y = \frac{2t^2}{1 - t^2}$.

Exercise 67 (Page 219)

1. Tangent $3x - y = 4$, normal $x + 3y = 28$.
2. Tangent $2x + y = -4$, normal $x - 2y = 3$.
3. Tangent $x + e^2y = 2e$, normal $e^2x - y = e^3 - e^{-1}$.
4. Tangent $x - 2y = 3$, normal $2x + y = 1$.
5. Tangent $3x + y = 3$, normal $x - 3y = 1$.
6. Tangent $x + y = 3$, normal $x - y = 0$.
7. $8x = y^3 - 48y$, vertical at $(\pm 16, \mp 4)$.
8. $y = x^2 - 1$, horizontal at $(0, -1)$.
9. $x^2 + y^2 = 25$, horizontal at $(0, \pm 5)$, vertical at $(\pm 5, 0)$.
10. $\frac{x^2}{4} + \frac{y^2}{16} = 1$, horizontal at $(0, \pm 4)$, vertical at $(\pm 2, 0)$.
11. $x = 2y^2 - 1$, $-1 \leqq x \leqq 1$, $-1 \leqq y \leqq 1$, vertical at $(-1, 0)$.
12. $y = 1 - 2x^2$, $-1 \leqq x \leqq 1$, $-1 \leqq y \leqq 1$, horizontal at $(0, 1)$.
13. $x^{\frac{2}{3}} + y^{\frac{2}{3}} = 1$, horizontal at $(\pm 1, 0)$, vertical at $(0, \pm 1)$.
14. $x = \cos(y/2)$, vertical at $[(-1)^k, 2k\pi]$.

Exercise 68 (Pages 221 to 222)

1. $\frac{d^2y}{dx^2} = 8t^3 = \frac{8}{x^3}$.

2. $\frac{d^2y}{dx^2} = -\frac{1}{t^3} = -\frac{1}{x^{\frac{3}{2}}}$.

3. $\frac{d^2y}{dx^2} = \frac{3}{4t} = \frac{3}{4} x^{-\frac{1}{2}}$.

4. $\frac{d^2y}{dx^2} = -\csc^3 t = -(1 - x^2)^{-\frac{3}{2}}$.

5. $\frac{d^2y}{dx^2} = 2e^{-3t} = \frac{2}{x^3}$.

6. $\dfrac{d^2y}{dx^2} = \dfrac{3}{4}e^{-t} = \dfrac{3}{4}x^{-\frac{1}{2}}.$

7. $-\dfrac{1}{a(1-\cos\phi)^2} = -\dfrac{1}{4a}\csc^4\dfrac{\phi}{2}.$

8. $\dfrac{a-2b}{4b(a-b)\sin\dfrac{a}{2b}\phi\cos^3\left(\dfrac{a-2b}{2b}\phi\right)}.$

9. $-\dfrac{b}{a^2}\csc^3\phi.$

10. $\dfrac{b(a\cos\phi - b)}{(a - b\cos\phi)^3}.$

11. $\dfrac{\sec^3\phi}{a\phi}.$

12. $\dfrac{n-2}{an}\tan^{n-4}\phi\sec^{n+2}\phi.$

13. $\dfrac{d^2y}{dx^2} = 2\sin^4 t\,(3\cos^2 t - \sin^2 t),\ \left(\pm\dfrac{1}{\sqrt{3}},\dfrac{3}{4}\right).$

14. $\dfrac{d^2y}{dx^2} = \dfrac{2(t-2)}{3(4t-t^2)^3},\ (16,6).$

15. $\dfrac{d^2y}{dx^2} = \dfrac{3(t^2-1)}{8(3t-t^3)^3},\ (5,\pm 2).$

16. $\dfrac{d^2y}{dx^2} = \dfrac{2(1-t)}{9(t+1)^3},\ (4,4).$

17. $\dfrac{d^2y}{dx^2} = \dfrac{48t(t^2+3)^3}{(3-t^2)^3},\ (0,0).$

18. $\dfrac{d^2y}{dx^2} = 2 + 2t^3,\ (-1,0).$

19. $\dfrac{d^2y}{dx^2} = 6(t^4-1),\ (\pm 1,-2).$

20. $\dfrac{d^2y}{dx^2} = \dfrac{6t-2}{(2t-1)^3t^3},\ (\tfrac{2}{9},3).$

Exercise 69 (Pages 231 to 232)

1. $\frac{13}{2}\sqrt{65}.$

2. $\frac{5}{4}\sqrt{5}.$

3. $2\sqrt{2}.$

4. $\frac{5}{4}\sqrt{5}.$

5. For "(2,0)" read "(0,2)." $\frac{1}{8}.$

6. $-5\sqrt{5}.$

7. $2x(1+x^4)^{-\frac{3}{2}}.$

8. $(1+x^2)^{-\frac{3}{2}}.$

9. $(2xy^2 - x^4)(y^2 + x^2y)^{-\frac{3}{2}}.$

10. $-\cos x\sin x.$

11. For "$-e^{-x/a}$" read "$+e^{-x/a}$." $\dfrac{1}{a(e^{x/a}+e^{-x/a})^2}.$

12. $\dfrac{-a^4b^4}{(a^4y^2+b^4x^2)^{\frac{3}{2}}}.$

13. $\frac{13}{6}\sqrt{13}.$

14. $-\frac{5}{4}\sqrt{5}.$

15. $-\sqrt{2}.$

16. $-\frac{37}{2}\sqrt{37}.$

17. $1/a.$

18. $ab(a^2\sin^2 t + b^2\cos^2 t)^{-\frac{3}{2}}.$

19. $-\dfrac{1}{3a\cos t\sin t} = -\dfrac{2}{3a}\csc 2t.$

20. $-\dfrac{1}{2a}(\cos^4 t + \sin^4 t)^{-\frac{3}{2}}.$

21. $1/at.$

22. $\dfrac{-1}{2a\sqrt{2-2\cos t}} = -\dfrac{1}{4a}\csc\dfrac{t}{2}.$

23. $x = \left(\dfrac{n-2}{2n^3-n^2}\right)^{1/(2n-2)}$ if $n > 2.$

Exercise 70 (Page 238)

1. $X = \dfrac{2x - x^7}{3},\ Y = \dfrac{7x^6+4}{12x^2}.$

2. $X = \dfrac{x-x^5}{2},\ Y = \dfrac{5x^4+3}{6x}.$

3. $X = 4x + \dfrac{2x}{y},\ Y = -y - 3y^2,\ 2y^3 = 3x^2.$

4. $X = \frac{3}{2}y^2 + 1,\ Y = -y^3.$

5. $X = x - \tan x,\ Y = \ln\cos x + 1.$

6. $X = \dfrac{12x^4+a^4}{8x^3},\ Y = \dfrac{4x^4+3a^4}{4a^2x}.$

7. For "$-e^{-x}$" read "$+e^{-x}$." $X = x - \frac{1}{4}(e^{2x}-e^{-2x}),\ Y = e^x + e^{-x}.$

8. $X = x + 3y^{\frac{2}{3}}x^{\frac{1}{3}},\ Y = y + 3y^{\frac{1}{3}}x^{\frac{2}{3}},\ x^{\frac{2}{3}} + y^{\frac{2}{3}} = 1.$

9. $X = x + \frac{2y^{\frac{1}{2}}}{a^{\frac{1}{2}}}(x + y)$, $Y = y + \frac{2x^{\frac{1}{2}}}{a^{\frac{1}{2}}}(x + y)$, $x^{\frac{1}{2}} + y^{\frac{1}{2}} = a^{\frac{1}{2}}$. So that $a(Y - X)^2 = \frac{1}{27}(2X + 2Y - 3a)^3$.

10. $X = \frac{2x^3}{a^2}$, $Y = -\frac{2y^3}{a^2}$, $x^2 - y^2 = a^2$.

11. $X = 6t^2 + 1$, $Y = -8t^3$.

12. $X = \frac{12 + a^4t^4}{8t}$, $Y = \frac{4 + 3a^4t^4}{4a^2t^3}$.

13. $X = \frac{a^2 - b^2}{a}\cos^3 t$, $Y = -\frac{a^2 - b^2}{b}\sin^3 t$.

14. $X = \frac{a^2 + b^2}{a}\sec^3 t$, $Y = -\frac{a^2 + b^2}{b}\tan^3 t$.

15. $X = a\cos^3 t + 3a\sin^2 t\cos t$, $Y = a\sin^3 t + 3a\sin t\cos^2 t$.

16. $X = a\cos^4 t + 2a\sin^2 t(\cos^4 t + \sin^4 t)$,
$Y = a\sin^4 t + 2a\cos^2 t(\cos^4 t + \sin^4 t)$.

17. $X = a\cos t$, $Y = a\sin t$.

18. $X = a(t - \sin t)$, $Y = a(1 - \cos t)$.

19. $X^2 + Y^2 = a^2$.

20. $(aX)^{\frac{2}{3}} + (bY)^{\frac{2}{3}} = (a^2 - b^2)^{\frac{2}{3}}$.

21. $(aX)^{\frac{2}{3}} - (bY)^{\frac{2}{3}} = (a^2 + b^2)^{\frac{2}{3}}$.

22. $Y = -[\frac{2}{3}(X - 1)]^{\frac{3}{2}} = -\frac{2}{9}\sqrt{6}(X - 1)^{\frac{3}{2}}$.

23. $(X + Y)^{\frac{2}{3}} - (X - Y)^{\frac{2}{3}} = 2a^{\frac{2}{3}}$.

24. $(X + Y)^{\frac{2}{3}} + (X - Y)^{\frac{2}{3}} = 2a^{\frac{2}{3}}$.

Exercise 71 (Pages 242 to 243)

1. $v_x = 3t^2 - 2t$, $v_y = 2$, $a_x = 6t - 2$, $a_y = 0$. At $t = 3$, $v_x = 21$, $v_y = 2$, $v = \sqrt{445}$, $a_x = 16$, $a_y = 0$, $a = 16$.

2. $v_x = 2$, $v_y = 4t^3$, $a_x = 0$, $a_y = 12t^2$. At $t = 2$, $v_x = 2$, $v_y = 32$, $v = \sqrt{1{,}028}$, $a_x = 0$, $a_y = 48$, $a = 48$.

3. $v_x = 2t + 1$, $v_y = 2t - 1$, $a_x = 2$, $a_y = 2$. At $t = 1$, $v_x = 3$, $v_y = 1$, $v = \sqrt{10}$, $a_x = 2$, $a_y = 2$, $a = \sqrt{8}$.

4. $v_x = 2$, $v_y = -e^{-t}$ $a_x = 0$, $a_y = e^{-t}$. At $t = 0$, $v_x = 2$, $y_y = -1$, $v = \sqrt{5}$, $a_x = 0$, $a_y = 1$, $a = 1$.

5. $v_x = \cos t$, $v_y = -3\sin 3t$, $a_x = -\sin t$, $a_y = -9\cos 3t$. At $t = \pi$, $v_x = -1$, $v_y = 0$, $v = 1$, $a_x = 0$, $a_y = 9$, $a = 9$.

6. $v_x = -\sin t$, $v_y = -2\sin 2t$, $a_x = -\cos t$, $a_y = -4\cos 2t$.
At $t = \pi/2$, $v_x = -1$, $v_y = 0$, $v = 1$, $a_x = 0$, $a_y = 4$, $a = 4$.

12. $a_t = \tan t\sec t$, $a_n = \sec t$.

13. $a_t = 0$, $a_n = 2$.

14. $a_t = 2$, $a_n = 4t^2$.

15. $a_t = e^t$, $a_n = e^{2t}$.

16. $a_t = \sqrt{2}e^t$, $a_n = \sqrt{2}e^t$.

17. $a_t = 2\tan t\sec^2 t$, $a_n = -\sec^2 t$.

18. $a_t = b\omega^2\cos(\omega t/2)$, $a_n = -b\omega^2\sin(\omega t/2)$.

19. $a_t = b\omega^2$, $a_n = b\omega^3 t$.

Exercise 72 (Page 249)

4. $\left(4, \frac{\pi}{4} + 2k\pi\right), \left(-4, \frac{5\pi}{4} + 2k\pi\right)$.

5. $(5, 2k\pi)$, $(-5, \pi + 2k\pi)$.

6. $(0, A)$.

Exercise 73 (Pages 253 to 254)

1. $r^2\sin 2\theta = 8$.

2. $r\sin^2\theta = 4\cos\theta$.

3. $r^2\cos 2\theta = 4$.

4. $r = 4\cos\theta$.

5. $r^2 = \cos 2\theta$.

6. $r^2 = \sin 2\theta$.

7. $x^2(x^2 + y^2) = y^2$.

8. $y^2(x^2 + y^2) = x^2$.

9. $(x^2 + y^2)^3 = y^2$.

10. $(x^2 + y^2)^3 = x^2$.

11. $(x^2 + y^2 - ax)^2 = b^2(x^2 + y^2)$.

12. $(x^2 + y^2 - ay)^2 = b^2(x^2 + y^2)$.

13. $r(\sqrt{3}\cos\theta + \sin\theta) = 8$.

14. $r(\sin\theta - \cos\theta) = 2.$ 15. $r\cos\theta = 3.$
16. $r = 4\sin\theta.$. 17. $r = 12\cos\theta.$ 18. $r = \cos\theta + \sin\theta.$
19. $r = \dfrac{4}{1 + \sin\theta}\cdot$ 20. $r = \dfrac{10}{1 + \cos\theta}\cdot$ 21. $r = \dfrac{8}{1 - \sin\theta}\cdot$
22. $x = 3$, straight line. 23. $x^2 + y^2 = 6y$, circle.
24. $x^2 + y^2 - 2y = 3$, circle. 25. $y = -2$, straight line.
26. $y^2 = 1 - 2x$, parabola. 27. $3x^2 + 4y^2 - 2x = 1$, ellipse.
28. $xy = 2$, hyperbola. 29. $x^2 - y^2 = 4$, hyperbola.

Exercise 74 (Page 257)

9. $-\frac{1}{3}\cot 3\theta,\ \tan^{-1}[1/3\,)\sqrt{3})].$ 10. $\frac{1}{4}\tan 4\theta,\ \tan^{-1}\frac{1}{4}.$
11. $-\dfrac{2 + \cos 2\theta}{\sin 2\theta},\ \dfrac{\pi}{2}$ 12. $\dfrac{2 + \sin\theta}{\cos\theta},\ \tan^{-1} 2.$ 13. $\theta,\ \tan^{-1}\pi.$
14. For "$+\dfrac{1}{\theta}$," read "$=\dfrac{1}{\theta}$." $-\theta,\ \pi - \tan^{-1}\pi.$

Exercise 75 (Pages 259 to 260)

10. $[1 + \sqrt{2},\ \pi \pm \cos^{-1}(\sqrt{2} - 1)].$
11. $\left(\dfrac{3 - \sqrt{33}}{4},\ \pm\cos^{-1}\dfrac{1 + \sqrt{33}}{8}\right),\ \left(\dfrac{3 + \sqrt{33}}{4},\ \pi \pm \cos^{-1}\dfrac{\sqrt{33} - 1}{8}\right).$
12. $\left(\dfrac{2\sqrt{2}}{3},\ \pm\cos^{-1}\dfrac{1}{\sqrt{3}}\right),\ \left(\dfrac{2\sqrt{2}}{3},\ \pi \pm \cos^{-1}\dfrac{1}{\sqrt{3}}\right),\ (0,0).$
13. $\left(\dfrac{2}{3},\ \pm\sin^{-1}\dfrac{1}{\sqrt{6}}\right),\ \left(\dfrac{2}{3},\ \pi \pm \sin^{-1}\dfrac{1}{\sqrt{6}}\right),\ \left(1, \pm\dfrac{\pi}{2}\right).$
15. $(1,0),\ (3,\pi).$ 16. $(1,\pi),\ (3,\pi),\ (\frac{1}{2},\ \pm\cos^{-1}\frac{1}{4}).$
17. $\left(\dfrac{2\sqrt{2}}{3},\ \pm\sin^{-1}\dfrac{1}{\sqrt{3}}\right),\ \left(\dfrac{2\sqrt{2}}{3},\ \pi \pm \sin^{-1}\dfrac{1}{\sqrt{3}}\right),\ \left(0, \dfrac{\pi}{2}\right).$
18. $\left(\dfrac{2}{3},\ \pm\cos^{-1}\dfrac{1}{\sqrt{6}}\right),\ \left(\dfrac{2}{3},\ \pi \pm \cos^{-1}\dfrac{1}{\sqrt{6}}\right),\ (1, \pm\pi).$

Exercise 76 (Page 263)

1. $(3\sqrt{13},\ \tan^{-1}\frac{3}{2}),\ (2\sqrt{13},\ \tan^{-1}\frac{2}{3})$ at $\tan^{-1}\frac{5}{12}.$
2. $\left(\dfrac{1 + \sqrt{17}}{2},\ \sin^{-1}\dfrac{\sqrt{17} - 1}{4}\right)$ at $\sin^{-1}\dfrac{\sqrt{17} - 1}{4} + \tan^{-1}\dfrac{1 + \sqrt{17}}{8}.$
3. $\left(\dfrac{3}{2}, \pm\dfrac{\pi}{3}\right)$ at $\dfrac{\pi}{6}\cdot$
4. $(0,0)$ at 0 and $\dfrac{\pi}{2}$; $\left(\dfrac{1}{2}\sqrt{3}, \dfrac{\pi}{3}\right),\ \left(\dfrac{1}{2}\sqrt{3}, \dfrac{2\pi}{3}\right)$ at $\tan^{-1}\dfrac{3\sqrt{3}}{5}.$
5. $(0,0)$ at 0; $\left(1, \pm\dfrac{\pi}{2}\right)$ at $\dfrac{\pi}{2}\cdot$ 6. Same as Answer 2 above.
7. $(0,0)$ at $\dfrac{\pi}{2}$; $\left(1, \pm\dfrac{\pi}{3}\right)$ at $\dfrac{\pi}{3}\cdot$ 8. $\left(\sqrt{8}, \dfrac{\pi}{12}\right),\ \left(\sqrt{8}, \dfrac{5\pi}{12}\right)$ at $\dfrac{\pi}{2}\cdot$

Exercise 77 (Page 266)

1. $TP = a\tan\theta,\ PN = a.$
2. $TP = -a\cot 2\theta\sqrt{\sec 2\theta},\ PN = a\sqrt{\sec 2\theta}.$
3. $TP = 2a\sin\dfrac{\theta}{2}\tan\dfrac{\theta}{2},\ PN = 2a\sin\dfrac{\theta}{2}\cdot$

4. $TP = -\frac{a}{2}\tan\frac{\theta}{2}\csc^3\frac{\theta}{2}$, $PN = \frac{a}{2}\csc^3\frac{\theta}{2}$.

5. $TP = a\tan\frac{\theta}{n}\sin^{n-1}\frac{\theta}{n}$, $PN = a\sin^{n-1}\frac{\theta}{n}$.

6. $TP = -a\tan\frac{\theta}{n}\csc^{n+1}\frac{\theta}{n}$, $PN = a\csc^{n+1}\frac{\theta}{n}$.

7. $ON = a$.

8. $OT = -1$.

9. $TP = \frac{\sqrt{1+b^2}}{b}r$, $OT = \frac{r}{b}$, $PN = \sqrt{1+b^2}\,r$, $ON = br$.

10. $r\cos\theta = 2$, $r\sin\theta = 3$.

11. $r\cos\theta = \pm 1$.

12. $r\cos\theta = 0$, $r\sin\theta = 0$.

13. $\theta = \pm\frac{\pi}{4}$.

15. $r\sin\left(\theta - \frac{\pi}{3}\right) = \frac{20}{3}$, $r\sin\left(\theta + \frac{\pi}{3}\right) = -\frac{20}{3}$.

16. $r\cos\theta = 0$.

17. $r\cos\left(\theta + \frac{\pi}{4}\right) = 3$, $r\sin\left(\theta + \frac{\pi}{4}\right) = -3$

18. $r\cos\theta = 2$, $r\sin\theta = 2$.

Exercise 79 (Page 273)

1. $v_r = 4$, $v_\theta = 8t$.

2. $v_r = 3\sin 3t$, $v_\theta = 3 - 3\cos 3t$.

3. $v_r = 6\cos 6t$, $v_\theta = 2\sin 6t$.

4. $v_r = -2\sin 2t$, $v_\theta = \cos 2t$.

5. $v_r = -\frac{2a\sin 4t}{\sqrt{\cos 4t}}$, $v_\theta = 2a\sqrt{\cos 4t}$.

6. $v_r = an\sin^{n-1}t\cos t$, $v_\theta = an\sin^n t$.

7. $v_r = a\sin t$, $v_\theta = b - a\cos t$.

8. $v_r = -an\cos nt$, $v_\theta = b - a\sin nt$.

9. $v = a\omega$.

10. $v = a\omega$.

11. $v = 1$.

12. $v = b$.

13. $v_r = br$, $v_\theta = \omega r$, $a_r = (b^2 - \omega^2)r$, $a_\theta = 2b\omega r$.

15. $a_r = -(b^2 + \omega^2)r$.

16. $a_r = (b^2 - \omega^2)r$.

Exercise 81 (Pages 280 to 281)

1. $\frac{dy}{dx} = -\frac{e^x\sin y + e^{-y}\sin x}{e^x\cos y + e^{-y}\cos x}$.

2. $\frac{dy}{dx} = \frac{\sec^2(x-y) - y}{\sec^2(x-y) + x}$.

3. $\frac{dy}{dx} = \frac{2x+y}{x-2y}$.

4. $\frac{dy}{dx} = \frac{x^2+y}{x-xy}$.

5. $\frac{dy}{dx} = \frac{\cos(2x-4y) + \sin(2x+4y)}{2[\cos(2x-4y) - \sin(2x+4y)]}$.

6. $\frac{dy}{dx} = -\frac{y(y + x\ln y)}{x(x + y\ln x)}$.

7. $\frac{dy}{dx} = -\frac{y(4x+1)}{2x(2x+1)}$.

8. $\frac{dy}{dx} = \frac{e^{-y} - e^y}{e^y + e^{-y}} = \frac{-2e^{-x}}{e^y + e^{-y}}$.

9. $\frac{dy}{dx} = \frac{2t}{3}$, $\frac{d^2y}{dx^2} = \frac{2}{9}$.

10. $\frac{dy}{dx} = \frac{2}{3t^2}$, $\frac{d^2y}{dx^2} = -\frac{2}{9t^5}$.

11. $\frac{dy}{dx} = 3t$, $\frac{d^2y}{dx^2} = \frac{3}{t}$.

12. $\frac{dy}{dx} = \frac{2}{t}$, $\frac{d^2y}{dx^2} = -\frac{2}{t^4}$.

13. $\frac{dy}{dx} = -\frac{3}{2}\cot t$, $\frac{d^2y}{dx^2} = -\frac{3}{4}\csc^3 t$.

14. $\frac{dy}{dx} = \frac{3}{2}\sin t$, $\frac{d^2y}{dx^2} = \frac{3}{4}\cos^3 t$.

15. $\frac{dy}{dx} = -4\sin t$, $\frac{d^2y}{dx^2} = -4$.

16. $\frac{dy}{dx} = -\frac{2\cos 2t}{\sin t}$, $\frac{d^2y}{dx^2} = -\frac{4\sin 2t\sin t + 2\cos 2t\cos t}{\sin^3 t}$.

17. $\frac{dy}{dx} = -\cos^2 t, \frac{d^2y}{dx^2} = 2\cos^3 t.$

18. $\frac{dy}{dx} = \frac{1}{2}e^{-t}, \frac{d^2y}{dx^2} = -\frac{1}{4}e^{-3t}.$

19. $\frac{dy}{dx} = -\frac{y}{x}, \frac{d^2y}{dx^2} = \frac{2y}{x^2}.$

20. $\frac{dy}{dx} = -\frac{y}{3x}, \frac{d^2y}{dx^2} = \frac{4y}{9x^2}.$

Exercise 82 (Page 283)

1. $\Delta y = 0.882, dy = 0.8.$
2. $\Delta y = 0.0175, dy = 0.0176.$
3. $\Delta y = 7.027, dy = 6.027.$
4. $\Delta y = 0.1, dy = 0.101.$
5. $\Delta y = 0.00860, dy = 0.00869.$
6. $\Delta y = -0.064, dy = -0.066.$
7. $dA = 2ah.$
8. $dA = 2\pi ah.$
9. $dA = \frac{ah}{2}\sqrt{3}.$
10. $dS = 8\pi ah.$
11. $dV = 4\pi a^2h.$
12. $dS = 2\pi bh.$
13. $dV = 2\pi abh.$
14. $dS = 8\pi ah.$
15. $dV = 6\pi a^2h.$
17. 10.1.
18. 0.0204.
19. 0.1414.
20. 4.96.
21. 8.944.
22. 0.09933.
23. 7.537.
24. 3.006.
25. 2.0043.
26. 0.8747.
27. For "$= 2$" read "$= 2/\sqrt{3}$." 0.5541
28. 0.7194.

Exercise 83 (Page 285)

1. $8 \pm 0.6.$
2. $0.5 \pm 0.12.$
3. $5 \pm 0.16.$
4. $0.4343 \pm 0.02.$
5. $0.9273 \pm 0.067.$
6. $0.7854 \pm 0.025.$
10. $\frac{dS}{S} = 0.04, S = 36\pi(1 \pm 0.04).$
11. $\frac{dV}{V} = 0.06, V = 36\pi(1 \pm 0.06).$
19. 14.
20. 22.
21. 16.
22. 5.

Exercise 84 (Page 287)

1. 1.4296.
2. 2.996.
3. 1.5571.
4. 0.7898.
5. 7.298.
6. 7.703.
7. 1.027.
8. 0.7390.
9. 1.0306.
10. 2.2789.
11. 1.4044.
12. 0.6346.
13. 3.0333.
14. For "$= 1.2$" read "$= 2.5$." 2.4746.
15. 0.5110.
16. 0.5671.
17. 1.2927.
18. 1.4546.
19. 0.7920.
20. 4.0955.

Exercise 86 (Page 297)

8. $\frac{11}{5}.$
9. 78.
10. $6 + \ln 2.$
11. $2(e^2 - 1).$
12. 1.
13. 2.
14. $e + 1/e - 2.$
15. $2 - \cos 1 - \sin 1.$
17. $\pi^2/2.$

Exercise 87 (Page 299)

1. $\frac{1}{4}(e^\pi - 1).$
2. $\pi^3/48.$
3. $\pi^5/320.$
4. $\pi^2/16.$
5. $\pi/20.$
6. $\pi/20.$
7. $\frac{1}{2}.$
8. $\frac{1}{3}.$
9. $3\pi/2.$
10. $19\pi/2.$
11. $\pi/2 - \frac{3}{4}\sqrt{3}.$
12. $\pi + \frac{3}{4}\sqrt{3}.$
13. $3\pi.$
14. $\pi - \frac{3}{2}\sqrt{3}.$
15. $2\pi + \frac{3}{2}\sqrt{3}.$
16. $\pi + 3\sqrt{3}.$

Exercise 88 (Page 304)

1. $\frac{1}{3}.$
2. $\frac{26}{3}.$
3. $\ln 3.$
4. $100\frac{13}{15}.$
5. $\frac{4}{15}.$
6. $\frac{2}{15}a^5.$
7. $\pi/16.$
8. $\frac{4}{3}.$
9. $3\pi a^2.$

Exercise 89 (Pages 310 to 311)

1. $\frac{1}{27}(13\sqrt{13} - 8).$
2. $\frac{61}{27}.$
3. $\frac{10}{3}.$
4. $\frac{3}{2} + \frac{1}{2}\ln 2.$
5. $\frac{17}{12}.$
6. $(a/2)(e^{h/a} - e^{-h/a}).$

11. 8. 12. 4. 13. 6.
14. 6. 15. $(\sqrt{5}/2)(1 - e^{-2\pi})$. 16. $\sqrt{5}\,(e^{\pi/2} - 1)$.
17. $\sqrt{2}\,(e^2 - 1)$. 18. 4π. 19. 80.
20. $\frac{3}{2}$. 21. $a^4 + a^2b^2 + b^4$. 22. $\dfrac{4n}{n+1}$.

Exercise 90 (Pages 317 to 318)

1. $(\pi/6)(17\sqrt{17} - 1)$. 2. $(\pi/18)(2\sqrt{2} - 1)$.
3. $(\pi/4)(e^2 - e^{-2} + 4)$. 4. 20π. 5. $4\pi\sqrt{2}$.
6. $(\pi/6)(5\sqrt{5} - 1)$. 7. 15π. 8. $10\pi/3$.
9. $\pi(\frac{15}{4} + \ln 2)$. 10. $4\pi\sqrt{2}$.
11. Use form 9 of Table 1 with $u = \sqrt{2}\cos\theta$, $A = -1$. $\sqrt{2}\,\pi \ln(1 + \sqrt{2})$.
12. 16π. 13. $8\pi^2$. 14. 64π. 15. $6\pi/5$.
16. 40π. 17. 55π. 18. $64\pi^2$. 19. $96\pi^2$.

Exercise 91 (Page 321)

1. 2. 2. Diverges. 3. 4. 4. Diverges.
5. Diverges. 6. 3. 7. $\frac{1}{8}$. 8. Diverges.
9. 1. 10. Diverges. 11. Diverges. 12. 1.
13. $\frac{9}{2}$. 14. 15. 15. π. 16. $\pi/2$.
17. 1. 18. $\frac{4}{3}$. 19. $\pi/2$. 20. 2π. 21. $\sqrt{2}$.

Exercise 92 (Page 325)

1. $-\dfrac{1}{3x^3} + \dfrac{3}{4x^4} - \dfrac{4}{5x^5} + C$. 2. $\ln x + \dfrac{3-4x}{2x^2} + C$.
3. $2x^{\frac{3}{2}}(x + 2) + C$. 4. $2x^{\frac{1}{2}}(2x - 5) + C$. 5. $\ln x + \dfrac{2}{x} - \dfrac{2}{x^2} + C$.
6. $\dfrac{x^2}{2} + 2x + \ln x + C$. 7. $\frac{1}{6}(3x + 4)^{\frac{3}{2}} + C$.
8. $-\frac{1}{2}(3 - 4x)^{\frac{1}{2}} + C$. 9. $\dfrac{1}{10 - 4x} + C$.
10. $\frac{1}{3}\ln(2 + 3x) + C$. 11. $\frac{1}{4}\ln(2x^2 + 1) + C$. 12. $\frac{1}{3}(4 + 3x^2)^{\frac{1}{2}} + C$.
13. $\frac{1}{3}(x^2 - 1)^{\frac{3}{2}} + C$. 14. $-\frac{1}{40}(1 - 2x^2)^{10} + C$. 15. $\sqrt{e^{2x} - 1} + C$.
16. $1/(e^{-x} + 1) + C$. 17. $\frac{2}{3}(\ln x)^{\frac{3}{2}} + C$. 18. $\frac{1}{2}(\ln x)^2 + C$.
19. $\frac{1}{10}\sin^5 2x + C$. 20. $-\frac{1}{8}\cos^4 2x + C$.
21. $\sec^3(x/3) + C$. 22. $\tan^4(x/4) + C$.

Exercise 93 (Page 329)

1. $\ln 2$. 2. $-\ln 2$. 3. $\frac{2}{5}\ln\sec(5x/2) + C$.
4. $-\frac{1}{3}\ln\csc(3x + 2) + C$. 5. $3\ln 2$.
6. $-\ln 2$. 7. $-\frac{1}{2}\ln(\csc 2x + \cot 2x) + C$.
8. $\frac{1}{2}\ln[\sec(2x - 3) + \tan(2x - 3)] + C$.
9. $\ln(1 + \sqrt{2})$. 10. $-\frac{1}{2}\ln 3 + C$.
11. $\ln(x + \sqrt{x^2 + 2}) + C$. 12. $\ln(x + \sqrt{x^2 - 3}) + C$.
13. $\dfrac{1}{4}\ln\dfrac{x-2}{x+2} + C$. 14. $\dfrac{1}{6}\ln\dfrac{x+3}{x-3} + C$.
15. $\frac{1}{3}\ln(3x + \sqrt{9x^2 - 5}) + C$. 16. $\frac{1}{10}\ln\dfrac{2x+5}{2x-5} + C$.

17. $\frac{1}{4}\ln(4x + \sqrt{16x^2 - 8}) + C.$

18. $\dfrac{1}{3\sqrt{2}}\ln\dfrac{x - \sqrt{2}}{x + \sqrt{2}} + C.$

19. $\ln 3.$

20. $\frac{1}{2}\ln\frac{4}{3}.$

Exercise 94 (Page 331)

1. $\sin^{-1}\dfrac{x}{3} + C.$
2. $\dfrac{1}{2}\sin^{-1}\dfrac{2x}{3} + C.$
3. $\dfrac{1}{5}\tan^{-1}\dfrac{x}{5} + C.$
4. $\dfrac{1}{6}\tan^{-1}\dfrac{3x}{2} + C.$
5. $\dfrac{1}{4}\sec^{-1}\dfrac{x}{4} + C.$
6. $\dfrac{1}{3}\sec^{-1}\dfrac{2x}{3} + C.$
7. $\dfrac{1}{3}\sin^{-1}\dfrac{3x}{4} + C.$
8. $\dfrac{1}{\sqrt{3}}\sin^{-1}\dfrac{\sqrt{3}\,x}{2} + C.$
9. $\dfrac{1}{6}\tan^{-1}\dfrac{2x}{3} + C.$
10. $\dfrac{1}{\sqrt{3}}\tan^{-1}\dfrac{\sqrt{3}\,x}{\sqrt{5}} + C.$
11. $\dfrac{1}{2}\sec^{-1}\dfrac{5x}{4} + C.$
12. $\dfrac{1}{2}\sec^{-1}\dfrac{\sqrt{3}\,x}{2} + C.$
13. $\pi/6.$
14. $\pi/18.$
15. $\pi/3.$
16. $\pi/6.$
17. $(\pi/12)\sqrt{3}.$
18. $\pi/16.$

Exercise 95 (Pages 333 to 334)

1. $\dfrac{1}{3}\tan^{-1}\dfrac{x - 2}{3} + C.$
2. $\sin^{-1}\dfrac{x + 3}{4} + C.$
3. $\dfrac{1}{2}\ln\dfrac{x + 2}{x + 4} + C.$
4. $\dfrac{1}{\sqrt{2}}\ln\left[(x - 1) + \sqrt{x^2 - 2x + \frac{5}{2}}\right] + C.$
5. $-\dfrac{1}{3}\tan^{-1}\dfrac{x - 4}{3} + C.$
6. $\ln\dfrac{x - 2}{x - 1} + C.$
7. $\sin^{-1}(x + 2) + C.$
8. $\dfrac{1}{\sqrt{5}}\ln\left[(x - 1) + \sqrt{x^2 - 2x - \frac{4}{5}}\right] + C.$
9. $-\dfrac{1}{\sqrt{8}}\ln\dfrac{8x - 1 + \sqrt{64 - 16x - 8x^2}}{x} + C.$
10. $\dfrac{1}{\sqrt{2}}\ln\dfrac{2 + \sqrt{2x^2 + 4x + 6}}{x + 1} + C.$
11. $\pi.$
12. $\pi/12.$
13. $-\frac{1}{4}\ln 3.$
14. $\ln(-1).$ For "$\sqrt{x^2 - 3x + 2}$" read "$\sqrt{3x - 2 - x^2}$." $\pi.$
15. $2\ln(x^2 + 2x + 5) + 3\tan^{-1}\dfrac{x + 1}{2} + C.$
16. $\ln(4x^2 - 4x - 3) - \dfrac{1}{2}\ln\dfrac{2x - 3}{2x + 1} + C = \dfrac{1}{2}\ln(2x - 3) + \dfrac{3}{2}\ln(2x + 1) + C.$
17. $\sqrt{x^2 + 2x} + 2\ln\left[(x + 1) + \sqrt{x^2 + 2x}\right] + C.$
18. $-\sqrt{4x - x^2} + 4\sin^{-1}\dfrac{x - 2}{2} + C.$
19. $-\sqrt{27 + 6x - x^2} + 3\sin^{-1}\dfrac{x - 3}{6} + C.$
20. $\frac{1}{2}\sqrt{4x^2 + 4x + 2} - \ln\left[(2x + 1) + \sqrt{4x^2 + 4x + 2}\right] + C.$

Exercise 96 (Page 335)

1. $-\frac{1}{4}\cos(4x - 7) + C.$
2. $-\frac{1}{3}\sin(4 - 3x) + C.$
3. $\frac{1}{3}\sec(3x - 2) + C.$
4. $-5\csc(x/5) + C.$
5. $\frac{2}{3}\tan(3x/2) + C.$
6. $-\frac{1}{2}\cot(2x - 4) + C.$
7. $\frac{1}{5}\tan 5x - x + C.$
8. $-\frac{4}{3}\cot 5x - x + C.$
9. $\frac{3}{2}\ln\sec(2x/3) + C.$
10. $\frac{5}{4}\ln[\sec(4x/5) + \tan(4x/5)] + C.$
11. $-\cot x + 2\ln\csc x + C.$
12. $x - \cot x + 2\ln(\cot x + \csc x) + C.$
13. $\dfrac{x}{2} - \dfrac{\sin 10x}{20} + C.$
14. $\dfrac{x}{2} + \dfrac{2}{5}\sin\dfrac{5x}{4} + C.$
15. $2\tan x - 2\sec x - x + C.$
16. $x - \sin^2 x + C.$
17. $-2\sqrt{2}\cos(x/2) + C.$
18. $\sqrt{2}\ln[\sec(x/2) + \tan(x/2)] + C.$
19. $\frac{3}{2}.$
20. $4.$
21. $\pi/8 + \frac{1}{4}.$
22. $\ln(2/\sqrt{3}).$

Exercise 97 (Pages 336 to 337)

1. $\frac{1}{3}e^{3x-4} + C.$
2. $\frac{1}{2}.$
3. $\frac{5}{7}e^{7x} + C.$
4. $\frac{1}{2}e^{x^2} + C.$
5. $\frac{1}{2}.$
6. $e^{\sin x} + C.$
7. $e^{\tan x} + C.$
8. $e^{e^x} + C.$
9. $\dfrac{9}{\ln 10}.$
10. $\dfrac{n^x}{\ln n} + \dfrac{x^{n+1}}{n+1} + C.$
11. $e^x - \dfrac{x^{e+1}}{e+1} + C.$
12. $\frac{1}{3}(e^2 - 1).$
13. $\dfrac{a^x e^x}{1 + \ln a} + C.$
14. $\dfrac{7^{8x}8^{7x}}{8\ln 7 + 7\ln 8} + C.$
15. $4(e^{x/4} - e^{-x/4}) + C.$
16. $\ln(e^x + 4) + C.$
17. $\frac{1}{2}(e^{2x} - e^{-2x} - 4x) + C.$
18. $\frac{1}{2}\ln(e^{2x} - e^{-2x}) + C.$
19. $-x + \ln(e^{2x} + 1).$
20. $-(x/2) + \frac{1}{2}\ln(e^{3x} + 2) + C.$

Exercise 98 (Pages 338 to 339)

1. $-\cos x + \dfrac{\cos^3 x}{3} + C.$
2. $\sin x - \dfrac{\sin^3 x}{3} + C.$
3. $\frac{1}{14}\sin^2 7x + C.$
4. $-\dfrac{\cos^3 x}{3} + \dfrac{\cos^5 x}{5} + C.$
5. $\dfrac{x}{8} - \dfrac{1}{2}\sin x + C.$
6. $\dfrac{3x}{8} - \dfrac{1}{4}\sin 2x + \dfrac{1}{32}\sin 4x + C.$
7. $2\sin\dfrac{x}{2} - \dfrac{2}{3}\sin^3\dfrac{x}{2} + C.$
8. $-\frac{1}{2}\cos 2x + \frac{1}{3}\cos^3 2x - \frac{1}{10}\cos^5 2x + C.$
9. $\sin^6(x/6) + C.$
10. $-2\cos^4(x/8) + C.$
11. $\frac{3}{5}\sin^{5/3} x + C.$
12. $-\frac{2}{3}\cos^{3/2} x + C.$
13. $\ln\sec x - (\sin^2 x)/2 + C.$
14. $2\sqrt{\sin x}\,(1 - \frac{1}{5}\sin^2 x) + C.$
15. $\frac{1}{4}\sin^4 x - \frac{1}{6}\sin^6 x + C.$
16. $\frac{5}{16}x + \frac{1}{4}\sin 2x + \frac{3}{64}\sin 4x - \frac{1}{48}\sin^3 2x + C.$
17. $\frac{5}{8}.$
18. For "$\dfrac{\pi}{6}$" read "$\dfrac{x}{6}$." $\dfrac{5}{4}.$
19. $3\pi/32.$
20. $\pi/64.$

Exercise 99 (Page 340)

1. $-\frac{1}{12}\sin 6x + \frac{1}{4}\sin 2x + C.$
2. $-\frac{1}{20}\cos 10x - \frac{1}{4}\cos 2x + C.$
3. $\frac{1}{14}\sin 7x + \frac{1}{6}\sin 3x + C.$
4. $-\frac{1}{16}\sin 8x + \frac{1}{12}$ in $6x + C.$
5. $-\frac{1}{6}\cos 3x - \frac{1}{2}\cos x + C.$
6. $\frac{1}{10}\sin 5x + \frac{1}{2}\sin x + C.$
7. $-\frac{1}{6}\cos(3x + 3) + \frac{1}{2}\cos(x + 3) + C.$

8. $\frac{1}{10} \sin (5x - 4) + \frac{1}{2} \sin (x - 4) + C.$
9. $-\frac{1}{14} \sin (7x - 3) + \frac{1}{6} \sin (3x - 3) + C.$
10. $-\frac{1}{4} \cos 2x - \frac{1}{4}x + C.$
11. $(x/4) + \frac{1}{8} \sin 2x + \frac{1}{16} \sin 4x + \frac{1}{24} \sin 6x + C.$
12. $-\frac{1}{4} \cos x - \frac{1}{14} \cos 7x + \frac{1}{52} \cos 13x + C.$
13. $\frac{1}{4} \sin x + \frac{1}{10} \sin 5x + \frac{1}{44} \sin 11x + C.$
14. $\frac{1}{4}x - \frac{1}{16} \sin 4x - \frac{1}{24} \sin 6x + \frac{1}{80} \sin 10x + \frac{1}{16} \sin 2x + C.$

Exercise 100 (Page 342)

1. $-\frac{1}{15} \cot^5 3x + C.$
2. $2 \tan^3 (x/6) + C.$
3. $\frac{1}{6} \sec^3 2x + C.$
4. $- \csc^5 (x/5) + C.$
5. $\frac{1}{2} \tan^2 x - \ln \sec x + C.$
6. $\frac{1}{6} \sec^3 2x - \frac{1}{2} \sec 2x + C.$
7. $- \cot^3 (x/3) - 3 \cot (x/3) + C.$
8. $- \cot^2 (x/2) + 2 \ln \csc (x/2) + C.$
9. $\frac{1}{4} \sec^4 x + C.$
10. $\frac{1}{5} \sec^5 x - \frac{1}{3} \sec^3 x + C.$
11. $\frac{3}{5} \tan^{\frac{5}{3}} x + C.$
12. $-2 \csc^{\frac{1}{2}} x + C.$
13. $-2 \cot^{\frac{1}{2}} x - \frac{2}{5} \cot^{\frac{5}{2}} x + C.$
14. $\sec x + \cos x + C.$
15. $\frac{1}{3} \tan^3 x - \tan x + x + C.$
16. $\frac{1}{3} \sec^2 x \tan x + \frac{2}{3} \tan x + C.$
17. $-\frac{1}{2} \cot^2 x + C.$
18. $\frac{1}{2} \sec x \tan x + \frac{1}{2} \ln (\sec x + \tan x) + C.$
19. 4.
20. Diverges. For "$\int_0^{\pi}$" read "$\int_{\pi}^{2\pi}$." $\frac{22}{15}$.
21. Diverges. For "$\int_0^{\pi/6}$" read "$\int_{\pi/6}^{\pi/2}$." $-\dfrac{1}{6} + \ln \dfrac{2}{\sqrt{3}}$.
22. Diverges. For "$\int_0^{\pi/4}$" read "$\int_{\pi}^{\pi/2}$," $\frac{2}{3}$.

Exercise 101 (Pages 344 to 345)

1. $\frac{2}{15} \sqrt{x - 1}\, (3x^2 + 4x + 8) + C.$
2. $x - \ln (2x + 3) + C.$
3. $\frac{4}{3}[x^{\frac{1}{4}} - \ln (1 + x^{\frac{3}{4}})] + C.$
4. $2 \tan^{-1} \sqrt{x} + C.$
5. $(x^2/2) - 4x + 12 \ln (x + 2) + 8/(x + 2) + C.$
6. $\frac{1}{180} \sqrt{2x - 3}\, (40x^4 - 240x^3 + 702x^2 - 621x + 567) + C.$
7. $\dfrac{x^2 + 8}{\sqrt{x^2 + 4}} + C.$
8. $\sqrt{x^2 - 4} - 2 \tan^{-1} \dfrac{\sqrt{x^2 - 4}}{2} + C.$
9. $\frac{4}{15} \sqrt{4 - \sqrt{x}}\, (3x - 4\sqrt{x} - 44) + C.$
10. $\dfrac{1 + x}{\sqrt{1 + 2x}} + C.$
11. $\frac{2}{9} \sqrt{x^3 + 8}\, (x^3 - 16) + C.$
12. $\frac{1}{3} \sqrt{x^2 - 4}\, (x^2 + 4) + C.$
13. $4 - 2 \ln 3.$
14. 3.
15. For "$\int_0^9$" read "$\int_0^3$." 9.
16. $\frac{172}{3}$.
17. $\frac{12}{5}$.
18. $\frac{44}{3}$.
19. $\frac{2}{15}(1 + \sqrt{2}).$
20. $\frac{40}{9}$.

Exercise 102 (Page 347)

1. $-\dfrac{1}{4x} \sqrt{4 - x^2} + C.$
2. $-\dfrac{1}{3x} \sqrt{3 + x^2} + C.$
3. $\dfrac{1}{16}\left(\tan^{-1} \dfrac{x}{2} + \dfrac{2x}{x^2 + 4}\right) + C.$
4. $\dfrac{1}{6}\left(\tan^{-1} \dfrac{x}{3} - \dfrac{3x}{x^2 + 9}\right) + C.$
5. $\dfrac{1}{125}\left(\dfrac{\sqrt{x^2 - 25}}{x}\right)^5 + C.$
6. $-\dfrac{x}{2\sqrt{x^2 - 2}} + C.$

7. $-\frac{1}{384}\frac{\sqrt{16-x^2}}{x^2}(x^2+8)+C.$

8. $\frac{1}{1875}\frac{\sqrt{25+x^2}}{x^3}(2x^2-25)+C.$

9. $\frac{1}{24}\frac{\sqrt{x^2-4}}{x^3}(x^2+2)+C.$

10. $\frac{x(4x^2-27)}{3(9-x^2)^{\frac{3}{2}}}+\sin^{-1}\frac{x}{3}+C.$

11. $\frac{1}{20}\left(\frac{x}{\sqrt{4-x^2}}\right)^5+C.$

12. $\frac{x}{\sqrt{25-x^2}}-\sin^{-1}\frac{x}{5}+C.$

13. $\frac{1}{27}\left(\frac{x}{\sqrt{x^2+9}}\right)^3+C.$

14. $\frac{1}{1215}\frac{x^3}{(9-x^2)^{\frac{5}{2}}}(8x^2-45)+C.$

15. $\frac{1}{1920}\frac{x^3(x^2+8)}{(x^2+16)^{\frac{5}{2}}}+C.$

16. $9\pi.$

17. $\frac{4}{45}.$ 18. $\pi.$ 19. $\frac{4}{75}.$ 20. $\frac{3}{80}.$

Exercise 103 (Pages 350 to 351)

1. $\frac{1}{2}xe^{2x}-\frac{1}{4}e^{2x}+C.$
2. $x\cos^{-1}x-\sqrt{1-x^2}+C.$
3. $x\ln x-x+C.$
4. $x\sin x+\cos x+C.$
5. $\frac{1}{2}x^2\sec^{-1}x-\frac{1}{2}\sqrt{x^2-1}+C.$
6. $-x^2e^{-x}-2xe^{-x}-2e^{-x}+C.$
7. $\frac{1}{4}x^4\ln x-\frac{1}{16}x^4+C.$
8. $-\frac{1}{3}x\cos 3x+\frac{1}{9}\sin 3x+C.$
9. $x^2\sin x+2x\cos x-2\sin x+C.$
10. $\frac{1}{3}x^3\tan^{-1}x-\frac{x^2}{6}+\frac{1}{6}\ln(1+x^2)+C.$
11. $\frac{1}{2}x^2\cot^{-1}x+\frac{x}{2}-\frac{1}{2}\tan^{-1}x+C.$
12. $\frac{x^2}{4}-\frac{x\sin x}{2}+\frac{\cos x}{2}+C.$
13. $\pi/8-\frac{1}{4}.$
14. $\frac{1}{4}.$
15. $\frac{1}{4}(e^2+1).$
16. $\frac{1}{2}e^{-\pi/2}.$

Exercise 104 (Pages 354 to 355)

1. $-\ln(x-1)-2\ln(x-2)+3\ln(x-3)+C.$
2. $-4\ln x+3\ln(x-1)+\ln(x+1)+C.$
3. $\ln(x-1)-5\ln(x-3)+4\ln(x-4)+C.$
4. $\frac{1}{5}\ln(x+1)+\frac{2}{5}\ln(x^2+4)-\frac{2}{5}\tan^{-1}(x/4)+C.$
5. $-\frac{4}{x}+3\ln x-2\ln(x-1)+C.$
6. $\ln(x-1)-\frac{1}{2}\ln(x^2+1)+\frac{1}{2}\tan^{-1}x+C.$
7. $4\ln x-2\ln(x^2+1)+C.$
8. $-\frac{5}{x-1}+12\ln(x-1)+4\ln(x^2+4)+14\tan^{-1}\frac{x}{2}+C.$
9. $\ln(x-1)+2\ln(x-2)-\ln(x-3)+C.$
10. $-2\ln(x+2)+\ln(x+1)+\ln(x-1)+C.$
11. $\ln x-\ln(x-1)+\ln(2x+1)+C.$
12. $\ln(x-1)+2\ln(x+2)+3\ln(x-3)+C.$
13. $-\frac{1}{x-1}-\frac{1}{x+1}-\ln(x-1)+\ln(x+1)+C.$
14. $4\ln(x-2)+\ln(x+2)-4\ln(x-1)-\ln(x+1)+C.$
15. $\ln(x-1)+\tan^{-1}x+C.$
16. $\ln(x-1)-\ln(x+1)-2\tan^{-1}x+C.$
17. $2\ln(x-2)+\ln(x^2+2x+4)+C.$
18. $2\ln x-\ln(x^2+4)+C.$
19. $\frac{3x+1}{2(x^2+1)}+\ln(x^2+1)+\frac{3}{2}\tan^{-1}x+C.$
20. $-\frac{1}{2(x^2+4)}+\tan^{-1}\frac{x}{2}+C.$

Exercise 105 (Page 356)

1. $\dfrac{1}{6}\ln\dfrac{x+3}{x-3}+C.$ **2.** $\dfrac{1}{3}\tan^{-1}\dfrac{x}{3}+C.$ **3.** $\dfrac{1}{2}\sin^{-1}\dfrac{2x}{3}+C.$

4. $\frac{1}{2}\ln\,(2x+\sqrt{4x^2-9})+C.$ **5.** $\dfrac{x}{2}\sqrt{25-4x^2}+\dfrac{25}{4}\sin^{-1}\dfrac{2x}{5}+C.$

6. $\dfrac{x}{2}\sqrt{4x^2-25}-\dfrac{25}{4}\ln\,(2x+\sqrt{4x^2-25})+C.$

7. $\frac{1}{4}(x^3+2)^{\frac{4}{3}}+C.$ **8.** $\frac{1}{3}(1+e^{2x})^{\frac{3}{2}}+C.$ **9.** $\frac{1}{2}\tan^{-1}x^2+C.$

10. $\frac{1}{2}\ln\,(x^2+\sqrt{x^2+1})+C.$ **11.** $\dfrac{2x-1}{(1-x)^2}+C.$

12. $\dfrac{1}{\ln 5}\,5^x+C.$ **13.** $2x+\dfrac{\sin 8x}{16}+C.$

14. $\dfrac{3x}{2}-\dfrac{\sin 6x}{12}+C.$ **15.** $-\frac{1}{13}e^{-2x}\,(2\sin 3x+3\cos 3x)+C.$

16. $\frac{1}{25}e^{-3x}\,(4\sin 4x-3\cos 4x)+C.$ **17.** $\dfrac{3}{8}x-\dfrac{\sin 2x}{4}+\dfrac{\sin 4x}{32}+C.$

18. $\frac{1}{3}$. **19.** $\pi/32$. **20.** $35\pi/256$.

Exercise 106 (Pages 361 to 362)

1. $\pi/4$. **2.** 1. **3.** $\ln\,(1+\sqrt{2})$. **4.** $\frac{2}{3}$.
5. $2\ln 2+1$. **6.** $\frac{1}{2}$. **7.** $\sqrt{3}$. **8.** $\ln\,(2+\sqrt{3})$.
9. $4-3\ln 3$. **10.** $\ln 5-\frac{4}{5}$. **11.** $2\pi-\frac{4}{3}$. **12.** 4π.
13. $\frac{4}{15}$. **14.** $\frac{4}{15}$. **15.** $\frac{16}{105}$. **16.** $\pi/16$. **17.** 9. **18.** π.
19. $a^2\,[\tan\,(\pi/8)+\frac{1}{3}\tan^3\,(\pi/8)]$. **20.** $\frac{1}{8}(4-\pi)$.
21. $\frac{1}{3}\sqrt{2}$. **22.** $3\pi/16$. **23.** $(\pi/4)(e^2-1)$. **24.** $\pi^2/4$.
25. Diverges to $+\infty$. **26.** $2\,(\ln 2)^2-4\ln 2+3$.
27. $7\pi/24$. **28.** Diverges to $+\infty$.
29. $2\pi^2$. **30.** $\frac{9}{2}\pi^2-8\pi$. **31.** $5\pi^2a^3$. **32.** $6\pi^3a^3$.

Exercise 107 (Page 365)

1. $2\sqrt{2}+\ln\,(2+\sqrt{2})$. **2.** $\ln\,(e^2+1)-1$.

3. $\ln\,(2+\sqrt{3})$. **4.** $\sqrt{2}\,\pi$. **5.** $4\sqrt{3}$.

6. $\ln\dfrac{\sqrt{1+e^2}-1}{\sqrt{2}-1}+\sqrt{1+e^2}-\sqrt{2}-1.$

7. $m[\sqrt{2}+\ln\,(1+\sqrt{2})]$.

8. $\pi\sqrt{1+4\pi^2}+\frac{1}{2}\ln\,(2\pi+\sqrt{1+4\pi^2})$.

9. $\pi\sqrt{1+4\pi^2}+\frac{1}{2}\ln\,(2\pi+\sqrt{1+4\pi^2})$.

10. $24+2\ln\,(3+2\sqrt{2})$.

11. $\pi[e\sqrt{1+e^2}+\ln\,(e+\sqrt{1+e^2})-\sqrt{2}-\ln\,(1+\sqrt{2})]$.

12. $(2\pi/3^9)(47\cdot 85\sqrt{85}+\frac{64}{5})$. **13.** $2\pi[\sqrt{2}+\ln\,(1+\sqrt{2})]$.

14. $\pi[\sqrt{2}+\ln\,(1+\sqrt{2})]$. **15.** $\frac{4}{5}\pi\sqrt{2}$.

16. $(8\pi/5)(3\sqrt{2}-4)$. **17.** $2\pi\sqrt{2}$.

18. $\pi ab+\dfrac{\pi a^2b}{\sqrt{a^2-b^2}}\sin^{-1}\dfrac{\sqrt{a^2-b^2}}{a}.$ **19.** $\pi a^2+\dfrac{\pi ab^2}{\sqrt{a^2-b^2}}\ln\dfrac{\sqrt{a^2-b^2}+a}{b}.$

20. $\pi a^2\left(\sqrt{6}-1-\frac{1}{\sqrt{2}}\ln\frac{2+\sqrt{3}}{1+\sqrt{2}}\right)$.

21. $\frac{32}{3}\pi a^2$.

Exercise 108 (Pages 368 to 369)

3. $\pi a/4$. 4. $2a/\pi$. 5. $2a/\pi$. 6. $2a^2/3$. 7. 9.
8. $\frac{12}{5}$. 9. $\frac{6}{7}\sqrt{2}$. 10. $\frac{1}{4}$. 11. $\frac{4}{7}$. 12. $2/\pi$.
13. $\frac{1}{2}$. 14. 0. 15. $\frac{1}{4}(\pi+\sqrt{3})$. 16. $\frac{1}{4}mA^2\omega^2$.
17. 0. 18. 6. 19. 24. 20. 18. 21. 1.
22. Diverges to $+\infty$. For "$\frac{1}{x^2+1}$" read "$\frac{1}{(x^2+1)^2}$." $\frac{\pi}{2}$.

Exercise 109 (Page 373)

2. $\left(\frac{4a}{3\pi},\frac{4b}{3\pi}\right)$. 3. $\left(\frac{a}{3},\frac{b}{3}\right)$. 4. $\left(\frac{\pi}{2}-1,\frac{\pi}{8}\right)$. 5. $(\frac{4}{5},\frac{3}{4})$.
6. $\left(\frac{\pi}{2},\frac{\pi}{8}\right)$. 7. $(\frac{8}{15},\frac{16}{105})$. 8. $(0,\frac{8}{5})$. 9. $(0,\frac{16}{45})$.
10. $(\frac{4}{5},\frac{4}{7})$ 11. $(\frac{12}{5},0)$. 12. $\left(0,\frac{3b}{5}\right)$. 13. (9,9).
14. (4,5) 15. $(\frac{8}{15},\frac{8}{21})$. 16. $(-\frac{5}{6}a,0)$. 17. $(0,\frac{20}{11})$.
18. $(\frac{17}{18},0)$. 19. $(0,-\frac{5}{6})$. 21. $\left(\pi a,\frac{5a}{6}\right)$.

Exercise 110 (Page 377)

2. $(\frac{2}{5},\frac{2}{5})$. 3. $\left(\frac{2}{5}\,\frac{\sqrt{2}+1}{2\sqrt{2}-1},0\right)$. 4. $(\frac{2}{5},0)$.
5. $\left[\frac{2a}{e+1},\frac{a(e^4+4e^2-1)}{4e(e^2-1)}\right]$. 6. $\left(\pi a,\frac{4a}{3}\right)$. 7. $\left[\frac{e^\pi-2}{5(e^{\pi/2}-1)},\frac{2e^\pi+1}{5(e^{\pi/2}-1)}\right]$.
8. $(4a/5,0)$. 10. $2m/3$. 11. $m/3$. 12. $5a/6$.
13. $\frac{1-3e^{-2}}{2(1-e^{-2})}$. 14. $\frac{\pi}{4}+\frac{1}{\pi}$. 15. $\frac{4}{3}\ln 4$. 16. $3a/8$.
17. $27a/16$. 18. $9a/16$. 19. $\frac{3}{8}$. 20. $15\pi/256$.
21. $\frac{1}{2}(x_1+x_2)$. 22. $\frac{e^2+4-3e^{-2}}{e^2+4-e^{-2}}$. 23. $\frac{2m}{5}\,\frac{\sqrt{2}-1}{2\sqrt{2}-1}$.

Exercise 111 (Pages 379 to 380)

1. $\left(\frac{7a}{5},\frac{7a}{5}\right)$. 2. $(\frac{41}{11},\frac{41}{11})$. 3. $\left(a,\frac{6\pi+28}{3\pi+24}\,a\right)$.
4. $\left(a,\frac{6+2\pi}{6+\pi}\,a\right)$. 5. $\left(\frac{3a}{5},0\right)$. 6. $(a,0)$.
7. $\left(-\frac{a}{3},0\right)$. 8. 4.6 in. from the base. 18. $24\pi a^3$.
19. $48\pi a^2$. 20. $30\pi a^3$. 21. $9\pi a^2\sqrt{10}$.

Exercise 112 (Page 385)

1. $\frac{64}{15}=\frac{4}{5}A$. 2. $\frac{128}{105}\sqrt{2}=\frac{32}{35}A$.
3. $\frac{1}{30}=\frac{2}{5}A$. 4. $\frac{8}{5}=\frac{6}{5}A$.

5. $\dfrac{e^3 - 1}{9} = \dfrac{e^2 + e + 1}{3} A.$

6. $\frac{4}{3} = \frac{4}{3}A.$

7. $\frac{4092}{15} = \frac{8184}{225}A.$

8. $\pi = A.$

15. $I_y = \dfrac{a^3b}{12} = \dfrac{a^2}{6} A,\ I_x = \dfrac{ab^3}{4} = \dfrac{b^2}{2} A,\ I_0 = \dfrac{a^2 + 3b^2}{6} A.$

16. $I_y = \dfrac{a^3b}{12} = \dfrac{a^2}{6} A,\ I_x = \dfrac{11}{12} ab^3 = \dfrac{11}{6} b^2A,\ I_0 = \dfrac{a^2 + 11b^2}{6} A.$

17. $I_y = \frac{1}{20} = \frac{3}{10}A,\ I_x = \frac{1}{28} = \frac{3}{14}A,\ I_0 = \frac{18}{35}A.$

18. $I_y = \frac{1}{36} = \frac{5}{18}A,\ I_x = \frac{1}{44} = \frac{5}{22}A,\ I_0 = \frac{50}{99}A.$

19. $I_y = \frac{3}{28} = \frac{5}{14}A,\ I_x = \frac{3}{52} = \frac{5}{26}A,\ I_0 = \frac{50}{91}A.$

20. $\frac{3}{2}\pi a^4 = \frac{3}{2}a^2A.$

21. $\dfrac{ab}{12} (a^2 + b^2) = \dfrac{1}{6} (a^2 + b^2)A.$

22. $\dfrac{\pi}{2} (b^4 - a^4) = \dfrac{1}{2} (a^2 + b^2)A.$

23. $\dfrac{3\pi}{4} = \dfrac{3\pi}{4} A.$

24. $\dfrac{3\pi}{64} = \dfrac{3}{8} A.$

25. $\dfrac{\pi}{32} = \dfrac{3}{8} A.$

Exercise 113 (Pages 388 to 389)

1. $\dfrac{\pi}{16} \rho(e^8 - 1) = \dfrac{1}{8} \dfrac{e^8 - 1}{e^2 - 1} M.$

2. $\dfrac{\pi}{14} \rho = \dfrac{2}{7} M.$

3. $\frac{3}{32}\pi^2\rho = \frac{3}{8}M.$

4. $\frac{2}{3}\pi\rho = \frac{2}{3}M.$

5. $\frac{4}{105}\pi\rho = \frac{16}{105}M.$

6. $\frac{8}{3}\pi\rho = (8/3\pi)M.$

7. $\frac{4096}{15}\pi\rho = \frac{32}{15}M.$

8. $\frac{1}{21}\pi\rho = \frac{10}{21}M.$

9. $\frac{64}{315}\pi\rho = \frac{8}{21}M.$

10. $\frac{8}{15}\pi\rho = \frac{2}{5}M.$

14. $\frac{7}{5}Ma^2.$

15. $\frac{13}{10}Ma^2.$

16. $\frac{3}{2}Ma^2.$

17. $\frac{1}{2}M(3a^2 + c^2).$

18. $\frac{1377}{4}a^4 = \frac{51}{4}a^2A.$

19. $1178\frac{2}{3}$ in.^{4} = $\frac{884}{33}A$ in.^{2}

20. $\frac{75}{4}\pi a^4 = \frac{15}{4}a^2A.$

Exercise 114 (Pages 395 to 396)

1. $F = 192w = 6$ tons, $x_c = 12\frac{1}{9}$ ft.

2. $F = 40\pi w = \dfrac{5\pi}{4}$ tons, $x_c = 10.1$ ft.

3. $F = w\left(20\pi - \dfrac{16}{3}\right) = \left(\dfrac{5\pi}{8} - \dfrac{1}{6}\right)$ tons, $x_c = \dfrac{303\pi - 160}{30\pi - 8}$ ft.

4. $F = w\left(20\pi + \dfrac{16}{3}\right) = \left(\dfrac{5\pi}{8} + \dfrac{1}{6}\right)$ tons, $x_c = \dfrac{303\pi + 160}{30\pi + 8}$ ft.

5. $F = \frac{8}{15}w = \frac{1}{60}$ ton, $x_c = \frac{4}{7}$ ft.

6. $F = \dfrac{8}{3} w_1(8\pi + 9\sqrt{3}) = \dfrac{1}{15} (8\pi + 9\sqrt{3})$ tons, $x_c = \dfrac{32\pi + 27\sqrt{3}}{8\pi + 9\sqrt{3}}.$

7. 2 ft.-lb.

8. 4×10^{-7} ft.-lb.

9. $2\pi I.$

10. 5,250 ft.-lb.

11. 1,100 ft.-lb.

12. $200\pi w = \frac{25}{4}\pi$ tons.

13. $12\pi w = \frac{3}{8}\pi$ tons.

14. $108\pi w = \frac{27}{8}\pi$ tons.

15. $4\pi w = \pi/8$ tons.

16. 346.6 ft.-lb.

17. 463.1 ft.-lb.

18. 670.6 ft.-lb.

19. 639.3 ft.-lb.

20. 605.1 ft.-lb.

21. 39.38 ft.-lb.

22. $\dfrac{Km_0M}{a(a + L)}.$

23. $\dfrac{2Km_0M}{a\sqrt{4a^2 + L^2}}.$

24. $\dfrac{2Km_0M}{\pi a^2}.$

25. $\dfrac{Km_0M}{\pi a^2} \ln (1 + \sqrt{2}).$

Exercise 115 (Page 400)

1. $3 + 9 + 27 = 39.$

2. $1 + \frac{1}{2} + \frac{1}{3} = 1\frac{5}{6}.$

3. $-2 + 4 - 8 = -6.$

4. $1 - \frac{1}{2} + \frac{1}{3} = \frac{5}{6}.$

5. $-\frac{1}{3} + 0 + \frac{1}{5} = -\frac{2}{15}.$

6. $1 + \frac{1}{4} + \frac{1}{9} = 1\frac{13}{16}.$

7. $-1 + \frac{1}{2} - \frac{1}{6} = -\frac{2}{3}.$

8. $2 + 2 + \frac{4}{3} = 5\frac{1}{3}$.

9. $2 + 5 + 8 = 15$, $S_n = \frac{n}{2}(3n + 1)$.

10. $4 + 5 + 6 = 15$, $S_n = \frac{n}{2}(n + 7)$.

11. $10 + 8 + 6 = 24$, $S_n = n(11 - n)$.

12. $6 + 3 + 0 = 9$, $S_n = \frac{3}{2}n(5 - n)$.

13. $1 + 2 + 4 = 7$, $S_n = 2^n - 1$.

14. $1 - \frac{1}{3} + \frac{1}{9} = \frac{7}{9}$, $S_n = \frac{1}{4}[3 + (-\frac{1}{3})^{n-1}]$.

15. $5 - 5 + 5 = 5$, $S_n = \frac{5}{2}[1 + (-1)^{n-1}]$.

16. $9 + 3 + 1 = 13$, $S_n = \frac{27}{2}[1 - (\frac{1}{3})^n]$.

Exercise 118 (Pages 411 to 412)

21. Divergent. 22. Convergent. 23. Divergent. 24. Convergent.
25. Divergent. 26. Convergent. 27. Convergent. 28. Convergent.
29. Convergent. 30. Convergent. 31. Convergent. 32. Convergent.

Exercise 119 (Pages 414 to 415)

1. Convergent. 2. Divergent. 3. Convergent. 4. Convergent.
5. Convergent. 6. Convergent. 7. Divergent. 8. Convergent.
9. Convergent. 10. Convergent. 11. Convergent. 12. Convergent.
13. Convergent. 14. Divergent.

Exercise 120 (Pages 419 to 420)

11. In Prob. 1 for "4^4" read "3^3." 0.2500.

12. 0.3494.

13. 0.7762.

Exercise 123 (Pages 429 to 430)

8. 0.004902. 9. 0.002577. 10. 5.099. 11. 0.06742.
12. 2.962. 13. 2.031. 14. 1.414. 15. 1.260.

Exercise 126 (Pages 441 to 442)

18. $\ln 3 = 1.09861$, $\ln 9 = 2.19722$, $\ln 10 = 2.30259$.

Exercise 127 (Pages 446 to 447)

1. $e^2\left[1 + (x - 2) + \frac{1}{2!}(x - 2)^2 + \frac{1}{3!}(x - 2)^3 + \cdots\right]$.

2. $\frac{1}{2}\sqrt{3} + \frac{1}{2}\left(x - \frac{\pi}{3}\right) - \frac{\sqrt{3}}{2 \cdot 2!}\left(x - \frac{\pi}{3}\right)^2 - \frac{1}{2 \cdot 3!}\left(x - \frac{\pi}{3}\right)^3 + \cdots$.

3. $\frac{\sqrt{2}}{2}\left[1 - \left(x - \frac{\pi}{4}\right) - \frac{1}{2!}\left(x - \frac{\pi}{4}\right)^2 + \frac{1}{3!}\left(x - \frac{\pi}{4}\right)^3 + \cdots\right]$.

4. $1 - (x - 1) + (x - 1)^2 - (x - 1)^3 + \cdots$.

5. $\frac{1}{3} - \frac{1}{3^2}(x - 2) + \frac{1}{3^3}(x - 2)^2 - \frac{1}{3^4}(x - 2)^3 + \cdots$.

6. $2 + \frac{1}{4}(x - 3) - \frac{1}{64}(x - 3)^2 + \frac{1}{512}(x - 3)^3 - \cdots$.

7. $(x - 1) - \frac{1}{2}(x - 1)^2 + \frac{1}{3}(x - 1)^3 - \frac{1}{4}(x - 1)^4 + \cdots$.

8. $\frac{\pi}{4} + \frac{1}{2}(x - 1) - \frac{1}{4}(x - 1)^2 + \frac{1}{12}(x - 1)^3 - \cdots$.

9. $1 - \frac{1}{2}(x - 1) + \frac{3}{8}(x - 1)^2 - \frac{5}{16}(x - 1)^3 + \cdots$.

10. $1 + \frac{5}{2}(x - 1) + \frac{15}{8}(x - 1)^2 + \frac{15}{6}(x - 1)^3 - \cdots$.

16. 26,903. 17. 0.006097. 18. 0.71934. 19. 0.68200.
20. 0.74194. 21. 2.031. 22. 2.9625. 23. 0.010204.

Exercise 129 (Page 455)

1. 32. **2.** e^3. **3.** 1. **4.** -1. **5.** 1.
6. $-\pi$. **7.** 2. **8.** $-\frac{1}{2}\sqrt{3}$. **9.** -1. **10.** $\frac{1}{3}$.
11. 2. **12.** 1. **15.** $-\frac{1}{2}$. **16.** $-\frac{3}{2}$.
17. $\frac{1}{3}$. **18.** -1. **19.** $\frac{1}{6}$. **20.** $\frac{1}{2}$.

Exercise 130 (Page 458)

1. 0. **2.** 0. **3.** $-\frac{1}{4}$. **4.** 3. **5.** $-\frac{1}{3}$. **6.** $\frac{5}{3}$.
7. 1. **8.** $-\frac{5}{3}$. **9.** 0. **10.** 1. **11.** 5. **12.** 7.
13. 0. **14.** 0. **15.** $\frac{1}{2}$. **16.** 0. **17.** e^{ab}.
18. e^{ab}. **19.** 1. **20.** 1. **21.** $e^{2/\pi}$. **22.** 1.

Exercise 131 (Pages 465 to 466)

1. $S_4 = 1.100, \ln 3 = 1.099$. **2.** $S_4 = 0.5236 = \pi/6$.
3. $S_4 = 0.7854 = \pi/4$. **4.** $S_4 = 0.1484 = \ln 1.16$.
5. $S_4 = 1.2111 = 2\sqrt{13} - 6$. **6.** $S_4 = 0.5526, \frac{1}{2}\tan^{-1} 2 = 0.5536$.
7. $S_4 = 32.4949 = \frac{16}{3}(17\sqrt{17} - 64)$.
8. $S_4 = 0.8042, \frac{1}{2}\ln 5 = 0.8047$. **9.** $S_4 = 0.6321 = 1 - e^{-1}$.
10. $S_4 = 0.6381, 2/\pi = 0.6367$.
16. $S_4 = 3.239, S_8 = 3.2411, I = 3.2412$.
17. $S_4 = 46.26, S_8 = 46.156, I = 46.150$.
18. $S_4 = 54.59, S_8 = 55.47, I = 55.52$. **19.** $S_4 = 8.7152, S_8 = 8.7155 = I$.
20. $S_4 = 0.6432, S_8 = 0.6431 = I$.
21. $S_4 = 0.4485, S_8 = 0.4470, I = 0.4469$.

Exercise 132 (Pages 469 to 470)

11. $x - \frac{1}{4}\sin 4x$. **12.** $\frac{1}{4}\sin 4x + 2\sin 2x + 3x$.
13. $\frac{1}{4}\sin 4x - 2\sin 2x + 3x$.
21. $$\int e^{ax}\cos bx\,dx = \frac{1}{a^2 + b^2} e^{ax}(a\cos bx + b\sin bx) + C,$$
$$\int e^{ax}\sin bx\,dx = \frac{1}{a^2 + b^2} e^{ax}(a\sin bx - b\cos bx) + C.$$

Exercise 133 (Pages 422 to 473)

15. $\cosh x = \frac{5}{4}, \tanh x = \frac{3}{5}, \coth x = \frac{5}{3}, \operatorname{sech} x = \frac{4}{5}, \operatorname{csch} x = \frac{4}{3}$.
16. $\sinh x = \pm\frac{12}{5}, \tanh x = \pm\frac{13}{5}, \coth x = \pm\frac{13}{12}, \operatorname{sech} x = \frac{5}{13}, \operatorname{csch} x = \pm\frac{5}{12}$.
17. $\sinh x = \frac{4}{3}, \cosh x = \frac{5}{3}, \coth x = \frac{5}{4}, \operatorname{sech} x = \frac{3}{5}, \operatorname{csch} x = \frac{3}{4}$.
18. $\sinh x = \pm\frac{3}{4}, \cosh x = \frac{5}{4}, \tanh x = \pm\frac{3}{5}, \coth x = \pm\frac{5}{3}, \operatorname{csch} x = \pm\frac{4}{3}$.
19. $\frac{77}{36}$. **20.** $\frac{24}{25}$.

Exercise 135 (Pages 479 to 480)

1. $\dfrac{\partial u}{\partial x} = 3x^2 + 2y, \dfrac{\partial u}{\partial y} = 2x + 6y$.

2. $\dfrac{\partial u}{\partial x} = x(x^2 + y^2)^{-\frac{1}{2}}, \dfrac{\partial u}{\partial y} = y(x^2 + y^2)^{-\frac{1}{2}}$.

3. $\dfrac{\partial u}{\partial x} = \dfrac{x}{x^2 + y^2}, \dfrac{\partial u}{\partial y} = \dfrac{y}{x^2 + y^2}$. **4.** $\dfrac{\partial u}{\partial x} = 2e^{2x}\sin 3y, \dfrac{\partial u}{\partial y} = 3e^{2x}\cos 3y$.

5. $\dfrac{\partial u}{\partial x} = -2xe^{-x^2-y^2}, \dfrac{\partial u}{\partial y} = -2ye^{-x^2-y^2}$.

6. $\dfrac{\partial u}{\partial x} = -4e^{-2y} \sin 4x, \dfrac{\partial u}{\partial y} = -2e^{-2y} \cos 4x.$

7. $\dfrac{\partial u}{\partial x} = 3ye^{3xy}, \dfrac{\partial u}{\partial y} = 3xe^{3xy}.$

8. $\dfrac{\partial u}{\partial x} = 2 \cos (2x + 3y), \dfrac{\partial u}{\partial y} = 3 \cos (2x + 3y).$

9. $\dfrac{\partial u}{\partial x} = -(2x + y) \sin (x^2 + xy), \dfrac{\partial u}{\partial y} = -x \sin (x^2 + xy).$

10. $\dfrac{\partial u}{\partial x} = \dfrac{y}{1 + x^2y^2}, \dfrac{\partial u}{\partial y} = \dfrac{x}{1 + x^2y^2}.$

11. $\dfrac{\partial u}{\partial x} = y^2z^2, \dfrac{\partial u}{\partial y} = 2xyz^2, \dfrac{\partial u}{\partial z} = 2xy^2z.$

12. $\dfrac{\partial u}{\partial x} = 2y - 4z, \dfrac{\partial u}{\partial y} = 2x + 3z, \dfrac{\partial u}{\partial z} = 3y - 4x.$

13. $\dfrac{\partial u}{\partial x} = -x(x^2 + y^2 + z^2)^{-\frac{3}{2}}, \dfrac{\partial u}{\partial y} = -y(x^2 + y^2 + z^2)^{-\frac{3}{2}},$

$\dfrac{\partial u}{\partial z} = -z(x^2 + y^2 + z^2)^{-\frac{3}{2}}.$

14. $\dfrac{\partial u}{\partial x} = \sin (2y + z^2), \dfrac{\partial u}{\partial y} = 2x \cos (2y + z^2), \dfrac{\partial u}{\partial z} = 2xz \cos (2y + z^2).$

15. $\dfrac{\partial u}{\partial x} = ye^{2z}, \dfrac{\partial u}{\partial y} = xe^{2z}, \dfrac{\partial u}{\partial z} = 2xye^{2z}.$

16. $\dfrac{\partial u}{\partial x} = z \ln y, \dfrac{\partial u}{\partial y} = \dfrac{xz}{y}, \dfrac{\partial u}{\partial z} = x \ln y.$

17. $\dfrac{\partial K}{\partial a} = \dfrac{b \sin C}{2}, \dfrac{\partial K}{\partial C} = -\dfrac{ab \cos C}{2}.$

18. $\dfrac{\partial c}{\partial a} = \dfrac{\sin C}{\sin A}, \dfrac{\partial c}{\partial C} = \dfrac{a \cos C}{\sin A}.$

19. $\dfrac{\partial c}{\partial a} = \dfrac{a - b \cos C}{\sqrt{a^2 + b^2 - 2ab \cos C}}, \dfrac{\partial c}{\partial C} = \dfrac{ab \sin C}{\sqrt{a^2 + b^2 - 2ab \cos C}}.$

20. $\dfrac{\partial K}{\partial a} = \dfrac{2a \sin B \sin C}{\sin (B + C)}, \dfrac{\partial K}{\partial C} = \dfrac{a^2 \sin^2 B}{\sin^2 (B + C)}.$

Exercise 136 (Page 483)

1. $8x^2 + 32xy + 6y^2 = 384t^2.$
2. $(2ty + 3t^2x) \cos xy = 5t^4 \cos t^5.$
3. $20xye^{2t} - 10x^2e^{-2t} = 10e^{2t}.$
4. $2y \cos t - 2x \sin t = 2 \cos 2t.$
5. $-\dfrac{2y}{x^2} e^{2t} + \dfrac{4}{x} e^{4t} = 2e^{2t}.$
6. $\dfrac{4x + 6y}{x^2 + y^2} = \dfrac{2}{t}.$
7. $30\pi.$
8. $-\frac{2}{5}.$
9. $44\pi/5.$
10. $-\frac{14}{9}.$
11. 8.
12. $-\frac{1}{6}.$
13. $\frac{11}{2}.$
14. $24/\sqrt{29}.$
15. $64/\sqrt{229}.$

Exercise 137 (Pages 484 to 485)

1. $\dfrac{\partial u}{\partial s} = 8s, \dfrac{\partial u}{\partial t} = -2t.$

2. $\dfrac{\partial u}{\partial s} = \dfrac{15}{2x + 3y}, \dfrac{\partial u}{\partial t} = \dfrac{-2}{2x + 3y}.$

3. $\dfrac{\partial u}{\partial s} = 6s^2 + 6t^2, \dfrac{\partial u}{\partial t} = 12st.$

4. $\dfrac{\partial u}{\partial s} = 2xy + x^2 + 6y, \dfrac{\partial u}{\partial t} = 2xy.$

5. $\dfrac{\partial u}{\partial s} = 4e^{2s}, \dfrac{\partial u}{\partial t} = 4e^{2t}.$

6. $\dfrac{\partial u}{\partial s} = 4 \cos (2x + 3y), \dfrac{\partial u}{\partial t} = \cos (2x + 3y).$

7. $\dfrac{\partial u}{\partial s} = 0, \dfrac{\partial u}{\partial t} = -12 \sin 12t.$

8. $\dfrac{\partial u}{\partial s} = 2te^{2st}, \dfrac{\partial u}{\partial t} = 2se^{2st}.$

10. $(x\,dy + y\,dx) \cos (xy).$

11. $-(x\,dy + y\,dx) \sin (xy).$

12. $(x\,dy + y\,dx)e^{xy}$. **14.** $\dfrac{x\,dy - y\,dx}{x^2}\cos\dfrac{y}{x}$.

15. $\dfrac{x\,dy - y\,dx}{x^2}e^{y/x}$. **16.** $\dfrac{x\,dy - y\,dx}{x\sqrt{x^2 - y^2}}$. **18.** $\dfrac{2\,dx + 3\,dy}{2x + 3y}$.

19. $5(4x - 5y)^4(4\,dx - 5\,dy)$. **20.** $(6\,dx - 3\,dy)\sec^2(6x - 3y)$.

Exercise 138 (Page 486)

1. $36t^5$. **2.** $4e^t$. **3.** $2\cos 4t + 2\sin 4t$.

4. $6\cos 6t$. **5.** $18t$. **6.** $\dfrac{3}{1 + 9t^2}$.

7. $4x^3$. **8.** $21x^2$. **9.** $2\tan^{-1}x + \dfrac{2x}{1 + x^2}$.

10. $-6\sin 6x$. **11.** $8\cos 8x$. **12.** $2e^{2x}$.

Exercise 139 (Page 489)

1. $\dfrac{ye^{xy} - 1}{2 - xe^{xy}}$. **2.** $\dfrac{2 - y^2(2x + 3)}{2xy(2x + 3) - 3}$. **3.** $\dfrac{4 + 4x^2y^2 - y}{5 + 5x^2y^2 + x}$.

4. $\dfrac{2xy + 2\cos(2x - y)}{\cos(2x - y) - x^2}$. **5.** $\dfrac{e^x\cos y - y}{x + e^x\sin y}$.

6. $\dfrac{2x + y}{x - 2y}$. **7.** $\dfrac{\partial z}{\partial x} = \dfrac{4x}{3z}, \dfrac{\partial z}{\partial y} = \dfrac{5y}{3z}$.

8. $\dfrac{\partial z}{\partial x} = \dfrac{1}{10z}, \dfrac{\partial z}{\partial y} = -\dfrac{4y}{5z}$. **9.** $\dfrac{\partial z}{\partial x} = -\dfrac{2x}{5z}, \dfrac{\partial z}{\partial y} = -\dfrac{3y}{5z}$.

10. $\dfrac{\partial z}{\partial x} = \dfrac{4x}{10z}, \dfrac{\partial z}{\partial y} = -\dfrac{1}{10z}$. **11.** $\dfrac{\partial z}{\partial x} = \dfrac{e^{x+2y+z}}{1 - e^{x+2y+z}}, \dfrac{\partial z}{\partial y} = \dfrac{2e^{x+2y+z}}{1 - e^{x+2y+z}}$.

12. $\dfrac{\partial z}{\partial x} = \dfrac{\partial z}{\partial y} = \dfrac{e^z\cos(x + y)}{1 - e^z\sin(x + y)}$. **13.** $\dfrac{\partial z}{\partial x} = \dfrac{\partial z}{\partial y} = \dfrac{1}{x + y + z - 1}$.

14. $\dfrac{\partial z}{\partial x} = \dfrac{-2\sin(2x - y + 3z)}{3\sin(2x - y + 3z) + 1}, \dfrac{\partial z}{\partial y} = \dfrac{\sin(2x - y + 3z)}{3\sin(2x - y + 3z) + 1}$.

15. For "= 2" read "= 4." $\dfrac{dx}{dt} = 0, \dfrac{dy}{dt} = 4$.

16. $\dfrac{dy}{dt} = -\dfrac{48}{5}, \dfrac{dz}{dt} = \dfrac{27}{5}$. **17.** $\dfrac{dx}{dt} = -6, \dfrac{dz}{dt} = -9$.

18. $\dfrac{dy}{dt} = -\dfrac{15}{4}, \dfrac{dz}{dt} = 0$. **19.** $\dfrac{dx}{dt} = 4, \dfrac{dz}{dt} = 16$.

Exercise 140 (Page 491)

1. 0.035, 0.27 per cent. **2.** 0.20, 0.65 per cent. **3.** 0.0023, 0.20 per cent.
4. 0.055. **5.** 0.069. **6.** 0.061. **7.** 0.013. **13.** 0.035.
14. 0.035. **15.** 0.06. **16.** 0.05. **17.** 0.0125. **18.** 0.02.

Exercise 141 (Pages 493 to 494)

1. $\dfrac{\partial^2 u}{\partial x^2} = 4, \dfrac{\partial^2 u}{\partial x\,\partial y} = 4, \dfrac{\partial^2 u}{\partial y^2} = 10$.

2. $\dfrac{\partial^2 u}{\partial x^2} = 24(2x + 3y), \dfrac{\partial^2 u}{\partial x\,\partial y} = 36(2x + 3y), \dfrac{\partial^2 u}{\partial y^2} = 54(2x + 3y)$.

3. $\dfrac{\partial^2 u}{\partial x^2} = y^2e^{xy}, \dfrac{\partial^2 u}{\partial x\,\partial y} = (xy + 1)e^{xy}, \dfrac{\partial^2 u}{\partial y^2} = x^2e^{xy}$.

4. $\dfrac{\partial^2 u}{\partial x^2} = -9 \sin (3x - y), \dfrac{\partial^2 u}{\partial x \, \partial y} = 3 \sin (3x - y), \dfrac{\partial^2 u}{\partial y^2} = - \sin (3x - y).$

5. $\dfrac{\partial u^2}{\partial x^2} = \dfrac{-2x^2 - 2y^2}{(x^2 - y^2)^2}, \dfrac{\partial^2 u}{\partial x \, \partial y} = \dfrac{-4xy}{(x^2 - y^2)^2}, \dfrac{\partial^2 u}{\partial y^2} = \dfrac{-2x^2 - 2y^2}{(x^2 - y^2)^2}.$

6. $\dfrac{\partial^2 u}{\partial x^2} = e^x \cos y, \dfrac{\partial^2 u}{\partial x \, \partial y} = -e^x \sin y, \dfrac{\partial^2 u}{\partial y^2} = -e^x \cos y.$

Exercise 142 (Page 500)

1. $2x - y - \dfrac{4}{3}x^3 + 2x^2y - xy^2 + \dfrac{y^3}{6} + \cdots.$

2. $1 - \dfrac{x^2}{2} - 2xy - 2y^2 + \cdots.$

3. $1 + xy + \frac{1}{2}x^2y^2 + \frac{1}{6}x^3y^3 + \cdots.$

4. $1 + x + \frac{1}{2}x^2 - \frac{1}{2}y^2 + \frac{1}{6}x^3 - \frac{1}{2}xy^2 + \cdots.$

5. $x - xy - \dfrac{x^3}{6} + \dfrac{xy^2}{2} - \cdots.$

6. $1 + \dfrac{x^2y^2}{2} - \cdots.$

7. $y + xy - \dfrac{y^2}{2} + \dfrac{x^2y}{2} - \dfrac{xy^2}{2} + \dfrac{y^3}{3} - \cdots.$

8. $1 + \dfrac{x}{2} - \dfrac{y}{2} - \dfrac{x^2}{8} - \dfrac{xy}{4} + \dfrac{3y^2}{8} + \dfrac{x^3}{16} + \dfrac{x^2y}{16} + \dfrac{3xy^2}{16} - \dfrac{5y^3}{16} - \cdots.$

9. (0,2), Min.

10. None.

11. (0,0) Min.

12. (0,0) Min, $(-\frac{2}{3},\frac{2}{3})$ Max.

13. $(-7,5)$ Min.

14. $(-1,-1)$ Max.

15. $(\frac{4}{3},\frac{9}{2})$ Min.

16. $\left(\dfrac{\pi}{3} + 2k\pi, \dfrac{\pi}{3} + 2k\pi\right)$ Max, $\left(\dfrac{5\pi}{3} + 2k\pi, \dfrac{5\pi}{3} + 2k\pi\right)$ Min.

17. $\left(\dfrac{\pi}{3} + k\pi, \dfrac{\pi}{3} + k\pi\right)$ Max, $\left(\dfrac{2\pi}{3} + k\pi, \dfrac{2\pi}{3} + k\pi\right)$ Min.

18. $\left(x, -\dfrac{4 + 2x}{3}\right)$ Min, (2,3) Max.

19. $(\frac{4}{9},\frac{8}{9},\frac{8}{9})$.

20. $(\frac{1}{2},1,1)$.

21. $(\frac{18}{7},\frac{12}{7},\frac{6}{7})$.

22. $(2^{-\frac{1}{5}},2,{}^{\frac{3}{10}}2^{\frac{3}{10}})$.

23. (12,12,12).

24. (6,12,18).

Exercise 143 (Pages 502 to 503)

1. $x^2 + y^2 = 4.$

2. $(x - y)^2 = 2x + 2y - 1$ or $x^{\frac{1}{2}} + y^{\frac{1}{2}} = 1.$

3. $x^{\frac{2}{3}} + y^{\frac{2}{3}} = 1.$

4. $4xy = 1.$

5. $4x^2 + 4y = 1.$

6. $y = (1 - n)(- x/n)^{n/(n-1)}.$

7. $x^2 - y^2 = a^2.$

8. $y^2 = 16x.$

9. $x^2 + 2xy + 2y^2 = 1.$

10. $2xy = 1$ and $2xy = -1.$

11. $x^{\frac{2}{3}} + y^{\frac{2}{3}} = 1.$

12. $y^2 = (1 - n)(- x/n)^{n/(n-1)}.$

13. $x = 0$ and $y = 0.$

14. $4x^2 - 4y - 1 = 0.$

15. $(x + y)^{\frac{2}{3}} + (x - y)^{\frac{2}{3}} = 2.$

16. $x = c^2 - \frac{9}{2}c^4, y = \frac{4}{3}c + 4c^2.$

17. $x = c + \sin c, y = -1 + \cos c.$

18. $y^2 = \frac{4}{27}(x - 2)^3.$

19. $x = \dfrac{1}{2}\left(\dfrac{1}{c^3} + 3c\right), y = \dfrac{1}{2}\left(\dfrac{3}{c} + c^3\right)$ or $(x + y)^{\frac{2}{3}} - (x - y)^{\frac{2}{3}} = 2^{\frac{4}{3}}.$

20. $x = \cos^6 c + 3 \cos^4 c \sin^2 c + 2 \sin^6 c,$
$y = 2 \cos^6 c + 3 \cos^2 c \sin^4 c + \sin^6 c,$ or $(y - x)^2 = \frac{1}{27}(2x + 2y - 3)^3.$

21. $x = -e^c \sin c, y = e^c \cos c.$

Exercise 144 (Pages 509 to 510)

1. $(3,2,-5)$, $(-3,2,\pm 5)$, $(-3,-2,\pm 5)$, $(3,-2,\pm 5)$.
2. $60°, 60°, 45°$. **3.** $90°, 60°, 30°$. **4.** $45°, 60°, 120°$.
5. $s = 7; \frac{2}{7}, \frac{3}{7}, \frac{6}{7}$. **6.** $s = 3; \frac{1}{3}, \frac{2}{3}, \frac{2}{3}$. **7.** $s = 15; \frac{2}{15}, \frac{1}{3}, \frac{14}{15}$.
8. $s = 9; \frac{1}{9}, \frac{4}{9}, \frac{8}{9}$. **9.** $s = 5; \frac{3}{5}, -\frac{4}{5}, 0$. **10.** $s = 13; \frac{12}{13}, \frac{5}{13}, 0$.
11. $(\sqrt{5})^2 = (\sqrt{3})^2 + (\sqrt{2})^2$. **12.** $2\sqrt{2} = 2\sqrt{2} = 2\sqrt{2}$.
13. $3^2 + 3^2 = (\sqrt{18})^2$. **14.** $\frac{x-2}{2} = \frac{y-1}{1} = \frac{z+1}{-1}$.
15. $\frac{x}{2} = \frac{y}{4} = \frac{z}{-1}$. **16.** $\frac{x}{5} = \frac{y}{1} = \frac{z-2}{0}$ or $x = 5y, z = 2$.
17. $\frac{x+2}{2} = \frac{y-3}{-2} = \frac{z+1}{1}$. **18.** $\frac{x-5}{6} = \frac{y}{0} = \frac{z}{0}$ or $y = 0, z = 0$.
19. $\frac{x-3}{0} = \frac{y-2}{1} = \frac{z-1}{-1}$ or $x = 3, y + z = 3$.
20. $\frac{x-3}{3} = \frac{y-4}{2} = \frac{z-2}{1}$. **21.** $\frac{x-2}{-1} = \frac{y-4}{4} = \frac{z-5}{5}$.
22. $\frac{x+4}{1} = \frac{y+2}{1} = \frac{z-5}{-3}$. **23.** $(0,-6,-6)$.
24. $(4,2,6)$. **25.** $(3,0,3)$.

Exercise 145 (Page 514)

1. $-2\mathbf{i} + 5\mathbf{j} - 14\mathbf{k}$, 15. **2.** $30\mathbf{i} + 15\mathbf{j} - 10\mathbf{k}$, 35.
3. $-2\mathbf{i} - 2\mathbf{j} - 2\mathbf{k}$, $2\sqrt{3}$. **4.** $8\mathbf{i} - 2\mathbf{j} + 12\mathbf{k}$, $2\sqrt{53}$.
5. $-6\mathbf{i} - 8\mathbf{j} + 16\mathbf{k}$, $2\sqrt{89}$. **6.** $9\mathbf{i} + 3\mathbf{j} - 7\mathbf{k}$, $\sqrt{139}$.
7. $7\mathbf{i} - 3\mathbf{j} + 11\mathbf{k}$, $\sqrt{179}$. **8.** $8\mathbf{i} + 5\mathbf{j} - 4\mathbf{k}$, $\sqrt{105}$.
9. $30\mathbf{i} + 8\mathbf{j} + 2\mathbf{k}$, $2\sqrt{242}$. **10.** $-2\mathbf{i} - \mathbf{j} + 14\mathbf{k}$.
11. $\mathbf{j} + 18\mathbf{k}$. **12.** $-5\mathbf{i} - 4\mathbf{j} + 8\mathbf{k}$. **13.** $-4\mathbf{i} - 3\mathbf{j} + 10\mathbf{k}$.
14. $-20\mathbf{i} - 19\mathbf{j} - 22\mathbf{k}$. **15.** $16\mathbf{i} + 17\mathbf{j} + 50\mathbf{k}$.
16. $(24,-24,48)$. **17.** $(4,-1,8)$. **18.** $(13,-10,26)$.
19. $(15,-8,30)$. **20.** $(0,5,0)$. **21.** $(0,2,2)$.

Exercise 146 (Page 517)

5. -4. **6.** -12. **7.** 3. **8.** 3.
9. 2. **10.** 4. **11.** $A = 90°$. **12.** $B = 30°$.
13. $C = 60°$. **14.** $A = 90°$. **15.** $B = 45°$. **16.** $C = 45°$.
19. 0. **20.** -1. **21.** -15. **22.** 3.

Exercise 147 (Pages 519 to 520)

1. $\frac{x}{2} + \frac{y}{3} + \frac{z}{4} = 1$. **2.** $x + y = 4$. **3.** $x + y + z = 6$.
4. $y = 2z$. **5.** $2x - y + 2z = 1$. **6.** $3x - y + z = 16$.
7. $x + y + z = 0$. **8.** $2x = z$. **9.** $2x + y + 2z = 6$.
10. $2x + y + 2z = -3$. **11.** $2x + y + 2z = 24$. **12.** $2x + y + 2z = 15$.
13. 0. **14.** $90°$. **15.** $45°$. **16.** $\cos^{-1}\frac{4}{9}$.
17. $\frac{10}{7}$. **18.** $\frac{1}{3}$. **19.** 1. **20.** $\frac{8}{15}$.

Exercise 148 (Pages 523 to 524)

1. $-4\mathbf{i} - 6\mathbf{j} + \mathbf{k}$.
2. $2\mathbf{i} + 3\mathbf{j} + 4\mathbf{k}$.
3. $2\mathbf{i} + 10\mathbf{j} + 3\mathbf{k}$.
4. $-4\mathbf{i} - 5\mathbf{j} - 6\mathbf{k}$.
5. $\frac{1}{2}\sqrt{11}$.
6. $\frac{1}{2}\sqrt{206}$.
7. $\frac{5}{2}\sqrt{2}$.
8. $\frac{3}{2}\sqrt{5}$.
9. $3x + y - z = 4$.
10. $-6x + 7y + 11z = 19$.
11. $3x - 5y - 4z = -11$.
12. $y + 2z = 7$.
13. $3x + y - z = 4$.
14. $-3x + 8y + 10z = 23$.
15. $-3x + 4y + 5z = 10$.
17. 8.
20. $\frac{11}{3}$.

Exercise 149 (Pages 525 to 526)

1. $\dfrac{x-6}{3} = \dfrac{y-16}{16} = \dfrac{z-40}{60}$.
2. $\dfrac{x-1}{2} = \dfrac{y-1}{4} = \dfrac{z-1}{5}$.
3. $\dfrac{x - \frac{1}{2}\sqrt{2}}{-1} = \dfrac{y - \frac{1}{2}\sqrt{2}}{1} = \dfrac{z-\pi}{4\sqrt{2}}$.
4. $\dfrac{x-e^2}{2e^2} = \dfrac{y-e^3}{3e^3} = \dfrac{z-2}{2}$.
5. $\dfrac{x}{1} = \dfrac{y}{0} = \dfrac{z}{1}$ or $z = x,\ y = 0$.
6. $\dfrac{x - \ln 2}{1} = \dfrac{y}{2} = \dfrac{z-4}{8}$.
7. 57.
8. $\frac{3}{2}$.
9. 5π.
10. $\frac{1}{2}\sqrt{34} + \frac{25}{6}\ln\dfrac{3+\sqrt{34}}{5}$.
11. $\ln(1+\sqrt{2})$.
12. $\sqrt{3}\,(e-1)$.

Exercise 150 (Page 527)

1. $\mathbf{r}' = 4\mathbf{i} + 2\mathbf{j} + 12\mathbf{k},\ \mathbf{r}'' = 2\mathbf{i} + 12\mathbf{k}$.
2. $\mathbf{r}' = 27\mathbf{i} - 108\mathbf{j} + 2\mathbf{k},\ \mathbf{r}'' = 18\mathbf{i} - 108\mathbf{j}$.
3. $\mathbf{r}' = -2\mathbf{i} - \frac{3}{2}\sqrt{2}\,\mathbf{k},\ \mathbf{r}'' = 4\mathbf{j} - \frac{9}{2}\sqrt{2}\,\mathbf{k}$.
4. $\mathbf{r}' = -\mathbf{i} + 3\mathbf{j},\ \mathbf{r}'' = 9\mathbf{k}$.
5. $\mathbf{r}' = e\mathbf{i} - e^{-1}\mathbf{j} + 2e^2\mathbf{k},\ \mathbf{r}'' = e\mathbf{i} + e^{-1}\mathbf{j} + 4e^2\mathbf{k}$.

Exercise 151 (Page 531)

15. For "(1,2,0)" read "(1,0,2)." $z = 2$.
16. $6z = y + 11$.
17. $2x = z$.
18. $5x = 3y + 4z$.
19. $x + y - 2z = 2$.
20. $2x + y - 2z = 1$.
21. $2x + 2y - z = 6$.
22. $-x + 4y + 4z = 20$.

Exercise 152 (Page 535)

13. $\dfrac{x}{3} + \dfrac{y}{4} - \dfrac{z}{5} = 1$.
14. $2x + \dfrac{2y}{3} - z = 5$.
15. $\dfrac{4x}{3} - y - z = 3$.
16. $x + 2y = 5z$.
17. $2x + 3y - 6z = 6$.
18. $x + 2y - z = 2$.

Exercise 153 (Page 538)

1. $4\sqrt{3}$.
2. $\frac{4}{11}\sqrt{22}$.
3. $2\sqrt{x^2+y^2+z^2}$.
4. 0.
5. $(1 - 2\sqrt{3})/10$.
6. $\frac{5}{26}\sqrt{2}$.
7. 0.
8. $-\frac{1}{8}$.
9. $e^x\cos(y+\alpha)$.
10. $e^{-x}\sin(\alpha - y)$.

Exercise 154 (Page 541)

1. $8\frac{67}{120}$. **2.** $\frac{2}{3}$. **3.** $\frac{1}{6}\ln 3$. **4.** $4 - \ln 3$. **5.** $\frac{1}{2}\ln 2$.
6. $\frac{16}{5}$. **7.** 160. **8.** 2. **9.** 4. **10.** $\frac{1}{12}$.
11. $1/e$. **12.** $e - 2$.
13. $\frac{3}{2}$. **14.** For "$\int^{e^y}$" read "$\int_1^{e^y}$." 9.
15. $\frac{\pi^2}{8} - \frac{\pi}{2} - \frac{1}{2} + \cos 1 + \sin 1$. **16.** $\pi - 1$.

Exercise 155 (Pages 548 to 549)

1. 3. **2.** $\frac{1}{6}$. **3.** $\frac{1}{10}$. **4.** π. **5.** $\frac{1}{4}$. **6.** $\pi/2$.
7. $\frac{16}{3}$. **8.** $\pi/2$. **9.** $\int_0^2 dy \int_0^{\sqrt{4-y^2}} dx$.
10. $\int_0^3 dx \int_0^{3-x} dy$. **11.** $\int_1^2 dy \int_1^y dx + \int_2^4 dy \int_{y/2}^2 dx$.
12. $\int_0^4 dx \int_0^x dy$. **13.** $\int_0^1 dy \int_y^{y^{\frac{1}{3}}} dx$. **14.** $\int_0^1 dx \int_{x^2}^x dy$.
15. $\pi/2$. **16.** 12π. **17.** $\frac{8}{3}$. **18.** 6. **19.** 3π. **20.** $\pi/2$.

Exercise 156 (Pages 553 to 554)

1. $\pi/4$. **2.** 4π. **3.** $9\pi/2$. **4.** $9\pi/2$.
5. $3\pi/16 + \frac{1}{8} + \frac{1}{2}\sqrt{2}$. **6.** $\frac{2}{3}$.
7. 1. **8.** π. **13.** 8π. **14.** 9π.
15. $\pi(\frac{16}{3} - 2\sqrt{3})$. **16.** $9\pi/2$.
17. π. **18.** 2π.
19. $\int_0^{\pi/2} d\theta \int_0^2 r\,dr = \pi$. **20.** $\int_0^{\pi/2} d\theta \int_0^{4\sin\theta} r^3\,dr = 6\pi^2$.
21. $\int_0^{\pi/4} d\theta \int_0^{\cot\theta\csc\theta} dr$ diverges to $+\infty$.
22. $\int_0^{\pi/2} d\theta \int_0^{\infty} e^{-r^2} r\,dr = \frac{\pi}{4}$.

Exercise 157 (Pages 557 to 558)

1. $(\frac{2}{5},\frac{4}{7})$. **2.** $(1,\frac{2}{3})$. **3.** $(0,\frac{3}{5})$. **4.** $(\frac{9}{20},\frac{9}{20})$.
5. $(\frac{1}{2},\frac{2}{5})$. **6.** $(\frac{5}{9},\frac{10}{27})$. **9.** $(\frac{5}{6},0)$. **10.** $(0,-\frac{17}{18})$.
11. $\left(\frac{128\sqrt{2}}{105\pi},0\right)$. **12.** $\left(\frac{\pi}{8}\sqrt{2},0\right)$.
14. $I_x = \frac{4}{15} = \frac{1}{5}A,\ I_y = \frac{4}{7} = \frac{3}{7}A$.
15. $I_x = \frac{128}{15} = \frac{4}{5}A,\ I_y = \frac{4096}{105} = \frac{128}{35}A$.
16. $I_x = \frac{8}{13} = \frac{5}{13}A,\ I_y = \frac{8}{21} = \frac{5}{21}A$. **17.** $I_x = \frac{1}{20} = \frac{3}{10}A,\ I_y = \frac{1}{28} = \frac{3}{14}A$.
18. $\frac{\pi}{4} = \frac{\pi}{4}A$. **19.** $\frac{3\pi}{64} = \frac{3}{8}A$. **20.** $\frac{\pi}{32} = \frac{3}{8}A$.
21. $I_x = I_y = \frac{\pi a^4}{4} = \frac{Aa^2}{4},\ I_0 = \frac{\pi a^2}{2} = \frac{Aa^2}{2}$.
22. $I_x = \frac{\pi a^4}{64} = \frac{Aa^2}{16},\ I_y = \frac{5\pi a^4}{16} = \frac{5Aa^2}{16},\ I_0 = \frac{3\pi a^4}{32} = \frac{3Aa^2}{8}$.

Exercise 158 (Pages 560 to 561)

7. $\frac{1}{12}[(1 + 4a^2)^{\frac{3}{2}} - 1]$. **8.** $\frac{1}{12}[(1 + 4a^2)^{\frac{3}{2}} - 1]$.
10. $6\sqrt{41}$. **11.** $\sqrt{2}$. **14.** $\frac{4}{3}\sqrt{2}$. **15.** $(\pi/4)\sqrt{2}$.

Exercise 159 (Pages 565 to 566)

1. 2. **2.** 2. **3.** 2. **4.** $\frac{17}{6}$. **5.** 1. **6.** 9.

11. $8\pi - 2\int_{2\pi/3}^{\pi} d\theta \int_{-\sec\theta}^{2} (2r\cos\theta + 2)r\,dr = \frac{16\pi}{3} + 6\sqrt{3}$.

12. $4\pi - 2\int_{2\pi/3}^{\pi} d\theta \int_{-\sec\theta}^{2} (1 + r^3\cos^3\theta)r\,dr = \frac{8\pi}{3} + \frac{27}{5}\sqrt{3}$.

13. $28\pi/3$. **14.** 16π. **15.** $\frac{\pi}{3}(10\sqrt{5} - 19)$.

16. $\frac{\pi}{3}(2 - \sqrt{3})a^3$. **17.** $\frac{1}{9}(6\pi - 8)a^3$. **18.** $\frac{1}{9}(3\pi + 20 - 16\sqrt{3})$.

Exercise 160 (Pages 571 to 572)

1. $(\frac{2}{3},\frac{1}{2},2)$. **2.** $(\frac{1}{2},3,\frac{2}{5})$. **3.** $\left(\frac{9\pi}{64}a,\frac{9\pi}{64}a,\frac{3}{8}a\right)$.

4. $(0,0,\frac{2}{3}c)$. **5.** $(\frac{3}{8}a,\frac{3}{8}b,\frac{3}{8}c)$. **6.** $(0,0,\frac{3}{8})$.

7. $(0,0,\frac{1}{3})$. **8.** $(0,0,\frac{1}{4})$. **9.** $(0,0,\frac{7}{20})$.

10. $(0,0,0)$. For part above $z = 0$, $\left(0,0,\frac{21a}{64 - 8\sqrt{3}}\right)$.

11. $(0,0,\frac{9}{10}a)$. **12.** $I_x = 19M$, $I_y = \frac{131}{6}M$, $I_z = \frac{5}{6}M$.

13. $I_x = I_y = \frac{7}{15}Ma^2$, $I_z = \frac{8}{15}Ma^2$. **14.** $I_x = I_y = I_z = \frac{1}{3}M$.

15. $I_x = I_y = \frac{1}{4}M$, $I_z = \frac{3}{10}M$. **16.** $\frac{1}{6}(a^2 + b^2)M$.

17. $\frac{53}{50}M$. **18.** $\frac{64 - 33\sqrt{3}}{5(8 - 3\sqrt{3})}a^2$. **19.** $\frac{8}{25}Ma^2$.

20. $M\left(\frac{a^2}{4} + \frac{h^2}{3}\right)$. **21.** $\frac{3}{2}Ma^2$. **22.** $\frac{3}{20}M(a^2 + 4h^2)$.

Exercise 161 (Pages 576 to 577)

19. $\frac{2}{3}a^2$. **20.** a^2. **21.** $32a/9\pi$. **22.** $\frac{3}{2}a^2$. **23.** $4a$.

Exercise 163 (Page 583)

1. $x^2 + y^2 = c$. **2.** $x^2 + 2y^2 + 3x - 5y = c$.

3. $x^2 + y^2 = cx^2y^2$. **4.** $\tan x = \sec y + c$. **5.** $e^{2x} = e^{2y} + c$.

6. $x^2 - 1 = c(1 + y^2)$. **7.** $xy = c$. **8.** $(3 + 2y)(4 - 2x) = c$.

9. $y^2 = c(1 + x^2)$. **10.** $y = cx(x + 2)$. **11.** $x = 2\sin(y + c)$.

12. $y^2(\cot 2x + \csc 2x) = c$. **13.** $y^2 = 4\sin(2x + c)$.

14. $xy = c(y - 1)$. **15.** $y = \frac{c - x}{1 + cx}$.

16. $y = c\sqrt{1 - x^2} - x\sqrt{1 - c^2}$. **17.** $y^2 - x^2 = c$.

18. $x^3y = c$. **19.** $x^2 + y^2 = c$. **20.** $x^2 + 2y^2 = c$.

21. $y^2 + 2\ln\sec x = c$. **22.** $x^2 + y^2 = 2\ln x + c$.

Exercise 164 (Page 586)

1. $xy = x^3 + c$. **2.** $y = \sin x + c\cos x$. **3.** $y = -2x + cx^2$.

4. $y = -3x + cx^3$. **5.** $y = x + c\sin x$. **6.** $y = -e^x + cx$.

7. $y = (x + 1)^2(\frac{1}{2}x^2 + x + c)$. **8.** $y = \frac{1}{3}(x^2 + 5) + c(x^2 - 1)^{-\frac{1}{2}}$.

9. $x^2 = y^2(\frac{1}{3} + cx^6)$. **10.** $x = y\ln x + cy$. **11.** $y^2(2x + cx^2) = 1$.

12. $x^2y(c - 2\ln x) = 1$. **13.** $y^5(x + cx^2) = 1$. **14.** $y^2e^x(c - x) = 1$.

15. $x = y^4 + cy$. **16.** $x = -6y + cy^3$.
17. $x = c \sin^2 x - \cos x - \sin^2 x \ln (\csc x + \cot x)$.
18. $x = -\frac{4}{3}y + cy^4$. **19.** $xy = -\cos y + c$. **20.** $xy^2 = y + c$.

Exercise 165 (Pages 588 to 589)

1. $\frac{1}{2} \ln (x^2 + y^2) + \tan^{-1} \frac{y}{x} = c$. **2.** $y^2 = cx - x^2$.
3. $(x + y)^2(x + 2y) = c$. **4.** $x^2(x^2 + 2y^2) = c$. **5.** $(x + y)^2(2x + y)^3 = c$.
6. $\ln x + 2\sqrt{\frac{y}{x}} = c$. **7.** $3x^2 + 2xy + y^2 = c$. **8.** $x(5x + 9y)^2 = c$.
9. Not homogeneous, but $y = vx$ leads to $x + \tan^{-1} (y/x) = c$. For "$= (x^2 + y^2)$" read "$= \sqrt{x^2 + y^2}$." $y + \sqrt{x^2 + y^2} = c$ or $x^2 = c^2 - 2cy$.
10. $x = y(2 \ln y + c)$.
11. For "$(y - x)$" read "$(y + x)$." $x^2 - 6xy - 3y^2 = c$.
12. $3xy^2 - x^3 = c$. **13.** $4x^3y + y^4 = c$. **14.** $xy + \cos y = c$.
15. $x^3y - y^3x = c$. **16.** $xy^2 - e^x = c$. **17.** $y + x^3 = cx$.
18. $x + y^4 = cy$. **19.** $x + \tan^{-1} (y/x) = c$. **20.** $y = x(x^2 + 2y^2 + c)$.

Exercise 166 (Pages 591 to 592)

1. $2y + x^2 = c,\ 2 \ln y + x^2 = c$. **2.** $2y + x^2 = c,\ 2y - 3x^2 = c$.
3. $2y - x^2 = c,\ \ln y - x = c$. **4.** $(x - c)^2 - y^2 = 1$.
5. $y = cx,\ y^2 - x^2 = c$. **6.** $y = cx,\ x^2y = c$.
7. $y = \sqrt{2}\, x^{\frac{3}{2}} + cx^{\frac{1}{2}}$.
8. $x = \pm\frac{1}{2}(e^{y-c} - e^{c-y}) = \pm \sinh (y - c)$.
9. $y = cx + 1 + c^2,\ 4y = 1 - x^2$. **10.** $y = cx - c^2,\ 4y = x^2$.
11. $(y - cx)^2 = c^3,\ 27y = 4x^3$. **12.** $y = cx - \ln c,\ y = 1 + \ln x$.
13. $x = cy - c^2,\ y^2 = 4x$. **14.** $(y - cx)^2 = 1 + c^2,\ x^2 + y^2 = 1$.
15. $y^2 = 2cx + c^2$, singular solution degenerates to $x^2 + y^2 = 0$.
16. $y^2 = cx - c^2,\ 2y = \pm x$.
17. $c^2x^2 = 2cy + 1$, singular solution degenerates to $x^2 + y^2 = 0$.
18. $cy = c^2x^2 + 1,\ y = \pm 2x$.
19. $y^2 = cx^2 - \frac{c}{c + 1}$, singular solution degenerates to $(x - 1)^2 + y^2 = 0$.
20. $y^2 = cx^2 + c^2$, singular solution degenerates to $x^4 + 4y^2 = 0$.

Exercise 167 (Pages 595 to 596)

1. $xy = c$. **2.** $x^2y = c$. **3.** $xy^2 = c$. **4.** $xy = c$.
5. $x^2 - y^2 = c$. **6.** $y = cx^2$. **7.** $x = cy^2$. **8.** $x^2 - y^2 = c$.
9. $x^2 - y^2 = c$. **10.** $x = cy^2$. **11.** $y = cx^2$. **12.** $y = cx$.
13. $r = ce^{\theta \cot A}$. **14.** $r = \frac{2k}{c - \theta}$.
15. $r = k \sin (\theta - c),\ r = k$. **16.** $r = ce^{\theta/\sqrt{k^2-1}}$.
17. $r \sin (\theta - c) = 2k,\ r = 2k$. **18.** $4xy = a^2$.
19. For "OP" read "OS." $x^{\frac{1}{2}} + y^{\frac{1}{2}} = a^{\frac{1}{2}}$.
20. $x^{\frac{2}{3}} + y^{\frac{2}{3}} = a^{\frac{2}{3}}$. **21.** $x^2 + y^2 = a^2$.

Exercise 168 (Pages 598 to 600)

10. 19 gram-moles. **18.** $t = 5\frac{\ln 8.2}{\ln 1.4} = 31.27$ min.

Exercise 169 (Page 603)

1. $y = \cos 2x + c_1x + c_2.$

2. $y = \dfrac{x^3}{6} + \dfrac{x^2}{2} + c_1x + c_2.$

3. $y = \dfrac{x^4}{2} - x \ln x + c_1x + c_2.$

4. $y = e^x - e^{-x} + c_1x + c_2.$

5. $y = \ln (x^2 + c_1) + c_2.$

6. $y = \dfrac{2}{c_1} \ln \dfrac{x}{c_1 - x} + c_2.$

7. $y = -x^2 + c_1x^3 + c_2.$

8. $y = -\cos x + c_1 \sin x + c_2.$

9. $y = x^2 + c_1x^4 + c_2.$

10. $y = x^3 + c_1 \ln x + c_2.$

11. $y = 2x^5 + c_1x^2 + c_2.$

12. $y = \displaystyle\int_{c_2}^{x} \frac{x\,dx}{c_1 + \ln x}.$ For "$= x\dfrac{dy}{dx} -$" read "$+ x\dfrac{dy}{dx} =$."

$c_1y = \ln (1 + c_1x^2) + c_2.$

13. $y = (c_1x + c_2)^2.$

14. $y^2 = c_1x + c_2.$

15. $4(x - c_2)^2 + y^2 = c_1.$

16. $y = c_2 - \ln (x + c_1).$

17. $y = c_2 - \ln \sec (3x + c_1).$

18. $(x - c_1)^2 + (y - c_2)^2 = 1.$

19. $x = c_1 \ln (e^y + \sqrt{e^{2y} - c_1^2}) + c_2.$

20. $c_1y^2 - c_1^2(x - c_2)^2 = 2.$

21. $9(x + c_2)^2 = 16(\sqrt{y} + c_1)(\sqrt{y} - 2c_1)^2.$

22. $$x + c_2 = \frac{9}{4c_1^5}\left[\frac{c_1^2}{(u - c_1)^2} - \frac{c_1^2}{(u + c_1)^2} - \frac{3c_1}{u - c_1} - \frac{3c_1}{u + c_1} - 3 \ln (u - c_1) + 3 \ln (u + c_1)\right], \text{ where } u = \sqrt{3y^{-\frac{1}{2}} + c_1^2}.$$

Exercise 170 (Page 608)

1. $y = ce^{7x}.$

2. $y = ce^{-3x/2}.$

3. $y = c_1e^{-x} \cos x + c_2e^{-x} \sin x.$

4. $y = c_1e^{-x} + c_2e^{2x}.$

5. $y = c_1e^x + c_2e^{3x}.$

6. $y = c_1e^x + c_2xe^x.$

7. $y = c_1e^x \cos 2x + c_2e^x \sin 2x.$

8. $y = c_1 \cos 4x + c_2 \sin 4x.$

9. $y = c_1e^{-3x} + c_2e^{3x}.$

10. $y = c_1x + c_2e^{9x}.$

11. $y = c_1x + c_2e^x + c_3e^{4x}.$

12. $y = c_1x + c_2e^{3x} + c_3xe^{3x}.$

13. $y = c_1x + c_2e^{4x} \cos 3x + c_3e^{4x} \sin 3x.$

14. $y = c_1e^{2x} + c_2e^{-x} \cos \sqrt{3}\, x + c_3e^{-x} \sin \sqrt{3}\, x.$

15. $y = c_1e^x + c_2xe^x + c_3x^2e^x.$

16. $y = c_1e^{-2x} + c_2e^{2x} + c_3e^{-3x} + c_4e^{3x}.$

17. $y = c_1 + c_2x + c_3x^2 + c_4e^{-5x}.$

18. $y = (c_1 + c_2x) \cos 3x + (c_3 + c_4x) \sin 3x.$

19. $y = c_1 + c_2x + c_3e^{4x} + c_4xe^{4x}.$

20. $c_1e^{2x} + c_2e^{-2x} + c_3 \cos x + c_4 \sin x.$

Exercise 171 (Page 612)

2. $y = \frac{2}{5}e^{3x} + ce^{-2x}.$

3. $y = \frac{1}{4} \sin x + ce^{-3x}.$

4. $y = x + ce^{2x}.$

5. $y = xe^x + ce^x.$

6. $y = 4x + 2e^x + c_1 \cos x + c_2 \sin x.$

7. $y = xe^{2x} + c_1e^{2x} + c_2e^{-2x}.$

8. $y = 2 \cos x - \sin x + c_1e^x \cos 2x + c_2e^x \sin 2x.$

9. $y = x^2 + 2x + c_1 + c_2e^{-x}.$

10. $y = (2x^2 + 6x + 7)e^{3x} + c_1e^{4x} + c_2e^{5x}.$

11. $y = -\frac{3}{2}x \cos 3x + c_1 \cos 3x + c_2 \sin 3x.$

12. $y = e^x + 2x^2e^{-x} + c_1e^{-x} + c_2xe^{-x}.$

13. $y = -e^{2x} \cos x - 3e^{2x} \sin x + c_1 + c_2e^{3x}.$

14. $y = \frac{1}{6}x^3e^x + c_1e^x + c_2xe^x + c_3x^2e^x$. **15.** $y = -\frac{3}{2}x^2 + c_1x + c_2e^{2x} + c_3e^{-2x}$.

16. $y = \frac{3}{4}(x - 1)e^{-2x} + c_1e^{3.36x} + e^{-0.18x}(c_2 \cos 1.085x + c_3 \sin 1.085x)$. For "$- 4y$" read "$+ 4y$." $y = \frac{9}{8}xe^{-2x} - \frac{27}{16}e^{-2x} + c_1e^{-x} + c_2e^{2x} + c_3xe^{2x}$.

17. $y = -2x^4 - 8x^3 - 24x^2 + c_1 + c_2x + c_3e^x$.

18. $y = -36e^x + c_1e^{-0.532x} + c_2e^{0.653x} + c_3e^{2.879x}$. For "$+ y$" read "$+ 2y$." $y = 6x^2e^x + c_1e^{-2x} + c_2e^x + c_3xe^x$.

19. $y = -x^3 - 3x + c_1e^{-x} + c_2e^x + c_3 \cos x + c_4 \sin x$.

20. $y = 3xe^{2x} + c_1e^{2x} + c_2e^{-2x} + c_3 \cos x + c_4 \sin x$.

21. $y = -x^3 + c_1x + c_2 + c_3e^x + c_4e^{-x}$.

24. $y = c_1x^{-3} + c_2x^{-4}$. **25.** $y = c_1x + c_2x^2$.

26. $y = \frac{12}{5}x + c_1x \cos (\sqrt{5} \ln x) + c_2x \sin (\sqrt{5} \ln x)$.

27. $y = c_1x^{-1} + c_2x^3$. **28.** $y = x^2 + c_1x + c_2x^{-2}$.

29. $y = \frac{1}{2}x(\ln x)^2 + c_1x + c_2x \ln x$.

30. $y = \frac{1}{24}x(\ln x)^4 + c_1x + c_2x \ln x + c_3x(\ln x)^2$.

31. $y = c_1 \sqrt{x} \cos (\frac{1}{2}\sqrt{19} \ln x) + c_2 \sqrt{x} \sin (\frac{1}{2}\sqrt{19} \ln x)$.

32. $y = c_1 + c_2x^{-1} + c_3x^{-2}$. **33.** $y = -2x + x^2 \ln x + c_1 + c_2x^2$.

INDEX

SOME DOVER SCIENCE BOOKS

SOME DOVER SCIENCE BOOKS

WHAT IS SCIENCE?,
Norman Campbell

This excellent introduction explains scientific method, role of mathematics, types of scientific laws. Contents: 2 aspects of science, science & nature, laws of science, discovery of laws, explanation of laws, measurement & numerical laws, applications of science. 192pp. 5⅜ x 8. Paperbound $1.25

FADS AND FALLACIES IN THE NAME OF SCIENCE,
Martin Gardner

Examines various cults, quack systems, frauds, delusions which at various times have masqueraded as science. Accounts of hollow-earth fanatics like Symmes; Velikovsky and wandering planets; Hoerbiger; Bellamy and the theory of multiple moons; Charles Fort; dowsing, pseudoscientific methods for finding water, ores, oil. Sections on naturopathy, iridiagnosis, zone therapy, food fads, etc. Analytical accounts of Wilhelm Reich and orgone sex energy; L. Ron Hubbard and Dianetics; A. Korzybski and General Semantics; many others. Brought up to date to include Bridey Murphy, others. Not just a collection of anecdotes, but a fair, reasoned appraisal of eccentric theory. Formerly titled *In the Name of Science*. Preface. Index. x + 384pp. 5⅜ x 8.
Paperbound $1.85

PHYSICS, THE PIONEER SCIENCE,
L. W. Taylor

First thorough text to place all important physical phenomena in cultural-historical framework; remains best work of its kind. Exposition of physical laws, theories developed chronologically, with great historical, illustrative experiments diagrammed, described, worked out mathematically. Excellent physics text for self-study as well as class work. Vol. 1: Heat, Sound: motion, acceleration, gravitation, conservation of energy, heat engines, rotation, heat, mechanical energy, etc. 211 illus. 407pp. 5⅜ x 8. Vol. 2: Light, Electricity: images, lenses, prisms, magnetism, Ohm's law, dynamos, telegraph, quantum theory, decline of mechanical view of nature, etc. Bibliography. 13 table appendix. Index. 551 illus. 2 color plates. 508pp. 5⅜ x 8.
Vol. 1 Paperbound $2.25, Vol. 2 Paperbound $2.25,
The set $4.50

THE EVOLUTION OF SCIENTIFIC THOUGHT FROM NEWTON TO EINSTEIN,
A. d'Abro

Einstein's special and general theories of relativity, with their historical implications, are analyzed in non-technical terms. Excellent accounts of the contributions of Newton, Riemann, Weyl, Planck, Eddington, Maxwell, Lorentz and others are treated in terms of space and time, equations of electromagnetics, finiteness of the universe, methodology of science. 21 diagrams. 482pp. 5⅜ x 8.
Paperbound $2.50

CHANCE, LUCK AND STATISTICS: THE SCIENCE OF CHANCE,
Horace C. Levinson
Theory of probability and science of statistics in simple, non-technical language. Part I deals with theory of probability, covering odd superstitions in regard to "luck," the meaning of betting odds, the law of mathematical expectation, gambling, and applications in poker, roulette, lotteries, dice, bridge, and other games of chance. Part II discusses the misuse of statistics, the concept of statistical probabilities, normal and skew frequency distributions, and statistics applied to various fields—birth rates, stock speculation, insurance rates, advertising, etc. "Presented in an easy humorous style which I consider the best kind of expository writing," Prof. A. C. Cohen, Industry Quality Control. Enlarged revised edition. Formerly titled *The Science of Chance*. Preface and two new appendices by the author. Index. xiv + 365pp. 5⅜ x 8. Paperbound $2.00

BASIC ELECTRONICS,
prepared by the U.S. Navy Training Publications Center
A thorough and comprehensive manual on the fundamentals of electronics. Written clearly, it is equally useful for self-study or course work for those with a knowledge of the principles of basic electricity. Partial contents: Operating Principles of the Electron Tube; Introduction to Transistors; Power Supplies for Electronic Equipment; Tuned Circuits; Electron-Tube Amplifiers; Audio Power Amplifiers; Oscillators; Transmitters; Transmission Lines; Antennas and Propagation; Introduction to Computers; and related topics. Appendix. Index. Hundreds of illustrations and diagrams. vi + 471pp. 6½ x 9¼.
Paperbound $2.75

BASIC THEORY AND APPLICATION OF TRANSISTORS,
prepared by the U.S. Department of the Army
An introductory manual prepared for an army training program. One of the finest available surveys of theory and application of transistor design and operation. Minimal knowledge of physics and theory of electron tubes required. Suitable for textbook use, course supplement, or home study. Chapters: Introduction; fundamental theory of transistors; transistor amplifier fundamentals; parameters, equivalent circuits, and characteristic curves; bias stabilization; transistor analysis and comparison using characteristic curves and charts; audio amplifiers; tuned amplifiers; wide-band amplifiers; oscillators; pulse and switching circuits; modulation, mixing, and demodulation; and additional semiconductor devices. Unabridged, corrected edition. 240 schematic drawings, photographs, wiring diagrams, etc. 2 Appendices. Glossary. Index. 263pp. 6½ x 9¼. Paperbound $1.25

GUIDE TO THE LITERATURE OF MATHEMATICS AND PHYSICS,
N. G. Parke III
Over 5000 entries included under approximately 120 major subject headings of selected most important books, monographs, periodicals, articles in English, plus important works in German, French, Italian, Spanish, Russian (many recently available works). Covers every branch of physics, math, related engineering. Includes author, title, edition, publisher, place, date, number of volumes, number of pages. A 40-page introduction on the basic problems of research and study provides useful information on the organization and use of libraries, the psychology of learning, etc. This reference work will save you hours of time. 2nd revised edition. Indices of authors, subjects, 464pp. 5⅜ x 8.
Paperbound $2.75

THE RISE OF THE NEW PHYSICS (formerly THE DECLINE OF MECHANISM), *A. d'Abro*

This authoritative and comprehensive 2-volume exposition is unique in scientific publishing. Written for intelligent readers not familiar with higher mathematics, it is the only thorough explanation in non-technical language of modern mathematical-physical theory. Combining both history and exposition, it ranges from classical Newtonian concepts up through the electronic theories of Dirac and Heisenberg, the statistical mechanics of Fermi, and Einstein's relativity theories. "A must for anyone doing serious study in the physical sciences," *J. of Franklin Inst.* 97 illustrations. 991pp. 2 volumes.

T3, T4 Two volume set, paperbound $5.50

THE STRANGE STORY OF THE QUANTUM, AN ACCOUNT FOR THE GENERAL READER OF THE GROWTH OF IDEAS UNDERLYING OUR PRESENT ATOMIC KNOWLEDGE, *B. Hoffmann*

Presents lucidly and expertly, with barest amount of mathematics, the problems and theories which led to modern quantum physics. Dr. Hoffmann begins with the closing years of the 19th century, when certain trifling discrepancies were noticed, and with illuminating analogies and examples takes you through the brilliant concepts of Planck, Einstein, Pauli, de Broglie, Bohr, Schroedinger, Heisenberg, Dirac, Sommerfeld, Feynman, etc. This edition includes a new, long postscript carrying the story through 1958. "Of the books attempting an account of the history and contents of our modern atomic physics which have come to my attention, this is the best," H. Margenau, Yale University, in *American Journal of Physics.* 32 tables and line illustrations. Index. 275pp. 5⅜ x 8.

T518 Paperbound $2.00

GREAT IDEAS AND THEORIES OF MODERN COSMOLOGY, *Jagjit Singh*

The theories of Jeans, Eddington, Milne, Kant, Bondi, Gold, Newton, Einstein, Gamow, Hoyle, Dirac, Kuiper, Hubble, Weizsäcker and many others on such cosmological questions as the origin of the universe, space and time, planet formation, "continuous creation," the birth, life, and death of the stars, the origin of the galaxies, etc. By the author of the popular *Great Ideas of Modern Mathematics.* A gifted popularizer of science, he makes the most difficult abstractions crystal-clear even to the most non-mathematical reader. Index. xii + 276pp. 5⅜ x 8½

T925 Paperbound $2.00

GREAT IDEAS OF MODERN MATHEMATICS: THEIR NATURE AND USE, *Jagjit Singh*

Reader with only high school math will understand main mathematical ideas of modern physics, astronomy, genetics, psychology, evolution, etc., better than many who use them as tools, but comprehend little of their basic structure. Author uses his wide knowledge of non-mathematical fields in brilliant exposition of differential equations, matrices, group theory, logic, statistics, problems of mathematical foundations, imaginary numbers, vectors, etc. Original publications, appendices. indexes. 65 illustr. 322pp. 5⅜ x 8. T587 Paperbound $2.00

THE MATHEMATICS OF GREAT AMATEURS, *Julian L. Coolidge*

Great discoveries made by poets, theologians, philosophers, artists and other non-mathematicians: Omar Khayyam, Leonardo da Vinci, Albrecht Dürer, John Napier, Pascal, Diderot, Bolzano, etc. Surprising accounts of what can result from a non-professional preoccupation with the oldest of sciences. 56 figures. viii + 211pp. 5⅜ x 8½.

S1009 Paperbound $2.00

COLLEGE ALGEBRA, *H. B. Fine*

Standard college text that gives a systematic and deductive structure to algebra; comprehensive, connected, with emphasis on theory. Discusses the commutative, associative, and distributive laws of number in unusual detail, and goes on with undetermined coefficients, quadratic equations, progressions, logarithms, permutations, probability, power series, and much more. Still most valuable elementary-intermediate text on the science and structure of algebra. Index. 1560 problems, all with answers. x + 631pp. 5⅜ x 8. Paperbound $2.75

HIGHER MATHEMATICS FOR STUDENTS OF CHEMISTRY AND PHYSICS, *J. W. Mellor*

Not abstract, but practical, building its problems out of familiar laboratory material, this covers differential calculus, coordinate, analytical geometry, functions, integral calculus, infinite series, numerical equations, differential equations, Fourier's theorem, probability, theory of errors, calculus of variations, determinants. "If the reader is not familiar with this book, it will repay him to examine it," *Chem. & Engineering News.* 800 problems. 189 figures. Bibliography. xxi + 641pp. 5⅜ x 8. Paperbound $2.50

TRIGONOMETRY REFRESHER FOR TECHNICAL MEN, *A. A. Klaf*

A modern question and answer text on plane and spherical trigonometry. Part I covers plane trigonometry: angles, quadrants, trigonometrical functions, graphical representation, interpolation, equations, logarithms, solution of triangles, slide rules, etc. Part II discusses applications to navigation, surveying, elasticity, architecture, and engineering. Small angles, periodic functions, vectors, polar coordinates, De Moivre's theorem, fully covered. Part III is devoted to spherical trigonometry and the solution of spherical triangles, with applications to terrestrial and astronomical problems. Special time-savers for numerical calculation. 913 questions answered for you! 1738 problems; answers to odd numbers. 494 figures. 14 pages of functions, formulae. Index. x + 629pp. 5⅜ x 8. Paperbound $2.00

CALCULUS REFRESHER FOR TECHNICAL MEN, *A. A. Klaf*

Not an ordinary textbook but a unique refresher for engineers, technicians, and students. An examination of the most important aspects of differential and integral calculus by means of 756 key questions. Part I covers simple differential calculus: constants, variables, functions, increments, derivatives, logarithms, curvature, etc. Part II treats fundamental concepts of integration: inspection, substitution, transformation, reduction, areas and volumes, mean value, successive and partial integration, double and triple integration. Stresses practical aspects! A 50 page section gives applications to civil and nautical engineering, electricity, stress and strain, elasticity, industrial engineering, and similar fields. 756 questions answered. 556 problems; solutions to odd numbers. 36 pages of constants, formulae. Index. v + 431pp. 5⅜ x 8. Paperbound $2.00

INTRODUCTION TO THE THEORY OF GROUPS OF FINITE ORDER, *R. Carmichael*

Examines fundamental theorems and their application. Beginning with sets, systems, permutations, etc., it progresses in easy stages through important types of groups: Abelian, prime power, permutation, etc. Except 1 chapter where matrices are desirable, no higher math needed. 783 exercises, problems. Index. xvi + 447pp. 5⅜ x 8. Paperbound $3.00

FIVE VOLUME "THEORY OF FUNCTIONS" SET BY KONRAD KNOPP

This five-volume set, prepared by Konrad Knopp, provides a complete and readily followed account of theory of functions. Proofs are given concisely, yet without sacrifice of completeness or rigor. These volumes are used as texts by such universities as M.I.T., University of Chicago, N. Y. City College, and many others. "Excellent introduction . . . remarkably readable, concise, clear, rigorous," *Journal of the American Statistical Association.*

ELEMENTS OF THE THEORY OF FUNCTIONS,
Konrad Knopp
This book provides the student with background for further volumes in this set, or texts on a similar level. Partial contents: foundations, system of complex numbers and the Gaussian plane of numbers, Riemann sphere of numbers, mapping by linear functions, normal forms, the logarithm, the cyclometric functions and binomial series. "Not only for the young student, but also for the student who knows all about what is in it," *Mathematical Journal.* Bibliography. Index. 140pp. 5⅜ x 8. Paperbound $1.50

THEORY OF FUNCTIONS, PART I,
Konrad Knopp
With volume II, this book provides coverage of basic concepts and theorems. Partial contents: numbers and points, functions of a complex variable, integral of a continuous function, Cauchy's integral theorem, Cauchy's integral formulae, series with variable terms, expansion of analytic functions in power series, analytic continuation and complete definition of analytic functions, entire transcendental functions, Laurent expansion, types of singularities. Bibliography. Index. vii + 146pp. 5⅜ x 8. Paperbound $1.35

THEORY OF FUNCTIONS, PART II,
Konrad Knopp
Application and further development of general theory, special topics. Single valued functions. Entire, Weierstrass, Meromorphic functions. Riemann surfaces. Algebraic functions. Analytical configuration, Riemann surface. Bibliography. Index. x + 150pp. 5⅜ x 8. Paperbound $1.35

PROBLEM BOOK IN THE THEORY OF FUNCTIONS, VOLUME 1.
Konrad Knopp
Problems in elementary theory, for use with Knopp's *Theory of Functions,* or any other text, arranged according to increasing difficulty. Fundamental concepts, sequences of numbers and infinite series, complex variable, integral theorems, development in series, conformal mapping. 182 problems. Answers. viii + 126pp. 5⅜ x 8. Paperbound $1.35

PROBLEM BOOK IN THE THEORY OF FUNCTIONS, VOLUME 2,
Konrad Knopp
Advanced theory of functions, to be used either with Knopp's *Theory of Functions,* or any other comparable text. Singularities, entire & meromorphic functions, periodic, analytic, continuation, multiple-valued functions, Riemann surfaces, conformal mapping. Includes a section of additional elementary problems. "The difficult task of selecting from the immense material of the modern theory of functions the problems just within the reach of the beginner is here masterfully accomplished," *Am. Math. Soc.* Answers. 138pp. 5⅜ x 8. Paperbound $1.50

NUMERICAL SOLUTIONS OF DIFFERENTIAL EQUATIONS,
H. Levy & E. A. Baggott

Comprehensive collection of methods for solving ordinary differential equations of first and higher order. All must pass 2 requirements: easy to grasp and practical, more rapid than school methods. Partial contents: graphical integration of differential equations, graphical methods for detailed solution. Numerical solution. Simultaneous equations and equations of 2nd and higher orders. "Should be in the hands of all in research in applied mathematics, teaching," *Nature*. 21 figures. viii + 238pp. 5⅜ x 8. Paperbound $1.85

ELEMENTARY STATISTICS, WITH APPLICATIONS IN MEDICINE AND THE BIOLOGICAL SCIENCES, *F. E. Croxton*

A sound introduction to statistics for anyone in the physical sciences, assuming no prior acquaintance and requiring only a modest knowledge of math. All basic formulas carefully explained and illustrated; all necessary reference tables included. From basic terms and concepts, the study proceeds to frequency distribution, linear, non-linear, and multiple correlation, skewness, kurtosis, etc. A large section deals with reliability and significance of statistical methods. Containing concrete examples from medicine and biology, this book will prove unusually helpful to workers in those fields who increasingly must evaluate, check, and interpret statistics. Formerly titled "Elementary Statistics with Applications in Medicine." 101 charts. 57 tables. 14 appendices. Index. vi + 376pp. 5⅜ x 8. Paperbound $2.00

INTRODUCTION TO SYMBOLIC LOGIC,
S. Langer

No special knowledge of math required — probably the clearest book ever written on symbolic logic, suitable for the layman, general scientist, and philosopher. You start with simple symbols and advance to a knowledge of the Boole-Schroeder and Russell-Whitehead systems. Forms, logical structure, classes, the calculus of propositions, logic of the syllogism, etc. are all covered. "One of the clearest and simplest introductions," *Mathematics Gazette*. Second enlarged, revised edition. 368pp. 5⅜ x 8. Paperbound $2.00

A SHORT ACCOUNT OF THE HISTORY OF MATHEMATICS,
W. W. R. Ball

Most readable non-technical history of mathematics treats lives, discoveries of every important figure from Egyptian, Phoenician, mathematicians to late 19th century. Discusses schools of Ionia, Pythagoras, Athens, Cyzicus, Alexandria, Byzantium, systems of numeration; primitive arithmetic; Middle Ages, Renaissance, including Arabs, Bacon, Regiomontanus, Tartaglia, Cardan, Stevinus, Galileo, Kepler; modern mathematics of Descartes, Pascal, Wallis, Huygens, Newton, Leibnitz, d'Alembert, Euler, Lambert, Laplace, Legendre, Gauss, Hermite, Weierstrass, scores more. Index. 25 figures. 546pp. 5⅜ x 8. Paperbound $2.25

INTRODUCTION TO NONLINEAR DIFFERENTIAL AND INTEGRAL EQUATIONS,
Harold T. Davis

Aspects of the problem of nonlinear equations, transformations that lead to equations solvable by classical means, results in special cases, and useful generalizations. Thorough, but easily followed by mathematically sophisticated reader who knows little about non-linear equations. 137 problems for student to solve. xv + 566pp. 5⅜ x 8½. Paperbound $2.00

AN INTRODUCTION TO THE GEOMETRY OF N DIMENSIONS,
D. H. Y. Sommerville

An introduction presupposing no prior knowledge of the field, the only book in English devoted exclusively to higher dimensional geometry. Discusses fundamental ideas of incidence, parallelism, perpendicularity, angles between linear space; enumerative geometry; analytical geometry from projective and metric points of view; polytopes; elementary ideas in analysis situs; content of hyper-spacial figures. Bibliography. Index. 60 diagrams. 196pp. 5⅜ x 8.
Paperbound $1.50

ELEMENTARY CONCEPTS OF TOPOLOGY, *P. Alexandroff*

First English translation of the famous brief introduction to topology for the beginner or for the mathematician not undertaking extensive study. This unusually useful intuitive approach deals primarily with the concepts of complex, cycle, and homology, and is wholly consistent with current investigations. Ranges from basic concepts of set-theoretic topology to the concept of Betti groups. "Glowing example of harmony between intuition and thought," David Hilbert. Translated by A. E. Farley. Introduction by D. Hilbert. Index. 25 figures. 73pp. 5⅜ x 8.
Paperbound $1.00

ELEMENTS OF NON-EUCLIDEAN GEOMETRY,
D. M. Y. Sommerville

Unique in proceeding step-by-step, in the manner of traditional geometry. Enables the student with only a good knowledge of high school algebra and geometry to grasp elementary hyperbolic, elliptic, analytic non-Euclidean geometries; space curvature and its philosophical implications; theory of radical axes; homothetic centres and systems of circles; parataxy and parallelism; absolute measure; Gauss' proof of the defect area theorem; geodesic representation; much more, all with exceptional clarity. 126 problems at chapter endings provide progressive practice and familiarity. 133 figures. Index. xvi + 274pp. 5⅜ x 8.
Paperbound $2.00

INTRODUCTION TO THE THEORY OF NUMBERS, *L. E. Dickson*

Thorough, comprehensive approach with adequate coverage of classical literature, an introductory volume beginners can follow. Chapters on divisibility, congruences, quadratic residues & reciprocity. Diophantine equations, etc. Full treatment of binary quadratic forms without usual restriction to integral coefficients. Covers infinitude of primes, least residues. Fermat's theorem. Euler's phi function, Legendre's symbol, Gauss's lemma, automorphs, reduced forms, recent theorems of Thue & Siegel, many more. Much material not readily available elsewhere. 239 problems. Index. I figure. viii + 183pp. 5⅜ x 8.
Paperbound $1.75

MATHEMATICAL TABLES AND FORMULAS,
compiled by Robert D. Carmichael and Edwin R. Smith

Valuable collection for students, etc. Contains all tables necessary in college algebra and trigonometry, such as five-place common logarithms, logarithmic sines and tangents of small angles, logarithmic trigonometric functions, natural trigonometric functions, four-place antilogarithms, tables for changing from sexagesimal to circular and from circular to sexagesimal measure of angles, etc. Also many tables and formulas not ordinarily accessible, including powers, roots, and reciprocals, exponential and hyperbolic functions, ten-place logarithms of prime numbers, and formulas and theorems from analytical and elementary geometry and from calculus. Explanatory introduction. viii + 269pp. 5⅜ x 8½.
Paperbound $1.25

A SOURCE BOOK IN MATHEMATICS,
D. E. Smith

Great discoveries in math, from Renaissance to end of 19th century, in English translation. Read announcements by Dedekind, Gauss, Delamain, Pascal, Fermat, Newton, Abel, Lobachevsky, Bolyai, Riemann, De Moivre, Legendre, Laplace, others of discoveries about imaginary numbers, number congruence, slide rule, equations, symbolism, cubic algebraic equations, non-Euclidean forms of geometry, calculus, function theory, quaternions, etc. Succinct selections from 125 different treatises, articles, most unavailable elsewhere in English. Each article preceded by biographical introduction. Vol. I: Fields of Number, Algebra. Index. 32 illus. 338pp. 5⅜ x 8. Vol. II: Fields of Geometry, Probability, Calculus, Functions, Quaternions. 83 illus. 432pp. 5⅜ x 8.

Vol. 1 Paperbound $2.00, Vol. 2 Paperbound $2.00,
The set $4.00

FOUNDATIONS OF PHYSICS,
R. B. Lindsay & H. Margenau

Excellent bridge between semi-popular works & technical treatises. A discussion of methods of physical description, construction of theory; valuable for physicist with elementary calculus who is interested in ideas that give meaning to data, tools of modern physics. Contents include symbolism; mathematical equations; space & time foundations of mechanics; probability; physics & continua; electron theory; special & general relativity; quantum mechanics; causality. "Thorough and yet not overdetailed. Unreservedly recommended," *Nature* (London). Unabridged, corrected edition. List of recommended readings. 35 illustrations. xi + 537pp. 5⅜ x 8. Paperbound $3.00

FUNDAMENTAL FORMULAS OF PHYSICS,
ed. by D. H. Menzel

High useful, full, inexpensive reference and study text, ranging from simple to highly sophisticated operations. Mathematics integrated into text—each chapter stands as short textbook of field represented. Vol. 1: Statistics, Physical Constants, Special Theory of Relativity, Hydrodynamics, Aerodynamics, Boundary Value Problems in Math, Physics, Viscosity, Electromagnetic Theory, etc. Vol. 2: Sound, Acoustics, Geometrical Optics, Electron Optics, High-Energy Phenomena, Magnetism, Biophysics, much more. Index. Total of 800pp. 5⅜ x 8.

Vol. 1 Paperbound $2.25, Vol. 2 Paperbound $2.25,
The set $4.50

THEORETICAL PHYSICS,
A. S. Kompaneyets

One of the very few thorough studies of the subject in this price range. Provides advanced students with a comprehensive theoretical background. Especially strong on recent experimentation and developments in quantum theory. Contents: Mechanics (Generalized Coordinates, Lagrange's Equation, Collision of Particles, etc.), Electrodynamics (Vector Analysis, Maxwell's equations, Transmission of Signals, Theory of Relativity, etc.), Quantum Mechanics (the Inadequacy of Classical Mechanics, the Wave Equation, Motion in a Central Field, Quantum Theory of Radiation, Quantum Theories of Dispersion and Scattering, etc.), and Statistical Physics (Equilibrium Distribution of Molecules in an Ideal Gas, Boltzmann Statistics, Bose and Fermi Distribution. Thermodynamic Quantities, etc.). Revised to 1961. Translated by George Yankovsky, authorized by Kompaneyets. 137 exercises. 56 figures. 529pp. 5⅜ x 8½.

Paperbound $2.50

MATHEMATICAL PHYSICS, *D. H. Menzel*
Thorough one-volume treatment of the mathematical techniques vital for classical mechanics, electromagnetic theory, quantum theory, and relativity. Written by the Harvard Professor of Astrophysics for junior, senior, and graduate courses, it gives clear explanations of all those aspects of function theory, vectors, matrices, dyadics, tensors, partial differential equations, etc., necessary for the understanding of the various physical theories. Electron theory, relativity, and other topics seldom presented appear here in considerable detail. Scores of definition, conversion factors, dimensional constants, etc. "More detailed than normal for an advanced text . . . excellent set of sections on Dyadics, Matrices, and Tensors," *Journal of the Franklin Institute*. Index. 193 problems, with answers. x + 412pp. 5⅜ x 8. Paperbound $2.50

THE THEORY OF SOUND, *Lord Rayleigh*
Most vibrating systems likely to be encountered in practice can be tackled successfully by the methods set forth by the great Nobel laureate, Lord Rayleigh. Complete coverage of experimental, mathematical aspects of sound theory. Partial contents: Harmonic motions, vibrating systems in general, lateral vibrations of bars, curved plates or shells, applications of Laplace's functions to acoustical problems, fluid friction, plane vortex-sheet, vibrations of solid bodies, etc. This is the first inexpensive edition of this great reference and study work. Bibliography, Historical introduction by R. B. Lindsay. Total of 1040pp. 97 figures. 5⅜ x 8. Vol. 1 Paperbound $2.50, Vol. 2 Paperbound $2.50, The set $5.00

HYDRODYNAMICS, *Horace Lamb*
Internationally famous complete coverage of standard reference work on dynamics of liquids & gases. Fundamental theorems, equations, methods, solutions, background, for classical hydrodynamics. Chapters include Equations of Motion, Integration of Equations in Special Gases, Irrotational Motion, Motion of Liquid in 2 Dimensions, Motion of Solids through Liquid-Dynamical Theory, Vortex Motion, Tidal Waves, Surface Waves, Waves of Expansion, Viscosity, Rotating Masses of Liquids. Excellently planned, arranged; clear, lucid presentation. 6th enlarged, revised edition. Index. Over 900 footnotes, mostly bibliographical. 119 figures. xv + 738pp. 6⅛ x 9¼. Paperbound $4.00

DYNAMICAL THEORY OF GASES, *James Jeans*
Divided into mathematical and physical chapters for the convenience of those not expert in mathematics, this volume discusses the mathematical theory of gas in a steady state, thermodynamics, Boltzmann and Maxwell, kinetic theory, quantum theory, exponentials, etc. 4th enlarged edition, with new material on quantum theory, quantum dynamics, etc. Indexes. 28 figures. 444pp. 6⅛ x 9¼. Paperbound $2.75

THERMODYNAMICS, *Enrico Fermi*
Unabridged reproduction of 1937 edition. Elementary in treatment; remarkable for clarity, organization. Requires no knowledge of advanced math beyond calculus, only familiarity with fundamentals of thermometry, calorimetry. Partial Contents: Thermodynamic systems; First & Second laws of thermodynamics; Entropy; Thermodynamic potentials: phase rule, reversible electric cell; Gaseous reactions: van't Hoff reaction box, principle of LeChatelier; Thermodynamics of dilute solutions: osmotic & vapor pressures, boiling & freezing points; Entropy constant. Index. 25 problems. 24 illustrations. x + 160pp. 5⅜ x 8. Paperbound $1.75